Introductory Statistics
A Problem-Solving Approach

Third Edition

Stephen Kokoska

Bloomsburg University

Austin • Boston • New York • Plymouth

Vice President, STEM: Daryl Fox
Program Director: Andrew Dunaway
Program Manager: Sarah Seymour
Senior Marketing Manager: Nancy Bradshaw
Marketing Assistant: Madeleine Inskeep
Executive Development Editor: Katrina Mangold
Development Editor: Leslie Lahr
Executive Media Editor: Catriona Kaplan
Associate Editor: Andy Newton
Assistant Editor: Justin Jones
Director of Content Management Enhancement: Tracey Kuehn
Senior Managing Editor: Lisa Kinne
Senior Content Project Manager: Vivien Weiss
Director of Design, Content Management: Diana Blume
Design Services Manager: Natasha Wolfe
Cover Designer: John Callahan
Text Designer: Gary Hespenheide
Director of Digital Production: Keri deManigold
Senior Media Project Manager: Elton Carter
Media Project Manager: Hanna Squire
Senior Workflow Project Manager: Paul W. Rohloff
Executive Permissions Editor: Robin Fadool
Composition: Lumina Datamatics, Inc.
Printing and Binding: LSC Communications
Cover Images: H. Mark Weidman Photography/Alamy

Library of Congress Control Number: 2019911302

Student Edition Hardcover:
ISBN-13: 978-1-319-04962-1
ISBN-10: 1-319-04962-1

Student Edition Loose-leaf:
ISBN-13: 978-1-319-05610-0
ISBN-10: 1-319-05610-5

Instructor Complimentary Copy:
ISBN-13: 978-1-319-05608-7
ISBN-10: 1-319-05608-3

© 2020, 2015, 2011 by W. H. Freeman and Company

All rights reserved.

Printed in the United States of America

1 2 3 4 5 6   24 23 22 21 20 19

Macmillan Learning
One New York Plaza
Suite 4600
New York, NY 10004-1562
www.macmillanlearning.com

In 1946, William Freeman founded W. H. Freeman and Company and published Linus Pauling's *General Chemistry*, which revolutionized the chemistry curriculum and established the prototype for a Freeman text. W. H. Freeman quickly became a publishing house where leading researchers can make significant contributions to mathematics and science. In 1996, W. H. Freeman joined Macmillan and we have since proudly continued the legacy of providing revolutionary, quality educational tools for teaching and learning in STEM.

Brief Contents

Chapter 0	Why Study Statistics	1
Chapter 1	An Introduction to Statistics and Statistical Inference	9
Chapter 2	Tables and Graphs for Summarizing Data	27
Chapter 3	Numerical Summary Measures	69
Chapter 4	Probability	119
Chapter 5	Random Variables and Discrete Probability Distributions	189
Chapter 6	Continuous Probability Distributions	247
Chapter 7	Sampling Distributions	299
Chapter 8	Confidence Intervals Based on a Single Sample	335
Chapter 9	Hypothesis Tests Based on a Single Sample	389
Chapter 10	Confidence Intervals and Hypothesis Tests Based on Two Samples or Treatments	451
Chapter 11	The Analysis of Variance	515
Chapter 12	Correlation and Linear Regression	557
Chapter 13	Categorical Data and Frequency Tables	629
Chapter 14	Nonparametric Statistics	657

Optional Sections (available online at macmillanlearning.com/introstats3e)

Section 6.5 The Normal Approximation to the Binomial Distribution

Section 12.6 The Polynomial and Qualitative Predictor Models

Section 12.7 Model Selection Procedures

Contents

Chapter 0
Why Study Statistics — 1

Chapter App: The Science of Intuition — 1
The Statistical Inference Procedure — 2
Problem Solving — 3
With a Little Help from Technology — 4
Chapter 0 Exercises — 6

Chapter 1
An Introduction to Statistics and Statistical Inference — 9

Chapter App: Lake Winnebago Fish — 9
1.1 Statistics Today — 10
1.2 Populations, Samples, Probability, and Statistics — 11
    Section 1.2 Exercises — 16
1.3 Experiments and Random Samples — 19
    Section 1.3 Exercises — 22
Chapter 1 Summary — 24
    Chapter 1 Exercises — 25

Chapter 2
Tables and Graphs for Summarizing Data — 27

Chapter App: Near-Earth Objects — 27
2.1 Types of Data — 28
    Section 2.1 Exercises — 31
2.2 Bar Charts and Pie Charts — 33
    Section 2.2 Exercises — 38
2.3 Stem-and-Leaf Plots — 42
    Section 2.3 Exercises — 47
2.4 Frequency Distributions and Histograms — 49
    Section 2.4 Exercises — 60
Chapter 2 Summary — 64
    Chapter 2 Exercises — 64

Chapter 3
Numerical Summary Measures — 69

Chapter App: SpaceX Payloads — 69
3.1 Measures of Central Tendency — 70
    Section 3.1 Exercises — 78
3.2 Measures of Variability — 82
    Section 3.2 Exercises — 89
3.3 The Empirical Rule and Measures of Relative Standing — 94
    Section 3.3 Exercises — 103
3.4 Five-Number Summary and Box Plots — 106
    Section 3.4 Exercises — 110
Chapter 3 Summary — 114
    Chapter 3 Exercises — 115

Chapter 4
Probability — 119

Chapter App: Dream Home — 119
4.1 Experiments, Sample Spaces, and Events — 120
    Section 4.1 Exercises — 128
4.2 An Introduction to Probability — 132
    Section 4.2 Exercises — 143
4.3 Counting Techniques — 147
    Section 4.3 Exercises — 154
4.4 Conditional Probability — 158
    Section 4.4 Exercises — 165
4.5 Independence — 170
    Section 4.5 Exercises — 179
Chapter 4 Summary — 184
    Chapter 4 Exercises — 185

Chapter 5
Random Variables and Discrete Probability Distributions — 189

Chapter App: Hooked on Vitamins — 189
- **5.1** Random Variables — 190
  - Section 5.1 Exercises — 194
- **5.2** Probability Distributions for Discrete Random Variables — 195
  - Section 5.2 Exercises — 203
- **5.3** Mean, Variance, and Standard Deviation for a Discrete Random Variable — 206
  - Section 5.3 Exercises — 212
- **5.4** The Binomial Distribution — 216
  - Section 5.4 Exercises — 226
- **5.5** Other Discrete Distributions — 230
  - Section 5.5 Exercises — 237
- Chapter 5 Summary — 241
  - Chapter 5 Exercises — 242

Chapter 6
Continuous Probability Distributions — 247

Chapter App: EMV Chip Technology — 247
- **6.1** Probability Distributions for a Continuous Random Variable — 248
  - Section 6.1 Exercises — 258
- **6.2** The Normal Distribution — 261
  - Section 6.2 Exercises — 274
- **6.3** Checking the Normality Assumption — 278
  - Section 6.3 Exercises — 284
- **6.4** The Exponential Distribution — 286
  - Section 6.4 Exercises — 291
- Chapter 6 Summary — 293
  - Chapter 6 Exercises — 294

Chapter 7
Sampling Distributions — 299

Chapter App: The Longest Zip Line — 299
- **7.1** Statistics, Parameters, and Sampling Distributions — 300
  - Section 7.1 Exercises — 305
- **7.2** The Sampling Distribution of the Sample Mean and the Central Limit Theorem — 309
  - Section 7.2 Exercises — 319
- **7.3** The Distribution of the Sample Proportion — 323
  - Section 7.3 Exercises — 328
- Chapter 7 Summary — 331
  - Chapter 7 Exercises — 331

Chapter 8
Confidence Intervals Based on a Single Sample — 335

Chapter App: Autonomous Cars — 335
- **8.1** Point Estimation — 336
  - Section 8.1 Exercises — 339
- **8.2** A Confidence Interval for a Population Mean When σ Is Known — 341
  - Section 8.2 Exercises — 350
- **8.3** A Confidence Interval for a Population Mean When σ Is Unknown — 355
  - Section 8.3 Exercises — 361
- **8.4** A Large-Sample Confidence Interval for a Population Proportion — 366
  - Section 8.4 Exercises — 370
- **8.5** A Confidence Interval for a Population Variance or Standard Deviation — 375
  - Section 8.5 Exercises — 380
- Chapter 8 Summary — 385
  - Chapter 8 Exercises — 385

Chapter 9
Hypothesis Tests Based on a Single Sample — 389

Chapter App: Heavy Metal — 389
- **9.1** The Parts of a Hypothesis Test and Choosing the Alternative Hypothesis — 390
  - Section 9.1 Exercises — 394
- **9.2** Hypothesis Test Errors — 396
  - Section 9.2 Exercises — 400
- **9.3** Hypothesis Tests Concerning a Population Mean When σ Is Known — 403
  - Section 9.3 Exercises — 410
- **9.4** *p* Values — 414
  - Section 9.4 Exercises — 419
- **9.5** Hypothesis Tests Concerning a Population Mean When σ Is Unknown — 422
  - Section 9.5 Exercises — 427

9.6	Large-Sample Hypothesis Tests Concerning a Population Proportion	432
	Section 9.6 Exercises	436
9.7	Hypothesis Tests Concerning a Population Variance or Standard Deviation	439
	Section 9.7 Exercises	443
Chapter 9 Summary		446
	Chapter 9 Exercises	447

Chapter 10
Confidence Intervals and Hypothesis Tests Based on Two Samples or Treatments — 451

Chapter App: Prescription Errors		451
Notation		452
10.1	Inference: Two Independent Samples, Population Variances Known	453
	Section 10.1 Exercises	460
10.2	Inference: Two Independent Samples, Normal Populations	463
	Section 10.2 Exercises	472
10.3	Paired Data	477
	Section 10.3 Exercises	483
10.4	Comparing Two Population Proportions Using Large Samples	487
	Section 10.4 Exercises	495
10.5	Comparing Two Population Variances or Standard Deviations	499
	Section 10.5 Exercises	506
Chapter 10 Summary		509
	Chapter 10 Exercises	510

Chapter 11
The Analysis of Variance — 515

Chapter App: Selenium Concentration		515
11.1	One-Way ANOVA	516
	Section 11.1 Exercises	523
11.2	Isolating Differences	528
	Section 11.2 Exercises	535
11.3	Two-Way ANOVA	539
	Section 11.3 Exercises	547
Chapter 11 Summary		550
	Chapter 11 Exercises	553

Chapter 12
Correlation and Linear Regression — 557

Chapter App: Wind Turbine Noise		557
12.1	Simple Linear Regression	558
	Section 12.1 Exercises	570
12.2	Hypothesis Tests and Correlation	574
	Section 12.2 Exercises	582
12.3	Inferences Concerning the Mean Value and an Observed Value of Y for $x = x^*$	586
	Section 12.3 Exercises	591
12.4	Regression Diagnostics	594
	Section 12.4 Exercises	599
12.5	Multiple Linear Regression	603
	Section 12.5 Exercises	615
Chapter 12 Summary		620
	Chapter 12 Exercises	624

Chapter 13
Categorical Data and Frequency Tables — 629

Chapter App: Traffic Composition		629
13.1	Univariate Categorical Data, Goodness-of-Fit Tests	630
	Section 13.1 Exercises	634
13.2	Bivariate Categorical Data: Tests for Homogeneity and Independence	639
	Section 13.2 Exercises	648
Chapter 13 Summary		652
	Chapter 13 Exercises	653

Chapter 14
Nonparametric Statistics — 657

Chapter App: Keyboard Angles		657
14.1	The Sign Test	658
	Section 14.1 Exercises	663
14.2	The Signed-Rank Test	666
	Section 14.2 Exercises	671
14.3	The Rank-Sum Test	674
	Section 14.3 Exercises	678
14.4	The Kruskal–Wallis Test	681
	Section 14.4 Exercises	684
14.5	The Runs Test	686
	Section 14.5 Exercises	690

14.6 Spearman's Rank Correlation	692	Table 10 Critical Values for the Wilcoxon Rank-Sum Statistic	T-22
Section 14.6 Exercises	694		
Chapter 14 Summary	697	Table 11 Critical Values for the Runs Test	T-25
Chapter 14 Exercises	698	Table 12 Greek Alphabet	T-27

Tables Appendix T-1

Table 1 Binomial Distribution Cumulative Probabilities — T-2

Table 2 Poisson Distribution Cumulative Probabilities — T-4

Table 3 Standard Normal Distribution Cumulative Probabilities — T-7

Table 4 Standardized Normal Scores — T-9

Table 5 Critical Values for the t Distribution — T-10

Table 6 Critical Values for the Chi-Square Distribution — T-11

Table 7 Critical Values for the F Distribution — T-13

Table 8 Critical Values for the Studentized Range Distribution — T-16

Table 9 Critical Values for the Wilcoxon Signed-Rank Statistic — T-19

Answers to Odd-Numbered Exercises A-1

Index I-1

Notes and Data Sources (available online at macmillanlearning.com/introstats3e) N-1

Optional Sections
(available online at www.macmillanlearning.com/introstats3e)

Section 6.5 The Normal Approximation to the Binomial Distribution

Section 12.6 The Polynomial and Qualitative Predictor Models

Section 12.7 Model Selection Procedures

Preface

Students frequently ask me why they need to take an introductory statistics course. My answer is simple: In almost every occupation and in ordinary daily life, you will have to make data-driven decisions and inferences, as well as assess risk. In addition, you must be able to translate complex problems into manageable pieces, recognize patterns, and, most importantly, solve problems. This text helps students develop the fundamental lifelong skill of solving problems and interpreting solutions in real-world terms.

One of my goals was to make this problem-solving approach accessible and easy to apply in many situations. I certainly want students to appreciate the beauty of statistics and connections to so many other disciplines. However, it is even more important for students to be able to apply problem-solving skills to a wide range of academic and career pursuits, including business, science and technology, and education.

Introductory Statistics: A Problem-Solving Approach, Third Edition, presents long-term, universal skills for students taking a one- or two-semester introductory-level statistics course. Examples include guided, explanatory solutions that emphasize problem-solving techniques. Example solutions are presented in a numbered, step-by-step format. The generous collection and variety of exercises provide ample opportunities for practice and review in a variety of contexts. Concepts, examples, and exercises are presented from a practical, realistic perspective; real and realistic data sets are current and relevant. The text uses mathematically correct notation and symbols and precise definitions to clearly illustrate statistical procedures and proper communication.

This book is designed to help students fully understand the steps in basic statistical arguments, emphasizing the importance of assumptions to follow valid arguments or identify inaccurate conclusions. Most importantly, students will understand the process of statistical inference. A four-step process (Claim, Experiment, Likelihood, Conclusion) is used throughout the text to present the smaller pieces of introductory statistics upon which the large, essential statistical inference puzzle is built.

New to This Edition

In this completely redesigned third edition, Steve Kokoska combines his animated, interactive teaching style and friendly writing with contemporary real-world examples, modern pedagogical features, and the use of technology. The exposition presents precise mathematics with clear, often humorous writing and a distinctive, organized solution process with reasons for every step. The following new features and enhancements build on the strong foundation set in previous editions.

Focus on R Software Wherever possible, a technology solution using R is presented at the end of each text example. This allows students to focus on concepts and interpretation and to gain a basic understanding of and familiarity with this widely used open-source statistical program.

Examples and Exercises More than 30% of the examples and exercises have been updated with an even greater liberal arts character; a wide variety of interesting and contemporary topics are included. New applications include the chance of discovering a pearl in an oyster, air quality, trace minerals, and pickleball.

CHAPTER APP Each chapter begins with a unique, real-world problem, providing an interesting introduction to new concepts and an application to begin discussion. Throughout the chapter, where relevant, Chapter App references provide questions and context that build up to the concluding Chapter App exercise.

Looking Back and Looking Forward At the beginning of every chapter, the revised Looking Back feature includes reminders of specific concepts linked to earlier sections that will be used to develop new skills. The Looking Forward list includes the learning objectives for the chapter now linked to each section.

New Subsections Each section has been carefully broken down into smaller subsections to better facilitate comprehension, retention, and recalling of information. Students now have a natural break to pause and reflect on what they have just learned, and to assess their learning by trying subsection-specific exercises.

Common Error This new marginal feature identifies and corrects common conceptual and computational errors in a just-in-time fashion.

Rapid Review Icons in the text indicate key content where Rapid Review assessment and videos can be found in your course in SaplingPlus, the new digital platform available with this edition. Rapid Reviews provide a math refresher with short review quizzes and videos to help students remember basic math skills.

Check It Out This marginal feature offers students a chance to stop and consider an important concept further in an informal way.

Features

Focus on Statistical Inference The main theme of this text is statistical inference and decision making through the interpretation of numerical results. The process of statistical inference is introduced in a variety of applications and statistical settings, all using a similar, carefully delineated, four-step approach: Claim, Experiment, Likelihood, and Conclusion.

Step-by-Step Solutions The solutions to selected examples are presented in logical, systematic steps. Each step is explained thoroughly, often graphically. This approach helps guide the reader through the necessary calculations to find a solution and interpret results.

TRY IT NOW **Try It Now References** Most examples conclude with a reference to a specific related exercise in the end-of-chapter set. With this, students can test their understanding of the example's concepts and techniques immediately.

A Closer Look The details provided in these sections offer straightforward explanations of various definitions and concepts. The itemized specifics, including hints, tips, and reminders, make it easier for the reader to interpret, comprehend, and learn important statistical ideas.

▶ **Theory Symbols** Only when appropriate to the discussion, more advanced material is offset with a blue triangle. This material can be skipped by the typical reader, but it provides more complete explanations of various topics.

How To Boxes This feature provides clear steps for constructing basic graphs or performing essential calculations.

Definition/Formula Boxes Definitions and formulas are clearly explained and easy to find within the chapter content.

Data Sets All data sets presented in the examples and exercises are available on the book's website (www.macmillanlearning.com/introstats3e) in various formats. This makes it easy for students and instructors to use their most familiar and comfortable technology.

Grouped Exercises A wide variety of interesting, engaging exercises on relevant topics, based on current data, appear at the end of each section. These problems provide plenty of opportunities for practice, review, and application of concepts. Section exercises are grouped according to level and purpose:

- **Concept Check:** True/False, Fill in the Blank, and Short Answer exercises designed to reinforce the basic concepts presented in the section.
- **Practice:** Basic, introductory problems to familiarize students with the relevant applications and solution methods.
- **Applications:** Realistic, appealing exercises to build confidence and promote routine understanding.
- **Extended Applications:** Applied problems that require extra creativity and thought.
- **Challenge Problems:** Additional exercises and technology projects that allow students to discover more advanced concepts and connections.

At the end of each chapter, **Chapter Exercises** test students' overall understanding of that chapter's concepts and enable them to practice for assessments.

Answers to odd-numbered section and chapter exercises are provided at the back of the book.

Chapter Summary A table at the end of each chapter provides a short summary and page reference for each key concept, relevant notation, and formula.

SaplingPlus for Statistics

Assessment—The "Office Hours" experience, while doing Homework

Tutorial-Style Formative Assessment

For select questions in our formative assessment environment, students' incorrect answers receive full solutions and feedback to guide their study. Many exercises are also designed to deliver error-specific feedback based on their common misconceptions about topics/learning objectives. This socratic feedback mechanism emulates the office hours experience, encouraging students to think critically about their identified misconception. Students are also provided with the fully-worked solutions to reinforce concepts and provide an in-product study guide for every problem.

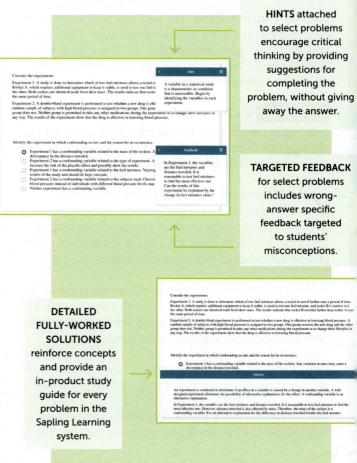

HINTS attached to select problems encourage critical thinking by providing suggestions for completing the problem, without giving away the answer.

TARGETED FEEDBACK for select problems includes wrong-answer specific feedback targeted to students' misconceptions.

DETAILED FULLY-WORKED SOLUTIONS reinforce concepts and provide an in-product study guide for every problem in the Sapling Learning system.

Adaptive Assessment Focused on Fundamental Concepts

LEARNINGCURVE

LearningCurve's adaptive quizzing encourages students to learn through practice. Through gamified elements, students are challenged to gain points by submitting correct answers to meet or exceed an instructor-determined score. Faculty and students are then provided detailed analytics on their performance via their own personal study plan with pathways to additional learning tools and insights. This tool focuses on getting students prepared for their upcoming class.

nalytics

SAPLING GRADEBOOK

STUDENT ANALYSIS

Through our heatmap dashboard instructors can get a quick visual representation of student performance on a given question. They can then dive deeper into each individual student to view and evaluate every submission they've made on a problem, getting to see a student's "work" on the question.

ITEM ANALYSIS

Instructors can also evaluate an aggregate view of their class's performance on a problem (item) through our **"Item Analysis"** tab. They can quickly see what percentage of students got this question correct, incorrect, or unanswered through the progress bar at the top. They can also see all of the student responses rolled up; giving them immediate insight into what their students' most common misconceptions are on this problem.

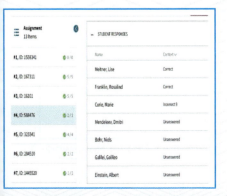

Interactive e-book

The **e-book** provides powerful study tools for students, multimedia content, and easy customization for instructors. Students can search, highlight, and bookmark specific information, making it easier to study and access key content. Media assets such as data sets, glossary terms, select exercise answers, and videos are linked within the e-book in **SaplingPlus,** allowing ease-of-use for students.

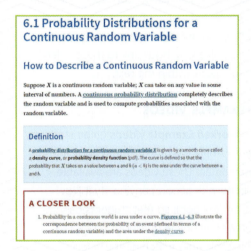

Integration

With **Deep Asset Linking,** instructors can organize individual pieces of Macmillan content into LMS Content folders. Students complete single-sign-on through the link to connect their Macmillan and LMS accounts. Once this occurs, students will have seamless access to the Macmillan content through their campus LMS using only the LMS username and password.

- **Multiple gradebook columns** generate automatically for each Macmillan assignment deployed to the LMS, allowing for assignments to be combined into gradebook categories within the LMS for gradebook calculations like dropping lowest scores or weighted percentages.

- **Automatic grade sync** occurs with Deep integration.

Media/Learning Objects

NEW INTRODUCTORY STATISTICS-SPECIFIC RESOURCES

RAPID REVIEW VIDEOS AND ASSESSMENT

Rapid Review videos discuss prerequisite or corequisite topics essential to students' understanding of key statistical concepts. These brief videos are presented in an engaging manner and include built-in, assignable assessment questions. Icons in the book and e-book link to these videos at relevant locations within the text.

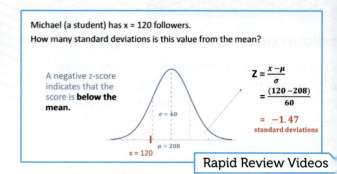
Rapid Review Videos

WORKED EXAMPLE VIDEOS

These new **Worked Example videos** guide students through an Example directly from the book in step-by-step detail. These videos are available to assign along with accompanying Try it Now Exercises from the book so that students can practice answering online questions related to the Example topic after watching the video.

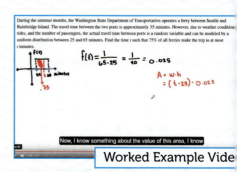
Worked Example Videos

R MARKDOWN FILES

R Markdown files demonstrate how to solve selected data-driven exercises from the book using **R**.

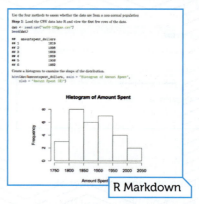
R Markdown

INTRODUCTORY STATISTICS-SPECIFIC ASSESSMENT

PRE-BUILT CHAPTER HOMEWORK ASSIGNMENTS are curated assignments containing questions directly from the book, with select questions containing error specific feedback to guide students to concept mastery.

Pre-Built Chapter Homework

aired with Assessment:

TRIED AND TRUE

STATISTICAL APPLETS give students hands-on opportunities to familiarize themselves with important statistical concepts and procedures, in an interactive setting that allows them to manipulate variables and see the results graphically. Icons in the textbook indicate when an applet is available for the material being covered.

STATBOARDS VIDEOS are brief whiteboard videos that illustrate difficult topics through additional examples, written and explained by a select group of statistics educators.

STATISTICAL VIDEO SERIES consists of StatClips, StatClips Examples, and Statistically Speaking "Snapshots." View animated lecture videos, whiteboard lessons, and documentary-style footage that illustrate key statistical concepts and help students visualize statistics in real-world scenarios.

EESEE CASE STUDIES (Electronic Encyclopedia of Statistical Examples and Exercises), developed by The Ohio State University Statistics Department, teach students to apply their statistical skills by exploring actual case studies using real data.

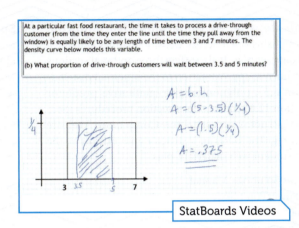
StatBoards Videos

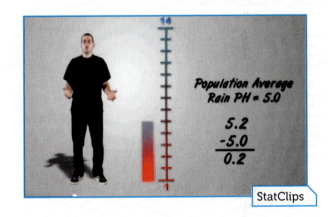
StatClips

ata Tools

DATA FILES are available in JMP, ASCII, Excel, TI, Minitab, SPSS, R, and CSV formats.

VIDEO TECHNOLOGY MANUALS available for TI-83/84 calculators, Minitab, Excel, JMP, SPSS, R, RStudio, Rcmdr, and CrunchIt!® provide brief instructions for using specific statistical software.

CRUNCHIT!® is Macmillan Learning's own web-based statistical software that allows users to perform all the statistical operations and graphing needed for an introductory statistics course and more. It saves users time by automatically loading data from IntroStats, and it provides the flexibility to edit and import additional data.

 JMP STUDENT EDITION (developed by SAS) is easy to learn and contains all the capabilities required for introductory statistics. JMP is the leading commercial data analysis software of choice for scientists, engineers and analysts at companies throughout the globe (for Windows and Mac).

Additional Resources Available with *Introductory Statistics: A Problem-Solving Approach*

Companion Website: www.macmillanlearning.com /introstats3e. This open-access website includes statistical applets and data set files. It also offers three optional sections covering the normal approximation to the binomial distribution (Section 6.5), polynomial and qualitative predictor models (Section 12.6), and model selection procedures (Section 12.7). Instructor access to the Companion Website requires user registration as an instructor and features all the open-access student web materials, plus:

- Instructor's Solutions Manual
- Lecture PowerPoint Slides
- Test Bank
- Clicker Questions
- Image PowerPoint Slides

Special Software Packages Student versions of JMP are available for packaging with the text. JMP is available inside SaplingPlus at no additional cost. Contact your Macmillan Learning representative for information or visit www.macmillanlearning.com.

With a course adoption, instructors may provide CrunchIt!® statistical software access at no cost to students. Please contact your Macmillan Learning representative to receive a course access code for CrunchIt!®

iClicker iClicker is a two-way radio-frequency classroom response solution developed by educators for educators. Each step of iClicker's development has been informed by teaching and learning. To learn more about packaging iClicker with this book, please contact your local Macmillan sales representative or visit www.iclicker.com.

Acknowledgments

I would like to thank the following colleagues who offered specific comments and suggestions on the third edition manuscript throughout various stages of development.

Achut Adhikari, *Miami University*
Eric Agyekum, *Vancouver Island University*
T. Jonathan Bayer, *Virginia Western Community College*
Julie M. Clark, *Hollins University*
Keith Coates, *Drury University*
Kossi D. Edoh, *North Carolina Agricultural and Technical State University*
Bree Ettinger, *Emory University*
Leonore A. Findsen, *Purdue University*
Beata I. Gebuza, *Gateway Community College*
Joel Haack, *University of Northern Iowa*
Susan Kay Herring, *Sonoma State University*
Joshua Himmelsbach, *Catholic University*
Daniel Inghram, *University of Central Florida*
Inyang Inyang, *Northern Virginia Community College Manassas Campus*
Yvette Janecek, *Blinn College*
Elizabeth Johnson, *George Mason University*
Pramod Kanwar, *Ohio University–Zanesville*
Lisa W. Kay, *Eastern Kentucky University*
Beverly Kludy, *Muskegon Community College*
Henryk Kolacz, *University of Alberta*
Tammi Kostos, *McHenry County College*
Michael P. Kowalski, *University of Alberta*
Subrata Kundu, *George Washington University*
Kurtis Lemmert, *Frostburg State University*
Sergio Loch, *Grand View University*
Ana Clare Mello, *Shasta College*
Nutan Mishra, *University of South Alabama*
Zahra Montazerti, *Carleton University*
Linda Myers, *Harrisburg Area Community College*
Michael C. Osborne, *Eastern Kentucky University*
Sergio Perez-Melo, *Florida International University*
Laurie Poe, *Santa Clara University*
Elizabeth J. Reed, *Michigan Technological University*
Sabrina Ripp, *Tulsa Community College*
Catherine Robinson, *University of Rhode Island*
Nursel Ruzgar, *Ryerson University*
John Samons, *Florida State College at Jacksonville*
Thobias Sando, *University of North Florida*
Yixun Shi, *Bloomsburg University of Pennsylvania*
Suryakala Srinivasan, *Rutgers University*
Wanhua Su, *MacEwan University*
Pamini Thangarajah, *Mount Royal University*
Kanapathi Thiru, *University of Alaska Anchorage*
Daniel Wang, *Central Michigan University*
Lisa Wellinghoff, *Wright State University*
Augustine Wong, *York University*
Karen Huynh Wong, *University of Toronto*
Casandra Wright, *Sam Houston State University*
Dong Zhang, *Bloomsburg University*
Yichuan Zhao, *Georgia State University*

My sincere appreciation and thanks go to the following instructors who created much of the ancillary package that accompanies the third edition: Robert Avakian, Oklahoma State University Institute of Technology, created the test bank; Julie Clark, Hollins University, authored the lecture slides; Nicole Dalzell, Wake Forest University, accuracy reviewed the test bank and Clicker questions; Mark Gebert, University of Kentucky, authored the Clicker questions and accuracy reviewed the solutions manual and back-of-book answers; Jennifer Kokoska, The Hill School, authored the solutions manual and back-of-book answers. Jean-Marie Magnier, Springfield Technical Community College, accuracy reviewed the practice quizzes; and Mark McKibben, West Chester University, authored the practice quizzes.

In addition, I'd like to thank John Samons, whose keen eye and solid statistical and instructional intuition benefit the accuracy of the book content.

A special thanks to the team at W. H. Freeman for their hard work. In particular, many thanks are due to program director Andrew Dunaway; senior content project manager Vivien Weiss; senior workflow project manager Paul Rohloff; copyeditor Jill Hobbs; design services manager Natasha Wolfe; and art manager Matthew McAdams. Additionally, I'd like to thank Daniel Lauve, director of content; Catriona Kaplan, executive media editor; Robin Fadool, executive permissions editor; Andy Newton, associate editor; Doug Newman, assistant media editor; Aaron Gladish, lead content developer; Ava Cas, content development manager; and Justin Jones, assistant editor. Finally, I especially appreciate the marketing and sales efforts of Nancy Bradshaw, senior marketing manager, and the entire sales force.

I am forever grateful to Karen Carson for her enthusiasm, sound advice, and exceptional editorial talent throughout the production of this third edition. Her confidence and support for this project has been unwavering, and I will miss her steady, guiding voice. And I could not have completed this project without Leslie Lahr. She has superb editing skills, a keen eye for style, and a knack for eliciting the best from an author, and she somehow always manages to find a splendid compromise between author and publisher.

It has also been a source of great pride and joy to work on this project with my daughter Jen. Her AP® Statistics experience, expertise in the classroom, and frank, honest review of material have been invaluable.

I am very grateful to the entire Antoniewicz family for providing the foundation for a wide variety of problems, including those that involve nephelometric turbidity units, floor slip testers, street hockey, and crazy crawler fishing lures.

I continue to learn a great deal with every day of writing. I believe this kind of exposition has made me a better teacher.

To Joan, thank you for your patience, understanding, inspiration, tasty treats, and ultra-caffeinated iced coffee.

About the Author

Eric Foster/Bloomsburg University

Steve Kokoska received his undergraduate degree from Boston College and his MS and PhD from the University of New Hampshire. His initial research interests included the statistical analysis of cancer chemoprevention experiments. He has published a number of research papers in mathematics journals, including *Biometrics*, *Anticancer Research*, and *Computer Methods and Programs in Biomedicine*; presented results at national conferences; and written several books. He has been awarded grants from the National Science Foundation, the Center for Rural Pennsylvania, and the Ben Franklin Program.

Steve is a longtime consultant for the College Board and has conducted workshops in Brazil, the Dominican Republic, Singapore, and China. He was the AP® Calculus Chief Reader for four years, has been involved with calculus reform and the use of technology in the classroom, and recently published an AP® Calculus text with James Stewart. He has been teaching at Bloomsburg University for 29 years and has served as director of the Honors Program.

Steve has been teaching introductory statistics classes throughout his academic career, and there is no doubt that this is his favorite course. This class (and book) provides students with basic, lifelong, quantitative skills that they will use in almost any job and teaches them how to think and reason logically. Steve believes very strongly in data-driven decisions and conceptual understanding through problem solving.

Steve's uncle, Fr. Stanley Bezuszka, a Jesuit and professor at Boston College, was one of the original architects of the so-called new math in the 1950s and 1960s. He had a huge influence on Steve's career. Steve helped Fr. B. with text accuracy checks, as a teaching assistant, and even with writing projects through high school and college. Along the way, Steve learned about the precision, order, and elegance of mathematics and developed an unbounded enthusiasm to teach.

Why Study Statistics

Mint Images/AGE Fotostock

CHAPTER APP

The Science of Intuition

Each year, some snowy owls migrate thousands of miles from the Arctic to southern Canada and the northern part of the United States. These owls are a favorite of bird watchers, spotted in fields and dunes, and have eerie pale yellow eyes. In December 2017, wildlife experts at the Raptor Center at the University of Minnesota reported that they were treating an unusually high number of injured snowy owls.

Near the end of December, the Raptor Center had treated 33 snowy owls. Julie Ponder, a veterinarian and the executive director of the Raptor Center, stated that this number was "significantly above normal."[1] The Raptor Center usually treats between 0 and 5 snowy owls each year.

There were so many snowy owls being treated for injuries that wildlife experts concluded it could not be a coincidence. This was a very natural, intuitive conclusion and is the essence of statistical inference. Wildlife experts observed an occurrence so rare and extraordinary that they instinctively concluded that it could not be due to pure chance or luck. There had to be another reason. And their logic was correct.

An investigation revealed that during the previous year, there had been an extra large food source in the Arctic for snowy owls—lemmings. This abundant food source enabled more snowy owls to survive, and more were able to leave the Arctic to find their own hunting territories. There were simply a lot more snowy owls migrating south and, therefore, more injuries.

We all have this same natural instinctive reaction when we see something extraordinary. Sometimes we think, "Wow, that's incredibly lucky." But more often, we question the observed outcome: "There must be some other explanation."

This natural reaction is the foundation of statistical inference. We make these kinds of decisions every single day. We gather evidence, make an observation, and conclude that the outcome is either reasonable or extraordinary. The purpose of statistics is to simply quantify this normal, everyday, deductive process. We need to learn probability so that we know for sure when an outcome is really rare. And we need to study the concepts of randomness and uncertainty.

The most important point here is that this process is not unusual or exceptional. The purpose of this text is to translate this normal, intuitive practice into statistical terms and models. This will make you better prepared to interpret outcomes, draw appropriate conclusions, and assess risk.

Here is another example of an extraordinary event involving icebergs. In April 2017, more than 400 icebergs floated into North Atlantic shipping lanes in one week.[2] This was an unusually large number for a seven day period at this time of year; the average is about 80. Due to the large number of icebergs, ships could no longer cut straight across the ocean. Instead, detours added extra travel time and increased costs. The Coast Guard commander who leads the ice patrol indicated that this was a drastic increase in icebergs in a very short amount of time.

Most of the icebergs that flow into the North Atlantic break off from the Greenland ice sheet. However, this extraordinary occurrence, and similar iceberg counts in 2014, 2015, and 2016, suggest that such an increase could not be due to pure chance alone. Although scientists are still searching for a definitive reason, many believe that climate change and wind patterns are contributing to this rare occurrence.

The Statistical Inference Procedure

The most important, fundamental theme in this text is statistical inference and decision making through problem solving. Computation is important and is shown throughout the text. However, calculators and computers remove the drudgery of hand calculations and allow us to concentrate more on interpretation and drawing conclusions. Many problems in this text contain a part asking you to interpret the numerical result or to draw a conclusion.

The process of questioning a rare occurrence or claim involves four steps.

Claim: This is the status quo, the ordinary, normal reasonable course of events, what we assume to be true.

Experiment: To check a claim, we conduct a relevant experiment or make an appropriate observation.

Likelihood: Here, we consider the likelihood of occurrence of the observed experimental outcome, assuming the claim is true. We will use many techniques to determine whether the experimental outcome is a reasonable observation (subject to some variability) or whether it is an exceptionally rare occurrence. We need to carefully consider and quantify our natural reaction to the relevant experiment. Using probability rules and concepts, we will convert our natural reaction to an experimental outcome into a precise measurement.

Conclusion: There are always only two possible conclusions.
1. If the outcome is reasonable, then we cannot doubt the original claim. The natural conclusion is that nothing out of the ordinary is occurring. More formally, there is no evidence to suggest that the claim is false.
2. If the experimental outcome is rare or extraordinary, we usually disregard the lucky alternative, and we think something is wrong. A rare outcome is a contradiction. Strange occurrences naturally make us question a claim. In this case, we believe there is evidence to suggest that the claim is false.

Let's try to apply these four steps to the snowy owl example.

The claim or status quo was that the Raptor Center treats between 0 and 5 injured snowy owls each calendar year. This assertion was the result of examination of historical data at the Raptor Center.

The experiment or observed outcome was the 33 injured snowy owls treated at the Raptor Center in December 2017.

Wildlife experts determined that the likelihood, or probability, of observing that many injured snowy owls was extremely low. Subject to normal variability, we should not see so many injured snowy owls in this location.

The conclusion was that this rare event was not due to pure chance or luck. Instead, some other reason explained this rare observation. The implication was that an unusually large number of snowy owls were migrating, due to an abundance of food. And this led to more injured snowy owls.

Problem Solving

Perhaps one of the most difficult concepts to teach is problem solving. We all struggle to solve problems: thinking about where to begin, what assumptions we can make, and which rules and techniques to use. One reason many students consider statistics a difficult course is that almost every problem is a word problem. These word problems have to be translated into mathematics.

To decipher a word problem, you might start by identifying the keywords and phrases, consider the relevant concepts, and develop a vision for the solution.

Here are some steps to help solve many of the problems in this text.

1. Find the *keywords*.
2. Correctly *translate* these words in statistics.
3. Determine the applicable *concepts*.
4. Develop a *vision*, or strategy, for the solution.

Many of the examples presented in this text can be solved by using this four-step process. The keywords in the problem lead to a translation into statistics. The statistics question is then solved by applying the appropriate, specific concepts. The keywords, translation, and concepts are used to develop a grand vision for solving the problem. This solution technique is not applicable to every problem. It is most appropriate for solving problems that involve probability, random variables, confidence intervals, and hypothesis testing, the foundation of most introductory statistics courses. As you become accustomed to this solution style, it will become routine, natural, and helpful.

With a Little Help from Technology

Although it is important to know and understand the underlying formulas, their derivation, and how to apply them, we will use and present the results from several different technology tools to supplement problem solving. Your focus should be on the interpretation of results, not the actual numerical calculations.

Most technology results in this text will be presented using R, a free powerful software tool for analyzing data. Several software programs provide an interface to R, such as RStudio and R Commander, and many free R packages are available that enhance the base distribution. **Figure 0.1** is a bar chart that shows the number of pardons by six past presidents.

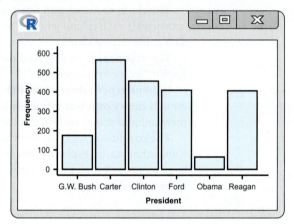

Figure 0.1 This is a bar chart created using R and RStudio.

The Texas Instruments TI-84 Plus CE graphing calculator has many common statistical features, including confidence intervals, hypothesis tests, and probability distribution functions. Data are entered and edited in the stat list editor, as shown in **Figure 0.2**. Meanwhile, **Figure 0.3** shows the results from a one-sample t test, and **Figure 0.4** provides a visualization of this hypothesis test.

Figure 0.2 The stat list editor.

Figure 0.3 One-sample t test output.

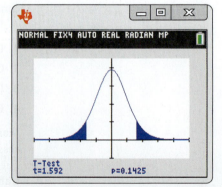

Figure 0.4 One-sample t test visualization.

CrunchIt! is available in the Macmillan Learning online homework system, and is accessed under the Resources tab. The opening screen (**Figure 0.5**) looks like a spreadsheet with pull-down menus at the top. You can enter data in columns, Var1, Var2, and so on; import data from a file; and export and save data.

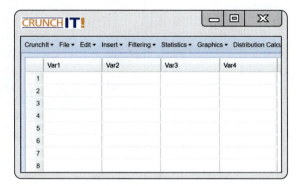

Figure 0.5 CrunchIt! opening screen.

Most Statistics, Graphics, and Distribution Calculator functions start with input screens. Output is displayed in a new screen. **Figure 0.6** shows the input screen for a bar chart with summarized data, and **Figure 0.7** shows the resulting graph.

Figure 0.6 Bar chart input screen.

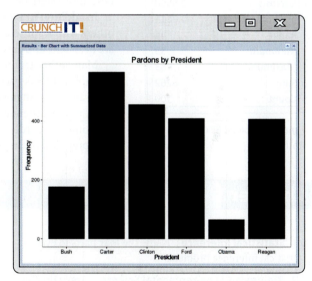

Figure 0.7 CrunchIt! bar chart.

Minitab is a powerful software tool for analyzing data. It has a logical interface, including a worksheet screen similar to a common spreadsheet. Data, graph, and statistics tools can be accessed through pull-down menus, and most commands can also be entered in a session window. **Figure 0.8** shows a bar chart of the number of movies for each genre from 1995 to 2018.

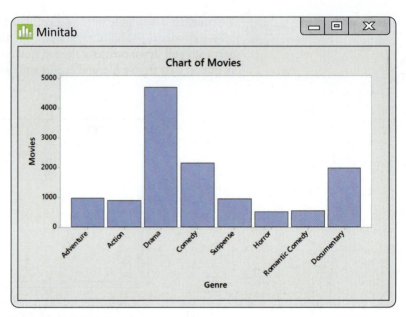

Figure 0.8 Minitab bar graph.

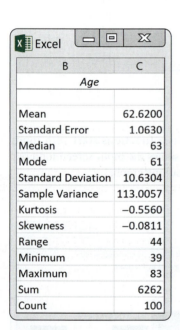

Figure 0.9 Excel descriptive statistics.

Excel 2019 includes many commonly used chart features accessible under the Insert tab. There are also probability distribution functions that allow the user to build templates for confidence intervals, hypothesis tests, and other statistical procedures. The Data Analysis tool pack provides additional statistical functions. **Figure 0.9** shows some descriptive statistics associated with the ages of current U.S. senators.

In addition to these tools, JMP statistical software is used by scientists, engineers, and others who want to explore or mine data. Various statistical tools and dynamic graphics are available, and this software features a friendly interactive interface.

Many other technology tools and statistical software packages are also available. For example, SPSS is used primarily in the social sciences, SAS incorporates a proprietary programming language, the TI-Nspire graphing calculator has a special Data & Statistics page, and Mathematica is a computer algebra system with many standard statistical functions. Regardless of your technology choice, remember that careful and thorough interpretation of the results is an essential part of using software properly.

Chapter Exercises

0.1 Statistical Inference Name the four parts to every statistical inference problem.

0.2 Biology and Environmental Science Apply the four statistical inference steps to the iceberg example in which an extraordinary number of icebergs floated into North Atlantic shipping lanes in one week.

0.3 Biology and Environmental Science In April 2017, the National Oceanic and Atmospheric Administration (NOAA) launched an investigation into the large number of humpback whale deaths along the East Coast of the United States. A total of 41 humpback whales perished in 15 months, all between Maine and North Carolina.[3] Historical data indicate that in an average year, only two humpback whales perish in this area, usually due to ship collisions. Explain why NOAA believes that there must be other factors contributing to these deaths.

0.4 Biology and Environmental Science It had been very rare for Southern California to experience a wildfire in the month of December. From 2000 to 2015, a total of seven wildfires in December burned an estimated 300 acres.[4] However, in December 2017, at least 29 wildfires occurred in Southern California. State two possible explanations for this rare rash of wildfires. Which explanation do you think is more plausible? Why?

0.5 Psychology and Human Behavior In October 2017, a North Carolina woman won the lottery twice in one day. Kimberly Morris first bought a $4 million Diamond Dazzler ticket from a supermarket and won $10,000. After claiming her prize at the North Carolina Education Lottery headquarters, on the way home she bought a second ticket. She became the first person to win a $1 million prize from this scratch-off game.[5] State two possible explanations for this rare occurrence. Which explanation do you think is more reasonable? Why?

0.6 Public Health and Nutrition When Universal Studios opened the Harry Potter roller coaster, an unusually high number of test subjects became sick (vomiting) after going on the ride.[6] Engineers were called in to examine the ride. One of the reasons for the increased illnesses may actually be the 3D glasses worn by riders.
(a) Explain why Universal Studios investigated the source of these illnesses.
(b) Apply the four statistical inference steps to this situation.

0.7 Biology and Environmental Science What do you think it means when a meteorologist says, "There is a 50% chance of rain today." Contact a meteorologist to ask what this statement means. Does the explanation agree with yours?

0.8 Biology and Environmental Science In April 2017, agencies in Ventura and Santa Barbara counties were inundated with calls from beachgoers who reported stranded sea lions.[7] There are usually a few reports of stranded sea lions every week, but the number of reports in April was extraordinarily high. Officials immediately began to investigate the sea lion deaths and conducted several tests to determine the possible cause. Explain why veterinary officials believed the sea lions did not die as a result of natural causes and investigated these deaths.

0.9 Public Health and Nutrition Recently, more than 130 people became sick after dining at the Chipotle restaurant in Sterling, Virginia.[8] The illnesses occurred over a seven-day period and the restaurant was closed for two days due to this incident.
(a) Explain why Chipotle officials investigated the source of these illnesses.
(b) Apply the four statistical inference steps to this situation.

0.10 Manufacturing and Product Development In October 2017, Fisher-Price issued a product recall for more than 63,000 Soothing Motions Seats due to overheating.[9] There had been 36 reports of the product overheating, including one report of a fire in the motor housing, but no injuries were reported.
(a) State two possible reasons for this observed high number of seats overheating.
(b) Why do you think Fisher-Price issued this recall?

0.11 Psychology and Human Behavior Suppose 15 home burglaries occurred in a small town during a one-year period. None occurred on a Thursday. Do you think there is evidence to suggest that something very unusual is happening in this town to prevent burglaries on that day of the week? Why or why not?

0.12 Demographics and Population Statistics The village of Kodinhi, India, has at least 400 pairs of twins, yet is home to only 2000 families.[10] Various organizations have sent teams to this village to search for an explanation for this phenomenon. Explain why researchers are investigating the source of this occurrence.

0.13 Marketing and Consumer Behavior According to the National Retail Federation (NRF), shoppers planned to pick up 15 Christmas gifts for relatives and friends in 2017.[11] Suppose a newspaper reporter surveyed 30 shoppers at the King of Prussia Mall on a Friday evening and found that 17 planned to buy 12 or fewer presents. Do you have any reason to doubt the claim made by the NRF? Why or Why not?

0.14 Medicine and Clinical Studies Amy Griffin, the assistant head coach of the women's soccer team at the University of Washington, became concerned when more than 50 players were diagnosed with cancer.[12] The women who developed this disease were mainly goalkeepers. Do you think that health officials should be concerned about the number of players who were diagnosed with cancer? Why or why not?

0.15 Prerequisites To understand the definitions and formulas in this text, you should feel comfortable with mathematical notation. To review and prepare for the notation used, make sure you are familiar with the following.
(a) Subscript notation. For example, $x_1, x_2, \ldots$.
(b) Superscript notation. For example, 2^3, x^2.
(c) Summation notation. For example, $\sum_{i=1}^{n} x_i$.
(d) The definition of a function.

An Introduction to Statistics and Statistical Inference

1

Looking Forward ▶

- Recognize that data and statistics are pervasive and that statistics are used to describe typical values and variability, as well as to make decisions that affect everyone: **Section 1.1**.
- Understand the relationships between a population, a sample, probability, and statistics: **Section 1.2**.
- Learn the basic steps in a statistical inference procedure: **Section 1.3**.

CHAPTER APP

thitiwat_t1980/Shutterstock

Lake Winnebago Fish

Microbeads are tiny pieces of plastic that are found in a variety of common products, including health and beauty items, especially soap and body scrubs. These microbeads are not always filtered by water treatment facilities, so they often end up in lakes and oceans. In December 2015, Congress amended the federal Food, Drug and Cosmetic Act. The Microbead-Free Waters Act of 2015 prohibits companies from manufacturing, packaging, and distributing rinse-off cosmetics that contain plastic microbeads. The purpose of this law was to address health concerns about microbeads in the water supply.

In 2017, researchers at the University of Wisconsin–Superior found that small pieces of plastic had passed through the local wastewater treatment facility and flowed into

Lake Winnebago.[1] These plastic pieces may contain toxins but also absorb additional toxins (microbeads) in the water from cosmetics, plastic litter, and clothing. If these plastic pieces are then consumed by fish, and people then consume the fish, this chain of events creates a potential public health concern.

Suppose researchers intend to conduct an extensive study to determine whether to issue any warnings about consuming fish from Lake Winnebago. One hundred random fish from this lake are obtained, and each is carefully examined for evidence of microbeads.

The methods presented in this chapter will enable us to identify the population of interest and the sample, as well as to understand the definition and importance of a random sample. Most importantly, we will characterize the deductive process used when an extraordinary event is observed and cannot be attributed to luck.

1.1 Statistics Today

Statistics All Around

Statistics data are everywhere: in newspapers, magazines, the Internet, the evening weather forecast, medical studies, and even sports reports. They are used to describe typical values and variability and to make decisions that affect every one of us. It is important to be able to read and understand statistical summaries and arguments with a critical eye. This chapter presents the basic elements of every statistics problem: a population and a sample and their connection to probability and statistics. Two common methods for data collection—observational sampling and experimentation—are also introduced.

Statistics data are used by professionals in many different disciplines. Actuaries are probably the biggest users of statistics. They conduct statistical analyses, assess risk, and estimate financial outcomes. An actuary helped compute your last car insurance bill.

Statistical analyses are used in a variety of settings. The National Agricultural Statistics Service publishes statistics on food production and supply, prices, farm labor, and even the price of land. Pollsters use statistical methods to predict a candidate's chances of winning an election. Using complex statistical analyses, companies make decisions about new products.

Both traditional statistical techniques and sophisticated new methods are used every day in making decisions that affect our lives directly. Pharmaceutical companies use a battery of standard statistical tests to determine a new drug's efficacy and possible side effects. Data mining, a combination of computer science and statistics, is a new technique used for constructing theoretical models and detecting patterns. Many companies apply this technique to understand their customers better and to respond quickly to their needs. Predictive microbiology is used to ensure that our food is not contaminated and is safe to consume. Given certain food properties and environmental parameters, a mathematical model can be used to predict a specific food's safety and shelf life.

Statistics is the science of collecting and interpreting data, as well as drawing logical conclusions from available information to solve real-world problems. This text presents several numerical and graphical procedures for organizing and summarizing data. However, the constant theme throughout the course is statistical inference using a four-step approach: claim, experiment, likelihood, and conclusion.

Numbers in the News

Here are some examples of statistics in the news.

1. **Statistical inference:** A large review was conducted by researchers from Reading University, United Kingdom; University of Copenhagen, Denmark; and Wageningen University and Research Center, the Netherlands, of 29 observational studies concerning a possible link between dairy consumption and cardiovascular disease.[2] Diary products often contain a lot of saturated fat, which has been linked to an

increased risk of heart attacks and strokes. However, this study involved almost 1 million people, and researchers concluded there was no increased risk of these health factors related to dairy consumption.

2. **Summary statistics:** In November 2017, the European Surveillance of Antimicrobial Consumption Network reported national antibiotic consumption data by country and hospital sector.[3] This document summarized the antibiotic consumption in defined daily doses (DDD) per 1000 inhabitants per day. The purpose of collecting and analyzing these data is to track and understand antibiotic resistance in Europe.

3. **Probability and odds:** According to the *Morbidity and Mortality Weekly Report* published by the Centers for Disease Control and Prevention (CDC), approximately 1 in 10 adults eats enough fruits or vegetables.[4] The CDC recommends that adults eat at least $1\frac{1}{2}$ to 2 cups of fruit and 2 to 3 cups of vegetables each day as part of a healthy eating pattern. This report indicates that very few Americans eat these recommended amounts. As a result, many Americans do not get the necessary vitamins and minerals. The CDC report suggested that fruit and vegetable consumption was lowest among men, young adults, and those living in poverty.

4. **Likelihood and inference:** Recent research suggests a link between dog ownership and heart disease.[5] Researchers in Sweden showed that single dog owners had a smaller risk of a heart attack compared to single non-owners. The probability of this happening by pure chance was so small that the researchers concluded that owning a dog significantly affects the risk of a heart attack.

5. **Relative frequency and probability:** In 2017, hackers were able to access sensitive data at Equifax, including names, Social Security numbers, birth dates, addresses, and even driver's license numbers.[6] Equifax Canada reported that approximately 100,000 Canadian consumers may have also had personal information stolen. If all 35 million Canadians are equally likely to have had their information stolen, then the probability that a randomly selected individual lost personal data to hackers is $0.0028448 = 100{,}000/35{,}151{,}728$. The relative frequency of occurrence is a good estimate of probability and is often used to develop statistical models and make predictions.

There has been an explosion of numerical information in stories like those just described, in business, in consumer reports, and even in casual conversation. Interpretation of graphs and evaluation of statistical arguments are no longer reserved for academics and researchers. It is essential for all of us to be able to understand arguments based on acquired data. This numerical, or quantitative, literacy is a vital life long tool.

No matter how you are employed or where you live, you will have to make decisions based on available information or data. Here are some questions you may have to consider.

1. Do you have enough information (data) to make a confident decision? How were the data obtained? If more information is necessary, how will these data be gathered?
2. How are the data summarized? Are the graphical and/or numerical techniques used appropriate? Does the summary represent the data accurately?
3. What is the appropriate statistical technique for analyzing the data? Are the conclusions reasonable and reliable?

1.2 Populations, Samples, Probability, and Statistics

Applications of Statistics

There are two very general applications of statistics: descriptive statistics and inferential statistics. Descriptive statistics involves summarizing and organizing the given information graphically and/or numerically. The focus of this text is statistical inference. The procedures of inferential statistics allow us to use the given data to draw conclusions and assess risk.

Here's a dictionary definition for inference: a deduction or logical conclusion.

Definition
Descriptive statistics: Graphical and numerical methods used to describe, organize, and summarize data.

Inferential statistics: Techniques and methods used to analyze a small, specific set of data to draw a conclusion about a large, more general collection of data.

EXAMPLE 1.1 Mishandled Baggage

The U.S. Department of Transportation publishes information concerning automobiles, public transportation, railroads, and waterways. Much of this information can be summarized or organized—in tables or charts, with a variety of graphs, and numerically—to describe typical values and variability. These summary descriptive procedures might be used to indicate preference for a certain airline or to promote safety records. **Figure 1.1** shows a bar graph of the total mishandled baggage reports for certain airlines in September 2018.

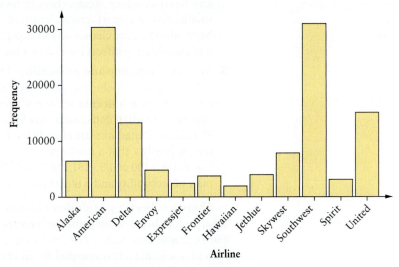

Figure 1.1 Total baggage reports, September 2018. (Source: Air Travel Consumer Report, U.S. Department of Transportation)

EXAMPLE 1.2 Summary Statistics

Motor Intelligence maintains a large database of information associated with automobile sales. For example, subscribers have access to reports involving marketing programs and lease price points, as well as statistical reports summarizing production and retail sales. This information can be neatly presented using tables, bar charts, pie charts, histograms, or stem-and-leaf plots, or it can be summarized numerically using the mean, median, quartiles, percentiles, variance, or standard deviation. These simple descriptive statistics reveal characteristics of the entire data set. Part of a table is shown in Table 1.1 on the facing page.[7]

EXAMPLE 1.3 Smartphone Ownership

According to a recent study, approximately 75% of all Americans own a smartphone. There has been continued growth in ownership of smartphones among Americans

Table 1.1 A portion of a table summarizing automobile sales.

	Sales			YTD Sales		
	December 2017	December 2016	% Chg	2017	2016	% Chg
General Motors Corp.	308,112	318,859	−3.4	2,999,605	3,042,421	−1.4
Total cars	58,151	89,348	−34.9	709,350	890,137	−20.3
Domestic car	57,683	88,501	−34.8	699,007	878,950	−20.5
Import car	468	847	−44.7	10,343	11,187	−7.5
Total light trucks	249,961	229,511	8.9	2,290,255	2,152,284	6.4
Domestic truck	243,905	225,580	8.1	2,249,215	2,138,091	5.2
Toyota Motor Sales USA Inc.	222,985	243,229	−8.3	2,434,518	2,449,630	−0.6
Total cars	88,078	100,263	−12.2	1,015,082	1,146,968	−11.5
Domestic car	62,425	74,326	−16.0	765,495	863,061	−11.3
Import car	25,653	25,937	−1.1	249,587	283,907	−12.1
Total light trucks	134,907	142,966	−5.6	1,419,436	1,302,662	9.0
Domestic truck	87,627	99,301	−11.8	945,271	928,966	1.8
Import truck	47,280	43,665	8.3	474,165	373,696	26.9
Nissan North America Inc.	138,226	152,743	−9.5	1,593,464	1,564,423	1.9
Total cars	54,719	67,006	−18.3	731,687	811,090	−9.8
Domestic car	47,276	57,910	−18.4	659,076	730,979	−9.8
Import car	7,443	9,096	−18.2	72,611	80,111	−9.4
Total light trucks	83,507	85,737	−2.6	861,777	753,333	14.4
Domestic truck	80,600	81,885	−1.6	823,949	727,840	13.2
Import truck	2,907	3,852	−24.5	37,828	25,493	48.4

age 50 and older, and nearly all younger adults own a smartphone. To check this ownership claim, several Americans are selected at random and each is asked if he or she owns a smartphone. The proportion of observed ownership is used to determine whether there is any evidence to suggest that the claim (of 75% ownership) is false. The collected data are used in inferential statistics to draw a conclusion regarding a claim.

EXAMPLE 1.4 Magnetic Fuel Savers

Many companies sell magnetic fuel savers for stoves, which are designed to condition liquefied propane gas (LPG) prior to combustion to increase power output, reduce emissions, and save gas. An independent agency tests these devices by recording the amount of gas necessary to boil a specific volume of water. Each test boil is classified by stove brand, shape and thickness of the pot, and burner size. The data collected are used to determine whether there is a difference in efficiency. If there is a difference in the amount of LPG used, further inferential statistical techniques will be used to isolate this difference. Such a difference may be due to the stove brand, type of pot, or burner size.

TRY IT NOW Go to Exercise 1.7

Population Versus Sample

Whether we are summarizing data or making an inference, every statistics problem involves a *population* and a *sample*. Consider the definitions that follow.

Definition

A **population** is the entire collection of individuals or objects to be considered or studied.

A **sample** is a subset of the entire population, a small selection of individuals or objects taken from the entire collection.

A **variable** is a characteristic of an individual or object in a population of interest.

A CLOSER LOOK

1. A population consists of all objects of a particular type. There are usually infinitely many objects in a population, or at least so many that we cannot look at all of them.
2. A sample is simply a handful from the population, usually a small part of a population.
3. A variable may be a *qualitative* (categorical) or a *quantitative* (numerical) attribute of each individual in a population. It is a characteristic of each item, or object, in the population.

The following examples illustrate the relationships among populations, samples, and variables.

EXAMPLE 1.5 High Anxiety

Various research studies suggest that whole-grain foods may be a natural help for those people who suffer from anxiety, or maybe even statistics anxiety![8] Whole grains generally contain high levels of magnesium, and magnesium deficiency can lead to anxiety. A new study is concerned with the magnesium level in a slice of whole-grain bread. One hundred slices of whole-grain bread are selected at random from various markets, and the magnesium level in each is carefully measured and recorded. Describe the population, sample, and variable in this problem.

Solution

STEP 1 The population consists of all slices of whole-grain bread in the entire world. Although this population is not infinite, we certainly could not examine every single slice.
STEP 2 The sample is the 100 slices selected at random. This is a subset of, or selection from, the population.
STEP 3 The variable in this problem is the magnesium level. This characteristic will be carefully measured for each slice, and the data will be summarized or used to draw a conclusion.

EXAMPLE 1.6 Asleep at the Wheel

The CDC released results from a study indicating that approximately 4% of all U.S. adults said they had fallen asleep at least once while driving in the last month. Nodding off while driving seems to be more common in men, and some officials claim the percentage of all U.S. adults who have fallen asleep while driving is greater than 4%. To check this claim, 10 adult drivers were selected from across the country. Each person was asked if he or she had fallen asleep while driving, and the results were recorded. Describe the population, sample, and variable in this problem.

Solution

STEP 1 The population consists of all adult drivers in the United States. This population is not infinite, but it is so large that it would be impossible to contact every adult driver.

STEP 2 The sample consists of the 10 adult drivers selected.

STEP 3 The variable in the problem is whether the driver has fallen asleep at the wheel, a yes/no response.

TRY IT NOW Go to Exercises 1.9 and 1.11

CHAPTER APP **Lake Winnebago Fish**

Describe the population, sample, and variable in this problem.

A CLOSER LOOK

Example 1.6 raises some important issues regarding the sample of 10 adult drivers.

1. How large a sample is necessary for us to be confident in our conclusion? Ten adult drivers may not seem like enough. But how many do we need? 100? 1000? We will consider the problem of sample size in Chapter 7 and beyond.

2. This problem does not say how the sample was obtained. Perhaps the first 10 drivers who recently renewed their licenses were selected. Or maybe only those from one state were included. To draw a valid conclusion, we need to be certain the sample is representative of the entire population. The formal definition of a representative sample is presented in Section 1.3.

Probability Versus Statistics

Statistical inference is based on, and follows from, basic probability concepts. *Probability* and inferential *statistics* are both related to a population and a sample, but from different perspectives. For the rest of this chapter, *statistics* really means *inferential statistics*.

Definition

To solve a **probability** problem, certain characteristics of a population are assumed to be known. We then answer questions concerning a sample from that population.

In a **statistics** problem, we assume very little about a population. We use the information about a sample to answer questions concerning the population.

Figure 1.2 illustrates this definition. Picture an entire population of individuals or objects. Suppose we know everything about the population and we select a sample from this population. A probability problem would involve answering a question concerning the sample. In a typical statistics problem, we assume very little about the population. We select a sample, analyze it completely, and use this information to draw a conclusion about the population.

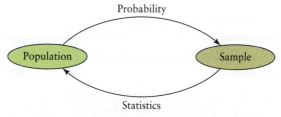

Figure 1.2 Relationships amoung probability, statistics, population, and sample.

In Figure 1.2, it might seem as if we can start our study anywhere in this circular diagram. However, we need to understand probability before we can learn statistics. A solid background in probability is necessary before we can actually do statistical inference.

EXAMPLE 1.7 Passport Please

According to *Paste Magazine*, only people from Finland travel abroad more than Americans do.[9] However, according to the State Department, only 36% of Americans hold a valid passport. There are many reasons more Americans do not travel abroad. For example, America itself is a diverse nation, with beaches, mountains, deserts, and all kinds of ethnic foods. And leaving the country can be very expensive and time-consuming. Consider the entire population of U.S. citizens and U.S. naturals (who can apply for a passport) and a sample of 20 from this population.

Population: All Americans eligible for a U.S. passport
Sample: The 20 Americans selected from this population.

Here is a probability question. The State Department reported that 36% of all Americans have a passport. What is the probability that 10 or more (of the 20) selected Americans have a passport? We know something about the population and try to answer a question about the sample.

Here is a statistics question. Suppose we interview the 20 Americans in the sample and find that 8 of the 20 hold a passport. What can we conclude about the percentage of *all* Americans who hold a passport? We know something about the sample and try to answer a question about the (whole or general) population.

Charles Taylor/Shutterstock

EXAMPLE 1.8 Apple Pay

Apple Pay is a method for making secure purchases and for sending and receiving money in stores, in apps, and on the web. According to a recent article, Apple Pay is now used by 35% of U.S. retail stores.[10] Consider the population consisting of all U.S. retail stores and a random sample of 100 from this population.

A probability question: Suppose 35% of all U.S. retail stores use Apple Pay. What is the probability that at most 30 (of the 100) retail stores in the sample use Apple Pay?

A statistics question: Of the 100 U.S. retail stores selected, 45 use Apple Pay. What does this suggest about the proportion of all U.S. retail stores that use Apple Pay?

TRY IT NOW Go to Exercises 1.19 and 1.21

thitiwat_t1980/Shutterstock

CHAPTER APP Lake Winnebago Fish

Write a probability question and a statistics question in the context of this problem.

Section 1.2 Exercises

Concept Check

1.1 True or False Inferential statistics are used to draw a conclusion about a population.

1.2 True or False Descriptive statistics are used to indicate how the data were collected.

1.3 True or False To answer a probability question, certain characteristics of a population are assumed to be known.

1.4 Fill in the Blank
(a) The entire collection of objects being studied is called the _____.

(b) A small subset from the set of all 2018 minivans is called a _____.

(c) Consider the amount of sugar in breakfast cereals. This characteristic of breakfast cereal (objects) is called a _____.

Practice

1.5 Probability/Statistics In each of the following problems, write a probability question and a statistics question associated with the given information.
(a) It has been reported that 81% of U.S. Americans have a social media profile.
(b) In general, 36% of all people use the same password for everything.
(c) In Norway, 52% of all passenger cars are plug-in electric vehicles.
(d) Fifty-four percent of all injuries due to snow shoveling fall into the category of pulled muscles.

Applications

1.6 Descriptive or Inferential Statistics Determine whether each of the following is a descriptive statistics problem or an inferential statistics problem.
(a) The Georgia Department of Transportation maintains records concerning all trucks stopped for inspection. A report of these inspections lists the proportion of all trucks stopped, by cargo carried.
(b) Eric Knudsen, a researcher at Stanford University Medical Center, obtains a random sample of wild owls and measures how far each can turn its neck. The data are used to conclude that an owl can turn its neck more than 120° from the forward position.
(c) A Navy research facility runs several tests to check the structural integrity of a new submarine. A laboratory report states the vessel can withstand pressure at depths of at most 800 ft.
(d) A safety inspector in Atlanta selects a sample of apartment buildings and checks the fire ladders on each. The proportion of broken ladders in the sample is used to estimate the proportion of apartment buildings with broken fire ladders in the entire city.
(e) We use electricity to power many different items in our homes. The U.S. Energy Information Administration suggested that the residential sector in the United States used approximately 1410 billion kilowatt-hours (kWh) of electricity in 2016.[11] The pie chart in **Figure 1.3** shows the percentage of consumption by end use.

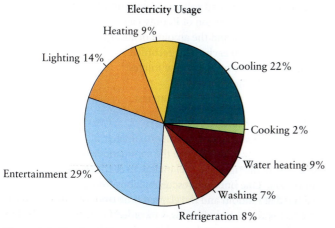

Figure 1.3 Pie chart illustrating how electricity is used in homes.

(f) A report from the CDC lists the percentage of people seeing their health care provider for influenza-like illness by state.

1.7 Descriptive or Inferential Statistics Determine whether each of the following is a descriptive statistics problem or an inferential statistics problem.
(a) The bar chart in **Figure 1.4** shows the number of branches for certain banks in Pennsylvania as of June 30, 2017.[12]

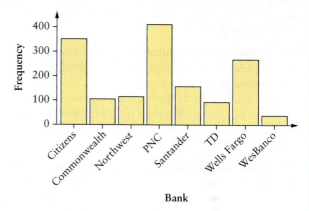

Figure 1.4 Total offices in Pennsylvania.

(b) The resting heart rate was measured for adult males from two separate groups: those who exercise at least three days per week, and those who do not exercise regularly. The resulting data are used to suggest that regular exercise decreases the resting heart rate in adult males.
(c) Interior Exterior Remodeling, in Northridge, California, maintains a comprehensive list of each home constructed by type, size, exterior color, and other descriptors.
(d) Researchers at the Center for Food Safety selected a sample of frozen toaster apple strudel sold in grocery stores. Measurements indicated the producer was baking each piece of strudel with less apple than advertised on the box.
(e) A report issued by the athletic department at Boston College listed each person who worked with a strength and conditioning coach, the number of minutes, and the type of exercises.
(f) The manager of the Early Bird Diner in South Carolina surveyed patrons and summarized breakfast orders by type of food.

1.8 Medicine and Clinical Studies Managers at Physicians Regional Healthcare System in Florida are interested in the length of stay (in days) of patients admitted for open-heart surgery. Hospital managers have decided to limit their investigation to open-heart patients who underwent surgery within the last year. Thirty open-heart surgery patients admitted to the hospital within the last year are selected. What is the population of interest, the sample, and the variable in this problem?

1.9 Marketing and Consumer Behavior T-shirt labels irritate many people's skin, so Calico Graphics of Wolfeboro, New Hampshire, would like to produce shirts without a label. The company wants to know whether there is an advantage to producing this type of T-shirt. Fifty people are surveyed about whether they cut the tags off their T-shirts. What is the population of interest, the sample, and the variable in this problem?

1.10 Psychology and Human Behavior Managers at Citigroup, Inc., in New York are concerned about the number of employees who eat and/or drink at their desks while working. Some managers believe this is an unnecessary distraction, and spills can cause computer failures and ruin documents. Thirty-five employees are selected, and each is questioned about eating/drinking while working. Describe the population and the sample in this problem.

1.11 Public Health and Nutrition Senator Martha McSally of Arizona is unsure of her vote on an emotional and controversial issue. Before voting, she would like to know what her constituents think about the proposed bill. An aide for the senator selects 500 people from Arizona and asks them whether they believe the bill should become law. Describe the population and the sample in this problem.

1.12 Economics and Finance In 2017, Hurricane Irma was a catastrophic Atlantic Ocean storm that caused an estimated $300 billion in damage in the Caribbean, Puerto Rico, and Florida. Individuals affected by this storm filed insurance claims for damage to their home and/or car. Nationwide Insurance has customers in southwest Florida, so the company is interested in the typical amount of a claim as a result of this storm. Eighty affected families are selected and their total claims are recorded. Describe the population and the sample in this problem.

1.13 Probability and Statistics In each of the following problems, identify the population and the sample, and determine whether the question involves probability or statistics.
  (a) Seventy-five percent of all people who buy a dining room table purchase matching chairs. Five people who purchased a dining room table within the last month are selected at random. What is the probability that all five purchased matching chairs?
  (b) Twenty-five people entering a Walgreens store in Oklahoma are selected at random. Of these 25 people, 7 bought some kind of dental product. Estimate the true proportion of people shopping at this Walgreens who buy dental products.
  (c) Historical records indicate 1 out of every 500 people using a particularly steep water slide suffer some kind of injury. Fifty people using the slide are selected at random. How many do you expect to be injured?
  (d) A building inspector in Henderson, Nevada, is checking public buildings with doors that open automatically. One hundred doors are randomly selected. Careful inspection reveals that 12 doors are broken. Use this information to estimate the percentage of automatic doors in Henderson that are broken.
  (e) One thousand people entering Los Angeles International Airport (LAX) are selected at random to complete a short survey regarding travel. The survey results show that 637 carry a frequent-flier card. Is there evidence to suggest the true proportion of travelers entering LAX who carry frequent flier cards is greater than 0.60?
  (f) A recent Redfin survey indicated that 60% of all people who sold a home in the past year saved money on commissions paid to the real estate agent. Agents gave sellers a refund, gave them a rebate, or contributed to the closing costs. Thirty-five people who sold a home in the past year are selected at random. Is it likely that more than 20 of these people received a discount on their commission?
  (g) Representatives from the Occupational Safety and Health Administration inspected several for-profit and Medicare-funded nursing homes for any violations. The resulting data will be used to determine whether there is any evidence to suggest that the quality of treatment differs in the two types of nursing homes.

1.14 Psychology and Human Behavior During each summer, many families spend part of their vacation time at a beach along the East or West Coast. Due to the popularity of movies like *Jaws* and *The Shallows*, and publicity about recent shark attacks on surfers, swimmers, snorkelers, and spearfishermen, Americans have become increasingly concerned about water activities. Research suggests that 58% of all shark attacks involve surfers.[13] One thousand records of shark attacks are selected, and each is categorized by victim group.
  (a) What is the population of interest?
  (b) What is the sample?
  (c) Describe the variable of interest.

1.15 Public Health and Nutrition Zinc is an important mineral that is linked to many enzymatic reactions in the body and is needed to maintain normal health.[14] This trace element is found in common pumpkin seeds. Fifty pumpkin seeds were selected and the amount of zinc was measured in each. Describe the population, sample, and variable in this problem.

1.16 Medicine and Clinical Studies Some evidence suggests that people with chronic hepatitis C have a liver enzyme level that fluctuates between normal and abnormal. Forty patients diagnosed with hepatitis C are selected and their liver enzyme levels are recorded each day for one month. Describe the population, sample, and variable in this problem.

1.17 Manufacturing and Product Development Paper towel manufacturers constantly advertise their products' strength, amount of stretch, and softness. A consumer group is interested in testing the absorption of Bounty paper towels. Thirty-five rolls are selected, and the amount of absorption for a single paper towel from each roll is recorded.
  (a) What is the population of interest?
  (b) What is the sample?
  (c) Describe the variable of interest.

1.18 Fuel Consumption and Cars U.S. Coast Guard Search and Rescue teams undertake approximately 20,000 rescue missions every year. Suppose the commander is interested in decreasing the time and fuel it takes to reach an individual in trouble. One hundred rescue missions from 2018 were selected at random and the elapsed time from the initial call to contact with the individual was recorded for each. Describe the population, sample, and variable in this problem.

Extended Applications

1.19 Manufacturing and Product Development While much of the cheddar cheese consumed around the world is processed, some is still produced in the traditional manner: made in small batches, wrapped in cloth to breathe, and allowed to age. Most traditional cheddar is aged one to two years; like fine wines, older cheddars assume their own character and flavor. Suppose 75% of all cheddars are aged less than two years, and a sample of 20 cheddar cheeses from around the world is obtained.
(a) Describe the population and the sample in this problem.
(b) Write a probability question and a statistics question involving this population and sample.

1.20 Marketing and Consumer Behavior Even though magazines, newspapers, and books have become more readily available in digital format, new research suggests that old-fashioned printed books are making a comeback. The Pew Research Center reported that 65% of Americans read a printed book in the past year.[15] Suppose a sample of 500 adults in the United States is obtained.
(a) Describe the population and the sample in this problem.
(b) Write a probability question and a statistics question involving this population and sample.

1.21 Technology and the Internet In 2017, concern arose that the Kaspersky antivirus software could be exploited by the Russian government. A survey of U.S. government agencies indicated that 15% had some trace of Kaspersky's software on their systems.[16] One hundred U.S. government agencies are selected, and their computer systems are carefully examined to determine if there is any trace of Kaspersky software.
(a) What is the population of interest?
(b) What is the sample?
(c) Describe the variable of interest.
(d) Write a probability question and a statistics question involving this population and sample.

1.22 Public Health and Nutrition The number of falls in nursing homes in Canada appeared to drop dramatically recently. However, health authorities have changed the way they count critical incidents. As a result, they no longer count most of the serious falls that result in injury.[17] Eighty nursing homes in Canada were selected and the number of all patient falls during the last year were recorded for each. Describe the population, sample, and variable in this problem.

1.3 Experiments and Random Samples

Experiments and Random Samples

Statisticians analyze data from two types of experiments: observational studies and experimental studies. The definitions for both types of studies are given below.

Definition
In an **observational study**, we observe the response for a specific variable for each individual or object.

In an **experimental study**, we investigate the effects of certain conditions on individuals or objects in the sample.

The data collected in an observational study may be summarized in a variety of ways or used to draw a conclusion about the entire population. The following is an example of an observational study.

EXAMPLE 1.9 Time for Breakfast?

A guidance counselor at Kerr Elementary School in Allen, Texas, is interested in the amount of time each student spends in the morning eating breakfast. Some students wake up an hour before the bus arrives, have a leisurely breakfast, read the comics, and complete last-minute homework. Others roll out of bed and onto the school bus. The guidance counselor decides to measure the amount of time from wake-up to school bus arrival. A random sample of students is selected, and each is asked for the amount of school-day preparation time. The data are summarized graphically and numerically in this observational study. ∎

In almost all statistical applications, it is important for the data to be *representative* of the relevant population. A representative sample has characteristics similar to those of the entire population and therefore can be used to draw a conclusion about the (general) population. The following definition describes a method for obtaining data in an *observational study* to ensure the resulting sample is representative of the corresponding population.

> **Definition**
> A **(simple) random sample** (SRS) of size n is a sample selected in such a way that every possible sample of size n has the same chance of being selected.

A CLOSER LOOK

1. In practice, a random sample may be very difficult to achieve. Statisticians employ various techniques, including random number tables and random number generators, to select a random sample.

2. If a sample is not random, then it is *biased*. Many different kinds of bias are possible, and many different factors may contribute to a biased sample.

3. *Nonresponse bias* is very common when data are collected using surveys. The majority of people who receive a survey in the mail simply discard it. The original collection of people receiving the survey may be random, but the final sample of completed surveys is not. Because the sample is biased, it is impossible to draw a valid conclusion.

4. *Self-selection bias* occurs when the individuals (or objects) choose to be included in the sample, as opposed to being randomly selected. For example, a television news program may ask viewers to respond to a yes/no question by dialing one of two phone numbers to cast their vote. Viewers *choose* to participate, and usually those with strong opinions (either way) vote. Many more did not have the opportunity to respond, so every single sample is *not* equally likely. Certainly this sample is biased, and hence no valid conclusion is possible.

5. If the population is infinite, then the number of simple random samples is also infinite. For finite populations, the formula for the number of possible random samples is presented in Chapter 7.

6. A simple random sample is vital to draw a reliable, confident conclusion. Before doing any analysis, you should always ask how the data were obtained. If any evidence of a pattern in selection is found, if the observations are associated or linked in some way, or if the observations share some connection, then the sample is not random. There is simply no way to transform bad data into good statistics.

EXAMPLE 1.10 People on the Go

Wawa convenience stores are located in Pennsylvania, New Jersey, Delaware, Maryland, Virginia, and Florida. They offer coffee, made-to-order foods, and gasoline. Howard Stoeckel, a member of the Wawa board of directors, is conducting a study to determine if changes should be made to the product offerings. He plans to choose 35 Wawa stores from the current 750 and will ask each store manager to complete a brief questionnaire concerning customer buying habits.

Stoeckel would like to obtain a simple random sample of size 35, a representative sample of the entire population of Wawas. One basic selection procedure would be to write each store location on a piece of paper, place all of them in a hat, thoroughly mix the papers, and then select 35 stores at random.

Although this procedure is straightforward and uncomplicated, it can be very tedious if the number of individuals or objects in the population is large. In addition, it is hard to guarantee a thorough mixing of the slips of paper.

More practical methods for selecting a simple random sample include the use of a random number table or a random number generator (available in most statistical software packages). In this example, we might assign each store a number, ranging from 1 to 750, and use a random number generator to produce a list of 35 numbers in this range. The stores associated with these 35 numbers would make up the random sample. ∎

TRY IT NOW Go to Exercises 1.31 and 1.33

Researchers often investigate the effects of certain conditions on individuals or objects. The data obtained are from an *experimental* study. Individuals are randomly assigned to specific groups, and certain factors are systematically controlled, or imposed, to investigate and isolate specific effects. The following example is of an experimental study.

EXAMPLE 1.11 Gardening Advice

The manager of Gardener's Supply Company claims that a new organic fertilizer, in comparison with the leading brand, increases the yield and size of tomatoes. To test this claim, tomato plants are randomly assigned to one of two groups. One group is grown using the leading fertilizer; the other is cultivated using the new product. At harvest time, the size and weight of each tomato are recorded, along with the total yield per plant. The data collected during this experiment are used to compare the two fertilizers. ∎

TRY IT NOW Go to Exercises 1.39 and 1.41

In an experimental study, researchers must be careful to ensure that significant effects are indeed due to an imposed treatment, or controlled factor. Confounding occurs when several factors together contribute to an effect, but no single cause can be isolated. Suppose the tomato plants in one of the groups in Example 1.11 are watered more and/or exposed to more sunlight and warmer temperatures. If the tomato plants that received the new fertilizer were subject to these different (favorable) growing conditions, a difference in yield could not be attributed to the new product.

The focus of this text is statistical inference, most of which is based on determining the likelihood of an observed experimental outcome. This strategy will be used informally in the early chapters of this book. Formal procedures will be presented beginning in Chapter 9. For now, we will follow the four-step process presented in the next subsection.

CHAPTER APP Lake Winnebago Fish

Is this an observational study or an experimental study? Describe a method for obtaining a simple random sample of fish from this lake.

Statistical Inference Procedure

The process of checking a claim can be divided into four parts.

Claim: This is a statement of what we assume to be true.

Experiment: To check the claim, we conduct a relevant experiment.

Likelihood: In this step, we consider the likelihood of occurrence of the observed experimental outcome assuming the claim is true. We will use many techniques to determine whether the experimental outcome is a reasonable observation (subject to reasonable variability) or a rare occurrence.

Conclusion: There are only two possible conclusions. (1) If the outcome is reasonable, then we cannot doubt the claim. We usually write, "There is no evidence to suggest the claim is false." (2) If the outcome is rare, we disregard the lucky alternative and question the claim. A rare outcome is a contradiction. It shouldn't happen (often) if the claim is true. In this case, we write, "There is evidence to suggest the claim is false."

EXAMPLE 1.12 Speaker Clarity

Bose Corporation ships a box containing 1000 SoundLink Micro speakers and claims that 999 are in perfect condition and only 1 is defective (due to variability in manufacturing). Upon receipt of the shipment, a quality control inspector reaches into the box, mixes the speakers around a bit, selects one at random, tests the sound, and finds that it's defective!

Claim: There were 999 good speakers and 1 defective speaker in the box.

Experiment: The quality control inspector selected one speaker from the box, tested it, and found it to be defective.

Likelihood: One of two things has happened.
1. The quality control inspector could be incredibly lucky. Intuitively, the chance of selecting the one defective speaker from among the 1000 total speakers is very small. It is possible to select the one defective speaker, but it is very unlikely.
2. The claim (999 perfect speakers, 1 defective) is false. Because the chance of selecting the single defective speaker is so small, it is more likely the manufacturer (Bose Corporation) was mistaken about the number of defective speakers in the shipment. (Perhaps there are really 999 defective speakers and only one good speaker in the box.)

We have found evidence that the claim is false by showing that the observed experimental outcome is unreasonable, an outcome so rare that it should almost never happen if the claim is really true.

Conclusion: Typically, statistical inference discounts the lucky alternative. Selecting the single defective speaker is an extremely rare occurrence. Therefore, there is evidence to suggest the manufacturer's claim is false because this outcome is very rare.

We will use this four-step process to check a claim in many different contexts. The method for determining likelihood is the key to this valuable tool for logical reasoning.

Section 1.3 Exercises

Concept Check

1.23 True or False In an observational study, we record the response for a specific variable for each individual or object.

1.24 True or False In an experimental study, we investigate the effects of certain conditions on at least three different groups.

1.25 True or False The number of simple random samples is always infinite.

1.26 True or False A simple random sample is representative of the entire population of interest.

1.27 True or False A simple random sample is a systematic pattern for selecting objects from a population.

1.28 Fill in the Blank
(a) If a sample is not random, then it is _____.
(b) It is very common to experience _____ when data are collected using surveys.
(c) _____ occurs when individuals ask to be included in a survey.

1.29 Statistical Inference Name the four parts of every statistical inference problem.

1.30 Liar, Liar Suppose an experimental outcome is very rare. What two things could have happened?

Applications

1.31 Fuel Consumption and Cars The administration at the University of Nebraska in Lincoln is interested in student reaction to a planned parking garage on campus. A dormitory near the proposed site is selected and several Student Senate members volunteer to solicit responses. One Thursday evening, the volunteers each take a specific dorm wing, knock on doors, and record student answers to several prepared questions.
(a) Is this an observational study or an experimental study?
(b) Describe the sample in this problem.
(c) Is this a random sample? Justify your answer.

1.32 Demographics and Population Statistics State Farm Insurance Company would like to estimate the proportion of volunteer firefighters across the country who are full-time teachers. The 25 largest volunteer fire companies in the United States are identified. Each is contacted and asked to complete a short survey regarding the number of volunteers and the occupation of each volunteer.
(a) Is this an observational study or an experimental study?
(b) Describe the sample in this problem.
(c) Is this a random sample? Justify your answer.

1.33 Manufacturing and Product Development The Hershey Company of Derry Township, Pennsylvania, has been accused of systematically underfilling 4-oz packages of Reese's

Pieces. An inspection team enters the manufacturing facility one afternoon and selects packages ready for shipment from various locations within the plant. The contents of each selected package are carefully measured.
   (a) Describe the population and the sample in this problem.
   (b) Is this a random sample? Justify your answer.

1.34 Manufacturing and Product Development SharkNinja has come under suspicion recently for shipping a large number of defective Ninja Coffee Bars. Several customers have complained about a faulty frother. The State Attorney General's office in Maryland would like to estimate the proportion of defective products shipped by the company. Describe a method for obtaining a simple random sample of shipped Ninja Coffee Bars.

1.35 Fuel Consumption and Cars Executives at Lyft are interested in the number of miles traveled by each driver during a typical 8-hour shift. Twenty-five drivers are selected from Atlanta, and the number of miles traveled by each driver is recorded.
   (a) Is this an observational study or an experimental study?
   (b) Describe the population and the sample in this problem.
   (c) Is this a random sample? Justify your answer.

1.36 Manufacturing and Product Development Gillette claims a new disposable razor provides a closer shave than any other brand currently on the market. One hundred men who are observed buying a disposable razor are selected and asked to participate in a shaving study.
   (a) Describe the population and the sample in this problem.
   (b) Is this a random sample? Justify your answer.

1.37 Technology and the Internet Smart TVs allow users to stream content services, browse the Internet, and check social media, all right on the TV set without additional hardware. However, the TVs can be tedious and confusing to set up. The owners' manual for a Samsung Smart TV claims it can be set up in less than 30 minutes. Describe a method for obtaining a simple random sample of customers who set up this Smart TV.

1.38 Sports and Leisure A National Football League (NFL) coach is permitted to initiate two challenges to referee calls per game (outside of the final 2 minutes in each half). If both challenges are successful, then the coach is given a third. During a challenge, the referee reviews the play in question on a replay monitor on the field, and the call is either confirmed or the challenge is upheld. Suppose the NFL reports the time required to resolve a coach's challenge is less than 5 minutes. A sports statistician would like to check this claim. Describe a method for obtaining a simple random sample of challenges during NFL games.

1.39 Public Health and Nutrition There is still a lot of concern about cell phone safety. The National Toxicology Program studied animals exposed to radiation levels equal to and higher than those currently allowed for cell phones. The researchers divided the rodents into two groups. One group was exposed to low levels of radiation, and the other was exposed to high levels. After two years, each animal was carefully examined for tumors. The data will be used to determine whether cell phone radiation is harmful.
   (a) Is this an observational study or an experimental study?
   (b) What is the variable of interest?

1.40 Travel and Transportation Transport Canada would like to test the fire extinguishers in certain Gulf Stream Coach RVs. Some evidence indicates that these devices may become clogged or require excessive force to operate. Each selected fire extinguisher will be carefully tested, and the data will then be used to determine whether a product recall should be issued.
   (a) Describe a method for obtaining a simple random sample of fire extinguishers in these vehicles.
   (b) Is this an observational study or an experimental study?

1.41 Biology and Environmental Science The Family Flowers Shop in Raleigh, North Carolina, claims to have developed a special spray for roses that causes the blossom to last longer than it does in an untreated flower. Fifty long-stemmed roses are obtained and randomly assigned to one of two groups: treated versus untreated. The treated roses are sprayed, and the lifetime of each blossom is carefully recorded.
   (a) Is this an observational study or an experimental study?
   (b) What is the variable of interest?
   (c) Describe a technique to randomly assign each rose to a group.

1.42 Fuel Consumption and Cars Electric and plug-in electric cars are designed to save gasoline and help the environment. The Tesla Model S and Model 3 are designed to have extra safety features, a convenient interior design, and up to an 8-year warranty. Although owning an electric car certainly has some benefits, many people complain about the slow acceleration, repair expense, and overall comfort. Thirty-five passengers are randomly selected. Each is blindfolded and taken for a ride in a traditional combustion-engine automobile and in a comparably sized electric car (over the same route). The passenger is then asked to select the car with the most comfortable ride.
   (a) Is this an observational study or an experimental study?
   (b) What is the variable of interest?
   (c) Describe possible sources of bias in these results.

1.43 Manufacturing and Product Development The ceramic tile used to construct the floors in a mall must be sturdy, easy to clean, and long-lasting. Before installing a specific tile, a construction firm orders a box of 25 tiles and uses a standard strength test on each. The results are used to determine whether the tiles will be used throughout the new mall.
   (a) Describe the population and the sample in this problem.
   (b) Is this a random sample? If so, justify your answer. If not, describe a technique for obtaining a random sample.

1.44 Manufacturing and Product Development Many comforters contain both white feathers and down as a means to provide a warm, soft cover. A bed-and-bath company would like to expand its line of products and sell comforters for queen- and king-size beds. Before manufacturing begins, a random sample of comforters is obtained from other companies and the proportions

of white feathers, down, and other components are measured and recorded. These data will be used to determine the exact mixture of feathers and down for the new line of comforters.
(a) Is this an observational study or an experimental study?
(b) What are the variables of interest?
(c) Describe a method for obtaining a random sample of comforters from current manufacturers.

1.45 Sports and Leisure BASF is using a new material to produce seat-back cushions for rail cars in the San Francisco Bay area. These new cushions help to reduce the overall weight of the seats and contribute to increased energy efficiency and better performance. Suppose a random sample of the new seat cushions is obtained and the weight is recorded for each.
(a) Is this an observational study or an experimental study?
(b) What is the variable of interest?
(c) Describe a method for determining whether the new cushions increase energy efficiency.

1.46 Manufacturing and Product Development A recent study considered the size of wine glasses in England from 1700 to 2017.[18] The research paper suggests that there had been a gradual increase in the size of these glasses until the 1990s, followed by a sharp, steady increase. Suppose a random sample of wine glasses from department stores and glassware manufacturers was obtained, and the size (in mL) was measured for each. The data will be used to determine whether one source markets larger glasses.
(a) Is this an observational study or an experimental study?
(b) What is the variable of interest?
(c) Describe a method for obtaining a random sample of wine glasses from both sources.

Chapter 1 Summary

Concept	Page	Notation / Formula / Description
Descriptive statistics	12	Graphical and numerical methods used to describe, organize, and summarize data.
Inferential statistics	12	Techniques and methods used to draw a conclusion or make an inference.
Population	14	The entire collection of individuals or objects to be considered or studied.
Sample	14	A subset of the entire population.
Variable	14	A characteristic of an individual or object in a population of interest.
Probability problem	15	Certain properties of a population are assumed or known. Questions involve a sample taken from this population.
Statistics problem	15	Information about a sample is used to answer questions concerning a population.
Observational study	19	We observe the response for a specific variable for each individual or object in the sample.
Experimental study	19	We investigate the effects of certain conditions on individuals or objects in the sample.
Simple random sample of size n	20	A sample selected in such a way that every possible sample of size n has the same chance of being selected.
Statistical inference procedure	21	Four-step process: claim, experiment, likelihood, and conclusion.

Chapter 1 Exercises

1.47 Descriptive or Inferential Statistics Determine whether each of the following is a descriptive statistics problem or an inferential statistics problem.
  (a) The Society of Government Economists conducted a salary and working conditions survey of top technology executives in the United States. A report issued by this group included a table that listed the number of bank executives in each state with salaries exceeding $1 million.
  (b) The Flowers Canada Growers obtained a sample of people who sent roses for Valentine's Day and recorded the color of the roses purchased. This information was used to construct a table listing the proportion of each color of rose purchased on Valentine's Day.
  (c) The Intergovernmental Panel of Climate Change collected data associated with global warming and predicted the extinction of up to 30% of plant and animal species in the world.
  (d) American Express conducted a survey of travelers at Los Angeles International Airport. The information was used to estimate the proportion of all travelers who make a purchase in an airport duty-free shop.
  (e) The International Olympic Committee presented a medal table for the Pyeongchang 2018 Winter Olympics. The table listed the number of gold, silver, and bronze medals won by each country.

1.48 Descriptive or Inferential Statistics Determine whether each of the following is a descriptive statistics problem or an inferential statistics problem.
  (a) A report by NASA listed each weather satellite orbiting the Earth and the number of years each has been in service.
  (b) The Agricultural Research Service obtained samples of natural cocoa from a variety of sources and measured the total antioxidant capacity in each sample. The resulting data were used to suggest that eating a moderate amount of chocolate may help prevent cancer, heart disease, and stroke.
  (c) The U.S. Patent Office issued a report listing every company that was granted a patent in 2018 and the number of patents awarded to each company.
  (d) A researcher at Emory University used brain scans to conclude that zen meditation may help treat disorders characterized by distracting thoughts.
  (e) Researchers from King's College in London compared the effects of e-cigarettes versus smoking. Scientists studied long-term users of both types of cigarettes and concluded that e-cigarettes are not as harmful as smoking.

1.49 Descriptive or Inferential Statistics Determine whether each of the following is a descriptive statistics problem or an inferential statistics problem.
  (a) The Food Channel conducted a blind taste test to determine the best chocolate for baking. A random sample of adults was obtained, and each was asked to select the best chocolate from among 10 varieties. The final report listed each chocolate along with the number of people who rated it the best.
  (b) After an extensive survey, the Association of Realtors in Chicago concluded that the mean price of a single-family home was less than $500,000.
  (c) After conducting several measurements, the Beijing Municipal Environmental Monitoring Center issued a warning that indicated the density of PM2.5 (fine particulate matter, a measure of air pollution) was over the safe limit.
  (d) International Living issued a report listing percentages of Americans who were retired and living in each foreign country.
  (e) Researchers at the Massachusetts Institute of Technology conducted a study of students to determine the effect of using laptops or digital devices during lectures. The study involved more than 800 undergraduate students, and researchers concluded that students who did not use any digital device actually performed better on their exams than those who used laptops.

1.50 Public Policy and Political Science A Parents Association would like to determine the proportion of teenagers who have the ability to prepare an entire meal. A sample of teenagers was obtained and each teen was asked if they could cook. Describe the population of interest, the sample, and the variable of interest in this problem.

1.51 Marketing and Consumer Behavior Hallmark is interested in the proportion of adults who sent a greeting card on Mother-in-Law Day. A sample of 400 adults was obtained and each adult was asked whether they sent a greeting card on this holiday, which started in 2002. Describe the population and the sample in this problem.

1.52 Medicine and Clinical Studies Researchers at St. Michael's Hospital in Vermont suggest that people who play contact sports show definitive changes in their brain structure and function. One thousand individuals who played contact sports were selected and each was evaluated for changes to brain structure and function. Describe the population, the sample, and the variable in this problem.

1.53 Public Policy and Political Science The Office of the Privacy Commissioner of Canada's (OPC) Contributions Program is interested in reaction to a proposal that would allow police to obtain cell phone records without a subpoena. One thousand people in British Columbia were called and each was asked to respond to several questions.
  (a) Is this an observational study or an experimental study?
  (b) Describe the sample in this problem.
  (c) Is this a random sample? Justify your answer.

1.54 Travel and Transportation Amtrak would like to estimate the proportion of travelers on the Auto Train from Lorton, Virginia, to Sanford, Florida, who utilize the free

Wi-Fi en route. At the end of the trip on January 3, an Amtrak representative stopped every third person getting off the train and asked them if they used the free Wi-Fi.
(a) Is this an observational study or an experimental study?
(b) Describe the sample in this problem.
(c) Is this a random sample? If not, suggest a method for obtaining a random sample.

1.55 Physical Sciences The Air Liquide Company has developed a new de-icing chemical for airplanes, consisting of glycol and several proprietary additives. The new chemical was designed to keep aircraft wings ice-free for a longer period of time. Ten typical Embraer commuter airplanes were obtained and randomly assigned to one of two groups: new chemical versus old chemical. Each plane was subject to constant icing conditions in a controlled environment and treated with one of the chemicals. The length of time until ice formed on the wings was recorded for each plane.
(a) Is this an observational study or an experimental study?
(b) What is the variable of interest?
(c) Describe a technique to randomly assign each plane to a chemical group.

Extended Applications

1.56 Vancouver, British Columbia, has some of the highest rents in Canada. The vacancy rate is reported to be 5%, which is forcing millennials to find some very creative living solutions. Five hundred apartments in Vancouver are selected at random, and each is classified as vacant or occupied.
(a) What is the population of interest?
(b) What is the sample?
(c) Describe the variable of interest.
(d) Write a probability question and a statistics question involving this population and sample.

1.57 Travel and Transportation The Channel Tunnel, or Chunnel, is a 31.4-mi railroad tunnel beneath the English Channel between Folkstone, Kent, in England and Coquelles in France. To ensure passenger safety, engineers selected the 35 deepest areas in the tunnel and measured the pressure on each section.
(a) Is this an observational study or an experimental study?
(b) What is the variable of interest?
(c) Is this a random sample? If so, justify your answer. If not, describe a technique to obtain a random sample.

1.58 Public Health and Nutrition ALDI reported that some of its Choceur Dark Chocolate Bars might contain almond pieces, which were not listed on the packaging. People who have a nut allergy could experience an allergic reaction if they ate the nut-containing bars. Even though no illnesses were reported related to consumption of this product, the FDA selected 100 of the chocolate bars from ALDI Food Stores in Pennsylvania and carefully checked each for almond pieces.
(a) Is this an observational study or an experimental study?
(b) What is the variable of interest?
(c) Is this a random sample? If so, justify your answer. If not, describe a technique to obtain a random sample.

1.59 Economics and Finance The new U.S. tax code (2018) caps the deduction for state and local taxes at $10,000. Some economists believe that this limit will cause people to move out of high-tax states like New York, New Jersey, and California. One thousand residents from high-tax states are selected and each is asked if they plan to move out of the state due to the new tax code. Describe the population, sample, and variable in this problem.

1.60 Business and Management Order picking is the process of finding an item in the current inventory in the warehouse in order to fill a customer order. It is a straightforward but extremely time-consuming task. A professor of logistics and operations management has developed a new method for order picking that involves wave picking and following the most efficient route through the warehouse. Two groups of sample orders were obtained. Each order in the first group was filled using the traditional order picking process; the orders in the second group were filled using the new method. The time to fill each order was measured.
(a) Is this an observational study or an experimental study?
(b) What is the variable of interest?

1.61 Manufacturing and Product Development PyroLance is a firefighting tool that uses water to blast through steel, brick or concrete walls, and even bullet-resistant glass. This tool uses ultrahigh-pressure technology consisting of millions of micro water droplets. Fifty PyroLance tools were obtained from fire companies in California, and the operating pressure was measured for each.
(a) Is this an observational study or an experimental study?
(b) What is the variable of interest?
(c) Is this a random sample? If so, justify your answer. If not, describe a technique to obtain a random sample.

thitiwat_t1980/ Shutterstock

CHAPTER APP

1.62 Lake Winnebago Fish There is some concern that plastic pieces are flowing into Lake Winnebago and then being consumed by fish. People then consume the fish, which in turn creates a potential public health concern. One hundred random fish are obtained from this lake, and each is carefully examined for evidence of microbeads.
(a) Describe the population, sample, and variable in this problem.
(b) Write a probability question and a statistics question in the context of this problem.
(c) Is this an observational study or an experimental study? Describe a method for obtaining a simple random sample of fish from this lake.
(d) Suppose an extraordinary number of fish (out of the 100 obtained) exhibit evidence of microbeads. Apply the statistical inference procedure to draw a conclusion when this rare event is observed.

2

Tables and Graphs for Summarizing Data

◀ **Looking Back**
- Understand the difference between descriptive and inferential statistics: **Section 1.2**.
- Understand the difference between a sample and the population: **Section 1.2**.
- Recognize the importance of a simple random sample in the statistical inference procedure: **Section 1.3**.

Looking Forward ▶
- Be able to classify a data set as categorical or numerical, discrete or continuous: **Section 2.1**.
- Learn how to construct graphical summaries for categorical data: **Section 2.2**.
- Learn how to construct tabular and graphical summaries for numerical data: **Sections 2.3 and 2.4**.

Dotted Yeti/Shutterstock

CHAPTER APP

Near-Earth Objects

NEO A near-Earth object is a comet or asteroid that passes close to our planet. Every so often, a nearby planet pulls one of these objects slightly and its new orbit brings it into the Earth's neighborhood. In December 2017, a needle-shaped asteroid that looked like a rocket stage passed close to Earth.[1] This object, named Oumuamua by astronomers, had lots of people thinking about alien visits and spaceships. However, the object did not have a heat signature and it followed a normal gravitational path.

The JPL Center for Near-Earth Object Studies maintains a list of all near-Earth objects (NEO), computes high-precision orbits and impact risks of these NEOs, and conducts long-term analyses of possible future orbits of dangerous asteroids. A random sample of near-Earth objects was obtained and the minimum possible close-approach distance in lunar distance (LD) was obtained for each. The data are given in the table.

60.8	3.7	52.7	29.2	55.3	42.4	34.7	43.6	45.7	21.0
21.9	16.4	19.7	50.6	42.8	77.2	71.3	36.1	34.5	42.2
71.5	18.3	42.1	37.1	39.0	46.0	21.0	60.0	55.1	64.7
68.3	71.0	34.4	21.9	44.3	68.5	40.3	46.7	72.8	53.6
47.3	51.1	38.5	65.5	15.2	64.4	3.1	53.8	57.2	31.3

The tabular and graphical techniques presented in this chapter will be used to describe the shape, center, and spread of this distribution of distances and to identify any outliers.

2.1 Types of Data

Categorical Versus Numerical Data

As members of an information society, we have access to all kinds of descriptive statistics: in newspapers, in research journals, and even through social media. Whether the information is obtained from a carefully designed experiment or an observational study, the first step is to organize and summarize the data. Tables, charts, and graphs reveal characteristics about the shape, center, and variability of a data set, or distribution. As an example, **Figure 2.1** shows a stacked bar chart of the number of property crimes by type in Philadelphia for five weeks in 2018.

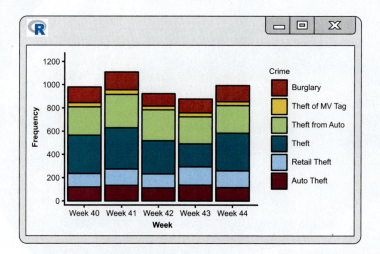

Figure 2.1 Total property crimes in Philadelphia over a five-week period.

The shape of a distribution may be symmetric or skewed. The center of a distribution refers to the position of the majority of the data, and measures of variability indicate the spread of the data. The variability (or dispersion) of a distribution describes how much the measurements vary, as well as how compact or how spread out the data are. Although they are not suitable for making inferences, the tabular and graphical techniques introduced in this chapter can help us describe the distribution of data and identify unusual characteristics.

The summary table or graph to be used, and later the statistical analysis to be performed, depends on the type of data. Consider cars entering the parking lot for a Saturday afternoon college football game. Here are several characteristics we could record: make of the car, number of people in the car, miles driven to the game, and parking location.

We'll do more with bivariate data in Chapter 12.

> **Definition**
> A data set consisting of observations on only a single characteristic, or attribute, is a **univariate** data set.
>
> If we measure, or record, two observations on each individual or object, the data set is **bivariate**.
>
> If there are more than two observations on the same individual or object, the data set is **multivariate**.

Suppose we record the type of coffee drink purchased by each person who enters a Starbucks—a univariate data set. The observations, such as latte, cappuccino, or frappuccino, are categorical. There is no natural ordering of the data, and each observation falls into only one category or class. We might also record how long each person must wait in line to place an order. This time, the responses, such as 1, 5, or 10 minutes, are numerical.

> **Definition**
> A **categorical**, or **qualitative**, univariate data set consists of non-numerical observations that may be placed in categories.
>
> A **numerical**, or **quantitative**, univariate data set consists of observations that are expressed with digits.

Discrete Versus Continuous Data

The following examples illustrate the two basic types of data sets.

EXAMPLE 2.1 Shreddin' the Slopes

A random sample of skiers at the Loon Mountain Ski Resort was obtained, and the ski brand used by each skier was recorded. The responses are given in the following table.

K2	Rossignol	Fischer	Volkl	Fischer	Rossignol
Volkl	Salomon	Dynaster	K2	Fischer	

Each response is non-numerical because there is no natural ordering. This is a (univariate) categorical data set. ∎

EXAMPLE 2.2 Priority Mail

The U.S. Postal Service offers Priority Mail Flat Rate boxes with which customers can expect delivery within 1–3 days for packages weighing up to 70 pounds. A random sample of small boxes shipped from post offices in Oklahoma was obtained, and each box was weighed. The resulting weights (in pounds) are given in the following table.

| 2.0 | 6.1 | 7.3 | 6.4 | 8.0 | 8.1 | 9.9 | 5.1 | 7.7 | 6.7 |
| 6.9 | 4.8 | 10.8 | 9.2 | 3.2 | 7.9 | 8.5 | 8.9 | 6.6 | 8.1 |

Because each observation is numerical, this is a (univariate) numerical data set. ∎

We can classify numerical data even further. Consider the following examples.

On a hot summer day in the southeast, suppose we record the number of lightning strikes within a specified county during the next 24 hours. The possible values are 0, 1, 2, 3 up to, say, 10. There are only a finite number of possible numerical values, and these values are discrete, isolated points on a number line (**Figure 2.2**). Now suppose we record the barometric pressure, in millibars, at 4:00 P.M. The possible values in this case are not discrete and isolated. Instead, the barometric pressure can (theoretically) be any number in the continuous interval 960 to 1070 millibars (**Figure 2.3**).

The number of lightning strikes is discrete. For instance, the number of possible lightning strikes can be 0, or 1, or 2, or 3, and so on, but it cannot be, for example, 2.5. However, if we had an instrument that could measure barometric pressure accurately enough, any number between 960 and 1070 millibars is possible (e.g., 995.466347789).

Figure 2.2 Possible values for a discrete data set: a finite number of values, isolated on a number line.

Figure 2.3 Possible values for a continuous data set: numerical values in some interval.

Definition

Countably infinite means you can count the set of possible values, but it will take you forever. See remark 2 in A Closer Look.

A numerical data set is **discrete** if the set of all possible values is finite, or countably infinite. Discrete data sets are usually associated with counting.

A numerical data set is **continuous** if the set of all possible values is an interval of numbers. Continuous data sets are usually associated with measuring.

A CLOSER LOOK

1. To decide whether a data set is discrete or continuous, consider all the possible values. Finite or countably infinite means discrete. An interval of possible values means continuous.

Mathematically, a set is countably infinite if it can be put into one-to-one correspondence with the counting numbers (1, 2, 3, 4, …). The three dots, …, mean the list continues in the same manner.

2. Countably infinite means there are infinitely many possible values, but they are countable. You may not ever be able to finish counting all of the possible values, but there exists a method for actually counting them. Think about counting all the grains of sand on your favorite beach: You could theoretically count them, but you would never finish.

3. The interval for a continuous data set can be any interval, of any length, open or closed. The exact interval may not be known, only that there is some interval of possible values.

4. In practice, we really can't observe any number in an interval because our measurement devices are limited. For example, we may only be able to achieve up to 10 digits of accuracy. So a continuous data set may contain any number in some interval in theory, but not in reality.

Figure 2.4 illustrates the classifications of univariate data.

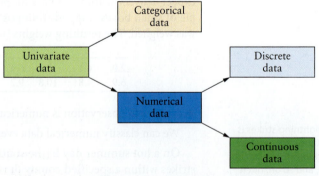

Figure 2.4 Classifications of univariate data.

Here is an example to illustrate these classifications.

EXAMPLE 2.3 Univariate Data Classificiations

A researcher obtained the following observations. Classify each resulting data set as categorical or numerical. If the data set is numerical, determine whether it is discrete or continuous.

(a) The number of books read by middle school students during the academic year.
(b) The duty status of crew members aboard a U.S. aircraft carrier.
(c) The length of time (in minutes) it takes to get a haircut.
(d) The number of garage sales advertised in a local newspaper.
(e) The types of candy received from different houses on Halloween.
(f) The weight of steel beams used in the construction of a high-rise apartment building.
(g) The type of hardwood floor purchased by customers at Green Tree Flooring Services.

Solution

(a) The observations are numbers, so the data set is numerical. The set of possible values is finite. We don't know the maximum number of books read, but the possible numbers in the data set represent counts. The data set is discrete.
(b) The observations are categorical—for example, off duty, on duty, or limited duty. There is no natural ordering; the possible responses fall into groups or classes. This data set is categorical.
(c) The observations are numbers and the set of possible values is some interval, perhaps 5 to 45 minutes. This is a numerical continuous data set.
(d) The observations are numbers and the set of possible values is finite. We can count the number of advertised garage sales. The minimum number may be 0 and the maximum may be 25. This is a numerical discrete data set.
(e) The observations may be Milky Way, Snickers, Nestle Crunch, etc. Although there may be some personal preference and an individual ranking, this is a categorical data set.
(f) The observations are numbers and the set of possible values is some interval, say, 13 to 20 pounds per foot. This is a numerical continuous data set.
(g) The observations are oak, maple, bamboo, walnut, etc. There may be some preference for the floor company, but this is a categorical data set.

TRY IT NOW Go to Exercises 2.7 and 2.11

CHAPTER APP Near-Earth Objects

Classify the data set as categorical or numerical. If the data set is numerical, determine whether it is discrete or continuous.

Methods for summarizing and displaying categorical data are discussed in Section 2.2, and tables and graphs for numerical data are presented in Section 2.3 and Section 2.4.

Section 2.1 Exercises

Concept Check

2.1 True or False A data set obtained by recording the height and weight of every person entering a doctor's office is univariate.

2.2 True or False Every data set is multivariate.

2.3 True or False A data set consisting of 37 times, in seconds, for pedestrians to cross a certain city street is univariate.

2.4 True or False Categorical data can be classified as discrete or continuous.

2.5 Fill in the Blank
(a) A _____ univariate data set consists of observations that are numbers.
(b) A _____ univariate data set consists of non-numerical observations.
(c) If the set of all possible values for a numerical data set is finite, then the data set is _____.
(d) If the set of all possible values for a numerical data set is some interval of numbers, then the data set is _____.

Practice

2.6 Univariate Data Classifications A set of observations is obtained as indicated in parts (a) through (h). In each case, classify the resulting data set as categorical or numerical. If the data set is numerical, determine whether it is discrete or continuous.
(a) The weights of several reams of paper.
(b) The number of cars towed from the Pennsylvania Turnpike during a given 24-hour period.
(c) The first ingredient in the product listing of boxes of cereal.
(d) The number of visitors to the Grand Canyon on certain days.
(e) The number of games that the Red Sox win during several seasons.
(f) The amount of sand (in tons) used on roads during winters in a small town.
(g) The diagnoses of patients in an emergency ward.
(h) The weight of various Girl Scout cookies.

2.7 Univariate Data Classifications A set of observations is obtained as indicated in parts (a) through (h). In each case, classify the resulting data set as categorical or numerical. If the data set is numerical, determine whether it is discrete or continuous.
(a) The lengths of the spans of bridges in New York State.
(b) The number of people hired by a company during certain weeks.
(c) The cloud ceiling (i.e., height of the lowest clouds) at airports around the country.
(d) The temperature of the coffee purchased at several fast-food restaurants.
(e) The number of errors found by the IRS in audited federal tax returns.
(f) The number of cases won by the Department of Justice during selected months.
(g) The type of notebook used by students in a statistics class.
(h) The classifications of Forward Operating Air Force bases (Main Air Base, Air Facility, Air Site, or Air Point).

2.8 Univariate Data Classifications A set of observations is obtained as indicated in parts (a) through (h). In each case, classify the resulting data set as categorical or numerical. If the data set is numerical, determine whether it is discrete or continuous.
(a) The number of steps on apartment fire escapes.
(b) The number of leaves on maple trees.
(c) The reason several automobiles fail inspection.
(d) The weight of fully loaded tractor trailers.
(e) The sneaker brand purchased from Zappos by several customers.
(f) The area of several Nebraska farms.
(g) The number of dandelions in the backyards of homes in certain developments.
(h) The cellular calling plan selected by customers.

2.9 Univariate Data Classifications A set of observations is obtained as indicated in parts (a) through (h). In each case, classify the resulting data set as categorical or numerical. If the data set is numerical, determine whether it is discrete or continuous.
(a) The number of engine revolutions per minute in automobiles.
(b) The thickness of the polar ice cap in several locations.
(c) The state in which families vacationed last summer.
(d) The type of Internet connection in county households.
(e) The duration of Olympic figure skating routines.
(f) The make of watch worn by people entering a certain department store.
(g) The number of raisins in 24-oz boxes.
(h) Whether or not selected computer passwords include a capital letter.

2.10 Numerical Observations A set of numerical observations is obtained as described in parts (a) through (h). Classify each resulting data set as discrete or continuous.
(a) The widths of posters at an art gallery.
(b) The time it takes to compile computer programs.
(c) The number of radioactive particles that escape from special containers during a one-hour period.
(d) The time it takes to bake batches of banana muffins.
(e) The concentration of carbon monoxide (in ppm) in homes during the winter.
(f) The number of pages in best-selling murder-mystery novels.
(g) The number of dents in automobiles that have service at Honda dealerships.
(h) The diameter of the aorta of patients in an emergency room.

2.11 Numerical Observations A set of numerical observations is obtained as described in parts (a) through (h). Classify each resulting data set as discrete or continuous.
(a) The weight of baseball bats.
(b) The area of selected dorm rooms.
(c) The number of bees in hives.
(d) The height of a storm surge during hurricanes.
(e) The amount of ink used in office printers during a week.
(f) The number of fish in office aquariums.
(g) The weight of boxes of chocolate-covered pretzels.
(h) The number of residents in assisted living centers.

2.12 Numerical Observations A set of numerical observations is obtained as described in parts (a) through (h). Classify each resulting data set as discrete or continuous.
(a) The time it takes giant slalom skiers to cover a race course.
(b) The number of magazines available for sale at newsstands.
(c) The number of black squares in crossword puzzles.

(d) The length of time spent waiting in line at grocery-store checkout lanes.
(e) The number of french fries in a small order from fast-food restaurants.
(f) The number of words in email messages received.
(g) The number of photos on iPhones.
(h) The length of time for police chases.

2.13 Univariate Data Classifications Classify each data set as categorical, discrete, or continuous.
(a) A random sample of mature Eastern tent caterpillars is obtained from a tree branch in a neighborhood yard. The length of each caterpillar is recorded.
(b) A random sample of prime-time television shows is obtained and the number of violent acts is recorded for each show.
(c) A representative sample of employees from a large company is obtained, and the overtime hours for the past month are recorded for each employee.
(d) An HMO selects a random sample of subscribers and records the number of office visits over the past year for each patient.
(e) Thirty-six apples are randomly selected from an orchard. Each is graded for quality of appearance: excellent, good, fair, or poor.
(f) A random sample of mattresses is obtained and the firmness (medium, medium firm, firm, or extra firm) of each is recorded.

2.14 Univariate Data Classifications Classify each data set as categorical, discrete, or continuous.
(a) A random sample of cheeses is obtained and the number of months each is allowed to age before sale is recorded.
(b) Sixteen universities are selected and each computer network system is carefully analyzed. The computer virus threat is assessed for each campus: low, medium, or high.
(c) A random sample of Waterford Normandy dinner plates is selected and the weight of each plate is recorded.
(d) Thirty-five new customers at a health club are selected and the body-fat percentage of each member is computed and recorded.
(e) A random sample of CDs is obtained from a local music store. The company that produced each CD is noted.
(f) A collection of pens is obtained from employees at a large company. For each pen, the outside diameter of the barrel at its widest point is measured and recorded.

2.15 Univariate Data Classifications Classify each data set as categorical, discrete, or continuous.
(a) A random sample of Hudson River ferry trips is obtained and the number of riders on each trip is recorded.
(b) A random sample of military helicopters is obtained and the weight of each is recorded.
(c) A random sample of communities in Canada is selected and the number of full-time police officers employed is recorded.
(d) A random sample of locations in the United States is selected. Temperature data are used to determine whether or not a new record-high temperature was set during the past year.
(e) A random sample of stock analysts is obtained and each is asked to rate a specific stock as buy, sell, or hold.
(f) A random sample of cross-country flights is obtained and the number of controllers each pilot talks to during the flight is recorded.

2.16 Univariate Data Classifications Classify each data set as categorical or numerical. If the data set is numerical, determine whether it is discrete or continuous.
(a) A random sample of homes with live streaming services is obtained and the company serving each home is recorded.
(b) A random sample of Supreme Court cases is obtained and the type of decision is recorded for each.
(c) A random sample of homes in Rockville, Maryland, is obtained and the distance from the nearest Metro stop is recorded for each.
(d) A random sample of returns to L. L. Bean is obtained and the reason for each return is recorded.
(e) A random sample of DMV offices in California is obtained and the number of people who renewed their driver's license on a certain day is recorded.
(f) A random sample of small businesses across the United States is obtained and each is classified by comparing the number of employees to the previous year (less, same, more).

2.2 Bar Charts and Pie Charts

Frequency Distributions

The natural summary measures for a categorical data set are the number of times each category occurred and the proportion of times each category occurred. These values are usually displayed in a table as in **Table 2.1**.

 Finding proportions.

Table 2.1 A frequency distribution summarizing the number of critically endangered species.

Class	Frequency	Relative frequency
Mammals	202	0.101
Birds	222	0.111
Reptiles	266	0.133
Amphibians	552	0.276
Fishes	468	0.234
Insect	290	0.145
Total	2000	1.000

Definition

A **frequency distribution** for categorical data is a summary table that presents categories, counts, and proportions.

(a) Each unique value in a categorical data set is a label, or **class**. In Table 2.1, the classes are mammals, birds, reptiles, etc.

(b) The **frequency** is the count for each class. In Table 2.1, the frequency for the mammals class is 202 (i.e., 202 mammals were on the critically endangered species list).

(c) The **relative frequency**, or sample proportion, for each class is the frequency of the class divided by the total number of observations. In Table 2.1, the relative frequency for the amphibians class is $552/2000 = 0.276$.

A frequency distribution for a categorical data set is illustrated in the next example.

EXAMPLE 2.4 Online Shopping

ECOMM

A random sample of e-commerce companies in the United States was obtained and the category for each company is given in the following table.

Pet products	Health	Sporting goods	Artisanal	Sporting goods
Sporting goods	Artisanal	Sporting goods	Sporting goods	Grocery
Health	Sporting goods	Pet products	Sporting goods	Artisanal
Artisanal	Grocery	Health	Pet products	Sporting goods
Sporting goods	Sporting goods	Artisanal	Health	Artisanal

Construct a frequency distribution to describe these data. What proportion of e-commerce companies are not classified as Health?

Solution

STEP 1 Each unique e-commerce category is a label, or **class**. This is a categorical data set. There are five unique classes and 25 observations in total.

STEP 2 Draw a table and list each unique class in the left-hand column. Find the **frequency** and **relative frequency** for each class. For example, because Health appears four times in the sample, the frequency for this class is 4. The relative frequency for Health is $4/25 = 0.16$.

Class	Frequency	Relative frequency	
Artisanal	6	0.24	(= 6/25)
Grocery	2	0.08	(= 2/25)
Health	4	0.16	(= 4/25)
Pet products	3	0.12	(= 3/25)
Sporting goods	10	0.40	(= 10/25)
Total	25	1.00	

We could also add the relative frequencies for all classes that are not Health related: $0.24 + 0.08 + 0.12 + 0.40 = 0.84$.

The proportion of e-commerce categories that are Health related is $4/25 = 0.16$. The total proportion is always 1.00. Therefore, the proportion of e-commerce categories that are not Health related is $1.00 - 0.16 = 0.84$.

TRY IT NOW Go to Exercise 2.23

A CLOSER LOOK

1. If you have to construct a frequency distribution by hand, an additional tally column is helpful. Insert this after the class column, and use a tally mark or tick mark to count observations as you read them from the table.

A tally mark is a short line drawn for each count up to four. On number five, draw a diagonal line across the other four. Count in sets of five.

2. The last (total) row is optional, but it is a good check of your calculations. The frequencies should sum to the total number of observations, and the relative frequencies should sum to 1.00 (subject to round-off error).

3. There is no rule for ordering the classes. In Example 2.4, the classes happen to be presented in alphabetical order.

Bar Charts

A **bar chart** is a graphical representation of a frequency distribution for categorical data. An example of a bar chart is shown in **Figure 2.5**.

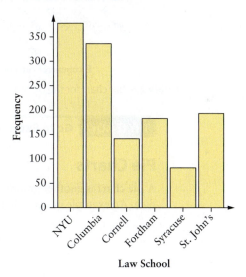

Figure 2.5 Bar chart showing the number of first-time candidates in July 2017 for the bar exam from certain law schools.

How to Construct a Bar Chart

1. Draw a horizontal axis with equally spaced tick marks, one for each class.
2. Draw a vertical axis for the frequency (or relative frequency) and use appropriate tick marks. Label each axis.
3. Draw a rectangle centered at each tick mark (class) with height equal to, or proportional to, the frequency of each class (also called the class frequency). The bars should be of equal width, but do not necessarily have to abut one another. There are often spaces between the bars to distinguish a bar chart from a histogram.

EXAMPLE 2.5 Online Shopping, Continued

Construct a bar chart for the online shopping data in Example 2.4.

Solution

STEP 1 Use the frequency distribution for the online shopping data. There are five classes, and the frequencies range from 2 to 10.

STEP 2 Draw a horizontal and a vertical axis. On the horizontal axis, draw five ticks for the five classes and label them with the class names. Because the greatest frequency is 10, draw and label tick marks from 0 to at least 10 on the vertical axis.

STEP 3 The height of each vertical bar is determined by the frequency of the class. For example, the frequency of Health is 4, so the height of the bar representing Health is 4. The resulting bar chart is shown in **Figure 2.6**. A technology solution is shown in **Figure 2.7**.

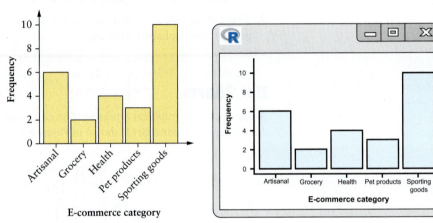

Figure 2.6 Bar chart for the online shopping data.

Figure 2.7 R bar chart.

TRY IT NOW Go to Exercise 2.25

Pie Charts

A **pie chart** is another graphical representation of a frequency distribution for categorical data. An example of a pie chart is shown in **Figure 2.8**.

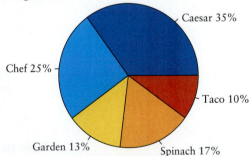

Figure 2.8 Type of salad ordered at the Spice Symphony restaurant.

How to Construct a Pie Chart

1. Divide a circle (or pie) into slices or wedges so that each slice corresponds to a class.
2. The size of each slice is measured by the angle of the slice. To compute the angle of each slice, multiply the relative frequency by 360° (the number of degrees in a whole or complete circle).
3. The first slice of a pie chart is usually drawn with an edge horizontal and to the right (0°). The angle is measured counterclockwise. Each successive slice is added counterclockwise with the appropriate angle.

 Finding percentages.

2.2 Bar Charts and Pie Charts

EXAMPLE 2.6 Online Shopping, Another View

Construct a pie chart for the online shopping data in Example 2.4.

Solution

STEP 1 Add a column to the frequency distribution for slice angle. Use the relative frequency of each class to find the slice angle.

Class	Relative frequency	Angle	
Artisanal	0.24	86.4°	$(= 0.24 \times 360°)$
Grocery	0.08	28.8°	$(= 0.08 \times 360°)$
Health	0.16	57.6°	$(= 0.16 \times 360°)$
Pet products	0.12	43.2°	$(= 0.12 \times 360°)$
Sporting goods	0.40	144.0°	$(= 0.40 \times 360°)$
Total	1.00	360.0°	

STEP 2 Draw a circle and mark slices using the angles in the frequency distribution. Draw the first slice with an edge extending from the center of the circle to the right. The remaining slices are drawn moving around the pie counterclockwise. It may be helpful to use a protractor and compass to draw the circle and measure the angles. See **Figure 2.9**. A technology solution is shown in **Figure 2.10**.

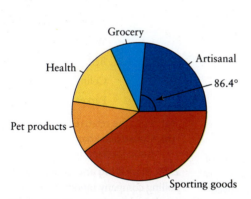

Figure 2.9 Pie chart for the online shopping data.

Figure 2.10 R pie chart.

Note: because Sporting goods corresponds to the largest slice of the pie (chart), this class has the greatest frequency (and relative frequency); it is the e-commerce category that occurred most often in the sample.

TRY IT NOW Go to Exercises 2.27 and 2.29

A CLOSER LOOK

1. A pie chart is hard to draw accurately by hand, even with a protractor and compass. A graphing calculator or computer software is quicker and more efficient for constructing this graph.
2. There are lots of pie-chart variations, such as exploding pie charts and 3D pie charts. Each is simply a visual representation of a frequency distribution for categorical data.

Section 2.2 Exercises

Concept Check

2.17 True or False A frequency distribution is a summary table for categorical data.

2.18 True or False The relative frequency for each class in a frequency distribution is a sample proportion.

2.19 True or False A bar chart is constructed using the frequency for each class.

2.20 True or False All the slices in a pie chart should have approximately the same angle.

2.21 True or False There should be no space between the bars in a bar chart.

Applications

2.22 Psychology and Human Behavior A random sample of TV viewers was obtained and each person was asked to select the entertainment category of his or her favorite show. The results are given in the table. TV

Comedy	Sports	Soap	Educational
Sports	Soap	Comedy	Sports
Soap	Drama	Drama	Soap
Drama	Comedy	Educational	Comedy
Drama	Drama	Soap	Drama
Reality	Comedy	Soap	Educational
Reality	Reality	Drama	Soap
Soap	Drama	Sports	Drama
Drama	Reality	Reality	Comedy
Reality	Soap	Sports	Soap
Reality	Comedy		

Construct a frequency distribution for these data.

2.23 Psychology and Human Behavior A random sample of patrons visiting the Rena Branston Gallery in San Francisco was obtained, and each was asked the type of art they most enjoyed viewing. The results are given in the table. ART

Abstract	Realist	Abstract	Expressionist
Surrealist	Expressionist	Abstract	Surrealist
Abstract	Abstract	Surrealist	Abstract
Surrealist	Abstract	Surrealist	Surrealist
Abstract	Abstract	Expressionist	Abstract
Surrealist	Surrealist	Realist	Abstract
Realist	Abstract	Abstract	Surrealist
Expressionist	Expressionist	Realist	Realist
Realist	Abstract	Realist	Realist
Realist	Realist	Expressionist	Abstract
Realist	Realist		

Construct a frequency distribution for these data.

2.24 Economics and Finance To help readers understand how retirement assets are taxed, AARP surveyed a random sample of members and asked each to name the largest source of their retirement income. A summary of their responses is given in the table. RETINC

Asset	Frequency
Tax-deferred account	60
Taxable account	45
Roth IRA	70
Pension	55
Annuity	30
Social Security	140

(a) Find the relative frequency for each asset.
(b) Construct a bar chart for these data using frequency on the vertical axis.

2.25 Travel and Transportation The table lists the number of speeding tickets issued for various counties in Texas for a one-week period during 2018. TXTIX

County	Frequency
Hidalgo	225
Cameron	165
Brazoria	145
Fort Bend	102
Harris	177
Montgomery	163
Williamson	120
Bell	118
Colin	102
Parker	102

(a) Find the relative frequency for each county.
(b) Construct a bar chart for these data using relative frequency on the vertical axis.

2.26 Marketing and Consumer Behavior An independent polling company randomly selected customers who recently purchased furniture from Overstock. Each person's reaction to the following statement was recorded: "Overall, I am very satisfied with the quality of the merchandise." Survey participants could answer very satisfied (VS), satisfied (S), neutral (N), unsatisfied (U), or very unsatisfied (VU). OVER

(a) Construct a frequency distribution for these data.
(b) Use the table in part (a) to construct a pie chart for these data.

2.27 Biology and Environmental Science The table lists the number of American adults in each category who responded to a survey regarding climate change beliefs and attitudes.[2]

Attitude	Frequency
Alarmed	228
Concerned	367
Cautious	304
Disengaged	76
Doubtful	152
Dismissive	127

(a) Find the relative frequency for each attitude.
(b) Construct a pie chart for these data.

2.28 Fuel Consumption and Cars There are three main types of distractions when driving an automobile that can endanger the driver, passengers, and other drivers.[3] A random sample of accidents in the United States was obtained and each was classified by distraction: visual (V), manual (M), cognitive (C), or none (N). VMCN

(a) Construct a frequency distribution for these data.
(b) Use the table in part (a) to construct a bar chart for these data.

2.29 Education and Child Development The grade distribution for a large calculus class at the University of New Hampshire is given in the table. GRADE

Grade	Frequency
A	10
B	43
C	56
D	26
F	15

(a) Find the relative frequency for each grade.
(b) Construct a bar chart using frequency on the vertical axis and a pie chart from the frequency distribution.
(c) How many students were in this calculus class? What proportion of students passed (i.e., received a D or better)?

2.30 Public Health and Nutrition A random survey of 200 customers who purchased a shake at Brennan's Big Chill showed the following proportions: SHAKE

Flavor	Relative frequency
Banana	0.100
Blue raspberry	0.185
Cherry	0.260
Chocolate	0.150
Lemon	0.080
Malt	0.225

(a) Find the frequency of each shake (class).
(b) Construct a bar chart using frequency on the vertical axis and a pie chart for these shake data.

2.31 Marketing and Consumer Behavior A random sample of customers who purchased a front-load washer at Home Depot was obtained and the brand of each washer was recorded. The results are given in the table. WASH

Samsung	Maytag	Whirlpool
GE	Whirlpool	LG
Samsung	GE	Samsung
LG	Maytag	LG
GE	GE	Maytag
LG	Samsung	Whirlpool
Whirlpool	Samsung	Samsung
Whirlpool	LG	GE
Whirlpool	Whirlpool	Samsung
Maytag	GE	Samsung

(a) Construct a frequency distribution for these data.
(b) Construct a bar chart using frequency on the vertical axis and a pie chart for these data.
(c) What proportion of people in this sample purchased an LG or a GE front-load washer?
(d) What proportion of people in this sample did not purchase a Samsung front-load washer?

2.32 Sports and Leisure Suppose each of the 350 exhibitors at the 2018 Comic Con convention in San Diego, California, was classified according to the type of entertainment genre. The proportions are given in the table. COMIC

Entertainment	Proportion
Comics	0.0714
Graphic novels	0.0857
Anime	0.0486
Manga	0.0771
Video games	0.3429
Toys	0.1029
Movies and television	0.2715

(a) Find the number of exhibitors in each entertainment classification.
(b) Carefully sketch a bar chart and a pie chart using the proportions for each class.

2.33 Psychology and Human Behavior Using the Library of Congress classification scheme, the Brookings Public Library in South Dakota recorded the type of book borrowed by 30 randomly selected patrons. The data are given in the table. BOOK

Medicine	Medicine	Education	Technology
Law	Science	Science	Technology
Science	Literature	Medicine	Education
Literature	Science	Law	Medicine
Law	Education	Literature	Technology
Science	Medicine	Technology	Medicine
Technology	Education	Technology	Medicine
Literature	Education		

(a) Construct a frequency distribution for these data.
(b) Carefully sketch a bar chart using relative frequency on the vertical axis and a pie chart for these data.
(c) Do you think the public library should try to purchase more books in one particular subject area? Why or why not?

2.34 Marketing and Consumer Behavior Cardinal Glass Industries produces several products for residential buildings, for vehicles, and for ordinary consumer use. The proportion of each type of manufactured product is given in the table. CARD

Building window	Vehicle window	Containers	Tableware	Lamps
0.35	0.15	0.10	0.25	0.15

Construct a bar chart and a pie chart for these data using the proportions in the table.

2.35 Business and Management The government of Canada uses the National Occupational Classification (NOC) system to classify occupations, based on job duties and work responsibilities.[4] A random sample of jobs in Edmonton was obtained. The classifications and the corresponding proportions are given in the table.

NOC	Proportion
Skill Type 0	0.064
Skill Level A	0.136
Skill Level B	0.252
Skill Level C	0.300
Skill Level D	0.248

(a) Construct a bar chart and a pie chart for these data using the proportions in the table.
(b) Suppose 250 occupations were selected for this study. Find the frequency, or the number of jobs, for each classification. **NOC**

2.36 Marketing and Consumer Behavior A survey of new homes built in the Sleepy Creek Mountains of West Virginia produced the following results for the type of siding used.

Siding	Frequency
Aluminum	20
Brick	15
Stucco	12
Vinyl	45
Wood	24

(a) Find the relative frequency for each siding classification.
(b) Construct a bar chart using frequency on the vertical axis and a pie chart for these data. **SIDING**

2.37 Technology and Internet A report by the research firm eMarketer indicated that more than 22 million U.S. adults no longer use a cable or satellite service and are instead using an Internet streaming service and/or free TV.[5] A random sample of adults using online streaming was obtained. The table lists the number of adults by service provider. **STREAM**

Service provider	Number of adults
Netflix	1267
Hulu	548
Amazon	498
Playstation Vue	606
Sling	844
Crackle	236
HBO	406

(a) Find the relative frequency associated with each service provider.
(b) Construct a bar chart using frequency on the vertical axis and a pie chart for these data.

2.38 Travel and Transportation Flight delays and cancellations are stressful and especially aggravating during major holiday times. The following table lists the ten worst airports by number of cancellations in 2017 (through December 20, 2017).[6] **CANCEL**

Airport	Number of cancellations
ATL: Atlanta	3918
ORD: Chicago	3505
IAH: Houston	3481
SFO: San Francisco	3057
LGA: New York	2886
EWR: Newark	2875
BOS: Boston	2541
MCO: Orlando	2510
FLL: Fort Lauderdale	2371
JFK: New York	2047

(a) Find the relative frequency associated with each airport.
(b) Construct a pie chart for these data.

Extended Applications

2.39 Sports and Leisure Complete the frequency distribution from a random sample of people visiting Atlantic City casinos. **ATLCITY**

Casino	Frequency	Relative frequency
Borgata	40	
Caesars	25	0.125
Tropicana	32	
Harrah's		0.110
Bally's	25	
Golden Nugget		0.280

(a) What is the size of the random sample?
(b) Which casino is most preferred by people in this survey? Justify your answer.

2.40 Travel and Transportation Families traveling to Walt Disney World in Florida often rent a car rather than use airport and hotel shuttle buses. A recent survey asked families to indicate the rental car agency used. The results are presented in the bar chart.

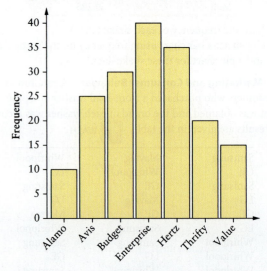

(a) Construct a frequency distribution for these survey results.

(b) How many observations were in this data set?
(c) What proportion of people did not use Hertz or Enterprise?
(d) Construct a pie chart for these data.

2.41 Marketing and Consumer Behavior A total of 1000 customers entering the Mall of America in Bloomington, Minnesota, were randomly selected and asked to rank the variety of stores. The results are given in the table.

Response	Frequency	Relative frequency
Excellent	50	
Very good	152	
Good	255	
Fair		0.4250
Poor		0.1180

(a) Complete the frequency distribution.
(b) Construct a bar chart using frequency on the vertical axis and a pie chart for these data.
(c) What proportion of customers did not rank the store variety as very good or excellent?

2.42 Biology and Environmental Science The table shows the number of tornadoes in the United States by month for 2016 and 2017.[7] **TRNDO**

Month	2016	2017
January	134	17
February	68	102
March	191	86
April	211	141
May	287	216
June	143	86
July	80	107
August	114	90
September	34	38
October	85	20
November	46	50
December	13	18

(a) Find the relative frequency for each month of 2016. Construct a bar chart using relative frequency on the vertical axis (and month on the horizontal axis).
(b) Find the relative frequency for each month of 2017. Construct a bar chart using relative frequency on the vertical axis (and month on the horizontal axis).
(c) How do these two bar charts compare? Describe any similarities or differences.

2.43 Economics and Finance A (tax-exempt) green bond is specifically issued for the development of land that is underutilized, such as areas with abandoned buildings or underdeveloped lots. The table shows the number of green bonds issued by qualified underwriters in the first half of 2017.[8] **GREEN**

Underwriter	Frequency
Morgan Stanley	20
Citi	29
HSBC	32
Barclays	22
BNP Paribas	29
JP Morgan	23
Natikis	22
CITIC	9

(a) Find the relative frequency associated with each underwriter.
(b) Construct a bar chart using frequency on the vertical axis, and a bar chart using relative frequency on the vertical axis. Which of these two graphs do you think is a better graphical description of underwriters and green bonds issued? Why?

2.44 Physical Sciences The table shows the number of regular producing oil wells in certain Texas counties as of February 2018.[9]

County	Frequency
Aransas	18
Brooks	44
Calhoun	29
Cherokee	97
Colorado	47
Cottle	41
Foard	73
Grimes	79
Kaufman	25
Lamb	71

(a) Construct a bar chart for these data. **WELLS**
(b) Construct a pie chart for these data.
(c) Is it reasonable to conclude that of these counties, Cherokee produces the most oil per month? Why or why not?

2.45 Sports and Leisure A side-by-side or a stacked bar chart may be used to compare categorical data obtained from two (or more) different sources or groups. **Figure 2.11** and **Figure 2.12** show an example of each—a comparison of test grades in two different sections of an introductory statistics course. The blue rectangles represent students from Section 01; the green rectangles represent students from Section 02. **TRAVPK**

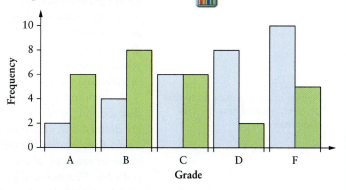

Figure 2.11 Side-by-side bar chart. Bars corresponding to the same category are placed side by side for easy comparison.

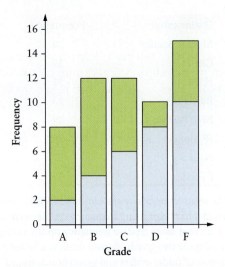

Figure 2.12 Stacked bar chart. Within each category, bars are stacked for comparison.

Marlin Travel has offices in several cities in Canada and last year offered two special summer vacation packages. Suppose the number of each package sold for each office is given in the following table.

City	Package A	Package B
Winnipeg	67	63
Vancouver	52	24
Toronto	86	55
Montreal	48	38
Calgary	50	36

(a) Compute the relative frequency for each city, for both vacation packages.
(b) Construct a side-by-side bar chart using the relative frequency of each class.
(c) Why should relative frequency be used for comparison in the side-by-side bar chart rather than frequency?

2.46 Sports and Leisure The table shows the number of hits and walks for the Boston Red Sox and the New York Yankees during the 2017 Major League Baseball season.[10] MLB2017

Batting	Boston Red Sox	New York Yankees
Singles	972	933
Doubles	302	266
Triples	19	23
Home runs	168	241
Walks	571	616

(a) Find the relative frequency for the Boston Red Sox for each batting category.
(b) Find the relative frequency for the New York Yankees for each batting category.
(c) Construct a side-by-side bar chart using the relative frequency for each batting category.

2.3 Stem-and-Leaf Plots

Background and Construction

This section introduces the stem-and-leaf plot, a graphical technique for describing numerical data. In Section 2.4, you will learn about some other tables and graphs for summarizing numerical data. The goal of all these techniques is the same: to get a quick idea of the distribution of the data in terms of shape, center, and variability. In addition, we are always watching for outliers, values that are very far away from the rest.

A stem-and-leaf plot is a graphical procedure used to describe numerical data. It is fairly easy to construct, even by hand, and most statistical software packages have options for drawing this graph. A stem-and-leaf plot is a combination of sorting and graphing. One advantage of this plot is that the actual data are used to create the graph; we do not lose the original data values as we do when summarizing by finding frequency or relative frequency.

A stem-and-leaf plot can be used to describe the shape, center, and variability of the distribution. In Section 2.4, some specific terms and expressions used to describe shape are defined and illustrated. To estimate the center of a distribution, or to find a typical value, first arrange the observations in increasing order. Simply approximate a middle value, or range of values, in this list. More precise definitions and computations are presented in Chapter 3. The variability refers to the spread or compactness of the data. In addition, we always check for outliers.

We will eventually need a specific quantitative measure for very far away.

The center of a distribution, or typical value, often occurs where the data are clustered.

How to Construct a Stem-and-Leaf Plot

There are, of course, exceptions to this two-digit rule.

To create a stem-and-leaf plot, each observation in the data set must have at least two digits. Think of each observation as consisting of two pieces (a stem and a leaf). For example, suppose we consider the number of people watching a movie, and in one theater there are 372 people. The number 372 could be split into the pieces 37 (the first two digits) and 2 (the last digit).

1. Split each observation into a

   Stem: one or more of the leading, or left-hand, digits; and a

   Leaf: the trailing, or remaining, digit(s) to the right.

   Each observation in the data set must be split at the same place, such as between the tens place and the ones place.

2. Write a sequence of stems in a column, from the smallest occurring stem to the largest. Include all stems between the smallest and largest, even if there are no corresponding leaves.

3. List all the digits of each leaf next to its corresponding stem. It is not necessary to put the leaves in increasing order, but make sure the leaves line up vertically.

4. Indicate the units for the stems and leaves in some manner.

Example: Stem-and-Leaf Plot

EXAMPLE 2.7   Waterfall Heights

FALLHT

Kerepakupai Meru, or Angel Falls, is the highest waterfall in the world.[11] Because the falls are so high (979 meters), by the time water reaches the canyon below, it has vaporized into a giant mist cloud. Suppose the table lists the total height, in meters, of several waterfalls in the world.

693	745	631	635	625	629	739	738	732	725
720	719	715	715	707	707	706	705	700	680
674	671	671	665	660	660	650	646	645	640
640	638	620	620	612	610	610	610	610	610
610	610	610	610	610	600	600	600	651	727

Construct a stem-and-leaf plot for these data.

Solution

STEP 1 There are only two options for splitting each observation in this example:

  (a) Split between the hundreds and the tens place (e.g., split 693 as 6 and 93); or

  (b) Split between the tens place and the ones place (e.g., split 693 as 69 and 3).

  If we split between the hundreds and the tens place, there will be only two stems because the only numbers in the hundreds place are 6 and 7. The resulting plot will not reveal much about the distribution of the data. The better split is between the tens place and the ones place.

STEP 2 Scan the data to find the smallest and the largest stems, and list all of the stems in a vertical column. Write each leaf next to its corresponding stem. For example,

  693 ⇒ 69 | 3   A 3 is placed in the 69 stem row.
           ↑   ↑
          stem leaf

  For 745, a 5 is placed in the 74 stem row.

  For 631, a 1 is placed in the 63 stem row.

  For 635, a 5 is placed in the 63 stem row.

44 CHAPTER 2 Tables and Graphs for Summarizing Data

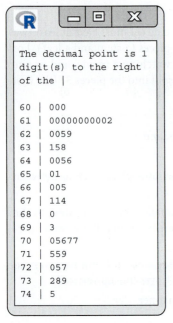

Figure 2.13 R stem-and-leaf plot.

STEP 3 Continue in this manner, to produce the following stem-and-leaf plot.

60	0 0 0
61	2 0 0 0 0 0 0 0 0 0 0
62	5 9 0 0
63	1 5 8
64	6 5 0 0
65	0 1
66	5 0 0
67	4 1 1
68	0
69	3
70	7 7 6 5 0
71	9 5 5
72	5 0 7
73	9 8 2
74	5

Stem = 10, Leaf = 1

Note that Stem = 10 means the rightmost digit in the stem is in the tens place and Leaf = 1 means the leftmost digit in each leaf is in the ones place. Reading from the graph, the smallest waterfall height is 600 meters and the largest is 745 meters. The center of a data set is a typical value or values near the middle of the observations when they are arranged in (increasing) order. For these data, the center appears to be in the 64 or 65 stem row. There are no outlying values. **Figure 2.13** shows a technology solution. ∎

TRY IT NOW Go to Exercises 2.63 and 2.65

A CLOSER LOOK

1. As a general rule of thumb, try to construct the plot with 5 to 20 stems. With fewer than 5, the graph is too compact; with more than 20, the observations are too spread out. Neither extreme reveals much about the distribution.

2. Sometimes, to help us find the center of the data, we put the leaves in increasing order, to make an ordered stem-and-leaf plot.

3. Some advantages of a stem-and-leaf plot: Each observation is a visible part of the graph and (in an ordered stem-and-leaf plot, as when using a computer) the data are sorted. However, a stem-and-leaf plot can get very big, very fast.

If a stem-and-leaf plot is made for a very large data set, the stems may be divided, usually in half or fifths. Consider the following example.

Example: Stems Divided

EXAMPLE 2.8 Fitness Professionals

FITPRO

Personal trainers provide one-on-one fitness and wellness exercises to individuals and to some small groups. The length and frequency of sessions varies, and the cost of a trainer is often determined by the trainer's area of specialization and experience. Suppose a random sample of personal trainer fees (per hour) was obtained, and the data are given in the table.[12]

58	74	69	66	87	89	61	60	79	82
73	66	71	85	97	63	51	80	73	71
77	57	66	67	89	56	54	80	74	53
90	80	70	58	78	55	66	95	52	90
65	75	67	71	71	56	89	61	92	68
68	76	62	72	77	78	84	80	67	61
58	93	78	64	60					

Construct a stem-and-leaf plot for these data.

Solution

STEP 1 If we split each observation between the tens place and the ones place, there will be five stems. However, the leaves will extend far to the right and the shape, center, and spread of the distribution will be unclear.

STEP 2 Divide each stem in half. The first 5-stem row holds numbers 50–54, the second 5-stem row holds numbers 55–59, the first 6-stem row holds numbers 60–64, etc.

STEP 3 The resulting stem-and-leaf plot, with divided stems, offers a better graphical description of the distribution. Note that the leaves have been ordered. Here is the stem-and-leaf plot for the personal trainer fee data.

```
5 | 1 2 3 4
5 | 5 6 6 7 8 8 8
6 | 0 0 1 1 1 2 3 4
6 | 5 6 6 6 6 7 7 7 8 8 9
7 | 0 1 1 1 1 2 3 3 4 4
7 | 5 6 7 7 8 8 8 9
8 | 0 0 0 0 2 4
8 | 5 7 9 9 9
9 | 0 0 2 3
9 | 5 7
```
Stem = 10, Leaf = 1

STEP 4 Notice that the data appear to tail off more slowly toward the large numbers. Therefore, the distribution of the data is slightly skewed to the right. In addition, the distribution is compact and there are no outlying values.

Figure 2.14 shows a technology solution.

TRY IT NOW Go to Exercise 2.67

```
The decimal point is 1
digit(s) to the right
of the |

5 | 1234
5 | 5667888
6 | 00111234
6 | 56666777889
7 | 0111123344
7 | 56778889
8 | 000024
8 | 57999
9 | 0023
9 | 57
```

Figure 2.14 R stem-and-leaf plot.

CHAPTER APP **Near-Earth Objects**

Construct a stem-and-leaf plot for these data. Describe the distribution in terms of shape, center, and spread.

Back-to-Back Stem-and-Leaf Plots

Two sets of data can be compared graphically using a back-to-back stem-and-leaf plot. Two plots are constructed using the same stem column. List the leaves for one data set to the left, and those for the other to the right.

EXAMPLE 2.9 **Cholesterol Levels**

Your total cholesterol level is the sum of your low-density lipoproteins (LDLs) and high-density lipoproteins (HDLs). A total cholesterol level of less than 200 mg/dL (milligrams per deciliter) is desirable, whereas 240 mg/dL or higher is considered high risk.[13] According to the Centers for Disease Control and Prevention, the average total cholesterol for adult Americans is about 200 mg/dL. Suppose a random sample of total cholesterol levels was obtained for men and women. The data are given in the table.

Men					Women				
110	124	132	147	157	183	190	201	211	154
164	172	180	193	201	186	212	213	169	173
210	224	112	158	165	177	195	203	203	189
173	181	193	205	216	207	158	218	213	205
194	194	179	185	185	205	204	189	179	177

Construct a back-to-back ordered stem-and-leaf plot for these data.

Solution

STEP 1 The following graph is a back-to-back stem-and-leaf plot for these data.

Men		Women
2 0	11	
4	12	
2	13	
7	14	
8 7	15	4 8
5 4	16	9
9 3 2	17	3 7 7 9
5 5 1 0	18	3 6 9 9
4 4 3 3	19	0 5
5 1	20	1 3 3 4 5 5 7
6 0	21	1 2 3 3 8
4	22	

Stem = 10, Leaf = 1

STEP 2 The center column of numbers (11, 12, 13, …) represents the stems for both groups. The stem-and-leaf plot for the men's data is constructed to the left, while the plot for the women's data is constructed to the right. Note that the leaves have been placed in increasing order, starting from the stem and proceeding outward.

STEP 3 The distribution for the men seems more spread out (i.e., has more variability), while the distribution of women's HDL-cholesterol levels is more compact and seems centered at a slightly greater value. **Figure 2.15** shows a technology solution.

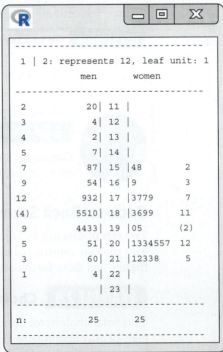

Figure 2.15 R back-to-back stem-and-leaf plot.

TRY IT NOW Go to Exercise 2.69

In mathematics, truncate means to discard the digits to the right of a specific place. To round a number to a certain position, consider the digit to the right of the rounding position. If this digit is 5 or greater, then round up. Otherwise, leave the rounding digit unchanged (and replace all digits to the right with 0).

 Rounding a number.

When constructing a stem-and-leaf plot, if there are two or more digits in each leaf, the trailing digits may be truncated or the entire leaf may be rounded. Suppose a data set includes the total yardage for randomly selected golf courses. Consider three observations, 6518, 6523, and 6576, and suppose each observation is split between the hundreds place and the tens place.

The following diagram shows the 65 stem row for three stem-and-leaf plots. The first is constructed with two-digit leaves. The second is constructed by simply truncating the ones, or last, digit. The third plot is constructed by rounding each leaf to the nearest 10.

Two-digit leaf
65 | 18 leaf = 18
65 | 23 leaf = 23
65 | 76 leaf = 76

Stem	Leaves
⋮	⋮
65	18 23 76
⋮	⋮

Stem = 100
Leaf = 10

Truncate each leaf
65 | 1̸8 leaf = 1
65 | 2̸3 leaf = 2
65 | 7̸6 leaf = 7

Stem	Leaves
⋮	⋮
65	1 2 7
⋮	⋮

Stem = 100
Leaf = 10

Round each leaf
65 | 18 rounds to 20, leaf = 2
65 | 23 rounds to 20, leaf = 2
65 | 76 rounds to 80, leaf = 8

Stem	Leaves
⋮	⋮
65	2 2 8
⋮	⋮

Stem = 100
Leaf = 10

Section 2.3 Exercises

Concept Check

2.47 True or False The stem in a stem-and-leaf plot must be only one digit.

2.48 True or False When constructing a stem-and-leaf plot, one can omit stem rows, between the smallest and largest stems, that contain no leaves.

2.49 True or False There may be more than one way to split each observation into a stem and a leaf.

2.50 True or False When constructing a stem-and-leaf plot, every observation must be split into a stem and a leaf in the same way.

2.51 Short Answer We try to construct a stem-and-leaf plot with 5–20 stems. What happens if we use fewer than 5 or more than 20 stems?

2.52 Short Answer Explain how you might identify an outlier in a stem-and-leaf plot.

Practice

2.53 Construct a stem-and-leaf plot for the following data. EX2.53

4.7	5.1	6.6	3.9	5.0	2.9	3.6	5.5	4.2	5.1
4.9	5.4	6.1	4.1	3.6	6.4	4.7	4.1	5.7	3.6
6.8	3.5	6.4	6.4	7.1	2.7	5.8	5.2	5.9	5.7

Determine a range of numbers to indicate the center of the data. Within this range, select one number that is a typical value for this data set.

2.54 Construct a stem-and-leaf plot for the given data. EX2.54

2.55 Construct a stem-and-leaf plot for the given data. Split each observation between the tens place and the ones place, and divide each stem in half. Determine a range of numbers to indicate the center of the data. Within this range, select one number that is a typical value for this data set. EX2.55

2.56 Construct a stem-and-leaf plot for the given data. Use the stem-and-leaf plot to identify any outliers in this distribution. EX2.56

2.57 Consider the following stem-and-leaf plot.

50	3
51	5
52	3 7
53	4 6
54	3 3 9
55	0 0 3 3 7
56	1 1 1 3 3 4 6 7 7
57	0 0 1 1 3 4 4 4 4 5 7 8 8
58	0 1 2 2 3 4 4 6 6 6 7 7
59	3 3 3 5 5 6 9
60	1 1 2

Stem = 10, Leaf = 1

(a) List the actual observations in the 54 stem row.
(b) What is a typical value for this data set?
(c) Do the data seem to be evenly distributed, or does one end tail off more slowly than the other?
(d) Does the stem-and-leaf plot suggest there are any outliers in this data set? If so, what are they?

2.58 Consider the data given in the table. EX2.58

1717	1719	1645	3739	3024	3664	3830
2991	2430	2730	3469	5086	2119	3021
3292	2844	3426	2067	3215	2767	3124
2573	2840	2449	2584	1505	1390	1645
2497	3466	3228	3192			

(a) Construct a stem-and-leaf plot by splitting each observation between the thousands place and the hundreds place.
(b) Construct a stem-and-leaf plot by splitting each observation between the hundreds place and the tens place (using two-digit leaves).
(c) Which plot presents a better picture of the distribution? Why?

Applications

2.59 Psychology and Human Behavior A random sample of patients involved in a psychology experiment was selected and the reaction time (in seconds) for each was recorded. REACT
(a) Construct a stem-and-leaf plot by splitting each observation between the ones place and the tenths place. Truncate the hundredths digit so that each leaf has a single digit.
(b) Construct a stem-and-leaf plot by splitting each observation between the ones place and the tenths place. Round each leaf to the nearest tenth so that each leaf has a single digit.
(c) Describe any differences between the two plots. What is a typical value?

2.60 Public Health and Nutrition The owner of Copperfield Racquet and Health Club randomly selected 50 people and recorded the number of calories burned after 20 minutes on a treadmill. CALBURN
(a) Construct a stem-and-leaf plot by splitting each observation between the tens place and the ones place.
(b) Construct a stem-and-leaf plot by splitting each observation between the tens place and the ones place, and by dividing each stem in half.
(c) Which stem-and-leaf plot is better? Why?

2.61 Physical Sciences A random sample of hot water temperatures (°F) on lower floors and upper floors in the Grand Hyatt San Antonio Hotel was obtained. WATEMP
(a) Construct a back-to-back stem-and-leaf plot to compare these two distributions.
(b) Using the plot in part (a), describe any similarities and/or differences between the distributions.

2.62 Physical Sciences The intensity of light is measured in foot-candles or in lux. In full daylight, the light intensity is approximately 10,700 lux; at twilight, the light intensity is about 11 lux. The recommended level of light in offices is 500 lux.[14] A random sample of 50 offices was obtained and the lux measurement at a typical work area was recorded for each. The data are given in the table. WORKLT

468	526	463	520	481	521	536	492	509	520
497	487	506	464	474	516	503	481	562	514
503	482	531	486	488	508	495	536	504	514
529	518	495	497	471	458	494	519	511	490
435	520	499	492	519	466	450	482	514	475

(a) Construct a stem-and-leaf plot for these light-intensity data.
(b) What is a typical light intensity? Are there any outliers? If so, what are they?

2.63 Public Health and Nutrition Some patients who visit an emergency room (ER) are admitted to the hospital for further observation or medical tests, whereas some are sent home after treatment. According to Hospital Review, patients who were admitted to the hospital spent an average of 96 minutes in the ER before being taken to a room.[15] Suppose a random sample of patients admitted to a hospital from the ER was obtained and the time in minutes until being taken to a room was recorded for each. ER
(a) Construct a stem-and-leaf plot for these data.
(b) What is a typical value? Are there any outliers.? If so, what are they?

2.64 Biology and Environmental Science South Carolina ports handled approximately 170,000 TEUs (Twenty-foot Equivalent Units) in January 2018.[16] A random sample of domestic containers was obtained and the volume of each (in TEUs) was recorded. SCPORT
(a) Construct a stem-and-leaf plot for these data. Split each observation between the tenths place and the hundredths place.
(b) Describe the container volume distribution in terms of shape, center, and spread. Are there any outliers? If so, what are they?

2.65 Business and Management A random sample of gasoline stations in Boston was obtained. The number of years each station has been in operation was recorded. GASSTA
(a) Construct a stem-and-leaf plot for these data.
(b) What is a typical number of years a station has been in operation? Are there any outliers? If so, what are they?

2.66 Manufacturing and Product Development Home Depot conducted a survey on the lifetime of dishwashers. Forty random users were contacted and asked to report the number of years their dishwasher lasted before needing replacement. DISH
(a) Construct a stem-and-leaf plot for these data. Split each observation between the ones place and the tenths place.
(b) Describe the distribution of dishwasher lifetimes in terms of shape, center, and spread.
(c) What is a typical lifetime? Are there any outliers? If so, what are they?

2.67 Biology and Environmental Science The Great Pumpkin Commonwealth promotes the hobby of growing giant pumpkins. This group establishes standards and regulations so that each pumpkin is of high quality and to ensure fairness in the competition for the largest pumpkin. A random sample of the largest pumpkins by weight (in pounds) from a recent competition was obtained.[17] PUMPKIN
(a) Construct a stem-and-leaf plot for these data.
(b) What is a typical weight for these giant pumpkins? Are there any outliers in the data set? If so, what are they?

Extended Applications

2.68 Sports and Leisure A greyhound race handicapper uses several factors to predict the winner, such as past performance, track condition, early speed, form, and competition. Races on a 5/16-mile track were randomly selected at the Naples–Fort Myers track, and the winning time (in seconds) was recorded for each. The data are given in the table.[18] **GREY**

30.59	30.97	31.01	31.51	31.57	31.29
30.93	30.81	30.66	30.89	30.80	31.05
31.25	32.13	31.22	30.90	30.73	30.85
30.94	31.23	31.42	30.59	30.78	30.96
31.07	31.06	31.14	31.19	31.41	31.02

(a) Construct a stem-and-leaf plot for these data. Split each observation between the tenths place and the hundredths place.
(b) What is a typical winning time? If a dog has never run better than 31.20 seconds in a 5/16-mile race, do you think it has a chance of winning? Justify your answer.
(c) Could a stem-and-leaf plot be constructed with the split between the ones place and the tenths place? How about between the tens place and the ones place? Explain your answer.

2.69 Biology and Environmental Science Many piano sellers recommend a special humidifier, especially for more expensive pianos. This device is installed inside the piano and works to keep the instrument in tune by maintaining a stable humidity. To test whether a humidifier really helps, several pianos with and without humidifiers were tuned and then checked six months later. Middle C was used as a measure of how well each piano stayed in tune. In a perfectly tuned piano, middle C has a frequency of 256 cycles per second. The frequency (in cycles per second) of middle C for each group, after six months, was obtained. **PIANO**

(a) Construct a back-to-back stem-and-leaf plot for these data.
(b) Use the plot in part (a) to describe any differences between the groups. Based on this plot, do you think a humidifier helps a piano stay in tune? Justify your answer.

2.70 Travel and Transportation National standards exist for every road sign, pavement marking, and traffic signal. However, there are no formal state policies regarding the duration of an amber light. According to Weiss & Associates, a New York law firm handling traffic tickets, the Administrative Code in New York City specifies that a traffic light remain amber for 2–3 seconds. In California, the amber light duration is based on the speed limit, and in Texas, the amber light must have a duration of 4.7 seconds on roads with a speed limit of 50 mph or higher.[19] Suppose a random sample of traffic signals for 50-mph roads in Cedar Hills, Texas, was selected, and the duration of the amber light was recorded for each. **AMBLT**

(a) Construct a stem-and-leaf plot for these data. Divide each stem into five parts.
(b) Based on the plot in part (a), do you believe this city has set the amber light duration to meet Texas regulations? Justify your answer.

2.71 Education and Child Development According to a recent study, approximately 70% of college graduates have student debt and more than 44 million Americans owe $1.4 trillion in student loan money.[20] Research suggests that most people with student loans expect to pay off the debt in their 40s. A random sample of student loan borrowers was obtained and the age at which each paid off their debt was recorded. **DEBT**

(a) Construct a stem-and-leaf plot for these data.
(b) Describe the distribution of ages in terms of shape, center, and spread.
(c) Estimate the proportion of students who were able to pay off their debt before the age of 40.

2.72 Marketing and Consumer Behavior Your credit score can have an effect on the amount of money you can borrow, the interest rate, and even your automobile insurance. The most widely used credit score, the FICO score, was created by the Fair Isaac Corporation and ranges in value from 300 to 850.[21] Suppose a random sample of Discover Card customers was obtained and the FICO score for each was recorded. **FICO**

(a) Construct a stem-and-leaf plot for these data.
(b) Describe the distribution of FICO scores in terms of shape, center, and spread.
(c) Consider the list of the different categories of FICO scores.

Category	FICO score
Excellent credit	750 +
Good credit	700–749
Fair credit	650–699
Poor credit	600–649
Bad credit	Less than 600

Estimate the proportion of Discover Card members with excellent credit. Use the credit categories to construct a bar chart for these data.

2.4 Frequency Distributions and Histograms

Background and Construction

MAIN Stem-and-leaf plots can be used to describe the shape, center, and variability of a numerical data set, but they can become huge and complex if the number of observations is large. A summary table like a frequency distribution for categorical data would be helpful. However, when the data set is numerical, there are no natural categories, as there are for qualitative data. The solution is to use intervals as categories, or classes. We can then construct a frequency distribution for continuous data (similar to the categorical case), as well as a histogram (analogous to a bar chart for categorical data). For a random sample of fishing vessels registered in the United Kingdom, **Figure 2.16** shows a histogram of the main power (in kW).[22]

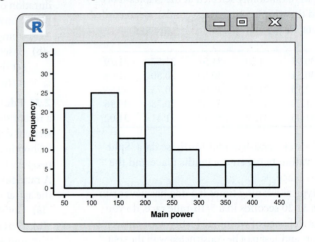

Figure 2.16 An example of a histogram.

Definition
A **frequency distribution** for numerical data is a summary table that displays classes, frequencies, relative frequencies, and cumulative relative frequencies.

How to Construct a Frequency Distribution for Numerical Data

In other words, partition the measurement axis into 5–20 subintervals.

1. Choose a range of values that captures all the data. Divide it into nonoverlapping (usually equal) intervals. Each interval is called a **class**, or **class interval**. The endpoints of each class are the class boundaries.
2. We use the left-endpoint convention. An observation equal to an endpoint is allocated to the class with that value as its lower endpoint. Hence, the lower class boundary is always included in the interval, and the upper class boundary is never included. This ensures that each observation falls into exactly one interval.
3. In practice, there should be 5–20 intervals. Use friendly numbers, for example, 10–20 and 20–30, rather than more complicated categories, such as 15.376–18.457 and 18.457–21.538.
4. Count the number of observations in each class interval. This count is called the **class frequency** or simply the **frequency**.
5. Compute the proportion of observations in each class. This ratio, the class frequency divided by the total number of observations, is called the **relative frequency**.
6. Find the **cumulative relative frequency** (CRF) for each class: the sum of all the relative frequencies of classes up to and including that class. This column is a running total, or accumulation, of relative frequency by row.

Example: Frequency Distribution

EXAMPLE 2.10 **Nuts and Bolts**

BOLT

Torque is a measure of the force needed to cause an object to rotate. It is usually measured in foot-pounds (ft-lb). As part of a quality-control program, Whirlpool inspectors measure the initial torque needed to loosen the balancing bolts on each leg of a clothes washer. A random sample of these measurements is given in the table.

20.4	41.3	13.0	24.1	11.0	44.4	28.4	37.5	16.9	53.4
36.4	14.9	62.1	25.6	63.7	31.7	43.5	57.2	23.1	45.7
24.2	38.1	35.5	51.1	26.4					

Construct a frequency distribution for these data.

Solution

STEP 1 The data set is numerical (continuous). The observations are measurements, and each can be any number in some interval. Scan the data to find the smallest and largest observations (11.0 and 63.7). Choose between 5 and 20 reasonable (equal) intervals that capture all the data.

STEP 2 The range of values 10–70 captures all the data. Divide this range using the friendly numbers 10, 20, 30, ... into the class intervals 10–20, 20–30, 30–40, etc.

Remember to use the left-hand endpoint convention.

STEP 3 Count the number of observations in each interval. For example, in the interval 10–20, there are four observations (16.9, 14.9, 11.0, and 13.0), so the frequency is 4.

STEP 4 Compute the proportion of observations in each class. For example, in the interval 10–20, the relative frequency is 4 (observations) divided by 25 (total number of observations).

RAPID REVIEW *Finding cumulative frequency.*

STEP 5 Find the CRF for each class. For example, for the class 30–40, the cumulative relative frequency is the sum of the relative frequencies of this class and of all those listed above it: $0.16 + 0.28 + 0.20 = 0.64$.

Class	Frequency	Relative frequency		Cumulative relative frequency	
10–20	4	0.16	(= 4/25)	0.16	(= 0.16)
20–30	7	0.28	(= 7/25)	0.44	(= 0.16 + 0.28)
30–40	5	0.20	(= 5/25)	0.64	(= 0.44 + 0.20)
40–50	4	0.16	(= 4/25)	0.80	(= 0.64 + 0.16)
50–60	3	0.12	(= 3/25)	0.92	(= 0.80 + 0.12)
60–70	2	0.08	(= 2/25)	1.00	(= 0.92 + 0.08)
Total	25	1.00			

```
     Var1   Freq  RelFreq  CRF
1  [10,20)    4     0.16   0.16
2  [20,30)    7     0.28   0.44
3  [30,40)    5     0.20   0.64
4  [40,50)    4     0.16   0.80
5  [50,60)    3     0.12   0.92
6  [60,70)    2     0.08   1.00
```

Figure 2.17 Frequency distribution using R.

STEP 6 As for categorical data, if you must construct a frequency distribution by hand, an additional tally column is helpful (as discussed in Section 2.2). Insert this after the class column, and use a tally mark or tick mark to count observations as you read them from the table. **Figure 2.17** shows a technology solution.

TRY IT NOW Go to Exercise 2.83

Common Error: Sum the cumulative relative frequency column in a frequency distribution.

Correction: The total CRF is not constant and should not be reported in the Total row of a frequency distribution.

The last (total) row in a frequency distribution is optional, but it is a good check of your calculations. The frequencies should sum to the total number of observations (25 in Example 2.10), and the relative frequencies should sum to 1.00 (subject to round-off error).

The CRF of the first class row is equal to the relative frequency of the first class. There are no other observations before the first class. The CRF of the last class should be 1.00 (subject to round-off error). You must accumulate all the data by the last class.

CRF gives the proportion of observations in that class and all previous classes. In Example 2.10, the CRF of the class 40–50 is 0.80. Interpretation: The proportion of torque measurements less than 50 is 0.80.

A CLOSER LOOK

1. Suppose you were given just the CRF for each class. To find the relative frequency for a class, take the class CRF and subtract the previous class CRF. In Example 2.10, to find the relative frequency for the class 50–60: 0.92 (CRF for the class 50–60) − 0.80 (CRF for the previous class 40–50) = 0.12.

2. If the data set is numerical and discrete, use the same procedure outlined above for constructing a frequency distribution. If the number of discrete observations is small, then each value may be a class, or category. In addition, certain understood liberties are sometimes acceptable in listing the classes. For example, suppose a discrete data set consists of integers from 1 to 30. One might use the classes 1–5, 6–10, 11–15, 16–20, 21–25, and 26–30. This is not a strict partition of the interval 1–30 even though these classes are disjoint, or do not overlap. These classes do not allow for all numbers between 1 and 30. For example, the value 5.5 is between 1 and 30 but does not fall into any of these classes. However, these classes work fine in this case because each observation is an integer. The resulting frequency distribution is perfectly valid.

CHAPTER APP **Near-Earth Objects**
Construct a frequency distribution for these data.

Histograms

A histogram is a graphical representation of a frequency distribution, a plot of frequency versus class interval. Given a frequency distribution, here is a procedure for constructing a histogram.

How to Construct a Histogram

1. Draw a horizontal (measurement) axis and place tick marks corresponding to the class boundaries.
2. Draw a vertical axis and place tick marks corresponding to frequency. Label each axis.
3. Draw a rectangle above each class with height equal to frequency.

EXAMPLE 2.11 Nuts and Bolts, Continued

Construct a frequency histogram for the torque data presented in Example 2.10. For reference, the technology solution for the frequency distribution from Example 2.10 is given in **Figure 2.18**.

Solution

STEP 1 Draw a horizontal axis and place tick marks corresponding to the class boundaries, or endpoints of each class: 10 through 70 by tens.

STEP 2 Draw a vertical axis for frequency and place appropriate tick marks by examining the frequency distribution. The frequencies range from 0 to 7, so draw tick marks at 0 to at least 7 on the vertical axis.

STEP 3 Draw a rectangle above each class with height equal to frequency. The resulting histogram is shown in **Figure 2.19**. **Figure 2.20** shows a technology solution.

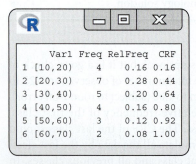

Figure 2.18 Frequency distribution from Example 2.10.

2.4 Frequency Distributions and Histograms

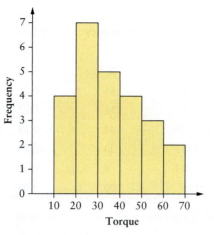

Figure 2.19 Frequency histogram for torque.

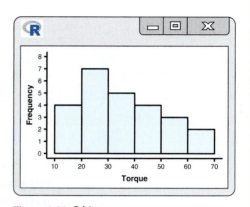

Figure 2.20 R histogram.

TRY IT NOW Go to Exercises 2.93 and 2.97

CHAPTER APP **Near-Earth Objects**

Construct a frequency histogram for these data.

A CLOSER LOOK

1. A histogram conveys information about the shape, center, and variability of the distribution. In addition, we can quickly identify any potential outliers.

2. If you must draw a histogram by hand, then you need to construct the frequency distribution first. However, calculators and computers construct histograms directly from the data. The frequency distribution is constructed in the background and is usually not displayed.

3. To construct a relative frequency histogram, plot relative frequency versus class interval. The only difference between a frequency histogram and a relative frequency histogram is the scale on the vertical axis. The two graphs are identical in appearance. Example 2.12 includes both a frequency histogram and a relative frequency histogram.

4. Histograms should not be used for inference. They provide a quick graphical summary of the distribution of data and merely suggest certain characteristics.

Histogram usually means frequency histogram.

Relative Frequency Histogram

EXAMPLE 2.12 **That Sinking Feeling**

A recent study suggests that parts of San Francisco Bay are sinking faster than the sea is rising. This phenomenon is due in part because much of the bay area is built on natural mud deposits. A random sample of locations in the San Francisco Bay area was obtained, and the yearly subsidence rate (in mm) was measured. The resulting frequency distribution is shown in the following table.

Class	Frequency	Relative frequency	Cumulative relative frequency
0–2	6	0.03	0.03
2–4	22	0.11	0.14
4–6	46	0.23	0.37
6–8	54	0.27	0.64
8–10	44	0.22	0.86
10–12	20	0.10	0.96
12–14	8	0.04	1.00
Total	200	1.00	

Use this table to construct a frequency histogram and a relative frequency histogram for these data.

Solution

STEP 1 For each graph, draw a horizontal axis and place tick marks at the class boundaries: 0, 2, 4, ..., 14.

STEP 2 For the frequency histogram:

(a) Draw a vertical axis for frequency. Since the largest frequency is 54, use tick marks at 0, 10, 20, ..., 60.

(b) Draw a rectangle above each class with height equal to frequency. The resulting frequency histogram is shown in **Figure 2.21**.

STEP 3 For the relative frequency histogram:

(a) Draw a vertical axis for relative frequency. Because the largest relative frequency is 0.27, use tick marks at 0, 0.05, 0.10, ..., 0.30.

(b) Draw a rectangle above each class with height equal to relative frequency. The resulting relative frequency histogram is shown in **Figure 2.22**.

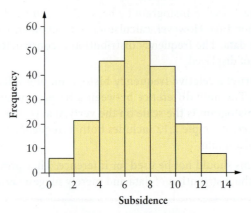

Figure 2.21 Frequency histogram for the San Francisco Bay area subsidence data.

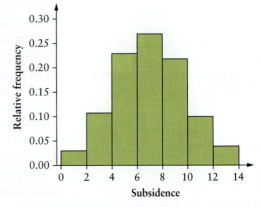

Figure 2.22 Relative frequency histogram for the San Francisco Bay area subsidence data.

TRY IT NOW Go to Exercise 2.95(a)

CHAPTER APP **Near-Earth Objects**

Construct a relative frequency histogram for these data.

Density Histograms

If the class widths are unequal in a frequency distribution, then neither the frequency nor the relative frequency should be used on the vertical axis of the corresponding histogram. To account for the unequal class widths, set the area of each rectangle equal to the relative frequency. In this case, the height of each rectangle is called the density, and it is equal to the relative frequency divided by the class width.

If two classes have the same frequency, but one class has double the width, then the corresponding rectangle in a traditional histogram would have double the area. This misrepresents the distribution.

 Area of a rectangle.

How to Find the Density
To find the density for each class:

1. Set the area of each rectangle equal to the relative frequency.
   The area of each rectangle is height times class width.
   Area of rectangle = Relative frequency
   $$= (\text{Height}) \times (\text{Class width})$$
2. Solve for height.
   Density = Height = (Relative frequency)/(Class width)

Example 2.13 shows an extended frequency distribution with the density of each class included, as well as the corresponding density histogram.

EXAMPLE 2.13 Accident Demographics

Younger drivers tend to be involved in more automobile accidents than older drivers are. This difference may be attributed to younger drivers' propensity for taking risks and inexperience. The following table shows the number of automobile accident fatalities in the United States in 2016 by age group.[23] The width of each class and the density calculations are also shown.

Class	Frequency	Relative frequency	Width	Density	
13–16	407	0.0129	3	0.0043	(= 0.0129/3)
16–20	2413	0.0765	4	0.0191	(= 0.0765/4)
20–25	4379	0.1389	5	0.0278	(= 0.1389/5)
25–30	3789	0.1202	5	0.0240	(= 0.1202/5)
30–40	5667	0.1798	10	0.0180	(= 0.1798/10)
40–50	4883	0.1549	10	0.0155	(= 0.1549/10)
50–60	5628	0.1785	10	0.0179	(= 0.1785/10)
60–70	4361	0.1383	10	0.0138	(= 0.1383/10)
Total	31527	1.0000			

Use this table to construct a density histogram for these data.

Solution

STEP 1 The class intervals are of unequal width, so the class density must be used as the height of each rectangle in a histogram.

STEP 2 Draw a horizontal axis corresponding to age. The classes range from 13 to 70. Use tick marks corresponding to the endpoints of each class: 13, 16, 20, ... , 70.

STEP 3 Add a vertical axis for density. The largest density is 0.0278, so use the tick marks 0, 0.005, 0.010, ... , 0.030.

STEP 4 Draw a rectangle above each class with height equal to density. The resulting density histogram is shown in **Figure 2.23**. A technology solution is shown in **Figure 2.24**.

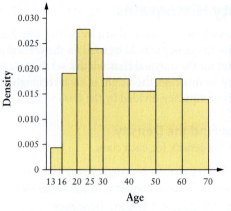
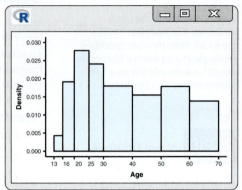

Figure 2.23 Histogram for unequal class widths: density histogram for the accident age data.

Figure 2.24 A technology solution: R density histogram.

TRY IT NOW Go to Exercise 2.101

Shape of a Distribution

Because the relative frequency is equal to the area of each rectangle in a density histogram, the sum of the areas of all the rectangles is 1. This is an important concept as we begin to associate area with probability.

The shape of a distribution, represented in a histogram, is an important characteristic. To help describe the various shapes, we draw a smooth curve along the tops of the rectangles that captures the general nature of the distribution (as shown in **Figure 2.25**). To help identify and describe distributions quickly, a smoothed histogram is often drawn on a graph without a vertical axis, without any tick marks on the measurement axis, and without any rectangles (as shown in **Figure 2.26**).

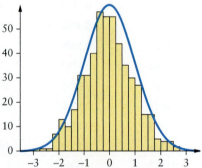
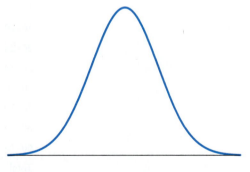

Figure 2.25 Smooth curve that captures the general shape of the distribution.

Figure 2.26 Typical smoothed histogram.

The first important characteristic of a distribution is the number of peaks.

Definition

1. A **unimodal** distribution has one peak. This is very common, as almost all distributions have a single peak.
2. A **bimodal** distribution has two peaks. This shape is not very common but may occur if data from two different populations are accidentally mixed.
3. A **multimodal** distribution has more than one peak. A distribution with more than two distinct peaks is very rare.

2.4 Frequency Distributions and Histograms

Examples of these three types of distributions are shown in **Figures 2.27–2.29**.

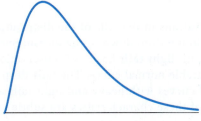

Figure 2.27 Unimodal distribution.

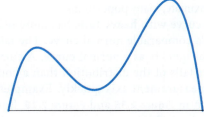

Figure 2.28 Bimodal distribution.

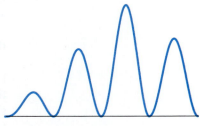

Figure 2.29 Multimodal distribution.

The following characteristics are used to further classify and identify unimodal distributions.

Definition

1. A unimodal distribution is **symmetric** if there is a vertical line of symmetry in the distribution.
2. The **lower tail** of a distribution is the leftmost portion of the distribution, and the **upper tail** is the rightmost portion of the distribution.
3. If a unimodal distribution is not symmetric, then it may be **skewed**.
   a. In a **positively skewed** distribution, or a distribution that is **skewed to the right**, the upper tail extends farther than the lower tail.
   b. In a **negatively skewed** distribution, or a distribution that is **skewed to the left**, the lower tail extends farther than the upper tail.

Figure 2.30 and **Figure 2.31** show examples of symmetric distributions. Each shows the (dashed) line of symmetry. The left half of the distribution is a mirror image of the right half. A bimodal or multimodal distribution may also be symmetric, and many distributions are approximately symmetric.

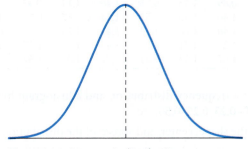

Figure 2.30 Symmetric distribution.

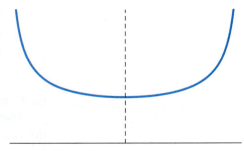

Figure 2.31 Symmetric distribution.

Examples of skewed distributions are shown in **Figure 2.32** and **Figure 2.33**. Positively skewed distributions are more common. The distribution of the lifetime of an electronics part might be positively skewed.

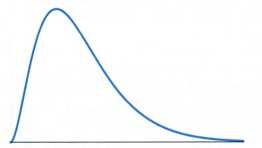

Figure 2.32 Positively skewed distribution.

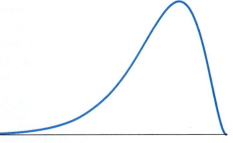

Figure 2.33 Negatively skewed distribution.

The most common unimodal distribution shape is a **normal curve** (as shown in **Figure 2.34**). This curve is symmetric and bell-shaped and can be used to model, or approximate, many populations.

A curve with **heavy tails** has more observations in the tails of the distribution than a comparable normal curve. The tails do not drop down to the measurement axis as quickly as a normal curve. A curve with **light tails** has fewer observations in the tails of the distribution than a comparable normal curve. The tails drop to the measurement axis quickly. Examples of curves with heavy and light tails are shown in **Figure 2.35** and **Figure 2.36**. Both of these characteristics are subtle and tricky to spot.

Figure 2.34 Normal curve.

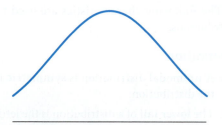

Figure 2.35 A distribution with heavy tails.

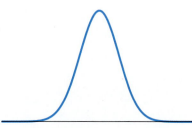

Figure 2.36 A distribution with light tails.

EXAMPLE 2.14 Solar Power

SOLAR

Solar panels have become more popular as their efficiency has increased and their prices have dropped. The amount of energy generated by a solar panel depends on the amount of direct sunlight and the panel's theoretical power production.[24] Suppose a sample of 50 solar panels located on the roofs of residential homes was obtained and a daily power output of each was measured (in kWh). The data are given in the following table.

0.78	0.89	1.85	0.75	1.46	1.12	0.95	1.38	1.31	1.48
0.83	1.40	1.05	1.52	1.34	0.67	0.71	1.66	1.42	1.43
1.47	1.40	1.60	1.53	1.41	1.75	1.24	0.45	0.24	1.57
1.13	1.20	1.71	1.76	1.45	1.06	1.35	1.80	1.63	1.62
1.51	1.67	1.34	1.41	1.77	1.04	1.52	1.71	1.66	1.64

(a) Construct a frequency distribution and a histogram for these data using the class intervals 0–0.25, 0.25–0.50, etc.

(b) Describe the shape, center, and spread of the distribution.

(c) What proportion of observations are less than 1 kWh?

(d) What proportion of observations are at least 1.5 kWh?

Solution

Construct the frequency distribution and use this table to sketch a histogram. Draw a smooth curve to approximate the distribution and use this to describe the shape, the middle of the distribution, and the variability. Use the relative frequencies or cumulative relative frequencies to find the proportion of observations in certain classes.

(a) The class intervals are given. Construct a frequency distribution and compute the frequency, relative frequency, and cumulative relative frequency for each class.

Class	Frequency	Relative frequency		Cumulative relative frequency	
0.00–0.25	1	0.02	(= 1/50)	0.02	(= 0.02)
0.25–0.50	1	0.02	(= 1/50)	0.04	(= 0.02 + 0.02)
0.50–0.75	2	0.04	(= 2/50)	0.08	(= 0.04 + 0.04)
0.75–1.00	5	0.10	(= 5/50)	0.18	(= 0.08 + 0.10)
1.00–1.25	7	0.14	(= 7/50)	0.32	(= 0.18 + 0.14)
1.25–1.50	15	0.30	(= 15/50)	0.62	(= 0.32 + 0.30)
1.50–1.75	14	0.28	(= 14/50)	0.90	(= 0.62 + 0.28)
1.75–2.00	5	0.10	(= 5/50)	1.00	(= 0.90 + 0.10)
Total	50	1.00			

Use the frequency distribution to sketch the histogram (**Figure 2.37**). A technology solution is shown in **Figure 2.38**.

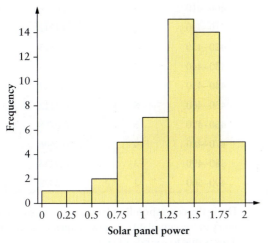

Figure 2.37 Histogram for the solar panel data.

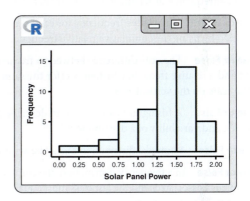

Figure 2.38 R histogram for the solar panel data.

(b) The distribution is negatively skewed. The observations are clustered in the upper tail, and the lower tail extends farther than the upper tail.

To estimate the center of the distribution, use the histogram to identify a value such that approximately half of the observations are below that number and half are above that number. A number between 1.25 and 1.50 appears to divide the ordered data in half. Typical values for this data set are in this range, and an estimate of the center is 1.37.

The variability is typically described as either compact (data that are compressed or squeezed together) or spread out (observations that extend over a wide range). Although this distinction is somewhat subjective for now, this data set seems fairly compact. All of the observations lie between 0.24 and 1.85 (even though the smallest class boundary is 0.00 and the largest class boundary is 2.00).

(c) Using the cumulative relative frequency column of the frequency distribution, the proportion of observations less than 1 is 0.18.

(d) There are two ways to find the proportion of observations that are at least 1.5.

(i) Add the relative frequencies that correspond to the classes that are at least 1.50.

$$\underset{1.50-1.75}{0.28} + \underset{1.75-2.00}{0.10} = 0.38$$

(ii) Find the cumulative relative frequency up to 1.50 and subtract this value from 1 (the total relative frequency).

Proportion of observations ≥ 1.50
$= 1 - ($proportion of observations $< 1.50)$
$= 1 - 0.62 = 0.38$

Section 2.4 Exercises

Concept Check

2.73 True or False The classes in a frequency distribution may overlap.

2.74 True or False The classes in a frequency distribution should have the same width.

2.75 True or False The cumulative relative frequency for each class in a frequency distribution may be greater than 1.

2.76 True or False The relative frequency for each class can be determined by using the cumulative relative frequencies.

2.77 True or False The only difference between a frequency histogram and a relative frequency histogram (for the same data) is the scale on the vertical axis.

2.78 True or False A histogram can be used to describe the shape, center, and variability of a distribution.

2.79 True or False A bimodal distribution cannot be symmetric.

2.80 True or False In a frequency distribution, the cumulative relative frequency associated with the last class must equal 1.

2.81 Short Answer
(a) When is a density histogram appropriate?
(b) In a density histogram, what is the sum of areas of all rectangles?

2.82 Fill in the Blank
(a) The most common unimodal distribution is _____.
(b) A unimodal distribution is _____ if there is a vertical line of symmetry.
(c) If a unimodal distribution is not symmetric, then it is _____.

Practice

2.83 Consider the data given in the table. EX2.83

87	91	91	89	81	86	81	85	86	86
89	86	90	87	89	90	88	88	83	90
85	85	90	89	79	92	83	78	91	85
80	91	87	87	90	83	82	86	80	92

Construct a frequency distribution to summarize these data using the class intervals 78–80, 80–82, 82–84, … .

2.84 Consider the data given and construct a frequency distribution to summarize these data. EX2.84

2.85 Consider the following frequency distribution.

Class	Frequency	Relative frequency	Cumulative relative frequency
400–410	5	0.0758	0.0758
410–420	8	0.1212	0.1970
420–430	10	0.1515	0.3485
430–440	12	0.1818	0.5303
440–450	9	0.1364	0.6667
450–460	8	0.1212	0.7879
460–470	5	0.0758	0.8637
470–480	4	0.0606	0.9243
480–490	3	0.0455	0.9698
490–500	2	0.0303	1.0001

Draw the corresponding frequency histogram. (Notice the last entry in the cumulative relative frequency column is not exactly 1. This is due to round-off error.)

2.86 Consider the following frequency distribution.

Class	Frequency	Relative frequency	Cumulative relative frequency
0.5–1.0	6	0.03	0.03
1.0–1.5	8	0.04	0.07
1.5–2.0	10	0.05	0.12
2.0–2.5	16	0.08	0.20
2.5–3.0	24	0.12	0.32
3.0–3.5	34	0.17	0.49
3.5–4.0	36	0.18	0.67
4.0–4.5	22	0.11	0.78
4.5–5.0	18	0.09	0.87
5.0–5.5	12	0.06	0.93
5.5–6.0	8	0.04	0.97
6.0–6.5	4	0.02	0.99
6.5–7.0	2	0.01	1.00

Draw the corresponding relative frequency histogram.

2.87 Complete the following frequency distribution.

Class	Frequency	Relative frequency	Cumulative relative frequency
100–150	155		
150–200	120		
200–250	130		
250–300	145		
300–350	150		
350–400	100		
Total			

2.88 Complete the frequency distribution.

Class	Frequency	Relative frequency	Cumulative relative frequency
1.0–1.1		0.05	
1.1–1.2	20		
1.2–1.3		0.15	
1.3–1.4	65		
1.4–1.5		0.25	
1.5–1.6	35		
1.6–1.7	25		
1.7–1.8			
Total	300		

2.89 Complete the frequency distribution and draw the corresponding histogram.

Class	Frequency	Relative frequency	Cumulative relative frequency
0–25			0.150
25–50			0.350
50–75			0.525
75–100			0.675
100–125			0.800
125–150			0.900
150–175			0.975
175–200			1.000
Total	1000		

2.90 Consider the given data. **EX2.90**
(a) Construct a frequency distribution to summarize these data using the class intervals 0–1, 1–2, 2–3, etc., and draw the corresponding histogram.
(b) Use the histogram to describe the shape of the distribution. Are there any outliers?

2.91 Consider the given data. **EX2.91**
(a) Construct a frequency distribution to summarize these data and draw the corresponding histogram.
(b) Use the histogram to describe the shape of the distribution.
(c) Use the frequency distribution to estimate the middle of the data: a number M such that 50% of the observations are below M and 50% are above M.
(d) Use the frequency distribution to estimate a number Q_1 such that 25% of the observations are below Q_1 and 75% are above Q_1.
(e) Use the frequency distribution to estimate a number Q_3 such that 75% of the observations are below Q_3 and 25% are above Q_3.

Applications

2.92 Biology and Environmental Science A weather station in Scranton, Pennsylvania, collects data on temperature, wind speed, and precipitation.[25] The average dewpoint for 50 randomly selected days in early 2018 was recorded. **DEWPT**
(a) Construct a frequency distribution to summarize these data, and draw the corresponding histogram.
(b) Describe the shape of the distribution. Are there any outliers?

2.93 Fuel Consumption and Cars The quality of an automobile battery is often measured by cold cranking amps (CCA), a measure of the current supplied at 0°F. Thirty automobile batteries were randomly selected and subjected to subfreezing temperatures. The resulting CCA data are given in the table. **BATTQ**

63	87	302	4	259	106	198	55	99	134
122	514	91	117	325	39	30	164	75	16
340	199	77	217	64	320	145	84	47	232

(a) Construct a frequency distribution to summarize these data, and draw the corresponding histogram.
(b) Describe the shape of the distribution.
(c) Estimate the middle of the distribution: a number M such that 50% of the data are below M and 50% are above M.

2.94 Marketing and Consumer Behavior The weights of diamonds and other precious stones are usually measured in carats. One carat is traditionally equal to 200 mg. A random sample of the weights (in carats) of conflict-free loose diamonds is given in the table.[26] **DIAMND**

0.31	1.16	0.50	0.50	1.01	1.53	1.42	1.38
1.55	1.70	1.74	1.44	0.52	1.05	1.30	1.45
1.76	0.40	1.00	1.51	1.82	0.71	1.91	0.50
2.68	1.00	1.51	1.30	1.66	1.95	1.85	0.61
1.91	2.48	1.10	2.03	1.63	2.11	1.59	1.08

(a) Construct a frequency distribution and a histogram for these data.
(b) Multiply each observation in the table by 200, to convert the weights into milligrams. Construct a frequency distribution and a histogram for these new, transformed data.
(c) Compare the two histograms. Are the shapes similar? Describe any differences.

2.95 Public Health and Nutrition Vitamin B_3 (niacin) helps detoxify the body, aids digestion, can ease the pain of migraine headaches, and helps promote healthy skin. A random sample of adults in the United States and in Europe was obtained and the daily intake of niacin (in milligrams) was recorded. The data are summarized in the table.

Class	United States frequency	Europe frequency
0–3	15	4
3–6	23	6
6–9	21	12
9–12	14	17
12–15	12	32
15–18	9	25
18–21	3	20
21–24	2	10

(a) Construct two relative frequency histograms, one for the United States and one for Europe.
(b) Describe the shape of each histogram. Does a comparison of the two histograms suggest any differences in niacin intake between the two samples? Explain.

2.96 Manufacturing and Product Development In the United States, yarn is often sold in hanks. For woolen yarn, one hank is approximately 1463 m. A quality control inspector uses a special machine to quickly measure each hank. A random sample was obtained during the manufacturing process, and the length (in meters) of each hank was recorded. YARN
(a) Construct a histogram for these data.
(b) Describe the distribution in terms of shape, center, and variability.

2.97 Sports and Leisure The National Hockey League (NHL) is concerned about the number of penalty minutes assessed to each player. While some people in attendance hope to see a lot of fighting (and penalty minutes), the League Office believes most fans are interested in good, clean hockey. A sample of total penalty minutes per player near the end of the 2017–2018 regular season was obtained, and the data are given in the table.[27] PENALTY

36	65	20	40	24	42	16	14	36	22
51	17	44	21	26	22	10	48	26	46
56	24	28	52	16	20	26	16	42	30
38	22	16	66	36	12	61	29	36	22
17	20	32	26	37	22	14	34	59	18

(a) Construct a histogram for these data. Describe the distribution in terms of shape, center, and variability.
(b) Find a value m for the number of minutes such that 90% of all players have fewer than m penalty minutes.

2.98 Economics and Finance Bitcoin is an electronic, or digital, currency, and transactions involving bitcoins are conducted without any banks. It is possible to buy goods and services using bitcoins, exchange bitcoins for other forms of currency, and acquire bitcoins by solving mathematical problems. A random sample of closing prices of a single bitcoin was obtained.[28] BITCOIN
(a) Construct a histogram for these data. Describe the distribution in terms of shape, center, and variability.

(b) Find two values, Q_1 and Q_3, such that the interval Q_1 to Q_3 contains the middle 50% of the data.

Extended Applications

2.99 Biology and Environmental Science Fruits such as cherries and grapes are harvested and placed in a shallow box or crate called a lug. The size of a lug varies, but one typically holds between 16 and 28 lb. A random sample of the weight (in pounds) of full lugs holding peaches was obtained, and the data are summarized in the table.

Class	Frequency	Relative frequency	Cumulative relative frequency
20.0–20.5	6		
20.5–21.0	12		
21.0–21.5	17		
21.5–22.0	21		
22.0–22.5	28		
22.5–23.0	25		
23.0–23.5	19		
23.5–24.0	15		
24.0–24.5	11		
24.5–25.0	10		

(a) Complete the frequency distribution.
(b) Construct a histogram corresponding to this frequency distribution.
(c) Estimate the weight w such that 90% of all full peach lugs weigh more than w.

2.100 Travel and Transportation The Shanghai Maglev train, connecting Pudong International Airport to Longyang Road Station, is the fastest-operating train in the world.[29] Magnets create a frictionless system in which the train operates at a distance of 100–150 mm from the rail. The size of this air gap is monitored constantly to ensure a safe ride. A random sample of the size of air gaps (in millimeters) at one specific location in the track was obtained. A partial frequency distribution for these data is shown in the table.

Class	Frequency	Relative frequency	Cumulative relative frequency
100–105			0.050
105–110			0.425
110–115			0.625
115–120			0.750
120–125			0.850
125–130			0.925
130–135			0.975
135–140			1.000
Total	200		

(a) Complete the frequency distribution.

(b) Draw a histogram corresponding to this frequency distribution.
(c) What proportion of air gaps were between 110 and 125 mm?

2.101 Biology and Environmental Science Many scientists have warned that global warming is causing the polar ice caps to melt and, therefore, sea levels around the world to rise. A random sample of linear relative sea level trends from random stations around the world was obtained (in mm/yr) and the data are summarized in the following table.[30]

Class	Frequency	Relative frequency	Width	Density
−20 to −10	4			
−10 to 0	14			
0–0.5	4			
0.5–1.0	6			
1.0–2.0	18			
2.0–3.0	42			
3.0–4.0	30			
4.0–10.0	24			
Total	142			

(a) Complete the frequency distribution.
(b) A traditional frequency histogram or relative frequency histogram is not appropriate in this case. Why not?
(c) Construct a density histogram corresponding to this frequency distribution.

2.102 Fuel Consumption and Cars The total cost of owning a car includes the amount spent on repairs. Before purchasing a new car, many consumers research the past quality of specific makes and models. The table lists the number of problems per 100 vehicles in 2017.[31] REPAIR

Lexus	110	Porsche	110	Toyota	12
Buick	126	Mercedes-Benz	131	Hyundai	133
BMW	139	Chevrolet	142	Honda	143
Jaguar	144	Kia	148	Lincoln	150
Mini	150	GMC	151	Cadillac	152
Audi	153	Volve	154	Chrysler	159
Subaru	164	Volkswagen	164	Mazda	166
Acura	167	Nissan	170	Land Rover	78
Mitsubishi	182	Ford	183	Ram	183
Dodge	187	Infiniti	203	Jeep	209
Fiat	298				

(a) Construct a histogram for these data.
(b) Describe the distribution in terms of shape, center, and variability.
(c) Find a number Q_1 such that 25% of the problem data are less than Q_1. Find a number Q_3 such that 25% of the problem data are greater than Q_3.
(d) How many values should be between Q_1 and Q_3? Find the actual number of values between Q_1 and Q_3. Explain any difference between these two values.

2.103 Public Health and Nutrition The Canadian Health Measures Survey is used to gather information to help improve health programs and services in Canada.[32] Suppose a random sample of males and females ages 20 to 39 who responded to the survey was obtained and the total grip strength was measured (in kilograms) for each. GRIP
(a) Use the same class intervals to construct a frequency distribution for both data sets.
(b) Construct two relative frequency histograms, one for the females and one for the males.
(c) Describe the shape of each histogram. Explain any differences in total grip strength between the two distributions.
(d) Why is it more appropriate to compare the two distributions using a relative frequency histogram rather than a frequency histogram?

Chapter 2 Summary

Concept	Page	Notation / Formula / Description
Categorical data set	29	Consists of observations that may be placed into categories.
Numerical data set	29	Consists of observations that are digits.
Discrete data set	30	The set of all possible values is finite, or countably infinite.
Continuous data set	30	The set of all possible values is an interval of numbers.
Frequency distribution	34	A table used to describe a data set. It includes the class, frequency, and relative frequency (and cumulative relative frequency, if the data set is numerical).
Class frequency	34	The number of observations within a class.
Class relative frequency	34	The proportion of observations within a class: class frequency divided by total number of observations.
Bar chart	35	A graphical representation of a frequency distribution for categorical data with a vertical bar for each class.
Pie chart	36	A graphical representation of a frequency distribution for categorical data that uses a slice, or wedge, for each class.
Stem-and-leaf plot	42	A graph used to describe numerical data. Each observation is split into a stem and a leaf.
Class cumulative relative frequency	50	The proportion of observations within a class and every class before it: the sum of all the relative frequencies up to and including the class.
Histogram	52	A graphical representation of a frequency distribution for numerical data.
Density histogram	55	A graphical representation of a frequency distribution for numerical data in which the class intervals are of unequal width.
Unimodal distribution	56	A distribution with one peak.
Bimodal distribution	56	A distribution with two peaks.
Multimodal distribution	56	A distribution with more than one peak.
Symmetric distribution	57	A distribution with a vertical line of symmetry.
Positively skewed distribution	57	A distribution in which the upper tail extends farther than the lower tail.
Negatively skewed distribution	57	A distribution in which the lower tail extends farther than the upper tail.
Normal curve	58	The most common distribution, a bell-shaped curve.

Chapter 2 Exercises

Applications

2.104 Business and Management There are several reasons why individuals may have no earned income during a given tax year. A random sample of local tax returns in Pennsylvania was obtained and the reasons for no earned income are summarized in the table. PATAX

Reason	Frequency
Disabled	15
Homemaker	35
Unemployed	75
Student	125
Military	65
Retired	85

(a) Add a relative frequency column to this table.
(b) Construct a bar chart and a pie chart for these data.

2.105 Fuel Consumption and Cars The coefficient of drag (C_d) is a measure of a car's aerodynamics. This unitless number is related directly to the speed of the car, overall performance, and miles per gallon. A low coefficient of drag indicates good performance. A random sample of new automobiles was examined, and the coefficient of drag was computed. AERO

(a) Construct a stem-and-leaf plot for these data.
(b) Use the plot in part (a) to describe the distribution in terms of shape, center, and variability.

2.106 Psychology and Human Behavior In spring 2017, Harris Interactive released the results of a survey in which adults were asked to name their favorite TV personality.[33] Ellen DeGeneres captured the top spot, with Mark Harmon second.

Jon Stewart, Jim Parsons, and Jay Leno rounded out the top five, and Tyra Banks, Jimmy Fallon, and Oprah Winfrey also received strong support. Suppose the results from this survey are given in the following table. **FAVETV**

TV personality	Frequency
Ellen DeGeneres	640
Mark Harmon	575
Jon Stewart	420
Jim Parsons	385
Jay Leno	330
Other	300

(a) Find the relative frequency for each category.
(b) Construct a pie chart for these data.
(c) What proportion of adults selected Jon Stewart or Jim Parsons?
(d) What proportion of adults did not select Jay Leno?

2.107 Physical Sciences Construction equipment used to build homes, businesses, and roads (e.g., cranes, backhoes, and front loaders) can be exceptionally loud. The noise level in dBA (A-weighted decibels) measured 50 ft away from several construction-related machines was recorded.[34] **NOISE**

(a) Construct a frequency distribution for these data.
(b) Draw the corresponding histogram.
(c) What proportion of construction equipment had a peak noise level below 80 dBA?
(d) What proportion of construction equipment had a peak noise level of at least 90 dBA?

2.108 Business and Management Many companies maintain technical support lines for customers. A random sample of the duration (in minutes) of customer support calls to a cable company was obtained, and the resulting stem-and-leaf plot is given here.

Stem	Leaf
0	1122334455556667888899
1	00012222223356689999
2	012334556678
3	000123478
4	334468
5	125
6	15
7	7

Stem = 10, Leaf = 1

(a) Describe the shape of this distribution of the duration of customer support calls.
(b) Use the plot to construct a frequency distribution using the class intervals 0–5, 5–10, 10–15, etc.
(c) What proportion of support calls lasted less than 15 minutes?
(d) If a call lasts at least 25 minutes, a supervisor monitors the conversation. What proportion of calls were monitored?

2.109 Technology and Internet Many police departments have been experimenting with and implementing state-of-the-art emergency 911 equipment, which is designed to allow a faster response time without voice contact. Caller information is displayed on a monitor, printed, and then processed. To compare the two procedures (old and new), a random sample of police response times (in minutes) was obtained. **POLICE**

(a) Construct a back-to-back stem-and-leaf plot for these data.
(b) Use the plot in part (a) to describe any similarities and/or differences between the distributions.
(c) Based on the plot in part (a), which procedure is better? Justify your answer.

2.110 Manufacturing and Product Development Microwave ovens are often rated by their output power, such as 900 watts. However, the actual output of a microwave oven tends to decrease with age. If the actual output is more than 400 watts below the rated output, then service is recommended. A random sample of five-year-old, 1000-watt-rated microwave ovens was obtained and tested for output. **MICRO**

(a) Construct a frequency distribution for these data and draw the corresponding histogram.
(b) Based on this random sample, what proportion of five-year-old, 1000-watt microwave ovens need service?
(c) Suppose the performance of these microwave ovens is graded by actual output power, according to the chart.

Power	Grade
900–1000	Excellent
800–900	Very good
700–800	Good
600–700	Fair
500–600	Poor
0–500	Not serviceable

Classify each power output, construct a frequency distribution by grade, and draw the resulting pie chart.

2.111 Sports and Leisure The NFL Scouting Combine is held every February and is a chance for college football players to exhibit their physical and mental abilities for coaches, general managers, and scouts. A random sample of 40-yard dash times (in seconds) for defensive backs (DB) and for safeties (S) at the 2018 Combine was obtained.[35] **DASH**

(a) Construct a back-to-back stem-and-leaf plot for these data.
(b) Use the plot to describe any similarities and/or differences between the distributions.
(c) Based on the plot in part (a), which group appears to run faster in the 40-yard dash? Justify your answer.

Extended Applications

2.112 Economics and Finance The World Bank released a study about paying taxes around the world. The report includes measures, such as time to comply, of the world's tax systems associated with a standardized business. For each of the 232

countries in the study, measures of time to comply (in hours) in 2017 were obtained.[36] **TAXSYS**
(a) Use the class intervals 0–200, 200–400, etc., to construct a frequency distribution and draw the corresponding histogram.
(b) Describe the distribution in terms of shape, center, and variability.
(c) What is a typical time to comply? Are there any outliers?
(d) What proportion of countries had a time to comply of at least 800 hours?

2.113 Fuel Consumption and Cars Remanufactured parts are common in the automotive industry. To ensure quality, Car Part Kings routinely checks the maximum output of rebuilt alternators. Each day, a random sample is obtained and the output delivered (in amps) at 2500 rpm is recorded. The results from a recent day are presented in the table.

Class	Frequency
30.0–32.0	8
32.0–33.0	7
33.0–34.0	10
34.0–34.5	25
34.5–35.0	30
35.0–35.5	40
35.5–36.0	45
36.0–50.0	5
Total	170

(a) Find the width and the density for each class.
(b) Construct a density histogram for these data.

2.114 Medicine and Clinical Studies A common cold usually lasts from 3 to 14 days. Some studies suggest echinacea, zinc, or vitamin C can prevent colds and/or shorten their duration. In a new study of the effectiveness of vitamin C for this application, patients with colds were randomly assigned to a placebo group or a vitamin C group. The duration of each cold (in days) was recorded, and the data are summarized in the following table.

Duration	Placebo frequency	Vitamin C frequency
3	0	3
4	0	6
5	8	7
6	7	10
7	21	18
8	10	15
9	26	17
10	15	10
11	8	9
12	3	2
13	1	3
14	1	0

(a) Use appropriate graphical procedures to compare the placebo and vitamin C data sets.
(b) Do the graphs suggest any differences in shape, center, or variability?
(c) Is there any graphical evidence to suggest that vitamin C reduced the duration of a cold?

2.115 Fuel Consumption and Cars Several different measurements can be used to evaluate the efficiency of a heat pump. However, all these calculations are meant to compare the amount of energy delivered by the heat pump to the amount of energy consumed by the unit.[37] A random sample of heat pumps was obtained and the seasonal energy efficiency ratio (SEER) was measured for each. **HEAT**
(a) Construct a stem-and-leaf plot for these data.
(b) Construct a frequency distribution for these data and draw the corresponding histogram.
(c) Describe the distribution in terms of shape, center, and variability. Are there any outliers? If so, what are they?
(d) Using the frequency distribution in part (a), approximate the proportion of heat pumps with a good SEER rating, that is, with a rating of 21 or greater.
(e) Suppose Consumer Reports classifies SEER ratings according to the following scheme: 35 or greater, excellent; at least 30 but less than 35, very good; at least 20 but less than 30, good; at least 15 but less than 20, fair; and less than 15, poor. Classify each SEER using this scheme, and construct a bar chart for these classification data.

2.116 Physical Sciences The Juno space probe has provided scientists with a wealth of information about the planet Jupiter. New measurements suggest the mysterious wind-sculpted bands on Jupiter extend as far as 3000 km down.[38] Suppose a random sample of these bands was obtained and the depth of each band (in km) was recorded from the information sent to Earth from Juno. **JUNO**
(a) Construct a frequency distribution for these data and draw the corresponding histogram.
(b) Describe the distribution in terms of shape, center, and variability. Are there any outliers? If so, what are they?

Challenge Problems

2.117 Sports and Leisure An ogive, or cumulative relative frequency polygon, is another type of visual representation of a frequency distribution. To construct an ogive:
(a) Plot each point (upper endpoint of class interval, cumulative relative frequency).
(b) Connect the points with line segments.

Figure 2.39 and Figure 2.40 show a frequency distribution and the corresponding ogive, respectively. The observations are ages. The values to be used in the plot are indicated in bold in the table.

Class	Frequency	Relative frequency	Cumulative relative frequency
12–16	8	0.08	0.08
16–20	10	0.10	0.18
20–24	20	0.20	0.38
24–28	30	0.30	0.68
28–32	15	0.15	0.83
32–36	10	0.10	0.93
36–40	7	0.07	1.00
Total	100	1.00	

Figure 2.39 Frequency distribution.

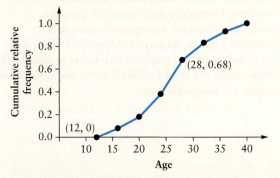

Figure 2.40 Corresponding ogive.

A random sample of game scores from Abby Sciuto's evening bowling league with Sister Rosita was obtained. **BOWL**
(a) Construct a frequency distribution for these data.
(b) Draw the resulting ogive for these data.

2.118 Public Health and Nutrition A doughnut graph is another graphical representation of a frequency distribution for categorical data. To construct a doughnut graph:
(a) Divide a (flat) doughnut (or washer) into pieces, so that each piece (bite of the doughnut) corresponds to a class.
(b) The size of each piece is measured by the angle made at the center of the doughnut. To compute the angle of each piece, multiply the relative frequency by 360° (the number of degrees in a whole, or complete, circle).

The manager at a Whole Foods Market obtained a random sample of customers who purchased at least one popular herb (for cooking or medicinal purposes). **Figure 2.41** and **Figure 2.42** show a frequency distribution and the corresponding doughnut graph, respectively.

Herb	Frequency	Relative frequency
Echinacea	25	0.125
Ephedra	15	0.075
Feverfew	20	0.100
Garlic	35	0.175
Ginkgo	40	0.200
Kava	30	0.150
Saw palmetto	20	0.100
St. John's wort	15	0.075
Total	200	1.000

Figure 2.41 Frequency distribution.

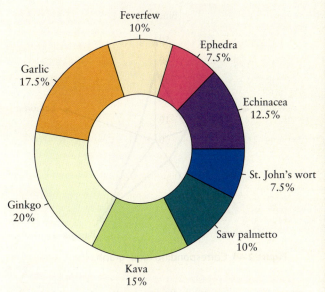

Figure 2.42 Corresponding doughnut graph.

A random sample of failed home inspection reports related to real estate sales in Austin, Texas, was selected and the cause of each was recorded. The resulting data are shown in the following table. **HERB**

Class	Frequency
Moisture in the basement	90
HVAC problems	75
Roofing problems	65
Electrical issues	55
Rotting wood	30
Security issues	45
Defective masonry	40

(a) Find the relative frequency for each class.
(b) Draw a doughnut graph for these data.

2.119 Sports and Leisure A spider graph, or radar chart, is yet another graphical representation of a frequency distribution for categorical data. To construct a radar graph, plot the frequency of each category along a separate axis that starts at the center of the chart and ends on the outer ring.

Suppose the manager at a Fresh Market records the type of quick bread mix purchased by randomly selected customers. **Figure 2.43** and **Figure 2.44** show a frequency distribution and the corresponding radar graph, respectively.

Class	Frequency
Blueberry	35
Walnut	17
Cranberry	12
Banana	42
Lemon	10

Figure 2.43 Frequency distribution.

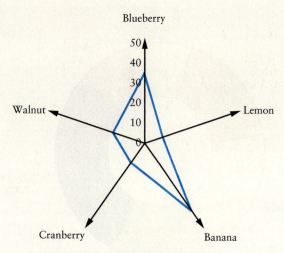

Figure 2.44 Corresponding radar graph.

The following frequency distribution lists the number and type of ground cover perennials purchased at a home and garden center over a two-week period. COVER

Class	Frequency
Creeping flox	10
Lamb's ear	31
Violet dark freckles	23
Corsican mint	37
Mazus	12
Blue star creeper	16

Draw a radar graph for these data.

CHAPTER APP

2.120 Near-Earth Objects The JPL Center for Near-Earth Object Studies maintains a list of all near-Earth objects (NEO), computes high-precision orbits and impact risks for these NEOs, and conducts long-term analyses of possible future orbits of dangerous asteroids. A random sample of NEOs was obtained and the minimum possible close-approach distance in lunar distance (LD) was obtained for each. NEO

(a) Classify the data set as categorical or numerical. If the data set is numerical, determine whether it is discrete or continuous.
(b) Construct a stem-and-leaf plot for these data.
(c) Construct a frequency distribution for these data.
(d) Construct a frequency histogram for these data.
(e) Construct a relative frequency histogram for these data.
(f) Use these tabular and graphical techniques to describe the shape, center, and spread of this distribution, as well as to identify any outlying values.

Numerical Summary Measures

NASA

◀ Looking Back
- Be able to classify a data set as categorical or numerical, discrete or continuous: **Section 2.1**.
- Remember how to construct a bar chart and a pie chart for categorical data: **Section 2.2**.
- Be able to construct a frequency distribution, stem-and-leaf plot, and histogram for numerical data: **Sections 2.3 and 2.4**.

Looking Forward ▶
- Learn how to compute and interpret common numerical summary measures that describe central tendency and variability: **Sections 3.1 and 3.2**.
- Learn how to use Chebyshev's Rule and the Empirical Rule to describe a distribution and how to apply these rules to the inference procedure: **Section 3.3**.
- Find a five-number summary and construct and interpret box plots: **Section 3.4**.

CHAPTER APP

SpaceX Payloads

Elon Musk is the founder and CEO of Space Exploration Technologies, or SpaceX, a private company that designs, builds, and launches rockets and spacecraft. The company's mission is to transport people to Mars and build a self-sustaining city there. Currently, SpaceX is a global leader in commercial launch services. The company has launched and recovered a spacecraft from orbit, has developed a fully reusable rocket, and is reducing the cost of access to space.

The company's spacecraft have delivered cargo to and from the international space station and have successfully completed missions for the U.S. Air Force, AsiaSat, and the Turkmenistan National Space Agency. The following table shows the weight (in kilograms, kg) of the payload for 40 randomly selected SpaceX missions.[1]

525	500	677	500	3170	3325	2296	1316	4535	4428
2216	2395	570	4159	1898	4707	1952	2034	553	5271
3136	4696	3100	3600	2257	4600	9600	2490	5600	5300
6070	2708	3669	6716	3310	475	4990	5200	3500	2205

The procedures presented in this chapter will be used to describe the center and variability of these data, as well as to search for any unusual observations.

3.1 Measures of Central Tendency

Notation

As we learned in Chapter 2, tabular and graphical procedures provide some very useful summaries of data. However, these techniques are not sufficient for statistical inference. For example, because there are no definite rules for constructing a histogram, two people may construct very different-looking displays for the same data, which could lead to different conclusions. The numerical summary measures presented in this chapter are more precise, combine information from the data into a single number, and allow us to draw a conclusion about an entire population. The two most common types of numerical summary measures describe the center and the variability of the data.

A numerical summary measure is a single number computed from a sample that conveys a specific characteristic of the entire sample. Measures of central tendency indicate where the majority of the data are centered, bunched, or clustered. There are many different measures of central tendency. They all combine information from a sample into a single number, and each has advantages and disadvantages.

To properly define and understand numerical summary measures, the following notation will be used.

A capital, or uppercase, *X* has a very different meaning (introduced in Chapter 5).

x

This stands for a specific, fixed observation on a variable. In general, lowercase letters are used to represent observations on a variable; y and z are also commonly used.

n

 Subscript notation.

This is generally used to denote the number of observations in a data set, or the sample size. If there are two relevant data sets, then m and n may be used to denote their sample sizes. Or, if there are two (or more) relevant data sets, then $n_1, n_2, n_3, \ldots$ may be used to denote their sample sizes.

The three dots, ..., mean the list continues in the same manner.

$x_1, x_2, x_3, \ldots, x_n$

This refers to a set of fixed observations on a variable. The subscripts indicate the order in which the observations were selected, not magnitude. For example, x_5 is the fifth observation drawn from a population, not the fifth largest observation.

 Summation notation.

$$\sum_{i=1}^{n} x_i = x_1 + x_2 + \cdots + x_n$$

This is an example of summation notation, often used to write long mathematical expressions more concisely. Here, the sum of n observations can be written more compactly by using the notation on the left side. Σ is the Greek capital letter sigma; i is the index of summation; 1 is the lower bound; and n is the upper bound. To make the notation more compact and less threatening, we will usually omit reference to the lower and upper bounds, $i = 1$ and n. Unless specifically indicated, each summation applies to all values of the variable. For example, the following notation is used to represent the sum of each squared observation:

$$\sum x_i^2 = x_1^2 + x_2^2 + \cdots + x_n^2$$

 Superscript notation.

The following example illustrates the use of this notation and some of the computations used throughout this text.

EXAMPLE 3.1 Sum Practice

Suppose $x_1 = 5$, $x_2 = 9$, $x_3 = 12$, $x_4 = -6$, $x_5 = 17$, and $x_6 = -2$. Compute the following sums.

EG3.1

(a) $(\sum x_i)^2$

(b) $\sum x_i^2$

(c) $\sum (x_i - 7)^2$

Solution

In each case, i is the index of summation, 1 is the lower bound, and 6 is the upper bound. Apply the definition of summation notation to each expression.

(a) In words, expression (a) says add all of the observations, and square the result.

$$\begin{aligned}
(\sum x_i)^2 &= (x_1 + x_2 + x_3 + x_4 + x_5 + x_6)^2 & \text{Expand summation notation.} \\
&= [5 + 9 + 12 + (-6) + 17 + (-2)]^2 & \text{Use the given data.} \\
&= (35)^2 = 1225 & \text{Add; square the sum.}
\end{aligned}$$

(b) In words, expression (b) says square each observation, and add the resulting values.

$$\begin{aligned}
\sum x_i^2 &= x_1^2 + x_2^2 + x_3^2 + x_4^2 + x_5^2 + x_6^2 & \text{Expand summation notation.} \\
&= (5)^2 + (9)^2 + (12)^2 + (-6)^2 + (17)^2 + (-2)^2 & \text{Use the given data.} \\
&= 25 + 81 + 144 + 36 + 289 + 4 & \text{Square each observation.} \\
&= 579 & \text{Add.}
\end{aligned}$$

(c) In words, expression (c) says subtract 7 from each observation, square each difference, and add the resulting values.

$$\begin{aligned}
\sum (x_i - 7)^2 &= (x_1 - 7)^2 + (x_2 - 7)^2 + (x_3 - 7)^2 + (x_4 - 7)^2 + (x_5 - 7)^2 + (x_6 - 7)^2 \\
& \hspace{8cm} \text{Expand summation notation.} \\
&= (5 - 7)^2 + (9 - 7)^2 + (12 - 7)^2 + (-6 - 7)^2 + (17 - 7)^2 + (-2 - 7)^2 \\
& \hspace{8cm} \text{Use the given data.} \\
&= (-2)^2 + (2)^2 + (5)^2 + (-13)^2 + (10)^2 + (-9)^2 & \text{Compute each difference.} \\
&= 4 + 4 + 25 + 169 + 100 + 81 = 383 & \text{Square each difference; add.}
\end{aligned}$$

TRY IT NOW Go to Exercises 3.3 and 3.4

The Sample Mean

The most commonly used measure of central tendency is the sample, or arithmetic, mean.

Definition

The **sample (arithmetic) mean**, denoted $\bar{x}$, of the n observations $x_1, x_2, \ldots, x_n$ is the sum of the observations divided by n. Written mathematically:

$$\bar{x} = \frac{1}{n} \sum x_i = \frac{x_1 + x_2 + \cdots + x_n}{n} \tag{3.1}$$

$\bar{x}$ is read as "x bar."

A CLOSER LOOK

1. The notation $\bar{x}$ is used to represent the sample mean for a set of observations denoted by $x_1, x_2, \ldots, x_n$. Similarly, $\bar{y}$ would represent the sample mean for a set of observations denoted by $y_1, y_2, \ldots, y_n$.
2. The **population mean** is denoted by μ, the Greek letter mu.

EXAMPLE 3.2 Bicycle Rides

BIKE1

Divvy is a bicycle sharing company in Chicago with more than 580 stations and 5800 bikes. People use bike sharing for quick errands, the commute to school or work, or leisurely rides to explore the city. The duration (in minutes) for 12 randomly selected trips is given in the following table.[2]

| 10 | 14 | 20 | 22 | 4 | 17 | 13 | 25 | 8 | 11 | 14 | 28 |

Find the sample mean bicycle trip duration.

Solution

Use Equation 3.1 to find the sample mean.

$$\bar{x} = \frac{1}{12}\sum x_i = \frac{1}{12}(x_1 + x_2 + \cdots + x_{12}) \quad \text{Add all the numbers; divide by } n = 12.$$

$$= \frac{1}{12}(10 + 14 + 20 + \cdots + 28)$$

$$= \frac{1}{12}(186) = 15.5$$

Figure 3.1 shows the sample mean using R. ∎

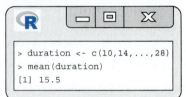

Figure 3.1 The sample mean using R.

A CLOSER LOOK

1. $\bar{x}$ is a sample characteristic. It describes the center of a fixed collection of data. There is no set rule to determine the number of included decimal places. Often, at least one extra decimal place to the right is used to write the result; then the sample mean has at least one more decimal place than the original data values.
2. The sample mean is an average. There are many other averages—for example, the geometric mean, the harmonic mean, a weighted mean, the median, and the mode. The general public usually associates the average with the sample mean.
3. μ is a population characteristic. It describes the center of an entire population. If the population happens to be of finite size N, then μ is the sum of all the values divided by N. Most populations of interest are infinite, or at least very large, so μ is an unknown constant that cannot be measured. It seems reasonable, however, to use $\bar{x}$ to estimate and draw conclusions about μ.
4. The population mean μ is a fixed constant. $\bar{x}$ varies from sample to sample. It is reasonable to think that two sample means associated with two samples from the same population should be *close*, but different.

If a data set contains outliers, meaning observations very far away from the rest, then the sample mean may not be a very good measure of central tendency. An outlier greatly influences the sample mean and tends to pull the mean in its direction. Example 3.3 shows how an outlier can affect the sample mean.

CHAPTER APP **SpaceX Payloads**

Find the sample mean payload weight.

The Sample Median

EXAMPLE 3.3 **One Long Bicycle Ride**

BIKE2

Modify the data in Example 3.2: Suppose one bicycle trip duration was 88 minutes, not 28. So the data set is now

| 10 | 14 | 20 | 22 | 4 | 17 | 13 | 25 | 8 | 11 | 14 | 88 |

The observation 88 is an obvious outlier; it is very far away from the rest of the data. The new sample mean is

$$\bar{y} = \frac{1}{12}(10 + 14 + 20 + \cdots + 88)$$
$$= \frac{1}{12}(246) = 20.5$$

Because $\bar{x} = 15.5$, $\bar{y} > \bar{x}$. The sample mean is pulled in the direction of the outlier and is therefore not necessarily a good measure of central tendency. The sample median is another measure of central tendency that is not as sensitive to outlying values. ∎

Definition

$\tilde{x}$ is read as "x tilde."

The **sample median**, denoted $\tilde{x}$, of the n observations $x_1, x_2, \ldots, x_n$ is the middle number when the observations are arranged in order from smallest to largest.

1. If n is odd, the sample median is the single middle value.
2. If n is even, the sample median is the mean of the two middle values.

A CLOSER LOOK

1. The median divides the data set into two parts, such that half of the observations lie below the median and half lie above it.
2. Only one calculation is necessary to find the median (no calculations are needed if n is odd). Put the observations in ascending order of magnitude (not the order in which the observations were selected), and find the middle value.
3. Similarly, $\tilde{y}$ represents the sample median for a set of observations denoted by $y_1, y_2, \ldots, y_n$.
4. The **population median** is denoted by $\tilde{\mu}$.

EXAMPLE 3.4 **Median Calculations**

The following three examples show how to find the median under various circumstances and the effect of an outlying value. The observations are already arranged in order from smallest to largest.

(a) 10 11 14 16 17

There are $n = 5$ observations. The middle number is in the third position. $\tilde{x} = 14$.

(b) 10 11 14 16 57

There are still $n = 5$ observations. The middle number is in the third position, and $\tilde{x} = 14$. Notice that the outlier 57 does not affect the median.

(c) 10 11 14 16 17 20

There are $n = 6$ observations. There is no single middle value. The median is the mean of the observations in the third and fourth positions. $\tilde{x} = \frac{1}{2}(14 + 16) = 15$. ∎

EXAMPLE 3.5 Heavy Metal

VANAD Vanadium is a trace metal found in water, soil, and the air. Trace amounts are believed to be important for good health, contributing to strong bones and helping the body's metabolism. A lack of vanadium may cause poor cartilage formation and too much may contribute to digestive problems. A random sample of soil was obtained from farms in Tabernacle, New Jersey. The amount of vanadium was measured in each (in micrograms per gram, μg/g) and the data are given in the table. Find the median amount of vanadium in the soil samples.

128	153	151	87	106	116	136
57	158	39	148	101	149	155

Solution

STEP 1 Arrange the observations in order. The position of each observation in the ordered list is given in the second row.

Ordered data:	39	57	87	101	106	116	128	136	148	149	151	153	155	158
Position:	1	2	3	4	5	6	7	8	9	10	11	12	13	14

STEP 2 There are $n = 14$ observations. The median is the mean of the two middle values (in the seventh and eighth positions).

$$\tilde{x} = \frac{1}{2}(128 + 136) = 132$$

Figure 3.2 shows a technology solution. ∎

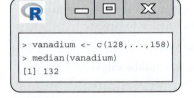

```
> vanadium <- c(128,...,158)
> median(vanadium)
[1] 132
```

Figure 3.2 The sample median using R.

TRY IT NOW Go to Exercises 3.13 and 3.15

CHAPTER APP **SpaceX Payloads**

Find the sample median payload weight.

A CLOSER LOOK

1. In general, the sample mean is not equal to the sample median, that is, $\bar{x} \neq \tilde{x}$. However, if the distribution of the sample is symmetric, then $\bar{x} = \tilde{x}$. If the sample distribution is approximately symmetric, then $\bar{x} \approx \tilde{x}$.

2. In general, the population mean is not equal to the population median, that is, $\mu \neq \tilde{\mu}$. However, if the distribution of the population is symmetric, then $\mu = \tilde{\mu}$.

3. The relative positions of $\bar{x}$ and $\tilde{x}$ suggest the shape of a distribution. The smoothed histograms in **Figure 3.3**, **Figure 3.4**, and **Figure 3.5** illustrate three possibilities.

(a) If $\bar{x} > \tilde{x}$, this suggests that the distribution of the sample is positively skewed, or skewed to the right (Figure 3.3).

(b) If $\bar{x} \approx \tilde{x}$, this suggests that the distribution of the sample is approximately symmetric (Figure 3.4).

(c) If $\bar{x} < \tilde{x}$, this suggests that the distribution of the sample is negatively skewed, or skewed to the left (Figure 3.5).

Recall: A histogram consists of rectangles drawn above each class with height proportional to frequency or relative frequency. We draw a curve along the tops of the rectangles to smooth out the histogram and display an enhanced graphical representation of the distribution.

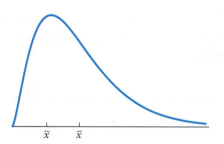

Figure 3.3 Positively skewed distribution.

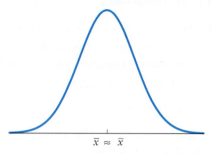

Figure 3.4 Approximately symmetric distribution.

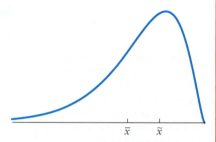

Figure 3.5 Negatively skewed distribution.

Other Measures of Central Tendency

Because the sample mean is extremely sensitive to outliers, and the sample median is very insensitive to outliers, it seems reasonable to search for a compromise measure of central tendency. A trimmed mean is moderately sensitive to outliers.

 Finding percentages.

Definition

A **$100p\%$ trimmed mean**, denoted $\bar{x}_{tr(p)}$, of the n observations $x_1, x_2, \ldots, x_n$ is the sample mean of the trimmed data set.

a. Order the observations from smallest to largest.

b. Delete, or trim, the smallest $100p\%$ and the largest $100p\%$ of the observations from the data set.

c. Compute the sample mean from the remaining data.

$100p$ is the **trimming percentage**, the percentage of observations deleted from each end of the ordered list.

A CLOSER LOOK

1. We compute a trimmed mean by deleting the smallest and largest values, which are possible outliers. Some statisticians believe that deleting any data is a bad idea because every observation contributes information about the distribution.

2. To compute a **$100p\%$ trimmed mean**, delete the smallest $100p\%$ and the largest $100p\%$ of the observations. Therefore, $2(100p)\%$ of the observations are removed.

3. There is no set rule for determining the value of p. It seems reasonable to delete only a few observations, and to select p so that np (the number of observations deleted from each end of the ordered data) is an integer.

4. Here is a specific example of the notation and an interpretation: $\bar{x}_{tr(0.05)}$ is a $(100)(0.05) = 5\%$ trimmed mean. In this example, 10% of the observations are discarded.

EXAMPLE 3.6 Salary Secrets

According to a survey by Aon Hewitt, employees can expect about a 3% raise in 2018.[3] However, variable pay (e.g., bonuses) is expected to be lower. Suppose a random sample of U.S. businesses was obtained and the raise as a percentage of salary was obtained for each. The data are given in the table.

5.8	1.3	3.5	4.4	2.3	1.4	1.1	3.5	2.7	3.1
4.8	2.6	2.8	3.8	0.9	6.0	3.3	2.2	4.1	1.8

Find a 10% trimmed mean.

Solution

STEP 1 The trimming percentage is 10%. $p = 10/100 = 0.10$. Find the number of observations to delete from each end of the ordered list.

There are $n = 20$ observations.

$np = (20)(0.10) = 2$ Trim 2 observations from each end of the ordered list.

Note that np may not be an integer. Computer software packages have algorithms for dealing with this issue.

STEP 2 The resulting data set is as follows.

~~0.9~~ ~~1.1~~ 1.3 1.4 1.8 2.2 2.3 2.6 2.7 2.8 3.1 3.3 3.5 3.5 3.8 4.1 4.4 4.8 ~~5.8~~ ~~6.0~~

STEP 3 Find the sample mean for the remaining data.

$$\bar{x}_{tr(0.10)} = \frac{1}{16}(1.3 + 1.4 + 1.8 + \cdots + 4.1 + 4.4 + 4.8) = 2.975$$

2.975 is the 10% trimmed mean. **Figure 3.6** shows a technology solution.

TRY IT NOW Go to Exercises 3.10 and 3.17

> **Common Error:** The original value of n is used in computing a trimmed mean.
> **Correction:** The sample mean of the trimmed data set is computed using the number of observations in the reduced data set.

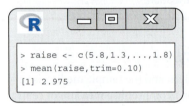

```
> raise <- c(5.8,1.3,...,1.8)
> mean(raise,trim=0.10)
[1] 2.975
```

Figure 3.6 Trimmed mean using R.

CHAPTER APP **SpaceX Payloads**

Find a 5% trimmed mean for the payload weight data.

Definition
The **mode**, denoted M, of a set of n observations $x_1, x_2, \ldots, x_n$ is the value that occurs most often, or with greatest frequency.

If all the observations occur with the same frequency, then the mode does not exist.

If two or more observations occur with the same greatest frequency, then the mode is not unique. If there are two modes, the distribution may be bimodal; if there are three modes, it may be trimodal, etc.

The mode is easy to compute and, intuitively, it does return a reasonable measure of central tendency. For example, consider a bell-shaped distribution. A random sample from this distribution should contain lots of values near the center. Therefore, the mode should suggest the middle of the distribution (**Figure 3.7**). For symmetric distributions, the mean, the median, and the mode will be about the same.

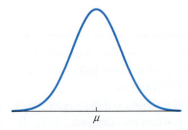

Figure 3.7 We expect the mode M of a sample from this distribution to be near the population mean μ.

A CLOSER LOOK

1. If a random sample is very large with many repeated values, then the grouped data may be presented in a summary table. Let $x_1, x_2, \ldots, x_k$ be a set of (representative) observations with corresponding frequencies $f_1, f_2, \ldots, f_k$. For example, x_7 occurs f_7 times in the random sample. The total number of observations in the data set is $n = \sum f_i$. If the data are grouped, there are other corresponding formulas for the measures of central tendency defined previously.

2. Remember, there are many other averages, including the weighted mean, the geometric mean, and the harmonic mean.

Summary Measures for Categorical Variables

The remainder of this section describes summary measures for qualitative data.

The natural summary measures for observations on a qualitative variable are simply the frequency and relative frequency of occurrence for each category. We have already done this! Recall Example 2.4, in which 25 e-commerce companies were randomly selected and each company type was recorded. Each response was categorical (type), and the data were summarized in a table listing only category, frequency of occurrence for each category, and relative frequency of occurrence for each category.

Suppose now that the commuter students at a small college are asked to complete a survey to identify the make of car they use to drive to school. Numerical summary measures for this categorical variable should include frequencies and relative frequencies, or proportions, as shown in the following table. (The cumulative relative frequency is used only for numerical data sets; it doesn't really make sense here because there is no natural ordering.)

Category	Frequency	Relative frequency
Buick	137	0.0938
Chevrolet	288	0.1973
Ford	202	0.1384
Honda	336	0.2301
Hyundai	322	0.2205
Kia	175	0.1199
Total	1460	1.0000

A dichotomous or Bernoulli variable is a special categorical variable that has only two possible responses. One response is often associated with, or called, a success (denoted S), and the other response is called a failure (denoted F). The two possible actual responses are ignored. For example, suppose a medical researcher selects children at random and asks them all whether they have had an ear infection within the past year. The response "had an ear infection" might be a success, and the response "had no ear infection" would be a failure. The same numerical measures are used to summarize observations on this kind of categorical variable: frequency and relative frequency of occurrence for each response. The relative frequency of successes has a special name.

Definition

For observations on a categorical variable with only two responses, the **sample proportion of successes**, denoted $\hat{p}$, is the relative frequency of occurrence of successes.

$\hat{p}$ is read as "p hat."

$$\hat{p} = \frac{\text{number of S's in the sample}}{\text{total number of responses}} = \frac{n(S)}{n} \tag{3.2}$$

Finding proportions.

A CLOSER LOOK

The symbol *p* is used in notation to represent several quantities: the population proportion of successes, the sample proportion of successes, and in the definition of the trimmed mean. The context in which the notation is used implies the appropriate concept.

1. The **population proportion of successes** is denoted p.
2. The success response is not necessarily associated with a good thing. For example, a researcher may be interested in the proportion of laboratory animals that die when exposed to a certain toxic chemical. A success may be associated with the death of an animal.
3. The sample proportion of successes $\hat{p}$ can be thought of as a sample mean in disguise. Suppose every S is changed to a 1, and every F to a 0. The sample mean for these new, transformed data is

$$\bar{x} = \frac{1}{n}(\text{a sum of 0's and 1's}) = \frac{n(S)}{n} = \hat{p}$$

EXAMPLE 3.7 Seat Belt Checkpoint

The target set by the U.S. federal government for seat belt usage is 92% by 2020. The National Occupant Protection Use Survey (NOPUS) indicated that 90.1% of drivers and front-seat passengers between 7:00 A.M. and 6:00 P.M. use seat belts.[4] Seat belt use is generally higher in the west, and recently use of seat belts among drivers and passengers of vans and SUVs has increased significantly. Suppose the State Police recently established a checkpoint along a heavily traveled rural road. A success was recorded for a driver wearing a seat belt, and a failure recorded otherwise. The observations from this checkpoint are given in the following table.

S	S	S	F	S	S	S	F	S	S	F	S	S	S	S
S	S	S	S	S	S	F	S	S	S	S	S	F	S	S

The sample contains 30 observations and 25 successes. The sample proportion of successes is

$$\hat{p} = \frac{n(S)}{n} = \frac{25}{30} = 0.8333$$

Approximately 83% of the drivers who were stopped at the checkpoint were wearing seat belts. It is reasonable to assume the value of $\hat{p}$ is close to the population proportion of successes; in this example, the true proportion of drivers who wear a seat belt. ∎

TRY IT NOW Go to Exercise 3.12

CHAPTER APP SpaceX Payloads

Suppose a successful payload weight is a weight of at least 4000 kg. Find the sample proportion of successes.

Section 3.1 Exercises

Concept Check

3.1 Fill in the Blank
(a) The two most common types of numerical summary measures describe the _____ and the _____ of the data.
(b) Measures of central tendency suggest where the data are _____.

3.2 True or False
(a) The sample mean and the population mean are always the same value.
(b) The sample mean and the sample median can be the same value.
(c) The sample mean is sensitive to outliers.

(d) In a random sample of n observations, suppose $n-1$ are positive and 1 is negative. The sample mean must be a positive number.
(e) When computing a trimmed mean, we discard the same number of observations from each end of the ordered list.
(f) The mode may not exist for a specific data set.
(g) It is reasonable to assume that the sample proportion of successes is close to the population proportion of successes.

Practice

3.3 Compute each summation using the following random sample. EX3.3

$x_1 = -20$, $x_2 = 12$, $x_3 = 32$, $x_4 = 18$, $x_5 = 28$

(a) $\sum x_i$ (b) $\sum x_i^2$ (c) $\sum (x_i - 10)$
(d) $\sum (x_i - 5)^2$ (e) $\sum (2x_i)$ (f) $2\sum x_i$

3.4 Suppose the following random sample is obtained.

50.9	50.0	52.5	47.3	48.0
47.3	52.0	46.9	45.7	54.1

Compute the following sums. EX3.4
(a) $\sum x_i^2$ (b) $\sum x_i^3$ (c) $\sum (x_i - 50)^2$
(d) $(\sum x_i)^2$ (e) $\sum (x_i - 49.5)$ (f) $\sum \dfrac{x_i}{10}$

3.5 Compute the mean for each sample with known sum.
(a) $\sum x_i = 1057$, $n = 10$
(b) $\sum x_i = 356$, $n = 27$
(c) $\sum x_i = 250.5$, $n = 36$
(d) $\sum x_i = 1.355$, $n = 11$
(e) $\sum x_i = -37.4$, $n = 15$
(f) $\sum x_i = 496.81$, $n = 28$

3.6 Find the position, or location, of the sample median in an ordered data set of size n.
(a) $n = 22$ (b) $n = 37$ (c) $n = 117$ (d) $n = 64$

3.7 Find the sample mean and the sample median for each data set.
(a) {5, 3, 7, 9, 11, 5, 6, 7, 7} EX3.7a
(b) {−7, 10, 25, 22, 36, −24, 0, 1, 12, 9, −11} EX3.7b
(c) {5.4, 3.3, 6.0, 10.1, 13.6, 7.7, 16.6, 28.9, 4.6} EX3.7c
(d) {−103.7, −110.35, −109.1, −99.7, −115.6} EX3.7d

3.8 Consider the data given in the following table.

5	7	8	27	3	15	7	6	4	5	5	1

Find the sample median. Note that this summary statistic is a better measure of central tendency than the sample mean for this data set. Why? EX3.8

3.9 Use the values of the sample mean and the sample median to suggest whether the distribution is symmetric, skewed to the left, or skewed to the right.
(a) $\bar{x} = 37$, $\tilde{x} = 49$
(b) $\bar{x} = 63.5$, $\tilde{x} = 62.75$
(c) $\bar{x} = -37$, $\tilde{x} = -16$
(d) $\bar{x} = -12.56$, $\tilde{x} = 12.56$

3.10 Compute the indicated trimmed mean for each data set.
(a) {24, 36, 26, 30, 28, 35, 33, 33, 34, 27}: $\bar{x}_{tr(0.10)}$ EX3.10a
(b) {72, 76, 76, 77, 85, 76, 80, 86, 62, 70}: $\bar{x}_{tr(0.20)}$ EX3.10b
(c) {182, 169, 180, 166, 173, 101, 188, 124, 182, 137, 100, 137, 118, 111, 137, 181, 189, 130, 168, 133}: $\bar{x}_{tr(0.20)}$ EX3.10c
(d) {5.5, 7.5, 7.3, 6.4, 5.3, 9.5, 7.2, 5.8, 7.0, 6.7, 9.0, 8.1, 8.4, 5.8, 5.4, 7.2, 7.4, 7.5, 5.9, 7.5}: $\bar{x}_{tr(0.15)}$ EX3.10d

3.11 Find the mode for each data set, if it exists.
(a) {3, 5, 6, 7, 3, 4, 6, 6, 8, 11, 13, 2, 1} EX3.11a
(b) {−17, −10, 0, 3, −5, 4.3, 12, 0, 5, −2.1, 1.7, −7} EX3.11b
(c) {6.6, 7.3, 5.2, 6.2, 8.3, 9.8, 4.1, 3.7} EX3.11c

3.12 Find the sample proportion of successes for each data set.
(a) {S, F, S, F, F, F, F, F, S, S, S, F, F, S} EX3.12a
(b) {F, S, S, F, S, F, F, S, S, S, S, S, S, S, F, S, S, S, S, S} EX3.12b
(c) {S, F, S, F, F, F, F, F, S, S, S, F, F, S, F, F, S, F, S, S, S, S, F, F, F, F, F, F, S, S, F, S, F} EX3.12c

Applications

3.13 Travel and Transportation Tractor trailers tend to exceed the speed limit (65 mph) on one downhill stretch of Route 80 in Pennsylvania. Using a radar gun, the following tractor trailer speeds (in mph) were observed. TRACTOR

81	66	67	69	79	62	70	73	67	60	61
67	74	65	77	74	64	71	64	67	61	

(a) Find the sample mean, $\bar{x}$.
(b) Find the sample median, $\tilde{x}$.
(c) What do your answers to parts (a) and (b) suggest about the shape of the distribution of speeds?

3.14 Biology and Environmental Science The air quality index (AQI) is a number used to convey the air pollution level. A larger AQI suggests that more people may experience adverse health effects. In spring 2018, a random sample of locations in mainland China was obtained and the AQI was measured at each.[5] AQI
(a) Find the sample mean and the sample median for these data.
(b) What do the summary statistics in part (a) suggest about the shape of the distribution of AQI?

3.15 Biology and Environmental Science The city of San Francisco regularly evaluates the conditions of all civic plazas, mini-parks, neighborhood parks, playgrounds, parkways, and regional parks. The results are presented to elected officials and the public and are used to improve park conditions, efficiently allocate resources, and improve maintenance. A random sample of San Francisco parks was obtained and the park score was recorded for each.[6] The data are given in the table. PARK

82.1	95.7	87.8	94.4	90.6	86.2	87.2	94.3
81.6	96.1	88.3	87.1	76.7	75.6	83.2	98.2
89.4	65.2	87.2	63.7				

(a) Find the sample mean and the sample median.
(b) Suppose the last observation had been 33.7 instead of 63.7. Find the sample mean and sample median for this revised data set. Explain how this change in the data affects the mean and median found in part (a).

3.16 Fuel Consumption and Cars Refer to the table that lists the atmospheric CO_2 concentration (in ppm) for 36 months ending in February 2018, recorded at the Mauna Loa Observatory, Hawaii.[7] 📊 CO2
(a) Find the sample mean and the sample median for these data.
(b) A certain group considers any monthly concentration less than 402 to be a success (not harmful to the environment). Find the sample proportion of successes.

3.17 Education and Child Development The Math SAT scores for all students in an introductory statistics class at Rio Salado College were obtained. 📊 SAT
(a) Find the sample mean and the sample median.
(b) Find a 5% trimmed mean.
(c) Using these three numerical summary measures, describe the shape of the distribution.

3.18 Manufacturing and Product Development A random sample of 12-oz cans of Dr. Pepper soda was obtained from Aldi's supermarket. The exact amount of soda (in ounces) in each can was measured. 📊 DRPEPP
(a) Find the sample mean and the sample median.
(b) What do the summary statistics in part (a) suggest about the shape of the distribution of the amount of soda in each can?
(c) Suppose any amount of 12 oz or greater is considered a success. Find the sample proportion of successes.

3.19 Biology and Environmental Science The rocky coast of Maine provides an ideal breeding ground for lobsters. In recent years, there has been increased demand for lobsters, especially from China, which has led to more commercial fishing harvesters in Maine. In fact, the past few years have seen a huge increase in the total pounds of lobsters landed. The number of pounds (in metric tons) caught by Maine fisheries for several recent years was obtained.[8] 📊 LOBS
(a) Find the sample mean and the sample median for this data set.
(b) Which statistic is a better measure of central tendency for these data? Justify your answer.

3.20 Sports and Leisure Some critics of Major League Baseball (MLB) believe the ball is "juiced" (livelier) because it is manufactured to give hitters an advantage. To investigate this claim, a sample of the earned run average (ERA) for American League starting pitchers for the 2017 season was obtained.[9] 📊 JUICED
(a) Find the sample mean and the sample median.
(b) Suppose the pitcher with the highest ERA plays in Colorado, where the air is thin and home runs are plentiful. To eliminate such outliers, find a 5% trimmed mean.
(c) Find the mode for the original data set, if it exists.

3.21 Biology and Environmental Science The water temperature (in degrees Fahrenheit) during the summer of 2017 at several locations off the coast of Florida is given in the table.[10] 📊 FL2017

82	84	79	81	80	86	85	84	86	84
84	79	79	72	77	79	70	74	81	84
86	80	83	83	87	87	86	87	85	86

(a) Find the sample mean and the sample median.
(b) Find a 10% trimmed mean for these data.
(c) Find the mode for the original data, if it exists.

3.22 Education and Child Development An educational study was designed to compare cooperative learning versus traditional lecture-style teaching. Two sections of an introductory statistics class were used. Seven students were randomly selected from each section. The scores on the second test (a 30-item exam) are given in the table. 📊 LEARN

| Traditional | 21 | 28 | 25 | 25 | 21 | 19 | 23 |
| Cooperative | 25 | 30 | 28 | 25 | 24 | 24 | 29 |

Which group of students did better, on average? Justify your answer.

3.23 Medicine and Clinical Studies Many of us cringe at the thought of a root canal because it is both painful and costly. A random sample of the cost of a root canal on a front tooth for city dentists and rural areas was obtained.[11] 📊 ROOT
(a) Find the sample mean and the sample median for each data set.
(b) Do you think it is more expensive to have a root canal in a city or rural area? Justify your answer.

Extended Applications

3.24 Sports and Leisure The 17th FINA World Championships were held in July 2017, in Budapest, Hungary. The women's 1-m springboard competition was won by Maddison Keeney from Austria. Many of the participants in this competition performed a back 1 1/2 somersault, as one of their dives. The scores from some of these dives were recorded for various participants.[12] 📊 FINA
(a) Find the sample mean and the sample median for this data set.
(b) Find the mode, if it exists.
(c) Multiply each score by the degree of difficulty, 2.3. Find the sample mean for this new data set. How does this sample mean compare with the sample mean found in part (a)?

3.25 Travel and Transportation The FAA air traffic control modernization plan calls for revised flight paths and procedures across the United States. These new procedures use very precise satellite-based navigation. The goals are to save time, to increase the number of planes landing and taking off at any given airport, and to reduce the amount of wasted fuel. However, these new flight paths have caused increased noise complaints in San Diego, Charlotte, and New York.[13] Suppose the noise level (in decibels, dB) of several aircraft was measured using one flight path in Charlotte, North Carolina. The data are given in the table. ▮▮ NOISE

71	64	68	66	66	72	72	70	74	67
69	63	68	74	66	62	73	67	62	66

(a) Find the sample mean and the sample median for these data.
(b) Based on the values found in part (a), describe the shape of this distribution.
(c) Some researchers believe that the wind velocity gradient can add as much as 4 dB to each reading. Add 4 dB to each observation in the data set. Compute the new sample mean. How does this compare with the sample mean found in part (a)?

3.26 Medicine and Clinical Studies Many physicians believe that the overuse of antibiotics has contributed to the emergence of multidrug-resistant bacteria that are responsible for health care–associated infections. The European Centre for Disease Prevention and Control recently released a report that presents the defined daily doses (DDD) per 1000 inhabitants per day for each country.[14] ▮▮ ANTIBIO
(a) Find the sample mean DDD for these data.
(b) It is possible that some physicians may be under-reporting antibiotic use. Multiply each observation by 1.10. Find the sample mean for this new data set. How does this sample mean compare with the sample mean found in part (a)?

3.27 Manufacturing and Product Development A new quality control program was recently started at a Hyundai manufacturing facility. Several times each day, randomly selected panels from a stamping press are inspected for defects. A nondefective panel is a success (S). A defective panel is a failure (F) and must be restamped at an additional cost. During a recent inspection, the following 32 observations were recorded. ▮▮ PANEL

S	S	S	S	F	S	F	F	S	S	S	S	S	S
S	S	F	S	S	S	S	S	S	S	S	F	S	S
S	S	S	S										

(a) Find the sample proportion of successes.
(b) Change each S to a 1, and each F to a 0. Find the sample mean for these new data. How does the mean compare with the sample proportion of successes found in part (a)?
(c) Suppose 8 additional panels were selected and inspected (for a total of 40 panels). Is it possible for the sample proportion of successes to be 0.9? Why or why not?

3.28 Sports and Leisure The playing time for rookies in the National Basketball Association (NBA) depends on many factors, including position and performance. A random sample of playing times per game (in minutes) for rookies in the NBA during the 2017–2018 season was obtained. The data are given in the table.[15] ▮▮ ROOKIE

19.9	16.3	15.4	15.1	15.0	14.9	13.3	11.8
11.3	10.5	13.3	10.1	9.2	8.0	6.7	7.0
5.8	5.5	5.6	10.3				

(a) Find the sample mean and the sample median.
(b) What do the summary statistics in part (a) suggest about the shape of the distribution of playing time for rookies?
(c) Can you change the maximum observation (19.9) so that the sample mean is equal to the sample median? Why or why not?

3.29 Medicine and Clinical Studies In a random sample of 13 patients with calcaneus bone fractures, the sample mean number of days until fracture healing was $\bar{x} = 37.85$ and the sample median was $\tilde{x} = 40$. Suppose an additional patient is added to the sample so that $x_{14} = 44.5$.
(a) Find the sample mean for all 14 patients.
(b) Is there any way to determine the sample median for all 14 patients? Explain your answer.

3.30 Fuel Consumption and Cars The estimated oil reserves (in millions of barrels) of four wells are given by ▮▮ OILRES

$$x_1 = 1078, \quad x_2 = 5833, \quad x_3 = 10{,}772, \quad x_4 = 7320$$

(a) Find x_5 so that the mean for all five observations is 6883.4.
(b) Find x_5 so that the sample mean is equal to the sample median.

3.31 Manufacturing and Product Development A consumer group has tested the drying time for 15 samples of exterior latex paint. The sample mean drying time is 83.8 minutes. What must the 16th drying time be if the 16th observation decreases the mean drying time by 30 seconds? By 1 minute?

3.32 Biology and Environmental Science The beaches along the coast of New Hampshire are famous for chilly waters, even during the hottest summer days. A recent sample of the water temperature on 24 randomly selected summer days was obtained. The following temperatures are in degrees Fahrenheit (°F). ▮▮ NHCHILL

58	58	53	53	59	57	54	61	56	60
57	61	56	55	59	60	55	53	55	58
59	53	59	63						

(a) Find the sample mean and the sample median.
(b) Convert each temperature to degrees Celsius (°C). Use the formula $C = (F - 32)/1.8$. Find the mean for all the water temperatures in degrees Celsius.
(c) What is the relationship between the sample means in parts (a) and (b)?

3.33 Economics and Finance Europeans take their coffee very seriously. For many, drinking coffee is part of their daily routine. The Belgians have many gourmet recipes involving coffee, and Lithuanians enjoy Kava, a strong, dark espresso. The cost (in euros) for a cup of coffee in each country in the European Union was obtained.[16] ▮▮ COFFEE

(a) Find the sample mean cost of a cup of coffee in the European Union.
(b) Convert each price to U.S. dollars. In March 2018, the exchange rate was 1 euro to U.S.$1.23. Find the sample mean cost of a cup of coffee in U.S. dollars.
(c) What is the relationship between the sample means in parts (a) and (b)?

3.34 Technology and Internet According to a recent study by the U.S. Energy Information Administration, the number of televisions in American homes is decreasing. Suppose a survey of U.S. residential homes revealed a count of televisions used per household. The (grouped) data are summarized in the table.[17] TV

Number of televisions	Frequency of occurrence
0	3
1	18
2	30
3	24
4	11
5	3

Find the sample mean and the sample median number of televisions per home.

3.2 Measures of Variability

The Sample Range

Measures of central tendency represent only one characteristic of a data set. However, these numerical summary measures alone are not sufficient to describe a sample completely. It is possible to have two very different data sets with (approximately) the same mean (and median). **Figure 3.8** and **Figure 3.9** show two smoothed histograms to illustrate the problem.

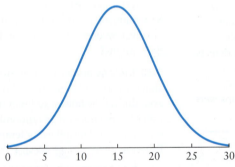

Figure 3.8 Sample 1: $x_1, x_2, \ldots, x_n$. The smoothed histogram suggests a compact distribution.

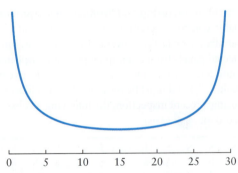

Figure 3.9 Sample 2: $y_1, y_2, \ldots, y_m$. The smoothed histogram suggest the data are more dispersed, or spread out.

The measures of central tendency (sample mean and sample median) are approximately the same ($\bar{x} \approx \bar{y} \approx 15$ and $\tilde{x} \approx \tilde{y} \approx 15$), but the data in Sample 1 are more compact because more of the data are clustered about the mean, $\bar{x} = 15$. To describe, or to distinguish, between the data sets, we need to consider their variability.

Definition
The **(sample) range**, denoted R, of the n observations $x_1, x_2, \ldots, x_n$ is the largest observation minus the smallest observation. Written mathematically,

$$R = x_{max} - x_{min} \tag{3.3}$$

where x_{max} denotes the maximum, or largest, observation, and x_{min} stands for the minimum, or smallest, observation.

A CLOSER LOOK

1. In theory, the sample range does measure, or describe, variability. A data set with a small range has little variability and is compact. A data set with a large range has lots of variability and is spread out.

2. The sample range is used in many quality control applications. For example, a production supervisor may want to maintain small variability in a manufacturing process. The sample range may be used to determine whether the process is still well controlled or varies abnormally.

The Sample Variance and Sample Standard Deviation

Despite being a reasonable measure that is very easy to compute, the sample range is not adequate for describing variability. For example, it may not accurately represent the variability of a distribution if the maximum and minimum values are outliers.

The sample ranges for the data sets summarized by the smoothed histograms in Figure 3.8 and Figure 3.9 are approximately the same: $R \approx 30 - 0 = 30$. Therefore, it is necessary to use a better, more sensitive measure of variability. To derive a more precise measure of variability, consider how far each observation lies from the mean.

A graph may be used to visualize the spread of data and to suggest another measurement. A dot plot is a graph that simply displays a dot corresponding to each observation along a number line. The stacked dot plot in **Figure 3.10** may be used to compare the variability in Sample 1 (x's) versus Sample 2 (y's).

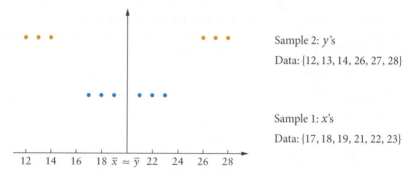

Figure 3.10 Stacked dot plot.

In Sample 1, the data set is compact; each observation is very close to the mean. In Sample 2, the data set is more spread out; each observation is far away from the mean. This analysis of Figure 3.10 suggests that a better measure of variability might include the distances from the mean.

Definition
Given a set of n observations $x_1, x_2, \ldots, x_n$, the **ith deviation about the mean** is $x_i - \bar{x}$.

A CLOSER LOOK

1. To calculate the ith deviation about the mean, find $\bar{x}$ in the given data set, then compute the difference $x_i - \bar{x}$. For example, the 7th deviation about the mean is the value $x_7 - \bar{x}$.

2. We usually do not need any one deviation about the mean; instead, all of the deviations about the mean are used together to find a suitable measure of variability.

3. If the ith deviation about the mean is positive, then the observation is to the right of the mean: If $x_i - \bar{x} > 0$, then $x_i > \bar{x}$.

   If the ith deviation about the mean is negative, then the observation is to the left of the mean: If $x_i - \bar{x} < 0$, then $x_i < \bar{x}$.

A data set with little variability should have small deviations about the mean, and the squares of the deviations should be small. Conversely, a data set with lots of variability should have large deviations about the mean, and the squares of the deviations should be large. This idea is used to define the **sample variance**.

Definition
The **sample variance**, denoted s^2, of the n observations $x_1, x_2, \ldots, x_n$ is the sum of the squared deviations about the mean divided by $n - 1$. Written mathematically,

$$s^2 = \frac{1}{n-1} \sum (x_i - \bar{x})^2$$
$$= \frac{1}{n-1}\left[(x_1 - \bar{x})^2 + (x_2 - \bar{x})^2 + \cdots + (x_n - \bar{x})^2\right] \tag{3.4}$$

The **sample standard deviation**, denoted s, is the positive square root of the sample variance. Written mathematically,

$$s = \sqrt{s^2} \tag{3.5}$$

 Finding square roots.

The sample variance s^2 is often called an average of the squared deviations about the mean, yet we divide the sum of the squared deviations by $n - 1$. Although this does not seem correct, dividing by $n - 1$ makes s^2 an unbiased estimator of σ^2. We will see later in the text that an unbiased statistic is, in some sense, a good thing. There are $n - 1$ degrees of freedom, a kind of dimension of variability, associated with the sample variance s^2.

A CLOSER LOOK

1. The **population variance**, a measure of variability for an entire population, is denoted by σ^2, and the **population standard deviation** is denoted by σ, the Greek letter sigma.

2. Just knowing s^2 doesn't seem to say much about variability. If $s^2 = 6$, for example, it is hard to infer anything about variability. However, the sample variance s^2 is a measure of variability, and it is useful in comparisons. For example, if Sample 1 and Sample 2 have similar units, and if $s_1^2 = 14$ and $s_2^2 = 10$, then the data in Sample 2 are more compact.

3. The sample standard deviation s is used (rather than s^2) in many statistical inference problems. So when finding s (by hand), we must compute s^2 first and then take the positive square root to find s.

4. The units for the sample standard deviation are the same as for the original data. A value of $s = 0$ means there is no variability in the data set.

5. The notation s_x^2 is used to represent the sample variance for a set of observations denoted by $x_1, x_2, \ldots, x_n$. Similarly, s_y^2 represents the sample variance for a set of observations $y_1, y_2, \ldots, y_n$.

EXAMPLE 3.8 Zucchini Weight

Farmer Moofy's Produce in Bloomsburg, Pennsylvania, sells a wide variety of fruits and vegetables and frequently donates crates of corn and zucchini to the local food cupboard. Five of the donated zucchini were randomly selected, and each of these squash was carefully weighed. The weights, in ounces, were 6.2, 4.5, 6.6, 7.0, and 8.2. Find the sample variance and the sample standard deviation for these data.

Solution

STEP 1 Find the sample mean:

$$\bar{x} = \frac{1}{5}(6.2 + 4.5 + 6.6 + 7.0 + 8.2) = \frac{1}{5}(32.5) = 6.5$$

STEP 2 Use Equation 3.4 to find the sample variance.

$$s^2 = \frac{1}{4}[(6.2-6.5)^2 + (4.5-6.5)^2 + (6.6-6.5)^2 + (7.0-6.5)^2 + (8.2-6.5)^2]$$

Use data and $\bar{x}$.

$$= \frac{1}{4}[(-0.3)^2 + (-2.0)^2 + (0.1)^2 + (0.5)^2 + (1.7)^2]$$

Compute the differences.

$$= \frac{1}{4}[0.09 + 4.0 + 0.01 + 0.25 + 2.89]$$

Square each difference.

$$= \frac{1}{4}(7.24) = 1.81$$

Add, divide by 4.

STEP 3 Take the positive square root of the variance to find the sample standard deviation.

$$s = \sqrt{1.81} \approx 1.3454$$

A technology solution is shown in **Figure 3.11**.

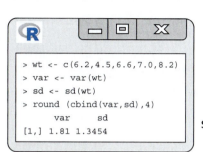

Figure 3.11 Variance and standard deviation using R.

A Computational Formula

Equation 3.4 is the definition of the sample variance and may be used to find s^2, but there is actually a more efficient technique for computing s^2.

Definition
The computational formula for the sample variance is

$$s^2 = \frac{1}{n-1}\left[\sum x_i^2 - \frac{1}{n}(\sum x_i)^2\right] \tag{3.6}$$

RAPID REVIEW Order of operations.

This is a convenient shortcut method for calculating s^2 without having to find all the deviations about the mean. Suppose $x_1, x_2, \ldots, x_n$ is a set of observations. To find s^2, Equation 3.6 says:

1. Find the sum of the squared observations, $\sum x_i^2$.
2. Find the sum of the observations, $\sum x_i$.
3. Square the sum of the observations, $(\sum x_i)^2$.
4. Multiply the square of the sum of the observations by $\frac{1}{n}$, $\frac{1}{n}(\sum x_i)^2$.
5. Subtract the two quantities, and multiply the difference by $\frac{1}{n-1}$.

$$s^2 = \frac{1}{n-1}\left[\sum x_i^2 - \frac{1}{n}(\sum x_i)^2\right]$$

EXAMPLE 3.9 Zucchini Weight (Continued)

Use the computational formula for s^2 to find the sample variance for the data in Example 3.8. The zucchini weights (in ounces) are 6.2, 4.5, 6.6, 7.0, and 8.2.

Solution

STEP 1 Find the sum of the squared observations.

$$\sum x_i^2 = 6.2^2 + 4.5^2 + 6.6^2 + 7.0^2 + 8.2^2$$
$$= 38.44 + 20.25 + 43.56 + 49.0 + 67.24 = 218.49$$

STEP 2 Find the sum of the observations.

$$\sum x_i = 6.2 + 4.5 + 6.6 + 7.0 + 8.2 = 32.5$$

STEP 3 Square the sum and multiply by $\frac{1}{n}$:

$$\frac{1}{5}(\sum x_i)^2 = \frac{1}{5}(32.5)^2 = 211.25$$

STEP 4 Subtract the two quantities, and multiply by $\frac{1}{n-1}$:

$$s^2 = \frac{1}{4}(218.49 - 211.25) = \frac{1}{4}(7.24) = 1.81$$

(The same answer as found in Example 3.8.)

TRY IT NOW Go to Exercises 3.41 and 3.51

CHAPTER APP **SpaceX Payloads**

Find the sample variance s^2 and the sample standard deviation s for the payload weight data.

It can be shown that Equation 3.4 and Equation 3.6 are equivalent. Exercise 3.64 at the end of this section asks for a proof. If you must find a sample variance by hand, then use the computational formula. It requires fewer calculations (is more efficient) and is usually more accurate (has less round-off error). In fact, most calculator and computer programs that find the sample variance use the computational formula.

The sample variance is always greater than or equal to zero: $s^2 \geq 0$. This is easy to see by looking at the definition in Equation 3.4. We sum squared deviations about the mean (always greater than or equal to zero) and divide by a positive number ($n-1$). There are two special cases.

1. $s^2 = 0$: This occurs if all the observations are the same. If all the observations are equal to some constant c, the mean is c, and all the deviations about the mean are zero. Hence, $s^2 = 0$. This makes sense intuitively as well: If all the observations are the same, there is no variability.

2. $n = 1$: This is a strange case, but it can occur. If $n = 1$, there is no variability; another way to think of it is that we cannot measure variability. The denominator in Equation 3.4 is zero, and anything divided by zero is undefined.

Quartiles

The sample variance (and the sample standard deviation) can be greatly influenced by outliers. An observation very far away from the rest has a large deviation about the mean, and in turn a large squared deviation about the mean, so it makes a large contribution to the sum (in the definition of the sample variance). The interquartile range is another measure of variability, and it is resistant to outliers.

Definition

Let $x_1, x_2, \ldots, x_n$ be a set of observations. The **quartiles** divide the data into four parts.

1. The **first (lower) quartile**, denoted Q_1 (or Q_L), is the median of the lower half of the observations when they are arranged in ascending order.
2. The **second quartile** is the median $\tilde{x} = Q_2$.
3. The **third (upper) quartile**, denoted Q_3 (or Q_U), is the median of the upper half of the observations when they are arranged in ascending order.
4. The **interquartile range**, denoted IQR, is the difference $\text{IQR} = Q_3 - Q_1$.

In smoothed histograms, the area under the curve between two values corresponds to the proportion of observations between those values. Therefore, here is a way to interpret part of Figure 3.12: 25% of the observations are between Q_1 and $\tilde{x}$.

The quartiles are illustrated in **Figure 3.12**.

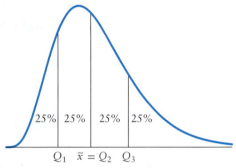

Figure 3.12 Smoothed histogram and quartiles.

There is a very intuitive method for finding quartiles. Arrange the data in order from smallest to largest. The median, $\tilde{x} = Q_2$, is the middle value. The first quartile Q_1 is the median of the lower half, and the third quartile Q_3 is the median of the upper half. In practice, a more general method is used for locating the position, or depth, of the first and third quartiles (in the ordered data set).

How to Compute Quartiles

Suppose $x_1, x_2, \ldots, x_n$ is a set of n observations.

1. Arrange the observations in ascending order, from smallest to largest.
2. To find Q_1, first compute $d_1 = \dfrac{n}{4}$.
   a. If d_1 is a whole number, then the depth of Q_1 (position in the ordered list) is $d_1 + 0.5$. Q_1 is the mean of the observations in positions d_1 and $d_1 + 1$.
   b. If d_1 is not a whole number, round up to the next whole number for the depth of Q_1.
3. To find Q_3, first compute $d_3 = \dfrac{3}{4}n$.
   a. If d_3 is a whole number, then the depth of Q_3 (position in the ordered list) is $d_3 + 0.5$. Q_3 is the mean of the observations in positions d_3 and $d_3 + 1$.
   b. If d_3 is not a whole number, round up to the next whole number for the depth of Q_3.

EXAMPLE 3.10   Pulse Rates

The following 10 observations represent the resting pulse rate for patients involved in an exercise study.

68   71   64   58   61   76   73   62   72   66

(a) Find the first quartile, the third quartile, and the interquartile range.

(b) Suppose there are 12 patients in the study, with $x_{11} = 78$ and $x_{12} = 81$. Find the first quartile, the third quartile, and the interquartile range for this modified data set.

Solution

(a)

STEP 1 Arrange the observations in order from smallest to largest.

Observation	58	61	62	64	66	68	71	72	73	76
Position	1	2	3	4	5	6	7	8	9	10

STEP 2 Find the depth of the first quartile.

$$d_1 = \frac{n}{4} = \frac{10}{4} = 2.5$$

Because d_1 is not a whole number, round up. The depth of the first quartile is 3.

Q_1 is in the third position in the ordered list.

Using the table of ordered data, $Q_1 = 62$.

STEP 3 Find the depth of the third quartile.

$$d_1 = \frac{3n}{4} = \frac{(3)(10)}{4} = 7.5$$

Because d_3 is not a whole number, round up. The depth of the third quartile is 8.

Q_3 is in the eighth position in the ordered list.

Using the table of ordered data, $Q_3 = 72$.

STEP 4 Find the interquartile range $IQR = Q_3 - Q_1$.

$IQR = 72 - 62 = 10$

A technology solution is shown in **Figure 3.13**.

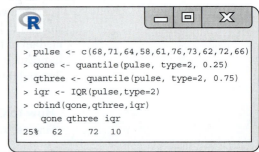

Figure 3.13 First quartile, third quartile, and interquartile range using R. The `type=2` option indicates the quantile algorithm.

(b)

STEP 1 If there are 12 patients in the study, arrange the observations in order from smallest to largest in the modified data set.

Observation	58	61	62	64	66	68	71	72	73	76	78	81
Position	1	2	3	4	5	6	7	8	9	10	11	12

STEP 2 Find the depth of the first quartile.

$$d_1 = \frac{n}{4} = \frac{12}{4} = 3$$

Because d_1 is a whole number, add 0.5. The depth of the first quartile is 3.5.

Q_1 is the mean of the observations in the third and fourth positions in the ordered list.

$$Q_1 = \frac{1}{2}(62 + 64) = 63$$

STEP 3 Find the depth of the third quartile.

$$d_3 = \frac{3n}{4} = \frac{(3)(12)}{4} = 9$$

Because d_3 is a whole number, add 0.5. The depth of the third quartile is 9.5.

Q_3 is the mean of the observations in the ninth and tenth positions in the ordered list.

$$Q_3 = \frac{1}{2}(73 + 76) = 74.5$$

STEP 4 Find the interquartile range.

$IQR = 74.5 - 63 = 11.5$

A technology solution is shown in **Figure 3.14**.

Figure 3.14 First quartile, third quartile, and interquartile range using R for the modified pulse data.

TRY IT NOW Go to Exercises 3.53 and 3.55

CHAPTER APP **SpaceX Payloads**

Find Q_1, Q_3, and IQR for the payload weight data.

A CLOSER LOOK

1. The interquartile range is the length of the interval that includes the middle half (middle 50%) of the data.
2. The interquartile range is not sensitive to outlying values. The lower and/or upper 25% of the distribution can be extreme without affecting Q_1 and/or Q_3.

Section 3.2 Exercises

Concept Check

3.35 True or False Every deviation about the mean is nonnegative.

3.36 True or False The sample standard deviation is always greater than or equal to 0.

3.37 True or False The sample standard deviation and the population standard deviation are always the same value.

3.38 True or False The computational formula for the sample variance is used only for large data sets.

3.39 True or False Quartiles divide the data into four parts.

3.40 True or False The first quartile is always less than or equal to the third quartile.

Practice

3.41 Find the sample range, sample variance, and sample standard deviation for each data set.

(a) {2.7, 6.0, 5.7, 5.4, 4.0, 3.1, 6.6, 5.7, 6.1, 3.0} EX3.41a
(b) {18.5, 23.5, 15.7, 15.7, 36.3, 20.8, 21.1, 20.2, 26.8, 19.9, 17.6, 17.5, 21.5, 22.4, 25.7} EX3.41b
(c) {23.94, −31.04, 37.09, 22.64, −61.23, 1.59, 23.09, 1.14} EX3.41c
(d) {0.13, 0.96, −0.50, 0.10, −1.65, −0.14, 1.43, −2.57, −1.28, −0.24, −0.90, −1.27, 1.53, 3.00, −1.28, 1.04, −0.90, 2.44, 1.70, 3.13} EX3.41d

3.42 Compute the sample variance and the sample standard deviation for each sample with known sum(s).

(a) $\sum x_i = 1219.29$ $\sum x_i^2 = 58945.1$ $n = 30$
(b) $\sum x_i = 35.2918$ $\sum x_i^2 = 7748.98$ $n = 17$
(c) $\sum x_i = 218.291$ $\sum x_i^2 = 3615.96$ $n = 15$
(d) $\sum (x_i - \bar{x})^2 = 49.784$ $n = 21$

3.43 Find the depth of the first quartile and the third quartile in an ordered data set of size n.

(a) $n = 60$
(b) $n = 37$
(c) $n = 100$
(d) $n = 48$

3.44 Find the first quartile, the third quartile, and the interquartile range for each data set.

(a) {20, 17, 37, 33, 29, 50, 20, 33} EX3.44a
(b) {13.1, 7.8, 11.9, 2.3, 6.7, 2.3, 7.4, 2.7, 8.9, 6.6, 6.8, 5.1, 2.2, 5.6, 5.5, 2.1, 7.7, 13.9, 1.6, 1.7} EX3.44b

(c) $\{-15, -13, -7, -15, -22, -12, -21, -21, -26, -17\}$
(d) $\{43.6, 44.1, 59.5, 52.3, 50.9, 39.7, 42.4, 58.5, 40.9, 38.5,$
$44.2, 60.3, 72.2, 34.8, 46.0, 54.7, 51.0, 54.3, 49.7, 62.9,$ EX3.44c
$44.6, 61.3, 52.4, 43.9, 68.8, 59.2, 57.1, 70.5, 52.3, 49.5\}$ EX3.44d

3.45 Consider the following data set. EX3.45

| 21 | 28 | 38 | 12 | 33 | 47 | 51 | 11 | 81 | 36 |

(a) Find the sample variance and the sample standard deviation.
(b) If 20 is subtracted from each observation in part (a), a new data set is formed.

| 1 | 8 | 18 | −8 | 13 | 27 | 31 | −9 | 61 | 16 |

Find the sample variance and the sample standard deviation for this new data set. How are these values related to the sample variance and the sample standard deviation found in part (a)?

(c) If each observation in part (a) is multiplied by 20, the following data set is formed.

| 420 | 560 | 760 | 240 | 660 |
| 940 | 1020 | 220 | 1620 | 720 |

Find the sample variance and the sample standard deviation for this new data set. How are these values related to the sample variance and the sample standard deviation found in part (a)?

3.46 How does an outlier affect each of the following?

(a) The sample variance
(b) The sample standard deviation
(c) The first quartile and the third quartile
(d) The interquartile range

Applications

3.47 Biology and Environmental Science The following turbidity readings (NTU) are from Lake Erie at weather station WE2 during 2017.[18] WE2

| 10.09 | 6.91 | 7.76 | 9.08 | 10.56 | 5.21 |
| 3.25 | 9.83 | 8.29 | 6.37 | 3.31 | 4.78 |

(a) Find the sample range R.
(b) Find the sample variance s^2 and the sample standard deviation s.
(c) Find the first quartile Q_1, the third quartile Q_3, and the interquartile range IQR.

3.48 Physical Sciences Feldspar is a raw material, used in the manufacture of glass, ceramic tiles, and insulation. The data set contains the 2017 feldspar mine production (in tons) for countries around the world.[19] FELD

(a) Find the sample standard deviation.
(b) Find the interquartile range.
(c) Which statistic, s or IQR, do you think is a better measure of variability for this data set? Justify your answer.

3.49 Fuel Consumption and Cars The gross vehicle weight rating (in pounds) for several 2018 automobiles is given in the table.[20] AUTOWT

| 4696 | 4790 | 5578 | 5831 | 4515 | 5586 | 5562 | 8590 |
| 4400 | 4756 | 6695 | 5434 | 5724 | 5830 | 5130 | 4222 |

(a) Find the sample variance and the sample standard deviation.
(b) Find the first and third quartiles.
(c) Find the interquartile range and the quartile deviation (another measure of variability), $QD = (Q_3 - Q_1)/2$.

3.50 Education and Child Development Many educators believe that success in school is related directly to the amount of time spent completing homework assignments. A research study compared the academic ability of 17-year-olds who spend less than one hour on homework every day and those who spend more than two hours on homework every day. Suppose the National Assessment of Educational Progress (NAEP) scores for each student in each group are given in the following table. NAEP

Less than one hour

| 290 | 289 | 291 | 289 | 289 | 294 | 288 | 291 |
| 293 | 290 | 290 | 291 | 290 | 290 | 296 | 292 |

More than two hours

| 303 | 305 | 302 | 297 | 294 | 303 | 299 | 297 | 303 | 299 |
| 300 | 295 | 297 | 297 | 297 | 293 | 296 | 297 | 302 | 294 |

(a) Find the sample variance, sample standard deviation, and interquartile range of the progress scores for students who spend less than one hour on homework.
(b) Find the sample variance, sample standard deviation, and interquartile range of the progress scores for students who spend more than two hours on homework.
(c) Use your answers to parts (a) and (b) to determine which data set has more variability.

3.51 Travel and Transportation In April 2018, Air Canada changed the operational aircraft on its Toronto to Barbados route to a Boeing 787-9. A random sample of these flights was obtained and the number of passengers on each flight is given in the table. TRTOBB

| 252 | 229 | 235 | 254 | 244 | 242 | 260 | 251 |

(a) Compute s^2 using the definition in Equation 3.4.
(b) Compute s^2 using the computational formula in Equation 3.6.
(c) How do your answers in parts (a) and (b) compare?

3.52 Public Policy and Political Science The president of the United States has the authority to grant clemencies, pardons, and commutations of sentences to criminals convicted of federal offenses. A sample of U.S. presidents was obtained, and the number of presidential clemency actions for each was recorded. The data are given in the table.[21] CLEM

President	Clemency actions
Calvin Coolidge	1691
Jimmy Carter	566
Woodrow Wilson	2827
John F. Kennedy	575
Barack Obama	1927
George W. Bush	200
William Clinton	459
Ronald Reagan	406
Gerald Ford	409
Richard Nixon	926
Lyndon Johnson	1187
Dwight Eisenhower	1157
Harry Truman	2044
Warren Taft	831
William McKinley	446

(a) Find Q_1, Q_3, and IQR for the clemency actions data.
(b) Find s^2 and s.
(c) Franklin D. Roosevelt had the highest number of clemency actions of any president, 3796. Add this value to the data set. Find IQR and s^2 for this expanded data set.
(d) How do IQR and s^2 compare in these two data sets? Explain why these values are the same/different.

3.53 Physical Sciences The following operating temperatures (°F) for a certain steam turbine were measured on 10 randomly selected days. **STEAM**

| 298 | 313 | 305 | 292 | 283 | 348 | 291 | 286 | 346 | 304 |

(a) Find Q_1, Q_3, and IQR.
(b) Find s^2 and s.
(c) Suppose the smallest observation (283) is changed to 226. Find IQR and s^2 for this modified data set.
(d) How do IQR and s^2 compare in these two data sets? Which measurement is more sensitive to outliers?

3.54 Marketing and Consumer Behavior Two measures designed to give a relative measure of variability are the **coefficient of variation**, denoted CV, and the **coefficient of quartile variation**, denoted CQV. These measures are defined by

$$CV = 100 \cdot \frac{s}{\bar{x}} \qquad CQV = 100 \cdot \frac{Q_3 - Q_1}{Q_3 + Q_1}$$

The areas (in square feet) for homes constructed in two new residential developments in San Antonio (one in the north central area and one on the city's west side) were recorded and are given in the table. **RESDEV**

North central development

| 2038 | 1939 | 2024 | 1990 | 2109 | 2102 | 1918 | 2022 |

West-side development

| 2061 | 2383 | 2638 | 2142 | 2382 | 1489 | 2070 | 2340 |
| 1725 | 2368 | 1674 | 1877 | | | | |

(a) Compute CV and CQV for each development.
(b) Compare the coefficient of variation and the coefficient of quartile variation for each development. Which data set has more variability?

3.55 Biology and Environmental Science The snowpack in California is the primary source of water that flows into the state's reservoirs, rivers, and streams. A random sample of streams in California was obtained and the flow rate (in cubic feet per second, cfs) was obtained for each.[22] The data are given in the table. **FLOW**

| 2730 | 187 | 30 | 263 | 111 | 200 | 355 | 4850 | 5 | 294 |
| 33 | 7 | 5 | 29 | 12 | 3 | 11 | 18 | 2 | 1100 |

(a) Find the sample variance and the sample standard deviation.
(b) Find Q_1, Q_3, and IQR for these data.
(c) Remove the two largest flow rates from the data set. Answer parts (a) and (b) for this reduced data set. Compare the sample standard deviation and IQR in these two data sets and explain how these values have changed.

3.56 Public Health and Nutrition The Center for Science in the Public Interest (CSPI), a consumer group concerned about nutrition labeling, has defined a new measure of breakfast cereal called the nutritional index (NI), which is based on calories, vitamins, minerals, and sugar content per serving. A larger NI indicates greater nutritional value. The NI was measured for randomly selected cereals sold by Kellogg's and General Mills. The results are given in the following table. **NI**

Kellogg's

| 86 | 70 | 77 | 79 | 71 | 80 | 88 | 62 | 81 | 82 |
| 75 | 83 | 70 | 67 | 72 | 68 | 74 | 80 | 62 | 74 |

General Mills

| 54 | 49 | 50 | 31 | 46 | 29 | 81 | 63 | 41 | 60 |
| 66 | 68 | 93 | 59 | 47 | 80 | 41 | 91 | 41 | 33 |

(a) Find s^2, s, and IQR for Kellogg's cereals.
(b) Find s^2, s, and IQR for General Mills cereals.
(c) Use the results in parts (a) and (b) to compare the variability in NI for the two companies' breakfast cereals.

3.57 Biology and Environmental Science It was unusually cold in Toronto, Canada, during January 2018. The weather was influenced by the polar jet stream and it led to one of the coldest December/January periods in the last 30 years. A random sample of heating degree days (in °C) from this period was obtained.[23] **POLJET**

(a) Find Q_1, Q_3, and IQR for these heating degree day data.
(b) How large could the minimum observation be without changing IQR?
(c) Find the coefficient of quartile variation (CQV, defined in Exercise 3.54) for the original data set.

3.58 Travel and Transportation Los Angeles Airport (LAX) has several Transportation Security Administration (TSA) checkpoints, but the wait time in these security lines can be more than 20 minutes. A random sample of travelers at LAX was obtained and the time (in minutes) required for each to pass through the TSA checkpoint was recorded for each. LAXTSA

(a) Find Q_1, Q_3, and IQR for these TSA wait time data.
(b) How many observations should be greater than Q_3? How many are actually greater than Q_3?
(c) Anyone who waits in line 20 minutes or longer usually files a formal complaint. Use these data to estimate the proportion of travelers who file complaints.

Extended Applications

3.59 Biology and Environmental Science The majority of wildland fires in the United States are caused by humans leaving campfires unattended, burning debris, or discarding cigarettes. The remaining fires are caused by lightning or lava. The following table lists the number of wildland fires in 2017 in selected states.[24] FIRES

State	AK	CA	CO	CT	FL	GA
Fires	364	9560	967	97	3280	3929

State	HI	IA	ID	KS	LA	MA
Fires	3	427	1598	71	1064	1216

State	MN	MO	MT	NH	NJ	PA
Fires	1036	3398	2422	36	735	537

(a) Find the sample variance and the sample standard deviation.
(b) Find Q_1, Q_3, and IQR for these data.
(c) Which measure, s or IQR, is a better measure of variability in this case? Why?
(d) Verify that the sum of the deviations about the mean is 0 (subject to round-off error).

3.60 Biology and Environmental Science Many live HD nest cams have been installed around the United States, including many that are positioned over wild bald eagle nests. The following table lists the weight (in grams) of several bald eagle eggs. NEST

122	111	124	126	110	135
123	125	124	121	122	114

(a) Find Q_1, Q_3, and IQR for these data.
(b) Suppose the largest weight (135) is changed to 127. Find Q_1, Q_3, and IQR for this modified data set.
(c) How large could the maximum weight be without changing IQR?
(d) How much could the minimum weight be raised before Q_1 changes?

3.61 Travel and Transportation A typical road bridge is constructed to last approximately 50 years. According to a recent report, almost 9% of all U.S. bridges are structurally deficient, including the Arlington Memorial Bridge and the Brooklyn Bridge. The number of structurally deficient bridges in each state and the District of Columbia as of 2017 was obtained.[25] DEFBRG

(a) Find the sample variance and the sample standard deviation of the number of structurally deficient bridges.
(b) Suppose each state is able to repair 10% of its structurally deficient bridges. Adjust each observation to reflect the reduced number of structurally deficient bridges (don't do any rounding), and find the sample variance and the sample standard deviation for this new data set.
(c) How do your answers to parts (a) and (b) compare?

3.62 Manufacturing and Product Development Satantango is one of the longest movies ever produced, lasting 450 minutes. Six other long movies were randomly selected and their length (in minutes) was recorded. MOVLEN

366	345	188	148	287	219

(a) Find each deviation about the mean.
(b) Verify that the sum of the deviations about the mean is 0 (subject to round-off error).
(c) Show that, in general, $\sum(x_i - \bar{x}) = 0$. *Hint*: Write this calculation as two separate sums, and use the definition of the sample mean.

3.63 Biology and Environmental Science The Virginia Estuarine and Coastal Observing System monitors the Chesapeake Bay and records values of several variables related to the water in the bay, including salinity, temperature, and turbidity. Suppose the wind speed (in miles per hour) at selected locations on a specific day are given in the table. BAY

2	5	11	17	9	7	11	10
6	9	24	27	8	8	10	28
30	25	14	24	10	18	18	25

(a) Find the sample variance and the sample standard deviation for these wind-speed data.
(b) Convert each observation to meters per second (multiply each observation by 0.44704). Find the sample variance and the sample standard deviation for the wind-speed data in meters per second.
(c) How do your answers to parts (a) and (b) compare?

3.64 Proof Prove that Equation 3.4 (definition of the sample variance) can be written as Equation 3.6. That is, show that

$$\frac{1}{n-1}\sum(x_i - \bar{x})^2 = \frac{1}{n-1}\left[\sum x_i^2 - \frac{1}{n}(\sum x_i)^2\right]$$

3.65 Public Health and Nutrition A nutritional study recently found the following number of calories in one slice of ten popular pizzas.[26] PIZZA

272	310	294	250	400	330	298	320	192	260

(a) Find the sample variance and the sample standard deviation.
(b) Add 15 (calories) to each observation. Find the sample variance and the sample standard deviation for this modified data set.
(c) How do your answers to parts (a) and (b) compare?
(d) Suppose a data set (x's) has variance s_x^2 and standard deviation s_x. A new (transformed) data set is created using the equation $y_i = x_i + b$, where b is a constant. How are the variance and the standard deviation of the new data set (s_y^2 and s_y) related to s_x^2 and s_x?

3.66 Medicine and Clinical Studies Some medical professionals suggest that bystanders should be prepared to provide first aid during an emergency because it can take 8–15 minutes before medical services arrive on a scene. Suppose eight emergencies in a large city were selected at random. The time from receiving notification to arrival on the scene is given in the table. **AID**

| 18 | 12 | 3 | 15 | 7 | 6 | 9 | 25 |

(a) Find the sample variance and the sample standard deviation.
(b) Multiply each observation by 3. (During the winter months, it generally takes three times as long for emergency services to arrive.) Find the sample variance and the sample standard deviation for this modified data set.
(c) How do your answers in parts (a) and (b) compare?
(d) Suppose a data set (x's) has variance s_x^2 and standard deviation s_x. A new (transformed) data set is created using the equation $y_i = ax_i$, where a is a constant. How are the variance and standard deviation of the new data set (s_y^2 and s_y) related to s_x^2 and s_x?

3.67 Transformed Data Combine the results obtained in the previous two exercises. Suppose a data set (x's) has variance s_x^2 and standard deviation s_x. A new (transformed) data set is created using the equation $y_i = ax_i + b$, where a and b are constants. How are the variance and the standard deviation of the new data set (s_y^2 and s_y) related to s_x^2 and s_x?

3.68 Anything Is Possible? Consider the following set of observations.

| 5 | 7 | 3 | 2 | 4 | 6 | 9 | 11 | 13 |

Can you find a subset of size $n = 7$ with $\bar{x} = 5$? If not, why not?

3.69 Marketing and Consumer Behavior Many homeowners pave their driveways with asphalt to enhance the driveway's appeal, to present a clean, finished look, and to make it easier to shovel snow in the winter (in the northeast). A random sample of driveways paved by Robert Young and Company was obtained. The area of each paved driveway (in square feet) is given in the following table. **PAVE**

| 316 | 318 | 271 | 340 | 439 | 415 | 537 | 423 | 538 | 393 |
| 527 | 427 | 398 | 417 | 402 | 244 | 553 | 524 | 430 | 343 |

(a) Find the sample variance and the sample standard deviation for this data set (x's).

(b) Since the cost for paving is $1000 plus $4 per square foot, let $y_i = 4x_i + 1000$. Find the sample variance and the sample standard deviation for the y's.
(c) How do your answers in parts (a) and (b) compare?
(d) Find Q_1, Q_3, and IQR for each data set. How do these values compare?

Challenge Problems

3.70 Biology and Environmental Science A whale-watching tour off the coast of Maine is considered a success if at least one whale is sighted. Thirty-two randomly selected summer tours are classified in the table.

S	S	S	S	F	S	S	S	S	S	S	S
S	S	S	S	S	S	S	F	S	S	S	S
S	F	S	S	S	S	S	S				

(a) Find the sample proportion of successes.
(b) Change each S to a 1 and each F to a 0. Find the sample variance for these new data. Write the sample variance in terms of the sample proportion.
(c) If a population happens to be of finite size N, then the population mean and the population variance, respectively, are defined by

$$\mu = \frac{1}{N}\sum x_i \qquad \sigma^2 = \frac{1}{N}\sum(x_i - \mu)^2$$

Suppose the table represents an entire population. Find the population variance for the data (consisting of 0's and 1's). Write the population variance in terms of the sample proportion.

3.71 Other Summary Statistics Many other summary statistics can be used to describe various characteristics of a numerical data set. Suppose $x_1, x_2, \ldots, x_n$ is a set of observations. For $r = 1, 2, 3, \ldots$, the **rth moment about the mean** $\bar{x}$ is defined as

$$m_r = \frac{1}{n}\sum(x_i - \bar{x})^r$$

For example, the second moment about the mean is

$$m_2 = \frac{1}{n}\sum(x_i - \bar{x})^2$$

Certain moments about the mean are used to define the **coefficient of skewness** (g_1) and the **coefficient of kurtosis** (g_2):

$$g_1 = \frac{m_3}{m_2^{3/2}} \qquad g_2 = \frac{m_4}{m_2^2}$$

The statistic g_1 is a measure of the lack of symmetry, and g_2 is a measure of the extent of the peak in a distribution.

Use technology to compute the values g_1 and g_2 for various distributions: skewed, symmetric, unimodal, uniform. Use your results to determine the values of g_1 that suggest more skewness in the distribution, and the values of g_2 that indicate a flatter, more uniform distribution.

3.3 The Empirical Rule and Measures of Relative Standing

Chebyshev's Rule

Measures of central tendency and measures of variability are used to describe the general nature of a data set. These two types of measures may be combined to describe the distribution of a data set more precisely. In addition, these values may be used to define measures of relative standing, to determine quantities used to compare observations from different data sets (with different units), or even to draw a conclusion or make an inference.

The first result combines the mean and the standard deviation to describe a distribution.

Chebyshev's Rule

What happens if $k = 1$?

Let $k > 1$. For any set of observations, the proportion of observations within k standard deviations of the mean [that is, lying in the interval $(\bar{x} - ks, \bar{x} + ks)$, where s is the standard deviation] is at least $1 - (1/k^2)$.

Recall interval notation: (a, b) denotes an open interval, with the endpoints not included, from a to b. Therefore, $(\bar{x} - ks, \bar{x} + ks)$ means the set of all x's such that $\bar{x} - ks < x < \bar{x} + ks$.

The graph in **Figure 3.15** and the accompanying table illustrate this concept. For any set of observations, the smoothed histogram shows that the proportion of observations captured in the interval $(\bar{x} - ks, \bar{x} + ks)$ is at least $1 - (1/k^2)$. For example, the proportion of observations within 1.5 standard deviations of the mean is at least 0.56 (or 56%). The proportion of observations within 3 standard deviations of the mean is at least 0.89 (or 89%).

In Figure 3.15, recall that we associate area under the smoothed histogram with the proportion of observations in the interval.

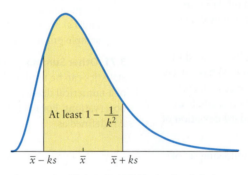

Figure 3.15 Illustration of Chebyshev's Rule.

Table 3.1 Chebyshev's Rule calculations for various values of k.

k	$1 - \dfrac{1}{k^2}$
1.5	$1 - \dfrac{1}{1.5^2} \approx 0.56$
2.0	$1 - \dfrac{1}{2.0^2} \approx 0.75$
2.3	$1 - \dfrac{1}{2.3^2} \approx 0.81$
3.0	$1 - \dfrac{1}{3.0^2} \approx 0.89$

A CLOSER LOOK

A symmetric interval about the mean is centered at the mean and has endpoints that are the same distance from the mean in each direction.

1. **Chebyshev's Rule** simply helps to describe a set of observations using symmetric intervals about the mean. If we move k standard deviations from the mean in both directions, then the proportion of observations captured is at least $1 - (1/k^2)$.

2. The total area under the curve (the sum of all the proportions or relative frequencies) is 1. Hence, Chebyshev's Rule also implies that the proportion of observations in the tails of the distribution, outside the interval $(\bar{x} - ks, \bar{x} + ks)$, is at most $1/k^2$.

3. As indicated in the statement of Chebyshev's Rule and as suggested in the table, you may use any value of k greater than 1, including decimals. The two most common values for k are $k = 2$ and $k = 3$. The actual proportions of observations within 2 and within 3 standard deviations must satisfy the conclusions in Chebyshev's Rule and the Empirical Rule (discussed later in this section). In addition, $k = 2$ and $k = 3$ provide the fundamental background to statistical inference.

Symmetric intervals.

4. Chebyshev's Rule is very conservative because it applies to any set of observations. Usually, the proportion of observations within k standard deviations of the mean is more than $1 - (1/k^2)$.
5. Chebyshev's Rule may also be used to describe a population. If the mean and the standard deviation are known, then μ and σ may be used in place of $\bar{x}$ and s, respectively. For any population, the proportion of observations that lie in the interval $(\mu - k\sigma, \mu + k\sigma)$ is at least $1 - (1/k^2)$.

EXAMPLE 3.11 Automobile Battery Lifetime

In a random sample of the lifetime (in months) of an Optima RedTop automobile battery, $\bar{x} = 54$ and $s = 5.3$. Use Chebyshev's Rule with $k = 2$ and $k = 3$ to describe this distribution of battery lifetimes.

Solution

STEP 1 For $k = 2$: $1 - \dfrac{1}{k^2} = 1 - \dfrac{1}{2^2} = 1 - \dfrac{1}{4} = \dfrac{3}{4} = 0.75$

At least 3/4 (or 75%) of the observations lie in the interval
$(\bar{x} - 2s, \bar{x} + 2s) = (54 - 2(5.3), 54 + 2(5.3)) = (43.4, 64.6)$

STEP 2 For $k = 3$: $1 - \dfrac{1}{k^2} = 1 - \dfrac{1}{3^2} = 1 - \dfrac{1}{9} = \dfrac{8}{9} \approx 0.89$

At least 8/9 (or 89%) of the observations lie in the interval
$(\bar{x} - 3s, \bar{x} + 3s) = (54 - 3(5.3), 54 + 3(5.3)) = (38.1, 69.9)$

STEP 3 Note also:

At most 1/4 (or 25%) of the observations lie outside the interval (43.4, 64.6).

At most 1/9 (or 11%) of the observations lie outside the interval (38.1, 69.9). ∎

EXAMPLE 3.12 Alice's Restaurant

In 1967, singer-songwriter Arlo Guthrie released an 18-minute song called "Alice's Restaurant," based on an incident in his life on Thanksgiving Day 1965 that began with a citation for littering. Most popular songs are much shorter. As an example, Adele's "When We Were Young" is approximately 5 minutes long. Suppose that in a random sample of the length (in minutes) of songs produced by classic rock bands, $\bar{x} = 3.35$ and $s = 0.5$.

(a) Find the approximate proportion of observations between 2.35 and 4.35 minutes.

(b) Find the approximate proportion of observations less than 1.85 or greater than 4.85 minutes.

(c) Approximately what proportion of songs last more than 5 minutes?

Solution

We don't know anything about the shape of the distribution of the length of songs. However, Chebyshev's Rule applies to any distribution, tells us about the proportion of observations captured by certain intervals, and may be used here if the questions involve symmetric intervals about the mean.

(a) Consider the interval (2.35, 4.35) and recognize that

$(\bar{x} - 2s, \bar{x} + 2s) = (3.35 - 2(0.5), 3.35 + 2(0.5)) = (2.35, 4.35)$

Therefore, (2.35, 4.35) is a symmetric interval about the mean, $k = 2$ standard deviations in each direction.

Using Chebyshev's Rule, at least $1 - (1/4) = 3/4$ (or 75%) of the observations lie between 2.35 and 4.35 minutes.

(b) Similarly, consider the interval (1.85, 4.85) and recognize that

$$(\bar{x} - 3s, \bar{x} + 3s) = (3.35 - 3(0.5), 3.35 + 3(0.5)) = (1.85, 4.85)$$

Therefore, (1.85, 4.85) is a symmetric interval about the mean, $k = 3$ standard deviations in each direction.

Using Chebyshev's Rule, at least $1 - (1/9) = 8/9$ (or 89%) of the observations lie between 1.85 and 4.85 minutes.

At most 1/9 (or 11%) of the observations are less than 1.85 or greater than 4.85 minutes.

(c) Since Chebyshev's Rule involves intervals in terms of the number of standard deviations from $\bar{x}$, find out how far 5 is from $\bar{x}$ in standard deviations.

$$\bar{x} + ks = 3.35 + k(0.5) = 5 \Rightarrow k = 3.3$$

We cannot assume anything about the shape of the distribution.

Use $k = 3.3$ in Chebyshev's Rule:

$$1 - \frac{1}{k^2} = 1 - \frac{1}{3.3^2} \approx 0.91$$

At least 0.91 (or 91%) of the observations lie in the interval

$$(\bar{x} - 3.3s, \bar{x} + 3.3s) = (3.35 - 3.3(0.5), 3.35 + 3.3(0.5)) = (1.7, 5.0)$$

Therefore, at most $1 - 0.91 = 0.09$ (or 9%) of the observations are outside this interval, either less than 1.7 or greater than 5.0 minutes. We cannot assume that the distribution is symmetric, so we do not know which part of the 9% is less than 1.7 and which part is more than 5.0 minutes. To be conservative, the best we can say is that at most 9% of the observations are more than 5 minutes long. ∎

TRY IT NOW Go to Exercise 3.89

CHAPTER APP **SpaceX Payloads**

Without assuming anything about the shape of the distribution of payload weights, use Chebyshev's Rule to describe the distribution (for $k = 2$ and for $k = 3$).

The Empirical Rule

A normal curve is bell-shaped and symmetric, centered at the mean.

If a set of observations can be reasonably modeled by a normal curve, then we can describe this distribution more precisely. The Empirical Rule involves the mean and standard deviation also, and the results apply to three specific symmetric intervals about the mean.

 Symmetric intervals.

Remember that we can associate area under the smoothed histogram, or area under the curve, with the proportion of observations in the interval.

The Empirical Rule
If the shape of the distribution of a set of observations is approximately normal, then

1. The proportion of observations within 1 standard deviation of the mean is approximately 0.68.
2. The proportion of observations within 2 standard deviations of the mean is approximately 0.95.
3. The proportion of observations within 3 standard deviations of the mean is approximately 0.997.

3.3 The Empirical Rule and Measures of Relative Standing

Figure 3.16 illustrates the Empirical Rule, the symmetric intervals about the mean, and the proportions. The Empirical Rule conclusions are more accurate than those from Chebyshev's Rule because we now (assume) more about the shape of the distribution (normality).

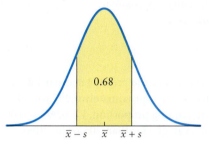
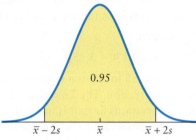
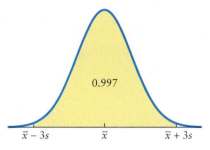

Figure 3.16 Symmetric intervals and proportions associated with the Empirical Rule.

For now, the reaons for the proportions 0.68, 0.95, and 0.997 will remain a mystery. We will discover where these numbers come from in Chapter 6.

A CLOSER LOOK

1. Given a set of observations, the Empirical Rule may be used to check normality. To test for normality numerically, find the mean, the standard deviation, and the three symmetric intervals about the mean $(\bar{x} - ks, \bar{x} + ks)$, $k = 1, 2, 3$. Compute the actual proportion of observations in each interval. If the actual proportions are close to 0.68, 0.95, and 0.997, then there is no evidence to suggest non-normality. Otherwise, there is evidence to suggest that the shape of the distribution is not normal. This process is sort of a Backward Empirical Rule.

2. The Empirical Rule may also be used to describe a population. If the distribution of the population is approximately normal, and the mean and the standard deviation are known, then μ and σ may be used in place of $\bar{x}$ and s, respectively.

3. The proportion of observations beyond 3 standard deviations from the mean is $1 - 0.997 = 0.003$ (pretty small). Therefore, if the shape of a (population) distribution is approximately normal, it would be unusual to have an observation more than 3 standard deviations from the mean. What if there is one? (See Example 3.14.)

EXAMPLE 3.13 World Record Speeding Ticket

In some parts of the world, the fine for speeding is partly determined by your income level. This ability-to-pay system led to the world's most expensive speeding ticket in 2010 of $290,000 to a Swedish man traveling about 180 mph in a Mercedes. In a random sample of more traditional ticket fines in Alberta, Canada, for the month of July 2018, suppose the shape of the distribution is approximately normal, with $\bar{x} = 130$ and $s = 25$ (in Canadian dollars). Approximately what proportion of observations is

(a) between 80 and 180?
(b) greater than 205 or less than 55?
(c) greater than 205?
(d) between 105 and 180?

Solution

Since the shape of the distribution is approximately normal, the Emprical Rule may be used to determine the proportion of observations captured by certain intervals, related in some way to three special symmetric intervals about the mean.

(a) Find the values 1, 2, and 3 standard deviations about the mean in each direction. See **Figure 3.17**. Notice that

$$(130 - 2(25), 130 + 2(25)) = (80, 180)$$

Therefore, 80 to 180 is a symmetric interval about the mean, 2 standard deviations in each direction. The Empirical Rule states that approximately 0.95 (or 95%) of the observations lie in this interval. See Figure 3.17.

(b) Notice that

$$(130 - 3(25), 130 + 3(25)) = (55, 205)$$

Therefore, 55 to 205 is a symmetric interval about the mean, 3 standard deviations in each direction. The Empirical Rule states that approximately 0.997 (or 99.7%) of the observations lie in this interval. The remaining proportion, $1 - 0.997 = 0.003$ (or 0.3%), of observations lie outside this interval, greater than 205 or less than 55. See **Figure 3.18**.

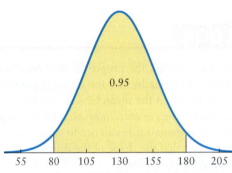

Figure 3.17 Approximately 0.95 (or 95%) of the observations lie within 2 standard deviations of the mean, in the interval (80, 105).

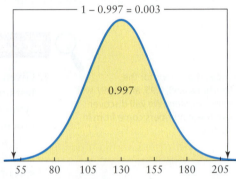

Figure 3.18 Approximately $1 - 0.997 = 0.003$ (or 0.3%) of the observations lie outside the interval (55, 205).

(c) Because a normal distribution is symmetric about the mean, the remaining proportion outside 3 standard deviations from the mean ($1 - 0.997 = 0.003$) is divided evenly between the two tails. Therefore approximately $0.003/2 = 0.0015$ (or 0.15%) of the observations are greater than 205. See **Figure 3.19**.

(d) (105, 180) is not a symmetric interval about the mean. However, approximately 0.68 of the observations lie in the interval (105, 155) (1 standard deviation from the mean). Approximately 0.95 of the observations lie in the interval (80, 180) (2 standard deviations from the mean). This means that $0.95 - 0.68 = 0.27$ of the observations lie in the intervals (80, 105) and (155, 180).

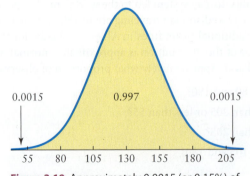

Figure 3.19 Approximately 0.0015 (or 0.15%) of the observations are greater than 205.

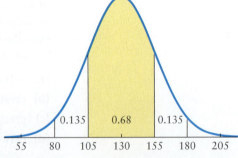

Figure 3.20 Approximately 0.27 (or 27%) of the observations lie in the interval (80, 105) and (155, 180).

Because a normal distribution is symmetric, $0.27/2 = 0.135$ of the observations lie between 155 and 180. Therefore, a total of approximately $0.68 + 0.135 = 0.815$ (or 81.5%) of the observations lie between 105 and 180. See **Figure 3.20**. ∎

TRY IT NOW Go to Exercises 3.87 and 3.91

CHAPTER APP SpaceX Payloads

Assume the shape of the distribution of payload weights is approximately normal. Approximately what proportion of payload weights is between 1250 and 7268?

EXAMPLE 3.14 Pain Management

First Horizon Pharmaceutical has just developed a new medicine for treating routine aches and pains. The company claims that the distribution of pain-relief times (in hours) is approximately normal, with mean $\mu = 8$ and standard deviation $\sigma = 0.2$. A patient with a typical muscle ache is randomly selected and the medicine is administered. The patient reports pain relief for only 7 hours. Is there any evidence to refute the manufacturer's claim?

Solution

Because the distribution of pain-relief times is approximately normal, the Empirical Rule may be used to determine how often observed times in certain intervals occur. If the observed pain-relief time is rare, then we should question the manufacturer's claim.

STEP 1 The shape of the distribution is approximately normal with $\mu = 8$ and $\sigma = 0.2$. Therefore,

- Approximately 0.68 of the population lies in the interval (7.8, 8.2).
- Approximately 0.95 of the population lies in the interval (7.6, 8.4).
- Approximately 0.997 of the population lies in the interval (7.4, 8.6).

STEP 2 The observation $x = 7$ hours lies outside the largest interval (7.4, 8.6). Only $1 - 0.997 = 0.003$ of the population lies outside this interval. More precisely (because of symmetry), only $0.003/2 = 0.0015$ of the population lies below 7.4. Seven hours is a very rare observation. Two things may have occurred:

- Seven hours is an incredibly lucky observation. Even though the proportion of observations below 7.4 is small, it is still possible for the manufacturer's claim to be true and for the pain reliever to last only 7 hours in this patient.
- The manufacturer's claim is false. Because an observation of 7 hours is so rare, it is more likely that one of the assumptions is wrong. The shape of the distribution may not be normal, the mean may be different from 8, and/or the standard deviation might be different from 0.2.

STEP 3 Typically, statistical inference discounts the lucky alternative. Therefore, because 7 hours is such an unlikely observation, there is evidence to suggest that the manufacturer's claim is false. Something is awry. We would rarely see pain relief of only 7 hours if the claim is true.

Note: We may be too quick to make an inference based on only a single observation. We will learn how to use more observations (information) to reach a more confident conclusion.

z-Scores

One method for comparing observations from different samples (with different units) is to use a standardized score. For a given observation, this relative measure is used to determine the distance from the mean in standard deviations.

Definition

Suppose $x_1, x_2, \ldots, x_n$ is a set of n observations with mean $\bar{x}$ and standard deviation s. The **z-score** corresponding to the ith observation x_i is given by

$$z_i = \frac{x_i - \bar{x}}{s} \tag{3.7}$$

z_i is a measure associated with x_i that indicates the distance from $\bar{x}$ in standard deviations.

Statisticians tend to measure distances in standard deviations, not miles, feet, inches, or meters. We often ask, "How many standard deviations from the mean is a given observation?"

> ### A CLOSER LOOK
>
> 1. z_i may be positive or negative (or zero). A positive z-score indicates the observation is to the right of the mean. A negative z-score indicates the observation is to the left of the mean.
> 2. A z-score is a measure of relative standing; it indicates where an observation lies in relation to the rest of the values in the data set. There are other methods of standardization, but this is the most commonly used.
> 3. Given a set of n observations, the sum of all the z-scores is 0: that is, $\sum z_i = 0$. Can you prove this?

EXAMPLE 3.15 SAT Versus ACT

Many colleges still require and strongly consider SAT or ACT scores in the admission process. These two admissions tests are scored on very different scales. Suppose the summary information for the math portion of each test is given in the table.

Test	Mean	Standard deviation
SAT Math	540	87
ACT Math	20.7	5.0

One applicant to a college scored 670 on the SAT Math test and another scored 27 on the ACT Math test. Which score is better, in terms of statistics?

Solution

Compute and compare the z-scores for each test. This will allow us to determine how many statistical steps each observation is from the mean. The higher the z-score, the better the test score.

STEP 1 Consider the z-scores to answer this question in terms of statistics. A larger z-score indicates a raw score that is farther from the mean in standard deviations, which is a more extraordinary observation.

STEP 2 First applicant: $z = \dfrac{670 - 540}{87} \approx 1.49$

670 is approximately 1.49 standard deviations to the right of the mean.

Second applicant: $z = \dfrac{27 - 20.7}{5.0} = 1.26$

27 is approximately 1.26 standard deviations to the right of the mean.

STEP 3 The first applicant's score is actually better, because the test score is farther from the mean to the right, indicating, in this case, a better performance.

EXAMPLE 3.16 Microwave Return Policy

The owner of Brooklyn Appliances is trying to establish a policy for the return of microwave ovens. In a random sample of the lifetime (in months) of microwaves, $\bar{x} = 72$ and $s = 12$. A customer recently returned with a microwave that had failed after 62 months. Is this a reasonable lifetime, or should the store provide some sort of refund (or even a new microwave)?

Solution

For any distribution, most observations are within 3 standard deviations of the mean or, equivalently, have a z-score between -3 and $+3$. Compute the z-score for this microwave lifetime, the number of statistical steps from the mean.

STEP 1 To determine whether 62 months is a reasonable microwave lifetime, consider the z-score corresponding to this observation.

STEP 2 $z = \dfrac{62-72}{12} = -0.83$

The observation (62 months) is only 0.83 standard deviation to the left of the mean. Because 62 months is within 1 standard deviation of the mean (regardless of the shape of the distribution), this is a very conservative, reasonable observation.

STEP 3 The microwave functioned for a reasonable amount of time subject to normal variability. No refund is necessary.

TRY IT NOW Go to Exercises 3.95 and 3.97

CHAPTER APP **SpaceX Payloads**

Suppose the population mean payload weight is 3200 kg and the population standard deviation payload weight is 2000 kg. A new SpaceX customer has a payload that weighs 5500 kg. Is there any evidence that this is an unusual weight? Justify your answer.

Percentiles

Another indication of relative standing is a percentile. Do you remember all of those standardized tests you took in grade school? The results were usually reported in terms of percentiles. The 90th percentile was a good score and the 25th percentile meant more homework in your future.

Definition
Let $x_1, x_2, \ldots, x_n$ be a set of observations. The **percentiles** divide the data set into 100 parts. For any integer r ($0 < r < 100$), the **rth percentile**, denoted p_r, is a value such that r percent of the observations lie at or below p_r (and $100 - r$ percent lie above p_r).

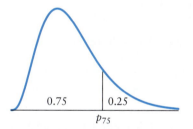

Figure 3.21 The 75th percentile is illustrated using a smoothed histogram.

The rth percentile has the same units as the observations; it is not a percent. **Figure 3.21** shows a smoothed histogram and illustrates the location of the 75th percentile on the measurement axis.

A CLOSER LOOK

1. The 50th percentile is the median, $p_{50} = \tilde{x}$.
2. The 25th percentile is the first quartile and the 75th percentile is the third quartile: $p_{25} = Q_1$, $p_{75} = Q_3$.

Remember: The area under the curve (smoothed histogram) between a and b corresponds to the proportion of observations between a and b. So the total area under the curve is 1.

How to Compute Percentiles
Suppose $x_1, x_2, \ldots, x_n$ is a set of n observations.
1. Arrange the observations in ascending order, from smallest to largest.
2. To find p_r, compute $d_r = \dfrac{r}{100} n$.
   a. If d_r is a whole number, then the depth of p_r (position in the ordered list) is $d_r + 0.5$. p_r is the mean of the observations in positions d_r and $d_r + 1$.
   b. If d_r is not a whole number, round up to the next whole number for the depth of p_r.

EXAMPLE 3.17 Visibility

The Clean Air Act and the National Park Service Organic Act protect the air quality in national parks. These laws help to provide visitors to Carlsbad Caverns with beautiful views of the Guadalupe Mountains and the Chihuahuan Desert. The visibility at Carlsbad Caverns is measured in terms of a haze index in deciviews (dv). Suppose historical records suggest that a haze index of 12 dv lies in the 80th percentile. Interpret this value.

Solution

Here, 12 is a single observation from the population of haze index records, and percentiles divide the observations into 100 parts. Because 12 lies at the 80th percentile, 80% of all haze index scores were 12 or less, and $100 - 80 = 20\%$ of all haze index scores are greater than 12.

Note: We do not know anything about the shape of the distribution of haze index scores, nor do we know the mean or standard deviation. There is no way of telling how far 12 is from the mean in standard deviations.

EXAMPLE 3.18 A Walk Across the Brooklyn Bridge

The Brooklyn Bridge in New York City is a popular tourist attraction, and many people enjoy a walk along the pedestrian walkway. A walk across the bridge takes approximately 25–60 minutes.[27] A random sample of people walking across the bridge was obtained, and their times are given in the table.

44	51	43	31	50	53	59	49	55	25
28	30	60	42	36	54	31	33	48	39
37	44	48	51	34	38	58	59	53	59

Find the time at which it took 20% of the walkers to make it across the bridge.

Solution

Find p_{20} (in minutes) so that 20% of the observations lie at or below and 80% lie above. Follow the steps for computing percentiles.

STEP 1 Order the data from smallest to largest. A portion of this ordered list is given in the following table.

Observation	25	28	30	31	31	33	34	36	37	38
Position	1	2	3	4	5	6	7	8	9	10

STEP 2 Find d_{20}: $d_{20} = \dfrac{r}{100} \cdot n = \dfrac{20}{100} \cdot 30 = 6$

STEP 3 Because d_{20} is a whole number, add 0.5 to find the depth. The depth of p_{20} is $d_{20} + 0.5 = 6 + 0.5 = 6.5$.

STEP 4 The 20th percentile, p_{20}, is the mean of the sixth and seventh observations.

$$p_{20} = \frac{1}{2}(33 + 34) = 33.5$$

Figure 3.22 shows a technology solution.

STEP 5 Twenty percent of the walkers made it across the bridge within 33.5 minutes, and 80% took longer than 33.5 minutes.

```
> walk <- c(44,...,59)
> quantile(walk,0.20,type=2)
20%
33.5
```

Figure 3.22 Use the R command `quantile()` to compute percentiles. *Note*: Use the option `type=2` to compute percentiles using the algorithm presented.

TRY IT NOW Go to Exercise 3.96

CHAPTER APP SpaceX Payloads

Find the 35th percentile for the payload weight data.

Section 3.3 Exercises

Concept Check

3.72 True or False Chebyshev's Rule applies to any set of data.

3.73 True or False The conclusion in Chebyshev's Rule applies to a symmetric interval about the mean.

3.74 True or False In a smoothed histogram, the area under the curve between two values a and b corresponds to the proportion of observations between a and b.

3.75 Short Answer Why is Chebyshev's Rule conservative?

3.76 True or False The Empirical Rule applies to any set of observations.

3.77 Short Answer If the shape of a distribution is approximately normal, what does it mean if an observation is more than 3 standard deviations from the mean?

3.78 True or False A z-score is a measure of relative standing.

3.79 True or False All quartiles are percentiles.

3.80 True or False A negative z-score means the observation is to the left of the mean.

Practice

3.81 For each data set with $\bar{x}$ and s given, find a symmetric interval k standard deviations about the mean, and use Chebyshev's Rule to compute the approximate proportion of observations with this interval.
(a) $\bar{x} = 50$, $s = 5$, $k = 2$
(b) $\bar{x} = 352$, $s = 10.5$, $k = 3$
(c) $\bar{x} = 17$, $s = 3.5$, $k = 1.6$
(d) $\bar{x} = 36.5$, $s = 10.45$, $k = 1.75$
(e) $\bar{x} = 158$, $s = 25$, $k = 2.5$
(f) $\bar{x} = -55$, $s = 0.125$, $k = 2.8$
(g) $\bar{x} = 1.7$, $s = 25.8$, $k = 2.25$

3.82 Assume the distribution of each data set is approximately normal, with $\bar{x}$ and s given. Find the intervals (referred to by the Empirical Rule) that are 1, 2, and 3 standard deviations about the mean. Carefully sketch the corresponding normal curve for each data set, indicating the endpoints of each interval.
(a) $\bar{x} = 20$, $s = 5$
(b) $\bar{x} = 37$, $s = 0.2$
(c) $\bar{x} = 675$, $s = 250$
(d) $\bar{x} = -5.5$, $s = 12$
(e) $\bar{x} = 98.6$, $s = 1.7$
(f) $\bar{x} = 5280$, $s = 150$

3.83 For each data set with $\bar{x}$ and s given, find the z-score corresponding to the given observation x.
(a) $\bar{x} = 8$, $s = 3$, $x = 17$
(b) $\bar{x} = 100$, $s = 16$, $x = 80$
(c) $\bar{x} = 15$, $s = 3$, $x = 17.5$
(d) $\bar{x} = 27$, $s = 4.5$, $x = 22$
(e) $\bar{x} = 122$, $s = 32$, $x = 175$
(f) $\bar{x} = -105$, $s = 33$, $x = -90$
(g) $\bar{x} = 6.55$, $s = 0.25$, $x = 6$
(h) $\bar{x} = 64$, $s = 8.75$, $x = 100$
(i) $\bar{x} = 0.025$, $s = 0.0018$, $x = 0.027$
(j) $\bar{x} = 407$, $s = 16$, $x = 500$

3.84 For each data set with $\bar{x}$ and s given, find an observation corresponding to the z-score given.
(a) $\bar{x} = 25$, $s = 5$, $z = 2.3$
(b) $\bar{x} = 9.8$, $s = 1.2$, $z = -0.7$
(c) $\bar{x} = -456$, $s = 37$, $z = 1.25$
(d) $\bar{x} = 37.6$, $s = 5.9$, $z = -1.96$
(e) $\bar{x} = 55$, $s = 0.05$, $z = 3.5$
(f) $\bar{x} = 3.14$, $s = 0.5$, $z = 1.28$
(g) $\bar{x} = 2.35$, $s = 0.94$, $z = -2.5$
(h) $\bar{x} = 0.529$, $s = 1.9$, $z = 0.55$

3.85 Find the position, or depth, of the indicated percentile in an ordered data set of size n.
(a) $n = 150$, P_{80}
(b) $n = 257$, P_{35}
(c) $n = 36$, P_{60}
(d) $n = 75$, P_{40}
(e) $n = 100$, P_{20}
(f) $n = 5035$, P_{70}

Applications

3.86 Demographics and Population Statistics The FBI uses public assistance in tracking criminals by maintaining the "Ten Most Wanted Fugitives" list. A fugitive is removed from this list if he or she is captured, the charges are dropped, or he or she no longer fits a certain profile. In a random sample of fugitives, the mean time on the list was 26.5 months, with a standard deviation of 4.3 months.
(a) What values are 1 standard deviation away from the mean? What values are 2 standard deviations away from the mean?
(b) Without assuming anything about the shape of the distribution of times, approximately what proportion of times are between 17.9 months and 35.1 months?

3.87 Travel and Transportation Royal Caribbean recently took delivery of the largest cruise ship in the world, the Symphony of the Seas. This Oasis-class vessel took 36 months to build, and it includes surf simulators, a zip line, and twin 40-foot rock-climbing walls. Suppose a random sample of large cruise ships was obtained, and the cruising speed of each was recorded. The sample mean was 25.6 knots and the standard deviation was 3.4 knots. Assume the shape of the speed distribution is approximately normal.

(a) What values are 2 standard deviations away from the mean? What values are 3 standard deviations away from the mean?
(b) Approximately what proportion of speeds are between 22.2 and 29.0 knots?

3.88 Fuel Consumption and Cars Companies that use business aircraft often purchase jets based on weight and total range, the maximum distance the plane can fly on one full tank of jet fuel. The Gulfstream G450 has a range of 8060 km, which lies in the 65th percentile of ranges for all business aircraft. Interpret this value.

3.89 Public Health and Nutrition The World Health Organization has indicated that vaccines are a powerful and cost-effective health intervention.[28] However, only 22% of vaccines that are developed are successful, and, from start to finish, the mean time to develop a vaccine is 7.6 years. Suppose the standard deviation is 1.2 years.
(a) What values are 1 standard deviation away from the mean? What values are 2 standard deviations away from the mean?
(b) Without assuming anything about the shape of the distribution of times, approximately what proportion of development times are between 5.2 and 10 years?
(c) Without assuming anything about the shape of the distribution of times, approximately what proportion of development times are either less than 4 years or greater than 11.2 years?
(d) Assuming the distribution of times is normal, what proportion of development times are between 5.2 and 10 years? Either less than 4 years or greater than 11.2 years?

3.90 Biology and Environmental Science The Commonwealth of Pennsylvania is concerned about the dwindling number of family-owned farms and the number of smaller, less efficient farms. For a random sample, the total acreage of each farm was recorded. The mean was 1125 acres, with a standard deviation of 250. The shape of the distribution of areas is not normal.
(a) Approximately what proportion of areas are between 625 and 1625 acres?
(b) Approximately what proportion of areas are between 375 and 1875 acres?
(c) Approximately what proportion of areas are less than 375 acres?
(d) Approximately what proportion of areas are between 750 and 1500 acres?

3.91 Physical Sciences During the spring, many rivers are monitored very carefully in case residents need to be warned about an impending flood. The depth (in feet) of the Susquehanna River at the Bloomsburg Bridge is measured and reported daily. In a random sample of depths, $\bar{x} = 16.7$, $s = 2.1$ ft, and the shape of the distribution is approximately normal.
(a) Approximately what proportion of depths are between 14.6 and 18.8 ft?
(b) Approximately what proportion of depths are less than 14.6 ft?
(c) Approximately what proportion of depths are between 14.6 and 23 ft?

3.92 Biology and Environmental Science Many farmers use the height of their corn on July 4 as an indicator for the entire crop. In a random sample of corn-stalk heights on July 4 in Columbia County, $\bar{x} = 25.6$, $s = 0.9$ in., and a histogram of the observations is bell-shaped.
(a) Approximately what proportion of observations are between 23.8 and 27.4 in.?
(b) Approximately what proportion of observations are between 22.9 and 26.5 in.?
(c) Approximately what proportion of observations are less than 27.4 in.?

3.93 Education and Child Development The Medical College Admission Test (MCAT) is a mutltiple-choice exam taken by pre-med majors that assesses problem solving, critical thinking, and general science knowledge. From May 2017 to April 2018, more than 150,000 students took the exam. The mean score was 500.2 and the standard deviation was 10.5, and a score of 513 was in the 89th percentile.
(a) Interpret this value, $p_{89} = 513$.
(b) The 50th percentile is the national average, approximately 500. Explain the meaning of average in this context.
(c) Suppose a pre-med major scored at the 99th percentile on the MCAT. Interpret this result.

3.94 Travel and Transportation Bicycle delivery services are utilized in many metropolitan areas because they can provide rush deliveries and are not subject to traffic jams or parking restrictions. Suppose an architectural firm would like to evaluate two bicycle delivery services in New York City. The first service has a mean and a standard deviation for delivery (in minutes) of 37 and 5, respectively. The second service has a mean of 42 with a standard deviation of 7. The company sent two test packages to the same location, one with each delivery service. The times to delivery were 33 and 35 minutes, respectively. Use z-scores to determine which service performed better.

3.95 Business and Management The Brewer's Fork in Boston, Massachusetts, is advertising quick lunches with a mean waiting time of 11 minutes and a standard deviation of 2.5 minutes. The general manager (Ron Fisher, a former statistician) also claims that the distribution of waiting times is approximately normal.
(a) Suppose your waiting time is 13 minutes. Is there any reason to that believe that the general manager's claim is false? Use a z-score to justify your answer.
(b) Suppose your waiting time is 20 minutes. Now is there any evidence to refute the general manager's claim?

3.96 Marketing and Consumer Behavior The time spent in a grocery store is an important issue for shoppers and for companies trying to market new products. Men tend to spend less time in a grocery store than women do, and people spend more time in the store on weekends. A random sample of shoppers at a local grocery store was obtained and the shopping time (in minutes) for each was recorded. GROCER

(a) Construct a histogram for these data.
(b) Use your histogram in part (a) to approximate the following percentiles: (i) 45th, (ii) 80th, (iii) 10th.
(c) Compute the exact percentiles in part (b) and compare your results.

3.97 Medicine and Clinical Studies Many physicians consider the time spent with patients the most rewarding and the most important in order to properly diagnose an illness. Suppose a general physician claims the mean amount of time she spends with patients is 15 minutes, with a standard deviation of 3 minutes, and the distribution of times is approximately normal.
(a) Suppose the general physician spends 11 minutes with a randomly selected patient. Is there any reason to refute the general physician's claim? Use a z-score to justify your answer.
(b) The health organization becomes concerned if the general physician spends more than 15 minutes with each patient. Suppose the general physician spends 22 minutes with a randomly selected patient. Is there any evidence to suggest the general physician is spending too much time with patients?

Extended Applications

3.98 Manufacturing and Product Development The engine in a tractor trailer is designed to last 1,000,000 miles before a rebuild or overhaul. The engines are also designed to run nonstop and have between 400 and 600 horsepower. A random sample of tractor trailers was obtained and the horsepower was measured for each engine. **HP**
(a) Find the mean and the standard deviation of these horsepower measurements.
(b) Find the actual proportion of observations within 1 standard deviation of the mean, within 2 standard deviations of the mean, and within 3 standard deviations of the mean.
(c) Using the results in part (b), do you think the shape of the distribution of horsepower measurements is normal? Why or why not?

3.99 Public Health and Nutrition The nutritional value of breakfast cereal is determined in part by the amount of added sugar, fiber content, vitamins, and salt. A random sample of breakfast cereals was obtained, and the amount of sodium in one serving, generally 1 cup, was measured for each.[29] **CEREAL**
(a) Find the actual proportion of observations within 1 standard deviation of the mean, within 2 standard deviations of the mean, and within 3 standard deviations of the mean.

(b) Using the results in part (a), do you think the shape of the distribution of sodium amounts is normal? Why or why not?
(c) Construct a histogram for these data. Describe the shape of the distribution.

3.100 Manufacturing and Product Development Paint viscosity is a measure of thickness that helps determine whether the paint will completely cover the existing color in a single coat. A random sample of latex paint viscosities (in Krebs units, KU) was obtained, and the data are given in the table. **PAINT**

| 113 | 124 | 141 | 115 | 115 | 129 | 113 | 129 | 112 | 112 |

(a) Find the mean and the standard deviation for these data.
(b) Find the z-score for each observation.
(c) Find the mean and the standard deviation for all of the z-scores.
(d) For any set of observations, can you predict the mean and the standard deviation of the corresponding z-scores? Try to prove this result.

3.101 Manufacturing and Product Development The Samsung Gear IconX headset is water resistant, is designed for runners, and has a battery life of 5–7 hours. A random sample of these headsets was obtained and the battery life of each was recorded, in hours. **HEADSET**
(a) Find the mean and the standard deviation for these data.
(b) Find the z-score for each observation.
(c) Find the 16th percentile and the 84th percentile of the z-scores. Explain the connection between these percentiles and the Empirical Rule.

Challenge Problems

3.102 Travel and Transportation According to the Massachusetts Bay Transportation Authority (MBTA), the ride from Chestnut Hill to Boston's Logan Airport on the MBTA takes less than 45 minutes. A random sample of travel times (in minutes) was obtained, and the results are given in the table.

| 46.5 | 38.3 | 39.1 | 41.1 | 42.0 | 37.6 | 41.6 | 45.5 |
| 39.0 | 34.8 | 36.5 | 38.6 | 38.4 | 44.4 | 42.4 | |

(a) Find the mean $\bar{x}$ and the standard deviation s for this data set.
(b) Compute the z-score for each observation and find $\sum z_i^2$.
(c) Find a general formula for $\sum z_i^2$ for any data set of size n.

3.103 Manufacturing and Product Development Reconsider Example 3.16. Find a good minimum guaranteed microwave oven lifetime. That is, if a microwave oven fails before a certain lifetime, then the store would provide a refund. Explain your reasoning.

3.4 Five-Number Summary and Box Plots

Standard Box Plot

A box plot, or box-and-whisker plot, is a compact graphical summary that conveys information about central tendency, symmetry, skewness, variability, and outliers. A standard box plot is constructed using the minimum and maximum values in the data set, the first and third quartiles, and the median. This collection of values is called the five-number summary.

Definition
The **five-number summary** for a set of observations $x_1, x_2, \ldots, x_n$ consists of the minimum value, the maximum value, the first and third quartiles, and the median.

Recall: The range of a data set is the largest observation (maximum value) minus the smallest observation (minimum value). This descriptive statistic was our first attempt at measuring variability in a data set.

These five numbers provide a glimpse of the symmetry, central tendency, and variability in a data set. For example, minimum and maximum values that are very far apart suggest lots of variability. A median that is approximately halfway between the minimum and maximum values and approximately halfway between the first and third quartiles suggests that the distribution is symmetric. A box plot is constructed as described below.

How to Construct a Standard Box Plot
Given a set of n observations $x_1, x_2, \ldots, x_n$:

1. Find the five-number summary: $x_{\min}, Q_1, \tilde{x}, Q_3, x_{\max}$.
2. Draw a (horizontal) measurement axis. Carefully sketch a box with edges at the quartiles: left edge at Q_1, right edge at Q_3. (The height of the box is irrelevant.)
3. Draw a vertical line in the box at the median.
4. Draw a horizontal line (whisker) from the left edge of the box to the minimum value (from Q_1 to $x_{\min}$). Draw a horizontal line (whisker) from the right edge of the box to the maximum value (from Q_3 to $x_{\max}$).

Recall: $x_{\min}$ denotes the minimum value and $x_{\max}$ denotes the maximum value.

Figure 3.23 illustrates this step-by-step procedure and shows a standard box plot with the five numbers indicated on a measurement axis. Note that the length of the box is the interquartile range. The box contains the middle 50% of the values.

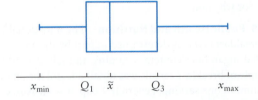

Figure 3.23 Standard box plot.

The position of the vertical line in the box (median) and the lengths of the horizontal lines (whiskers) indicate symmetry or skewness as well as variability. **Figure 3.24** shows a standard box plot for a distribution of data that is skewed to the right. The lower half of the data are in the interval from 3 to 4.5, while the upper half of the data are much more spread out, from 4.5 to 11. **Figure 3.25** shows a standard box plot for a fairly symmetric distribution with lots of variability. The lower and upper half of the data are evenly distributed, but the whiskers extend far from each edge of the box. That is, 25% of the data are between 0 and 4, and 25% are between 7 and 11.

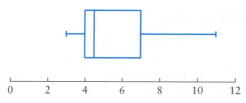

Figure 3.24 Standard box plot for data skewed to the right.

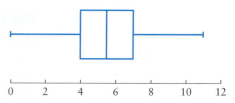

Figure 3.25 Standard box plot for a symmetric distribution.

Example: Standard Box Plot

EXAMPLE 3.19 Red Wine and Blood Pressure

Some evidence suggests that consumption of nonalcoholic red wine may decrease systolic and diastolic blood pressure.[30] Suppose the systolic blood pressure for 30 randomly selected participants involved in this research study is given in the table. Construct a standard box plot for these data.

177	167	148	122	138	175	128	107	169	191
188	203	180	102	135	142	116	142	197	138
168	196	114	181	67	188	168	160	176	150

Solution

STEP 1 Find the five-number summary:

$$x_{\min} = 67 \quad Q_1 = 135 \quad \tilde{x} = 163.5 \quad Q_3 = 180 \quad x_{\max} = 203$$

STEP 2 Draw a measurement axis and sketch a box with edges at $Q_1 = 135$ and $Q_3 = 180$.

STEP 3 Draw a vertical line inside the box at the median, $\tilde{x} = 163.5$.

STEP 4 Draw a horizontal line from $Q_1 = 135$ to $x_{\min} = 67$, and another horizontal line from $Q_3 = 180$ to $x_{\max} = 203$. The resulting box plot is shown in **Figure 3.26**.

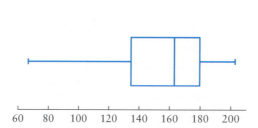

Figure 3.26 Standard box plot for the systolic blood pressure.

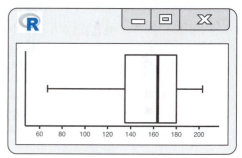

Figure 3.27 R box plot.

STEP 5 The box plot suggests that the data are negatively skewed, or skewed to the left. The lower half of the data are much more spread out than the upper half are. A technology solution is shown in **Figure 3.27**.

A CLOSER LOOK

1. A box plot has only one measurement axis, and it may be horizontal or vertical. Many software packages, including CrunchIt! and R, draw box plots with a vertical measurement axis by default. The construction and interpretations are the same. The horizontal=TRUE option in R produces a box plot with a horizontal measurement axis.

2. The software does not usually include tick marks on the measurement axis for the five-number summary. The tick marks and scale are selected simply for convenience.

Modified Box Plot

Using a standard box plot based on the five-number summary to describe a data set does have some disadvantages. For instance, an examination of the graph cannot reveal how many observations are between each quartile and the extreme. Each whisker is drawn from the quartile to the extreme, regardless of the number of observations in between. In addition, there are no provisions for identifying outliers. A standard (graphical) technique for distinguishing outliers is important because these values play an important role in statistical inference. Therefore, many statisticians prefer to use a modified box plot to describe a data set graphically. This type of graph still conveys information about center, variability, symmetry, and skewness, but it is more precise and also plots outliers, if present in the data set.

How to Construct a Modified Box Plot

Given a set of n observations $x_1, x_2, \ldots, x_n$:

1. Find the quartiles, the median, and the interquartile range:

   $Q_1, \tilde{x}, Q_3, \text{IQR} = Q_3 - Q_1$

2. Compute the two inner fences (low and high) and the two outer fences (low and high) using the following formulas:

   $\text{IF}_L = Q_1 - 1.5(\text{IQR}) \qquad \text{IF}_H = Q_3 + 1.5(\text{IQR})$
   $\text{OF}_L = Q_1 - 3(\text{IQR}) \qquad \text{OF}_H = Q_3 + 3(\text{IQR})$

   Think of the interquartile range as a step. The inner fences are 1.5 steps away from the quartiles, and the outer fences are 3 steps away from the quartiles.

3. Draw a (horizontal) measurement axis. Carefully sketch a box with edges at the quartiles: left edge at Q_1, right edge at Q_3. Draw a vertical line in the box at the median.

4. Draw a horizontal line (whisker) from the left edge of the box to the most extreme observation within the low inner fence. This line will extend from Q_1 to at most IF_L. Draw a horizontal line (whisker) from the right edge of the box to the most extreme observation within the high inner fence. This line will extend from Q_3 to at most IF_H.

5. Any observations between the inner and outer fences (between IF_L and OF_L, or between IF_H and OF_H) are classified as mild outliers and are plotted separately with shaded circles. Any observations outside the outer fences (less than OF_L, or greater than OF_H) are classified as extreme outliers and are plotted separately with open circles. *Note*: Some statistical packages will use other symbols for outliers and may not distinguish between mild and extreme outliers.

Figure 3.28 shows the relationship between construction points for a modified box plot and the location of any outliers.

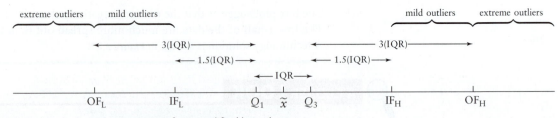

Figure 3.28 Construction points for a modified box plot.

EXAMPLE 3.20 Sled Dog Trips

SLED

The Ignace, Ontario, fishing and hunting resort Agimac River Outfitters offers guided sled dog trips on wooded trails and along beautiful lakes.[31] Most trips last approximately 2 1/2 hours, but there are all-day trips and the weather conditions may affect the length of a scheduled trip. A random sample of sled dog trips was

Winterdance Dogsled Tours

obtained and the length (in hours) of each was recorded. The data are given in the table. Construct a modified box plot for these data.

0.7	9.5	2.2	1.7	2.6	3.7	0.8	2.7	0.5	2.7
11.9	1.1	1.4	3.1	1.4	0.8	7.9	3.1	2.6	0.6
0.1	3.5	6.1	2.6	1.3	5.2	4.3	0.6	1.9	4.5

Solution

STEP 1 Find the quartiles, the median, and the interquartile range.
$Q_1 = 1.1 \quad \tilde{x} = 2.6 \quad Q_3 = 3.7 \quad \text{IQR} = 3.7 - 1.1 = 2.6$

STEP 2 Find the inner and outer fences.
$\text{IF}_L = 1.1 - (1.5)(2.6) = 1.1 - 3.9 = -2.8$
$\text{IF}_H = 3.7 + (1.5)(2.6) = 3.7 + 3.9 = 7.6$
$\text{OF}_L = 1.1 - (3)(2.6) = 1.1 - 7.8 = -6.7$
$\text{OF}_H = 3.7 + (3)(2.6) = 3.7 + 7.8 = 11.5$

STEP 3 Draw a (horizontal) measurement axis. Carefully sketch a box with edges at the quartiles: left edge at Q_1, right edge at Q_3. Draw a vertical line in the box at the median.

STEP 4 Draw a horizontal line (whisker) from the left edge of the box to the most extreme observation within the inner fence IF_L (0.1).

Draw a horizontal line (whisker) from the right edge of the box to the most extreme observation within the inner fence IF_H (6.1).

STEP 5 Plot any mild outliers, observations between -6.7 and -2.8, or between 7.6 and 11.5. There are two mild outliers, 7.9 and 9.5.

Plot any extreme outliers, observations less than -6.7 or greater than 11.5. There is one extreme outlier, 11.9.

STEP 6 The resulting modified box plot is shown in **Figure 3.29**. The box plot suggests the data are positively skewed, or skewed to the right. The upper half of the data are much more spread out than the lower half. There are two mild outliers and one extreme outlier.

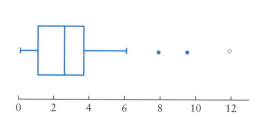

Figure 3.29 Modified box plot for the sled dog trip data.

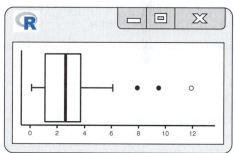

Figure 3.30 R box plot for the sled dog trip data.

Note: IF_L and OF_L are negative even though an observed trip time cannot be less than 0 hours. That's OK. This is a correct statistical calculation, not a contradiction, even though it seems odd. **Figure 3.30** shows a technology solution. ∎

TRY IT NOW Go to Exercises 3.115 and 3.121

CHAPTER APP **SpaceX Payloads**

Construct a modified box plot for the payload weight data. Describe the distribution in terms of symmetry, skewness, variability, and outliers.

A CLOSER LOOK

When we compare two (or more) data sets graphically, the corresponding box plots may be placed on the same measurement axis (one above the other using a horizontal axis, or side-by-side with a vertical axis). **Figure 3.31** shows three box plots on the same measurement axis, representing the number of gallons of gasoline pumped in randomly selected vehicles at three different stations.

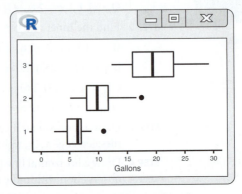

Figure 3.31 R box plot for gasoline data.

Section 3.4 Exercises

Concept Check

3.104 True or False The five-number summary for a set of observations is determined by finding the five most extreme observations.

3.105 True or False In a standard box plot, a line is drawn in the box at the sample mean.

3.106 True or False A box plot may reveal whether a distribution is symmetric.

3.107 True or False A modified box plot always has markers for mild and extreme outliers.

3.108 True or False The median line in a box plot could be at one of the edges of the box.

3.109 True or False If a modified box plot has an extreme outlier, then there must also be at least one mild outlier.

Practice

3.110 Find the five-number summary for each data set.

(a) {34, 40, 34, 32, 32, 40, 35, 35, 28, 35}   EX3.110a
(b) {57, 65, 70, 71, 67, 56, 52, 66, 74, 57, 67, 78}   EX3.110b
(c) {94, 80, 91, 94, 83, 92, 83, 93, 96, 80, 87, 98, 81, 93}
(d) {2.3, 1.8, 2.1, 1.0, 2.4, 2.3, 0.4, 9.8, 0.6, 1.4, 3.1, 10.9, 3.8, 0.5, 0.9, 2.2, 1.3, 1.3}   EX3.110c EX3.110d
(e) {166.8, 103.1, 119.9, 141.9, 110.6, 189.8, 121.6, 141.6, 133.6, 178.2, 158.9, 145.9, 139.1, 148.6, 135.0, 174.0, 152.4, 119.7, 196.9, 118.7, 159.7, 150.3, 113.8, 108.9, 163.2}   EX3.110e
(f) {−33.8, −9.8, −18.5, −11.5, −36.3, −33.1, −21.1, −26.2, −25.4, −32.1, −35.9, −28.0, −38.2, −12.0, −29.2, −40.1, −13.1}   EX3.110f

3.111 Construct a standard box plot for each five-number summary.

(a) $x_{min} = 15.3, Q_1 = 21.8, \tilde{x} = 25.3, Q_3 = 28.2, x_{max} = 34.2$
(b) $x_{min} = 70.9, Q_1 = 167.8, \tilde{x} = 187.1, Q_3 = 225.3, x_{max} = 329.3$
(c) $x_{min} = 0.06, Q_1 = 5.3, \tilde{x} = 13.7, Q_3 = 30.8, x_{max} = 122.3$
(d) $x_{min} = 10.1, Q_1 = 10.7, \tilde{x} = 11.3, Q_3 = 12.5, x_{max} = 26.7$

3.112 For each data set with Q_1 and Q_3 given, find the interquartile range and the inner and outer fences.

(a) $Q_1 = 22,$ $Q_3 = 46$
(b) $Q_1 = 1255,$ $Q_3 = 1306$
(c) $Q_1 = 65.75,$ $Q_3 = 75.21$
(d) $Q_1 = 914.9,$ $Q_3 = 1140.5$
(e) $Q_1 = 1.275,$ $Q_3 = 4.07$
(f) $Q_1 = 0.265,$ $Q_3 = 2.51$
(g) $Q_1 = -33.67,$ $Q_3 = -23.90$
(h) $Q_1 = 98.43,$ $Q_3 = 98.81$

3.113 For each data set with Q_1 and Q_3 given, determine whether the observation x is a mild outlier, an extreme outlier, or neither.

(a) $Q_1 = 20,$ $Q_3 = 29,$ $x = 35$
(b) $Q_1 = 486.1,$ $Q_3 = 510.9,$ $x = 440$
(c) $Q_1 = 5.18,$ $Q_3 = 6.32,$ $x = 4.2$

(d) $Q_1 = 96.3$, $Q_3 = 101.1$, $x = 116.5$
(e) $Q_1 = 68.92$, $Q_3 = 69.07$, $x = 68.4$
(f) $Q_1 = 101.26$, $Q_3 = 144.59$, $x = 132.6$

3.114 For each of the following box plots, find the five-number summary. (Estimate these numbers as best you can by using the tick marks on each graph.)

(a)
(b)
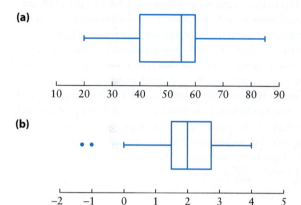

(c)
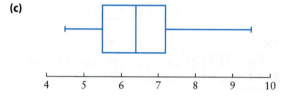

(d)
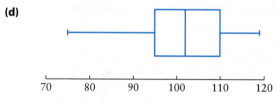

(e)
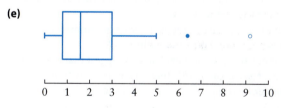

(f)
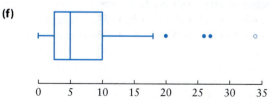

(g)
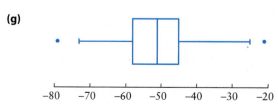

Applications

3.115 Business and Management A Roth's supermarket in Salem, Oregon, is using a new statistical tool to help in ordering bottles of raspberry iced tea. A random sample of the number of bottles sold per day is given in the table. Construct a modified box plot for these data. Describe the distribution of the number of bottles sold. **BOTTLE**

48	52	46	58	50	46	59	51	46	48
45	47	50	48	49	49	49	48	48	45

3.116 Travel and Transportation Major airlines compete for customers by advertising on-time arrival. A random sample of flights arriving at Boston's Logan International Airport was obtained and the actual arrival time was compared to the scheduled arrival time. The differences (in minutes) are given in the table (negative numbers indicate the flight arrived before the scheduled arrival time).[32] Construct a modified box plot for these data. Describe the distribution in terms of symmetry, skewness, and variability. Are there any outliers? If so, are they mild or extreme? **ARRIVAL**

3	−16	−1	−24	−30	23	−15	1	−2	−41
53	−26	−26	24	53	49	−13	−18	−14	17
18	52	−8	−36	−25	31	−14	−4	−18	−25

3.117 Public Policy and Political Science A recent study reported the property tax bills (in dollars) for randomly selected condos in Bonita Springs, Florida. Use the following modified box plot to describe the distribution of the data.

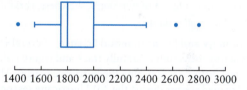

3.118 Education and Child Development A psychologist in Memphis, Tennessee, randomly selected a group of 6-year-olds and recorded the time (in minutes) each child needed to complete a 20-piece picture puzzle. The data were used to predict their readiness for first grade. The standard box plot for the data is shown here. Describe the distribution of completion times.

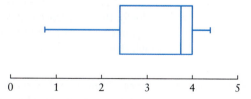

3.119 Public Health and Nutrition As part of a new physical fitness program, the Kromrey Middle School in Middleton, Wisconsin, records the number of sit-ups each sixth-grade student can complete in one minute. A random sample for males and females was obtained and the modified box plots are shown here. Describe the male and female data separately. What similarities and/or differences do the box plots suggest?

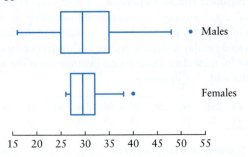

3.120 Economics and Finance Many government program budgets are determined by the annual inflation rate. The inflation rate (as a percent change in the consumer price index since 1913) for the United States for the years 1965–2017 was recorded.[33] Construct a modified box plot for these data. Describe the distribution of the inflation rate over the past 53 years in terms of symmetry, skewness, variability, and outliers. **INFLAT**

3.121 Sports and Leisure The salaries for head football coaches at NCAA Division I schools have increased dramatically, driven by television revenue and pressure to compete for a national championship. Nick Saban, the head coach at Alabama, is one of the highest-paid coaches, earning more than $11 million per year. A sample of Division I head football coaches was obtained, and their salaries were recorded (in millions of dollars).[34] Construct a modified box plot for these data. Describe the distribution of salaries for these football coaches in terms of symmetry, skewness, variability, and outliers. **COACH**

3.122 Biology and Environmental Science Several weather centers across the country carefully track and record data for tropical disturbances during the hurricane season. There were 17 named storms during the 2017 hurricane season, 10 hurricanes, and 6 major hurricanes. The following table contains the minimum central pressure (in millibars) for each named tropical storm and hurricane.[35] **HURRI**

| 993 | 1007 | 992 | 1009 | 1005 | 981 | 967 | 941 | 914 |
| 938 | 972 | 962 | 909 | 981 | 958 | 995 | 995 | |

Construct a modified box plot for the hurricane data. Describe the distribution in terms of symmetry, skewness, variability, and outliers. How would the graph change if a standard box plot were used?

3.123 Economics and Finance Many of us have various forms of debt, including mortgages, cars, medical bills, and even student loans. In 2017, the mean balance on credit cards was approximately $6300.[36] Suppose a random sample of Americans was obtained and the credit card balance for each was recorded. Construct a modified box plot for these credit card balance data. Describe the distribution in terms of symmetry, skewness, variability, and outliers. Does the box plot support the claim about the mean credit card balance? Justify your answer. **CREDIT**

Extended Applications

3.124 Public Policy and Political Science There is very little legislation regulating yellow-light times. However, the National Motorists Association Foundation has developed some general guidelines for appropriate yellow-light duration. It suggests that 3 seconds should be the minimum time at any intersection and that the yellow-light time should depend on the speed limit.[37] A random sample of yellow-light times (in seconds) at intersections with different speed limits is given below. Construct a modified box plot for each set of data on the same measurement axis. Describe and compare the distributions. Do the box plots suggest that yellow-light times are longer at intersections with a higher speed limit? **YELLOW**

30 mph

| 3.7 | 3.7 | 3.4 | 2.2 | 3.4 | 3.6 | 3.5 | 3.8 | 2.6 | 2.1 |
| 3.1 | 3.9 | 4.3 | 1.8 | 3.0 | 1.6 | 4.5 | 2.7 | 2.2 | 2.4 |

50 mph

| 4.1 | 4.7 | 4.3 | 4.1 | 4.8 | 5.4 | 2.9 | 3.9 | 2.1 | 6.9 |
| 4.6 | 3.3 | 6.6 | 5.0 | 5.5 | 3.6 | 3.7 | 5.9 | | |

3.125 Manufacturing and Product Development Many people enjoy sampling unique beers from micro-breweries, and some people even brew beer in their own home. A random sample of home brew recipes for American IPA and American Lager was obtained and the alcohol by volume (ABV) was measured for each.[38] **BREW**

(a) Construct a modified box plot for each data set on the same measurement axis.

(b) Compare the distributions in terms of symmetry, skewness, variability, and outliers.

(c) Is there any graphical evidence to suggest that one home brew has a greater ABV than the other? Justify your answer.

3.126 Public Health and Nutrition Despite little empirical evidence, many people believe that vitamin C helps prevent the common cold. One of the best-selling supplements is Doctor's Best Vitamin C, a 1000-mg capsule containing a proprietary ascorbic acid. A random sample of these vitamin C tables was obtained and analyzed by an independent laboratory for the precise vitamin C content, measured in milligrams. **VITC**

(a) Construct a modified box plot for the vitamin C data.

(b) Describe the distribution in terms of symmetry, skewness, variability, and outliers.

(c) Does the box plot suggest any graphical evidence that the 1000-mg claim is wrong? Justify your answer.

3.127 Public Policy and Political Science If you receive a jury summons, you are obligated to appear in court. There are, however, several general (honest) instant, temporary, and hardship excuses to avoid jury duty. For example, firefighters and physicians may be automatically excused from serving. The compensation for serving on a jury varies by state; some states reimburse child-care expenses and/or transportation costs; some states do not compensate jurors for the first few days of service. A sample of states was obtained, and the juror pay per day as of April 1, 2018, for each was recorded.[39] The data are given in the table. **JURYPAY**

State	Pay	State	Pay
Arizona	12.00	Arkansas	5.00
California	15.00	Colorado	50.00
Connecticut	50.00	Delaware	20.00
Florida	30.00	Georgia	5.00
Hawaii	30.00	Idaho	10.00
Indiana	40.00	Iowa	10.00
Louisiana	25.00	Maryland	15.00
Minnesota	20.00	Missouri	6.00
Montana	13.00	Nebraska	35.00
Oklahoma	20.00	Rhode Island	15.00
South Dakota	50.00	Virginia	30.00

(a) Construct a modified box plot for the jury pay per day data.

(b) Describe the distribution in terms of symmetry, skewness, variability, and outliers.

(c) Suppose that in January 2019, the $5.00 rate in Georgia was raised to $65.00 but all other rates given remained the same. Change the value for Georgia to $65.00 and construct a new modified box plot. How does this new box plot compare with the one in part (a)? Describe any similarities and/or differences.

3.128 Travel and Transportation The Federal Aviation Administration (FAA) maintains an extensive aircraft database, including fields for aircraft type, engine type, number of seats, and cruising speed. A random sample of turbo-jets from this database was obtained and the engine thrust (in pounds) for each was recorded.[40] **THRUST**

(a) Construct a modified box plot for the engine thrust data.

(b) Describe the distribution in terms of symmetry, skewness, variability, and outliers.

(c) The FAA claims that the mean engine thrust for turbo-jets is 10,000 pounds. Does the box plot suggest any graphical evidence that this claim is wrong? Justify your answer.

3.129 Biology and Environmental Science Landslides and mudslides are usually triggered by storms and rain events. Areas in which there have been wildfires or increased development are often vulnerable to landslides during heavy rains. A random sample of landslides from around the world was obtained and the distance traveled (in km) was recorded for each.[41]

(a) Construct a modified box plot for the landslide data.

(b) Describe the distribution in terms of symmetry, skewness, variability, and outliers. **SLIDES**

Chapter 3 Summary

Concept	Page	Notation / Formula / Description
Sample mean	71	$\bar{x} = \dfrac{1}{n}\sum x_i$
Population mean	72	μ: the mean of an entire population.
Sample median	73	$\tilde{x}$: the middle value of the ordered data.
Population median	73	$\tilde{\mu}$: the middle value of an entire population.
Trimmed mean	75	$\bar{x}_{\text{tr}(p)}$: the sample mean of a *trimmed* data set.
Mode	76	The value that occurs most often.
Sample proportion of successes	77	$\hat{p} = \dfrac{n(S)}{n}$: the relative frequency of occurrence of successes.
Population proportion of successes	78	p: the true proportion of successes in an entire population.
Sample range	82	R: the largest observation ($x_{\max}$) minus the smallest observation ($x_{\min}$).
Deviation about the mean	83	$x_i - \bar{x}$
Sample variance	84	$s^2 = \dfrac{1}{n-1}\sum(x_i - \bar{x})^2$
Sample standard deviation	84	$s = \sqrt{s^2}$: the positive square root of the sample variance.
Population variance	84	σ^2: the variance for an entire population.
Population standard deviation	84	$\sigma = \sqrt{\sigma^2}$: the positive square root of the population variance.
Quartiles	87	The quartiles divide the data into four parts. Q_1 is the first quartile and Q_3 is the third quartile. Q_2 is the median.
Interquartile range	87	$\text{IQR} = Q_3 - Q_1$
Chebyshev's Rule	94	For any set of observations, the proportion of observations within k standard deviations of the mean is at least $1 - \dfrac{1}{k^2}$.
Empirical Rule	96	If a distribution is approximately normal, the proportion of observations within 1, 2, and 3 standard deviations about the mean is approximately 0.68, 0.95, and 0.997, respectively.
z-Score	99	$z_i = \dfrac{x_i - \bar{x}}{s}$, an observation's (signed) distance from the mean in standard deviations.
Percentiles	101	The percentiles divide a data set into 100 parts.
Five-number summary	106	$x_{\min}, Q_1, \tilde{x}, Q_3, x_{\max}$
Box plot	106	A graphical description of a data set, constructed using the five-number summary. The graph conveys information about central tendency, symmetry, skewness, and variability.
Modified box plot	108	A graphical description of a data set, constructed using $\tilde{x}, Q_1, Q_3$, IQR, and the inner and outer fences. This box plot also indicates any outliers.

Chapter 3 Exercises

Applications

3.130 Public Health and Nutrition Most multivitamins contain calcium for building strong bones and lowering the risk of heart disease. A random sample of multivitamins was obtained, and the calcium content for each (in milligrams) is given in the table. **CALCONT**

156	151	173	201	182	166	173	180	174	185
160	178	173	169	203	190	187	202	173	171

(a) Find the mean, the variance, and the standard deviation.
(b) Find the proportion of observations within 1, 2, and 3 standard deviations about the mean.
(c) Using the proportions obtained in part (b), do you think the distribution of observations is normal? Why or why not?

3.131 Manufacturing and Product Development Many boxed cake mixes include special high-altitude baking instructions. To determine any difference between baking times at low and high altitudes, the Compass Marketing Research team made several similar cakes in 9-in. round pans in Miami and Denver and carefully recorded the baking time (in minutes). The data are given in the table. **CAKE**

Low-altitude times (Miami)

25.1	25.6	24.9	23.7	25.5	22.4	24.7	24.2	25.6
24.8	23.9	24.4	24.7	24.4	26.4	24.7	24.7	26.8
24.9	24.3							

High-altitude times (Denver)

22.8	30.0	27.3	30.3	28.3	31.1	27.0	26.8	26.3
29.1	23.5	26.2	29.2	23.0				

(a) Construct a modified box plot for each data set on the same measurement axis.
(b) Describe each box plot in terms of center, shape, variability, and outliers.
(c) Describe the similarities and differences between the two distributions.

3.132 Sports and Leisure The longest-running sitcom, by episodes, in U.S. television history is *The Adventures of Ozzie and Harriet*. Other long-running sitcoms include *Cheers*, *Frasier*, and *Happy Days*. The number of episodes for some of the top 70 sitcoms are given in the table.[42] **SITCOM**

435	380	351	275	263	264	255	247	249	249
236	209	210	209	233	234	222	275	254	274
172	173	184	201	202	200	163	170	172	157

(a) Find the median, the first and third quartiles, and the interquartile range.
(b) Find the 30th and the 95th percentiles.
(c) The sitcom *Night Court* was set in a Manhattan court presided over by judge Harold T. Stone. It ran for nine seasons and 193 episodes. Using the data in the table, in what percentile does this episode count lie?

3.133 Manufacturing and Product Development General Electric manufactures and sells the largest offshore wind turbine in the world, the Haliade-X. This wind turbine has a 220-m rotor, a 107-m blade, and a rated mean electrical generating capacity of 12 megawatts (MW)[43] with a standard deviation of 0.07 MW. Suppose the distribution of electrical generating capacity is approximately normal and a quality control inspector is trying to develop a plan to perform routine maintenance based on z-scores.

(a) Suppose a randomly selected wind turbine is inspected and found to have a generating capacity of 12.04 MW. Is there any reason to believe this generating capacity is unusual? Why or why not?
(b) Suppose another randomly inspected wind turbine has a generating capacity of 11.8 MW. Is there any reason to believe this generating capacity is unusual? Why or why not?

3.134 Sports and Leisure String tension in tennis rackets is usually measured in pounds. Recommended string tensions are usually in the mid-60s (pounds) for oversize rackets, and in the high 50s to low 60s for mid-overs. Higher tensions tend to decrease the size of the "sweet spot" and reduce power, but increase control. A random sample of string tension was obtained from tennis rackets of players on the professional tour. **TENNIS**

(a) Find the range, sample variance, interquartile range, coefficient of variation, and coefficient of quartile variation for each type of racket. (CV and CQV were defined in Exercise 3.54.)
(b) Using the results from part (a), compare the variability in string tension for the two types of rackets.
(c) Construct a modified box plot for each type of racket on the same measurement axis. Does this graphical comparison support your numerical comparison in part (b)?

3.135 Manufacturing and Product Development Many homes that use forced hot air for heat have air ducts installed in every room. A system using galvanized pipe is constructed to distribute heat throughout the house. A random sample of 6-in-diameter, 5-ft-long, 28-gauge galvanized pipe was obtained from various manufacturers and the weights (in pounds) recorded. **PIPE**

(a) Find the sample mean and the sample median.
(b) Use your results from part (a) to describe the symmetry of the distribution.
(c) Find a 10% trimmed mean. Is the use of a trimmed mean to measure central tendency justified (or necessary) in this case? Why or why not?

3.136 Medicine and Clinical Studies Although caffeine is believed to be safe when consumed in moderate amounts, some health experts suggest that intake of 400 mg of caffeine (the amount in about two cups of coffee) is harmless.[44] The amount of caffeine in a cup of coffee varies according to coffee bean, brewing technique, filter, etc. A random sample of 8-oz cups of coffee was obtained and the caffeine content (in milligrams) was measured. The data are given in the table. CAFF

| 208 | 250 | 188 | 217 | 201 | 147 | 220 | 204 | 164 | 203 |
| 204 | 194 | 190 | 233 | 247 | 218 | 220 | 239 | 186 | 208 |

(a) Find the mean, median, variance, and standard deviation.
(b) Construct a modified box plot for these data.
(c) Use your results from parts (a) and (b) to describe the data.
(d) Based on your results in parts (a) and (b), do you believe a person who drinks two cups of coffee ingests a harmful amount of caffeine (more than 400 mg)? Justify your answer.

3.137 Public Health and Nutrition The Centers for Disease Control and Prevention recommends that children between 2 and 15 months should receive the DTP vaccine for diphtheria, tetanus, and pertussis (whopping cough). The percentage of children ages 12–23 months who have received the DTP immunization as of 2016 in various countries was obtained.[45] DTP
(a) Find the mean, variance, and standard deviation.
(b) Find the proportion of observations within 1, 2, and 3 standard deviations about the mean. Use these proportions to determine whether this distribution is approximately normal.
(c) Construct a modified box plot for these data. Does this graph support your conclusion in part (b)? Why or why not?

3.138 Physical Sciences The tallest dam in the world is the 1001-ft-high Jinping-I Dam constructed on the River Yalong in China. Its primary purpose is to generate hydroelectric power. There are approximately 3800 public utility dams in the United States. A random sample of these dams was obtained and the height (in feet) of each was recorded.[46] DAMS
(a) Find the mean, variance, and standard deviation.
(b) Find the proportion of observations within 1, 2, and 3 standard deviations about the mean. Use these proportions to determine whether the distribution of dam heights is approximately normal.
(c) Construct a modified box plot for these data. Does this graph support your conclusion in part (b)? Why or why not?

3.139 Physical Sciences A standard often used for measuring brightness is lux. For example, bright moonlight has 0.1 lux and bright sunshine has 100,000 lux. The light required for general office work is approximately 400 lux. A random sample of the brightness in office cubicles was obtained. OFFICE
(a) Find the mean, variance, and standard deviation.
(b) Construct a modified box plot for these data. Classify any outliers as mild or extreme.
(c) Using the data in the table, in what percentile does 400 lux lie?
(d) Use Chebyshev's Rule to describe this data set ($k = 2, 3$).

3.140 Manufacturing and Product Development The density of tires is an important selling point for serious mountain bike riders. The tire industry uses a type A durometer to measure the indentation hardness for mountain bike tires. Suppose the distribution of tire hardness is approximately normal, with mean 45 and standard deviation 7.
(a) Carefully sketch the normal curve for tire hardness.
(b) Is a tire hardness of 30 unusually soft? Justify your answer.
(c) A certain bicycle shop claims the hardness of all its tires is at the 84th percentile. If this is true, what is the minimum hardness of any tire in the store?

3.141 Physical Sciences There were approximately 1550 earthquakes around the world in 2017 of magnitude 5.0 or greater.[47] A random sample of the magnitudes (on the Richter scale) of earthquakes during a seven-day period in March 2018 was obtained. QUAKE
(a) Find the mean, variance, and standard deviation.
(b) Find the proportion of observations within 1, 2, and 3 standard deviations about the mean. Use these proportions to determine whether the distribution of magnitudes is approximately normal.
(c) Construct a modified box plot for these data. Does this graph support your conclusion in part (b)? Why or why not?
(d) Find the 40th and the 80th percentiles.
(e) How likely is an earthquake of magnitude of 5 or greater? Justify your answer.

3.142 Public Policy and Political Science The World Happiness report is a survey of global happiness, first published in 2012. The 2017 report ranks 155 countries based on the Cantril ladder, a measure of life satisfaction. The happiness scores were obtained.[48] HAPPY
(a) Find the mean, variance, and standard deviation.
(b) Find the median, quartiles, and interquartile range.
(c) Construct a modified box plot for these data.
(d) Which measure do you think is a better measure of central tendency: the mean or the median? Justify your answer.
(e) Suppose countries with a happiness score greater than 7 are classified as deliriously happy. What proportion of countries fall into this category?

Extended Applications

3.143 Physical Sciences The process of extracting oil from the ground can be divided into four main activities: exploration, well development, production, and site abandonment. In 2018, the mean depth of crude oil developmental wells was approximately 4938 ft.[49] Assume the standard deviation is 450 ft and the distribution of development well depths is approximately normal.
(a) What proportion of development wells have depths between 4038 and 5838 ft?
(b) What proportion of development wells have depths less than 3588 ft?

(c) What proportion of development wells have depths between 4488 and 6288 ft?
(d) Suppose a new development well was drilled in 2019 to a depth of 5300 ft. Is there any evidence to suggest that the mean depth of wells has changed? Justify your answer.

3.144 Physical Sciences A building code officer inspected random home fire extinguishers for pressure (in psi), and the data were recorded. **FIREX**
(a) Construct a modified box plot for these data.
(b) Use the empirical rule to decide whether this distribution of pressures is approximately normal.
(c) Create a new set of observations, $y_i = \ln(x_i)$, where ln is the natural logarithm function. Construct a modified box plot for this new set of data. Use this graph and the Empirical Rule to decide whether the distribution of the transformed data is approximately normal.

3.145 Manufacturing and Product Development Some of our favorite Halloween candies include Snickers, Twix, Kit Kat bars, and M&Ms. However, one of the most popular Halloween candies is Reese's Cups. Suppose the manufacturer (Hershey) claims that the mean weight of a king-size package is 2.8 oz with standard deviation 0.02 oz.
(a) Without assuming anything about the shape of the distribution of the weights of these packages, what proportion of king-size packages have weights between 2.76 and 2.84 oz.
(b) Suppose a random king-size package weighs 2.78 oz. Do you believe the manufacturer's claim about the mean weight? Justify your answer.

3.146 Manufacturing and Product Development The actual width of a 2×4 piece of lumber is approximately 1 3/4 in. but can vary considerably. Larry's Lumber advertises consistent dimensions for better building, and claims all 2×4s sold have a mean width of 1 3/4 in. with a standard deviation of 0.02 in.
(a) Assume the distribution of widths is approximately normal. Find a symmetric interval about the mean that contains almost all of the 2×4 widths.
(b) Suppose a random 2×4 has width 1.78 in. Is there any evidence to suggest the Larry's Lumber claim is wrong? Justify your answer.
(c) Suppose a random 2×4 has width 1.68 in. Is there any evidence to suggest the Larry's Lumber claim is wrong? Justify your answer.

3.147 Biology and Environmental Science Some fish have been found to have mercury levels greater than 1 ppm (parts per million), a level considered safe by the U.S. Food and Drug Administration. Suppose the mean mercury level for smallmouth bass in the Susquehanna River is 0.7 ppm with standard deviation 0.1 ppm, and the distribution of mercury level is approximately normal.
(a) Is it likely that a fisherman will catch a smallmouth bass with a mercury level greater than 1 ppm? Justify your answer.
(b) Suppose the standard deviation is 0.05 ppm. Now, is it likely that a fisherman will catch a smallmouth bass with a mercury level greater than 1 ppm? Justify your answer.
(c) Carefully sketch the normal curves for parts (a) and (b) on the same measurement axis.

3.148 Sports and Leisure The longest-running Broadway show, with more than 12,000 performances, is *Phantom of the Opera*. A sample of Broadway shows was obtained, and the number of performances of each was recorded.[50] **PHANTOM**
(a) Find the sample mean and the sample median number of performances. What do these values suggest about the shape of the distribution?
(b) Find the sample variance and the sample standard deviation. Find the proportion of observations within 1 standard deviation of the mean, within 2 standard deviations of the mean, and within 3 standard deviations of the mean. What do these proportions suggest about the shape of the distribution?
(c) Find the first quartile, the third quartile, and the interquartile range. Construct a modified box plot for the performance data. Use this graph to describe the distribution in terms of symmetry, skewness, variability, and outliers. Does your description based on the box plot agree with your answers to parts (a) and (b)? Why or why not?
(d) Find out how many performances there have been for *Phantom of the Opera* and modify this value in the data set. How will this value affect the sample mean, sample median, sample variance, and quartiles? Find these values and verify your predictions.

3.149 Sports and Leisure Powerlifting is a very competitive, individualized sport in which participants try to lift as much weight as they can in one attempt of the squat, bench press, and deadlift. The total weight (in kg) of female powerlifting competitors in the open division from various meets was obtained.[51] **POWER**
(a) Find the sample mean and the sample median for these data. What do these values suggest about the shape of the distribution?
(b) Find the sample variance and the sample standard deviation. Find the proportion of observations within 1 standard deviation of the mean, within 2 standard deviations of the mean, and within 3 standard deviations of the mean. Use these proportions to determine whether the distribution of total weights is approximately normal.
(c) Construct a modified box plot for these data. Does this graph support your conclusion in part (b)? Why or why not?
(d) Suppose a competitor lifts a total of 500 kg in a meet. Do you believe this woman has a good chance of winning the meet (based on highest total weight)? Justify your answer.

CHAPTER APP

3.150 SpaceX Payloads SpaceX spacecraft have delivered cargo to and from the international space station and have successfully completed missions for the U.S. Air Force, AsiaSat, and the Turkmenistan National Space Agency. The table shows the weight (in kg) of the payload for 40 randomly selected SpaceX missions. **SPACEX**

525	500	677	500	3170	3325	2296	1316	4535	4428
2216	2395	570	4159	1898	4707	1952	2034	553	5271
3136	4696	3100	3600	2257	4600	9600	2490	5600	5300
6070	2708	3669	6716	3310	475	4990	5200	3500	2205

(a) Find the sample mean payload weight.
(b) Find the sample median payload weight.
(c) Use the sample mean and the sample median to describe the shape of the distribution of payload weight.
(d) Find a 5% trimmed mean for the payload weight data.
(e) Suppose a successful payload weight is a weight of at least 4000 kg. Find the sample proportion of successes.
(f) Find the sample variance s^2 and the sample standard deviation s for the payload weight data.
(g) Find Q_1, Q_3, and IQR for the payload weight data.
(h) Without assuming anything about the shape of the distribution of payload weights, use Chebyshev's Rule to describe the distribution (for $k = 2$ and for $k = 3$).
(i) Assume the shape of the distribution of payload weights is approximately normal. Approximately what proportion of payload weights is between 1250 and 7268?
(j) Suppose the population mean payload weight is 3200 kg, the population standard deviation payload weight is 2000 kg, and the distribution of payload weights is approximately normal. A new SpaceX customer has a payload that weighs 5500 kg. Is there any evidence that this is an unusual weight? Justify your answer.
(k) Find the 35th percentile for the payload weight data.
(l) Construct a modified box plot for these data. Describe the distribution in terms of symmetry, skewness, variability, and outliers.

Probability

korisbo/Shutterstock

◀ Looking Back
- Understand the relationships among a population, a sample, probability, and statistics: **Section 1.2**.

Looking Forward ▶
- Learn the basic concepts associated with an experiment: **Section 4.1**.
- Understand the intuitive definition of the probability of an event and several useful probability rules: **Section 4.2**.
- Learn how to use counting techniques to find the probability of an event: **Section 4.3**.
- Understand and use the definition of conditional probability in a variety of contexts: **Section 4.4**.
- Understand and use the definition of independent events to compute probabilities: **Section 4.5**.

CHAPTER APP

Dream Home

In 1997, HGTV started a contest to give away a Dream Home. Every year, HGTV produces a special television series to showcase the featured home and the experts who helped design and build the home. The first home was a log house overlooking the Grand Tetons in Jackson Hole, Wyoming. Since then, the homes have become more stunning and extraordinary each year. Past homes have been located on St. Simons Island, Georgia; Martha's Vineyard, Massachusetts; and Lake Tahoe, California.

The 2018 prize was a spectacular home overlooking Puget Sound in Gig Harbor, Washington. The winner also received $250,000 in cash and a new car. Entry in this sweepstakes is open to residents of the United States and participants may enter every day online or by mail. The contest usually runs from the end of December to the middle of February.

The probability of winning the Dream Home sweepstakes depends on the number of entries you submit and the total number of entries. Suppose you submit 50 entries, or one entry every day the contest is open. There have been approximately 130 million entries in this sweepstakes the past few years. Therefore, the probability of winning the Dream Home is 0.0000003846. Well, at least we can dream about the home.

The techniques presented in this chapter can be used to determine the probability of certain events, such as winning the Dream Home. We'll start with a very intuitive definition of probability and develop several rules to help compute more complex probabilities.

4.1 Experiments, Sample Spaces, and Events

Experiments

To understand probability concepts, we need to think carefully about **experiments**. Consider the activity, or act, of tossing a coin, selecting a card from a standard poker deck, counting the number of people standing on a city bus, or even testing a cell phone for defects before shipment. In each of these activities, the outcome is uncertain. For example, when we test a new cell phone, we do not know (for sure) whether it will be defect-free. This idea of uncertainty leads to the definition of an experiment.

> **Definition**
> An **experiment** is an activity in which there are at least two possible outcomes and the result of the activity cannot be predicted with absolute certainty.

Here are some examples of experiments:

1. Roll a six-sided die and record the number that lands face up.

   We cannot say with certainty that the number face up will be a 1, a 2, etc., so this activity is an experiment.

2. Using a radar gun, record the speed of a pitch at a Red Sox baseball game.

   We're not sure whether the pitch will be a fastball, curveball, slider, etc. And even if we steal the signal from the catcher, we cannot predict the speed of the pitch with certainty.

3. Count the number of patients who arrive at the emergency room of a city hospital during a 24-hour period.

   Although past records might help us estimate the patient volume, there is no way of predicting the exact number of patients who visit the emergency room during a 24-hour period.

4. Select two Ninja Blenders and inspect each for flaws in materials and workmanship.

   Even though a careful quality control process might be in place, there is no way of knowing whether both blenders will be flawless, one will contain a flaw, or both will have flaws.

CHAPTER APP **Dream Home**

Explain why the selection of the Dream Home winner is an experiment.

Outcomes

Because we don't know for sure what will happen when we conduct an experiment, we need to consider all possible outcomes. This sounds easy (just think about all the things that can happen), but it can be tricky. Sometimes it involves a lot of counting, but often outcomes can be visualized using a tree diagram. Consider the following examples.

EXAMPLE 4.1 Zip Codes

Suppose an outgoing letter at a New York City post office is selected at random and the first digit in the address zip code is recorded. How many possible outcomes are there, and what are they?

Solution

STEP 1 The first digit of a zip code can be any integer from 0 to 9.

STEP 2 There are 10 possible outcomes.

The outcomes are 0, 1, 2, 3, 4, 5, 6, 7, 8, and 9.

This is an experiment because we cannot predict the first digit in a zip code with certainty.

EXAMPLE 4.2 Drone Pilot

According to the Federal Aviation Administration (FAA), anyone who flies a drone (outside) weighing more than 0.55 pound must register the device. Suppose two drone pilots flying devices that weigh more than 0.55 pound are selected at random and their drones are checked for registration. How many possible outcomes are there, and what are they?

Solution

STEP 1 If the drone is registered, denote this observation by R (for registered). If the drone is not registered, use U (for unregistered).

STEP 2 Each outcome is a pair of observations, one on each drone. There are four possible outcomes, and here is one way to write them: RR, RU, UR, UU.

The first letter indicates the observation on the first drone, and the second letter indicates the observation on the second drone.

RU is a different outcome from UR. RU means the first drone was registered and the second drone was not. UR means the first drone was not registered and the second drone was.

There are lots of other ways to denote these four outcomes. There is no single correct, or best, notation. Write the outcomes so that others can understand and interpret your list.

All of the outcomes from the experiment in Example 4.2 can be determined by constructing a **tree diagram**, a visual road map of possible outcomes. **Figure 4.1** is a tree diagram associated with this experiment.

Tree diagrams will also be extremely useful for determining probabilities in problems involving the Probability Multiplication Rule and Bayes' Rule. Problems of this type are presented in Section 4.5.

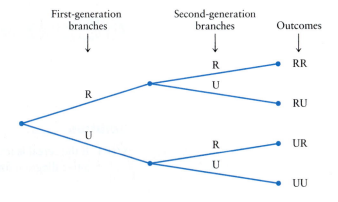

Figure 4.1 Tree diagram for Example 4.2.

The first-generation branches indicate the possible choices associated with the first drone, and the second-generation branches represent the choices for the second drone. A path from left to right represents a possible experimental outcome.

EXAMPLE 4.3 Drone Pilot (Continued)

Extend the previous example. How many outcomes are there if we select three drones and record their registration status?

Solution

Now there are eight possible outcomes: RRR, RRU, RUR, RUU, URR, URU, UUR, UUU.

Figure 4.2 is a tree diagram for this extended experiment. Again, every path from left to right represents a possible outcome.

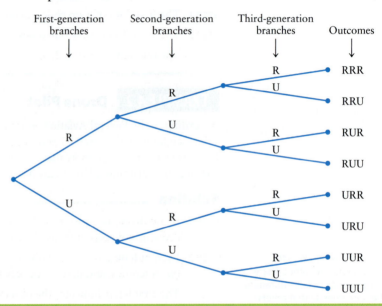

Figure 4.2 Tree diagram for Example 4.3.

Tree diagrams are also used to prove the Multiplication Rule (Section 4.3), an arithmetic technique used to count the number of possible outcomes in certain experiments.

A CLOSER LOOK

1. Tree diagrams are a fine technique for finding all the possible outcomes for an experiment. However, they can get very big, very fast.
2. A tree diagram does not have to be symmetric, as the trees are in Figure 4.1 and Figure 4.2. The branches and paths depend on the experiment. Consider the next example.

EXAMPLE 4.4 Breakfast of Champions

A consumer in Blakesburg, Iowa, is searching for a box of his favorite breakfast cereal. He will check all three grocery stores in town if necessary but will stop if the cereal is found. The experiment consists of recording the cereal search result. How many possible outcomes are there, and what are they?

Solution

STEP 1 If the cereal is in stock, use the letter I; if it is out of stock, use O. **Figure 4.3** shows a tree diagram for this experiment.

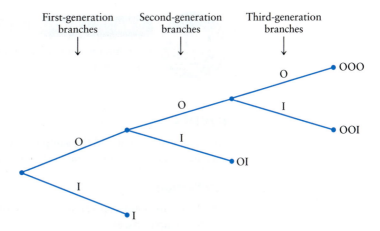

Figure 4.3 Tree diagram for Example 4.4.

STEP 2 The tree diagram shows four possible paths from left to right. The outcomes are as follows:

Outcome	Experiment result
I	The cereal is in stock at store 1.
OI	The cereal is not in stock at store 1, but it is in stock at store 2.
OOI	The cereal is not in stock at stores 1 and 2, but it is in stock at store 3.
OOO	The cereal is not in stock at any store.

This tree diagram is not symmetric, but all possible outcomes are represented by left-to-right paths.

The paths representing the outcomes have different lengths. Some of the outcomes are represented by a shorter sequence of letters because the experiment ends early if the consumer finds the cereal at the first or second store.

The Sample Space

Definition
The **sample space** associated with an experiment is a listing of all possible outcomes using set notation. It is the collection of all outcomes written mathematically, in set notation (using curly braces), and denoted by S.

The symbol S is used to denote several different objects, or quantities, in probability and statistics; a small s is also used in many different situations. The context of the problem reveals the relevant meaning of the symbol.

EXAMPLE 4.5 **Sample Spaces**

Find the sample space for each of the four experiments in Examples 4.1 through 4.4.

Solution

We determined the outcomes for each experiment. Simply write the sample space mathematically, using set notation.

STEP 1 First digit of zip code numbers: $S = \{0, 1, 2, 3, 4, 5, 6, 7, 8, 9\}$.

STEP 2 Drone pilot experiment: $S = \{RR, RU, UR, UU\}$.

STEP 3 Extended drone pilot experiment:
$S = \{RRR, RRU, RUR, RUU, URR, URU, UUR, UUU\}$.

STEP 4 Cereal experiment: $S = \{I, OI, OOI, OOO\}$.

> **CHAPTER APP** **Dream Home**
>
> Describe the sample space in the selection of the Dream Home winner experiment.

Events

Given an experiment and the sample space, we usually study and find the probability of specific collections of outcomes, called **events**.

> ### Definition
> 1. An **event** is any collection (or set) of outcomes from an experiment (any subset of the sample space).
> 2. A **simple event** is an event consisting of exactly one outcome.
> 3. An event has **occurred** if the resulting outcome is contained in the event.

> ### A CLOSER LOOK
>
> 1. An event may be given in standard set notation, or it may be defined in words. If a written definition is given, we usually need to translate the words into mathematics to identify the event outcomes.
> 2. Notation:
>   (a) Events are denoted with capital letters, for example, $A, B, C, \ldots$.
>   (b) Simple events are often denoted by $E_1, E_2, E_3, \ldots$.
> 3. It is possible for an event to be empty. An event containing no outcomes is denoted by $\{\}$ or $\emptyset$ (the empty set).

EXAMPLE 4.6 College Dining

Two resident students at Bucknell University are selected and asked if they purchased a meal plan (M) or cook for themselves (C). The experiment consists of recording the response from both students.

There are four possible outcomes. A tree diagram is shown in **Figure 4.4**. The sample space is $S = \{MM, MC, CM, CC\}$.

There are four relevant simple events:

$E_1 = \{MM\}, E_2 = \{MC\}, E_3 = \{CM\}, E_4 = \{CC\}$

Here are some other events, in words and in set notation.

Let A be the event that both students made the same choice. $A = \{MM, CC\}$.

Let B be the event that at most one student purchased a meal plan.

$B = \{CC, MC, CM\}$ contains observations with at most one M.

Let D be the event that at least one student cooks for himself.

$D = \{CM, MC, CC\}$ contains observations with one or more Cs. ∎

Translate *at most* and *at least* carefully. These expressions appear frequently in probability and statistics questions.

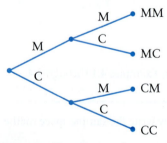

Figure 4.4 Tree diagram for Example 4.6.

EXAMPLE 4.7 On-Time Delivery

A UPS driver may deliver packages and letters to floors 2 through 6 in an office building and use one of three elevators (labeled A, B, and C). The experiment consists of recording the floor and elevator used.

There are 15 possible outcomes because there are three elevators for each of the five floors. A tree diagram works again. The sample space is the following set:

S = {2A, 3A, 4A, 5A, 6A, 2B, 3B, 4B, 5B, 6B, 2C, 3C, 4C, 5C, 6C}

The number in each outcome represents the floor, and the letter represents the elevator.

Let E be the event that the delivery is made on an odd floor using elevator B.

$E = \{3B, 5B\}$

Let F be the event that the delivery is made on an even floor.

This definition says nothing about the elevator used. Therefore, there are no restrictions on the elevator in this event.

$F = \{2A, 4A, 6A, 2B, 4B, 6B, 2C, 4C, 6C\}$

Let G be the event that the delivery is made using elevator C.

$G = \{2C, 3C, 4C, 5C, 6C\}$

■

TRY IT NOW Go to Exercises 4.21 and 4.23

When an experiment is conducted, only one outcome can occur. For example, if the UPS driver used elevator B to deliver to the third floor, the experimental outcome is 3B. The observed outcome may be included in several relevant events. In the delivery example just presented, if the outcome 4C is observed, then the events F and G have occurred. The event E did not occur.

New Events from Old

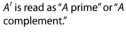

RAPID REVIEW: Set notation, complement, union, intersection and disjoint events.

Given an experiment, the sample space, and some relevant events, we often combine events in various ways to create and study new events. Events are really sets, so the methods of combining events are set operations.

Definition

Let A and B denote two events associated with a sample space S.

A' is read as "A prime" or "A complement."

1. The event **A complement**, denoted A', consists of all outcomes in the sample space S that are not in A.
2. The event **A union B**, denoted $A \cup B$, consists of all outcomes that are in A or B or both.
3. The event **A intersection B,** denoted $A \cap B$, consists of all outcomes that are in both A and B.
4. If A and B have no elements in common, they are **disjoint** or **mutually exclusive**, written $A \cap B = \{\}$.

A CLOSER LOOK

1. The event A' is also called **not A**. The word *not* in the context of a probability question usually means you need to find the complement of an event.
2. *Or* usually means **union**; A or B means $A \cup B$.
3. *And* usually means **intersection**; A and B means $A \cap B$.
4. Any outcome in both A and B is included only once in the event $A \cup B$.
5. The three events defined previously could be denoted using any new symbols. A', $A \cup B$, and $A \cap B$ are customary mathematical symbols to denote complement, union, and intersection, respectively.
6. It is possible for one of these new events to contain all the outcomes in the sample space.

EXAMPLE 4.8 Top Secret Information

U.S. government and nongovernmental repositories store records that may be classified as confidential (C), secret (S), or top secret (T). A national security information officer is interested in the types of documents stored in a Washington, D.C., location and their possible declassification. She conducts an experiment by selecting two documents at random and recording the security classification of each.

The sample space for this experiment has nine outcomes. See **Figure 4.5**.

$S = \{CC, CS, CT, SC, SS, ST, TC, TS, TT\}$

Consider the following events:

$A = \{CC, SS, TT\}$	Both documents have the same classification.
$B = \{CC, CS, SC, SS\}$	Neither document is classified as top secret.
$C = \{CC, CS, CT, SC, TC\}$	At least one document is classified as confidential.
$D = \{CS, SC, ST, TS\}$	Exactly one document is classified as secret.

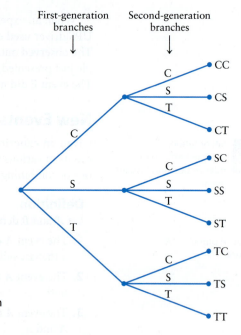

Figure 4.5 Tree diagram for Example 4.8.

Here are some new events created from the four given events.

$D' = \{CC, CT, SS, TC, TT\}$

  $= D$ complement: Neither document is classified as secret or both are.

  $=$ All outcomes in S not in D.

$A \cup C = \{CC, SS, TT, CS, CT, SC, TC\ \}$

  $=$ Both documents have the same classification or at least one is confidential.

  $=$ All outcomes in A or C or both.

$A \cap D = \{\ \}$

  $=$ Both documents have the same classification and exactly one is secret.

  $=$ All outcomes in A and D. The events A and D are disjoint.

$(A \cup C)' = \{ST, TS\}$

  $= A$ union C, complement.

  $=$ All outcomes in S not in $A \cup C$.

$(A \cap D)' = S$
$ = A \text{ intersection } D, \text{ complement.}$
$ = \text{All outcomes in } S \text{ not in } A \cap D.$

TRY IT NOW Go to Exercises 4.29 and 4.33

Venn Diagrams and Extended Definitions

A **Venn diagram** may be used to visualize a sample space and events, to determine outcomes in combinations of events, and to answer probability questions in later sections. To construct a Venn diagram, draw a rectangle to represent the sample space. Various figures (often circles) are drawn inside the rectangle to represent events. The Venn diagrams in **Figure 4.6** illustrate various combinations of events.

In a Venn diagram, plane regions represent events. We often add labeled points to denote outcomes. Later, probabilities assigned to events will be added to these diagrams.

The definitions of union, intersection, and disjoint events can be extended to a collection consisting of more than two events.

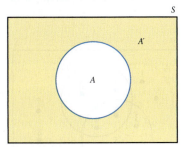
A complement: A'

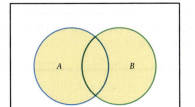
A union B: $A \cup B$

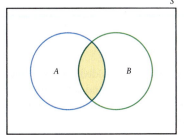
A intersection B: $A \cap B$

Definition
Let $A_1, A_2, A_3, \ldots, A_k$ be a collection of k events.
1. The event $A_1 \cup A_2 \cup \cdots \cup A_k$ is a **generalized union** and consists of all outcomes in at least one of the events $A_1, A_2, A_3, \ldots, A_k$.
2. The event $A_1 \cap A_2, \cap \cdots \cap A_k$ is a **generalized intersection** and consists of all outcomes in each of the events $A_1, A_2, A_3, \ldots, A_k$.
3. The k events $A_1, A_2, A_3, \ldots, A_k$ are **disjoint** if no two have any element in common.

EXAMPLE 4.9 Consumer Ratings

Kelley Blue Book is a resource for consumers buying new or used cars. Each 2018 car receives an overall rating based on consumer reviews. Suppose the rating range is 0 to 9, with 0 as the lowest rating and 9 as the highest rating. Consider an experimenter in which a random 2018 automobile is selected and the Kelley Blue Book rating is recorded. The sample space is

$S = \{0, 1, 2, 3, 4, 5, 6, 7, 8, 9\}$

Consider the following events:

$A = \{0, 1, 2, 3, 4\} \qquad B = \{3, 4, 5, 6\}$
$C = \{7, 8\} \qquad D = \{2, 4, 6, 9\}$

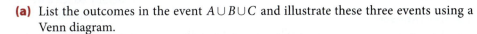
A and B are disjoint: $A \cap B = \{\ \}$

Figure 4.6 Venn diagrams.

(a) List the outcomes in the event $A \cup B \cup C$ and illustrate these three events using a Venn diagram.

(b) List the outcomes in the event $A \cup B \cup D$ and illustrate these three events using a Venn diagram.

(c) List the outcomes in each of the following events:
  (i) $A \cap B \cap C$
  (ii) $A \cap B \cap D$
  (iii) $(A \cup B)'$
  (iv) $(A \cup B \cup D)'$

Solution

(a) The event $A \cup B \cup C$ consists of all the outcomes in at least one of the events A, B, or C.

$A \cup B \cup C = \{0, 1, 2, 3, 4, 5, 6, 7, 8\}$

Figure 4.7 shows the relationship among the events A, B, and C, and the sample space S.

(b) The event $A \cup B \cup D$ consists of all the outcomes in at least one of the events A, B, or D.

$A \cup B \cup D = \{0, 1, 2, 3, 4, 5, 6, 9\}$

Figure 4.8 shows the relationship among the events A, B, and D, and the sample space S.

(c) (i) $A \cap B \cap C = \{\}$ There are no outcomes in all three events.
(ii) $A \cap B \cap D = \{4\}$ 4 is the only outcome in all three events.
(iii) $(A \cup B)' = \{7, 8, 9\}$ All outcomes in S not in $A \cup B$.
(iv) $(A \cup B \cup D)' = \{7, 8\} = C$ All outcomes in S not in $A \cup B \cup D$.

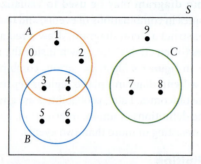

Figure 4.7 The events A, B, and C in Example 4.9.

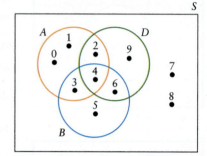

Figure 4.8 The events A, B, and D in Example 4.9.

TRY IT NOW Go to Exercises 4.13 and 4.15

Section 4.1 Exercises

Concept Check

4.1 True or False In an experiment, the result of the activity is always fairly certain.

4.2 True or False A tree diagram is always symmetric.

4.3 True or False A sample space consists of all possible outcomes.

4.4 True or False A simple event is one in which the outcome is very rare.

4.5 Short Answer
(a) The word _____ is usually associated with the complement of an event.
(b) The word _____ is usually associated with union.
(c) The word _____ is usually associated with intersection.

4.6 Suppose A, B, and C are events. Classify each expression as always true, sometimes true, or never true.
(a) $A \cup B = B \cup A$
(b) $A \cap B \cap C = C \cap B \cap A$
(c) $(A \cap B)' = (B \cap A)'$
(d) $A = A'$
(e) $A \cup C = A$
(f) $A \cup B = B \cup C$

Practice

4.7 An experiment consists of rolling a six-sided die, recording the number face up, and then tossing a coin and recording either a head or a tail. Carefully sketch a tree diagram and find the sample space for this experiment.

4.8 A basketball player is going to select a sneaker with red, blue, green, or black stripes and in either low- or high-top style. An experiment consists of recording the color and style. Carefully sketch a tree diagram and find the sample space for this experiment.

4.9 An experiment consists of selecting one letter from B, I, N, G, and O, and selecting one of five rows. How many possible outcomes are there in this experiment? Carefully sketch the corresponding tree diagram.

4.10 One playing card is selected from a regular 52-card deck. An experiment consists of recording the denomination (ace, 2, 3, 4, 5, 6, 7, 8, 9, 10, jack, queen, king) and suit (club, diamond, heart, or spade). How many possible outcomes are there in this experiment?

4.11 Consider an experiment with the sample space $S = \{0, 1, 2, 3, 4, 5, 6, 7, 8, 9\}$

and the events
$A = \{0, 2, 4, 6, 8\}$; $\quad B = \{1, 3, 5, 7, 9\}$;
$C = \{0, 1, 2, 3, 4\}$; $\quad D = \{5, 6, 7, 8, 9\}$
Find the outcomes in the following events.
(a) A' (b) C' (c) D'
(d) $A \cup B$ (e) $A \cup C$ (f) $A \cup D$

4.12 Use the sample space and events defined in Exercise 4.11 to find the outcomes in each of the following events.
(a) $B \cap C$ (b) $B \cap D$ (c) $A \cap B$
(d) $A \cap C$ (e) $(B \cap C)'$ (f) $B' \cup C'$

4.13 Consider an experiment with sample space
$S = \{a, b, c, d, e, f, g, h, i, j, k\}$
and the events
$A = \{a, c, e, g\}$; $\quad B = \{b, c, f, j, k\}$;
$C = \{c, f, g, h, i\}$; $\quad D = \{a, b, d, e, g, h, j, k\}$
Find the outcomes in each of the following events.
(a) A' (b) C' (c) D'
(d) $A \cap B$ (e) $A \cap C$ (f) $C \cap D$

4.14 Use the sample space and the events defined in Exercise 4.13 to find the outcomes in each of the following events.
(a) $A \cup B \cup D$ (b) $B \cup C \cup D$
(c) $B \cap C \cap D$ (d) $A \cap B \cap C$

4.15 Use the sample space and the events defined in Exercise 4.13 to find the outcomes in each of the following events.
(a) $(A \cap B \cap C)'$ (b) $A \cup B \cup C \cup D$
(c) $(B \cup C \cup D)'$ (d) $B' \cap C' \cap D'$

4.16 The Venn diagram shows the relationship between two events.

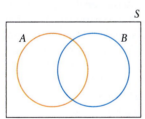

Redraw the Venn diagram for each part of this problem and carefully shade in the region corresponding to each new event.
(a) $(A \cup B)'$ (b) $(A \cap B)'$ (c) $A' \cap B$
(d) $A \cap B'$ (e) $A' \cap B'$ (f) $A' \cup B'$

4.17 The Venn diagram shows the relationship among three events.

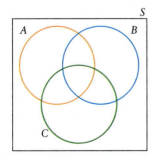

Redraw the Venn diagram for each part of this problem and carefully shade in the region corresponding to each new event.
(a) $A \cup B \cup C$ (b) $A \cap B \cap C$ (c) $A \cup C$
(d) $B \cap C$ (e) $B \cap C'$ (f) $(A \cup B)' \cap C$
(g) $(A \cup B \cup C)'$ (h) $A' \cap B' \cap C'$ (i) $B \cap C \cap A'$

4.18 Consider an experiment with sample space
$S = \{YYY, YYN, YNY, YNN, NYY, NYN, NNY, NNN\}$
(a) Find the outcomes in each of the following events.
$A =$ Exactly one Y
$B =$ Exactly two Ns
$C =$ At least one Y
$D =$ At most one N

Find the outcomes in each of the following events and write each as a combination of the events $A, B, C,$ and D.
(b) Exactly one Y or at most one N
(c) Two or more Ns
(d) Exactly two Ns and at least one Y
(e) Two or more Ys

4.19 Consider an experiment with sample space
$S = \{0, 1, 2, 3, 4, 5, 6, 7, 8, 9\}$ and the events
$A = \{0, 1, 2, 7, 8, 9\}$; $\quad B = \{0, 1, 2, 4, 8\}$;
$C = \{0, 1, 3, 9\}$; $\quad D = \{1, 4, 9\}$

Draw a separate Venn diagram to illustrate the relationship among each collection of events and the sample space S.
(a) B and C (b) A and D
(c) $A, B,$ and C (d) $A, C,$ and D

Applications

4.20 Physical Sciences An experiment consists of recording the time zone (E, C, M, P) and strength (L, M, H) of the next earthquake in the 48 contiguous states of the United States.
(a) Carefully sketch a tree diagram to illustrate the possible outcomes for this experiment.
(b) Find the sample space S for this experiment.

4.21 Economics and Finance Three taxpayers are selected at random and asked whether they contributed to an individual retirement account (IRA) last year. An experiment consists of recording each response. Construct a tree diagram to represent this experiment and find the outcomes in the sample space.

4.22 Travel and Transportation Two people who work in New York City are selected at random and asked how they get to work: drive, take a train, or take a bus. An experiment consists of recording each response. Construct a tree diagram to represent this experiment and find the outcomes in the sample space.

4.23 Sports and Leisure Snowmobiles are popular recreational, off-road vehicles in Strathcona County, Alberta, Canada. Snowmobile riders must wear a helmet, the operator must be able to produce a valid registration and proof of insurance, and their snowmobile must have a working muffler.

Suppose a peace officer stops a snowmobile operator at random and records the following data:
(a) Whether the operator has a valid registration
(b) Whether the snowmobile is properly insured
(c) The condition of the snowmobile muffler: none, standard, or modified

Construct a tree diagram to represent this experiment and find the outcomes in the sample space.

4.24 Physical Sciences A construction crew excavating a site for a building foundation must remove the rock and prepare a trench for concrete footers. An experiment consists of recording the type of rock present (I, igneous; S, sedimentary; M, metamorphic) and the number of days needed to prepare the site (1 to 5).
(a) Carefully sketch a tree diagram to illustrate the possible outcomes for this experiment.
(b) Find the sample space S for this experiment.

4.25 Marketing and Consumer Behavior Suppose a Whitman's Sampler box of chocolates has four candies left, but only one of them is your favorite, a strawberry cream. An experiment consists of testing (well, tasting) each chocolate until the strawberry cream is found.
(a) How many possible outcomes are there for this experiment?
(b) Is the outcome FSFF (Failure, Strawberry cream, Failure, Failure) possible? Why or why not?

4.26 Sports and Leisure An experiment consists of recording the number of pins knocked down on each roll during a frame of a bowling game. A bowler may take a maximum of two rolls per frame. How many outcomes are in the sample space for this experiment? *Hint*: If the first roll is a 10 (a strike), the experiment is over.

Now suppose this experiment involves candlepin bowling. In this case, a bowler may take a maximum of three rolls per frame. If the bowler knocks down all 10 pins, the experiment is over. How many outcomes are in the sample space for this experiment?

4.27 Sports and Leisure A sports statistician must carefully chart opposition football plays in preparation for the next game. An experiment consists of recording the type of play (pass or rush) and the yards gained $(-99, -98, -97, \ldots, -2, -1, 0, 1, 2, \ldots, 97, 98, 99)$ on a randomly selected first down. How many outcomes are in the sample space for this experiment?

4.28 Medicine and Clinical Studies The emergency room (ER) in a rural hospital is staffed in four 6-hour shifts (1, 2, 3, 4). During any shift, an ER patient is attended to by either a general physician (G), a surgeon (R), or an intern (I). An experiment consists of coding the next ER patient by shift and the attending doctor. Consider the following events.
A = The attending doctor is the general physician.
B = The patient is admitted during the second shift.
C = The patient is admitted during shift 3 or is seen by the intern.
D = The patient is admitted during shift 4 and is seen by the general physician.
(a) Find the sample space S for this experiment.
(b) List the outcomes in each of the events A, B, C, and D.
(c) List the outcomes in the events $A \cup B$ and $A \cap B$.

4.29 Psychology and Human Behavior Drivers entering the Coastland Center Mall parking lot at the main entrance may turn left, right, or go straight. An experiment consists of recording the direction of the next car entering the mall and the vehicle style (sedan, SUV, van, or pickup). Consider the following events.
A = The next vehicle is a van.
B = The next vehicle is a sedan or pickup.
C = The next vehicle turns left.
D = The next vehicle goes straight or turns right.
(a) Find the sample space S for this experiment.
(b) List the outcomes in each of the events A, B, C, and D.
(c) List the outcomes in the events $C \cup D$ and $C \cap D$.

4.30 Public Health and Nutrition Each patient with a regular appointment at Park Family Dentistry is classified by the number of cavities found (assume four is the maximum) and as late (L) or on time (T) for the appointment.
(a) Find the sample space S for this experiment.
(b) Describe the following events in words.
$A = \{0L, 1L, 2L, 3L, 4L\}$
$B = \{3L, 4L, 3T, 4T\}$
$C = \{1L, 3L, 1T, 3T\}$
$D = \{0L, 0T\}$
$E = \{0L, 0T, 1L, 2L, 3L, 4L\}$
$F = \{4T\}$

4.31 Travel and Transportation Every passenger arriving at the San Francisco International Airport is classified as traveling on a domestic (D) or international (I) flight and by the number of checked bags (assume five is the maximum).
(a) Find the sample space S for this experiment.
(b) Describe the following events in words.
$A = \{D0, I0\}$
$B = \{I0, I1, I2, I3, I4, I5\}$
$C = \{D1, I1, D2, I2\}$
$D = \{I0, I5\}$
$E = \{D1, I1, D3, I3, D5, I5\}$

4.32 Marketing and Consumer Behavior A researcher working for a Five Guys fast-food restaurant in Savannah, Georgia, selects random customers and classifies each according to order type [hot dog (D), burger (B), or sandwich (S)], one topping [lettuce (L), pickles (P), or tomatoes (T)], and type of fries [Five Guys style (F) or cajun style (C)].
(a) Find the sample space S for this experiment.
(b) Describe the following events in words.
$A = \{BLF, BLC, BPF, BPC, BTF, BTC\}$
$B = \{BLF, BLC\}$
$C = \{SLF, SPF, STF, SLC, SPC, STC\}$
$D = \{DPF, DPC, BPF, BPC, SPF, SPC\}$

4.33 Education and Child Development The Villa Maria music school in Montreal was scheduled to close in January due to low enrollment. However, protesters were fighting to keep

the school open, as many alumni have had successful careers in music. Suppose a student at the school is selected at random and classified by extracurricular activity [choir (C), music shows (M), or ensemble (E)], type of instrument played [string (S), woodwind (W), brass (B), or keyboard (K)], and intention to pursue a career in music [yes (Y) or no (n)].
(a) Find the sample space S for this experiment.
(b) List the outcomes in each event.
   (i) The student plays a woodwind instrument.
   (ii) The student plays a brass instrument and intends to pursue a career in music.
   (iii) The student is in the choir or does not intend to pursue a career in music.
   (iv) The student is in an ensemble and plays the keyboard.
   (v) The student does not play the keyboard or participate in music shows.

Extended Applications

4.34 Sports and Leisure A single six-sided die is rolled. If the number face up is even, then the experiment is over. If the number face up is odd, then the die is rolled again. The experiment continues until the number face up is even.
(a) Carefully sketch (part of) a tree diagram to illustrate the possible outcomes for this experiment.
(b) Find the sample space for this experiment.

4.35 Business and Management Centrelink is an agency in the Australian Government that controls payments to retirees. Recently, staff reductions have led to higher-than-normal busy signals when individuals call with questions. Suppose that if a person calling Centrelink receives a busy signal, she will hang up and try again later, and will stop calling as soon as she reaches an official. An experiment consists of recording the calling pattern. A possible outcome is BBH: a busy signal (B) on the first two calls, and (finally) help (H) on the third call.
(a) How many possible outcomes are there in this experiment?
(b) List some of the outcomes for this experiment.

4.36 Marketing and Consumer Behavior Musicnotes sells sheet music online in the following genres: rock, jazz, New Age, and country. An experiment consists of recording the preferred genre for the next customer and the number of songs purchased (assume five is the maximum). Consider the following events.
A = The next customer prefers rock.
B = The next customer prefers jazz and buys at least three songs.
C = The next customer buys at most two songs.
D = The next customer prefers country and buys one song.
(a) Find the sample space S for this experiment.
(b) Find the outcomes in each of the following events.
   (i) A' (ii) $A \cup C$ (iii) $A \cap D$
   (iv) $C \cap D$ (v) $A \cap C \cap D$ (vi) $(A \cap B)'$

4.37 Marketing and Consumer Behavior Starwood SPG hotels are categorized by the number of points required to reserve a free night, 1–7. Categories 1 and 2 represent relatively inexpensive hotels, whereas category 7 hotels are some of the most luxurious in spectacular locations. An experiment consists of selecting a Starwood SPG member and recording the hotel category corresponding to the most recent hotel stay and whether the guest used points. Consider the following events.
A = The guest did not use points.
B = The guest stayed at a category 1, 2, or 3 hotel.
C = The guest stayed at a category 5, 6, or 7 hotel and used points.
D = The guest stayed at a category 1, 3, 5, or 7 hotel.
(a) Find the sample space S for this experiment.
(b) Find the outcomes in each of the following events.
   (i) B' (ii) $A \cup B$ (iii) $A \cap B$
   (iv) $C \cap D$ (v) $A \cap B \cap D$ (vi) $(A \cap D)'$

4.38 Travel and Transportation An experiment consists of selecting a random passenger on a train from Washington, D.C.'s Union Station to Trenton, New Jersey, and recording the purpose of travel (business or pleasure) and the number of pieces of luggage (zero to four). Consider the following events.
A = The passenger is traveling on business.
B = The passenger has no luggage.
C = The passenger has at most one piece of luggage.
D = The passenger has three pieces of luggage or is traveling for pleasure.
(a) Find the sample space S for this experiment.
(b) Find the outcomes in each of the following events.
   (i) $A \cup B$ (ii) $A \cap B$
   (iii) $B \cup C$ (iv) $B \cap C$
   (v) $A \cap D$ (vi) $A \cap B \cap C \cap D$

4.39 Marketing and Consumer Behavior Bubble tea is made with tea (of course), milk or fruit flavors, a sweetener, and QQ (the ingredient that gives the impression of bubbles, often tapioca pearls). At Boba Guys in the Canal Street Market, customers can purchase bubble tea consisting of black, white, or green tea, with fresh milk or soy milk, and with tapioca pearls or boba (large tapioca pearls). An experiment consists of recording these three options for the next customer.
(a) Carefully sketch a tree diagram to illustrate the possible outcomes for this experiment.
(b) Find the sample space S for this experiment.
(c) Consider the following events.
A = The next customer purchase uses black tea.
B = The next customer purchase uses soy milk.
C = The next customer purchase uses black tea or fresh milk.
Find the outcomes in each of the following events.
   (i) $A \cup B$ (ii) $B \cup C$
   (iii) $B \cap C$ (iv) C'

4.40 Demographics and Population Statistics There are more than 2000 homeowner/condominium associations (HOAs) in south Florida with more than 500,000 managed dwellings. Florida law stipulates that each HOA board must have a minimum of three members, and suppose board bylaws contain a clause limiting the size to six. An experiment consists of selecting an HOA at random and recording the number of HOA board members, whether the HOA has a clubhouse, and

whether the HOA levied a special assessment within the last year.
(a) Find the sample space S for this experiment.
(b) Consider the following events
A = The HOA has a clubhouse and levied a special assessment last year.
B = The HOA has five or six board members.
C = The HOA did not levy a special assessment last year.

Find the outcomes in each of the following events.
(i) $A \cap B$ (ii) $(A \cup B)'$ (iii) $B \cap C$
(iv) $B' \cap C$ (v) $(A \cap B \cap C)'$ (vi) $(A \cup C)' \cap B$

(c) Draw a Venn diagram to illustrate the relationship among the three events A, B, and C.

4.2 An Introduction to Probability

The Probability of an Event

P is really just a function. In the expression P(A), the input is the event A and the output is a probability.

Given an experiment, some events are more likely to occur than others. For any event A, we need to assign a number to A that corresponds to this intuitive likelihood of occurrence. The likelihood that A will occur is simply the probability of the event A. For example, the probability that an amateur golfer hits a hole-in-one is approximately 0.00008 (a pretty unlikely event). The probability that a randomly selected home in America has a barbecue grill is 0.75 (a much more likely event). The notation P(A) is used to denote this likelihood, the probability of an event A. To begin our discussion of probability, consider the following working definition.

Definition
The probability of an event A is a number between 0 and 1 (including those endpoints) that measures, or conveys, the likelihood that A will occur.

1. If the probability of an event is close to 1, then the event is likely to occur.
2. If the probability of an event is close to 0, then the event is not likely to occur.

If the probability of an event A is 1, then the event is a certainty: It will occur. If the probability of an event B is 0, then B is definitely not going to occur. What about events with probabilities in between? How do we decide to assign a probability of 0.3, for example, to an event C? We need a reasonable, all-purpose rule for linking an event to its likelihood of occurrence. The natural (theoretical) definition for assigning a probability to an event is very intuitive.

Definition
The **relative frequency of occurrence of an event** is the number of times the event occurs divided by the total number of times the experiment is conducted.

EXAMPLE 4.10 Pick a Card, Any Card

A regular 52-card deck contains 13 clubs, 13 diamonds, 13 hearts, and 13 spades. Suppose an experiment consists of selecting one card from the deck and recording the suit. What is the probability of selecting a club?

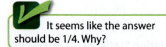
It seems like the answer should be 1/4. Why?

Solution

STEP 1 Let C be the event that a club is selected. We want the probability of the event C, which is denoted by P(C).

STEP 2 To estimate the probability of C, it seems reasonable to conduct the experiment several times and see how often a club is selected. If C occurs often (we get a club a lot of the time), then the likelihood (probability) should be high. If C occurs rarely, then the probability should be close to 0.

STEP 3 To estimate the likelihood of selecting a club, we use the relative frequency of occurrence of a club, which is the frequency divided by total trials, or

$$\text{Relative frequency} = \frac{\text{number of times a club is selected}}{\text{total number of selections}}$$

Relative frequency was defined in Chapter 2 in the context of frequency distributions.

STEP 4 Suppose that after 10 tries, a club was selected only twice. The relative frequency is $2/10 = 0.2$. This is an estimate of P(C). It's quick and easy, but it doesn't seem very accurate.

After every selection, the observed card is placed back in the deck. The deck is shuffled, and another selection is made.

STEP 5 Suppose we try the experiment a few more times. With more observations, we should be able to make a better guess at P(C). The table shows values for N, the number of trials, and $\hat{p}$, the relative frequency of occurrence of a club.

N	10	50	100	200	300	400	500	600	700
$\hat{p}$	0.2	0.3	0.29	0.23	0.223	0.205	0.228	0.252	0.267

N	800	900	1000	1100	1200	1300	1400	1500
$\hat{p}$	0.245	0.243	0.227	0.254	0.260	0.256	0.261	0.249

For example, after 300 selections, the relative frequency of occurrence of the event C was 0.223.

STEP 6 Figure 4.9 shows a plot of relative frequency versus number of trials. The graph shows a remarkable pattern. As N increases, the points come noticeably closer to the dashed line. The relative frequencies seem to be homing in on one number (around 0.25); it seems reasonable to conclude that this relative frequency, whatever it is, should be the probability of the event C.

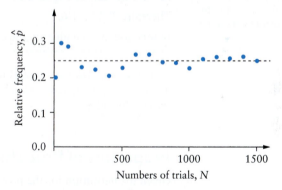

Figure 4.9 Scatter plot of relative frequency versus number of trials.

STEP 7 In the long run, the relative frequencies tend to stabilize, or even out, and become almost constant. They close in on one number, the limiting relative frequency. The probability of the event C is the limiting relative frequency.

If an experiment is conducted N times and an event occurs n times, then the probability of the event is approximately n/N (the relative frequency of occurrence). The **probability of an event A**, P(A), is the limiting relative frequency, the proportion of time that the event A will occur in the long run. This is a basic and sensible definition, a rule for assigning probability to an event. Given an event, all we need to do is find the limiting relative frequency.

Although this definition makes sense, and Example 4.10 and Figure 4.9 support and illustrate our intuition, it presents a real practical problem. We cannot conduct experiments over and over, compute relative frequencies, and only then estimate the true probability. How will we ever know the true limiting relative frequency? How large should N be? When are we close enough? Will we ever hit the limiting relative frequency exactly? The definition is nice, but it seems to offer little hope of ever finding the true probability of an event. Fortunately, there is another way to determine the exact probability in some cases. Consider the next two examples.

EXAMPLE 4.11 Call It in the Air

Suppose an experiment consists of tossing a fair coin and recording the side that lands face up. The event H is the coin landing with heads face up. Find $P(H)$.

Solution

There are only two possible outcomes on each flip of the coin, and both are equally likely to occur. In the long run, we expect heads to occur half of the time.

Therefore, $P(H) = 1/2$.

Without flipping the coin thousands of times, making estimates, or guessing at the limiting relative frequency, we are certain the probability is 1/2.

If we were to conduct this experiment over and over, the relative frequency of occurrence of H would close in on 1/2.

EXAMPLE 4.12 Roll the Die

An experiment consists of tossing a fair six-sided die and recording the number that lands face up. Consider the event $E = \{1\}$, rolling a one. Find $P(E)$.

Solution

There are six possible outcomes for each roll of the die, and all are equally likely to occur. In the long run, we expect a 1 to occur one-sixth of the time.

Therefore, $P(E) = 1/6$.

In this case, we can identify the exact limiting relative frequency.

The relative frequency of occurrence of a 1 would get closer and closer to 1/6 as the number of rolls gets larger and larger.

These two examples suggest that it is indeed possible to find the limiting relative frequency! They are special cases, however, because in each experiment, all of the outcomes are equally likely.

Properties of Probability

Given the definition for the probability of an event A, here are some reasonable properties of probability.

Properties of Probability

1. For any event A, $0 \leq P(A) \leq 1$.

    The probability of any event is a limiting relative frequency, and a relative frequency is a number between 0 and 1. An event with probability close to 0 is very unlikely to occur, and an event with probability close to 1 is very likely to occur.

2. For any event A, $P(A)$ is the sum of the probabilities of all the outcomes in A.

    To compute $P(A)$, just add up the probability of each outcome or simple event in A.

3. The sum of the probabilities of all possible outcomes in a sample space is 1: $P(S) = 1$.

    The sample space is an event. If an experiment is conducted, S is guaranteed to occur.

4. The probability of the empty set, or empty event, is 0: $P(\{\,\}) = P(\emptyset) = 0$. This event contains no outcomes.

The word chance is also used to express likelihood. A 10% chance means the probability is 0.10.

In the next example, the probability (limiting relative frequency) of each simple event is assumed to be known. We will use the properties just provided and some earlier definitions to develop some common tools and strategies for solving similar probability questions.

EXAMPLE 4.13   Bread by Mail

Zingerman's sells various kinds of bread online. The store uses minimal ingredients, bakes in a stone hearth oven, and offers traditional as well as specialty loaves. Suppose the store sells six types of bread and an experiment consists of classifying the next customer's single bread purchase. The probability of each simple event (bread purchase) is given in the table.

Maglido Photography/Shutterstock

	Rye	Sourdough	Raisin	Pumpernickel	Dinkelbrot	Chocolate
Simple event	1	2	3	4	5	6
Probability	0.08	0.12	0.10	0.25	0.15	0.30

Consider the following events.

$A = \{1, 2\}$
 = The next customer buys rye or sourdough bread.

$B = \{3, 4, 5\}$
 = The next customer buys raisin, pumpernickel, or dinkelbrot bread.

$C = \{6\}$
 = The next customer buys chocolate bread.

$D = \{1, 4\}$
 = The next customer buys rye or pumpernickel bread.

Find $P(A)$, $P(C)$, $P(B \cup D)$, $P(A \cap D)$, and $P(A \cap B)$.

Solution

STEP 1 $P(A) = P(1) + P(2)$ *Add the probabilities of each simple event in A.*
$= 0.08 + 0.12 = 0.20$

STEP 2 $P(C) = P(6) = 0.30$ *There is only one outcome in C.*

STEP 3 $P(B \cup D) = P(\{1, 3, 4, 5\})$ *Find the outcomes in the event $B \cup D$. Add the probabilities of each simple event.*
$= P(1) + P(3) + P(4) + P(5)$
$= 0.08 + 0.10 + 0.25 + 0.15 = 0.58$

STEP 4 $P(A \cap D) = P(1) = 0.08$ *The intersection is one simple event. Check the probability given in the table.*

STEP 5 $P(A \cap B) = P(\{\}) = 0$ *The intersection is empty, so the probability is 0.* ∎

TRY IT NOW   Go to Exercises 4.61 and 4.63

Equally Likely Outcomes

To find probabilities in the previous example, we looked at each event piece by piece; that is, we broke down each event into simple events. Let's apply the same properties in an **equally likely outcome experiment**.

Suppose an experiment has n equally like outcomes, $S = \{e_1, e_2, e_3, \ldots, e_n\}$. Each simple event has the same chance of occurring, so the probability of each is $1/n$; $P(e_i) = 1/n$. The limiting relative frequency of the simple event e_i is $1/n$. This is exactly what we found in Examples 4.11 and 4.12. Now, consider an event $A = \{e_1, e_2, e_3, e_4, e_5\}$. To find $P(A)$, add up the probabilities of each simple event in A.

Think about tossing a fair coin, or rolling a fair die, or randomly selecting a student in a class to answer a question.

$$P(A) = P(e_1) + P(e_2) + P(e_3) + P(e_4) + P(e_5)$$
$$= \frac{1}{n} + \frac{1}{n} + \frac{1}{n} + \frac{1}{n} + \frac{1}{n} = \frac{5}{n}$$
$$= \frac{\text{number of outcomes in } A}{\text{number of outcomes in the sample space } S} = \frac{N(A)}{N(S)}$$

Finding Probabilities in an Equally Likely Outcome Experiment

In an equally likely outcome experiment, the probability of any event A is the number of outcomes in A divided by the total number of outcomes in the sample space S. Finding the probability of any event, in this case, means counting the number of outcomes in A, counting the number of outcomes in the sample space S, and dividing.

$$P(A) = \frac{N(A)}{N(S)}$$

> You will not always see the phrase "equally likely outcomes" in these types of probability questions. We will identify some keywords, develop some problem-solving strategies, and work with familiar experiments that imply equally likely outcomes.

Section 4.3 presents some special counting rules to help us compute probabilities associated with common experiments and events. However, we can solve some of these problems already and can even use our results to make a statistical inference.

EXAMPLE 4.14 Bank Teller Jobs

The PNC Bank on Radio Road in Naples, Florida, has five tellers. Tellers 1 and 2 are trainees, whereas tellers 3, 4, and 5 are experienced. Tellers 2, 3, and 4 are female, and tellers 1 and 5 are male. At the end of the day, two tellers will be randomly selected and all of their transactions for the day will be audited.

(a) What is the probability that both trainees will be selected for the audit?

(b) What is the probability that one male and one female will be selected for the audit?

(c) What is the probability that two females will be selected for the audit?

Solution

Since the tellers are randomly selected, each outcome is equally likely. To find the probability of each event, count the number of outcomes in that event and divide by the total number of outcomes in the sample space.

The experiment consists of selecting two tellers at random. The outcomes consist of two tellers who can be represented by their numbers. Therefore, 12 represents the outcome that tellers 1 and 2 were selected. The order of selection does not matter. For example, 12 and 21 both represent the event that tellers 1 and 2 were selected. We can (a) list all possible outcomes systematically, (b) sketch a tree diagram, or (c) use combinations (to be presented in Section 4.3). There are 10 outcomes in the sample space.

$$S = \{12, 13, 14, 15, 23, 24, 25, 34, 35, 45\}$$

(a) Let A = both trainees are selected for the audit. Because the trainees are tellers 1 and 2, there is only one outcome in the event A: $A = \{12\}$.

$$P(A) = \frac{\text{number of outcomes in } A}{\text{number of outcomes in } S} = \frac{N(A)}{N(S)} = \frac{1}{10} = 0.10$$

(b) Let B = one male and one female teller are selected. Tellers 2, 3, and 4 are female, and tellers 1 and 5 are male. Examine the sample space carefully to list the outcomes in B.

$$B = \{12, 13, 14, 25, 35, 45\} \Rightarrow P(B) = \frac{N(B)}{N(S)} = \frac{6}{10} = 0.60$$

(c) Let $C =$ two females are selected. Tellers 2, 3, and 4 are female. Examine the sample space again, and pick out the matching outcomes.

$$C = \{23, 24, 34\} \Rightarrow P(C) = \frac{N(C)}{N(S)} = \frac{3}{10} = 0.30$$

The next example involves an equally likely outcome experiment and an inference question. We'll need to compute the likelihood of the observed event to help us draw a conclusion.

EXAMPLE 4.15 Better Breakfast Sandwich

A Panera Bread restaurant in Groton, Connecticut, sells an egg, cheese, and bacon breakfast sandwich on either a bagel (B) or an English muffin (E). The owner believes the demand for each kind of bread is the same, so the shop should continue to stock bagels and English muffins in equal numbers. Five customers are selected at random. Each customer buys only a breakfast sandwich and the bread choice for each is recorded.

(a) Find the probability that exactly one person buys a breakfast sandwich on a bagel.

(b) Suppose all five customers purchase a breakfast sandwich on a bagel. Is there any evidence to suggest that demand is weighted more toward one bread choice?

Solution

Since the demand for each bread choice is assumed to be the same and the customers are selected at random, all of the outcomes in this experiment are equally likely. Therefore, to compute probability we need to count the number of outcomes in the event and divide by the total number of outcomes in the sample space.

The experiment consists of selecting five customers at random and recording their breakfast sandwich bread choice. Each outcome is a sequence of five letters: Bs and/or Es. For example, the outcome BBEBE indicates that the first customer's bread choice is a bagel, the second customer's bread choice is a bagel, the third customer's bread choice is an English muffin, the fourth customer chooses a bagel, and the fifth customer chooses an English muffin. There are 32 possible outcomes; a systematic listing helps, and a tree diagram works (but is big). (The Multiplication Rule also works here; this very useful counting technique is presented in Section 4.3.) Here is the sample space:

$S =$ {BBBBB, BBBBE, BBBEB, BBBEE, BBEBB, BBEBE, BBEEB, BBEEE, BEBBB,
BEBBE, BEBEB, BEBEE, BEEBB, BEEBE, BEEEB, BEEEE, EBBBB, EBBBE, EBBEB,
EBBEE, EBEBB, EBEBE, EBEEB, EBEEE, EEBBB, EEBBE, EEBEB, EEBEE, EEEBB,
EEEBE, EEEEB, EEEEE}

(a) Let $A =$ exactly one person buys a breakfast sandwich with a bagel. Examine the sample space and carefully list all the outcomes in A.

$A =$ {BEEEE, EBEEE, EEBEE, EEEBE, EEEEB}

$$P(A) = \frac{N(A)}{N(S)} = \frac{5}{32} = 0.15625 \qquad \text{Equally likely outcomes.}$$

(b) The key phrase *Is there any evidence to suggest* signifies a statistical inference problem. It seems reasonable to determine the probability of the experimental, or observed, outcome that all five customers selected a bagel, and use this value to draw a conclusion about the claim.

The claim is that the demand for each bread choice is equal. If this is true, then all of the outcomes in the sample space S are equally likely.

The experiment consists of observing the bread choice for the next five customers. Let G = the observed outcome, everyone selected a bagel.

Find the likelihood of the event G occurring. There is only one outcome in G, so the probability of the event G is

$$P(G) = \frac{N(G)}{N(S)} = \frac{1}{32} = 0.03125 \qquad \text{Count and divide.}$$

The conclusion: Because this probability is so small, all five people selecting a bagel is a rare event. But it happened! Under the assumption of equal demand, this is an extraordinary occurrence. It shouldn't happen (often), but it did. This suggests the restaurant owner's assumption is wrong; there is evidence to suggest that the demand for each bread type is not equal.

Note: There is really evidence to suggest that *some* assumption is wrong. It could be, for example, that the five customers were not selected at random. To draw a conclusion about the demand for these two bread types, we must accept all other assumptions are true. ■

TRY IT NOW   Go to Exercises 4.59 and 4.73

CHAPTER APP   **Dream Home**

Suppose you submit 50 entries and there are approximately 130 million entries in this sweepstakes. Find the probability of winning the Dream Home.

The Complement Rule

Consider an experiment, two events A and B, and known probabilities $P(A)$ and $P(B)$. Suppose we use A and B to create a new event using complement, union, or intersection. Sometimes we can use the known probabilities $P(A)$ and $P(B)$ to calculate the probability of the new event quickly. We may not have to break down the new event into simple events, or even count all the outcomes in the new event (if it is an equally likely outcome experiment). The **Complement Rule** and the **Addition Rule for two events** are two rules that help us perform probability calculations.

The Complement Rule
For any event A, $P(A) = 1 - P(A')$.

 Complement of a set.

$(A')'$ is read as "A complement, complement." What is $(A')'$? All outcomes not in A', which is A!

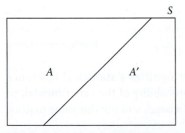

Figure 4.10 Venn diagram for visualizing the Complement Rule.

A CLOSER LOOK

1. The Complement Rule is easy to visualize and justify by looking at a Venn diagram. **Figure 4.10** shows an event A and its complement A'.

   Remember, the area of a region represents probability. So $P(A) + P(A') = P(S) = 1$, which can be written as $P(A) = 1 - P(A')$ or $P(A') = 1 - P(A)$.

2. The Complement Rule is incredibly handy; it is used in various contexts throughout probability and statistics. The problem is, how do you know when to use it? Look for keywords such as *not*, *at least*, or *at most*. A rule of thumb: If you face a very long probability calculation involving many simple events, or one that may require lots of counting, try looking at the complement.

EXAMPLE 4.16 Law and Order

Three public defenders are assigned to cases randomly. An experiment consists of recording the lawyer (by number) assigned to the next three cases. For example, the outcome 132 means lawyer 1 was assigned case 1, lawyer 3 was assigned case 2, and lawyer 2 was assigned case 3.

(a) Find the probability that all three cases are assigned to different lawyers.
(b) Find the probability that lawyer 2 is not assigned to any of the three cases.
(c) Find the probability that lawyer 2 is assigned to at least one case.

Solution

There are 27 possible outcomes; a tree diagram helps to identify all 27.

Here's another way to establish that this sample space has 27 outcomes.

Each case can be assigned to one of three lawyers.

Number of possible assignments for each case

$$\underset{\text{Case 1}}{3} \times \underset{\text{Case 2}}{3} \times \underset{\text{Case 3}}{3} = 27$$

Here's the sample space.

$S = \{111, 112, 113, 121, 122, 123, 131, 132, 133,$
$\quad\quad 211, 212, 213, 221, 222, 223, 231, 232, 233,$
$\quad\quad 311, 312, 313, 321, 322, 323, 331, 332, 333\}$

(a) Let A = all three cases are assigned to different lawyers. Find all the outcomes in S with a 1, a 2, and a 3.

$A = \{123, 132, 213, 231, 312, 321\}$

$$P(A) = \frac{N(A)}{N(S)} = \frac{6}{27} = 0.2222 \quad\quad \text{Equally likely outcomes.}$$

(b) Let B = lawyer 2 is not assigned to any of the three cases. Find all the outcomes without a 2.

$B = \{111, 113, 131, 133, 311, 313, 331, 333\}$

$$P(B) = \frac{N(B)}{N(S)} = \frac{8}{27} = 0.2963$$

(c) The keywords *at least one* suggest we might be able to use the Complement Rule. Consider the complement C', the event that lawyer 2 has no cases. Count the number of outcomes that have no 2s and use the Complement Rule.

Let C be the event that lawyer 2 has at least one case. The outcomes in C include those with one 2, two 2s, and three 2s. That seems like a lot of counting. This is a good opportunity to use the Complement Rule.

$P(C) = 1 - P(C')$ \quad\quad Complement Rule.
$\quad\quad = 1 - P(\text{lawyer 2 is assigned 0 cases})$ \quad\quad Interpretation of C'.
$\quad\quad = 1 - P(B)$ \quad\quad $C' = B$ in this example.
$\quad\quad = 1 - \frac{8}{27} = 1 - 0.2963 = 0.7037$

TRY IT NOW Go to Exercises 4.55 and 4.65

CHAPTER APP Dream Home

Suppose 10 million entries are from residents of Nebraska. What is the probability that the winner is not from Nebraska?

The Addition Rule

Often we can find the probability of the union of two events by using known probabilities, rather than resorting to more counting.

The Addition Rule for Two Events

1. For any two events A and B, $P(A \cup B) = P(A) + P(B) - P(A \cap B)$.
2. For any two *disjoint* events A and B, $P(A \cup B) = P(A) + P(B)$.

Disjoint events.

A CLOSER LOOK

1. **Figure 4.11** helps illustrate and justify this rule.

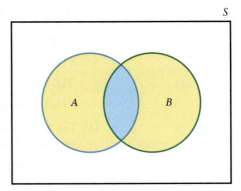

Figure 4.11 Venn diagram for illustrating the Addition Rule.

To find the probability of the union, start by adding $P(A) + P(B)$. This sum includes the region of intersection, $P(A \cap B)$, twice. Adjust this total by subtracting the intersection area, once.

2. $A \cup B = B \cup A$; order doesn't matter here. So, $P(A \cup B) = P(B \cup A)$.
3. The first, more general, formula always works.
   If A and B are disjoint, then $P(A \cap B) = 0$.
4. ▶ The Addition Rule can be extended to more than two events.

   For any three events A, B, and C:
   $$P(A \cup B \cup C) = P(A) + P(B) + P(C)$$
   $$- P(A \cap B) - P(A \cap C) - P(B \cap C)$$
   $$+ P(A \cap B \cap C)$$

   You can also visualize and derive this formula by using a Venn diagram. In this case, the sum $P(A) + P(B) + P(C)$ includes the double intersections twice and the triple intersection three times. We therefore need to adjust the total accordingly. ◀

5. Let $A_1, A_2, A_3, \ldots, A_k$ be a collection of k *disjoint* events.

   $$P(A_1 \cup A_2 \cup \cdots \cup A_k) = P(A_1) + P(A_2) + \cdots + P(A_k)$$

   If the events are disjoint, to find the probability of a union, just add up the corresponding probabilities. This approach is especially useful when questions ask about

4.2 An Introduction to Probability

Common Error
$P(A+B)$
Correction: A and B are events, or sets. You cannot add two events.
$P(A) \cup P(B)$
Correction: $P(A)$ and $P(B)$ are numbers. You cannot take the union of two numbers.

the number of individuals or objects with a specific attribute. For example, suppose 10 people are asked whether they received a flu shot this winter. The probability of at least 3 of those people having received a flu shot is the probability of 0, plus the probability of 1, plus the probability of 2, plus the probability of 3.

6. Complement, union, and intersection are operations applied to events. It doesn't make sense to take the union of probabilities (which are numbers). Similarly, addition and subtraction are operations on real numbers. You shouldn't try to add or subtract events.

EXAMPLE 4.17 Super Bowl Ads

The percentages in Example 4.17 represent likelihood of occurrence. Divide each by 100 to convert to a probability.

The 2018 Super Bowl ads included Morgan Freeman lip syncing and Eli Manning and O'Dell Beckham Jr. dancing. According to *Variety*,[1] two of the best commercials were for Amazon Alexa and Tide laundry detergent. A survey indicated that 70% of all people watching the Super Bowl saw the ad for Alexa, 35% saw the ad for Tide, and 25% saw both. Suppose a person who watched the Super Bowl is selected at random.

(a) Draw a Venn diagram to illustrate the events in this problem.
(b) What is the probability that the person saw at least one of these two ads?
(c) What is the probability that the person saw neither ad?
(d) What is the probability that the person saw just the ad for Alexa?
(e) What is the probability that the person saw just one of these two ads?

Solution

Write the probability of each event, construct a Venn diagram to visualize this experiment, and use the probability rules.

(a) Define the events given in this problem.

Let A = the person saw the Alexa ad; $P(A) = 0.70$.

Let T = the person saw the Tide ad; $P(T) = 0.35$.

Saw both means saw the ads for Alexa *and* Tide, which means intersection. Therefore, $P(A \cap T) = 0.25$

Note that these three probabilities add up to more than 1. That's OK because the events A and T intersect.

Remember that area of a region corresponds to probability. To construct a Venn diagram, start on the inside and work your way toward the outside.

(i) The shaded area represents the probability that the person saw both the Alexa and Tide ads, $P(A \cap T)$. We know that $P(A \cap T) = 0.25$. Because $P(A) = 0.70$, the remaining area representing A corresponds to $0.70 - 0.25 = 0.45$.

(ii) Similarly, because $P(A \cap T) = 0.25$ and $P(T) = 0.35$, the remaining area representing T corresponds to $0.35 - 0.25 = 0.10$.

(iii) The total probability in the entire sample space must sum to 1; the remaining probability is $1 - (0.45 + 0.25 + 0.10) = 0.20$.

Figure 4.12 is the Venn diagram that corresponds to this experiment.

(b) The probability of seeing at least one ad means seeing the ad for Alexa, or the ad for Tide, or both. That's a union of two events.

$$P(A \cup T) = P(A) + P(T) - P(A \cap T)$$ Addition Rule for two events.
$$= 0.70 + 0.35 - 0.25 = 0.80$$ Use the known probabilities.

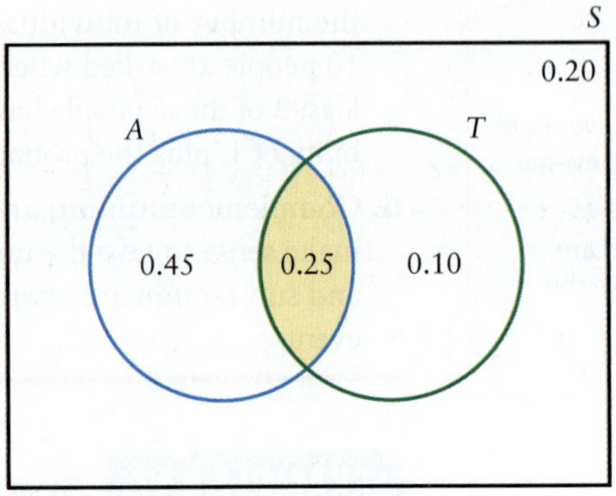

Figure 4.12 Venn diagram for Example 4.17.

The Venn diagram supports this answer. Look at the region that represents $P(A \cup T)$, and add up the corresponding probabilities.

(c) *Saw neither* means did not see either the Alexa ad or the Tide ad. Because the event $A \cup T$ means saw the Alexa ad or the Tide ad, *neither* suggests the complement of $A \cup T$.

$$P[(A \cup T)'] = 1 - P(A \cup T)$$ Complement Rule applied to the event $A \cup T$.
$$= 1 - 0.80 = 0.20$$ Use the previous answer.

(d) *Saw just the ad for Alexa* means saw the Alexa ad but not both ads. This is not simply $P(A)$, because this probability includes more than just the Alexa ad. Start with the probability the person saw the Alexa ad, and subtract the probability of seeing both.

$$P(\text{just Alexa}) = P(A) - P(A \cap T)$$ Use the Venn diagram to write a probability expression.
$$= 0.70 - 0.25 = 0.45$$ Use the known probabilities.

(e) *Saw just one* of these two ads means saw the Alexa ad or the Tide ad, but not both. Start with the union and then subtract the intersection.

$$P(\text{Exactly one}) = P(A \cup T) - P(A \cap T)$$ Use the Venn diagram to write a probability expression.
$$= 0.80 - 0.25 = 0.55$$ Use the known probabilities.

TRY IT NOW Go to Exercises 4.69 and 4.71

EXAMPLE 4.18 Trail Blazing

Suppose the Franconia Notch State Park in Grafton County, New Hampshire, has six trails for hiking. Observation reports from park officials were used to compile the following table, showing the probability of visitors hiking each trail. Assume a person can hike only one trial in a day.

Trail	T_1	T_2	T_3	T_4	T_5	T_6
Probability	0.10	0.25	0.20	0.30	0.10	0.05

Consider the following events.

$A = \{T_1, T_2\}$ (easy trails)
$B = \{T_2, T_3, T_6\}$ (waterfall trails)
$C = \{T_4, T_5\}$ (long trails)
$D = \{T_6\}$ (dog-friendly trial)

Suppose a hiker is selected at random. Find the probability he
(a) hiked an easy trail or one with a waterfall.
(b) hiked an easy trail or a long trail.
(c) hiked an easy trail, or a long trail, or a dog-friendly trail.

Solution

Find the probability of A, B, C, and D. Break down each event and consider the individual outcomes, or simple events.

$P(A) = P(T_1) + P(T_2) = 0.10 + 0.25 = 0.35$
$P(B) = P(T_2) + P(T_3) + P(T_6) = 0.25 + 0.20 + 0.05 = 0.50$
$P(C) = P(T_4) + P(T_5) = 0.30 + 0.10 = 0.40$
$P(D) = P(T_6) = 0.05$

(a) *Or* means union. Find the corresponding events and translate everything into a probability expression.

$$P(A \cup B) = P(A) + P(B) - P(A \cap B) \quad \text{(General) Addition Rule.}$$
$$= P(A) + P(B) - P(T_2) \quad \text{Find the outcomes in } A \cap B.$$
$$= 0.35 + 0.50 - 0.25 = 0.60 \quad \text{Use known probabilities.}$$

(b) Part (b) is the same kind of a question; *or* means union.

$$P(A \cup C) = P(A) + P(C) \quad \text{A and C are disjoint.}$$
$$= 0.35 + 0.40 = 0.75 \quad \text{Use known probabilities.}$$

(c) *Or* means union again in part (c), but now with three events.

$$P(A \cup C \cup D) = P(A) + P(C) + P(D) \quad \text{Three disjoint events.}$$
$$= 0.35 + 0.40 + 0.05 = 0.80 \quad \text{Use known probabilities.}$$

TRY IT NOW Go to Exercise 4.67

Section 4.2 Exercises

Concept Check

4.41 True or False The probability of any event is always a number between 0 and 1, inclusive.

4.42 Fill in the Blank
(a) If the probability of an event is close to 1, then the event is _____.
(b) If the probability of an event is close to 0, then the event is _____.
(c) Although it could be difficult to find, the intuitive definition of the probability of an event is _____.

4.43 True or False There is no way to determine the sum of the probabilities of all possible outcomes in a sample space.

4.44 True or False In an equally likely outcome experiment, the probability of an event A is the number of outcomes in A.

4.45 True or False For any event A, $P(A) + P(A') = 1$.

4.46 Fill in the Blank For any two events A and B, $P(A \cup B) = P(A) + P(B) - $ _____.

4.47 True or False If an outcome in an equally likely outcome experiment is rare, then the probability of the entire sample space can be less than 1, $P(S) < 1$.

4.48 True or False If there are more than 30 outcomes in an experiment, then we can consider all of the outcomes in the sample space to be equally likely.

4.49 Short Answer Consider the following expression: $P(A \cap B) = P(A \cup B)$. Explain whether this is always true, sometimes true, or never true. Justify your answer.

Practice

4.50 Consider an experiment with the probability of each simple event given in the following table.

Simple event	e_1	e_2	e_3	e_4
Probability	0.07	0.09	0.13	0.18

Simple event	e_5	e_6	e_7
Probability	0.22	0.15	0.16

The events A, B, C, and D are defined by
$A = \{e_1, e_2, e_3\}$ $B = \{e_2, e_4, e_6, e_7\}$
$C = \{e_1, e_5, e_7\}$ $D = \{e_3, e_4, e_5, e_6, e_7\}$

Find the following probabilities.
(a) $P(A)$
(b) $P(C)$
(c) $P(D)$
(d) $P(A \cup B)$
(e) $P(A \cap C)$
(f) $P(B \cap D)$
(g) $P(A')$
(h) $P(A \cap C')$
(i) $P(A' \cap D)$
(j) $P(C')$
(k) $P(B \cap C \cap D)$
(l) $P[(B \cup C)']$

How do you know there is no other possible simple event in this experiment?

4.51 An experiment consists of rolling a special 18-sided die and recording the number face up. All of the numbers, 1 through 18, are equally likely. Find the probability of each event.
(a) A = rolling an even number
(b) B = rolling a number divisible by 3
(c) C = rolling a number less than 7
(d) D = rolling a number at least 10

4.52 An experiment consists of rolling a special 22-sided die and recording the number face up. All of the numbers, 1 through 22, are equally likely. Find the probability of each event.
(a) A = rolling a number greater than 10 and even
(b) B = rolling a prime number or a number divisible by 5
(c) C = rolling at most an 11
(d) D = rolling a number divisible by 2 and 3

4.53 Consider an experiment, the events A and B, and probabilities $P(A) = 0.55$, $P(B) = 0.45$, and $P(A \cap B) = 0.15$. Find the probability of each event.
(a) A or B occurring
(b) A and B occurring
(c) Just A occurring
(d) Just A or just B occurring

4.54 Consider an experiment, the events A and B, and probabilities $P(A) = 0.26$, $P(B) = 0.68$, and $P(A \cup B) = 0.80$. Find each probability.
(a) $P(A \cap B)$
(b) $P(A')$
(c) $P[(A \cap B)']$
(d) $P[(A \cup B)']$

4.55 Consider an experiment, the events A and B, and probabilities $P(A) = 0.355$, $P(B) = 0.406$, and $P(A \cap B) = 0.229$. Find each probability.
(a) $P(A \cup B)$
(b) $P[(A \cup B)']$
(c) $P(B')$
(d) $P[(A \cap B)']$

4.56 Carefully sketch a Venn diagram showing the relationship between two events. Add probabilities to the appropriate regions so that the following statements are true: $P(A \cap B) = 0.31$, $P(A) = 0.57$, and $P(B) = 0.48$.

4.57 Carefully sketch a Venn diagram showing the relationships among three events. Add probabilities to the appropriate regions so that the following statements are true.
$P(A) = 0.46$ $P(B) = 0.35$ $P(C) = 0.44$
$P(A \cap B) = 0.05$ $P(A \cap C) = 0.18$
$P(B \cap C) = 0.14$ $P(A \cap B \cap C) = 0.03$

4.58 An experiment consists of rolling a special 10-sided die and recording the number face up. All of the numbers, 1 through 10, are equally likely. Consider the event $A = \{1, 2, 3, 4\}$.
(a) Use technology to simulate this experiment for $n = 100, 200, 500, 1000, 1500, 2000, \ldots, 10,000$ times. Compute the relative frequency of occurrence of A at each value of n.
(b) Sketch a scatter plot of the relative frequency of occurrence of A versus n.
(c) Explain the pattern in this scatter plot.
(d) Find $P(A)$. Justify your answer.

Applications

4.59 Manufacturing and Product Development Valassis, a marketing services company, offers a cafeteria-style benefit program; an employee may select three benefits from five choices. The five possible benefits are health insurance, life insurance, a prescription plan, dental insurance, and vision insurance.
(a) How many different benefit packages can an employee select? List them.
(b) If all benefit packages are equally likely, what is the probability that an employee selects a package that includes health insurance?
(c) If all benefit packages are equally likely, what is the probability that an employee selects a package that includes life insurance and a prescription plan?

4.60 Psychology and Human Behavior Suppose a person preparing for work has 10 basic blue ties to choose from: Four are silk, two are wool, one is linen, and the rest are cotton. During the morning rush to work, the person often reaches for one of the blue ties without regard to type. Therefore, suppose the person selects a tie at random.
(a) What is the probability that a wool tie is selected?
(b) What is the probability that a cotton tie or the linen tie is selected?
(c) What is the probability that a silk tie is not selected?
(d) What is the probability that a linen tie is selected?

4.61 Economics and Finance In 2018, the federal income tax brackets were adjusted for inflation and the top marginal income tax decreased slightly to 37%. The table lists the tax bracket and the approximate proportion of American households in each bracket.

Tax bracket	Proportion
0%	0.2673
10%	0.1987
12%	0.3060
22%	0.1741
24%	0.0334
32%	0.0127
35%	0.0013
37%	0.0065

Suppose an American household is selected at random.
(a) What is the probability that the household is in the 0%, 10%, 12%, or 22% tax bracket?
(b) What is the probability that the household is in the 35% or 37% tax bracket?
(c) What is the probability that the household is not in the 24% tax bracket?

4.62 Marketing and Consumer Behavior Delorenzo's Pizza offers five different toppings on its pizzas: pepperoni, sausage, olives, mushrooms, and anchovies. A large pizza comes with any two different toppings.
(a) How many different two-topping pizzas are possible?
(b) Suppose that all of the pizzas are equally likely. What is the probability that the next pizza ordered has at least one meat topping?
(c) What is the probability that the next pizza ordered does not have anchovies?
(d) Suppose one more large pizza choice is added: plain cheese with no toppings. Answer parts (b) and (c) with this added assumption.

4.63 Demographics and Population Statistics The table lists the proportion of Canadians who live in each province and territory.[2]

Geographic name	Proportion
Newfoundland and Labrador	0.0148
Prince Edward Island	0.0041
Nova Scotia	0.0263
New Brunswick	0.0213
Quebec	0.2323
Ontario	0.3826
Manitoba	0.0364
Saskatchewan	0.0312
Alberta	0.1156
British Columbia	0.1322
Yukon	0.0010
Northwest Territories	0.0012
Nunavut	0.0010

Suppose a Canadian citizen is selected at random.
(a) What is the probability that the person lives in Nova Scotia or New Brunswick?
(b) What is the probability that the person does not live in British Columbia or Yukon?
(c) What is the probability that the person does not live in one of the three most highly populated provinces or territories?
(d) If the total population in Canada is 35,151,728 people, how many people live in Nunavut?

4.64 Psychology and Human Behavior Thieves use many different methods to obtain personal information. The table lists the delivery method and the number of reports of identity theft scams in Australia for 2017.[3]

Delivery method	Number of reports
Phone	31,299
Email	14,194
Internet	2257
In person	258
Social networking	407
Mobile applications	178
Text message	6965
Mail	374

Suppose an Australian identity theft scam report is selected at random.
(a) What is the probability that the delivery method is email?
(b) What is the probability that the delivery method is the Internet or social networking?
(c) What is the probability that the delivery method is not a text message?

4.65 Marketing and Consumer Behavior A marketing firm can place an advertisement using several forms of media. The table shows the probability that a randomly selected person in a targeted region will see the advertisement in the given medium.

Medium	Newspaper	Radio	Magazine	TV
Probability	0.15	0.10	0.08	0.30
Medium	Internet	Billboard	Not seen	
Probability	0.12	0.05	0.20	

Consider the following events.
$A = \{$magazine, newspaper$\}$
$B = \{$TV, radio, Internet$\}$
$C = \{$magazine, newspaper, Internet, billboard$\}$

Find the following probabilities.
(a) $P(A)$, $P(B)$, $P(C)$
(b) $P(A \cup B)$, $P(A \cap B)$, $P(B \cap C)$
(c) $P(A')$, $P(A' \cap C)$, $P(A \cap B \cap C)$
(d) $P(B' \cap C')$, $P[(B \cup C)']$

4.66 Sports and Leisure The Pennsylvania Pick 3 day lottery number consists of three digits, each 0–9.
(a) How many possible Pick 3 day numbers are there?
(b) If all of the Pick 3 day numbers are equally likely, find the probability that all three digits are the same.
(c) If all of the Pick 3 day numbers are equally likely, find the probability that all three digits are either 8s or 9s.

(d) There is also an evening number, consisting of three digits, 0–9. If all of the day and evening numbers are equally likely, what is the probability that the two numbers are the same?

4.67 Marketing and Consumer Behavior The table lists the most popular convention centers in the United States.[4] Suppose the probability given represents the likelihood that a randomly selected U.S. convention will be held at that site.

Site	Probability
Chicago	0.288
Las Vegas	0.225
Washington, D.C.	0.105
Orlando	0.075
Atlanta	0.064
Dallas	0.055
San Francisco	0.048
Other	0.140

Suppose a convention is randomly selected. Consider the events
$A = \{$Convention is in Chicago or Las Vegas$\}$
$B = \{$Convention is not in Orlando$\}$
$C = \{$Convention is in Washington, D.C.$\}$

Find the following probabilities.
- **(a)** $P(A)$, $P(B)$, $P(C)$
- **(b)** $P(A \cap B)$, $P(A \cup C)$, $P(A \cap C)$
- **(c)** $P(A' \cup C)$, $P(A \cup B \cup C')$

4.68 Economics and Finance According to a recent Gallup poll, 52% of Americans own stocks and 63% of Americans own real estate.[5] Suppose 27% of Americans own both (that is, own stock and real estate) and suppose an American is selected at random.
- **(a)** What is the probability that the American owns either stock or real estate?
- **(b)** What is the probability that the American does not own either stock or real estate?
- **(c)** What is the probability that the American owns only stock? Only real estate?

4.69 Business and Management The Occupational Safety and Health Administration (OSHA) routinely inspects worksites in an effort to protect workers from injuries on the job. For example, OSHA inspectors may examine ladders and electrical wiring on a construction site.[6] Suppose 23% of all worksites are cited for fall protection violations, 14% for hazard communication standards violations, and 5% for both violations, and suppose a random worksite is inspected.
- **(a)** What is the probability that the worksite is cited for violating at least one of these standards?
- **(b)** What is the probability that the worksite is not cited for violating either standard?
- **(c)** What is the probability that the worksite is cited for violating just the fall protection standard?

4.70 Travel and Transportation Tourists traveling to Edmonton, Alberta, often visit popular points of interest and landmarks. For example, Sir Winston Churchill Square, the Government House Alberta, and La Cite Francophone are common tourist stops. Suppose 35% of all Edmonton tourists visit Whyte Avenue, 50% visit the Ice Castles, and 65% visit at least one of these points of interest. An Edmonton tourist is selected at random.
- **(a)** What is the probability that the tourist visits both points of interest?
- **(b)** What is the probability that the tourist visits neither point of interest?
- **(c)** What is the probability that the tourist visits just the Ice Castles?

Extended Applications

4.71 Marketing and Consumer Behavior Of all those people who enter Uncle's Stereo, a discount electronics store in New York City, 28% purchase a digital camera, 5% buy a home theater receiver, and 4% buy both. Suppose a customer is selected at random.
- **(a)** Sketch a Venn diagram with probabilities to illustrate the relationship between the events $D = $ purchase a digital camera and $H = $ purchase a home theater receiver.
- **(b)** What is the probability that the customer buys a digital camera or a home theater receiver?
- **(c)** What is the probability that the customer buys either a digital camera or a home theater receiver, but not both?
- **(d)** What is the probability that the customer buys only a digital camera?
- **(e)** What is the probability that the customer does not buy a home theater receiver?

4.72 Demographics and Population Statistics The following table shows the ABO and Rh blood-type probabilities for people in the United States.[7] (This table is called a *joint probability table*. Each number in the table can be thought of as the probability of an intersection; for example, the probability of blood type A and negative Rh is 0.063.)

		ABO type			
		O	A	B	AB
Rh type	Positive	0.374	0.357	0.085	0.034
	Negative	0.066	0.063	0.015	0.006

Suppose a U.S. resident is selected at random. Find the following probabilities.
- **(a)** The person has Rh-positive blood.
- **(b)** The person has type B blood.
- **(c)** The person does not have type O blood.
- **(d)** The person has type AB or Rh-negative blood.

4.73 Manufacturing and Product Development A tire manufacturer has started a program to monitor production. In every batch of eight tires, two will be randomly selected and tested for defects electronically. An experiment consists of recording the condition of these two tires: defect-free (G) or reject (B). Suppose two of the eight tires in a batch actually have serious defects.
- **(a)** List the outcomes in this experiment.
- **(b)** What is the probability that both tires selected will be defect-free?
- **(c)** What is the probability that at least one of the tires selected will have a defect?
- **(d)** What is the probability that both tires selected will have a defect?

4.74 Medicine and Clinical Studies The number of ER visits has increased over the past several years in the United States. One reason may be the increased cost of health insurance and especially deductibles. People tend to postpone routine health care, which results in more visits to the ER. Of all patients who visit an ER, suppose 32% are seen in less than 15 minutes, 8% are admitted to the hospital, and 5% are seen in less than 15 minutes and admitted to the hospital.[8] Suppose a patient who made an ER visit is selected at random.
(a) Sketch a Venn diagram showing the relationship between the events seen in less than 15 minutes and admitted to the hospital, and add probabilities to the appropriate regions.
(b) What is the probability that the patient was seen in less than 15 minutes or admitted to the hospital?
(c) What is the probability that the patient was seen in less than 15 minutes but not admitted to the hospital?
(d) What is the probability that the patient was neither seen in less than 15 minutes nor admitted to the hospital?

4.75 Psychology and Human Behavior According to a recent study, five of the top 10 states people are moving out of are in the northeast.[9] Suppose a survey was conducted in which people moving out of northeast states were asked to list the reasons for leaving. Here is a summary of the responses:
33% were moving due to the cost of living.
31% were moving due to the weather.
45% were moving due to the job market.
5% were moving due to the cost of living and the weather.
3% were moving due to the weather and the job market.
13% were moving due to the cost of living and the job market.
2% were moving due to all three factors.

Suppose a person leaving the northeast is selected at random.
(a) Sketch a Venn diagram showing the relationships among these three events: moving due to the cost of living (C), moving due to weather (W), and moving due to the job market (J).
(b) What is the probability that the person is leaving the northeast due to the cost of living or the weather?
(c) What is the probability the person is leaving the northeast due to the weather or the job market?
(d) What is the probability that the person is leaving the northeast due to the cost of living, the weather, or the job market?
(e) What is the probability that the person is leaving the northeast for some other reason?

4.76 Marketing and Consumer Behavior The Westin Kansas City at Crown Center hotel recently conducted a guest survey. One question asked guests to list the reasons for choosing the Westin while staying in Kansas City. The results are summarized in the Venn diagram showing the relationship among four responses: reviews (R); age of guests (A); location (L); and value (V).

Suppose a Westin guest is randomly selected. Find the probability the guest is staying at the Westin
(a) due to value.
(b) due to location.
(c) due to age or value.
(d) for all four reasons.
(e) for some other reason.
(f) just for location.
(g) due to just reviews and age.

Challenge Problems

4.77 Reconsider Example 4.15. Suppose the restaurant owner records the type of breakfast sandwich for the next 10 customers. Find the probability that everyone buys a breakfast sandwich on a bagel. What do you think about the assumption of equal demand now? Justify your answer.

4.78 Five Events Sketch a Venn diagram that shows all possible intersections of five events. How many regions are represented in this diagram?

4.3 Counting Techniques

The Multiplication Rule

In an equally likely outcome experiment, computing probabilities means counting. To find the probability of an event A, count the number of outcomes in the event A and divide by the number of outcomes in the entire sample space S: $P(A) = N(A)/N(S)$. If $N(S)$ is large, drawing a tree diagram or listing all of the possible outcomes is impractical. For certain experiments, the following rules may be used instead to count outcomes in an event and/or a sample space.

The Multiplication Rule

Suppose an outcome in an experiment consists of an ordered list of k items selected using the following procedure:

1. There are n_1 choices for the first item.
2. There are n_2 choices for the second item, no matter which first item was selected.
3. The process continues until there are n_k choices for the kth item, regardless of the previous items selected.

There are $N(S) = n_1 \cdot n_2 \cdot n_3 \cdot \ldots \cdot n_k$ outcomes in the sample space S.

A CLOSER LOOK

1. You can picture (and even prove) this rule by drawing a tree diagram and counting the number of paths from left to right.
2. To use this rule, think of each choice as a slot, or a position, to fill.

$$\underbrace{n_1}_{\text{Item 1}} \times \underbrace{n_2}_{\text{Item 2}} \times \cdots \times \underbrace{n_k}_{\text{Item }k} = n_1 \cdot n_2 \cdot \ldots \cdot n_k$$

Number of choices for each slot

3. This counting technique can also be used for events, not just for sample spaces.

EXAMPLE 4.19 Louisville Slugger

A Little League Louisville Slugger bat can be customized on the barrel, handle, grip, and the end cap. There are 14 barrel colors, 16 handle colors, 20 grip designs, and 5 end cap styles. How many possible Louisville sluggers can be manufactured?

Solution

STEP 1 This is a counting problem, and there are four slots to fill: barrel, handle, grip, and end cap. We'll assume that all choices are compatible, and that the choice of any one item does not depend on any other item.

STEP 2 Here's how to apply the Multiplication Rule.

$$\underbrace{14}_{\text{Barrel}} \times \underbrace{16}_{\text{Handle}} \times \underbrace{20}_{\text{Grip}} \times \underbrace{5}_{\text{End cap}} = 22{,}400$$

There are 22,400 possible Louisville Sluggers.

EXAMPLE 4.20 Specialty Plates

Texas drivers can purchase a special Adopt-a-Beach license plate so that a portion of the fee goes to beach cleanup efforts. This license plate consists of two letters, two numbers, and a letter.

(a) How many different Adopt-a-Beach license plates are possible?
(b) How many Adopt-a-Beach plates begin with BB?

Solution

(a) This is a counting problem. There are five slots to fill: two letters, then two numbers, and one letter. There are 26 possible letters for the first, second, and fifth slots, and 10 possible numbers for the third and fourth slots. Use the Multiplication Rule.

$$\underbrace{26}_{\text{Letter}} \times \underbrace{26}_{\text{Letter}} \times \underbrace{10}_{\text{Number}} \times \underbrace{10}_{\text{Number}} \times \underbrace{26}_{\text{Letter}} = 1{,}757{,}600$$

There are 1,757,600 possible different Adopt-a-Beach license plates.

(b) If the license plate begins with BB, then each of the first two letter positions is fixed; there is only one choice. We are still free to choose any number in the next two positions, and any letter in the last position. The Multiplication Rule still works.

$$\underbrace{1}_{\text{Letter}} \times \underbrace{1}_{\text{Letter}} \times \underbrace{10}_{\text{Number}} \times \underbrace{10}_{\text{Number}} \times \underbrace{26}_{\text{Letter}} = 2600$$

There are 2600 Adopt-a-Beach license plates that begin with BB.

EXAMPLE 4.21   Five-of-a-Kind

In the game of Yahtzee, five fair dice are rolled and the numbers that land face up are recorded.

(a) How many different rolls are possible?

(b) What is the probability of rolling a Yahtzee (all five dice show the same number face up)?

Solution

Keeweeboy/Dreamstime.com

(a) There are five slots to fill, one for each die. Use the Multiplication Rule.

$$\underbrace{6}_{\text{Die 1}} \times \underbrace{6}_{\text{Die 2}} \times \underbrace{6}_{\text{Die 3}} \times \underbrace{6}_{\text{Die 4}} \times \underbrace{6}_{\text{Die 5}} = 7776$$

There are 7776 possible rolls, or outcomes, in the sample space.

(b) There are only six possible Yahtzees: 11111, 22222, 33333, 44444, 55555, and 66666. Because all the outcomes are equally likely (fair dice), the probability of rolling a Yahtzee is

$$P(\text{Yahtzee}) = \frac{\text{number of Yahtzees}}{\text{number of different rolls}} = \frac{6}{7776} = 0.0007716$$

EXAMPLE 4.22   Win, Place, or Show

Suppose there are 12 entries in the Preakness Stakes horse race. An experiment consists of recording the finish: the first-, second-, and third-place horses. For example, the outcome (7, 9, 2) means horse 7 came in first, horse 9 came in second, and horse 2 came in third.

(a) How many different finishes are possible?

(b) What is the probability of a finish with horse 4 or 5 in first place?

(c) What is the probability that horse 7 will not finish first, second, or third?

Solution

(a) There are three positions to fill, but the number of choices in the second slot depends on the first choice and the number of choices in the third slot depends on the first two choices. Even though we are drawing from the same reduced collection, we can still use the Multiplication Rule.

There are 12 horses that could finish first. Once a first-place horse is selected, there are only 11 left that could come in second. After the first- and second-place horses are selected, there are only 10 possible horses for third place. The Multiplication Rule is used here to count the number of finishes.

$$\underbrace{12}_{\text{First}} \times \underbrace{11}_{\text{Second}} \times \underbrace{10}_{\text{Third}} = 1320$$

There are 1320 possible different finishes. The number of outcomes in the sample space associated with this experiment is $N(S) = 1320$.

(b) Let A be the event that horse 4 or 5 wins the race. We'll assume that all of the outcomes are equally likely so that $P(A) = N(A)/N(S) = N(A)/1320$.

There are two choices for first place (horse 4 or 5). There are then 11 choices for second place (the horse not selected for first, plus the remaining 10), and 10 choices for third place.

The Multiplication Rule is used to find the number of outcomes in A:

$$\underbrace{2}_{\text{First}} \times \underbrace{11}_{\text{Second}} \times \underbrace{10}_{\text{Third}} = 220$$

Finally, $P(A) = 220/1320 = 0.1667$.

(c) The word *not* suggests the use of a complement, but a direct approach may be easier here. Let the event $B =$ horse 7 does not finish first, second, or third. $P(B) = N(B)/N(S) = N(B)/1320$.

Use the Multiplication Rule again to count the number of outcomes in the event B. We do not want horse 7 in the top three. That leaves 11 possible horses for first place, 10 for second, and 9 for third.

$$\underbrace{11}_{\text{First}} \times \underbrace{10}_{\text{Second}} \times \underbrace{9}_{\text{Third}} = 990$$

$P(B) = 990/1320 = 0.75$. ■

TRY IT NOW   Go to Exercises 4.99 and 4.113

Permutations

The following concise notation is often used to write large numbers associated with counting problems.

Definition
For any positive whole number n, the symbol $n!$ (read "***n* factorial**") is defined as

$$n! = n(n-1)(n-2)\cdots(3)(2)(1)$$

In addition, $0! = 1$ (0 factorial is 1).

A CLOSER LOOK

1. To find $n!$, just start with n, multiply by $(n-1)$, then $(n-2)$, ..., down to 1. For example,
   $7! = (7)(6)(5)(4)(3)(2)(1) = 5040$
   $10! = (10)(9)(8)(7)(6)(5)(4)(3)(2)(1) = 3,628,800$

2. Factorials get really big, really fast: Just try finding 50!. If you absolutely have to find a large factorial, then you should probably use technology.

4.3 Counting Techniques

Consider a generalization of the horse-racing problem. Suppose there are n items to choose from, there are r positions to fill, and the order of selection matters. There are n choices for the first position, $n-1$ choices for the second position, and $n-2$ choices for the third position. This process continues until there are $n-(r-1)$ choices for the rth position. The product of these numbers is the total number of **permutations**.

Definition

Given a collection of n different items, an ordered arrangement, or subset, of these items is called a **permutation**. The number of permutations of n items, taken r at a time, is given by

$$_nP_r = n(n-1)(n-2)\cdots[n-(r-1)]$$

> $_nP_r$ is also referred to as n items permuted r at a time.

Using the definition of factorial

$$_nP_r = \frac{n!}{(n-r)!}$$

In the denominator, do the subtraction first, then the factorial.

A CLOSER LOOK

1. The formula for the number of permutations is valid only if all n items are different.

2. A distinguishing characteristic of a permutation is that order matters. For example, if outcome AB is different from outcome BA, that suggests a permutation. Suppose an experiment consists of selecting two students from a class of 35. The first one selected will be the president and the second will be the vice president. Order certainly matters here, so we will be counting permutations. If the two students selected will form a committee, however, then the order of selection does not matter. Counting in this case involves a *combination*, which will be introduced a little later.

3. Here is an example to illustrate the calculation of $_nP_r$ and to show that the number of permutations is really just a special case of the Multiplication Rule.

$$_{12}P_3 = \frac{12!}{(12-3)!} = \frac{12!}{9!}$$

Definition of $_{12}P_3$, $n=12$, $r=3$.

$$= \frac{(12)(11)(10)(\not{9})(\not{8})(\not{7})(\not{6})(\not{5})(\not{4})(\not{3})(\not{2})(\not{1})}{(\not{9})(\not{8})(\not{7})(\not{6})(\not{5})(\not{4})(\not{3})(\not{2})(\not{1})}$$

Definition of factorial.

$$= (12)(11)(10)$$

Cancel like terms.

EXAMPLE 4.23 Consumer Electronics

A columnist reporting from the Consumer Electronics Show has found 10 memorable products. However, she has enough room in her column to write about only six of them, and suppose the order in which she writes about these products affects their publicity and demand. How many different product arrangements are possible?

Solution

STEP 1 There are $n=10$ items, we need to choose $r=6$, and the order in which the products are arranged in the column matters. For example, if capital letters represent products, then the product arrangement ABCDEF is different from ABCEDF. We must count the number of permutations of 10 items, taken 6 at a time.

> If you compute $_nP_r$ by hand, there is always a lot of canceling. $_nP_r$ is a count, so the answer must be an integer.

STEP 2

$$_{10}P_6 = \frac{10!}{(10-6)!} = \frac{10!}{4!}$$

Definition of $_nP_r$ using factorials.

$$= \frac{(10)(9)(8)(7)(6)(5)(\not{4})(\not{3})(\not{2})(\not{1})}{(\not{4})(\not{3})(\not{2})(\not{1})}$$

Definition of factorial.

$$= (10)(9)(8)(7)(6)(5) = 151,200$$

Cancel; multiply.

There are 151,200 ordered product arrangements in the column. **Figure 4.13** shows a technology solution.

Figure 4.13 $_{10}P_6$ calculated using R. Note that R does not have a built-in function for computing permutations.

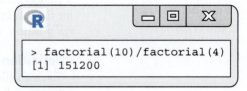

```
> factorial(10)/factorial(4)
[1] 151200
```

EXAMPLE 4.24   Intelligence Division

A fan of *Chicago P.D.* has recorded nine episodes from the most recent season of this show. However, he has time to watch only four episodes. Suppose he selects four shows at random.

(a) How many different ordered arrangements of episodes are possible?

(b) If the season finale is recorded, what is the probability that he will select and watch this episode last?

Solution

(a) There are $n = 9$ episodes to choose from. We need to count the number of ordered arrangements of $r = 4$ recordings.

$$_9P_4 = \frac{9!}{(9-4)!} = \frac{9!}{5!}$$ Definition of $_nP_r$ using factorials.

$$= \frac{(9)(8)(7)(6)(\cancel{5})(\cancel{4})(\cancel{3})(\cancel{2})(\cancel{1})}{(\cancel{5})(\cancel{4})(\cancel{3})(\cancel{2})(\cancel{1})}$$ Definition of factorial.

$$= (9)(8)(7)(6) = 3024$$ Cancel; multiply.

There are 3024 different ordered arrangements of four episodes.

(b) This is an equally likely outcome experiment. Count the number of arrangements in which the final recording is selected last, and divide this count by the total number of ordered arrangements.

Let A = the last recording selected is the season finale. There are four positions to fill, but the last slot is fixed (with the season finale). The first three positions can be filled by any of the remaining eight recordings, in any order.

$$\underbrace{\underset{\text{Rec 1}}{8} \times \underset{\text{Rec 2}}{7} \times \underset{\text{Rec 3}}{6}}_{_8P_3} \times \underset{\text{Rec 4}}{1} = 336$$

$$P(A) = \frac{336}{3024} = 0.1111$$

Figure 4.14 shows a technology solution.

Figure 4.14 Find the total number of permutations, the number of outcomes in the event A, and divide.

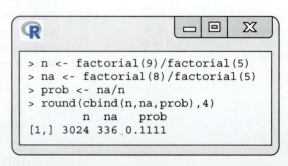

```
> n <- factorial(9)/factorial(5)
> na <- factorial(8)/factorial(5)
> prob <- na/n
> round(cbind(n,na,prob),4)
        n   na   prob
[1,] 3024  336 0.1111
```

TRY IT NOW   Go to Exercises 4.97 and 4.101

Combinations

In many experiments, the order in which the items are selected does not matter: for example, selecting five manufactured items from a batch of 50 for inspection, choosing nine people from 35 for a search committee, or picking three tax returns from 100 for a federal audit. In each case, the order of selection is not important; the collection, or group selected, is a single outcome. These unordered arrangements are called **combinations**.

Definition
Given a collection of n different items, an unordered arrangement, or subset, of these items is called a **combination**. The number of combinations of n items, taken r at a time, is given by

$$_nC_r = \binom{n}{r} = \frac{n!}{r!(n-r)!} = \frac{_nP_r}{r!}$$

A CLOSER LOOK

1. $\binom{n}{r}$ is read as "n choose r."

2. To find $_nC_r$ from $_nP_r$, we need to collapse all ordered arrangements of the same r items into one possible outcome. Dividing by $r!$ does this because every unordered set of r distinct items can be arranged in $r!$ ways.

3. If you have to calculate $_nC_r$ by hand, there is always a lot of cancellation. The final answer must be an integer because it is a count.

EXAMPLE 4.25 Jury Duty

How many different ways are there to select a jury of 12 people from a pool of 20 candidates?

Solution

STEP 1 There are $n = 20$ prospective jurors, and we need to choose $r = 12$, without regard to order. A jury is an unordered arrangement of 12 people. We need to count the number of combinations of 20 items, taken 12 at a time.

STEP 2 $\binom{20}{12} = \frac{20!}{12!(20-12)!} = \frac{20!}{12!\,8!}$ *Definition of $\binom{20}{12}$.*

$= \frac{(20)(19)(18)(17)(16)(15)(14)(13)(12!)}{12!\,8!} = 125{,}970$ *Cancellation; computation.*

There are 125,970 ways to select a jury of 12 from a pool of 20 candidates. **Figure 4.15** shows a technology solution.

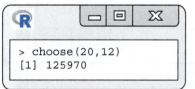

Figure 4.15 Use the R command choose() to compute combinations.

EXAMPLE 4.26 An Apple a Day

A produce bin at Giant Supermarket contains 18 apples; 12 are McIntosh and 6 are Cortland. A shopper randomly selects four apples from the bin.

(a) What is the probability that all of the apples selected are McIntosh?

(b) What is the probability that the shopper selects three McIntosh apples?

Solution

Since the shopper is selecting apples at random, all of the outcomes are equally likely. The order in which the apples are selected does not matter. To find the number of outcomes in the sample space, count combinations. To count the number of outcomes in each event, use the Multiplication Rule and the formula for $_nC_r$.

There are $n = 18$ apples in the bin, and we need to choose $r = 4$, without regard to order.

$$_{18}C_4 = \binom{18}{4} = \frac{18!}{4!(18-4)!} = \frac{18!}{4!14!} = \frac{(18)(17)(16)(15)}{4!} = 3060$$

There are 3060 outcomes in the sample space, all equally likely.

(a) Let A = select all McIntosh apples. Count the number of ways to select all McIntosh apples. There are 12 McIntosh apples, so we count the number of ways to select 4 apples from the 12 McIntosh apples, without regard to order.

$$N(A) = \binom{12}{4} = \frac{12!}{4!(12-4)!} = \frac{12!}{4!8!} = \frac{(12)(11)(10)(9)}{4!} = 495$$

Because this is an equally likely outcome experiment,

$$P(A) = \frac{N(A)}{N(S)} = \frac{495}{3060} = 0.1618$$

(b) Let B = select three McIntosh apples (and, therefore, 1 Cortland apple). To find the number of outcomes in B, there are two cases to consider: the number of ways to select three McIntosh apples and the number of ways to select one Cortland apple.

$$\binom{12}{3} \times \binom{6}{1} = 220 \times 6 = 1320$$

↑   ↑

The number of ways to select 3 McIntosh apples from 12, without regard to order.   The number of ways to select 1 Cortland apple from 6, without regard to order.

$$P(B) = \frac{N(B)}{N(S)} = \frac{1320}{3060} = 0.4314$$

Figure 4.16 shows a technology solution.

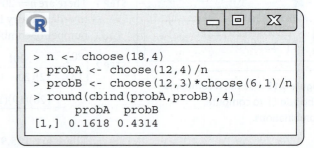

Figure 4.16 Probability calculations.

TRY IT NOW   Go to Exercises 4.103 and 4.107

Section 4.3 Exercises

Concept Check

4.79 True or False For fixed values of n and r, $_nP_r$ is always greater than or equal to $_nC_r$.

4.80 True or False The value of $_nC_r$ can be negative.

4.81 True or False The value of $_nP_r$ must be an integer.

4.82 True or False The expression for $_nC_r$ is valid only if all n items are different.

4.83 Fill in the Blank A _____ is a visualization of the multiplication rule.

4.84 Fill in the Blank An ordered arrangement is called a _____.

4.85 Fill in the Blank An unordered arrangement is called a _____.

4.86 Fill in the Blank The expression for $_nP_r$ is a special case of _____.

4.87 Short Answer Counting rules are most helpful in what kind of an experiment?

Practice

4.88 Find the number of permutations indicated.
(a) $_8P_4$ (b) $_{11}P_7$ (c) $_{12}P_4$
(d) $_{10}P_{10}$ (e) $_{10}P_1$ (f) $_{10}P_0$
(g) $_9P_2$ (h) $_{20}P_2$ (i) $_{100}P_2$

4.89 Find the number of combinations indicated.
(a) $\binom{9}{5}$ (b) $\binom{9}{4}$ (c) $\binom{14}{7}$
(d) $\binom{10}{10}$ (e) $\binom{10}{1}$ (f) $\binom{10}{0}$
(g) $\binom{12}{3}$ (h) $\binom{16}{7}$ (i) $\binom{20}{18}$

4.90 How many permutations of the letters in the word HISTOGRAM are possible?

4.91 A businessman's outfit consists of a pair of pants, a shirt, and a tie. Suppose he can choose from among 5 pairs of pants, 8 shirts, and 15 ties.
(a) How many different outfits are possible?
(b) Suppose a winter outfit includes a sweater and he can select one of 7 sweaters. Now how many different winter outfits are possible?

4.92 Earle Bailey has 20 songs to choose from but can play only 7 in the next half-hour on SiriusXM Classic Vinyl. How many different (ordered) playlists are possible?

4.93 A grocery store has 6 cashiers on duty, 10 baggers, and 4 people who will help customers load their groceries into a car. How many different checkout crews are possible?

4.94 A television station is developing a new identifying three-note theme. How many different three-note themes are possible if there are 20 notes to choose from and no note can be repeated?

4.95 A small hardware store has twenty 60-W equivalent LED light bulbs on the shelf, of which three are defective.
(a) How many different handfuls of six light bulbs are possible?
(b) How many different handfuls of five good light bulbs and one defective light bulb are possible?
(c) How many different handfuls of three good light bulbs and three defective light bulbs are possible?

4.96 A limousine service has 15 cars available for rides; 8 are Lincoln Town Cars and the remainder are Cadillac models. Suppose an event organizer requests five limousines from this company.
(a) How many different collections of five limousines are possible?
(b) How many different collections of two Lincoln Town Cars and three Cadillacs are possible?
(c) How many different collections of at least four Lincoln Town Cars are possible?

Applications

4.97 Manufacturing and Product Development Suppose Target sells a certain combination lock, which is really a permutation lock, with 40 numbers, 0 to 39. The combination for each lock is set at the factory and consists of three numbers.
(a) How many lock combinations are possible if numbers can be repeated?
(b) If all lock combinations are equally likely, what is the probability of selecting a lock with only single-digit numbers in the combination?
(c) Answer parts (a) and (b) if the lock combination must be three different numbers.

4.98 Fuel Consumption and Cars Suppose GEICO offers automobile insurance with specific levels of coverage according to the following table.

Coverage	Levels
Medical:	$10,000; $20,000; $50,000; $100,000
Bodily injury liability:	$50,000; $100,000
Property damage liability:	$25,000; $50,000; $100,000
Uninsured motorists:	$50,000; $100,000; $200,000
Comprehensive:	$250,000; $500,000; $1,000,000

Suppose an automobile policy must have coverage in all five categories.
(a) How many different automobile policies are possible?
(b) How many policies have comprehensive coverage of at least $500,000?
(c) How many policies have bodily injury liability and property damage liability of $100,000?

4.99 Manufacturing and Product Development L.L. Bean sells 10 different canoes, 5 types of paddles, and 8 suggested (canoe) accessories. A package sale consists of a canoe, a set of paddles, and an accessory.
(a) How many different package sales does the company offer?
(b) The Old Town Discovery Sport and the Saranac 146 canoes are the two most popular models. How many package sales contain these two canoes?
(c) The West Branch Canoe is the most expensive. If L.L. Bean eliminates this canoe style, how many package sales are possible?

4.100 Fuel Consumption and Cars A small tool-and-die shop manufactures kneuter valves. A shipment of 15 valves to a Swedish automobile assembly plant contains three defective

valves. Suppose the assembly plant randomly selects four valves from the shipment.
  (a) What is the probability that all four valves will be defect-free?
  (b) What is the probability that the plant will select all three defective valves?
  (c) What is the probability that the plant will select at least one defective valve?

4.101 Medicine and Clinical Studies A physician routinely visits a local nursing home on Thursday mornings to examine patients. Suppose the facility has 20 residents, but the physician has time to check only 8. The supervisor places 8 random patients on an ordered list and presents the schedule to the physician.
  (a) How many different schedules are possible?
  (b) If there are 15 women and 5 men in the facility, what is the probability that all appointments will be with women?

4.102 Economics and Finance A financial consultant has 12 people on her contact list. Suppose she will randomly select 8 people to call to discuss a certain stock purchase before the market opens.
  (a) How many different calling schedules are possible?
  (b) Suppose only 2 of the 12 will definitely purchase the stock when contacted. What is the probability that these 2 people will be the first 2 called?
  (c) Suppose another 2 of the 12 will ask to sell certain shares when contacted. What is the probability that these 2 people will not be called?

4.103 Sports and Leisure In preparation for the coming season, a bass fisherman decides to buy 5 random lures out of the 10 new ones in the local tackle shop.
  (a) How many different collections of 5 new lures are possible?
  (b) Suppose 1 of the 10 lures is a Crazy Crawler. What is the probability that the fisherman will not select this lure?
  (c) Suppose 3 of the 10 are Excalibur lures. What is the probability that at least 1 of the 5 selected will be an Excalibur lure?

4.104 Psychology and Human Behavior A music collector has 15 unopened, mint-condition classic rock albums in her collection. Suppose she decides to select three to sell at an upcoming auction.
  (a) How many different ways are there to select three albums from her collection?
  (b) Suppose the albums are selected at random and five are by the Beatles. What is the probability that all three selected will be by the Beatles?
  (c) What is the probability that none of the three will be by the Beatles?
  (d) Suppose that one album is by the Doors. What is the probability that this album is not selected?
  (e) What is the probability that the Doors album and two of the Beatles albums are selected?

4.105 Business and Management The Gagosian art gallery in New York City has 20 stored paintings but has just made room to display several of them. Seven paintings will be randomly selected and offered to the public for sale.
  (a) How many different collections of 7 paintings are possible?
  (b) Suppose 10 of the 20 stored works are by the same local artist. What is the probability that all 7 of the selected paintings will be by this artist?
  (c) The featured room in the gallery receives the most attention, and the order in which the paintings are displayed in this room is related to buyer interest. Suppose the 7 selected paintings will be placed in this featured room. How many different arrangements are possible?

4.106 Economics and Finance The purchasing agent for a state office building placed a call for bids on painting all the walls and ceilings. Suppose that eight sealed bids are received by the deadline. The bids will be opened in random order.
  (a) In how many different ways can the bids be opened?
  (b) What is the probability that the lowest bid will be opened first?

4.107 Marketing and Consumer Behavior A builder remodeling a residential kitchen decides to place a splashguard behind the sink consisting of 8 six-inch-square ceramic tiles decorated with different botanical herbs. The tiles will be installed in a custom-made wooden panel. The tile supplier has 12 different herb designs to choose from, and the builder selects 8 of these 12 at random. Suppose the order in which the tiles are arranged on the splashguard does not matter.
  (a) Two of the 12 herb tiles contain a blue tint that matches the kitchen color scheme. What is the probability that these 2 tiles will be included in the splashguard?
  (b) The family actually grows 5 of the 12 herbs in a backyard garden. What is the probability that all 5 of these will be included on the splashguard?

4.108 Travel and Transportation A PennDOT road line-painting crew consists of a foreman, a driver, and a painter. Suppose a supervisor is preparing the schedule to paint lines on roads in Edinboro and 10 foremen, 15 drivers, and 17 painters are available.
  (a) How many different crews are possible?
  (b) Suppose the crews are selected at random, and there is one foreman who has a severe personality conflict with one driver. What is the probability that neither of these individuals will be on the road painting crew?
  (c) Eight of the painters have been cited by a supervisor for improper painting. What is the probability that the crew will include one of these painters?
  (d) Two foremen, 4 drivers, and 5 of the painters are women. What is the probability that the crew will consist of all women?

4.109 Education and Child Development A university library is preparing a display case of books written by faculty members. There are 25 new faculty books, but there is room for only 10 in the display case. Suppose 10 books are selected at random.

(a) How many different faculty book collections can be displayed?
(b) If 3 of the books are written by members of the Department of Mathematics, what is the probability that none of the displayed books is written by a faculty member from this department?
(c) If 15 of the new books are written by faculty members from the College of Science and Technology, what is the probability that all 10 displayed books are written by faculty members from this college?
(d) If none of the 10 displayed books is written by faculty members from the College of Science and Technology, is there any evidence to suggest the selection process was not random? Justify your answer.

4.110 Psychology and Human Behavior In a family with five children, two of the five are selected at random each evening to do the dishes. The first one selected washes, and the second one dries.
(a) How many different wash–dry crews are possible?
(b) Suppose there are two girls and three boys in the family. If the two girls are selected to wash and dry, is there any evidence to suggest the selection process was not random? Justify your answer.

4.111 Demographics and Population Statistics A physician at Geisinger Medical Center is conducting a behavioral study concerning twins. There are 20 potential sets of twins for the study; 14 sets are fraternal twins and the remainder are identical sets of twins. The physician will select 8 sets of twins at random for the study.
(a) What is the probability that the physician will select all fraternal sets of twins?
(b) What is the probability that the physician will select all of the identical sets of twins?
(c) What is the probability that the physician will select the same number of fraternal and identical sets of twins?
(d) What is the probability that the physician will select at most 2 identical sets of twins?

4.112 Public Health and Nutrition There are 12 types of cookies available on the buffet at Golden Corral. Five contain naturally occurring fat and the remainder contain trans fat. A customer is going to select four cookies at random for dessert, each a different type.
(a) What is the probability that none of the cookies contains trans fat?
(b) What is the probability that one of the cookies contains trans fat?
(c) What is the probability that at least two of the cookies contain trans fat?

Extended Applications

4.113 Manufacturing and Product Development A remote-control garage door opener has a series of 10 two-position (0 or 1) switches used to set the access code. The code is initially set at the factory, and the switch sequence on the remote control and the opener must match for the customer to use the system.
(a) How many different access codes are possible?
(b) If all access codes are equally likely, what is the probability that a randomly selected system will have a code with exactly one 0?
(c) To increase security and ensure that customers will have different access codes, new systems have 10 three-position switches (0, 1, or 2). Answer parts (a) and (b) using the new system.

4.114 Psychology and Human Behavior An annual family picture following Thanksgiving dinner is arranged with all 10 family members in a row in front of a fireplace.
(a) How many different arrangements of family members are possible?
(b) Suppose the family includes one set of twins, and all arrangements are equally likely. What is the probability that the twins will be in the middle two places (positions 5 and 6)?
(c) What is the probability that the twins will be side by side in the picture?
(d) Suppose the family includes five males and five females. What is the probability that the picture arrangement will alternate male, female, male, female, etc., or female, male, female, male, etc.?

4.115 Public Policy and Political Science A special committee on community development has 4 members from the town council. The full town council has 14 members: 6 Democrats and 8 Republicans.
(a) How many different committees on community development are possible?
(b) Suppose the committee members are selected at random. What is the probability of a committee consisting of all Republicans?
(c) Suppose every member of the committee selected is a Democrat. Do you believe the selection process was random? Justify your answer.

4.116 Sports and Leisure Texas Hold 'em poker has become very popular in gambling casinos and is seen on ESPN and Poker Central. In July 2017, Scott Blumstein won the World Series of Poker in Las Vegas and a cool $8.15 million. The game is played with a standard 52-card deck and starts with each player being dealt two (random) cards face down (hole cards). There is a round of betting, the dealer then flips three cards face up (the flop), betting, one card is flipped (the turn), betting, a fifth card is flipped (the river), and more betting. Let's focus on the two hole cards, called a (pre-flop) hand, in this problem.
(a) How many (two-card, pre-flop) hands are possible in Texas Hold 'em?
(b) What is the probability that a pre-flop hand consists of two aces?
(c) What is the probability that a pre-flop hand consists of a pair, meaning two cards of the same rank?
(d) What is the probability that a pre-flop hand consists of two cards of the same suit?
(e) What is the probability that a pre-flop hand consists of two consecutive cards, such as an ace–two, or two–three, etc., or king–ace?

4.117 Public Policy and Political Science The Robert F. Kennedy Justice Building in Washington, D.C., contains

68 murals and 35 sculptures. Suppose that during a visitor tour, you are able to see 20 total items (murals and sculptures) at random.
(a) What is the probability that a visitor will see 10 murals and 10 sculptures?
(b) What is the probability that the visitor will see only murals?
(c) What is the probability that the visitor will see all 18 murals depicting the codifiers of law?
(d) Suppose it is possible to see some of the 12 physiographic and industrial regions of the national domain panels but still see only 20 total items. What is the probability that the visitor will see all of these panels and murals but no sculptures?

Challenge Problems

4.118 The Complement Rule Reconsider Example 4.22. Verify the probability in part (c) using the Complement Rule.

4.119 Combination Patterns Find the sum.
(a) $\binom{2}{0}+\binom{2}{1}+\binom{2}{2}$ (b) $\binom{3}{0}+\binom{3}{1}+\binom{3}{2}+\binom{3}{3}$
(c) $\binom{4}{0}+\binom{4}{1}+\binom{4}{2}+\binom{4}{3}+\binom{4}{4}$ (d) $\binom{n}{0}+\binom{n}{1}+\binom{n}{2}+\cdots+\binom{n}{n}$

4.120 Straight Poker Consider a regular deck of 52 playing cards. For a five-card poker hand, find the probability of each event.
(a) One pair.
(b) Two pairs.
(c) Three of a kind: three cards of the same rank and two others of different ranks, such as JJJ74.
(d) A straight: five cards in sequence; the ace can be either high or low.
(e) A flush: five cards of the same suit.

4.121 Roundtable Discussion How many different ways are there to arrange n people at a round table? (*Hint*: A simple rotation of a seating plan, shifting each person around the table but keeping the order the same, is not a different arrangement.)

4.122 Travel and Transportation Suppose there are n items, of which n_1 are of one type, n_2 are of a second type,..., and n_k are of the kth type, and $n_1 + n_2 + \cdots + n_k = n$. The number of unordered arrangements of the n items is a generalized combination given by

$$\binom{n}{n_1\ n_2\ \cdots\ n_n} = \frac{n!}{n_1!\ n_2!\ \cdots\ n_k!}$$

(Think about grabbing a handful of different-colored M&Ms.) Suppose the Amtrak Auto Train from Washington, D.C., to Florida has 10 sleeper cars, 2 diner cars, and 14 car carriers. Discounting the engine and caboose, how many different unordered arrangements of cars in the train are there?

4.123 Public Policy and Political Science The U.S. Senate Appropriations Subcommittee on Energy and Water Development has 18 members. The full Senate has 47 Democrats, 51 Republicans, and 2 Independents.
(a) How many different 18-member Senate committees are possible?
(b) If the committee members are selected at random, what is the probability of a committee consisting of all Democrats?
(c) What is the probability that the committee consists of exactly 6 Democrats?

4.124 Three Tiers The Dockland Stadium in Melbourne, Austrialia, has three tiers and a retractable roof; it is used mostly for Aussie rules football. Suppose a fan wants to see a game and will buy a ticket at random. There are 65 tickets left in Tier 2, 40 in Tier 3, and the remainder in Tier 1. If the probability of selecting a Tier 1 ticket is 0.125, what is the probability of selecting a Tier 3 ticket?

4.4 Conditional Probability

Background and Definition

The probability questions we have considered so far have all been examples of unconditional probability. No special conditions were imposed, nor was any extra information given. However, sometimes two events are related such that the probability of one depends on whether the other has occurred. In this case, knowing something extra may affect the probability assignment. This type of situation usually involves two events. The extra information may be expressed as an event separate from the event whose probability is desired.

EXAMPLE 4.27 **Morning Commute**

Consider a banker who commutes 30 miles to work every day. Because of several factors (e.g., weather, road construction, family obligations), the probability that she makes it to work on time on any random day is 0.5. If the event T is

T = the banker makes it to work on time,

dan_prat/iStock/Getty Images

The vertical bar, |, in the probability statement is read as "given."

then $P(T) = 0.5$. This is an *unconditional* probability statement: No extra information related to the event T is known or given.

Suppose a random day is selected, and the road conditions are terrible because of a snowstorm. The probability that the banker arrives at work on time is surely lower, perhaps around 0.1. Knowing the extra information (a snowstorm) changes the probability assignment for T.

The statement "What is the probability that the banker arrives at work on time if it is snowing?" is a *conditional* probability question. The extra information is that it's snowing outside. If the event F is defined as

$F =$ a snowstorm,

then this conditional probability is written as $P(T \mid F) = 0.1$; the probability that the banker arrives at work on time, given that it is snowing, is 0.1.

Suppose another random day is selected, but this time the banker wakes up before the alarm goes off and leaves the house early. The probability that she makes it to work on time is certainly higher, say, close to 0.95. Once again, knowing some extra information changes the probability assignment for T. If the event E is

$E =$ the banker leaves her house early,

then $P(T \mid E) = 0.95$.

Knowing something extra may change the probability assignment. We need to determine how to use any additional information to compute the (possibly) new probability. Consider the next example.

EXAMPLE 4.28 A Single Roll

Consider an experiment in which a fair, six-sided die is rolled and the number that lands face up is recorded. The sample space is $S = \{1, 2, 3, 4, 5, 6\}$. Consider the following events.

$$A = \{1\} = \text{roll a 1} \quad \text{and} \quad B = \{1, 3, 5\} = \text{roll an odd number}$$

Finding $P(A)$ is an unconditional probability question because no extra information is known. Because all of the outcomes in the experiment are equally likely, and because there is one outcome in A and six outcomes in the sample space, $P(A) = 1/6$.

Suppose someone rolls the die, covers it with her hands, peeks at the number, and reports, "I rolled an odd number." With this added information, the probability of the number being a 1 is now $P(A \mid B) = 1/3$. This conditional probability is reasonable because now we have to consider only three possibilities. In other words, we have reduced the sample space from six outcomes to three, and the number of outcomes in A (and B) is 1.

The idea of reducing, or shrinking, the sample space is key to calculating conditional probabilities. The definition of **conditional probability**, and some justification for it, are given next.

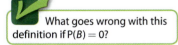
What goes wrong with this definition if $P(B) = 0$?

Definition
Suppose A and B are events with $P(B) > 0$. The **conditional probability of the event A given that the event B has occurred**, $P(A \mid B)$, is

$$P(A \mid B) = \frac{P(A \cap B)}{P(B)}$$

A CLOSER LOOK

1. The unconditional probability of an event A can be written as

$$P(A) = \frac{P(A)}{1} = \frac{P(A)}{P(S)} = \frac{\text{probability of the event } A}{\text{probability of the relevant sample space}}$$

We use this same reasoning to find $P(A \mid B)$.

2. Given that B has occurred, the relevant sample space has changed. It is reduced from S to B. (See **Figure 4.17**.)

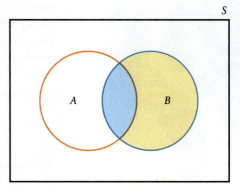

Figure 4.17 An illustration for calculating conditional probability.

3. Given that B has occurred, the only way A can occur is if $A \cap B$ has occurred, because the sample space has been reduced to B.
4. $P(A \mid B)$ is the probability that A has occurred, $P(A \cap B)$, divided by the probability of the relevant sample space, $P(B)$.

Definition Check

EXAMPLE 4.29 A Single Roll (Continued)

The experiment consists of rolling a fair, six-sided die and recording the number that lands face up. The sample space is $S = \{1, 2, 3, 4, 5, 6\}$. Consider the following events.

$A = \{1\} = $ roll a 1   and $B = \{1, 3, 5\} = $ roll an odd number

Find $P(A \mid B)$, the probability of rolling a 1 given that an odd number has occurred, or was rolled.

Solution

STEP 1 We will need the following probabilities.

$P(B) = 3/6$ and $P(A \cap B) = P(1) = 1/6$

STEP 2 Use the definition of conditional probability.

$$P(A \mid B) = \frac{P(A \cap B)}{P(B)} = \frac{1/6}{3/6} = \frac{1}{6} \cdot \frac{6}{3} = \frac{1}{3}$$

This answer agrees with the intuitive result (thank goodness). ■

Common Error

$P(A \mid B) = \dfrac{P(A \cap B)}{P(A)}$

Correction: In the definition of conditional probability, the given (or extra known) information is always in the probability statement in the denominator:

$P(A \mid B) = \dfrac{P(A \cap B)}{P(B)}$

A CLOSER LOOK

Here are some facts about union, intersection, and conditional probability to help you translate and solve many of the problems that follow.

1. $P(A \cup B) = P(B \cup A)$

   This is always true, because the two events contain the same outcomes:

   $A \cup B = B \cup A$ (all the outcomes in A or B or both)

2. $P(A \cap B) = P(B \cap A)$

   This is also always true, because these two events also contain the same outcomes:

   $A \cap B = B \cap A$.

> When are these two conditional probabilities equal?

3. $P(A|B) \neq P(B|A)$

These two probabilities could be the same, but in general they are different.

It's OK to switch A and B with union and intersection, but not with conditional probability.

4. The keywords *given* and *suppose* often signal partial information and, therefore, indicate a conditional probability question.

Contingency Tables and Conditional Probability

EXAMPLE 4.30 Coffee and Cookies

In 2018, Dunkin® started to offer Girl Scout Cookie flavorings that could be added to any kind of coffee.[10] Suppose 550 people who purchased iced coffee at Dunkin® were selected at random and each sale was categorized by the size of the drink (small, medium, large) and the flavoring (Thin Mints, Coconut Caramel, Peanut Butter, other). The results of this study are presented in the **two-way table**, also called a **contingency table**. The numbers in the table represent frequencies. For example, in the third row and fourth column, 40 people purchased a large iced coffee with some other flavoring. The last column contains the sum of each row; similarly, the bottom row contains the sum for each column. These sums are often called *marginal totals*.

> You can think of this table as representing all of the simple events in an equally likely outcome experiment. For example, let the outcome $A \cap T$ mean a person purchased a small iced coffee and Thin Mints flavoring. The probability of $A \cap T$, $P(A \cap T)$, is the number of outcomes in $A \cap T$ divided by the number of outcomes in the sample space: $N(A \cap T)/N(S) = 40/550$.

Reading a contingency table.

		Flavoring				
		Thin Mints (T)	Coconut Caramel (M)	Peanut Butter (E)	Other (O)	
Size	Small (A)	40	35	50	65	190
	Medium (B)	80	25	15	50	170
	Large (C)	65	75	10	40	190
		185	135	75	155	550

Assume that these results are representative of the entire population of people who purchase iced coffee at Dunkin®, so the relative frequency of occurrence is the true probability of the event. A person purchasing Dunkin® iced coffee is selected at random.

(a) Find the probability that the person buys a medium iced coffee.

(b) Find the probability that the person adds Peanut Butter flavoring.

(c) Suppose the person buys a large iced coffee. What is the probability that he adds Thin Mints flavoring?

(d) Suppose the person adds one of the three Girl Scout Cookie flavorings. What is the probability that she buys a size large iced coffee?

Solution

(a) This is an unconditional probability question, asking only about the event B. Compute the relative frequency of occurrence of B, which is the proportion of people who purchased size medium.

$$P(B) = \frac{80 + 25 + 15 + 50}{550} = \frac{170}{550} = 0.3091$$

(b) This is also an unconditional probability question. Find the relative frequency of occurrence of Peanut Butter flavoring.

$$P(E) = \frac{50 + 15 + 10}{550} = \frac{75}{550} = 0.1364$$

(c) This is a conditional probability question; the key word is *suppose*. The given information is *buys a large iced coffee*. We need the probability of the event T given that the event C has occurred. Use the definition for conditional probability.

$$P(T\,|\,C) = \frac{P(T \cap C)}{P(C)} = \frac{65/550}{190/550} = \frac{65}{550} \cdot \frac{550}{190} = \frac{65}{190} = 0.3421$$

This probability can also be obtained directly by reducing the sample space in the two-way table. The shaded row is the reduced sample space.

	Thin Mints (T)	Coconut Caramel (M)	Peanut Butter (E)	Other (O)	
Small (A)	40	35	50	65	190
Medium (B)	80	25	15	50	170
Large (C)	65	75	10	40	190
	185	135	75	155	550

In the reduced sample space, 190 outcomes are in the event C. Therefore,

$$P(T\,|\,C) = \frac{\text{number of outcomes in } T \text{ and in the reduced sample space}}{\text{number of outcomes in the reduced sample space}}$$

$$= \frac{65}{190} = 0.3421$$

(d) Solve this conditional probability question by reducing the sample space via the two-way table.

	Thin Mints (T)	Coconut Caramel (M)	Peanut Butter (E)	Other (O)	
Small (A)	40	35	50	65	190
Medium (B)	80	25	15	50	170
Large (C)	65	75	10	40	190
	185	135	75	155	550

There are 395 ($= 185 + 135 + 75$) outcomes in the reduced sample space, and 150 ($= 65 + 75 + 10$) people purchased size large in this reduced sample space. Note that purchasing one of the three flavorings can be denoted as $T \cup M \cup E$ or O'.

$$P(C\,|\,O') = \frac{150}{395} = 0.3797$$

TRY IT NOW Go to Exercises 4.145 and 4.155

CHAPTER APP **Dream Home**

Suppose the following contingency table lists the number of entries (in millions) in the United States by region and time when the entry was submitted.

		Entry time		
		Early	Middle	Late
Region	East	8.1	5.6	3.4
	Southeast	2.3	5.9	8.9
	Midwest	6.5	7.7	9.6
	West	5.8	6.1	7.1

Suppose an entry is selected at random.
(a) If the entry is from the southeast, what is the probability that it was submitted early?
(b) If the entry was submitted late, what is the probability that it was from the west?

Example: Use the Definition

EXAMPLE 4.31 Attitudes and Ideology

The Cooperative Congressional Election Study Survey asks individuals to rate various institutions on a scale from 1 to 7, with 1 being "very liberal" and 7 being "very conservative."[11] A recent Gallup poll indicates that 29% of all Americans consider themselves to be Democrats, and the survey suggests that 10% of Americans are Democrats and rate the Supreme Court as middle of the road (in ideology). Suppose that an American is selected at random. If the person is a Democrat, what is the probability that he or she rates the Supreme Court as middle of the road?

Solution

This is a conditional probability question. Translate the question into an appropriate probability statement and use the definition for conditional probability.

STEP 1 Consider the following events.

D = the American is a Democrat.

M = the individual rates the Supreme Court as middle of the road.

The statement of the problem includes two probabilities involving these two events.

$P(D) = 0.29$ *Percentage converted to unconditional probability.*

$P(M \cap D) = 0.10$ *The word and means intersection.*

STEP 2 $P(M|D) = \dfrac{P(M \cap D)}{P(D)}$ *Question translated into a conditional probability expression; definition of conditional probability.*

$= \dfrac{0.10}{0.29} = 0.3448$ *Use known probabilities.*

If the individual is a Democrat, the probability that he or she rates the Supreme Court as middle of the road is 0.3448.

TRY IT NOW Go to Exercises 4.141 and 4.143

Joint Probability Table

Steps for Calculating a Conditional Probability

To find the conditional probability of the event A given that the event B has occurred:

(a) Calculate $P(B)$ and $P(A \cap B)$.

(b) Find $P(A|B) = \dfrac{P(A \cap B)}{P(B)}$.

The body of the table contains intersection probabilities: the probability of a row event *and* a column event. For example, the probability that a person is male and 13–24 years old is 0.21, the entry at the intersection of the first row and the first column. The probabilities obtained by summing across rows and down columns are called marginal probabilities. The total probability in the table is 1.00.

EXAMPLE 4.32 It's a Digital World

The 2018 Global Digital suite of reports suggests that more than 4 billion people around the world use the Internet. The number of people using social media, such as Facebook, Twitter, and Instagram, has also increased dramatically. The following *joint probability table* lists the probabilities of people using Facebook corresponding to age (in years) and gender.[12]

		Age				
		13–24 (A)	25–44 (B)	45–64 (C)	65+ (D)	
Gender	Male (M)	0.21	0.26	0.07	0.02	0.56
	Female (F)	0.15	0.19	0.08	0.02	0.44
		0.36	0.45	0.15	0.04	1.00

Suppose a Facebook user is selected at random.

(a) Find the probability the person is female and 45–64 years old.

(b) Suppose the person is male. What is the probability that he is 13–24 years old?

(c) Suppose the person is 25–44 years old. What is the probability the person is female?

Solution

(a) The keyword in part (a) is *and*, which means intersection.

The probability of female (F) and age 45–64 (C) is found by reading the appropriate cell in the joint probability table.

	13–24 (A)	25–44 (B)	45–64 (C)	65+ (D)	
Male (M)	0.21	0.26	0.07	0.02	0.56
Female (F)	0.15	0.19	0.08	0.02	0.44
	0.36	0.45	0.15	0.04	1.00

Therefore, $P(F \cap C) = 0.08$.

(b) The keyword here is *suppose*. That suggests conditional probability. The given, or extra known, information is male.

$$P(A \mid M) = \frac{P(A \cap M)}{P(M)}$$ Question translated into a conditional probability expression; definition of conditional probability.

$$= \frac{0.21}{0.56} = 0.375$$ Use known probabilities.

(c) This is another conditional probability question. This time, the event B is given.

$$P(F \mid B) = \frac{P(F \cap B)}{P(B)} = \frac{0.19}{0.45} = 0.4222$$

TRY IT NOW Go to Exercises 4.135 and 4.153

Mutually Exclusive Exhaustive Events

A CLOSER LOOK

1. ▶ Notice in Example 4.32,

$P(A) = P(A \cap M) + P(A \cap F)$
$ = P(A \cap M) + P(A \cap M')$

In general, for any two events A and B,

$P(A) = P(A \cap B) + P(A \cap B')$

This decomposition technique is often needed to find $P(A)$. The Venn diagram in **Figure 4.18** illustrates this equation.

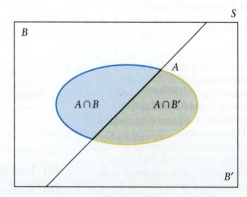

Figure 4.18 Venn diagram showing the decomposition of the event A.

Note that the events B and B' make up the entire sample space: $S = B \cup B'$.

2. Suppose B_1, B_2, and B_3 are mutually exclusive and exhaustive: $B_1 \cup B_2 \cup B_3 = S$. For any event A,
$$P(A) = P(A \cap B_1) + P(A \cap B_2) + P(A \cap B_3).$$

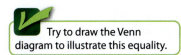
Try to draw the Venn diagram to illustrate this equality.

Section 4.4 Exercises

Concept Check

4.125 True or False In an unconditional probability statement, no extra relevant information related to the question is given.

4.126 True or False Extra information always changes a probability assignment.

4.127 True or False In an equally likely outcome experiment, for any two events A and B, $P(A|B) = P(B|A)$.

4.128 True or False For any two events A and B, $P(A'|B) = 1 - P(A|B)$.

4.129 Fill in the Blank In the conditional probability statement $P(A|B)$, the relevant sample space is _____.

4.130 True or False
(a) $P(A \cup B) = P(B \cup A)$
(b) $P(A \cap B) = P(B \cap A)$
(c) $P(A|B) = P(B|A)$

4.131 Short Answer Suppose B_1, B_2, B_3, and B_4 are mutually exclusive and exhaustive events. For any other event A, write $P(A)$ as a sum of probabilities involving the events B_1, B_2, B_3, and B_4.

Practice

4.132 Identify each of the following statements as a conditional or unconditional probability question.
(a) The probability that a randomly selected car will start in the morning.
(b) The probability that a person will remember to bring home a loaf of bread after work if he leaves a Post-It note reminder on the steering wheel.
(c) The probability that the next batter will get a hit in a baseball game.
(d) The probability that a randomly selected heart transplant operation will be successful.
(e) For all one-way streets in a large city, the probability that the street has more than two lanes.
(f) The probability that a certain flight will take off on time given that there is a severe thunderstorm warning in the area.

4.133 Identify each of the following statements as a conditional or unconditional probability question.
(a) The probability that a randomly selected circuit board will be defective, given that it was manufactured during the third shift.
(b) The probability that a waitress receives a tip of more than 18% of the cost of the meal.
(c) The probability that the next customer in a bookstore will buy a magazine.
(d) The probability that a company's sales will increase, given that more money is spent on advertising.
(e) The probability that a bowler will make three strikes in a row.
(f) The probability that an order at a fast-food restaurant will be correct given that it was ordered and obtained at the drive-through.

4.134 Consider the following joint probability table describing the events A, B, C, D, E, F, and G.

	F	G
A	0.12	0.05
B	0.15	0.07
C	0.17	0.04
D	0.19	0.02
E	0.11	0.08

(a) Verify that this is a valid joint probability table; that is, each probability must be greater than or equal to 0, and the sum of all probabilities must equal 1.
(b) Compute the marginal probabilities.
(c) Find $P(A \cap F)$, $P(B \cap G)$, and $P(D \cap G)$.
(d) Find $P(A|G)$, $P(F|D)$, and $P(E|C)$.
(e) Verify that $P(C) = P(C \cap F) + P(C \cap G)$.

4.135 Consider the following joint probability table.

	B_1	B_2	B_3
A_1	0.095	0.016	0.007
A_2	0.205	0.188	0.003
A_3	0.155	0.238	0.093

(a) Find $P(A_1)$, $P(A_2)$, and $P(A_3)$.
(b) Find $P(B_1)$, $P(B_2)$, and $P(B_3)$.
(c) Find $P(A_1 \cap B_1)$, $P(A_2 \cap B_2)$, and $P(A_3 \cap B_3)$.
(d) Find $P(A_1|B_1)$, $P(B_1|A_1)$, and $P(A_1'|B_1')$.
(e) Find $P(B_2|A_2)$ and $P(B_3|A_3)$.

4.136 Consider the following joint probability table.

	C_1	C_2	C_3	
A	0.135	0.125	0.206	0.466
B	0.145	0.174	0.215	0.534
	0.280	0.299	0.421	1.000

(a) Find $P(A)$ and $P(C_2)$.
(b) Find $P(A \cap C_1)$ and $P(B \cap C_3)$.
(c) Find $P(C_2 \mid B)$, $P(A \mid C_3)$, and $P(A \mid C_3')$.
(d) Verify that $P(B) = P(B \cap C_1) + P(B \cap C_2) + P(B \cap C_3)$. Carefully sketch a Venn diagram to illustrate this equality.

4.137 A recent survey classified each person according to the following two-way table.

	B_1	B_2	B_3	
A_1	178		815	
A_2		150	244	
A_3	165	202		
	466	583	985	

(a) Complete the two-way table.
(b) How many people participated in this survey?

Assume that the results from this survey are representative of the entire population, and that one person from this population is selected at random.

(c) Find $P(A_1)$, $P(A_2)$, and $P(A_3)$.
(d) Find $P(B_1 \cap A_1)$, $P(B_2 \cap A_2)$, and $P(B_3 \cap A_3)$.
(e) Find $P(A_3 \mid B_1)$, $P(B_2 \mid A_2)$, and $P(A_3 \mid B_1')$.

4.138 Consider an experiment and three events A, B, and C defined in the Venn diagram.

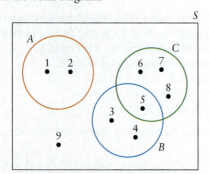

The table gives the probability of each outcome.

Outcome	1	2	3	4	5
Probability	0.01	0.12	0.11	0.10	0.15

Outcome	6	7	8	9
Probability	0.25	0.14	0.08	0.04

Find the following probabilities.
(a) $P(A)$, $P(B)$, and $P(C)$. Why don't these three probabilities sum to 1?
(b) $P(A \cap B)$ and $P(B \cap C)$.
(c) $P(B \mid C)$ and $P(C \mid B)$.

(d) $P(A \mid B')$, $P(C \mid A')$, and $P[1 \mid (A \cup B)']$.
(e) $P(3 \mid B)$, $P(4 \mid B)$, and $P(5 \mid B)$. Why do these three probabilities sum to 1?

4.139 Consider an experiment and three events A, B, and C defined in the Venn diagram.

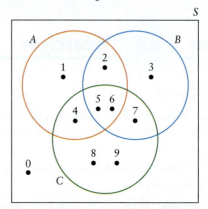

The table gives the probability of each outcome.

Outcome	0	1	2	3	4
Probability	0.135	0.130	0.142	0.128	0.147

Outcome	5	6	7	8	9
Probability	0.083	0.072	0.063	0.055	0.045

Find the following probabilities.
(a) $P(A)$, $P(B)$, and $P(C)$
(b) $P(A \cap B)$ and $P(B \cap C)$
(c) $P(A \mid B)$, $P(B \mid C)$, and $P[(A \cap B) \mid C]$
(d) $P(0 \mid C')$, $P(7 \mid C)$, and $P[(A \cup B) \mid C']$
(e) $P(2 \mid B)$, $P(3 \mid B)$, and $P(7 \mid B)$

Applications

4.140 Sports and Leisure Consider a regular 52-card deck of playing cards. Suppose two cards are drawn at random from the deck without replacement.
(a) What is the probability that the second card is an ace, given that the first card is a king?
(b) What is the probability that the second card is an ace, given that the first card is an ace?
(c) What is the probability that the second card is a heart, given that the first card is a heart?
(d) Suppose two cards are drawn at random from the deck with replacement. What is the probability that the second card is a heart, given that the first card is a heart?

4.141 Economics and Finance In the United States, some evidence suggests a link between people who participate in an office football pool and those who cheat on their income taxes. Suppose 25% of all people participate in an office football pool. The IRS estimates that 15% of all people participate in an office football pool and cheat on their income tax return. Suppose a person is randomly selected. If the person is known to participate in an office football pool, what is the probability that she cheats on her income tax return?

4.142 Marketing and Consumer Behavior According to a recent study, approximately 31% of U.S. consumers purchase groceries online.[13] Suppose 24% purchase food online and regularly use Amazon for this purpose, and 9% purchase food online and regularly use Kroger's online service. Suppose an online grocery shopper is selected at random.
(a) Given that the person shops for groceries online, what is the probability that he regularly uses Amazon?
(b) Given that the person shops for groceries online, what is the probability that he regularly uses Kroger?
(c) If the person shops online for groceries, what is the probability that he does not regularly use Amazon?

4.143 Travel and Transportation In Bangor, Maine, 80% of all people use chains on their car tires (for winter driving), 60% carry a snow shovel in their car and use chains, and 15% carry a shovel but do not use chains. Suppose a person from this area is selected at random.
(a) If the person uses chains, what is the probability that she carries a shovel?
(b) Given that the person does not use chains, what is the probability that she carries a shovel?

4.144 Business and Management The Statistics Canada Annual Greenhouse, Sod, and Nursery Survey includes information about the type of labor in each agricultural sector. The two-way table shows the number of employees in each sector in 2017.[14]

	Greenhouse	Sod	Nursery
Seasonal	18741	1140	7268
Permanent	14348	428	3332

Suppose a Canadian agricultural worker is selected at random.
(a) What is the probability that the worker is seasonal?
(b) Suppose the person works in a greenhouse. What is the probability that the worker is permanent?
(c) Suppose the person does not work in a nursery. What is the probability that the worker is seasonal?
(d) Suppose the person is permanent. What is the probability that he or she works in a nursery?

4.145 Sports and Leisure Amateur radio operators, or hams, have been communicating and providing a public service since 1909, when the first amateur radio club was organized in New York. Even though the number of hams is decreasing, there are still an estimated 6 million amateur radio users around the world. The two-way table shows the number of active U.S. FCC licenses held by individuals by type and by state.[15]

		State			
		AR	CA	FL	MI
Type	Novice	152	1145	628	258
	Tech	10352	64590	17648	10005
	General	4475	20060	11193	4998
	Advanced	1178	4486	3120	1207
	Extra	4117	15606	9199	4438

Suppose a ham radio operator from one of these four states is selected at random.
(a) What is the probability that the ham is from Arizona?
(b) What is the probability that the ham is from California and has a Tech license?
(c) Suppose the ham has an Advanced license. What is the probability that she is from Florida?
(d) Suppose the ham is from Michigan. What is the probability that she has a Novice license?
(e) Suppose the ham does not have an Advanced or Extra license. What is the probability that she is from Arizona or California?

4.146 Biology and Environmental Science The Fish Passage center in Oregon provides information on the Smolt Monitoring Program, which collects data on salmon, steelhead, trout, and lamprey; oversees hydrosystem operations; and monitors the number and type of fish passing by certain dams. The two-way table shows the number of fish by location and by type for one day in early May 2018.[16]

		Fish		
		Chinook adult	Chinook jack	Steelhead
Location	Willamette Falls	424	27	79
	Bonneville Dam	5392	153	26
	Ice Harbor Dam	112	3	3
	Little Goose Dam	49	2	4

Assume this table is representative of each spring day and that a fish passing one of these locations is selected at random.
(a) What is the probability that the fish is a Chinook adult?
(b) Suppose the fish is a Chinook jack. What is the probability that it was selected at Ice Harbor Dam?
(c) Suppose the fish is selected at Willamette Falls. What is the probability that it is a steelhead?
(d) Suppose the fish is not a steelhead. What is the probability that it was selected at Bonneville Dam?
(e) If the fish is selected at Ice Harbor Dam or Little Goose Dam, what is the probability that it is a Chinook jack?

4.147 Psychology and Human Behavior The Fremont Bridge Bicycle Counter in Seattle records the number of bicycles that cross the bridge using the sidewalks. There is an east and west sidewalk, and bicycles are counted traveling in either direction. Recent records suggest that 45.1% of bicycles cross the bridge in the morning, 33.1% in the afternoon, and 21.8% in the evening. In addition, 22.1% cross the bridge in the morning and use the east sidewalk, 24.2% cross the bridge in the afternoon and use the east sidewalk, and 15.4% cross the bridge in the evening and use the east sidewalk.[17] Suppose a bicycle rider crossing this bridge is selected at random.
(a) Suppose the person crosses the bridge in the afternoon. What is the probability that he used the east sidewalk?
(b) Suppose the person uses the west sidewalk. What is the probability that he crosses the bridge in the evening?
(c) Suppose the person does not cross the bridge in the evening. What is the probability that he uses the east sidewalk?

4.148 Marketing and Consumer Behavior The manager at Geek Window Cleaning in Houston, Texas, is looking at many ways to increase business for next year. The probability that she will advertise more is 0.75, the probability of advertising more and increasing revenue is 0.35, and the probability that revenue will not increase given no additional advertising is 0.80.
(a) Suppose the store manager decides to advertise more. What is the probability that revenue will increase?
(b) If the store manager does advertise more, what is the probability that revenue will not increase?
(c) What is the probability that the store manager will not advertise more and revenue will not increase?

4.149 Marketing and Consumer Behavior A random sample of adults who dine out regularly was obtained, consisting of 1200 women and 800 men. A survey showed that 965 of the women and 535 of the men rely on food photos on social media to select a restaurant.[18] Suppose a person included in this survey is randomly selected.
(a) What is the probability that the person selected is a woman and relies on food photos?
(b) Suppose the person selected is a man. What is the probability that he relies on food photos?
(c) Suppose the person selected relies on food photos. What is the probability that the person is a woman?

4.150 Travel and Transportation Most roadside assistance companies will help start your car, change a flat tire, or provide towing to the nearest service station. In addition, many make a claim about the amount of time they take to respond to an incident. A random sample of drivers who needed roadside assistance was obtained, and each was categorized by roadside assistance company and time until arrival. Suppose the resulting two-way table is representative of all drivers who need assistance from these companies.

		Arrival time	
		At most 1 hour	More than 1 hour
Company	Good Sam	125	21
	AAA	202	35
	Allstate	191	16
	AARP	133	42

A driver needing assistance from one of these roadside assistance companies is selected at random.
(a) What is the probability that the driver uses AAA and it takes more than 1 hour until arrival?
(b) Suppose the driver uses Allstate. What is the probability that it will take at most 1 hour until arrival?
(c) Suppose it takes more than 1 hour for arrival. What is the probability that the driver uses AARP?
(d) Suppose the driver uses Good Sam or AAA. What is the probability that it takes at most 1 hour until arrival?

Extended Applications

4.151 Demographics and Population Statistics According to a recent English Housing Survey, 14.9% of all households in England are in London and 85.1% are outside of London. In addition, 7.1% of households are owner occupied and in London, and 55.5% are owner occupied and outside of London.[19] Suppose a household in England is selected at random.
(a) Suppose the household is in London. What is the probability that it is owner occupied?
(b) Suppose the household is outside of London. What is the probability that it is owner occupied?
(c) Suppose the household is outside of London. What is the probability that it is not owner occupied?

4.152 Biology and Environmental Science Homeowners who cultivate small backyard gardens are often worried about pests (e.g., rabbits and groundhogs) ruining plants. Some gardeners protect their gardens with a fence, others spread chemicals around the perimeter of the garden to keep animals away, and some do nothing. The joint probability table shows the relationships among these garden protection methods and their results.

		Garden defense		
		Fence	Chemicals	Nothing
Result	Pests	0.05	0.08	0.34
	No pests	0.30	0.20	0.03

Suppose a backyard gardener is selected at random.
(a) Suppose the garden had pests. What is the probability that the gardener did not use a garden defense method?
(b) Suppose the gardener used chemicals. What is the probability that there were pests?
(c) Given that the garden had no pests, which method of defense did the gardener most likely use? Justify your answer.

4.153 Business and Management According to a recent survey, Bank of America, Wells Fargo, and Sprint are some of the companies ranked worst in customer service.[20] Suppose the table represents the results of a survey of customer satisfaction for Comcast, the company that received the highest proportion of negative responses regarding its customer service.

		Customer service			
		Excellent	Good	Fair	Poor
Region	North	0.102	0.059	0.062	0.004
	Midwest	0.105	0.105	0.144	0.007
	South	0.075	0.084	0.213	0.040

Suppose this table is representative of all Comcast user customer satisfaction and a customer is selected at random.
(a) What is the probability that the customer is from the midwest and customer service is fair?

(b) If the customer is from the north, what is the probability that customer service is excellent?
(c) Suppose customer service is poor. What is the probability that the customer is from the south?
(d) If the customer service is good or fair, which region is the customer most likely from? Justify your answer.

4.154 Public Health and Nutrition The following partial two-way table lists the number of residents admitted into continuing care facilities in Canada by province and source.[21]

		Source of admission		
		Hospital	Residential care	Home
Province	Newfoundland	622	437		1297
	Ontario	13,331		15,796	37,472
	Manitoba	1367	463	415	2245
	Saskatchewan		1276	850	4352
	Alberta	5520	865	1324	7709
	British Columbia	4300	3835	3285	11,420
		27,366		21,908	

(a) Complete this table.
(b) Suppose the person admitted was from Newfoundland. What is the probability that the source of admission was residential care?
(c) Suppose the source of admission was a home. What is the probability that the person admitted was from Alberta?
(d) Suppose the person admitted was from Manitoba. What is the probability that the source of admission was a hospital?

4.155 Public Health and Nutrition The Clayton Middle School in Salt Lake City, Utah, is set to examine its school lunch program. A survey of 2200 students asked students about their lunch type and how they got to school in the morning. The following (partial) two-way table is assumed to represent the entire student body.

		Arrival mode		
		Bus	Car	Walk
Lunch	Carries		466	142	
	Buys	345		500	967
		970			

(a) Complete this table.
(b) Suppose a student at the school is randomly selected. What is the probability that the student carries a lunch and gets to school by car?
(c) Suppose the student takes the bus to school. What is the probability that the student buys lunch?
(d) Suppose the student does not walk to school. What is the probability that the student carries a lunch?
(e) If the student buys lunch, how did he or she most likely get to school?

4.156 Demographics and Population Statistics The Bureau of Labor Statistics maintains case and demographic characteristics for all work-related injuries. The two-way table (**Table 4.1**) shows the number of nonfatal occupational injuries and illnesses involving days away from work by industry and by nature of the illness or injury in 2016.[22] Suppose a person who suffered one of these injuries in one of these industries is selected at random.
(a) What is the probability that the person suffered a cut?
(b) What is the probability that the person was in accommodation and food services?
(c) What is the probability that the person suffered a bruise and was in real estate?
(d) Suppose the person was in construction. What is the probability that he or she suffered a sprain?
(e) Suppose the person suffered a heat burn. What is the probability that he or she was in agriculture?
(f) Suppose the person was in professional and technical services or education services. What is the probability that he or she suffered a puncture?
(g) Suppose the person suffered a fracture or cut. In which industry did he or she most likely work?

Table 4.1 The number of injuries by industry and by illness or injury

		Illness or injury						
		Sprains	Fractures	Cuts	Punctures	Bruises	Heat burns	Chemical burns
Industry	Agriculture	4500	2020	1250	260	1370	60	30
	Construction	22,880	11,880	9470	2340	4660	830	180
	Manufacturing	35,110	12,560	13,170	1790	8750	1940	900
	Wholesale trade	20,230	4890	5180	650	4280	240	310
	Retail trade	48,190	10,080	13,070	2070	12,650	1110	680
	Transportation and warehousing	43,320	7740	3480	690	9580	400	190
	Utilities	1460	450	290	30	200	30	0
	Real estate	5470	1270	1520	330	970	110	230
	Professional and technical services	4050	1700	800	1450	1010	40	30
	Administrative and waste services	15,340	5040	4390	560	4540	270	150
	Educational services	3960	1340	430	170	980	60	30
	Arts entertainment and recreation	5690	1170	1340	140	1400	210	20
	Accommodation and food services	19,990	5020	15,120	490	6290	6400	300

Table 4.2 The number of movies released by studio and by genre

		Movie genre					
		Action	Adventure	Drama	Comedy	Thriller	Other
Studio	23	12	28	19	32	18	52
	Warner Bros.	14	9	16	12	10	26
	20th Century Fox	34	6	25	32	37	18
	Sony Pictures	27	17	34	26	22	16
	Universal	21	28	12	14	31	29
	Other	56	102	137	64	44	98

4.157 Marketing and Consumer Behavior The action movie genre captured the largest share of gross ticket sales in 2017.[23] Suppose the two-way table (**Table 4.2**) shows the number of movies released in the past few years by studio and by genre. Suppose a recently released movie is selected at random.
(a) What is the probability that the movie is an adventure?
(b) What is the probability that the movie is from Universal and a thriller?
(c) Suppose the movie is from Sony Pictures. What is the probability that it is a drama?
(d) Suppose the movie is a comedy. What is the probability that it is from Warner Brothers?
(e) Suppose the movie is not from some other studio. What is the probability that it is some other genre?

Challenge Problems

4.158 Public Policy and Political Science A survey of voters in a certain district asked if they favored a return to stronger isolationism. The following three-way table classifies each response by gender, political party, and response.

	Male			Female		
	Dem	Rep	Ind	Dem	Rep	Ind
Yes	202	126	105	234	101	95
No	124	288	85	312	66	150

Suppose a random voter is selected from this district.
(a) What is the probability that the voter is in favor of isolationism, a female, and a Republican?
(b) What is the probability that the voter is not in favor of isolationism?
(c) Suppose the voter is female. What is the probability that she is a Democrat?
(d) Suppose the voter is not in favor of isolationism. What is the probability that the voter is a Republican and male?
(e) Suppose the voter is not an Independent. What is the probability that he or she is in favor of isolationism?

Mike Wulf/CSM/Shutterstock

4.5 Independence

Background and Definition

In the last section, we learned about conditional probability, which considers whether knowing extra information may change a probability assignment. Often, however, additional information has no effect on the probability assignment. Consider the following examples.

EXAMPLE 4.33 Market Movers

Financial consultants and casual investors are always looking for the best stocks to buy and important stock market indicators. Some people rely on monetary policy, trading volume, or the inflation rate. However, stock prices are not related to the weather, a crystal ball, the political party of the president, general economic news, or even the Super Bowl winner.

Let R = the stock market will rise on a random day. Suppose the (unconditional) probability is 0.35: $P(R) = 0.35$.

If the New England Patriots win the Super Bowl, this extra information has no effect on the Dow Jones Industrial Index. What is the probability that the stock market will rise, given the event N = the New England Patriots win the Super Bowl?

$P(R \mid N) = 0.35$

If extra information is given, sometimes we simply say, "So what?"

Knowing extra information here does not change the conditional probability assignment. Intuitively, the events R (stock market rise) and N (Patriots win the Super Bowl) are unrelated, or *independent*.

EXAMPLE 4.34 Whale Watching

Whale watching tours are popular attractions off the coast of Puerto Vallarta, Mexico. Anxious visitors often ask boat captains about the best time of day for whale watching. However, from years of experience and thousands of trips, records indicate that the time of day (i.e., morning or afternoon) does not change the probability of seeing a whale.[24]

Suppose the events W and M are defined as $W =$ seeing, or encountering, a whale on a boat trip, and $M =$ morning boat trip.

$P(W \mid M) = P(W)$ and $P(W \mid M') = P(W)$

Whether a tourist takes a morning boat trip or whale watching tour at some other time has no effect on the chance of seeing a whale. The events seeing a whale and boat trip time are independent.

In these two examples, the occurrence or nonoccurrence of one event has no effect on the occurrence of the other. In this case, the two events are **independent**.

Definition
Two events A and B are **independent** if and only if

$$P(A \mid B) = P(A)$$

If A and B are not independent, they are said to be **dependent** events.

A CLOSER LOOK

One way to verify independent events: Is $P(A \mid B) = P(A)$? If so, then A and B are independent; if not, they are dependent.

1. Here's how the *if and only if* part of this definition works.

   If we know the events A and B are independent, then

   $P(A \mid B) = P(A)$ and $P(B \mid A) = P(B)$

   Or, if either one of these equations is true (the other is also true), then the events are independent.

2. If A and B are independent events, then so are all the combinations of these two events and their complements.

   Mathematical translation: If $P(A \mid B) = P(A)$, then $P(A \mid B') = P(A)$, $P(A' \mid B) = P(A')$, and $P(A' \mid B') = P(A')$.

3. Unfortunately, independent events cannot be represented on a Venn diagram. There is no good way to visualize independent events. In problems that involve independent events, we'll have to translate the words into a probability question and then use an appropriate formula.

4. It is reasonable to think of independent events as unrelated. One might conclude that they are therefore disjoint. However, this is not true!

   Suppose A and B are mutually exclusive and $P(A) \neq 0$ (there is some positive probability associated with the event A, or some chance that the event A will occur). Then

   $P(A \mid B) = 0 \neq P(A)$

   The probability of A given B has to be 0, because A and B are disjoint. Once B occurs, A cannot occur. Therefore, disjoint events are **dependent**.

The Probability Multiplication Rule

In Section 4.4, we learned an expression for finding conditional probability:

$$P(A|B) = \frac{P(A \cap B)}{P(B)} \quad \text{or} \quad P(B|A) = \frac{P(A \cap B)}{P(A)}$$

We can solve both of these equations for $P(A \cap B)$ to obtain the following **Probability Multiplication Rule**.

The Probability Multiplication Rule
For any two events A and B,

$$\left.\begin{array}{l} P(A \cap B) = P(B) \cdot P(A|B) \\ = P(A) \cdot P(B|A) \end{array}\right\} \text{Always true}$$

$$= P(A) \cdot P(B) \quad \text{Only true if } A \text{ and } B \text{ are independent.}$$

A CLOSER LOOK

1. The real skill in applying this rule is knowing which equality to use. The first two equalities are always true. Use one of these only if A and B are dependent and you need to find $P(A \cap B)$. Read the problem carefully to determine which conditional and unconditional probabilities are given.

   If A and B are independent, use the third equality to compute the probability of intersection. The word *independent* will not always appear in the problem. It may be implied or can be inferred from the type of experiment described.

2. If events are dependent, you can use a modified tree diagram to visualize and apply the Probability Multiplication Rule. In **Figure 4.19**, the probability of traveling along any branch is written along the appropriate leg. Second-generation branch probabilities are conditional. For example, $P(C|A)$ (the probability of C given A) is the probability of taking path C, given path A.

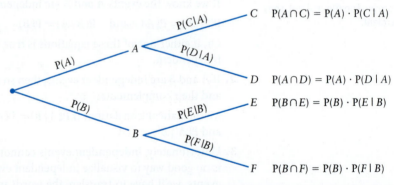

Figure 4.19 The Probability Multiplication Rule visualized on a tree diagram.

On this road map, to determine a final probability, we multiply probabilities along the way.

A modified tree diagram is useful here as well. Try drawing one to illustrate this extended rule.

All probabilities coming from a single node must sum to 1. To find the probability of traveling along a complete path from left to right (equivalent to the probability of an intersection), we multiply probabilities along the path.

3. The Probability Multiplication Rule can be extended. For any three events A, B, and C:
$$P(A \cap B \cap C) = P(A) \cdot P(B|A) \cdot P(C|A \cap B).$$

4. If the events $A_1, A_2, \ldots, A_k$ are mutually independent, then
$$P(A_1 \cap A_2 \cap \cdots \cap A_k) = P(A_1) \cdot P(A_2) \cdots P(A_k).$$
In words, if the events are mutually independent, the probability of an intersection is the product of the corresponding probabilities.

Examples

There are lots of probability rules, formulas, and diagrams in this section. These examples and problem-solving strategies are intended to help you translate the questions into mathematics and to illustrate these concepts.

EXAMPLE 4.35 Mobile Shoppers

Cyber Monday is a huge online shopping day. More people are now using their mobile devices, smartphones, or tablets to shop and make purchases on this Monday after Thanksgiving. In 2017, approximately 21% of Cyber Monday shoppers used their mobile device to make a purchase.[25] If a person used a mobile device to make a purchase, the probability of making the purchase from Amazon was 0.55. Suppose a Cyber Monday purchase is selected at random. What is the probability the purchase was made with a mobile device and was from Amazon?

Solution

We need to find the probability of using a mobile device *and* buying from Amazon. The word *and* suggests intersection. Determine whether the events are independent, determine which probabilities are given, and use the appropriate form of the Probability Multiplication Rule.

STEP 1 Define the following events.

M = used a mobile device to make a purchase.

A = made a purchase from Amazon.

We are given the probability of using a mobile device to make a purchase: $P(M) = 0.21$. In addition, we have the probability of making a purchase from Amazon given that a mobile device was used: This conditional probability statement is written as $P(A \mid M) = 0.55$.

STEP 2 We need the probability that the purchase was made with a mobile device and was from Amazon: $P(M \cap A)$.

$$P(M \cap A) = P(M) \cdot P(A \mid M) \quad \text{Probability Multiplication Rule.}$$
$$= (0.21)(0.55) = 0.1155 \quad \text{Use the given probabilities.}$$

Note: Using the Probability Multiplication Rule, we can also write:

$$P(M \cap A) = P(A) \cdot P(M \mid A)$$

This is a correct application of the rule, but it doesn't help in this problem because the probabilities on the right-hand side are not given. An accurate but inappropriate use of the Probability Multiplication Rule is evident because the probabilities given in the problem and in the equality are mismatched. Simply try the other equality. ∎

TRY IT NOW Go to Exercises 4.173 and 4.191

EXAMPLE 4.36 Residue Theory

According to the latest U.S. Food and Drug Administration pesticide residue report, 62% of all domestic vegetables examined by researchers contained pesticide residues.[26] Although most of these residues were compliant with federal standards, many researchers believe that the standards are outdated and too weak. Assume that the report used a representative sample, and suppose that two domestic vegetables are selected at random.

(a) What is the probability that both vegetables contain pesticide residues?

(b) What is the probability that both vegetables do not contain pesticide residues?

(c) What is the probability that exactly one vegetable contains pesticide residues?

Solution

Define some relevant events, determine whether the events are independent, determine which probabilities are given, and use the appropriate form of the Probability Multiplication Rule.

(a) Let V_i = vegetable i contains pesticide residues; $P(V_i) = 0.62$ (given) for $i = 1, 2$.

Since the vegetables are selected at random, the events V_1 and V_2 are independent.

Both vegetables contain pesticide residues means vegetable 1 contains pesticide residues *and* vegetable 2 contains pesticide residues.

$P(V_1 \cap V_2) = P(V_1) \cdot P(V_2)$ Both vegetables contain pesticide residues; independent events.

$= (0.62)(0.62)$ Probability of each vegetable containing pesticide residues.

$= 0.3844$

The probability that both vegetables contain pesticide residues is 0.3844.

(b) *Both vegetables do not contain pesticide residues* means vegetable 1 does not contain pesticide residues *and* vegetable 2 does not contain pesticide residues. "Does not contain pesticide residues" is the complement of "does contain pesticide residues."

$P(V_1' \cap V_2') = P(V_1') \cdot P(V_2')$ Independent events.

$= [1 - P(V_1)] \cdot [1 - P(V_2)]$ Complement Rule.

$= (0.38)(0.38)$

$= 0.1444$

The probability that both vegetables do not contain pesticide residues is 0.1444.

(c) *Exactly one vegetable contains pesticide residues* means

Vegetable 1 contains pesticide residues *and* vegetable 2 does not, *or*

Vegetable 1 does not contain pesticide residues *and* vegetable 2 does.

We can translate this into a probability expression:

P(Exactly 1 vegetable contains pesticide residues)

$= P[(V_1 \cap V_2') \cup (V_1' \cap V_2)]$

We don't usually see this step, but there really is a union of two events in the background. Notice the word *or* separating the two statements just given. These two events, separated by union, are disjoint, and the probability of the union of disjoint events is the sum of the corresponding probabilities.

$= P(V_1 \cap V_2') + P(V_1' \cap V_2)$

$= P(V_1) \cdot P(V_2') + P(V_1') \cdot P(V_2)$ Independent events.

$= (0.62)(0.38) + (0.38)(0.62)$ Use known probabilities.

$= 0.4712$

The probability that exactly one vegetable contains pesticide residues is 0.4712.

Note: There is another way to interpret and solve part (c). If we inspect two vegetables for pesticide residues, one of three things must happen: 0 vegetables have pesticide residues, 1 has pesticide residues, or 2 have pesticide residues. The probability of these three events must sum to 1. (Why?)

From part (a), P(2 contain pesticide residues) = 0.3844

From part (b), P(0 contain pesticide residues) = 0.1444

> In Chapter 5, we will convert all (symbolic) outcomes into real numbers, and use the probabilities of experimental outcomes to find the probabilities associated with real numbers.

P(1 vegetable contains residues)

$= 1 - [\text{P}(0 \text{ vegetables contain residues}) + \text{P}(2 \text{ vegetables contain residues})]$ *Complement Rule.*

$= 1 - (0.1444 + 0.3844)$ *Use known probabilities.*

$= 1 - 0.5288$

$= 0.4712$

TRY IT NOW   Go to Exercises 4.175 and 4.177

EXAMPLE 4.37   Who Needs a Car?

In some parts of the United States, owning a car is truly necessary. In other places, such as New York City, you probably don't need to own a car. Owning a car is expensive and some cities have plenty of public transportation. One way to measure the auto-dependence of an area is to consider the proportion of households with vehicles. According to a recent study, the percentage of households with vehicles in Boston, Massachusetts, is 66.2%; in Cambridge, Massachusetts, 63.2%; and in Worcester, Massachusetts, 80.7%.[27] Suppose a household from each city is randomly selected.

(a) Find the probability that all three households have vehicles.

(b) Find the probability that none of the three households has a vehicle.

(c) Find the probability that exactly one of the three households has a vehicle.

Solution

(a) Consider the following events.

$B =$ household B from Boston has a vehicle.

$C =$ household C from Cambridge has a vehicle.

$W =$ household W from Worcester has a vehicle.

Assume these three events are independent.

$\text{P}(B \cap C \cap W)$ *All three means intersection.*

$= \text{P}(B) \cdot \text{P}(C) \cdot \text{P}(W)$ *Independent events.*

$= (0.662)(0.632)(0.807) = 0.3376$

The probability that all three households have a vehicle is 0.3376.

(b) *None of the three households has a vehicle* means the household in Boston does *not* have a vehicle *and* the household in Cambridge does *not* have a vehicle *and* the household in Worcester does *not* have a vehicle. Translate this sentence into mathematics using intersection and complement.

$\text{P}(B' \cap C' \cap W')$ *Math translation; intersection.*

$= \text{P}(B') \cdot \text{P}(C') \cdot \text{P}(W')$ *Independent events.*

$= [1 - \text{P}(B)] \cdot [1 - \text{P}(C)] \cdot [1 - \text{P}(W)]$ *Complement Rule.*

$= (1 - 0.662)(1 - 0.632)(1 - 0.807)$ *Use given probabilities.*

$= (0.338)(0.368)(0.193)$

$= 0.0240$

The probability that none of the three households has a vehicle is 0.0240.

(c) To write a probability statement that corresponds to *exactly one household has a vehicle*, think about how this can happen. Household B has a vehicle and households C and W do not, or household C has a vehicle and households B and W do not, or household W has a vehicle and households B and C do not. Translate this sentence into probability using intersection and complement.

$P(B \cap C' \cap W') + P(B' \cap C \cap W') + P(B' \cap C' \cap W)$ *Three ways exactly one household has a vehicle.*

$= P(B) \cdot P(C') \cdot P(W') + P(B') \cdot P(C) \cdot P(W') + P(B') \cdot P(C') \cdot P(W)$ *Independent events.*

$= (0.662)(0.368)(0.193) + (0.338)(0.632)(0.193) + (0.338)(0.368)(0.807)$ *Known probabilities; Complement Rule.*

$= 0.0470 + 0.0412 + 0.1004$

$= 0.1886$

The probability that exactly one household has a vehicle is 0.1886.

TRY IT NOW Go to Exercise 4.187

EXAMPLE 4.38 Winter Tires

Winter tires are designed to reduce automobile crashes and improve driver safety in a wide range of winter weather conditions. Although using winter tires has some clear advantages, it also has some notable drawbacks: increased cost, decreased fuel economy, and the aggravation of mounting and installation. Despite these disadvantages, the number of Canadian drivers using winter tires has increased significantly. According to a recent survey, 55% of drivers in Alberta use winter tires (event A), 65% in Ontario do (event O), and 49% in British Columbia do (event B).[28] Suppose one driver from each region is randomly selected. Find the probability that at least one driver uses winter tires.

Solution

The words *at least one* mean the event *one, two, or three* drivers use winter tires. The complement of this event is *none* of the drivers uses winter tires. Use the Complement Rule to find the probability that at least one driver uses winter tires.

P(at least one driver uses winter tires)

$= 1 - P(0 \text{ drivers use winter tires})$ *Complement Rule.*

$= 1 - P(A' \cap O' \cap B')$ *All three do not use winter tires.*

$= 1 - P(A') \cdot P(O') \cdot P(B')$ *Independent events.*

$= 1 - [1 - P(A)] \cdot [1 - P(O)] \cdot [1 - P(B)]$ *Complement Rule.*

$= 1 - (1 - 0.55)(1 - 0.65)(1 - 0.49)$ *Use given probabilities.*

$= 1 - (0.45)(0.35)(0.51)$

$= 1 - 0.0803 = 0.9197$

The probability that at least one driver uses winter tires is 0.9197.

Challenge: Find this probability using a direct approach, without using the Complement Rule.

TRY IT NOW Go to Exercise 4.179

Bayes' Rule

EXAMPLE 4.39 A Traveling Salesperson

During his frequent trips to Dallas, Texas, a traveling salesperson stays at hotel A 50% of the time, at hotel B 30% of the time, and at hotel C 20% of the time. When checking in, there is some problem with the reservation 3% of the time at hotel A, 6% of the time at hotel B, and 10% of the time at hotel C. Suppose the salesperson is on one of his trips to Dallas.

(a) Find the probability that the salesperson stays at hotel A and has a problem with the reservation.

(b) Find the probability that the salesperson has a problem with the reservation.

(c) Suppose the salesperson has a problem with the reservation. What is the probability that the salesperson is staying at hotel A?

Solution

Consider the following events: A = stays at hotel A; B = stays at hotel B; C = stays at hotel C; and R = problem with the reservation.

Use these events and convert all the given percentages into probabilities. The phrase *of the time* indicates conditional probability.

$P(A) = 0.50 \quad P(B) = 0.30 \quad P(C) = 0.20$

$P(R|A) = 0.03 \quad P(R|B) = 0.06 \quad P(R|C) = 0.10$

To find $(R'|A)$, apply the Complement Rule to a conditional probability statement: $P(R'|A) = 1 - P(R|A)$.

This experiment can be represented with a modified tree diagram (**Figure 4.20**). Remember, the probabilities along all paths coming from a node must sum to 1, and second-generation branch probabilities are conditional.

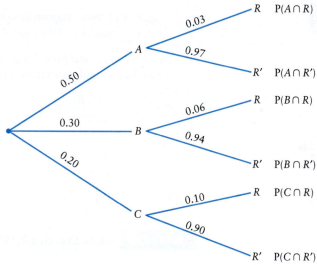

Figure 4.20 A visualization of the experiment and the probabilities presented in Example 4.39 using a modified tree diagram.

(a) The word *and* here means intersection. Determine whether the events are independent, determine which probabilities are given, and use the appropriate form of the Probability Multiplication Rule.

The events A and R are dependent. The likelihood of a problem with a reservation depends on the hotel.

$P(A \cap R) = P(A) \cdot P(R|A)$ Probability Multiplication Rule.

$\quad\quad\quad\quad = (0.50)(0.03)$ Use known probabilities.

$\quad\quad\quad\quad = 0.0150$

The probability of staying at hotel A and having a problem with the reservation is 0.0150.

(b) To find P(R), think about how this event can occur or how this can happen. Which paths from left to right in the tree diagram involve the event R? The tree diagram suggests that three compound events (paths) involve a problem with the reservation. This is an unconditional probability question, but we need to consider all the paths from left to right in the tree diagram in which the salesperson has a problem with the reservation.

$$P(R) = P(A \cap R) + P(B \cap R) + P(C \cap R) \qquad \text{Decomposition of } R.$$
$$= P(A) \cdot P(R|A) + P(B) \cdot P(R|B) + P(C) \cdot P(R|C) \qquad \text{Probability Multiplication Rule.}$$
$$= (0.50)(0.03) + (0.30)(0.06) + (0.20)(0.10) \qquad \text{Use known probabilities.}$$
$$= 0.0150 + 0.0180 + 0.0200 = 0.0530$$

The probability of a problem with the reservation (regardless of the hotel) is 0.0530. **Figure 4.21** shows this decomposition of R using a Venn diagram.

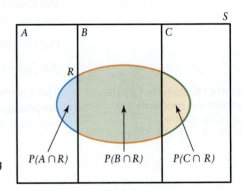

Figure 4.21 Venn diagram showing the decomposition of the event R.

(c) This is a conditional probability question, and the given information is that the salesperson has a problem with the reservation.

$$P(A|R) = \frac{P(A \cap R)}{P(R)} \qquad \text{Definition of conditional probability.}$$
$$= \frac{0.0150}{0.0530} \qquad \text{Use results from parts (a) and (b).}$$
$$= 0.2830$$

The probability that the salesperson stayed at hotel A, given a problem with the reservation, is 0.2830. ∎

TRY IT NOW Go to Exercise 4.193

A CLOSER LOOK

1. Part (c) of the hotel example illustrates **Bayes' rule**. This theorem loosely states: Given certain conditional probabilities (and other unconditional probabilities), we are able to solve for a new conditional probability where the events are inverted, or swapped.

   In the hotel example, we were given the conditional probabilities $P(R|A), P(R|B)$, and $P(R|C)$. Using these probabilities and the unconditional probabilities $P(A), P(B)$, and $P(C)$, we were able to find $P(A|R)$, a conditional probability with the events A and R switched.

2. Suppose $P(A), P(B)$, and $P(A \cap B)$ are known. To decide whether A and B are independent, check the equation $P(A \cap B) = P(A) \cdot P(B)$. If the probability of the intersection is equal to the product of the probabilities, then the events are independent. If not, they are dependent.

3. Many applications in probability and statistics involve repeated sampling from a population *with replacement*. In this case, each draw is independent of any other draw.

Other applications involve sampling *without replacement*, for example, exit polls and telephone surveys. Consider each individual response as an event. These events are definitely dependent. However, if the population is large enough and the sample is small relative to the size of the population, then the events are *almost independent*. In this case, calculating probabilities assuming independence results in little loss of accuracy. Exercise 4.198 illustrates this idea.

Section 4.5 Exercises

Concept Check

4.159 Fill in the Blank Two events A and B are independent if and only if $P(A\,|\,B) = $ _____.

4.160 Fill in the Blank If A and B are independent events, $P(A \cap B) = $ _____.

4.161 Fill in the Blank If the events A, B, and C are independent, $P(A \cap B \cap C) = $ _____.

4.162 True or False For any two events A and B, $P(A \cap B) = P(A) \cdot P(B\,|\,A)$.

4.163 True or False For any two events A and B, $P(A) \cdot P(B\,|\,A) = P(B) \cdot P(A\,|\,B)$.

4.164 True or False When sampling from a population with replacement, each draw is independent of any other draw.

4.165 Short Answer In your own words, describe what it means for two events A and B to be independent.

Practice

4.166 Decide whether each pair of events is independent or dependent.
(a) $A = $ make an error on your income tax return, and $B = $ file Form 1040 long.
(b) $C = $ put together a swing set correctly, and $D = $ read the directions.
(c) $E = $ run out of milk, and $F = $ the refrigerator breaks down.
(d) $G = $ break your pencil lead while writing, and $H = $ feel overly stressed.

4.167 Decide whether each pair of events is independent or dependent.
(a) $A = $ a random email has an attachment, and $B = $ a random email message is spam.
(b) $C = $ one paper towel is enough to completely clean a spill, and $D = $ you use a generic paper towel.
(c) $E = $ no accidents are reported in 24 hours in a county, and $F = $ there are no storms in the area.
(d) $G = $ your automobile insurance bill increases, and $H = $ you had one speeding ticket within the last year.

4.168 Decide whether each pair of events is independent or dependent.
(a) $A = $ it rains during the day, and $B = $ there is water in your basement.
(b) $C = $ take a city bus to work, and $D = $ watch the latest episode of *Blue Bloods*.
(c) $E = $ have a salad with dinner, and $F = $ work on a home project.
(d) $G = $ receive approval for a home mortgage, and $H = $ credit score is above 750.

4.169 Suppose the following probabilities are known: $P(A) = 0.25$, $P(B) = 0.30$, $P(B\,|\,A) = 0.34$, and $P(C\,|\,A \cap B) = 0.62$.
(a) Find $P(A \cap B)$, $P(B'\,|\,A)$, and $P(A \cap B')$.
(b) Find $P(A \cap B \cap C)$, $P(C'\,|\,(A \cap B))$, and $P(A \cap B \cap C')$.
(c) Are the events A and B independent? Justify your answer.

4.170 Suppose the events A, B, and C are independent and $P(A) = 0.55$, $P(B) = 0.45$, and $P(C) = 0.35$. Find the following probabilities.
(a) $P(A \cap B)$, $P(A \cap C)$, and $P(B \cap C)$
(b) $P(A \cap B \cap C)$ and $P(A' \cap B' \cap C')$
(c) $P(A \cap B' \cap C')$ and $P(A' \cap B \cap C)$

4.171 Suppose the following probabilities are known: $P(A) = 0.40$, $P(B\,|\,A) = 0.25$, $P(C\,|\,A) = 0.45$, and $P(D\,|\,A) = 0.30$.
(a) Find $P(A \cap B)$, $P(A \cap C)$, and $P(A \cap D)$.
(b) Are the events A and B independent? Justify your answer.
(c) Find $P(B'\,|\,A)$. If the event A occurs, are there any other events in addition to B, C, and D that can occur? Justify your answer.

4.172 Suppose the probability that an individual has blue eyes is 0.41. Four people are randomly selected.
(a) Find the probability that all four have blue eyes.
(b) Find the probability that none of the four has blue eyes.
(c) Find the probability that exactly two have blue eyes.

4.173 Suppose an experiment can be visualized in the modified tree diagram.

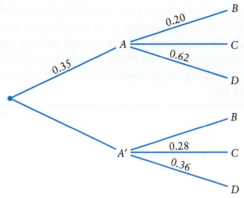

(a) Identify and determine each missing path probability.
(b) Find P(A∩C) and P(A'∩B).
(c) Find P(D).

4.174 Suppose an experiment can be visualized in the modified tree diagram.

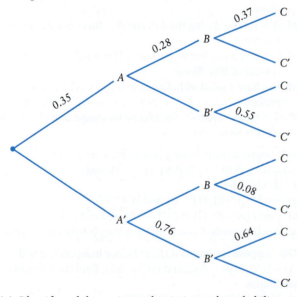

(a) Identify and determine each missing path probability.
(b) Find P(A∩B∩C) and P(A'∩B∩C').
(c) Find P(C). Are the events B and C independent? Justify your answer.

Applications

4.175 Demographics and Population Statistics Fishing is often considered a quiet, serene pastime, but the job of commercial fishermen is actually very dangerous. For example, *The Deadliest Catch* on Discovery Channel chronicles the risky lives of fishermen on the Bering Sea. According to recent data, the (yearly) fatality rate for fishermen is 0.0009.[29] Suppose two fishermen are selected at random.
(a) What is the probability that both fishermen will be fatally injured during the year?
(b) What is the probability that neither will be fatally injured during the year?
(c) What is the probability that exactly one will be fatally injured during the year?

4.176 Public Policy and Political Science The port of South Louisiana is the largest-tonnage port in the United States. Inspectors randomly select ships at one of the facilities and check for safety violations. Past records indicate that 90% of all ships inspected have no safety violations. Suppose two ships are selected at random.
(a) What is the probability that both ships have safety violations?
(b) What is the probability that neither ship has a safety violation?
(c) What is the probability that exactly one ship has a safety violation?

4.177 Marketing and Consumer Behavior Approximately 16% of Americans own a smart speaker, such as Amazon Alexa or Google Home.[30] These devices can control a TV or other smart devices in a home, or even tell jokes. Suppose two Americans are selected at random.
(a) What is the probability that both people own smart speakers?
(b) What is the probability that exactly one person owns a smart speaker?
(c) What is the probability that at most one person owns a smart speaker?

4.178 Public Health and Nutrition If a smoothie is made the right way, it can provide many nutritional benefits, adding fiber, protein, vitamins, and minerals to your diet. In a recent survey of Canadian social and lifestyle choices, 33% of respondents reported that they use nondairy milk as an ingredient in homemade smoothies.[31] Suppose three Canadians who make homemade smoothies are selected at random.
(a) What is the probability that all three use nondairy milk in the smoothies?
(b) What is the probability that none of the three uses nondairy milk in the smoothies?
(c) What is the probability that at least one uses nondairy milk in the smoothie?

4.179 Physical Sciences The San Francisco Bay Area is near several geological fault lines and subject to the constant threat of earthquakes. According to a study by the U.S. Geological Survey, the probability of a magnitude 6.7 or greater earthquake in the Greater Bay Area in the next 30 years is 0.72. The probability of a large earthquake within the next 30 years along four major fault lines is given in the table.[32]

Fault line	Probability
North San Andreas	0.22
Hayward	0.33
San Gregorio	0.06
Green Valley Concord	0.16

Suppose earthquakes occur in this area independent of one another.
(a) What is the probability that there will be a major earthquake within the next 30 years in all four fault regions?

(b) What is the probability that there will be no major earthquake within the next 30 years in any of the four regions?
(c) What is the probability that there will be a major earthquake within the next 30 years in at least one of the four regions?
(d) Suppose each probability is doubled if we consider the next 40 years. What is the probability that there will be a major earthquake within the next 40 years in at least one of the four regions?

4.180 Public Policy and Political Science Recent national elections suggest that the political ideology of U.S. adults is very evenly divided. In a 2018 Gallup survey, approximately 25% were Republicans, 30% were Democrats, and 45% were Independents.[33] In addition, suppose 51% of all Republicans, 16% of all Democrats, and 28% of all Independents describe their political views as conservative. Suppose a U.S. adult is selected at random.
(a) What is the probability that the adult is a Republican and describes her political views as conservative?
(b) What is the probability that the adult is a Democrat and describes her political views as conservative?
(c) Suppose the adult described her views as conservative. What is the probability that she is an Independent?

4.181 Medicine and Clinical Studies According to the Alzheimer's Association, approximately 17% of Americans aged 65–74 have Alzheimer's disease.[34] This disease is the sixth leading cause of death in the United States, and there is no treatment to prevent, cure, or even slow it. Suppose four Americans in this age group are selected at random.
(a) What is the probability that all four have Alzheimer's disease?
(b) What is the probability that exactly one has Alzheimer's disease?
(c) What is the probability that at least two have Alzheimer's disease?

4.182 Public Policy and Political Science The rental vacancy is a component of the index of leading economic indicators and is used by organizations to determine the need for new housing programs and initiatives. The first-quarter 2018 rental vacancies for four U.S. regions are given in the table.[35]

Region	Rental vacancy rate (percent)
Northeast	5.7
Midwest	7.8
South	8.8
West	4.7

Suppose a rental unit from each region is selected at random.
(a) What is the probability that all four units are vacant?
(b) What is the probability that just the unit in the northeast is vacant?
(c) What is the probability that exactly one unit is vacant?
(d) What is the probability that at least one unit is vacant?

4.183 Economics and Finance Detailed analysis of two technology stocks indicates that over the next six months, the probability that the price of stock 1 will rise is 0.42 and the probability that the price of stock 2 will rise is 0.63. Suppose the stock prices react independently.
(a) What is the probability that both stock prices will rise over the next six months?
(b) What is the probability that stock 1 will rise and stock 2 will fall?
(c) Suppose both stocks are in the technology sector, and stock 2 tends to follow stock 1. If stock 1 rises over the next six months, the chance of stock 2 rising is 81%. Now what is the probability of both stock prices rising over the next six months?

4.184 Sports and Leisure The PGA Tour maintains statistical reports on variables such as money leaders, driving distance, and driving accuracy. The table lists the probability that selected players were able to hit the fairway off the tee.[36] This is the probability a tee shot comes to rest in the fairway or on the green.

Golfer	Probability
Ryan Armour	0.7243
Jim Furyk	0.6694
Henrik Stenson	0.7681

Suppose these three golfers are playing a round at Sawgrass and they tee up on number 11, one of the most difficult holes to play on the professional tour.
(a) What is the probability that all three players will hit the fairway?
(b) What is the probability that none of the three players will hit the fairway?
(c) What is the probability that exactly one of the players will hit the fairway?
(d) What is the probability that all three players will hit the fairway on all four rounds of the tournament?

4.185 Sports and Leisure As of May 2018, Larry Bird had the twelfth highest career free-throw percentage in NBA history, 88.6%. Steve Nash was number one.[37] Suppose Larry is still playing and he steps up to the free-throw line for two shots. It is unlikely that the two shots are independent. If he misses the first shot, the probability that he makes the second is 0.95, and if he makes the first shot, the probability that he makes the second is 0.85.
(a) What is the probability that he makes both shots?
(b) What is the probability that he misses both shots?
(c) What is the probability that he makes only one shot?

4.186 Demographics and Population Statistics Most of us look forward to retirement, kicking back, and enjoying life in general. However, a Nationwide Retirement Institute study found that 27% of recent retirees say their lives are worse now than when they were working.[38] Most retirees blame their negative views on the lack of income. Suppose three recent retirees are selected at random.
(a) What is the probability that exactly two of the three say their lives are worse since retiring?

(b) What is the probability that all three say their lives are worse since retiring?
(c) Suppose another random sample of three recent retirees is obtained. What is the probability that exactly one retiree from each sample says that his or her life is worse since retiring?

4.187 Psychology and Human Behavior A survey conducted by YouGov showed that more than half of all Americans, British, and Germans believe that alien life exists. In addition, 30% of Americans believe that alien life has already contacted us but that the government has kept this information secret.[39] Suppose three Americans are selected at random.
(a) What is the probability that all three believe aliens have already contacted us?
(b) What is the probability that none of the three believes that aliens have contacted us?
(c) What is the probability that exactly one of the three believes that aliens have contacted us?

4.188 Demographics and Population Statistics Truck driving as an occupation is more of a lifestyle than a job. It involves long hours of solitude, different routines, and time away from family. According to the president and CEO of the Women in Trucking Association, 7% of truck drivers are women.[40] Suppose four truck drivers are selected at random.
(a) What is the probability that at least three are women?
(b) What is the probability that exactly two are women?
(c) Suppose another random sample of four truck drivers is obtained.
  (i) What is the probability that there is exactly one woman in each group?
  (ii) What is the probability that there is at least one woman in each group?

Extended Applications

4.189 Biology and Environmental Science Beachcombers enjoy searching the shore for shells and other interesting items. Some hope they may even find a real treasure, an oyster with a pearl inside. Alas, the chance of finding an oyster with a pearl inside is approximately 1 in 12,000.[41] Suppose four oysters are selected at random from various beaches.
(a) What is the probability that none of the four oysters contains a pearl?
(b) What is the probability that at least one contains a pearl?
(c) How many oysters would one have to search for the probability of finding at least one pearl to be 0.50?

4.190 Travel and Transportation A family trying to arrange a vacation is using the Internet to name their own price for a rental car. The software reports that 50% of all people name a price of $30 per day, 40% bid $25 per day, and 10% bid $20 per day. The Internet company also reports that 90% of all $30 bids are accepted, 60% of all $25 bids are accepted, and only 5% of all $20 bids are accepted.
(a) What is the probability that the family will submit a bid of $25 and have it accepted?
(b) What is the probability that their bid will be accepted?
(c) Suppose their bid is accepted. What is the probability that it is for $20?

4.191 Medicine and Clinical Studies Penicillin was one of the word's first antibiotics, and it is commonly used to treat a wide range of bacterial infections. However, many people have allergic reactions to penicillin, including nausea, hives, and rash. Some people outgrow these allergic reactions, and some mistakenly attribute such reactions to penicillin. Consider the following statements concerning patients' medical records and reactions to penicillin.[42]
- The probability that a patient has a penicillin allergy listed in his or her medical records is 0.10.
- If a patient has a penicillin allergy listed in the medical record, the probability that he or she will actually have a reaction to penicillin is 0.10.
- The probability of a patient not having a penicillin allergy listed and having a reaction is 0.045.
(a) Define events and write a probability statement for each statement listed.
(b) Construct a tree diagram to illustrate these events and probabilities.
(c) What is the probability that a randomly selected patient has a penicillin allergy listed in his or her medical record and has a reaction?
(d) Suppose a patient has a reaction. What is the probability that a penicillin allergy was listed in his or her medical record?

4.192 Medicine and Clinical Studies A tine test is commonly used to determine whether a person has been exposed to tuberculosis. Approximately 4% of people in the United States have been exposed to tuberculosis.[43] When the tine test is administered, 98% of all people who have been exposed test positive and 95% of those not exposed test negative. Suppose a person is randomly selected and given the tine test.
(a) What is the probability that the person tests positive and has been exposed to tuberculosis?
(b) What is the probability that the person tests positive?
(c) Suppose the test is positive; what is the probability that the person actually has been exposed?

4.193 Travel and Transportation There are five major air carriers with flights from Boston to Los Angeles; 40.5% of all passengers take American Airlines, 30.5% take Jet Blue, 12% take Delta, 8.1% take United, and 8.9% take Virgin America. Data from 2018 indicate that 6.3% of all American Airlines flights from Boston to LA are late, 14.2% of Jet Blue flights from Boston to LA are late, 19.2% of Delta flights from Boston to LA are late, 12.5% of United flights from Boston to LA are late, and 5.7% of Virgin America flights from Boston to LA are late.[44] Suppose a passenger taking a flight from Boston to Los Angeles is randomly selected.
(a) What is the probability that the passenger takes American Airlines and is late?
(b) What is the probability that the passenger is late? On time?
(c) Suppose the passenger arrives late. Which airline did the passenger most likely fly?

4.194 Manufacturing and Product Development Very often people engage in metal detecting at the beach. Some of these people hope that this surprisingly addictive hobby might lead to an important discovery or even real treasure. The Teknetics Delta 4000 metal detector is 82% effective; that is, the probability of this device detecting buried metal is 0.82.[45] Suppose a piece of precious metal is buried in the sand and three people with this metal detector scan the area.
(a) What is the probability that no one will detect this metal?
(b) What is the probability that at least one person will detect this metal?
(c) How many people would have to scan the area to be 99.99% sure the precious metal would be discovered?

4.195 Public Health and Nutrition Many more adults in the United States have celiac disease today than a decade ago. A new study suggests that approximately 1% of Americans have celiac disease and should avoid eating foods with gluten.[46] Suppose five U.S. adults are selected at random.
(a) What is the probability that exactly one of the five has celiac disease?
(b) What is the probability that only the first adult and the fifth adult selected have celiac disease?
(c) Suppose all five adults have celiac disease. Do you believe the claim concerning the percentage of adults with celiac disease? Justify your answer.

4.196 Fuel Consumption and Cars After a minor collision, a driver must take his car to one of two body shops in the area. Consider the following events.

D = driver takes his car to shop D.
L = driver takes his car to shop L.
T = the work is complete on time.
B = the cost is less than or equal to the estimate (under budget).

The modified tree diagram provides a visualization of the relationships among these events.

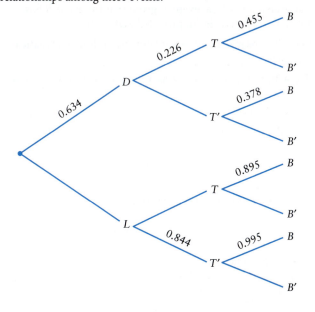

(a) Complete the tree diagram by filling in the missing path probabilities.
(b) What is the probability that the car is repaired under budget, on time, and with company D?
(c) What is the probability that the cost of the repair exceeds the estimate?
(d) What is the probability that the car is repaired under budget, given that it is ready on time?

4.197 Demographics and Population Statistics Toronto is Canada's largest city and is one of the world's most expensive real estate markets. Because the cost of housing is so high, 42% of the city's young professionals are considering leaving.[47] Suppose six young professionals in Toronto are selected at random.
(a) What is the probability that all six are considering leaving Toronto?
(b) What is the probability that exactly one young professional is considering leaving Toronto?
(c) Suppose all six are considering leaving Toronto. Do you believe the claim concerning the percentage of young professionals considering leaving the city? Justify your answer.

4.198 Sports and Leisure Suppose two cards are drawn without replacement from a regular deck of 52 playing cards. Consider these events:

A_1 = an ace is selected on the first draw.
A_2 = an ace is selected on the second draw.

(a) Find $P(A_2 \mid A_1)$ and $P(A_2)$. Are the events A_1 and A_2 independent? Justify your answer.
(b) Suppose the two cards are drawn without replacement from six regular 52-card decks shuffled together. Find $P(A_2 \mid A_1)$ and $P(A_2)$ for this experiment. Are the events A_1 and A_2 independent? Justify your answer.
(c) In part (b), the events are *almost independent*. For six decks, find $P(A_1 \cap A_2)$ exactly, and then find the same probability assuming the two events are independent (with the probability of an ace on any draw being 24/312).

Challenge Problems

4.199 The Traveling Salesperson Reconsider Example 4.39. Suppose the salesperson has a problem with the reservation. In which hotel did the salesperson most likely stay?

4.200 Manufacturing and Product Development In February 2018, BMW of North America recalled certain vehicles with possibly defective head air bags. BMW claimed that 4% of the recalled vehicles actually had defective air bags. Suppose six of the recalled vehicles are selected at random.
(a) What is the probability that none of the six vehicles has a defective air bag?
(b) What is the probability that at least one of the vehicles has a defective air bag?
(c) Suppose all six of the vehicles have defective air bags. Do you believe the claim made by BMW? Justify your answer.

Chapter 4 Summary

Concept	Page	Notation / Formula / Description
Experiment	120	An activity in which there are at least two possible outcomes and the result cannot be predicted with certainty.
Tree diagram	121	A visual road map of possible outcomes in an experiment.
Sample space	123	S: a listing of all possible outcomes, using set notation.
Event	124	Any collection of outcomes from an experiment.
Simple event	124	An event consisting of exactly one outcome.
Complement	125	A': all outcomes in the sample space S not in A.
Union	125	$A \cup B$: all outcomes in A or B or both.
Intersection	125	$A \cap B$: all outcomes in both A and B.
Disjoint events	125	Two events are disjoint if their intersection is empty: $A \cap B = \{\}$.
Venn diagram	127	Geometric representation of a sample space and events.
Probability of an event	132	The limiting relative frequency of occurrence.
Equally likely outcome experiment	135	All outcomes in the experiment have the same chance of occurring.
Probability of an event A in an equally likely outcome experiment	136	$P(A) = N(A)/N(S)$
Complement Rule	138	$P(A) = 1 - P(A')$
Addition Rule	138	$P(A \cup B) = P(A) + P(B) - P(A \cap B)$ $P(A \cup B) = P(A) + P(B)$ if A and B are disjoint.
Multiplication Rule	148	$N(S) = n_1 \cdot n_2 \cdot n_3 \cdots n_k$
n factorial	150	$n! = n(n-1)(n-2)\cdots(3)(2)(1); 0! = 1$
Permutation	151	An ordered arrangement. $_nP_r = n(n-1)(n-2)\cdots[n-(r-1)] = \dfrac{n!}{(n-r)!}$
Combination	153	An unordered arrangement. $_nC_r = \binom{n}{r} = \dfrac{n!}{r!(n-r)!}$
Conditional probability	159	The conditional probability of the event A given that the event B has occurred is $P(A|B) = P(A \cap B)/P(B)$, provided $P(B) \neq 0$.
Two-way (contingency) table	161	A two-way table with observed frequencies corresponding to classifications of two variables.
Joint probability table	163	A two-way table with probabilities corresponding to the intersection of two events.
Independent events	171	Two events A and B are independent if $P(A|B) = P(A)$, meaning that the occurrence of the event B does not affect the occurrence or non-occurrence of the event A.
Dependent events	171	If two events A and B are not independent, then they are dependent.
Probability Multiplication Rule	172	$P(A \cap B) = P(B) \cdot P(A|B) = P(A) \cdot P(B|A)$ $P(A \cap B) = P(A) \cdot P(B)$ if A and B are independent.

Chapter 4 Exercises

Applications

4.201 Marketing and Consumer Behavior A home decorating store received a shipment of 20 different large copper mirrors, and the store manager selects three mirrors at random for display.
(a) How many different displays are possible?
(b) Suppose three of the mirrors were damaged during shipping and contain hairline cracks. What is the probability that two of the mirrors selected will be damaged?
(c) What is the probability that at least one of the mirrors selected will be damaged?

4.202 Physical Sciences At a state middle-school science fair, students launch bottle rockets designed and built from plastic two-liter beverage containers. An experiment consists of recording the general appearance of a rocket [bad (B), good (G), or excellent (E)] and the maximum altitude [low (L), medium (M), or high (H)]. Consider the following events:
A = the rocket is rated as excellent.
B = the rocket flies to a high altitude.
C = the rocket is rated as bad or flies low.
D = the rocket is good and flies to a medium altitude.
(a) Find the sample space S for this experiment.
(b) List the outcomes in each of the events A, B, C, and D.
(c) List the outcomes in $A \cup B, B \cup C$, and D'.
(d) List the outcomes in $A \cap B, C \cap D$, and $(B \cap D)'$.

4.203 Travel and Transportation Four high-speed, or bullet, trains run each day between Shanghai Hongqiao Railway Station and Chengdu East Railway Station, China (trains 1, 2, 3, and 4). On each train, a passenger may select a first-class seat (A), second-class seat (B), or business-class seat (C). An experiment consists of recording the train and seat for a passenger. Consider the following events:
E = the passenger travels on train 2.
F = the passenger selects a second-class seat.
G = the passenger travels on train 1 or selects a first-class seat.
H = the passenger selects a business-class seat and travels on train 3.
(a) Carefully sketch a tree diagram to illustrate the possible outcomes for this experiment.
(b) Find the sample space for this experiment.
(c) Find the outcomes in each of the events E, F, G, and H.
(d) List the outcomes in $E \cup F, F \cap G$, and H'.
(e) List the outcomes in $E \cap H', E \cup F \cup G$, and $F \cup G'$.

4.204 Sports and Leisure There have been 24 James Bond movies. The first, *Dr. No*, was released in 1962, and the 25th is scheduled for release in April 2020. A recent survey asked participants to name the actor who is the best James Bond (in their opinion).[48] The table represents the proportion of people who selected each actor.

Actor	Proportion
Sean Connery	0.43
Roger Moore	0.14
Pierce Brosnan	0.10
Daniel Craig	0.07
George Lazenby	0.02
Timothy Dalton	0.02
Not sure	0.22

Assume this table is representative of the American movie-going public. One person is selected at random and asked which actor is the best James Bond. Consider the events:
A = best actor is Daniel Craig, George Lazenby, or Timothy Dalton.
B = best actor is not Sean Connery.
C = not sure of the best actor.
Find the following probabilities.
(a) $P(A), P(B)$, and $P(C)$
(b) $P(A \cap B), P(A \cup C)$, and $P(B \cup C)$
(c) $P(C'), P(A' \cup B)$, and $P(B' \cap C)$

4.205 Biology and Environmental Science The germination rate for pumpkin seeds is directly related to the prevailing weather conditions. The Autumn Gold is a popular medium-sized pumpkin that ripens to a deep orange. If conditions are seasonable, the probability of germination is 0.90.[49] If it is dry, suppose the probability that a random seed will germinate is 0.65. Recent weather history suggests there is a 40% chance of a dry start to the growing season. Suppose an Autumn Gold pumpkin seed is randomly selected.
(a) What is the probability that the growing season will be dry and the seed will germinate?
(b) What is the probability that the seed will germinate?
(c) Suppose the seed does not germinate. What is the probability that the growing season had a dry start?

4.206 Economics and Finance Many Americans use savings bonds to supplement retirement funds or to pay for qualified higher-education expenses. Savings bonds are a very safe investment, but they have declined in popularity for many reasons. According to a recent study, 8.6% of all American households still own savings bonds.[50] Suppose four American households are randomly selected.
(a) What is the probability that all own savings bonds?
(b) What is the probability that none of the four owns savings bonds?
(c) What is the probability that exactly two of the four own savings bonds?

4.207 Marketing and Consumer Behavior At Helms, an old-fashioned barber shop in Wooster, Ohio, 70% of all customers get a haircut, 40% get a shave, and 15% get both.

(a) What is the probability that a randomly selected customer gets a shave or a haircut?
(b) What is the probability that a randomly selected customer gets neither?
(c) What is the probability that a randomly selected customer gets only a shave?
(d) What is the probability that a randomly selected customer gets a shave, given that he gets a haircut?
(e) Suppose two customers are selected at random. What is the probability that both get only a haircut?

4.208 Medicine and Clinical Studies More and more people are trying herbal remedies, including tumeric tea, oil of oregano, and devil's claw, to reduce pain and inflammation. The joint probability table shows the relationship between having tried an herbal remedy and highest academic degree earned.

	Highest academic degree earned			
	Vocational	High school	College degree	Graduate degree
Tried	0.23	0.17	0.06	0.05
Not tried	0.04	0.12	0.15	0.18

Suppose one person is randomly selected.
(a) What is the probability that the person has tried an herbal remedy, given that the highest degree earned is from college?
(b) If the person has not tried an herbal remedy, what is the probability that the highest degree earned is from high school?
(c) Suppose the person has not earned a graduate degree. What is the probability that the person has tried an herbal remedy?
(d) Suppose two people are selected at random. What is the probability that exactly one has tried an herbal remedy?

4.209 Psychology and Human Behavior Do you believe in ghosts? According to a recent survey, 45% of Americans believe in ghosts.[51] Suppose that of those who believe in ghosts, 20% say they have seen a ghost. Suppose an American is selected at random.
(a) If the person believes in ghosts, what is the probability that she has never seen a ghost?
(b) What is the probability that the person believes in ghosts and has seen a ghost?
(c) What is the probability that the person believes in ghosts and has not seen a ghost?

4.210 Travel and Transportation A super-commuter is a person who commutes to work from one large metro area to another by car, rail, bus, or even air. Super-commuters are not necessarily elite business travelers, but rather middle-income individuals who are willing to commute long distances to secure affordable housing or better schools for their children. According to a recent study, 10% of workers in Stockton, California, are super-commuters, as are 4.8% in San Francisco and 3.8% in Los Angeles.[52] Suppose three workers are selected at random, one from each city.
(a) What is the probability that all three are super-commuters?
(b) What is the probability that none of the three is a super-commuter?
(c) What is the probability that only the worker from Stockton is a super-commuter?
(d) What is the probability that exactly two of the workers are super-commuters?

4.211 Rustic Chic Weddings According to an annual survey by The Knot, weddings in traditional venues (e.g., banquet halls, country clubs, and hotels) have declined over the past 10 years. Instead, couples are opting for rustic chic weddings. In 2017, approximately 15% of couples were married on farms, on ranches, or in a barn.[53] Suppose two weddings are selected at random.
(a) What is the probability that both take place in rustic venues?
(b) What is the probability that neither takes place in a rustic venue?
(c) Suppose at least one wedding takes place in a rustic venue; what is the probability that both take place in a rustic venue?

4.212 Marketing and Consumer Behavior According to a recent survey by *USA Today*, 54% of shoppers regularly give a charitable donation at the cash register.[54] Suppose four shoppers are selected at random.
(a) What is the probability that all four will give a charitable donation at the cash register?
(b) What is the probability that none of the four will give a charitable donation at the cash register?
(c) What is the probability that exactly two of the four will give a charitable donation at the cash register?
(d) Suppose there are two cash registers open, and four shoppers in each line. What is the probability that at least one shopper gives a charitable donation at the cash register?

4.213 Marketing and Consumer Behavior Many home buyers are now making offers without ever actually seeing the property first. This may be due to very tight housing markets and buyers' increased use of online resources. In Los Angeles, 57% of home buyers make a sight-unseen offer.[55] Of those buyers in LA who make a sight-unseen offer, suppose 35% of the offers are accepted by the seller. Suppose a Los Angeles home buyer is selected at random.
(a) Suppose the buyer makes a sight-unseen offer. What is the probability that it will be rejected?
(b) What is the probability that the buyer makes a sight-unseen offer and it is accepted?
(c) Suppose three Los Angeles home buyers are selected at random. What is the probability that exactly one makes a sight-unseen offer?

4.214 Economics and Finance Many families establish trusts for the children and relatives. There are two basic types of trusts: living trusts and testamentary trusts. Suppose a financial

consultant has created 15 trusts for clients, and 8 are living trusts. A federal accountant is going to select 4 of the 15 trusts at random to be audited.
(a) How many different ways are there to select 4 of the 15 trusts?
(b) What is the probability that the federal accountant will select none of the living trusts?
(c) What is the probability that the federal accountant will select at least one living trust?
(d) What is the probability that the federal accountant will select at least three living trusts?

Extended Applications

4.215 Travel and Transportation Despite the many modes of transportation and their widespread availability, a recent survey indicated that 11% of Americans have never traveled outside of the state where they were born (home).[56] Suppose that of those Americans who have never left home, 32% own luggage, and of those people who have traveled outside the state where they were born, 78% own luggage. Suppose an American is selected at random.
(a) What is the probability that the person has never left home and does not own luggage?
(b) What is the probability that the person owns luggage?
(c) Suppose the person owns luggage. What is the probability that he or she has never traveled away from home?

4.216 Public Health and Nutrition Most people believe that football and soccer players are very susceptible to head injuries. However, a recent study indicated that 67% of people in the performing arts have experienced at least one theater-related head impact.[57] Suppose four people in the performing arts are selected at random.
(a) What is the probability that all four have experienced a head-impact injury?
(b) What is the probability that none of the four has experienced a head-impact injury?
(c) What is the probability that exactly one of the four has experienced a head-impact injury?

4.217 Economics and Finance Customers at a Publix grocery store in Charleston, South Carolina, can pay for purchases with cash, a debit card, or a credit card. Fifty-five percent of all customers use cash and 38% use a debit card. Careful research has shown that, of those paying with cash, 75% use coupons; of those using a debit card, 35% use coupons; and of those using a credit card, only 10% use coupons. Suppose a customer is randomly selected.
(a) What is the probability that the customer pays with a credit card and does not use coupons?
(b) What is the probability that the customer does not use coupons?
(c) If the customer does not use coupons, what is the probability that he paid with a debit card?

4.218 Sports and Leisure Cycling has become very popular in the United Kingdom (U.K.), and there are now more than 2 million people who cycle at least once a week. A recent study indicated that 12% of cyclists planned to take a cycling vacation abroad within the next year.[58] Suppose five U.K. cyclists are selected at random.
(a) What is the probability that all five plan to take a cycling vacation abroad?
(b) What is the probability that exactly one of the five plans to take a cycling vacation abroad?
(c) Suppose none of the five plans to take a cycling vacation abroad. Is there any evidence to suggest that the study's claim is false? Justify your answer.

4.219 Fuel Consumption and Cars Auto Parts Warehouse offers a wide variety of parts and accessories for cars. Consider the following events:
A = a randomly selected customer purchases a manual.
B = a randomly selected customer purchases trim accessories.
C = a randomly selected customer purchases a car-care product.
Suppose the following probabilities are known:
$P(A) = 0.44$, $P(B) = 0.52$, $P(C) = 0.39$,
$P(A \cap B) = 0.19$, $P(A \cap C) = 0.10$, $P(B \cap C) = 0.23$,
$P(A \cap B \cap C) = 0.08$
(a) Carefully sketch a Venn diagram illustrating the relationship among these three events and label each region with the corresponding probability.
(b) Find the probability of just event A occurring.
(c) Find the probability of none of the events (A, B, or C) occurring.
(d) Find $P(A|C)$, $P(B|A \cap C)$, and $P(A \cap B \cap C|A)$.

4.220 Travel and Transportation Florida's Pasco County has special evacuation plans in the event of a hurricane. Suppose residents can take one of five different major highways out of the county. Department of Transportation officials have produced the following table indicating the probability that a resident will use a selected road.

Road	A	B	C	D	E
Probability	0.20	0.18	0.26	0.32	0.04

Suppose three Pasco County residents are selected at random and a hurricane strikes.
(a) What is the probability that all three will take the same escape route?
(b) What is the probability that exactly one will take escape route E?
(c) What is the probability that two will take escape route C?
(d) What is the probability that all three will take route B?

4.221 Economics and Finance Most consumers think of ATMs as a quick, convenient way to access cash from their bank accounts. However, ATMs also provide other important banking services. Suppose the joint probability table represents the results of a survey conducted by the National Credit Union Administration corresponding to United States region and ATM service.

		ATM service			
		View balance	Deposit	Transfer money	Withdraw cash
Region	North	0.03	0.04	0.05	0.11
	South	0.08	0.06	0.03	0.04
	Midwest	0.09	0.04	0.09	0.07
	West	0.06	0.03	0.12	0.06

Suppose this table is representative of all people in the United States with an ATM card, and one transaction is selected at random.
(a) What is the probability that the transaction involves a transfer?
(b) Suppose the transaction is in the south. What is the probability that it involves a deposit?
(c) Suppose the transaction is to withdraw cash. What is the probability that it is in the west?
(d) Suppose the transaction is not to withdraw cash. What is the probability that it is in the midwest?
(e) Suppose the transaction is not in the north. What is the probability that it is to view a balance?
(f) Are the events South and Withdraw cash independent? Justify your answer.
(g) Suppose two transactions are selected at random. What is the probability that both were from the west and involved a transfer of funds?

4.222 Manufacturing and Product Development Automated quality testing using specialized machines has helped to improve and increase production of semiconductors. A company claims that a new quality-testing machine is 90% effective; that is, it will detect a defective semiconductor 90% of the time.[59] Suppose a defective semiconductor is *inspected* by three quality-testing machines.
(a) What is the probability that at least one machine will detect the defective semiconductor?
(b) Suppose none of the three machines detects the defective semiconductor. Is there any evidence to suggest the company's claim (90% effective) is wrong? Justify your answer.
(c) How many quality-testing machines would be necessary to be 99.999% sure that a defective semiconductor is identified?

Challenge Problems

4.223 Free Nights During the month of August, one guest at the Golden Nugget in Atlantic City will be selected at random to participate in a contest to win free lodging. A fair quarter will be tossed until the first head is recorded. If the first head occurs on toss x, the contestant will win x free nights' stay at the Golden Nugget.
So, if a head is obtained on the first coin toss, the contest is over, and the guest wins one free night. If the first head appears on the fourteenth toss (i.e., 13 tails and then a head), the guest wins 14 free nights. Theoretically, a guest could win any number of free nights, 1, 2, 3, 4,..., although it seems unlikely someone could win, for example, 100 free nights.
(a) Use technology to model this contest. Try your simulation 10 times and record the number of free nights awarded each time. Did anyone win five or more free nights' stay?
(b) Consider event A = the guest wins five or more free nights at the Golden Nugget. Simulate the contest $n = 50$ times and compute the relative frequency of occurrence of event A. Repeat this process for $n = 100, 150, 200, \ldots, 2000$.
(c) Construct a plot of the relative frequency versus the number of simulations. Describe any patterns.
(d) Use your results in parts (b) and (c) to estimate the probability of winning five or more free nights at the Golden Nugget.
(e) Find the exact probability of winning five or more free nights at the Golden Nugget. *Hint*: Consider the complement of event A.

CHAPTER APP

korisbo/ Shutterstock

4.224 Dream Home Suppose you submit 50 entries in the Dream Home sweepstakes, one entry every day the contest is open. There were approximately 130 million entries in this sweepstakes last year.
(a) Explain why the selection of the Dream Home winner is an experiment.
(b) Describe the sample space in the selection of the Dream Home winner experiment.
(c) Suppose you submit 50 entries and there are approximately 130 million entries in this sweepstakes. Find the probability of winning the Dream Home.
(d) Suppose 10 million entries are from residents of Nebraska. What is the probability that the winner is not from Nebraska?
(e) Suppose the following contingency table lists the number of entries (in millions) in the Unites States by region and when the entry was submitted.

		Entry time		
		Early	Middle	Late
Region	East	8.1	5.6	3.4
	Southeast	2.3	5.9	8.9
	Midwest	6.5	7.7	9.6
	West	5.8	6.1	7.1

Suppose an entry is selected at random.
(i) If the entry is from the southeast, what is the probability that it was submitted early?
(ii) If the entry was submitted late, what is the probability that it was from the west?

Random Variables and Discrete Probability Distributions

5

◀ Looking Back
- Recall the definition of an experiment, a sample space, and operations on events: **Section 4.1**.
- Remember the properties and rules used to compute the probability of various events: **Sections 4.2, 4.4, and 4.5**.

Looking Forward ▶
- Explore the concept of a random variable, which serves as a bridge between the experimenter's world and the statistician's world, and determine how information is transferred between worlds: **Section 5.1**.
- Understand the connection between an experimental outcome and the number associated with that outcome, and describe how to construct and represent a probability distribution for a discrete random variable: **Section 5.2**.
- Learn how to compute and interpret the mean, variance, and standard deviation for a discrete random variable: **Section 5.3**.
- Understand a binomial experiment, and compute probabilities and solve inference problems associated with a binomial random variable: **Section 5.4**.
- Become familiar with and utilize other important discrete probability distributions: **Section 5.5**.

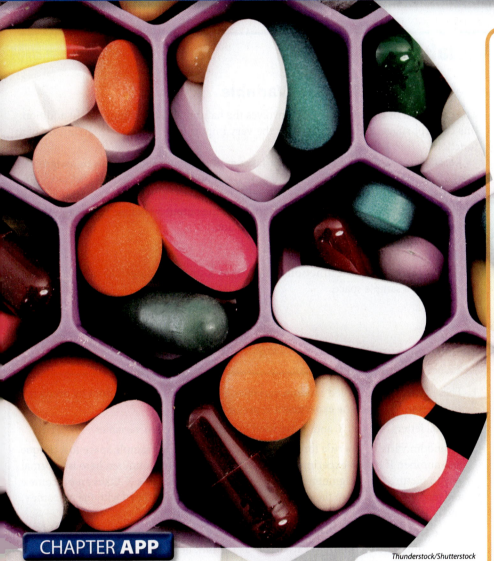

Thunderstock/Shutterstock

CHAPTER APP

Hooked on Vitamins

Each year, Americans spend billions of dollars on vitamins and dietary supplements. An estimated 90,000 vitamin and dietary supplement products are currently sold in the United States, in the form of pills, powders, drinks, and bars. A recent survey suggested that more than half of all Americans take vitamin supplements. However, older Americans are really hooked on vitamins. A Gallup poll indicated that 68% of those age 65 and older take vitamin supplements.[1]

Unfortunately, many of these products do not contain the actual ingredients advertised. Others lack conclusive proof that they provide the stated cure or relief.

A recent study found that only 48% of supplements in a sample contained what was actually on the labels: The rest were either fake, counterfeit, or contaminated.

To check this claim (48% real), a random sample of 50 dietary supplements was obtained. Each was tested using DNA barcoding, and 21 were found to contain the ingredients listed on the label. The techniques presented in this chapter will allow us to compute the likelihood of at most 21 of the supplements (out of the 50) being authentic. This result will be used to determine whether there is any evidence that the claim is false.

5.1 Random Variables

Definition of a Random Variable

The definition of a random variable involves the natural idea of assignment. In addition, this concept suggests that we consider the very important mathematical definition of a function.

Recall that a function f is a rule that takes an input value and returns, or assigns, an output value (according to the rule). Suppose the function f is defined by $f(x) = x^2 + 4$. This rule indicates that f takes an input x and assigns, or maps, x to the value $x^2 + 4$. For example, the function f assigns the input 1 to the output 5 because $f(1) = 1^2 + 4 = 5$. A random variable is just a special kind of function.

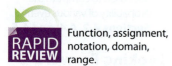
Function, assignment, notation, domain, range.

Definition
A **random variable** is a function that assigns a unique numerical value to each outcome in a sample space.

A CLOSER LOOK

1. These functions are called random variables because their values cannot be predicted with certainty before the experiment is performed.
2. Capital letters, such as X and Y, are used to represent random variables.
3. A random variable is a rule for assigning each outcome in a sample space to a unique real number. If e is an experimental outcome and x is a real number, here is a formal way to picture this assignment: $X(e) = x$. The random variable X takes an outcome e and maps, or assigns, it to the number x. The number x is associated with the outcome e, and is a value that the random variable can take on, or assume.
4. **Figure 5.1** and **Figure 5.2** help us understand the concept of a random variable. These figures illustrate the random variable X as the link between experimental outcomes and numerical values.

$X: S \to R$
A random variable maps elements of a sample space to the real numbers.

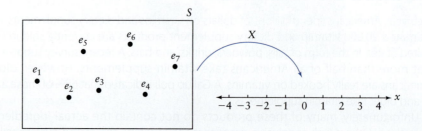

Figure 5.1 A random variable assigns a numerical value to each outcome.

5.1 Random Variables

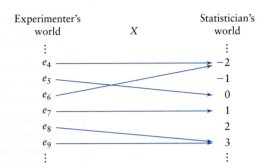

Figure 5.2 This visualization of a random variable illustrates the connection between each experimental outcome and the associated real number.

The rule for a random variable may be given by a formula, as a table, or even in words. Note that several outcomes may be assigned to the same number, but each outcome is assigned to only one number.

The next example shows how a specific random variable maps outcomes to numbers. The notation will get shorter and more concise as the concept of assignment becomes clearer.

EXAMPLE 5.1 That Sinking Feeling

In late April 2018, several sinkholes opened up in an Ocala, Florida, neighborhood. At least eight families were evacuated, and officials conducted tests to determine the cause of the sinkholes and the risk that additional holes might develop.[2] The Florida Department of Environmental Protection maintains a database of subsidence incident reports, but many sinkholes go unreported and some occur in rural areas and are unobserved. Suppose three sinkholes in Florida are selected at random, and each is classified as a collapse sinkhole (C) or some other type (O). Let the random variable X be defined as the number of collapse sinkholes out of the three selected.

Solution

Find all the outcomes in the sample space S. Use the definition of the random variable to find the numerical value associated with each outcome. To find the probability that X takes on a specific value, consider all the outcomes that are mapped to that value.

STEP 1 The experiment consists of recording the type of each sinkhole. Each outcome consists of a sequence of three letters, with each letter being either a C or an O. There are eight possible outcomes (from the Multiplication Rule). Here is the sample space:

$S = \{OOO, OOC, OCO, OCC, COO, COC, CCO, CCC\}$

STEP 2 The random variable X takes each outcome and returns the number of collapse sinkholes (indicated by C). The following table illustrates this mapping and the values that the random variable X can assume.

Outcome	Value of X
OOO	0
OOC	1
OCO	1
OCC	2
COO	1
COC	2
CCO	2
CCC	3

More formally, one can write

$X(OOO) = 0, X(OOC) = 1, X(OCO) = 1, \ldots$

A number is assigned to each outcome. Note that the outcomes OOC, OCO, and COO are all mapped to the same number, 1, and the outcomes OCC, COC, and CCO are all mapped to 2.

STEP 3 Here's the important concept: We are no longer interested in the sequence of letters, or outcomes, but rather focus on the numbers associated with the outcomes. We need to consider the number of possible values that X can assume and the probability that X assumes each value.

The expression $X = 1$ is an event defined in terms of a random variable.

STEP 4 To find, say, the probability that X takes on the value 1, think about which outcomes are assigned to 1 and sum the probabilities of those outcomes. The probability that the random variable X equals 1 is, in this example,

$P(X = 1) = P(OOC) + P(OCO) + P(COO)$

because these three outcomes are mapped to 1. As shown in **Figure 5.3**, the random variable links these three outcomes and their associated probabilities to the number 1.

Figure 5.3 The random variable X maps three outcomes to the number 1. The symbol x represents a possible value of the random variable X.

Outcome	X	x
OOO		0
OOC		1
OCO		2
OCC		3
COO		
COC		
CCO		
CCC		

TRY IT NOW Go to Exercise 5.15

Random Variable Types

Two types of random variables are possible, depending on the number of possible values that the random variable can assume.

RAPID REVIEW: Finite and countably infinite sets, intervals.

Definition
A random variable is **discrete** if the set of all possible values is finite, or countably infinite. A random variable is **continuous** if the set of all possible values is an interval of numbers.

These definitions are analogous to those for discrete and continuous data sets. The following points about random variables are also similar.

A CLOSER LOOK

1. Discrete random variables are usually associated with counting, and continuous random variables are usually associated with measuring.
2. To decide whether a random variable is discrete or continuous, consider all the possible values that the random variable could assume. Finite or countably infinite means discrete. An interval of possible values means continuous.
3. Recall that countably infinite means there are infinitely many possible values, but they are countable. You may not ever be able to finish counting all of the possible values, but there exists a theoretical method for actually counting them.

4. The interval of possible values for a continuous random variable can be any interval of any length, open or closed. The exact interval may not be known, only that some interval of possible values exists.

5. In practice, no measurement device is precise enough to return any number in some interval. In theory, a continuous random variable may assume any value in some interval (but not in reality).

Recall that an experiment may sometimes result in a numerical value right away, not a symbol or a token. In this case, we do not need any extra link or connection to the real numbers. The description of the experiment is the same as the definition of the random variable. The values that the random variable can assume are the possible distinct experimental outcomes. In the following example, several experiments are described and each associated random variable is identified.

EXAMPLE 5.2 Discrete or Continuous

For each of the following experiments, determine whether the associated random variable is discrete or continuous.

(a) The Nassau Inn located in downtown Princeton, New Jersey, offers 157 guest rooms. Each room is unique. The inn caters to tourists and also offers accommodations for business meetings. Let the random variable X be the number of occupied rooms on a randomly selected day.

(b) One way to travel between parks at the Walt Disney World Resort is to use the world-famous monorail system. Let the random variable Y be the amount of time it takes to travel from the Magic Kingdom to Epcot via monorail.

(c) Ring manufactures and sells a video doorbell for homes. At the facility where the doorbells are made and assembled, finished doorbells are randomly selected, tested, and carefully checked for defects. If a defective doorbell is found, the assembly line is shut down. An experiment consists of recording whether the selected doorbell is good (G) or defective (B). The sample space is $S = \{B, GB, GGB, GGGB, GGGGB, \ldots\}$. Let the random variable X be the number of doorbells inspected until a defect is found.

(d) Let the random variable Y be the length of the largest fish caught on the next party boat arriving back to the dock in Belmar, New Jersey.

Solution

(a) There is no need to use a collection of symbols to represent experimental outcomes for the rooms at the inn. The possible values for X (and the distinct experimental outcomes) are finite: $0, 1, 2, 3, \ldots, 157$. These values are distinct, disconnected points on a number line. The random variable X is discrete.

(b) Y is a measurement, the amount of time it takes to travel from the Magic Kingdom to Epcot via monorail. The possible values for Y are any number in some interval, say, 20 to 60 minutes. The random variable Y is continuous.

(c) The values X can assume are $1, 2, 3, 4, \ldots$. The number of possible values is countably infinite. If we plot these values on a number line, they are separate, distinct, and disconnected. The random variable X is discrete.

(d) Y is a measurement, and it can (theoretically) take on any value in some interval. The possible values for Y are any number in some interval, say, 5 to 25 inches. The random variable Y is continuous.

TRY IT NOW Go to Exercises 5.11 and 5.13

CHAPTER APP Hooked on Vitamins

A random sample of 50 dietary supplements was obtained. Let X be the number tested that actually contained the supplement shown on the label. Is X discrete or continuous?

Section 5.1 Exercises

Concept Check

5.1 True or False The set of all possible values for a random variable can be infinite.

5.2 True or False A random variable may assign more than one numerical value to an outcome.

5.3 True or False A random variable can be both discrete and continuous.

5.4 True or False If the set of possible values for a random variable is greater than 500, then the random variable is continuous.

5.5 Fill in the Blank A random variable is a special kind of _____.

5.6 Fill in the Blank A random variable maps elements of the _____ to the _____.

5.7 Fill in the Blank Discrete random variables are usually associated with _____.

5.8 Fill in the Blank Continuous random variables are usually associated with _____.

5.9 Short Answer If X is a discrete random variable, explain how to find $P(X = 2)$.

Practice

5.10 Classify each random variable as discrete or continuous.
(a) The number of lightning bugs in the backyard of a rural home
(b) The volume of ice cream in one scoop
(c) The area of a randomly selected baseball field including the foul territory
(d) The number of late deliveries in one month by a package delivery service
(e) The number of girls born in a rural hospital during the next year
(f) The interest rate on a savings account at a randomly selected bank in Philadelphia
(g) The number of tickets sold in the next Powerball lottery
(h) The number of oil tankers registered to a certain country at a given time

5.11 Classify each random variable as discrete or continuous.
(a) The number of visitors to the Museum of Science in Boston on a randomly selected day
(b) The camber-angle adjustment necessary for a front-end alignment
(c) The total number of pixels in a photograph produced by a digital camera
(d) The number of days until a rose begins to wilt after purchase from a flower shop
(e) The running time for the latest *Star Wars* movie
(f) The volume of soft drink that the next customer pours from a self-serve beverage dispenser at a Five Guys restaurant

5.12 Classify each random variable as discrete or continuous.
(a) The number of people requesting vegetarian meals on a flight from New York to London
(b) The exact thickness (in millimeters) of a paper towel
(c) The time it takes a driver to react after the car in front stops suddenly
(d) The number of escapees in the next prison breakout
(e) The length of time that a deep-space probe remains in contact with Earth
(f) The number of points on a randomly selected buck (The definition of a point is an antler projection at least 1 in. in length from the base to tip. The brow tine and the main beam tip are counted as points regardless of length.)

5.13 Classify each random variable as discrete or continuous.
(a) The number of votes necessary to elect a new pope
(b) The amount of sugar in a 16-oz sweetened bottled drink purchased in a New York City cafe
(c) The total number of riders on all forms of public transportation in the United States during the year
(d) The number of residents in an assisted-living center who have hardening of the arteries
(e) The amount of lead measured in the soil of a children's playground
(f) The time it takes an automobile to pass through the George Massey Tunnel in Vancouver, British Columbia

5.14 Classify each random variable as discrete or continuous.
(a) The total weight of all food collected during a food drive
(b) The number of guests at a wedding reception
(c) The number of people who are hospitalized following an automobile accident
(d) The total number of steps taken during a regular workday by a hotel valet
(e) The amount of time it takes a robot vacuum to clean a room
(f) The weight of a grizzly bear in British Columbia

Applications

5.15 Marketing and Consumer Behavior Zappos is an online retailer that sells shoes for men, women, boys, and girls. An experiment consists of classifying the next two items purchased as either men's, women's, boys', or girls' shoes. Let the random variable X be the number of sales of women's or girls' shoes.
(a) List the outcomes in the sample space.
(b) What are the possible values for X? Is X discrete or continuous? Justify your answer.

5.16 Education and Child Development An experiment consists of showing a 4-year-old child an interactive instructional video and then asking that the child to tie his shoelaces. The random variable Y is the length of time that the child takes to tie the first shoelace. Is Y discrete or continuous? Justify your answer.

5.17 Biology and Environmental Science The Waynesburg Lions Club receives a shipment of 300 Christmas trees from Wending Creek Farms in Coudersport, Pennsylvania, to sell as a fundraiser. Classify each of the following random variables as discrete or continuous.
(a) The number of trees that are more than 6 ft tall
(b) The moisture content (expressed as a percentage) of a randomly selected tree
(c) The number of Douglas fir trees in the shipment
(d) The diameter of the trunk at the bottom of a randomly selected tree

5.18 Biology and Environmental Science To map the current of bottom water in a certain part of the Atlantic Ocean, a dye is released and used to trace the water flow. Let the random variable X be the maximum distance (in meters) from release at which the dye is detected after one day. Is X discrete or continuous? Justify your answer.

5.19 Psychology and Human Behavior An experiment consists of recording the behavior of a randomly selected San Diego cab driver as a traffic signal changes from red to green. Let the random variable X be the acceleration (in ft/s^2) of the cab 1 second after the light changes. Is X discrete or continuous? Justify your answer.

5.20 Technology and the Internet Many computers are now being equipped with a solid-state drive (SSD) instead of a hard-disk drive. SSDs use flash memory to store data, so they have no moving parts. Therefore, they provide much better performance and tend to be more reliable. Suppose the laptop computers for Nationwide Insurance's regional sales force are examined. Let X be the number of laptops with a SSD. Is X discrete or continuous? Justify your answer.

5.21 Sports and Leisure The Boston Red Sox play their spring training games in JetBlue Park, Fort Myers, Florida. Suppose a game against the Tampa Bay Rays is selected. Classify each of the following random variables as discrete or continuous.
(a) The price of a randomly selected ticket
(b) The time it takes to complete the game
(c) Whether the game is postponed due to rain (0) or is completed (1)
(d) The speed of the first pitch in the bottom of the third inning
(e) The number of fans in attendance
(f) The number of hot dogs sold during the entire game
(g) The total number of errors in the game
(h) The weight (in ounces) of the bat used by the third hitter in the fifth inning

5.22 Travel and Transportation Key West Express offers ferry trips to Key West from two locations in southwest Florida. Suppose a trip from Marco Island to Key West is selected. Classify each of the following random variables as discrete or continuous.
(a) The number of passengers on board
(b) The number of cars on board
(c) The time it takes to reach Key West
(d) Whether the ferry leaves Marco Island on time (0) or is late (1)
(e) The speed of the ferry at mid-voyage
(f) The direction of the wind (in degrees) as the ferry leaves Marco Island
(g) The number of hot chocolates sold during the trip
(h) The temperature of the ocean water near Key West
(i) The height of the largest wave encountered on the trip
(j) The number of people who wear a sweater during the trip

5.2 Probability Distributions for Discrete Random Variables

Definition and Example

A random variable is a rule that assigns each experimental outcome to a real number. To complete the description of a discrete random variable so that we can understand and answer questions involving the random variable, we need to know all possible values that the random variable can assume and all the associated probabilities. This collection of values and probabilities is called a *probability distribution*. Because random variables are used to model populations, a probability distribution is a theoretical description of a population.

A random variable provides the link between experimental outcomes and real numbers. An experimental outcome and the probability assigned to that outcome are both associated with exactly the same value of the random variable. This connection determines probability assignments for a random variable.

Definition
The **probability distribution for a discrete random variable** X is a method for specifying all of the possible values of X and the probability associated with each value.

A CLOSER LOOK

1. A probability distribution for a discrete random variable may be presented in the form of an itemized listing, a table, a graph, or a function.
2. A probability mass function (pmf), denoted p, is the probability that a discrete random variable is equal to some specific value. In symbols, it is defined by

$$p(x) = \underbrace{P(X = x)}_{\text{Rule}}.$$

In words, the rule for the function p evaluated at an input x is the probability of an event, the probability that the random variable X takes on the specific value x. The function p and its probability rule are used interchangeably.

Suppose X is a discrete random variable. Then $p(7)$ means find the probability that the random variable X equals 7, or $P(X = 7)$.

EXAMPLE 5.3 Probability Distribution Construction

Suppose an experiment has eight possible outcomes, each denoted by a sequence of three letters, where each letter is either an N or a D. The probability of each outcome is given in the following table.

Outcome	NNN	NND	NDN	DNN	NDD	DND	DDN	DDD
Probability	0.336	0.224	0.144	0.084	0.096	0.056	0.036	0.024

The random variable X is defined to be the number of Ds in an outcome. Find the probability distribution for X.

Solution

STEP 1 The probability distribution for X consists of all possible values that X can assume, along with the associated probabilities. The following table shows the random variable assignment and a technique for calculating the probability of each value.

Experiment			Probability distribution	
Probability	Outcome	X	Value, x	Probability
0.336	NNN →		0	$P(X = 0) = 0.336$
0.224	NND			
0.144	NDN →		1	$P(X = 1) = 0.224 + 0.144 + 0.084 = 0.452$
0.084	DNN			
0.096	NDD			
0.056	DND →		2	$P(X = 2) = 0.096 + 0.056 + 0.036 = 0.188$
0.036	DDN			
0.024	DDD →		3	$P(X = 3) = 0.024$

5.2 Probability Distributions for Discrete Random Variables

To find the probability that X takes on a specific value x, find all the outcomes that are mapped to x, and add the probabilities of these outcomes.

STEP 2 The random variable X takes on the values 0, 1, 2, and 3. The probability distribution can be presented in a table like the one shown here.

x	0	1	2	3
$p(x)$	0.336	0.452	0.188	0.024

> ✓ Looking at just this (probability distribution) table, how do you know X cannot assume any other value?

TRY IT NOW Go to Exercise 5.35

A CLOSER LOOK

1. Think carefully about this process of constructing a probability distribution. To find the probability that X takes on the value x, look back at the experiment and find all the outcomes that are mapped to x. Add the probabilities associated with these outcomes.

2. The probability distribution for a random variable X is a reference for use in answering probability questions about the random variable. For example, we'll need to answer probability questions such as "Find $P(X = 3)$." Think of $X = 3$ as an event stated in terms of a random variable. The details needed for answering this question are in the probability distribution.

Probability Distribution Presentations

The next example illustrates various methods for presenting a probability distribution.

EXAMPLE 5.4 Words of Wisdom (Teeth)

With age comes wisdom, and so do wisdom teeth. Most people experience the emergence of their wisdom teeth between the ages of 17 and 21. However, not everyone has all four wisdom teeth, or even three. Suppose the random variable Y represents the number of wisdom teeth with which a person is born. The probabilities of Y taking on various values are as follows: 5/15 for 0 wisdom teeth; 4/15 for one; 3/15 for two; 2/15 for three; and 1/15 for all four. Here are several ways to represent the probability distribution for Y.

> Don't worry about where these actual probabilities came from here. In this example, focus only on the methods for conveying all the values and probabilities. In fact, approximately 35% of all people really are born without any wisdom teeth.

Solution

STEP 1 A complete listing of all possible values and associated probabilities (use either the probability mass function p or the Assignment Rule).

$P(Y = 0) = 5/15$
$P(Y = 1) = 4/15$
$P(Y = 2) = 3/15$
$P(Y = 3) = 2/15$
$P(Y = 4) = 1/15$

> The random variable Y can take on the values 0, 1, 2, 3, or 4, and the probability of each value is given. There can be no other value of Y because the probabilities sum to 1.

STEP 2 A table of values and probabilities.

y	0	1	2	3	4
$p(y)$	5/15	4/15	3/15	2/15	1/15

This kind of table is the most common way to present a probability distribution for a discrete random variable. It concisely lists all the values that Y can assume and the associated probabilities.

STEP 3 A probability histogram.

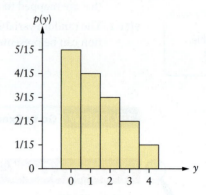

In this example, the sum of the areas of all the rectangles is 1.

The distribution of Y is represented graphically. A rectangle is drawn for each value y, centered at y, with height equal to $p(y)$.

STEP 4 A point representation.

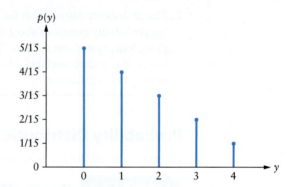

Plot the points $(y, p(y))$ and draw a line from $(y, 0)$ to $(y, p(y))$.

STEP 5 A formula.

$$p(y) = \frac{5-y}{15} \quad y = 0, 1, 2, 3, 4$$

This shows the rule for the probability mass function. For example, to find $p(2)$, which is the probability that $Y = 2$, let $y = 2$ in the formula to find

$$p(2) = \frac{5-2}{15} = \frac{3}{15}$$

Probability Questions Involving a Random Variable

All of the techniques presented in the preceding example are valid methods for presenting a probability distribution. Use the style that is most convenient or appropriate, or what is called for in the question. Often, a graphical representation of the distribution will be helpful. Sometimes, having a formula for the probability distribution is more useful. In the next example, we'll construct another probability distribution and consider some probability questions involving a random variable.

EXAMPLE 5.5 Who Wants Coffee?

The Oddly Correct Coffee Bar in Kansas City carefully monitors customer orders and has found that 70% of all customers ask for some kind of coffee (C), whereas the remainder order a specialized tea (T). Suppose four customers are selected at random. Let the random variable X be the number of customers who order coffee.

(a) Find the probability distribution for X.
(b) Find the probability that more than two customers order coffee.
(c) Suppose at least two customers order coffee. What is the probability that all four customers order coffee?

Solution

The experiment consists of observing four customer choices. Each outcome consists of a sequence of four letters, with each letter being either a C or a T. Using the Multiplication Rule, there are 16 possible outcomes: CCCC, CCCT, CCTC,....

Because the customers are selected at random, each choice is independent, and the probability of each outcome is obtained by multiplying the corresponding probabilities. For example,

$P(CTCT) = P(C \cap T \cap C \cap T)$ *First customer buys coffee and second customer buys tea and ...*

$= P(C) \cdot P(T) \cdot P(C) \cdot P(T)$ *Events are independent; multiply corresponding probabilities.*

$= (0.70)(0.30)(0.70)(0.30)$ $P(T) = 1 - P(C)$

$= 0.0441$

The following table lists all of the possible experimental outcomes, the probability of each outcome (computed as just shown), and the value of the random variable assigned to each outcome.

Outcome	Probability	x	Outcome	Probability	x
TTTT	0.0081	0	CTTC	0.0441	2
TTTC	0.0189	1	CTCT	0.0441	2
TTCT	0.0189	1	CCTT	0.0441	2
TCTT	0.0189	1	TCCC	0.1029	3
CTTT	0.0189	1	CTCC	0.1029	3
TTCC	0.0441	2	CCTC	0.1029	3
TCTC	0.0441	2	CCCT	0.1029	3
TCCT	0.0441	2	CCCC	0.2401	4

This table shows that the values of X are 0, 1, 2, 3, and 4.

(a) We know all of the values that X can assume. Summarize all of the probability assignments in the table by each value of the random variable and construct the probability distribution for X.

$p(0) = P(X = 0) = P(TTTT) = 0.0081$

There is only one outcome assigned to the value 0, and the probability of that outcome is 0.0081.

$p(1) = P(X = 1)$ *Definition of a probability mass function.*

$= P(TTTC \text{ or } TTCT \text{ or } TCTT \text{ or } CTTT)$ *These outcomes are mapped to 1.*

$= P(TTTC) + P(TTCT) + P(TCTT) + P(CTTT)$ *Or means union; the outcomes are disjoint.*

$= 0.0189 + 0.0189 + 0.0189 + 0.0189 = 0.0756$

Continue in this manner to obtain the probability distribution for X.

x	0	1	2	3	4
$p(x)$	0.0081	0.0756	0.2646	0.4116	0.2401

(b) Use the probability distribution to determine the values of X that are greater than 2, and add the associated probabilities.

$$P(X > 2) = P(X = 3) + P(X = 4)$$ Only values of X greater than 2.
$$= 0.4116 + 0.2401 = 0.6517$$ Use the probability distribution table.

The probability of more than two customers ordering coffee is 0.6517.

Note: How would this probability change if the question asked for the probability that two or more customers order coffee?

(c) The phrase "Given that at least two customers order coffee" suggests conditional probability, and the number of values that X can assume is now reduced. Use the definition of conditional probability with events involving the random variable.

Given that X is at least 2, find the probability that X is exactly 4.

$$P(X = 4 \mid X \geq 2) = \frac{P(X = 4 \cap X \geq 2)}{P(X \geq 2)}$$ Definition of conditional probability.

$$= \frac{P(X = 4)}{P(X \geq 2)}$$ Intersection of $X = 4$ and $X \geq 2$.

$$= \frac{0.2401}{0.2646 + 0.4116 + 0.2401}$$ Use the probability distribution.

$$= \frac{0.2401}{0.9163} = 0.2620$$

Given that at least two people order coffee, the probability that exactly four order coffee is 0.2620.

Here is a way to visualize this conditional probability using the probability distribution table.

x	0	1	2	3	4
p(x)	0.0081	0.0756	0.2646	0.4116	0.2401

Given that X is either 2, 3, or 4, the reduced, or relevant, probability is 0.9163. The proportion of time that X is equal to 4, given X is 2, 3, or 4, is 0.2401/0.9163. ∎

TRY IT NOW Go to Exercises 5.37 and 5.39

Inference and Random Variables

The probability distribution of a random variable reveals which values of the random variable are most likely to occur. This information is extremely helpful in making a statistical inference. Consider the following example.

EXAMPLE 5.6 Burglar Alarms

Doberman Security sells a motion detector burglar alarm designed especially for home use. To feel very secure, a home owner installs three of these burglar alarms. Each device operates independently, and the manufacturer claims that the probability that each burglar alarm detects motion during a burglary is 0.95. Suppose a burglary occurs and all three devices fail to detect the motion. Do you have reason to doubt the manufacturer's claim? Justify your answer.

Solution

The phrase "Do you have reason to doubt the manufacturer's claim," suggests statistical inference. Consider the claim, the experiment, and the likelihood of the experimental outcome. Use this result to draw a conclusion. Define a random variable, and use the probability distribution to determine the probability that all three burglar alarms fail.

STEP 1 Let S represent a success, in which a burglar alarm functions properly and detects motion. Let F represent a failure, in which a burglar alarm fails to detect motion. There are eight possible experimental outcomes. Let X be the number of successes.

STEP 2 The table lists each outcome, the probability of each outcome, and the value of the random variable associated with each outcome.

Outcome	Probability	x
FFF	0.0001	0
SFF	0.0024	1
FSF	0.0024	1
FFS	0.0024	1
SSF	0.0451	2
SFS	0.0451	2
FSS	0.0451	2
SSS	0.8574	3

Note: The probabilities are rounded to four places to the right of the decimal.

Because each burglar alarm operates independently, the probability of each outcome is the product of the corresponding probabilities. For example,

$$P(SFS) = P(S \cap F \cap S)$$
$$= P(S) \cdot P(F) \cdot P(S)$$
$$= (0.95)(0.05)(0.95) = 0.0451$$

STEP 3 Use the links in the table to construct the probability distribution for X.

x	0	1	2	3
$p(x)$	0.0001	0.0072	0.1353	0.8574

STEP 4 Use the four-step inference procedure.

Claim: The probability that each burglar alarm will function properly is 0.95.
Experiment: The value of the random variable observed is $x = 0$.
Likelihood: The likelihood of the observed outcome is $P(X = 0) = 0.0001$.
Conclusion: Because this probability is so small, the outcome of observing zero successes is very rare. But it happened! This small probability suggests the assumption is wrong. There is evidence to suggest that the claim of 0.95 (motion detection probability) is wrong. ∎

It seems very intuitive and reasonable to consider the probability of the observed outcome as a measure of likelihood. Later we'll modify this likelihood slightly to be more conservative, to consider results at least as extreme, and to account for continuous random variables.

TRY IT NOW Go to Exercise 5.50

Properties of a Probability Distribution

As the previous examples suggest, the following properties must be true for every probability distribution for a discrete random variable X.

Properties of a Valid Probability Distribution for a Discrete Random Variable

1. $0 \le p(x) \le 1$

The probability that X takes on any value, $p(x) = P(X = x)$, must be between 0 and 1.

2. $\sum_{\text{all } x} p(x) = 1$

The sum of all the probabilities in a probability distribution for a discrete random variable must equal 1.

The following example involves a probability distribution for a discrete random variable and illustrates these two properties.

EXAMPLE 5.7 Coney Island Skeeball

Neville Elder/Corbis/Getty Images

Skeeball is a nostalgic, historical arcade game that is played by rolling balls up an inclined plane. The object is to score as many points as possible by landing the balls in holes with certain point values. The number of points earned on one roll is a random variable Y. Suppose Y has the following probability distribution.

y	0	10	20	30	40	50	100
$p(y)$	0.30	0.24	0.18	?	0.08	0.04	0.02

(a) Find $p(30)$.

(b) Find $P(10 \leq Y \leq 40)$ and $P(10 < Y < 40)$.

(c) Construct the corresponding probability histogram.

Solution

(a) The sum of all the probabilities must equal 1.

$$p(30) = 1 - [p(0) + p(10) + p(20) + p(40) + p(50) + p(100)]$$
$$= 1 - (0.30 + 0.24 + 0.18 + 0.08 + 0.04 + 0.02)$$
$$= 1 - 0.86 = 0.14$$

(b) The values that Y takes on between 10 and 40 inclusive are 10, 20, 30, and 40.

$$P(10 \leq Y \leq 40) = p(10) + p(20) + p(30) + p(40)$$
$$= 0.24 + 0.18 + 0.14 + 0.08 = 0.64$$

The values that Y takes on strictly between 10 and 40 are 20 and 30.

$$P(10 < Y < 40) = p(20) + p(30)$$
$$= 0.18 + 0.14 = 0.32$$

In this example, including (or excluding) an endpoint (a single value) changes the probability assignment. It is important to remember that a single value may make a difference in a probability assignment for a discrete random variable.

(c) To construct the probability histogram, draw a rectangle for each value y, centered at y, with height equal to $p(y)$.

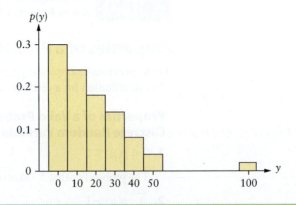

Section 5.2 Exercises

Concept Check

5.23 True or False A probability distribution is a theoretical model of a population.

5.24 True or False The sum of all the probabilities in a probability distribution for a discrete random variable must equal 1.

5.25 True or False For a discrete random variable, under certain circumstances $p(x)$ could be less than 0.

5.26 True or False The values of a discrete random variable must be evenly spaced, or at regular intervals, on a number line.

5.27 Fill in the Blank The probability distribution for a discrete random variable is a method for specifying _____ and _____.

5.28 Short Answer Suppose X is a discrete random variable. Is the expression $P(X \leq 6) = P(X < 6)$ always true, sometimes true, or never true? Justify your answer.

5.29 Short Answer Briefly describe several methods to represent a probability distribution for a discrete random variable.

Practice

5.30 The probability distribution for the random variable X is given in the table.

x	1	2	3	4	5	6	7
$p(x)$	0.35	0.20	0.15	0.12	?	0.08	0.03

(a) Find $p(5)$.
(b) Find $P(2 \leq X \leq 6)$ and $P(2 < X \leq 6)$.
(c) Find $P(X < 4)$.
(d) Find the probability that X takes on the value 1 or 7.

5.31 The probability distribution for the random variable Y is given in the table.

y	10	20	25	30	45	50
$p(y)$	0.155	0.237	0.184	0.122	?	0.258

(a) Find $p(45)$.
(b) Find $P(Y \geq 25)$ and $P(Y > 25)$.
(c) Find the probability that Y is divisible by 10.
(d) Construct the corresponding probability histogram.

5.32 The probability distribution for the random variable X is given in the table.

x	-3	-2	-1	0	1	2	3
$p(x)$	0.20	0.10	0.05	0.30	0.05	0.10	0.20

(a) Find $P(X \geq 0)$ and $P(X > 0)$.
(b) Find $P(X^2 > 1)$.
(c) Find $P(X \geq 2 \mid X \geq 0)$.
(d) Construct the corresponding probability histogram.

5.33 Determine whether the probability distribution is valid. Justify your answer.

(a)
x	2	4	6	8	10	12
$p(x)$	0.15	0.16	0.17	0.19	0.19	0.20

(b)
x	2	4	6	8	10	12
$p(x)$	0.25	0.25	0.25	-0.25	0.25	0.25

(c)
x	2	4	6	8	10	12
$p(x)$	0.05	0.20	0.25	0.25	0.20	0.05

(d)
x	2	4	6	8	10	12
$p(x)$	0.06	0.25	0.30	0.12	0.18	0.10

5.34 The table lists all of the possible outcomes for an experiment, the probability of each outcome, and the value of a random variable assigned to each outcome. Use this table to construct the probability distribution for X. Construct the corresponding probability histogram.

Outcome	Probability	x	Outcome	Probability	x
AA	0.01	1	CA	0.03	3
AB	0.02	2	CB	0.06	3
AC	0.03	3	CC	0.09	3
AD	0.04	4	CD	0.12	4
BA	0.02	2	DA	0.04	4
BB	0.04	2	DB	0.08	4
BC	0.06	3	DC	0.12	4
BD	0.08	4	DD	0.16	4

5.35 The table lists all of the possible outcomes for an experiment, the probability of each outcome, and the value of a random variable assigned to each outcome. Use this table to construct the probability distribution for Y. Construct the corresponding probability histogram. EX5.35

5.36 The probability distribution for a discrete random variable X is given by the following formula:

$$p(x) = \frac{x(x+1)}{112} \quad x = 1, 2, \ldots, 6$$

(a) Verify that this is a valid probability distribution.
(b) Find $P(X = 4)$.
(c) Find $P(X > 2)$.
(d) Find the probability that X takes on the value 3 or 4.
(e) Construct the corresponding probability histogram.

Applications

5.37 Manufacturing and Product Development A wooden kitchen cabinet is carefully inspected at the manufacturing facility before it is sent to a retailer. The random variable X is the number of defects found in a randomly selected cabinet. The probability distribution for X is given in the table.

x	0	1	2	3	4	5
$p(x)$	0.900	0.050	0.025	0.020	0.004	0.001

Suppose a cabinet is selected at random.
(a) What is the probability that the cabinet is defect free?
(b) What is the probability that the cabinet has at most two defects?
(c) What is the probability that two randomly selected cabinets both have at least three defects?
(d) Find $P(2 \leq X \leq 4)$ and $P(2 < X < 4)$.

5.38 Fuel Consumption and Cars The cost of an automobile insurance policy depends on many factors, including marital status, where you live, credit score, age of the car, and driving history.[3] Suppose that for some driver category, the probability distribution for the random variable Y, the amount (in dollars) of a collision deductible on a random automobile policy, is given in the table.

y	0	100	200	500	1000
$p(y)$	0.05	0.11	0.21	0.25	0.38

(a) Find $P(Y > 0)$.
(b) Find $P(Y \leq 200)$.
(c) What is the probability that a randomly selected policy has a $500 collision deductible?
(d) Suppose two policies are selected at random. What is the probability that both have $200 collision deductibles?
(e) Suppose two policies are selected at random. What is the probability that at least one has a collision deductible of $500 or less?

5.39 Sports and Leisure Ice fishing has gained in popularity even though it often involves cold-weather apparel, a portable heater, and an ice shelter. The New York Freshwater Fishing Regulations stipulate that no more than seven ice fishing lines may be used at one time.[4] Suppose the number of lines that an ice fisherman uses is a random variable X, with the probability distribution given in the table.

x	1	2	3	4	5	6	7
$p(x)$	0.371	0.247	0.158	0.096	0.063	0.042	0.023

Suppose an ice fisherman is selected at random.
(a) What is the probability that the fisherman has exactly one line?
(b) What is the probability that the fisherman has at least two lines?
(c) What is the probability that the fisherman has at most three lines?
(d) Suppose the fisherman has at least three lines. What is the probability that he has at least five lines?
(e) Suppose three fishermen are selected at random. What is the probability all three will have more than five lines?

5.40 Biology and Environmental Science To care for the environment, many grocery stores no longer use plastic bags, and many customers bring their own reusable bags. However, some grocery stores sell plastic bags, and a recent survey indicated that 30% of all grocery store shoppers buy plastic bags.[5] Suppose three grocery store shoppers are selected at random. Let X be the number who buy plastic bags.
(a) Construct the probability distribution for X. Draw the corresponding probability histogram.
(b) What is the probability that all three shoppers buy plastic bags?
(c) What is the probability that at least one of the three shoppers buys plastic bags?

5.41 Marketing and Consumer Behavior Twinkies are an American tradition, sold for more than 83 years, originating in a Chicago bakery, and forever linked to legal lingo. (You may have read about the "Twinkie defense," associated with a 1979 murder trial in San Francisco.) According to an eating habits survey, approximately 34% of Americans say Twinkies are delicious.[6] Suppose four Americans are selected at random. Let Y be the total number of Americans who say Twinkies are delicious.
(a) Construct the probability distribution for Y.
(b) What is the probability that at least one American says Twinkies are delicious?
(c) Suppose at least two Americans say Twinkies are delicious. What is the probability that all four say they are delicious?

5.42 Manufacturing and Product Development Staples has six special drafting pencils for sale, two of which are defective. A student buys two of these six drafting pencils, selected at random. Let the random variable X be the number of defective pencils purchased. Construct the probability distribution for X.

5.43 Business and Management Two packages are independently shipped from Princeton, New Jersey, to the Convention Center in Kansas City, Missouri, and each is guaranteed to arrive within four days. The probability that a package arrives on the first day is 0.10, on the second day is 0.15, on the third day is 0.25, and on the fourth day is 0.50. Let the random variable X be the total number of days for both packages to arrive. Construct the probability distribution for X.

5.44 Marketing and Consumer Behavior Alvin, Simon, and Theodore work in the same office, and each routinely stops for a bagel on the way to work. The probability that Alvin stops for a bagel is 0.70, Simon 0.80, and Theodore, 0.90; and each stops independently of the others. Let the random variable X be the total (of the three) who stop for a bagel on the way to work on a randomly selected day. Find the probability distribution for the random variable X.

Extended Applications

5.45 Biology and Environment Science Agway, a farm and garden supply store, sells winter fertilizer in 50-lb bags. For customers who purchase this product, the probability distribution for the random variable X, the number of bags sold, is given in the table.

x	1	2	3	4	5
$p(x)$	0.55	0.35	0.07	0.02	0.01

Suppose a person buying winter fertilizer is randomly selected.
(a) What is the probability that the customer buys more than two bags?
(b) What is the probability that the customer does not buy two bags [$P(X \neq 2)$]?
(c) Find the probability that two randomly selected customers each buy one bag.
(d) Suppose two customers are randomly selected. What is the probability that the total number of bags purchased will be at least eight?
(e) Let the random variable Y be the number of pounds sold to a randomly selected customer buying winter fertilizer. Find the probability distribution for Y.

5.46 Sports and Leisure Suppose the probability that a person claims he or she was at the Woodstock Festival and Concert is 0.20. An experiment consists of randomly selecting people in Berkeley, California, and asking them whether they were at Woodstock. The experiment stops as soon as one person says he or she was there. The random variable X is the number of people stopped and questioned (until one person says he or she was there). Let Y and N stand for Yes and No responses, respectively.
(a) List the first several outcomes in the sample space.
(b) Find the probability of each outcome in part (a).
(c) Find the value of the random variable associated with each outcome in (a).
(d) What are the possible values for the random variable X?
(e) Find a formula for the probability distribution of X.

5.47 Sports and Leisure A game show contestant on *Let's Make a Deal* selects two envelopes from a total of six envelopes holding prize money. Two of the envelopes contain $100, one envelope contains $250, two envelopes contain $500, and the last envelope contains $1000. Let the random variable M be the maximum of the two prizes.
(a) Find the probability distribution for M.
(b) Suppose two contestants independently select prize envelopes on two different days. What is the probability that both win the top prize?

5.48 Economics and Finance The Cornbread Cafe in Eugene, Oregon, recently installed a new computer system that allows customers to order their meals electronically from their tables. The manager of the restaurant claims that this new system will decrease the waiting time, and that the probability of getting a meal in less than 7 minutes (with this system in place) is 0.75. Suppose four customers are selected at random. Let the random variable X be the number of customers who get their meal in less than 7 minutes.
(a) Find the probability distribution for X.
(b) Suppose none of the four customers gets the meal in less than 7 minutes. Is there any evidence to suggest that the manager's claim is false? Justify your answer.

5.49 Demographics and Population Statistics Most physicians in Canada graduated from a medical school in Canada. However, some attended medical school in the United States or another foreign country. The probability that a physician in Alberta attended a Canadian medical school is 0.656; in British Columbia, 0.708; in Quebec, 0.885; and in Ontario, 0.731.[7] Suppose one physician is independently selected from each province. Let the random variable Y be the number of physicians who attended a Canadian medical school.
(a) Construct a probability distribution for Y.
(b) What is the probability that at least one physician attended a Canadian medical school?
(c) Suppose another group of physicians is selected, one from each province. What is the probability that at least three physicians from each group attended a Canadian medical school?

5.50 Psychology and Human Behavior Memorial Day is a wonderful holiday that traditionally marks the beginning of summer. It provides a chance to travel and try out that new backyard barbecue recipe. According to a recent survey, 60% of American families barbecue during Memorial Day weekend.[8] Suppose four American families are selected at random. Let the random variable X be the number of families who barbecue during Memorial Day weekend.
(a) Find the probability distribution for X.
(b) What is the probability that at least three families barbecue during Memorial Day weekend?
(c) Suppose none of the four families barbecues during Memorial Day weekend. Is there any evidence to suggest that the survey claim is wrong? Justify your answer.

Challenge Problems

5.51 Uneven Steps Suppose X is a random variable with the probability distribution given in the table.

x	0	1	2	3	4	5
$p(x)$	0.21	0.18	0.15	0.12	0.10	0.08

x	6	7	8	9	10
$p(x)$	0.06	0.04	0.03	0.02	0.01

Consider the function $F(x) = P(X \leq x)$, which is the cumulative distribution function (CDF) of the random variable X. This function accumulates probability up to and including x.
(a) Find $F(x)$ for $x = 0, 1, 2, \ldots, 10$.
(b) Find $F(-1)$ and $F(25)$.
(c) Find $F(x)$ for any value $x < 0$. Justify your answer.
(d) Find $F(x)$ for any value $x > 10$. Justify your answer.
(e) Find $F(2.1)$, $F(2.5)$, and $F(2.99)$.
(f) Sketch a graph of the function F. Describe the shape of this graph.

5.3 Mean, Variance, and Standard Deviation for a Discrete Random Variable

The Mean

Just as there are descriptive measures of a sample (for example, $\bar{x}$, s^2, and s), there are corresponding descriptive measures of a population (μ, σ^2, and σ). As we learned in Chapter 3, these population parameters describe the center and variability of the entire population. They are usually unknown values that we would like to estimate.

However, because a random variable may be used to model a population, these (population) descriptive measures are inherent in and determined by the probability distribution. This section presents the methods used to compute the mean, variance, and standard deviation of a discrete random variable (or population). The next example suggests a definition of *expected value*.

EXAMPLE 5.8 Amateur Umpiring

Baseball is America's national pastime, but baseball umpires often deal with a lot of grief from players and fans. Suppose an amateur umpire works high school baseball games. Every day on which there are games in his region, his name is entered into an umpire pool. The probability of being selected to umpire on any day is 4/5. If he is selected, he will earn $50. On days he is not selected, he earns nothing. How much money does this amateur umpire earn per day, *on average*? Or, in the long run, how much does the umpire earn each day?

Solution

STEP 1 This question focuses on the amount earned each day, *on average*, not on any one particular day. Consider the probabilities given and consider five typical days.

On four of five days, the umpire earns $50. On the fifth day, he earns $0. The total earned for the five typical days is $200. To find the average amount earned each day, divide by 5: $200/5 = $40.

STEP 2 Consider a random variable X that takes on only two values, 0 and 50, with probabilities 0.20 and 0.80, respectively. Another way to compute the *average* earned each day is to use this probability distribution.

$$40 = \underset{\text{Value}}{0} \times \underset{\text{Probability}}{0.20} + \underset{\text{Value}}{50} \times \underset{\text{Probability}}{0.80}$$

The long-run average earnings per day can be found by using a probability distribution. Multiply each value by its corresponding probability, and sum these products. ∎

Definition

Let X be a discrete random variable with probability mass function $p(x)$. The **mean**, or **expected value**, of X is

$$\underset{\text{Notation}}{E(X) = \mu = \mu_X} = \underset{\text{Calculation}}{\sum_{\text{all } x} [x \cdot p(x)]} \tag{5.1}$$

5.3 Mean, Variance, and Standard Deviation for a Discrete Random Variable 207

A CLOSER LOOK

1. The capital E stands for *expected value* and is a function. The function E takes a random variable as an input and returns the expected value. More generally, E accepts as an input any function of a random variable. For example, suppose $f(X)$ is a function of a discrete random variable X. The expected value of $f(X)$ is

$$E[f(x)] = \sum_{\text{all } x}[f(x) \cdot p(x)] \qquad (5.2)$$

2. μ is the mean, or expected value, of a random variable (which may model a population). If necessary, the associated random variable is used as a subscript for identification–for example, μ_X or μ_Y.

3. The mean is easy to compute: Multiply each value of the random variable by its corresponding probability, and add the products.

4. The mean of a random variable is a *weighted average* and is what happens on average. The mean may not be any of the possible values of the random variable.

RAPID REVIEW Weighted average.

EXAMPLE 5.9 Mortgage Shopping

The First Columbia Bank & Trust Company offers four different fixed-rate home mortgages. The interest rate on each is related to the length of the loan. Suppose the length (in years) of a random mortgage is a discrete random variable X, with probability distribution given in the table.

x	10	15	20	30
$p(x)$	0.10	0.25	0.20	0.45

Find the mean of X.

Solution

STEP 1 X is a discrete random variable. To find the mean, use Equation 5.1.

STEP 2 $\mu = \sum_{\text{all } x}[x \cdot p(x)]$ *Equation 5.1.*

$= (10)(0.10) + (15)(0.25) + (20)(0.20) + (30)(0.45)$ *Multiply each value by its corresponding probability, and sum.*

$= 1.00 + 3.75 + 4.00 + 13.50 = 22.25$

STEP 3 The mean, or long-run value, of X is 22.25. On average, the length of a mortgage is 22.25 years. In this example, the mean is not a possible value of X. **Figure 5.4** shows a technology solution. ■

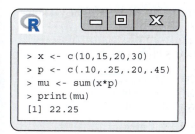

Figure 5.4 The mean of the random variable X.

Common Error

$\mu = \dfrac{1}{n} \sum_{\text{all } x} x$

where n is the number of possible values of the random variable.

Correction: This is an attempt to find a population mean using the concept of a sample mean. The mean of a discrete random variable is
$\mu = \sum_{\text{all } x}[x \cdot p(x)]$.

EXAMPLE 5.10 Preventive Medicine

Many physicians recommend essential blood tests for their patients every year, including complete blood count, liver function tests, and cholesterol level.[9] Suppose X is a random variable that represents the number of essential blood tests ordered for a randomly selected patient following a routine yearly physical exam. The probability distribution for X is given in the table.

x	0	1	2	3	4	5
$p(x)$	0.05	0.07	0.11	0.13	0.24	0.40

Find the expected number of essential blood tests ordered for a randomly selected patient.

Solution

STEP 1 X is a discrete random variable; there is a finite number of values of X. Find the mean using Equation 5.1.

$$\mu = \sum_{\text{all } x}[x \cdot p(x)] \qquad \text{Equation 5.1.}$$

$$= (0)(0.05) + (1)(0.07) + (2)(0.11) + (3)(0.13)$$
$$+ (4)(0.24) + (5)(0.40)$$

Multiply each value by its corresponding probability, and sum.

$$= 0.00 + 0.07 + 0.22 + 0.39 + 0.96 + 2.00 = 3.64$$

STEP 2 The mean number of blood tests ordered is 3.64. **Figure 5.5** shows a technology solution.

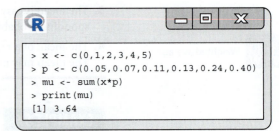

Figure 5.5 The mean number of essential blood tests ordered for a randomly selected patient.

```
> x <- c(0,1,2,3,4,5)
> p <- c(0.05,0.07,0.11,0.13,0.24,0.40)
> mu <- sum(x*p)
> print(mu)
[1] 3.64
```

The Variance and Standard Deviation

The **variance** and **standard deviation** of a random variable measure the spread of the distribution. The variance is computed using the expected value function, and the standard deviation together with the mean can be used to determine the most likely values of the random variable.

Definition

Let X be a discrete random variable with probability mass function $p(x)$. The **variance** of X is

$$\underbrace{\text{Var}(X) = \sigma^2 = \sigma_X^2}_{\text{Notation}} = \underbrace{\sum_{\text{all } x}[(x-\mu)^2 \cdot p(x)]}_{\text{Calculation}} = \underbrace{E[(X-\mu)^2]}_{\substack{\text{Definition in terms} \\ \text{of expected value}}} \qquad (5.3)$$

The **standard deviation** of X is the positive square root of the variance.

$$\underbrace{\sigma = \sigma_X}_{\text{Notation}} = \underbrace{\sqrt{\sigma^2}}_{\text{Calculation}} \qquad (5.4)$$

A CLOSER LOOK

1. In words, the variance is the expected value of the squared deviations about the mean.
2. The symbol Var stands for variance and is also a function. The function Var takes a random variable, or function of a random variable, as an input, and returns the variance.
3. To compute the variance using Equation 5.3:
   (a) Find the mean μ of X using Equation 5.1.
   (b) Find each difference: $(x - \mu)$.
   (c) Square each difference: $(x - \mu)^2$.
   (d) Multiply each squared difference by the associated probability.
   (e) Sum the products.
4. There is a computational formula for the variance of a random variable.

Computational Formula for σ^2

$$\sigma^2 = E(X^2) - E(X)^2 = E(X^2) - \mu^2 \qquad (5.5)$$

In words, the variance is the expected value of X squared minus the expected value of X, squared. In theory, the computational formula for the variance is faster and more accurate than using the definition.

Examples Involving μ, σ^2, and σ

EXAMPLE 5.11   Children in Day Care

Suppose the discrete random variable X, the age of a randomly selected child at the Precious Angels Child Care Center in Cleveland, Ohio, has the probability distribution given in the table.

x	1	2	3	4	5	6	7
$p(x)$	0.05	0.10	0.15	0.25	0.20	0.15	0.10

(a) Find the expected value, variance, and standard deviation of X.

(b) Find the probability that the random variable X takes on a value within 1 standard deviation of the mean.

Solution

(a) Expected value: Find the expected value of X using Equation 5.1.

$$E(X) = \sum_{\text{all } x}[x \cdot p(x)]$$

$$= (1)(0.05) + (2)(0.10) + (3)(0.15) + (4)(0.25) + (5)(0.20)$$
$$\quad + (6)(0.15) + (7)(0.10)$$
$$= 0.05 + 0.20 + 0.45 + 1.00 + 1.00 + 0.90 + 0.70$$
$$= 4.30 = \mu$$

Variance: Find the variance of X using Equation 5.3.

$$\text{Var}(X) = \sum_{\text{all } x}[(x - \mu)^2 \cdot p(x)] \qquad \text{Equation 5.3.}$$

$$= (1 - 4.30)^2(0.05) + (2 - 4.30)^2(0.10) + (3 - 4.30)^2(0.15) \quad \text{Sum over all values of } x.$$
$$\quad + (4 - 4.30)^2(0.25) + (5 - 4.30)^2(0.20) + (6 - 4.30)^2(0.15)$$
$$\quad + (7 - 4.30)^2(0.10)$$

$$= (10.89)(0.05) + (5.29)(0.10) + (1.69)(0.15) \qquad \text{Square each difference.}$$
$$\quad + (0.09)(0.25) + (0.49)(0.20) + (2.89)(0.15)$$
$$\quad + (7.29)(0.10)$$

$$= 0.5445 + 0.5290 + 0.2535 + 0.0225 \qquad \text{Compute each product.}$$
$$\quad + 0.0980 + 0.4335 + 0.7290$$
$$= 2.61 = \sigma^2$$

Here is a tabular method for visualizing the steps in computing the variance using the definition. Sum the last column to obtain σ^2.

x	$x-\mu$	$(x-\mu)^2$	$p(x)$	$(x-\mu)^2 \cdot p(x)$
1	−3.30	10.89	0.05	0.5445
2	−2.30	5.29	0.10	0.5290
3	−1.30	1.69	0.15	0.2535
4	−0.30	0.09	0.25	0.0225
5	0.70	0.49	0.20	0.0980
6	1.70	2.89	0.15	0.4335
7	2.70	7.29	0.10	0.7290
				2.6100 $\leftarrow \sigma^2$

Sum this column

Variance: Using the computational formula.

Find $E(X^2)$, the expected value of X^2.

$$E(X^2) = \sum_{\text{all } x}[x^2 \cdot p(x)]$$ Equation 5.2.

$= 1^2(0.05) + 2^2(0.10) + 3^2(0.15) + 4^2(0.25) + 5^2(0.20)$ Sum over all values of x.
$\quad + 6^2(0.15) + 7^2(0.10)$

$= 1(0.05) + 4(0.10) + 9(0.15) + 16(0.25) + 25(0.20)$ Square each x.
$\quad + 36(0.15) + 49(0.10)$

$= 0.05 + 0.40 + 1.35 + 4.00 + 5.00 + 5.40 + 4.90$ Compute each product.

$= 21.10$

Use this result to find the variance.

$\sigma^2 = E(X^2) - \mu^2$ Equation 5.5.

$= 21.10 - (4.30)^2$ Use previous results.

$= 21.10 - 18.49$ Find μ^2.

$= 2.61$ Find the difference.

Here is a tabular method for visualizing the steps in computing the variance using the computational formula

$x^2 \cdot p(x)$	x^2	$p(x)$	x	$x \cdot p(x)$
0.05	1	0.05	1	0.05
0.40	4	0.10	2	0.20
1.35	9	0.15	3	0.45
4.00	16	0.25	4	1.00
5.00	25	0.20	5	1.00
5.40	36	0.15	6	0.90
4.90	49	0.10	7	0.70

Sum this column for $E(X^2)$.

Sum this column for μ.

$E(X^2) \rightarrow 21.10 \qquad\qquad 4.30 \leftarrow \mu$

$\sigma^2 = E(X^2) - \mu^2 = 21.10 - (4.30)^2 = 2.61$

Standard deviation: The positive square root of the variance.

$\sigma = \sqrt{\sigma^2} = \sqrt{2.61} \approx 1.6155$

Figure 5.6 shows a technology solution.

(b) The phrase "within 1 standard deviation of the mean" (of X) means in the interval $(\mu - \sigma, \mu + \sigma)$. Write a probability statement, find the value(s) of X that lie in this interval, and add the corresponding probabilities.

5.3 Mean, Variance, and Standard Deviation for a Discrete Random Variable 211

```
> x <- c(1,2,3,4,5,6,7)
> p <- c(0.05,0.10,0.15,0.25,0.20,0.15,0.10)
> mu <- sum(x*p)
> var <- sum(x^2*p) - mu^2
> sd <- sqrt(var)
> round(cbind(mu,var,sd),4)
       mu  var     sd
[1,]  4.3 2.61 1.6155
```

Figure 5.6 Mean, variance, and standard deviation of the random variable.

$$
\begin{aligned}
P(\mu - \sigma \leq X \leq \mu + \sigma) & & \text{Translation to a probability statement.} \\
= P(4.30 - 1.6155 \leq X \leq 4.30 + 1.6155) & & \text{Use values for } \mu \text{ and } \sigma. \\
= P(2.6845 \leq X \leq 5.9155) & & \text{Compute the difference and sum.} \\
= P(X = 3) + P(X = 4) + P(X = 5) & & \text{Find values of } X \text{ in the interval.} \\
= 0.15 + 0.25 + 0.20 & & \text{Use corresponding probabilities.} \\
= 0.60 & & \text{Compute the sum.}
\end{aligned}
$$

The probability that X takes on a value within 1 standard deviation of the mean is 0.60. In the context of the problem, the probability that the age of a randomly selected child is within 1 standard deviation of the mean is 0.60. ∎

TRY IT NOW Go to Exercises 5.63 and 5.67

A CLOSER LOOK

1. The computational formula for the variance is quicker and produces less round-off error. Therefore, you should use Equation 5.5 to find the variance of a discrete random variable.

2. Because Example 5.11 (b) involves an interval that is 1 standard deviation from the mean, it might seem as if the solution involves the Empirical Rule. However, the random variable X is not (approximately) normal; the Empirical Rule does not apply. In addition, even though Chebyshev's Rule applies to any distribution, it should not be used if the probability distribution is known. Chebyshev's Rule provides only an estimate, a lower bound for the probability that X is within k standard deviations of the mean. The exact probability can be determined by using the known probability distribution. (Actually, Chebyshev's Rule can't help at all here, because k must be greater than 1.)

EXAMPLE 5.12 Lady Luck

One of the instant scratch-off games in the North Carolina Education Lottery is Lady Luck. The payout (in dollars) is a discrete random variable with the probability distribution shown in the table.[10]

x	1000	400	100	50	30	25
$p(x)$	0.00000098	0.0000015	0.000283	0.000883	0.000354	0.004000
x	10	5	4	2	1	0
$p(x)$	0.008667	0.007334	0.013333	0.073314	0.100000	0.791830

(a) Find the expected payout, as well as the variance and standard deviation of the payout.

(b) Interpret the values found in part (a). In particular, if it costs $1.00 to purchase a ticket, what happens in the long run?

Solution

(a) We can use the tabular method and the computational formula to find the mean, the variance, and then the standard deviation. However, **Figure 5.7** shows a technology solution.

```
> x <- c(1000,400,100,50,30,25,10,5,4,2,1,0)
> p <- c(0.00000098,0.0000015,0.000283,0.000883,0.000354,0.004000,
+        0.008667,0.007334,0.013333,0.073314,0.100000,0.79183)
> mu  <- sum(x*p)
> ex2 <- sum(x^2*p)
> var <- ex2 - mu^2
> sd  <- sqrt(var)
> round(cbind(mu,ex2,var,sd),4)
         mu     ex2     var     sd
[1,] 0.608 10.7327 10.3631 3.2192
```

Figure 5.7 Mean, variance, and standard deviation of the payout.

$$\mu = \sum_{\text{all } x}[x \cdot p(x)] = (1000)(0.00000098) + \cdots + (0)(0.79183) = 0.608$$

$$E(X^2) = \sum_{\text{all } x}[x^2 \cdot p(x)] = (1000)^2(0.00000098) + \cdots + (0)^2(0.79183) = 10.7327$$

$$\sigma^2 = E(X^2) - \mu^2 = 10.7327 - (0.608)^2 = 10.3631$$

$$\sigma = \sqrt{\sigma^2} = \sqrt{10.3631} = 3.2192$$

(b) The mean payout from this instant scratch-off game is approximately $0.61, with variance $10.36 and standard deviation $3.22. If it costs $1.00 to play (purchase a ticket), in the long run the player loses $1.00 - 0.61 = 0.39$, or 39 cents, on average. This doesn't seem like much. However, from the state's point of view, every time someone buys a ticket, North Carolina makes 39 cents (on average). ■

TRY IT NOW Go to Exercises 5.69 and 5.71

Section 5.3 Exercises

Concept Check

5.52 True or False The mean of a discrete random variable must be a possible value of the random variable.

5.53 True or False The expected value of a discrete random variable can be negative.

5.54 True or False The variance of a discrete random variable can be negative.

5.55 True or False The standard deviation of a discrete random variable is always non-negative.

5.56 True or False The computational formula for σ^2 should be used only when the number of possible values for the random variable is small.

5.57 True or False For any discrete random variable X, the probability that X is within 2 standard deviations of the mean is approximately 0.68.

5.58 Fill in the Blank Suppose $f(X)$ is a function of a discrete random variable. $E[f(X)] = $ _____.

5.59 Fill in the Blank The variance of a discrete random variable is the expected value of _____.

5.60 Short Answer Explain in your own words why σ^2 is a measure of variability.

Practice

5.61 Suppose X is a discrete random variable. Complete the table to find the mean, variance, and standard deviation of X. Use technology to check your answers.

$x^2 \cdot p(x)$	x^2	$p(x)$	x	$x \cdot p(x)$
		0.10	2	
		0.16	4	
		0.20	6	
		0.24	8	
		0.18	10	
		0.12	12	

5.62 The probability distribution for a random variable X is given in the table.

x	5	10	15	20
$p(x)$	0.10	0.15	0.70	0.05

(a) Find the mean, variance, and standard deviation of X.
(b) Find the probability that X takes on a value smaller than the mean.
(c) Using the probability distribution, explain why the value of the mean of X makes sense.

5.63 Suppose Y is a discrete random variable with the probability distribution given in the table.

y	-20	-10	0	10	20
$p(y)$	0.30	0.15	0.10	0.15	0.30

(a) Find μ, σ^2, and σ.
(b) Find $P(\mu - 2\sigma \leq Y \leq \mu + 2\sigma)$.
(c) Find $P(Y \geq \mu)$ and $P(Y > \mu)$.

5.64 Suppose the random variable X has the probability distribution given in the table.

x	2	3	5	7	11	13
$p(x)$	0.15	0.25	0.15	0.10	0.30	0.05

(a) Find the mean, variance, and standard deviation of X.
(b) Suppose the random variable Y is defined by $Y = 2X + 1$. Find the mean, variance, and standard deviation of Y.
(c) Suppose the random variable W is defined by $W = X^2 + 1$. Find the mean, variance, and standard deviation of W.

5.65 Suppose X is a discrete random variable with the probability distribution given in the table.

x	1	2	3	5	8	13	21
$p(x)$	0.05	0.10	0.15	0.20	0.25	0.20	0.05

(a) Find the mean, variance, and standard deviation of X.
(b) Find the probability that X is more than 1 standard deviation from the mean.
(c) Find $P(X \leq \mu + 2\sigma)$.

5.66 Suppose X is a discrete random variable with the probability distribution given in the table.

x	-2	-1	0	1	2
$p(x)$	0.05	0.10	0.20	0.30	0.35

(a) Find the mean, variance, and standard deviation of X.
(b) Find $P(X = 2 \mid X \geq 0)$.
(c) Suppose the random variable Y is defined by $Y = X^2$. Find the mean, variance, and standard deviation of Y.
(d) Suppose the random variables X_1 and X_2 are independent and both have the same probability distribution as given in the table. Find $P(X_1 > 0 \cap X_2 > 0)$.

Applications

5.67 Manufacturing and Product Development In a 0.25-lb bag of red pistachio nuts, some shells are too difficult to pry open by hand. Suppose the random variable X, the number of pistachios in a randomly selected bag that cannot be opened by hand, has the probability distribution given in the table.

x	0	1	2	3	4	5
$p(x)$	0.500	0.250	0.100	0.050	0.075	0.025

(a) Is this a valid probability distribution? Justify your answer.
(b) Find the mean, variance, and standard deviation of X.
(c) Find the probability that X takes on a value less than the mean.
(d) Suppose two bags of pistachios are selected at random. What is the probability that both bags have four or more pistachios that are too difficult to open by hand?

5.68 Sports and Leisure Suppose the number of rides that a visitor enjoys at Disney World during a day is a random variable with the probability distribution given in the table.

x	5	6	7	8	9	10	11	12
$p(x)$	0.04	0.07	0.09	0.12	0.20	0.30	0.13	0.05

(a) Find the mean number of rides for a Disney World visitor.
(b) Find the variance and standard deviation of the number of rides for a Disney World visitor.
(c) Find the probability that the number of rides for a randomly selected Disney World visitor is within 1 standard deviation of the mean.
(d) Find the probability that the number of rides for a randomly selected Disney World visitor is less than 1 standard deviation above the mean.

5.69 Public Policy and Political Science In 2017, recalls of 93 children's products involved approximately a total of 11.8 million items.[11] In particular, children's clothing is recalled for a variety of reasons, such as drawstrings that are too long and pose a hazard, small buttons that may break off and cause choking, and material that fails to meet federal flammability standards. Suppose the number of recalls of children's clothing during a given month is a random variable with the probability distribution given in the table.

x	0	1	2	3	4	5	6
p(x)	0.005	0.185	0.275	0.305	0.200	0.020	0.010

(a) Find the mean, variance, and standard deviation of the number of recalls of children's clothing during a given month.
(b) Suppose the number of recalls in a given month is at least three. What is the probability that the number of recalls that month will be at least five?
(c) If the number of recalls in a given month is greater than 1 standard deviation from the mean, that is, more than $\mu + \sigma$, the federal government issues a special warning directed toward parents. What is the probability that a special warning will be issued during a given month?

5.70 Manufacturing and Product Development Pringles, a product invented in the 1960s, are uniform potato chips sold in distinctive tubes. Each chip is fried in vegetable oil for 11 seconds, and there are approximately 90 chips in each tube. Suppose the number of chips in each tube is actually a random variable with the probability distribution given in the table.

x	88	89	90	91	92	93
p(x)	0.03	0.07	0.30	0.25	0.20	0.15

(a) Find the mean, variance, and standard deviation of the number of chips in a tube.
(b) Find the probability that the number of chips in a randomly selected tube is within 1 standard deviation of the mean.
(c) Find the probability that the number of chips in a randomly selected tube is within 2 standard deviations of the mean.
(d) Suppose three tubes are selected at random. Find the probability that all three have at least 90 chips.

5.71 Public Health and Nutrition In Pennsylvania, the Indoor Tanning Regulation Act requires all tanning facilities to register with the Department of Health. Indoor tanning devices usually operate on a timer, but the exposure to ultraviolet radiation can vary based on the age and type of bulbs.[12] Suppose that past records at the Darkside Tanning Company indicate that most sessions range from 10 to 30 minutes. Suppose the duration of a tanning session (in minutes) is a discrete random variable with the probability distribution given in the table.

x	10	12	15	20	25	30
p(x)	0.30	0.25	0.15	0.12	0.10	0.08

(a) Find the mean, variance, and standard deviation of the duration of a tanning session time.
(b) Find the probability that a randomly selected session has a duration within 1 standard deviation of the mean.
(c) Find the probability that a randomly selected session has a duration within 2 standard deviations of the mean.
(d) Suppose a sunlamp lasts for 100 hours. After approximately how many tanning sessions will the sunlamp have to be replaced?

5.72 Public Health and Nutrition For nurses, additional patients contribute heavily to increased stress and job burnout. In recognition of this issue, the Massachusetts Nurse–Patient Assignment Limits Initiative appeared on the ballot in November 2018.[13] Suppose the number of patients assigned to each nurse at the Massachusetts General Hospital in Boston, Massachusetts, is a random variable with the probability distribution given in the table.

x	3	4	5	6	7	8
p(x)	0.07	0.12	0.18	0.37	0.17	0.09

(a) Find the mean, variance, and standard deviation of the number of patients assigned to each nurse.
(b) Find the probability that the number of assigned patients is greater than 1 standard deviation to the right of the mean.
(c) Suppose three nurses are selected at random. What is the probability that exactly two of the three have five assigned patients?

5.73 Psychology and Human Behavior A certain elevator in Tampa's tallest office building, 100 North Tampa, is used heavily between 8:00 A.M. and 9:00 A.M. as employees arrive for work. Suppose the number of people who board the elevator on the ground floor going up is a random variable with the probability distribution given in the table.

x	1	2	3	4	5	6
p(x)	0.002	0.010	0.050	0.060	0.080	0.090
x	7	8	9	10	11	12
p(x)	0.100	0.120	0.140	0.150	0.150	0.048

(a) Find μ, σ^2, and σ.
(b) For a randomly selected elevator ride from the ground floor going up, what is the probability that the number of riders is within 1 standard deviation of the mean?
(c) For two randomly selected elevator rides from the ground floor going up, what is the probability that both trips have a number of riders more than 2 standard deviations from the mean?
(d) For two randomly selected elevator rides from the ground floor going up, what is the probability that both trips have the same number of riders?

5.74 Manufacturing and Product Development Tim Hortons is a chain that first opened in Hamilton, Ontario. The first two stores offered only coffee and doughnuts, but the selection has grown to muffins, cakes, bagels, and the very successful Timbit (a bite-sized doughnut hole).[14] Despite its success and growth, Tim Hortons does not offer free coffee refills. However, suppose the company is exploring a policy to offer free coffee refills, and the number of refills for a randomly selected customer is a random variable with the probability distribution given in the table.

x	0	1	2	3	4
p(x)	0.14	0.50	0.23	0.11	0.02

(a) Find the mean, variance, and standard deviation for the number of coffee refills.
(b) Suppose each refill costs Tim Hortons 37 cents. Find the mean, variance, and standard deviation of the refill cost.
(c) Suppose four customers who order a cup of coffee are selected at random. Find the probability that exactly one customer has two refills.

Extended Applications

5.75 Manufacturing and Product Development A cordless drill has several torque settings for driving different screws into different materials. A manufacturer models the torque setting required for a randomly selected task with a probability distribution given by

$$p(x) = \frac{(x-12)^2}{247} \quad x = 1, 5, 10, 15, 20$$

(a) Verify that this is a valid probability distribution.
(b) Find the mean, variance, and standard deviation of X.
(c) A torque setting is classified as rare if it is more than 1 standard deviation from the mean. Find the probability of a task requiring a rare torque setting.

5.76 Six Degrees of Kevin Bacon The actor Kevin Bacon has been in so many movies that almost everyone in Hollywood can be connected to him within six degrees. Let $n = 2{,}341{,}194$, the number of actors linked to Kevin Bacon. The probability distribution for the Bacon Number of a randomly selected Hollywood personality is given in the table.[15]

x	p(x)	x	p(x)	x	p(x)
0	1/n	4	390,201/n	8	119/n
1	3452/n	5	34,150/n	9	7/n
2	403,921/n	6	4181/n	10	1/n
3	1,504,560/n	7	601/n		

(a) Find the mean, variance, and standard deviation for the Bacon Number. Interpret the mean in the context of the problem.
(b) Find $P(X \geq \mu - \sigma)$.
(c) The number of movies Y in which a Hollywood personality has appeared is related to the Bacon Number by the formula $Y = 2X + 5$. Find the mean, variance, and standard deviation of the number of movies in which a Hollywood personality has appeared.

5.77 Marketing and Consumer Behavior While the temperature range of most household ovens is approximately 200–600°F, most consumers use only four or five common settings. Suppose the probability distribution for the oven-temperature setting for a randomly selected use is given in the table.

x	300	325	350	375	400	500
p(x)	0.040	0.205	0.400	0.075	0.200	0.080

(a) Find the mean, variance, and standard deviation of the oven-temperature settings.
(b) Suppose three different oven uses are randomly selected. Find the probability that the temperature settings for all three are at least 400°F.
(c) Suppose three different oven uses are randomly selected. Find the probability that exactly one use is for 350°F.

5.78 Education and Child Development An elementary class rarely remains the same size from the beginning of the school year until the end. Families move in and out of the district, some students are reassigned, and scheduling conflicts necessitate changes. Suppose the change in the number of students in a class is a random variable with the probability distribution given by

$$p(x) = \frac{|x|+1}{19} \quad x = -3, -2, -1, 0, 1, 2, 3$$

(a) Verify that this is a valid probability distribution.
(b) Find the mean, variance, and standard deviation of the change in class size.
(c) Suppose two classes are selected at random. Find the probability that both classes remain the same size for the entire year.

5.79 Psychology and Human Behavior Many organizations publish wedding guidelines with specific suggestions regarding the reception, wedding cake, flowers, and even themes and styles. However, the number of bridesmaids and groomsmen is usually a very personal decision made by the bride and groom, and it can even say something about your personality. For semiformal weddings, three to five bridesmaids are typical, with a possible flower girl and/or ring bearer.[16] Suppose the probability distribution for the number of bridesmaids at a semiformal wedding is given in the table.

x	1	2	3	4	5	6
p(x)	0.05	0.21	0.34	0.27	0.08	0.05

(a) Find the mean, variance, and standard deviation for the number of bridesmaids at a semiformal wedding.
(b) Suppose a randomly selected semiformal wedding has at least four bridesmaids. What is the probability that it has exactly six bridesmaids?
(c) Suppose four semiformal weddings are randomly selected. What is the probability that all four have at least three bridesmaids?

5.80 Medicine and Clinical Studies In a recent study at the University of Southern California Schaeffer Center for Health Policy and Economics, researchers concluded that customers overpaid for prescriptions 23% of the time.[17] This overpayment often occurs because the copay is higher than the full cost of a drug. Suppose four prescriptions are selected at random. Let X be a random variable defined to be the number of prescriptions for which the customer overpaid.
(a) Find the probability distribution for X.
(b) Find the mean, variance, and standard deviation of X.
(c) Suppose all four customers overpaid for their prescriptions. Is there any evidence to suggest that the percentage reported in this study is wrong? Justify your answer.

Challenge Problems

5.81 Dichotomous Random Variable Suppose the random variable X takes on only two values, according to the probability distribution in the table.

x	0	1
$p(x)$	0.4	0.6

(a) Find the mean, variance, and standard deviation of X.
(b) Suppose $P(X = 1) = 0.7$, and therefore $P(X = 0) = 1 - 0.7 = 0.3$. Find the mean, variance, and standard deviation of X.
(c) Suppose $P(X = 1) = 0.8$. Find the mean, variance, and standard deviation of X.
(d) Suppose $P(X = 1) = p$, and $P(X = 0) = 1 - p = q$. Find the mean, variance, and standard deviation of X in terms of p and q.
(e) For what values of p (and q) is the variance of X greatest?

5.82 Linear Function Suppose X is a discrete random variable with mean μ_X and variance σ_X^2. Let Y be a linear function of X, such that $Y = aX + b$, where a and b are constants. Find the mean and variance of Y in terms of μ_X and σ_X^2.

5.83 Variance Computation Formula Suppose X is a discrete random variable that takes on a finite number of values. Prove the variance computation formula. That is, show that

$$E[(X - \mu)^2] = E(X^2) - \mu^2$$

Hint: Write $E[(X - \mu)^2]$ as a sum using the probability mass function $p(x)$. Expand and simplify.

5.84 Standardization Suppose X is a discrete random variable with mean μ_X and variance σ_X^2. Let Y be defined in terms of X by

$$Y = \frac{X - \mu_X}{\sigma_X}$$

Find the mean, variance, and standard deviation of Y.

5.4 The Binomial Distribution

Binomial Experiment

The previous sections introduced the general definition and probability distribution for a discrete random variable. This section presents a specific discrete random variable that is both common and very important. The binomial random variable can be used to model many real-world populations and to do more formal inference.

As with any random variable, there is a related experiment in the background. Consider the following experiments (and look for similarities):

1. Toss a coin 50 times and record the sequence of heads and tails.
2. Identify 40 volcanoes around the world and record whether each one erupts during the year.
3. Select a random sample of 25 customers at a fast-food restaurant and record whether each pays with exact change.
4. Drill a series of randomly selected test oil wells. Each well will either yield oil worth drilling or be classified as dry. Record the result for each well.

All of these experiments share four properties in common. These properties are used to describe a **binomial experiment**, and they are necessary to define a **binomial random variable**.

Properties of a Binomial Experiment

1. The experiment consists of n identical trials.
2. Each trial can result in only one of two possible (mutually exclusive) outcomes. One outcome is usually designated a success (S) and the other a failure (F).
3. The outcomes of the trials are independent.
4. The probability of a success p remains constant from trial to trial.

A CLOSER LOOK

1. A trial is a small part of the larger experiment. A trial results in a single occurrence of either a success or a failure. For example, flipping a coin once, or drilling one test oil well, is a single trial. A typical binomial experiment might consist of $n = 50$ trials.
2. A success does not have to be a good thing. For example, an experiment may consist of carefully monitoring tornadoes in a selected area. A success might be a tornado that causes property damage. Success and failure could stand for heads and tails, acceptable and not acceptable, or even dead and alive.
3. Trials are independent if whatever happens on one trial has no effect on any other trial. For example, any one voter response has no effect on any other voter response.
4. The probability of a success on every trial is exactly the same. For example, the probability of the tossed (fair) coin landing with a head face up is always 1/2.

In a binomial experiment, outcomes consist of sequences of Ss and Fs. For example, SSFSFSFS is a possible outcome in a binomial experiment with $n = 8$ trials.

Binomial Random Variable

The Binomial Random Variable
The binomial random variable maps each outcome in a binomial experiment to a real number; it is defined to be the number of successes in n trials.

Notation

Why is $P(F) = 1 - p$?

1. The probability of a success is denoted by p. Therefore, $P(S) = p$ and $P(F) = 1 - p = q$.
2. A binomial random variable X is completely determined by the number of trials n and the probability of a success p. If we know those two values, then we will be able to answer any probability question involving X.

The shorthand notation $X \sim B(n, p)$ means X is (distributed as) a binomial random variable with n trials and probability of a success p.

For example, $X \sim B(25, 0.4)$ means X is a binomial random variable with 25 trials and probability of success 0.4.

Our goal now is to find the probability distribution for a binomial random variable. Given n and p, we want to find the probability of obtaining x successes in n trials, $P(X = x) = p(x)$. We will solve this problem by first considering a simple case with $n = 5$.

EXAMPLE 5.13 Binomial Experiment with $n = 5$

Consider a binomial experiment with $n = 5$ trials and probability of success p.

For example, suppose five people are selected at random. Let p be the probability that a randomly selected person snores.

(a) A typical outcome with two successes is SFFSF. Find the probability of this outcome.
(b) Another possible outcome with two successes is FSFSF. Find the probability of this outcome.
(c) Compare your results from parts (a) and (b).
(d) Find the probability that $X = 2$ successes.

Solution

(a) The probability of this outcome is

$$P(SFFSF) = P(S \cap F \cap F \cap S \cap F)$$

Probability of a success on the first trial and a failure on the second trial and ....

$$= P(S) \cdot P(F) \cdot P(F) \cdot P(S) \cdot P(F)$$

Trials are independent (property of a binomial experiment).

$$= p \cdot (1-p) \cdot (1-p) \cdot p \cdot (1-p)$$

$P(S) = p, P(F) = 1-p$

$$= p^2(1-p)^3$$

Multiplication is commutative.

(b) The probability of this outcome is

$$P(FSFSF) = P(F) \cdot P(S) \cdot P(F) \cdot P(S) \cdot P(F)$$

Trials are independent.

$$= (1-p) \cdot p \cdot (1-p) \cdot p \cdot (1-p)$$

$P(S) = p, P(F) = 1-p$

$$= p^2(1-p)^3$$

Multiplication is commutative.

(c) The results are identical. Every other outcome with two successes, and therefore three failures, has exactly the sample probability $p^2(1-p)^3$. Therefore, the probability of an outcome depends on the number of successes (and failures), not on the order in which they appear.

(d) To compute the probability that $X = 2$ successes, find all the outcomes mapped to a 2, and add the corresponding probabilities. However, every outcome that is mapped to a 2 has the same probability, so all we need to know is how many outcomes are mapped to a 2.

$$P(X=2) = (\text{number of outcomes with 2 successes}) \cdot p^2 \cdot (1-p)^3$$

▶ Generalizing, suppose $X \sim B(n, p)$. We want to find the probability of obtaining x successes in n trials. The probability of any single outcome with x successes, and therefore $n-x$ failures, is $p^x(1-p)^{n-x}$. The probability of obtaining x successes is

$$P(X=x) = (\text{number of outcomes with } x \text{ successes}) \cdot p^x \cdot (1-p)^{n-x}$$

The number of successes and the number of failures must sum to n, the total number of trials.

We need a method for quickly counting the number of outcomes with x successes. Recall that for any positive whole number n, the symbol $n!$ (read as "n factorial") is defined as

$$n! = n(n-1)(n-2)\cdots(3)(2)(1)$$

In addition, $0! = 1$ (0 factorial is 1). Given a collection of n items, the number of combinations of size x is given by

$$_nC_x = \binom{n}{x} = \frac{n!}{x!(n-x)!}$$

The number of outcomes with x successes is determined using combinations. Suppose $X \sim B(n, p)$. The number of outcomes with x successes is $\binom{n}{x}$. We can now write an expression for the probability of obtaining x successes in n trials. ◀

✓ *Can you figure out why this is true?*

The Binomial Probability Distribution

Suppose X is a binomial random variable with n trials and probability of success $p: X \sim B(n, p)$. Then

$$p(x) = P(X=x) = \underbrace{\binom{n}{x}}_{\substack{\text{Number of outcomes} \\ \text{with } x \text{ successes}}} \underbrace{p^x(1-p)^{n-x}}_{\substack{\text{Probability of } x \text{ successes and } n-x \\ \text{failures in any single outcome}}}, \quad x = 0, 1, 2, 3, 4, \ldots, n \qquad (5.6)$$

Example: Use the Probability Mass Function

EXAMPLE 5.14 **Earth Hour**

Earth Hour is an annual global event organized by the World Wildlife Fund in which individuals are encouraged to turn off their lights for 1 hour to highlight the dangers of climate change. It usually takes place in late March each year. BC Hydro reported that 70% of British Columbians believe that Earth Hour is relevant and that they intend to participate by turning off their lights at the specified time.[18] Suppose 10 British Columbians are selected at random.

(a) Find the probability that exactly six intend to participate in Earth Hour.

(b) Find the probability that at least seven intend to participate in Earth Hour.

Solution

Let X be the number of British Columbians (out of the 10 selected) who intend to participate in Earth Hour. The experiment exhibits the properties of a binomial experiment, so $X \sim B(10, 0.70)$ ($n = 10$, $p = 0.70$).

(a) Translate the question into a probability statement involving the random variable X. Exactly six means $X = 6$. Use Equation 5.6 to find the relevant probability.

$$P(X = 6) = \binom{10}{6}(0.70)^6(1-0.70)^{10-6}$$ Equation 5.6.

$$= (210)(0.70)^6(0.30)^4$$ Compute $_{10}C_6$.

$$= 0.2001$$

The probability that exactly 6 of 10 randomly selected British Columbians plan to participate in Earth Hour is 0.2001.

(b) The phrase "at least seven" means seven or more. So we need to find the probability that X is greater than or equal to 7. In this experiment that means the probability that X is 7 or 8 or 9 or 10.

$$P(X \geq 7) = P(X = 7 \text{ or } X = 8 \text{ or } X = 9 \text{ or } X = 10)$$

$$= P(X = 7) + P(X = 8) + P(X = 9) + P(X = 10)$$ *Or* means union; the outcomes are disjoint.

$$= \binom{10}{7}(0.70)^7(0.30)^3 + \binom{10}{8}(0.70)^8(0.30)^2$$ Equation 5.6 four times.

$$+ \binom{10}{9}(0.70)^9(0.30)^1 + \binom{10}{10}(0.70)^{10}(0.30)^0$$

$$= 0.2668 + 0.2335 + 0.1211 + 0.0282$$ Compute combinations and powers, and multiply.

$$= 0.6496$$

The probability that at least seven British Columbians will participate in Earth Hour is 0.6496.

Note:

1. The two most important elements for solving this type of problem are (1) the probability distribution and (2) the probability statement.

2. Often, the properties of a binomial experiment will not be stated explicitly in the problem. Usually, we must decipher the problem to see the n trials, to identify a success, to recognize independence, and to presume that the probability of a success remains constant from trial to trial.

Jason Doucette/Alamy

A CLOSER LOOK

1. Even for small values of n, many of the probabilities associated with a binomial random variable are a little tedious to calculate and are subject to lots of round-off error. Technology helps, and Appendix Table 1 presents **cumulative probabilities** for a binomial random variable, for various values of n and p.

2. Cumulative probability is an important concept. If $X \sim B(n, p)$, the probability that X takes on a value less than or equal to x is a cumulative probability. Add all the probabilities associated with values up to and including x. Symbolically, cumulative probability is

$$P(X \leq x) = \sum_{k=0}^{x} P(X = k) \tag{5.7}$$

$$= P(X = 0) + P(X = 1) + P(X = 2) + \cdots + P(X = x)$$

3. Graphically, cumulative probability is like standing on a special staircase, looking down (or back), and measuring the height. The steps are labeled $0, 1, 2, \ldots, n$; the height of step x is $P(X = x)$; and the total height of the staircase is 1. **Figure 5.8** illustrates the staircase and $P(X \leq 3)$. The number of steps is $n+1$, and the height of each step depends on n and p. In this example, $n = 10$ and $p = 0.25$. The largest steps (the highest probabilities) are associated with $X = 1, 2$, and 3. Steps 7, 8, 9, and 10 are hard to see because $P(X = 7)$, $P(X = 8)$, $P(X = 9)$, and $P(X = 10)$ are so small.

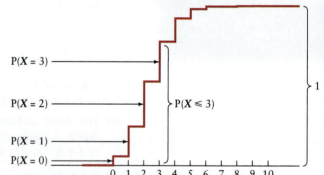

Figure 5.8 Staircase analogy for cumulative probability.

Here's another way to visualize cumulative probability for a discrete random variable. Sketch the probability histogram for the random variable X such that each rectangle has width 1. Then, the area of the rectangle centered at x is $P(X = x)$, and the cumulative probability $P(X \leq x)$ is the sum of the areas of the rectangles centered at x and to the left of x. **Figure 5.9** illustrates the case of $P(X \leq 3)$.

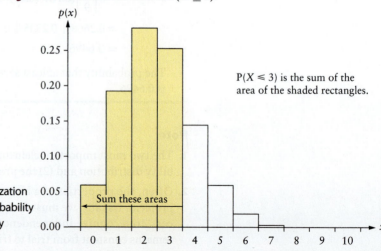

Figure 5.9 Visualization of cumulative probability using a probability histogram.

Every probability question about a binomial random variable can be answered using cumulative probability. Other, faster methods may also be available, but cumulative probability always works. The following example illustrates some of the techniques for converting to and using cumulative probability.

Example: Use Cumulative Probability

EXAMPLE 5.15 **Digital Pizza**

Online orders now make up more than half of all orders for some big chain pizza stores. According to a recent survey, approximately 60% of all pizza stores offer online ordering.[19] Suppose 20 pizza stores are randomly selected.

(a) Find the probability that at most 12 offer online orders.

(b) Find the probability that exactly 14 offer online orders.

(c) Find the probability that at least 9 offer online orders.

(d) Find the probability that between 10 and 15 (inclusive) offer online orders.

Solution

Let X be the number of pizza stores (out of the 20 selected) that offer online orders. X is a binomial random variable with $n = 20$ and $p = 0.60$: $X \sim B(20, 0.60)$.

(a) "At most 12 offer online orders" means that X is less than or equal to 12.

$$P(X \leq 12) = 0.5841 \qquad \text{Cumulative probability; use Appendix Table 1.}$$

(b) "Exactly 14 offer online orders" means that X is equal to 14.

$$P(X = 14) = P(X \leq 14) - P(X \leq 13) \qquad \text{Use cumulative probability.}$$
$$= 0.8744 - 0.7500 \qquad \text{Use Appendix Table 1.}$$
$$= 0.1244$$

This solution may also be found using the probability mass function for a binomial random variable (Equation 5.6). In addition, most statistical software has a built-in function to compute binomial probabilities for single values. See **Figure 5.10**.

```
> # Part (a)
> # Probability X is less than or equal to 12
>
> p <- pbinom(12,20,0.60)
> round(p,4)
[1] 0.5841

> # Part (b)
> # Probability that X equals 14
>
> p <- dbinom(14,20,0.60)
> round(p,4)
[1] 0.1244
```

Figure 5.10 $P(X \leq 12)$, cumulative probability; $P(X = 14)$, probability that X equals a single value.

(c) "At least 9 offer online orders" means 9 or more; X is greater than or equal to 9.

$$P(X \geq 9) = 1 - P(X < 9) \qquad \text{The Complement Rule.}$$
$$= 1 - P(X \leq 8) \qquad \text{The first value } X \text{ takes on less than 9 is 8.}$$
$$= 1 - 0.0565 \qquad \text{Use Appendix Table 1.}$$
$$= 0.9435$$

(d) "Between 10 and 15 (inclusive)" means that X can be any value from 10 to 15, including 10 and 15.

$P(10 \leq X \leq 15) = P(X \leq 15) - P(X \leq 9)$ Use cumulative probability.
$\qquad\qquad\qquad\quad = 0.9490 - 0.1275$ Use Appendix Table 1.
$\qquad\qquad\qquad\quad = 0.8215$

Figures 5.10 and **5.11** show technology solutions.

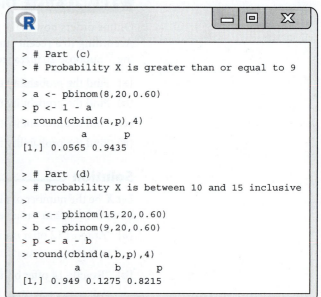

Figure 5.11 $P(X \geq 9)$, using cumulative probability; $P(10 \leq X \leq 15)$, using cumulative probability.

TRY IT NOW Go to Exercises 5.103 and 5.105

CHAPTER APP **Hooked on Vitamins**

Find the probability that at least 25 bottles contained the supplement shown on the label.

Example: Inference

EXAMPLE 5.16   Lower Cholesterol

The cholesterol-lowering drugs known as statins may cause several painful side effects, including blurred vision and fever. In a recent study, researchers claimed that 10% of people taking a statin suffer from some form of muscle pain. Unfortunately, people who experience this muscle pain may stop taking the statin and, therefore, are at greater risk of a heart attack or stroke.[20] Suppose 25 people who need a statin are selected at random. Each is given a 40-mg dose (per day), and the number of people who experience muscle pain is recorded.

(a) Find the probability that at most one person will experience muscle pain.

(b) Suppose seven people experience muscle pain. Is there any evidence to suggest that the researcher's claim is wrong? Justify your answer.

Solution

This example implicitly describes a binomial experiment. There are $n = 25$ trials with two outcomes (muscle pain or no muscle pain), the trials are independent, (random sample), and the probability of a success (experience muscle pain) remains constant from trial to trial.

Let X be the number of people (out of the 25 selected) who experience muscle pain after taking a statin. X is a binomial random variable with $n = 25$ and $p = 0.10$: $X \sim B(25, 0.10)$.

Translate the question into a probability statement involving the random variable X, convert to cumulative probability if necessary, and use Appendix Table 1.

(a) "At most one" means that X is any value up to and including 1.

$P(X \leq 1) = 0.2712$ Cumulative probability; use Appendix Table 1.

Figure 5.12 shows a technology solution.

(b) Use the experimental outcome to draw a conclusion concerning the claim. To decide whether seven people experiencing muscle pain is reasonable, follow the four-step inference procedure. This process now involves a random variable.

The researcher's claim is $p = 0.10$. This implies that the random variable X has a binomial distribution with $n = 25$ and $p = 0.10$.

Claim: $p = 0.10 \Rightarrow X \sim B(25, 0.10)$

The experimental outcome is that seven people experienced muscle pain. Write this outcome as an observed value of the random variable.

Experiment: $x = 7$

It seems reasonable to consider $P(X = 7)$, the probability of observing seven people with muscle pain, and draw a conclusion based on this probability. However, to be conservative (to give the person making the claim the benefit of the doubt), we always consider a tail probability. We accumulate the probability in a tail of the distribution; if it is small, then there is evidence to suggest the claim is false.

Which tail do we use? The answer depends on the mean of the distribution (and later on, the alternative hypothesis). Formulas for the mean, variance, and standard deviation of a binomial random variable are given next. Intuitively, however, the mean of a binomial random variable is $\mu = np$. If $n = 25$ and $p = 0.10$, we expect to see $\mu = (25)(0.10) = 2.5$ people experience joint pain. Because $x = 7$ is to the right of the mean, we'll consider a right-tail probability. See **Figure 5.13**.

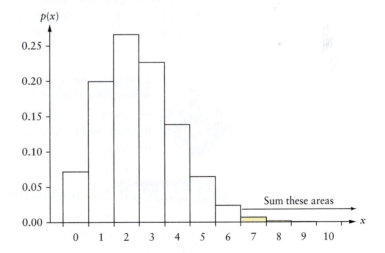

Figure 5.13 A portion of the probability histogram for the random variable X in Example 5.16. The right-tail probability, $P(X \geq 7)$, is the sum of the heights (or areas) of the rectangles above 7, 8, 9, ..., 25.

Likelihood:

$P(X \geq 7) = 1 - P(X < 7)$ The Complement Rule.

$ = 1 - P(X \leq 6)$ The first value X takes on less than 7 is 6.

$ = 1 - 0.9905$ Cumulative probability; use Appendix Table 1.

$ = 0.0095$

Conclusion: Because this tail probability is so small (less than 0.05), it is very unusual to observe seven or more people with muscle pain (if the claim is true). But it happened! Either this is an incredibly lucky occurrence or someone is lying.

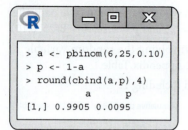

```
> a <- pbinom(6,25,0.10)
> p <- 1-a
> round(cbind(a,p),4)
          a      p
[1,] 0.9905 0.0095
```

Figure 5.14 Tail probability calculations using R.

We usually discount the lucky possibility and conclude that there is evidence to suggest the claim is false.

Figure 5.14 shows a technology solution.

TRY IT NOW Go to Exercises 5.107 and 5.109

μ, σ^2, and σ for a Binomial Random Variable

A random variable is often described, or characterized, by its mean and variance (or standard deviation): μ and σ^2 (or σ). If we know μ and σ, we can use Chebyshev's Rule to determine the most likely values of the random variable. For any population (random variable), most (at least 89%) of the values are within 3 standard deviations of the mean. This fact provides another approach to statistical inference for determining the likelihood of an experimental outcome.

To find the mean and variance of a binomial random variable, we could use the mathematical definitions (Equations 5.1 and 5.3). These formulas are used to produce the general results given here. However, the mean is intuitive. Consider a binomial random variable $X \sim B(10, 0.5)$. We expect to see $(10)(0.5) = 5 = np$ successes in 10 trials. (Think about tossing a fair coin 10 times.) Similarly, if $X \sim B(100, 0.75)$, we expect to see $(100)(0.75) = 75 = np$ successes. The mean of a binomial random variable with n trials and probability of a success p is $\mu = np$.

Mean, Variance, and Standard Deviation of a Binomial Random Variable

If X is a binomial random variable with n trials and probability of success p, $X \sim B(n, p)$, then

$$\mu = np \quad \sigma^2 = np(1-p) \quad \sigma = \sqrt{np(1-p)} \tag{5.8}$$

Given a binomial random variable n and probability p, we know the mean, variance, and standard deviation immediately. There is no need to use the formulas or the tabular method to find μ and σ^2. Here is an example to illustrate the use of this concept.

CHAPTER APP **Hooked on Vitamins**

Find the mean, variance, and standard deviation of the random variable X, the number of supplements that contained the ingredients shown on the label.

EXAMPLE 5.17 **Burnout in Tech**

Increased workload, few pay increases, and technological advancements have all contributed to employee burnout in many organizations. Occupational burnout is especially prevalent among tech workers, causing health concerns including insomnia, depression, and heart disease. A recent survey indicated that approximately 57% of all tech workers currently suffer from workplace burnout.[21] Suppose 100 tech workers are selected at random.

(a) Find the mean, variance, and standard deviation of the number of tech workers who suffer from workplace burnout.

(b) Suppose 61 of the 100 tech workers suffer from workplace burnout. Is there any evidence to suggest the survey's claim is false? Justify your answer.

Solution

(a) Let X be the number of tech workers (out of the 100 selected) who suffer from workplace burnout. The information given suggests that X is a binomial random variable with $n = 100$ and $p = 0.57$: $X \sim B(100, 0.57)$.

Use Equation 5.8 to find the mean, variance, and standard deviation.

$\mu = np = (100)(0.57) = 57$

$\sigma^2 = np(1-p) = (100)(0.57)(1-0.57) = (100)(0.57)(0.43) = 24.51$

$\sigma = \sqrt{\sigma^2} = \sqrt{24.51} = 4.95$

The expected number of tech workers who suffer from burnout is 57, with a variance of 24.51 and a standard deviation of 4.95.

(b) The key phrase, "Is there any evidence," suggests this problem involves statistical inference. Use the experimental outcome and the four-step inference procedure to draw a conclusion concerning the survey's claim.

Identify all four parts and use the mean and standard deviation to determine the most likely values of the relevant random variable.

The survey claims $p = 0.57$. This implies that the random variable X has a binomial distribution with $n = 100$ and $p = 0.57$.

Claim: $p = 0.57 \Rightarrow X \sim B(100, 0.57)$

The experimental outcome is that 61 tech workers suffer from workplace burnout.

Experiment: $x = 61$

Likelihood: From part (a), $\mu = 57$ and $\sigma = 4.95$. Most observations are within 3 standard deviations of the mean. Therefore, most values of X are in the interval

$(\mu - 3\sigma, \mu + 3\sigma) = (57 - 3(4.95), 57 + 3(4.95))$
$= (57 - 14.85, 57 + 14.85)$
$= (42.15, 71.85)$

Conclusion: Because 61 lies in this interval, or the interval includes 61, it is a reasonable observation. There is no evidence to suggest that the claim of $p = 0.57$ is false. ∎

TRY IT NOW Go to Exercises 5.113 and 5.117

CHAPTER APP **Hooked on Vitamins**

Suppose 21 dietary supplements contained the ingredients shown on the label. Is there any evidence to suggest that the claim ($p = 0.48$) is false?

A CLOSER LOOK

1. In statistics, we usually measure distance in standard deviations, not in miles, feet, inches, meters, or other units. We often want to know how many standard deviations from the mean is a given observation.

2. The inference problem in part (b) of Example 5.17 can also be solved using the tail probability approach. (Try it!) This method leads to the same conclusion and is generally more precise than constructing an interval about the mean. Chapter 9 introduces a more formal process for checking claims (hypothesis tests).

3. Whenever we test a claim, there are only two possible conclusions:
   (a) There is evidence to suggest the claim is false.
   (b) There is no evidence to suggest the claim is false.

   In either case, we never state with absolute certainty that the claim is true or the claim is false. This is because we never look at the entire population, only at a sample. With a large, random (representative) sample, we can be relatively confident in our conclusion, but never absolutely sure.

Section 5.4 Exercises

Concept Check

5.85 True or False A binomial random variable is completely described by the number of trials, n.

5.86 True or False There can be three or more outcomes in each trial of a binomial experiment.

5.87 True or False For any binomial random variable, $P(F) = 1 - P(S)$.

5.88 True or False Every probability question about a binomial random variable can be answered using cumulative probability.

5.89 True or False Suppose $X \sim B(n, p)$. Then $E(X) = np$.

5.90 True or False The most common values for a binomial random variable are less than the mean.

5.91 Fill in the Blank The binomial random variable is a count of _____.

5.92 Fill in the Blank Suppose $X \sim B(n, p)$. The number of outcomes with x successes is _____.

5.93 Short Answer Suppose $X \sim B(n, p)$. What are all the possible values of X?

5.94 Short Answer Identify the four properties of a binomial experiment.

5.95 Short Answer Write a probability expression involving a binomial random variable X that represents cumulative probability.

Practice

5.96 Suppose $X \sim B(15, 0.25)$. Find the probability.
(a) $P(X \leq 2)$ (b) $P(X < 2)$ (c) $P(X = 7)$
(d) $P(X > 6)$ (e) $P(X \geq 6)$ (f) $P(3 \leq X \leq 10)$

5.97 Suppose $X \sim B(20, 0.40)$. Find the probability.
(a) $P(X \geq 12)$ (b) $P(X \neq 10)$
(c) $P(X \leq 15)$ (d) $P(2 < X \leq 8)$

5.98 Suppose $X \sim B(25, 0.70)$. Find the probability.
(a) $P(X \geq 1)$ (b) $P(X \geq 10)$
(c) $P(X \geq 17.5)$ (d) $P(10.1 \leq X \leq 19)$

5.99 Suppose X is a binomial random variable with $n = 25$ and $p = 0.80$.
(a) Find the mean, variance, and standard deviation of X.
(b) Find the probability that X is within 1 standard deviation of the mean.
(c) Find the probability that X is more than 2 standard deviations from the mean.

5.100 Suppose X is a binomial random variable with $n = 30$ and $p = 0.40$.
(a) Find the mean, variance, and standard deviation of X.
(b) Find the intervals $\mu \pm \sigma$, $\mu \pm 2\sigma$, and $\mu \pm 3\sigma$.
(c) Find $P(X > \mu + 3\sigma)$.
(d) Find $P(X \leq \mu - 2\sigma)$.

5.101 Suppose X is a binomial random variable with $n = 20$ and $p = 0.45$.
(a) Find the mean, variance, and standard deviation of X.
(b) Find $P(X > \mu + \sigma \mid X > \mu)$.
(c) Find $P(X < \mu - \sigma \cup X > \mu + \sigma)$.
(d) Find a value m such that $P(X \leq m) \approx 0.5$.

5.102 Suppose X is a binomial random variable with $n = 10$ and $p = 0.50$.
(a) Create a table of values of X and associated probabilities. (*Hint*: This is quick and easy using technology.)
(b) Use the table in part (a) and the definitions of expected value and variance (Equations 5.1 and 5.3) to find μ, σ^2, and σ.
(c) Use Equation 5.8 to find μ, σ^2, and σ. Check these answers against those you found in part (b).

Applications

5.103 Technology and the Internet Even though many of us worry that someone will steal one of our passwords, a recent report revealed that only 10% of active gmail accounts use two-factor authentication (2FA).[22] This security tool involves a password and a second form of authorization, such as a code that is sent to your cell phone. Suppose 20 gmail accounts are selected at random.
(a) Find the probability that at most two gmail accounts use 2FA.
(b) Find the probability that at least four gmail accounts use 2FA.
(c) Find the expected number of gmail accounts that use 2FA.
(d) Suppose fewer than four gmail accounts use 2FA. What is the probability that none uses 2FA?

5.104 Fuel Consumption and Cars The battery manufacturer Varta sells a car battery with 800 cold-cranking amps and advertises great performance even in bitterly cold weather. Varta claims that after sitting on a frozen Minnesota lake for 10 days at temperatures below 32°F, this battery will still have enough power to start a car. Suppose the actual probability of starting a car following this experiment is 0.75, and 15 randomly selected cars (equipped with this battery) are subjected to these grueling conditions.
(a) Find the probability that fewer than 10 cars will start.
(b) Find the probability that more than 12 cars will start.
(c) Suppose 9 cars actually start. Is there any evidence to suggest that the probability of starting a car is different from 0.75? Justify your answer.

5.105 Marketing and Consumer Behavior Amy's Bread, in Chelsea Market, New York City, is trying to determine how many loaves of raisin bread to make each day. Over the past few months, the store has baked 50 loaves each day and has sold out with probability 0.80. Suppose the owner continues this practice and 30 days are selected at random.
(a) What is the expected number of days on which all 50 loaves will be sold?
(b) Find the probability of selling all 50 loaves on at least 20 days.
(c) Find the probability of selling all 50 loaves on at most 18 days.

5.106 Sports and Leisure Wild Animal Safari in Pine Mountain, Georgia, offers a drive-through, 3.5-mile experience in which animals come right up to your car. Suppose park officials claim that the probability of some car damage by an animal during a safari drive-through is 0.60. Suppose 20 cars are selected at random.
(a) Find the probability that exactly 10 cars will be damaged.
(b) Find the probability that at least 15 cars will be damaged.
(c) Find the probability that no more than 12 cars will be damaged.
(d) Suppose 19 cars are damaged. Is there any evidence to suggest the claim of 0.60 is false? Justify your answer.

5.107 Public Health and Nutrition Parents tend to be very good at diagnosing their children's routine medical problems, such as an ear infection, sinus infection, or strep throat. If an ailment is identified correctly, a trip to the doctor's office may be avoided. A physician may confer with a parent by telephone and simply call a pharmacy with a prescription for an antibiotic. Suppose parents are correct 90% of the time, and 50 families with a child suffering from some minor illness are selected at random.
(a) Find the mean, variance, and standard deviation of the number of parents who identify their child's illness correctly.
(b) Find the probability that at least 42 parents are correct.
(c) Find the probability that between 42 and 47 (inclusive) parents are correct.
(d) Suppose 41 parents are actually correct. Is there any evidence to suggest that less than 90% of parents are correct? Justify your answer.

5.108 Public Health and Nutrition A building inspector enforces building, electrical, mechanical, plumbing, and energy code requirements for the safety and health of people in a certain city, county, or state. In Colusa County, California, suppose the probability that a building inspector will find at least one code violation at a commercial building is 0.25. Suppose 30 commercial buildings are selected at random.
(a) Find the mean, variance, and standard deviation of the number of commercial buildings with at least one violation.
(b) Find the probability that the number of commercial buildings with at least one violation will be within 1 standard deviation of the mean.
(c) Find the probability that the number of commercial buildings with at least one violation will be more than 2 standard deviations from the mean.
(d) Suppose the actual number of commercial buildings with at least one violation is 10. Is there any evidence to suggest that code violations are found in more than 25% of commercial buildings? Justify your answer.

5.109 Education and Child Development Some school districts are trying various approaches to decrease truancy since graduation requirements have increased in those districts. In addition, some evidence suggests that consistent attendance improves learning. However, *The Baltimore Sun* recently reported that approximately 18% of Maryland students are chronically absent from school during the academic year.[23] Suppose 35 Maryland students are selected at random.
(a) What is the probability that exactly five students are chronically absent?
(b) What is the probability that at least eight students are chronically absent?
(c) Suppose four students are chronically absent. Is there any evidence to suggest that the proportion of students who are chronically absent is different from 0.18? Justify your answer.

5.110 Sports and Leisure The very familiar cry of "Are we there yet?" is commonly heard from the back seat during a family road trip. According to a report in *NewsUSA*, 85% of American families who planned to take a vacation during the summer of 2018 planned to travel by car.[24] Suppose 40 American families who took a vacation during the summer of 2018 are selected at random.
(a) What is the probability that at least 33 families traveled by car?
(b) What is the probability that between 30 and 35 (inclusive) traveled by car?
(c) Suppose 28 families actually traveled by car. Is there any evidence to suggest that the proportion of families who traveled by car for vacation changed? Justify your answer.

5.111 Business and Management Many different kinds of companies are adopting data-intensive tools to improve the services and the products they offer. In turn, the demand for data-science jobs has increased, and this demand is expected to continue. According to a recent report, 12% of open job

listings in Washington were for data-science positions, and 8% of job openings in Maryland were for these kinds of positions.[25] Suppose 20 open job listings in Washington and 20 open job listings from Maryland are selected at random.
(a) Find the probability that at most two job listings in Washington are for data-science positions.
(b) Find the probability that none of the job listings in Maryland is for a data-science position.
(c) Find the probability that at least one job listing from each state is for a data-science position.

5.112 Economics and Finance Self-employment offers greater work–life flexibility and the chance to supplement a regular income. In Canada, approximately 45% of the workforce is composed of freelancers, independent contractors, and on-demand workers.[26] Suppose 40 Canadian workers are selected at random.
(a) Find the mean, variance, and standard deviation of the number of Canadian workers (out of the 40 selected) who are self-employed.
(b) Construct intervals 1, 2, and 3 standard deviations from the mean.
(c) Suppose 27 workers (out of the 40 selected) are self-employed. How many standard deviations from the mean is this observation? What does this (statistical) distance measure indicate about the likelihood of observing 27 workers who are self-employed?

5.113 Technology and the Internet Even though robots can now vacuum floors, build cars, and make candy bars, the Organization for Economic Cooperation and Development claims that only 14% of jobs could be taken over by robots.[27] To check this claim, 50 jobs from around the world were randomly selected and each was assessed to determine if it could be completed by a robot.
(a) If the claim is true, find the probability that exactly seven jobs could be completed by a robot.
(b) If the claim is true, find the probability that at most four jobs could be completed by a robot.
(c) Suppose 12 jobs could be completed by a robot. Is there any evidence to suggest that the claim is false? Justify your answer.

5.114 Psychology and Human Behavior For those people with pets, it is often very tempting to feed a dog or cat table food. Table food is usually too fatty for pets and can cause severe stomach problems. Even so, according to a recent survey, 68% of pet owners feed their pets table food or scraps.[28] Suppose 50 pet owners are selected at random.
(a) Find the mean, variance, and standard deviation of the number of pet owners who feed them table scraps.
(b) Suppose 28 (out of the 50 pet owners) feed their pets table scraps. Is there any evidence to suggest that the claim (of 68%) is false?
(c) How many standard deviations away from the mean is this observation (of 28)? What does this distance suggest about the likelihood of observing 28 pet owners who feed their pets table scraps?
(d) Explain the relationship between your answers in parts (b) and (c).

Extended Applications

5.115 Business and Management In the *Lethal Weapon* movies, the character played by Joe Pesci is concerned about the accuracy of orders at fast-food restaurant drive-through windows. Suppose the probability that an order at a drive-through window at a fast-food restaurant will be filled correctly is 0.75. Twenty orders are selected at random.
(a) What is the probability that exactly 15 orders will be filled correctly?
(b) What is the probability that at most 12 orders will be filled correctly?
(c) What is the probability that between 10 and 14 (inclusive) orders will be filled correctly?
(d) Suppose two groups of 20 random orders are independently selected. What is the probability that at least 16 orders will be filled correctly in both groups?

5.116 Economics and Finance In many U.S. cities, such as San Francisco and Seattle, there is fierce competition in the housing market. As a consequence, buyers may not win a home with a first offer and sellers often receive more than the listed price. According to Zillow, 24% of U.S. homebuyers paid more than the asking price in 2017.[29] Suppose 20 U.S. home sales from 2017 are selected at random.
(a) Find the probability that at least six of the home sales were for more than the asking price.
(b) Find the expected number of home sales made for more than the asking price. Find the probability that the number of home sales is less than the mean.
(c) Suppose that at most four of the home sales were for more than the asking price. What is the probability that none of the home sales was for more than the asking price?

5.117 Marketing and Consumer Behavior More children are being rushed to the hospital because they were able to open the protective cap on a medication bottle and were poisoned by a common drug. A recent research study suggested that 25% of all preschool children can open a medication bottle. Suppose 10 preschool children are selected at random. Let the random variable X be the number of children who can open the bottle.
(a) Construct a probability histogram for the random variable X.
(b) Find the mean, variance, and standard deviation of X. Indicate the mean on the graph from part (a).
(c) Find $P(\mu - \sigma \leq X \leq \mu + \sigma)$ and illustrate this probability on the graph from part (a).
(d) Suppose that when these 10 children try to open a medicine bottle with a new design for the cap, only one child is able to open the bottle. Is there any evidence to suggest that the new cap is more effective in stopping children from opening the bottle? Justify your answer.

5.118 Manufacturing and Product Development A company has developed a very inexpensive explosive-detection machine for use at airports. However, if an explosive is actually concealed in a suitcase, the probability of it being detected by this machine is only 0.60. Therefore, several of these machines will be used simultaneously to screen each piece of luggage

independently. Suppose a piece of luggage actually contains an explosive.
(a) If three machines screen this luggage, what is the probability that exactly one will detect the explosive? What is the probability that none of the three will detect the explosive?
(b) If four machines screen this luggage, what is the probability that at least one device will detect the explosive?
(c) If five machines screen this luggage, what is the probability that at least one device will detect the explosive?
(d) How many machines are necessary for screening to be certain that at least one device will detect the explosive with probability 0.999 or greater?

5.119 Marketing and Consumer Behavior Forever 21 sells women's flip-flops in (oddly enough) 21 different colors. Despite this vast array of colors, 50% of all flip-flop purchases are in white. Suppose 30 buyers are selected at random.
(a) Find the mean, variance, and standard deviation of the number of buyers who purchase white flip-flops.
(b) Find the probability that the number of white flip-flops purchased will be within 2 standard deviations of the mean. Compare this with the predicted result from Chebyshev's Rule.
(c) Suppose two groups of 30 customers are independently selected. What is the probability of at least one group having exactly 15 people who buy white flip-flops?

5.120 Manufacturing and Product Development Sales reports indicate that approximately 1.2 billion watches are sold around the world each year. The demand for watches continues to increase, especially for minimalist watches. This huge demand has caused some manufacturers to cut corners. Timex, a leading watch producer, claims the proportion of its minimalist watches that are defective is 0.02. Tourneau, a leading watch retailer in the United States, receives a shipment of 50,000 minimalist watches. Before accepting the entire lot, Tourneau engineers select a random sample of 25 watches and thoroughly test each one. If four or more watches are found to be defective, the entire shipment will be sent back. Otherwise, the shipment will be accepted.
(a) Suppose the claim is true: The actual proportion of defectives is $p = 0.02$. What is the probability that the shipment will be rejected? (This is one type of error probability. The company, Tourneau, would be making a mistake if this event occurred: It would reject the shipment when the proportion of defective watches is as claimed.)
(b) Suppose the actual proportion of defective watches is $p = 0.05$. What is the probability that the shipment will be accepted? (This is another type of error probability. In this case, the company would also be making a mistake: It would accept the shipment when the proportion of defective watches is too high.)
(c) Suppose the actual proportion of defective watches is $p = 0.07$. What is the probability that the shipment will be accepted?
(d) Compare and explain the relationship between the probabilities in parts (b) and (c).

5.121 Psychology and Human Behavior As a result of stricter training requirements, fewer big fires, higher-paying jobs in cities, and changes in society, the number of volunteer firefighters is declining. Approximately 70% of all U.S. firefighters are volunteers, and the total number of volunteer firefighters has decreased steadily over the last two decades.[30] Suppose 30 U.S. firefighters are selected at random.
(a) Find the probability that exactly 22 of the firefighters are volunteers.
(b) Find the probability that more than 25 of the firefighters are volunteers.
(c) Suppose 17 of the firefighters are volunteers. Is there any evidence to suggest that the proportion of volunteer firefighters has decreased? Justify your answer.
(d) Suppose 50 firefighters are selected at random from the west and 50 firefighters are selected from the northeast. What is the probability that at least 40 of the firefighters will be volunteers in both groups?

5.122 Sports and Leisure Most ski resorts operate beginner, intermediate, and advanced terrains to appeal to people with varying abilities. The table lists several ski areas in Canada and the proportion of skiers who attempt the advanced terrain during their visit.

Ski area	Probability
Big White	0.28
Kicking Horse	0.60
Norquay	0.44

Suppose 20 skiers are randomly selected from each area.
(a) Find the probability that exactly five skiers at Big White will attempt the advanced terrain.
(b) Find the probability that more than eight skiers at Kicking Horse will attempt the advanced terrain.
(c) Find the probability that between 12 and 16 (inclusive) skiers at Norquay will attempt the advanced terrain.
(d) Find the probability that at most five skiers at all three locations will attempt the advanced terrain.

5.123 Manufacturing and Product Development In June 2018, Sanders issued a recall of its Milk Chocolate Covered Fudge Mini Bites because some might contain undeclared almonds. It was discovered that an almond-containing product was mistakenly added to a production line, and anyone allergic to peanuts could experience a life-threatening reaction after eating one of these Mini Bites. Suppose a package contains 20 random Fudge Mini Bites and production records indicate the probability that a Mini Bite contains undeclared almonds is 0.008.
(a) Find the probability that a package contains at least one Mini Bite with undeclared almonds.
(b) Suppose two packages are selected independently. Find the probability that at least one package contains at least one Mini Bite with undeclared almonds.
(c) How many packages would need to be selected for the probability that at least one package contains at least one Mini Bite with undeclared almonds to be greater than 0.5?

5.5 Other Discrete Distributions

Geometric Random Variable

There are many other discrete probability distributions. This section presents three commonly used distributions, along with brief background information, properties, and examples. Many of the problems involving these distributions are solved using the same general technique:

1. Define a random variable and identify its probability distribution (distribution statement).
2. Translate the question into a probability expression in which the event is stated in terms of the random variable (probability statement).
3. If necessary, try to convert the probability statement into an equivalent expression involving cumulative probability. Use tables and technology wherever possible.

The **geometric distribution** is closely related to the binomial distribution. In a binomial experiment, n (the number of trials) is fixed and the number of successes varies. The binomial random variable is the number of successes in n trials. In a geometric experiment, the number of successes is fixed at 1, and the number of trials varies.

Properties of a Geometric Experiment
1. The experiment consists of identical trials.
2. Each trial can result in only one of two possible outcomes: a success (S) or a failure (F).
3. The trials are independent.
4. The probability of a success p is constant from trial to trial.

A geometric experiment ends when the first success is obtained.

The Geometric Random Variable
The **geometric random variable** is the number of trials necessary until the first success occurs.

Think of an experiment in which you continue to phone a friend until you get through. The number of calls necessary until the first success (reaching your friend) is the value of a geometric random variable.

The derivation of the probability distribution involves the properties just given. Let X be a geometric random variable, the number of trials until the first success (including the trial in which the success is obtained). Given p, the probability of a success, find the probability of needing x trials, $P(X = x) = p(x)$.

$$P(X=1) = P(S) = p$$

$X = 1$ means the first trial results in a success, and the experiment is over. The probability of a success is simply p.

$$P(X=2) = P(F \cap S) = P(F) \cdot P(S) = (1-p)p$$

$X = 2$ means the first trial is a failure and the second trial is a success. Because trials are independent, we multiply the corresponding probabilities.

$$P(X=3) = P(F \cap F \cap S) = P(F) \cdot P(F) \cdot P(S)$$
$$= (1-p)(1-p)p = (1-p)^2 p$$

$X = 3$ means the first two trials are failures and the third trial is a success. We use independence again, and multiply the corresponding probabilities.

$$P(X = 4) = P(F \cap F \cap F \cap S) = P(F) \cdot P(F) \cdot P(F) \cdot P(S)$$
$$= (1-p)(1-p)(1-p)p = (1-p)^3 p$$

$X = 4$ means the first three trials are failures and the fourth trial is a success. We use independence again, and multiply the corresponding probabilities. In general,

$$P(X = x) = \underbrace{P(F) \cdot P(F) \cdots P(F)}_{x-1 \text{ failures}} \cdot P(S) = \underbrace{(1-p)(1-p)\cdots(1-p)}_{x-1 \text{ terms}} \cdot p$$
$$= (1-p)^{x-1} p$$

The event $X = x$ means the first $x - 1$ trials are failures and the xth trial is the first success. This generalization is the formula for the probability distribution.

The Geometric Probability Distribution
Suppose X is a geometric random variable with probability of a success p. Then

$$p(x) = P(X = x) = (1-p)^{x-1} p \qquad x = 1, 2, 3, \ldots \qquad (5.9)$$

$$\mu = \frac{1}{p} \qquad \sigma^2 = \frac{1-p}{p^2} \qquad \sigma = \sqrt{\frac{1-p}{p^2}} \qquad (5.10)$$

A CLOSER LOOK

1. The geometric random variable is discrete. The number of possible values is countably infinite: 1, 2, 3, ...

2. The geometric distribution is completely characterized, or defined, by one parameter, p.

3. We do not need a table to find cumulative probabilities associated with a geometric random variable because there is an easy formula for computing these values. If X is a geometric random variable with probability of success p, then

$$P(X \leq x) = 1 - (1-p)^x \qquad (5.11)$$

We can also use technology to find the probabilities associated with a geometric random variable.

4. Equation 5.9 is indeed a valid probability distribution.
   ▶ Each probability is between 0 and 1, and the sum of all the probabilities is an infinite series. The sum

$$\sum_{x=1}^{\infty} P(X = x) = \sum_{x=1}^{\infty} (1-p)^{x-1} p$$

is called a geometric series and it does sum to 1! ◀

✓ There is a formula for the sum of a geometric series. Can you use it to show that this sum is 1?

EXAMPLE 5.18 Bekins Men Are Careful, Quick, and Kind

Several years ago, the Bekins Moving Company used a clever jingle in many of its advertisements. The song started, "Bekins men are careful, quick, and kind; Bekins takes a load off of your mind...."

The number of people who change residences in the United States has declined steadily since 1985. According to the U.S. Census Bureau, approximately 11% of all people changed residences in 2016.[31] This is the lowest one-year move rate since the Census Bureau began collecting these data in 1948. Suppose researchers at the Bekins Moving Company randomly call people in the United States and ask if they have moved in the last year.

(a) What is the probability that the fourth person called will be the first to have moved in the past year?

(b) What is the probability that it will take at least six calls before the researchers speak to someone who has moved in the past year?

Solution

This experiment ends when the first mover, or success, is obtained. Consider a geometric random variable and an appropriate probability expression for each part.

(a) Let X be the number of calls necessary until the first mover is found. X is a geometric random variable with $P(S) = 0.11 = p$.

The probability that the first mover (success) is found on the fourth call:

$$P(X = 4) = (1-p)^{4-1} p \qquad \text{Equation 5.9}$$
$$= (1-0.11)^3 (0.11) = (0.89)^3 (0.11) = 0.0775 \qquad \text{Use } p = 0.11.$$

The probability that the first mover is found on the fourth call is 0.0775.

(b) "At least six calls before the researchers speak to someone who has moved in the past year" means the first success will occur on the sixth call or later.

The probability that at least six calls will be needed is

$$P(X \geq 6) = 1 - P(X < 6) \qquad \text{The Complement Rule.}$$
$$= 1 - P(X \leq 5) \qquad \text{The first value } X \text{ takes on that is less than 6 is 5.}$$
$$= 1 - [1 - (1-p)^5] \qquad \text{Use Equation 5.11.}$$
$$= 1 - [1 - (0.89)^5] \qquad \text{Use } p = 0.11.$$
$$= 1 - 0.4416 = 0.5584 \qquad \text{Expand and simplify.}$$

The probability that it will take six or more calls to find the first mover is 0.5584.

Figure 5.15 shows technology solutions.

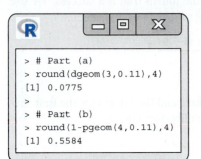

```
> # Part (a)
> round(dgeom(3,0.11),4)
[1] 0.0775
>
> # Part (b)
> round(1-pgeom(4,0.11),4)
[1] 0.5584
```

Figure 5.15 Probability calculations using built-in R functions. Note that the input values for the R probability mass function and for cumulative probability associated with a geometric random variable are $x - 1$ and p, respectively.

TRY IT NOW Go to Exercises 5.141 and 5.145

CHAPTER APP **Hooked on Vitamins**

Suppose dietary supplements are selected at random and tested. What is the probability that the first bottle that contains the supplement shown on the label is the fifth selected?

The Poisson Random Variable

The distribution is named after the French mathematician Simeon Denis Poisson (1781–1840).

The **Poisson probability distribution** has many practical applications and is often associated with rare events. A **Poisson random variable** is a count of the number of occurrences of a certain event in a given unit of time, space, volume, distance, etc., such as the number of arrivals to a hospital emergency room in a certain 30-minute period, the number of asteroids that pass through Earth's orbit during a given year, or the number of bacteria in a milliliter of drinking water.

Properties of a Poisson Experiment

1. The probability that a single event occurs in a given interval of time (specific area, region, volume, etc.) is the same for all intervals.
2. The number of events that occur in any interval is independent of the number that occur in any other interval.

These properties are often referred to as a Poisson process and can be difficult to verify.

The Poisson Random Variable

The **Poisson random variable** is a count of the number of times the specific event occurs during a given interval.

The Poisson distribution is completely determined by the mean, denoted by the Greek letter lambda, λ. Because the Poisson distribution is often used to count rare events, the mean number of events per interval is usually small. The probability distribution is given here.

The Poisson Probability Distribution
Suppose X is a Poisson random variable with mean λ. Then

$$p(x) = P(X = x) = \frac{e^{-\lambda}\lambda^x}{x!} \qquad x = 0, 1, 2, 3, \ldots \qquad (5.12)$$

$$\mu = \lambda \qquad \sigma^2 = \lambda \qquad \sigma = \sqrt{\lambda} \qquad (5.13)$$

A CLOSER LOOK

1. The Poisson random variable is discrete. The number of possible values is countably infinite: $0, 1, 2, 3, \ldots$.
2. The Poisson distribution is completely characterized by only one parameter, λ. The mean and the variance are both equal to the same value, λ.
3. Equation 5.12 is a valid probability distribution. All of the probabilities are between 0 and 1, and the sum of all the probabilities is 1 (another infinite series).
4. The e in Equation 5.12 is the base of the natural logarithm. $e \approx 2.71828$ is an irrational number, and most calculators have this special constant built in.
5. The denominator of Equation 5.12 contains $x!$ (x factorial).
   Recall that $x! = x(x-1)(x-2)\cdots(3)(2)(1)$ and $0! = 1$.
6. Appendix Table 2 contains values for $P(X \leq x)$ (cumulative probability) for various values of λ. We can also use technology to find values of the probability mass function and the cumulative probability.

EXAMPLE 5.19 Monthly Marine Occurrences

According to the Transportation Safety Board of Canada, approximately six fishing vessels are involved in accidents each month.[32] Some of the accident types are collision, fire, grounding, and sinking. Suppose six is the mean number of fishing vessel accidents per month and a random month is selected.

(a) Find the probability that there will be exactly four accidents involving fishing vessels.

(b) Find the probability that there will be at least eight accidents involving fishing vessels.

(c) Find the probability that the number of accidents involving fishing vessels will be within 1 standard deviation of the mean.

Solution

Since a fishing vessel accident is rare, and the mean number of fishing vessel accidents is small in a fixed amount of time (a month), consider a Poisson random variable. Write an appropriate probability expression for each part and convert to cumulative probability if necessary.

Let X be the number of fishing vessel accidents per month. X has a Poisson distribution with $\lambda = 6$.

(a) The probability of *exactly four* means $P(X = 4)$.

$$P(X = 4) = \frac{e^{-6} 6^4}{4!} = 0.1339 \qquad \text{Use Equation 5.12.}$$

$$= P(X \leq 4) - P(X \leq 3) \qquad \text{Or, convert to cumulative probability.}$$

$$= 0.2851 - 0.1512 = 0.1339 \qquad \text{Use Appendix Table 2.}$$

(b) "At least eight" means eight or more: $X \geq 8$.

$$\begin{aligned} P(X \geq 8) &= 1 - P(X < 8) & \text{The Complement Rule.} \\ &= 1 - P(X \leq 7) & \text{The first value } X \text{ takes on less than 8 is 7.} \\ &= 1 - 0.7440 & \text{Use Appendix Table 2.} \\ &= 0.2560 \end{aligned}$$

(c) "Within one standard deviation of the mean" describes the interval $(\mu - \sigma, \mu + \sigma)$.

$$\mu = 6 = \sigma^2 \Rightarrow \sigma = \sqrt{6} = 2.4495$$

$$\begin{aligned} P(\mu - \sigma \leq X \leq \mu + \sigma) \\ &= P(6 - 2.4495 \leq X \leq 6 + 2.4495) & \text{Use values for } \mu \text{ and } \sigma. \\ &= P(3.5505 \leq X \leq 8.4495) & \text{Compute the difference and sum.} \\ &= P(4 \leq X \leq 8) & \text{Use properties of the Poisson distribution.} \\ &= P(X \leq 8) - P(X \leq 3) & \text{Convert to cumulative probability.} \\ &= 0.8472 - 0.1512 & \text{Use Appendix Table 2.} \\ &= 0.6960 & \text{Compute the difference.} \end{aligned}$$

Figure 5.16 shows technology solutions.

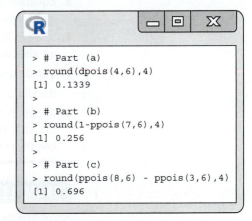

Figure 5.16 Probability calculations using built-in R functions for the probability mass function and for cumulative probability.

TRY IT NOW Go to Exercises 5.139 and 5.143

The Hypergeometric Random Variable

The **hypergeometric probability distribution** arises from an experiment involving sampling without replacement from a finite population. Each element in the population is labeled a success or failure. The **hypergeometric random variable** is a count of the number of successes in the sample. For example, consider a shipment of 12 automobile tires, of which two are defective, and a random sample of four tires. A hypergeometric random variable may be defined as a count of the number of good tires selected.

Properties of a Hypergeometric Experiment

1. The population consists of N objects, of which M are successes and $N - M$ are failures.
2. A sample of n objects is selected without replacement.
3. Each sample of size n is equally likely.

The Hypergeometric Random Variable

The **hypergeometric random variable** is a count of the number of successes in a random sample of size n.

The hypergeometric probability distribution is completely determined by n, N, and M. The probability of obtaining x successes is derived using many concepts introduced earlier: independence, the Multiplication Rule, equally likely outcomes, and combinations.

The Hypergeometric Probability Distribution

Suppose X is a hypergeometric random variable characterized by sample size n, population size N, and number of successes M. Then

$$p(x) = P(X = x) = \frac{\binom{M}{x}\binom{N-M}{n-x}}{\binom{N}{n}} \tag{5.14}$$

$$\max(0, n-N+M) \leq x \leq \min(n, M)$$

$$\mu = n\frac{M}{N} \qquad \sigma^2 = \left(\frac{N-n}{N-1}\right) n \frac{M}{N}\left(1 - \frac{M}{N}\right) \tag{5.15}$$

A CLOSER LOOK

1. ▶ Here is an explanation for the strange restriction on the possible values for the random variable X.

    $\max(0, n-N+M) \leq x$: x must be at least 0 or $n-N+M$, whichever is bigger. If $n-N+M$ is positive, it is impossible to obtain fewer than $n-N+M$ successes.

    $x \leq \min(n, M)$: x can be at most n or M, whichever is smaller. The greatest number of successes possible is either n or the total number of successes in the population.

    Suppose $n = 5$, $N = 10$, and $M = 6$. Then

    $\max(0, n-N+M) = \max(0, 5-10+6) = \max(0, 1) = 1$ and
    $\min(n, N) = \min(5, 10) = 5 \Rightarrow 1 \leq x \leq 5$.

    It is impossible to obtain less than 1 success. Also, the greatest number of successes possible is 5. ◀

2. The hypergeometric random variable is discrete. All of the probabilities are between 0 and 1, and the probabilities do sum to 1.

3. Recall that $\binom{n}{r}$ is a combination. The number of combinations of n items taken r at a time is $_nC_r = \binom{n}{r} = \frac{n!}{r!(n-r)!}$.

EXAMPLE 5.20 Apple Keyboards

In May 2018, a class action lawsuit was filed in Northern California district court alleging that certain Apple keyboards in MacBook and MacBook Pro models were defective—that is, they were prone to fail, which resulted in nonresponsive keys and other issues.[33]

Suppose an Apple Store has 10 MacBook Pro models for sale. Two of the 10 MacBooks have defective keyboards and will fail soon after purchasers begin using them. Suppose four of the MacBooks are randomly selected.

(a) What is the probability that exactly two MacBooks will have good, or working, keyboards?

(b) What is the probability that at least three MacBooks will have working keyboards?

Solution

The population is finite, $N = 10$ MacBooks; there are $M = 8$ working keyboards, or successes; the sample size is $n = 4$; and all four MacBooks are selected at random without replacement. Consider a hypergeometric distribution.

Let X be the number of working keyboards, or successes, in the sample. X has a hypergeometric distribution with $n = 4$, $N = 10$, and $M = 8$. Transform each question into a probability expression, convert to cumulative probability if necessary, and use Equation 5.14 and/or technology.

(a) "Exactly two" means $X = 2$.

$$P(X=2) = \frac{\binom{M}{x}\binom{N-M}{n-x}}{\binom{N}{n}} = \frac{\binom{8}{2}\binom{10-8}{4-2}}{\binom{10}{4}} \quad \text{Use Equation 5.14.}$$

$$= \frac{\binom{8}{2}\binom{2}{2}}{\binom{10}{4}} \quad \text{In the numerator, from the 8 good keyboards, choose 2; from the 2 bad keyboards, choose 2. In the denominator, } \binom{10}{4} \text{ is the total number of ways to choose 4 keyboards from 10.}$$

$$= \frac{(28)(1)}{210} = 0.1333 \quad \text{Use the formula for combination.}$$

The probability of selecting exactly two Macbooks with working keyboards is 0.1333.

(b) "At least three" means three or more. The maximum number of successes is $\min(n, M) = \min(4, 8) = 4$. In this case, three or more means 3 or 4.

$$P(X \geq 3) = P(X=3) + P(X=4) \quad \text{Consider the values } X \text{ can assume that are greater than or equal to 3.}$$

$$= \frac{\binom{8}{3}\binom{2}{1}}{\binom{10}{4}} + \frac{\binom{8}{4}\binom{2}{0}}{\binom{10}{4}} \quad \text{Use Equation 5.14.}$$

$$= \frac{(56)(2)}{210} + \frac{(70)(1)}{210} \quad \text{Use the formula for a combination.}$$

$$= 0.5333 + 0.3333 = 0.8667$$

Note: This problem can also be solved using cumulative probability:
$P(X \geq 3) = 1 - P(X \leq 2)$.

Figure 5.17 shows technology solutions. ∎

TRY IT NOW Go to Exercise 5.157

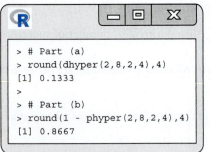

```
> # Part (a)
> round(dhyper(2,8,2,4),4)
[1] 0.1333
>
> # Part (b)
> round(1 - phyper(2,8,2,4),4)
[1] 0.8667
```

Figure 5.17 Probability calculations using built-in R functions for the probability mass function and for cumulative probability. The input values are x, M, $N - M$, and n.

Section 5.5 Exercises

Concept Check

5.124 True or False In a geometric experiment, the probability of a success varies from trial to trial.

5.125 True or False A geometric experiment ends when the first success is observed.

5.126 True or False The number of possible values for a geometric random variable is infinite.

5.127 True or False For a Poisson random variable, the mean is equal to the variance.

5.128 Fill in the Blank A Poisson random variable is often used to count _____.

5.129 Short Answer Explain why it would be unusual to use a Poisson random variable with $\lambda = 705$ to model a population.

5.130 Short Answer Explain the difference between a hypergeometric experiment and a binomial experiment.

5.131 Short Answer Suppose X is a hypergeometric random variable with sample size 5, population size 25, and number of successes 10. Find the possible values for X.

Practice

5.132 Suppose X is a geometric random variable with probability of success 0.35. Find the following probabilities.
(a) $P(X = 4)$
(b) $P(X \geq 3)$
(c) $P(X \leq 2)$
(d) $P(X \geq \mu)$

5.133 Suppose X is a geometric random variable with mean $\mu = 4$. Find the following probabilities.
(a) $P(X = 1)$
(b) $P(3 \leq X \leq 7)$
(c) $P(X > \mu + 2\sigma)$
(d) $P(X < 3 \mid X < 10)$

5.134 Suppose X is a Poisson random variable with $\lambda = 2$. Find the following probabilities.
(a) $P(X = 0)$
(b) $P(2 \leq X \leq 8)$
(c) $P(X > 5)$
(d) $P(X \leq 6)$

5.135 Suppose X is a Poisson random variable with $\lambda = 4.5$. Find the following probabilities.
(a) $P(X > \mu)$
(b) $P(X = 2)$
(c) The probability X is either 4 or 5.
(d) $P(X \leq \mu + 2\sigma)$

5.136 Suppose X is a hypergeometric random variable with $n = 5$, $N = 12$, and $M = 6$.
(a) Find $P(X = 2)$.
(b) Find $P(X = 5)$.
(c) Find the mean, variance, and standard deviation of X.
(d) Find $P(X = 6)$. Explain the meaning of this answer in the context of this random variable.

5.137 Suppose X is a hypergeometric random variable with sample size 8, population size 16, and number of successes in the population 12.
(a) List the possible values for X.
(b) Find the mean, variance, and standard deviation of X.
(c) Find $P(X = 5)$.
(d) Find $P(X = 8)$.

Applications

5.138 Psychology and Human Behavior According to the Anxiety and Depression Association of America (ADAA), approximately 8.7% of all adults suffer from a specific phobia, such as high bridges or old elevators.[34] An experiment consists of selecting adults at random and asking them if they suffer from any kind of phobia.
(a) What is the probability that the fifth adult selected will be the first with a specific phobia?
(b) What is the probability that at least eight adults will be selected before identifying a person with a specific phobia?
(c) What is the mean number of adults who must be selected before identifying a person with a specific phobia?
(d) Suppose the 35th adult is the first with a specific phobia. Is there any evidence to suggest the ADAA claim is false? Justify your answer.

5.139 Public Policy and Political Science Kate Middleton, the Duchess of Cambridge, has, on average, 4.75 official engagements per week.[35] Suppose a random week is selected.
(a) What is the probability that the Duchess will have exactly three official engagements?
(b) What is the probability that the Duchess will have more than seven official engagements?
(c) Suppose two weeks are selected at random. What is the probability that the Duchess will have at most two official engagements in both weeks?

5.140 Economics and Finance Women in Dubai represent a growing share of investors and are very active in the real estate market. During 2018, women accounted for 30% of Dubai realty transactions.[36] Suppose an auditor randomly examines Dubai real estate transactions made in 2018.
(a) What is the probability that the second transaction examined is the first made by a woman?
(b) What is the probability that the tenth transaction examined is the first made by a woman?
(c) What is the mean number of transactions that will be examined before one will be made by a woman?
(d) What is the probability that at least five transactions will be examined before one is made by a woman?

5.141 Technology and the Internet Despite advances in technology, burglar alarms that detect motion or sound are still not very reliable. Across the United States, false (burglar) alarms account for approximately 17% of all calls (about a burglary) to police. Suppose burglar alarm calls are selected at random.

(a) What is the probability that the third call will be the first false alarm?
(b) What is the probability that the twelfth call will be the first false alarm?
(c) What is the mean number of calls before one will be a false alarm?
(d) Suppose the first false alarm occurs after the fifth call. What is the probability that the first false alarm will occur after the ninth call?
(e) Suppose burglar alarm calls from two cities are selected at random. What is the probability that the first false alarm will occur before the third call in both cities?

5.142 Psychology and Human Behavior Various types of crimes, including property crimes, car thefts, and even violent crimes, occur on college campuses. The University of Denver reported that there are, on average, 2.4 bicycle thefts per month.[37] Suppose a random month is selected.

(a) What is the probability that there will be exactly three bicycle thefts?
(b) What is the probability that there will be fewer than six bicycle thefts?
(c) What is the probability that there will be at least five bicycle thefts?
(d) Suppose there are between 2 and 10 (inclusive) bicycle thefts. What is the probability of more than five bicycle thefts?

5.143 Travel and Transportation Bad weather is to blame for some of the worst highway crashes in Canada. In February 2013, a 50-car pile-up shut down Highway 401 near Woodstock, Ontario; in January 2017, nearly 100 vehicles were involved in a pile-up near Toronto. Highway 63 in Alberta has a notorious reputation. Suppose approximately four accidents occur every week on this road, and a week is randomly selected.

(a) Find the probability that there are no more than four crashes.
(b) Find the probability that the number of crashes is more than $\mu + 2\sigma$.
(c) To obtain government funding for safety improvements, a road must have five weeks in a row with six or more crashes. What is the probability of this happening?

5.144 Biology and Environmental Science In January 2018, a blue moon combined with a lunar eclipse as the moon was at its closest point to Earth, creating a *super blue blood moon*. The best place in the United States to see this celestial event was in the west. Suppose that in a group of 25 people who work in the same Palo Alto, California, office, 15 actually saw the super blue blood moon. Five people from this group are selected at random.

(a) What is the probability that none of the five people saw the super blue blood moon?
(b) What is the probability that at least four people saw the super blue blood moon?
(c) What is the probability that at most two people saw the super blue blood moon?

5.145 Psychology and Human Behavior American families don't get together for a home-cooked meal these days as often as they did in the past. However, a recent survey indicated that 27% of Americans actually cook every day. Suppose Americans are selected at random.

(a) What is the probability that the first person who cooks every day will be the fifth selected?
(b) What is the probability that the first person who cooks every day will be selected after the tenth person?
(c) What is the mean number of Americans selected before a person who cooks every day is identified?
(d) What is the probability that the first person who cooks every day will be identified before selecting 15 people?

5.146 Manufacturing and Product Development Flat-panel displays in televisions and computer monitors often develop dead pixels, or pixels that become locked in one state (e.g., red) at all times. Manufacturers maintain that dead pixels are a natural defect, and they have developed various policies for returning the devices after the discovery of a dead pixel. Suppose the mean number of dead pixels in a new LG OLED TV is 2.5. One of these TVs is randomly selected and inspected for dead pixels.

(a) What is the probability that there will be no dead pixels?
(b) If the number of dead pixels is more than $\mu + 3\sigma$, the assembly line is automatically stopped and examined. What is the probability that the assembly line will be stopped?
(c) What is the probability that the number of dead pixels will be within 2 standard deviations of the mean?

5.147 Public Health and Nutrition Approximately 12,000 people in Britain become ill each year from eating contaminated oysters, with most cases being due to norovirus. Two studies found that approximately 70% of all oysters on sale in the United Kingdom are infected with norovirus.[38] Suppose oysters in the United Kingdom are randomly selected and tested for norovirus.

(a) What is the probability that the second oyster selected will be the first infected with norovirus?
(b) What is the probability that the first oyster infected with norovirus will be one of the first three selected?
(c) Suppose the first oyster infected with norovirus is the 10th selected. Is there any evidence to suggest that the claim (70%) is false? Justify your answer.

5.148 Business and Management Fifteen lobstermen have their boats anchored at a small pier along the New Hampshire coast. Five of these lobstermen have been fined within the past year for commercial lobster-size violations. Suppose four lobstermen are selected at random.

(a) What is the probability that exactly two have been fined for violations within the past year?

(b) What is the probability that all four have been fined for violations within the past year?

(c) What is the probability that at least one has been fined for violations within the past year?

5.149 Demographics and Population Statistics Buchtal, a manufacturer of ceramic tiles, reports 3.9 job-related accidents per year. Accident categories include trip, fall, struck by equipment, transportation, and handling. Suppose a year is selected at random.

(a) What is the probability that there will be no job-related accidents?

(b) What is the probability that the number of accidents that year will be between two and five (inclusive)?

(c) If the number of accidents is more than 3 standard deviations above the mean, the company's insurance carrier will raise its rates. What is the probability of an increase in the company's insurance bill?

5.150 Sports and Leisure Amusement park rides are great family fun, but in 2016 more than 30,000 injuries were associated with amusement attractions nationwide. According to a recent study, on average, 12 children are treated in a hospital ER every day as a result of an injury from an amusement park ride.[39] Suppose a day is selected at random.

(a) What is the probability that no children will be treated in an ER as a result of an injury from an amusement park ride?

(b) What is the probability that at most 15 children will be treated in an ER as a result of an injury from an amusement park ride?

(c) Suppose that 20 children are treated in an ER as a result of an injury on an amusement park ride. Is there any evidence to suggest that the claim (of 12 every day) is wrong? Justify your answer.

5.151 Fuel Consumption and Cars Twenty people are in the Danville Department of Motor Vehicles waiting area. Eight people are waiting to have their license renewed, and the remaining people are there for other reasons, such as a road test or an automobile registration. Suppose six people in the waiting area are selected at random.

(a) What is the probability that exactly two are there to have their license renewed?

(b) What is the probability that all six are there to have their license renewed?

(c) What is the probability that at most four are there to have their license renewed?

Extended Applications

5.152 Marketing and Consumer Behavior The Sweet Leaf Iced Teas Company is sponsoring a conventional bottle-cap sweepstakes game. Under each bottle cap is a note either saying, "You are not a winner," or identifying the prize awarded. Suppose 20 of the game bottles are placed on a shelf in the supermarket, and two of them are winners. A customer randomly selects six bottles from the shelf.

(a) What is the probability of selecting no winning bottles?

(b) What is the probability of selecting both winning bottles?

(c) What is the mean number of winning bottles selected?

(d) How many bottles would the customer have to purchase to expect one winning bottle?

5.153 Physical Sciences More than 1 million earthquakes occur worldwide each year. However, major earthquakes, those with magnitude greater than 7, occur, on average, once per month.[40] Suppose a random month is selected.

(a) What is the probability of exactly two earthquakes of magnitude 7 or higher?

(b) What is the probability of at most four earthquakes of magnitude 7 or higher?

(c) Suppose there are four earthquakes of magnitude 7 or higher. Is there any evidence to suggest that the mean is different from 1? Justify your answer.

5.154 Business and Management Managers at Brooks Gourmet Burgers acknowledge that a variety of errors may occur in customer orders received via telephone. A recent audit revealed that the probability of some type of error in a telephone order is 0.20. In an attempt to correct these errors, a supervisor randomly selects telephone orders and carefully inspects each one.

(a) What is the probability that the third telephone order selected will be the first to contain an error?

(b) What is the probability that the supervisor will inspect between two and six (inclusive) telephone orders before finding an error?

(c) What is the probability that the inspector will examine at least seven orders before finding an error?

(d) What is the probability that the first error will occur on the fourth telephone order or later?

(e) Suppose the first four telephone orders contain no errors. What is the probability that the first error will occur on the eighth order or later?

5.155 Economics and Finance The manager of Capitol Park Plaza, an apartment complex in Washington, D.C., collects the rent from each tenant on the first day of every month. Past records indicate that the mean number of tenants who do not pay the rent on time in any given month is 4.7. Consider the rent collection for the next month.

(a) Find the probability that every tenant will pay the rent on time.

(b) Find the probability that at least seven tenants will be late with their rent.

(c) Suppose the number of delinquent rent payments in a month is independent of the number in every other month. What is the probability that at most three tenants will be late with their rent in two consecutive months?

5.156 Psychology and Human Behavior The mattress company Amerisleep recently conducted a survey and concluded that more than half of all Americans sleep on the job, but the type of work and salary affect how often people grab some shut eye.[41] Suppose that the mean number of naps per month on the job by a randomly selected American worker is four.

(a) What is the probability that a randomly selected American worker does not take a single nap during a month? One nap? Two naps?

(b) Suppose two American workers are selected at random. What is the probability that the total number of naps for the two Americans during a month is zero? One? Two?

(c) Suppose the mean number of times per month an American worker naps on the job is eight. What is the probability that a randomly selected American worker does not take a single nap during a month? One nap? Two naps?

(d) How do your answers in parts (b) and (c) compare? What property does this suggest about a Poisson random variable?

5.157 Psychology and Human Behavior The Travelers Risk Index is an annual study of driver distractions. The 2018 survey indicated that 25% of people multitask when driving—for example, by responding to text messages, emails, or calls. Those people who do so believe they can perform multiple tasks safely while driving.[42] In a group of 30 drivers, 6 multitask while driving. Suppose 4 people from this group are selected at random.

(a) What is the probability that exactly 1 person multitasks while driving?

(b) What is the probability that at most 2 people multitask while driving?

(c) Suppose the group of consists of 50 people, 10 of whom multitask while driving. Find the probabilities in parts (a) and (b) given this new, larger group.

(d) Suppose 4 people are selected at random from across the country. What is the probability that exactly 1 person multitasks while driving? That at most 2 multitask while driving?

(e) Compare all of these probabilities. Explain how the hypergeometric distribution is related to the binomial distribution.

5.158 Travel and Transportation We would all like to fly first class, with extra leg room, plush seats, and endless snacks. Frequent flier programs increase the chance of an upgrade, but airlines would, of course, prefer to sell these premium seats. Suppose a flight from New York to Los Angeles has 16 first-class (occupied) seats. The airline claims four people are seated in first class as a result of an upgrade. Six first-class passengers are selected at random.

(a) Find the mean, variance, and standard deviation of the number of people who received an upgrade.

(b) Find the probability that at least three people received an upgrade.

(c) Suppose none of the six received an upgrade. Is there any evidence to suggest that the airline's claim (four people received an upgrade) is false? Justify your answer.

Challenge Problems

5.159 Approaching Poisson Suppose X is a Poisson random variable with $\lambda = 2$. Let the random variable Y have the probability distribution in the following table.

y	$P(Y = y)$
0	$P(X = 0) = 0.1353$
1	$P(X = 1) = 0.2707$
2	$P(X = 2) = 0.2707$
3	$P(X \geq 3) = 0.3233$

Find the expected value of Y.
Suppose the distribution of Y is changed slightly, at the right tail, as given in the following table.

y	$P(Y = y)$
0	$P(X = 0) = 0.1353$
1	$P(X = 1) = 0.2707$
2	$P(X = 2) = 0.2707$
3	$P(X = 3) = 0.1804$
4	$P(X \geq 4) = 0.1429$

Find the expected value of Y.
Suppose the distribution of Y is changed again, once more at the right tail.

y	$P(Y = y)$
0	$P(X = 0) = 0.1353$
1	$P(X = 1) = 0.2707$
2	$P(X = 2) = 0.2707$
3	$P(X = 3) = 0.1804$
4	$P(X = 4) = 0.0902$
5	$P(X \geq 5) = 0.0527$

Find the expected value of Y.
Continue in this manner. To what number is $E(Y)$ converging, and why does this make sense?

5.160 A Committed Relationship Suppose X is a geometric random variable with probability of success $p = 0.40$ and Y is a binomial random variable with the same probability of success $p = 0.40$. For $a = 1, 2, 3, \ldots, 10$, construct a table with the following probabilities.

(a) $P(X = a)$
(b) $P(Y = 1)/a$, where $Y \sim B(a, 0.40)$
(c) $P(X \leq a)$
(d) $1 - P(Y = 0)$, where $Y \sim B(a, 0.40)$

Carefully examine the table and write a general formula to explain each equality. Can you prove these results?

5.161 When One Success Is Not Enough Suppose a geometric experiment is conducted with probability of success p. Let the random variable X be the number of trials necessary until two successes occur.

(a) What are the possible values for X?
(b) Find $P(X = 2)$.
(c) Find $P(X = 3)$ and $P(X = 4)$.
(d) Find $P(X = x)$.

Chapter 5 Summary

Concept	Page	Notation / Formula / Description
Random variable	190	A function that assigns a unique numerical value to each outcome in a sample space.
Discrete random variable	192	The set of all possible values is finite, or countably infinite.
Continuous random variable	192	The set of all possible values is an interval of numbers.
Probability distribution for a discrete random variable	195	A method for conveying all the possible values of the random variable and the probability associated with each value.
Mean, or expected value, of a discrete random variable X	206	$\mu = E(X) = \sum_{\text{all } x}[x \cdot p(x)]$.
Variance of a discrete random variable X	208	$\sigma^2 = \text{Var}(X) = \sum_{\text{all } x}[(x-\mu)^2 \cdot p(x)]$.
Properties of a binomial experiment	216	1. n identical trials. 2. Each trial can result in only a success (S) or a failure (F). 3. Trials are independent. 4. Probability of a success is constant from trial to trial.
Binomial random variable	217	The number of successes in n trials.
Binomial probability distribution	218	If $X \sim B(n, p)$ then $$p(x) = \binom{n}{x} p^x (1-p)^{n-x}, \ x = 0, 1, 2, 3, \ldots, n$$ where $\mu = np$, $\sigma^2 = np(1-p)$, and $\sigma = \sqrt{np(1-p)}$.
Cumulative probability	220	$P(X \leq x)$
Geometric random variable	230	The number of trials necessary to realize the first success.
Geometric probability distribution	231	$p(x) = (1-p)^{x-1} p, \ x = 1, 2, 3, \ldots$ where $\mu = \dfrac{1}{p}$, $\sigma^2 = \dfrac{1-p}{p^2}$, and $\sigma = \sqrt{\dfrac{1-p}{p^2}}$.
Poisson random variable	232	A count of the number of times a specific event occurs during a given interval.
Poisson probability distribution	233	$p(x) = \dfrac{e^{-\lambda} \lambda^x}{x!}, \ x = 0, 1, 2, 3, \ldots$ where $\mu = \lambda$, $\sigma^2 = \lambda$, and $\sigma = \sqrt{\lambda}$.
Hypergeometric random variable	234	A count of the number of successes in a random sample of size n from a population of size N.
Hypergeometric probability distribution	234	$p(x) = \dfrac{\binom{M}{x}\binom{N-M}{n-x}}{\binom{N}{n}}, \ \max(0, n-N+M) \leq x \leq \min(n, M)$ where $\mu = n\dfrac{M}{N}$, $\sigma^2 = \left(\dfrac{N-n}{N-1}\right) n \dfrac{M}{N}\left(1-\dfrac{M}{N}\right)$, and $\sigma = \sqrt{\left(\dfrac{N-n}{N-1}\right) n \dfrac{M}{N}\left(1-\dfrac{M}{N}\right)}$.

Chapter 5 Exercises

Applications

5.162 Biology and Environmental Science Hard water is high in mineral content, often calcium and magnesium. These minerals cause scaling, which may support bacteria growth, can build up and clog water lines, and cause certain appliances, such as clothes washers, to run less efficiently. A recent report claimed that 90% of all American homes have hard water.[43] Suppose 30 American homes are selected at random and tested for hard water.
(a) Find the probability that exactly 28 homes have hard water.
(b) Find the probability that at least 25 homes have hard water.
(c) Suppose only 20 homes have hard water. Is there any evidence to suggest that the proportion of American homes with hard water is less than 0.90? Justify your answer.

5.163 Business and Management IKEA is a Swedish company that sells ready-to-assemble furniture. Shoppers contact customer service to find a nearby store, for online shipping questions, and even for help with assembly. IKEA classifies all telephone calls to its customer support staff by the amount of time the customer is on hold. If the customer is on hold for no more than 60 seconds, then the call is classified as successful (actually, this sounds like a miracle). The supervisor in technical support claims 80% of all calls are successful. Suppose 25 calls to technical support are selected at random.
(a) Find the mean, variance, and standard deviation of the number of successful calls.
(b) Find the probability that at least 18 calls will be successful.
(c) Suppose 21 calls are successful. Is there any evidence to suggest the supervisor's claim is false? Justify your answer.

5.164 Economics and Finance Overdraft fees from some of the largest banks range from $9 to $39. Consumers believe these charges are annoying and excessive, but the fees generate huge revenue for banks, $34.3 billion in 2017.[44] The Overdraft Protection Act of 2013 is designed to limit overdraft fees in a variety of ways and to require fees to be reasonable and proportional to the amount of the overdraft. Let X be the amount of an overdraft fee for a randomly selected bank. The probability distribution for X is given in the table.

x	10	12	15	20	25	27	30	35	38
$p(x)$	0.02	0.06	0.08	0.10	0.16	0.28	0.15	0.07	0.08

(a) Find the mean, variance, and standard deviation of the overdraft amount.
(b) Find the probability that a randomly selected bank has an overdraft fee greater than $25.
(c) Find the probability that a randomly selected bank has an overdraft fee less than $\mu - \sigma$.
(d) Suppose three banks are selected at random. What is the probability that at least one bank has an overdraft fee less than $20?

5.165 Manufacturing and Product Development Thales Alenia Space is a European company that manufactures communications satellites. Researchers at the company have determined that the most common reason for a satellite to fail once it is in orbit is a problem related to opening and initiating the solar panels. Suppose the probability of a failure related to the solar panels is 0.08.
(a) What is the probability that the fifth satellite launched will be the first to fail due to a solar-panel problem?
(b) What is the number of satellites expected to be launched before the first one fails due to a solar-panel problem?
(c) Thales Alenia Space is preparing an advertising campaign in which it claims to have had 20 successful launches in a row. What is the probability that the first failure due to a solar-panel problem will occur after the 20th launch?

5.166 Business and Management An easy-assembly, tools-provided, four-drawer, wood filing cabinet comes with detailed step-by-step instructions. Even though each filing cabinet is carefully packaged, pieces are often missing. This can aggravate the customer and increase the cost to the producer, which must provide phone support and ship the missing parts. Suppose the mean number of missing pieces per packaged filing cabinet is 0.7, and one filing cabinet is randomly selected from the stockroom.
(a) What is the probability that no pieces will be missing from the package?
(b) If there are more than five missing pieces, the manufacturer identifies the packager and issues a warning. What is the probability of a warning being issued at the packaging plant?
(c) Suppose three filing cabinets are randomly selected. What is the probability that each will have no more than one missing piece?

5.167 Sports and Leisure Backyard barbecue enthusiasts enjoy the convenience and clean, steady heat of gas grills. Gas grills are usually fueled by liquid propane, which avoids the need to deal with briquettes or lighter fluid. However, each year, many Americans end up in the ER due to injuries associated with using a gas grill. The Gas Grill Safety Commission (GGSC) claims that 8% of all injuries related to a gas grill are caused by starting the grill with the cover closed. Suppose people who suffered an injury involving a gas grill are selected at random.
(a) What is the probability that the third person selected is the first who suffered an injury by starting the grill with the cover closed?
(b) What is the probability that at least 10 people will be selected before identifying the first who suffered an injury by starting the grill with the cover closed?
(c) What is the mean number of people who must be selected before identifying a person who suffered an injury by starting the grill with the cover closed?
(d) Suppose the 30th person is the first who suffered an injury by starting the grill with the cover closed. Is there any evidence to suggest the GGSC claim is false? Justify your answer.

5.168 Public Health and Nutrition Lead paint is still a leading cause of lead poisoning in the United States. The presence of lead paint is associated with the age of the building. According to Conservation Consultants, homes built before 1940 are more likely to contain lead paint.[45] In a group of 25 New York City apartments, 16 contain lead paint. Suppose 8 of the 25 will be selected at random for inspection.
(a) What is the probability none of the eight apartments contains lead paint?
(b) What is the probability that at most four of the apartments contain lead paint?
(c) What is the probability that at least six apartments contain lead paint?

5.169 Sports and Leisure Over the past decade, approximately 3.3 shark attacks occured each year at North Carolina beaches.[46] Consider the number of shark attacks during a random year.
(a) What is the probability that there will be no shark attacks?
(b) What is the probability that between two and five shark attacks (inclusive) will occur?
(c) If there is evidence to suggest that the mean number of shark attacks per year has increased, the Coast Guard will begin more patrols to adequately protect the public. Suppose eight shark attacks occur in year 13. Is there evidence to suggest the need for more patrols? Justify your answer.
(d) Find out exactly how many shark attacks occurred in a recent year in North Carolina. Determine the probability of this happening.
(e) Florida has the highest number of shark attacks per year, approximately 24.4, followed by Hawaii (6.5 per year) and South Carolina (3.9 per year). What is the probability that there will be no attacks in these three states plus North Carolina in a given year?

5.170 Public Policy and Political Science A survey conducted by the Annenberg Public Policy Center (APPC) showed that only 63% of Americans could name any of the five rights protected by the First Amendment (freedom of religion, speech, the press, to assemble peaceably, and to petition the government for a redress of grievances).[47] Suppose 40 Americans are selected at random, and each is asked to name the five rights protected by the First Amendment.
(a) Find the probability that exactly 23 can name any of the five rights in the First Amendment.
(b) Find the probability that at most 30 can name any of the five rights in the First Amendment.
(c) Find the probability that more than 33 can name any of the five rights in the First Amendment.
(d) Suppose the number of people who can name any of the five rights in the First Amendment is within 2 standard deviations of the mean. What is the probability that the actual number who can name any of the five rights in the First Amendment is within 1 standard deviation of the mean?

5.171 Sports and Leisure The cruise ship industry continues to build larger-capacity ships, stop at more local ports, and increase the number of on-board activities. In 2017, more than 25 million people took a cruise; moreover, approximately 24% of all people in the United States have taken a cruise at some point in their lives.[48] Suppose 25 people from the United States are selected at random.
(a) Find the probability that exactly seven people have been on a cruise.
(b) Find the probability that at most five people have been on a cruise.
(c) Suppose 11 people have taken a cruise. Is there any evidence to suggest that the percentage of people who have taken a cruise has increased? Justify your answer.

5.172 Psychology and Human Behavior The army emphasizes cleanliness and neatness in its military barracks. Each cadet is responsible for maintaining his or her area in top condition. Periodic inspections are held, and those receiving top scores are rewarded. Suppose the mean number of violations discovered per cadet during a barracks inspection is 2.7.
(a) What is the probability that a randomly selected cadet will have exactly three violations during an inspection?
(b) If a cadet has six or more violations, he or she is assigned to KP (kitchen) duty for one week. What is the probability that a randomly selected cadet will be assigned to KP duty following a barracks inspection?
(c) If every member of a 10-cadet unit has no violations, then each will receive a weekend pass. What is the probability of this happening following a barracks inspection?

5.173 Travel and Transportation The Toronto Transit Commission (TTC) is conducting a study to determine whether to install safety barriers on subway platforms.[49] These automatic barriers are already used in parts of Europe and Asia, but the cost to retrofit stations in Toronto is estimated at $1 billion. Suppose only 20% of Toronto residents are in favor of installing these barriers. Fifty people from Toronto are selected at random, and each is asked whether the barriers should be installed.
(a) Find the mean, variance, and standard deviation of the number of people who believe the barriers should be installed.
(b) What is the probability that at most 10 people believe the barriers should be installed?
(c) Find the probability that the number of people who believe the barriers should be installed is within 2 standard deviations of the mean.

5.174 Sports and Leisure *Star Wars* fans were delighted with the release of *Solo: A Star Wars Story*. The critics proclaimed the film to be flawed yet fun, and the Flixster movie site, Rotten Tomatoes, rated the movie at 71% on the Tomatometer.[50] However, only 64% of all people who saw the movie liked it. Suppose 30 people who saw the movie are selected at random.
(a) What is the probability that exactly 20 people liked the movie?
(b) What is the probability that at least 22 people liked the movie?

(c) Suppose 16 people liked the movie. Is there any evidence to suggest that the claim (64%) is wrong? Justify your answer.

5.175 Technology and the Internet The Creepy and Cool survey conducted by RichRelevance provides insight into consumers' opinions related to cutting-edge technologies, including augmented reality (AR) and artificial intelligence (AI). The fourth annual survey revealed that approximately 80% of consumers believe companies are obligated to disclose whether they are using AI, and how they are using it.[51] Suppose 50 consumers are selected at random.
(a) What is the probability that at least 45 consumers believe companies should disclose the use of AI?
(b) What is the probability that between 38 and 43 (inclusive) consumers believe companies should disclose the use of AI?
(c) Suppose 36 consumers believe companies should disclose the use of AI. Is there any evidence to suggest the poll results are wrong? Justify your answer.

5.176 Biology and Environmental Science Three species of prairie dogs live in Colorado. Prairie dogs, whose burrows are used by other animals, can damage rangeland and carry diseases. During the winter months, there are approximately five black-tailed prairie dogs per acre living in Colorado's eastern plains.[52] Suppose an acre in the eastern plains is selected at random.
(a) What is the probability that there are no black-tailed prairie dogs in this acre?
(b) What is the probability that there are at most six black-tailed prairie dogs in this acre?
(c) Suppose there are actually 10 black-tailed prairie dogs in this acre. Is there any evidence to suggest that the number of black-tailed prairie dogs per acre has changed? Justify your answer.

5.177 Public Health and Nutrition Approximately 29 million people in the United States have diabetes. However, in a report published by UpWell Health, 45% of patients with diabetes indicated that they had skipped medical care at times due to the cost.[53] Suppose U.S. residents living with diabetes are contacted, and each is asked if they have skipped medical care due to cost.
(a) Find the probability that the third person contacted will be the first who has skipped medical care.
(b) Find the probability that the first person who has skipped medical care will be within the first six people contacted.
(c) Suppose the first person who has skipped medical care is the 12th person contacted. Is there any evidence to suggest that the proportion of people skipping medical care due to cost has changed? If so, how? Justify your answer.

Extended Applications

5.178 Discrete Uniform Random Variable Suppose X is a random variable with the probability distribution given by

$$p(x) = \frac{1}{5} \quad x = 1, 2, 3, 4, 5$$

(a) Find the mean, variance, and standard deviation of X.
(b) Suppose $p(x) = 1/6$, $x = 1, 2, 3, 4, 5, 6$. Find the mean, variance, and standard deviation of X.
(c) Suppose $p(x) = 1/n$, $x = 1, 2, 3, \ldots, n$. Find the mean, variance, and standard deviation of X in terms of n.

5.179 Medicine and Clinical Studies According to the National Institutes of Health, approximately 35% of U.S. adults attempt to diagnose a medical condition online.[54] Highmark Insurance Company is concerned about the rising number of online diagnosers and the resulting failure to consult a physician. It has decided to select 25 policyholders at random. If the number of online diagnosers is 11 or fewer, then no action will be taken. Otherwise, the company will begin a campaign to remind policyholders that they should always consult a physician to confirm a medical condition.
(a) Suppose the true proportion of online diagnosers is 0.35. What is the probability that Highmark will begin a new reminder campaign?
(b) Suppose the true proportion of online diagnosers is 0.40. What is the probability that no action will be taken? What if the true proportion is 0.50?
(c) Suppose the decision rule is changed such that if the number of online diagnosers is 12 or fewer, then no action will be taken. Answer parts (a) and (b) using this rule.

5.180 Marketing and Consumer Behavior Kohl's is running a sale in which customers may save as much as 40% on any purchase. Once a customer decides to make a purchase, he or she selects two sales prize tickets at random from a large bin placed at the front of the store. Each ticket has a percentage marked on it, and the probability of selecting each ticket is given in the table.

Percentage	10%	20%	30%	40%
Probability	0.50	0.35	0.10	0.05

The larger of the two percentages selected is used for the purchase.
(a) Let X be the maximum of the two prize ticket percentages. Find the probability distribution for X.
(b) Find the mean, variance, and standard deviation of X.
(c) What is the probability that a customer will receive at least 20% off on his or her purchase?

5.181 Psychology and Human Behavior All parents are concerned about the whereabouts of their children and their safety. Yet children should be allowed personal space and the chance to build trust from parents. To be certain about safety, 30% of parents set up Global Positioning System (GPS) tracking systems on their children's phones without the children's consent.[55] Suppose 40 parents with children who have cell phones are selected at random.
(a) What is the probability that exactly 14 parents have secretly set up GPS tracking systems?
(b) Find the largest value n such that the probability of n or fewer parents who have secretly set up GPS tracking systems is at most 0.20.

(c) Suppose 16 parents have secretly set up GPS tracking systems. Is there any evidence to suggest that the proportion of parents who do this has changed? Justify your answer.

5.182 Biology and Environmental Science Recent satellite data indicate that 24% of the world's sandy beaches are eroding.[56] Suppose 30 sandy beaches from around the world are selected at random.
(a) What is the probability that exactly seven beaches are eroding?
(b) What is the probability that at least 10 beaches are eroding?
(c) Suppose at most eight beaches are eroding. What is the probability that at most four beaches are eroding?

Challenge Problems

5.183 A Day on the Dock Two crews work on a receiving dock at a fabric manufacturing plant. The first crew unloads four shipments every day and the second crew unloads seven shipments every day. A supervisor records whether each shipment is complete (a success) or missing items (a failure).

Suppose X_1 is a binomial random variable, representing the number of complete shipments for crew 1, with parameters $n_1 = 4$ and $p = 0.6$. Similarly, let X_2 be a binomial random variable, representing the number of complete shipments for crew 2, with parameters $n_2 = 7$ and $p = 0.6$. Assume X_1 and X_2 are independent.

(a) Use technology to generate a random observation for X_1 (the number of complete shipments for crew 1) and a random observation for X_2 (the number of complete shipments for crew 2). Add these two values to compute a random total number of complete shipments for crews 1 and 2.

Repeat this process to generate a total of 1000 complete shipments for crews 1 and 2. Compute the relative frequency of occurrence of each observation.

Suppose Y is a binomial random variable with $n = 11$ and $p = 0.6$. Use technology to construct a table of probabilities for $y = 0, 1, 2, 3, \ldots, 11$. Compare these probabilities with the relative frequencies obtained in the preceding step.

(b) Suppose a new receiving crew is added that unloads five shipments each day. Let X_3 be a binomial random variable, representing the number of complete shipments for crew 3, with parameters $n_3 = 5$ and $p = 0.6$.

Use technology to generate random observations for X_1, X_2, and X_3. Add these three values to compute a random number of complete shipments for crews 1, 2, and 3.

Repeat this process to generate a total of 1000 complete shipments for crews 1, 2, and 3. Compute the relative frequency of occurrence of each observation.

Suppose Y is a binomial random variable with $n = 16$ and $p = 0.6$. Use technology to construct a table of probabilities for $Y = 0, 1, 2, 3, \ldots, 16$. Compare these probabilities with the relative frequencies obtained in the preceding step.

(c) Suppose another receiving crew is added that unloads nine shipments each day. Let X_4 be a binomial random variable, representing the number of complete shipments for crew 4, with parameters $n_2 = 9$ and $p = 0.6$. Let Y represent the total number of complete shipments for all four crews.
  (i) Find $P(Y = 15)$, the probability of exactly 15 total complete shipments.
  (ii) Find $P(Y \leq 12)$.
  (iii) Find $P(Y > 16)$.
  (iv) How many total complete shipments can be expected?

CHAPTER APP

5.184 Hooked on Vitamins A recent study claimed that only 48% of vitamin supplements actually contained the ingredients indicated on the labels. The remainder were either fake, counterfeit, or contaminated. A random sample of 50 dietary supplements was obtained. Each was tested using DNA barcoding to determine whether it contained the supplement shown on the label.
(a) Let X be the number tested that actually contained the supplement shown on the label. Is X discrete or continuous?
(b) Find the probability that at least 25 bottles contained the supplement shown on the label.
(c) Find the mean, variance, and standard deviation of the random variable X, the number of supplements that contained the ingredients shown on the label.
(d) Suppose dietary supplements are selected at random and tested. What is the probability that the first bottle that actually contains the ingredients shown on the label is the fifth selected?

Continuous Probability Distributions

simpson33/Deposit Photos

CHAPTER APP

EMV Chip Technology

Europay, MasterCard, and Visa (EMV) is a global standard associated with credit card chip technology. It is used to authenticate credit and debit cards and is designed to be more secure than the magnetic strip found on many cards. In theory, an EMV card is difficult to duplicate, which makes it much harder for someone to commit fraud using a counterfeit card.

Despite the added security, some merchants have been slow to adopt this new technology. There have been delays in obtaining EMV terminals, there is a large expense

◀ Looking Back

- Remember how to completely describe and compute probabilities associated with a discrete random variable: **Section 5.2**.
- Recall the characteristics of and probability computations associated with the binomial, geometric, Poisson, and hypergeometric random variables: **Sections 5.4 and 5.5**.

Looking Forward ▶

- Learn how to completely describe a continuous random variable and how to compute probabilities associated with a continuous random variable: **Section 6.1**.
- Understand the characteristics of the normal distribution and compute probabilities involving a normal random variable: **Section 6.2**.
- Learn several methods to check for non-normality: **Section 6.3**.
- Understand the characteristics of the exponential distribution and compute probabilities involving an exponential random variable: **Section 6.4**.

associated with this upgrade, and these cards tend to slow the in-store check-out process. Due to the increased transaction time, some Aldi supermarkets with EMV terminals have taped over the card insertion slot because card swipes are faster.[1]

According to Bluefin, a payment security firm, the mean processing time for an EMV chip card is 13 seconds, approximately double the time needed for a swipe card.[2] Suppose the standard deviation is 2.5 seconds and the distribution of EMV chip card process times is approximately normal. The concepts presented in this chapter will allow us to determine the most reasonable process times for customers and to decide when a merchant or customer has a legitimate complaint about the time it takes to process a transaction.

6.1 Probability Distributions for a Continuous Random Variable

How to Describe a Continuous Random Variable

Suppose X is a continuous random variable; X can take on any value in some interval of numbers. A **continuous probability distribution** completely describes the random variable and is used to compute probabilities associated with the random variable.

> **Definition**
> A **probability distribution for a continuous random variable** X is given by a smooth curve called a **density curve**, or **probability density function** (pdf). The curve is defined so that the probability that X takes on a value between a and b $(a < b)$ is the area under the curve between a and b.

A CLOSER LOOK

1. Probability in a continuous world is area under a curve. **Figures 6.1–6.3** illustrate the correspondence between the probability of an event (defined in terms of a continuous random variable) and the area under the density curve.

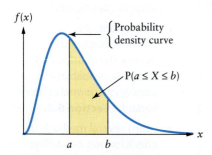

Figure 6.1 The shaded area is $P(a \leq X \leq b)$.

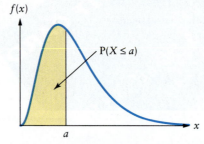

Figure 6.2 The shaded area is $P(X \leq a)$.

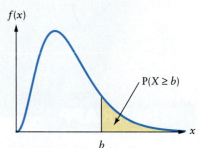

Figure 6.3 The shaded area is $P(X \geq b)$.

2. The density curve, or **probability density function**, is usually denoted by f. It is a function, defined for all real numbers. $f(x)$ is not the probability that the random variable X equals the specific value x. Rather, the function f leads to, or conveys, probability through area.

3. The shape of the graph of a density function can vary considerably. However, a density function must satisfy the following two properties.

   (a) The density function f must be defined so that the total area under the curve is 1. The total probability associated with any random variable must be 1. A specific value of the density function, $f(x)$, may be greater than 1 (while the total area under the curve is still exactly 1).

   (b) Values of the density function must be greater than or equal to 0; that is, $f(x) \geq 0$ for all x. Therefore, the complete graph of a density function lies on or above the x-axis. See **Figure 6.4**.

> Remember that $f(x)$ is not a probability. The density function is used to compute probability.

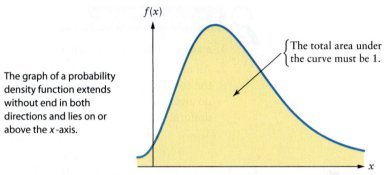

The graph of a probability density function extends without end in both directions and lies on or above the x-axis.

The total area under the curve must be 1.

Figure 6.4 A valid probability density function.

4. If X is a continuous random variable with density function f, the probability that X equals any one specific value is 0. That is, $P(X = a) = 0$ for any a. The reason: There is no area under a single point.

> $P(X = a)$ translated: Find the area under the curve between a and a. This probability expression represents the area of a line segment. There is no second dimension, so the area is 0.

▶ This seems like a contradiction. Certainly we can observe specific values of X, yet the probability of observing any single value is 0. Recall: Probability is a limiting relative frequency. There are (uncountably) infinite possible values for any continuous random variable. Therefore, the limiting relative frequency of occurrence of any single value is 0.

Because no probability is associated with a single point, the following four probabilities are all the same.

> This is not necessarily true for a discrete random variable.

$$P(a \leq X \leq b) = P(a < X \leq b) = P(a \leq X < b) = P(a < X < b) \quad (6.1)$$

In fact, we can remove as many single points as we want from any interval, and the probability will stay the same. The only reasonable probability questions concerning continuous random variables involve intervals. And we can almost always sketch a graph to visualize these probabilities, or regions. ◀

So how do we find area under a curve, and therefore probability? In general, this is a calculus question. Don't panic. We'll use a little geometry, tables, and technology to find the necessary area (probability).

The Uniform Distribution

The (continuous) uniform distribution provides a good opportunity to illustrate the connection between area under the curve and probability. For this random variable, the total probability 1 is distributed evenly, or uniformly, between two values. Computing probabilities associated with this random variable, therefore, can be simplified to finding the area of a rectangle.

 Piecewise defined functions.

Definition
The random variable X has a **uniform distribution** on the interval $[a, b]$ if

$$f(x) = \begin{cases} \dfrac{1}{b-a} & \text{if } a \leq x \leq b \\ 0 & \text{otherwise} \end{cases} \quad -\infty < a < b < \infty \quad (6.2)$$

$$\mu = \frac{a+b}{2} \qquad \sigma^2 = \frac{(b-a)^2}{12} \quad (6.3)$$

 A CLOSER LOOK

1. a and b can be any real numbers, as long as a is less than b ($a < b$).
2. All of the probability (action) is between a and b. The probability density function is the constant $1/(b-a)$ between a and b, and 0 outside of this interval. Hence, there is no area and no probability outside the interval $[a, b]$. **Figure 6.5** shows a graph of the uniform probability density function.

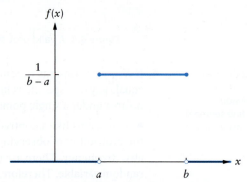

Figure 6.5 The graph of the probability density function for a uniform random variable.

3. Equation 6.2 is a valid probability density function because $f(x) \geq 0$ for all x, and the total area under the curve is 1. The area under the curve for $x < a$ is 0, and the area under the curve for $x > b$ is 0. Between a and b, the area under the curve is the area of a rectangle (area = width × height). See **Figure 6.6**.

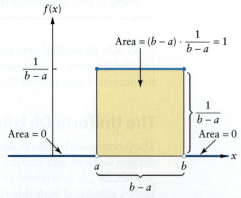

Figure 6.6 The total area under the curve is 1. Here, the density curve consists of three line segments.

Uniform Distribution Example

The following example involves a uniform distribution and illustrates visualizing and calculating probabilities associated with a continuous random variable.

EXAMPLE 6.1   Reef Dives

Bonaire, an island in the Dutch Caribbean, is considered one of the top 10 diving destinations in the world. There is generally 60–100 feet (ft) of visibility, the current is mild, and at least 58 dive sites can be reached from shore. Guides take tourist groups on commercial boats to scuba dive and snorkel at selected locations. A careful examination of boat records has shown that the time it takes to reach a randomly selected dive site has a uniform distribution between 5 and 25 minutes. Suppose a dive site is selected at random.

(a) Carefully sketch a graph of the probability density function.
(b) Find the probability that it takes at most 10 minutes to reach the dive site.
(c) Find the probability that it takes between 10 and 20 minutes to reach the dive site.
(d) Find the mean time it takes to reach a dive site, as well as the variance and standard deviation.

Solution

The time it takes to reach a dive site has a uniform distribution between 5 and 25 minutes. For a continuous random variable, probability is an appropriate area under the density curve. Translate each question into a probability statement, sketch the corresponding region, and compute the area under the curve.

(a) Let X be the time it takes to reach a dive site. The random variable X has a uniform distribution between the times $a = 5$ and $b = 25$. Use Equation 6.2 to find

$$\frac{1}{b-a} = \frac{1}{25-5} = \frac{1}{20} = 0.05$$

The probability density function is

$$f(x) = \begin{cases} 0.05 & \text{if } 5 \leq x \leq 25 \\ 0 & \text{otherwise} \end{cases}$$

Figure 6.7 shows the graph of the probability density function.

(b) We have the distribution of X. Translate the question in part (b) into a probability statement, sketch the region corresponding to the probability statement, and find the area of that region.

At most means up to and including 10. We need the probability that X is less than or equal to 10: $P(X \leq 10)$. See **Figure 6.8**.

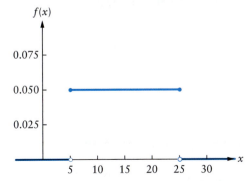

Figure 6.7 The graph of the probability density function for a uniform random variable on the interval $a = 5$ to $b = 25$.

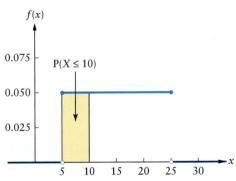

Figure 6.8 The area of the shaded region is $P(X \leq 10)$.

The probability statement P(X ≤ 10) simplifies to P(5 ≤ X ≤ 10) in this case because there is no probability (area) for X less than 5.

$P(X \le 10) = P(5 \le X \le 10)$
   = area under the density curve between 5 and 10
   = area of a rectangle
   = width × height
   = (5)(0.05) = 0.25

The probability that it takes at most 10 minutes is 0.25.

If X is a continuous random variable:

P(10 ≤ X ≤ 20)
= P(10 < X ≤ 20)
= P(10 ≤ X < 20)
= P(10 < X < 20)

(c) The probability that it takes between 10 and 20 minutes to reach a dive site in terms of the random variable X is P(10 ≤ X ≤ 20). Even though the word *inclusive* is not used in the question, we chose to write the interval including the endpoints. It doesn't really matter! Remember: In a continuous world, single values contribute no probability and do not change the probability calculation.

$P(10 \le X \le 20)$ = area under the density curve between 10 and 20
   = area of a rectangle
   = width × height
   = (10)(0.05) = 0.50

The probability that it takes between 10 and 20 minutes is 0.50. See **Figure 6.9**.

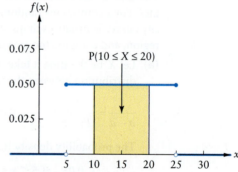

Figure 6.9 The area of the shaded region is P(10 ≤ X ≤ 20).

(d) Use Equation 6.3 to find the mean and variance.

$$\mu = \frac{a+b}{2} = \frac{5+25}{2} = \frac{30}{2} = 15$$

The mean time it takes to reach a dive site is 15 minutes. Because the uniform distribution is symmetric, the mean is the middle of the distribution, and the mean is equal to the median.

> Find the length of time t such that 90% of all dive sites are reached within t minutes.

$$\sigma^2 = \frac{(b-a)^2}{12} = \frac{(25-5)^2}{12} = \frac{20^2}{12} = \frac{400}{12} \approx 33.3$$

$$\sigma = \sqrt{\sigma^2} = \sqrt{33.3} \approx 5.8$$

The standard deviation is approximately 5.8 minutes.

TRY IT NOW Go to Exercises 6.17 and 6.19

Cumulative Probability

To find a probability associated with any continuous random variable, the probability statement is often rewritten to use cumulative probability. From Chapter 5, cumulative probability means accumulated probability up to and including a fixed value. Cumulative probability is defined in the same way for a continuous random variable. The cumulative probability up to and including x is $P(X \leq x)$ and is the area under the density curve to the left of the fixed value x. **Figure 6.10** illustrates this cumulative probability.

Suppose X is a continuous random variable, and a and b are constants. Here are some typical probability statements involving X, and equivalent expressions using cumulative probability.

> It doesn't really matter whether we use $\leq$ or $<$, because one point contributes no probability. However, for consistency and accuracy throughout this text, cumulative probability will mean up to and including x; we use $\leq$ (not $<$).

$$P(X \geq b) = 1 - P(X < b) \qquad \text{(Figure 6.11)} \qquad \text{The Complement Rule.}$$
$$= 1 - \underbrace{P(X \leq b)}_{\text{Cumulative probability}} \qquad \text{A single value contributes no probability.}$$

$$P(a \leq X \leq b) = P(X \leq b) - P(X < a) \qquad \text{(Figure 6.12)}$$
$$= P(X \leq b) - P(X \leq a) \qquad \text{A single value contributes no probability.}$$

Find all the probability up to b, find all the probability up to a, and subtract. The difference is the probability that X lies in the interval from a to b.

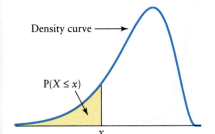

Figure 6.10 The shaded area is the cumulative probability $P(X \leq x)$.

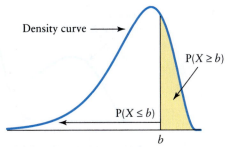

Figure 6.11 Use the Complement Rule to convert to cumulative probability.

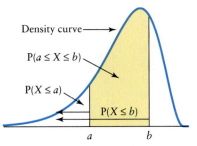

Figure 6.12 The shaded area is $P(a \leq X \leq b)$.

Here is one more way to visualize cumulative probability. As x moves from left to right, we accumulate more and more probability. Therefore, as x increases, cumulative probability also increases. Imagine starting at an altitude (or probability) of 0 and walking up a (smooth) hill. At any point along the walk, measure the altitude. This distance is the cumulative probability. **Figure 6.13** shows the relationship between the area under the density curve and the altitude, or cumulative probability.

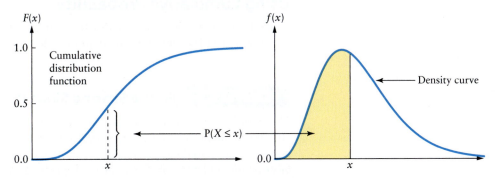

Figure 6.13 Visualizing cumulative probability: The altitude, or height, is equal to the shaded area, the cumulative probability.

A CLOSER LOOK

> Why is 1 the maximum value of the cumulative distribution function?

1. The drawing on the left in Figure 6.13 is a graph of the **cumulative distribution function**, F, defined by $F(x) = P(X \leq x)$. This function starts at 0 and is always increasing, until it reaches a maximum value of 1.

2. The mean μ and the variance σ^2 for a continuous random variable are computed using calculus. Although we will not consider any of these calculations, we will interpret and use these values as usual. μ is a measure of the center of the distribution, and σ^2 (or σ) is a measure of the spread, or variability, of the distribution.

   Figure 6.14 shows the graphs of the density functions for the random variables X and Y.

   (a) The mean of X is less than the mean of Y, $\mu_X < \mu_Y$, because the center of the distribution of X is to the left of the center of the distribution of Y.

   (b) The standard deviation of X is greater than the standard deviation of Y, $\sigma_X > \sigma_Y$, because the distribution of X is more spread out, and thus has more variability, than the distribution of Y.

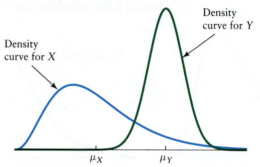

Figure 6.14 Graphs of the density functions for X and Y. The mean and the variance (and standard deviation) convey the same information (as for a discrete random variable) about the center and variability of the distribution.

Using Cumulative Probability

The following example illustrates the use of cumulative probability to compute probability associated with a continuous random variable.

EXAMPLE 6.2 Figure Skating Short Program

The rules for the women's figure skating short program stipulate the length as 2 minutes, 50 seconds. Suppose the time inconsistency (in seconds) of a randomly selected short program in relation to the stipulated length is a random variable X. The graph of the probability density function for X is shown in **Figure 6.15**. A negative value of X indicates that a short program ended before 2 minutes, 50 seconds, and a positive value indicates that a short program went longer than 2 minutes, 50 seconds. The general cumulative probability expression $P(X \leq x)$ is illustrated in **Figure 6.16**.

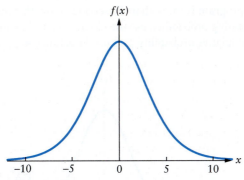

Figure 6.15 Graph of the probability density function for the time inconsistency of a short program.

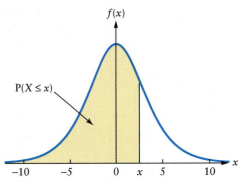

Figure 6.16 Visualization of cumulative probability for the time inconsistency of a short program.

For this random variable, cumulative probability (the area of the shaded region in Figure 6.16) can be computed using this equation:

Recall: e is the base of the natural logarithm; $e \approx 2.71828$. Most calculators have a specific key for e.

$$P(X \leq x) = \frac{1}{1 + e^{-x/2}} \quad \text{for all } x \quad (6.4)$$

Suppose a short program is randomly selected.

(a) What is the probability that the short program ends 5 seconds or more before the stipulated time?

 Exponents.

(b) What is the probability that the short program is more than 10 seconds long?

(c) What is the probability that the short program is within 3 seconds of the stipulated time?

Solution

(a) If the short program ends 5 seconds or more before the stipulated time, this means $X \leq -5$. The expression $P(X \leq -5)$ is a specific cumulative probability; there is no need to convert this expression. Use Equation 6.4. This probability is illustrated in **Figure 6.17**.

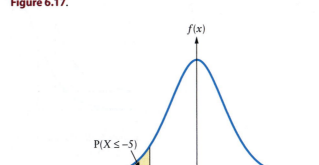

Figure 6.17 Visualization of the cumulative probability $P(X \leq -5)$.

$$P(X \leq -5) = \frac{1}{1 + e^{-(-5)/2}} = 0.0759 \quad \text{Use Equation 6.4.}$$

The probability that a randomly selected short program ends 5 seconds or more before the stipulated time is 0.0759.

(b) If the short program is more than 10 seconds long, this means $X > 10$. To compute the corresponding probability, use the Complement Rule to convert to an expression involving cumulative probability, and use Equation 6.4. See **Figure 6.18**.

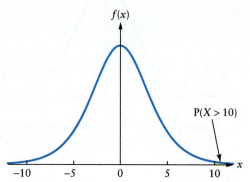

Figure 6.18 Visualization of the probability $P(X > 10)$.

$$P(X > 10) = 1 - P(X \leq 10) \qquad \text{The Complement Rule.}$$
$$= 1 - \frac{1}{1 + e^{-10/2}} \qquad \text{Use Equation 6.4.}$$
$$= 1 - 0.9933 = 0.0067 \qquad \text{Simplify.}$$

The probability that a randomly selected short program is more than 10 seconds long is 0.0067.

(c) If the short program is within 3 seconds of the stipulated time, this means $-3 \leq X \leq 3$. To compute the corresponding probability, find the difference between two cumulative probabilities. See **Figure 6.19**.

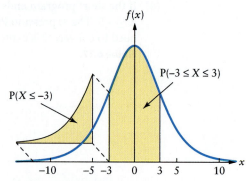

Figure 6.19 Visualization of the probability $P(-3 \leq X \leq 3)$.

$$P(-3 \leq X \leq 3) = P(X \leq 3) - P(X < -3)$$
$$= P(X \leq 3) - P(X \leq -3) \qquad \text{A single value contributes no probability.}$$
$$= \left(\frac{1}{1 + e^{-3/2}}\right) - \left(\frac{1}{1 + e^{-(-3)/2}}\right) \qquad \text{Use Equation 6.4.}$$
$$= 0.8176 - 0.1824 = 0.6352 \qquad \text{Simplify.}$$

The probability that a randomly selected short program is within 3 seconds of the stipulated time is 0.6352.

TRY IT NOW Go to Exercise 6.27

6.1 Probability Distributions for a Continuous Random Variable

Working Backward

In some problems, a probability is given and we need to work backward to find a solution. Consider the following example.

Ceri Breeze/Shutterstock

EXAMPLE 6.3 Summer Sailing Schedule

During the summer months, the Washington State Department of Transportation operates a ferry between Seattle and Bainbridge Island. The travel time between the two ports is approximately 35 minutes.[3] However, due to weather conditions, tides, and the number of passengers, the actual travel time between ports is a random variable and can be modeled by a uniform distribution between 25 and 65 minutes. Find the time t such that 75% of all ferries make the trip in at most t minutes.

Solution

STEP 1 Because X has a uniform distribution with $a = 25$ and $b = 65$, the probability density function is

$$f(x) = \begin{cases} \dfrac{1}{40} = 0.025 & 25 \leq x \leq 65 \\ 0 & \text{otherwise} \end{cases}$$

We need to find the value t such that $P(X \leq t) = 0.75$.

This is a *backward* problem, because we know the probability (0.75) and need to find a starting value (t)—that is, a value of X that produces this probability.

STEP 2 Use **Figure 6.20** to visualize this backward probability question and to solve for t.

$P(X \leq t)$ = area under the density curve from 25 to t
= area of a rectangle
= width × height
= $(t - 25)(0.025)$

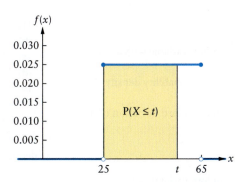

Figure 6.20 The area of the shaded region is $P(X \leq t)$.

STEP 3 Set the expression for probability equal to 0.75, and solve for t.

$(t - 25)(0.025) = 0.75$

$t - 25 = \dfrac{0.75}{0.025} = 30$ Divide both sides by 0.025.

$t = 30 + 25 = 55$ Add 25 to both sides.

Seventy-five percent of all ferry rides make the trip within 55 minutes. ∎

TRY IT NOW Go to Exercise 6.25

Section 6.1 Exercises

Concept Check

6.1 True or False The graph of a probability density function may extend below the x-axis.

6.2 True or False For a continuous random variable X with probability density function f, $P(X = x) = f(x)$.

6.3 True or False For a continuous random variable, there is no probability associated with a single value.

6.4 True or False The mean μ and the variance σ^2 of a continuous random variable describe the center and spread of the distribution, respectively.

6.5 True or False For a continuous random variable X with probability density function f, it is possible that for some value a, $f(a) = 2$.

6.6 True or False For a continuous random variable X and any value a, $P(X \leq a) = P(X < a)$.

6.7 Short Answer Explain how to compute probabilities associated with a continuous random variable.

6.8 Short Answer Explain why a cumulative distribution function can never have a value greater than 1.

6.9 Short Answer Suppose X has a uniform distribution on the interval $[a, b]$. Explain why the mean is $\frac{a+b}{2}$ and why the mean is equal to the median.

Practice

6.10 Suppose X is a uniform random variable with $a = 0$ and $b = 16$.
(a) Carefully sketch a graph of the probability density function for X.
(b) Find the mean, variance, and standard deviation of X.
(c) Find $P(X \geq 4)$.
(d) Find $P(2 \leq X < 12)$.
(e) Find $P(X \leq 7)$.

6.11 Suppose X is a uniform random variable with $a = -5$ and $b = 25$.
(a) Carefully sketch a graph of the probability density function for X.
(b) Find the mean, variance, and standard deviation of X.
(c) Find $P(-10 < X < -1)$.
(d) Find $P(X > 0)$ and $P(X \geq 0)$.
(e) Find $P(X \geq 20 \mid X \geq 10)$.

6.12 Suppose X is a uniform random variable with $a = 50$ and $b = 100$.
(a) Find the mean, variance, and standard deviation of X.
(b) Find $P(\mu - \sigma \leq X \leq \mu + \sigma)$.
(c) Find $P(X \geq \mu + 2\sigma)$.
(d) Find a value c such that $P(X \leq c) = 0.20$.

6.13 Suppose X is a uniform random variable with $a = 25$ and $b = 75$.
(a) Find the mean, variance, and standard deviation of X.
(b) Find the probability that X is more than 2 standard deviations from the mean.
(c) Find a value c such that $P(X \geq c) = 0.40$.
(d) Suppose two values of X are selected at random. What is the probability that both values are between 30 and 40?

6.14 Suppose X is a continuous random variable with probability density function given by

$$f(x) = \begin{cases} \dfrac{x}{8} & \text{if } 0 \leq x \leq 4 \\ 0 & \text{otherwise} \end{cases}$$

The graph of f is shown in the figure.

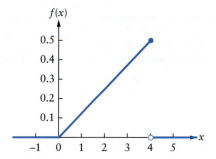

(a) Find $P(X \leq 1)$.
(b) Find $P(X > 3)$.
(c) Find $P(X > 4)$.
(d) Find $P(2 \leq X \leq 3)$.
(e) Find $P(X \leq 2 \mid X \leq 3)$.
(f) Find a value c such that $P(X \leq c) = 0.5$. Explain why c is not equal to 2, halfway between 0 and 4.

6.15 Suppose X is a continuous random variable with probability density function f. The graph of f is shown in the figure.

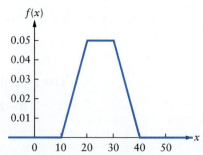

(a) Verify that this is a valid probability density function.
(b) Find a formula for the probability density function.
(c) Find $P(X \leq 15)$.
(d) Find $P(X > 27)$.
(e) Explain why the mean of X is 25.

Applications

6.16 Manufacturing and Product Development A Gold Canyon candle is designed to last 9 hours. However, depending on the wind, air bubbles in the wax, the quality of the wax, and the number of times the candle is re-lit, the actual burning time (in hours) is a uniform random variable with $a = 6.5$ and $b = 10.5$. Suppose one of these candles is randomly selected.
(a) Find the probability that the candle burns at least 7 hours.
(b) Find the probability that the candle burns at most 8 hours.
(c) Find the mean burning time and the probability that the burning time of a randomly selected candle will be within 1 standard deviation of the mean.
(d) Find a time t such that 25% of all candles burn longer than t hours.

6.17 Sports and Leisure According to Major League Baseball rules, a baseball should weigh between 5 and 5.25 ounces (oz) and have a circumference of between 9 and 9.25 inches (in.). Suppose the weight of a baseball (in ounces) has a uniform distribution with $a = 5.085$ and $b = 5.155$, and the circumference (in inches) has a uniform distribution with $a = 9.0$ and $b = 9.1$.
(a) Find the probability that a randomly selected baseball has a weight greater than 5.14 oz.
(b) Find the probability that a randomly selected baseball has a circumference less than 9.03 in.
(c) Suppose the weight and the circumference are independent. Find the probability that a randomly selected baseball will have a weight between 5.11 and 5.13 oz and a circumference between 9.04 and 9.06 in.

6.18 Manufacturing and Product Development Pre-manufactured wooden roof trusses allow builders to complete projects faster and with lower on-site labor costs. The connector plates for trusses are made from Grade A steel and are hot-dip galvanized. The thickness of a truss connector (in inches) varies slightly and has a uniform distribution with $a = 0.036$ and $b = 0.050$.
(a) If the manufacturer will use only connectors with a minimum thickness of 0.04 in., what proportion of connectors is rejected?
(b) Suppose a truss connector is selected at random. Find the probability that the truss connector has a thickness between 0.042 and 0.045 in.
(c) Find the mean, variance, and standard deviation of the thickness of a truss connector.

6.19 Travel and Transportation When the Department of Transportation (DOT) repaints the center lines, edge lines, or no-passing-zone lines on a highway, epoxy paint is sometimes applied. This paint is more expensive than latex but lasts longer. If this paint splashes onto a vehicle, it has to be completely sanded off, and that area of the vehicle has to be repainted. The DOT has warned motorists that the drying time for this epoxy paint (in minutes) has a uniform distribution with $a = 30$ and $b = 60$. Suppose epoxy paint is applied to a small section of center line.
(a) What is the probability that the paint will be dry within 45 minutes?
(b) What is the probability that the paint will be dry in between 40 and 50 minutes?
(c) Find a value t such that the probability of the paint taking at least t minutes to dry is 0.75.
(d) If the DOT road crew removes all of the cones on the center line 55 minutes after painting, what is the probability that the paint will still be wet?

6.20 Public Health and Nutrition A group of researchers recently analyzed the salt content in breads sold around the world and concluded that more than one-third of all loaves exceed the maximum salt target for bread.[4] Suppose the amount of salt per 100 g in a loaf of bread has a uniform distribution with $a = 0.95$ and $b = 1.31$. A loaf of bread is selected at random.
(a) Find the probability that the loaf of bread contains more than 1.2 g of salt per 100 g.
(b) Suppose the loaf of bread contains more than 1.1 g of salt per 100 g. What is the probability that it contains more than 1.2 g of salt per 100 g?
(c) If a loaf of bread contains at most 1 g of salt per 100 g, it meets the voluntary target established by the Food and Drug Administration. Suppose 5 loaves of bread are selected at random. Find the probability that at most one exceeds this target.

6.21 Manufacturing and Product Development The length (in inches) of a Wamsutta Extra-Firm Standard pillow is a random variable with a uniform distribution on the interval [23.75, 25.00]. Suppose one of these pillows is selected at random.
(a) Find the mean, variance, and standard deviation of the length of the pillow.
(b) Find the probability that the length of the pillow is within 1 standard deviation of the mean.
(c) Find a value L such that the probability that the length of the pillow is greater than L is 0.25.

Extended Applications

6.22 Psychology and Human Behavior Some of the common over-the-counter medications to help people relieve headaches include Advil, Aleve, Excedrin, and aspirin. Suppose Excedrin is formulated so that an individual will feel headache pain relief within 30 minutes. The probability density function for X, the time (in minutes) it takes to feel headache relief after taking an Excedrin tablet, is given by

$$f(x) = \begin{cases} 0.05 & \text{if } 0 \leq x \leq 10 \\ -0.0025(x - 30) & \text{if } 10 < x \leq 30 \\ 0 & \text{otherwise} \end{cases}$$

The graph of f is shown in the figure.

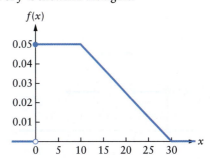

(a) Verify that this is a valid probability density function.
(b) If a randomly selected person takes an Excedrin tablet, what is the probability that he will experience relief within 5 minutes?
(c) What is the probability that the person will experience relief between 20 and 30 minutes after taking the tablet?
(d) Find a value t such that the probability of experiencing relief within t minutes after taking the tablet is 0.75.
(e) If it takes less than 15 minutes to experience relief after taking a tablet, people consider the medication a success. Suppose 20 people, each of whom has a headache, are selected at random. What is the probability that exactly 14 of them experience relief successfully? What is the probability that at least 16 people experience relief successfully? What is the probability that at most 10 people experience relief successfully?

6.23 Marketing and Consumer Behavior The city of Ventura, California, recently installed new parking meters that are easier and faster to use and that offer users the option of refilling the meter remotely.[5] The graph of the probability density function for the length of time a car is parked (in hours) at a metered spot in Ventura is shown in the figure. Suppose a car parked at a metered spot in Ventura is selected at random.

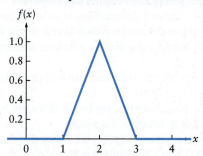

(a) What is the probability that the car is parked for less than 2 hours?
(b) What is the probability that the car is parked for less than 1.4 hours?
(c) What is the probability that the car is parked for more than 2.6 hours?
(d) What is the probability that the car is parked for between 1.4 and 2.6 hours?

6.24 Marketing and Consumer Behavior Marini's candy store on the beach boardwalk in Santa Cruz sells candy in bulk. Customers can mix products from more than 100 barrels. The graph of the probability distribution function for the number of pounds (lb) of candy purchased by a randomly selected customer is shown in the figure.

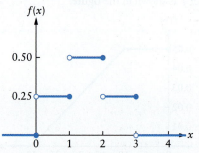

(a) Verify that this is a valid probability density function.
(b) Find the probability that the next customer buys at most 2 lb of candy.
(c) Find the probability that the next customer buys more than 1 lb of candy.
(d) Suppose the next customer buys at most 1.5 lb of candy. What is the probability that she buys at most 0.5 lb of candy?

6.25 Economics and Finance On any given trading day, the fluctuation, or change, in the price (in dollars) of General Electric stock, listed on the New York Stock Exchange, is between -2.00 and 2.00. Suppose the change in price is a random variable and the graph of the probability density function is shown in the figure.

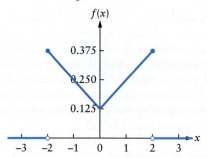

(a) Verify that this is a valid probability density function.
(b) What is the probability that the stock price increases by at least $1.00 on a randomly selected day?
(c) What is the probability that the change in stock price is between -1.00 and 1.00?
(d) Find a value c such that $P(-c \leq X \leq c) = 0.90$.

6.26 Public Health and Nutrition The health benefits from eating yogurt include preventing hypertension, regulating cholesterol levels, and strengthening the immune system. However, there are also varying amounts of sodium, carbohydrates, fat, and caffeine in each serving depending on the yogurt type. Suppose the amount of fat (in grams, g) in a cup of chocolate frozen yogurt is a random variable X with probability density function given by

$$f(x) = \begin{cases} -0.08(x-5) & \text{if } 0 \leq x \leq 5 \\ 0 & \text{otherwise} \end{cases}$$

The graph of f is shown in the figure.

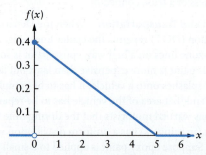

(a) Verify that this is a valid probability density function.
(b) What is the probability that a randomly selected cup of chocolate frozen yogurt will have fat content less than 2.5 g?

(c) What is the probability that a randomly selected cup of chocolate frozen yogurt will have fat content between 2 and 3 g?
(d) Find a value c such that 5% of all cups of chocolate frozen yogurt have fat content of at least c.
(e) What is the probability that a randomly selected cup of chocolate frozen yogurt will have fat content of at least 4 g?
(f) What is the probability that the fifth cup of chocolate frozen yogurt will be the first to have fat content of at least 4 g?

6.27 Biology and Environmental Science The Jasper SkyTram transports passengers to the top of Whistlers Mountain in the Canadian Rockies, where there are spectacular views, boardwalks, and hiking trails.[6] The SkyTram travel time (in minutes) up the mountain is a random variable X with probability density function given by

$$f(x) = \begin{cases} -0.75x^2 + 11.25x - 41.4375 & \text{if } 6.5 \leq x \leq 8.5 \\ 0 & \text{otherwise} \end{cases}$$

The cumulative distribution function for X is given by

$$P(X \leq x) = \begin{cases} 0 & \text{if } x < 6.5 \\ -0.25(6.5-x)^2(x-9.5) & \text{if } 6.5 \leq x < 8.5 \\ 1 & \text{if } x \geq 8.5 \end{cases}$$

(a) Sketch the graph of the probability density function f.
(b) Sketch the graph of the cumulative distribution function.
(c) Find the probability that a SkyTram ride up the mountain takes less than 7 minutes.
(d) Find the probability that a SkyTram ride up the mountain takes more than 8.25 minutes.
(e) Find the probability that five Skytram rides up the mountain all take more than 8 minutes each.

Challenge Problems

6.28 Marketing and Consumer Behavior Dinner customers at the Primanti Brothers restaurant in Pittsburgh, Pennsylvania, often experience a long wait for a table. For a randomly selected customer who arrives at the restaurant between 6:00 P.M. and 7:00 P.M., the waiting time (in minutes) is a continuous random variable such that

$$P(X \leq x) = \begin{cases} 1 - e^{-0.05x} & \text{if } x \geq 0 \\ 0 & \text{otherwise} \end{cases}$$

Suppose a dinner customer is randomly selected.
(a) What is the probability that the person must wait for a table for at most 20 minutes?
(b) What is the probability that the person must wait for a table for more than 30 minutes?
(c) What is the probability that the person must wait for a table for between 15 and 30 minutes?

6.29 Psychology and Human Behavior Parents with children younger than age 16 often spend a lot of time during the day driving their kids to various places—for example, to and from after-school activities, music lessons, sports practices and games, the library, and a friend's home. Suppose a family has k child(ren) younger than 16 ($k = 1, 2, 3, 4, 5$), and let the random variable X_k be the time (in hours) spent taxiing them during the day. X_k has a uniform distribution with $a = 0$ and $b = k$. For example, for a family with two children, X_2 has a uniform distribution with $a = 0$ and $b = 2$.
(a) For a family with three children, what is the probability that the parents will spend less than 1 hour driving kids on a randomly selected day?
(b) For a family of four children, what is the mean number of hours spent driving kids? What is the probability that the driving time will be greater than 2 standard deviations from the mean?
(c) For a family with five children younger than 16, find a time t such that the probability of driving kids more than t hours is 0.25.
(d) Suppose five families are selected at random, the first with one child younger than 16, the second with two children younger than 16, and so on. What is the probability that all five families drive less than 30 minutes on a randomly selected day? What is the probability that all five families drive more than 90 minutes on a randomly selected day?

6.30 Using Cumulative Probability Suppose X is a continuous random variable with cumulative distribution function given by

$$F(x) = P(X \leq x) = \begin{cases} 1 - e^{-x^2/8} & \text{if } x \geq 0 \\ 0 & \text{otherwise} \end{cases}$$

(a) Sketch the graph of F.
(b) Find $P(X \leq 4)$.
(c) Find $P(X > 2)$.
(d) Find $P(1 \leq X \leq 3)$.
(e) Find $P(X \leq 2 \mid X \leq 4)$.

6.2 The Normal Distribution

The Bell-Shaped Curve

The normal probability distribution is very common and is the most important distribution in all of statistics. This bell-shaped density curve can be used to model many natural phenomena, and the normal distribution is used extensively in statistical inference.

Recall that a random variable is completely described by certain parameters—for example, a binomial random variable by n and p, and a Poisson random variable by λ. A normal distribution is completely characterized, or determined, by its mean μ and variance σ^2 (or by its mean μ and standard deviation σ).

The Normal Probability Distribution

Suppose X is a normal random variable with mean μ and variance σ^2. The probability density function is given by

$$f(x) = \frac{1}{\sigma\sqrt{2\pi}} e^{-(x-\mu)^2/(2\sigma^2)} \tag{6.5}$$

and

$$-\infty < x < \infty \quad -\infty < \mu < \infty \quad \sigma^2 > 0 \tag{6.6}$$

A CLOSER LOOK

We've seen the number e before, associated with the Poisson distribution.

1. In this probability density function, e is the base of the natural logarithm; $e \approx 2.71828$. π is another constant, commonly used in trigonometry; $\pi \approx 3.14159$.

2. We use the shorthand notation $X \sim N(\mu, \sigma^2)$ to indicate that X is (distributed as) a normal random variable with mean μ and variance σ^2. For example, $X \sim N(5, 36)$ means that X is a normal random variable with mean $\mu = 5$ and variance $\sigma^2 = 36$ (and $\sigma = 6$).

3. Equation 6.6 means that x can be any real number (the density curve continues forever in both directions), the mean μ can be any real number (positive or negative), and the variance can be any positive real number.

Bell-shaped: Imagine placing a bell on a table and passing a plane (a piece of paper) through the bell perpendicular to the table. The intersection of the plane and the bell is a bell-shaped curve.

4. For any mean μ and variance σ^2, the density curve is symmetric about the mean μ, unimodal, and bell-shaped, as shown in **Figure 6.21**.

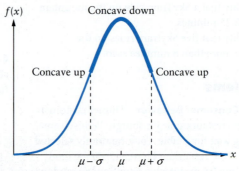

Figure 6.21 Graph of the probability density function for a normal random variable with mean μ and variance σ^2.

▶ The graph of the probability density function changes concavity at $x = \mu - \sigma$ and again at $x = \mu + \sigma$. ◀

The mean is equal to the median because the normal distribution is symmetric.

▶ It can be shown (using calculus) that the total area under this density curve is 1 (even though it extends forever in both directions, getting closer and closer to the x-axis but never touching it). ◀

5. The mean μ is a location parameter, and the variance σ^2 determines the spread of the distribution. As the variance increases, the total area under the probability density function (1) is rearranged. The graph is compressed down and pushed out (on the tails). **Figure 6.22** and **Figure 6.23** show the effects of μ and σ^2 on the location (center) and spread of the density curve.

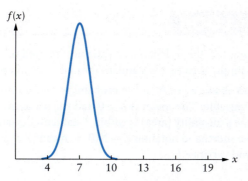

Figure 6.22 Graph of the probability density function for a normal distribution with $\mu = 7$ and small σ^2.

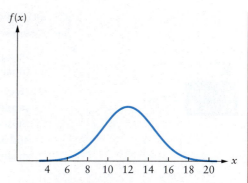

Figure 6.23 Graph of the probability density function for a normal distribution with $\mu = 12$ and large σ^2.

The Standard Normal Random Variable

Note: In many of the figures that involve the graph of the probabiity density function for a normal random variable, we will not include the vertical, or density, axis. We will focus on the area of specific regions that correspond to probability expressions.

Suppose X is a normal random variable with mean μ and variance σ^2: $X \sim N(\mu, \sigma^2)$. The probability X lies in some interval; for example $[a, b]$, is the area under the density curve between a and b (**Figure 6.24**).

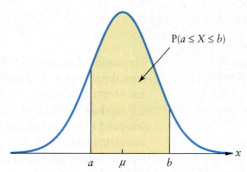

Figure 6.24 The shaded region corresponds to $P(a \leq X \leq b)$.

The shaded region in Figure 6.24 is not a simple geometric figure; it's bounded by a curve! Consequently, there is no nice formula for the area of this region, corresponding to $P(a \leq X \leq b)$. However, a probability statement associated with any normal random variable can be transformed into an equivalent expression involving a **standard normal random variable**. Cumulative probabilities associated with this distribution are provided in the Appendix Table 3.

The Standard Normal Random Variable

The normal distribution with $\mu = 0$ and $\sigma^2 = 1$ (and $\sigma = 1$) is called the **standard normal distribution**. A random variable that has a standard normal distribution is called a **standard normal random variable**, usually denoted by Z. The probability density function for Z is given by

Let $\mu = 0$ and $\sigma = 1$ in Equation 6.5.

$$f(z) = \frac{1}{\sqrt{2\pi}} e^{-z^2/2} \qquad -\infty < z < \infty \tag{6.7}$$

 Z scores.

> ## A CLOSER LOOK
>
> 1. In Equation 6.7, the independent variable z is used to define the probability density function simply because the standard normal random variable is usually denoted by Z.
>
> 2. **Figure 6.25** shows a graph of the probability density function for a standard normal random variable. The mean is $\mu = 0$ and the standard deviation is $\sigma = 1$. Note that most of the probability (area) is within 3 standard deviations of the mean, between -3 and 3. The shorthand notation $Z \sim N(0, 1)$ means Z is a normal random variable with mean 0 and variance 1.
>
>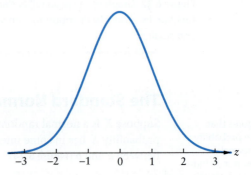
>
> **Figure 6.25** Graph of the probability density function for a standard normal random variable.
>
> 3. The standard normal distribution is not common, but it is used extensively as a reference distribution. Any probability statement involving any normal random variable can be transformed into an equivalent expression (with the same probability) involving a Z random variable. We will learn how to standardize expressions shortly. Therefore, you need to become an expert at computing probabilities in the Z world. Probabilities associated with Z are computed using cumulative probability, as described in the next section. **Figure 6.26** shows the steps for computing probabilities associated with a normal random variable.
>
>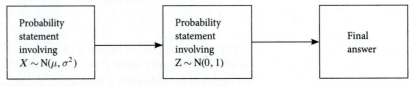
>
> **Figure 6.26** Strategy for computing a probability associated with any normal random variable.
>
> ## Use Cumulative Probability
>
> Probabilities associated with a standard normal random variable Z are computed using cumulative probability. Appendix Table 3 contains values for $P(Z \leq z)$ for selected values of z. **Figure 6.27** shows the geometric region corresponding to $P(Z \leq z)$, and **Figure 6.28** illustrates the use of Appendix Table 3. Locate the units and tenths digits in z along the left side of the table. Find the hundredths digit in z across the top row. The intersection of this row and column, in the body of the table, contains the cumulative probability. Note that Appendix Table 3 is limited; various technology solutions are more versatile and accurate.

We will often refer to a standard normal distribution as a *Z world*.

6.2 The Normal Distribution

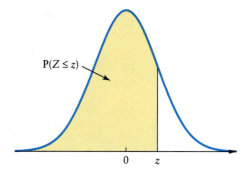

Figure 6.27 The shaded area under the graph of the standard normal density function corresponds to P(Z ≤ z).

z	0.00	0.01	0.02	0.03	0.04	0.05	0.06	0.07	0.08	0.09
⋮	⋮	⋮	⋮	⋮	⋮	⋮	⋮	⋮	⋮	⋮
1.0	0.8413	0.8438	0.8461	0.8485	0.8508	0.8531	0.8554	0.8577	0.8599	0.8621
1.1	0.8643	0.8665	0.8686	0.8708	0.8729	0.8749	0.8770	0.8790	0.8810	0.8830
1.2	0.8849	0.8869	0.8888	0.8907	0.8925	0.8944	0.8962	0.8980	0.8997	0.9015
1.3	0.9032	0.9049	0.9066	0.9082	0.9099	0.9115	0.9131	0.9147	0.9162	0.9177
1.4	0.9192	0.9207	0.9222	0.9236	0.9251	0.9265	0.9279	0.9292	0.9306	0.9319
⋮	⋮	⋮	⋮	⋮	⋮	⋮	⋮	⋮	⋮	⋮

Figure 6.28 The entry in Table 3 at the intersection of the 1.2 row and the 0.03 column is $P(Z \leq 1.23) = 0.8907$.

The following example illustrates the use of Appendix Table 3 to find probabilities associated with Z.

EXAMPLE 6.4 Probability Calculations Associated with the Standard Normal Distribution

Use Appendix Table 3 to find each probability associated with the standard normal distribution.

(a) $P(Z \leq 1.45)$

(b) $P(Z \geq -0.6)$

(c) $P(-1.25 \leq Z \leq 2.13)$

(d) Find the value b such that $P(Z \leq b) = 0.90$.

Solution

(a) This probability expression is already a cumulative probability. Go directly to Appendix Table 3. Find the intersection of the 1.4 row and the 0.05 column.

$P(Z \leq 1.45) = 0.9265$ Cumulative probability; use Appendix Table 3.

Figure 6.29 is a visualization of this probability, and **Figure 6.30** shows a technology solution.

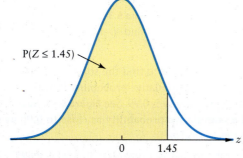

Figure 6.29 The area of the shaded region is $P(Z \leq 1.45)$.

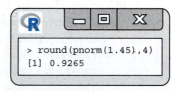

Figure 6.30 Use the R command `pnorm()` to find cumulative probability associated with a normal random variable.

(b) This is a right-tail probability. Convert to a cumulative probability and use Appendix Table 3.

$$P(Z \geq -0.6) = 1 - P(Z < -0.6) \quad \text{The Complement Rule.}$$
$$= 1 - P(Z \leq -0.6) \quad \text{One value doesn't matter.}$$
$$= 1 - 0.2743 = 0.7257 \quad \text{Use Appendix Table 3.}$$

Figure 6.31 is a visualization of this probability, and **Figure 6.32** shows a technology solution.

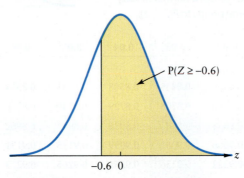
Figure 6.31 The area of the shaded region is $P(Z \geq -0.6)$.

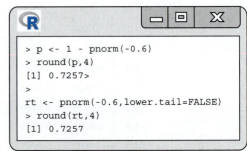
Figure 6.32 Use cumulative probability or set `lower.tail=FALSE` to find the right tail probability.

(c) Find all the probability up to 2.13, find all the probability up to -1.25, and subtract. The difference is the probability that Z lies in this interval.

$$P(-1.25 \leq Z \leq 2.13)$$
$$= P(Z \leq 2.13) - P(Z < -1.25) \quad \text{Use cumulative probability.}$$
$$= P(Z \leq 2.13) - P(Z \leq -1.25) \quad \text{One value doesn't matter.}$$
$$= 0.9834 - 0.1056 = 0.8778 \quad \text{Use Appendix Table 3.}$$

Figure 6.33 is a visualization of this probability, and **Figure 6.34** shows a technology solution.

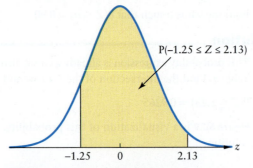
Figure 6.33 The area of the shaded region is $P(-1.25 \leq Z \leq 2.13)$.

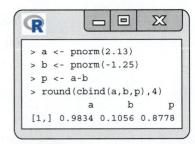
Figure 6.34 Use cumulative probability to find $P(-1.25 \leq Z \leq 2.13)$.

(d) In this problem, we need to work backward to find the solution. This is an **inverse cumulative probability** problem. The cumulative probability is given. We need the value b such that the cumulative probability is 0.90. See **Figure 6.35**. Search the body of Appendix Table 3 to find a cumulative probability as close to 0.90 as possible. Read the row and column entries to find b.

In the body of Table 3, the closest cumulative probability to 0.90 is 0.8997. This corresponds to $1.28 \approx b$. **Figure 6.36** shows a more accurate technology solution.

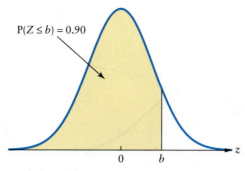

Figure 6.35 The area of the shaded region is $0.90 = P(Z \leq b)$.

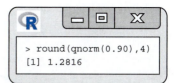

Figure 6.36 Use the R command `qnorm()` to find inverse cumulative probability associated with a normal random variable.

Note: Linear interpolation can be used to find a more exact answer. A technology solution like the one presented in Figure 6.36, which uses a special inverse cumulative probability function, is preferable.

TRY IT NOW Go to Exercise 6.45

Interpolation

Interpolation is a method of approximation. It is often used to estimate a value at a position between two given values in a table. Linear interpolation assumes that the two known values lie on a straight line.

Linear interpolation.

In Example 6.4(d), 0.90 is between the Appendix Table 3 known cumulative probabilities 0.8997 and 0.9015. Suppose the two points (1.28, 0.8997) and (1.29, 0.9015) lie on a straight line. The approximate z value corresponding to the cumulative probability 0.90 is

$$1.28 + (0.01)(0.90 - 0.8997)/(0.9015 - 0.8997) = 1.2817$$

Standardization

The following rule provides the connection between any normal random variable and the standard normal random variable.

Standardization Rule

If X is a normal random variable with mean μ and variance σ^2, then a standard normal random variable is given by

$$Z = \frac{X - \mu}{\sigma} \tag{6.8}$$

A CLOSER LOOK

There are other types of standardization, but $Z = \frac{X - \mu}{\sigma}$ is the most common.

1. The process of converting from X to Z is called **standardization**. Z is a standardized random variable.

2. Using this rule, any probability involving a normal random variable can be transformed into an equivalent expression involving a Z random variable. We can then convert to cumulative probability if necessary, and use Appendix Table 3.

3. The Standardization Rule given previously is illustrated in **Figure 6.37**, using cumulative probability.

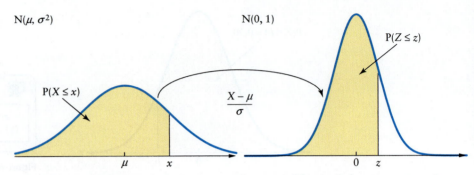

Figure 6.37 An illustration of standardization. The areas of the shaded regions are equal.

The following calculation shows why the two shaded regions in Figure 6.37 have the same area, as well as how to use the rule to compute probabilities involving any normal random variable.

Assume $X \sim N(\mu, \sigma^2)$.

$P(X \leq x)$ The original (cumulative) probability statement.

Remember the phrase: Whatever you do to one side of the inequality, you have to do to the other side.

$$= P\left(\frac{X-\mu}{\sigma} \leq \frac{x-\mu}{\sigma}\right)$$

Work within the probability statement. Subtract the mean of X and divide by the standard deviation of X, on both sides of the inequality (standardize).

$$= P(Z \leq z)$$

Apply the Standardization Rule within the probability statement. The expression with X is transformed into Z. The expression with x becomes some fixed value z. Use Appendix Table 3 to find this probability.

Standardization Examples

The following examples involve normal random variables and standardization. The hardest part of these types of problems is (as before) (1) to define and identify the probability distribution, and (2) to write a probability statement. Given a probability statement involving a normal random variable, all we have to do is standardize and use cumulative probability. Even for backward problems (with a known probability), we still standardize and still use cumulative probability. Note that the technology solutions presented do not require standardization.

EXAMPLE 6.5 **Probability Calculations Associated with a Normal Random Variable**

Suppose X is a normal random variable with mean 10 and variance 4: $X \sim N(10, 4)$, and $\sigma = \sqrt{4} = 2$.

(a) Find $P(X > 12.5)$.

(b) Find $P(9 \leq X \leq 10)$.

(c) Find the value b such that $P(X \leq b) = 0.75$.

Solution

(a) X is a normal random variable. We know the mean and standard deviation. Standardize and use cumulative probability associated with Z.

$$P(X > 12.5) = P\left(\frac{X-10}{2} > \frac{12.5-10}{2}\right) \quad \text{Standardize.}$$

$$= P(Z > 1.25) \quad \text{Equation 6.8; simplify.}$$

$$= 1 - P(Z \leq 1.25) \quad \text{The Complement Rule.}$$

$$= 1 - 0.8944 = 0.1056 \quad \text{Use Appendix Table 3.}$$

Figure 6.38 illustrates this standardization and solution.

6.2 The Normal Distribution   269

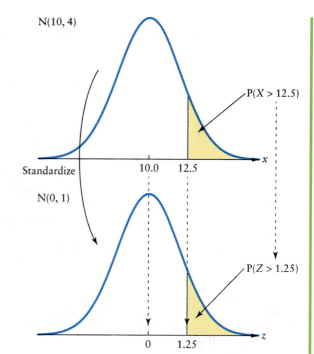

Figure 6.38 A visualization of the standardization in Example 6.5 (a). The value $X=10$ is transformed to $Z=0$, and $X=12.5$ is transformed to $Z=1.25$. The areas of the shaded regions are the same.

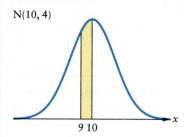

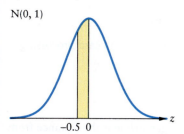

Figure 6.39 A visualization of the standardization in Example 6.5 (b).

(b) Here, we need to standardize again. Work within the probability statement to write an equivalent expression involving Z.

$$P(9 \leq X \leq 10) = P\left(\frac{9-10}{2} \leq \frac{X-10}{2} \leq \frac{10-10}{2}\right) \quad \text{Standardize.}$$
$$= P(-0.5 \leq Z \leq 0) \quad \text{Equation 6.8; simplify.}$$
$$= P(Z \leq 0) - P(Z < -0.5) \quad \text{Use cumulative probability.}$$
$$= P(Z \leq 0) - P(Z \leq -0.5) \quad \text{One value doesn't matter.}$$
$$= 0.5000 - 0.3085 = 0.1915 \quad \text{Use Appendix Table 3.}$$

Figure 6.39 illustrates this standardization and solution.

(c) Convert the given expression into a cumulative probability involving Z. Because the probability is already given, this is an inverse, or backward, cumulative probability problem. So we will work backward in Appendix Table 3.

$$P(X \leq b) = P\left(\frac{X-10}{2} \leq \frac{b-10}{2}\right) \quad \text{Standardize.}$$
$$= P\left(Z \leq \frac{b-10}{2}\right) = 0.75 \quad \text{Equation 6.8.}$$

There is no other simplification within the probability statement. However, the resulting probability statement involves Z and is a cumulative probability. Find a value in the body of Appendix Table 3 as close to 0.75 as possible. Set the corresponding z equal to $\left(\frac{b-10}{2}\right)$, and solve for b.

$$\frac{b-10}{2} = 0.6745 \quad \text{Appendix Table 3; interpolation.}$$
$$b - 10 = 1.349 \quad \text{Multiply both sides by 2.}$$
$$b = 11.349 \quad \text{Add 10 to both sides.}$$

Therefore, $P(X \leq 11.349) = 0.75$ and $b = 11.349$. **Figure 6.40** illustrates this standardization and solution.

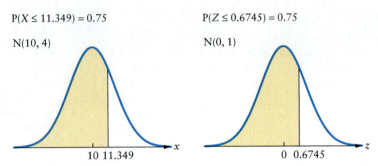

Figure 6.40 A visualization of the standardization in Example 6.5 (c).

Figures 6.41–6.43 show technology solutions. Note that there is no need to standardize using R to evaluate these probability expressions.

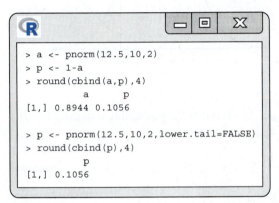

Figure 6.41 $P(X \geq 12.5)$

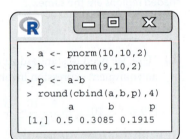

Figure 6.42 $P(9 \leq X \leq 10)$

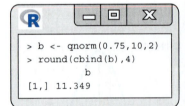

Figure 6.43 Inverse cumulative probability solution using R.

TRY IT NOW Go to Exercises 6.49 and 6.53

CHAPTER APP **EMV Chip Technology**

Find the probability that a randomly selected EMV chip card processing time is less than 10 seconds.

EXAMPLE 6.6 Seat Pitch

Seat pitch, more commonly called leg room, on a passenger airline is the distance from the back of one seat to the front of the one directly behind it. The greater the seat pitch, the more comfortable the seat and the less likely you are to travel with your knees against your chest. Some passengers pay a premium for extra leg room rather than sit in a regular coach seat. Suppose the seat pitch for all coach seats is normally distributed, with mean 31 in. and standard deviation 0.5 in.[7]

(a) For a randomly selected coach seat, find the probability that the seat pitch is between 30.5 and 32 in. (considered barely comfortable).

(b) Any seat pitch less than 30 in. is considered constricted. Find the probability that a randomly selected coach seat is constricted.

TomCarpenter/Shutterstock

Solution

Define a normal random variable and translate each question into a probability statement. Standardize and use cumulative probability associated with Z if necessary.

(a) Let X be the seat pitch in inches. The keywords in the problem suggest $X \sim N(31, 0.25)$.

Between 30.5 and 32 means in the interval [30.5, 32] (whether it is closed or open doesn't matter). Find the probability that X lies in this interval.

$P(30.5 \leq X \leq 32)$

$= P\left(\dfrac{30.5-31}{0.5} \leq \dfrac{X-31}{0.5} \leq \dfrac{32-31}{0.5}\right)$ Standardize.

$= P(-1.00 \leq Z \leq 2.00)$ Equation 6.8; simplify.

$= P(Z \leq 2.00) - P(Z \leq -1.00)$ Use cumulative probability.

$= 0.9772 - 0.1587 = 0.8185$ Use Appendix Table 3.

The probability that a randomly selected coach seat has seat pitch between 30.5 and 32 in. is 0.8185.

(b) A seat is constricted if the value of X is less than 30 in.
Find $P(X < 30)$.

$P(X < 30) = P\left(\dfrac{X-31}{0.5} < \dfrac{30-31}{0.5}\right)$ Standardize.

$= P(Z < -2.00) = P(Z \leq -2.00)$ Equation 6.8; simplify.

$= 0.0228$ Cumulative probability; use Appendix Table 3.

The probability that a randomly selected coach seat is constricted is 0.0228.

Figure 6.44 shows technology solutions to parts (a) and (b). **Figure 6.45** illustrates the standardization in part (b).

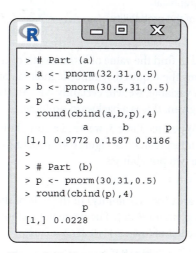

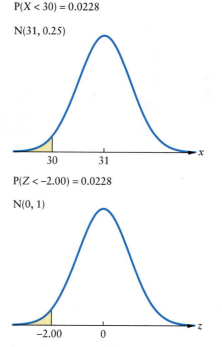

Figure 6.44 Normal probability calculations using R.

Figure 6.45 A visualization of the standardization in Example 6.6 (b).

TRY IT NOW Go to Exercises 6.55 and 6.59

The Normal Distribution and Inference

EXAMPLE 6.7 Stoppage Time

During a soccer game, stoppage time is an allowance made at the discretion of the referee for time lost through substitutions, player injury, or a team simply wasting time. This extra time is added on to the end of the match. During the World Cup soccer games in 2018, the mean stoppage time per match was 6.4 minutes.[8] Assume the distribution of stoppage time is normal with standard deviation 1.6 minutes.

(a) Find a symmetric interval about the mean time, $[\mu - b, \mu + b]$, such that 95% of all stoppage times lie in this interval.

(b) A random soccer game was selected following the World Cup and the stoppage time was 10.1 minutes. Is there any evidence to suggest that the mean stoppage time has changed? Justify your answer.

Solution

(a)

STEP 1 Let X be the stoppage time in minutes of a randomly selected World Cup soccer match. The information given indicates that X is a normal random variable with mean $\mu = 6.4$ and standard deviation $\sigma = 1.6$: $X \sim N(6.4, 2.56)$.

"Find a symmetric interval about the mean such that 95% of all stoppage times lie in this interval" translates as "Find a value of b such that $P(6.4 - b \leq X \leq 6.4 + b) = 0.95$." **Figure 6.46** illustrates this probability statement.

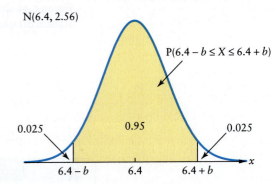

Figure 6.46 A graphical representation of the probability statement.

STEP 2 We need to find the value of b, the length of time in each direction from the mean that forms the required symmetric interval. Since this question involves a normal random variable, we will certainly have to standardize. And, because the probability is given, this is a backward problem.

To use Appendix Table 3, we need a cumulative probability statement. So we need another interpretation of Figure 6.46 involving cumulative probability and the value b. Here are two possibilities.

i. $P(X \leq 6.4 - b) = 0.025$

The area (or probability) in the tails of the distribution is $1 - 0.95 = 0.05$ (the Complement Rule). The normal distribution is symmetric, so the probability to the left of $(6.4 - b)$ is $0.05/2 = 0.025$.

ii. $P(X \leq 6.4 + b) = 0.975$

The probability to the left of $(6.4 + b)$ is $0.95 + 0.025 = 0.975$.

STEP 3 We'll use the expression in (i).

$$P(X \leq 6.4 - b) = P\left(\frac{X - 6.4}{1.6} \leq \frac{(6.4 - b) - 6.4}{1.6}\right) = 0.025 \quad \text{Standardize.}$$

$$= P\left(Z \leq \frac{-b}{1.6}\right) \quad \text{Equation 6.8; simplify.}$$

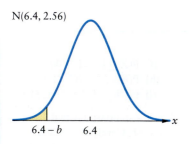

$P(X < 6.4 - b) = 0.025$
N(6.4, 2.56)

$P(Z < -1.96) = 0.025$
N(0, 1)

Figure 6.47 A visualization of the standardization in Example 6.7 (a).

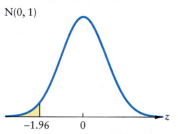

Figure 6.48 A technology solution using R. Use inverse cumulative probability to find each endpoint.

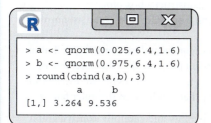

N(6.4, 2.56)

Figure 6.49 $P(X \geq 10.1)$ is a right-tail probability and a measure of the likelihood in this statistical inference.

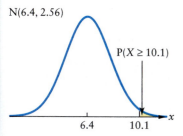

Figure 6.50 Tail probability calculation using R.

There is no further simplification within the probability statement. The resulting expression involves Z and is a cumulative probability. Find a value in the body of Appendix Table 3 as close to 0.025 as possible. **Figure 6.47** illustrates this standardization.

Set the corresponding z value equal to $\frac{-b}{1.6}$, and solve for b.

$\frac{-b}{1.6} = -1.96$ \hfill Appendix Table 3.

$-b = -3.136$ \hfill Multiply both sides by 1.6.

$b = 3.136$ \hfill Multiply both sides by −1.

The value of b is 3.136 minutes and the symmetric interval about the mean is

$P(6.4 - b \leq X \leq 6.4 + b) = P(6.4 - 3.136 \leq X \leq 6.4 + 3.136)$
$= P(3.264 \leq X \leq 9.536) = 0.95$

Thus, 95% of all stoppage times are between 3.264 and 9.536 minutes.

Technology can be used to find the endpoints of the interval without solving for b. See **Figure 6.48**.

(b) Part (b) is a statistical inference problem. We see the phrase, "Is there any evidence to suggest," which signals the four-step inference process.

The claim or status quo is $\mu = 6.4$ minutes. This implies, as in part (a), that the random variable X has a normal distribution, with $\mu = 6.4$ and $\sigma = 1.6$.

Claim: $\mu = 6.4 \Rightarrow X \sim N(6.4, 2.56)$

The experimental outcome is that the stoppage time in a random soccer game was 10.1 minutes. Write this outcome as an observed value of the random variable.

Experiment: $x = 10.1$

It seems reasonable to consider $P(X = 10.1)$, the probability of observing a stoppage time of 10.1 minutes, and then draw a conclusion based on this probability. However, since X is a continuous random variable, the probability of a single value is 0. So $P(X = 10.1) = 0$.

Therefore, as before, we consider a tail probability as a measure of likelihood. And, since 10.1 is to the right of the mean, we'll consider a right-tail probability. See **Figure 6.49**.

Likelihood:

$P(X \geq 10.1) = P\left(\frac{X - 6.4}{1.6} \geq \frac{10.1 - 6.4}{1.6}\right)$ \hfill Standardize.

$= P(Z \geq 2.31)$ \hfill Equation 6.8; simplify.

$= 1 - P(Z \leq 2.31)$ \hfill The Complement Rule, in a continuous world.

$= 1 - 0.9896 = 0.0104$ \hfill Use Appendix Table 3; simplify.

Conclusion: Because this tail probability is so small (less than 0.05), it is very unusual to observe a stoppage time of 10.1 minutes or greater (if the claim of $\mu = 6.4$ minutes is true). But it happened! Therefore, there is evidence to suggest the claim is false. In the context of this problem, there is evidence to suggest that the mean stoppage time per soccer match is different from (well, greater than) 6.4 minutes.

Figure 6.50 shows a technology solution. ■

TRY IT NOW Go to Exercises 6.57 and 6.61

CHAPTER APP **EMV Chip Technology**

Suppose a EMV chip card transaction is selected at random and the processing time is 20 seconds. Is there any evidence to suggest that the mean processing time is greater than 13 seconds?

simpson33/Deposit Photos

Section 6.2 Exercises

Concept Check

6.31 True or False The graph of the probability density function for any normal random variable is bell-shaped.

6.32 True or False The mean and variance of a normal random variable determine the location and spread of the distribution.

6.33 Fill in the Blank The standard normal random variable has mean _____ and variance _____.

6.34 Fill in the Blank Any probability statement involving a normal random variable can be converted to an equivalent statement involving a standard normal random variable through the process of _____.

6.35 Multiple Choice For any normal random variable X, the statement $P(X \leq x)$ is (a) a cumulative probability; (b) an inverse cumulative probability; or (c) standardized.

6.36 Short Answer Suppose X is a normal random variable and a and b are constants. Rewrite each expression in terms of cumulative probability.
(a) $P(X \geq b)$
(b) $P(a \leq X \leq b)$
(c) $P(X \leq a \cup X \geq b)$

Practice

6.37 Let the random variable Z have a standard normal distribution. Find each of the following probabilities and carefully sketch a graph corresponding to each expression.
(a) $P(Z \leq 2.16)$
(b) $P(Z < 2.16)$
(c) $P(Z \leq -0.47)$
(d) $P(0.73 > Z)$
(e) $P(-1.75 \geq Z)$
(f) $P(-0.35 \leq Z \leq 0.65)$
(g) $P(Z < 5)$
(h) $P(Z \leq -4)$
(i) $P(Z \leq 4)$
(j) $P(Z \geq -5)$

6.38 Let the random variable Z have a standard normal distribution. Find each of the following probabilities and carefully sketch a graph corresponding to each expression.
(a) $P(-1.33 > Z)$
(b) $P(Z < 2.35)$
(c) $P(Z > 2.59)$
(d) $P(-1.56 < Z < -0.56)$
(e) $P(0.13 < Z < 2.44)$
(f) $P(-0.05 < Z < 0.76)$
(g) $P(Z \geq 2.67)$
(h) $P(Z \leq 1.42)$
(i) $P(Z \leq -2 \cup Z \geq 2)$
(j) $P(-1.82 < Z \leq -0.94)$

6.39 Let the random variable Z have a standard normal distribution. Find each of the following probabilities.
(a) $P(-1.00 \leq Z \leq 1.00)$
(b) $P(-2.00 \leq Z \leq 2.00)$
(c) $P(-3.00 \leq Z \leq 3.00)$

Do you recognize these probabilities (from Chapter 3). What rule are they associated with?

6.40 Let the random variable Z have a standard normal distribution. Solve each expression for b. Carefully sketch a graph corresponding to each probability statement.
(a) $P(Z \leq b) = 0.8686$
(b) $P(Z < b) = 0.1867$
(c) $P(Z < b) = 0.0016$
(d) $P(Z \geq b) = 0.2643$
(e) $P(Z > b) = 0.9382$
(f) $P(Z \geq b) = 0.5000$
(g) $P(b < Z) = 0.0192$
(h) $P(b > Z) = 0.9938$
(i) $P(-b < Z < b) = 0.7995$
(j) $P(-b \leq Z \leq b) = 0.5527$

6.41 Let the random variable Z have a standard normal distribution. Solve each expression for b. Carefully sketch a graph corresponding to each probability statement.
(a) $P(Z \leq b) = 0.5100$
(b) $P(Z > b) = 0.1080$
(c) $P(Z \geq b) = 0.0500$
(d) $P(Z \leq b) = 0.0100$
(e) $P(-b \leq Z \leq b) = 0.8000$
(f) $P(-b < Z < b) = 0.6535$

6.42 Let the random variable Z have a standard normal distribution. Recall the definition of percentiles. $P(Z \leq 1.0364) = 0.85$, so 1.0364 is the 85th percentile. Find each of the following percentiles for a standard normal distribution.
(a) 10th
(b) 27th
(c) 85th
(d) 40th
(e) 49th
(f) 61st

6.43 Let Z be a standard normal random variable and recall the calculations necessary to construct a box plot.
(a) Find the first and third quartiles for a standard normal distribution.
(b) Find the inner fences for a standard normal distribution.
(c) Find the probability that Z is beyond the inner fences.
(d) Find the outer fences for a standard normal distribution.
(e) Find the probability that Z is beyond the outer fences.

6.44 Compute each probability and carefully sketch a graph corresponding to each expression.
(a) $X \sim N(3, 0.0225)$, $P(X \leq 3.25)$
(b) $X \sim N(52, 49)$, $P(X > 60)$
(c) $X \sim N(-7, 1)$, $P(X \leq -4.5)$
(d) $X \sim N(235, 121)$, $P(X > 200)$
(e) $X \sim N(242, 132)$, $P(X \geq 350)$
(f) $X \sim N(1.17, 3.94)$, $P(X < -1.45)$

6.45 Use technology to compute each probability and to carefully sketch a graph corresponding to each expression.
(a) $X \sim N(3.7, 4.55)$, $P(3.0 \leq X \leq 4.0)$
(b) $X \sim N(62, 100)$, $P(50 < X < 70)$
(c) $X \sim N(32, 30)$, $P(X \geq 45)$
(d) $X \sim N(77, 0.01)$, $P(X < 76.95)$
(e) $X \sim N(-50, 16)$, $P(X < -55 \cup X > -45)$
(f) $X \sim N(7.6, 12)$, $P(8 \leq X \leq 9)$

6.46 Use technology to solve each expression for b.
(a) $X \sim N(17, 28)$, $P(X < b) = 0.75$
(b) $X \sim N(303, 70)$, $P(X \leq b) = 0.05$
(c) $X \sim N(0, 25)$, $P(-b \leq X \leq b) = 0.90$
(d) $X \sim N(-12, 2)$, $P(X > b) = 0.35$
(e) $X \sim N(37, 2.25)$, $P(\mu - b \leq X \leq \mu + b) = 0.68$
(f) $X \sim N(26.35, 7.21)$, $P(X < b) = 0.11$

6.47 Suppose X is a normal random variable with mean 25 and standard deviation 6: $X \sim N(25, 36)$.
(a) Find the first and third quartiles for X.
(b) Find the inner fences for X.

(c) Find the probability that X is beyond the inner fences.
(d) Find the outer fences for X.
(e) Find the probability that X is beyond the outer fences.

Applications

6.48 Economics and Finance San Francisco is one of the most expensive U.S. cities in which to live. As of April 2018, the mean rent for a one-bedroom apartment in the Mission District was $3442.[9] Assume that the distribution of rents is approximately normal and the standard deviation is $300. A one-bedroom apartment in the Mission District is selected at random.
(a) Find the probability that the rent is less than $3000.
(b) Find the probability that the rent is between $3200 and $3600.
(c) Find a rent r such that 90% of all rents are less than r dollars per month.

6.49 Marketing and Consumer Behavior According to an annual survey conducted by TheKnot.com, the mean cost of a wedding in 2017 was $33,391.[10] This is actually $2000 less than the mean in 2016, but still daunting for parents whose children are thinking about marriage. Suppose the cost for a wedding is normally distributed, with a standard deviation of $1500, and a wedding is selected at random.
(a) Find the probability that the wedding costs more than $35,000.
(b) Find the probability that the wedding costs between $30,000 and $34,000.
(c) Find the probability that the wedding costs less than $29,000.

6.50 Public Policy and Political Science The president of the United States gives a State of the Union Address every year in late January or early February. Since Lyndon Johnson's address in 1966, Richard Nixon gave some of the shortest messages, and Bill Clinton presented an address lasting 1 hour, 28 minutes, in 2000.[11] The mean length of these addresses is approximately 53 minutes and the standard deviation is 14.4 minutes. Assume the length of a State of the Union Address is normally distributed.
(a) What is the probability that the next State of the Union Address will be between 45 and 55 minutes long?
(b) What is the probability that the next State of the Union Address will be more than 90 minutes long?
(c) What is the probability that the next two State of the Union Addresses will be less than 30 minutes long?

6.51 Manufacturing and Product Development A standard Versa-Lok block used in residential and commercial retaining wall systems has mean weight 37.19 kg (kilograms).[12] Assume the standard deviation is 0.8 kg and the distribution is approximately normal. A standard block unit is selected at random.
(a) What is the probability that the block weighs more than 38 kg?
(b) What is the probability that the block weighs between 36 and 37 kg?
(c) If the block weighs less than 35.5 kg, it cannot be used in certain commercial construction projects. What is the probability that the block cannot be used?

6.52 Marketing and Consumer Behavior Movie trailers are designed to entice audiences by showing scenes from coming attractions. Several trailers are usually shown in a theater before the start of the main feature, and most are available via the Internet. The duration of a movie trailer is approximately normal, with mean 150 seconds and standard deviation 30 seconds.
(a) What is the probability that a randomly selected trailer lasts less than 1 minute?
(b) Find the probability that a randomly selected trailer lasts between 2 minutes and 3 minutes, 15 seconds.
(c) Any movie trailer that lasts beyond 4 minutes, 30 seconds is considered too long. What proportion of movie trailers is too long?
(d) Find a symmetric interval about the mean such that 99% of all movie trailer durations lie in this interval.

6.53 Biology and Environmental Science The salinity, or salt content, in the ocean is expressed in parts per thousand (ppt). This number varies with depth, rainfall, evaporation, river runoff, and ice formation. The mean salinity of the oceans is 35 ppt.[13] Suppose the distribution of salinity is normal and the standard deviation is 0.52 ppt, and suppose a random sample of ocean water from a region in the tropical Pacific Ocean is obtained.
(a) What is the probability that the salinity is more than 36 ppt?
(b) What is the probability that the salinity is less than 33.5 ppt?
(c) A certain species of fish can survive only if the salinity is between 33 and 35 ppt. What is the probability that this species can survive in a randomly selected area?
(d) Find a symmetric interval about the mean salinity such that 50% of all salinity levels lie in this interval. What are the endpoints of this interval called?

6.54 Public Health and Nutrition Many people grab an energy bar for breakfast or for a snack to make it through the afternoon slump at work. A Clif Chocolate Chip Energy Bar weighs 68 g, and the mean amount of protein in each bar is 9 g.[14] Suppose the amount of protein in a bar is normally distributed and the standard deviation is 0.15 g, and a random Clif Chocolate Chip Energy Bar is selected.
(a) What is the probability that the amount of protein is less than 8.75 g?
(b) What is the probability that the amount of protein is between 8.8 and 9.2 g?
(c) Suppose the amount of protein is at least 9.1 g. What is the probability that it is more than 9.3 g?
(d) Suppose three bars are selected at random. What is the probability that all three will have protein content between 8.7 and 9.3 g?

6.55 Public Health and Nutrition Many typical household cleaners contain toxic chemicals. 2-Butoxyethanol is found in multipurpose cleaners and is a very powerful solvent.

The Environmental Protection Agency (EPA) has a safety standard for this chemical when used in the workplace, but cleaning at home in a confined area can cause levels to rise well above this standard. The mean percentage of 2-butoxyethanol in Rain-X Glass Cleaner is 3%.[15] Suppose the distribution is approximately normal and the standard deviation is 1%, and a random bottle of Rain-X Glass Cleaner is selected.

(a) What is the probability that the percentage of 2-butoxyethanol is less than 2.5%?
(b) What is the probability that the percentage of 2-butoxyethanol is between 2.2% and 3.5%?
(c) Suppose the EPA has established a limit of 5% 2-butoxyethanol in all consumer products. What is the probability that a bottle exceeds this limit?

6.56 Psychology and Human Behavior In many U.S. families, both parents work outside the home, while children spend time at daycare centers or are cared for by other relatives. The mean amount of time fathers spend on child care is 8 hours per week.[16] Suppose this time distribution is approximately normal with standard deviation 0.75 hour, and suppose a father is randomly selected.

(a) What is the probability that the father spends at least 8.5 hours on child care in a given week?
(b) What is the probability that the father spends between 6.5 and 7.5 hours on child care in a given week?
(c) What is the probability that the father spends less than 6 hours on child care in a given week?
(d) What is the probability that the father will spend at least 9 hours on child care in each of five randomly selected weeks?

6.57 Sports and Leisure The base running ability of a Major League Baseball (MLB) player is often measured as sprint speed, in feet per second (ft/sec), during a player's fastest 1-second window. The elite speedsters steal second base easily, make it from first to home on a long double, and are generally outfielders. The mean sprint speed for all MLB players is 27 ft/sec.[17] Suppose the distribution of sprint speeds is normal with standard deviation 1.5 ft/sec, and suppose an MLB player is selected at random.

(a) What is the probability that he will have a sprint speed faster than Mookie Betts (28.2 ft/sec)?
(b) What is the probability that he will have a sprint speed between 25 and 30 ft/sec?
(c) Suppose the player has a sprint speed of 23 ft/sec. Is there any evidence to suggest that the claim (the mean is 27 ft/sec) is false? Justify your answer.

Extended Applications

6.58 Sports and Leisure People who ride in hot-air balloons usually fly just above the treetops at 200–500 ft. In populated areas, however, they usually stay at an altitude of at least 1000 ft. The amount of flying time possible in a hot-air balloon depends on many factors, including the number of propane burners, the number of people in the basket, and the weather. Assume the time spent aloft is normally distributed, with mean 1.5 hours and standard deviation 0.45 hour. Suppose a hot-air balloon flight is selected at random.

(a) What is the probability that the flight time is between 1 and 2 hours?
(b) What is the probability that the flight time is more than 1 hour, 15 minutes?
(c) Find a value t such that 10% of all flights last less than t hours.
(d) Suppose a person offering hot-air balloon rides charges $50 for each ride of at least 1 hour, and $1.00 for every minute after 1 hour. What proportion of rides costs more than $100?

6.59 Sports and Leisure The Daytona 500, often referred to as the Great American Race, is a spectacular sporting event, complete with an impressive pre-race show. Austin Dillon won this race in 2018, when the mean speed per lap for all racers was 150.545 mph.[18] Assume the speed is normally distributed with a standard deviation of 16 miles per hour (mph), and a driver and a lap are selected at random.

(a) What is the probability that the speed on this lap is less than 135 mph?
(b) What is the probability that the speed on this lap is between 150 and 160 mph?
(c) The fastest recorded speed at Daytona International Speedway is 210.364 mph, achieved in 1987. What is the probability that the speed on this lap will set a new record?
(d) What is the probability that the four leaders will all have a speed of at least 170 mph on this lap?

6.60 Biology and Environmental Science Crystal Dam on the Gunnison River in Colorado is a double-curvature thin-arch type. Regulating gates in the powerplant control the total flow from the reservoir. Historical data suggest the mean total release from the reservoir per day is 1454 cubic feet per second (cfs).[19] Suppose the total release distribution is normal with standard deviation 340 cfs, and a random day is selected.

(a) What is the probability that the total release is between 1300 and 1600 cfs?
(b) What is the probability that the total release is less than 800 cfs?
(c) Suppose the release is at least 1700 cfs. What is the probability that it will be at least 2000 cfs?
(d) If the release is more than 2100 cfs for three consecutive days, then areas near Grizzly Ridge will flood. What is the probability of flooding near Grizzly Ridge?

6.61 Economics and Finance A tariff is a border tax paid by an importer and collected by customs agents. During the 2018 G7 summit in Canada, President Donald Trump indicated he wanted to eliminate all tariffs. At that time, the World Trade Organization (WTO) claimed that the mean tariff for imported goods in the United States was 2.4%.[20] Suppose the tariff distribution is normal with standard deviation 0.6%, and a random imported product is selected.

(a) What is the probability that the tariff is more than 3.5%?
(b) What is the probability that the tariff is between 2% and 3%?

(c) Suppose the tariff on the product is 1.5%. Is there any evidence to suggest the claim made by the WTO is false? Justify your answer.

6.62 Biology and Environmental Science Many backyard gardeners prefer Silver Queen Hybrid corn. This late-season variety is very sweet and has tender, white kernels. In some locations in the Northeast, gardeners have trouble harvesting this variety because of its longer growing time. The temperature of the soil should be at least 65°F before planting, and the growing time is approximately normal, with mean 92 days and standard deviation 5 days.
(a) What is the probability that a randomly selected seed will mature in less than 90 days?
(b) What is the probability that a randomly selected seed will mature in between 95 and 100 days?
(c) Suppose a row in a backyard garden contains 12 plants. What is the probability that four will be ready for dinner by the 95th day?
(d) Find a value h such that 99% of all plants are ready to be harvested within h days.

6.63 Manufacturing and Product Development Violin bows are made from various woods to accommodate musicians' preferences and demands. Some commonly used woods include snakewood, ironwood, hakia, and pernambuco. While the bows are carefully handcrafted, they vary slightly in weight. Suppose a bowmaker claims the weight of his bows is normally distributed, with mean 60 g and standard deviation 3.2 g.
(a) What is the probability that the weight of a randomly selected ironwood bow is between 58 and 62 g?
(b) Good musicians can detect an unacceptable bow weight—that is, a weight that differs from the mean by more than 2 standard deviations. What is the probability that a bow weight is unacceptable?
(c) Any manufactured bow that weighs more than 66 g is reworked to decrease the weight. What is the probability that a randomly selected ironwood bow will need rework?
(d) Suppose the weight of a randomly selected bow is 55 g. Is there any evidence to suggest the mean weight is less than 60 g? Justify your answer.

6.64 Medicine and Clinical Studies Repeated industrial tasks often cause work-related muscle disorders. Measurements of joint angles (of the shoulder and elbow, for example) required to complete a certain task can be used to predict future injuries. The shoulder joint angle required to fasten an aluminum door frame on an assembly line varies according to the worker's height, arm length, and location. The shoulder joint angle for this task is normally distributed, with mean 23.7 degrees and standard deviation 1.9 degrees. Suppose an employee is randomly selected.
(a) What is the probability that the shoulder joint angle will be between 20 and 25 degrees?
(b) What is the probability that the joint angle will be less than 18 degrees?
(c) If the joint angle is more than 28 degrees, there is a good chance the employee will suffer from a muscle disorder. What is the probability that the employee will suffer from a muscle disorder?
(d) If the joint angle is between 21.7 and 25.7 degrees, then management believes the ergonomics of the task are adequate. If five employees are randomly selected, what is the probability that four of the five have adequate ergonomics?

6.65 Biology and Environmental Science Many lakes are carefully monitored for pH concentration, total phosphorus, chlorophyll, nitrogen, and total suspended solids. These data are used to characterize the condition of the lake and to chart year-to-year variability. Based on the State of the Lake and Ecosystem Indicators Report, the Port Henry region of Lake Champlain has a mean total phosphorus concentration of 14 milligrams per liter (mg/L).[21] Suppose the total phosphorus concentration distribution is normal, with standard deviation 5.5 mg/L, and a random water sample from Port Henry is obtained.
(a) What is the probability that the total phosphorus is less than 12 mg/L?
(b) What is the probability that the total phosphorus differs from the mean by more than 5 mg/L?
(c) Suppose the total phosphorus is less than 20 mg/L. What is the probability that it is less than 10 mg/L?
(d) If the total phosphorus measurement is 25 mg/L, is there any evidence to suggest the mean has increased?

6.66 Manufacturing and Product Development High-pressure washers have become popular for cleaning siding, decks, and windows. This equipment is available in various engine types and horsepower. Suppose the power rating (in horsepower, hp) for a residential pressure washer is normally distributed, with mean 20 hp and standard deviation σ.
(a) The probability that a randomly selected power rating is within 2.5 hp of the mean is 0.7229. Find the value of σ.
(b) A leading consumer magazine advised its readers to purchase pressure washers with a power rating of 15 hp or more. What proportion of pressure washers have this rating?
(c) If the power rating is more than 26.5 hp, the pressure washer will crack, or even break, certain windows. What is the probability that a pressure washer could break a window?

6.67 Physical Sciences Hydroelectric projects are carefully monitored, and their energy capability is predicted for several years into the future. Suppose the Klamath Hydro Project, located on the upper Klamath River in south-central Oregon, generates electricity according to a normal distribution. The Pacific Northwest Utilities Conference Committee claims the mean electricity generated per year is 35 megawatts (MW).
(a) The probability that the Klamath Hydro Project generates less than 34 MW during any randomly selected year is 0.3540. Find the standard deviation.
(b) Suppose the years are independent, and the hydro project will record a profit in a given year if it is able to generate at least 37.8 MW that year. What is the probability that the project will record a profit for four consecutive years?

(c) Suppose the electricity generated during a certain year is 33.5 MW. Is there any evidence to suggest that the claim by the Pacific Northwest Utilities Conference Committee is false? Justify your answer.

6.68 Manufacturing and Product Development Dining-room chairs come in many different woods, styles, and shapes. The height of the seat of a randomly selected oak dining-room chair is approximately normal, with mean 85 centimeters (cm) and standard deviation 1.88 cm.
(a) Find a value h such that 99% of all dining-room chairs have height less than h.
(b) Consumer testing indicates that any chair seat higher than 90 cm is uncomfortable to use when eating. What is the probability that a randomly selected dining-room chair is uncomfortable?
(c) Find the first and third quartiles of the dining-room chair height distribution.
(d) There is some evidence to suggest that, after five years of use, the mean height of these chairs has decreased, due to wear, erosion, and humidity. Suppose that after five years, the probability that the height is more than 86 cm is 0.0718. Find the mean height after five years.

6.69 Biology and Environmental Science Sand dollars are a favorite of serious conchologists (people who collect sea shells). They can be found on the beaches of Fort Myers and Sanibel Island in Florida. Found on shore, a sand dollar is white, with mean diameter 3 in.[22] Suppose the diameter of a sand dollar is normally distributed, with standard deviation 0.55 in.
(a) Find a value d such that 90% of all sand dollars have diameter greater than d.
(b) Large sand dollars, those with diameter greater than 4 in., can be sold to Kelly's Shell Shack. What is the probability that a randomly selected sand dollar can be sold?
(c) Suppose 10 sand dollars are selected at random. What is the probability that exactly 7 of the 10 are between 2.5 and 3.5 in. in diameter?
(d) There is some speculation that environmental hazards have reduced the diameter of sand dollars. Suppose a randomly selected sand dollar is 2 in. in diameter. Is there any evidence to suggest that the mean diameter has decreased? Justify your answer.

6.70 Biology and Environmental Science One measure of air quality is the level of nitrogen dioxide (NO_2). This nasty-smelling gas is formed naturally in the atmosphere, but the major source of NO_2 in the atmosphere is due to burning of fossil fuels. Although air pollution is becoming worse in many parts of the world, Canada has some of the best air quality of any country.[23] Suppose the amount of NO_2 in the atmosphere in Windsor, Ontario, is a normal random variable with mean 10 mg/m^3 and standard deviation 2.5 mg/m^3.[24]
(a) What is the probability that a randomly selected day has an NO_2 between 7 and 13 mg/m^3?
(b) If the NO_2 level is greater than 13 mg/m^3, then an Air Quality Alert is issued. What is the probability that an alert will be issued on a randomly selected day?
(c) What is the probability that 5 days out of 7 will have an NO_2 level less than 6 mg/m^3?

Challenge Problems

6.71 Sports and Leisure The International Tennis Federation (ITF) establishes the specifications for tennis balls. The diameter of a tennis ball used in any tournament must be between 2.5 and 2.625 in. Suppose the diameter of a tennis ball is approximately normal, with mean 2.5625 in. and standard deviation 0.04 in.
(a) What is the probability that a randomly selected tennis ball will meet the ITF diameter specifications?
(b) Suppose six tennis balls will be used in a tournament game. What is the probability that exactly one will not meet the ITF diameter specifications? Assume independence.
(c) Suppose two tennis balls are selected at random. What is the probability that one will be less than 2.5 in. in diameter and one will be greater than 2.63 in. in diameter?

6.72 Fuel Consumption and Cars The EPA reports that the miles per gallon (mpg) for a 2018 Dodge Challenger under controlled conditions is a normal random variable X, with mean μ and standard deviation σ. Suppose independent tests have shown that for a randomly selected Challenger, the probability that the mpg is less than 14.5 is 0.025, and the probability that the mpg is greater than 23.5 is 0.10. Find the mean μ and the standard deviation σ.

6.3 Checking the Normality Assumption

Four Methods

Almost every inferential statistics procedure requires certain assumptions—for example, that observations are selected independently or that variances are equal (for analysis of variance). And many statistical techniques are valid only if the observations are from a normal distribution. If an inference procedure requires normality, and the population distribution is not normal, then the conclusions are worthless. Therefore, it seems reasonable to be able to perform some kind of check for normality, to make sure there is no evidence to refute this assumption.

Until now, we have been using the normal distribution as a model for describing the variability of a random variable X, and we have been assuming that we know the values of the population mean μ and the population variance σ^2. If those values are not known, the sample mean $\bar{x}$ and the sample standard deviation s can be used as estimates of the unknown parameters μ and σ. However, we still cannot be sure that the normal distribution is an appropriate model to describe a particular set of observations. We need a way to check whether a set of observations does seem to come from a population with a normal distribution. We can use four different methods to look for any evidence of non-normality. Three of them rely on techniques that we have seen before; the fourth one is a new method.

Given a set of observations, $x_1, x_2, x_3, \ldots, x_n$, the following four methods may be used to check for any evidence of non-normality, such as a distribution that is not bell-shaped, a skewed distribution, or a distribution with heavy tails.

(1) Graphs

Construct a histogram, a stem-and-leaf plot, and/or a dot plot. Examine the shape of the distribution for any indications that the distribution is not bell-shaped. In a random sample, the distribution of the sample should be similar to the distribution of the population.

(2) Backward Empirical Rule

To use the Empirical Rule to test for normality, find the mean, the standard deviation, and the three symmetric intervals about the mean $(\bar{x} - ks, \bar{x} + ks)$, $k = 1, 2, 3$. Compute the actual proportion of observations in each interval. If the actual proportions are close to 0.68, 0.95, and 0.997, then normality seems reasonable. Otherwise, there is evidence to suggest that the shape of the distribution is not normal.

(3) IQR/s

Find the interquartile range IQR and standard deviation s for the sample, and compute the ratio IQR/s. If the data are approximately normal, then IQR/$s \approx 1.3$.

$P(Z \leq -0.6745) = 0.25$, and
$P(Z \leq 0.6745) = 0.75$

Here is some justification for this ratio. Consider a standard normal random variable, Z ($\mu = 0, \sigma = 1$). The first quartile for Z is -0.6745 and the third quartile is 0.6745. The interquartile range divided by the standard deviation is $[0.6745 - (-0.6745)]/1 = 1.349$. In a random sample, the interquartile range should be close to the population interquartile range, and the standard deviation should be close to the population standard deviation. Any normal distribution can be standardized, or compared to Z, so IQR/$s \approx 1.3$.

(4) Normal Probability Plot

 Scatter plot.

A normal probability plot is a scatter plot of each observation versus its corresponding standardized normal score. For a normal distribution, the points will fall close to a straight line.

Normal Probability Plot

How to Construct a Normal Probability Plot

Suppose $x_1, x_2, \ldots, x_n$ is a set of observations.

1. Order the observations from smallest to largest, and let $x_{(1)}, x_{(2)}, \ldots, x_{(n)}$ represent the set of ordered observations.

The standardized normal scores are expected values. For example, in repeated samples of size n from the Z distribution, on average the smallest value is z_1, on average the next largest value is $z_2, \ldots$, on average the largest value is z_n.

2. Find the standardized normal scores for a sample of size n in Appendix Table 4, $z_1, z_2, \ldots, z_n$.

3. Plot the ordered pairs $(z_i, x_{(i)})$.

Most of the standardized normal scores are always between −2.0 and +2.0, because approximately 95% of all observations lie within 2 standard deviations of the mean.

If the scatter plot is nonlinear, there is evidence to suggest the data did not come from a normal distribution. Most statistical software (for example, R, Minitab, and the TI-84 Plus CE) automatically computes the expected Z values. Appendix Table 4 provides standardized normal scores for some values of n.

Figures 6.51–6.54 are examples of normal probability plots.

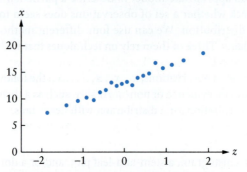

Figure 6.51 A normal probability plot. The points lie along an approximate straight line. There is no evidence of non-normality.

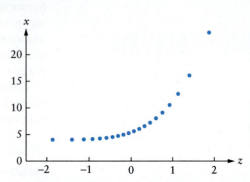

Figure 6.52 A normal probability plot. The curved graph suggests that the distribution is not normal and is skewed.

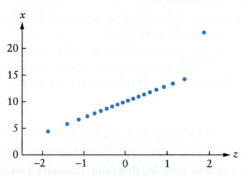

Figure 6.53 A normal probability plot. The plot suggests that the distribution is not normal and that the data set contains an outlier.

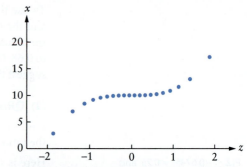

Figure 6.54 A normal probability plot. The plot suggests that the distribution is not normal and has heavy tails.

The data axis can be horizontal or vertical. To use a horizontal data axis, plot the points $(x_{(i)}, z_i)$. **Figure 6.55** shows a normal probability plot with the data plotted on the vertical axis, and **Figure 6.56** shows a normal probability plot (using the same data and standardized normal scores) with the data plotted on the horizontal axis.

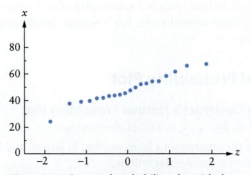

Figure 6.55 A normal probability plot with the data plotted on the vertical axis.

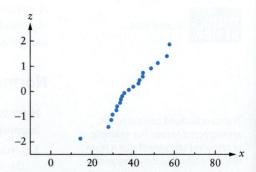

Figure 6.56 A normal probability plot with the data plotted on the horizontal axis.

Interpretation of a normal probability plot is very subjective. Even if the axes are reversed, we are still looking for the points to lie along, or close to, a straight line.

All four methods can be used to check the normality assumption, and any one (or several) may suggest the data did not come from a normal distribution. Because we are searching for evidence of non-normality, even if we fail to reject the normality assumption in each test, we still cannot say with absolute certainty that the data came from a normal distribution.

Examples

EXAMPLE 6.8   Arctic Ice Extent

NOAA/National Oceanic and Atmospheric Administration/ Department of Commerce

Weather data from the Arctic can help scientists construct better climate models. However, the harsh environment makes it difficult to collect information. Recently, automated aircraft and instrument-bearing tethered balloons have been used to gather data in regions that are difficult for scientists to reach. A random sample of spring days was selected and the sea ice extent ($\times 10^6$ km^2) was recorded for each. The 20 observations are given in the table.

| 13.1 | 13.3 | 14.2 | 11.7 | 12.9 | 12.0 | 13.3 | 14.3 | 10.7 | 12.2 |
| 11.3 | 13.5 | 15.1 | 13.5 | 13.8 | 13.5 | 11.7 | 12.8 | 14.6 | 12.7 |

Is there any evidence to suggest that this distribution is not normally distributed?

Solution

STEP 1 **Figure 6.57** shows a frequency histogram for these data. There are no obvious outliers, and the distribution seems approximately normal.

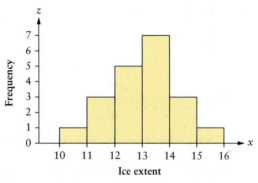

Figure 6.57 Frequency histogram for the ice extent data.

STEP 2 The sample mean and the sample standard deviation are $\bar{x} = 13.01$ and $s = 1.15$, respectively. The following table lists three symmetric intervals about the mean, the number of observations in each interval, and the proportion of observations in each interval (recall that $n = 20$).

Interval	Frequency	Proportion
$(\bar{x} - s, \bar{x} + s) = (11.86, 14.16)$	12	0.60
$(\bar{x} - 2s, \bar{x} + 2s) = (10.71, 15.31)$	19	0.95
$(\bar{x} - 3s, \bar{x} + 3s) = (9.56, 16.46)$	20	1.00

The actual proportions are close to those given by the Empirical Rule (0.68, 0.95, and 0.997).

STEP 3 The quartiles are $Q_1 = 12.1$, $Q_3 = 13.65$.

$$IQR/s = (13.65 - 12.1)/1.15 = 1.3508$$

This ratio is close to 1.3.

STEP 4 The following table lists each observation, along with the corresponding normal score from Appendix Table 4.

Observation	Normal score	Observation	Normal score
10.7	−1.87	13.3	0.06
11.3	−1.40	13.3	0.19
11.7	−1.13	13.5	0.31
11.7	−0.92	13.5	0.45
12.0	−0.74	13.5	0.59
12.2	−0.59	13.8	0.74
12.7	−0.45	14.2	0.92
12.8	−0.31	14.3	1.13
12.9	−0.19	14.6	1.40
13.1	−0.06	15.1	1.87

Plot these points to obtain the normal probability plot, as shown in **Figure 6.58**. **Figure 6.59** shows a technology solution. The points lie along an approximately straight line.

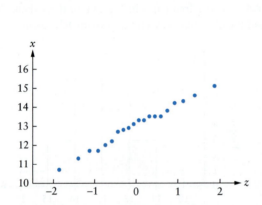

Figure 6.58 Normal probability plot for the sea extent data.

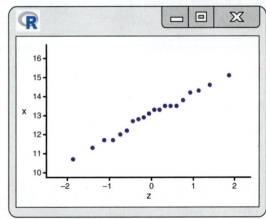

Figure 6.59 Normal probability plot using R.

The histogram, backward Empirical Rule, IQR/s, and normal probability plot show no significant evidence of non-normality. Remember, however, that this decision is subjective. ∎

TRY IT NOW Go to Exercises 6.87 and 6.89

EXAMPLE 6.9 Chemotherapy Protocol

EG6.9 A certain protocol for chemotherapy states that the total dose for patients younger than age 12 is no greater than 450 mg/m^2 within six months. A random sample of 30 patients undergoing this form of chemotherapy was obtained, and their medical records were examined to determine the total dose of the drug over the past six months. The data are given in the table.

350	351	352	353	354	358	361	364	371	376
377	378	387	396	399	402	406	408	412	424
427	430	432	437	440	441	443	446	447	449

Is there any evidence to suggest the distribution of the six-month total dosage is not normally distributed?

Solution

STEP 1 **Figure 6.60** shows a frequency histogram for these data. Although the graph appears symmetric, it is not bell-shaped. Most of the data are concentrated in the tails of the distribution. This suggests the data are not from a normal distribution.

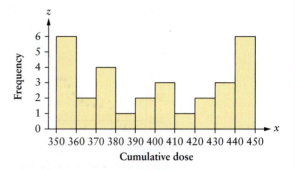

Figure 6.60 Frequency histogram for the cumulative chemotherapy dose data.

STEP 2 The sample mean is $\bar{x} = 399.03$ and the sample standard deviation is $s = 34.94$. The table lists three symmetric intervals about the mean, the number of observations in each interval, and the proportion of observations in each interval (computed using $n = 30$).

Interval	Frequency	Proportion
$(\bar{x} - s, \bar{x} + s) = (364.09, 433.97)$	15	0.50
$(\bar{x} - 2s, \bar{x} + 2s) = (329.15, 468.91)$	30	1.00
$(\bar{x} - 3s, \bar{x} + 3s) = (294.21, 503.85)$	30	1.00

The first two proportions (0.50 and 1.00) are significantly different from those given by the Empirical Rule (0.68 and 0.95). This suggests the population of total chemotherapy doses is not normal.

STEP 3 The quartiles are $Q_1 = 364.00$ and $Q_3 = 432.00$.

$$IQR/s = (432.00 - 364.00)/34.94 = 1.9463$$

This ratio is significantly different from 1.3, so there is more evidence to suggest the underlying population is not normal.

STEP 4 The following table lists each observation, along with the corresponding normal score.

Observation	Normal score	Observation	Normal score	Observation	Normal score
350	−2.04	377	−0.38	427	0.47
351	−1.61	378	−0.29	430	0.57
352	−1.36	387	−0.21	432	0.67
353	−1.18	396	−0.12	437	0.78
354	−1.02	399	−0.04	440	0.89
358	−0.89	402	0.04	441	1.02
361	−0.78	406	0.12	443	1.18
364	−0.67	408	0.21	446	1.36
371	−0.57	412	0.29	447	1.61
376	−0.47	424	0.38	449	2.04

The normal probability plot is shown in **Figure 6.61**. The points do not lie along a straight line. Each tail is flat, which makes the graph look S-shaped. This suggests that the underlying population is not normal. **Figure 6.62** shows a technology solution.

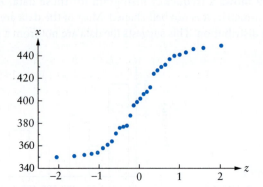

Figure 6.61 Normal probability plot for the cumulative chemotherapy dose data.

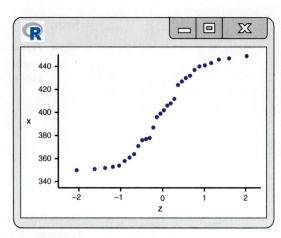

Figure 6.62 Normal probability plot using R.

The histogram, backward Empirical Rule, IQR/s, and the normal probability plot all suggest that this sample did not come from a normal population.

TRY IT NOW Go to Exercise 6.93

Section 6.3 Exercises

Concept Check

6.73 Short Answer Name four methods to search for evidence of non-normality.

6.74 True or False In a normal probability plot, the data axis can be horizontal or vertical.

6.75 True or False For data from a normal distribution, the points in the corresponding normal probability plot will fall close to a straight line or a bell-shaped curve.

6.76 True or False Most standardized normal scores are between -2.0 and $+2.0$.

6.77 True or False In a random sample, the distribution of the sample should be similar to the distribution of the population.

6.78 Fill in the Blank For data from a normal distribution, IQR/s ≈ _____.

6.79 Short Answer Explain a method to approximate the normal scores for $n = 25$.

Practice

6.80 Consider the following 20 observations. EX6.80

15.4	13.9	14.9	16.2	16.6	15.4	17.2	18.5
19.3	13.0	16.5	20.2	16.4	15.3	18.5	17.9
15.5	17.4	16.3	14.3				

Construct a normal probability plot. Is there any evidence to suggest the data are from a non-normal population? Justify your answer.

6.81 Consider the following 20 observations. EX6.81

52.0	52.1	58.8	88.0	49.9	18.7	43.1	47.6
90.0	49.8	54.8	35.1	56.1	53.2	76.5	45.4
34.1	19.5	58.7	25.7				

Construct a normal probability plot. Is there any evidence to suggest the data are from a non-normal population? Justify your answer.

6.82 Examine each normal probability plot. Is there any evidence to suggest the data are from a non-normal population? Justify your answer.

(a)
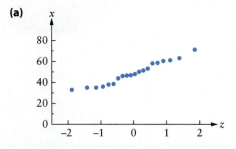

(b)
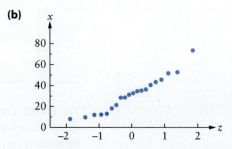

(c)
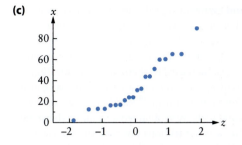

(d)
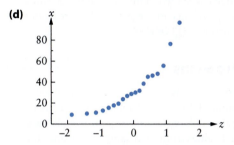

6.83 Use the four methods presented in this section to determine whether there is any evidence to suggest the data are from a non-normal population. EX6.83

6.84 Consider the data in the table. EX6.84

5.32	9.87	11.25	10.94	5.58
6.29	7.47	10.75	6.22	8.00

Use the four methods presented in this section to determine whether there is any evidence to suggest the data are from a non-normal population.

Applications

6.85 Public Health and Nutrition All across the United States, there are abandoned factories that were once used to melt lead. The total soil lead concentration is a health concern in these areas, and the EPA suggests that a total concentration greater than 400 parts per million (ppm) can be harmful to children playing there.[25] Hedley Street near the Delaware River in Philadelphia is the site of two old lead factories. Suppose samples of soil around homes in the area were obtained and the lead level in each was carefully measured (in ppm). The data are given in the table. LEAD

2223	641	1275	1796	802	691	1716	1340
1504	1006	778	1476	914	1108	1704	2106
1069	1398	1680	1705	2376	1577	1601	2201
1809	1847	1721	1069	1157	2110		

(a) Construct a normal probability plot for these data. Is there any evidence to suggest non-normality?
(b) Use the backward Empirical Rule to check for evidence of non-normality.

6.86 Education and Child Development Some people claim children who practice yoga are more physically fit, self-confident, and self-aware. A random sample of pre-teens (ages 10–12) practicing yoga was obtained and their meditation (or quiet breathing) times (in minutes) per day were recorded. Use the four methods presented in this section to determine whether there is any evidence to suggest the data are from a non-normal population. YOGA

6.87 Sports and Leisure Many NBA players express themselves on the court through their sneakers. High school basketball players often insist on wearing the same expensive sneakers worn by their favorite player. A random sample of sneakers worn by Notre Dame Prep (in Massachusetts) basketball players was obtained, and the retail price (in dollars) for each pair of sneakers is given in the table.

96.70	112.05	120.70	106.40	86.60
126.40	134.75	76.75	142.20	116.70

Use the four methods presented in this section to determine whether there is any evidence to suggest the data are from a non-normal population. SNEAKER

6.88 Public Policy and Political Science Bicycle paths are usually planned and constructed according to certain guidelines (for example, the American Association of State Highway and Transportation Officials guidelines for construction). There are construction standards for width, offset from the road, maximum grade, and horizontal and vertical clearances. A random sample of bicycle-path widths (in feet) was obtained.
(a) Find the sample mean and the sample standard deviation for these data. PATHS
(b) Find the intervals $(\bar{x} - s, \bar{x} + s), (\bar{x} - 2s, \bar{x} + 2s)$, and $(\bar{x} - 3s, \bar{x} + 3s)$.
(c) Find the proportion of observations in each interval in part (b). Is there any evidence to suggest the data are from a non-normal population?

6.89 Biology and Environmental Science A New York City antipollution law prohibits commercial drivers from leaving their engines running while parked at a curb for more than three minutes, or more than one minute in a school zone. A citizen who notices a commercial vehicle idling for a long period of time may file a complaint with the city's Department of Environmental Protection. If a summons is issued, the reward is 25% of the fine.[26] A random sample of idling times (in minutes) that resulted in fines was obtained. Use the methods presented in this section to determine whether there is any evidence to suggest the data are from a non-normal population. IDLING

6.90 Sports and Leisure A typical round of golf takes approximately 3.5 hours (walking, without an electric cart). Many weekend golfers take more time as a result of lost golf balls, thinking about certain shots, and talking to other players. The manager at the Westchester Golf Course in Los Angeles, California, obtained a random sample of round times (in hours), and the data are given in the table. GOLF

3.86	4.36	4.66	4.92	4.09	4.40	4.15	4.30
4.45	3.83	4.23	4.42	4.34	4.28	4.56	4.63
4.24	4.34	4.33	3.73				

(a) Construct a stem-and-leaf plot for these data.
(b) Compute the ratio IQR/s.

(c) Construct a normal probability plot for these data.
(d) Is there any evidence to suggest these data are from a non-normal population? Justify your answer.

6.91 Marketing and Consumer Behavior The predominant acid in frozen concentrated orange juice (FCOJ) is citric acid, and the amount is usually given as a percentage. Degrees Brix is a measure of the total soluble solids in FCOJ and is also a percentage. The Brix/acid ratio is computed by simple division, and 12 is considered an ideal ratio. A random sample of FCOJ was obtained from different sellers, and the Brix/acid ratio was measured for each. These measurements were used to construct the normal probability plot.

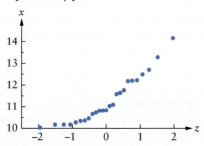

Does this plot suggest the data are from a non-normal distribution? Justify your answer.

6.92 Biology and Environmental Science There are approximately 1500 black bears in Great Smoky Mountains National Park. Park visitors are warned that these animals are dangerous and unpredictable, and that it is illegal to disturb or displace a black bear.[27] To track and protect these animals, suppose a random sample of male black bears was obtained and the weight of each (in pounds) was recorded. Use the four methods presented in this section to determine whether there is any evidence to suggest the data are from a non-normal population. ▌▍▎ BEARS

6.93 Travel and Transportation The Calgary Airport Tunnel is 620 meters long, goes under one runway and three taxiways, and accommodates approximately 13,000 vehicles each day.[28] A random sample of cars traveling through the tunnel was obtained and the time (in seconds) to drive through the tunnel was recorded for each. Use the four methods presented in this section to determine whether there is any evidence to suggest the data are from a non-normal population. ▌▍▎ TUNNEL

6.94 Biology and Environmental Science Hawaii was the first state to ban suncreens that contain oxybenzone and octinoxate. These two chemicals are believed to be harmful to marine life, including algae, sea urchins, fish, and especially coral reefs. Increased oxybenzone concentrations due to swimmers using sunscreen near coral reefs may cause significant environmental damage.[29] Random water samples near Hawaiian coral reefs were obtained and the oxybenzone concentration (in parts per trillion) was measured in each. Use the four methods presented in this section to determine whether there is any evidence to suggest the data are from a non-normal population. ▌▍▎ OXYBEN

Challenge Problems

6.95 Normal Scores Generate 500 random samples of size 10 from a standard normal distribution. Order each sample from smallest to largest. Find $\bar{x}_{(1)}$, the sample mean of the 500 smallest values from each sample. Consider the next largest value in each sample. Find $\bar{x}_{(2)}$, the sample mean of these 500 values. Continue in this manner to find $\bar{x}_{(3)}, \bar{x}_{(4)}, \ldots,$ and $\bar{x}_{(10)}$, the mean of the 500 largest values from each sample. Compare these 10 sample means, $\bar{x}_{(1)}, \bar{x}_{(2)}, \ldots, \bar{x}_{(10)}$, with the standardized normal scores for $n = 10$ in Appendix Table 4.

Try a similar procedure for $n = 20$. Generate 500 random samples of size 20 from a standard normal distribution. Order each sample from smallest to largest. Find $\bar{x}_{(1)}, \bar{x}_{(2)}, \ldots, \bar{x}_{(20)}$ and compare these values with the standardized normal scores for $n = 20$ in Appendix Table 4.

Explain why these sample means should be good estimates of the standardized normal scores.

Generate 500 random observations from a normal distribution with mean 50 and standard deviation 10. Arrange the observations in order from smallest to largest and denote this ordered list $x_{(1)}, x_{(2)}, \ldots, x_{(500)}$.

Form the ordered pairs $(x_{(1)}, 1/500), (x_{(2)}, 2/500), \ldots, (x_{(i)}, i/500), \ldots, (x_{(500)}, 1)$. Plot these points in a rectangular coordinate system and describe the shape of the graph. Which curve does this graph approximate? Why?

6.4 The Exponential Distribution

The Probability Density Function

There are many other common continuous distributions, such as the t distribution, chi-square distribution, and F distribution. We will learn a little about each of these distributions in Chapter 8 and Chapter 9 as we study confidence intervals and hypothesis tests. This section presents the exponential distribution, which is related to several continuous distributions. Remember that probability in a continuous world is the area under the curve, and if necessary, try to convert any probability statement into an equivalent expression involving cumulative probability.

The **exponential probability distribution** is often used to model the time to failure of an electronic part, or the waiting time between events. This distribution is completely characterized by one parameter, λ.

 Exponents.

The Exponential Probability Distribution
Suppose X is an exponential random variable with parameter λ (with $\lambda > 0$). The probability density function is given by

$$f(x) = \begin{cases} \lambda e^{-\lambda x} & \text{if } x \geq 0 \\ 0 & \text{otherwise} \end{cases} \tag{6.9}$$

A CLOSER LOOK

1. The symbol e in Equation 6.9 is the base of the natural logarithm ($e \approx 2.71828$). The constant e is also used in the Poisson distribution and the normal distribution.

2. If the exponential distribution is used to model the lifetime of a light bulb, a machine, or even a human being, then λ represents the failure rate.

3. The exponential distribution has positive probability only for $x \geq 0$. **Figure 6.63** and **Figure 6.64** show the graphs of a general probability density function for an exponential random variable and a probability density function with $\lambda = 2$.

Notice that $f(x) = \lambda$ when $x = 0$, because $e^0 = 1$.

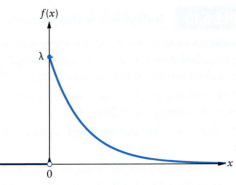

Figure 6.63 A graph of the probability density function for an exponential random variable with parameter λ.

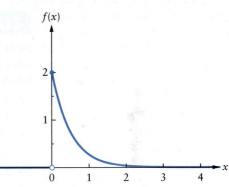

Figure 6.64 A graph of the probability density function for an exponential random variable with $\lambda = 2$.

4. The mean and variance for an exponential random variable X with parameter λ are

$$E(X) = \mu = \frac{1}{\lambda} \qquad \sigma^2 = \frac{1}{\lambda^2} \tag{6.10}$$

Cumulative Probability

Probabilities associated with an exponential random variable with parameter λ are computed using cumulative probability. We do not need a table for these calculations! **Figure 6.65** illustrates cumulative probability associated with an exponential random variable. Remember: There is no area (or probability) for $x < 0$.

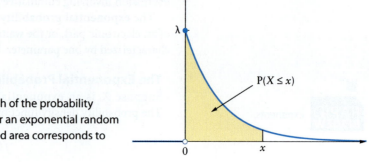

Figure 6.65 A graph of the probability density function for an exponential random variable. The shaded area corresponds to $P(X \leq x)$.

The formula for cumulative probability is given by

$$P(X \leq x) = \begin{cases} 0 & \text{if } x < 0 \\ 1 - e^{-\lambda x} & \text{if } x \geq 0 \end{cases} \quad (6.11)$$

If $x \geq 0$, the probability that X assumes a value greater than x is a right-tail probability, and is given by

$$P(X > x) = 1 - P(X \leq x) = 1 - (1 - e^{-\lambda x}) = 1 - 1 + e^{-\lambda x} = e^{-\lambda x} \quad (6.12)$$

The following example illustrates the use of cumulative probability and the formula for right-tail probability to find probabilities associated with an exponential random variable.

Example

EXAMPLE 6.10 Relief from the Common Cold

Some pharmacists recommend Zicam to relieve many symptoms due to the common cold. After the prescribed dose is taken, suppose the length of time (in hours) until symptoms return is a random variable X that has an exponential distribution with parameter $\lambda = 0.1$.

(a) Carefully sketch a graph of the probability density function for X. Find the mean, variance, and standard deviation of X.

(b) What is the probability that the length of time until symptoms return is less than the mean?

(c) What is the probability that the length of time until symptoms return is at least 12 hours?

(d) What is the probability that the length of time until symptoms return is between 8 and 16 hours?

Solution

(a) Use Equation 6.9 to sketch the graph (**Figure 6.66**) and Equation 6.10 to compute the mean, variance, and standard deviation.

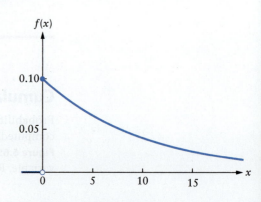

Figure 6.66 A graph of the probability density function for an exponential random variable with $\lambda = 0.1$.

$$\mu = \frac{1}{\lambda} = \frac{1}{0.1} = 10$$

$$\sigma^2 = \frac{1}{\lambda^2} = \frac{1}{0.1^2} = 100 \quad \sigma = \sqrt{\sigma^2} = \sqrt{100} = 10$$

(b) The length of time until symptoms return is modeled by an exponential random variable ($\lambda = 0.1$). The mean is 10. Translate the question into a probability statement, and use cumulative probability where appropriate.

$P(X < 10)$ Translation to a probability statement.

$= 1 - e^{-0.1(10)}$ Use the formula for cumulative probability.

$= 1 - e^{-1}$ Simplify.

$= 1 - 0.3679 = 0.6321$

The probability that the length of time until symptoms return is less than the mean is 0.6321. **Figure 6.67** illustrates this probability.

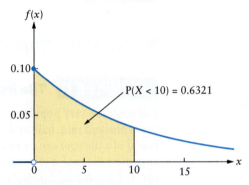

Figure 6.67 The shaded region represents the probability that the symptoms return in less than 10 hours.

(c) Translate the question into a probability statement, and use cumulative probability where appropriate. At least 12 hours means 12 hours or more.

$P(X \geq 12)$ Translation to a probability statement.

$= e^{-0.1(12)}$ Use the formula for right-tail probability.

$= e^{-1.2} = 0.3012$ Simplify.

The probability that it will take at least 12 hours for symptoms to return is 0.3012.

(d) Translate the question into a probability statement, and use cumulative probability where appropriate. Translation to a probability statement.

$P(8 \leq X \leq 16)$ Use cumulative probability.

$= P(X \leq 16) - P(X < 8)$ Formula for cumulative probability.

$= [1 - e^{-0.1(16)}] - [1 - e^{-0.1(8)}]$ Simplify

$= 0.7981 - 0.5507 = 0.2474$

The probability that the symptoms will return in between 8 and 16 hours is 0.2474. **Figures 6.68** and **6.69** show technology solutions.

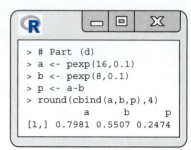

Figure 6.68 $P(X < 10)$: Use cumulative probability. $P(X \geq 12)$: Use the option to compute right-tail probability.

Figure 6.69 $P(8 \leq X \leq 16)$: Compute each cumulative probability and subtract.

TRY IT NOW Go to Exercises 6.105 and 6.107

Memoryless Property

Kent Weakley/Shutterstock

EXAMPLE 6.11 Tile Roof Lifetime

Tile roofs are very popular in Florida. These roofs are stylish; have low long-term costs; can withstand rain, hail, and high winds; and last, on average, 20 years. The lifetime (in years) of a tile roof can be modeled by an exponential random variable X with $\lambda = 0.05$. Suppose a tile roof is selected at random.

(a) What is the probability that the tile roof will last for at least eight years?

(b) Suppose the tile roof lasts for eight years. What is the probability that it will last for at least another eight years?

Solution

(a) Translate the question into a probability statement. Use cumulative probability or right-tail probability where appropriate. At least eight years means eight years or longer.

$P(X \geq 8)$ *Translation to a probability statement.*

$= e^{-0.05(8)} = 0.6703$ *Right-tail probability.*

(b) Since we are given that the tile roof lasts for eight years, this is a conditional probability question. Write the events in terms of X, and use the definition of conditional probability and cumulative or right-tail probability where appropriate.

$P(X \geq 8 + 8 \mid X \geq 8)$ *Translation to a probability statement.*

$= \dfrac{P[(X \geq 16) \cap (X \geq 8)]}{P(X \geq 8)}$ *Definition of conditional probability.*

$= \dfrac{P(X \geq 16)}{P(X \geq 8)}$ $(X \geq 16) \cap (X \geq 8) = (X \geq 16)$.

$= \dfrac{e^{-0.05(16)}}{e^{-0.05(8)}} = \dfrac{0.4493}{0.6703}$ *Right-tail probabilities.*

$= 0.6703$

This conditional probability of lasting an additional eight years is the same as the unconditional probability of lasting the initial eight years. According to this model, the fact that the tile roof has lasted eight years does not affect the probability of it lasting an additional eight years. This unrealistic result is called the memoryless property of an exponential random variable. At any point in time, the exponential random variable forgets or ignores what happened earlier: "The future is independent of the past."

TRY IT NOW Go to Exercises 6.109 and 6.113

Section 6.4 Exercises

Concept Check

6.96 True or False For any exponential random variable, the mean is equal to the standard deviation.

6.97 True or False An exponential random variable can take on only values greater than or equal to 0.

6.98 Short Answer Suppose X is an exponential random variable with parameter λ.
(a) $P(X \leq x) =$
(b) $P(X > x) =$

6.99 Short Answer Give two examples of populations that an exponential distribution might be used to model.

Practice

6.100 Suppose X is an exponential random variable with $\lambda = 0.2$.
(a) Carefully sketch a graph of the density curve for X.
(b) Find the mean, variance, and standard deviation of X.
(c) Find $P(X < 3)$.
(d) Find $P(1 < X < 9)$.

6.101 Suppose X is an exponential random variable with $\lambda = 0.25$.
(a) Carefully sketch a graph of the density curve for X.
(b) Find $P(0.01 \leq X \leq 0.05)$.
(c) Find $P(X > 0.06)$.

6.102 Suppose X is an exponential random variable with $\lambda = 0.025$.
(a) Find $P(X > 30)$ and illustrate this probability using an appropriate density curve.
(b) Find $P(X > 20)$.
(c) Find $P(X > 50 \mid X \geq 30)$.

6.103 Suppose X is an exponential random variable and $P(X \leq 20) = 0.7981$. Find the value of λ.

6.104 Suppose X_i, for $i = 1, 2, 3, 4$, is an exponential random variable with $\lambda = i/10$. If the X_i's are independent, find the probability that the values of all four random variables are less than 1.

Applications

6.105 Public Health and Nutrition Heavy cream is used in pasta sauce, soup, and biscuits and is whipped into a topping for cakes and pies. According to food experts, heavy cream holds up in the freezer, on average, for approximately 3 months. Suppose the distribution of time (in months) until a random carton of heavy cream stored in the freezer spoils is an exponential random variable with $\lambda = 0.325$, and a random container of heavy cream stored in the freezer is selected at random.
(a) Find the probability that the heavy cream spoils in less than 4 months.
(b) Find the probability that the heavy cream spoils in between 2 and 3 months.
(c) If the heavy cream has been in the freezer for more than 5 months, most chefs will assume it has spoiled. What is the probability that the cream is still good?

6.106 Fuel Consumption and Cars The purpose of an automobile's timing belt is to provide a connection between the camshaft and the crankshaft. This allows the valves to open and close in sync with the pistons. Suppose the lifetime of a timing belt (in miles, mi) can be modeled by an exponential random variable with mean 100,000 mi.
(a) Find the value of λ.
(b) Find the probability that a randomly selected timing belt lasts for more than 120,000 mi.
(c) Find the probability that a randomly selected timing belt lasts for between 75,000 and 125,000 mi.
(d) If the timing belt on a new car breaks within 70,000 mi, the dealer will install a new belt free of charge. What is the probability the dealer will be forced to install a new timing belt free of charge on a randomly selected car?

6.107 Business and Management Suppose the dentists at You Can't Handle the Tooth in Nashua, New Hampshire, never see a patient on time. The amount of time (in minutes) past the appointment time that a random patient must wait to see a dentist has an exponential distribution with $\lambda = 0.01$. Suppose a patient is selected at random, and the appointment time has passed.
(a) What is the probability that the dentist will see the patient within 5 minutes after the appointment time?
(b) Find the probability that the dentist will see the patient between 10 and 20 minutes after the appointment time.
(c) Find the probability that the patient will have to wait at least 30 minutes past the appointment time.

6.108 Psychology and Human Behavior In Columbia, South Carolina, on a randomly selected Friday night between 9:00 P.M. and 2:00 A.M., the time (in minutes) between calls to a 911 dispatcher has an exponential distribution with $\lambda = 0.125$. Suppose a Friday evening in Columbia is selected at random.
(a) If a 911 call comes in at 10:07 P.M., what is the probability that the next call will occur before 10:30 P.M.?

(b) If a 911 call comes in at 11:30 P.M., what is the probability that the next call will occur after 11:45 P.M.?

(c) Find the mean, variance, and standard deviation for the time between 911 calls.

(d) Find a value t such that if a 911 call is received between 9:00 P.M. and 2:00 A.M., then the probability that the next 911 call will occur more than t minutes later is 0.75.

6.109 Public Policy and Political Science During the Cold War, there were frequent radio tests of alert warnings. The regular program was interrupted and replaced by a long high-pitched signal. An announcer would break in with a message similar to "This has been a test of the emergency broadcasting system." Suppose the time (in hours) between these tests had an exponential distribution with mean 24 hours.

(a) If a test occurred at 6:00 A.M., what is the probability that another one would occur before 6:00 P.M.?

(b) If a test occurred at 10:00 P.M., what is the probability that another one would not occur until after 6:00 A.M.?

(c) If a test occurred at 9:00 A.M., what is the probability that the next test would occur between 12:00 P.M. and 1:00 P.M.?

6.110 Manufacturing and Product Development L.L. Bean recently changed its lifetime product guarantee to allow customers to return purchases for only one year from the date of purchase, with a receipt. Some customers had been taking advantage of the lifetime guarantee and were returning heavily worn products many years, and even decades, after purchase.[30] L.L. Bean outdoor gear is designed to be durable and long lasting, and the Bean Boot is one of the company's signature products. Suppose the lifetime (in years) of a Bean Boot can be modeled by an exponential random variable X with parameter λ.

(a) Suppose $P(X > 5) = 0.05$. Find the value of λ.

(b) Find the probability that a randomly selected Bean Boot will wear out in less than 1 year.

(c) Suppose a family of four avid hikers all get Bean Boots. What is the probability that all four pairs of boots will last at least three years?

(d) Find the probability that a randomly selected Bean Boot lasts for at least a decade.

Extended Applications

6.111 Public Health and Nutrition The U.S. Public Health Service's Advisory Committee on Immunization Practices recommends a tetanus booster every 10 years. Suppose the lifetime (in years) of a tetanus booster shot has an exponential distribution with $\lambda = 0.05$.

(a) Find the probability that a randomly selected tetanus booster shot is still protecting against tetanus after more than 10 years.

(b) Suppose a physician recommends a booster shot after a period when only 10% of all tetanus shots still have an effect. How long should you wait before getting another booster shot?

(c) Suppose two people independently receive tetanus shots. What is the probability that both shots still have an effect after more than five years?

6.112 Medicine and Clinical Studies A patient's blood cholesterol level is often checked during a routine physical examination. A blood sample is taken (from the finger or arm) and tested for total cholesterol and HDL (high-density lipoprotein, the "good cholesterol") cholesterol levels, measured in milligrams per deciliter (mg/dL). If total cholesterol is less than 200 mg/dL, then no action is taken. If a patient's total cholesterol is approximately 300 mg/dL, a new drug, together with a strict diet, is prescribed to reduce this number to a safe level. The amount of time (in days) it takes to reduce total cholesterol to a safe level has an exponential distribution with $\lambda = 1/15$. A patient with total cholesterol of 300 mg/dL is randomly selected and placed on this drug-and-diet regimen.

(a) Find the mean number of days until the total cholesterol level is safe (less than 200 mg/dL).

(b) Find the probability that the total cholesterol level will be safe in a number of days within 2 standard deviations of the mean.

(c) Find a value d such that 75% of all such patients have a safe cholesterol level within d days.

(d) Suppose two patients are selected at random. Find the probability that this drug-and-diet regimen works for at least one of the patients within 10 days.

6.113 Marketing and Consumer Behavior A toy manufacturer routinely tests new toys in a controlled environment before deciding whether to actually market a toy. Research has shown that the amount of time (in minutes) a randomly selected child plays with a new toy has an exponential distribution with $\lambda = 0.05$. Suppose a new toy is presented for study.

(a) What is the probability that a randomly selected child will play with the toy for at most 10 minutes?

(b) What is the probability that a randomly selected child will play with the toy for between 5 and 20 minutes?

(c) Suppose a child plays with the toy for at least 15 minutes. What is the probability that the child will play with the toy for another 20 minutes?

(d) If four different children each play independently with a new toy for at least 25 minutes, then the toy is immediately brought to market. What is the probability of this happening for a newly designed toy?

6.114 Physical Sciences Four large water pumps supply Bellingham, Washington, with water. If water pump i, for $i = 1, 2, 3, 4$, breaks down, then the time to repair it has an exponential distribution with $\lambda = 1/(2i)$ hours. Suppose the water pumps operate independently, and when breakdowns occur, the repair times are also independent.

(a) Suppose water pump 1 breaks down. What is the probability that it will take more than 30 minutes to repair?

(b) Answer part (a) for water pumps 2, 3, and 4.

(c) Suppose all four water pumps break simultaneously. What is the probability that at least one of the four water pumps will not be repaired within one hour?

6.115 Public Health and Nutrition From the instant a fresh-baked chocolate-chip cookie is taken out of the oven, the time (in minutes) the wonderful aroma lasts has an exponential distribution with $\lambda = 1/30$. Suppose a chocolate-chip cookie is done baking and is taken out of the oven.

(a) Find the mean, variance, and standard deviation for the time the aroma lasts.

(b) What is the probability that the aroma lasts for at least 40 minutes?
(c) What is the probability that the aroma lasts for between 30 and 50 minutes?
(d) Find a value t such that the probability that the aroma lasts for at most t minutes is 0.90.
(e) Suppose a second batch of cookies is taken out of the oven 10 minutes after the first. What is the probability that there will be no aroma from either batch 35 minutes after the first batch was done?

6.116 Psychology and Human Behavior Sensory memory in humans is very short-term and lasts approximately 200–500 milliseconds after an individual perceives an object.[31] Suppose the length of time (in milliseconds) an object remains in sensory memory for a randomly selected adult has an exponential distribution with $\lambda = 0.003$. A flashing grid of letters is displayed to a randomly selected adult and the sensory memory is measured.
(a) What is the probability that the sensory memory will last for at most 200 milliseconds?
(b) Find the median time that the sensory memory will last.
(c) Find a time t such that only 5% of all adults' sensory memory lasts at least t milliseconds.

6.117 Manufacturing and Product Development The Aurora LED Smart Lighting System consists of nine interlocking panels and is sold with a generous 3-year free replacement warranty. The length of time, in years, until a randomly selected LED panel needs to be replaced can be modeled by an exponential random variable X with parameter λ.
(a) Find the smallest value of λ such that $P(4 \leq X \leq 5) = 0.03$.
(b) Find the probability that a randomly selected panel needs to be replaced within 3 years.
(c) Assume the LED panels operate independently. Find the probability that at least one panel needs to be replaced within 3 years.

Challenge Problems

6.118 Memoryless Property Suppose X is an exponential random variable with parameter λ. If a and b are constants (> 0), confirm the memoryless property by showing that $P(X \geq a) = P(X \geq a+b \mid X \geq b)$.

Chapter 6 Summary

Concept	Page	Notation / Formula / Description
Probability distribution for a continuous random variable	248	A smooth curve (density curve) defined such that the probability X takes on a value between a and b is the area under the curve between a and b.
Uniform distribution	250	If X has a uniform distribution on the interval $[a,b]$, then $$f(x) = \begin{cases} \frac{1}{b-a} & \text{if } a \leq x \leq b \\ 0 & \text{otherwise} \end{cases} \quad \mu = \frac{a+b}{2}, \quad \sigma^2 = \frac{(b-a)^2}{12}$$
Normal distribution	262	If X is a normal random variable with mean μ and variance σ^2, then the probability density function is given by $f(x) = \frac{1}{\sigma\sqrt{2\pi}} e^{-(x-\mu)^2/(2\sigma^2)}$
Standard normal distribution	263	A normal distribution, with $\mu = 0$ and $\sigma^2 = 1$ is the standard normal distribution. The standard normal random variable is usually denoted by Z: $Z \sim N(0,1)$.
Standardization	267	If X is a normal random variable with mean μ and variance σ^2, then $Z = \frac{X-\mu}{\sigma}$ is a standard normal random variable.
Normal probability plot	279	A scatter plot of each observation versus its corresponding expected value from a Z distribution. For a normal distribution, the points will fall close to a straight line.
Exponential distribution	287	If X is an exponential random variable with parameter λ, then the probability density function is given by $$f(x) = \begin{cases} \lambda e^{-\lambda x} & \text{if } x \geq 0 \\ 0 & \text{otherwise} \end{cases} \quad \mu = \frac{1}{\lambda}, \quad \sigma^2 = \frac{1}{\lambda^2}$$

Chapter 6 Exercises

Applications

6.119 Technology and Internet In the Gensler Architectural firm, all of the computers are part of a local area network, connected to one main printer. When an architect prints a document, the job is placed in a queue. It may take several minutes before the document begins to print, due to the number of other print jobs and the complexity of the document. The time (in minutes) until the document starts to print has an exponential distribution with $\lambda = 0.40$. Suppose a randomly selected document is sent to the main printer.
 (a) What is the mean time until the document begins to print?
 (b) What is the probability that the document will begin to print within 30 seconds?
 (c) What is the probability that the document will need more than 5 minutes before starting to print?
 (d) Find a value t such that only 2% of all documents take at least t minutes before starting to print.

6.120 Public Health and Nutrition The amount of sodium in a randomly selected 8-oz serving of chicken noodle soup has a normal distribution, with mean $\mu = 343$ mg and standard deviation $\sigma = 21$.[32] Suppose an 8-oz serving is randomly selected.
 (a) What is the probability that the amount of sodium is more than 325 mg?
 (b) What is the probability that the amount of sodium is between 330 and 360 mg?
 (c) Find a symmetric interval about the mean such that 90% of all 8-oz servings have amounts of sodium in that interval.
 (d) Suppose a randomly selected 8-oz serving has a sodium level of 390 mg. Is there any evidence to suggest the mean sodium level reported in the question description is false?

6.121 Public Health and Nutrition The Food and Drug Administration (FDA) reviews all advertisements for drugs to check for omissions regarding a drug's risk; inadequate, incorrect, or inconsistent labeling; misleading claims; unsupported comparative claims; and unapproved purposes. Lengthy legal reviews of advertisements have increased the total review time. Suppose the length of time between a request for a review and final approval has a normal distribution, with mean $\mu = 21$ days and variance $\sigma = 4$ days. Consider a randomly selected advertisement submitted to the FDA for review.
 (a) What is the probability that the advertisement will be reviewed in less than 14 days?
 (b) What is the probability that the advertisement will be reviewed in between 15 and 19 days?
 (c) Suppose the advertisement takes at least 20 days for review. What is the probability that it will take less than 30 days for review?
 (d) Suppose two independent advertisements are submitted for review simultaneously. What is the probability that both will take more than 30 days for review?

6.122 Marketing and Consumer Behavior Canister vacuums are often rated according to their ease of use, noise level, emissions, cleaning ability, and length of the power cord. The length (in feet) of the electric cord on a canister vacuum is a random variable X with a uniform distribution on the interval [20, 30]. Suppose a canister vacuum is selected at random.
 (a) Carefully sketch a graph of the probability density function for the random variable X.
 (b) What is the probability that the power cord is shorter than 22 ft?
 (c) What is the probability that the power cord is longer than 26 ft?
 (d) Find a value f such that 75% of all power cords are longer than f ft.

6.123 Physical Sciences A random sample of the weekly U.S. field production of crude oil, in thousands of barrels per day, is given on the website.[33] Is there any evidence to suggest that the data are from a non-normal population? Justify your answer. CRUDE

6.124 Business and Management The manager for Marriott's Crystal Shores claims that the time it takes to make a room reservation over the phone is approximately normally distributed, with mean 4 minutes and standard deviation 45 seconds. Suppose a call placed to the inn to make a room reservation is selected at random.
 (a) What is the probability that it will take less than 3 minutes to make the reservation?
 (b) What is the probability that it will take between 3 1/2 and 4 1/2 minutes to make the reservation?
 (c) Find the first and the third quartile times.
 (d) Suppose it takes 7 minutes to make the reservation. Is there any evidence to suggest the manager's claim ($\mu = 4$ minutes) is false? Justify your answer.

6.125 Fuel Consumption and Cars The Tesla Model X is an electric car that can travel approximately 238–295 mi on a full charge.[34] Tesla dealers claim the time it takes to fully recharge depends on the percent depleted, but is approximately normal with mean 2.5 hours and standard deviation 0.75 hour (on a 240-volt line). Suppose a random Tesla Model X is connected to a home recharging station.
 (a) What is the probability that the amount of time to fully recharge is between 1 and 2 hours?
 (b) What is the probability that the amount of time to fully recharge is more than 4.5 hours?
 (c) If the amount of time to fully recharge is less than 30 minutes, the car could still travel another 100 mi. What is the probability that the car could travel another 100 mi?
 (d) Suppose three Tesla Model X cars are selected at random. What is the probability that all three will have a time to fully recharge greater than 3 hours?
 (e) Suppose it takes 3 1/2 hours for a random Tesla Model X to recharge. Is there any evidence to suggest that the claim is false? Justify your answer.

6.126 Manufacturing and Product Development Oriental rugs are made from various wools and woven in several different countries. The pile height (in millimeters, mm) of an Oriental rug varies slightly and can be modeled by a uniform random variable on the interval [6, 10].
(a) Carefully sketch a graph of the probability density function for pile height.
(b) What is the probability that a randomly selected Oriental rug will have pile height less than 7 mm?
(c) What is the probability that a randomly selected Oriental rug will have pile height between 8.5 and 9.5 mm?
(d) Find a value h such that 90% of all Oriental rugs have pile height less than h.

6.127 Technology and Internet The life expectancy of a laser-printer toner cartridge varies considerably, according to the toner and drum type and how the cartridge is used. Printed pages containing lots of graphics require more toner, whereas pages with mostly text require considerably less toner. Page coverage is usually measured as a proportion (or percentage). For example, a typical text page has approximately 0.05 coverage. A random sample of printed pages from an office printer was obtained, and the page coverage was carefully measured. Is there any evidence to suggest the data are from a non-normal population? **TONER**

6.128 Sports and Leisure Jockeys in the United States and England work very hard to keep their weight down. Many participate in weight-loss programs, carefully monitor their diet, and exercise regularly. The weight of a male jockey is approximately normal, with mean 52 kg and standard deviation 1.2 kg. Suppose a male jockey is randomly selected.
(a) What is the probability that the jockey weighs more than 53 kg?
(b) What is the probability that the jockey weighs between 50 and 54 kg?
(c) Find a value w such that 80% of all male jockeys weigh more than w.
(d) Suppose the jockey selected weighs 57 kg. Is there any evidence to suggest the claimed mean (52 kg) is wrong? Justify your answer.

6.129 Public Health and Nutrition The amount of caffeine in a cup of coffee varies considerably, even if it is brewed by the same person, using the same brewing method and ingredients. Suppose the amount of caffeine in an 8-oz cup of Dunkin' coffee is approximately normally distributed, with mean 180 mg and standard deviation 15 mg.[35]
(a) What is the probability that a randomly selected cup of Dunkin' coffee will have less than 170 mg of caffeine?
(b) What is the probability that a randomly selected cup of Dunkin' coffee will have between 160 and 190 mg of caffeine?
(c) A recent article in a medical journal suggests that an 8-oz cup of coffee with more than 200 mg of caffeine could cause a person's heart to race. What is the probability that a randomly selected cup will have more than 200 mg of caffeine?
(d) Suppose a random cup of Dunkin' coffee has 157 mg of caffeine. Is there any evidence to suggest the claimed mean (180 mg) is wrong? Justify your answer.

6.130 Manufacturing and Product Development The quality of a kitchen knife is often measured by the sharpness and total lifetime of the blade. One test for sharpness involves mounting the knife with the blade vertical and lowering a specially designed pack of paper onto the blade. The sharpness is measured by the depth of the cut. A greater depth indicates a sharper knife. Suppose the depth of the cut for a randomly selected knife has a normal distribution, with mean 92 mm and standard deviation 21 mm.
(a) What is the probability that a randomly selected kitchen knife has a sharpness measure less than 75 mm?
(b) A kitchen knife with a sharpness measure of at least 100 mm qualifies as a steak knife. What proportion of kitchen knives are steak knives?
(c) The Kitchen Gadgets Association would like to set a maximum sharpness for butter knives. Find a value c such that 15% of all knives have sharpness less than c.
(d) Suppose a randomly selected kitchen knife has sharpness greater than 90 mm. What is the probability that it has sharpness greater than 100 mm?

6.131 Economics and Finance Some of the variables that affect the monthly payment of a new-car loan are the total amount borrowed, the interest rate, and the length of the loan. In 2018, the mean length of a new-car loan was 68 months, with loans of 72 and 84 months becoming more common.[36] Suppose the length of a new-car loan is approximately normal, with standard deviation 9 months.
(a) What is the probability that a new-car loan is for at most 60 months?
(b) What is the probability that a new-car loan length is between 50 and 70 months?
(c) Find a symmetric interval about the mean such that 95% of all new-car loan lengths fall in this interval.
(d) Suppose the amount borrowed on a new-car loan is also approximately normal, with mean $20,000 and standard deviation $5000. If the length of the loan and the amount borrowed are independent, what is the probability that the loan will be for more than $27,000 and for less than 72 months?

6.132 Fuel Consumption and Cars Even though automobiles are becoming more fuel-efficient, the cost of gasoline is still taking a huge chunk of our personal income. According to GasBuddy, the mean amount spent on gasoline in 2018 was $1900. A random sample of U.S. households was obtained, and each was asked for the amount (in dollars) spent on gasoline during the last year. Is there any evidence to suggest that the data are from a non-normal distribution? **GAS**

6.133 Physical Sciences Large reservoirs of oil found underground are under very high pressure, which allows the oil to be pumped to the surface. All oil fields contain some water, and as water is pumped back into a well to maintain high pressure, the water content increases. Suppose the proportion of water in a randomly selected barrel of oil pumped to the surface has a normal distribution, with mean 0.12 and standard deviation 0.025.
(a) What is the probability that a randomly selected barrel of oil has a proportion of water less than 0.12?

(b) What is the probability that a randomly selected barrel of oil has a proportion of water between 0.15 and 0.17?

(c) The higher the proportion of water in oil, the more expensive it is to separate the oil from the water. If the proportion of water is greater than 0.20, then the well is too expensive to operate and maintain. What is the probability that a randomly selected well is too expensive?

6.134 Travel and Transportation The Washington, D.C., Metro has some long and deep escalators. The longest continuous escalator is at the Wheaton (Red Line) stop. It is 230 ft long and approximately 140 ft deep.[37] For those who simply ride the escalator (without extra steps), it takes a little more than 2.5 minutes to complete the journey. However, because many people walk while on the escalator, the amount of time to travel the 230 ft is approximately normal, with mean 1.9 minutes and standard deviation 0.3 minute. Suppose an escalator rider is selected at random.

(a) What is the probability that it takes less than 2 minutes to ride the escalator?

(b) What is the probability that it takes between 1.5 and 2.5 minutes to ride the escalator?

(c) Suppose that five escalator riders are selected at random. What is the probability that three of the five take at least 2 minutes, 20 seconds to ride the escalator?

(d) The D.C. Metro Maintenance Crew claims to have increased the speed of the Wheaton escalator. Following this adjustment, the amount of time for a random escalator rider was 1.75 minutes. Is there any evidence to suggest that the mean time to ride the escalator has decreased? Justify your answer.

Extended Applications

6.135 Travel and Transportation The U.S. Customs and Border Protection Agency maintains a table of the estimated wait times for reaching the primary inspection booth when crossing the Mexican/U.S. border. The processing goal for passenger vehicles is 15 minutes.[38] Suppose the time to cross the border at El Paso, Texas, is a random variable X with probability density function given by

$$f(x) = \begin{cases} 0.02x & \text{if } 0 \leq x \leq 10 \\ 0 & \text{otherwise} \end{cases}$$

(a) Carefully sketch a graph of the probability density function.

(b) Find the probability that it takes less than 5 minutes for a randomly selected passenger vehicle to cross the border at El Paso.

(c) Find the probability that it takes more than 8 minutes for a randomly selected passenger vehicle to cross the border at El Paso.

(d) Find the probability that it takes between 2 and 6 minutes for a randomly selected passenger vehicle to cross the border at El Paso.

(e) Suppose it takes less than 2 minutes for a randomly selected passenger vehicle to cross the border at El Paso. What is the probability that it takes less than 1 minute?

6.136 Public Health and Nutrition Meat or poultry classified as lean has less than 4 g of saturated fat. Suppose the amount of saturated fat in a randomly selected piece of lean meat or poultry is a random variable X with probability density function given by

$$f(x) = \begin{cases} 0.1 & \text{if } 0 \leq x < 1 \\ 0.2 & \text{if } 1 \leq x < 2 \\ 0.3 & \text{if } 2 \leq x < 3 \\ 0.4 & \text{if } 3 \leq x < 4 \\ 0 & \text{otherwise} \end{cases}$$

(a) Carefully sketch a graph of the probability density function.

(b) What is the probability that a randomly selected piece of lean meat or poultry has less than 1.5 g of saturated fat?

(c) What is the probability that a randomly selected piece of lean meat or poultry has more than 3 g of saturated fat?

(d) What is the probability that a randomly selected piece of lean meat or poultry has between 2 and 4 g of saturated fat?

(e) Suppose a randomly selected piece of lean meat or poultry has at most 3 g of saturated fat. What is the probability that it has at most 1 g of saturated fat?

6.137 Manufacturing and Product Development Four people working independently on an assembly line all perform the same task. The time (in minutes) to complete this task for person i, for $i = 1, 2, 3, 4$, has a uniform distribution on the interval $[0, i]$. Suppose each person begins the task at the same time.

(a) What is the probability that person 2 takes less than 90 seconds to complete the task?

(b) What is the mean completion time for each person?

(c) What is the probability that all four people complete the task in less than 30 seconds?

(d) What is the probability that exactly one person completes the task in less than 1 minute?

6.138 Probability Calculations Using a Density Function The probability density function for a random variable X is given by

$$f(x) = \begin{cases} -\dfrac{1}{4}x + \dfrac{1}{2} & \text{if } 0 \leq x < 2 \\ -\dfrac{1}{4}x + 1 & \text{if } 2 \leq x < 4 \\ 0 & \text{otherwise} \end{cases}$$

(a) Carefully sketch a graph of the probability density function.

(b) Find $P(X < 1)$.

(c) Find $P(X \geq 3)$.

(d) Find $P(1 < X < 3)$.

6.139 Greek Yogurt Chobani yogurt is made using a special straining technique to remove excess liquid. This process yields a thicker, creamier yogurt with more protein per serving than a regular yogurt.[39] Suppose the amount of protein in a 5.3-oz

serving of Chobani Flip salted caramel crunch is normally distributed, with mean 12 g and standard deviation 0.7 g. A 5.3-oz serving is selected at random from the assembly line.
(a) What is the probability that there is more than 13.5 g of protein in the cup of yogurt?
(b) If the amount of protein in a cup of yogurt is between 11 and 13 g, the manufacturing process is considered to be in control. What is the probability that the process is considered in control? Out of control?
(c) Suppose the cup of yogurt has less than 12.5 g of protein. What is the probability that it has more than 11.5 g?
(d) Suppose that six cups of yogurt are selected at random. What is the probability that at least three cups have at least 11.5 g of protein?

6.140 Sports and Leisure During a Major League Baseball (MLB) game, there is a lot of time spent between batters, between innings, and between pitches. Consequently, the mean *action time* in a MLB game is only 18 minutes.[40] Suppose MLB action time can be modeled by a normal distribution, with standard deviation 1.2 minutes.
(a) Find the probability that a randomly selected MLB game has at most 17 minutes of action time.
(b) Find the probability that a randomly selected MLB game has between 16 and 19 minutes of action time.
(c) Suppose the mean action time μ for a MLB game is unknown, and the probability of five MLB games with action time less than 20 minutes is 0.5722. Find the value of μ.

simpson33/Deposit Photos

CHAPTER APP

6.141 EMV Chip Technology According to Bluefin, a payment security firm, the mean processing time for an EMV chip card is 13 seconds, approximately double the time needed for a swipe card. Suppose the standard deviation is 2.5 seconds and the distribution of EMV chip card processing times is approximately normal.
(a) Find the probability that a randomly selected EMV chip card processing time is less than 10 seconds.
(b) Suppose an EMV chip card transaction is selected at random and the processing time is 20 seconds. Is there any evidence to suggest that the mean processing time is greater than 13 seconds?
(c) A random sample of 20 EMV chip card processing times was obtained. Is there any evidence to suggest that this distribution is not normally distributed? **CHIP**
(d) The number c is the customer irritation index and is defined to be a value such that 1% of all EMV chip processing times take longer than c seconds. Find the value of c.

7 Sampling Distributions

◀ Looking Back
- Recall that parameters such as μ, σ^2, and p may completely characterize, or describe, a random variable and, therefore, a population: **Chapter 5 and Chapter 6**.
- These parameters, μ, σ^2, and p, are constant and usually unknown, but we would like to estimate these values, or draw a conclusion about these parameters: **Chapter 5 and Chapter 6**.

Looking Forward ▶
- Understand and utilize methods to find a sampling distribution: **Section 7.1**.
- Discover the Central Limit Theorem, investigate the distribution of $\overline{X}$, and solve probability and inference problems using this result: **Section 7.2**.
- Use the distribution of $\hat{P}$ to solve probability and inference problems: **Section 7.3**.

CHAPTER APP

Ras Al Khaimah/United Arab Emirates/Andrew Fosker/PinPep/WENN.com/Newscom

The Longest Zip Line

The Jebel Jais Flight is the world's longest zip line. Opened in 2018, it allows thrill seekers to jump off Jebel Jais, the highest mountain in the United Arab Emirates (UAE). The launch point is 1680 meters above sea level, with riders then dropping 400 meters; the total length of the zip line is 2831 meters, and the zip line cable weighs more than 6 tons.

There are no age restrictions for riders as they travel at speeds up to 150 kph. A spokesperson for the ride claimed that the mean ride time is 1.75 min with a standard deviation of 0.2 min.

A random sample of 40 Jebel Jais Flight riders was obtained. The time for each ride was recorded, and the sample mean was 1.80 min. This mean time was longer than expected, and could provide sufficient evidence that the claimed mean is wrong. The concepts presented in this chapter will allow us to find the distribution of the sample mean, to answer probability questions about this random variable, and to make inferences about a population mean—in this case, 1.75 min. These ideas all rely on the Central Limit Theorem, the most important result in probability and statistics.

7.1 Statistics, Parameters, and Sampling Distributions

Parameters and Statistics

The terms *parameter* and *statistic* have been used in previous chapters in intuitive contexts. The following definitions distinguish between measures represented by symbols such as μ, σ^2, and p, and the quantities used to estimate these values.

> **Definition**
> A **parameter** is a numerical descriptive measure of a population.
> A **statistic** is any quantity computed from values in a sample.

A CLOSER LOOK

1. A parameter is a population quantity. It is used to describe some characteristic of a population. Usually we cannot measure a parameter; it is an unknown constant that we would like to estimate.
2. A statistic is any sample quantity. There are infinitely many quantities one could compute using the data in a sample. For example, $\bar{x}$ and $\tilde{x}$ are statistics, as is the sum of the smallest and the largest values divided by 2.

EXAMPLE 7.1 Parameter Versus Statistic

In each of the following statements, identify the **boldface** number as the value of a population parameter or a sample statistic.

(a) In a recent survey of young adults, **66%** of millennials said they have "always believed the world is round."

(b) A spokesperson for Google reported that the proportion of all people working for the company who are women is **0.31**.

(c) The U.S. Department of Transportation recently reported that the mean age of all highway bridges in the United States is **42** years.

(d) The manager of a large hotel located near Disney World indicated that 20 selected guests had a mean length of stay equal to **5.6** days.

(e) In Canada's World Survey 2018, **21%** of those who responded said that the environment was the most important world issue.

Solution

(a) 66% is a statistic. This number describes a characteristic of a sample of millennials.

(b) 0.31 is a parameter. This number describes a characteristic of the entire population of Google employees.

(c) 42 (years) is a parameter. This number is a characteristic of all the highway bridges in the United States.

(d) 5.6 (days) is a statistic. This number describes a characteristic of the sample of 20 guests.

(e) 21% is a statistic. This number describes a characteristic of the sample of individuals who responded to the survey.

TRY IT NOW Go to Exercises 7.11 and 7.12

CHAPTER APP **The Longest Zip Line**

The mean ride time is 1.75 min with standard deviation 0.2 min. Are these values parameters or statistics?

Sampling Distribution of a Statistic

Suppose the mean is computed for a sample of size n from a population of interest. Denote this value as $\bar{x}_1$. In a second sample of size n from the same population, let $\bar{x}_2$ be the mean. It is reasonable to expect $\bar{x}_1$ and $\bar{x}_2$ to be close to each other but not necessarily equal. The important realization is that $\bar{x}_1$ will be different from $\bar{x}_2$. In fact, the sample mean will most likely differ from sample to sample. This statistic is subject to sampling variability. Therefore, the sample mean, $\bar{X}$, is a random variable, and has a mean, a variance, a standard deviation, and a probability distribution. This distribution is called a sampling distribution.

Statistics are random variables!

Any statistic is a random variable because the value of the statistic differs from sample to sample. One cannot predict the value of a statistic with absolute certainty. To make a reliable inference based on a specific statistic, we need to know the properties of the distribution of the statistic.

Definition
The **sampling distribution** of a statistic is the probability distribution of the statistic taken from all possible random samples of a specific size (n).

 A CLOSER LOOK

1. A sampling distribution (like any random variable) describes the long-run behavior of the statistic.

2. Here is a technique to approximate a sampling distribution and a method for finding the exact sampling distribution.

(a) Recall that to approximate the distribution of a population, we construct a histogram (or stem-and-leaf plot) using values from the population. If the sample is representative, then the histogram should be similar in shape, center, and spread to the population distribution.

Similarly, to approximate the distribution of a statistic, we obtain (many) values of the statistic and construct a histogram. The resulting graph approximates the sampling distribution of the statistic.

For example, to approximate the sampling distribution of the mean of a sample of size $n = 10$ from a population: (i) obtain several samples of size 10 from the population; (ii) compute the sample mean for each sample; (iii) construct a histogram using all the sample means. The histogram approximates the sampling distribution of the sample mean.

(b) In some cases, we can obtain the exact sampling distribution of a statistic. If the statistic is a discrete random variable, the sampling distribution includes all the values that the statistic assumes and the associated probabilities. If the statistic is a continuous random variable, the sampling distribution consists of a probability density curve.

Sampling Distribution Examples

EXAMPLE 7.2 Font Properties

Ogg is a popular font used by designers and calligraphers. This font looks like hand lettering and is seen on book covers, in magazines, and posters. Ogg is available in five font weights: 100, 123, 200, 300, and 321.

Suppose a sample of three Ogg font weights is selected (without replacement) and the sample median weight is computed. Find the sampling distribution for the sample median.

Solution

> The order of selection does not matter here. Therefore, this is a combination.

STEP 1 There are $\binom{5}{3} = 10$ ways to select three font weights from the five in the population. List all the possible samples, the computed value of the statistic for each sample, and the probability of selecting each sample. Summarize the resulting table to construct the sample distribution.

STEP 2 Here is the table showing each sample, the sample median, and the probability of obtaining each sample.

Sample	$\tilde{x}$	Probability	Sample	$\tilde{x}$	Probability
100, 123, 200	123	0.1	100, 300, 321	300	0.1
100, 123, 300	123	0.1	123, 200, 300	200	0.1
100, 123, 321	123	0.1	123, 200, 321	200	0.1
100, 200, 300	200	0.1	123, 300, 321	300	0.1
100, 200, 321	200	0.1	200, 300, 321	300	0.1

The median for each sample is the middle value. The probability of each sample is $1/10 = 0.1$ because we assume that each sample is equally likely.

STEP 3 Of all the possible samples, only three values are possible for the sample median in this case. Sum the probabilities associated with each value. The probability distribution for the random variable $\tilde{X}$ lists the possible values (of $\tilde{X}$) and the associated probabilities.

$\tilde{x}$	123	200	300
$p(\tilde{x})$	0.3	0.4	0.3

TRY IT NOW Go to Exercise 7.17

A CLOSER LOOK

1. In Example 7.2, $\tilde{X}$ is a discrete random variable. Using equations from Chapter 5, the mean, variance, and standard deviation for $\tilde{X}$ are $\mu = 206.9$, $\sigma^2 = 4731.09$, and $\sigma = 68.78$.

2. The font weights were selected without replacement. Suppose three font weights are selected with replacement. That is, select a font weight, record the value, and place it back in the collection. The same font weight could be selected two or three times. This sampling scheme changes the probability distribution for $\tilde{X}$.

✓ Find the sampling distribution for $\tilde{X}$ if the sampling is done with replacement. *Hint:* There are $5 \times 5 \times 5 = 125$ possible samples.

EXAMPLE 7.3   Sick Days

Employees at Cerner and Hallmark Cards in Kansas City are allowed up to three sick days during a calendar year. Suppose the probability distribution for X, the number of sick days used by an employee during a year, is given in the following table.

x	0	1	2	3
$p(x)$	0.40	0.35	0.20	0.05

Two employees are independently selected at random, and the number of sick days is recorded for each. Consider the statistic M, the maximum number of sick days taken by either employee. Find the probability distribution for M.

Solution

STEP 1 A sample consists of two numbers: The first represents the sick days for employee 1, and the second denotes the sick days for employee 2. There are 16 possible samples. Using the Multiplication Rule, there are two slots to fill: $4 \times 4 = 16$. List all the possible samples, the computed value of the statistic for each sample, and the probability of selecting each sample.

STEP 2 The probability associated with each sample is computed using the independence assumption. For example, the probability that the first employee used one sick day and the second employee used two sick days is given by

$P(X = 1 \cap X = 2)$
$= P(X=1) \cdot P(X=2)$ *Independent events.*
$= (0.35)(0.20) = 0.07$ *Use the probability distribution for X.*

STEP 3 Construct a table listing each sample, the sample maximum, and the probability associated with the sample.

Sample	m	Probability	Sample	m	Probability
0, 0	0	0.1600	2, 0	2	0.0800
0, 1	1	0.1400	2, 1	2	0.0700
0, 2	2	0.0800	2, 2	2	0.0400
0, 3	3	0.0200	2, 3	3	0.0100
1, 0	1	0.1400	3, 0	3	0.0200
1, 1	1	0.1225	3, 1	3	0.0175
1, 2	2	0.0700	3, 2	3	0.0100
1, 3	3	0.0175	3, 3	3	0.0025

The maximum for each sample is the largest of the two values.

STEP 4 There are four possible values for the discrete random variable M. Sum the probabilities associated with each value. The probability distribution is given in the table.

m	0	1	2	3
$p(m)$	0.1600	0.4025	0.3400	0.0975

> Recall that a capital letter, such as M, represents a random variable. The corresponding lowercase letter m denotes a specific value that the random variable can assume.

> ✓ Find the mean, variance, and standard deviation for the random variable M.

TRY IT NOW   Go to Exercises 7.15 and 7.19

Random Samples

In almost all observational studies, it is assumed that the data are obtained from a simple random sample. Usually, the sampling is done without replacement. Consider an exit poll or a study of the time spent each week on lawn care by homeowners. An individual is selected from the population and an observation is recorded. The individual is not placed back into the population. There is no chance the individual will be selected again.

If sampling is done without replacement, individual responses are dependent. However, if the population is large enough and the sample is small relative to the size of the population, then the responses are *almost* independent. Calculating probabilities with an assumption of independence results in little loss of accuracy. As a rule of thumb, if the sample size is at most 5% or 10% of the total population, then successive observations can be considered independent. Even though sampling is done without replacement, the data are assumed to be part of a simple random sample.

Recall the following definition from Chapter 1.

Definition
A **(simple) random sample** (SRS) of size n is a sample selected in such a way that every possible sample of size n has the same chance of being selected.

A CLOSER LOOK

1. Suppose the population is finite, of size N, and the sample is of size n. The number of possible simple random samples is $\binom{N}{n}$.

2. A random sample consists of individuals or objects, and a variable is a characteristic of an individual or object. A value of the variable is obtained for each member of the random sample. We often refer to the values of the variable as the random sample rather than the individuals in the sample. For example, consider a study in which the amount of the trace element chromium is measured in coal from around the world. The random sample consists of pieces of coal (objects) and the values are the chromium measurements. However, it is common practice to say, "Consider the random sample of chromium measurements."

3. Unless stated otherwise, all data presented in this text are obtained from a simple random sample.

EXAMPLE 7.4 Corn Maturity

CORN

The table lists the number of days to maturity for the entire population of 22 different varieties of corn sold by Burpee.[1]

| 62 | 75 | 67 | 72 | 63 | 92 | 72 | 89 | 75 | 30 | 71 | 85 |
| 80 | 70 | 77 | 92 | 72 | 75 | 77 | 100 | 78 | 80 | 72 | 100 |

Note: The population mean is the sum of all the observations divided by $n = 24$: $\mu = 76.08$ days.

Find an approximate sampling distribution for the sample mean of five observations from this population.

Solution

STEP 1 There are $\binom{24}{5} = 42{,}504$ possible samples of size 5. Instead of considering every one of these samples, select some (say, 100) samples of size 5, compute the mean for each sample, and construct a histogram of the sample means.

STEP 2 The table lists the first few samples of size 5 and the mean for each sample.

Sample					$\bar{x}$	Sample					$\bar{x}$
67	92	80	92	75	81.2	100	92	75	72	62	80.2
70	72	75	72	89	75.6	100	75	100	85	72	86.4
30	72	100	77	100	75.8	72	100	89	100	70	86.2
⋮					⋮	⋮					⋮

STEP 3 **Figure 7.1** shows a histogram of the sample means for 100 samples of size 5.

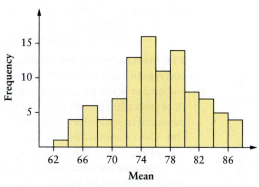

Figure 7.1 Histogram of the sample means.

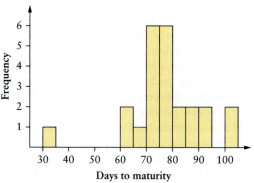

Figure 7.2 Histogram of the original population.

STEP 4 There are some very interesting, curious results here. Even though the population is not normally distributed (the population is finite; also consider the graph of the original population of corn varieties in **Figure 7.2**), the shape of the sampling distribution appears to be approximately normal! In addition, the center of the sampling distribution of the mean is approximately the population mean ($\mu = 76.08$ days). Although the relationship between the original population parameters and the sampling distribution parameters is not clear, there is certainly less variability in the sampling distribution than in the population distribution ($\sigma = 13.91$). These three observations suggest the exact distribution of the sample mean, as discussed in the next section.

Section 7.1 Exercises

Concept Check

7.1 Fill in the Blank A parameter describes a _____. A statistic describes a _____.

7.2 True or False Statistics are used to estimate parameters.

7.3 True or False Any statistic is a random variable.

7.4 True or False An exact sampling distribution can never be obtained.

7.5 Short Answer Describe a method to approximate a sampling distribution.

7.6 Fill in the Blank If sampling is done without replacement, individual responses are _____.

7.7 Short Answer Suppose a population is finite, of size N, and the sample is of size n. How many simple random samples are possible?

7.8 Short Answer Suppose two sample means of size n, $\bar{x}_1$ and $\bar{x}_2$, are obtained from the same population. Describe the relationship between $\bar{x}_1$ and $\bar{x}_2$.

7.9 Short Answer Describe three possible relationships between the distribution of the sample mean and the distribution of the original population. *Hint:* Describe their shape, center, and variability.

Applications

7.10 Parameter Versus Statistic In each of the following statements, identify the **boldface** number as the value of a population parameter or a sample statistic.
(a) A researcher in Boston conducted a study to investigate the prevalence of Alzheimer's disease in people older than the age of 85. The results indicated that **47.2%** of the patients studied experienced symptoms consistent with the disease.
(b) After an extensive audit, the Metropolitan Transportation Authority reported that **90%** of the speedometers on the D line do not work.
(c) A consumer magazine tested a random sample of 19 green teas for taste and health benefits. The mean cost per cup of tea was reported to be **14** cents.
(d) A computer manufacturer claims that the mean lifetime of laptop batteries is **6.7** hours.
(e) A random sample of people visiting the San Diego Zoo was obtained and the mean time spent at the zoo was reported to be **3.54** hours.

7.11 Parameter Versus Statistic In each of the following statements, identify the **boldface** number as the value of a population parameter or a sample statistic.
(a) A half-ton pickup truck is designed to safely carry a maximum of 1000 pounds. The Department of Transportation is concerned that drivers are hauling much heavier loads. The State Police randomly stopped 50 pickup trucks on an interstate highway and carefully weighed the contents in the truck bed. The mean weight was **1037** pounds.
(b) A toy manufacturer issued a recall for a Spin Art Kit because the battery compartment in the device could overheat and cause a burn hazard. The proportion of buyers who took advantage of the recall was reported to be **0.65**.
(c) A recent study indicated that more people today are rising before 6:00 A.M. each weekday to prepare for work and to help children get ready for school. In a random sample of 500 adults, **42%** said they get up each weekday before 6:00 A.M.
(d) In a random sample of dentists, **80%** recommended a certain product to help whiten teeth.
(e) During a recent winter, the month of January was particularly cold. A power company in Pennsylvania reported that the mean number of kilowatts used by each customer during January was **1346**.

7.12 Parameter Versus Statistic In each of the following statements, identify the **boldface** number as the value of a population parameter or a sample statistic.
(a) The maker of a new, enhanced water product conducted a survey to determine if people are drinking enough liquids each day. In a random sample of Americans, **90%** were chronically dehydrated.
(b) Last year, **62%** of all home robberies in Washington, D.C., took place between 10:00 A.M. and 3:00 P.M.
(c) NASA reported that the mean time for all spacewalks at the International Space Station was **45** minutes.
(d) The International Energy Agency reported that **93%** of Canada's renewable energy comes from hydroelectric sources.
(e) In a random sample of publishing companies, **23%** of sales were e-books.

7.13 Sports and Leisure Dick's Sporting Goods sells five different exercise balls with weights (in kilograms) given in the table. **EXBALL**

| 10 | 12 | 15 | 18 | 25 |

(a) Find the population mean and the population median.
(b) Suppose a random sample of size 3 is selected from this population without replacement. Find the sampling distribution of the sample mean. Find the mean, variance, and standard deviation for the sample mean.
(c) Suppose a random sample of size 3 is selected from this population without replacement. Find the sampling distribution of the sample median. Find the mean, variance, and standard deviation for the sample median.
(d) Compare the mean of the sample mean with the population mean. Compare the mean of the sample median with the population median.

7.14 Manufacturing and Product Development The coverage of a gallon of paint depends on the surface, the type and quality of the paint, and the applicator. A local hardware store stocks 30 different paints, and the coverage for each is measured in square feet. **PAINT**
(a) Use a computer or calculator to draw 50 random samples of size 5 from this population without replacement, and compute the mean for each sample.
(b) Construct a histogram of the sample means.
(c) Use the histogram to approximate the sampling distribution of the mean. What is the approximate shape of the distribution? What is the approximate value of the mean of the sample mean?
(d) Find the population mean. How does this compare with the approximate mean of the sampling distribution?

7.15 Demographics and Population Statistics For planning purposes, General Electric has determined the probability distribution for the retirement age X of employees in the mortgage work group. The probability distribution for X is given in the table.

x	64	65	66
$p(x)$	0.1	0.7	0.2

(a) Find the mean of X.
(b) Suppose two employees from this work group are selected at random. Find the exact probability distribution for the sample mean, $\overline{X}$.
(c) Find the mean of $\overline{X}$. How does this compare with your answer in part (a)?

7.16 Public Health and Nutrition The U.S. Food and Drug Administration regulates the use of certain terms used on food labels, such as reduced fat, low fat, and light. Lean meat must contain less than 4.5 g of saturated fat per serving.[2] Suppose the probability distribution for the amount (in grams) of saturated fat in a randomly selected serving of lean meat is given in the table.

x	0	1	2	3	4
$p(x)$	0.50	0.25	0.10	0.10	0.05

(a) Suppose two servings of lean meat are selected at random. Find the exact probability distribution for the sample median amount of saturated fat $\tilde{X}$.
(b) Find the mean, variance, and standard deviation of $\tilde{X}$.

7.17 Biology and Environmental Science In July 2018, a jaguar escaped overnight from its regular habitat at the Audubon Zoo in New Orleans.[3] The animal was captured the next day and no one was injured. Suppose there are five jaguars in the Audobon Zoo and the weight (in pounds) of each is given in the table. **JAG**

| 170 | 160 | 200 | 150 | 180 |

A random sample of three jaguars is selected without replacement.
(a) Find the sampling distribution of the sample mean $\overline{X}$.
(b) Find the sampling distribution of the total weight for all three jaguars, T.

7.18 Education and Child Development The number of copies sold (in millions) for the all-time best-selling children's books are given in the table. **KIDSBK**

| 9.9 | 9.9 | 7.1 | 7.5 | 6.4 |

Suppose two of these books are selected at random without replacement.
(a) Find the sampling distribution of the sample mean.
(b) Find the sampling distribution of the total number of books sold.

7.19 Public Policy and Political Science The Rules, Procedures and the Rights of Parliament Committee of the Senate of Canada has 14 members.[4] This committee often meets with fewer than 14 members due to senators' schedules and other issues. Suppose the number of senators present at a committee meeting is a random variable X with the probability distribution given in the table.

x	10	11	12	13	14
$p(x)$	0.20	0.30	0.30	0.15	0.05

(a) Find the mean of X.
(b) Suppose two committee meetings are selected at random. Find the exact probability distribution of the sample mean $\overline{X}$.
(c) Find the exact probability distribution for the minimum number of committee members present at two committee meetings M.

Extended Applications

7.20 Fuel Consumption and Cars An automobile manufacturer lists several specifications for every one of its cars. For example, the length, width, wheelbase, turning circle, curb weight, and interior room measurements are readily available to customers. The acceleration time from 0 to 60 miles per hour (in seconds) for 10 four-cylinder cars is given in the table.[5] **CARSPEC**

| 7.8 | 6.9 | 8.8 | 7.2 | 6.1 | 6.0 | 7.4 | 6.3 | 7.1 | 7.8 |

(a) Use a computer or calculator to draw 30 random samples of size 3 without replacement, and compute the standard deviation for each sample.
(b) Construct a histogram of the sample standard deviations.

(c) Use the histogram to describe the shape of the sampling distribution of the standard deviation.
(d) Compute the population standard deviation. Find an approximate mean of the sampling distribution. How do these two numbers compare?

7.21 Travel and Transportation American Airlines offers limited first-class seating to passengers on a flight from Newark to Los Angeles. The probability distribution for the number of passengers in first class on a randomly selected flight is a random variable X. The probability distribution for X is given in the table.

x	5	6	7	8
$p(x)$	0.50	0.30	0.15	0.05

(a) Find the variance of X.
(b) Suppose two American Airlines cross-country flights are randomly selected. Find the exact probability distribution for the sample variance S^2.
(c) Find the mean of S^2. How does this compare with your answer in part (a)?

7.22 Sports and Leisure The number of championship round appearances for several members of the Professional Bowlers Association is given in the table.[6] **BOWL**

| 11 | 8 | 12 | 6 |

Suppose a random sample of two bowlers from this population is selected with replacement.
(a) Find the sampling distribution of the minimum number of appearances.
(b) Find the sample distribution of the maximum number of appearances.
(c) Find the sample mean for both distributions.

7.23 Sports and Leisure Although the Golden State Warriors won the 2017–2018 NBA championship, some of the players' individual performances during the season weren't so stellar. The table lists some of the players with the most technical fouls during the season.[7] **FOULS**

Player	Technical fouls
Dwight Howard	17
Draymond Green	15
Kevin Durant	14
Russell Westbrook	14
Chris Paul	11

Suppose a random sample of two of these players is selected without replacement.
(a) Find the sampling distribution for the maximum number of technical fouls.
(b) Find the sampling distribution for the median number of technical fouls.

7.24 Psychology and Human Behavior Five people own and operate Botanical Interests in Denver, Colorado. Each person is married, and the number of years each has been married is given in the table. **MARRIED**

5	3	7	2	12

Suppose a random sample of two of the owners is selected with replacement. Let D be a statistic defined to be the absolute value of the difference in the number of years each has been married. For example, if 2 and 7 were selected, the value of D would be $|2-7|=|-5|=5$. Find the sampling distribution of D.

7.25 Simple Random Sample Consider the population consisting of the numbers 1 through 50.
(a) Find the population mean μ.
(b) Use technology to select 50 random samples of size 10 from this population. Find the sample mean for each sample.
(c) Construct a histogram of the 50 sample means. Where is the histogram centered?

7.26 Biology and Environmental Science The number of coal power plants has been decreasing due to cheaper energy sources, lower demand for electricity, and stricter environmental regulations. However, coal is still used to produce energy in many regions of the world. The number of coal power plants operating in various countries is given in the table. **COAL**

868	137	54	70	2	44	133	12	7	117
10	4	9	16	22	17	5	3	0	170
106	58	13	340	128	85	17	14	12	5

(a) Find the population mean number of coal power plants operating μ.
(b) Find the population standard deviation of coal power plants operating σ.
(c) Use technology to select 100 random samples of size 10 from this population. Find the sample mean for each sample.
(d) Construct a histogram of the 100 sample means. Where is the histogram centered? How does this value compare to μ?
(e) Find the standard deviation for the 100 sample means. How does this value compare to σ?

7.27 Biology and Environmental Science Cane toads are prevalent during the rainy season in Florida and expel a toxic substance that can severely harm pets. Recently, a 10-year-old boy in southwest Florida established a business to trap the invasive creatures at residences. It takes about an hour to check a typical yard, and Landen Grey, the Toad Trapper, had more than 50 customers.[8] Suppose the number of cane toads trapped at a residence is a random variable X, with the probability distribution given in the table.

x	0	1	2	3	4
$p(x)$	0.05	0.15	0.25	0.30	0.25

(a) Find the mean of X, μ_X, and the standard deviation of X, σ_X.
(b) Suppose two residences are selected at random. Find the exact probability distribution for the sample mean $\overline{X}$.
(c) Find the mean of $\overline{X}$, $\mu_{\overline{X}}$, and the standard deviation of $\overline{X}$, $\sigma_{\overline{X}}$.
(d) How is μ_X related to $\mu_{\overline{X}}$? How is σ_X related to $\sigma_{\overline{X}}$?

Challenge Problems

7.28 Which Statistic Is Better? Each of the following graphs shows the probability distribution for two statistics that could be used to estimate a parameter θ (describing an underlying population). Select the statistic that would be a better estimator of θ and justify your answer.

(a)
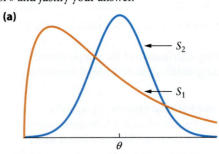

(b)
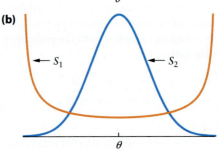

(c)
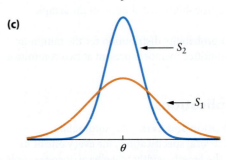

(d)
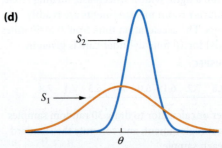

7.2 The Sampling Distribution of the Sample Mean and the Central Limit Theorem

Sampling Variability

It seems reasonable to use a value of the sample mean $\bar{x}$ to make an inference about the population mean μ. However, as discussed in the previous section, the sample mean varies from sample to sample. This **sampling variability** makes it difficult to know how far a specific $\bar{x}$ is from μ, or even whether the sample mean is an overestimate or underestimate. In this section, we explore the sampling variability of the sample mean. The probability distribution of the sample mean $\overline{X}$ can be used to make a sensible guess about the true value of the population mean.

To develop a reliable estimate, we need to know the exact probability distribution of the sample mean. As the next few examples will show, the distribution of $\overline{X}$ is related to n, the sample size, and to the parameters of the original, or underlying, population.

EXAMPLE 7.5 Approximate Distribution of $\overline{X}$

Consider a population consisting of the numbers $1, 2, 3, \ldots, 20$. The population mean is $\mu = (1 + 2 + 3 + \cdots + 20)/20 = 10.5$. Use frequency histograms to approximate the distribution of the mean, $\overline{X}$.

Solution

STEP 1 Consider a random sample of n observations selected with replacement. For $n = 5$, five numbers are selected at random from the population, and the sample mean is computed. This procedure is repeated 500 times. A histogram of the resulting 500 sample means is shown in **Figure 7.3**. For $n = 10$, a similar procedure is followed. The resulting histogram of the sample means is shown in **Figure 7.4**.

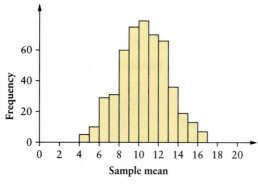
Figure 7.3 Histogram of sample means for $n = 5$.

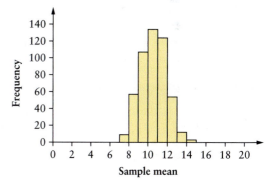
Figure 7.4 Histogram of sample means for $n = 10$.

The original population is certainly not normally distributed; it is finite, and a smoothed probability histogram is a horizontal line segment. It is amazing and perhaps counterintuitive that $\overline{X}$ has a normal distribution.

STEP 2 These histograms suggest some very surprising results. Each distribution appears to be centered near the population mean, 10.5. In addition, the shape of each distribution is approximately normal! Also notice that the variability of the sampling distribution decreases as n increases. The sampling distribution for $\overline{X}$ when $n = 10$ is more compact than that when $n = 5$.

STEP 3 This example implies that the distribution of the sample mean is approximately normal, with the mean being equal to the underlying (original) population mean and the variance being related to the sample size, n.

In the next example, we obtain the exact sampling distribution of the mean.

EXAMPLE 7.6 CD Terms

A certificate of deposit (CD) is really just a special savings account with a fixed interest rate and a fixed date of withdrawal. Goldman Sachs Bank offers CDs with various rates, minimum deposit amounts, and either a 2-, 3-, or 4-year term. Research indicates that the term selected by a customer is a random variable X, with probability distribution given in the table.

x	2	3	4
$p(x)$	0.5	0.3	0.2

Suppose a random sample of three Goldman Sachs Bank CDs is selected.

(a) Find the sampling distribution of $\overline{X}$, the sample mean term.

(b) Find the mean and the variance of the random variable X.

(c) Find the mean and the variance of the random variable $\overline{X}$.

(d) How do the results from (b) compare with the results from (c)?

Solution

(a) Each sample consists of three numbers: The first represents the term for CD 1, the second for CD 2, and the third for CD 3. By the Multiplication Rule, there are $3 \times 3 \times 3 = 27$ possible samples. For each sample, list the sample mean and the probability of selecting that sample.

The probability associated with each sample is computed using independence. For example, the probability of observing the sample 2, 2, 4 is

$P(X = 2 \cap X = 2 \cap X = 4)$ *Intersection of three events.*

$= P(X = 2) \cdot P(X = 2) \cdot P(X = 4)$ *Independence.*

$= (0.5)(0.5)(0.2) = 0.05$ *Use the probability distribution for X.*

Use this method to help construct a table listing all possible samples, the value of the sample mean, and the probability of each sample.

Sample	$\overline{x}$	Probability	Sample	$\overline{x}$	Probability
2, 2, 2	2	0.125	3, 3, 4	10/3	0.018
2, 2, 3	7/3	0.075	3, 4, 2	3	0.030
2, 2, 4	8/3	0.050	3, 4, 3	10/3	0.018
2, 3, 2	7/3	0.075	3, 4, 4	11/3	0.012
2, 3, 3	8/3	0.045	4, 2, 2	8/3	0.050
2, 3, 4	3	0.030	4, 2, 3	3	0.030
2, 4, 2	8/3	0.050	4, 2, 4	10/3	0.020
2, 4, 3	3	0.030	4, 3, 2	3	0.030
2, 4, 4	10/3	0.020	4, 3, 3	10/3	0.018
3, 2, 2	7/3	0.075	4, 3, 4	11/3	0.012
3, 2, 3	8/3	0.045	4, 4, 2	10/3	0.020
3, 2, 4	3	0.030	4, 4, 3	11/3	0.012
3, 3, 2	8/3	0.045	4, 4, 4	4	0.008
3, 3, 3	3	0.027			

Summarize the results in this table to write the probability distribution for $\overline{X}$. We need to list all the possible values of $\overline{X}$ and the corresponding probabilities.

$\overline{x}$	2	7/3	8/3	3	10/3	11/3	4
$p(\overline{x})$	0.125	0.225	0.285	0.207	0.114	0.036	0.008

(b) The mean and the variance for (the discrete random variable) X are

$E(X) = (2)(0.5) + (3)(0.3) + (4)(0.2) = 2.7$

$Var(X) = (2 - 2.7)^2(0.5) + (3 - 2.7)^2(0.3) + (4 - 2.7)^2(0.2) = 0.61$

(c) The mean and the variance for (the discrete random variable) $\overline{X}$ are

$E(\overline{X}) = (2)(0.125) + (7/3)(0.225) + \cdots + (4)(0.008) = 2.7$

$Var(\overline{X}) = (2 - 2.7)^2(0.125) + \cdots + (4 - 2.7)^2(0.008) = 0.2033$

(d) Notice the following extraordinary relationships between the means and the variances.

$E(\overline{X}) = 2.7 = E(X)$

The mean of the sample mean is equal to the original population mean!

$Var(\overline{X}) = 0.2033 = \dfrac{0.61}{3} = \dfrac{Var(X)}{n}$

The variance of $\overline{X}$ is equal to the original variance divided by the sample size. ■

Distribution Connections

Here is one more approach to illustrate the very important connections between the distribution of the sample mean and the distribution of the original population.

Consider three original, or underlying, distributions: (1) the standard normal distribution, a normal distribution with mean 0 and standard deviation 1; (2) a uniform distribution with parameters $a = 0$ and $b = 1$; and (3) an exponential distribution with parameter $\lambda = 0.5$. The probability density function and the mean for each distribution are shown in the first row of **Figure 7.5**.

Now consider the following process for each distribution. Select 500 samples of size $n = 2$ and compute the mean for each sample. Construct a histogram of the sample means. The resulting smoothed histogram is shown in the second row of Figure 7.5. Repeat this procedure for $n = 5$, 10, and 20, for each underlying distribution. These smoothed histograms are also shown in Figure 7.5.

Notice the following incredible patterns.

1. If the underlying population is normal, the distribution of the sample mean appears to be normal, regardless of the sample size. See Figure 7.5(a).

2. Even if the underlying population is not normal, the distribution of the sample mean becomes more normal as n increases. See Figure 7.5(b) and Figure 7.5(c).

3. The sampling distribution of the mean is centered at the mean of the underlying population. See Figure 7.5(a), Figure 7.5(b), and Figure 7.5(c).

4. As the sample size n increases, the variance of the distribution of the sample mean decreases. See Figure 7.5(a), Figure 7.5(b), and Figure 7.5(c).

The previous examples and observations lead to the following properties concerning the sample mean $\overline{X}$.

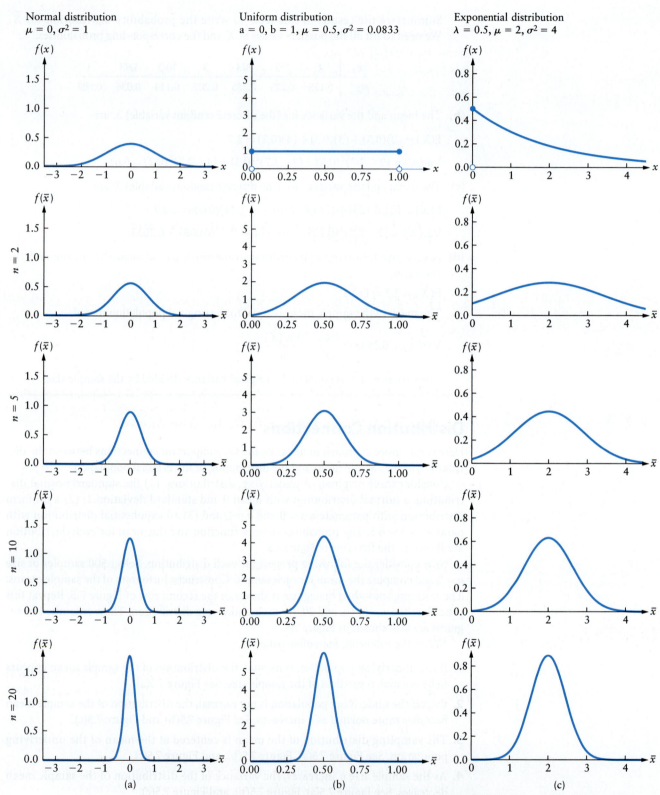

Figure 7.5 Three different original, or underlying, populations, and approximations to the distribution of the sample mean for various sample sizes.

Properties of the Sample Mean

Let $\overline{X}$ be the mean of observations in a random sample of size n drawn from a population with mean μ and variance σ^2.

1. The mean of $\overline{X}$ is equal to the mean of the underlying population.

    In symbols: $\mu_{\overline{X}} = \mu$

2. The variance of $\overline{X}$ is equal to the variance of the underlying population divided by the sample size.

    In symbols: $\sigma^2_{\overline{X}} = \dfrac{\sigma^2}{n}$

    The standard deviation, or standard error, of $\overline{X}$ is $\sigma_{\overline{X}} = \sqrt{\dfrac{\sigma^2}{n}} = \dfrac{\sigma}{\sqrt{n}}$.

3. If the underlying population is normally distributed, then the distribution of $\overline{X}$ is also exactly normal for any sample size.

    In symbols: $\overline{X} \sim N(\mu, \sigma^2/n)$

A statistic $\hat{\theta}$ is an **unbiased estimator** of a population parameter θ if $E(\hat{\theta}) = \theta$, and the mean of $\hat{\theta}$ is θ.

The mean of the sampling distribution of $\overline{X}$ is the same as the mean of the underlying distribution that we want to estimate, so the sample mean is an **unbiased estimator** of the population mean μ.

The Central Limit Theorem

Even if the underlying population is not normal, the previous examples suggest the distribution of $\overline{X}$ becomes more normal as the sample size increases. This amazing result is the most important idea in all of statistics, the **Central Limit Theorem**.

Central Limit Theorem (CLT)

Let $\overline{X}$ be the mean of observations in a random sample of size n drawn from a population with mean μ and finite variance σ^2. As the sample size n increases, the sampling distribution of

$$Z = \dfrac{\overline{X} - \mu}{\sigma/\sqrt{n}}$$

approaches the standard normal distribution. In practice, or informally, this means the sampling distribution of $\overline{X}$ will increasingly approximate a normal distribution, with mean μ and variance σ^2/n, regardless of the shape of the underlying population distribution.

In symbols: $\overline{X} \stackrel{\bullet}{\sim} N(\mu, \sigma^2/n)$

The symbol $\stackrel{\bullet}{\sim}$ means "is approximately distributed as."

The CLT is really a remarkable result. No matter what the shape of the underlying population (skewed, bimodal, lots of variability, etc.), the CLT says that the distribution of $\overline{X}$ is approximately normal, as long as n is large enough!

A CLOSER LOOK

1. ▶ A better name for this result might be the *normal convergence theorem*. The distribution of $\overline{X}$ converges, or gets closer and closer, to a normal distribution. ◀

2. If the original population is normally distributed, then the distribution of $\overline{X}$ is normal, no matter how large or small the sample size (n).

3. If the original population is not normal, the CLT says that the distribution of $\overline{X}$ approaches a normal distribution as n increases, and the approximation improves as n increases: The approximation gets better and better as the sample size n gets bigger and bigger.

There is no magical threshold value for n. However, in most cases, if $n \geq 30$, then the approximation is pretty good. Even for severely skewed populations, as long as n is at least 30, the approximation is reasonable.

Some people believe that if $n = 29$, the distribution of $\overline{X}$ will not be approximately normal, but if $n = 31$, it will be. This is simply not true. The CLT should be interpreted to mean that the distribution of $\overline{X}$ approaches a normal distribution—that is, looks more and more like a normal distribution—as n increases. In some cases, the approximation will be excellent for n as small as 5. In other cases, n might have to be at least 26 before the approximation is good.

4. To compute a probability involving the sample mean, we treat $\overline{X}$ just like any other normal random variable: Standardize and use cumulative probability where appropriate. We use the same method even if $\overline{X}$ is only approximately normal.

5. The expression for the variance of $\overline{X}$ mathematically confirms our observations regarding the variability of the sampling distribution. As n increases, the distribution of the sample mean becomes more compact. $\sigma_{\overline{X}}^2 = \sigma^2/n$ and σ^2 is constant. This fraction becomes smaller as n (the denominator) increases.

6. A more general version of the CLT includes a statement about the sum of independent observations T. If n is sufficiently large, the distribution of T approaches a normal distribution with mean $n\mu$ and variance $n\sigma^2$.

In symbols: $T \overset{\bullet}{\sim} N(n\mu, n\sigma^2)$.

CHAPTER APP: The Longest Zip Line

Suppose the population mean ride time is 1.75 min with a population standard deviation of 0.2 min. A random sample of 40 riders was obtained. Find the distribution of the sample mean $\overline{X}$.

An even more general version of the CLT concludes, essentially, that any statistic that is a sum or a mean tends toward a normal distribution as n increases. This is useful in many inference problems because the appropriate statistic is often a sum or a mean. It is easy to compute probabilities associated with a normal statistic (random variable). Therefore, the likelihood, or tail probability, associated with an observed value of the statistic is a straightforward calculation.

The CLT also helps explain why so many real-world distributions are approximately normal. Almost any statistic can be decomposed into a sum of other variables. For example, the height of a tomato plant after six weeks might be directly related to, or be the sum of, the effects of a large number of independent variables, including the amount of water, fertilizer, and sunlight it has received. Therefore, the distribution of the height should be approximately normal. This single theorem explains empirical evidence that suggests almost every measurement distribution is approximately normal.

CLT Examples

The following examples illustrate the properties of $\overline{X}$ and the Central Limit Theorem.

EXAMPLE 7.7 Green Line Time

The Massachusetts Bay Transportation Authority (MBTA) Green Line from the Boston College stop to Park Street in downtown Boston has trolleys leaving regularly throughout the day beginning at 5:01 A.M. Although the length of the trip and the number of stops are constant, the time taken for each trip varies due to weather, traffic, and time of day.

7.2 The Sampling Distribution of the Sample Mean and the Central Limit Theorem

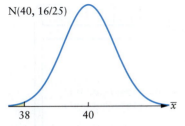

The shaded area represents $P(\overline{X} < 38)$.

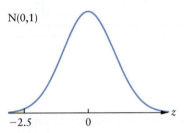

The shaded area represents $P(Z < -2.5)$.

Figure 7.6 A visualization of the standardization in part (a).

Common Error

$\sigma_{\overline{X}} = \sigma$

Correction: The standard deviation of the sample mean $\overline{X}$ is $\sigma/\sqrt{n}$.

Recall that one value in a continuous world (Z) contributes no probability:

$P(Z < -1.25) = P(Z \le -1.25)$.

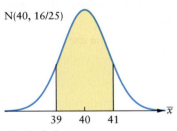

The shaded area represents $P(39 \le \overline{X} \le 41)$.

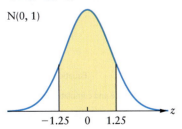

The shaded area represents $P(-1.25 \le Z \le 1.25)$.

Figure 7.8 A visualization of the standardization in part (b).

According to the MBTA, the mean time for the trip is approximately 40 min. Suppose the travel time is normally distributed with a standard deviation $\sigma = 4$ min. A random sample of 25 trips is obtained, and the time for each is recorded.

(a) Find the probability that the sample mean time will be less than 38 min.

(b) Find the probability that the sample mean will be within 1 min of the population mean (40 min).

Solution

The underlying distribution is normal, so the distribution of $\overline{X}$ is exactly normal. Find the mean, variance, and standard deviation of $\overline{X}$; translate each question into a probability statement; standardize; and use cumulative probability where appropriate.

(a) The underlying distribution is normal with $\mu = 40$ and $\sigma^2 = 16$. The sample size is $n = 25$. The sample mean is (exactly) normally distributed.

$$\overline{X} \sim N(\mu, \sigma^2/n) = N(40, 16/25); \quad \sigma_{\overline{X}} = \sqrt{16/25} = 4/5 = 0.8 \text{ min}$$

To solve part (a), begin with a probability statement involving the random variable $\overline{X}$. We need the probability that $\overline{X}$ will be less than 38 min.

$$P(\overline{X} < 38) = P\left(\frac{\overline{X} - 40}{0.8} < \frac{38 - 40}{0.8}\right) \quad \text{Standardize.}$$

$$= P(Z < -2.5) \quad \text{Equation 6.8; simplify.}$$

$$= 0.0062 \quad \text{Use Appendix Table 3.}$$

The probability that the sample mean time for the 25 trips will be less than 38 min is 0.0062. **Figure 7.6** is a visualization of the standardization, and **Figure 7.7** shows a technology solution.

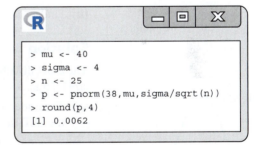

Figure 7.7 $P(\overline{X} < 38)$ using R.

(b) "Within 1 min of the population mean" means $39 \le \overline{X} \le 41$.

Write a probability statement, and standardize again.

$P(39 \le \overline{X} \le 41)$

$$= P\left(\frac{39 - 40}{0.8} \le \frac{\overline{X} - 40}{0.8} \le \frac{41 - 40}{0.8}\right) \quad \text{Standardize.}$$

$$= P(-1.25 \le Z \le 1.25) \quad \text{Equation 6.8; simplify.}$$

$$= P(Z \le 1.25) - P(Z \le -1.25) \quad \text{Use cumulative probability.}$$

$$= 0.8944 - 0.1056 \quad \text{Use Appendix Table 3.}$$

$$= 0.7888$$

Figure 7.8 is a visualization of the standardization. $P(39 \le \overline{X} \le 41)$, or the area of the shaded region in the $\overline{X}$ world, is the same as $P(-1.25 \le Z \le 1.25)$, or the area of the shaded region in the Z world. This is due to the Standardization Rule, Equation 6.8. **Figure 7.9** shows a technology solution.

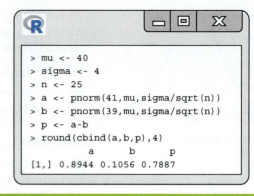

Figure 7.9 $P(39 \leq \overline{X} \leq 41)$ using R.

TRY IT NOW Go to Exercises 7.43 and 7.45

CHAPTER APP **The Longest Zip Line**
Find the probability that the sample mean ride time is less than 1.72 min.

EXAMPLE 7.8 Milk Deliveries

In upstate New York, milk tanker trucks follow a daily routine, stopping at the same dairy farms every day. Farm output, however, varies because of weather, time of year, number of cows, and other factors. From years of recorded data, the mean amount of milk collected by a truck for processing is 7750 liters (L), with a standard deviation of 150 L. Suppose 36 trucks are randomly selected.

(a) Find the probability that the sample mean amount of milk picked up by the 36 trucks is more than 7800 L.

(b) Find a value m such that the probability that the sample mean is less than m is 0.1.

Solution

The exact distribution of the underlying population is not known. However, the sample size is large: $n = 36$ (≥ 30). Therefore the CLT can be applied.

(a) The underlying distribution has $\mu = 7750$ and $\sigma = 150$. The sample size is $n = 36$. By the CLT, the distribution of $\overline{X}$ is approximately normal.

$$\overline{X} \stackrel{.}{\sim} N(\mu, \sigma^2/n) = N(7750, 150^2/36) = N(7750, 625)$$

The standard deviation of $\overline{X}$ is

$$\sigma_{\overline{X}} = \sqrt{\sigma^2/n} = \sqrt{625} = 25 \quad \text{(or } \sigma_{\overline{X}} = \sigma/\sqrt{n} = 150/6 = 25\text{)}$$

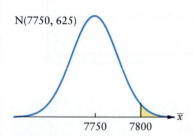

N(7750, 625)

The shaded area represents $P(\overline{X} > 7800)$.

To solve part (a), start with a probability statement involving $\overline{X}$. We need the probability that the sample mean will be more than 7800 L.

$$P(\overline{X} > 7800) = P\left(\frac{\overline{X} - 7750}{25} > \frac{7800 - 7750}{25}\right) \quad \text{Standardize.}$$
$$= P(Z > 2) \quad \text{Equation 6.8; simplify.}$$
$$= 1 - P(Z \leq 2) \quad \text{Use Complement Rule and cumulative probability.}$$
$$= 1 - 0.9772 \quad \text{Use Appendix Table 3.}$$
$$= 0.0228$$

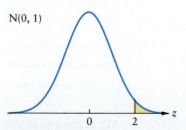

N(0, 1)

The shaded area represents $P(Z > 2)$.

Figure 7.10 A visualization of the standardization in part (a).

The probability that the sample mean amount of milk picked up by the 36 trucks will be greater than 7800 L is 0.0228. **Figure 7.10** helps to visualize this standardization process, and **Figure 7.11** shows a technology solution.

7.2 The Sampling Distribution of the Sample Mean and the Central Limit Theorem

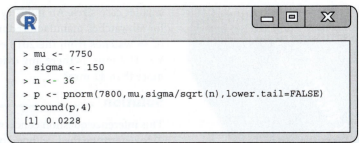

Figure 7.11 $P(\overline{X} > 7800)$ using R.

(b) To solve part (b), once again we need to translate the text of the question into a probability statement. Convert the expression into a cumulative probability statement involving Z. Because the probability is already given, this is an inverse cumulative probability problem. Work backward in Appendix Table 3.

$$P(\overline{X} < m) = P\left(\frac{\overline{X} - 7750}{25} < \frac{m - 7750}{25}\right) \quad \text{Standardize.}$$

$$= P\left(Z < \frac{m - 7750}{25}\right) = 0.1 \quad \text{Equation 6.8.}$$

There is no further simplification within the probability statement. However, the resulting probability statement involves Z and is a cumulative probability. Find a value in the body of Appendix Table 3 as close to 0.1 as possible. Set the corresponding z value equal to $\left(\frac{m - 7750}{25}\right)$, and solve for m.

$$\frac{m - 7750}{25} = -1.28 \quad \text{Appendix Table 3.}$$

$$m - 7750 = -32 \quad \text{Multiply both sides by 25.}$$

$$m = 7718 \quad \text{Add 7750 to both sides.}$$

The probability that the sample mean will be less than $m = 7718$ is (approximately) 0.1.

Figure 7.12 helps to visualize this standardization and solution. **Figure 7.13** shows a technology solution.

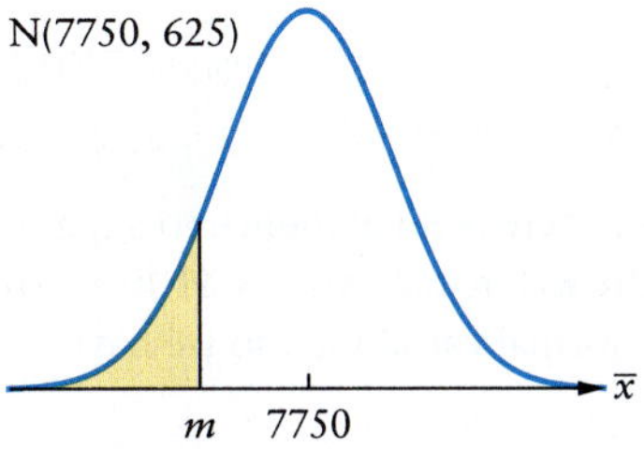

N(7750, 625)

This graph shows a value of m such that $P(\overline{X} < m) = 0.1$.

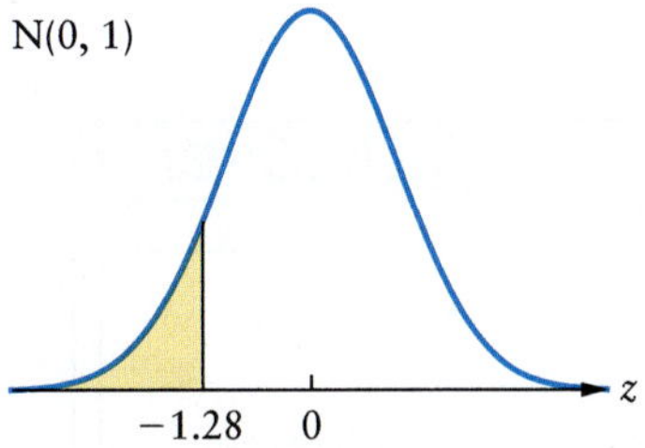

N(0, 1)

Using Appendix Table 3, $P(Z < -1.28) = 0.1$.

Figure 7.12 A visualization of the standardization in part (a).

Figure 7.13 R calculations to find the value of m.

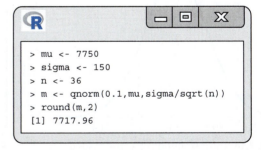

TRY IT NOW Go to Exercises 7.57 and 7.61

The CLT and Inference

EXAMPLE 7.9 **Baselworld**

The most important watch exposition takes place each year in Basel, Switzerland. Baselworld 2018 featured more than 600 exhibitors and stunning new watches from Rolex and Patek Philippe. The general trend was the use of high-tech materials and

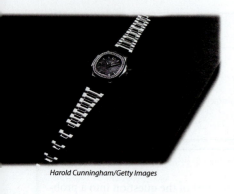

Harold Cunningham/Getty Images

smaller case sizes. An industry spokesperson claimed the mean case size (in diameter) for all watches manufactured in 2018 is 41 mm.[9] Suppose $\sigma = 1$ mm, a random sample of 35 watches from the 2018 exposition is selected, and the sample mean case size is $\bar{x} = 41.2$ mm. Is there any evidence to suggest that the population mean case size is more than 41 mm?

Solution

This inference question concerns a population mean. Use the CLT and the distribution of $\bar{X}$ to determine the likelihood of an observed sample mean of $\bar{x} = 41.2$ or greater.

STEP 1 Assume the underlying distribution has $\mu = 41$ mm and $\sigma = 1$ mm. We do not know the shape of the underlying distribution of case sizes, but the sample size is large: $n = 35\ (\geq 30)$. By the CLT, the distribution of $\bar{X}$ is approximately normal.

STEP 2 Claim: $\mu = 41 \Rightarrow \bar{X} \stackrel{\cdot}{\sim} N(41, 1^2/35)$, $\sigma_{\bar{X}} = 1/\sqrt{35}$

Experiment: $\bar{x} = 41.2$

Likelihood: We are looking for any evidence that the mean case size is more than 41 mm, so we will find the right-tail probability.

> Remember that $P(\bar{X} = 41.2) = 0$ and we always consider a tail probability to be conservative.

$$P(\bar{X} \geq 41.2) = P\left(\frac{\bar{X} - 41}{1/\sqrt{35}} \geq \frac{41.2 - 41}{1/\sqrt{35}}\right) \quad \text{Standardize.}$$
$$= P(Z \geq 1.18) \quad \text{Equation 6.8; simplify.}$$
$$= 1 - P(Z \leq 1.18) \quad \text{Use cumulative probability.}$$
$$= 1 - 0.8810 \quad \text{Appendix Table 3.}$$
$$= 0.1190$$

Conclusion: This probability is large (>0.05), so there is no evidence to suggest the mean case size is greater than 41 mm. If the mean case size is 41 mm, an observation of $\bar{x} = 41.2$ is reasonable, subject to normal variability, so we have no reason to doubt the claim.

Figure 7.14 shows the distribution of $\bar{X}$ and the right-tail probability; Figure 7.15 shows a technology solution.

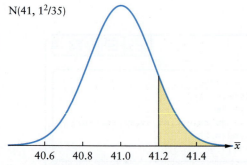

Figure 7.14 The distribution of $\bar{X}$. The shaded region represents the right-tail probability $P(\bar{X} \geq 41.2)$.

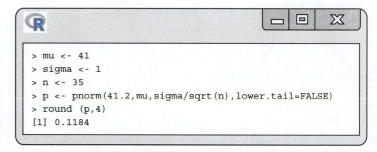

```
> mu <- 41
> sigma <- 1
> n <- 35
> p <- pnorm(41.2,mu,sigma/sqrt(n),lower.tail=FALSE)
> round (p,4)
[1] 0.1184
```

Figure 7.15 Likelihood calculation using R.

TRY IT NOW Go to Exercises 7.47 and 7.53

Section 7.2 Exercises

Concept Check

7.29 True or False The sample mean varies from sample to sample.

7.30 Fill in the Blank Let $\overline{X}$ be the mean of observations in a random sample of size n from a population with mean μ and variance σ^2.
(a) The mean of $\overline{X}$ is _____.
(b) The variance of $\overline{X}$ is _____.

7.31 True or False The distribution of $\overline{X}$ is never exactly normal.

7.32 True or False If the underlying population is not normal, the CLT says the distribution of $\overline{X}$ approaches a normal distribution as n increases.

7.33 True or False As n increases, the distribution of the sum of the observations approaches a normal distribution.

7.34 True or False As n increases, the variance of $\overline{X}$ also increases.

7.35 Short Answer Explain how the CLT suggests that so many real-world distributions are approximately normal.

Note: To find the distribution of $\overline{X}$, or any random variable, you need to describe the distribution, by name if possible, and provide the values of the parameters that characterize the distribution. For example, the statement "$\overline{X}$ has a normal distribution with mean $\mu = 27$ and variance $\sigma^2_{\overline{X}} = 0.35$, or $\overline{X} \sim N(27, 0.35)$," completely describes the distribution of $\overline{X}$.

Practice

7.36 Consider a normally distributed population with mean $\mu = 10$ and standard deviation $\sigma = 2.5$. Suppose a random sample of size n is selected from this population. Find the distribution of $\overline{X}$ and the indicated probability in each of the following cases.
(a) $n = 7$, $P(\overline{X} \leq 9)$
(b) $n = 12$, $P(\overline{X} > 11.5)$
(c) $n = 15$, $P(9.5 \leq \overline{X} \leq 10.5)$
(d) $n = 25$, $P(\overline{X} \geq 10.25)$
(e) $n = 100$, $P(\overline{X} \leq 9.8 \cup \overline{X} \geq 10.2)$

7.37 Suppose X is a normal random variable with mean $\mu = 17.5$ and standard deviation $\sigma = 6$. A random sample of size $n = 24$ is selected from this population.
(a) Find the distribution of $\overline{X}$.
(b) Carefully sketch a graph of the probability density functions for X and $\overline{X}$ on the same coordinate axes.
(c) Find $P(X \leq 14)$ and $P(\overline{X} \leq 14)$.
(d) Find $P(15 < X < 19)$ and $P(15 < \overline{X} < 19)$.

7.38 Suppose X is a random variable with mean $\mu = 50$ and standard deviation $\sigma = 49$. A random sample of size $n = 38$ is selected from this population.
(a) Find the approximate distribution of $\overline{X}$. Why is the CLT necessary here?
(b) Find $P(\overline{X} < 49)$.
(c) Find $P(\overline{X} \geq 52)$.
(d) Find $P(49.5 \leq \overline{X} \leq 51.5)$.
(e) Find a value c such that $P(\overline{X} > c) = 0.15$.

7.39 Suppose X is a random variable with mean $\mu = 1000$ and standard deviation $\sigma = 100$. A random sample of size $n = 36$ is selected from this population.
(a) Find the approximate distribution of $\overline{X}$. Carefully sketch a graph of the probability density function.
(b) Find $P(\overline{X} > 975)$.
(c) Find $P(\overline{X} \leq 1030)$.
(d) Find $P(\mu_{\overline{X}} - \sigma_{\overline{X}} \leq \overline{X} \leq \mu_{\overline{X}} + \sigma_{\overline{X}})$.
(e) Find a symmetric interval about the mean $\mu_{\overline{X}}$ such that $P(\mu_{\overline{X}} - c \leq \overline{X} \leq \mu_{\overline{X}} + c) = 0.95$.

7.40 Suppose X is a random variable with mean $\mu = 30$ and standard deviation $\sigma = 50$. A random sample of size $n = 40$ is selected from this population.
(a) Find the approximate distribution of $\overline{X}$. Carefully sketch a graph of the probability density function.
(b) Find $P(\overline{X} \geq 38)$.
(c) Find $P(20 \leq \overline{X} \leq 40)$.
(d) Find $P(\overline{X} < 15)$.
(e) Find a value c such that $P(\overline{X} \leq c) = 0.001$.

7.41 The figure shows the graphs of the probability density functions for the random variable X, the random variable $\overline{X}$ for $n = 5$, and the random variable $\overline{X}$ for $n = 15$.

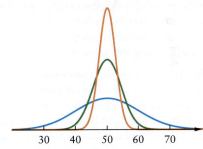

Identify the graph of each probability density function.

7.42 The figure shows the graphs of the probability density function for the random variable X, the approximate density functions for the random variable $\overline{X}$ for $n = 5$, and the random variable $\overline{X}$ for $n = 15$.

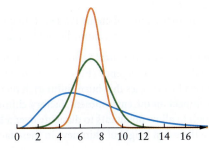

Identify the graph of each probability density function.

Applications

7.43 Public Health and Nutrition One measure of general health is body mass index (BMI). Adults with a BMI between 18.5 and 25 are generally considered to have an ideal body weight for their height.[10] A recent study indicated that Canadians are among the most overweight people in the world as measured by BMI (despite all the snow shoveling exercise).[11] Suppose the BMI for male adults in Canada is normally distributed, with mean 27.4 and standard deviation 1.5.
  (a) Suppose one male adult in Canada is selected at random. What is the probability that his BMI is more than 28?
  (b) Suppose 10 male adults from Canada are selected at random. What is the probability that the sample mean BMI is greater than 28?
  (c) For $n = 10$, what is the probability that the sample mean BMI is less than 26.5?
  (d) Bliss YogaSpa in Edmonton claims that its clients have BMI measurements between 26 and 27. Suppose ten Bliss clients are selected at random. What is the probability that the sample mean will be in this interval?

7.44 Public Policy and Political Science In certain hurricane-prone areas of the United States, concrete columns used in construction must meet specific building codes. The minimum diameter for a cylindrical column is 8 in. Suppose the mean diameter for all columns is 8.25 in., with a standard deviation of 0.1 in. A building inspector randomly selects 35 columns and measures the diameter of each.
  (a) Find the approximate distribution of $\overline{X}$. Carefully sketch a graph of the probability density function.
  (b) What is the probability that the sample mean diameter for the 35 columns will be greater than 8 in.?
  (c) What is the probability that the sample mean diameter for the 35 columns will be between 8.2 and 8.4 in.?
  (d) Suppose the standard deviation is 0.15 in. Answer parts (a), (b), and (c) using this value of σ.

7.45 Manufacturing and Product Development A large part of the luggage market is made up of overnight bags. These bags vary by weight, exterior appearance, material, and size. Suppose the volume of overnight bags is normally distributed, with mean $\mu = 1750$ in^3 and standard deviation $\sigma = 250$ in^3. A random sample of 15 overnight bags is selected, and the volume of each is found.
  (a) Find the distribution of $\overline{X}$.
  (b) What is the probability that the sample mean volume is more than 1800 in^3?
  (c) What is the probability that the sample mean volume is within 100 in^3 of 1750?
  (d) Find a symmetric interval about 1750 such that 95% of all values of the sample mean volume lie in this interval.

7.46 Biology and Environmental Science The U.S. Environmental Protection Agency (EPA) is concerned about pollution caused by factories that burn sulfur-rich fuel. To decrease the impact on the environment, factory chimneys must be high enough to allow pollutants to dissipate over a larger area. Assume the mean height of chimneys in these factories is 100 m (an EPA-acceptable height) with a standard deviation of 12 m. A random sample of 40 chimney heights is obtained.
  (a) What is the probability that the sample mean height for the 40 chimneys is greater than 102 m?
  (b) What is the probability that the sample mean height is between 101 and 103 m?
  (c) Suppose the sample mean is 98.5 m. Is there any evidence to suggest that the true mean height for chimneys is less than 100 m? Justify your answer.

7.47 Manufacturing and Product Development The manager at the Aldi's grocery store in Manahawkin, New Jersey, suspects a supplier is systematically underfilling 12-oz bags of potato chips. To check the manufacturer's claim, a random sample of 100 bags of potato chips is obtained, and each bag is carefully weighed. The sample mean is 11.9 oz. Assume $\sigma = 0.3$ oz.
  (a) Find the probability that the sample mean is 11.9 oz or less.
  (b) How can this probability in part (a) be so small when 11.9 is so close to the population mean $\mu = 12$?
  (c) From your answer to part (a), is there any evidence to suggest that the mean weight of bags of potato chips is less than 12 oz?

7.48 Psychology and Human Behavior In attempting to flee from police, criminals sometimes barricade themselves inside a building and create a police standoff. The criminal is usually armed with a dangerous weapon and may hold hostages. Suppose the mean length of a police standoff, ending with some sort of resolution, is 6.5 hr, with a standard deviation of 4 hr. A random sample of 35 police standoffs is selected.
  (a) Find the distribution of the sample mean.
  (b) What is the probability that the sample mean for the 35 police standoffs will be greater than 7 hr?
  (c) A new psychological technique was used in negotiations with the criminals involved in the 35 police standoffs. Suppose the sample mean for the 35 police standoffs is $\bar{x} = 5.1$ hr. Is there any evidence to suggest that the mean police standoff time is shorter when this new technique is used?

7.49 Biology and Environmental Science During August, the hottest month of the year in Houston, Texas, the mean amount of rainfall is 3.54 in.[12] Suppose the standard deviation is 1.1 in., and 30 months (all August) are selected at random.
  (a) What is the probability that the sample mean rainfall is less than 3.3 in.?
  (b) What is the probability that the sample mean rainfall is greater than 3.9 in.?
  (c) Find a symmetric interval about the mean, 3.54, such that the probability the sample mean lies in this interval is 0.90.

7.50 Biology and Environmental Science Carbon dioxide (CO_2) is one of the primary gases contributing to the greenhouse effect and global warming. The mean amount of CO_2 in the atmosphere for June 2018 was 410.79 parts per million (ppm).[13] Suppose 40 atmospheric samples were selected

at random in July 2018 and the standard deviation for CO_2 in the atmosphere was found to be $\sigma = 20$ ppm.
(a) Find the probability that the sample mean CO_2 level is less than 405 ppm.
(b) Find the probability that the sample mean CO_2 level is between 408 and 415.
(c) Suppose the sample mean CO_2 level is 418 ppm. Is there any evidence to suggest that the population mean CO_2 level has increased? Justify your answer.

7.51 Sports and Leisure The sport of pickleball is gaining in popularity, not only among older adults but also in school gym classes. In this game, the court is smaller than a tennis court; the paddle is like a giant table tennis paddle; and the ball is a wiffle ball, light and perforated. The mean weight of an Onix Pure 2 Outdoor Pickleball is 25 g (with diameter 2.875 in. and 40 holes).[14] Suppose the weight of this Onix pickleball is normally distributed, with a standard deviation of 0.9 g, and 10 balls are selected at random.
(a) Find the probability that the sample mean weight is less than 24.5 g.
(b) Find the probability that the sample mean weight is greater than 25.25 g.
(c) These 10 pickleballs can be used in a tournament if the sample mean weight is between 24.4 and 25.6 g. What is the probability that these 10 balls cannot be used in a tournament?

7.52 Biology and Environmental Science There are many regulations for catching lobsters off the coast of New England, including required permits, allowable gear, and size prohibitions. The Massachusetts Division of Marine Fisheries requires a minimum carapace (shell) length measured from a rear eye socket to the center line of the body shell. Any lobster measuring less than 3.25 in. must be returned to the ocean. For all lobsters caught, suppose the carapace length is normally distributed, with a mean of 4.125 in. and a standard deviation of 1 in. A random sample of 15 lobsters is obtained.
(a) Find the distribution of the sample mean carapace length.
(b) What is the probability that the sample mean carapace length is more than 4.5 in.?
(c) What is the probability that the sample mean carapace length is between 3.8 and 4.2 in.?
(d) If the sample mean carapace length is less than 3.5 in., a lobsterman will look for other places to set his traps. What is the probability that a lobsterman will be looking for a different location?

7.53 Sports and Leisure One measure of an athlete's ability is the height of his or her vertical leap. Many professional basketball players are known for their remarkable vertical leaps, which lead to amazing dunks. At the 2018 NBA Draft Combine, Donte DiVincenzo and Josh Okogie had the highest vertical leap at 42 in.[15] However, the mean vertical leap of all NBA players is reported to be 28 in. Suppose the standard deviation is 7 in., and 36 NBA players are selected at random.
(a) What is the probability that the mean vertical leap for the 36 players will be less than 26 in.?
(b) What is the probability that the mean vertical leap for the 36 players will be between 27.5 and 28.5 in.?
(c) A high-priced athletic trainer has been hired to work with a group of NBA players to improve their hip flexibility, which should improve their vertical leap. After one month of training, the mean vertical leap for 50 of these players selected at random is 29.75 in. Is there any evidence to suggest this flexibility program has increased the mean vertical leap (from $\mu = 28$ in.)?

7.54 Physical Sciences The ozone hole is a region in the Southern Hemisphere that has increased in area fairly regularly since 1979, but decreased in 2017 to its smallest size since 1988. The mean ozone hole area (in million km^2) in 2017 was 17.4,[16] with standard deviation 5.5. Suppose 33 days are selected at random and the ozone hole is measured each day.
(a) What is the probability that the sample mean ozone hole area is less than 16 million km^2?
(b) What is the probability that the sample mean ozone hole area is between 15.5 and 18.5 million km^2?
(c) Some researchers suggest that the decreased use of certain chemicals has caused the size of the ozone hole area to decrease. Suppose the sample mean for the 33 days is 14.75 million km^2. Is there any evidence to suggest that the mean ozone hole area has decreased? Justify your answer.

7.55 Physical Sciences A recent study of the Kitchener (Ontario) Fire Department revealed response times that are longer than standards set by the National Fire Protection Association. The mean turnout time, which is defined to be the length of time between a fire station receiving a call and trucks leaving the station, was 101 sec.[17] Suppose the standard deviation is 18 sec and 35 calls are selected at random.
(a) Find the probability that the sample mean turnout time is greater than 99 sec.
(b) Find the probability that the sample mean turnout time is between 95 and 105 sec.
(c) Technology upgrades are expected to decrease the turnout time. After these upgrades, suppose the sample mean turnout time for the 35 calls is 94 sec. Is there any evidence to suggest that the population mean turnout time has decreased? Justify your answer.

Extended Applications

7.56 Public Health and Nutrition Recent research studies suggest that lycopene, found in tomatoes, may reduce the risk of certain cancers. According to the U.S. Department of Agriculture/Nutrition Coordinating Center's Carotenoid Database for U.S. Foods, the mean amount of lycopene in a medium-sized fresh tomato is 3.7 mg. Suppose the amount of lycopene is a normal random variable with standard deviation 1 mg. Consider the sample mean based on 5 observations, $\overline{X}_5$, and the sample mean based on 20 observations, $\overline{X}_{20}$.
(a) Find the distributions for $\overline{X}_5$ and $\overline{X}_{20}$.
(b) Carefully sketch the probability density functions for X, $\overline{X}_5$, and $\overline{X}_{20}$ on the same coordinate axes.

(c) Find $P(X < 3)$, $P(\bar{X}_5 < 3)$, and $P(\bar{X}_{20} < 3)$. Explain why these values are different.
(d) Find $P(3.6 \le X \le 3.8)$, $P(3.6 \le \bar{X}_5 \le 3.8)$, and $P(3.6 \le \bar{X}_{20} \le 3.8)$. Explain why these values are different.

7.57 Sports and Leisure A health club recently added tanning booths as well as a weight room, pool, and racquetball courts. For insurance purposes, an employee maintains careful records of the length of time each member spends in a tanning booth. The mean length of time spent tanning is 15 min, with a standard deviation of 2 min. Consider a random sample of 35 members who use a tanning booth.
(a) Find the distribution of the total time spent tanning by the 35 members T.
(b) Find the probability that the total time spent tanning is between 8 and 9 hr.
(c) If the total time spent tanning is more than 9.2 hr, an employee must work overtime. What is the probability that the employee works overtime on a day when 35 members tan?
(d) Find a value t such that $P(T \ge t) = 0.01$.

7.58 Sports and Leisure *The Tonight Show* with Johnny Carson lasted for 30 years and featured movie stars, comedians, animal acts, and Carnac the Magnificent. The first part of the show was reserved for Johnny's monologue. The mean length of a monologue was 12 min, with a standard deviation of 45 sec. Suppose 40 shows are selected at random, and the length of each monologue is recorded.
(a) Find the distribution of the total monologue time for the 40 shows T.
(b) Find $P(T > 500)$.
(c) Find $P(470 \le T \le 490)$.
(d) After his retirement, Carson's production company started to sell DVDs of his monologues. Suppose one DVD holds 7.9 hr of recording. What is the probability that the company will be able to fit 40 randomly selected monologues onto one DVD?

7.59 Manufacturing and Product Development A (destructive) tensile test is standard procedure for testing a cross-wire weld. Suppose a certain weld is designed to withstand a force with a mean of 0.8 kilonewton (kN). To check the quality of welds, a manufacturer randomly selects 25 welds (every hour) and performs a tensile test on each weld. If the mean force required (for the 25 welds) is between 0.75 and 0.85 kN, the process is allowed to continue. Otherwise, it is shut down. Suppose the force distribution is normal, with a standard deviation of 0.1 kN.
(a) If the true mean is 0.8 kN, what is the probability that the process will be shut down? This probability represents the chance of making one kind of error.
(b) If the true mean is 0.82 kN, what is the probability that the process will be allowed to continue? What if $\mu = 0.84$ kN? These probabilities represent the chance of making a different kind of error.
(c) Suppose the process is allowed to continue if the mean force required is between 0.76 and 0.84 kN. Answer parts (a) and (b) with this new interval.

7.60 Manufacturing and Product Development Shetland wool is considered some of the finest in the world because it is soft, durable, and easy to spin. The mean fleece weight from a typical sheep is 3.25 lb, with a standard deviation of 0.4 lb. Suppose a farmer has 100 sheep ready to be sheared.
(a) Find the distribution for the total weight of fleece from the 100 sheep T.
(b) What is the probability that the total fleece weight is less than 323 lb?
(c) What is the probability that the total fleece weight is between 330 and 340 lb?
(d) Find a value t such that $P(T \ge t) = 0.15$.

7.61 Physical Sciences A wingsuit is a specially designed jumpsuit with added surface area to provide more lift. Extreme-sports enthusiasts wear this suit and jump from planes or a fixed object (BASE jumping, where BASE stands for Building, Antenna, Span, Earth). One of the most popular BASE environments is Troll Wall in Norway. Suppose the mean flight time for a wingsuit jump at Troll Wall is 22.5 sec, with a standard deviation of 2.3 sec. Thirty-two jumpers are selected at random to test a new wingsuit at Troll Wall.
(a) What is the probability that the sample mean flight time is less than 22 sec?
(b) Suppose the sample mean flight time for these jumpers is 23.15 sec. Is there any evidence to suggest that this new wingsuit increases the mean flight time?
(c) Find a value w such that $P(\bar{X} \le w) = 0.005$.

7.62 Travel and Transportation The Airbus Beluga XL transport aircraft, named after the whale, is oddly shaped but specially designed to help companies move large parts from production facilities. The cargo door is in the forward section, and the cargo bay can accommodate payloads of approximately 53 tons.[18] Suppose shipping records indicate the mean cargo weight is 42 tons, with a standard deviation of 4 tons. Forty flights are selected at random, and the cargo weight for each is recorded.
(a) Find the probability that the sample mean cargo weight $\bar{X}$ is less than 41 tons.
(b) Find the probability that the sample total cargo weight T is less than 1640 tons.
(c) Compare your answers in part (a) and part (b). Explain this result.
(d) Suppose the sample mean cargo weight is $\bar{x} = 42.75$ tons. Is there any evidence to suggest that the population mean cargo weight has increased? Justify your answer.

Challenge Problems

7.63 Fuel Consumption and Cars Companies receiving large shipments of raw materials of any product often use a specific plan for accepting the entire shipment. An acceptance sampling plan usually includes the sample size for close inspection, the acceptance criterion, and the rejection criterion. For a given plan, the operating characteristic (OC) curve shows the probability of accepting the entire lot as a function of the actual quality level.

Standard clip-on weights for steel rims on automobiles (used when tires are balanced) are available in 0.25-oz to 6-oz sizes. Suppose an automobile garage receives a large shipment of 2-oz weights. A garage mechanic will select a random sample of 30 weights and weigh each one on a precise scale. If the sample mean is within 0.05 oz of the printed weight (of 2 oz), then the shipment is accepted. Otherwise, the entire shipment is rejected and returned to the manufacturer. Suppose the population standard deviation is 0.13 oz.

(a) Find the probability of accepting the entire shipment if the true population mean weight is 1.86, 1.88, 1.90, 1.92, 1.94, 1.96, 1.98, 2.00, 2.02, 2.04, 2.06, 2.08, 2.10, 2.12, or 2.14 oz.
(b) Plot the probability of accepting the entire shipment versus the true population mean weight. The resulting graph is the OC curve for the given acceptance sampling plan.

7.64 Normal Approximation to the Binomial Distribution Consider a binomial experiment with n trials and probability of success p. If we assign a 0 to each failure and a 1 to each success, then the binomial random variable can be defined as a sum. By the CLT, as n increases, the distribution of X (the sum) approaches a normal distribution with mean np and variance $np(1-p)$. Suppose X is a binomial random variable with $n = 30$ and probability of success $p = 0.5$.

(a) Construct a probability histogram for the binomial random variable X. Find $P(12 \leq X \leq 16)$ using the binomial distribution.
(b) Find the approximate normal distribution for X. Find $P(12 \leq X \leq 16)$ using the normal distribution for X.
(c) Compare the probabilities found in parts (a) and (b).
(d) Find $P(11.5 \leq X \leq 16.5)$ using the normal distribution for X. Compare this answer with the probability in part (a). Why do you think this is a much better approximation?

7.65 Sports and Leisure The AMC Hamilton theater is showing 20 different movies over the weekend. The runtime (in minutes) for each is given in the table. **AMC**

| 126 | 112 | 124 | 115 | 124 | 143 | 130 | 112 | 136 | 121 |
| 141 | 117 | 115 | 136 | 133 | 130 | 134 | 118 | 120 | 110 |

(a) Find the (population) mean runtime for the 20 movies.
(b) Use technology to generate at least 100 random samples of size 5 (without replacement). Find the sample mean runtime for each sample.
(c) Construct a histogram of the sample means found in part (a). Describe the distribution.

7.3 The Distribution of the Sample Proportion

The Sampling Distribution of $\hat{P}$

The sampling distribution of the sample mean was introduced in Section 7.2, and the CLT was used if the underlying population is not normal. Knowing the distribution of $\overline{X}$, we can use the sample mean to make an inference about the population mean μ. Similarly, we are often interested in drawing a conclusion about the population proportion p (the probability of a success). For example, a politician might want to estimate the proportion of voters in a district in favor of a certain highway bill, or a quality-control supervisor might need to estimate the true proportion of defective parts in a large shipment.

It seems reasonable to use a value of the sample proportion $\hat{P}$ to make an inference concerning the population proportion p. Therefore, knowledge of the **sampling distribution of the sample proportion** is necessary. We need to completely characterize the variability of this statistic.

Consider a sample of n individuals or objects (or trials), and let X be the number of successes in the sample. The sample proportion (introduced in Section 3.1) is defined to be

$$\hat{P} = \frac{X}{n} = \frac{\text{the number of successes in the sample}}{\text{the sample size}} \tag{7.1}$$

The sample proportion is simply the proportion of successes in the sample, or a relative frequency.

An approach similar to the one applied in Section 7.2 can be used to approximate the distribution of $\hat{P}$: Generate lots of sample proportions, construct a histogram, and try to characterize the distribution in terms of shape, center, and variability. This sampling distribution is summarized next.

The Sampling Distribution of $\hat{P}$

Let $\hat{P}$ be the sample proportion of successes in a sample of size n from a population with true proportion of success p.

1. The mean of $\hat{P}$ is the true population proportion.
   In symbols: $\mu_{\hat{P}} = p$

2. The variance of $\hat{P}$ is $\sigma^2_{\hat{P}} = \dfrac{p(1-p)}{n}$.

   The standard deviation of $\hat{P}$ is $\sigma_{\hat{P}} = \sqrt{\dfrac{p(1-p)}{n}}$.

3. If n is large and both $np \geq 5$ and $n(1-p) \geq 5$, then the distribution of $\hat{P}$ is approximately normal.
   In symbols: $\hat{P} \overset{\cdot}{\sim} N(p, p(1-p)/n)$

As n increases, the distribution of $\hat{P}$ approaches a normal distribution. There is no threshold value for n. The larger the value of n and the closer p is to 0.5, the better is the approximation.

A CLOSER LOOK

1. It may be a little surprising to learn that $\hat{P}$ is approximately normal. However, the sample proportion can be written as a sample mean. Assign 0 to each failure and 1 to each success. X, the total number of successes, is a sum (of 0s and 1s). Therefore, the sample proportion X/n is really a sample mean. Remember that the CLT says that the sample mean is approximately normal for n sufficiently large.

2. Because the mean of $\hat{P}$ is the true population proportion, $\hat{P}$ is an unbiased estimator for p.

3. A large sample isn't enough for normality. The two products np and $n(1-p)$ must both be greater than or equal to 5. This is called the nonskewness criterion. It guarantees that the distribution of $\hat{P}$ is approximately symmetric, that is, centered far enough away from 0 or 1.

4. To compute a probability involving the sample proportion, treat $\hat{P}$ like any other normal random variable: Standardize and use cumulative probability where appropriate.

The next example illustrates the properties of $\hat{P}$ and the technique for computing probabilities associated with this random variable.

Probabilities Associated with $\hat{P}$

EXAMPLE 7.10 Comic-Con Spending

Comic-Con International: San Diego is a nonprofit, educational corporation that creates awareness of, and appreciation for, comics and similar artforms. The 2018 convention featured approximately 700 exhibitors and an estimated 130,000 attendees. Most of these fans buy lots of swag, such as collectible comic items, T-shirts, toys, and figurines, and approximately 65% of all attendees spend more than $500.[19] Suppose 130 attendees of the San Diego Comic-Con convention are selected at random and the number who actually spend more than $500 is determined.

Mike Blake/Reuters/Newscom

(a) Find the distribution of the sample proportion of attendees who spend more than $500, $\hat{P}$. Carefully sketch the probability density function for this random variable.

(b) What is the probability that the sample proportion (for the 130 attendees selected) is greater than 0.70?

(c) Find the probability that the sample proportion will be between 0.57 and 0.67.

Solution

This problem involves the sample proportion and the random variable $\hat{P}$. Translate each question into a probability statement, use the approximate distribution of $\hat{P}$, standardize, and use cumulative probability where appropriate.

(a) For $n = 130$ and $p = 0.65$, check the nonskewness criterion.

$$np = (130)(0.65) = 84.5 \geq 5 \quad \text{and} \quad n(1-p) = (130)(0.35) = 45.5 \geq 5$$

Both inequalities are satisfied. The distribution of $\hat{P}$ is approximately normal with

$$\mu_{\hat{P}} = p = 0.65 \quad \text{and} \quad \sigma_{\hat{P}}^2 = \frac{p(1-p)}{n} = \frac{(0.65)(0.35)}{130} = 0.00175$$

In symbols: $\hat{P} \overset{\cdot}{\sim} N(0.65, 0.00175) \quad \sigma_{\hat{P}} = \sqrt{0.00175}$

Figure 7.16 shows a graph of the probability density function and a visualization of the probability in part (b).

> **Common Error**
> $\sigma_{\hat{P}} = \frac{p(1-p)}{n}$
>
> **Correction:** The variance of $\hat{P}$ is $\sigma_{\hat{P}}^2 = \frac{p(1-p)}{n}$. The standard deviation of $\hat{P}$ is $\sigma_{\hat{P}} = \sqrt{\frac{p(1-p)}{n}}$.

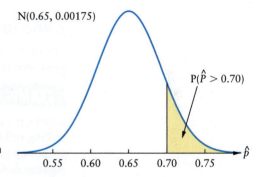

Figure 7.16 A graph of the probability density function for $\hat{P}$. The shaded area represents $P(\hat{P} > 0.70)$.

(b) To find the probability that the sample proportion is greater than 0.70, write a corresponding probability statement, and standardize.

$P(\hat{P} > 0.70)$ — Probability that the sample proportion is greater than 0.70.

$= P\left(\dfrac{\hat{P} - 0.65}{\sqrt{0.00175}} > \dfrac{0.70 - 0.65}{\sqrt{0.00175}} \right)$ — Standardize.

$= P(Z > 1.20)$ — Equation 6.8; simplify.

$= 1 - P(Z \leq 1.20)$ — Use cumulative probability.

$= 1 - 0.8849$ — Use Appendix Table 3.

$= 0.1151$

The probability that the sample proportion is greater than 0.70 is 0.1151.

(c) Write an appropriate probability statement and standardize.

$P(0.57 \leq \hat{P} \leq 0.67)$ — Probability the sample proportion is between 0.57 and 0.67.

$= P\left(\dfrac{0.57 - 0.65}{\sqrt{0.00175}} \leq \dfrac{\hat{P} - 0.65}{\sqrt{0.00175}} \leq \dfrac{0.67 - 0.65}{\sqrt{0.00175}} \right)$ — Standardize.

$= P(-1.91 \leq Z \leq 0.48)$ — Equation 6.8; simplify.

$= P(Z \leq 0.48) - P(Z \leq -1.91)$ — Use cumulative probability.

$= 0.6844 - 0.0281$ — Use Appendix Table 3.

$= 0.6563$

The probability that the sample proportion will be between 0.57 and 0.67 is 0.6563.

Figure 7.17 helps to visualize this standardization, and **Figures 7.18** and **7.19** show technology solutions to parts (b) and (c).

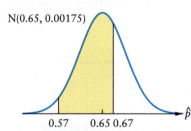

The shaded area represents $P(0.57 \leq \hat{P} \leq 0.67)$.

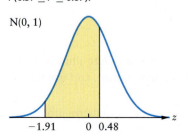

The shaded area represents $P(-1.91 \leq Z \leq 0.48)$.

Figure 7.17 A visualization of the standardization in part (c).

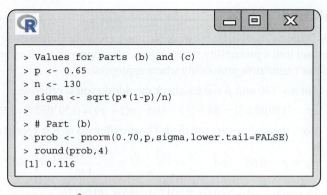

Figure 7.18 $P(\hat{P} > 0.70)$: right-tail probability.

Figure 7.19 $P(0.57 \leq \hat{P} \leq 0.67)$.

TRY IT NOW Go to Exercises 7.77 and 7.79

$\hat{p}$ and Inference

The next example illustrates an inverse cumulative probability problem and an inference associated with the random variable $\hat{P}$.

EXAMPLE 7.11   Too Many Regulations

Company executives often complain that there are too many government regulations. Stifling rules and endless bureaucracy may limit creativity, new-product research, and corporate profits. Suppose 60% of all CEOs believe there are too many government regulations for business. A random sample of 150 CEOs is obtained, and each participant is asked whether he or she believes there are too many government regulations.

(a) Find a value r such that the probability that the sample proportion is greater than r is 0.25.

(b) In recent years, big business has lobbied politicians to relax regulations to stimulate the economy. Suppose the sample proportion for the 150 CEOs is 0.56. Is there any evidence to suggest that the true proportion of CEOs who believe there are too many regulations has decreased?

Solution

For $n = 150$ and $p = 0.60$, check the nonskewness criterion.

$$np = (150)(0.60) = 90 \geq 5 \quad \text{and} \quad n(1-p) = (150)(0.40) = 60 \geq 5$$

Another way to ask this question: Find the third quartile of the $\hat{P}$ distribution.

Both inequalities are satisfied. The distribution of $\hat{P}$ is approximately normal with

$$\mu_{\hat{P}} = p = 0.60 \quad \text{and} \quad \sigma^2 = \frac{p(1-p)}{n} = \frac{(0.60)(0.40)}{150} = 0.0016$$

In symbols: $\hat{P} \overset{.}{\sim} N(0.60, 0.0016)$ $\sigma_{\hat{P}} = \sqrt{0.0016} = 0.04$

(a) We need to find a value of r such that $P(\hat{P} > r) = 0.25$. **Figure 7.20** illustrates this probability statement. This is a backward problem: Standardize, and use inverse cumulative probability.

To find the value r, write an equivalent expression involving cumulative probability. Standardize and work backward in Appendix Table 3.

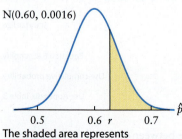

N(0.60, 0.0016)

The shaded area represents $P(\hat{P} > r) = 0.25$.

Figure 7.20 A visualization of the backward probability statement in part (a).

$$P(\hat{P} \leq r) = P\left(\frac{\hat{P} - 0.60}{0.04} \leq \frac{r - 0.60}{0.04}\right) \quad \text{Standardize.}$$

$$= P\left(Z \leq \frac{r - 0.60}{0.04}\right) = 0.75 \quad \text{Equation 6.8.}$$

7.3 The Distribution of the Sample Proportion

There is no further simplification, but the resulting probability statement involves Z and cumulative probability. Find a value in the body of Appendix Table 3 that is as close to 0.75 as possible; use interpolation if necessary or desirable. Set the corresponding z value equal to $\left(\dfrac{r - 0.60}{0.04}\right)$, and solve for r.

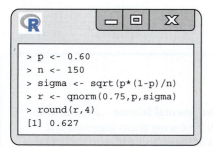

$$\dfrac{r - 0.60}{0.04} = 0.675 \qquad \text{Appendix Table 3; interpolation.}$$

$$r - 0.60 = 0.027 \qquad \text{Multiply both sides by 0.04.}$$

$$r = 0.627 \qquad \text{Add 0.60 to both sides.}$$

Figure 7.21 Use the R function `qnorm` to solve this inverse cumulative probability problem.

The probability that the sample proportion is greater than 0.627 is 0.25. **Figure 7.21** shows a technology solution.

(b) To decide whether the observed sample proportion is reasonable, we need to use the usual four-step inference procedure. Consider the claim, the experiment, and the likelihood of the experimental outcome, and then draw a conclusion.

Claim: $p = 0.60$ (60% of all CEOs believe there are too many government regulations.) $\Rightarrow \hat{P} \stackrel{\cdot}{\sim} N(0.60, 0.0016)$

Experiment: The proportion of CEOs who believe there are too many government regulations is $\hat{p} = 0.56$.

Likelihood: Because $\hat{p} = 0.56$ is to the left of the claimed mean ($p = 0.60$), and because we are looking for evidence that the mean has decreased, consider a left-tail probability as a measure of likelihood.

$$P(\hat{P} \leq 0.56) = P\left(\dfrac{\hat{P} - 0.60}{0.04} \leq \dfrac{0.56 - 0.60}{0.04}\right) \qquad \text{Standardize.}$$

$$= P(Z \leq -1.00) \qquad \text{Equation 6.8; simplify.}$$

$$= 0.1587 \qquad \text{Appendix Table 3.}$$

Figure 7.22 helps to visualize this standardization, and **Figure 7.23** shows a technology solution for this probability calculation. Recall that in statistical inference problems, any probability less than or equal to 0.05 is considered small. So 0.05 can be thought of as a cutoff value.

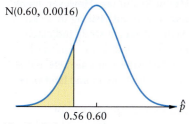

The shaded area represents $P(\hat{P} \leq 0.56)$.

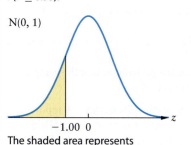

The shaded area represents $P(Z \leq -1.00)$.

Figure 7.22 A visualization of the standardization in part (b).

Conclusion: This likelihood (tail) probability is larger than 0.05, so it is reasonable to observe a sample proportion of 0.56 or smaller. There is no evidence to suggest that the claim of $p = 0.60$ is wrong.

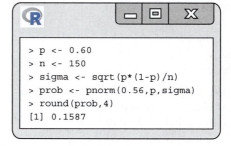

Figure 7.23 Since $\hat{P}$ is approximately normal, use the R function `pnorm`.

TRY IT NOW Go to Exercises 7.83 and 7.87

Section 7.3 Exercises

Concept Check

7.66 Fill in the Blank The mean of $\hat{P}$ is _____ and the variance is _____.

7.67 True or False The distribution of $\hat{P}$ is approximately normal for any sample size n.

7.68 True or False The sample proportion can be characterized as a sample mean.

7.69 True or False The sample proportion is an unbiased estimator for p.

7.70 Short Answer Explain why the inequalities $np \geq 5$ and $n(1-p) \geq 5$ must be true for the distribution of $\hat{P}$ to be approximately normal.

7.71 Short Answer Explain why the standard deviation of $\hat{P}$ decreases as the sample size n increases.

Practice

7.72 Suppose a random sample of size n is obtained. Check the nonskewness criterion, and find the distribution of the sample proportion $\hat{P}$.
(a) $n = 100$, $p = 0.25$
(b) $n = 150$, $p = 0.90$
(c) $n = 100$, $p = 0.75$
(d) $n = 1000$, $p = 0.85$
(e) $n = 5000$, $p = 0.006$

7.73 Suppose a random sample of size $n = 200$ is obtained from a population with probability of success $p = 0.40$. Find each of the following probabilities.
(a) $P(\hat{P} \leq 0.37)$
(b) $P(\hat{P} > 0.45)$
(c) $P(0.38 \leq \hat{P} \leq 0.42)$
(d) $P(\hat{P} < 0.33 \text{ or } \hat{P} > 0.47)$

7.74 Suppose a random sample of size $n = 500$ is obtained from a population with probability of success $p = 0.50$. Find each of the following probabilities.
(a) $P(0.52 > \hat{P})$
(b) $P(\hat{P} \geq 0.47)$
(c) $P(0.44 \leq \hat{P} < 0.49)$
(d) $P(0.45 \leq \hat{P} < 0.55)$

7.75 Suppose a random sample of size $n = 80$ is obtained from a population with probability of success $p = 0.35$.
(a) Find a value a such that $P(\hat{P} \leq a) = 0.10$.
(b) Find a value b such that $P(\hat{P} > b) = 0.01$.
(c) Find a value c such that $P(0.35 - c \leq \hat{P} \leq 0.35 + c) = 0.95$.

7.76 Suppose a random sample of size $n = 1000$ is obtained from a population with probability of success $p = 0.25$.
(a) Find a value a such that $P(\hat{P} \leq a) = 0.05$.
(b) Find a value b such that $P(\hat{P} > b) = 0.005$.
(c) Find a value c such that $P(0.25 - c \leq \hat{P} \leq 0.25 + c) = 0.99$.
(d) Find the quartiles of the distribution of $\hat{P}$.

Applications

7.77 Biology and Environmental Science A recent audit of British Columbia fish-processing plants suggested that 70% of all plants are out of compliance with environmental regulations.[20] Additional regulations are necessary, including an updated permit process, to protect the marine environment. Suppose this survey is representative of all fish-processing plants in Canada, and a random sample of 250 Canadian fish-processing plants is obtained. Each plant was inspected to determine compliance with current regulations.
(a) Find the distribution of the sample proportion of fish-processing plants that are out of compliance.
(b) Find the probability that the sample proportion is less than 0.66.
(c) Find the probability that the sample proportion is more than 0.71.
(d) Find the probability that the sample proportion is between 0.68 and 0.78.

7.78 Sports and Leisure There are certain iconic landmarks in the United States that everyone really needs to visit, such as the Empire State Building, Niagara Falls, the Golden Gate Bridge, and Mount Rushmore. However, many Americans have traveled abroad without visiting these kinds of locations in the United States. A recent study indicated that only 75% of Americans have visited iconic landmarks in their own cities.[21] Suppose 200 Americans are selected at random and asked if they have visited any iconic landmark in their own city.
(a) Find the distribution of the sample proportion of Americans who have visited an iconic landmark in their own city.
(b) Find the probability that the sample proportion is less than 0.71.
(c) Find the probability that the sample proportion is more than 0.78.
(d) Find a value m such that the probability that the sample proportion is less than m is 0.01.

7.79 Demographics and Population Statistics A report by the Inspector General at the Department of Justice (DOJ) revealed that only 16% of criminal investigators in the four law enforcement components were women.[22] Suppose 150 DOJ criminal investigators are selected at random.
(a) Find the distribution of the sample proportion of criminal investigators who are women.
(b) Find the probability that the sample proportion is more than 0.19.
(c) Find the probability that the sample proportion is less than 0.11.
(d) Find a symmetric interval about the mean ($p = 0.16$) such that the probability that the sample proportion is in this interval is 0.90.

7.80 Sports and Leisure Anyone who bets money in a casino is classified as a winner if he or she wins more than he or she loses. Casino operators in Atlantic City, New Jersey, believe that the proportion of all players who go home a winner is 0.46. Suppose 75 Atlantic City gamblers are selected at random.
(a) Find the sampling distribution of the proportion of gamblers who go home winners.
(b) Find the probability that the sample proportion is less than 0.40.
(c) Find the probability that the sample proportion is more than 0.45.
(d) Find a symmetric interval about the mean ($p = 0.46$) such that the probability that the sample proportion is in this interval is 0.99.

7.81 Public Policy and Political Science A Gallop poll suggested that 64% of American adults are in favor of legalizing the recreational use of marijuana.[23] Suppose 225 American adults are selected at random.
(a) Find the probability that the sample proportion of American adults in favor of legalizing marijuana is less than 0.60.
(b) Find the probability that the sample proportion of American adults in favor of legalizing marijuana is between 0.65 and 0.70.
(c) Find a value m such that $P(\hat{P} > m) = 0.95$.

7.82 Public Health and Nutrition Dermatologists suggest that 70% of all people are allergic to the oil produced by poison ivy.[24] The itchy rash associated with poison ivy can be unbearable; we just need to remember that adage, "Leaves of three, let them be." Suppose 125 people are selected at random, and each is tested for an allergic reaction to poison ivy.
(a) Find the probability that the sample proportion of people allergic to poison ivy is less than 0.65.
(b) Find the probability that the sample proportion of people allergic to poison ivy is between 0.62 and 0.72.
(c) Some dermatologists believe that the proportion of people who are allergic to poison ivy is increasing due to a variety of environmental and behavioral factors. Suppose 99 of the 125 people are allergic to poison ivy. Is there any evidence to suggest that the proportion of people who are allergic to poison ivy has increased? Justify your answer.

7.83 Education and Child Development Admissions offices at universities and colleges keep careful records of acceptance and yield rates. The yield rate is the percentage of admitted students who decide to accept an offer of admission. Historical data help admissions personnel plan for the incoming classes and guide some admission decisions. The yield rate for Harvard University is 82%.[25] Suppose 100 students who were accepted to Harvard are randomly selected.
(a) Find the probability that the sample proportion of those who enroll is greater than 0.87.
(b) Find the probability that the sample proportion of those who enroll is between 0.75 and 0.85.
(c) Suppose the actual sample proportion of those who enroll is 0.72. Is there any evidence to suggest that the yield has decreased? Justify your answer.

7.84 Business and Management A certain philanthropic organization funds one out of every 10 grant proposals. Suppose a random sample of 300 grant proposals is obtained.
(a) Find the probability that the sample proportion of funded grant proposals is less than 0.075.
(b) Find the probability that the sample proportion of funded grant proposals is between 0.11 and 0.15.
(c) If the funding rate increases, the board of directors may become concerned about resources being depleted. Suppose the actual sample proportion of funded grant proposals is 0.16. Is there any evidence to suggest that the funding rate has increased? Justify your answer.

7.85 Sports and Leisure The fielding percentage for a major league baseball team is a measure of how well defensive players handle a batted ball (without error) and is based on total chances (putouts + assists + errors). At the All-Star break of the 2018 baseball season, the Boston Red Sox's fielding percentage was 0.987.[26] Suppose 400 chances are randomly selected.
(a) Find the distribution of the sample proportion of chances handled without error.
(b) Find the probability that the sample proportion is less than 0.98.
(c) Find a value f such that $P(\hat{P} \leq f) = 0.05$.
(d) If the team fielding percentage for these random chances is greater than 0.995, then everyone on the team receives a bonus. What is the probability of a team bonus being awarded?

7.86 Psychology and Human Behavior According to a report by the Federal Reserve Board, 40% of Americans cannot cover a $400 emergency expense.[27] If they did not have the cash on hand, most people said they would borrow or sell something. The good news is that this percentage represents a decrease from a few years ago. Suppose 350 Americans are selected at random.
(a) Find the probability that the sample proportion of Americans who cannot cover a $400 emergency expense is less than 0.37.
(b) Find the probability that the sample proportion is between 0.41 and 0.42.
(c) Find a value c such that $P(\hat{P} \geq c) = 0.01$.

7.87 Biology and Environmental Science Ships dumping garbage and ordinary beachgoers contribute to the increasing amount of trash that washes onto the shore and collects in the oceans all over the world. In 2017, the International Coastal Cleanup collected more than 20 million pieces of trash from the world's beaches and waterways. Common debris items collected included food wrappers, plastic beverage bottles, and plastic grocery bags. Cigarettes accounted for 12.1% of all debris items.[28] A random sample of 450 debris items from a clean-up along a Texas beach was obtained.

(a) Find the probability that the proportion of cigarette debris items is greater than 0.14.
(b) Find the probability that the proportion of cigarette debris items is between 0.10 and 0.15.
(c) Suppose 36 of the debris items are cigarettes. Is there any evidence to suggest that the true proportion of cigarette debris items is different from 0.121? Justify your answer.

7.88 Economics and Finance In Canada, 18% of renter households spend more than 50% of their income on housing.[29] According to the Canada Mortgage and Housing Corporation, spending more than 50% of one's income on housing is considered to be a crisis level of spending, and could put families at risk. Suppose 200 Canadian renters are selected at random.
(a) Find the probability that the sample proportion of renters spending more than 50% of income on housing is between 0.15 and 0.20.
(b) Suppose the sample proportion is less than 0.16. Find the probability that it is less than 0.12.
(c) Suppose the sample proportion for the 200 Canadians selected is 0.22. Is there any evidence to suggest that the true proportion ($p = 0.18$) has increased? Justify your answer.

Extended Applications

7.89 Manufacturing and Product Development Low-quality coffee shipped from various locations in Europe tends to contain a high proportion of defective beans (beans composed of foreign matter; moldy, black, unripe, or fermented beans; or those known as stinkers). Suppose a shipper claims that the proportion of defective beans is 0.07. A U.S. packaging company receives a huge shipment of coffee beans and randomly selects 1000 beans. If the sample proportion of defective beans is more than 0.09, then the entire shipment will be returned to the supplier in Europe.
(a) If the true proportion of defective beans is 0.07 (as claimed), what is the probability that the shipment will be sent back?
(b) If the true proportion of defective beans is 0.08, what is the probability that the shipment will be accepted?

7.90 Manufacturing and Product Development Mueller's Hardware in Pittsburgh, Pennsylvania, routinely receives shipments of 4 ft × 8 ft sheets of 1/2-in.-thick plywood. A plywood sheet may contain defects, such as a knot, a split, or a deviation in wood structure. Defective sheets reduce profits because they are sold at a lower price. The manufacturer claims the proportion of all plywood sheets that are defective is 0.05. Suppose a large shipment of plywood sheets is received and 200 are randomly selected for inspection. If the sample proportion of defective sheets is more than 0.09, the entire shipment will be sent back to the supplier.

(a) If the true proportion of defective plywood sheets is 0.05, what is the probability that the entire shipment will be sent back?
(b) If the true proportion of defective plywood sheets is 0.03, what is the probability that the entire shipment will be sent back?
(c) If the true proportion of defective plywood sheets is 0.10, what is the probability that the shipment will be accepted?

7.91 Demographics and Population Statistics The high cost of living, increasing home prices, and earthquakes are forcing many San Francisco Bay Area residents to leave the area. Suppose the proportion of residents who are planning to leave the area is $p (< 0.50)$ and 200 San Francisco Bay Area residents are selected at random.
(a) Let $\hat{P}$ be the sample proportion of residents planning to leave the area. The variance of $\hat{P}$ is 0.001242. Find the value of p.
(b) Find the value of r such that $P(\hat{P} \leq r) = 0.70$.

7.92 Economics and Finance Many Americans worry about how much they should tip when dining out. However, a Zagat survey indicated that Americans, in general, tip well, and 43% are in favor of eliminating tipping completely, even if it would mean higher meal costs.[30] Suppose a random sample of n Americans is obtained.

Let $\hat{P}$ be the sample proportion of Americans who are in favor of eliminating tips. If $P(\hat{P} \leq 0.47) = 0.90$, find the value of n, the number of Americans in the random sample.

Challenge Problems

7.93 Manufacturing and Product Development Suppose a company is receiving a large shipment of peel-and-stick vinyl floor tiles. The manufacturer claims that the proportion of defective floor tiles is 0.05. The company will select a random sample of 200 floor tiles, carefully inspect each, and determine whether it is defective. The acceptance sampling plan states: Accept the entire shipment if the sample proportion of defective tiles is 0.08 or less; otherwise, reject the entire shipment.
(a) Find the probability of accepting the entire shipment if the true proportion of defective floor tiles is 0.01, 0.02, 0.03, 0.04, 0.05, 0.06, 0.07, 0.08, 0.09, 0.10, or 0.15.
(b) Plot the probability of accepting the entire shipment (on the y-axis) versus the true proportion of defective floor tiles (on the x-axis). The resulting graph is the operating characteristic (OC) curve for the given acceptance sampling plan.

7.94 Maximum Variance For a fixed sample size n, find the value of p that maximizes the variance of the sample proportion, $\sigma_{\hat{P}}^2 = \dfrac{p(1-p)}{n}$.

Hint: Compute the value of the variance for several different values of p. Plot the variance (on the y-axis) versus the value of p (on the x-axis).

Chapter 7 Summary

Concept	Page	Notation / Formula / Description
Parameter	300	A numerical descriptive measure of a population.
Statistic	300	Any quantity computed from values in a sample.
Sampling distribution	301	The probability distribution of a statistic based on all possible random samples of specific size (n).
Properties of the sample mean $\bar{X}$	313	$\mu_{\bar{X}} = \mu$, $\sigma^2_{\bar{X}} = \dfrac{\sigma^2}{n}$. If the underlying population is normal, then $\bar{X}$ is normal.
Central Limit Theorem	313	As the sample size n increases, the sampling distribution of $\bar{X}$ will increasingly approximate a normal distribution, with mean μ and variance σ^2/n, regardless of the shape of the underlying population distribution.
Distribution of the sample proportion of successes $\hat{P}$	323	$\mu_{\hat{P}} = p$, $\sigma^2_{\hat{P}} = \dfrac{p(1-p)}{n}$. If n is large and both $np \geq 5$ and $n(1-p) \geq 5$, then $\hat{P}$ is approximately normal.

Chapter 7 Exercises

Applications

7.95 Manufacturing and Product Development Until the early 1900s, people who colored their hair used only herbs and natural dyes. Today, hair-coloring products contain two main ingredients: hydrogen peroxide and ammonia. The makers of a Clairol hair-color product claim that the mean amount of hydrogen peroxide in each bottle is 0.10 mg/m^3. Assume the standard deviation is 0.05 mg/m^3. A random sample of these hair-color products was obtained, and the amount of hydrogen peroxide in each bottle was measured. **DYES**
(a) Suppose the manufacturer's claim is true. Find the distribution of the sample mean.
(b) Is there any evidence to suggest that the manufacturer is including too much hydrogen peroxide in the product? Justify your answer.

7.96 Physical Sciences A floor slip tester is used to measure the safety of a floor by comparing the measured coefficient of static friction with accepted standards and guidelines. Several factors can affect floor safety, such as dampness, polishes, and maintenance chemicals. A marble floor is considered safe if the coefficient of static friction is no greater than 0.5. A random sample of 50 rainy days was selected, and the coefficient of static friction for the marble floor was measured on each day. The resulting sample mean was 0.6. Is there any evidence to suggest that the marble floor is unsafe on rainy days? Assume the underlying population standard deviation is 0.2 and justify your answer.

7.97 Marketing and Consumer Behavior In the fresh fruits and vegetables section of a Kroger grocery store, customers can purchase any desired amount by placing the food in a plastic bag to be weighed and/or priced at the checkout line. For those people who purchase cucumbers, the probability distribution for the number purchased is given in the table.

x	1	2	3	4	5
$p(x)$	0.10	0.50	0.20	0.15	0.05

(a) Suppose two customers who purchase cucumbers are selected at random. Find the exact probability distribution for the sample mean number of cucumbers purchased $\bar{X}$.
(b) Find the mean, variance, and standard deviation of $\bar{X}$.

7.98 Marketing and Consumer Behavior A drive-in movie theater charges viewers by the carload but keeps careful records of the number of people in each car. The probability distribution for the number of people in each car entering the drive-in is given in the table.

x	1	2	3	4	5	6
$p(x)$	0.02	0.30	0.10	0.30	0.20	0.08

(a) Suppose two cars entering the drive-in are selected at random. Find the exact probability distribution for the maximum number of people in either one of the cars, M.
(b) Find the mean, variance, and standard deviation of M.

7.99 Demographics and Population Statistics Many clubs and companies around the country offer hot-air balloon rides. Most impose strict safety regulations and take at most four adults at one time, plus a pilot. The mean weight for an adult male in the United States is 195.7 lb.[31] Suppose the distribution is normal with a standard deviation of 30 lb.
(a) If a hot-air balloon pilot (an adult male) takes three adult males for a ride, what is the distribution for the total weight aboard T? Carefully sketch the probability distribution for T.
(b) If the total weight is less than 900 lb, the pilot will have enough fuel to extend the ride by a few minutes. What is the probability of an extended ride?
(c) If the total weight is more than 975 lb, then the balloon will not be able to take off. What is the probability that the balloon will not be able to take off?

7.100 Biology and Environmental Science A typical houseplant produces oxygen (from carbon dioxide) in varying amounts, depending on the amount of light and water. Suppose a medium Norfolk Island pine produces 7.5 mL of oxygen per hour when exposed to normal sunlight. Thirty-five Norfolk Island pines are selected at random, and the oxygen output is carefully measured for each. Assume $\sigma = 1.75$ mL/hr.
(a) What is the probability that the sample mean oxygen produced is less than 7 mL/hr?
(b) What is the probability that the sample mean oxygen produced is between 7.25 and 7.5 mL/hr?
(c) Suppose each plant is exposed to a new high-intensity grow light and the sample mean oxygen produced for the 35 plants is 8.1 mL/hr. Is there any evidence that the new lamp has increased oxygen output?
(d) Answer parts (a), (b), and (c) if $\sigma = 3.75$ mL/hr.

7.101 Marketing and Consumer Behavior Seafood restaurants along the coast of New England offer a variety of entrees, but lobster is the most popular meal. At Newick's Seafood Restaurant in Dover, New Hampshire, 37% of all diners order lobster. Suppose 120 customers are selected at random, and the meal ordered by each is recorded.
(a) Find the distribution of the sample proportion of diners who order lobster, and carefully sketch a graph of the probability density function.
(b) Find the probability that the sample proportion of diners who order lobster is less than 0.30.
(c) Find the probability that the sample proportion of diners who order lobster is between 0.35 and 0.40.
(d) The manager of Newick's is concerned that more customers might be ordering lobster. This would require a change in restaurant ordering and a shift in kitchen staff. Suppose the observed proportion of diners who order lobster is 0.42. Is there any evidence to suggest that the proportion of diners who order lobster has increased? Justify your answer.

7.102 Biology and Environmental Science Tropical rainforests are located in areas near the equator and cover approximately 6% of Earth's surface. The temperature in a typical rainforest ranges between 70°F and 85°F, and some rainforests receive almost 400 in. of rain per year.[32] The mean amount of rain per year in any one rainforest is $\mu = 110$ in., with standard deviation $\sigma = 20$ in. Suppose 30 rainforests are selected at random, and the amount of rain per year is recorded for each.
(a) What is the probability that the mean rainfall per year for the 30 rainforests is more than 115 in.?
(b) What is the probability that the mean rainfall is between 100 and 110 in.?
(c) Find a value r such that the probability that the mean rainfall is less than r is 0.001.

7.103 Technology and the Internet An office manager has several computers running distributed programs. Because of the demands on the system, the machines may crash at various times during the day and require a hard reset. The probability distribution for the number of times a randomly selected machine crashes during a day X is given in the table.

x	0	1	2	3	4	5
$p(x)$	0.50	0.30	0.10	0.07	0.02	0.01

(a) Find the mean, variance, and standard deviation for the number of crashes for a single machine during a day.
(b) Suppose $n = 2$ machines are selected at random. Find the sampling distribution of the statistic T, the total number of crashes for the two machines.
(c) Find the mean, variance, and standard deviation of T.
(d) Verify the relationships $\mu_T = 2\mu_X$ and $\sigma_T^2 = 2\sigma_X^2$.

7.104 Parameter Versus Statistic In each of the following statements, identify the **boldface** number as the value of a population parameter or a sample statistic.
(a) Some political observers claim that state senators spend too much time addressing colleagues about pending legislation. The mean length for a random sample of speeches was **23.7** min.
(b) A consulting firm prepared a report from census data and concluded that the proportion of single-family homes in a certain county is **0.52**.
(c) In a random sample of adults, the mean number of (eye) blinks per day was **22,037**.
(d) Federal Express found that the variance of the number of daily round-trip miles traveled by all trucks based in Atlanta was **1501.9**.
(e) A random sample of people who snowboard at least five times per year found the mean number of injuries per person per year to be **3.4**.
(f) In a survey conducted by the U.S. Postal Service, **80%** of respondents supported the proposed new six-day package, five-day letter delivery schedule.

7.105 Medicine and Clinical Studies The stapes in the middle ear is the smallest bone in the human body, with a mean length of 3 mm.[33] Suppose the standard deviation of the length is 0.16 mm. This bone, located in the ear, is part of a leverage system that can affect hearing. Some researchers have speculated that high noise levels can affect the development of this bone and inhibit hearing. In a random sample of adults

who have lived in a large city their entire lives, the length of the right-ear stapes bone was obtained. ▌▌ STAPE
(a) Find the distribution of the sample mean. Carefully sketch a graph of the probability density function.
(b) Is there any evidence to suggest that the length of the stapes bone for lifetime city adults is different from 3 mm? Justify your answer.

7.106 Sampling Distribution Each of the following graphs shows the probability distribution for an underlying distribution and for the sampling distribution of the sample mean of n observations (drawn from the underlying distribution). Identify each probability distribution function.

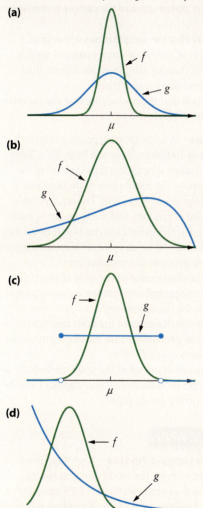

7.107 Economics and Finance Cryptocurrencies are exchanged using blockchain technology, a unique digital ledger that records and transmits transactions without the need for a bank. A cryptocurrency with an efficient, speedy blockchain technology tends to be more valuable. A recent study claimed that the mean transaction speed for Bitcoin is 78 min.[34] Suppose this transaction speed is normally distributed, with a standard deviation of 6.5 min. A random sample of Bitcoin transactions was obtained, and the transaction speed (in minutes) of each was recorded. The resulting data are given in the table. ▌▌ BITCOIN

| 86.2 | 71.0 | 85.8 | 68.7 | 75.9 |
| 73.6 | 72.2 | 74.0 | 75.4 | 81.9 |

(a) Suppose that the study's claim is true. Find the distribution of the mean transaction speed for Bitcoin.
(b) Use the data to determine whether there is any evidence to suggest that the study's claim is false.

7.108 Travel and Transportation The Greenline Taxi Company in New York City keeps careful records of items left behind by riders. Most items are claimed, but many remain in a lost-and-found area at the company's headquarters. Records indicate that the proportion of riders who leave an item in a taxi is 0.12. Suppose a random sample of 250 riders is obtained.
(a) Find the sampling distribution of the sample proportion of riders who leave an item behind. Carefully sketch a graph of the probability density function.
(b) What is the probability that the sample proportion of riders who leave an item behind is more than 0.15?
(c) What is the probability that the sample proportion of riders who leave an item behind is between 0.11 and 0.115?
(d) Some people speculate that in bad economic times, riders (and people in general) are more careful with their belongings. Suppose this sample was obtained during a recession, and the sample proportion of riders who left an item behind was 0.09. Is there any evidence to suggest that the true proportion is less than 0.12?

7.109 Manufacturing and Product Development Eyedrops are used by many people to soothe and relieve irritation and to lubricate their eyes. A manufacturer claims that the amount of dextran, which helps to prolong the effect of eyedrops, contained in its eyedrops product is 70%. Thirty-six randomly selected bottles of these eyedrops were obtained, and the sample mean amount of dextran was 68.25%. Assume that the population standard deviation is $\sigma = 5\%$.
(a) Find the sampling distribution of the sample mean.
(b) Find the probability that the amount of dextran is more than 71%.
(c) Does the observed sample mean suggest that the manufacturer's claim is wrong? Justify your answer.

7.110 Public Health and Nutrition Most health insurance providers require policy holders to share the cost of certain medical services, such as by paying $20 for an office visit. A recent government study indicated that the proportion of all people who have a copayment of more than $20 for an office visit is 0.65. Suppose 1000 insured individuals are randomly selected, and the office copayment amount for each person is recorded.
(a) Find the distribution of the sample proportion of people who have a copayment of more than $20 for an office visit. Verify the nonskewness criterion.
(b) Find the probability that the sample proportion is more than 0.66.
(c) Find the probability that the sample proportion is between 0.64 and 0.67.

(d) Find a value h such that the probability that the sample proportion is less than h is 0.01.

7.111 Public Health and Nutrition Mangoes contain the chemical compound urushiol, found in poison ivy, poison oak, and poison sumac. Approximately 50% of people exposed to urushiol in a mango experience an allergic reaction, such as itching or hives. Suppose 150 adults are selected at random, and each is tested for mango sensitivity.
(a) Find the distribution of the sample proportion of people who experience an allergic reaction to mangoes.
(b) Find the probability that the sample proportion is between 0.45 and 0.55.
(c) Suppose the observed sample proportion is 0.59. Is there any evidence to suggest that the true proportion has increased? Justify your answer.

Extended Applications

7.112 Physical Sciences Carlinville, Illinois, has just started an ambitious recycling program. Special trucks collect recyclable products once a week, with these items then being sorted into barrels of paper, glass, and plastic. The amount of glass recycled per week by a single household is normally distributed, with mean $\mu = 27$ lb and standard deviation $\sigma = 7$ lb. Suppose 12 households are randomly selected.
(a) What is the probability that the total glass collected for the 12 homes T will be less than 350 lb?
(b) Find a value g such that $P(T \geq g) = 0.05$.
(c) If the total glass collected for the 12 homes is more than 400 lb, the recycling plant will make a profit. Find a value of μ such that the probability of making a profit is 0.10.

7.113 Travel and Transportation Current travel trends suggest that people are looking for more personalized experiences and are willing to pay extra for these services. In addition, more people are considering impulse-based trips—that is, booking a trip at the last minute. A recent survey indicated that 53% of travel bookings are impulse bookings, made within a week of the trip.[35] Suppose 250 travel bookings are selected at random, and the number of impulse bookings is recorded.
(a) What is the probability that the sample proportion of impulse bookings is between 0.45 and 0.50?
(b) Find a value b such that the probability that the sample proportion is less than b is 0.01.
(c) Suppose 152 of the 250 travel bookings are classified as impulse-based trips. Is there any evidence to suggest that the true proportion of impulse bookings has increased? Justify your answer.

7.114 Manufacturing and Product Development A manufacturer of ice pops fills each plastic container with a fruity liquid, leaving enough room so that consumers can freeze the product. A filling machine is set so that the amount of liquid in each ice pop is normally distributed, with a mean of 8.00 oz and a standard deviation of 0.25 oz. Suppose 16 ice pops are randomly selected.
(a) Find the probability distribution of the sample mean number of ounces in each container, $\bar{X}$.
(b) Find the probability that the sample mean is less than 7.9 oz.
(c) Find the probability that the sample mean is more than 8.15 oz.

(d) Suppose the filling machine operator can fine-tune the process by controlling the standard deviation of the fill. Find a value for σ such that the probability that the sample mean is more than 8.05 oz is 0.05.

7.115 Biology and Environmental Science During summer 2018, a huge population of 17-year cicadas emerged from the ground in central New York. These prehistoric-looking bugs have red eyes and huge wings, and they make some pretty loud noises. Suppose the mean noise level is 90 decibels and the standard deviation is 15.6 decibels. During the cicada emergence in Virginia and other southern states, a random sample of 40 locations was selected and the noise level (in decibels) was carefully measured.
(a) Find the probability distribution of the sample mean noise level.
(b) Find the probability that the sample mean is less than 60 decibels, the level of moderate conversational speech.
(c) Find the probability that the sample mean is more than 110 decibels, the level of a passing train.
(d) Find a value of n such that the probability that the sample mean is less than 93 decibels is 0.95.

7.116 Sports and Leisure A large sporting goods company has just received a shipment of 100,000 table tennis balls. USA Table Tennis tournament regulations specify that the diameter of the ball must be 40 mm. Suppose the distribution of the diameter is normal, with a standard deviation of 0.4 mm. Twenty-five table tennis balls will be selected at random, and the diameter of each will be carefully measured. If the mean diameter is within 0.2 mm of 40 mm, then the shipment will be accepted. Otherwise, the entire shipment will be returned to the manufacturer.
(a) Suppose the true mean diameter of the table tennis balls is 40 mm. What is the probability that the entire shipment will be sent back to the manufacturer?
(b) Suppose the true mean diameter of the table tennis balls is 40.4 mm. What is the probability that the shipment will be accepted by the sporting goods store?
(c) Suppose the true mean diameter of the table tennis balls is 39.4 mm. What is the probability that the shipment will be accepted by the sporting goods store?

Ras Al Khaimah/ United Arab Emirates/ Andrew Fosker/ PinPep/WENN.com/ Newscom

CHAPTER APP

7.117 The Longest Zip Line A spokesperson for Jebel Jais Flight, the world's longest zip line, claims that the mean ride time is 1.75 min, with a standard deviation of 0.2 min. A random sample of 40 Jebel Jais zip line riders was obtained.
(a) The mean ride time is 1.75 min, with a standard deviation of 0.2 min. Are these values parameters or statistics?
(b) A random sample of 40 riders was obtained. Find the distribution of the sample mean $\bar{X}$.
(c) Find the probability that the sample mean ride time is less than 1.72 min.
(d) The time for each ride was recorded, and the sample mean was 1.80 min. Is there any evidence to suggest that the true mean ride time is greater than 1.75 min?

Confidence Intervals Based on a Single Sample

◀ Looking Back
- Remember the most common numerical summary measures, such as the sample mean, sample variance, and sample proportion: **Section 3.1 and Section 3.2**.
- Recall that the sampling distribution of a statistic is the probability distribution of the statistic: **Section 7.1**.

Looking Forward ▶
- Learn the properties of point estimators and understand what makes an estimator good: **Section 8.1**
- Construct confidence intervals (CIs) for a population mean: **Section 8.2 and Section 8.3**.
- Construct confidence intervals for a population proportion: **Section 8.4**.
- Construct confidence intervals for a population variance or standard deviation: **Section 8.5**.

Scharfsinn/Deposit Photos

CHAPTER APP

Autonomous Cars

Self-driving, or autonomous, cars are vehicles that can drive themselves, without any human interaction. Fully autonomous cars may be commercially available very soon; thousands of driverless taxis are planned to launch in Singapore, and GM intends to produce an autonomous car with no steering wheel in 2019.

Tesla and other companies already sell a limited form of autopilot, which requires the driver to intervene if something unexpected occurs. Waymo is testing cars on certain public roads in California and Arizona. One measure of success in testing autonomous cars is the number of miles driven between disengagements, which are situations in which a driver/passenger has to take control of the car. Companies would naturally like the number of miles between disengagements to be as large as possible.

The Department of Motor Vehicles in Sacramento, California, selected 15 Waymo autonomous cars at random and recorded the number of miles between the most recent disengagements. The summary statistics were $\bar{x} = 5595.3$ and $s = 376.8$.[1] The concepts presented in this chapter will be used to construct an interval in which we are fairly certain that the true mean miles between disengagements lies. This interval of numbers can then be used to make an inference concerning the population mean.

8.1 Point Estimation

Good Estimators

A **point estimate** of a population parameter is a single number computed from a sample, which serves as a guess for the parameter. Using the terminology introduced in Chapter 7:

1. An **estimator** is a statistic of interest and is therefore a random variable. So an estimator has a distribution, a mean, a variance, and a standard deviation.
2. An **estimate** is simply a specific value of an estimator.

> **Definition**
> An **estimator** (statistic) is a rule used to produce a point estimate of a population parameter.

Suppose we need to estimate a population parameter θ, for which there are many different statistics (rules) available. Which one should we use? **Figure 8.1** shows the sampling distributions of three different statistics for estimating θ. These graphs suggest some properties of a good statistic.

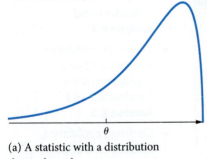
(a) A statistic with a distribution that is skewed.

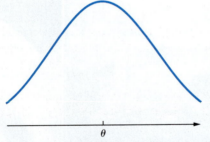
(b) A statistic with large variance.

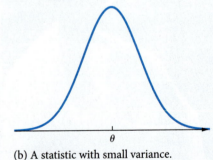
(b) A statistic with small variance.

Figure 8.1 The sampling distributions for three different statistics for estimating θ. The horizontal axis represents all possible values of $\hat{\theta}$, an estimator (statistic) for the parameter θ.

The statistic in Figure 8.1(a) is unlikely to produce a value close to θ. The sampling distribution is skewed to the left, and most of the values of the statistic are found to the right of θ. The statistic in Figure 8.1(b) is centered at θ. On average (in the long run), this statistic will produce θ (that's good!). However, this statistic has large variance (that's bad!). Even though the sampling distribution is centered at the true value of the population parameter, specific estimates will probably be far away from θ. The statistic in Figure 8.1(c) exhibits two very desirable properties: It is centered at the true value of the population parameter, and it has small variance. These observations suggest two rules for selecting a statistic.

Definition

A statistic $\hat{\theta}$ is an **unbiased estimator** of a population parameter θ if $E(\hat{\theta}) = \theta$, that is, if the mean of $\hat{\theta}$ is θ.

If $E(\hat{\theta}) \neq \theta$, then the statistic $\hat{\theta}$ is a **biased estimator** of θ.

Figure 8.2 illustrates the sampling distribution of an *unbiased* estimator for θ. The distribution of the statistic is centered at θ; the value of the statistic is, on average, θ. **Figure 8.3** shows the sampling distribution of a *biased* estimator for θ. On average, the value of this statistic is greater than θ.

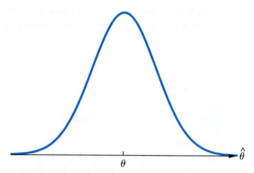

Figure 8.2 The sampling distribution for an unbiased estimator for θ.

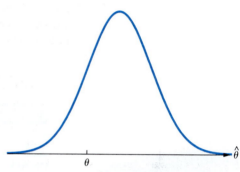

Figure 8.3 The sampling distribution for a biased estimator for θ.

We have already worked with several unbiased estimators in previous chapters.

1. The sample mean $\overline{X}$ is an unbiased statistic for estimating the population mean μ because $E(\overline{X}) = \mu$.
2. The sample proportion $\hat{P}$ is an unbiased statistic for estimating the population proportion p because $E(\hat{P}) = p$.
3. The sample variance S^2 is an unbiased statistic for estimating the population variance σ^2 because $E(S^2) = \sigma^2$.

Even though S^2 is an unbiased estimator for σ^2, the sample standard deviation S is a biased estimator for the population standard deviation σ.

$$E(S) = E(\sqrt{S^2}) \neq \sqrt{E(S^2)} = \sqrt{\sigma^2} = \sigma$$

The expected value operation does not pass freely through the square-root symbol.

Even though S is a biased estimator for σ, it is still very important in statistical inference.

If several statistics are available for estimating the same parameter, say θ, it seems reasonable to use one that is unbiased. Therefore, the first rule for choosing a statistic is that the sampling distribution should be centered at θ: The estimator should be unbiased.

The second rule for choosing a statistic is that, of all unbiased statistics, the best statistic to use is the one with the smallest variance. The point estimate produced using this statistic will, on average, be close to the true value of the population parameter. **Figure 8.4** illustrates this second rule for choosing a statistic.

Suppose we have two statistics to choose from for estimating θ, $\hat{\theta}_1$ and $\hat{\theta}_2$, with sampling distributions as shown in **Figure 8.5**. The statistic $\hat{\theta}_1$ is unbiased but has large variance; $\hat{\theta}_2$ is slightly biased but has small variance. The choice of an estimator is a difficult decision in this case, and there is no single, definitive right answer.

Note that for a given parameter θ, a MVUE may not exist.

Suppose we need to estimate the population parameter θ, and we have several unbiased statistics from which to choose. If one of these statistics has the smallest possible variance, it is called the minimum-variance unbiased estimator (MVUE). If the underlying population is normal, the sample mean $\overline{X}$ is the MVUE for estimating μ. So, if the population is normal, the sample mean is a really good statistic to use for estimating μ. $\overline{X}$ is unbiased, and it has the smallest variance of all possible unbiased estimators for μ.

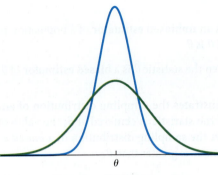
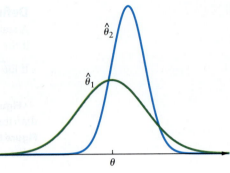

Figure 8.4 Sampling distributions of two unbiased statistics for estimating θ. Use the statistic with the smaller variance.

Figure 8.5 Sampling distributions of an unbiased statistic $\hat{\theta}_1$ with large variance and a biased statistic $\hat{\theta}_2$ with small variance.

Point Estimates

EXAMPLE 8.1   Colorado Snowpack

SNOWPK

Photo courtesy of USDA's Natural Resources Conservation Service (NRCS), Colorado Snow Survey Supervisor in Colorado, Brian Dominos

The spring snowpack in the Colorado mountains is a good predictor of water supply during the summer months. Due to the small snowpack during the spring of 2018, this geographic area faced expanded drought designations, slow stream flows, and concerns about wildfires and water supply for agricultural purposes. A random sample of sites in the upper Colorado River basin during July was obtained and the snow water equivalent at each site was measured (in inches).[2] The data are given in the table.

22.4	21.8	12.5	18.6	17.1	16.0
16.5	24.9	11.1	18.3	33.7	22.6

Find point estimates for the population mean snow water equivalent and for the population median snow water equivalent in July.

Solution

STEP 1 Use the sample mean to estimate the population mean.

$$\bar{x} = \frac{1}{12}(22.4 + 21.8 + \cdots + 33.7 + 22.6)$$

$$= \frac{1}{12}(235.5) = 19.625$$

A point estimate for the population mean snow water equivalent in July is 19.625 in.

STEP 2 Use the sample median to estimate the population median.

Order the observations from smallest to largest. The sample median is the middle value. There are $n = 12$ observations, so the middle value is in position 6.5, that is, the mean of the values in positions 6 and 7.

Ordered observations:

11.1   12.5   16.0   16.5   17.1   18.3   18.6   21.8   22.4   22.6   24.9   33.7

$$\tilde{x} = \frac{1}{2}(18.3 + 18.6) = 18.45$$

A point estimate for the population median snow water equivalent in July is 18.45 in.

Figure 8.6 shows a technology solution.

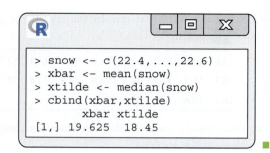

Figure 8.6 Use the R commands `mean()` and `median()` to find the sample mean and the sample median.

TRY IT NOW Go to Exercises 8.13 and 8.15

CHAPTER APP **Autonomous Cars**
Identify the point estimates and the statistics.

Section 8.1 Exercises

Concept Check

8.1 True or False An estimator is a random variable.

8.2 Fill in the Blank A statistic $\hat{\theta}$ is an unbiased estimator of θ if _____.

8.3 Short Answer Name two characteristics of a good estimator.

8.4 Short Answer If two statistics are unbiased, which one would you use?

8.5 Fill in the Blank An unbiased statistic with the smallest possible variance is called _____.

8.6 Short Answer Describe a case in which one would use a biased statistic, instead of an unbiased statistic, to estimate a parameter θ.

8.7 Short Answer Suppose $x_1, x_2, \ldots, x_n$ is a random sample of n observations from a population with mean μ. Give examples of several estimators for a population mean.

Practice

8.8 The figure shows the graphs of the probability density function for three different statistics that could be used to estimate a population parameter θ. Which statistic would you use, and why?

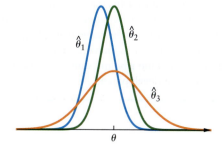

8.9 The figure shows the graphs of the probability density function for three different statistics that could be used to estimate a population parameter θ. Which statistic would you use, and why?

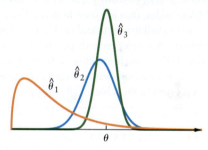

8.10 Why is an unbiased estimator for a parameter θ better than a biased estimator for θ?

8.11 Suppose there are several unbiased estimators for θ. What criterion would you use for selecting one of these estimators, and why?

Applications

8.12 Business and Management A recent survey asked employees in technology-related jobs what perks they would like their employer to provide. Out of 1200 randomly selected workers polled, 975 said they would like ongoing training paid for by their employer. Find a point estimate for the population proportion of all workers in technology-related jobs who would like ongoing training.

8.13 Biology and Environmental Science A hand is a traditional unit used to measure the height of a horse. One hand is 4 in., and the height of a horse is measured from the ground to the horse's shoulder. A random sample of horses sold at auctions around the country revealed the following heights (in hands). **HANDHT**

15.8	13.7	11.0	17.1	19.3	14.6	14.4	13.8
18.7	16.5	12.2	17.9	16.8	16.3	12.8	18.7
10.4	15.6	11.7	12.8				

(a) Find a point estimate for the population mean height of all horses sold at auctions.
(b) Find a point estimate for the population median height of all horses sold at auctions.
(c) Find a point estimate for the population variance of the height of all horses sold at auctions.
(d) Any horse with a height less than 14.5 hands is considered short. Find a point estimate for the true proportion of short horses sold at auctions.

8.14 Biology and Environmental Science During the first few weeks of January every year, biologists associated with the U.S. Fish and Wildlife Service and the Maryland Department of Natural Resources physically count ducks, geese, and other waterfowl birds near the Chesapeake Bay. In 2018, these scientists counted 641,200 Canada geese and 1,023,300 total waterfowl.[3] Find a point estimate for the proportion of Canada geese that were observed during the 2018 winter waterfowl count.

8.15 Biology and Environmental Science Cowrie shells have been used as money in many parts of the world, including China and Africa. Today, they are used in decorations and jewelry, and cowrie-shell bracelets and necklaces are popular near seaside resorts. A jeweler recently purchased several hundred cowrie shells to use in making earrings. A random sample of the finished earrings was obtained and the weight (in grams) of each is given in the table. **COWRIE**

7.3	7.4	7.4	7.2	7.2	7.8	7.2	7.5	7.7	7.9
7.7	7.4	7.3	7.1	7.5	7.3	7.0	7.7	7.0	7.2
7.5	7.0	7.2	7.3	7.4	7.7	7.6	7.4	6.9	7.3

(a) Find a point estimate for the first quartile and a point estimate for the third quartile.
(b) The smallest (in weight) 20% of all cowrie-shell earrings are sold at a discount. Find a point estimate for the 20th percentile of the cowrie-shell earring weight distribution.

8.16 Manufacturing and Product Development A company that manufactures a centrifugal pump for golf-course sprayers would like to rate the pressure (in psi) developed by this unit. Thirty pumps were randomly selected and tested. **SPRAY**
(a) Find a point estimate for the minimum pressure developed by this pump.
(b) Find a point estimate for the maximum pressure developed by this pump.
(c) Use your answers to parts (a) and (b) to construct an interval estimate for the pressure developed by this pump.

8.17 Business and Management The annual Canadian Lawyer Legal Fees survey received responses from 135 firms in Quebec and 107 firms in Nova Scotia. This survey examines fee structures associated with 10 different practice areas. The results of the survey indicated that 75 of the firms in Quebec and 50 of the firms in Nova Scotia handle civil litigation cases.
(a) Find a point estimate for the proportion of all legal firms in Quebec that handle civil litigation cases.
(b) Find a point estimate for the proportion of all legal firms in Nova Scotia that handle civil litigation cases.
(c) Let p_d denote the difference in population proportions between all the firms in Quebec that handle civil litigation cases and all the firms in Nova Scotia that handle civil litigation cases. Find a point estimate for p_d.

8.18 Medicine and Clinical Studies Biogen recently reported that a new drug, BAN2401, was statistically significant in slowing the progression of Alzheimer's disease. The results of this phase 1 trial are encouraging, but the drug only slows the cognitive decline; it does not cure the disease. Suppose a random sample of patients taking this new drug was obtained. The dosage (every two weeks) in milligrams was recorded for each patient. **BAN2401**
(a) Find a point estimate for the mean dosage.
(b) Find a point estimate for the variance of the dosage.
(c) Find a point estimate for the first quartile and a point estimate for the third quartile.

8.19 Biology and Environmental Science In May 2018, the Kilauea volcano erupted, sending lava crawling across Hawaii's Big Island, opening fissures, and triggering earthquakes. This volcano also emits noxious sulfur dioxide gas (SO_2) during periods of sustained eruption. A random sample of days was obtained, and the mass of the SO_2 plume from the volcano was recorded (in metric tons).[4] **LAVA**
(a) Find a point estimate for the minimum mass of the SO_2 plume.
(b) Find a point estimate for the maximum mass of the SO_2 plume.
(c) Find a point estimate for the interquartile range of the mass of the SO_2 plume.

8.20 Biology and Environmental Science A common merganser duck usually has one brood per year, laying approximately 6–17 eggs. However, in July 2018, an amateur photographer spotted a female duck leading 76 ducklings across Lake Bemidji. This sighting attracted tourists and scientists alike. A random sample of duck eggs in the area was obtained and the egg length of each was measured (in centimeters). **DUCK**
(a) Find a point estimate for the mean egg length.
(b) Find a point estimate for the variance in egg length.
(c) Find a point estimate for the range in egg length.
(d) Any egg with a length less than 6.2 cm is considered at risk for development. Find a point estimate for the proportion of eggs at risk for development.

8.2 A Confidence Interval for a Population Mean When σ Is Known

Confidence Interval Definitions

In Section 8.1, we discovered that a good estimator is unbiased and has small variance. However, an estimator produces only a single value that serves as a best guess for a population parameter. In this section, we use this single value to produce a **confidence interval**. Such an interval of values is constructed so that we can be reasonably sure that the true value of the population parameter lies within it.

> **Definition**
>
> A **confidence interval** (CI) for a population parameter is an interval of values constructed so that, with a specified degree of confidence, the value of the population parameter lies within it.
>
> The **confidence coefficient** is the probability that the CI includes the population parameter in repeated samplings.
>
> The **confidence level** is the confidence coefficient expressed as a percentage.

> **A CLOSER LOOK**
>
> 1. A confidence interval for a population parameter is usually expressed as an open interval, for example, (10.5, 15.8), where 10.5 is the left endpoint, or lower bound, and 15.8 is the right endpoint, or upper bound. The interval extends all the way to, but does not include, the endpoints.
> 2. Typical confidence coefficients are 0.90, 0.95, and 0.99.
> 3. Typical confidence levels are, therefore, 90%, 95%, and 99%.

Construction of a Confidence Interval

The following steps provide background information on the construction of a confidence interval for a population mean μ.

1. Suppose either (a) the underlying population is normal, or (b) the sample size n is large, or both, and the population standard deviation σ is known.
2. Using the properties of $\overline{X}$ and the Central Limit Theorem (if necessary): The sample mean $\overline{X}$ is (approximately) normal with mean μ and variance σ^2/n. In symbols: $\overline{X} \sim N(\mu, \sigma^2/n)$.
3. Using the Empirical Rule, approximately 95% of all values of the sample mean lie within 2 standard deviations of the population mean. **Figure 8.7** shows an example of a single value, or point estimate, within 2 standard deviations of the mean.
4. Even though we know that the distribution of $\overline{X}$ is centered at μ, we do not know the true value of μ. To capture μ, it seems reasonable to step 2 standard deviations away from an estimate $\overline{x}$ in both directions. The resulting (rough) 95% confidence interval $\left(\overline{x} - 2 \cdot \frac{\sigma}{\sqrt{n}}, \overline{x} + 2 \cdot \frac{\sigma}{\sqrt{n}}\right)$ is illustrated in **Figure 8.8**.

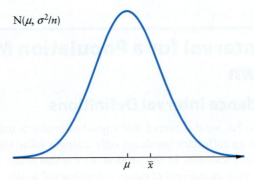

Figure 8.7 The sampling distribution of $\bar{X}$ and a typical value.

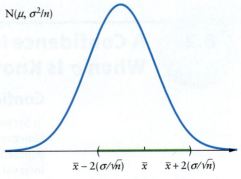

Figure 8.8 A rough 95% confidence interval that probably captures the true value μ.

A 95% Confidence Interval

To construct a more accurate 95% confidence interval for μ, using the same assumptions, begin with the standardized random variable Z.

1. $\bar{X} \sim N(\mu, \sigma^2/n) \Rightarrow Z = \dfrac{\bar{X} - \mu}{\sigma/\sqrt{n}} \sim N(0, 1)$

2. Use Appendix Table 3 to find a symmetric interval about 0 such that the probability that Z lies in this interval is 0.95. **Figure 8.9** helps to visualize this interval.

   $P(-1.96 < Z < 1.96) = 0.95$

Figure 8.9 A graph of the probability density function for a standard normal random variable and an illustration of a symmetric interval about 0 such that $P(-1.96 < Z < 1.96) = 0.95$.

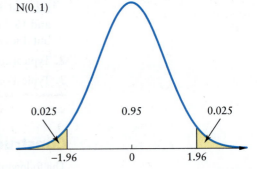

Solving compound inequalities.

3. Substitute for Z and rewrite the interval inside the probability statement so that μ is caught in the middle.

▶ $P\left(-1.96 < \dfrac{\bar{X} - \mu}{\sigma/\sqrt{n}} < 1.96\right) = 0.95$ Substitute for Z.

$P\left((-1.96)\dfrac{\sigma}{\sqrt{n}} < \bar{X} - \mu < (1.96)\dfrac{\sigma}{\sqrt{n}}\right) = 0.95$ Multiply all three parts by $\sigma/\sqrt{n}$.

$P\left(-\bar{X} - (1.96)\dfrac{\sigma}{\sqrt{n}} < -\mu < -\bar{X} + (1.96)\dfrac{\sigma}{\sqrt{n}}\right) = 0.95$ Subtract $\bar{X}$ from all three parts.

$P\left(\bar{X} + 1.96\dfrac{\sigma}{\sqrt{n}} > \mu > \bar{X} - 1.96\dfrac{\sigma}{\sqrt{n}}\right) = 0.95$ Multiply all three parts by -1. Multiplying by -1 changes the direction of the inequalities.

$P\left(\bar{X} - 1.96\dfrac{\sigma}{\sqrt{n}} < \mu < \bar{X} + 1.96\dfrac{\sigma}{\sqrt{n}}\right) = 0.95$ Rewrite the expression in increasing order. ◀

The last expression includes a formula for a 95% confidence interval for μ: Step exactly 1.96 (not 2) standard deviations away from a specific value $\bar{x}$ in both directions. This interval is shown in **Figure 8.10**.

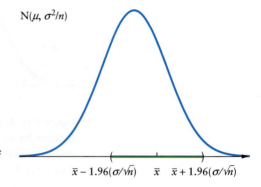

Figure 8.10 A graph of the probability density function for $\bar{X}$ and an illustration of an exact 95% confidence interval for μ.

A General $100(1-\alpha)$% Confidence Interval

The last step is to find a more general $100(1-\alpha)$% confidence interval for μ, using the same assumptions. In most cases, α is small. For example, if $\alpha = 0.05$, the resulting confidence level is $100(1-0.05)\% = 95\%$. Usually, the confidence level is given and we need to work backward to find α. The next definition is necessary to start with a probability statement involving Z (and then work to capture μ as before).

Definition

$z_{\alpha/2}$ is a **critical value**. It is a value on the measurement axis in a **standard normal distribution** such that $P(Z \geq z_{\alpha/2}) = \alpha/2$.

> In this definition, the subscript on z could be any, usually small, value, represented by any variable, or letter. For example,
> $$P(Z \geq z_c) = c$$
>

A CLOSER LOOK

1. $z_{\alpha/2}$ is simply a z value such that $\alpha/2$ of the area (probability) lies to the right of $z_{\alpha/2}$. The value $-z_{\alpha/2}$ is just the negative critical value.
2. Critical values are always defined in terms of right-tail probability.
3. The z critical values are easy to find by using the Complement Rule and working backward. For example,

   $P(Z \geq z_{\alpha/2}) = \alpha/2$ Definition of a critical value.
   $P(Z \leq z_{\alpha/2}) = 1 - \alpha/2$ The Complement Rule.

   Work backward in Appendix Table 3 to find $z_{\alpha/2}$.

To find a general confidence interval for μ, start once again in the Z world. Find a symmetric interval about 0 such that the probability that Z lies in this interval is $1 - \alpha$ (**Figure 8.11**).

$$P(-z_{\alpha/2} < Z < z_{\alpha/2}) = 1 - \alpha \qquad (8.1)$$

Rewrite Equation 8.1 to obtain the probability statement.

$$P\left(\bar{X} - z_{\alpha/2} \cdot \frac{\sigma}{\sqrt{n}} < \mu < \bar{X} + z_{\alpha/2} \cdot \frac{\sigma}{\sqrt{n}}\right) = 1 - \alpha$$

Figure 8.12 illustrates this interval for a specific value $\bar{x}$.
These derivations lead to the general result.

> Remember, a large sample usually means $n \geq 30$.

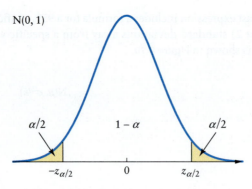
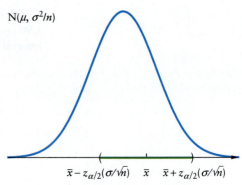

Figure 8.11 A graph of the probability density function for a standard normal random variable and an illustration of a symmetric interval about 0 such that $P(-z_{\alpha/2} < Z < z_{\alpha/2}) = 1 - \alpha$.

Figure 8.12 A graph of the probability density function for $\overline{X}$ and an illustration of a $100(1-\alpha)\%$ confidence interval for μ.

How to Find a $100(1-\alpha)\%$ Confidence Interval for a Population Mean When σ Is Known

Given a random sample of size n from a population with mean μ, if

1. the underlying population distribution is normal and/or n is large, and
2. the population standard deviation σ is known, then

the endpoints for a $100(1-\alpha)\%$ confidence interval for μ have the values

$$\overline{x} \pm z_{\alpha/2} \cdot \frac{\sigma}{\sqrt{n}} \tag{8.2}$$

A CLOSER LOOK

1. Equation 8.2 can be used only if σ is known. From a practical standpoint, it is unlikely that we will know the value of σ. Nevertheless, this is a very instructive procedure to learn and understand so that we can construct a confidence interval for an arbitrary parameter.

2. If n is large and σ is unknown, some statisticians suggest using the sample standard deviation s in Equation 8.2 instead of σ. This produces an approximate confidence interval for μ. The next section develops an exact confidence interval for μ when σ is unknown.

3. As the confidence coefficient increases (with σ and n being held constant), the critical value $z_{\alpha/2}$ increases. Therefore, the confidence interval is wider.

Example Confidence Interval

EXAMPLE 8.2 Cell Phone Weight

CELLWT

Cell phones have become an essential part of our everyday lives. We use them for communication, photography, Internet browsing, and banking, among many other functions. Therefore, choosing the right cell phone depends on many factors, including the size and resolution of the display, processing power, battery life, weight, and price. Suppose the weight of a typical cell phone is normally distributed with a standard deviation of 18 g. In a random sample of 15 cell phones, the sample mean weight was $\overline{x} = 155.7$ g. Find a 95% confidence interval for the true mean weight of cell phones.

Note: Appendix Table 3 includes a list of common critical values.

Common Error
The endpoints for a $100(1-\alpha)\%$ CI for μ are $\bar{x} \pm z_{\alpha/2} \cdot \sigma$.
Correction: Use the standard deviation of the random variable $\bar{X}$. The endpoints are $\bar{x} \pm z_{\alpha/2} \cdot \dfrac{\sigma}{\sqrt{n}}$.

Intermediate calculations are shown rounded, but final numerical answers are computed and presented using greater accuracy.

Solution

We need a 95% CI for μ. The population is normal and σ is known. Find the appropriate critical value and use Equation 8.2.

STEP 1 $\bar{x} = 155.7; \quad \sigma = 18; \quad n = 15$ *Given.*

$1 - \alpha = 0.95 \Rightarrow \alpha = 0.05 \Rightarrow \alpha/2 = 0.025$ *Find $\alpha/2$.*

$P(Z \geq z_{\alpha/2}) = P(Z \geq z_{0.025}) = 0.025$ *Definition of critical value.*

$P(Z \leq z_{0.025}) = 1 - 0.025 = 0.975$ *The Complement Rule.*

$z_{0.025} = 1.96$ *Use Appendix Table 3.*

STEP 2   Use Equation 8.2.

$\bar{x} \pm z_{\alpha/2} \cdot \dfrac{\sigma}{\sqrt{n}}$ *Equation 8.2.*

$= \bar{x} \pm z_{0.025} \cdot \dfrac{\sigma}{\sqrt{n}}$ *Use the value of α.*

$= 155.7 \pm (1.96) \dfrac{18}{\sqrt{15}}$ *Use summary statistics and values for σ and $z_{\alpha/2}$.*

$= 155.7 \pm 9.11$ *Simplify.*

$= (146.59, 164.81)$ *Compute endpoints.*

(146.59, 164.81) is a 95% confidence interval for the true mean weight (in grams) of a cell phone.

Figure 8.13 shows a technology solution.

Figure 8.13 Find the critical value, and then use the summary statistics and Equation 8.2 to find the confidence interval.

```
> xbar <- 155.7
> n <- 15
> sigma <- 18
> alpha <- 0.05
> zcrit <- qnorm(1 - alpha/2)
> 
> left <- xbar - zcrit * sigma/sqrt(n)
> right <- xbar + zcrit * sigma/sqrt(n)
> round(cbind(left,right),2)
      left  right
[1,] 146.59 164.81
```

TRY IT NOW   Go to Exercises 8.34 and 8.35

There are two important ideas to remember when constructing a confidence interval.

1. The population parameter, in this case μ, is fixed. The confidence interval varies from sample to sample. It is correct to say, "We are 95% confident that the interval captures the true mean μ." The statement, "We are 95% confident that μ lies in the interval," (incorrectly) implies that the interval is fixed and the parameter μ varies.

2. The confidence coefficient, a probability, is a long-run limiting relative frequency. In repeated samples, the proportion of confidence intervals that capture the true value of μ approaches 0.95. **Figure 8.14** illustrates this concept. We cannot be certain about any one specific confidence interval. The confidence pertains to the long-run process.

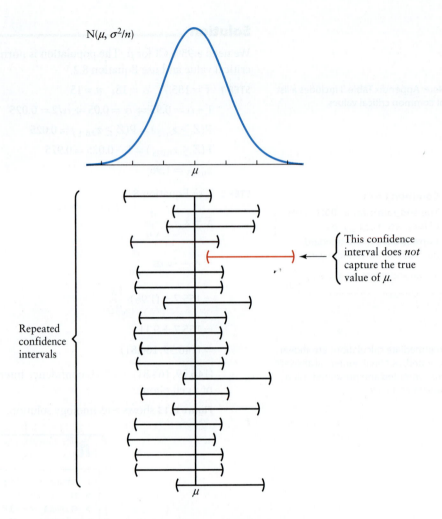

Figure 8.14 An illustration of the meaning of a confidence coefficient. In repeated CIs, the proportion of all 95% CIs that capture the true value of μ is 0.95.

Inference Using a Confidence Interval

In the next example, data are presented rather than summary statistics. Equation 8.2 is used again to construct a confidence interval for a population mean.

EXAMPLE 8.3 Vitality Snack

The Chia Squeeze Vitality Snack is available in six flavors; it is advertised as a good source of fiber, containing fruits and vegetables, gluten-free and vegan, and only 70 calories. The drink has a smoothie-like texture, and the manufacturing company claims each squeeze packet contains 1200 mg of omega-3 fatty acids.[5] A random sample of Chia Squeeze Vitality Snacks was obtained and the amount of omega-3 was carefully measured in each. The data are given in the table. Assume that $\sigma = 12$ mg.

1192	1200	1207	1185	1198	1194	1210	1197	1212	1209
1189	1202	1194	1196	1179	1191	1214	1197	1213	1213
1211	1193	1204	1187	1196	1194	1220	1193	1194	1194
1177	1181	1217	1213	1204	1197	1221	1198	1210	1183

(a) Find a 99% confidence interval for the mean amount of omega-3 in each Chia Squeeze Vitality Snack.

(b) Using the confidence interval from part (a), is there any evidence to suggest that the population mean amount of omega-3 is different from 1200 mg?

Solution

Construct a 99% CI for μ. No information is given about the shape of the underlying population distribution. Because $n = 40$ (≥ 30), the Central Limit Theorem applies, and we can use Equation 8.2.

(a) $\sigma = 12; \quad n = 40$ — Given.

$\bar{x} = \dfrac{1}{40}(1192 + \cdots + 1183) = 1199.475$ — Compute the sample mean.

$1 - \alpha = 0.99 \Rightarrow \alpha = 0.01 \Rightarrow \alpha/2 = 0.005$ — Find $\alpha/2$.

$z_{0.005} = 2.5758$ — Common critical value; Appendix Table 3.

Use Equation 8.2 to construct the confidence interval.

$\bar{x} \pm z_{\alpha/2} \cdot \dfrac{\sigma}{\sqrt{n}}$ — Equation 8.2.

$= \bar{x} \pm z_{0.005} \cdot \dfrac{\sigma}{\sqrt{n}}$ — Use the value of α.

$= 1199.475 \pm (2.5758)\dfrac{12}{\sqrt{40}}$ — Use summary statistics and values for σ and $z_{0.005}$.

$= 1199.475 \pm 4.887$ — Simplify.

$= (1194.59, 1204.36)$ — Compute endpoints.

$(1194.59, 1204.36)$ is a 99% confidence interval for the true mean amount of omega-3 in a Chia Squeeze Vitality Snack. **Figure 8.15** shows a technology solution.

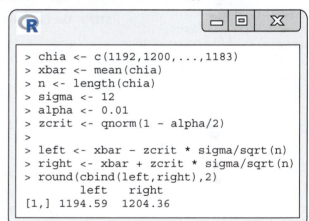

Figure 8.15 Find the sample mean and the critical value, and then use the summary statistics and Equation 8.2 to find the confidence interval.

(b) Use the four-step inference procedure and the CI as a measure of likelihood. If the CI in part (a) captures 1200, then there is no evidence to suggest that μ is different from 1200.

Claim: $\mu = 1200$

Experiment: $\bar{x} = 1199.475$

Likelihood: The likelihood in this case is expressed as a 99% confidence interval, an interval of likely values of μ: $(1194.59, 1204.36)$.

Conclusion: This CI includes 1200, so there is no evidence to suggest that μ is different from 1200. ∎

TRY IT NOW Go to Exercises 8.43 and 8.47

One More Example

Here is one more example involving a data set instead of summary statistics.

EXAMPLE 8.4 Chelsea Power Failures

Some business owners and residents in Chelsea, Quebec, are very angry about frequent and sustained power outages in the city. Hydro Quebec claims that most of the outages are caused by trees or branches falling on power lines, but because the power supply is so unreliable, many people in the area have invested in backup power generators.[6] A random sample of 50 power outage events was obtained and the length of time (in hours) until full power was restored was recorded for each event. The data are given in the table. Assume $\sigma = 3.5$.

15.6	22.0	13.4	14.1	6.3	3.8	14.9	14.3	0.8	6.0	6.9	15.6	9.9
2.8	15.4	7.4	4.6	7.7	15.4	8.1	14.7	8.7	8.7	8.9	8.3	9.1
11.7	11.8	10.2	15.5	11.0	15.1	11.9	10.8	8.6	13.3	13.3	8.6	2.6
11.3	0.6	20.9	3.4	9.9	8.4	6.4	1.3	3.8	7.2	8.1		

Find a 98% confidence interval for the true mean length of a power outage in Chelsea.

Solution

STEP 1 $\sigma = 3.5;\quad n = 50$ *Given.*

$\bar{x} = \dfrac{1}{50}(15.6 + \cdots + 8.1) = 9.782$ *Compute the sample mean.*

$1 - \alpha = 0.98 \Rightarrow \alpha = 0.02 \Rightarrow \alpha/2 = 0.01$ *Find $\alpha/2$.*

$z_{0.01} = 2.3263$ *Common critical value; Appendix Table 3.*

STEP 2 Use Equation 8.2 to construct the confidence interval.

$\bar{x} \pm z_{\alpha/2} \cdot \dfrac{\sigma}{\sqrt{n}}$ *Equation 8.2.*

$= \bar{x} \pm z_{0.01} \cdot \dfrac{\sigma}{\sqrt{n}}$ *Use the value of α.*

$= 9.782 \pm (2.3263)\dfrac{3.5}{\sqrt{50}}$ *Use summary statistics and values for σ and $z_{0.01}$.*

$= 9.782 \pm 1.151$ *Simplify.*

$= (8.63, 10.93)$ *Compute endpoints.*

$(8.63, 10.93)$ is a 98% confidence interval for the true mean length of a power outage in Chelsea. **Figure 8.16** shows a technology solution.

Figure 8.16 Find the sample mean and the critical value, and then use the summary statistics and Equation 8.2 to find the confidence interval.

```
> power <- c(15.6,22.0,...,8.1)
> xbar <- mean(power)
> n <- length(power)
> sigma <- 3.5
> alpha <- 0.02
> zcrit <- qnorm(1 - alpha/2)
>
> left <- xbar - zcrit * sigma/sqrt(n)
> right <- xbar + zcrit * sigma/sqrt(n)
> round(cbind(left,right),2)
     left right
[1,] 8.63 10.93
```

Bound on the Error of Estimation

Many confidence intervals have endpoint formulas that are of the same general form. Suppose the statistic $\hat{\theta}$ is used to estimate the population parameter θ. The general form of a confidence interval is

$$(\text{Point estimate using } \hat{\theta}) \pm (\text{critical value}) \cdot [(\text{estimate of}) \text{ standard deviation of } \hat{\theta}]$$

The critical value may be from the standard normal distribution or some other reference distribution. If the actual standard deviation of the statistic is not known (a likely situation), then an estimate is used.

The (two-sided) confidence interval derived earlier is the most common CI. However, other two-sided confidence intervals are possible, as well as one-sided confidence intervals consisting of either only a lower bound or only an upper bound. The derivation of several one-sided confidence intervals is outlined in the Challenge Problems.

Most of the time, as in the examples given earlier, there is no control over the sample size. The data or summary statistics are presented, and a confidence interval is constructed. However, a certain desired accuracy may be expressed by the width of a confidence interval. Given σ and the confidence level, a sample size n can be computed such that the resulting confidence interval has a certain desired width.

> The error of estimation is the step, the distance from the sample mean to the left endpoint and to the right endpoint of the confidence interval. We would like to find a sample size n so that the step is no bigger than B. Therefore, B is called a bound on the error of estimation.

Suppose n is large (but unknown), σ is known, and the confidence level is $100(1-\alpha)\%$. If the desired width is W, let $B = W/2$, called the **bound on the error of estimation**. Half the width of the confidence interval is given by the step (in each direction) from $\bar{x}$. The endpoints of the confidence interval for μ are (from Equation 8.2)

$$\bar{x} \pm \underbrace{z_{\alpha/2} \cdot \frac{\sigma}{\sqrt{n}}}_{B=\text{bound}}$$

 Solving equations.

Let $B = z_{\alpha/2} \cdot \frac{\sigma}{\sqrt{n}}$, and solve for n. The resulting formula for n is given by

$$n = \left(\frac{\sigma \cdot z_{\alpha/2}}{B}\right)^2 \tag{8.3}$$

A CLOSER LOOK

> In symbols,
> $\sigma \uparrow \Rightarrow n \uparrow$
> $z_{\alpha/2} \uparrow \Rightarrow n \uparrow$
> $B \downarrow \Rightarrow n \uparrow$

1. ▶ Consider the effect on n as one value (on the right-hand side of Equation 8.3) changes while the other two remain constant.

   As σ increases (with $z_{\alpha/2}$ and B constant), the fraction becomes larger, the square is larger, and therefore n is larger. This result makes sense, because a larger underlying variance would require a larger sample size to preserve a certain bound on the error of estimation.

   As $z_{\alpha/2}$ increases (equivalently, as the confidence level increases, with σ and B constant), again the fraction becomes larger, the square is larger, and therefore n is larger. More confidence requires a larger n to maintain the same bound on the error of estimation.

   As B decreases (with σ and $z_{\alpha/2}$ constant), the denominator becomes smaller, which makes the fraction larger, the square larger, and n larger. A smaller bound on the error of estimation (i.e., a smaller CI) means more information (a larger sample size) is needed to maintain the same confidence level. ◀

2. When Equation 8.3 is applied, it is likely that the value of n will not be an integer. Always round up to the nearest whole number. This guarantees a $100(1-\alpha)\%$ confidence interval with a bound on the error of estimation of at most B. Any larger sample size will also be sufficient.

3. It is unlikely that σ, which is needed in Equation 8.3, will be known. However, often an experimenter can make a very good guess at σ from previous experience or a small preliminary study.

The next example illustrates the use of Equation 8.3.

EXAMPLE 8.5 Geyser Height

Yellowstone National Park, which opened in 1872, includes magnificent scenery, historic sites, attractions of scientific interest, and recreational areas. The Castle Geyser is one of the oldest geysers in the basin, has a very large cone, and has changed eruption patterns over time.[7] A National Park Service researcher would like to find a 95% confidence interval for the mean height of the Castle Geyser eruption with a bound on the error of estimation of 5 ft. Previous experience suggests that the population standard deviation for the height is approximately 15 ft. How large a sample is necessary to achieve this accuracy?

Solution

STEP 1 $\sigma = 15$; $B = 5$ — Given.

$1 - \alpha = 0.95 \Rightarrow \alpha/2 = 0.025 \Rightarrow z_{0.025} = 1.96$ — Find the critical value.

STEP 2 Use Equation 8.3 to find the value of n.

$$n = \left(\frac{\sigma \cdot z_{\alpha/2}}{B}\right)^2$$ — Equation 8.3.

$$= \left[\frac{(15)(1.96)}{5}\right]^2$$ — Use known values.

$$= (5.88)^2 = 34.57$$ — Computation.

The necessary sample size is $n \geq 35$ (always round up). This will guarantee that a 95% confidence interval for the population mean height will have a bound on the error of estimation no greater than 5 ft.

TRY IT NOW Go to Exercises 8.39 and 8.41

Jochen Lambrechts/Shutterstock

Section 8.2 Exercises

Concept Check

8.21 True or False A confidence interval for a population mean is guaranteed to contain the true value of μ.

8.22 True or False A critical value is always defined in terms of right-tail probability.

8.23 True or False Many different two-sided confidence intervals and one-sided confidence intervals can be constructed based on the same information/data.

8.24 Fill in the Blank
(a) $P(Z \geq z_{\alpha/2}) =$ _____
(b) $P(Z \leq z_{\alpha/2}) =$ _____
(c) $P(Z \leq -z_{\alpha/2}) =$ _____

8.25 Fill in the Blank As the confidence coefficient increases, the confidence interval becomes _____.

8.26 Fill in the Blank To find a $100(1-\alpha)$% confidence interval for a population mean when σ is known, the underlying population must be either _____ or _____.

8.27 Short Answer Suppose the confidence coefficient and σ are constant. Explain why the bound on the error of estimation decreases as the sample size n increases.

8.28 Short Answer Explain the meaning of a confidence coefficient.

Practice

8.29 Find each critical value.
(a) $z_{0.10}$ (b) $z_{0.05}$ (c) $z_{0.025}$ (d) $z_{0.01}$
(e) $z_{0.005}$ (f) $z_{0.001}$ (g) $z_{0.0005}$ (h) $z_{0.0001}$

8.30 In each of the following cases, the sample mean, the sample size, the population standard deviation, and the

confidence level are given. Assume that the underlying population is normally distributed. Find the associated confidence interval for the population mean.
(a) $\bar{x} = 15.6$, $n = 12$, $\sigma = 3.7$,   95%
(b) $\bar{x} = 6322$, $n = 17$, $\sigma = 225$,   90%
(c) $\bar{x} = -45.78$, $n = 9$, $\sigma = 12.35$,   80%
(d) $\bar{x} = 0.0795$, $n = 24$, $\sigma = 0.006$,   99%
(e) $\bar{x} = 37.68$, $n = 27$, $\sigma = 2.2$,   99.9%

8.31 In each of the following cases, the sample mean, the sample size, the population standard deviation, and the confidence level are given. Find the associated confidence interval for the population mean.
(a) $\bar{x} = 17.6$, $n = 32$, $\sigma = 10.27$,   95%
(b) $\bar{x} = 136.8$, $n = 45$, $\sigma = 25.44$,   99%
(c) $\bar{x} = 335.7$, $n = 65$, $\sigma = 125.3$,   90%
(d) $\bar{x} = -6.7$, $n = 52$, $\sigma = 2.25$,   98%
(e) $\bar{x} = 20.11$, $n = 37$, $\sigma = 1.76$,   99.99%

8.32 Two statisticians are given the same data from an experiment. Each uses these data to construct a confidence interval for the true population mean μ. The resulting CIs are (8.55, 10.85) and (8.40, 11.0).
(a) What is the value of the sample mean $\bar{x}$?
(b) One CI has a confidence level of 95%, and the other has a confidence level of 99.9%. Match the confidence level with the confidence interval, and justify your answer.

8.33 In each of the following cases, the population standard deviation, the bound on the error of estimation, and the confidence level are given. Find a value for the sample size necessary to satisfy these requirements.
(a) $\sigma = 7.9$, $B = 2.5$,   95%
(b) $\sigma = 10.77$, $B = 5$,   99%
(c) $\sigma = 0.55$, $B = 0.001$,   98%
(d) $\sigma = 35.97$, $B = 3.5$,   95%
(e) $\sigma = 55$, $B = 2$,   99.9%

Applications

8.34 Physical Sciences Over the last 30 years, the Abandoned Mine Reclamation Program has closed more than 6000 of the approximately 17,000 abandoned mines in Utah. However, Utah's largest coal mine, the Sufco Mine on the Wasatch Plateau, is still open and recently received a reduction in its royalty rate, the percentage of the value of mine production paid to the state.[8] To help project the royalty savings, a random sample of 40 mining days was obtained, and the production was recorded for each. The sample mean was $\bar{x} = 15.07$ thousand tons. Assume $\sigma = 1.8$. Find a 95% confidence interval for the true mean production per day.

8.35 Public Health and Nutrition After a ham is cured, it may be smoked to add flavor or to ensure it lasts longer. Typical grocery-store hams are smoked for a short period of time, whereas gourmet hams are usually smoked for at least one month. A random sample of 36 grocery-store hams was obtained, and the length of the smoking time was recorded for each. The mean was $\bar{x} = 140$ hr. Assume $\sigma = 8$ hr.

(a) Find a 99% confidence interval for the mean amount of time a grocery-store ham is smoked.
(b) What assumptions did you make to construct the confidence interval in part (a)?

8.36 Sports and Leisure The London Eye is a giant Ferris wheel on the banks of the River Thames in London. This popular tourist attraction offers stunning views and has appeared in movies and on television. A random sample of London Eye riders was obtained and the length of each ride was measured (in minutes). Assume $\sigma = 5$ min. EYE
(a) Find a 95% confidence interval for the mean amount of time spent on this ride.
(b) Find a 99% confidence interval for the mean amount of time on this ride.
(c) Explain why the interval in part (b) is wider.

8.37 Marketing and Consumer Behavior The owner of a small tailoring shop keeps careful records of all alterations. Two common alterations to men's clothing are lengthening the inseam on pants and letting out a sports coat around the waist. A random sample of each type of alteration was obtained, and the summary statistics are given in the table. Measurements are in inches.

Alteration	Sample size	Sample mean	Assumed σ
Pants inseam	33	0.74	0.22
Sports coat waist	42	1.05	0.37

(a) Find a 95% confidence interval for the mean alteration of the inseams on men's pants.
(b) How large a sample is necessary for the bound on the error of estimation to be 0.05 for the confidence interval in part (a)?
(c) Find a 90% confidence interval for the mean alteration of the waists of sports coats.
(d) How large a sample is necessary for the bound on the error of estimation to be 0.07 for the confidence interval in part (c)?

8.38 Physical Sciences In July 2018, an iceberg with an estimated mass of 11 million tons floated dangerously close to shore near Innaarsuit, Greenland.[9] The 300-ft-tall iceberg caused some town residents to evacuate and many fishing boats were pulled to shore. Suppose a team of scientists aboard a research vessel operating in this area selects a random sample of 49 icebergs. The height is carefully measured for each, and the sample mean is $\bar{x} = 102$ m. Assume $\sigma = 25$ m.
(a) Find a 99% confidence interval for the mean height of icebergs in this area.
(b) Some experts suggest that global warming has led to more icebergs of greater height. Using the interval in part (a), is there any evidence to suggest that the mean height of icebergs in this area is more than 95 m? Justify your answer.

8.39 Travel and Transportation Lighthouses are constructed to guide ships traveling in rocky waters and to allow sailing

at night. Although the size of a lighthouse can be reported in many ways, the height is usually measured from the base of the tower to the top of the ventilator ball. A random sample of 18 lighthouses in France and England was obtained, and the height of each was recorded. The sample mean was $\bar{x} = 33.75$ m. Assume the distribution of lighthouse heights is normal and $\sigma = 5.4$ m.
(a) Find a 95% confidence interval for the mean height of all lighthouses in France and England.
(b) How large a sample is necessary to ensure that the width of the resulting 95% confidence interval is 2 m?
(c) Why is the confidence interval constructed in part (a) valid even though the sample size $n = 18$ is less than 30?

8.40 Biology and Environmental Science Koalas, which live in eastern Australia, are not really bears, but are related to the wombat and the kangaroo. These cuddly-looking creatures grow to a size of 27 to 36 in. and live 13–17 years.[10] A random sample of 35 koalas was obtained, and each was carefully weighed. The mean weight was $\bar{x} = 20.75$ lb. Assume $\sigma = 3.05$ lb.
(a) Find a 99% confidence interval for the mean weight of all koalas.
(b) Researchers believe that if the true mean weight is less than 23 lb, these animals may be suffering from malnutrition. Is there any evidence to suggest that the koala population is suffering from malnutrition? Justify your answer.

8.41 Technology and the Internet Many companies are increasingly concerned about computer security and employees using their computers for personal use. A random sample of 50 employees of Liberty Mutual Insurance Company was obtained and each was asked whether they use the Internet for nonbusiness (personal) purposes during the day. Assume $\sigma = 0.5$ hr. COMPSEC
(a) Find a 95% confidence interval for the mean number of hours all employees at Liberty Mutual use the Internet for personal reasons.
(b) Using the confidence interval in part (a), is there any evidence to suggest that the mean time spent using the Internet for personal reasons is more than 1 hr? Justify your answer.
(c) How large a sample is necessary for the bound on the error of estimation to be 0.1 hr?
(d) Do you think the underlying distribution of time spent on the Internet for personal use is normal? Explain your answer.

8.42 Sports and Leisure A random sample of professional wrestlers was obtained, and the annual salary (in dollars) for each was recorded. The summary statistics were $\bar{x} = 57,500$ and $n = 18$. Assume the distribution of annual salary is normal, with $\sigma = 9500$.
(a) Find a 90% confidence interval for the true mean annual salary for all professional wrestlers.
(b) How large a sample is necessary for the bound on the error of estimation to be 3000?

(c) How large a sample is necessary for the bound on the error of estimation to be 1000?

8.43 Travel and Transportation The Boeing 747-8i is one of the fastest commercial aircraft. Its top speed is Mach 0.86 (Mach 1 is the speed of sound), and the plane is used by only three commercial operators: Air China, Korean Air, and Lufthansa. A random sample of 12 747-8i flights was obtained and the cruising speed was measured for each (in mph). The sample mean was 560.4 mph. Assume the cruising speed distribution is normal and $\sigma = 52$ mph.
(a) Find a 99% confidence interval for the true mean cruising speed of all 747-8i planes.
(b) If the mean cruising speed is greater than 600 mph, then these planes will have to be inspected more often to ensure they are maintaining their structural integrity. Is there any evidence to suggest that the mean cruising speed is greater than 600 mph? Justify your answer.

8.44 Public Policy and Political Science Many chimney fires are caused by a buildup of creosote, a highly flammable material that forms in flues due to the condensation of certain gases. In an effort to promote fireplace and wood-stove safety, a town in Vermont has started a new inspection program. Forty homes with wood stoves were selected at random, and the amount of creosote buildup 1 ft from the top of the chimney was carefully measured. The resulting sample mean thickness was $\bar{x} = 0.131$ in. Assume $\sigma = 0.02$ in.
(a) Find a 95% confidence interval for the true mean thickness of creosote buildup 1 ft from the top of chimneys.
(b) Approximately 1/8 in. of creosote buildup is considered safe. If evidence suggests that the true mean thickness is greater than 0.125 in., the town will embark on an extensive safety program. Using the interval constructed in part (a), should the town stress greater safety? Justify your answer.

8.45 Physical Sciences According to the U.S. National Weather Service, at any given moment of any day, approximately 2000 thunderstorms are occurring worldwide. Many of these storms include lightning strikes. Sensitive electronic equipment is used to record the number of lightning strikes worldwide every day. Twelve days were selected at random, and the number of lightning strikes on each day was recorded. The sample mean was $\bar{x} = 8.6$ million. Assume the distribution of the number of lightning strikes per day is normal, with $\sigma = 0.35$ million.
(a) Find a 99% confidence interval for the mean number of lightning strikes per day worldwide.
(b) Do you think the normality assumption in this problem is reasonable? Why or why not?

8.46 Biology and Environmental Science During summer 2018, a large, persistent bloom of algae known as red tide occurred in southwest Florida. This algal bloom really consists of simple plants that live in the sea and freshwater; when they grow out of control, they produce microscopic toxins that affect people, fish, marine mammals, and birds. A random sample of water was obtained from each monitored beach and the toxins were measured in each (cell count/mL). Assume the distribution of red tide toxin count per milliliter is normal, with $\sigma = 625$. 📊 REDTIDE

(a) Find a 95% confidence interval for the mean number of toxins per milliliter.
(b) If the mean cell count is greater than 1000, a level considered high, people at the beach can experience severe respiratory irritation. Is there any evidence that the mean cell count is high? Justify your answer.
(c) Do you think the normality assumption in this problem is reasonable? Why or why not?

8.47 Biology and Environmental Science A few years ago, a disease infected and killed approximately 96% of sea stars (or starfish) off the central coast of British Columbia.[11] The cause of this wasting disease is still unknown. However, this event initiated new research that revealed how much sea stars actually enrich the British Columbia coast. To monitor the recovery of the sea star population, a random sample of 36 morning sun stars was obtained and each was measured (in inches). The sample mean was 7.1 in. Assume $\sigma = 2.1$ in.

(a) Find a 95% confidence interval for the true mean size of morning sun stars.
(b) The population mean size prior to the infection and die-off was $\mu = 8.2$ in. Is there any evidence to suggest that the population mean is less than 8.2 in.? Justify your answer.

Extended Applications

8.48 Biology and Environmental Science Peregrine falcons were placed on the U.S. Endangered Species list in the 1970s. Their population dwindled primarily as a result of the introduction of new chemicals and pesticides. Although it may seem like a peculiar nesting area, the New York City area is home to many falcon pairs. The tall buildings and bridges provide lots of open space for hunting. These birds have a wingspan of 3.3 to 3.6 ft and weigh between 18.8 and 56.5 oz. A random sample of 35 falcons was obtained and each was carefully weighed. The sample mean was $\bar{x} = 47.3$ oz. Assume $\sigma = 6.2$ oz.

(a) Find a 95% confidence interval for the true mean weight of peregrine falcons.
(b) Based on your answer from part (a), is there any evidence to suggest that the mean weight is less than 48 oz?

8.49 Economics and Finance Representative Bill Cassidy has become concerned about the disparity in home prices in Louisiana's 6th District (Monroe). A random sample of homes sold within the past year in each of two parishes was obtained. The summary statistics are given in the table. Prices are in dollars.

Parish	Sample size	Sample mean	Assumed σ
East Feliciana	30	125,200	5,750
Iberville	36	155,900	25,390

(a) Find a 99% confidence interval for the mean selling price of all homes in East Feliciana Parish.
(b) Find a 99% confidence interval for the mean selling price of all homes in Iberville Parish.
(c) Using the confidence intervals in parts (a) and (b), is there any evidence to suggest that the mean selling price is different for the two parishes? Justify your answer.

8.50 Sports and Leisure There are three main types of exercises: range-of-motion (flexibility), strengthening, and endurance. A random sample of people who exercise regularly was obtained. The type of exercise and length (in minutes) of each workout was recorded. The table summarizes the information obtained.

Exercise type	Sample size	Sample mean	Assumed σ
Range-of-motion	65	25.2	5.2
Strengthening	32	73.6	10.7
Endurance	40	82.2	12.5

(a) Construct a 99% confidence interval for the mean workout length for each of the three types of exercises.
(b) A group of exercise scientists claims that the length of a workout for range-of-motion exercises is the same as for the other two types. Based on the confidence intervals in part (a), is there any evidence to refute this claim?

8.51 Psychology and Human Behavior A random sample of male college athletes was obtained and their coping skills levels were measured by means of an extensive psychological profile. Each athlete was also classified by sport. The table summarizes the information obtained.

Sport	Sample size	Sample mean	Assumed σ
Football	35	65.77	14.07
Basketball	30	53.90	12.50
Hockey	32	68.45	10.25

(a) Find a 95% confidence interval for the true mean coping skills level for all male athletes in each sport.
(b) Use your answers to part (a) to determine whether there is any evidence to suggest that the mean coping skills level for male basketball players is different from that of male football players. Justify your answer.
(c) How large a sample is necessary for the bound on the error of estimation for each confidence interval in part (a) to be 2?

8.52 Sports and Leisure Two competing ski slopes in Colorado advertise their powder base each day in an effort to attract more skiers. A random sample of the powder base depth (in inches) was obtained for each ski resort on days during a recent winter. Assume $\sigma_B = 2.5$ and $\sigma_V = 2.7$. 📊 BRECK 📊 VAIL

(a) Find a 99% confidence interval for the true mean powder depth at Breckenridge Ski Resort.

(b) Find a 99% confidence interval for the true mean powder depth at Vail Ski Resort.
(c) Use the results in parts (a) and (b) to determine whether there is any evidence to suggest a difference in the true mean powder depths at these two ski resorts.
(d) What assumptions were necessary to construct the confidence intervals in parts (a) and (b)?

8.53 Public Health and Nutrition The nutritional value of every food product sold in the United States is listed on the package in terms of protein, fat, and other nutrients. A researcher randomly selected fifty 1-oz samples of various nuts and carefully measured the amount of protein (in grams) in each sample. The sample mean amount of protein and the assumed standard deviation for each type of nut are given in the table.

Nut	Sample mean	Assumed σ
Cashew	5.17	0.40
Filbert	4.24	0.60
Pecan	2.60	0.95

(a) Find a 95% confidence interval for the true mean amount of protein in each type of nut. Based on these intervals, is there any evidence to suggest that the mean amount of protein is different for cashews and pecans? How about filberts and pecans?
(b) Answer part (a) using a sample size of 18 for each nut.

8.54 Medicine and Clinical Studies A recent study suggests that adults older than age 60 with a high waist-to-hip ratio may have reduced cognitive function.[12] A random sample of 35 women older than age 60 was obtained and the waist-to-hip ratio was computed for each. The sample mean was $\bar{x} = 0.75$. Assume $\sigma = 0.19$.
(a) Find a 95% confidence interval for the true mean waist-to-hip ratio in women older than age 60.
(b) Use the result in part (a) to determine if there is any evidence to suggest that the mean waist-to-hip ratio in women older than 60 is greater than 0.80 (considered over weight). Justify your answer.

8.55 Fuel Consumption and Cars The Shell Airflow Starship semi truck is designed with a carbon fiber body and aluminum extenders to make the vehicle more aerodynamic. Other design features are included to ensure better fuel economy for these big rigs, which transport freight across the country.[13] A random sample of test trips for this new truck was obtained and the fuel economy was measured for each outing, in miles per gallon. Assume $\sigma = 1.25$ mpg. RIGS
(a) Find a 95% confidence interval for the true mean miles per gallon for the Shell semi truck.
(b) Most semi trucks get about 7 mpg. Is there any evidence to suggest that the mean mpg for the new truck is greater than 7 mpg? Justify your answer.

(c) A spokesperson for Shell indicated that the engineers hoped the new truck would get at least 10 mpg. Is there any evidence to suggest that the mean mpg for the new truck is different from 10 mpg? Justify your answer.

Challenge Problems

8.56 One-Sided Confidence Intervals Given a random sample of size n from a population with mean μ, assume the underlying population is normal or n is large, and the population standard deviation σ is known.
(a) Rewrite the probability statement

$$P\left(\frac{\bar{X} - \mu}{\sigma/\sqrt{n}} > -z_\alpha\right) = 1 - \alpha$$

to find a one-sided $100(1-\alpha)\%$ confidence interval for μ bounded above. That is, find an upper bound for the population mean μ.
(b) Write a similar probability statement to find a one-sided $100(1-\alpha)\%$ confidence interval for μ bounded below.
(c) First-class mail in the United States includes personal correspondence and all kinds of bills. Each piece of first-class mail must weigh less than 13 oz. In a random sample of 25 first-class letters, the sample mean weight was $\bar{x} = 2.2$ oz. Assume the population is normal and $\sigma = 0.75$ oz. Find a 95% one-sided confidence interval bounded above for the population mean weight of first-class letters.

8.57 Sports and Leisure Steeplechase horseraces are run on courses that include obstacles such as brush fences, stone walls, timber walls, and water jumps. A random sample of 17 winning times (in seconds) in the $2\frac{3}{8}$-mile Saratoga Steeplechase Race was obtained. The sample mean was $\bar{x} = 259.79$ sec. Assume that $\sigma = 7.5$ and that the population distribution of winning times is normal. Find a one-sided 99% confidence interval, bounded below, for the true mean winning time.

8.58 The "Best" Confidence Interval There are actually infinitely many different confidence intervals for a population mean μ. For example, suppose that, in a random sample of size n from a population with mean μ, the underlying population is normal or n is large, and the population standard deviation σ is known. One could start with a probability statement of the form

$$P\left(-z_{3\alpha/4} < \frac{\bar{X} - \mu}{\sigma/\sqrt{n}} < z_{\alpha/4}\right) = 1 - \alpha$$

to find a $100(1-\alpha)\%$ confidence interval for μ. Why is the traditional two-sided confidence interval (presented in this section) the best CI for a population mean μ?

8.3 A Confidence Interval for a Population Mean When σ Is Unknown

The t Distribution

In Section 8.2, we presented a confidence interval for a population mean μ. However, Equation 8.2 (based on a standard normal, or Z, distribution) is valid only if σ is known (and either the underlying population is normal or the sample size is large). Although this is a very instructive application, it is unrealistic to assume that the population standard deviation is known. This section introduces a more practical and useful approach to constructing a confidence interval for μ.

Recall that if σ is known and either the underlying population is normal or the sample size is large, then the expression

$$Z = \frac{\overline{X} - \mu}{\sigma/\sqrt{n}}$$

has a standard normal distribution. In Section 8.2, we used this expression to derive the confidence interval. Note that only one component of this expression, $\overline{X}$, contributes to the variability.

If σ is unknown, a similar standardization is used:

$$T = \frac{\overline{X} - \mu}{S/\sqrt{n}} \tag{8.4}$$

The symbol T is used in notation to represent both the sample total (in Chapter 7) and a random variable with a t distribution. The context in which the notation is used implies the appropriate concept.

However, the distribution of this random variable is not normal. It is reasonable to believe that the distribution of T is centered at 0, but there is more variability in this expression, with contributions from two sources, $\overline{X}$ and S. The most common confidence interval for a population mean μ is based on Equation 8.4, and the t distribution introduced next.

The t distribution is closely related to the standard normal distribution. Most continuous random variables are defined by a probability density function, but the important point here is to understand the properties of a t distribution.

Properties of a t Distribution

1. A t distribution is completely determined (characterized) by only one parameter ν, called the number of degrees of freedom (df). The symbol ν is the Greek letter "nu." ν must be a positive integer, $\nu = 1, 2, 3, 4, \ldots$, and there is a different t distribution corresponding to each value of ν.

The t distribution was derived by William Gosset in 1908. Working for Guinness Breweries, he published his result using the pseudonym Student. For this reason, the distribution is still often called "Student's t distribution."

2. If T (a random variable) has a t distribution with ν degrees of freedom, denoted $T \sim t_\nu$, then

$$\mu_T = 0 \quad \text{and} \quad \sigma_T^2 = \frac{\nu}{\nu - 2} \quad (\nu \geq 3) \tag{8.5}$$

3. The density curve for every t distribution is bell-shaped and centered at 0, but more spread out than the density curve for a standard normal random variable Z. As ν increases, the density curve for T becomes more compact and closer to the density curve for Z. See **Figure 8.17**.

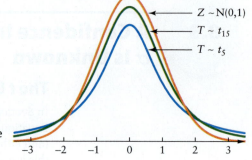

Figure 8.17 A comparison of density curves for two t distributions with the density curve for a standard normal random variable.

t Critical Values

To construct a confidence interval for μ based on a t distribution, we need a definition and notation for a t **critical value**.

Definition

$t_{\alpha,\nu}$ is a **critical value** related to a **t distribution** with ν degrees of freedom. If T has a t distribution with ν degrees of freedom, then $P(T \geq t_{\alpha,\nu}) = \alpha$.

A CLOSER LOOK

Remember, the symbol α in the notation $t_{\alpha,\nu}$ is simply a placeholder. Any symbol could be used here to represent a right-tail probability.

$$P(T \geq t_{\alpha,\nu}) = \alpha$$

1. For any t distribution, $t_{\alpha,\nu}$ is simply a t value (a value on the measurement axis) such that α of the area (probability) lies to the right of $t_{\alpha,\nu}$. The negative critical value is $-t_{\alpha,\nu}$. Because the t distribution is symmetric, $P(T \leq -t_{\alpha,\nu}) = P(T \geq t_{\alpha,\nu}) = \alpha$. See **Figure 8.18**.

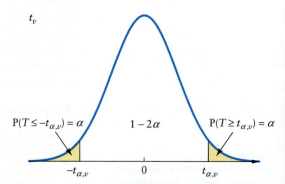

Figure 8.18 An illustration of the definition of a t critical value.

2. Critical values are always defined in terms of right-tail probability.

Appendix Table 5 presents selected critical values associated with various t distributions. Right-tail probabilities appear in the top row of this table, and degrees of freedom are listed in the left column. In the body of the table, $t_{\alpha,\nu}$ is at the intersection of the α column and the ν row. The following example illustrates the use of this table for finding critical values associated with a t distribution.

EXAMPLE 8.6 Critical Value Lookups

Find each critical value: (a) $t_{0.05,12}$, (b) $t_{0.01,21}$

Solution

(a) Use Appendix Table 5 to find the intersection of the $\alpha = 0.05$ column and the $\nu = 12$ row.

8.3 A Confidence Interval for a Population Mean When σ Is Unknown

ν	0.20	0.10	α 0.05	0.025	0.01	0.005	0.001	0.0005	0.0001
$\vdots$	$\vdots$	$\vdots$	$\vdots$	$\vdots$	$\vdots$	$\vdots$	$\vdots$	$\vdots$	$\vdots$
10	0.8791	1.3722	1.8125	2.2281	2.7638	3.1693	4.1437	4.5869	5.6938
11	0.8755	1.3634	1.7959	2.2010	2.7181	3.1058	4.0247	4.4370	5.4528
12	0.8726	1.3562	1.7823	2.1788	2.6810	3.0545	3.9296	4.3178	5.2633
13	0.8702	1.3502	1.7709	2.1604	2.6503	3.0123	3.8520	4.2208	5.1106
14	0.8681	1.3450	1.7613	2.1448	2.6245	2.9768	3.7874	4.1405	4.9850
$\vdots$	$\vdots$	$\vdots$	$\vdots$	$\vdots$	$\vdots$	$\vdots$	$\vdots$	$\vdots$	$\vdots$

Therefore, $t_{0.05,12} = 1.7823$. If T has a t distribution with $\nu = 12$, then $P(T \geq 1.7823) = 0.05$. See **Figure 8.19**.

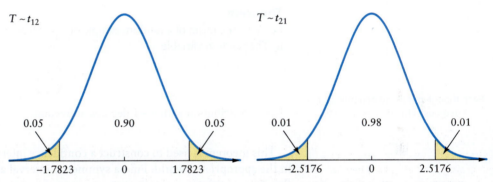

Figure 8.19 Visualization of $t_{0.05,12} = 1.7823$

Figure 8.20 Visualization of $t_{0.01,21} = 2.5176$

(b) Similarly, use Appendix Table 5 to find the intersection of the $\alpha = 0.01$ column and the $\nu = 21$ row. $t_{0.01,21} = 2.5176$, and if T has a t distribution with $\nu = 21$, then $P(T \geq 2.5176) = 0.01$. See **Figure 8.20**.

Figure 8.21 shows technology solutions.

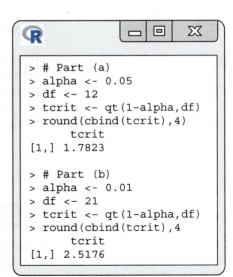

Figure 8.21 Use the R quantile function for the t distribution to find critical values.

TRY IT NOW Go to Exercise 8.67

A CLOSER LOOK

1. Appendix Table 5 is very limited. However, by using technology, we can find almost any critical value needed.
2. In Appendix Table 5, as ν increases, the t critical values approach the corresponding Z critical values. This numerical observation is analogous to the graphical comparison in Figure 8.17. The density curve for t_ν approaches the density curve for Z as ν increases. Notice that even for ν as small as 30, $t_{\alpha,30} \approx z_\alpha$.

One-Sample t Confidence Interval

▶ A confidence interval for a population mean when σ is unknown is based on the following theorem.

Theorem
Let $\overline{X}$ be the mean of a random sample of size n from a normal distribution with mean μ. The random variable

$$T = \frac{\overline{X} - \mu}{S/\sqrt{n}} \qquad (8.6)$$

has a t distribution with $n - 1$ degrees of freedom.

Even though n is in the denominator, $n - 1$ degrees of freedom is correct!

This theorem is used to construct a confidence interval for μ. As in Section 8.2, start in the appropriate t world. Find a symmetric interval about 0 such that the probability that T lies in this interval is $1 - \alpha$.

$$P(-t_{\alpha/2,n-1} < T < t_{\alpha/2,n-1}) = 1 - \alpha$$

$$P\left(-t_{\alpha/2,n-1} < \frac{\overline{X} - \mu}{S/\sqrt{n}} < t_{\alpha/2,n-1}\right) = 1 - \alpha \qquad (8.7)$$

Rewrite Equation 8.7 to obtain the probability statement

$$P\left(\overline{X} - t_{\alpha/2,n-1} \cdot \frac{S}{\sqrt{n}} < \mu < \overline{X} + t_{\alpha/2,n-1} \cdot \frac{S}{\sqrt{n}}\right) = 1 - \alpha \qquad (8.8)$$

This probability statement leads to the general result. ◀

How to Find a $100(1-\alpha)\%$ Confidence Interval for a Population Mean When σ Is Unknown

Given a random sample of size n with sample standard deviation s from a population with mean μ, if the underlying population distribution is normal, the endpoints of a $100(1-\alpha)\%$ confidence interval for μ have the values

$$\overline{x} \pm t_{\alpha/2,n-1} \cdot \frac{s}{\sqrt{n}} \qquad (8.9)$$

A CLOSER LOOK

1. Equation 8.9 can be used with any sample size n (≥ 2) and produces an exact (not an approximate) confidence interval for μ.
2. This confidence interval for μ (Equation 8.9) is valid only if the underlying population is normal.

Confidence Interval Example

EXAMPLE 8.7 **Orange Blossom Perfume**

Distillation is a process for separating and collecting substances according to their reaction to heat. When heat is applied to a mixture, the substance that evaporates and is collected as it cools is called the distillate. The unevaporated portion of the mixture is called the residue. Oil obtained from orange blossoms through distillation is used in perfume. Suppose the oil yield is normally distributed. In a random sample of 11 distillations, the sample mean oil yield was $\bar{x} = 980.2$ g with standard deviation $s = 27.6$ g. Find a 95% confidence interval for the true mean oil yield per batch.

Solution

We need a 95% CI for μ. The population is normal, and σ is unknown. Find the appropriate t critical value and use Equation 8.9.

STEP 1 $\bar{x} = 980.2$; $s = 27.6$; $n = 11$ *Given.*

$1 - \alpha = 0.95 \Rightarrow \alpha = 0.05 \Rightarrow \alpha/2 = 0.025$ *Find $\alpha/2$.*

$t_{\alpha/2,\, n-1} = t_{0.025,\, 10} = 2.2281$ *Use Appendix Table 5 with $\nu = 10$.*

STEP 2 Use Equation 8.9.

$$\bar{x} \pm t_{\alpha/2,\, n-1} \cdot \frac{s}{\sqrt{n}}$$ *Equation 8.9.*

$$= \bar{x} \pm t_{0.025,\, 10} \cdot \frac{s}{\sqrt{n}}$$ *Use the values of α and n.*

$$= 980.2 \pm (2.2281) \cdot \frac{27.6}{\sqrt{11}}$$ *Use summary statistics and critical value.*

$$= 980.2 \pm 18.54$$ *Simplify.*

$$= (961.66,\, 998.74)$$ *Compute endpoints.*

(961.66, 998.74) is a 95% confidence interval for the true mean oil yield (in grams) through distillation in a typical batch.

Figure 8.22 shows a technology solution.

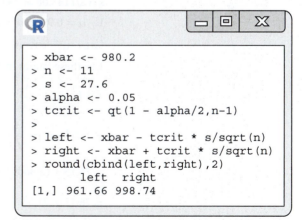

```
> xbar <- 980.2
> n <- 11
> s <- 27.6
> alpha <- 0.05
> tcrit <- qt(1 - alpha/2, n-1)
>
> left <- xbar - tcrit * s/sqrt(n)
> right <- xbar + tcrit * s/sqrt(n)
> round(cbind(left, right), 2)
      left  right
[1,] 961.66 998.74
```

Figure 8.22 Find the critical value, and then use the summary statistics and Equation 8.9 to find the confidence interval.

Common Error
Use of the critical value $t_{\alpha/2,\, n}$
Correction: The critical value is based on $n - 1$ degrees of freedom.

TRY IT NOW Go to Exercises 8.69 and 8.75

CHAPTER APP **Autonomous Cars**

Find a 95% confidence interval for the true mean number of miles between recent disengagements.

Inference Using a Confidence Interval

The next example illustrates the same process using actual data rather than summary statistics. The resulting confidence interval is used to draw a conclusion about the value of the population mean.

EXAMPLE 8.8 Rogue Waves

Rogue waves are greater than twice the size of surrounding waves and occur unpredictably, often from directions other than the prevailing wind and waves. Some of the known causes of such waves are constructive interference and focusing of wave energy. Some scientists now believe that conditions in the Bermuda Triangle are just right for massive rogue waves, which may explain the mysterious disappearances of planes and ships in this area.[14] A random sample of rogue waves was obtained and the height (in feet) of each is given in the table.[15]

133	50	70	112	92	80	72	66	28	154
84	95	61	20	98	91	69	60	26	62

(a) Find a 99% confidence interval for the true mean height of rogue waves. Assume that the underlying population is normal.

(b) Using the confidence interval in part (a), is there any evidence to suggest that the true mean height of rogue waves is different from 50 ft? Justify your answer.

Solution

(a) Construct a 99% confidence interval for μ. The underlying distribution is assumed to be normal and σ is unknown. Use Equation 8.9.

$n = 20$ — Sample size.

$\bar{x} = \dfrac{1}{20}(133 + 50 + \cdots + 62) = 76.15$ — Compute the sample mean.

$s^2 = \dfrac{1}{19}[(133 - 76.15)^2 + \cdots + (62 - 76.15)^2] = 1133.08$

$s = \sqrt{1133.08} = 33.66$ — Compute the sample standard deviation.

$1 - \alpha = 0.99 \Rightarrow \alpha = 0.01 \Rightarrow \alpha/2 = 0.005$ — Find $\alpha/2$.

$t_{\alpha/2,\,n-1} = t_{0.005,\,19} = 2.8609$ — Use Appendix Table 5 with $\nu = 19$.

Use these values in Equation 8.9.

$\bar{x} \pm t_{\alpha/2,\,n-1} \cdot \dfrac{s}{\sqrt{n}}$ — Equation 8.9.

$= \bar{x} \pm t_{0.005,\,19} \cdot \dfrac{s}{\sqrt{n}}$ — Use the values of α and n.

$= 76.15 \pm (2.8609) \cdot \dfrac{33.66}{\sqrt{20}}$ — Use summary statistics and critical value.

$= 76.15 \pm 21.53$ — Simplify.

$= (54.62, 97.68)$ — Compute endpoints.

$(54.62, 97.68)$ is a 99% confidence interval for the true mean height of rogue waves. **Figure 8.23** shows a technology solution.

(b) The CI constructed in part (a) is an interval in which we are 99% confident the true value of μ lies. If 50 lies in this interval, there is no evidence to suggest μ is different from 50. Use the four-step inference procedure.

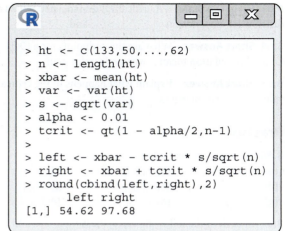

Figure 8.23 Compute the summary statistics and then find the critical value. Use Equation 8.9 to find the confidence interval.

Claim: $\mu = 50$

Experiment: $\bar{x} = 76.15$

Likelihood: The likelihood is expressed as a 99% confidence interval, an interval of likely values for μ: (54.62, 97.68), from part (a).

Conclusion: The claimed mean, 50 ft, is not included, or captured, by the confidence interval. There is evidence to suggest that μ is different from (greater than) 50 ft. ∎

TRY IT NOW Go to Exercises 8.77 and 8.79

CHAPTER APP **Autonomous Cars**

Is there any evidence to suggest that the mean number of miles between disengagements is different from 6000?

A CLOSER LOOK

1. We can use technology to find the appropriate t critical value for any sample size and any confidence level. If you must use Appendix Table 5 to find a critical value for a value of ν and/or a value of α not listed, use linear interpolation.

   Recall that interpolation is a method of approximation. It is often used to estimate a value at a position between two given values in a table. Linear interpolation (discussed in Chapter 6) assumes the two known values lie on a straight line.

2. Sample size calculation is more complicated in this case. σ is unknown and the critical value, $t_{\alpha/2, n-1}$, depends on n. Equation 8.3 is often used with an estimate for σ and the assumption that $z_{\alpha/2} \approx t_{\alpha/2, n-1}$.

Section 8.3 Exercises

Concept Check

8.59 True or False Every t distribution is symmetric and centered at 0.

8.60 True or False Every t distribution has more variability than does the Z distribution.

8.61 True or False The confidence interval for μ based on a t distribution is valid for any underlying distribution.

8.62 True or False The confidence interval for μ based on a t distribution is used only for sample sizes less than or equal to 30.

8.63 Fill in the Blank A critical value associated with a t distribution depends on _____ and _____.

8.64 Short Answer What are the endpoints for a $100(1-\alpha)\%$ CI for a population mean μ when σ is unknown?

8.65 Short Answer Explain what happens to the critical value $t_{\alpha,n-1}$ as n increases.

Practice

8.66 Find each critical value.
(a) $t_{0.10,5}$ (b) $t_{0.20,24}$ (c) $t_{0.005,19}$
(d) $t_{0.025,7}$ (e) $t_{0.005,15}$ (f) $t_{0.001,6}$
(g) $t_{0.0005,23}$ (h) $t_{0.0001,3}$ (i) $t_{0.05,11}$

8.67 In each of the following cases, the sample size and the confidence level are given. Assume σ is unknown. Find the appropriate t critical value for use in constructing a confidence interval for the population mean.
(a) $n=15$, 95%
(b) $n=21$, 98%
(c) $n=31$, 99%
(d) $n=17$, 99.9%
(e) $n=12$, 99.99%
(f) $n=4$,   98%

8.68 In each of the following cases, use Appendix Table 5 and linear interpolation (if necessary) to approximate the critical value. Verify each approximation using technology.
(a) $t_{0.15,10}$ (b) $t_{0.07,23}$ (c) $t_{0.0025,20}$
(d) $t_{0.01,35}$ (e) $t_{0.005,42}$ (f) $t_{0.025,75}$
(g) $t_{0.02,45}$ (h) $t_{0.003,52}$ (i) $t_{0.05,45}$

8.69 In each of the following cases, the sample mean, the sample standard deviation, the sample size, and the confidence level are given. Assume the underlying population is normally distributed. Find the associated confidence interval for the population mean.
(a) $\bar{x}=211.2$, $s=44.37$, $n=27$, 95%
(b) $\bar{x}=74.42$, $s=31.8$, $n=10$, 98%
(c) $\bar{x}=138.9$, $s=22.3$, $n=28$, 99%
(d) $\bar{x}=-28.3$, $s=41.33$, $n=20$, 95%
(e) $\bar{x}=1014.5$, $s=67.9$, $n=17$, 99.9%

8.70 In each of the following cases, the sample mean, the sample standard deviation, the sample size, and the confidence level are given. Assume the underlying population is normally distributed. Find the associated confidence interval for the population mean.
(a) $\bar{x}=0.234$, $s=0.081$, $n=16$, 95%
(b) $\bar{x}=259.6$, $s=76.9$, $n=26$, 99%
(c) $\bar{x}=22.85$, $s=7.19$, $n=27$, 99%
(d) $\bar{x}=380.9$, $s=28.4$, $n=21$, 95%
(e) $\bar{x}=88.1$, $s=17.45$, $n=19$, 99.9%

8.71 In each of the following problems, put the values in order from smallest to largest. (*Note:* There is no need to use a table or technology here. Use your knowledge of the t distribution and the Z distribution.)
(a) $t_{0.01,5}$; $t_{0.01,27}$; $z_{0.01}$; $t_{0.01,17}$
(b) $t_{0.025,13}$; $t_{0.025,11}$; $t_{0.025,45}$; $z_{0.025}$
(c) $t_{0.05,15}$; $t_{0.001,15}$; $t_{0.02,15}$; $t_{0.025,15}$
(d) $t_{0.0001,21}$; $t_{0.1,21}$; $t_{0.005,21}$; $t_{0.05,21}$
(e) $t_{0.10,6}$; $t_{0.001,26}$; $t_{0.05,17}$; $z_{0.0001}$

Applications

8.72 Manufacturing and Product Development During the manufacture of certain commercial windows and doors, hot steel ingots are passed through a rolling mill and flattened to a prescribed thickness. The machinery is set to produce a steel section 0.25 in. thick. Fourteen steel sections were selected at random and the thickness of each was recorded. The data are given in the table. STEEL

| 0.213 | 0.298 | 0.236 | 0.324 | 0.254 | 0.271 | 0.204 |
| 0.252 | 0.307 | 0.297 | 0.301 | 0.291 | 0.222 | 0.246 |

Assume the underlying distribution of section thickness is normal.
(a) Find a 99% confidence interval for the true mean steel section thickness.
(b) As the rollers erode, the machine begins to produce steel sections that are too thick. Using your answer to part (a), is there any evidence to suggest that the true mean steel section thickness is more than 0.25 in.? Justify your answer.
(c) Use the methods described in Section 6.3 to check for any evidence of non-normality.

8.73 Manufacturing and Product Development A typical washing machine has several different cycles, including soak, wash, and rinse. The energy consumption of a washing machine is linked to the length of each cycle. A random sample of 21 washing machines was obtained, and the length (in minutes) of each wash cycle was recorded. The summary statistics are $\bar{x}=37.8$ and $s=5.9$. Assume the underlying distribution of the main wash cycle times is normal.
(a) Find a 90% confidence interval for the true mean wash cycle time.
(b) Interpret the confidence interval found in part (a).
(c) Suppose a 95% confidence interval for the true mean wash cycle time is constructed. Would this interval be smaller or larger than the interval in part (a)? Justify your answer.

8.74 Biology and Environmental Science The planetary boundary layer (PBL) is the lowest layer of the troposphere; its characteristics are influenced by contact with the ground. Wind speed, temperature, and moisture in the PBL all affect weather patterns around the globe. A random sample of days was obtained and the height of the PBL (in meters) above the Great Basin Desert was measured using weather radar. Assume the underlying distribution of PBL heights is normal. PBL
(a) Find a 95% confidence interval for the true mean height of the PBL above the Great Basin Desert.
(b) Is there any evidence to suggest that the true mean height of the PBL above the Great Basin Desert is different from 700 m? Justify your answer.

8.75 Physical Sciences The Earth is structured in layers: crust, mantle, and core. A recent study was conducted to estimate the mean depth of the upper mantle in a specific farming region in California. Twenty-six sample sites were selected at random, and the depth to the upper mantle was measured using changes in seismic velocity and density. The summary statistics are $\bar{x} = 127.5$ km and $s = 21.3$ km. Suppose the depth of the upper mantle is normally distributed. Find a 95% confidence interval for the true mean depth of the upper mantle in this farming region, and interpret your result.

8.76 Fuel Consumption and Cars In many rural areas, newspaper carriers deliver morning papers using their automobiles because the length of the route prohibits walking. In a random sample of 28 carriers who use their automobiles, the sample mean route length was $\bar{x} = 16.7$ mi with $s = 3.4$ mi.
(a) If the distribution of route lengths is normal, find a 95% confidence interval for the true mean route length of newspaper carriers who use their automobiles.
(b) If the mean length of the routes is more than 20 mi, the circulation department becomes concerned that papers will not be delivered by 7:00 A.M. Using your answer to part (a), is there any evidence to suggest that the true mean route length is more than 20 mi? Justify your answer.

8.77 Sports and Leisure The popular sport of mountain biking involves riding bikes off-road, often over very rough terrain. Trek is a leading manufacturer of mountain bikes, which are specially designed for durability and performance. A random sample of Trek mountain bikes was obtained, and the weight (in kilograms) of each is given in the table.[16] **TREK**

| 10.35 | 11.93 | 15.22 | 9.17 | 12.46 | 12.31 | 23.18 | 14.06 |
| 15.88 | 23.59 | 11.34 | 23.16 | 10.70 | 12.70 | 10.93 | 13.24 |

(a) Assume the underlying distribution is normal. Find a 99% confidence interval for the true mean weight of Trek mountain bikes.
(b) Suppose Trek claims to have the lightest mountain bikes on the market, with a mean weight of 10 kg. Using your answer in part (a), is there any evidence to suggest that this claim is false? Justify your answer.

8.78 Technology and the Internet A computer supply store sells a wide variety of generic replacement ink cartridges for printers. A consumer group is concerned that the cartridges may not contain the specified amount of ink (30 mL). A random sample of 17 black replacement cartridges was obtained, and the amount of ink (in milliliters) in each is given in the table. **INKCART**

30.27	29.12	29.33	29.70	29.68	29.21	29.35
28.54	28.84	29.08	30.01	29.74	29.87	29.26
30.61	29.50	29.80

Assume the underlying distribution of ink amount is normal.
(a) Find a 95% confidence interval for the true mean amount of ink in each black cartridge.
(b) Is there any evidence to suggest that the cartridges are underfilled? Justify your answer.

8.79 Travel and Transportation Some municipal managers complain a city bus route is the most difficult service to fulfill. Traffic jams and roadwork are often the cause of unreliable service. In an effort to analyze current city bus transportation in Atlanta, Georgia, a random sample of routes and stops was obtained. The number of minutes the bus was late (compared with the posted route times) was recorded. The resulting data are given in the table. A value of 0.00 means the bus arrived on time. **BUS**

10.29	1.83	3.21	0.00	2.37	0.00	0.00
1.47	6.91	9.96	0.00	0.00	2.10	7.98
5.02	0.00	5.19	9.52	0.40		

(a) Assume the underlying distribution of late times is normal. Find a 90% confidence interval for the true mean number of minutes a city bus in Atlanta is late.
(b) Use the methods described in Section 6.3 to determine whether there is any evidence to suggest that the distribution of late times is non-normal.
(c) If city buses in Atlanta consistently arrive more than 5 min late, the mayor receives a large number of phone calls from irate citizens. Using your answer to part (a), is there any evidence to suggest that the mayor will be receiving nasty phone calls? Justify your answer.

8.80 Biology and Environmental Science In 2018, the United States decided to freeze the vehicle emissions standards as of 2021. Up until this time, the automobile industry had been required to reduce emissions with each new model year. A random sample of 2019 model automobiles was obtained, and the greenhouse gas (GHG) emissions were measured for each (in grams/mile). Assume the distribution of GHG emissions is normal. **GHG**
(a) Find a 95% confidence interval for the true mean vehicle GHG emissions for the 2019 model year.
(b) Suppose the mean industry target is 255 g/mi. Using the CI from part (a), is there any evidence to suggest that the automobile industry has not met this goal? Justify your answer.

8.81 Sports and Leisure One factor in rating a National Hockey League team is the mean weight of its players. A random sample of players from five teams was obtained. The weight (in pounds) of each player was carefully measured, and the resulting data are given in the table.[17]

Team	Sample size	Sample mean	Sample standard deviation
Boston	15	198.9	10.5
Dallas	15	207.1	14.2
Detroit	18	197.3	11.8
Edmonton	12	203.1	10.9
Philadelphia	10	195.8	12.1

(a) Assuming normality, find a 95% confidence interval for the true mean weight of players for each hockey team.

(b) The sample size for Boston and Dallas is the same. Why is the 95% confidence interval for the weight of Dallas players wider?
(c) If the true mean weight for two teams is different, then it is likely that there will be a more physical game when the two teams meet. Is there any evidence to suggest that the true mean player weights are different for Boston and Edmonton? Justify your answer.

8.82 Marketing and Consumer Behavior A recent study indicated that Americans are keeping their cars longer.[18] One reason may be vehicle quality. However, older vehicles do not include all of the current safety features. A random sample of new- and used-car owners was obtained, and the length of ownership (in months) was recorded for each. OWNCAR
(a) Find a 99% confidence interval for the true mean time for which people keep a new car.
(b) Find a 99% confidence interval for the true mean time for which people keep a used car.
(c) Using your answers in parts (a) and (b), do you think new-car owners and used-car owners keep their cars for different lengths of times? Justify your answer.

8.83 Economics and Finance Along with gold and silver, platinum is considered a precious metal and is traded on the commodities market. A random sample of the price of platinum (in dollars per troy ounce) was obtained from jewelers around the world. The resulting summary statistics were $\bar{x} = 830.60$, $s = 15.25$, and $n = 55$. Assume the underlying distribution of the price is normal.
(a) Find a 95% confidence interval for the current true mean price per troy ounce of platinum.
(b) One month ago, a brokerage firm reported the true mean price of platinum to be $825.75 per troy ounce, with a recommendation to buy. Is there any evidence to suggest that the true mean price has increased? Justify your answer.

8.84 Physical Sciences The Canadian Forces Snowbirds is an Air Demonstration Squadron that consists of pilots (of course), technicians, and various support officers from the Canadian Armed Forces and National Defense Public Service employees. The Snowbirds usually give about 60 performances each year, and the tour of duty for the pilots is limited to three years. A random sample of Snowbirds performances was obtained, and the speed (in knots) of the aircraft during a Tutor Formation was recorded for each. Assume the underlying distribution of the speed is normal. SNOWBIRD
(a) Find a 95% confidence interval for the true mean speed of the aircraft during a Tutor Formation.
(b) A true mean of 185 knots for the Tutor Formation is considered a safe speed. Is there any evidence to suggest that the true mean speed is different from 185 knots? Justify your answer.

Extended Applications

8.85 Psychology and Human Behavior The ambient temperature in which humans are comfortable varies with culture, activity, metabolic rate, psychological state, environment, and season. For most people in the United States, the comfort zone is 68°F to 78°F. During a recent winter, a random sample of homeowners was selected from two different parts of the country. The thermostat temperature setting (in °F) was recorded for each home, and the summary statistics are given in the table.

Region	Sample size	Sample mean	Sample standard deviation
New England	14	70.2	2.75
South	11	72.1	1.55

(a) Find a 95% confidence interval for the true mean thermostat setting for New England homeowners during winter.
(b) Find a 95% confidence interval for the true mean thermostat setting for southern homeowners during winter.
(c) What assumption(s) did you make in constructing these two confidence intervals?
(d) Using your answers to parts (a) and (b), is there any evidence to suggest the New England mean thermostat setting is different from the southern thermostat setting during winter? Justify your answer.

8.86 Public Health and Nutrition Juvenile courts in all states maintain careful records for cases, including demographics, charges, and dispositions. One variable of interest for repeat offenders is the number of days since the last offense (or arrest). A random sample of repeat offenders appearing in juvenile court was obtained for three states. The number of days since the last offense was recorded for each, and the summary statistics are given in the table.

State	Sample size	Sample mean	Sample standard deviation
Ohio	17	180.6	37.8
California	29	162.7	25.2
Massachusetts	22	115.3	17.6

(a) Assume the underlying distribution of days since the last offense is normal for each state. Find a 99% confidence interval for the true mean number of days since the last offense for repeat offenders in each state.
(b) Using your answers to part (a), is there any evidence to suggest that the true mean numbers of days since the last offense in Ohio and California are different? How about California and Massachusetts? Justify each answer.

8.87 Medicine and Clinical Studies Arrhythmia, or an irregular heartbeat, may be caused by heart disease or by environmental factors such as stress, caffeine, tobacco, or even cold medicine. One type of arrhythmia is atrial flutter, in which the heart beats very fast, at more than 250 beats per minute. In a recent research study, randomly selected patients identified to have atrial flutter were carefully monitored. The number of

beats per minute for the most recent flutter for each patient, by gender, was recorded. Assume the underlying distribution of beats per minute in patients with atrial flutter is normal. **HEART**
(a) Find a 98% confidence interval for the true mean beats per minute in male patients with atrial flutter.
(b) Find a 98% confidence interval for the true mean beats per minute in female patients with atrial flutter.
(c) Using your answers to parts (a) and (b), is there any evidence to suggest that the true mean beats per minute in atrial flutter patients are different for men and women? Justify your answer.
(d) Do you think the normality assumption is reasonable in this case? Why or why not?

8.88 Sports and Leisure The Wimbledon Championships tennis tournament, where women players curtsy to the Queen and fans eat strawberries and cream, finishes in the second week of July. Each match is played on a grass court, as opposed to a clay or hard court. Some people believe the grass court is more challenging and speeds up the game. Random samples of match lengths (in minutes) from Wimbledon and from the U.S. Clay Court Championships were obtained. The data are summarized in the table.

Court type	Sample size	Sample mean	Sample standard deviation
Grass	18	65.7	25.3
Clay	12	83.2	35.8

Assume the underlying distribution of match lengths is normal on grass and on clay.
(a) Find a 99% confidence interval for the true mean match length for grass courts at Wimbledon.
(b) Find a 99% confidence interval for the true mean match length for clay courts at the U.S. Clay Court Championships.
(c) Is there any evidence to suggest the mean match times for grass courts and clay courts are different? Justify your answer.

8.89 Public Health and Nutrition During a medical emergency, people who dial 911 expect an ambulance to arrive quickly and personnel to provide vital care. Health insurance companies believe there is a marked difference in the response time for rural areas versus cities because of differences in coverage area and number of qualified paramedics. Random samples of ambulance response times (in minutes) were obtained for rural areas and cities. **DIAL911**
(a) Assuming normality, find a 99% confidence interval for the true mean response time for rural-area ambulances.
(b) Assuming normality, find a 99% confidence interval for the true mean response time for city ambulances.
(c) Use the methods described in Section 6.3 to determine whether there is any evidence to suggest that either response time distribution is non-normal.
(d) Is there any evidence to suggest that the mean response time is different for rural areas and cities? Justify your answer.

8.90 Physical Sciences Researchers have discovered a new method for increasing the efficiency of solar panels. A new antireflective coating allows a panel to absorb more sunlight from any angle. A random sample of solar panels was obtained, and each was treated with the new coating. The rate of sunlight absorbed (as a percentage) was measured for each panel. **SOLAR**
(a) Assuming normality, find a 95% confidence interval for the true mean rate of sunlight absorbed.
(b) The mean rate of sunlight absorbed for untreated solar panels is 91.5. Is there any evidence to suggest that the mean rate of sunlight absorbed for treated solar panels is greater than 91.5? Justify your answer.

8.91 Sports and Leisure The Jackpine Gypsies Motorcycle Club holds the Sturgis Rally in the Black Hills of South Dakota every summer. There are motocross, hill climb, and short-track races, plus plenty of parties and bikes. In an attempt to learn more about the approximately 750,000 participants, a random sample of bike enthusiasts was obtained. Some of the summary data are given in the table.

Variable	Sample size	Sample mean	Sample standard deviation
Age (males, years)	60	38.9	7.9
Age (females, years)	40	35.6	4.5
Distance traveled (miles)	75	257.5	56.8

Assume the underlying distributions are normal.
(a) Find a 95% confidence interval for the true mean age of men attending the rally and a 95% confidence interval for the true mean age of women attending the rally.
(b) Is there any evidence to suggest that the mean age for men is different from the mean age for women? Justify your answer.
(c) Find a 99% confidence interval for the true mean distance traveled to the rally. Interpret this result.

8.92 Biology and Environmental Science Some researchers have suggested that climate change is likely to reduce the number of hailstorms, but the size of hail in the storms will probably be larger. A random sample of hail reports was obtained, and the size (diameter, in inches) of the hail was measured for each.[19] Assume the underlying distribution of hail size is normal. **HAIL**
(a) Find a 95% confidence interval for the true mean size of hail.
(b) Is there any evidence to suggest that the true mean size of hail is greater than 1 in.? Justify your answer.

Challenge Problems

8.93 Biology and Environmental Science Given a random sample of size n from a normal population with mean μ (and σ unknown):
(a) Use the method outlined in Exercise 8.56 to find a one-sided $100(1-\alpha)\%$ confidence interval for μ bounded above, and another bounded below.

(b) In August 2018, the Mendocino Complex Fire in California became the largest wildfire ever recorded in the state, burning through more than 283,000 acres. Some researchers have suggested that wildfires in California will continue to increase in size due to environmental laws and use of water resources. A random sample of wildfires in California was obtained, and the number of acres burned was recorded for each.[20] Assume the underlying population distribution is normal, and find a one-sided 99% confidence interval, bounded below, for the true mean acres burned during a California wildfire. **FIRES**

8.4 A Large-Sample Confidence Interval for a Population Proportion

Large-Sample Confidence Interval for p

Many surveys and experiments are conducted to estimate a population proportion, the true fraction of individuals or objects that exhibit a specific characteristic. For example, pollsters routinely estimate the proportion of Americans who favor a particular candidate for office. Food companies use randomly selected consumers to test-market new products, because the true proportion of shoppers who will purchase a product is important for predicting profit. Insurance agencies constantly analyze data to estimate the proportion of homeowners who will file a claim during the next year.

Let p = the true population proportion, the fraction of individuals or objects with a specific characteristic (the probability of a success). As in Section 7.3, it is reasonable to use a value of the sample proportion, $\hat{p}$, to construct a confidence interval for p. In a sample of n individuals or objects, let X be the number of individuals with the characteristic (or the number of successes in the sample). Recall that the sample proportion is the proportion of individuals with the specific characteristic in the sample, or a relative frequency of those individuals.

$$\hat{P} = \frac{X}{n} = \frac{\text{the number of individuals with the characteristic}}{\text{the sample size}} \tag{8.10}$$

From Section 7.3, if n is large and both $np \geq 5$ and $n(1-p) \geq 5$, then the random variable $\hat{P}$ is approximately normal with mean p and variance $p(1-p)/n$. Since $\hat{P}$ is approximately normal, we can standardize to obtain

$$Z = \frac{\hat{P} - p}{\sqrt{\frac{p(1-p)}{n}}} \sim N(0, 1) \tag{8.11}$$

▶ As in Section 8.2, start with an appropriate probability statement involving the random variable Z in Equation 8.11. Find a symmetric interval about 0 such that the probability that Z lies in this interval is $1 - \alpha$.

$$P(-z_{\alpha/2} < Z < z_{\alpha/2}) = 1 - \alpha$$

$$P\left(-z_{\alpha/2} < \frac{\hat{P} - p}{\sqrt{\frac{p(1-p)}{n}}} < z_{\alpha/2}\right) = 1 - \alpha \tag{8.12}$$

A Challenge Problem in this section asks you to find a confidence interval for p without using $\hat{p}$ in the denominator.

Trying to sandwich p in Equation 8.12 is a little tricky because p appears in both the numerator and the denominator (inside a square root). Instead, because n is large, we typically use the sample proportion, $\hat{p}$, as a good estimate of p in the denominator. Manipulating the inequality (inside the probability statement) leads to the following general result. ◀

How to Find a Large-Sample $100(1-\alpha)\%$ Confidence Interval for a Population Proportion

Use $\hat{p}$ as an estimate of p to check the nonskewness criterion.

Given a random sample of size n. If n is large and both $n p \geq 5$ and $n(1-p) \geq 5$, the endpoints for a large-sample $100(1-\alpha)\%$ confidence interval for p, the true proportion, have the values

$$\hat{p} \pm z_{\alpha/2} \cdot \sqrt{\frac{\hat{p}(1-\hat{p})}{n}} \tag{8.13}$$

Confidence Interval for p and Inference

EXAMPLE 8.9 Magnesium Levels

Recent research suggests that many Americans have deficient magnesium levels. Vitamin D, which increases calcium and phosphate levels, cannot be metabolized without magnesium, but this study suggests that many Americans are consuming only half the recommended daily allowance of magnesium.[21] A random sample of 1200 Americans was obtained, and each was tested for magnesium level. In total, 540 were found to have deficient magnesium levels.

(a) Find a 95% confidence interval for the true proportion of Americans who have deficient magnesium levels.

(b) The American Medical Association claims that 50% of Americans are magnesium deficient. Is there any evidence to suggest that the percentage of Americans who have deficient magnesium levels is different from this value? Justify your answer.

Solution

(a) We need a 95% CI for p. Check the nonskewness criterion to confirm that the distribution of $\hat{P}$ is approximately normal. Find the appropriate critical value, and use Equation 8.13.

$n = 1200$ Sample size.

$\hat{p} = \dfrac{x}{n} = \dfrac{540}{1200} = 0.45$ Compute the sample proportion.

$n\hat{p} = (1200)(0.45) = 540 \geq 5$ Check nonskewness criterion.

$n(1-\hat{p}) = (1200)(0.55) = 660 \geq 5$

Both inequalities are satisfied, so $\hat{P}$ is approximately normal, and Equation 8.13 can be used to construct a confidence interval for p.

$1 - \alpha = 0.95 \Rightarrow \alpha = 0.05 \Rightarrow \alpha/2 = 0.025$ Find $\alpha/2$.

$z_{\alpha/2} = z_{0.025} = 1.96$ Common critical value.

Use Equation 8.13.

$\hat{p} \pm z_{\alpha/2} \cdot \sqrt{\dfrac{\hat{p}(1-\hat{p})}{n}}$ Equation 8.13.

$= \hat{p} \pm z_{0.025} \cdot \sqrt{\dfrac{\hat{p}(1-\hat{p})}{n}}$ Use the value of α.

$= 0.45 \pm (1.96) \cdot \sqrt{\dfrac{(0.45)(0.55)}{1200}}$ Use the values for $z_{0.025}$ and $\hat{p}$.

$= 0.45 \pm 0.0281$ Simplify.

$= (0.4219, 0.4781)$ Compute endpoints.

$(0.4219, 0.4781)$ is a 95% confidence interval for the true proportion p of Americans who are magnesium deficient. **Figure 8.24** shows a technology solution.

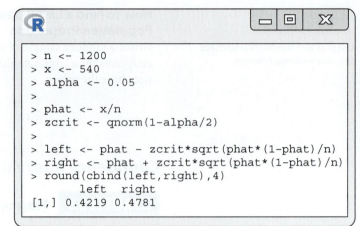

Figure 8.24 Use $\hat{p}$ and the critical value to find a confidence interval for p.

(b) The CI constructed in part (a) is an interval in which we are 95% confident that the true value of p lies. If 0.50 lies in this interval, there is no evidence to suggest p is different from 0.50. Use the four-step inference procedure.

Claim: $p = 0.50$

Experiment: $\hat{p} = 0.45$

Likelihood: The likelihood is expressed as a 95% confidence interval, an interval of likely values for p: (0.4219, 0.4781).

Conclusion: The claimed value, 0.50, does not lie in this confidence interval. There is evidence to suggest that p is different from (less than) 0.50. ∎

TRY IT NOW Go to Exercises 8.107 and 8.109

Here is another example involving a confidence interval for a population proportion.

EXAMPLE 8.10   Solar Flare Activity

Solar flares on the sun are violent explosions that release enormous amounts of energy and radiation. These emissions can harm spacecraft and satellites in Earth's orbit and disrupt communication on the ground. The year 2019 was near the end of an approximate 11-year solar flare activity cycle.[22] However, in an effort to plan for possible disruptions during the next solar cycle, a study was conducted over several years. On 86 of 350 randomly selected days, at least one solar flare occurred. Find a 99% confidence interval for the true proportion of days on which there is at least one solar flare.

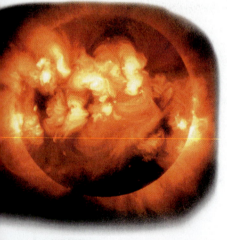

JAXA/NASA

Solution

STEP 1 $n = 350$; $x = 86$ *Given.*

STEP 2   The sample proportion: $\hat{p} = \dfrac{86}{350} = 0.2457$

Check the nonskewness criterion.

$n\hat{p} = (350)(0.2457) = 86 \geq 5$

$n(1-\hat{p}) = (350)(0.7543) = 264 \geq 5$

Both inequalities are satisfied. $\hat{P}$ is approximately normal.

STEP 3 $1 - \alpha = 0.99 \Rightarrow \alpha = 0.01 \Rightarrow \alpha/2 = 0.005$ *Find $\alpha/2$.*

$z_{\alpha/2} = z_{0.005} = 2.5758$ *Common critical value.*

STEP 4   Use Equation 8.13.

$$\hat{p} \pm z_{\alpha/2} \cdot \sqrt{\dfrac{\hat{p}(1-\hat{p})}{n}}$$ *Equation 8.13.*

$$= \hat{p} \pm z_{0.005} \cdot \sqrt{\frac{\hat{p}(1-\hat{p})}{n}}$$ Use the value of α.

$$= 0.2457 \pm (2.5758) \cdot \sqrt{\frac{(0.2457)(0.7543)}{350}}$$ Use the values for $z_{0.005}$ and $\hat{p}$.

$$= 0.2457 \pm 0.0593$$ Simplify.

$$= (0.1864, 0.3050)$$ Compute endpoints.

$(0.1864, 0.3050)$ is a 99% confidence interval for the true proportion p of days on which there is at least one solar flare. **Figure 8.25** shows a technology solution.

```
> n <- 350
> x <- 86
> alpha <- 0.01
>
> phat <- x/n
> zcrit <- qnorm(1-alpha/2)
>
> left <- phat - zcrit*sqrt(phat*(1-phat)/n)
> right <- phat + zcrit*sqrt(phat*(1-phat)/n)
> round(cbind(left,right),4)
        left right
[1,] 0.1864 0.305
```

Figure 8.25 Use $\hat{p}$ and the critical value to find a confidence interval for p.

TRY IT NOW Go to Exercises 8.117 and 8.119

Error of Estimation

Recall from Section 8.2 that a sample size can be determined such that the resulting confidence interval for μ has certain properties. Similarly, given the confidence level, a sample size n can be computed such that the resulting confidence interval for p has a desired width.

Suppose n is large, the confidence level is $100(1-\alpha)\%$, and B is the bound on the error of estimation. Half the width of the confidence interval is the step in each direction from $\hat{p}$. Therefore, the bound on the error of estimation is

$$B = z_{\alpha/2} \cdot \sqrt{\frac{\hat{p}(1-\hat{p})}{n}} \qquad (8.14)$$

Solving for n yields

$$n = \hat{p}(1-\hat{p})\left(\frac{z_{\alpha/2}}{B}\right)^2 \qquad (8.15)$$

Look carefully at this result. Although the mathematics is correct, there is a real issue with Equation 8.15. The critical value $z_{\alpha/2}$ and the bound on the error of estimation B can be specified, but $\hat{p}$ is unknown. We do not know $\hat{p}$ until we have the sample size n (and x).

There are two solutions to this problem.

1. Use a reasonable estimate for $\hat{p}$ from previous experience. Researchers often conduct many similar experiments over time and can make very realistic guesses for $\hat{p}$.
2. If no prior information is available, use $\hat{p} = 0.5$ in Equation 8.15. This produces a very conservative, large value for n. This result occurs because the product $\hat{p}(1-\hat{p})$ is largest when $\hat{p} = 0.5$.

For given $z_{\alpha/2}$ and B, Equation 8.15 is greatest (maximized) when $\hat{p} = 0.5$.

In either case, the value of n is unlikely to be an integer. Always round up, as this guarantees the resulting confidence interval will have a bound on the error of estimation of at most B. Any larger sample size will also suffice. The following example illustrates the use of Equation 8.15.

EXAMPLE 8.11 Negative Equity

During the housing crisis, more than 30% of homeowners owed lenders more than the actual value of their homes, meaning they had negative equity. Although an economic recovery occurred subsequent to the crisis, millions of Americans still have negative equity in their homes. To help banks set interest rates, a company plans to estimate the proportion of American homeowners with negative equity. A 95% confidence interval for p with bound on the error of estimation of 0.02 is needed. How large a sample size is necessary in each of the following cases?

(a) Prior experience suggests $\hat{p} \approx 0.2$.

(b) There is no prior information about the value of $\hat{p}$.

Solution

(a) Use $\hat{p} = 0.2$ and $B = 0.02$ in Equation 8.15.

$1 - \alpha = 0.95 \Rightarrow \alpha = 0.05 \Rightarrow \alpha/2 = 0.025 \Rightarrow z_{0.025} = 1.96$ Find the critical value.

$$n = \hat{p}(1-\hat{p})\left(\frac{z_{\alpha/2}}{B}\right)^2$$ Equation 8.15.

$$= (0.2)(0.8)\left(\frac{1.96}{0.02}\right)^2$$ Substitute given values.

$$= (0.16)(98)^2 = 1536.64$$ Computation.

The necessary sample size is $n \geq 1537$.

(b) With no prior information, use $\hat{p} = 0.5$, $B = 0.02$, and $z_{0.025} = 1.96$ in Equation 8.15.

$$n = \hat{p}(1-\hat{p})\left(\frac{z_{\alpha/2}}{B}\right)^2$$ Equation 8.15.

$$= (0.5)(0.5)\left(\frac{1.96}{0.02}\right)^2$$ Substitute given values.

$$= (0.25)(98)^2 = 2401$$ Computation.

The necessary sample size in this case is $n \geq 2401$.

TRY IT NOW Go to Exercise 8.113

Section 8.4 Exercises

Concept Check

8.94 Short Answer Which two conditions must be true to construct a large sample confidence interval for a population proportion?

8.95 Short Answer Explain how a large sample confidence interval for a population proportion can be used in statistical inference.

8.96 True or False A large sample confidence interval for a population proportion will always include 0.50.

8.97 True or False When computing the sample size necessary to construct a confidence interval for a population proportion of a certain width, round n to the nearest 100.

8.98 True or False When computing the sample size necessary to construct a confidence interval for a population proportion of a certain width, if no prior information is available, use a value for $\hat{p}$ close to 0 or close to 1.

8.99 Short Answer Using Equation 8.15, is it possible that a confidence interval for a population proportion may include values greater than 1 or values less than 0? If so, how? If not, why not?

Practice

8.100 In the following cases, the sample size and the number of individuals or objects with a specified characteristic are given. Check the nonskewness criterion to determine whether the distribution is approximately normal.
(a) $n = 105$, $x = 85$
(b) $n = 1750$, $x = 1645$
(c) $n = 225$, $x = 220$
(d) $n = 183$, $x = 3$
(e) $n = 377$, $x = 350$
(f) $n = 480$, $x = 478$

8.101 In the following cases, the sample size, the number of individuals or objects with a specified characteristic, and the confidence level are given. Find the associated confidence interval for the population proportion.
(a) $n = 150$, $x = 70$, 95%
(b) $n = 225$, $x = 65$, 98%
(c) $n = 500$, $x = 468$, 90%
(d) $n = 95$, $x = 63$, 99%
(e) $n = 2450$, $x = 986$, 99.9%

8.102 In the following cases, the sample size, the number of individuals or objects with a specified characteristic, and the confidence level are given. Find the associated confidence interval for the population proportion.
(a) $n = 1336$, $x = 1001$, 99%
(b) $n = 775$, $x = 680$, 95%
(c) $n = 85$, $x = 41$, 95%
(d) $n = 335$, $x = 290$, 98%
(e) $n = 566$, $x = 47$, 99.9%

8.103 In the following cases, the confidence level, the bound on the error of estimation, and an estimate for $\hat{p}$ are given. Find the sample size necessary to produce a confidence interval for p with this bound and confidence.
(a) 95%, $B = 0.05$, $\hat{p} \approx 0.45$
(b) 98%, $B = 0.07$, $\hat{p} \approx 0.32$
(c) 99%, $B = 0.10$, $\hat{p} \approx 0.14$
(d) 90%, $B = 0.001$, $\hat{p} \approx 0.057$
(e) 99.9%, $B = 0.03$, $\hat{p} \approx 0.22$

8.104 In the following cases, the confidence level and the bound on the error of estimation are given. Find the sample size necessary to produce a confidence interval for p with this bound and confidence level.
(a) 99%, $B = 0.06$
(b) 95%, $B = 0.10$
(c) 98%, $B = 0.002$
(d) 90%, $B = 0.05$
(e) 99.9%, $B = 0.20$

8.105 Three factors affect the width of a large-sample confidence interval for p: the confidence level, the sample size, and the sample proportion $\hat{p}$. Determine whether the width of the resulting confidence interval for p increases or decreases under each of the following conditions.
(a) Confidence level and $\hat{p}$ are constant, and n increases.
(b) Sample size and $\hat{p}$ are constant, and confidence level decreases.
(c) Sample size and confidence level are constant, and $\hat{p}(>0.5)$ increases.

8.106 Three factors affect the size of the sample necessary to produce a confidence interval for p with certain properties: the confidence level, the bound on the error of estimation, and the estimate of $\hat{p}$. Determine whether the necessary sample size increases or decreases under each of the following conditions.
(a) Confidence level and estimate of $\hat{p}$ are constant, and B is smaller.
(b) B and estimate of $\hat{p}$ are constant, and confidence level increases.
(c) B and confidence level are constant, and estimate of $\hat{p}$ is closer to 0.
(d) B and confidence level are constant, and estimate of $\hat{p}$ is closer to 1.

Applications

8.107 Education and Child Development In 2018, Chicago Public Schools officials were working to improve the cleanliness of the school facilities. However, blitz inspections in July showed that 25% of schools still failed to meet cleanliness standards.[23] Suppose a random sample of 230 Chicago Public Schools was obtained just before classes started in the fall. When each was thoroughly inspected, 55 schools failed inspection due to cleanliness problems.
(a) Find a 95% confidence interval for the true proportion of Chicago Public Schools that would fail inspection.
(b) Is there any evidence to suggest that the true proportion is less than 0.25? Justify your answer.

8.108 Public Policy and Political Science A new standard for 30-, 40-, and 50-gal residential gas water heaters includes a flame arrester. This device helps to prevent flashback fires from flammable liquid vapor nearby. The Consumer Product Safety Commission would like to estimate the proportion of homes affected by this safety standard.
(a) In a random sample of 575 homes, 235 had gas water heaters. Find a 90% confidence interval for the true proportion of homes with gas water heaters.
(b) Prior research suggests that the proportion of homes with gas water heaters is approximately 0.40. How large a sample is necessary for the bound on the error of estimation to be 0.03 for a 95% confidence interval?

8.109 Marketing and Consumer Behavior A successful company usually has high brand-name and logo recognition among consumers. For example, Coca-Cola's red-and-white logo is recognized by 94% of the world's population.[24] A software firm developing a product would like to estimate the proportion of people who recognize the Linux penguin logo. Of the 952 randomly selected consumers surveyed, 132 could identify the product associated with the penguin.
(a) Is the distribution of the sample proportion $\hat{P}$ approximately normal? Justify your answer.
(b) Find a 95% confidence interval for the true proportion of consumers who recognize the Linux penguin.
(c) The company will market a Linux version of its new software if the true proportion of people who recognize the logo is greater than 0.10. Is there any evidence to suggest that the true proportion of people who recognize the logo is greater than 0.10? Justify your answer.

8.110 Travel and Transportation To advertise appropriate vacation packages, Best Bets Travel would like to learn more about families planning overseas trips. In a random sample of 125 families planning a trip to Europe, 15 indicated France was their travel destination.
(a) For those families planning vacations to Europe, find a 98% confidence interval for the true proportion traveling to France.
(b) Suppose no prior estimate of $\hat{p}$ is known. How large a sample is necessary for the bound on the error of estimation to be 0.05 with the confidence level in part (a)?

8.111 Public Health and Nutrition Vegetarian and plant-based options are becoming more widely available in restaurants and grocery stores. Many people who stop eating meat do so due to health reasons. A random sample of adults in three countries was obtained, and the number who declared themselves to be vegetarian was recorded. The data are summarized in the table.

Country	Sample size	Vegetarian
United States	375	37
United Kingdom	406	60
Canada	525	57

(a) Find a 95% confidence interval for the true proportion of U.S. adults who are vegetarian.
(b) Find a 95% confidence interval for the true proportion of U.K. adults who are vegetarian.
(c) Find a 95% confidence interval for the true proportion of Canadian adults who are vegetarian.
(d) Which of the confidence intervals from parts (a), (b), and (c) is the narrowest? Why?

8.112 Economics and Finance A systematic risk survey was conducted in the United Kingdom to assess the nation's views of confidence in the stability of the financial system. A random sample of 500 adults was obtained; 230 cited Brexit as the greatest risk to the U.K. financial system, 178 cited a cyber attack, and 66 named regulation.[25]
(a) Find a 95% confidence interval for the true proportion of U.K. residents who believe Brexit is the greatest risk to the financial system.
(b) Find a 95% confidence interval for the true proportion of U.K. residents who believe a cyber attack is the greatest risk to the financial system.
(c) Find a 95% confidence interval for the true proportion of U.K. residents who believe regulation is the greatest risk to the financial system.

8.113 Technology and the Internet Many websites require users to change their passwords frequently, and some people routinely change their passwords for greater security. However, in a recent survey of 2500 U.S. consumers, 875 indicated they never change their passwords, except if prompted to do so.
(a) Find a 95% confidence interval for the true proportion of consumers who never change their passwords.
(b) Suppose no prior estimate of $\hat{p}$ is known. How large a sample is necessary for the bound on the error of estimation to be 0.04 with a confidence level of 90%?

8.114 Economics and Finance When unemployment rises, high school students face more competition from college students and adult workers for summer jobs. In a random sample of 188 high school students looking for summer work, 61 said they were able to find a job.
(a) Find a 95% confidence interval for the true proportion of high school students who were able to find a summer job.
(b) In a similar study conducted the previous year, the sample proportion of high school students who were able to find a job was 0.25. Use this estimate to find the sample size necessary for the bound on the error of estimation to be 0.025 with a confidence level of 95%.

8.115 Marketing and Consumer Behavior Most walk-behind lawn mowers have three options for disposal of grass clippings: by bagging, by mulching, or by side discharge. The manager at an Aubuchon Hardware store conducted a survey to determine which disposal method is most common. The results are given in the table, classified by area mowed. A small area is less than 1/2 acre, a medium area is 1/2 to 1 acre, and a large area is more than 1 acre.

Area	Sample size	Disposal method		
		Bagging	Mulching	Side discharge
Small	125	85	35	5
Medium	157	70	40	47
Large	144	42	45	57

(a) For people with small yards, find a 95% confidence interval for the true proportion who dispose of grass clippings by bagging.
(b) For people with medium yards, find a 95% confidence interval for the true proportion who dispose of grass clippings by mulching.
(c) For people with large yards, find a 95% confidence interval for the true proportion who dispose of grass clippings by side discharge.

8.116 Business and Management A U.S. textile company is interested in the proportion of orders shipped to another country. In a random sample of 1560 clothing orders placed to U.S. companies, 500 were exported.
(a) Find a 99% confidence interval for the true proportion of clothing orders shipped to other countries.
(b) In the previous year, the true proportion of clothing orders shipped to other countries was believed to be 0.30. Using your answer to part (a), is there any evidence to suggest the true proportion has changed? Justify your answer.

8.117 Business and Management Many dairy farms have experienced bankruptcy over the past decade as a result of wild fluctuations in conventional milk prices. However, organic farms, which do not treat cows with antibiotics or hormones and which use hay grown without chemicals, have remained solvent and even expanded. In a random sample of 1400 New England dairy farms, 90 are certified as organic.

(a) Find a 99% confidence interval for the true proportion of New England dairy farms certified as organic.

(b) Five years ago, an extensive census reported that 3% of all New England dairy farms were organic. Is there any evidence to suggest that this proportion has changed? Justify your answer.

8.118 Marketing and Consumer Behavior In a recent survey of home-buyer preferences, consumers were asked about desired characteristics, such as a wood-burning fireplace, a den/library, and flooring. In particular, each person was asked whether a separate dining room is essential. The sample size and the number who responded *Yes* to this question are given in the table by geographic region.

Geographic region	Sample size	Number who responded Yes
Northeast	225	180
Midwest	276	224
South Central	301	232
South Atlantic	454	377
West	366	304

(a) Find a 99% confidence interval for the true proportion of home-buyers who believe a separate dining room is essential in each geographic region.

(b) Which confidence interval in part (a) is the largest? Why?

8.119 Travel and Transportation The U.S. Transportation Security Administration (TSA) screens approximately 2.1 million travelers in the nation's airports each day. All carry-on and checked baggage is carefully screened, and every passenger must pass through a metal detector or an advanced imaging-technology scanner. The agency claims that approximately 93% of passengers who were in a TSA Pre-check lane waited less than 5 min.[26] In a random sample of 505 passengers in the TSA Pre-check lane, 458 waited less than 5 min.

(a) Find a 99% confidence interval for the true proportion of airline passengers who wait less than 5 min in the TSA Pre-check lane.

(b) Is there any evidence to suggest that the TSA claim is false? Justify your answer.

8.120 Economics and Finance For homeowners with an adjustable loan, a mortgage reset refers to a change in the interest rate and, therefore, the payment. In 2018, the Bank of Canada indicated that half of all mortgages were scheduled for a reset during the next year.[27] To check this claim, a random sample of 328 mortgages was obtained, and 145 were scheduled for a mortgage reset within the next year.

(a) Find a 95% confidence interval for the true proportion of mortgage resets that will occur within the next year.

(b) Is there any evidence to suggest that the Bank of Canada claim is wrong? Justify your answer.

Extended Applications

8.121 Sports and Leisure Deep-sea fishing takes place farther out to sea in waters at least 30 m deep. And because it is fishing, no one is guaranteed to catch anything. A random sample of people who went deep-sea fishing near Gloucester, Massachusetts, was obtained and each was asked if he or she caught a fish (a success!). The resulting data are given in the table, by company.

Party boat	Sample size	Number who caught a fish
Gloucester Fleet	260	160
Yankee Fleet	380	288
Patriot Wave	310	125

(a) Find a 95% confidence interval for the true proportion of people who catch a fish for each party boat.

(b) Is there any evidence to suggest that the true proportion of people who catch a fish on a Patriot Wave party boat is different from that for either of the other two boat companies? Justify your answer.

8.122 Psychology and Human Behavior The Drivers Technology Association in the United Kingdom recently studied the behavior of drivers with and without radar detectors. A random sample of 550 users and 562 non-users was obtained. In the past three years, 108 users and 68 non-users were found to have had an accident.

(a) Find a 99% confidence interval for the true proportion of radar-detector users who had an accident in the past three years.

(b) Find a 99% confidence interval for the true proportion of radar-detector non-users who had an accident in the past three years.

(c) Is there any evidence to suggest the two true proportions are different? Justify your answer.

8.123 Public Health and Nutrition Birth weight is often used as a measure of health and quality of life. Some research suggests that the proportion of babies born with a low birth weight is increasing, and data suggest that babies are more likely to be born at low birth weight in some communities and states than in others. A random sample of newborns was obtained from Nevada and California, and the birth weight of each was classified. The resulting data are given in the table.[28]

State	Sample size	Number with low birth weight
California	452	31
Nevada	402	35

(a) Find a 95% confidence interval for the true proportion of babies born in California with low birth weight.

(b) Find a 95% confidence interval for the true proportion of babies born in Nevada with low birth weight.

(c) Is there any evidence to suggest that the two true proportions are different? Justify your answer.

8.124 Economics and Finance The first question on the U.S. IRS Income Tax Form 1040 asks taxpayers whether they want $3 to go to the Presidential Election Campaign Fund. A large Washington, D.C., political-action committee obtained a random sample of registered voters and asked them if they checked *Yes* on this question on the past year's return. The results are given in the table, by party affiliation.

Political party	Sample size	Number who checked Yes
Democrat	237	70
Republican	388	184
Independent	155	23

(a) Find a 95% confidence interval for the true proportion of registered voters in each political party who checked Yes on the Presidential Election Campaign Fund question.
(b) Is there any evidence to suggest that the proportion of Independent voters who checked Yes is different from the proportions of either Democrat or Republican voters who checked Yes? Justify your answer.
(c) Suppose the results obtained in this study are preliminary and a larger survey is planned. How large a sample is necessary for each political party for the bound on the error of estimation to be 0.02 with a confidence level of 95%?

8.125 Medicine and Clinical Studies A new prescription medication is designed to ease the pain of arthritis. In a clinical trial, both treatment and placebo groups were studied to determine whether they had any adverse reactions to the drug. The table lists the number of patients in each group who experienced each adverse reaction. Assume each group represents a random sample.

Adverse reaction	Treatment group ($n = 465$)	Placebo group ($n = 154$)
Headache	61	15
Rash	31	9

(a) Find a 95% confidence interval for the true proportion of people who suffered a headache in each of the groups.
(b) Is there any evidence to suggest that the true proportion of people who suffered a headache is different for the two groups? Justify your answer.
(c) Find a 98% confidence interval for the true proportion of people who experienced a rash in each of the groups.
(d) Is there any evidence to suggest that the true proportion of people who experienced a rash is different for the two groups? Justify your answer.

8.126 Demographics and Population Statistics Since 1973, when the United States ended the draft for military service, the number of active-duty personnel has decreased from a little more than 2 million to 1.29 million. However, the percentage of active female enlisted military service members has increased. A random sample of enlisted personnel from each of the four military branches was obtained. The enlisted female representation, by service, is given in the table.

Service	Sample size	Number of females
Army	478	72
Navy	436	80
Marines	362	29
Air Force	425	71

(a) Find a 95% confidence interval for the true proportion of female enlisted service members for each military branch.
(b) A Pentagon spokesperson claims the proportion of female enlisted service members for each military branch is Army, 0.14; Navy, 0.19; Marines, 0.08; and Air Force, 0.19. Is there any evidence to suggest that any of these claimed proportions is false? Justify your answer.
(c) Is there any evidence to suggest that the true proportion of female enlisted service members is different for any two military branches? Justify your answer.

Challenge Problems

8.127 Technology and the Internet Given a random sample of size n and the sample proportion of individuals with a specific characteristic $\hat{p}$, suppose n is large and both $n\hat{p} \geq 5$ and $n(1-\hat{p}) \geq 5$.
(a) Use the method outlined in Exercise 8.56 to find a one-sided $100(1-\alpha)\%$ confidence interval for p bounded above, and another bounded below.
(b) Some Internet users download and share illegal copies of songs (and movies). A random sample of 260 Internet users who regularly download music was obtained, and 171 indicated they did not care if they were violating copyright laws. Find a one-sided 95% confidence interval, bounded above, for the population proportion of Internet users who download music and who do not care about violating copyright laws.

8.128 The Wilson Confidence Interval Consider the probability statement used to construct a confidence interval for a population proportion (Equation 8.12):

$$P\left(-z_{\alpha/2} < \frac{\hat{P} - p}{\sqrt{\frac{p(1-p)}{n}}} < z_{\alpha/2}\right) = 1 - \alpha$$

Rewrite the inequality (without substituting $\hat{p}$ for p in the denominator) to obtain a $100(1-\alpha)\%$ confidence interval for p (the Wilson interval) with endpoints

$$\frac{n\hat{p} + z_{\alpha/2}^2/2}{n + z_{\alpha/2}^2} \pm \frac{z_{\alpha/2}\sqrt{n}}{n + z_{\alpha/2}^2}\sqrt{\hat{p}(1-\hat{p}) + z_{\alpha/2}^2/(4n)} \quad (8.16)$$

(a) In a random sample of size 100, suppose the sample proportion is $\hat{p} = 0.60$. Find a 95% confidence interval for p using Equation 8.13. Find a 95% confidence interval for p using the Wilson interval (Equation 8.16). Which is wider? Why?
(b) Let $n = 120, 140, 160, \ldots, 500$. For each value of n, let $\hat{p} = 0.60$ and compute both CIs for p. What happens to the Wilson CI as n increases? Why?

8.129 Necessary Sample Size To find the sample size necessary to construct a $100(1-\alpha)\%$ confidence interval for a population proportion with a bound on the error of estimation B, Equation 8.15 is used:

$$n = \hat{p}(1-\hat{p})\left(\frac{z_{\alpha/2}}{B}\right)^2$$

For a 95% confidence interval and $B = 0.10$, let $\hat{p} = 0.05, 0.10, 0.15, \ldots, 0.95$, and find the sample size necessary for each value of $\hat{p}$. Plot the values of n versus $\hat{p}$ (n on the vertical axis and $\hat{p}$ on the horizontal axis). Describe the pattern. When is n largest?

8.5 A Confidence Interval for a Population Variance or Standard Deviation

The Chi-Square Distribution

Many real-world problems involve estimation of variability to find out whether measurements are clustered around a central value or spread out over a wide range. For example, quality-control specialists continuously monitor production processes for increases or changes in range or variability, and most scientists claim increased quantities of carbon dioxide in the atmosphere have contributed to greater climate variability. It seems reasonable to use the sample variance S^2 as an estimator for the population variance σ^2, a measure of variability. A confidence interval for a population variance σ^2 is based on S^2, a new standardization, and a **chi-square distribution**.

S^2 is an unbiased estimator for σ^2: $E(S^2) = \sigma^2$.

A **chi-square** (abbreviated χ^2) **distribution** has positive probability only for nonnegative values. The probability density function for a χ^2 random variable (details in the margin) is 0 for $x < 0$. We will focus on the properties of a chi-square distribution and the method for finding critical values associated with this distribution.

Properties of a Chi-Square Distribution

The probability density function for a chi-square random variable X, with ν degrees of freedom, is given by

$$f(x) = \begin{cases} \dfrac{e^{-x/2} x^{(\nu/2)-1}}{2^{\nu/2} \cdot (\nu/2)} & x \geq 0 \\ 0 & \text{otherwise} \end{cases}$$

1. A chi-square distribution is completely determined by one parameter, the number of degrees of freedom (ν). The degrees of freedom must be a positive integer, $\nu = 1, 2, 3, 4, \ldots$. There is a different chi-square distribution corresponding to each value of ν.

2. If X has a chi-square distribution with ν degrees of freedom, denoted $X \sim \chi^2_\nu$, then

$$\mu_X = \nu \quad \text{and} \quad \sigma^2_X = 2\nu \tag{8.17}$$

The mean of X is ν, the number of degrees of freedom, and the variance is 2ν, twice the number of degrees of freedom.

3. Suppose $X \sim \chi^2_\nu$. The density curve for X is positively skewed (not symmetric). As x increases, the density curve gets closer and closer to the x-axis but never touches it. As ν increases, the density curve becomes flatter and actually looks more like a normal distribution density curve. See **Figure 8.26**.

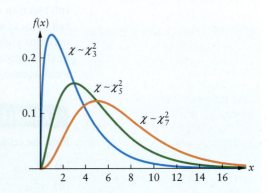

Figure 8.26 Density curves for several chi-square distributions.

χ^2 Critical Values

The definition and notation for a χ^2 **critical value** are analogous to those for a Z critical value and a t critical value.

Definition

$\chi^2_{\alpha,\nu}$ is a **critical value** related to a **chi-square distribution** (a χ^2 critical value) with ν degrees of freedom. If $X \sim \chi^2_\nu$, then $P(X \geq \chi^2_{\alpha,\nu}) = \alpha$.

A CLOSER LOOK

1. $\chi^2_{\alpha,\nu}$ is a value on the measurement axis in a chi-square world with ν degrees of freedom such that α of the area (probability) lies to the right of $\chi^2_{\alpha,\nu}$. There is no symmetry in chi-square critical values (as there are for Z and t critical values).

2. **Critical values** are always defined in terms of right-tail probability. The notation is just more general here because a chi-square distribution is not symmetric. It will be necessary to find critical values denoted $\chi^2_{1-\alpha,\nu}$ (with large values for $1-\alpha$). By definition, $P(X \geq \chi^2_{1-\alpha,\nu}) = 1-\alpha$, and by the Complement Rule, $P(X \leq \chi^2_{1-\alpha,\nu}) = \alpha$. See **Figure 8.27**.

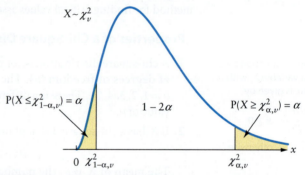

Figure 8.27 An illustration of chi-square critical values.

Appendix Table 6 presents selected critical values associated with various chi-square distributions. Right-tail probabilities appear in the top row, and degrees of freedom are listed in the left column. In the body of the table, $\chi^2_{\alpha,\nu}$ is at the intersection of the α column and the ν row. The first part of this table presents left-tail critical values corresponding to large right-tail probabilities. The second half contains right-tail critical values corresponding to small right-tail probabilities. The next example illustrates the use of this table for finding critical values associated with a chi-square distribution.

EXAMPLE 8.12 Critical Value Lookups

Find the critical value.

(a) $\chi^2_{0.05,10}$

(b) $\chi^2_{0.99,7}$

Solution

(a) The notation $\chi^2_{0.05,10}$ represents a critical value related to a chi-square distribution with 10 degrees of freedom. By definition, if $X \sim \chi^2_{10}$, then $P(X \geq \chi^2_{0.05,10}) = 0.05$. A portion of Appendix Table 6 shows the intersection of the $\alpha = 0.05$ column and the $\nu = 10$ row.

8.5 A Confidence Interval for a Population Variance or Standard Deviation   **377**

				α				
ν	0.10	0.05	0.025	0.01	0.005	0.001	0.0005	0.0001
$\vdots$	$\vdots$	$\vdots$	$\vdots$	$\vdots$	$\vdots$	$\vdots$	$\vdots$	$\vdots$
8	13.3616	15.5073	17.5345	20.0902	21.9550	26.1245	27.8680	31.8276
9	14.6837	16.9190	19.0228	21.6660	23.5894	27.8772	29.6658	33.7199
10	15.9872	18.3070	20.4832	23.2093	25.1882	29.5883	31.4198	35.5640
11	17.2750	19.6751	21.9200	24.7250	26.7568	31.2641	33.1366	37.3670
12	18.5493	21.0261	23.3367	26.2170	28.2995	32.9095	34.8213	39.1344
$\vdots$	$\vdots$	$\vdots$	$\vdots$	$\vdots$	$\vdots$	$\vdots$	$\vdots$	$\vdots$

Therefore, $\chi^2_{0.05,10} = 18.3070$. If $X \sim \chi^2_{10}$, then $(X \geq 18.3070) = 0.05$, as illustrated in **Figure 8.28**.

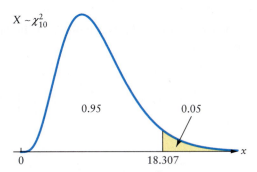
Figure 8.28 Visualization of $\chi^2_{0.05,10} = 18.3070$.

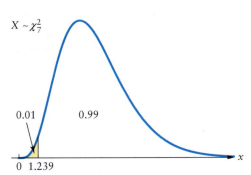
Figure 8.29 Visualization of $\chi^2_{0.99,7} = 1.2390$.

(b) The notation $\chi^2_{0.99,7}$ represents a critical value related to a chi-square distribution with 7 degrees of freedom. A portion of Appendix Table 6 shows the intersection of the $\alpha = 0.99$ column and the $\nu = 7$ row.

				α				
ν	0.9999	0.9995	0.999	0.995	0.99	0.975	0.95	0.90
$\vdots$	$\vdots$	$\vdots$	$\vdots$	$\vdots$	$\vdots$	$\vdots$	$\vdots$	$\vdots$
5	0.0822	0.1581	0.2102	0.4117	0.5543	0.8312	1.1455	1.6103
6	0.1724	0.2994	0.3811	0.6757	0.8721	1.2373	1.6354	2.2041
7	0.3000	0.4849	0.5985	0.9893	1.2390	1.6899	2.1673	2.8331
8	0.4636	0.7104	0.8571	1.3444	1.6465	2.1797	2.7326	3.4895
9	0.6608	0.9717	1.1519	1.7349	2.0879	2.7004	3.3251	4.1682

Therefore, $\chi^2_{0.99,7} = 1.2390$. If $X \sim \chi^2_7$, then $P(X \geq 1.2390) = 0.99$ and $P(X \leq 1.2390) = 0.01$, as illustrated in **Figure 8.29**.

Figures 8.30 and **8.31** show technology solutions.

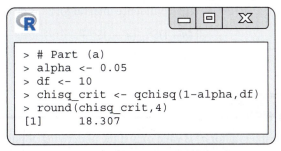

Figure 8.30 Use the R quantile function for the chi-square distribution to find the critical value.

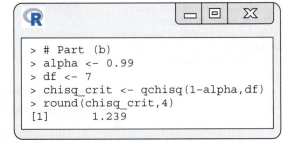

Figure 8.31 Use the R quantile function for the chi-square distribution to find the critical value.

TRY IT NOW   Go to Exercise 8.135

Appendix Table 6 is limited as it includes only a handful of values for α, and $\nu = 1 - 40$, 50, 60, 70, 80, 90, 100. However, using technology, we can find almost any critical value needed.

Confidence Interval for σ^2 or σ

A confidence interval for a population variance is based on the next theorem.

Theorem

Let S^2 be the sample variance of a random sample of size n from a normal distribution with variance σ^2. The random variable

$$X = \frac{(n-1)S^2}{\sigma^2} \tag{8.18}$$

has a chi-square distribution with $n-1$ degrees of freedom.

This is just another kind of standardization, a transformation to a chi-square distribution.

▶ Let $X \sim \chi^2_{n-1}$ and find an interval that captures $1 - \alpha$ in the middle of this chi-square distribution (see **Figure 8.32**.)

$$P(\chi^2_{1-\alpha/2, n-1} < X < \chi^2_{\alpha/2, n-1}) = 1 - \alpha \tag{8.19}$$

$$P\left(\chi^2_{1-\alpha/2, n-1} < \frac{(n-1)S^2}{\sigma^2} < \chi^2_{\alpha/2, n-1}\right) = 1 - \alpha \tag{8.20}$$

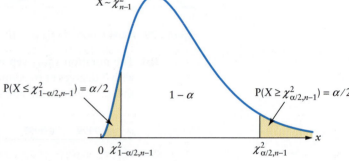

Figure 8.32 An illustration of the probability statement in Equation 8.19.

Inside the probablity statement in Equation 8.20, divide each term by $(n-1)S^2$ and then take the reciprocal of each term (change the direction of the inequality).

Rewrite Equation 8.20 to obtain the probability statement

$$P\left(\frac{(n-1)S^2}{\chi^2_{\alpha/2, n-1}} < \sigma^2 < \frac{(n-1)S^2}{\chi^2_{1-\alpha/2, n-1}}\right) = 1 - \alpha \tag{8.21}$$

This probability statement leads to the following general result. ◀

How to Find a 100(1−α)% Confidence Interval for a Population Variance

Given a random sample of size n from a population with variance σ^2, if the underlying population is normal, a $100(1-\alpha)\%$ confidence interval for σ^2 is given by

The critical value in the right tail of the chi-square distribution is part of the expression for the left endpoint, and the critical value in the left tail of the chi-square distribution is part of the expression for the right endpoint.

$$\left(\frac{(n-1)S^2}{\chi^2_{\alpha/2, n-1}}, \frac{(n-1)S^2}{\chi^2_{1-\alpha/2, n-1}}\right) \tag{8.22}$$

🔍 A CLOSER LOOK

1. This confidence interval for σ^2 is valid only if the underlying population is normal.
2. Take the square root of each endpoint of Equation 8.22 to find a $100(1-\alpha)\%$ confidence interval for the population standard deviation σ.

Confidence Interval for σ^2 and Inference

EXAMPLE 8.13 Kiln-Fired Dishes

Earthenware dishes are made from clay and are fired, or exposed to heat, in a large kiln. Large fluctuations in the kiln temperature can cause cracks, bumps, or other flaws in the dishes (and increase cost). With the kiln set at 800°C, a random sample of 19 temperature measurements (in °C) was obtained. The sample variance was $s^2 = 17.55$.

(a) Find a 95% confidence interval for the true population variance in the temperature of the kiln when set to 800°C. Assume the underlying distribution is normal.

(b) Quality-control engineers have determined that the maximum variance in temperature during firing should be 16°C². Based on the confidence interval constructed in part (a), is there any evidence to suggest the true temperature variance is greater than 16°C²? Justify your answer.

> **Common Error**
> Use of $\nu = n$ to find the critical values.
> **Correction:** The CI for σ^2 is based on a chi-square distribution with $\nu = n - 1$ degrees of freedom.

Solution

(a) The underlying distribution is assumed to be normal. Find the appropriate chi-square critical values, and use Equation 8.22 to construct a 95% CI for σ^2.

$s^2 = 17.55;\quad n = 19;\quad \nu = 19 - 1 = 18$ Given.

$1 - \alpha = 0.95 \Rightarrow \alpha = 0.05 \Rightarrow \alpha/2 = 0.025$ Find $\alpha/2$.

Use Appendix Table 6 with $\nu = 18$ to find the critical values.

$\chi^2_{\alpha/2, n-1} = \chi^2_{0.025, 18} = 31.5264$

$\chi^2_{1-\alpha/2, n-1} = \chi^2_{0.975, 18} = 8.2307$

Use Equation 8.22 to find the endpoints of the confidence interval.

$\dfrac{(n-1)s^2}{\chi^2_{\alpha/2, n-1}} = \dfrac{(n-1)s^2}{\chi^2_{0.025, 18}} = \dfrac{(18)(17.55)}{31.5264} = 10.02$ Left endpoint.

$\dfrac{(n-1)s^2}{\chi^2_{1-\alpha/2, n-1}} = \dfrac{(n-1)s^2}{\chi^2_{0.975, 18}} = \dfrac{(18)(17.55)}{8.2307} = 38.38$ Right endpoint.

(10.02, 38.38) is a 95% confidence interval for the true population variance in temperature when the kiln is set to 800°C.

$(\sqrt{10.02}, \sqrt{38.38}) = (3.17, 6.20)$ is a 95% confidence interval for the population standard deviation.

Figue 8.33 shows a technology solution.

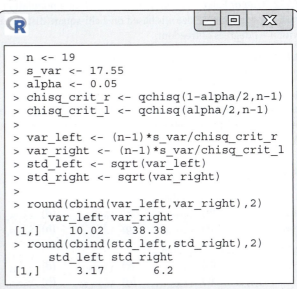

Figure 8.33 Use s^2, n, and α. Find the critical values, and use Equation 8.22.

(b) The CI constructed in part (a) is an interval in which we are 95% confident the true value of σ^2 lies. If 16 lies in this interval, there is no evidence to suggest that σ^2 is different from 16. Use the four-step inference procedure.

Claim: $\sigma^2 = 16$

Experiment: $s^2 = 17.55$

Likelihood: The likelihood is expressed as a 95% confidence interval, an interval of likely values for σ^2: (10.02, 38.38), from part (a).

Conclusion: The required kiln temperature variance, $16°C^2$, is included in this confidence interval. There is no evidence to suggest that σ^2 is different from 16.

Note: The confidence interval for σ^2 is not symmetric about the point estimate s^2 and is quite large (see **Figure 8.34**). The confidence interval will be narrow only for very large values of n.

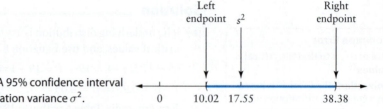

Figure 8.34 A 95% confidence interval for the population variance σ^2.

TRY IT NOW Go to Exercises 8.143 and 8.145

Section 8.5 Exercises

Concept Check

8.130 True or False A confidence interval for a population variance based on a chi-square distribution is valid only if the underlying population is normal.

8.131 True or False A confidence interval for a population variance based on a chi-square distribution is symmetric about the sample variance.

8.132 True or False A confidence interval for a population variance with sample size n is based on a chi-square distribution with $n-1$ degrees of freedom.

8.133 True or False Chi-square critical values satisfy the equation $\chi^2_{1-\alpha/2,\nu} = -\chi^2_{\alpha/2,\nu}$.

8.134 True or False A confidence interval for a population standard deviation can be constructed by taking the square root of each endpoint of the confidence interval for the population variance.

Practice

8.135 Find the critical value.
(a) $\chi^2_{0.10,5}$ (b) $\chi^2_{0.001,31}$ (c) $\chi^2_{0.05,16}$ (d) $\chi^2_{0.025,21}$
(e) $\chi^2_{0.99,11}$ (f) $\chi^2_{0.95,15}$ (g) $\chi^2_{0.975,23}$ (h) $\chi^2_{0.995,9}$

8.136 Find the critical value.
(a) $\chi^2_{0.90,12}$ (b) $\chi^2_{0.01,15}$ (c) $\chi^2_{0.9999,22}$ (d) $\chi^2_{0.005,3}$
(e) $\chi^2_{0.9995,19}$ (f) $\chi^2_{0.005,34}$ (g) $\chi^2_{0.999,26}$ (h) $\chi^2_{0.0001,40}$

8.137 In the following cases, the sample size and the confidence level are given. Find the appropriate chi-square critical values for use in constructing a confidence interval for the population variance.
(a) $n = 22$, 95% (b) $n = 37$, 99%
(c) $n = 11$, 98% (d) $n = 31$, 90%
(e) $n = 5$, 95% (f) $n = 37$, 99.9%

8.138 In the following cases, the sample variance, the sample size, and the confidence level are given. Assume the underlying population is normally distributed. Find the associated confidence interval for the population variance.
(a) $s^2 = 5.65$, $n = 35$, 95%
(b) $s^2 = 45.62$, $n = 26$, 98%
(c) $s^2 = 50.41$, $n = 6$, 80%
(d) $s^2 = 7.68$, $n = 37$, 90%
(e) $s^2 = 32.22$, $n = 5$, 99%
(f) $s^2 = 70.67$, $n = 28$, 99.9%

8.139 In the following cases, the sample variance, the sample size, and the confidence level are given. Assume the underlying population is normally distributed. Find the associated confidence interval for the population variance.
(a) $s^2 = 3.08$, $n = 14$, 99%
(b) $s^2 = 64.10$, $n = 11$, 95%
(c) $s^2 = 59.07$, $n = 6$, 80%
(d) $s^2 = 7.35$, $n = 27$, 99.98%
(e) $s^2 = 31.38$, $n = 22$, 95%
(f) $s^2 = 12.39$, $n = 18$, 99%

8.140 Use Appendix Table 6 and linear interpolation to approximate the critical value. Verify each approximation using technology.
(a) $\chi^2_{0.05, 45}$ (b) $\chi^2_{0.005, 65}$ (c) $\chi^2_{0.01, 72}$ (d) $\chi^2_{0.025, 56}$
(e) $\chi^2_{0.95, 85}$ (f) $\chi^2_{0.999, 52}$ (g) $\chi^2_{0.90, 66}$ (h) $\chi^2_{0.975, 75}$

Applications

8.141 Business and Management A personnel manager at Inserra Supermarkets is concerned about the pattern of overtime hours claimed by employees. Although a special budget is allocated to pay for overtime hours, large fluctuations may cause cashflow problems. In a random sample of 25 employees, the sample variance for the number of overtime hours claimed was 4.25. Assume the distribution of overtime hours is normal.
(a) Find a 95% confidence interval for the population variance of overtime hours.
(b) Find a 95% confidence interval for the population standard deviation of overtime hours.

8.142 Got Milk? One measure of milk production in the United States is the amount (in gallons) produced per cow. Milk production is affected by weather, feed, and other environmental factors. A random sample of the amount of milk produced per cow per quarter was obtained.[29] MILK
(a) Find a 95% confidence interval for the true population variance of milk produced per cow per quarter.
(b) What assumption(s) did you make in constructing this confidence interval?
(c) Find a 95% confidence interval for the true population standard deviation of milk produced per cow per quarter.
(d) The U.S. Department of Agriculture claims the standard deviation of milk produced per cow per quarter is 400 gal. Is there any evidence to refute this claim? Justify your answer.

8.143 Technology and the Internet Everyone who owns or has access to a computer probably has several passwords for various websites. These passwords are supposed to be secret, random combinations of characters. However, some psychologists believe passwords are predictable based on personality traits, and many computer hackers claim that any password can be acquired. To study computer security at Spectaguard Acquisition, a special program was developed and used to systematically try various passwords for randomly selected user accounts. The time (in hours) needed to obtain each password is given in the table. PASS

1.88	1.78	5.67	1.71	4.13	0.03	2.09	2.55
6.60	1.66	2.28	0.28	3.52	4.02	2.64	4.47
4.94	0.68						

Assume the distribution of times is normal.
(a) Find a 98% confidence interval for the true population variance of the time needed to acquire someone's password.
(b) Use the methods described in Section 6.3 to check for any evidence of non-normality.

8.144 Physical Science Commercial and recreational fishermen carefully monitor tides to help improve catches. Knowledge of tides is also important for ships navigating through shallow waters and for serious beach goers. A random sample of high-tide observations (in feet) for Boston, Massachusetts, from summer 2018 was obtained; the summary statistics were $\bar{x} = 10.34$, $s^2 = 1.2455$, and $n = 25$.[30]
(a) Assuming normality, find a 99% confidence interval for the true variance in high-tide level.
(b) Historical evidence suggests that the variance in high tides is 0.70. Is there any evidence to suggest that the variance in high tides has increased? Justify your answer.

8.145 Biology and Environmental Science Summer temperatures in Phoenix, Arizona, routinely soar above 100°F, but residents claim that it is a *dry* heat. The low humidity makes outside physical activity easier and attracts tourists, especially golfers, and retirees. The table contains relative humidity data (percentages) for several summer days in Phoenix.[31] HUMID

5	7	11	8	14	12	23	35
31	22	25	12	11	16	12	17
18	6	8	12	16	25	22	29

(a) Find a 95% confidence interval for the true variance in relative humidity in Phoenix during the summer.
(b) Is there any evidence to suggest the data are from a non-normal distribution? Justify your answer.

8.146 Manufacturing and Product Development In residential and commercial buildings, the thickness of window glass varies according to the width of the window. Wider windows require thicker window panes. As part of quality control, a manufacturer randomly samples windows and carefully measures the width of each (in millimeters). Consider the data in the table.

Window width	Sample size	Sample mean	Sample variance
< 250	15	6.02	0.003
250–1100	18	7.95	0.040
1100–2250	22	10.12	0.500

(a) Find a 99% confidence interval for the true variance of window thickness for each window width category. Assume normality.
(b) Suppose a process is considered *in control* if the variance is 0.02 mm² (or less). Is there any evidence to suggest that any of the three processes is out of control?

8.147 Education and Child Development The government in Ontario plans to implement an ambitious child care program that would fund daycare servcies for preschool-aged children and increase subsidies for middle- and lower-income families with small children. To examine current child care patterns, a random sample of children from two daycare centers was obtained. The number of hours spent at the daycare center during the week for each child was recorded. The summary data are given in the table.

Daycare center	Sample size	Sample variance
Sunrise Children's Center	14	6.28
Right Start Wondershool	22	37.64

(a) Find a 95% confidence interval for the true population variance in hours spent at Sunrise Children's Center.
(b) Find a 95% confidence interval for the true population variance in hours spent at Right Start Wonderschool.
(c) Is there any evidence to suggest that the two population variances are different? Justify your answer.

8.148 Sports and Leisure Some of the stages of the Tour de France bicycling championships cover more than 200 km. In a random sample of 17 riders on Stage 4 of the 2018 Tour de France, La Bauel to Sarzeau, a distance of 195 km, times were measured (in minutes), and the sample variance was 5.69 min^2.[32] Assuming normality, find a 99% confidence interval for the true population variance in times for Stage 4 of the Tour de France.

8.149 Sports and Leisure During the National Football League preseason, coaches appraise both new players and tested veterans. Suppose a rookie is trying to unseat a veteran for the starting tailback position. A random sample of preseason runs was selected for each player, and the number of yards gained on each play was recorded. TAILBACK
(a) Find a 99% confidence interval for the variance of yardage gained per run for the veteran tailback.
(b) Find a 99% confidence interval for the variance of yardage gained per run for the rookie tailback.
(c) Suppose the coach will start the more consistent and dependable tailback. Which player do you think will get the starting position, and why?

8.150 Travel and Transportation Airline travelers are often plagued by long delays on the tarmac waiting to take off. The U.S. Department of Transportation rule for flights departing from a U.S. airport states that airlines are required to move an airplane so that passengers can deplane before the plane has spent 3 hr on the tarmac.[33] A random sample of 25 planes at the Charlotte Douglas International Airport was obtained, and the tarmac time (in minutes) was recorded for each. The sample variance was 181.55.
(a) Find a 95% confidence interval for the population variance in tarmac times. Assume normality.
(b) Airlines would like to reduce tarmac times, of course, and lower the variance. Is there any evidence to suggest that the population variance is less than 100 min^2? Justify your answer.

8.151 Physical Sciences Following an earthquake in January in the Gulf of Alaska, there was actually a small water table level rise in some locations in Florida. According to the U.S. Geological Survey, water levels do respond to seismic waves, and temporary changes in groundwater levels can occur hundreds of miles away from the earthquake's epicenter.

A random sample of days was obtained and the groundwater level in Collier County was measured (in feet).[34] GRNDWTR
(a) Find a 98% confidence interval for the population variance in groundwater level at this site.
(b) Find a 98% confidence interval for the population standard deviation in groundwater level at this site.
(c) Use the methods described in Section 6.3 to test for any evidence of non-normality.

8.152 Fuel Consumption and Cars One measure of a vehicle's front-end alignment is the caster angle, the relationship between the upper and lower ball joints. The specifications for a certain sports car include a caster angle of 2.8 deg with variance 0.40 deg^2. Twenty-two new Maserati Quattroporte sedans were randomly selected, and the caster angle was measured for each. The sample variance was 0.55 deg^2.
(a) Find a 95% confidence interval for the population variance in caster angle. Assume normality.
(b) Is there any evidence to suggest that the population variance in caster angle is different from 0.40 deg^2? Justify your answer.

8.153 Business and Management Even though the budget for a movie may exceed $400 million, worldwide distribution can provide a huge net income for a studio. A sample of movies was obtained, and the budget (in dollars) and the U.S. gross income (in dollars) was recorded for each movie.[35] MOVIES
(a) Find a 95% confidence interval for the population variance in gross income.
(b) Compute the ratio of U.S. revenue to budget (ratio = gross/budget) for each movie. Find a 95% confidence interval for the population variance in ratio of revenue to budget.
(c) If the variance in the ratio is greater than 1, it is considered risky to invest in a new film. Is there any evidence to suggest that investing in a new movie is risky? Justify your answer.

8.154 Marketing and Consumer Behavior Silver bars used in trading on the bullion market are manufactured to certain height, width, and length specifications. In addition, the silver content of the bars varies. Five randomly selected silver bars were carefully analyzed for silver content (in ounces). The sample variance was 65.5 oz^2. Assume the distribution of silver content is normal, and find a 99% confidence interval for the population variance in silver content.

8.155 Psychology and Human Behavior The Great Wall of China was started in 214 B.C. and designed as a defense against nomadic tribes. This 1500-mi-long wall, built from earth and stone, is visible from space and is one of the most famous structures in the world. Suppose a random sample of midpoints between guard towers along the Great Wall was obtained, and the height of the wall at each point was measured (in feet). The data are given in the table.[36] GRWALL

26.9	22.2	21.8	26.6	19.2	29.3	21.9	19.0
23.5	26.7	18.3	27.7	18.8	25.0	24.9	21.1
24.1	24.3	19.5	20.8	27.1			

(a) Find a 95% confidence interval for the population variance in height at the midpoints between guard towers of the Great Wall.
(b) Using the confidence interval constructed in part (a), is there any evidence to suggest that the true population variance in the height at midpoints between guard towers is less than 12 ft^2? Justify your answer.

8.156 Travel and Transportation In 2018, the city of Ottawa planned to change bus route 12 to ease congestion downtown. However, residents of Vanier sharply criticized the plan, claiming the change would add substantial time to their commute.[37] A random sample of bus trips for each route (current and revised) was obtained, and the route completion time was measured (in minutes) for each. VANIER
(a) Find a 95% confidence interval for the population variance in completion time for the current bus 12 route.
(b) Find a 95% confidence interval for the population variance in completion time for the revised bus 12 route.
(c) City officials claim that less variability in completion time allows riders to plan their commute and arrival time more accurately. Is there any evidence to suggest that the population variance for the revised bus route is less than the current route? Justify your answer.

Extended Applications

8.157 Medicine and Clinical Studies Many researchers believe moderate physical activity, such as walking, will help prevent weight gain. To study this claim, doctors in Colorado asked randomly selected patients to wear a pedometer to measure the distance walked (in miles) per week. The table presents the data collected for men and women.

Patient	Sample Size	Sample variance
Men	32	5.75
Women	28	7.66

(a) Find a 95% confidence interval for the true variance in distance walked per week for men.
(b) Find a 95% confidence interval for the true variance in distance walked per week for women.
(c) Is there any evidence to suggest that the true variance is different for men and women? Justify your answer.
(d) What assumption(s) did you make in constructing the confidence intervals in parts (a) and (b)?

8.158 Biology and Environmental Science Rice is a staple food for much of the world's population. The height of a rice plant (usually 0.4–5 mm) depends on the rice variety and the environment. Genetic engineering is currently under way to protect rice plants from disease and to decrease the variability in plant height. In a controlled experiment, plant heights were measured (in millimeters); 30 natural rice plants had a sample variance in height of 1.5 mm^2, and 22 genetically engineered rice plants had a sample variance in height of 0.89 mm^2. Assume normality.

(a) Find a 95% confidence interval for the true variance in height for natural rice plants.
(b) Find a 95% confidence interval for the true variance in height for genetically engineered rice plants.
(c) Is there any evidence to suggest that the population variance is different for natural and genetically engineered rice plants?

8.159 Public Health and Nutrition The World Health Organization and the U.S. Food and Drug Administration suggest that adults limit their consumption of sugar. In 2018, the city of Baltimore prohibited soda and other sugary drinks from childen's menus in restaurants in an attempt to lower children's sugar consumption. A random sample of popular sugary drinks was obtained and the amount of sugar (grams) per fluid ounce was measured in each.[38] SUGAR
(a) Find a 95% confidence interval for the population variance in the amount of sugar per drink.
(b) Find a 95% confidence interval for the population standard deviation in the amount of sugar per drink.

8.160 Physical Sciences A microwave radiometer is used to measure the column water vapor and the infrared brightness (IB) temperatures in clouds. Clouds were randomly selected, and a weather station in Coffeyville, Kansas, collected summary statistics.

Cloud type	Column water vapor (cm)		IB temperature (°C)	
	Sample size	Sample variance	Sample size	Sample variance
Cirrus	11	0.06	17	201.7
Cumulus	21	0.08	28	225.6

(a) Find a 99% confidence interval for the population variance in column water vapor and in IB temperature for cirrus clouds.
(b) Find a 99% confidence interval for the population variance in column water vapor and in IB temperature for cumulus clouds.
(c) Is there any evidence to suggest that the variance in column water vapor or IB temperature is different in cirrus and cumulus clouds? Justify your answers.

8.161 Medicine and Clinical Studies A new medicinal spray has been developed to help ease the itch and burn associated with poison ivy and poison oak contact. People who suffer symptoms from contact with these poisons want and need immediate relief. A research study was conducted to measure the time (in minutes) from application of this spray to relief from itching. The summary statistics were reported for a random sample of children and adults.

Population	Sample size	Sample variance
Children	11	1.57
Adults	25	2.38

(a) Find a 95% confidence interval for the population variance in time to relief for children.
(b) Find a 95% confidence interval for the population variance in time to relief for adults.
(c) Is there any evidence to suggest that the variance in time to relief is different in children and adults? Justify your answer.

8.162 Biology and Environmental Science The agriculture industry is under pressure to produce more while minimizing the effect on the environment. Many corn farmers have turned to hybrids that produce greater yields when more fertilizer is applied. However, hybrid corn tends to be more variable because it is sensitive to environmental conditions. A random sample of corn farmers was obtained. Some planted hybrid corn and others non-hybrid corn, all under similar environmental conditions. The yield (bushels/acre) is summarized in the table.

Corn	Sample size	Sample mean	Sample variance
Hybrid	18	180.60	313.96
Non-hybrid	22	170.85	65.36

(a) Find a 95% confidence interval for the population variance in yield for hybrid corn.
(b) Find a 95% confidence interval for the population variance in yield for non-hybrid corn.
(c) Is there any evidence to suggest that the variance in hybrid corn yield is greater than the variance in non-hybrid corn yield? Justify your answer.

8.163 Biology and Environmental Science The Mississippi River flows from Lake Itasca to the Gulf of Mexico; it borders or passes through ten states. Some researchers suggest that greater variability in the flow rate along the Mississippi River is contributing to more frequent floods. A random sample of the flow rate of the Mississippi River (in thousands of cubic feet per second) at Baton Rouge, Louisiana, over two 10-year periods was obtained. The summary statistics are given in the table.

Years	Sample size	Sample variance
2008–2017	25	42,395.1
1980–1989	32	24,180.8

(a) Find a 95% confidence interval for the population variance in flow rate for the years 2008–2017.
(b) Find a 95% confidence interval for the population variance in flow rate for the years 1980–1989.
(c) Is there any evidence to suggest that the variance in flow rate increased from 1980–1989 to 2008–2017? Justify your answer.

Challenge Problems

8.164 Sports and Leisure Suppose we are given a random sample of size n from a normal population with variance σ^2.
(a) Use the method outlined in Exercise 8.56 to find a one-sided $100(1-\alpha)\%$ confidence interval for σ^2 bounded above, and another bounded below.
(b) A random sample of soccer stadiums from around the world was obtained, and the seating capacity for each was recorded. Find a one-sided 95% confidence interval, bounded above, for the population variance in seating capacity for soccer stadiums. **SEATING**

8.165 Normal Approximation to the Chi-Square Distribution Given a random sample of size n from a normal population with variance σ^2, a $100(1-\alpha)\%$ confidence interval for σ^2 (based on a chi-square distribution) is given by Equation 8.22:

$$\frac{(n-1)s^2}{\chi^2_{\alpha/2, n-1}} < \sigma^2 < \frac{(n-1)s^2}{\chi^2_{1-\alpha/2, n-1}}$$

If n is large ($n > 30$), then the chi-square random variable $(n-1)S^2/\sigma^2$ is approximately normal with mean $n-1$ and variance $2(n-1)$. Therefore, for large n,

$$P\left(-z_{\alpha/2} < \frac{\frac{(n-1)S^2}{\sigma^2} - (n-1)}{\sqrt{2(n-1)}} < z_{\alpha/2}\right) = 1-\alpha \quad (8.23)$$

(a) Rewrite Equation 8.23 to obtain an approximate $100(1-\alpha)\%$ confidence interval for a population variance.
(b) The thickness of pavement on roads and highways depends on the predicted weight and volume of vehicular traffic. The thickness of the pavement (in millimeters) was measured at 51 random locations along Route 95. The sample variance was $s^2 = 16.25$.
   (i) Find a 95% confidence interval for the population variance in pavement thickness using Equation 8.22.
   (ii) Find an approximate 95% confidence interval for the population variance in pavement thickness using the equation derived in part (a).
   (iii) Which of these two intervals is wider? Why?

Chapter 8 Summary

Concept	Page	Notation / Formula / Description
Estimator	336	A statistic, or rule, used to produce a point estimate of a population parameter.
Unbiased estimator	337	A statistic $\hat{\theta}$ is an unbiased estimator of θ if $E(\hat{\theta}) = \theta$.
Biased estimator	337	A statistic $\hat{\theta}$ is a biased estimator of θ if $E(\hat{\theta}) \neq \theta$.
MVUE	337	Minimum variance unbiased estimator.
Confidence interval (CI)	341	An interval of values constructed so that with a specified degree of confidence, the value of the population parameter lies in this interval.
Confidence coefficient	341	The probability that the confidence interval includes the population parameter in repeated samplings.
Confidence level	341	The confidence coefficient expressed as a percentage.
z_α	343	z critical value; a value such that $P(Z \geq z_\alpha) = \alpha$.
Probability density function for a random variable with a t distribution	355	$f(x) = \dfrac{1}{\sqrt{\pi \nu}} \dfrac{\Gamma\left(\dfrac{\nu+1}{2}\right)}{\Gamma\left(\dfrac{\nu}{2}\right)} \left(1 + \dfrac{x^2}{\nu}\right)^{-(\nu+1)/2}$ $-\infty < x < \infty$, $\nu = \text{df}$
$t_{\alpha, \nu}$	356	t critical value; a value such that $P(T \geq t_{\alpha, \nu}) = \alpha$, where T has a t distribution with ν degrees of freedom.
$\chi^2_{\alpha, \nu}$	376	Chi-square critical value; a value such that $P(X \geq \chi^2_{\alpha, \nu}) = \alpha$, where X has a chi-square distribution with ν degrees of freedom.

Chapter 8 Exercises

Applications

8.166 Medicine and Clinical Studies Most patients in need of a kidney transplant are placed on a dialysis machine. Some evidence suggests that the longer patients remain on dialysis, the worse they fare following a transplant. The mean waiting time for a kidney transplant is 5 years (60 months). Suppose a random sample of kidney transplant patients was obtained, and the wait time (in months) was recorded for each. The data are given in the table. KIDNEY

58.1	63.9	74.4	57.8	85.5	68.5	53.5	100.8
87.6	83.7	78.4	49.7	76.6	40.7	63.6	50.3
75.5	76.8	101.4	64.0	43.7	64.3	48.8	76.2

(a) Find a 95% confidence interval for the true mean wait time for a kidney transplant.
(b) Find a 95% confidence interval for the true variance in wait time for a kidney transplant.
(c) Construct a normal probability plot for the wait-time data. Is there any evidence to suggest non-normality?
(d) Is there any evidence to suggest that the waiting time is different from 60 months? Justify your answer.

8.167 Biology and Environmental Science Photosynthetic, microscopic plankton depend upon the concentration of chlorophyll in the ocean. In turn, plankton populations affect marine life, which can affect the availability of certain foods. Some of the highest chlorophyll concentrations are located near the Atlantic and Pacific coasts.[39] A random sample of 35 days was selected, and the chlorophyll near Beaufort, South Carolina, was measured (in $\mu g/L$) for each. The sample mean chlorophyll concentration was 2.93 $\mu g/L$. Assume the population standard deviation is $\sigma = 2.12$ $\mu g/L$.
(a) Find a 98% confidence interval for the true mean chlorophyll concentration near Beaufort.
(b) Find the sample size necessary to construct a 98% confidence interval for the true mean chlorophyll concentration with a bound on the error of estimation $B = 0.5$.

8.168 Marketing and Consumer Behavior Parrot Jungle Island is a roadside attraction in Miami, Florida, featuring tropical birds, crocodiles, and more than 2000 varieties of plants and flowers. A new advertising campaign was developed to attract more out-of-state visitors. In a random sample of 270 visitors, 189 were area residents.
(a) Find the sample proportion of visitors who were area residents. Check the nonskewness criterion.
(b) Find a 99% confidence interval for the true proportion of visitors who were area residents.

(c) Suppose no prior knowledge of the proportion of visitors who were area residents is available. Find the sample size necessary for a 99% confidence interval with a bound on the error of estimation of 0.05.

8.169 Sports and Leisure Most Major League Baseball (MLB) parks have a device to measure the speed of every pitch. The results are often displayed on a scoreboard, and tracked by fans and coaches. A random sample of pitches made during the first inning of baseball games around the country was selected. The speed of each pitch (in mph) is given in the table. PITCH

| 92 | 94 | 90 | 95 | 95 | 95 | 96 | 86 | 99 | 88 | 92 |
| 93 | 88 | 95 | 93 | 84 | 93 | 84 | 92 | 93 | 90 | 94 |

(a) Assume normality. Find a 95% confidence interval for the population mean speed of pitches in the first inning of MLB games.
(b) Find a 99% confidence interval for the population variance in speed of pitches in the first inning of MLB games.
(c) The mean speed of all pitches made in the first inning of games during the previous season was 91.225 mph. Is there any evidence to suggest that the mean speed has changed?

8.170 Economics and Finance Automated teller machines (ATMs) have made it easier and faster for customers to check their account balances, transfer money, and obtain cash. Many banks are now considering the next generation of ATMs, which will feature news headlines, full-motion video, and tickets to events. In a random sample of 500 customers, 280 said ATMs should offer postage stamps.
(a) Find the sample proportion of customers who believe ATMs should offer postage stamps, and check the nonskewness criterion.
(b) Find a 99% confidence interval for the true proportion of customers who believe ATMs should offer postage stamps.
(c) A bank official claims that the proportion of customers who believe ATMs should offer postage stamps is 0.60. Is there any evidence to refute this claim?

8.171 Physical Sciences Scientists have recently discovered that the shape of a galaxy is related to its age: A galaxy grows bigger and puffier as it ages. A random sample of known galaxies was obtained and the age of each (in gigayears) was recorded.[40] GALAXY
(a) Assume normality. Find a 95% confidence interval for the population mean age of galaxies.
(b) Is there any evidence to suggest that the population mean age is greater than 13.772 gigayears, the estimated age of the universe?

8.172 Medicine and Clinical Studies Passengers on long airline flights may develop deep vein thrombosis (DVT), potentially dangerous blood clots. Although most blood clots in the bloodstream dissolve naturally, travelers on long flights have three times the risk of developing a blood clot. In a research study of people who traveled frequently as part of their job, 22 of 8755 developed a blood clot within eight weeks of a long flight.
(a) Find a 90% confidence interval for the true proportion of passengers on long flights who develop DVT within eight weeks of a long flight.

(b) Is there any evidence to suggest that this proportion is greater than 0.5%? Justify your answer.

8.173 Psychology and Human Behavior People who watch multiple DVDs or several online TV episodes in a row are indulging in binge-watching. Most experts agree that this is mostly a harmless, enjoyable addiction. In a recent survey of 2693 U.S. adults with Internet access, 1670 watched multiple TV episodes back to back.
(a) Find a 99% confidence interval for the true proportion of adults with Internet access who watch multiple TV episodes back to back.
(b) Find the sample size necessary to construct a 99% confidence interval with a bound on the error of estimation of 0.02.

8.174 Sports and Leisure The Isle of Man Tourist Trophy (TT) is, perhaps, the world's craziest motorcycle race, held on public roads with no run-off and very few safety barriers. One lap is 60.7 km. In 2018, Michael Dunlop won the four-lap event. A random sample of lap speeds (in mph) was obtained.[41] RACE
(a) Find a 99% confidence interval for the population mean speed of Isle of Man TT racers.
(b) Find a 99% confidence interval for the population variance in speed of Isle of Man TT racers. Why do you suppose this interval is so small?
(c) What assumptions did you make in constructing these two confidence intervals?
(d) Race organizers are concerned that increasing racing speeds are contributing to more accidents. Is there any evidence to suggest that the true mean speed of racers is 133 mph or greater? Justify your answer.

8.175 Business and Management Many businesses rent a limousine to chauffeur important clients to and from an airport, hotel, or office. A random sample of the cost (in dollars) of renting a limousine for the entire day in Los Angeles was obtained. For $n = 25$, the mean was $\bar{x} = 982.75$ and the standard deviation was $s = 45.10$. Assume the cost distribution is normal.
(a) Find a 95% confidence interval for the population mean cost of renting a limousine for the entire day.
(b) Find a 95% confidence interval for the population variance in the cost of renting a limousine for the entire day.
(c) Find a 95% confidence interval for the standard deviation in the cost of renting a limousine for the entire day.

8.176 Economics and Finance According to a recent report, approximately 15 million people have a credit card, checking account, or savings that they keep hidden from their partner.[42] A random sample of millennials and Gen Xers involved in a live-in relationship was obtained, and participants were asked if they had a hidden account. The data are given in the table.

Generation	Sample size	Number with hidden finances
Millennials	500	150
Gen Xers	625	149

(a) Find a 95% confidence interval for the true proportion of millennials with a hidden financial account.

(b) Find a 95% confidence interval for the true proportion of Gen Xers with a hidden financial account.
(c) Is there any evidence to suggest that the true proportion of millennials with a hidden financial account is different from the true proportion of Gen Xers with a hidden financial account? Justify your answer.

8.177 Physical Sciences The Quabbin Reservoir in Massachusetts is 18 miles long and has a capacity of 412 billion gal. This reservoir serves approximately 2.5 million people, with a daily yield of 300 million gal of water. The depth of the reservoir is carefully monitored and measured at certain locations and times each day. In a random sample of 18 summer days, the depth (in feet) was measured at location S-1 at noon. The mean depth was 75.4 ft. Assume the population standard deviation is $\sigma = 7.58$ ft.
(a) Find a 99% confidence interval for the mean depth of the reservoir at this location.
(b) Historical records indicate that the mean depth at this location for the previous 10 years is 78.4 ft. Is there any evidence to suggest that this population mean has changed?

8.178 Public Health and Nutrition Rising costs have caused many Americans to live without health insurance. A random sample of nonelderly adults in various states was obtained and asked whether they had health insurance. The resulting data are given in the table.[43]

State	Sample size	Number without health insurance
Arizona	1220	170
California	1080	97
Florida	1156	169

(a) Find a 95% confidence interval for the proportion of nonelderly adults without health insurance for each state.
(b) Is there any evidence to suggest that the true proportion of nonelderly adults without health insurance is different for any pair of states? Justify your answer.

8.179 Public Health and Nutrition The Centers for Disease Control and prevention (CDC) recently reported that one in four Americans is living with a disability.[44] In a random sample of 520 American men, 115 had a disability, and of 678 women, 190 had a disability.
(a) Find a 99% confidence interval for the true proportion of American men who suffer from a disability.
(b) Find a 99% confidence interval for the true proportion of American women who suffer from a disability.
(c) Is there any evidence to suggest that the true proportion of American men who suffer from a disability is different from the true proportion of American women who suffer from a disability? Justify your answer.

Extended Applications

8.180 Medicine and Clinical Studies The drugs Ritalin and Adderall are designed to stimulate the central nervous system and are widely prescribed for children diagnosed with attention-deficit disorder (ADD). Suppose a random sample of girls and boys in 12th grade was obtained and the number taking Ritalin/Adderall was recorded. The data are given in the table.

Group	Sample size	Number taking Ritalin/Adderall
Girls	375	24
Boys	480	39

(a) Find a 95% confidence interval for the true proportion of 12th-grade girls who are taking Ritalin/Adderall.
(b) Find a 95% confidence interval for the true proportion of 12th-grade boys who are taking Ritalin/Adderall.
(c) Is there any evidence to suggest that the proportion of girls in this age group taking Ritalin/Adderall is different from the proportion of boys? Justify your answer.

8.181 Public Health and Nutrition Tannin is a general term for certain nonvolatile phenolic substances in many fruits that provide an astringent sensation, such as that observed in apple cider. Some evidence suggests that the tannin level in apples is affected by the fertilizer regimen. A random sample of apples was obtained from both fertilized and unfertilized trees. The percentage of tannin was measured in each apple, and the data (sample size, sample mean, and sample standard deviation) are reported in the table.

Apples	Sample size	Sample mean	Sample standard deviation
Fertilized trees	48	0.30	0.058
Unfertilized trees	55	0.35	0.077

(a) Find a 95% confidence interval for the true mean tannin percentage in apples from fertilized trees.
(b) Find a 95% confidence interval for the true mean tannin percentage in apples from unfertilized trees.
(c) Is there any evidence to suggest that the percentage of tannin is different in apples from fertilized and unfertilized trees?

8.182 Sports and Leisure The number of students participating in high school sports has increased every year for almost three decades. Recently, the number of girls participating in high school sports reached an all-time high.[45] A random sample of high school students from across the country was obtained, and the students were asked whether they participate in a high school sport. The results are given in the table.

Group	Sample size	Number participating
Boys	1250	725
Girls	1475	782

(a) Find a 95% confidence interval for the true proportion of boys participating in a high school sport.
(b) Find a 95% confidence interval for the true proportion of girls participating in a high school sport.
(c) Is there any evidence to suggest that the proportions of boys and girls participating in a high school sport are different? Justify your answer.

8.183 Public Health and Nutrition Elevated levels of mercury can cause hair loss, fatigue, and memory lapses. The concentration of mercury in a person's blood can be greatly affected by diet, especially the amount of fish consumed. Random samples of adults who eat fish two to three times per week and of adults who never eat fish were obtained. The mercury concentration in the blood of each person was measured (in micrograms per liter of blood), and the summary data are reported in the table.

Group	Sample size	Sample mean	Sample standard deviation
Fish	18	4.662	0.298
No fish	27	2.079	0.309

Assume both distributions are normal.
(a) Find a 95% confidence interval for the population mean mercury concentration in the blood of adults who eat fish regularly.
(b) Find a 95% confidence interval for the population mean mercury concentration in the blood of adults who never eat fish.
(c) Suppose the safe level of mercury (set by the U.S. Environmental Protection Agency) is 5 μg per liter of blood. Is there any evidence to suggest that either group is over the safe limit of mercury concentration?
(d) Use the confidence intervals in parts (a) and (b) to determine whether the two groups have different mean mercury concentration levels.

8.184 Biology and Environmental Science There is growing concern that Americans are generating too much trash. More zoning permits are being sought for landfills in rural areas, and trash haulers routinely move garbage out of state and even out to the ocean. In a recent study, 65 households were randomly selected, and the amount of trash (in pounds) generated by each in one week was recorded. The sample mean was $\bar{x} = 52.3$ with a standard deviation of $s = 10.75$. Assume normality.
(a) Find a 98% confidence interval for the true mean amount of trash generated by an American household per week.
(b) Find an approximate 98% confidence interval for the true mean amount of trash generated by an American household per week (based on a Z distribution rather than a t distribution). Compare your answers in parts (a) and (b). How do they differ?
(c) Find a 95% confidence interval for the variance in the amount of trash generated by an American household per week.

8.185 Medicine and Clinical Studies In a major health study, a random sample of adult males was obtained and each was tested for symptoms of heart disease yearly for a decade. Subjects were divided into those who regularly donated blood and those who did not. The results are given in the table.

Group	Sample size	Number with heart disease
Donate blood	145	51
Do not donate blood	527	210

(a) Find a 95% confidence interval for the population proportion of males who donate blood who have heart disease.
(b) Find a 95% confidence interval for the population proportion of males who do not donate blood who have heart disease.
(c) Is there any evidence to suggest the proportion of males with heart disease is different for those who donate blood and those who do not?

8.186 Psychology and Human Behavior Although much of fundraising is about building relationships, online fundraising sites such as Kickstarter, GoFundMe, and Crowdrise have made it easier for organizations to raise funds for a specific cause and for donors to contribute. A recent study compared the donation amount (in dollars) on weekdays versus weekends. A random sample of online donations was obtained for each group. Assume normality. 📊 DONATE
(a) Find a 95% confidence interval for the true mean online donation amount on weekdays.
(b) Find a 95% confidence interval for the true mean online donation amount during the weekend.
(c) Is there any evidence to suggest that these two population means are different? Justify your answer.

8.187 Economics and Finance Many job ads include a statement indicating that salary is negotiable. However, fewer than half of all job seekers attempt to secure a higher starting salary. A random sample of workers in three cities was obtained, and the participants were asked if they tried to negotiate their starting salary. The results are given in the table.[46]

City	Sample size	Number who negotiated
Boston	250	110
Dallas	205	105
Denver	180	52

(a) Find a 95% confidence interval for the true proportion of job seekers who negotiate a starting salary for each city.
(b) Is there any evidence to suggest that the true proportion of job seekers who negotiate a starting salary is different for any pair of cities? Justify your answer.

Scharfsinn/Deposit Photos

CHAPTER APP

8.188 Autonomous Cars The Department of Motor Vehicles in Sacramento, California, selected 15 Waymo autonomous cars at random and recorded the number of miles between the most recent disengagements. The summary statistics are $\bar{x} = 5595.3$ and $s = 376.8$.
(a) Identify the point estimates and the statistics.
(b) Find a 95% confidence interval for the true mean number of miles between disengagements.
(c) Is there any evidence to suggest that the mean number of miles between disengagements has reached 6000?

Hypothesis Tests Based on a Single Sample

apichart609@gmail.com/Deposit Photos

◀ Looking Back

- Recall that $\bar{x}$, $\hat{p}$, and s^2 are point estimates for the parameters μ, p, and σ^2: **Section 8.1**.
- Remember how to construct and interpret confidence intervals: **Chapter 8**.
- Think about the concept of a sampling distribution for a statistic and the process of standardization: **Section 7.1 and Section 6.2**.

Looking Forward ▶

- Understand the formal decision process and learn the four-step hypothesis test procedure: **Section 9.1**.
- Learn to identify in context the errors that might occur in a hypothesis test: **Section 9.2**.
- Learn how to conduct hypothesis tests concerning a population mean: **Section 9.3 and Section 9.5**.
- Understand the concept of p values and their use in hypothesis tests: **Section 9.4**.
- Learn how to conduct hypothesis tests concerning a population proportion: **Section 9.6**.
- Learn how to conduct hypothesis tests concerning a population variance: **Section 9.7**.

CHAPTER APP

Heavy Metal

In a study conducted by Consumer Reports, researchers concluded that common baby food products, such as packaged entrees, fruits, and vegetables, contained measurable levels of certain heavy metals. Of the foods tested, all purchased from a variety of U.S. grocery stores, 68% contained levels of toxic chemicals that food safety experts suggest are worrisome.[1]

Consumer Reports also marked 15 tested products as the worst, indicating that babies who eat one serving of those foods each day are at risk for health problems. Food products that contained rice and/or sweet potatoes were more likely to contain high levels of heavy metals. Moreover, organic baby foods were just as likely to contain toxic chemicals as were common baby food products.

In response to this study, suppose the Food and Drug Administration (FDA) issued new guidelines to manufacturers for processing baby foods. Several months after the release of the new guidelines, an independent agency conducted another research study examining common baby food products. Of the 125 baby food products inspected, 75 were found to contain high levels of toxic chemicals. The concepts presented in this chapter will be used to construct a formal hypothesis test to determine whether there is any evidence to suggest that less than 68% of baby foods contain toxic chemicals.

9.1 The Parts of a Hypothesis Test and Choosing the Alternative Hypothesis

Definition: Hypothesis

You have probably heard the word **hypothesis** used in many different contexts. An engineer might have a hypothesis concerning gas mileage on a certain car. She might claim that a new hybrid engine design will significantly improve the miles-per-gallon rating. Or a biologist might hypothesize that a special combination of nutrients will significantly increase the growth rate of yellow corn.

Definition
In statistics, a **hypothesis** is a declaration, or claim, in the form of a mathematical statement, about the value of a specific population parameter (or about the values of several population characteristics).

Here are some examples of statistical hypotheses.

1. $\mu = 14.5$

   where μ is the population mean time (in minutes) it takes for an adult's pupils to dilate after treatment with phenylephrine.

2. $p > 0.70$

   where p is the population proportion of all people older than age 65 who will need long-term care services at some point in their lifetime.

3. $\sigma^2 \neq 30.5$

   where σ^2 is the population variance in the amount (in gallons) of coal tar in a 5-gal bucket.

A hypothesis is a claim about a population parameter, not about a sample statistic. For example, $\mu = 5$ and $p = 0.27$ are valid hypotheses, but $\bar{x} = 27$ and $s = 32.5$ are not.

Hypothesis Test Parts

There are four parts to every hypothesis test, and it is important to identify each part in every test.

Four Parts of a Hypothesis Test

H_0 is read as "H sub zero," and H_a is read as "H sub a."

1. The **null hypothesis**, denoted H_0.

   This is the claim (about a population parameter) assumed to be true, what is believed to be true, or the hypothesis to be tested. Sometimes referred to as the no-change hypothesis, this claim usually represents the status quo or existing state. Although there is an implied inequality in H_0, the null hypothesis is written in terms of a single value (with an equal sign), for example, $\theta = 5$. Although it may seem strange, we usually try to reject the null hypothesis.

2. The **alternative hypothesis**, denoted H_a.

   This statement identifies other possible values of the population parameter, or simply a possibility not included in the null hypothesis. H_a indicates the possible values of the parameter if H_0 is false. Experiments are often designed to determine whether there is evidence in favor of H_a. The alternative hypothesis represents change in the current standard or existing state.

3. The **test statistic**, denoted TS.

   This statistic is a rule, related to the null hypothesis, involving the information in a sample. The value of the test statistic will be used to determine which hypothesis is more likely to be true, H_0 or H_a.

4. The **rejection region** or **critical region**, denoted RR or CR.

   This interval or set of numbers is specified such that if the value of the test statistic lies in the rejection region, then the null hypothesis is rejected. There is also a corresponding nonrejection region; if the value of the test statistic lies in this set, then we cannot reject H_0.

A CLOSER LOOK

1. The test of a statistical hypothesis is a procedure to decide whether there is evidence to suggest that the alternative hypothesis, H_a, is true. The ultimate objective of a hypothesis test is to use the information in a sample to decide which hypothesis is more likely to be true, H_0 or H_a. Usually, we are looking for evidence to reject the null hypothesis.

2. The rejection region and the nonrejection region divide the world (values of the test statistic) into parts. **Figure 9.1** illustrates this concept in terms of a parameter. The cutoff value, or dividing line, is determined by considering likely values for the test statistic if H_0 is true, and is included in one of the regions.

   The value of the parameter must lie in either the rejection region or the nonrejection region. In Section 9.3, we'll see how easy it is to specify a rejection region.

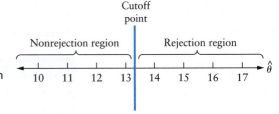

Figure 9.1 An illustration of a rejection region and a nonrejection region, where θ is a population parameter.

3. The hypothesis test procedure is very prescriptive. Once the four parts are identified, the sample data are used to compute a value of the test statistic. There are only two possible conclusions.

   (a) If the value of the test statistic lies in the rejection region, then we reject H_0. There is evidence to suggest that the alternative hypothesis is true.

   (b) If the value of the test statistic does not lie in the rejection region, then we cannot reject H_0. There is no evidence to suggest that the alternative hypothesis is true.

   We never say that we "accept H_0." A hypothesis test is designed to find evidence in favor of the alternative hypothesis. If we fail to find evidence in favor of H_a, this does not imply that H_0 is true.

A hypothesis test can only provide support in favor of H_a. If the value of the test statistic lies in the rejection region, we reject the null hypothesis. In such a case, there is evidence to suggest that H_a is true. If the value of the test statistic does not lie in the rejection region, do not reject H_0. In this case, there is no evidence to suggest that H_a is true. Watch the wording: We never accept the null hypothesis. Instead, we state that there is no evidence to suggest that the claim is false.

4. The formal hypothesis test procedure is analogous to the four-step inference procedure used in the previous chapters. In fact, many of the concepts presented in earlier chapters are combined here to produce this traditional, well-established hypothesis test procedure.

The **claim** corresponds to H_0, a claim about a population parameter. The **experiment** is equivalent to a value of the test statistic. **Likelihood** is expressed in terms of the non-rejection region (likely values of the test statistic) and the rejection region (unlikely values of the test statistic). The **conclusion** is completely determined by the region in which the value of the test statistic lies. If the value is in the rejection region, we reject H_0; otherwise, we cannot reject H_0.

Writing H_0 and H_a

Suppose a researcher conducts a hypothesis test concerning the population parameter θ, and θ_0 is a specific value of θ. The null hypothesis is always stated in terms of a single value. In this case, there are only three possible alternative hypotheses:

$$H_0: \theta = \theta_0$$
$$\left.\begin{array}{l}H_a: \theta > \theta_0 \\ \theta < \theta_0\end{array}\right\} \text{one-sided alternatives}$$
$$\theta \neq \theta_0 \} \text{two-sided alternative}$$

Translating inequalities to symbols.

Only one alternative hypothesis is selected. H_a answers the question, "What is the experimenter trying to prove, or detect, about θ?" It takes a little practice to decide which H_a is appropriate. The same specific value of the parameter (θ_0, above) always appears in H_0 and H_a.

A valid set of null and alternative hypotheses must include a statement similar to H_0 above and the relevant alternative, one of the three possibilities just given. For example, $H_0: \mu = 17$; $H_a: \mu < 17$ is a valid set of null and alternative hypotheses. $H_a: \tilde{\mu} \neq 25$; $H_a: \tilde{\mu} \neq 26$ is not. The next examples focus on identifying H_0 and the relevant alternative hypothesis.

EXAMPLE 9.1 Trust Me, I'm a Doctor

Trust in the workplace, in social settings, and in the media allows us to work and live with an expectation that conversations and decisions will be addressed in good faith. However, trust, in general, has been declining for many years; most concerning, trust has been decreasing in public health and safety institutions and workers. According to a recent survey, only 58% of Americans trust doctors.[2] Suppose a national advertising campaign is conducted to address confidence in doctors and medical leaders, and an experiment (new survey) is conducted to determine whether it has been effective. What null and alternative hypotheses should be used?

Solution

STEP 1 The claim assumed to be true involves the pre-campaign population proportion of Americans who trust doctors. It is assumed that 58% of all Americans trust their doctor. Therefore, the null hypothesis, stated in terms of a single value for p, is

$$H_0: p = 0.58$$

STEP 2 The existing proportion of Americans who trust doctors is 0.58. The experiment (new survey) is designed to detect an increase in this proportion, for example, to answer the question, "Do Americans now have greater trust in doctors?" Researchers hope to find evidence that the proportion of Americans who trust doctors is greater than 0.58. Therefore, the alternative hypothesis is

$H_a: p > 0.58$

TRY IT NOW Go to Exercises 9.19 and 9.21

EXAMPLE 9.2 Aussie Drivers

As in many other places around the world, motorists in Australia can easily be caught in rush-hour traffic. Even though the roads are crowded, it was reported that the mean number of kilometers driven per year by each Australian driver is 13,716.[3] Over the past year, more public transportation has been made available in most large cities to encourage people to drive less and to use buses and trains instead. An observational study is conducted to determine whether the mean number of kilometers driven each year has decreased. State the appropriate null and alternative hypotheses.

Solution

STEP 1 The claim assumed to be true involves the population mean number of kilometers driven by all Australians. The null hypothesis, the status quo, is given in terms of the parameter μ.

$H_0: \mu = 13,716$

STEP 2 We are searching for evidence in favor of a smaller mean number of kilometers driven. More public transportation is introduced with the hope that the mean number of miles driven is less than the current mean. Therefore,

$H_a: \mu < 13,716$

> **Common Error**
> **Interchanging H_0 and H_a**
> **Correction:** H_0 represents the claim, or status quo. We are looking for evidence in favor of H_a.

EXAMPLE 9.3 Recycled Paper

The thickness (measured in inches) of recycled printer paper is important, because sheets that are too thick will clog the machine, and paper that is too thin will rip and bleed toner. The variance in thickness for 20-lb printer paper at a manufacturing plant is known to be 0.0007. A new process is developed that uses more recycled fiber, and an experiment is conducted to detect any difference in the variance in paper thickness. State the appropriate null and alternative hypotheses.

Solution

STEP 1 The assumption, or existing state, involves the population variance in the thickness of recycled printer paper. The null hypothesis is given in terms of σ^2.

$H_0: \sigma^2 = 0.0007$

STEP 2 The experiment is designed to detect any difference in the population variance. This suggests a two-sided alternative.

$H_a: \sigma^2 \neq 0.0007$

TRY IT NOW Go to Exercises 9.13 and 9.25

CHAPTER APP Heavy Metal

State the null hypothesis and the alternative hypothesis in terms of p, the true proportion of baby foods that contain toxic chemicals.

Section 9.1 Exercises

Concept Check

9.1 True or False A statistical hypothesis is always stated in terms of a population parameter.

9.2 True or False In a test of a statistical hypothesis, we attempt to find evidence in favor of the null hypothesis.

9.3 True or False In a test of a statistical hypothesis, there may be more than one alternative hypothesis.

9.4 True or False If the value of the test statistic lies in the nonrejection region, then the null hypothesis is true.

9.5 Short Answer State the four parts to every hypothesis test.

9.6 Short Answer State the only two possible conclusions in a test of a statistical hypothesis.

Practice

9.7 Determine whether the statement is a valid hypothesis. Classify each valid hypothesis as a null hypothesis or an alternative hypothesis. Justify your answers.
(a) $\mu = 0.355$
(b) $\hat{p} < 0.42$
(c) $s > 3.5$
(d) $\bar{x} \neq 16$
(e) $\mu > 22.66$
(f) $p \neq 0.15$
(g) $\tilde{x} < 47.5$
(h) $\tilde{\mu} = 12$
(i) $\sigma = 400.5$

9.8 Determine whether the statement is a valid hypothesis. Classify each valid hypothesis as a null hypothesis or an alternative hypothesis. Justify your answers.
(a) $\sigma^2 = 49.55$
(b) IQR $= 25$
(c) $\mu \neq 17$
(d) $\bar{y} = 100.7$
(e) $Q_3 = 7.65$
(f) $\mu < 33.79$
(g) $\sigma \neq 8.95$
(h) $p = 0.77$
(i) $\bar{x}_{\text{tr}(0.025)} = 712.5$

9.9 Determine whether the pair of statements is a valid set of null and alternative hypotheses. Justify your answers.
(a) $H_0: p = 0.55$; $H_a: p < 0.55$
(b) $H_0: \mu = 9.7$; $H_a: \mu \geq 9.7$
(c) $H_0: \sigma^2 = 98.6$; $H_a: \sigma^2 = 101$
(d) $H_0: \tilde{\mu} = 38.9$; $H_a: \tilde{\mu} = 38.9$

9.10 Determine whether the pair of statements is a valid set of null and alternative hypotheses. Justify your answers.
(a) $H_0: \mu = 30$; $H_a: \mu \neq 30$
(b) $H_0: \sigma = 3.55$; $H_a: \sigma > 3.55$
(c) $H_0: p \leq 0.32$; $H_a: p > 0.32$
(d) $H_0: \bar{x} = 78.5$; $H_a: \bar{x} \neq 78.5$

9.11 Determine whether the pair of statements is a valid set of null and alternative hypotheses. Justify your answers.
(a) $H_0: p = 0.50$; $H_a: p \neq 0.50$
(b) $H_0: \mu = 25.6$; $H_a: \mu < 25.6$
(c) $H_0: \mu < 35.9$; $H_a: \mu \geq 35.9$
(d) $H_0: \sigma^2 = 95$; $H_a: \sigma^2 > 95$

9.12 Determine whether the statement is an acceptable conclusion to a hypothesis test.
(a) The value of the test statistic does not lie in the rejection region. Therefore, we accept the null hypothesis.
(b) The value of the test statistic lies in the rejection region. Therefore, there is evidence to suggest that the null hypothesis is not true.
(c) The value of the test statistic does not lie in the rejection region. Therefore, there is evidence to suggest that the null hypothesis is true.
(d) The value of the test statistic lies in the nonrejection region. Therefore, there is no evidence to suggest that the alternative hypothesis is true.
(e) The value of the test statistic does not lie in the rejection region. Therefore, there is no evidence to suggest that the alternative hypothesis is true.
(f) The value of the test statistic lies in the rejection region. Therefore, there is evidence to suggest that the alternative hypothesis is true.

Applications

9.13 Education and Child Development The College Board reported that the mean cumulative SAT score (Evidence-Based Reading and Writing and Mathematics) for 2017 high school graduates was 1060.[4] Officials from the Pennsylvania State System of Higher Education would like to know whether the students enrolled for fall classes have a mean cumulative SAT score greater than 1060. A random sample of students from across the system is obtained, each student's cumulative SAT score is recorded, and the mean cumulative score is recorded. State the null and alternative hypotheses.

9.14 Travel and Transportation A recent report by AAA estimated that 88 million Americans, and in particular 44% of millennials, are planning to take a family trip in 2018.[5] Liberty Travel would like to determine if a different proportion of baby boomers are planning to take a trip. State the null and the alternative hypotheses in terms of the population proportion of baby boomers who are planning to take a family trip.

9.15 Biology and Environmental Science During the summer months, wildfires in the western United States pose a great hazard to people, residential and commercial buildings, and animals. Previous records indicate that the mean number of acres burned during a wildfire is 17,060. The most recent summer was unusually wet, and firefighting officials would like to know whether the number of acres burned during wildfires was any less than the mean. State the null and alternative hypotheses.

9.16 Biology and Environmental Science Tourism officials in Palm Beach County, Florida, claim that the mean diameter of sand dollars found on Delray Beach is 4.25 cm. Some scientists claim that warmer water off the coast has inhibited growth of all sea life and that the diameter of sand dollars has decreased. State the null and alternative hypotheses.

9.17 Manufacturing and Product Development The stereotypical video-game player is a male teenager. However, recent studies suggest that 43% of video-game players are at least 36 years old.[6] A software company has decided to develop and market a sophisticated stock-market video game if the mean age of all video-game players is greater than 25. A random sample of players will be obtained, and the resulting data will be used to test the relevant hypothesis. Let μ represent the mean age of all video-game players. Which of the following sets of null and alternative hypotheses is appropriate? Justify your answer.

(a) $H_0: \mu = 25$ versus $H_a: \mu > 25$
(b) $H_0: \mu = 25$ versus $H_a: \mu < 25$
(c) $H_0: \mu = 25$ versus $H_a: \mu \neq 25$

9.18 Physical Sciences There are approximately 16 fiber-optic lines under the ocean off the Florida coast. These lines carry telephone and Internet communications between Florida and Europe, Latin America, and the Caribbean. During storms, these lines sway dramatically and often damage coral or get caught in anchors. Technical reports including the distance the line swayed indicate that the variance in sway during storms is 32 ft. A study will be conducted to determine whether a new, heavier cable housing will decrease the variance in sway. State the relevant null and alternative hypotheses in terms of σ^2, the variance in cable sway.

9.19 Travel and Transportation Lime recently began to offer dockless electric scooters in San Francisco. However, residents have complained that the discarded scooters are blocking the public right of way. Lime representatives claim that 70% of all scooters are picked up by juicers, independent contractors who recharge and/or repair each scooter. A study will be conducted to determine whether the juicers are picking up less than 70% of the scooters. State the null and alternative hypotheses in terms of p, the true proportion of scooters picked up by juicers.

9.20 Business and Management According to the Organization for Economic Cooperation and Development (OECD), the average employee in the 36 member countries works $\mu = 1776$ hr per year.[7] A random sample of employees in Canada will be selected, and the resulting hours worked per year will be used to determine whether Canadians work less than a typical OECD employee. State the relevant null and alternative hypotheses in terms of μ.

9.21 Travel and Transportation Bluebird, a school bus company in Fort Valley, Georgia, will install seat belts on all buses if more than 50% of all parents favor this change. A random sample of parents in certain school districts will be obtained, and the responses will be used to test the relevant hypothesis. If p is the true proportion of parents who favor seat-belt installation, which of the following sets of null and alternative hypotheses is appropriate? Justify your answer.

(a) $H_0: p = 0.50$ versus $H_a: p \neq 0.50$
(b) $H_0: p = 0.50$ versus $H_a: p < 0.50$
(c) $H_0: p = 0.50$ versus $H_a: p > 0.50$

9.22 Public Policy and Political Science Suppose U.S. Representative Chellie Pingree is considering a run for governor of Maine. She will enter the campaign if there is evidence to suggest that more than 65% of all state residents favor her candidacy. A random sample of likely voters is obtained, and the resulting survey data will be used to test the relevant hypothesis. Let p be the proportion of all voters who favor Pingree's candidacy. State the null and alternative hypotheses in terms of p.

9.23 Education and Child Development Students at Stetson University plan to ask administrators to build more on-campus housing on the DeLand campus if there is any evidence that the median monthly rent for off-campus housing is more than $400 per person. A random sample of students living off-campus will be obtained, and the resulting rent data will be used to test the relevant hypothesis. Let $\tilde{\mu}$ be the median monthly rent for students. State the null and alternative hypotheses in terms of the population median $\tilde{\mu}$.

9.24 Travel and Transportation Commercial airline pilots often begin as flight engineers or first officers for smaller, regional airlines. Historically, newly hired pilots at major airline companies have approximately 2000 hours of flight experience. The Federal Aviation Administration (FAA) is conducting a study to determine whether the mean flight experience of newly hired pilots has decreased. State the null and alternative hypotheses in terms of the population mean μ.

9.25 Medicine and Clinical Studies A new surgical procedure has been developed to remove cataracts; it involves a smaller incision. This more expensive procedure will be implemented at a major hospital only if there is evidence that the standard deviation in time to recovery is less than seven days. A random sample of patients will receive the new procedure, and the resulting recovery times will be used to test the relevant hypothesis. Let σ be the population standard deviation in recovery time. State the null and alternative hypotheses in terms of σ.

9.26 Public Policy and Political Science Officials in Dexter, a small town in upstate New York, have decided to install more fire hydrants in an effort to decrease insurance rates for many businesses. Prior to installing the new fire hydrants, the mean distance to a fire hydrant for downtown buildings was 525 ft. After the new installations, a random sample of downtown buildings will be obtained, and the distance to the nearest fire hydrant will be recorded. The data will be used to determine whether there is evidence that the mean distance to a fire hydrant has decreased. State the null and alternative hypotheses in terms of μ, the population mean distance to a fire hydrant.

9.27 Public Policy and Political Science The U.S. Health Insurance Portability and Accountability Act of 1996 (HIPAA) was designed to protect personal privacy. However, this law has created mountains of paperwork and may even increase the cost of medical care. The federal government has decided to consider repealing this law if more than 60% of all hospitals are experiencing increased costs due to this regulation. A random sample of hospitals will be obtained, and the resulting information will be used to test the relevant hypothesis. State the null and alternative hypotheses in terms of p, the population proportion of hospitals experiencing increased costs as a result of HIPAA.

9.28 Public Health and Nutrition Due to Ontario's aging-in-place strategy, the nature of long-term care has significantly changed. Only people with high-care needs are eligible, the admission criteria are stricter, and as of February 2018, there were 33,080 people on the waiting list for long-term care placements.[8] A new admission protocol has been implemented in an effort to reduce the mean time to placement from 161 days. A random sample of patients waiting for admission will be obtained and the information will be used to test the relevant hypotheses. State the null and alternative hypotheses in terms of μ, the mean time to placement.

9.29 Public Policy and Political Science The Ocean City Community Committee in New Jersey will renovate the fishing pier on South Philadelphia Avenue if more than 80% of the residents favor the plan. A random sample of residents will be obtained, and their responses will be used to test the relevant hypothesis. Let p be the population proportion of residents who favor renovating the fishing pier. Which of the following sets of null and alternative hypotheses is appropriate? Justify your answer.

(a) H_0: $p = 0.80$ versus H_a: $p \neq 0.80$
(b) H_0: $p = 0.80$ versus H_a: $p < 0.80$
(c) H_0: $p = 0.80$ versus H_a: $p > 0.80$

9.30 Business and Management The Savannah Sugar Refinery in Louisiana will invest in new energy-saving devices if there is evidence to suggest that the mean amount of energy (in kilowatts) used per day will decrease from 1925. Suppose μ is the population mean amount of energy used per day, and an experiment is designed to test the new equipment. State the null and alternative hypotheses in terms of μ.

9.31 Economics and Finance Suppose U.S. monetary policy will be set by the Federal Reserve Board by examining the median consumer price $\tilde{\mu}$ for a large fixed set of common commodities. The Fed will raise the interest rate if there is evidence that the median price is less than $125.50. A random sample of counties will be obtained, and the cost of these goods in each county will be used to test the relevant hypothesis. State H_0 and H_a in terms of $\tilde{\mu}$.

9.32 Economics and Finance The mean credit card debt in American households with a net worth of $500,000 or more is $8139.[9] Some research suggests that households with a smaller net worth have less credit card debt. A random sample of American households with net worth less than $500,000 was obtained, and the credit card debt of each was recorded. Let μ represent the mean credit card debt for American households with net worth less than $500,000. State the relevant null and alternative hypotheses in terms of μ.

9.33 Physical Sciences If an impact gun is used to mount automobile tires, the torque setting and incoming air pressure must be kept fairly constant to ensure consistent results. Suppose a Subaru dealership is testing a new air compressor and will install this device if the variance in air pressure at 50 psi is less than 1.25 psi. State the relevant null and alternative hypotheses in terms of σ^2, the variance in air pressure.

9.2 Hypothesis Test Errors

Human Error

To conduct a hypothesis test, we use the information in a sample to reach a decision about the value of a population parameter. The sample data lead to a value of the test statistic and the ultimate decision (reject or do not reject H_0). However, there is always a chance of making a mistake (the wrong decision). Even a simple random sample is only a (usually small) portion of the entire population. This limited information could lead to the wrong conclusion about the population parameter. To fully understand the structure of a hypothesis test, we need to examine what could possibly go wrong.

The water in Lake Jean, located in Ricketts Glen State Park, Pennsylvania, is clear and cool, sometimes very cool. Park officials have decided to post an advisory warning for swimmers if there is any evidence to suggest that the mean water temperature μ is less than 62°F. Otherwise, the lifeguards will post no signs and allow swimming as usual. Each day, lifeguards measure the temperature of the water (in °F) at 15 randomly selected locations around the lake. They use the value of the sample mean water temperature $\bar{x}$ to test the hypotheses

$$H_0: \mu = 62 \quad \text{versus} \quad H_a: \mu < 62$$

The lifeguards have decided to use a cutoff point of 61.5°F for their sample mean. They cannot measure the temperature of the water at every location in the lake (and hence compute the population mean). Also, recall that the sample mean varies around the population

Yarvin World Journeys/Alamy

In this hypothesis test, $\bar{X}$ is the test statistic.

mean. The lifeguards have decided that a sample mean value of $\bar{x} \leq 61.5$ is far enough away from 62 that it cannot be attributed to ordinary variation about the population mean.

The rejection region is any value $\bar{x}$ less than or equal to 61.5 (see **Figure 9.2**). If the value $\bar{x}$ is 61.5 or less, then H_0 is rejected, and an advisory warning is posted. If the value $\bar{x}$ is greater than 61.5, the lifeguards cannot reject H_0. There is no evidence to suggest that the mean water temperature is less than 62°F.

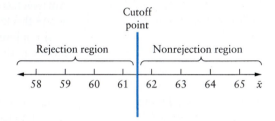

Figure 9.2 The rejection region and nonrejection region for the Lake Jean water-temperature hypothesis test.

Here is what can happen when the sample data are collected and the hypothesis is tested.

This is all very theoretical. You never really know whether you made a mistake in a hypothesis test.

1. Suppose the true population mean μ is 62°F or greater; the water is warm enough for swimming.
   (a) If $\bar{x} > 61.5$°F, then there is no evidence to reject H_0. This conclusion is correct.
   (b) If $\bar{x} \leq 61.5$°F, then there is evidence to reject H_0. An advisory warning is posted. This conclusion is incorrect, because the water really is warm enough for swimming.
2. Suppose the true population mean μ is less than 62°F; the water is too cold for swimming.
   (a) If $\bar{x} > 61.5$°F, then there is no evidence to reject H_0. This conclusion is incorrect, because the water is really too cold for swimming.
   (b) If $\bar{x} \leq 61.5$°F, then there is evidence to reject H_0. An advisory warning is posted. This conclusion is correct.

Error Definitions

The Lake Jean example illustrates the two possible errors in a hypothesis test.

Definition
1. The value of the test statistic may lie in the rejection region, but the null hypothesis is true. If we reject H_0 when H_0 is really true, this is called a **type I error**. The probability of a type I error is called the **significance level** of the hypothesis test and is denoted by α: P(type I error) $= \alpha$.
2. The value of the test statistic may not lie in the rejection region, but the alternative hypothesis is true. If we do not reject the null hypothesis when H_a is really true, this is called a **type II error**. The probability of a type II error is denoted by β: P(type II error) $= \beta$.

The table illustrates the decisions and errors in a hypothesis test.

		Decision	
		Reject H_0	Do not reject H_0
Truth	H_0	Type I error	Correct decision
	H_a	Correct decision	Type II error

Errors in Context

Each example in this section describes a specific hypothesis test. The type I error and the type II error are described in context, and the real-world consequences of a wrong decision are given.

EXAMPLE 9.4   Keep It Cool

Since some people play long, intense computer games, and both academics and business people prefer portability, laptop computers have become more popular than desktop models. To alleviate common overheating issues, many users purchase a cooling pad. This inexpensive accessory provides extra airflow to keep the laptop temperature down. One company that mass-produces cooling pads has a known defect rate of 0.07. Suppose Amazon has ordered a large number of cooling pads from this company and is willing to accept this known defect rate or lower.

A random sample of cooling pads will be obtained and carefully inspected. The information in the sample will be used to test the hypotheses

$$H_0: p = 0.07 \quad \text{versus} \quad H_a: p > 0.07$$

If H_0 is rejected, the entire shipment of cooling pads will be sent back to the manufacturer. Discuss the consequences of the decision to reject or not to reject for each truth assumption.

> There are two possible truth assumptions, as indicated in the table presented earlier: H_0 is true or H_a is true.

Solution

STEP 1   Suppose H_0 is true: The proportion of defective cooling pads is really 0.07 (or less).

   (a) If H_0 is rejected, the entire shipment of cooling pads will be sent back. This is a type I error, and in this case the manufacturer will not be happy. The cooling pads are of acceptable quality, but the entire shipment is returned.

   (b) If H_0 is not rejected, then the cooling pads shipment is accepted. Everyone is happy in this case. The hypothesis test indicated no evidence of a higher proportion of defective cooling pads.

STEP 2   Suppose H_a is true: The proportion of defective cooling pads is greater than 0.07.

   (a) If H_0 is rejected, the entire shipment of cooling pads will be returned. This is the correct conclusion. The hypothesis test indicated the cooling pad shipment contained a higher proportion of defectives, and Amazon will be glad to return a bad batch.

   (b) If H_0 is not rejected, then the cooling pad shipment will be accepted. This is a type II error, and in this case, Amazon will not be happy. Too many cooling pads are defective, yet Amazon will accept a shipment of poor quality. ■

TRY IT NOW   Go to Exercises 9.55 and 9.57

EXAMPLE 9.5   Take Two Aspirin

Approximately 1 billion people worldwide take an aspirin regularly to treat or prevent heart disease.[10] Suppose Walgreens receives a large shipment of Bayer aspirin, in which each tablet should contain 80 mg of acetylsalicylic acid, the active ingredient in aspirin. A random sample of tablets will be obtained and chemically analyzed. The information in the sample will be used to test the hypotheses

$$H_0: \mu = 80 \quad \text{versus} \quad H_a: \mu \neq 80$$

where μ is the mean amount of acetylsalicylic acid in each tablet. If H_0 is rejected, the entire aspirin shipment will be returned to Bayer. Discuss the consequences of the decision to reject or not to reject for each truth assumption.

Solution

STEP 1   Suppose H_0 is true: The mean amount of the active ingredient is 80 mg.

   (a) If H_0 is rejected, the aspirin shipment is returned to Bayer. This is a type I error, and Bayer will not like this decision. The aspirin has the specified amount of acetylsalicylic acid, but the entire shipment is returned.

(b) If H_0 is not rejected, then the aspirin bottles are placed on the shelves at Walgreens. Everyone is happy. The aspirin contains the correct amount of the active ingredient, and the hypothesis test did not indicate otherwise.

STEP 2 Suppose H_a is true: The mean amount of active ingredient is not 80 mg.

(a) If H_0 is rejected, the aspirin shipment is returned to Bayer. This is the correct conclusion. The hypothesis test indicated the mean amount of acetylsalicylic acid is different from 80 mg, and Walgreens is happy to return a bad batch of aspirin.

(b) If H_0 is not rejected, the aspirin bottles are placed on the shelves at Walgreens, ready for sale. This is a type II error, and in this case Walgreens may encounter some very unhappy customers. The mean amount of the active ingredient is not as specified, yet the aspirin is placed in store stock and sold.

TRY IT NOW Go to Exercises 9.47 and 9.49

EXAMPLE 9.6 Cold and Rolled

Nucor Corporation manufactures steel in cold-rolled sheets for use in a wide variety of products. The steel is produced to have a certain thickness, but the focus in the manufacturing process is on the variance in thickness. Each hour, a random sample of cold-rolled sheets of 10-gauge steel is obtained and the thickness of each sheet (in inches) is carefully measured. The sample variance is used to test the hypotheses

$$H_0: \sigma^2 = 0.04 \quad \text{versus} \quad H_a: \sigma^2 > 0.04$$

where σ^2 is the population variance in sheet metal thickness. If H_0 is rejected, the entire manufacturing process is shut down for inspection. Discuss the consequences of the decision to reject or not to reject for each truth assumption.

Remember: H_0 is stated in terms of an equality, but there is an implied inequality in the null hypothesis. The manufacturing process will be stopped only if there is evidence that the population variance is greater than 0.04. The status quo or no-change state is really $\sigma^2 \leq 0.04$.

Solution

STEP 1 Suppose H_0 is true: The population variance is 0.04 in. (or less).

(a) If H_0 is rejected, the manufacturing process is shut down. This is a type I error. The null hypothesis is true, but H_0 is rejected. This error is bad for Nucor. The process is stopped unnecessarily, and production time is lost.

(b) If H_0 is not rejected, then the facility continues to manufacture metal sheets. This is the correct decision.

STEP 2 Suppose H_a is true: The population variance in thickness is greater than 0.04 in.

(a) If H_0 is rejected, the correct decision has been made. The variance is too high and the hypothesis test procedure suggested that the manufacturing process should be shut down for inspection.

(b) If H_0 is not rejected, the facility continues to hum along. This is an incorrect decision, a type II error. H_a is true, but the null hypothesis is not rejected. In this case, there is a good chance that the metal sheets are being made with too much variability in thickness.

TRY IT NOW Go to Exercises 9.51 and 9.53

CHAPTER APP Heavy Metal

Discuss the consequences of the decision to reject or not to reject the null hypothesis for each truth assumption.

Hypothesis Test Efficiency

A perfect hypothesis test would have the probability of a type I error and the probability of a type II error both equal to 0.

The *efficiency*, or *goodness*, of a hypothesis test is often measured in terms of α and β. Intuitively, in an effective test, the probability of both types of errors should be small. In fact, because no one wants to make a mistake, it would be ideal to have $\alpha = \beta = 0$. However, this is impossible, because any decision in a hypothesis test is made using information in a sample. We can never examine the entire population, so there is always a chance of making a mistake.

In a hypothesis test, we usually control, or set, the value of α. We do not have any direct control over β, but we can often compute the probability of a type II error for a specific alternative value of the parameter.

▶ It seems reasonable to set α as small as possible, but for a fixed sample size, α and β are inversely related. This means that for a fixed sample size n, making α smaller forces β to be larger. The only way to decrease both α and β (simultaneously) is to increase the sample size. More information (larger n) means less of a chance of making a mistake. The most common values for α are 0.05 and 0.01. The probability of a type II error actually depends on the specific alternative value of the population parameter being tested. Therefore, the probability of a type II error is usually written as a function of the population parameter. For example, $\beta(\mu_a)$ represents the probability of a type II error if the true value of μ is μ_a.

Consider a hypothesis test with $H_0: \theta = \theta_0$ and $H_a: \theta \neq \theta_0$. As the alternative value of the parameter θ_a moves farther away from the hypothesized value θ_0 the probability of a type II error $\beta(\theta_a)$ decreases. This seems reasonable because for values of θ_a far away from θ_0, we have a better chance of detecting the difference between the hypothesized value and the alternative value of the population parameter.

Later in this chapter, we will visualize and learn how to compute the probability of a type II error for a specific alternative value of the population characteristic. ◀

TRY IT NOW Go to Exercise 9.59

Section 9.2 Exercises

Concept Check

9.34 Fill in the Blank
(a) Rejecting H_0 when H_0 is really true is called a _____.
(b) Rejecting the null hypothesis when H_a is true is called a _____.

9.35 True or False Under certain circumstances, the probability of a type I error and the probability of a type II error can be 0.

9.36 True or False In a hypothesis test, we always know if we made an error in the conclusion.

9.37 True or False In a hypothesis test, we never really know if we have made the correct decision.

9.38 True or False In a hypothesis test, for a fixed sample size, α and β are inversely related.

9.39 True or False In a hypothesis test, we usually specify a value for α.

9.40 Short Answer What are the most common values for the probability of a type I error?

Practice

9.41 Consider a hypothesis test with
$$H_0: \mu = 180 \quad \text{and} \quad H_a: \mu < 180$$
Determine whether the decision is correct or in error. Identify each error as type I or type II.
(a) The true value of μ is 180 and H_0 is rejected.
(b) The true value of μ is 179 and H_0 is rejected.
(c) The true value of μ is 160 and H_0 is not rejected.
(d) The true value of μ is 182 and H_0 is rejected.

9.42 Consider a hypothesis test with
$$H_0: p = 0.44 \quad \text{and} \quad H_a: p \neq 0.44$$
Determine whether the decision is correct or in error. Identify each error as type I or type II.
(a) The true value of p is 0.44 and H_0 is not rejected.
(b) The true value of p is 0.41 and H_0 is not rejected.
(c) The true value of p is 0.45 and H_0 is not rejected.
(d) The true value of p is 0.42 and H_0 is rejected.

9.43 Consider a hypothesis test with
$$H_0: \sigma^2 = 26.5 \quad \text{and} \quad H_a: \sigma^2 > 26.5$$

Determine whether the decision is correct or in error. Identify each error as type I or type II.
(a) The true value of σ^2 is 26.0 and H_0 is rejected.
(b) The true value of σ^2 is 27.0 and H_0 is not rejected.
(c) The true value of σ^2 is 26.4 and H_0 is not rejected.
(d) The true value of σ^2 is 26.5 and H_0 is not rejected.

9.44 Recall that the probability of a type II error depends on the alternative specific value of the population parameter. Consider a hypothesis test with

$$H_0: \mu = 10 \quad \text{and} \quad H_a: \mu > 10$$

(a) For a fixed sample size, how do $\beta(11)$ and $\beta(15)$ compare? Are these two values approximately the same or different? If they are different, which is smaller, and why?
(b) What happens to $\beta(\mu_a)$ as μ_a increases (gets farther and farther away from 10)?

9.45 Why is there always a chance of making a mistake (an incorrect decision) in any hypothesis test?

9.46 Because we usually control the value of α, we could simply set the probability of a type I error to a very small value, say 0.0001. What's wrong with this strategy?

Applications

9.47 Travel and Transportation Highway 405, which runs from northern California to southern California, is the busiest interstate road in the United States. An estimated 374,000 vehicles use this road every day, often referred to as Carmageddon.[11] Transportation officials have decided to conduct a hypothesis test and will raise tolls in California to fund planned repairs if there is evidence to suggest that the mean number of vehicles per day on highway 405 has increased.
(a) Write the null and alternative hypotheses about μ, the mean number of vehicles per day that use highway 405.
(b) For the hypotheses in part (a), describe the type I and type II errors.
(c) If a type I error is committed, who is more angry, the transportation officials or drivers, and why?
(d) If a type II error is committed, who is more angry, the transportation officials or drivers, and why?

9.48 Economics and Finance A recent analysis of farms in Nebraska suggests that most operations are under financial stress. Many Nebraska farms have a cash flow problem, which makes it difficult to pay debts.[12] A proposal has been made to award grants so that farmers can improve their working capital to gross revenue ratio. The state will begin the program only if there is evidence to suggest that the true mean debt per farm is greater than $1 million. A random sample of Nebraska farms will be obtained, and the information in the sample will be used to test the following hypotheses:

$$H_0: \mu = 1 \text{ million} \quad \text{and} \quad H_a: \mu > 1 \text{ million}$$

(a) Describe a type I error and a type II error in this context.
(b) From the farmer's perspective, which error is more serious? Why?
(c) From the state's point of view, which error is more serious? Why?

9.49 Physical Sciences The annual Waikiki Roughwater Swim contest is held over a 2.4-mile course and ends near the Hilton Rainbow Tower. The 2017 winner was Rhys Mainstone with a time of 51 minutes, 45 seconds.[13] Before the race, a random sample of the current velocity (in knots) along the race course is obtained, and the resulting information is used to determine whether the race should be canceled. A mean current velocity μ of more than 0.65 knot is considered unsafe.
(a) State the null and alternative hypotheses.
(b) Describe type I and type II errors in this context.
(c) Which error is more serious for the swimmers? Why?
(d) Which error is more serious for the race organizers? Why?

9.50 Education and Child Development A recent poll indicated that 8% of American students have missed a test because they overslept. Suppose officials at a state college decide to apply a new academic policy regarding exams. If a student arrives late for a test, then the student receives a 0 for the test and cannot take a make-up test at a later time.
(a) What hypotheses should be tested, in terms of p, the true proportion of students who have missed a test because they overslept, if college officials want to prove that the new academic policy is causing fewer students to be late for exams?
(b) Which error, type I or type II, is more serious for college officials? Why?
(c) Which error, type I or type II, is more serious for students? Why?

9.51 Medicine and Clinical Studies Some research suggests that eating a moderate amount of dark chocolate may increase the level of antioxidants, compounds that protect us against free radicals, which can cause heart disease and cancer.[14] A study was designed, and the antioxidant concentration in patients who ate dark chocolate regularly was measured. This information was used to test the hypotheses

$$H_0: \mu = 0.4 \quad \text{and} \quad H_a: \mu > 0.4$$

where μ is the true mean percentage of antioxidants in the bloodstream.
(a) Describe a type I error and a type II error in this context.
(b) The Hershey Company, maker of Hershey's chocolates, is very interested in the results of this study. Which error is more serious for this company? Why?

9.52 Public Policy and Political Science The Dallas, Texas, city council is going to consider a zoning variance for the Estates on Frankford apartment complex so that the developer may extend the structure closer to the nearest road. Some council members are concerned about safety, but they will vote for the measure if more than 60% of all city residents favor the variance. A random sample of residents will be obtained and asked whether they favor the zoning variance. Let p be the true proportion of residents who favor the extended structure.
(a) State the null and alternative hypotheses in terms of p.
(b) Describe a type I error and a type II error in this context.

(c) Which error is more serious for the developer? Why?
(d) Which error is more serious for the city council members? Why?

9.53 Manufacturing and Product Development The National Fire Protection Association maintains an extensive list of codes and standards that address equipment, extinguishing systems, and fire-protective gear.[15] Suppose a firefighter's helmet should withstand a temperature of 1200°F for 30 seconds. The Seattle Fire Department is ordering helmets from Kidde Fire Fighting. It will accept the shipment only if there is no evidence to suggest that the helmets fail to meet the standard. Let μ be the mean temperature that the helmets can withstand.
(a) State the null and alternative hypotheses in terms of μ.
(b) Describe a type I error and a type II error in the context of this problem.

9.54 Education and Child Development For most cities in Canada, the mean cost of child care per month is $1000.[16] Officials in Toronto believe childcare costs in their city are higher than this mean, and are forcing more parents to work extra hours or second jobs. A random sample of families who utilize child care services in Toronto will be obtained, and the child care costs incurred by each family will be recorded. These data will be used to determine whether there is any evidence that the mean child care cost in Toronto μ is greater than $1000.
(a) State the null and alternative hypotheses in terms of μ.
(b) Describe a type I error and a type II error in this context.

9.55 Technology and the Internet A new 3D scanner is being developed for use in airports that can reportedly detect liquids in amounts greater than 5 mL. If successful, passengers would no longer have to remove items from carry-on luggage for screening. The scanner manufacturer claims the detection rate, meaning successfully detecting a liquid, is 0.98. Airport officials in London have decided to conduct a test to determine if there is any evidence to suggest that the detection rate p is less than 0.98.
(a) State the null and alternative hypotheses.
(b) Describe a type I error and a type II error in this context.

9.56 Medicine and Clinical Studies An electroencephalogram (EEG) is often used to measure brain, or neural, activity expressed as electrical voltage. Research suggests that transcendental meditation (TM) produces a simplified state of rest and relaxation resulting in an EEG with smaller than normal variance. Suppose normal brain activity has variance $\sigma^2 = 15$ volts². A random sample of patients who practice TM will be selected, and their brain activity will be measured during TM. The resulting information will be used to test the research theory.
(a) State the null and alternative hypotheses in terms of σ^2.
(b) Describe a type I and a type II error in this context.
(c) If there is evidence to suggest that TM decreases the variance in brain activity, then the National Science Foundation (NSF) will commit more money for TM research. Which error is more serious for the NSF? Why? Which error is more serious for TM researchers? Why?

9.57 Business and Management A recent poll suggests that 20% of jobs in the United States are held by a worker under contract, that is, one without benefits or full-time employment.[17] Because business evolves quickly in response to competition and suppliers, labor leaders believe that the proportion of contract workers has increased. A random sample of American jobs will be obtained, and the information will be used to determine if the true proportion of contract workers p is greater than 0.20.
(a) State the null and alternative hypotheses in terms of p.
(b) Describe a type I and a type II error in this context.
(c) If there is evidence that the proportion of contract workers has increased, businesses will be forced to pay an additional tax to the federal government. Which error is more serious for the federal government? Why? Which error is more serious for businesses? Why?

Extended Applications

9.58 Marketing and Consumer Behavior Many families have decided to use their TVs for broadband-delivered video (for example, from Netflix, Hula, and Sling) instead of pay-TV (cable and satellite) services. A local cable TV provider in Kansas City, Missouri, Spectrum Cable, is concerned about losing market share and plans to conduct a hypothesis test to determine whether more advertising is needed. A random sample of homes in the city will be obtained, and the data will be used to determine whether there is any evidence that the true proportion of homes with broadband-delivered video is greater than 0.30.
(a) State the null and alternative hypotheses in terms of p.
(b) Describe type I and type II errors in this context.
(c) For a fixed sample size, which is smaller, $\beta(0.32)$ or $\beta(0.35)$? Justify your answer.

9.59 Sports and Leisure South Carolina Department of Natural Resources personnel enforce hunting and fishing regulations and conduct routine safety checks on recreational boats. Past experience indicates that 15% of all boats inspected have at least one safety violation. Recent accidents on lakes in South Carolina have prompted calls for more extensive inspections. A random sample of recreational boats will be obtained and inspected. If there is evidence that the true proportion of boats with safety violations p is more than 0.15, a methodical inspection of every boat launched at popular sites will be started.
(a) State the null and alternative hypotheses in terms of p.
(b) Describe type I and type II errors in this context.
(c) For a fixed cutoff value, what happens to the probability of a type I error as the value of p approaches 0.15 from the right (that is, 0.20, 0.19, …).

9.60 Physical Sciences Civil engineers are going to test a highway bridge outside of Washington, D.C., for structural integrity. A random sample of locations along the bridge will be selected, and the concrete stress (in pounds per square inch, psi) will be measured. These data will be used to determine the safety of the bridge.

(a) Suppose 6400 psi is considered a safe concrete stress level. Which pair of hypotheses should be tested?

$H_0: \mu = 6400$ and $H_a: \mu > 6400$

or

$H_0: \mu = 6400$ and $H_a: \mu < 6400$

(b) If you regularly drive over the bridge being tested, would you prefer $\alpha = 0.1$ or $\alpha = 0.01$? Why?

9.61 Medicine and Clinical Studies The failure for all metal hip implants is approximately $p = 0.62$; corrective surgery is needed, usually to replace a failed implant, within five years.[18] A new implant design by researchers at the Missouri Bone and Joint Research Foundation uses a stronger, thinner metal and is expected to cause fewer complications and further surgeries. A long-term study was conducted to examine the failure rate of the new implant design.

(a) State the null and alternative hypotheses in terms of p.
(b) Describe type I and type II errors in this context.
(c) For a fixed sample size, which is smaller, $\beta(0.60)$ or $\beta(0.50)$?

(d) If you have invested in the company that manufactures this new implant, would you prefer $\alpha = 0.1$ or $\alpha = 0.01$? Why?

9.62 Biology and Environmental Science Due to rising sea levels and the redirecting of freshwater flow, the salinity level in the Florida Everglades has become unstable. A permanent change in the salinity will severely affect plant and animal life. Suppose the historic mean salinity level in the Everglades is 11 ppt (parts per thousand). A study will be conducted to determine if the salinity level has risen above this historic mean.

(a) State the null and alternative hypotheses in terms of μ.
(b) Describe type I and type II errors in this context.
(c) For a fixed sample size, witch is smaller, $\beta(12)$ or $\beta(20)$?
(d) If there is evidence that the salinity level has increased, the state will work to redirect freshwater back into the Everglades. If you are working for an environmental group to save the plants and animals in the Everglades, which would your prefer, $\alpha = 0.1$ or $\alpha = 0.01$? Why?

9.3 Hypothesis Tests Concerning a Population Mean When σ Is Known

Connect Four

In the previous sections and chapters, all of the statistics tools have been provided to construct a formal hypothesis test. Recall that every hypothesis test consists of four parts, and the null hypothesis, an equality stated in terms of a population parameter, represents the current state, or what is assumed to be true. The following example is used to develop a hypothesis test about a population mean μ when σ is known. From a practical standpoint, it is unlikely that we will know the value of σ. However, this is a very instructive procedure to learn and understand so that we can conduct a hypothesis test for an arbitrary parameter.

EXAMPLE 9.7 Patient Triage

No one wants to wait long in the emergency room to see a doctor for treatment. Nevertheless, many constraints make it difficult to treat patients in a timely manner. The Permanente Medical Group in California decided to address this issue by using predictive analysis, new staffing models, and a system-wide change in culture.[19] Prior to this experiment, the mean time to treat very ill patients (as opposed to critically ill patients or those with a minor injury) entering the emergency room was 20 minutes (with standard deviation $\sigma = 5.0$ minutes). During the waiting-room experiment, a random sample of 36 very ill patients was selected and the time to treatment for each was recorded. The sample mean time was $\bar{x} = 16.1$ minutes. Conduct a hypothesis test to determine whether there is any evidence to suggest that the waiting-room experiment reduced the mean time to treatment for very ill patients. Use $\alpha = 0.05$.

Solution

STEP 1 Let μ be the mean time to treatment for very ill patients entering the emergency room. The current state, or what is assumed to be true, involves time to treatment prior to the waiting-room experiment. Therefore, the null hypothesis is

$H_0: \mu = 20$

STEP 2 Leaders at the Permanente Medical Group hope these new procedures will decrease time to treatment for very ill patients. They hope to find evidence in favor of the alternative hypothesis

$$H_a: \mu < 20$$

STEP 3 We need to know whether 16.1 minutes is a reasonable observation under the null hypothesis, or whether it is too far away from the assumed mean, 20 minutes. Because H_0 is assumed to be true, and $n = 36 \geq 30$, by the Central Limit Theorem, $\overline{X}$ is approximately normal with mean $\mu = 20$ and standard deviation $\sigma/\sqrt{n}$. Although $\overline{X}$ could be used as the test statistic, it's easier to identify unlikely values of a related standard normal random variable. Therefore, the appropriate test statistic is obtained by standardizing.

$$\text{TS: } Z = \frac{\overline{X} - 20}{\sigma/\sqrt{n}}$$

STEP 4 The alternative hypothesis is one-sided, left-tailed, so unusual values of the test statistic Z are in the left tail of the distribution. We should reject the null hypothesis if the value of Z is to the left of some cutoff value, or endpoint. The cutoff value is determined so that the probability of a type I error is 0.05. Select the cutoff value such that if H_0 is true,

$$P(Z \leq \text{cutoff value}) = 0.05$$

Using the definition of a Z critical value (from Section 8.2),

$$P(Z \leq -z_{0.05}) = 0.05 \quad \text{and} \quad -z_{0.05} = -1.6449 \quad \text{Appendix Table 3.}$$

In interval notation, the rejection region is $(-\infty, -1.6449]$.

the rejection region is written as

$$\text{RR: } Z \leq -z_{0.05} = -1.6449$$

If the value of Z is less than or equal to the critical value -1.6449, then the observed value of $\overline{X}$ is considered rare, and we reject the null hypothesis. **Figure 9.3** illustrates the critical value and the rejection region for this hypothesis test.

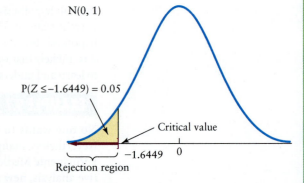

Figure 9.3 The critical value and the rejection region for the emergency-room hypothesis test. The critical value divides the z measurement axis into two parts: the rejection region and the nonrejection region.

STEP 5 The value of the test statistic is

The small letters here (z and x̄) represent specific values.

$$z = \frac{\overline{x} - 20}{\sigma/\sqrt{n}} = \frac{16.1 - 20}{5.0/\sqrt{36}} = -4.68$$

The value $\overline{x} = 16.1$ is 4.68 standard deviations to the left of the mean; it is a very unusual observation if the null hypothesis is really true. This means either 16.1 is an incredibly lucky observation or the assumption ($\mu = 20$) is wrong. As usual, we discount the lucky possibility.

More formally, because $z = -4.68$ (≤ -1.6449) lies in the rejection region, we reject the null hypothesis at the $\alpha = 0.05$ level of significance. There is evidence to suggest that the true population mean time to treatment μ is less than 20 minutes. ∎

General Procedure

In any hypothesis test, as in Example 9.7, we assume H_0 is true and consider the likelihood of the sample outcome (expressed as a single value of the test statistic).

1. If the value of the test statistic is reasonable under the null hypothesis, then we cannot reject H_0.
2. If the value of the test statistic is unlikely under the null hypothesis, then we reject H_0.

As presented in Section 9.1, there are three possible alternative hypotheses: one two-sided alternative and two one-sided alternatives. The rejection region depends on the alternative hypothesis. The general procedure for a hypothesis test concerning μ is summarized next.

Hypothesis Tests Concerning a Population Mean When σ Is Known

Use this as a template for a hypothesis test about a population mean when σ is known.

Given a random sample of size n from a population with mean μ, assume

1. The underlying population is normal or n is large.
2. The population standard deviation σ is known.

A hypothesis test about the population mean μ with significance level α has the form:

$H_0: \mu = \mu_0$

$H_a: \mu > \mu_0, \quad \mu < \mu_0, \quad \text{or} \quad \mu \neq \mu_0$

TS: $Z = \dfrac{\overline{X} - \mu_0}{\sigma/\sqrt{n}}$

The rejection region always includes the endpoint of the (infinite) interval(s).

RR: $Z \geq z_\alpha, \quad Z \leq -z_\alpha, \quad \text{or} \quad |Z| \geq z_{\alpha/2}$

A CLOSER LOOK

1. μ_0 is a fixed, hypothesized value of the population mean μ.
2. Use only one (appropriate) alternative hypothesis and the corresponding rejection region. The graphs in **Figure 9.4**, **Figure 9.5**, and **Figure 9.6** illustrate the rejection region for each alternative hypothesis.

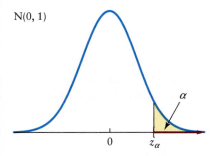

Figure 9.4 Rejection region for $H_a: \mu > \mu_0$.

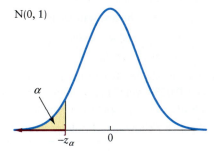

Figure 9.5 Rejection region for $H_a: \mu < \mu_0$.

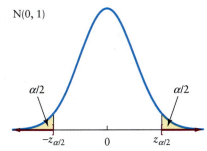

Figure 9.6 Rejection region for $H_a: \mu \neq \mu_0$.

3. For a two-sided alternative hypothesis, the rejection region $|Z| \geq z_{\alpha/2}$ (written using the absolute value of Z) is simply a shorthand way to write $Z \geq z_{\alpha/2}$ or $Z \leq -z_{\alpha/2}$.
4. For a given significance level, the corresponding critical value is found using Appendix Table 3, backward. This procedure was presented in Chapter 8. Common values for α are 0.05, 0.025, and 0.01.
5. The hypothesis test procedure described here can be used only if σ is known. If n is large and σ is unknown, some statisticians suggest using the sample standard deviation s instead of σ. This produces an approximate test statistic. Section 9.5 presents an exact test procedure concerning μ when σ is unknown.

Hypothesis Test Examples

The next examples illustrate this hypothesis test procedure.

EXAMPLE 9.8 The Dead Zone

The biological desert in the Gulf of Mexico called the Dead Zone is a region in which there is very little or no oxygen. Most marine life in the Dead Zone dies or leaves the region. The area of this region varies and is affected by agriculture, fertilizer runoff, and weather. The long-term mean area of the Dead Zone is 5460 square miles.[20] As a result of recent flooding in the midwest and subsequent runoff from the Mississippi River, researchers believe that the Dead Zone area will increase. A random sample of 36 days was obtained, and the sample mean area of the Dead Zone was 6258 mi^2. Is there any evidence to suggest that the current mean area of the Dead Zone is greater than the long-term mean? Assume that the population standard deviation is 1850 and use $\alpha = 0.025$.

Solution

This example involves a hypothesis test concerning a population mean when σ is known. Use the template for a one-sided, right-tailed test about μ. The underlying population distribution is unknown, but n is large and σ is known. Determine the appropriate alternative hypothesis and the corresponding rejection region, find the value of the test statistic, and draw a conclusion.

STEP 1 The current state, or assumed mean, is $\mu = 5460 \,(= \mu_0)$.

The sample size is $n = 36$; $\sigma = 1850$ and $\alpha = 0.025$.

We are looking for evidence that the current mean area of the Dead Zone is greater than the long-term mean. Therefore, the alternative hypothesis is one-sided, right-tailed.

STEP 2 The four parts of the hypothesis test are

$H_0: \mu = 5460$

$H_a: \mu > 5460$

TS: $Z = \dfrac{\overline{X} - \mu_0}{\sigma/\sqrt{n}}$

RR: $Z \geq z_\alpha = z_{0.025} = 1.96$

STEP 3 The value of the test statistic is

$$z = \frac{\overline{x} - \mu_0}{\sigma/\sqrt{n}} = \frac{6258 - 5460}{1850/\sqrt{36}} = 2.5881 \geq 1.96$$

STEP 4 Because 2.5881 lies in the rejection region, we reject the null hypothesis at the $\alpha = 0.025$ level. There is evidence to suggest that the current mean area of the Dead Zone is greater than 5460 mi^2.

Figure 9.7 shows a technology solution.

```
> mu_zero <- 5460
> xbar <- 6258
> n <- 36
> sigma <- 1850
>
> alpha <- 0.025
> zcrit <- qnorm(1 - alpha)
>
> ts <- (xbar - mu_zero)/(sigma / sqrt(n))
> round(cbind(zcrit,ts),4)
     zcrit   ts
[1,]  1.96 2.5881
```

Figure 9.7 R calculations to find the critical value and the value of the test statistic.

TRY IT NOW Go to Exercises 9.81 and 9.83

EXAMPLE 9.9 Natural Defense

White blood cells are the body's natural defense mechanism against disease and infection. The mean white blood cell count in healthy adults, measured as part of a CBC (complete blood count), is approximately $7.5 \times 10^3/\mu L$.[21] A company developing a new drug to treat arthritis pain must check for any side effects. A random sample of patients using the new drug was selected, and the white blood cell count of each patient was measured. The results are given in the table ($\times 10^3/\mu L$).

6.50	8.74	8.69	8.00	6.85	8.84	6.76	7.93	6.58	7.65	8.84
7.00	8.44	6.70	8.28	9.20	7.65	6.45	6.95	7.66	10.12	

Assume the distribution of white blood cell counts is normal and $\sigma = 1.1$. Conduct a hypothesis test to determine whether there is any change in the mean white blood cell count due to the arthritis drug. Use $\alpha = 0.01$.

Solution

STEP 1 The assumed mean is $\mu = 7.5$ ($= \mu_0$); the sample size is $n = 21$; $\sigma = 1.1$; and $\alpha = 0.01$.

The company is looking for any change in the mean white blood cell count. Therefore, the relevant alternative hypothesis is two-sided.

The sample size is small ($n < 30$), but the population is assumed to be normal. The hypothesis test concerning a population mean when σ is known can be used.

STEP 2 The four parts of the hypothesis test are

$H_0: \mu = 7.5$

$H_a: \mu \neq 7.5$

TS: $Z = \dfrac{\overline{X} - \mu_0}{\sigma/\sqrt{n}}$

RR: $|Z| \geq z_{\alpha/2} = z_{0.005} = 2.5758$ ($Z \leq -2.5758$ or $Z \geq 2.5758$)

STEP 3 The sample mean is

$$\overline{x} = \frac{1}{21}(6.50 + 8.74 + \cdots + 10.12) = 7.8014$$

The value of the test statistic is

$$z = \frac{\overline{x} - \mu_0}{\sigma/\sqrt{n}} = \frac{7.8014 - 7.5}{1.1/\sqrt{21}} = 1.2557$$

STEP 4 The value of the test statistic, $z = 1.2557$, does not lie in the rejection region. We do not reject the null hypothesis at the $\alpha = 0.01$ level of significance. There is no evidence to suggest that the new arthritis drug has changed the mean white blood cell count.

Figure 9.8 shows a technology solution.

```
> blood_cell <- c(6.50,8.69,...,10.12)
>
> mu_zero <- 7.5
> xbar <- mean(blood_cell)
> n <- length(blood_cell)
> sigma <- 1.1
>
> alpha <- 0.01
> zcrit <- qnorm(1 - alpha/2)
>
> ts <- (xbar - mu_zero)/(sigma / sqrt(n))
> round(cbind(zcrit,ts),4)
      zcrit     ts
[1,] 2.5758 1.2557
```

Figure 9.8 R calculations to find the sample mean, critical value, and the value of the test statistic.

TRY IT NOW Go to Exercises 9.85 and 9.93

Calculating the Probability of a Type II Error

▶ Recall that the probability of a type II error depends on the alternative value of the parameter under investigation. The next example presents a method for computing β in a hypothesis test about a population mean μ when σ is known.

EXAMPLE 9.10 Particulates Matter

Recently, China introduced a comprehensive plan for dealing with air pollution that mandated a nationwide cap on coal use, stopped new coal-burning plants from being constructed, and increased the use of filters and scrubbers at existing coal plants. One measure of air pollution is the concentration of PM2.5, the smallest polluting particles that pose the greatest health risks. In 2018, the mean density of PM2.5 in Beijing was 58 $\mu g/m^3$.[22] A random sample of 36 days is selected and the PM2.5 concentration is recorded for each. A hypothesis test is conducted to determine whether there is any evidence that the mean PM2.5 concentration is less than 58 $\mu g/m^3$. Let $\alpha = 0.025$, assume $\sigma = 15$, and find the probability of a type II error if the true mean PM2.5 concentration is now 50 $\mu g/m^3$.

Roman Pilipey/EPA/Shutterstock

Solution

STEP 1 The assumed mean is $\mu = 58\ (= \mu_0)$; $n = 36$, $\sigma = 15.0$, and $\alpha = 0.025$.

The Chinese government hopes the air pollution plan reduces the concentration of PM2.5. This is a one-sided, left-tailed test.

The underlying population distribution is unknown, but the sample size is large, ($n = 36 \geq 30$). The hypothesis test concerning a population mean μ when σ is known is appropriate.

STEP 2 The four parts of the hypothesis test are

$H_0: \mu = 58$

$H_a: \mu < 58$

TS: $Z = \dfrac{\overline{X} - \mu_0}{\sigma/\sqrt{n}}$

RR: $Z \leq -z_\alpha = -z_{0.025} = -1.96$

STEP 3 To compute the probability of a type II error, we need to write and visualize the rejection region in terms of $\overline{X}$. Start by writing the definition of a type I error, and work backward to a statement involving $\overline{X}$.

$P(Z \leq -1.96) = 0.025$ Definition of a type I error.

$P\left(\dfrac{\overline{X} - 58}{15/\sqrt{36}} \leq -1.96\right) = 0.025$ Use the definition of the test statistic.

$P(\overline{X} \leq 53.1) = 0.025$ Isolate $\overline{X}$. Within the probability expression, multiply both sides by $15/\sqrt{36}$, and add 58 to both sides.

$\overline{X}_c = 53.1$ is the critical value in the $\overline{X}$ world, or distribution (see **Figure 9.9**). If the sample mean PM2.5 concentration is less than or equal to 53.1, reject the null hypothesis. This is equivalent to saying if the value of the test statistic z is less than or equal to -1.96, reject the null hypothesis.

STEP 4 $\beta(50)$ is the probability of not rejecting the null hypothesis if the real mean is 50 (often denoted $\mu_a = 50$; the a in the subscript stands for **alternative mean**). Write a probability statement for $\beta(50)$ using the critical value 53.1, and standardize.

9.3 Hypothesis Tests Concerning a Population Mean When σ Is Known

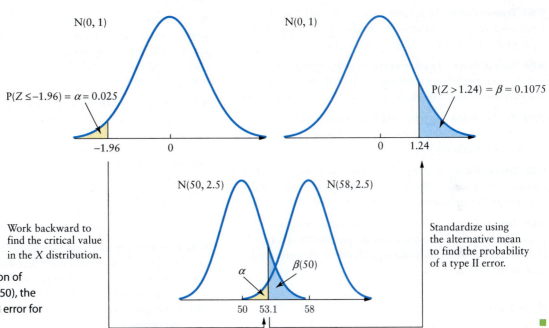

$$\beta(\mu_a) = P(\overline{X} > \overline{X}_c)$$ Definition of a type II error.

$$\beta(50) = P(\overline{X} > 53.1)$$ Use values for μ_a and $\overline{x}_c$.

$$= P\left(\frac{\overline{X} - 50}{15/\sqrt{36}} > \frac{53.1 - 50}{15/\sqrt{36}}\right)$$ Standardize using $\mu_a = 50$. Assume $\sigma = 15.0$ is unchanged even if the true mean is 50.

$$= P(Z \geq 1.24)$$ Equation 6.8

$$= 1 - P(Z \leq 1.24)$$ Use cumulative probability.

$$= 1 - 0.8925 = 0.1075$$ Use Appendix Table 3.

Figure 9.9 Visualization of the calculations for $\beta(50)$, the probability of a type II error for $\mu_a = 50$.

TRY IT NOW Go to Exercises 9.103 and 9.105

A CLOSER LOOK

1. By referring to Figure 9.9, we can visualize the inverse relationship between α and β. For example, if α increases, $\overline{X}_c$ moves to the right, and the area of the region corresponding to β decreases. Exercise 9.96 at the end of this section is designed to explore and confirm this concept numerically.

2. The logic and method for finding the probability of a type II error for the other alternative hypotheses are the same.
   (a) Work backward to find $\overline{X}_c$.
   (b) Write a probability statement for $\beta(\mu_a)$ in terms of $\overline{X}$ and $\overline{X}_c$.
   (c) Standardize to find the probability.

3. Rather than trying to memorize a formula for the probability of a type II error corresponding to each alternative hypothesis, use the definition, a drawing, and the three-step procedure just described. This conceptual approach will help when we consider the hypothesis test procedure concerning a population proportion.

Section 9.3 Exercises

Concept Check

9.63 True or False The hypothesis test concerning a population mean when σ is known can be used only when the underlying population is normal.

9.64 True or False In general, we reject the null hypothesis in a test concerning a population mean when σ is known and the value of the test statistic is large in magnitude.

9.65 True or False In a hypothesis test concerning a population mean when σ is known, the probability of a type II error does not depend on the true value of the population mean.

9.66 True or False In a hypothesis test concerning a population mean when σ is known, there are only three possible alternative hypotheses.

9.67 True or False In a hypothesis test concerning a population mean when σ is known, the conclusion is dependent on whether the value of the test statistic lies in the rejection region.

9.68 True or False In a hypothesis test concerning a population mean when σ is known, the probability of a type I error α and the probability of a type II error β always sum to 1. That is, $\alpha + \beta = 1$.

Practice

9.69 Consider a hypothesis test concerning a population mean with $H_0: \mu = 170$, $H_a: \mu < 170$, $n = 38$, and $\sigma = 15$.
(a) Write the appropriate test statistic.
(b) Write the rejection region corresponding to the value of α.

(i) $\alpha = 0.01$ (ii) $\alpha = 0.025$ (iii) $\alpha = 0.05$
(iv) $\alpha = 0.10$ (v) $\alpha = 0.001$ (vi) $\alpha = 0.0001$

9.70 Consider a hypothesis test concerning a population mean from a normal population with $H_0: \mu = 45.6$, $H_a: \mu > 45.6$, $n = 16$, and $\sigma = 15$.
(a) Write the appropriate test statistic.
(b) Write the rejection region corresponding to the value of α.

(i) $\alpha = 0.01$ (ii) $\alpha = 0.025$ (iii) $\alpha = 0.05$
(iv) $\alpha = 0.10$ (v) $\alpha = 0.005$ (vi) $\alpha = 0.0005$

9.71 Consider a hypothesis test concerning a population mean from a normal population with $H_0: \mu = -11$, $H_a: \mu \neq -11$, $n = 21$, and $\sigma = 4.5$.
(a) Write the appropriate test statistic.
(b) Write the rejection region corresponding to the value of α.

(i) $\alpha = 0.01$ (ii) $\alpha = 0.2$ (iii) $\alpha = 0.05$
(iv) $\alpha = 0.10$ (v) $\alpha = 0.001$ (vi) $\alpha = 0.0002$

9.72 Consider a hypothesis test concerning a population mean with $H_0: \mu = 3.55$, $H_a: \mu < 3.55$, $n = 49$, and $\sigma = 6.2$. Find the significance level α for each rejection region.
(a) $Z \leq -1.6449$ (b) $Z \leq -2.5758$ (c) $Z \leq -2.0537$
(d) $Z \leq -2.3263$ (e) $Z \leq -3.0902$ (f) $Z \leq -3.7190$

9.73 Consider a hypothesis test concerning a population mean with $H_0: \mu = 7.6$, $H_a: \mu \neq 7.6$, $n = 37$, and $\sigma = 4.506$. Find the significance level α for each rejection region.
(a) $|Z| \geq 1.96$ (b) $|Z| \geq 1.6449$ (c) $|Z| \geq 2.8070$
(d) $|Z| \geq 3.2905$ (e) $|Z| \geq 1.2816$ (f) $|Z| \geq 2.3263$

9.74 Consider a hypothesis test concerning a population mean from a normal population with $H_0: \mu = 98.6$, $H_a: \mu > 98.6$, $n = 10$, and $\sigma = 1.2$. Find the significance level α for each rejection region.
(a) $Z \geq 3.7190$ (b) $Z \geq 0.8416$ (c) $Z \geq 2.3263$
(d) $Z \geq 1.6449$ (e) $Z \geq 3.2905$ (f) $Z \geq 2.8782$

9.75 Consider a random sample of size 25 from a normal population with hypothesized mean 212 and $\sigma = 2.88$.
(a) Write the four parts for a one-sided, right-tailed hypothesis test concerning the population mean with $\alpha = 0.01$.
(b) What assumptions are made for the test in part (a) to be appropriate?
(c) Suppose the sample mean is $\bar{x} = 213.5$. Find the value of the test statistic, and draw a conclusion about the population mean.

9.76 Consider a random sample of size 32 from a population with hypothesized mean 3.14 and $\sigma = 6.8$. EX9.76
(a) Write the four parts for a one-sided, left-tailed hypothesis test concerning the population mean with $\alpha = 0.001$.
(b) What assumptions are made for the test in part (a) to be appropriate?
(c) Compute the sample mean, and find the value of the test statistic. Draw a conclusion about the population mean.

9.77 Consider a random sample of size 48 from a population with hypothesized mean 365.25 and $\sigma = 22.3$.
(a) Write the four parts for a two-sided hypothesis test concerning the population mean with $\alpha = 0.05$.
(b) What assumptions are necessary for the test in part (a) to be appropriate?
(c) Suppose the sample mean is $\bar{x} = 360.0$. Find the value of the test statistic, and draw a conclusion about the population mean.

9.78 Consider a one-sided, right-tailed hypothesis test concerning the mean μ_0 from a normal population with σ known, sample size n, and $\alpha = 0.05$. Explain the error in each of the following statements.
(a) The rejection region is RR: $Z \leq 1.96$.
(b) The test statistic is $Z = \dfrac{\mu_0 - \overline{X}}{\sigma/\sqrt{n}}$.
(c) The value of the test statistic does not lie in the rejection region. We accept the null hypothesis and conclude that $\mu = \mu_0$.
(d) The null hypothesis is $H_0: \mu > \mu_0$.
(e) The probability of a type II error when $\mu = \mu_a$ is $1 - \alpha = 0.05 = 0.95$.
(f) The probability of a type II error is high if the true population mean is very far away from μ_0.

Applications

9.79 Business and Management The mean income per year of employees who produce internal and external newsletters and magazines for corporations was reported to be $51,500. These editors and designers work on corporate publications but not on marketing materials. As a result of increased use of technology and oversupply, these corporate communications workers may be experiencing a decrease in salary. A random sample of the salaries of 38 corporate communications workers revealed that $\bar{x} = 49{,}762$. Conduct a hypothesis test to determine whether there is any evidence to suggest that the mean income per year of corporate communications workers has decreased. Assume $\sigma = 3750$ and use $\alpha = 0.01$.

9.80 Biology and Environmental Science Motor vehicles contribute approximately 20% of all U.S. emissions released into the atmosphere each year, mostly carbon dioxide but other global-warming gases as well. Therefore, carbon dioxide emissions are often used to determine whether a motor vehicle is efficient, or green. A random sample of 31 2018 automobiles was obtained and the CO_2 emissions for each was measured (in g/mi).[23] The mean CO_2 emissions level for all automobiles tested by the U.S. Environmental Protection Agency (EPA) in 2017 was 352 g/mi. Is there any evidence that the mean CO_2 emissions level has decreased? Assume $\sigma = 66$ and use $\alpha = 0.05$. CO2

9.81 Marketing and Consumer Behavior After analyzing a database of more than 2 million telephone calls, Phone Power World Unlimited reported that the mean length of all international calls was 295 seconds. Following an extensive advertising campaign, the company expected the length of international calls to increase. A few months after the ads first appeared, a random sample of 48 calls was obtained. The sample mean was $\bar{x} = 306.3$ seconds. Assume that the population standard deviation is 52 seconds. Is there any evidence to suggest that the advertising campaign was successful? Use $\alpha = 0.01$.

9.82 Physical Sciences The U.S. Geological Survey collects water-level measurement data from various locations. Data from La Pine Basin, Oregon, are being used to study changes in the water table. Twelve days were randomly selected, and the water table in feet below land surface on each day is given in the table.

12.30	12.38	12.29	12.37	12.30	12.38
12.35	12.42	12.37	12.46	12.40	12.52

Is there any evidence to suggest that the mean water table is less than 12.4 ft? Assume the underlying distribution is normal, $\sigma = 0.07$, and use $\alpha = 0.025$. WATLEV

9.83 Sports and Leisure The length of a motorboat is traditionally measured in two ways. The distance along the centerline from the outside of the front hull to the rear is the length overall (LOA). The length of waterline, or load waterline (LWL), is the length of the boat on the line where the boat meets the water. Members of the Community Association in the lakeside city of Vermilion, Ohio, are concerned that residents are using bigger boats, contributing to more noise and water pollution. Past registration records indicate the mean LOA for boats allowed on the lake is 35 ft. A random sample of 41 boats on the lake was obtained, and each LOA was carefully measured. The sample mean LOA was $\bar{x} = 36.22$ ft. Assume $\sigma^2 = 5.7$ ft^2.
(a) Is there any evidence to suggest that the mean LOA has increased? Use $\alpha = 0.01$.
(b) Does your answer to part (a) change if $\alpha = 0.1$? Why or why not?

9.84 Economics and Finance The price of coffee has been steadily declining since October 2012. The International Coffee Organization (ICO) reported that the mean price for coffee in July 2018 was 107.20 U.S. cents/lb.[24] A random sample of 34 days was obtained and the composite indicator for Brazilian Natural coffee was recorded for each. The sample mean was $\bar{x} = 120.29$. Assume the population standard deviation is 8.6 U.S. cents/lb. Is there any evidence to suggest that the mean composite indicator for Brazilian Natural is greater than 107.20? Use $\alpha = 0.05$.

9.85 Public Health and Nutrition The U.S. Department of Agriculture suggests that active eight-year-old boys need 2000 calories per day to maintain a healthy weight.[25] An education researcher believes many students in this group do poorly in school because they have an inadequate diet and, therefore, not enough energy. A random sample of academically at-risk eight-year-old boys was obtained, and their caloric intake was carefully measured for one day. The summary statistics were $n = 37$, $\bar{x} = 1889$. Assume $\sigma = 360$.
(a) Is there any evidence to suggest that this group of students has a mean caloric intake below the daily energy requirement? Use $\alpha = 0.05$.
(b) How would your answer to part (a) change if $\alpha = 0.01$?

9.86 Biology and Environmental Science An average lawn has a mean of 21 blades of grass per square inch. A garden store sells an expensive fertilizer designed to transform an average lawn into a lush, thick carpet within three weeks. To test the claim, a random sample of average lawn plots was obtained. Each was treated with the fertilizer according to the instructions on the package. Three weeks later, the density of each plot was measured by counting the blades of grass per square inch. The summary statistics were $n = 32$, $\bar{x} = 22.4$. Is there any evidence to suggest that the fertilizer improves the thickness of an average lawn? Assume $\sigma = 2.7$ and use $\alpha = 0.005$.

9.87 Technology and the Internet Based on test results, Speedtest.net publishes an Internet download index for countries all over the world. In November 2017, the mean fixed broadband download speed for the entire world was 40.11 Mbps. In August 2018, a random sample of countries was obtained and the download speed for each was recorded.[26] Assume that $\sigma = 32$. Is there any evidence to suggest that the mean world download speed has increased? Use $\alpha = 0.01$. SPEED

9.88 Public Policy and Political Science The U.S. Department of Health and Human Services defines "response time" as the time from receipt of a report of child neglect to the initial investigation. Once a call to the agency is received and logged in, the clock starts. Face-to-face contact with the alleged victim marks the initial investigation. There is some concern that heavy workloads and poor organization have contributed to a significant increase in the long-term mean response time of 14.0 hr. A random sample of cases was selected, and the response time for each was recorded. Assume $\sigma = 3.2$ and conduct a hypothesis test to determine whether there is any evidence that the mean response time has increased. Use $\alpha = 0.001$. CHILD

9.89 Manufacturing and Product Development A major cause of injuries in highway work zones is weak construction barriers. One federal government road-barrier specification involves the velocity of a front-seat passenger immediately following impact. This impact velocity must be less than 12 m/sec. TSS GmbH, a German company, has just developed a new high-impact road barrier with special absorbing material. In controlled tests, the impact velocity was measured for 12 randomly selected crashes. The sample mean was 11.85 m/sec. Assume the distribution of impact velocities is normal and $\sigma = 0.26$.
(a) Conduct a hypothesis test to determine whether there is sufficient evidence to suggest that the true mean impact velocity is less than 12 m/sec. Use $\alpha = 0.05$.
(b) What is your conclusion if $\alpha = 0.01$?

9.90 Psychology and Human Behavior Many people live under a lot of stress and various time constraints. Consequently, some people run out of the house without their keys and are locked out. Rather than break into their own home, most people in these circumstances call a locksmith. In 2018, the mean service charge for a locksmith was $181.00.[27] A random sample of 26 locksmith service charges in Black Springs, Arkansas, was obtained. The sample mean was $191.33. Assume that the underlying population distribution is normal and that $\sigma = 24$. Is there any evidence to suggest that the mean locksmith service charge in this city is greater than $181? Use $\alpha = 0.05$.

9.91 Travel and Transportation There have been huge advances in technology and manufacturing, but some evidence suggests that airplane flights have actually become longer.[28] One major reason is that most airlines have reduced cruising speeds to save on fuel and, therefore, fuel costs. The historical mean flight time from Madrid to Barcelona is 55 minutes. A random sample of flights on this route was obtained and the flight times (in minutes) were recorded. Is there any evidence to suggest that the mean flight time for this route has increased? Assume $\sigma = 7.1$ and use $\alpha = 0.025$. FLIGHT

9.92 Public Health and Nutrition Egg Beaters is an egg substitute with no fat or cholesterol. The Original Egg Beaters is produced to have 5 g of protein, one fewer gram than a regular egg.[29] To check the protein claim, 18 servings of Egg Beaters were randomly selected and the amount of protein in each was measured. The resulting data are given in the table. EGGBT

5.54	3.83	3.01	5.83	4.64	6.26	4.45	5.71	4.84
4.41	5.62	5.49	5.51	4.11	6.46	6.26	4.95	4.40

Is there any evidence to suggest that the mean amount of protein in a serving of Egg Beaters is less than 5 g? Assume the distribution of protein is normal and $\sigma = 0.95$. Use $\alpha = 0.02$.

9.93 Public Policy and Political Science Residential mailboxes in Des Moines, Iowa, should be installed such that the bottom of the mailbox is 42 in. above the ground. This rule is designed for safety and to accommodate short mail carriers. A random sample of 75 mailboxes in the city was selected. The height of each was carefully measured, and the sample mean was 43.22 in. Assume $\sigma = 7.6$ in. and use $\alpha = 0.05$. Is there any evidence to suggest that the true mean height of mailboxes in Des Moines is different from 42 in.?

9.94 Public Health and Nutrition Iodine is an important nutrient for the human body, especially in hormone development. Milk is a natural source of iodine, and the mean amount in cow's milk is 88 μg/250 mL.[30] Milk from organic farms is reported to have lower concentrations of elements such as zinc, iodine, and selenium. A random sample of 35 gal of organic milk from Humboldt County, California, was obtained and the iodine concentration was measured in each. The sample mean was $\bar{x} = 86.5$. Assume $\sigma = 3.4$. Is there any evidence to suggest that the mean iodine concentration in organic milk is less than 88 μg/mL? Use $\alpha = 0.01$.

9.95 Travel and Transportation The mean commuting time for a Canadian worker is 26.2 min.[31] The mean commuting time varies by city, with Regina commuters spending on average 18 min commuting and Montreal commuters 30 min. A random sample of Toronto commuters was obtained and the commuting time for each (in minutes) was recorded. Assume $\sigma = 6$ and use $\alpha = 0.025$. Is there any evidence to suggest that the mean commuting time for people living and working in Toronto is greater than 26.2 min? COMMUTE

Extended Applications

9.96 Type II Errors The four parts of a hypothesis test concerning a population mean from a normal population are as follows:

$H_0: \mu = 50$

$H_a: \mu > 50$

TS: $Z = \dfrac{\bar{X} - \mu_0}{\sigma/\sqrt{n}}$

RR: $Z \geq z_\alpha$

Assume the sample size is $n = 25$; $\sigma = 7.5$, and $\alpha = 0.01$.
(a) Find the probability of a type II error for the alternative mean $\mu_a = 54$; that is, find $\beta(54)$.
(b) Find $\beta(55)$ and $\beta(56)$.
(c) Repeat parts (a) and (b) for $\alpha = 0.025$.

9.97 Fuel Consumption and Cars Biodiesel fuels are made from vegetable oils or animal fats and may be used instead of conventional diesel fuel. One advantage of using biodiesel is a possible decrease in regulated emissions, specifically total hydrocarbons (HC). Using the heavy-duty transient Federal Test Procedure (FTP), the mean HC emission is

0.23 g/hp-hr (grams per horsepower-hour), with standard deviation $\sigma = 0.07$. A random sample of heavy-duty engines was obtained, and each was tested with biodiesel fuel. The resulting HC measurements are given in the table. 📊 **FUEL**

0.17	0.18	0.02	0.18	0.19	0.01	0.16	0.06
0.10	0.24	0.13	0.21	0.29	0.10	0.15	0.23
0.14	0.23	0.14	0.30	0.06	0.18	0.24	0.21
0.10	0.24						

(a) Suppose the underlying population is normal. Is there any evidence to suggest that the use of biodiesel has decreased the mean level of HC emissions? Use $\alpha = 0.01$.

(b) Suppose a company is thinking about building a $20 million biodiesel fuel production facility. The company will invest the money only if there is overwhelming evidence in favor of decreased HC emissions. Which error (type I or type II) is more important to the fuel company? Would the company prefer a smaller or a larger significance level? Justify your answer.

9.98 Manufacturing and Product Development The left outside panel of an apartment-size, frost-free refrigerator is designed to have width 23.625 in. Each hour, 10 such panels are randomly selected from the assembly line and carefully measured. If there is any evidence that the mean width is different from 23.625 in., the assembly line is shut down for cleaning and inspection. Suppose the distribution of panel widths is normal and $\sigma = 0.15$.

(a) During a specific hour of operation, $\bar{x} = 23.7$ in. Should the assembly line be shut down? Use $\alpha = 0.05$.

(b) Using $\alpha = 0.05$, find the critical values in the $\bar{X}$ distribution. (*Note:* There are two critical values, because this is a two-sided hypothesis test.)

9.99 Manufacturing and Product Development A manufacturer claims the weight of a package of its unsalted pretzels is (at least) 15.5 oz. To test this claim, a consumer group randomly selected 250 packages and carefully weighed the contents of each. The sample mean was $\bar{x} = 15.45$ oz.

(a) Conduct a one-sided, left-tailed hypothesis test to see whether there is any evidence that the true mean weight of the pretzel packages is less than 15.5 oz. Use $\alpha = 0.01$ and assume $\sigma = 0.26$.

(b) If your conclusion in part (a) is to reject the null hypothesis, explain how this conclusion is possible even though the sample mean, 15.45, is so close to the hypothesized mean, 15.5. If your conclusion in part (a) is to not reject H_0, check those calculations one more time.

9.100 Biology and Environmental Science Lake Vostok is Antarctica's largest and deepest subsurface lake. It is buried under approximately 2 mi of ice, and scientists believe it contains evidence of thousands of tiny organisms and fish. The mean depth of the lake is believed to be 432 m.[32] A random sample of the depth of this lake was obtained. Assume the underlying distribution is normal and that $\sigma = 30$ meters. 📊 **EX9.100**

(a) Conduct a hypothesis test to determine whether there is any evidence to suggest that the mean depth is less than 432 m. Use $\alpha = 0.01$.

(b) Find the probability of a type II error if the true mean depth of the lake is 403 m; that is, find $\beta(403)$.

(c) Carefully sketch a graph illustrating the probability found in part (b) using density curves for $\bar{X}$.

(d) Find $\beta(398)$.

9.101 Marketing and Consumer Behavior The water-treatment facility owners in Lake Havasu City, Arizona, recently proposed changes to all user fees. There is a treatment capacity fee, a connection fee, and a monthly minimum fee. All prices have been set assuming the mean monthly usage is 714 ft^3 per household. To check this claim, a random sample of 16 homes was obtained and the monthly usage for each home was recorded. The sample mean was 601.2 ft^3. Assume the distribution of monthly water usage per household is normal, with standard deviation 283 ft^3.

(a) Conduct a two-sided hypothesis test to determine whether there is any evidence that the mean monthly water usage is different from 714 ft^3. Use $\alpha = 0.05$.

(b) If your conclusion in part (a) is to not reject the null hypothesis, explain how this conclusion is possible even though the sample mean, 601.2, is so far away from the hypothesized mean, 714. If your conclusion in part (a) is to reject the null hypothesis, try checking your calculations one more time.

9.102 Marketing and Consumer Behavior The Carpet Corner in Gladstone, Missouri, offers a variety of carpets, wood flooring, and tiles. Sales records indicate that the mean amount of carpet installed in a wall-to-wall carpeted residential home by crews from this store is 1250 ft^2. The store manager believes that when there is a sale, customers translate the savings into carpeting a larger area. A random sample of 45 wall-to-wall carpet purchases was selected during a sale. The sample mean was $\bar{x} = 1305$ ft^2; assume that $\sigma = 155$ ft^2.

(a) Conduct a one-sided, right-tailed test of $H_0: \mu = 1250$ versus $H_a: \mu > 1250$. Is there any evidence to suggest that the true mean square footage of carpet purchased is larger during a sale? Use $\alpha = 0.01$.

(b) Find the probability of a type II error if the true mean square footage of carpet purchased during a sale is $\mu_a = 1330$; that is, find $\beta(1330)$.

(c) Carefully sketch a graph that includes the distribution of $\bar{X}$ if $\mu = 1250$ and if $\mu = 1330$. Shade in the areas that correspond to the probability of a type I error (0.01) and to the probability of a type II error found in part (b).

9.103 Physical Sciences Kiln-dried solid grade A teak wood should have a moisture content of no more than 12%. A furniture company recently purchased a large shipment of this wood for use in constructing dining room sets. Thirty-seven pieces of teak wood were randomly selected and carefully measured for moisture content. The sample mean moisture content was $\bar{x} = 12.3\%$. Assume $\sigma = 1.25$.

(a) Is there any evidence to suggest that the true population mean moisture content is greater than 12%? Use $\alpha = 0.01$.

(b) Find the probability of a type II error if the true population mean moisture content is 12.2%; that is, find $\beta(12.2)$.

9.104 Fuel Consumption and Cars TB Wood's produces many makes and models of automobile drive belts. One popular poly V-belt is designed to have a length of 1050 mm. To ensure that the belts meet this specification, 18 are randomly selected from the assembly line every hour and carefully measured. If there is any evidence that the population mean length is different from 1050 mm, the entire line is shut down for inspection.

(a) Assume the distribution of belt lengths is normal and $\sigma = 3.7$ mm. In a two-sided hypothesis test of $H_0: \mu = 1050$ versus $H_a: \mu \neq 1050$, find the critical values in the $\overline{X}$ distribution. Use $\alpha = 0.01$.

(b) In a sample of 18 belts, suppose $\bar{x} = 1049$. Should the assembly line be shut down? Justify your answer using the Z distribution and the appropriate $\overline{X}$ distribution.

9.105 Manufacturing and Product Development The 2013 Kawasaki Jet Ski STX-15F has strong acceleration, low emissions, as well as a 1498 cc engine with digital fuel injection, and is designed to have a load capacity of 496 lb.[33] If the passenger and gear weight exceed this capacity, the jet ski will not function properly and there will be a severe strain on the engine. To determine whether customers are following the load capacity recommendation, a random sample of these jet skis is obtained from marinas around the country and each load is carefully weighed as owners prepare for a ride. The summary statistics are $n = 33$ and $\bar{x} = 498.42$. Assume $\sigma = 46.7$.

(a) Is there any evidence to suggest that the true mean weight of passengers and gear is greater than 496 lb? Use $\alpha = 0.01$.

(b) Suppose the engineers have determined that the watercraft will overheat rapidly if the load is 525 lb. Find the probability of a type II error if the true mean weight of passengers and gear is 525 lb; that is, find $\beta(525)$.

Challenge Problems

9.106 The Power of a Test The probability of a type II error β represents the likelihood of accepting the null hypothesis when the alternative hypothesis is true. The power of a statistical test is $\pi = 1 - \beta$, the probability of (correctly) rejecting the null hypothesis, or detecting a difference in the hypothesized value of the population parameter.

A hot torsion test is used to determine the workability of a metal. Suppose a carbon steel rod is designed to fail (break) with mean axial load of 800 newtons (N) under certain temperature and speed conditions. A random sample of 25 rods is obtained, and the axial load failure is measured for each. Suppose the underlying distribution of axial load is normal and $\sigma = 50$.

(a) Consider a hypothesis test of $H_0: \mu = 800$ versus $H_a: \mu < 800$ with $\alpha = 0.025$. Find the probability of a type II error and the power of this test for $\mu = 775$; that is, find $\beta(775)$ and $\pi(775)$.

(b) Use technology to compute the power, $\pi(\mu_a)$, for $\mu_a = 730, 735, 740, \ldots, 800$.

(c) Use the values from part (b) to carefully sketch a plot of $\pi(\mu_a)$ versus μ_a. The resulting plot is called a power curve.

9.4 p Values

Significance Level Dilemma

The last piece of a hypothesis test, the rejection region, establishes a firm decision rule based on the value of the test statistic. If the value of the test statistic lies in the rejection region, we reject H_0. If not, we cannot reject the null hypothesis. Using a fixed rejection region associated with a specific value (of the significance level) can lead to a peculiar dilemma. Consider the following example.

EXAMPLE 9.11 Sleep Deprivation

Research suggests that most people need at least 7 hr of sleep each night to function well the next day. Some research suggests that lack of sleep can adversely affect the immune system, increase the risk of respiratory diseases, and affect body weight, blood pressure, and hormone production.[34] Suppose adults need 7.0 hr of sleep per night. A random sample of 32 adults was obtained, and the sleeping time for each was recorded. The sample mean was $\bar{x} = 6.8$ hr, and assume $\sigma = 0.67$ hr. Is there any evidence to suggest that the mean sleeping time for adults is less than 7 hr?

Solution

STEP 1 The assumed mean is $\mu = 7.0 \, (= \mu_0)$; $n = 32$ and $\sigma = 0.67$.

We are searching for any evidence to suggest that the mean sleeping time for adults is less than 7.0 hr. This is a one-sided, left-tailed test.

The underlying population distribution is unknown, but the sample size is large ($n = 32 \geq 30$). The hypothesis test concerning a population mean μ when σ is known is relevant.

STEP 2 No value of α is given. The first three parts of the hypothesis test are

$H_0: \mu = 7$

$H_a: \mu < 7$

TS: $Z = \dfrac{\overline{X} - \mu_0}{\sigma/\sqrt{n}}$

The value of the test statistic is

$$z = \dfrac{\overline{x} - 7.0}{\sigma/\sqrt{n}} = \dfrac{6.8 - 7.0}{0.67/\sqrt{32}} = -1.6886$$

STEP 3 The following table shows the resulting rejection region and the conclusion for various values of α.

α	Rejection region	Conclusion
0.10	$Z \leq -z_{0.10} = -1.2816$	$z = -1.6886$ lies in the rejection region. We reject the null hypothesis at the $\alpha = 0.10$ level.
0.05	$Z \leq -z_{0.05} = -1.6449$	$z = -1.6886$ lies in the rejection region. We reject the null hypothesis at the $\alpha = 0.05$ level.
0.025	$Z = -z_{0.025} = -1.9600$	$z = -1.6886$ does not lie in the rejection region. We cannot reject the null hypothesis at the $\alpha = 0.025$ level.
0.01	$Z \leq -z_{0.01} = -2.3263$	$z = -1.6886$ does not lie in the rejection region. We cannot reject the null hypothesis at the $\alpha = 0.01$ level.

STEP 4 In this example, the decision to reject or not reject H_0 depends on the value of α. To avoid this dilemma, an alternative method for reporting the result of a hypothesis test is often used. This technique involves computing a tail probability, or ***p* value**.

p Value Definition

Definition
The ***p* value**, denoted p, for a hypothesis test is the smallest significance level (value of α) for which the null hypothesis, H_0, can be rejected.

The symbol p is used to represent both the population proportion of successes and the p value. The context in which the notation is used implies the appropriate concept.

The p value is simply a tail probability, and the tail is determined by the alternative hypothesis. Consider a hypothesis test concerning a population mean μ when σ is known. Suppose either the underlying population is normal or n is large, and let z be the value of the test statistic. The table presents the probability definition of a p value for each alternative hypothesis.

Alternative hypothesis	Probability definition	Illustration
$H_a: \mu > \mu_0$	$p = P(Z \geq z)$	Figure 9.10
$H_a: \mu < \mu_0$	$p = P(Z \leq z)$	Figure 9.11
$H_a: \mu \neq \mu_0$	$p/2 = P(Z \geq z)$ if $z \geq 0$	Figure 9.12
	$p/2 = P(Z \leq z)$ if $z < 0$	Figure 9.13

The p value conveys the strength of the evidence in favor of the alternative hypothesis. For a small p value, the value of the test statistic lies far out in a tail of the distribution.

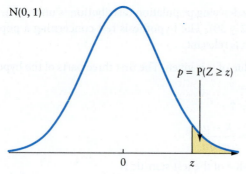

Figure 9.10 Illustration of the *p* value for $H_a: \mu > \mu_0$.

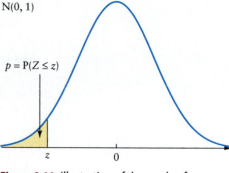

Figure 9.11 Illustration of the *p* value for $H_a: \mu < \mu_0$.

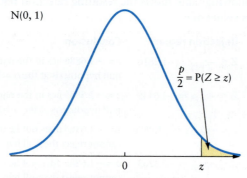

Figure 9.12 Illustration of the *p* value for $H_a: \mu \neq \mu_0, z \geq 0$.

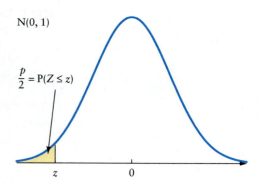

Figure 9.13 Illustration of the *p* value for $H_a: \mu \neq \mu_0, z < 0$.

That is, values of the test statistic in this tail of the distribution are very unlikely and represent more evidence in favor of H_a. Small values of *p* are usually $p \leq 0.05$.

If α is the significance level of the hypothesis test, the conclusion is usually written in one of two ways.

1. If $p \leq \alpha$: Reject the null hypothesis. There is evidence to suggest that H_0 is false at the $p =$ (observed *p* value) level of significance.
2. If $p > \alpha$: Do not reject the null hypothesis. There is no evidence to suggest that the null hypothesis is false.

Consider a one-sided, right-tailed hypothesis test concerning a population mean μ when σ is known ($H_a: \mu > \mu_0$). Suppose either the underlying population is normal or *n* is large, α is the significance level, and *z* is the value of the test statistic. Recall the definition of a critical value: $\alpha = P(Z \geq z)$. **Figure 9.14** and **Figure 9.15** demonstrate

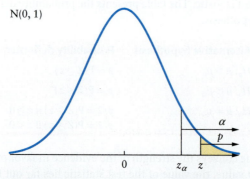

Figure 9.14 If $p \leq \alpha$, then $z \geq z_\alpha$. Conclusion: Reject H_0.

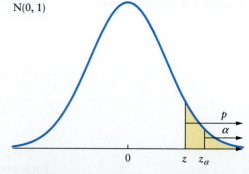

Figure 9.15 If $p > \alpha$, then $z < z_\alpha$. Conclusion: Do not reject H_0.

the relationship between the p value and a fixed significance level. These figures also illustrate the definition of a p value, that is, the smallest significance level for which H_0 can be rejected. If the significance level (α) is less than p, then $z < z_\alpha$, and we cannot reject H_0.

p Value Examples

EXAMPLE 9.12 Hyperloop One

Elon Musk, the billionaire founder of PayPal, Tesla, and SpaceX, has been developing a transportation system called the Hyperloop, an alternative to bullet trains. Passengers would ride in a capsule and travel at a speed of approximately 670 mph in a vacuum tube, or tunnel.[35] A 500-m test tunnel, known as DevLoop, has been constructed in the Nevada desert. Suppose 20 random tests are selected and the speed of the capsule is carefully measured for each. The sample mean is $\bar{x} = 660.1$. Assume the distribution of capsule speed is normal, with $\sigma = 25$. Conduct a one-sided, left-tailed hypothesis test with $\alpha = 0.05$ and compute the p value. Is there any evidence to suggest that the true mean speed is less than 670 mph?

Solution

STEP 1 The assumed mean is $\mu = 670\ (= \mu_0)$; $n = 20$ and $\sigma = 25$.

This is a one-sided, left-tailed test. The underlying population is assumed to be normal. The hypothesis test concerning a population mean μ when σ is known is relevant.

STEP 2 The value of the test statistic will be used to compute a p value. The first three parts of the hypothesis test are

$H_0: \mu = 670$

$H_a: \mu < 670$

TS: $Z = \dfrac{\overline{X} - \mu_0}{\sigma/\sqrt{n}}$

The value of the test statistic is

$$z = \dfrac{\bar{x} - 670}{\sigma/\sqrt{n}} = \dfrac{660.1 - 670}{25/\sqrt{20}} = -1.771$$

STEP 3 The p value is a left-tail probability, illustrated in **Figure 9.16**.

$p = P(Z \le -1.771) \approx P(Z \le -1.77)$ Definition of p value for $H_a: \mu < \mu_0$.

$= 0.0384$ Use Appendix Table 3.

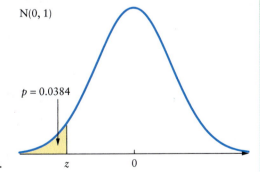

Figure 9.16 Illustration of the p value for the Hyperloop example.

Because $p = 0.0384 \le 0.05\ (= \alpha)$, we reject the null hypothesis. There is evidence to suggest that the true mean capsule speed is less than 670 mph at the $p = 0.0384$ level of significance. **Figure 9.17** shows a technology solution.

Figure 9.17 R calculations to find the value of the test statistic and the *p* value. The slight difference in the *p* values is because the results using technology are more accurate than rounding and referring to Appendix Table 3.

```
> mu_zero <- 670
> xbar <- 660.1
> n <- 20
> sigma <- 25
>
> z <- (xbar - mu_zero) / (sigma / sqrt(n))
> p_value <- pnorm(z)
>
> round(cbind(z,p_value),4)
           z p_value
[1,] -1.771  0.0383
```

TRY IT NOW Go to Exercises 9.123 and 9.131

EXAMPLE 9.13 Rotor Speed

ROTOR The La Grande-1 generating station on the La Grande Rivière in Quebec was constructed by 6000 workers using 640,000 m^3 of concrete. As water flows into James Bay, 12 generating unit rotors spin at a rate of 204 km/hr.[36] If there is any variation in the rotor spin rate, then a generating unit must be shut down for safety and inspection. A random sample of days was obtained and the rotor spin rate was recorded on each day at noon. The data are given in the table.

208	200	206	201	217	196	219	217	197
198	206	197	224	200	206	212	210	208
215	203	221	187	220	207	192		

Assume that the underlying population is normal and that $\sigma = 10$. Is there any evidence to suggest that the true mean rotor speed is different from 204 km/hr? Use $\alpha = 0.05$ and the *p* value associated with the test statistic to justify your answer.

Solution

STEP 1 The assumed mean is $\mu = 204 \, (= \mu_0)$; $n = 25$ and $\sigma = 10.0$.

The phrase *is different* means this is a two-sided test. The distribution of rotor speed is assumed to be normal. The hypothesis test concerning a population mean μ when σ is known is relevant.

STEP 2 The first three parts of the hypothesis test are

$H_0: \mu = 204$

$H_a: \mu \neq 204$

TS: $Z = \dfrac{\overline{X} - \mu_0}{\sigma/\sqrt{n}}$

STEP 3 The sample mean is

$$\overline{x} = \frac{1}{25}(208 + 200 + \cdots + 192) = 206.68$$

The value of the test statistic is

$$z = \frac{\overline{x} - \mu_0}{\sigma/\sqrt{n}} = \frac{206.68 - 204}{10.0/\sqrt{25}} = 1.34$$

STEP 4 Since this is a two-sided test and the value of the test statistic ($z = 1.34$) is positive, $p/2$ is a right-tail probability (see **Figure 9.18**).

$$p/2 = P(Z \geq 1.34)$$
$$= 0.0901$$
$$p = 2(0.0901) = 0.1802$$

Definition of a p value for $H_a: \mu \neq \mu_0$.

Use Appendix Table 3.

Solve for p.

Figure 9.19 shows a technology solution.

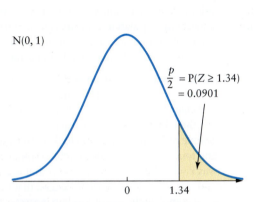

Figure 9.18 Illustration of the p value for the rotor speed example.

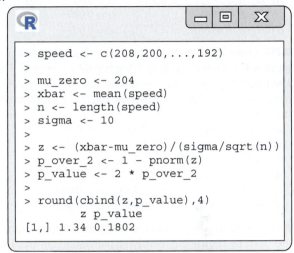

Figure 9.19 R calculations to find the value of the test statistic and the p value.

Because $p = 0.1802 > 0.05 (=\alpha)$, we do not reject the null hypothesis. There is no evidence to suggest that the mean rotor speed is different from 204 km/hr.

TRY IT NOW Go to Exercises 9.121 and 9.125

Section 9.4 Exercises

Concept Check

9.107 True or False In a two-sided hypothesis test, the p value may be greater than 1.

9.108 True or False In a one-sided hypothesis test, the p value may be greater than 0.5.

9.109 True or False A small p value suggests more evidence in favor of H_a.

9.110 True or False The p value is the smallest significance level for which the null hypothesis can be rejected.

9.111 Short Answer If α is the significance level of a hypothesis test and $p \leq \alpha$, what is the conclusion?

9.112 Short Answer Suppose the p value associated with a hypothesis test is $p = 0.0001$. What is the conclusion and why?

Practice

9.113 For each p value and significance level, determine whether the null hypothesis would be rejected or not rejected.
(a) $p = 0.067$, $\alpha = 0.05$
(b) $p = 0.0043$, $\alpha = 0.01$
(c) $p = 0.159$, $\alpha = 0.05$
(d) $p = 0.0260$, $\alpha = 0.025$
(e) $p = 0.001$, $\alpha = 0.05$
(f) $p = 0.1770$, $\alpha = 0.01$

9.114 Consider a hypothesis test concerning a population mean with σ known, n large, and alternative hypothesis $H_a: \mu > \mu_0$. Find the p value associated with each value of the test statistic.
(a) $z = 1.87$
(b) $z = 2.55$
(c) $z = 1.20$
(d) $z = 0.57$
(e) $z = 3.88$
(f) $z = -1.14$

9.115 Consider a hypothesis test concerning a population mean with σ known, underlying population normal, and alternative hypothesis $H_a: \mu < \mu_0$. Find the p value associated with each value of the test statistic.
(a) $z = -2.05$
(b) $z = -1.43$
(c) $z = -3.22$
(d) $z = -0.67$
(e) $z = -4.58$
(f) $z = 0.25$

9.116 Consider a hypothesis test concerning a population mean with σ known, underlying population normal, and alternative hypothesis $H_a: \mu \neq \mu_0$. Find the p value associated with each value of the test statistic.
(a) $z = -1.77$
(b) $z = 1.43$
(c) $z = 2.58$
(d) $z = -0.37$
(e) $z = 3.58$
(f) $z = 0.85$

9.117 Consider a hypothesis test concerning a population mean with σ known and n large. For the alternative hypothesis, value of the test statistic, and significance level, find the p value and determine whether H_0 is rejected or not rejected.
(a) $H_a: \mu > 12.5$, $z = 1.43$, $\alpha = 0.001$
(b) $H_a: \mu < -0.56$, $z = -2.05$, $\alpha = 0.05$
(c) $H_a: \mu \neq 1200$, $z = 1.75$, $\alpha = 0.10$
(d) $H_a: \mu > 37.7$, $z = 3.11$, $\alpha = 0.01$
(e) $H_a: \mu < 52.68$, $z = -1.16$, $\alpha = 0.025$
(f) $H_a: \mu \neq 46.68$, $z = -2.35$, $\alpha = 0.001$

9.118 Consider a hypothesis test concerning a population mean with σ known and underlying population normal. For the alternative hypothesis, value of the test statistic, and significance level, find the p value and determine whether H_0 is rejected or not rejected.
(a) $H_a: \mu > 3.14$, $z = 2.52$, $\alpha = 0.05$
(b) $H_a: \mu > 9.80$, $z = 1.39$, $\alpha = 0.05$
(c) $H_a: \mu < 186{,}000$, $z = -2.28$, $\alpha = 0.01$
(d) $H_a: \mu < 4.135$, $z = 0.17$, $\alpha = 0.01$
(e) $H_a: \mu \neq 1.62$, $z = -1.63$, $\alpha = 0.001$
(f) $H_a: \mu \neq 0.671$, $z = 2.96$, $\alpha = 0.001$

9.119 Consider a hypothesis test concerning a population mean with σ known and underlying population normal. For the hypothesis test indicated, find the p value and determine whether H_0 is rejected or not rejected.

	H_0	H_a	$\bar{x}$	σ	n	α
(a)	$\mu = 10$	$\mu > 10$	11.50	7.56	18	0.05
(b)	$\mu = 2.718$	$\mu < 2.718$	2.60	0.56	21	0.01
(c)	$\mu = 57.72$	$\mu \neq 57.72$	56.42	1.58	14	0.01
(d)	$\mu = -16.18$	$\mu > -16.18$	2.35	21.23	8	0.001
(e)	$\mu = 273$	$\mu < 273$	275.80	17.80	15	0.05
(f)	$\mu = 6.63$	$\mu \neq 6.63$	7.17	1.08	27	0.05

Applications

9.120 Public Health and Nutrition High levels of the chemical dioxin, which may be ingested through food and air, can cause diabetes, immune disorders, and cancer. The U.S. EPA has set a maximum acceptable level of dioxin in blood at 10 ppt (parts per trillion). Residents living near an old incinerator known to produce dioxin air pollution have complained of many health problems. A random sample of people living within a 10-mi radius of the facility was obtained, and the dioxin level (in ppt) in the blood of each was measured. Is there any evidence to suggest that the mean dioxin level in the blood of residents living near the incinerator is greater than 10 ppt? Assume $\sigma = 2.66$, use $\alpha = 0.05$, and compute the p value. DIOXIN

9.121 Sports and Leisure The National Football League rates each quarterback after every game according to a formula that involves completions, passing yards, touchdown passes, and interceptions. The mean quarterback rating in 2016 was 91.03[37]; assume $\sigma = 11.06$. There is some speculation that rule changes have affected quarterback ratings. A random sample of quarterbacks was obtained and their 2017 ratings were recorded.[38] Is there any evidence to suggest that the true mean quarterback rating has changed? Use $\alpha = 0.05$ and compute the p value. QBACK

9.122 Medicine and Clinical Studies The active ingredient in antiseptic liquid bandages is 8-hydroxyquinoline (8h). This chemical can be harmful when it is ingested through simple skin contact, by swallowing, or inhalation. Suppose the list of ingredients on a 3M Nexcare liquid bandage indicates that the volume of 8h is 1%. A random sample of eight bottles of this product was obtained, and each was analyzed to obtain the volume of 8h. The sample mean was $\bar{x} = 1.025\%$. Assume the underlying population is normal and $\sigma = 0.04$. Is there any evidence to suggest that the true mean percentage of 8h is different from 1%? Use $\alpha = 0.01$ and compute the p value.

9.123 Fuel Consumption and Cars Frontier Toyota in Valencia, California, offers a multi-point inspection for any car and claims that the service department will finish the job in less than 30 min. A local newspaper reporter selected a random sample of 58 customers who took advantage of the dealer's offer and found the sample mean time to complete the safety checkup was $\bar{x} = 32.2$ min. Assume $\sigma = 5.7$ and use $\alpha = 0.01$.
(a) Is there any evidence to suggest that the true mean time to complete the safety checkup is greater than 30 min?
(b) Find the p value for the hypothesis test in part (a).

9.124 Sports and Leisure The Der Dachstein mountain in Austria includes a suspension bridge, a sky walk that extends out from a ledge to offer a view of the Alps, and an ice palace. The panorama gondola takes passengers up to a glacier, and rises 1000 m in approximately 6 min.[39] Suppose a random sample of gondola rides was selected and the time (in minutes) to reach the top was recorded for each. The data are given in the table. GNDOLA

| 6.8 | 7.3 | 5.2 | 6.1 | 6.1 | 6.7 | 7.4 | 4.7 | 5.3 |
| 6.4 | 6.2 | 7.1 | 5.1 | 4.1 | 5.9 | 6.8 | 8.4 | 7.3 |

Assume the underlying population is normal and $\sigma = 1.3$. Is there any evidence to suggest that the mean cable car time is greater than 5.5 min? Use $\alpha = 0.05$ and compute the p value to draw a conclusion.

9.125 Manufacturing and Product Development Nitterhouse Masonry Products, in Chambersburg, Pennsylvania, produces architectural concrete masonry products. The Dover is the largest block in a certain product line, is used primarily for residential retaining walls, and is manufactured to weigh 40 lb. A quality control inspector for the company randomly selected eight blocks, and the weight (in pounds) of each is given in the table. DOVER

| 43.4 | 39.8 | 42.3 | 42.6 | 41.0 | 39.8 | 39.6 | 38.8 |

Assume the underlying population is normal and $\sigma = 2$ lb. Is there any evidence to suggest that the true mean weight of the blocks is different from 40 lb? Use $\alpha = 0.05$ and compute the p value to draw a conclusion.

9.126 Travel and Transportation For takeoff, Boeing engineers claim the 787-10 Dreamliner needs a runway of, on average, 6300 ft. To check this claim, officials from the FAA randomly selected 787-10 flights and measured the length (in feet) of the runway needed for takeoff. Assume $\sigma = 375$ and use $\alpha = 0.05$. Is there any evidence to suggest that the true mean runway length needed for takeoff of a Boeing 787-10 is different from 6300 ft? Use the p value to draw a conclusion. **RUNWAY**

9.127 Manufacturing and Product Development Rustic Shingle is a four-way interlocking aluminum roofing system designed to withstand winds up to 120 mph. Each square is manufactured to weigh 42 lb. A random sample of Rustic Shingles was obtained and the weight of each is given in the table. **SHINGLE**

41.6	41.1	40.0	42.8	43.3	38.3	41.4	40.2
45.7	43.4	42.1	42.6	39.9	42.8	40.8	38.2
42.0	37.4	36.4	43.4	39.8	43.0	41.5	44.9
44.3	42.4	41.0	39.8	41.7	38.0	41.7	40.5

Assume $\sigma = 2.5$. Is there any evidence to suggest that the true mean weight of a square is different from 42 lb? Use $\alpha = 0.05$ and compute the p value to draw a conclusion.

9.128 Manufacturing and Product Development The Wood Flooring Manufacturers Association claims a new product restores a high-gloss finish to old wood floors. A gloss meter is used to measure the shine of a wood floor, and a high-gloss finish should have a rating of at least 80 units. To check the manufacturer's claim, the sealer and the finish were applied to 35 randomly selected wood floors and the gloss was measured after each application dried. The sample mean gloss was $\bar{x} = 78.6$. Assume $\sigma = 3.45$. Is there any evidence to suggest that the manufacturer's claim is false? Use $\alpha = 0.01$ and compute the p value to draw a conclusion.

9.129 Physical Sciences Hydrothermal vents form as seawater percolates down through fissures in the ocean crust near subduction zones. The cold seawater is heated by hot magma and emerges from the vent full of fine-grained minerals. A random sample of vent fields was obtained and the depth of each (in meters below sea level) was recorded.[40] Vent fields deeper than 2500 m are costly and difficult to explore. Is there any evidence to suggest that the mean depth of vent fields is greater than 2500 m? Use $\alpha = 0.01$ and $\sigma = 860$, and compute the p value to draw a conclusion. **VENT**

Extended Applications

9.130 Manufacturing and Product Development Arizona state procurement specifications for instant-dry highway paint indicate that the paint must dry in less than 60 sec. A random sample of paint was obtained from a potential seller, and the drying time of each batch was carefully tested. Assume $\sigma = 8.0$ and use $\alpha = 0.05$. **AZHWY**

(a) Is there any evidence to suggest that the mean drying time for this highway paint is less than 60 sec? Use the p value to draw a conclusion.

(b) Using the results from part (a), do you believe there is overwhelming evidence to suggest that the paint dries in less than 60 sec? Why or why not?

(c) Carefully sketch a graph to illustrate the p value in this hypothesis test.

9.131 Biology and Environmental Science A large farm selling round fescue hay bales claims the mean weight per bale is 1600 lb. To check this claim, personnel from the local agricultural extension service randomly selected 25 round bales and carefully weighed each. The sample mean was $\bar{x} = 1595.6$ lb. Assume the underlying population is normal and $\sigma = 23$ lb.

(a) Is there any evidence to suggest that the true mean weight of these round hay bales is less than 1600 lb? Use $\alpha = 0.01$ and compute the p value to draw a conclusion.

(b) Carefully sketch a graph to illustrate the p value in this hypothesis test.

9.132 Public Policy and Political Science A technical panel investigating residential smoke alarms recommends that the sound emitted by the device (during a fire) should be at least 85 decibels (dB) at 10 ft away. In a random sample of 60 smoke alarms from homes in a large city, the sample mean decibel level (at 10 ft away) was $\bar{x} = 84.88$. Assume $\sigma = 5.6$.

(a) Is there any evidence that the true mean decibel level produced by smoke alarms in this city is less than 85? Use $\alpha = 0.01$ and compute the p value to draw a conclusion.

(b) What is the smallest significance level at which this hypothesis test would be significant?

9.133 Physical Sciences Jakarta, the capital of Indonesia, population 10 million, is one of the fastest sinking cities in the world. This isn't surprising since it is built on swampy land and 13 rivers run through the city. North Jakarta has been sinking, on average, 2 cm per month.[41] In late 2018, a random sample of locations in North Jakarta was obtained and the monthly sinking amount was recorded (in centimeters). Assume $\sigma = 0.5$.

(a) Is there any evidence that the true mean sinking amount per month is greater than 2 cm? Use $\alpha = 0.05$ and compute the p value to draw a conclusion. **JAKARTA**

(b) Carefully sketch a graph to illustrate the p value in this hypothesis test.

9.134 Economics and Finance According to the *Financial Post*, the average Canadian non-mortgage debt balance is $22,800 (CAD).[42] However, the credit reporting agency TransUnion claimed that residents of Vancouver have a higher-than-average consumer debt. A random sample of Vancouverites was obtained and the non-mortgage debt of each was recorded. Assume $\sigma = 10,000$. **DEBT**

(a) Is there any evidence to suggest that the true mean non-mortgage debt of Vancouverites is greater than $22,800? Use $\alpha = 0.01$ and compute the p value to draw a conclusion.

(b) Using your results from part (a), do you believe there is overwhelming evidence to suggest that the true mean non-mortgage debt of Vancouverites is greater than $22,800? Why or why not?

(c) The mean non-mortgage debt for people living in Calgary was reported to be $29,000. Is there any evidence to suggest that the true mean non-mortgage debt of Vancouverites is different from the mean of people living in Calgary? Use $\alpha = 0.01$, $\sigma = 10,000$, and compute the p value to draw a conclusion.

9.5 Hypothesis Tests Concerning a Population Mean When σ Is Unknown

One-Sample t Test

In Section 9.3, we examined hypothesis tests concerning a population mean μ. Those tests are based on a standard normal, or Z, distribution and are valid only if σ is known (and either the underlying population is normal or the sample size is large, or both). It is unrealistic to assume that the population standard deviation is known. However, as with confidence intervals, we can assume the underlying population is normal and develop a hypothesis test procedure that is based on the following result (from Section 8.3).

Theorem
Given a random sample of size n from a normal distribution with mean μ, the random variable

$$T = \frac{\overline{X} - \mu}{S/\sqrt{n}}$$

has a t distribution with $\nu = n - 1$ degrees of freedom.

This is simply another kind of standardization, and yes, the distribution of T is characterized by $n - 1$ (not n) degrees of freedom. The general procedure for a hypothesis test concerning μ is summarized next.

Hypothesis Tests Concerning a Population Mean When σ Is Unknown

This is the template for a hypothesis test about a population mean when σ is unknown.

Given a random sample of size n from a normal population with mean μ, a hypothesis test concerning the population mean μ with significance level α has the form

$H_0: \mu = \mu_0$

$H_a: \mu > \mu_0, \quad \mu < \mu_0, \quad$ or $\quad \mu \ne \mu_0$

Remember: Use only one (appropriate) alternative hypothesis and the corresponding rejection region.

TS: $T = \dfrac{\overline{X} - \mu_0}{S/\sqrt{n}}$

RR: $T \ge t_{\alpha, n-1}, \quad T \le -t_{\alpha, n-1}, \quad$ or $\quad |T| \ge t_{\alpha/2, n-1}$

A CLOSER LOOK

1. This procedure is often called a *small-sample* test (concerning a population mean), or simply a t test. As long as the underlying population is normal, this test is valid (and exact) for any sample size n (large or small).

2. Appendix Table 5 presents selected critical values associated with various t distributions. These critical values were used to construct confidence intervals in Section 8.3.

3. Figure 9.20, Figure 9.21, and Figure 9.22 illustrate the rejection region for each alternative hypothesis.

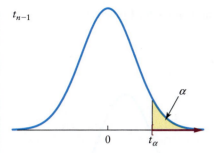

Figure 9.20 Rejection region for $H_a: \mu > \mu_0$.

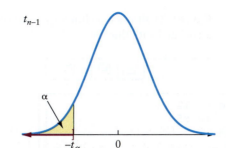

Figure 9.21 Rejection region for $H_a: \mu < \mu_0$.

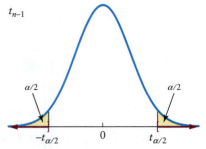

Figure 9.22 Rejection region for $H_a: \mu \neq \mu_0$.

t Test Examples

The next example illustrates this small sample hypothesis test procedure.

EXAMPLE 9.14   Attention Span

According to a study by Microsoft, the attention span of an adult has decreased from the mean of 12 sec found in 2000, some would say largely due to technology.[43] It appears that people no longer want to read through content but prefer short snippets of information. A random sample of 18 adults was obtained and the attention span of each was measured using a standardized test. The summary statistics were $\bar{x} = 10.85$ sec and $s = 2.1$ sec. Is there any evidence to suggest that the true mean attention span is less than 12 sec? Assume the underlying distribution is normal, and use $\alpha = 0.05$.

Solution

This example involves a hypothesis test about a population mean when σ is unknown. Use the template for a one-sided, left-tailed t test about μ. The underlying population is assumed to be normal, and σ is unknown. Determine the appropriate alternative hypothesis and the corresponding rejection region, find the value of the test statistic, and draw a conclusion.

STEP 1 The assumed mean is $\mu = 12$ ($= \mu_0$); $n = 18$, $s = 2.1$, and $\alpha = 0.05$.

We are looking for evidence to suggest that the mean attention span is less than the previous mean. The relevant alternative hypothesis is one-sided, left-tailed.

The underlying population is assumed to be normal, but σ is unknown. A t test is appropriate.

STEP 2 The four parts of the hypothesis test are

$H_0: \mu = 12$

$H_a: \mu < 12$

TS: $T = \dfrac{\bar{X} - \mu_0}{S/\sqrt{n}}$

RR: $T \leq -t_{\alpha, n-1} = t_{0.05, 17} = -1.7396$

STEP 3 The value of the test statistic is

$$t = \frac{\bar{x} - \mu_0}{s/\sqrt{n}} = \frac{10.85 - 12}{2.1/\sqrt{18}} = -2.3234$$

STEP 4 Because -2.3234 lies in the rejection region, we reject the null hypothesis at the $\alpha = 0.05$ level. There is evidence to suggest that the mean attention span of adults has decreased from 12 sec.

Figure 9.23 shows a technology solution and Figure 9.24 illustrates the p value associated with this test.

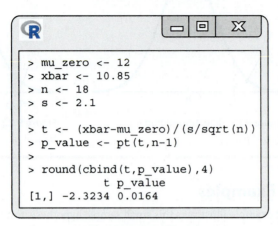

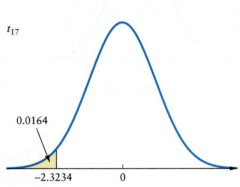

Figure 9.23 R calculations to find the critical value, the value of the test statistic, and the p value.

Figure 9.24 p value illustration: $p = P(T \leq -2.3234) = 0.0164 \leq 0.05 = \alpha$.

TRY IT NOW Go to Exercises 9.157 and 9.161

Here is another example of a hypothesis test concerning μ when the population variance is unknown.

EXAMPLE 9.15 Water Park Water

Schlitterbahn Waterparks & Resorts

Most water parks use a recycling system so that the only water loss is due to evaporation and splashing. Despite a sophisticated recycling system, the Schlitterbahn Waterpark in New Braunfels, Texas, has informed the city water department of its need for 250,000 gal of water per day. The city water department selected a random sample of days, and the park's water usage (in thousands of gallons) on each day is given in the table.

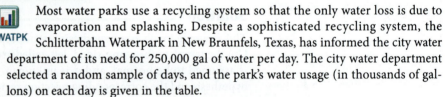

140.1	244.6	244.1	270.1	269.6	201.0	234.3	292.3
205.7	242.3	263.0	219.0	233.3	229.0	303.5	337.4
264.0	248.6	260.5	210.4	236.9			

Is there any evidence to suggest that the mean water usage is different from 250,000 gal? Assume the underlying population is normal and use $\alpha = 0.01$.

Solution

STEP 1 The assumed mean is 250 ($= \mu_0$) thousand gallons; $n = 21$ and $\alpha = 0.01$.

The water department is looking for water usage different from 250,000 gal. Therefore, the relevant alternative hypothesis is two-sided.

The underlying population is assumed to be normal, and σ is unknown. A t test is appropriate.

STEP 2 The four parts of the hypothesis test are

$H_0: \mu = 250$

$H_a: \mu \neq 250$

TS: $T = \dfrac{\overline{X} - \mu_0}{S/\sqrt{n}}$

RR: $|T| \geq t_{\alpha/2,\, n-1} = t_{0.005, 20} = 2.8453 \quad (T \leq -2.8453 \text{ or } T \geq 2.8453)$

STEP 3 The sample mean is

$$\bar{x} = \frac{1}{21}(140.1 + 244.6 + \cdots + 236.9) = 245.2238$$

Find the standard deviation.

$$s^2 = \frac{1}{20}\left[1{,}296{,}137.59 - \frac{1}{21}(5149.7)^2\right] = 1665.4269$$

$$s = \sqrt{1665.4269} = 40.8096$$

The value of the test statistic is

$$t = \frac{\bar{x} - \mu_0}{s/\sqrt{n}} = \frac{245.2238 - 250}{40.8096/\sqrt{21}} = -0.5363$$

STEP 4 The value of the test statistic, $t = -0.5363$, does not lie in the rejection region. We cannot reject the null hypothesis. There is no evidence to suggest park water usage is different from 250,000 gal per day.

Figure 9.25 shows a technology solution and **Figure 9.26** illustrates the p value associated with this test.

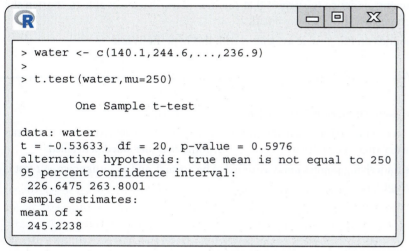

Figure 9.25 Use the R function `t.test()` to conduct a t test on vectors of data.

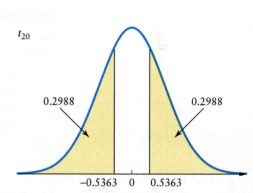

Figure 9.26 p value illustration:
$p = 2P(T \leq -0.5363) = 0.5976 \geq 0.01 = \alpha$. ∎

TRY IT NOW Go to Exercises 9.159 and 9.163

p Value Bounds

Most technology solutions provide the exact p value associated with this and other hypothesis tests. However, the table listing critical values for various t distributions is very limited. Therefore, when using this table, the best we can do is bound the p value associated with a hypothesis test.

How to Bound the *p* Value for a *t* Test

Suppose t is the value of the test statistic in a one-sided hypothesis test.

1. Select the row in Appendix Table 5 that corresponds to $n-1$, the number of degrees of freedom associated with the test.
2. Place $|t|$ in this ordered list of critical values.
3. To compute p:
   (a) If $|t|$ is between two critical values in the $n-1$ row, then the p value is bounded by the corresponding significance levels.
   (b) If $|t|$ is greater than the largest critical value in the $n-1$ row, then $p < 0.0001$ (the smallest significance level in the table).
   (c) If $|t|$ is less than the smallest critical value in the $n-1$ row, then $p > 0.20$ (the largest significance level in the table).

If $t < 0$ for a right-tailed test, or $t > 0$ for a left-tailed test, then $p > 0.5$. If the hypothesis test is two-sided, this method produces a bound on $p/2$.

RAPID REVIEW — Absolute value inequalities.

In the next example, we'll use Appendix Table 5 to find a bound on the p value, and then use technology to confirm the answer.

EXAMPLE 9.16 Delivery Time

A certain daily delivery route for Hostess breads and snack cakes includes eight grocery stores and four convenience stores. The historical mean time to complete these deliveries (to the 12 stores) and return to the distribution center is 6.5 hr. A new driver has been assigned to this route, and a random sample of his route completion times (in hours) was obtained. The data are given in the table.

| 6.61 | 6.25 | 6.40 | 6.57 | 6.35 | 5.95 | 6.53 | 6.29 |

Assume the underlying population is normal.

(a) Is there any evidence to suggest that the new driver has been able to shorten the route completion time? Use $\alpha = 0.01$.

(b) Find bounds on the p value associated with this hypothesis test.

Solution

(a)

STEP 1 The assumed mean is $6.5 (= \mu_0)$; $n = 8$ and $\alpha = 0.01$.

We would like to know whether the new driver has been able to shorten the mean delivery time. The relevant alternative hypothesis is one-sided, left-tailed.

The underlying population is assumed to be normal, and σ is unknown. A t test is appropriate.

STEP 2 $H_0: \mu = 6.5$

$H_a: \mu < 6.5$

TS: $T = \dfrac{\overline{X} - \mu_0}{S/\sqrt{n}}$

RR: $T \leq -t_{\alpha,\,n-1} = t_{0.01,\,7} = -2.9980$

STEP 3 The sample mean is

$$\overline{x} = \frac{1}{8}(6.61 + 6.25 + \cdots + 6.29) = 6.3688$$

Find the standard deviation.

$$s^2 = \frac{1}{7}\left[324.8095 - \frac{1}{8}(50.95)^2\right] = 0.0460$$

$$s = \sqrt{0.0460} = 0.2144$$

The value of the test statistic is

$$t = \frac{\overline{x} - \mu_0}{s/\sqrt{n}} = \frac{6.3688 - 6.5}{0.2144/\sqrt{8}} = -1.7317$$

The value of the test statistic, $t = -1.7317$, does not lie in the rejection region. We cannot reject the null hypothesis. There is no evidence to suggest that the new driver has been able to shorten the mean delivery time for this route.

(b)

$|t| = |-1.7317| = 1.7317$

In Appendix Table 5, row $n - 1 = 8 - 1 = 7$, place 1.7308 in the ordered list of critical values.

$$1.4149 \leq 1.7317 \leq 1.8946$$

$$t_{0.10,\,7} \leq 1.7317 \leq t_{0.05,\,7}$$

Therefore, $0.05 \leq p \leq 0.10$ (See **Figure 9.27**)

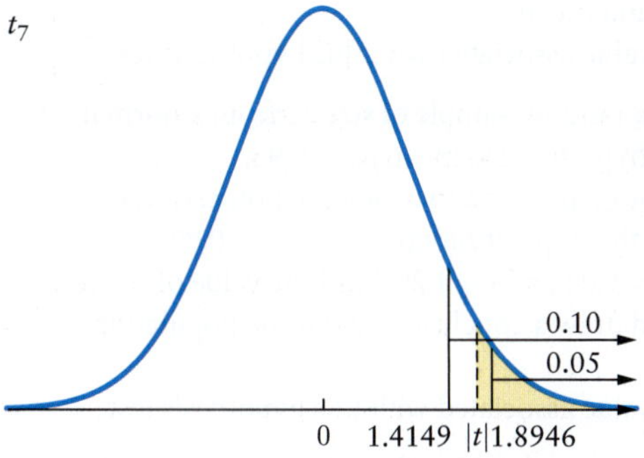

Figure 9.27 Visualization of the bounds on the *p* value. The shaded area in the figure corresponds to the *p* value.

```
> time <- c(6.61,6.25,...,6.29)
>
> t.test(time,mu=6.5,alternative="less",conf.level=0.99)

            One Sample t-test

data: time
t = -1.7317, df = 7, p-value = 0.06347
alternative hypothesis: true mean is less than 6.5
99 percent confidence interval:
     -Inf 6.595971
sample estimates:
mean of x
  6.36875
```

Figure 9.28 Use the R function t.test() with the appropriate options to conduct a *t* test on vectors of data.

The *p* value associated with this hypothesis test is between 0.05 and 0.10. Because the smallest possible value, 0.05, is greater than the significance level, 0.01, we cannot reject the null hypothesis. **Figure 9.28** shows a technology solution and confirms the bound on the *p* value.

TRY IT NOW Go to Exercises 9.165 and 9.167

Section 9.5 Exercises

Concept Check

9.135 True or False In a hypothesis test concerning a population mean when σ is unknown, the test statistic is based on a *t* distribution with $n-1$ degrees of freedom.

9.136 True or False The hypothesis test concerning a population mean when σ is unknown is valid only when the underlying population is normal.

9.137 True or False In a hypothesis test concerning a population mean when σ is unknown, there are only three possible alternative hypotheses.

9.138 True or False In a hypothesis test concerning a population mean when σ is unknown, the *p* value can never be determined.

9.139 True or False In a hypothesis test concerning a population mean when σ is unknown, the test statistic is based on the standard normal distribution when the sample size is greater than 30.

9.140 True or False In a hypothesis test concerning a population mean when σ is unknown with alternative hypothesis $H_a: \mu > \mu_0$, the value of the test statistic will always be greater than 0.

Practice

9.141 Consider a hypothesis test concerning a population mean from a normal population with $H_0: \mu = 10.5$ and $H_a: \mu > 10.5$, and σ unknown.
(a) Write the appropriate test statistic.
(b) Find the rejection region corresponding to the value of *n* and α.
  (i) $n = 6$, $\alpha = 0.01$
  (ii) $n = 23$, $\alpha = 0.025$
  (iii) $n = 17$, $\alpha = 0.05$
  (iv) $n = 29$, $\alpha = 0.10$
  (v) $n = 10$, $\alpha = 0.001$
  (vi) $n = 9$, $\alpha = 0.0001$

9.142 Consider a hypothesis test concerning a population mean from a normal population with $H_0: \mu = 22.41$ and $H_a: \mu < 22.41$, and σ unknown.
(a) Write the appropriate test statistic.
(b) Find the rejection region corresponding to the value of *n* and α.
  (i) $n = 15$, $\alpha = 0.01$
  (ii) $n = 11$, $\alpha = 0.0005$
  (iii) $n = 21$, $\alpha = 0.05$
  (iv) $n = 24$, $\alpha = 0.10$
  (v) $n = 5$, $\alpha = 0.001$
  (vi) $n = 31$, $\alpha = 0.0001$

9.143 Consider a hypothesis test concerning a population mean from a normal population with $H_0: \mu = 1.67$ and $H_a: \mu \neq 1.67$, and σ unknown.

(a) Write the appropriate test statistic.
(b) Find the rejection region corresponding to the value of n and α.

(i) $n = 12$, $\alpha = 0.01$ (ii) $n = 19$, $\alpha = 0.20$
(iii) $n = 26$, $\alpha = 0.05$ (iv) $n = 28$, $\alpha = 0.10$
(v) $n = 7$, $\alpha = 0.001$ (vi) $n = 56$, $\alpha = 0.02$

9.144 Consider a hypothesis test concerning the mean from a normal population with H_0: $\mu = 0.082$ and H_a: $\mu > 0.082$, and σ unknown. Find the significance level (α) corresponding to the value of n and the rejection region.

(a) $n = 27$, $T \geq 2.0555$ (b) $n = 9$, $T \geq 4.5008$
(c) $n = 16$, $T \geq 2.6025$ (d) $n = 20$, $T \geq 3.5794$

9.145 Consider a hypothesis test concerning the mean from a normal population with H_0: $\mu = -15.76$ and H_a: $\mu < -15.76$, and σ unknown. Find the significance level (α) corresponding to the value of n and the rejection region.

(a) $n = 23$, $T \leq -1.3212$ (b) $n = 4$, $T \leq -10.2145$
(c) $n = 31$, $T \leq -2.7500$ (d) $n = 18$, $T \leq -2.5524$

9.146 Consider a hypothesis test concerning the mean from a normal population with H_0: $\mu = 5128$ and H_a: $\mu \neq 5128$, and σ unknown. Find the significance level (α) corresponding to the value of n and the rejection region.

(a) $n = 30$, $|T| \geq 1.6991$ (b) $n = 6$, $|T| \geq 4.0321$
(c) $n = 17$, $|T| \geq 3.6862$ (d) $n = 14$, $|T| \geq 5.1106$

9.147 Consider a hypothesis test concerning the mean from a normal population with H_0: $\mu = 2.53$ and H_a: $\mu > 2.53$, and σ unknown. Find bounds on the p value for the value of n and the test statistic t.

(a) $n = 24$, $t = 2.35$ (b) $n = 3$, $t = 8.55$
(c) $n = 12$, $t = 5.68$ (d) $n = 8$, $t = 1.52$

9.148 Consider a hypothesis test concerning the mean from a normal population with H_0: $\mu = 6.28$ and H_a: $\mu < 6.28$, and σ unknown. Find bounds on the p value for the value of n and the test statistic t.

(a) $n = 25$, $t = -1.97$ (b) $n = 13$, $t = -0.63$
(c) $n = 18$, $t = -2.28$ (d) $n = 27$, $t = -3.58$

9.149 Consider a hypothesis test concerning the mean from a normal population with H_0: $\mu = 1.414$ and H_a: $\mu \neq 1.414$, and σ unknown. Find bounds on the p value for the value of n and the test statistic t.

(a) $n = 10$, $t = 2.04$ (b) $n = 23$, $t = -3.14$
(c) $n = 14$, $t = -5.52$ (d) $n = 11$, $t = 1.75$

9.150 Consider a random sample of size 20 from a normal population with hypothesized mean 1.618.
(a) Write the four parts for a one-sided, left-tailed hypothesis test concerning the population mean with $\alpha = 0.05$.
(b) Suppose $\bar{x} = 1.5$ and $s = 0.45$. Find the value of the test statistic, and draw a conclusion about the population mean.
(c) Find the p value associated with this hypothesis test.

9.151 Consider a random sample of 11 observations, given in the table, from a normal population with a hypothesized mean 57.71.

59.94	58.93	59.41	60.66	59.00	60.98
58.85	55.21	59.02	61.14	59.25	

EX9.151

(a) Write the four parts for a one-sided, right-tailed hypothesis test concerning the population mean with $\alpha = 0.01$.
(b) Compute the sample mean, the sample standard deviation, and the value of the test statistic. Draw a conclusion about the population mean.
(c) Find the p value associated with this hypothesis test.

9.152 Consider a random sample of size 28 from a normal population with hypothesized mean $\mu_0 = 9.96$.
(a) Write the four parts for a two-sided hypothesis test concerning the population mean with $\alpha = 0.002$.
(b) Suppose $\bar{x} = 9.04$ and $s = 1.20$. Find the value of the test statistic, and draw a conclusion about the population mean.
(c) Find the p value associated with this hypothesis test.

9.153 Consider a two-sided hypothesis test concerning the mean μ_0 from a normal population, with sample size $n = 25$, standard deviation s, and $\alpha = 0.05$. Explain the error in each statement.

(a) The test statistic is $T = \dfrac{\bar{X} - \mu_0}{\sigma / \sqrt{25}}$.

(b) The test statistic is $T = \dfrac{\bar{X} - \mu_0}{S / \sqrt{24}}$.

(c) The rejection region is RR: $T \geq 1.7109$.

(d) If the value of the test statistic is $t = 2.6732$, then $p \leq 0.005$.

Applications

9.154 Biology and Environmental Science The Atlantic bluefin tuna is the largest and most endangered of the tuna species. These fish are found throughout the North Atlantic Ocean. In 2017, the mean weight of bluefin tuna reported to New England and Mid-Atlantic National Oceanic and Atmospheric Administration (NOAA) fisheries was 414.7 lb.[44] The Center for Biological Diversity is concerned that this species has been overfished and that the mean weight has decreased. Suppose a random sample of 12 Atlantic bluefin tuna was obtained in June 2018 from commercial fishing boats and weighed. The summary statistics were $\bar{x} = 411.7$ and $s = 37.8$. Conduct a hypothesis test to determine whether there is any evidence that the mean weight of Atlantic bluefin tuna is less than 414.7 lb. Assume the distribution of weight is normal and use $\alpha = 0.05$.

9.155 Sports and Leisure The mean cost of a movie ticket during the first quarter of 2018 was $9.16.[45] During the second quarter of 2018, the domestic box office take was approximately $3.33 billion thanks to smash hits like *Avengers: Infinity War* and *The Incredibles 2*. A random sample of theaters was obtained and the cost of a movie ticket was recorded. Is there any evidence to suggest that the true mean movie ticket cost is greater than $9.16? Assume the underlying distribution is normal and use $\alpha = 0.025$. MOVIE

9.156 Biology and Environmental Science The historical mean yield of soybeans from farms in Indiana is 31.9 bushels per acre. Following a recent dry summer, a random sample of 26 farms across the state was obtained. The mean yield in bushels per acre was $\bar{x} = 30.088$, with $s = 4.433$. Is there any evidence to suggest that the lack of rain, or some other factor, adversely affected the soybean yield in Indiana? Assume the underlying population is normal and use $\alpha = 0.01$.

9.157 Manufacturing and Product Development The mean square footage for new single-family homes at the end of 2017 was 2560 ft^2. New home size had been decreasing over the two previous years as builders tried to construct more entry-level homes. However, some evidence suggests that the square footage is increasing again, especially in certain markets. A random sample of new single-family homes in Charlotte, North Carolina, was obtained and the square footage of each was recorded.[46] Is there any evidence to suggest that the true mean new single-family home square footage in Charlotte has increased? Assume the underlying distribution is normal and use $\alpha = 0.01$. NCHOME

9.158 Business and Management The 40-hour work week did not become a U.S. standard until 1940. Today, many white-collar employees work more than 40 hours per week because management demands longer hours or offers large monetary incentives for the extra time. A random sample of white-collar employees was obtained, and the number of hours each worked during the last week is given in the table.

44.7	42.0	45.8	43.0	42.8	50.9	47.0
41.9	49.3	45.6	45.7	39.4	39.0	44.4

Is there any evidence that the true mean number of hours worked by white-collar employees is greater than 40? Use $\alpha = 0.01$. What assumption(s) did you make to complete this hypothesis test? WORKWK

9.159 Marketing and Consumer Behavior Kennebunkport, Maine, is a popular tourist town, but businesses suffer during the winter months, especially January. The mean hotel room occupation rate per day, a measure of tourist activity, is 23.1% during winter months. A new advertising campaign was launched to attract more tourists to this town during the winter. Following the campaign, a random sample of nine winter days was selected. The mean hotel room occupation rate was $\bar{x} = 24.6\%$ and $s = 2.1\%$. Is there any evidence to suggest that the mean hotel room occupation rate has increased? Assume normality and use $\alpha = 0.01$. If this test is significant, can you conclude the ad campaign caused the increase?

9.160 Economics and Finance Economic growth in China has contributed to increased demand for crude oil and rising gasoline prices in several countries. In June 2018, the mean amount of crude oil imported per day in China was 8.36 million barrels.[47] A random sample of 17 days in the last quarter of 2018 was obtained, and the amount of crude oil imported was recorded for each (in million barrels per day, bpd). The summary statistics were $\bar{x} = 8.792$ million bpd and $s = 0.862$ million bpd. Is there any evidence to suggest that the mean amount of crude oil imported per day in China has increased? Use $\alpha = 0.05$ and assume normality.

9.161 Medicine and Clinical Studies The density of physicians is defined as the number of physicians per 1000 people. Suppose the World Health Organization suggests that a physician density of 2.3 is necessary to meet a population's primary health care needs. A random sample of countries was obtained and the physician density of each was recorded.[48] Is there any evidence to suggest that the world true mean physician density is less than 2.3? Assume normality and use $\alpha = 0.01$. PHYSDEN

9.162 Medicine and Clinical Studies Over time, certain metabolic changes can cause the lens of the eye to become opaque, which leads to loss of vision. In this case, many adults elect to have cataract surgery in which the natural lens of the eye is removed and an artificial lens implant is inserted. A random sample of cataract surgeries was obtained and the length of each surgery was recorded (in hours). Past records indicate that the mean time for this type of surgery is 0.75 hr. Assume the underlying population is normal. LENS
(a) Is there any evidence to suggest that the mean time has decreased? Use $\alpha = 0.01$.
(b) Find bounds on the p value associated with this test.

9.163 Economics and Finance Classical ballet dancers are both artists and athletes. They work exceptionally hard in a very demanding and prestigious form of dance, and their careers begin early and end early. According to the Bureau of Labor Statistics, the mean hourly wage for a ballet dancer in a performing arts company is $19.36.[49] A random sample of performing arts companies was obtained, and the hourly wage for each is given in the table. BALLET

32.64	24.50	17.96	17.46	16.64
16.37	16.25	15.85	14.74	14.23

Is there any evidence to suggest that the mean hourly wage has changed? Assume normality and use $\alpha = 0.05$.

9.164 Public Policy and Political Science While campaigning, Ontario Premier Doug Ford promised he would lower the minimum price of a bottle or can of beer to $1.00. Brewers are not required to accept the buck-a-beer challenge, and as of early September 2018, most large and craft brewers had not lowered their prices.[50] Suppose the mean price for a 0.5-L bottle of beer purchased in an Ontario supermarket was $2.82 before the buck-a-beer challenge. A random sample of supermarket beer prices was obtained after the challenge. Assume the underlying population is normal and use $\alpha = 0.05$. Is there any evidence to suggest that the true mean price for a bottle of beer in Ontario has decreased? BEER

Extended Applications

9.165 Demographics and Population Statistics Police reports and insurance claims indicate that the mean loss during a residential break-in is $1381. Actuaries working for Eagle Pacific Insurance Company need to check whether the mean loss has changed during the past year so that they can determine new policy rates. A random sample of 17 residential break-ins was obtained, and the loss due to each burglary was recorded. The sample mean loss (in dollars) was 1857, and the standard deviation was $s = 786$. Assume the loss distribution is normal and use $\alpha = 0.05$.

(a) Is there any evidence to suggest that the true mean loss as a result of a residential break-in has changed?
(b) Find bounds on the p value associated with this hypothesis test.
(c) Carefully sketch a graph illustrating the p value and the value of the test statistic.

9.166 Manufacturing and Product Development An engineered hardwood floor consists of an inner core of plywood glued and pressed together and a wear layer fused on top. A thicker wear layer can be sanded and refinished more often, and a good-quality engineered hardwood has a 4-mm top layer. A random sample of engineered hardwood was obtained from Hadinger Flooring, and the top layer thickness was measured for each, in millimeters. HWOOD

(a) Is there any evidence to suggest that the mean wear thickness is less than 4 mm? Assume the distribution is normal and use $\alpha = 0.05$.
(b) Find bounds on the p value associated with this hypothesis test.

9.167 Fuel Consumption and Cars Coalbed methane is an important source of natural gas in the United States. The Powder River coalfield has approximately 2500 wells, each producing, on average, 159,350 ft³ of gas per day. To maintain sufficient storage facilities, the mean well output is carefully monitored. A random sample of 11 wells was obtained, and the daily methane output for each was recorded. The sample mean was $\bar{x} = 163{,}288$ ft³ and the standard deviation was $s = 8792$ ft³ cubic feet. Assume the distribution of methane output per day is normal, and use $\alpha = 0.01$.

(a) Is there any evidence to suggest that the mean methane output per well per day has increased?
(b) Find bounds on the p value associated with this hypothesis test.

9.168 Marketing and Consumer Behavior The manager of a Piggly Wiggly grocery store claims a membership card will save consumers (through automatic discounts and extra coupons) at least $15.00 per week, on average. To check this claim, a random sample of 11 shoppers with membership cards was obtained and their weekly grocery bills were inspected. The sample mean savings was $\bar{x} = \$14.35$ and $s = \$3.75$. Assume the distribution of savings is normal.

(a) Is there any evidence to refute the manager's claim? Use $\alpha = 0.025$.
(b) If you rejected the null hypothesis, how can you explain this conclusion when $14.35 is so close to $15.00? If you did not reject the null hypothesis, how can you explain this conclusion when $14.35 is certainly less than $15.00?
(c) Find bounds on the p value associated with this hypothesis test.

9.169 Biology and Environmental Science Lionfish are native to Indo-Pacific waters but were introduced into the Atlantic in the 1990s. These spiny fish have huge appetites, feed on native species, and have a mean length of 14.5 in.[51] A random sample of 24 lionfish caught during the annual lionfish round-up near Tequesta, Florida, was obtained, and the length of each was recorded (in inches). The summary statistics were $\bar{x} = 15.005$ and $s = 4.039$.

(a) Is there any evidence to suggest that the true mean length of lionfish in this area near Tequesta is greater than 14.5 in.? Use $\alpha = 0.01$ and assume the underlying population is normal.
(b) Find bounds on the p value associated with this test.

9.170 Sports and Leisure Triathlons involve swimming, cycling, and running in succession over various distances. The August 2018 Moose Nugget Triathlon near Anchorage, Alaska, featured a 1500-m swim, 40 km-cycling course, and a 10-km run. A random sample of female racers was obtained, and their finish times were recorded (in hours).[52] Food vendors at the finish line planned to stay for 3.5 hr after the start of the race. Is there any evidence to suggest that the mean time to finish this triathlon was more than 3.5 hr? Assume the underlying distribution is normal and use $\alpha = 0.01$. TRIATH

9.171 Biology and Environmental Science A quahog is a chewy Atlantic hard-shell clam with a blue-gray shell. Quahogs have different names, depending on size; for example, the width of a littleneck clam is less than 5 cm (across the shell). The historical mean width of a littleneck is 4.75 cm. Following a recent oil spill off the coast of Maine, a random sample of 46 littlenecks was obtained, and the width of each was recorded. The summary statistics were $\bar{x} = 4.66$ cm and $s = 0.25$ cm.

(a) Use a one-sided, left-tailed hypothesis test to show that there is evidence to suggest that the mean width of littlenecks has decreased. Assume the population of littleneck widths is normal and use $\alpha = 0.01$.
(b) Explain why there is statistical evidence that the population mean has decreased even though the sample mean (4.66) is so close to the historical mean (4.75).
(c) Find bounds on the p value associated with this hypothesis test.

9.172 Physical Sciences The Palmer Drought Severity Index (PDSI) is a measure of prolonged abnormal dryness or wetness. It indicates general conditions and is not affected by local variations. An index value of -4 indicates extreme drought, and $+4$ represents very wet conditions. A random sample of the PDSI for selected regions in the United States for August 2018 was obtained, and the data are given in the table.[53] PDSI

0.98	0.61	0.94	0.95	−1.22	−3.21	−3.77
0.85	0.93	1.82	−1.66	−0.72	0.64	1.56
0.72	2.08	0.70	1.50	−2.03	−2.73	2.38

(a) Is there any evidence to suggest that the true mean PDSI is different from 0? Assume the underlying distribution is normal and use $\alpha = 0.01$. Find bounds on the p value associated with this test.
(b) Explain your conclusion in part (a) in terms of farming conditions and reservoir levels.

9.173 Physical Sciences Alaska registers the most earthquakes in the United States, and a recent increase in seismic activity has pushed Oklahoma into second place on the list of states with the most earthquakes, ahead of California. A random sample of earthquakes in the United States during

September 2018 was obtained, and the magnitude of each was recorded.[54] **QUAKE**

(a) An earthquake with magnitude of 1.0 to 4.0 is considered weak and generally causes minor or no damage. Is there any evidence to suggest that the true mean magnitude of earthquakes is greater than 1.0? Assume the underlying distribution is normal and use $\alpha = 0.05$.

(b) Find bounds on the p value associated with this test.

9.174 Public Health and Nutrition Fluoride is added to public water supplies and many dental products to help prevent tooth decay. Sodium fluoride is a common additive in toothpaste, and the concentration is usually measured in parts per million by weight (ppmF). However, some research suggests that high concentrations of fluoride can be toxic and may cause brain damage, immune disorders, and changes in bone structure and strength. The manufacturer of Pepsodent toothpaste claims the concentration of fluoride in every tube is 1000 ppmF. A random sample of toothpaste tubes was obtained, and the fluoride concentration in each was determined. **TPASTE**

(a) Test the relevant hypothesis concerning the mean fluoride concentration per tube of toothpaste. Assume normality and use $\alpha = 0.05$.

(b) Find bounds on the p value associated with this hypothesis test.

9.175 Public Health and Nutrition A recent study suggested that artificially sweetened soft drinks actually have a negative impact on weight and other health issues. Aspartame is the sweetener most commonly used in diet sodas. A random sample of diet sodas was obtained and the amount of aspartame in an 8-oz can was recorded for each (in milligrams). The data are given in the table. **SODA**

125	125	0	58	118	118	83
0	123	57	50	50	19	66

Assume the underlying population is normal.

(a) Is there any evidence to suggest that the mean amount of aspartame in an 8-oz can of diet soda is greater than 65 mg? Use $\alpha = 0.01$.

(b) Find bounds on the p value associated with this test.

9.176 Marketing and Consumer Behavior Red Rocks Park in Morrison, Colorado, is a naturally formed, rock, open-air amphitheater located 6450 ft above sea level. A random sample of events held at this site in summer 2018 was obtained, and the attendance for each was recorded. The summary statistics were $\bar{x} = 8722$, $s = 460.4$, and $n = 11$. Assume the distribution of attendance is normal and use $\alpha = 0.01$.

(a) The amphitheater operators believe that 9000 people is the optimal attendance in terms of comfort and profit. Is there any evidence to suggest that the true mean attendance is different from 9000?

(b) Find bounds on the p value associated with this test.

(c) How large a sample size would be necessary for this test to be significant (assuming $\bar{x}$ and s remain the same)?

Challenge Problems

9.177 Psychology and Human Behavior A small-sample (exact) hypothesis test concerning the mean of a Poisson random variable λ is constructed in the following manner. Suppose the random variable X is a count, modeled by a Poisson random variable. The four parts of the hypothesis test are

$H_0: \lambda = \lambda_0$
$H_a: \lambda > \lambda_0, \quad \lambda < \lambda_0, \quad \text{or} \quad \lambda \neq \lambda_0$
TS: $X =$ the number of events that occur in the specified interval
RR: $X \geq x_\alpha, \quad X \leq x'_\alpha, \quad \text{or} \quad X \leq x_{\alpha/2} \quad \text{or} \quad X \geq x'_{\alpha/2}$

The critical values x_α, x'_α, $x_{\alpha/2}$, and $x'_{\alpha/2}$ are obtained from the Poisson distribution with parameter λ_0 to yield the desired significance level α.

At a large hotel in Los Angeles, on average four people per day forget to take their room key and, therefore, lock themselves out. In an effort to lower the number of these service calls, the hotel staff has placed special signs on the back of every hotel room door reminding visitors to take their key. On a randomly selected day, two people were locked out of their rooms.

(a) Write the four parts of a small-sample, one-sided, left-tailed test concerning the mean number of people locked out of their room, based on a Poisson distribution. Use $\alpha = 0.05$.

(b) Is there any evidence to suggest that the mean has decreased?

(c) Find the p value associated with this hypothesis test.

9.178 Public Health and Nutrition A large-sample test concerning the mean of a Poisson random variable is based on a normal approximation. If X has a Poisson distribution with (large) mean λ, then X is approximately normal with mean λ and standard deviation $\sqrt{\lambda}$.

The four parts of the large-sample hypothesis test are

$H_0: \lambda = \lambda_0$
$H_a: \lambda > \lambda_0, \quad \lambda < \lambda_0, \quad \text{or} \quad \lambda \neq \lambda_0$
TS: $Z = \dfrac{X - \lambda_0}{\sqrt{\lambda_0}}$

$X =$ the number of events that occur in the specified interval
RR $Z \geq z_\alpha, \quad Z \leq -z_\alpha, \quad \text{or} \quad |Z| \geq z_{\alpha/2}$

Postal workers who deliver mail are at high risk for dog bites. The number of dog bites to postal employees in the United States per (fiscal) year peaked during the mid-1980s, but with increased public awareness and employee training, the number steadily decreased to fewer than 3000 dog bites per year. However, in fiscal year 2002, the number of dog bites began to rise again, especially during the "dog days" of summer.

In 2017, the mean number of dog bites to postal workers per day was 17. A random day in July 2018 was obtained, and the number of dog bites to postal workers was 23.

(a) Write the four parts of a large-sample, one-sided, right-tailed test concerning the mean number of dog bites to postal employees per day, based on a normal approximation to the Poisson distribution. Use $\alpha = 0.01$.

(b) Is there any evidence to suggest that the mean number of dog bites to postal employees per day has increased?

(c) Find the p value associated with this hypothesis test.

9.6 Large-Sample Hypothesis Tests Concerning a Population Proportion

Test for One Proportion

Many experiments and observational studies are conducted to draw a conclusion about a population proportion. For example, a quality control inspector may need to decide whether the proportion of defective products in a delivery is greater than 0.05. Based on a random sample, a decision is made either to accept the delivery or to send the entire shipment back. Medical researchers routinely assess the proportion of patients who recover from various illnesses. This information may be used to determine whether there is evidence that a new drug performs better than an existing treatment, or whether the proportion of patients who recover is greater than some threshold value.

These decisions involve the parameter p, the true population proportion, the fraction of individuals or objects with a specific characteristic, or the probability of a success. As in Section 7.3 and Section 8.4, it is reasonable to use the sample proportion, $\hat{p}$, as an estimate of the population proportion p. In a sample of n individuals or objects, let X be the number of individuals with the relevant characteristic. Recall the definition of the random variable $\hat{P}$, the sample proportion:

$$\hat{P} = \frac{X}{n} = \frac{\text{the number of individuals or objects with the characteristic}}{\text{the sample size}} \quad (9.1)$$

From Section 7.3, if n is large and both $np \geq 5$ and $n(1-p) \geq 5$, then the random variable $\hat{P}$ is approximately normal with mean p and variance $p(1-p)/n$, written as $\hat{P} \sim N(p, p(1-p)/n)$. These results about the distribution of $\hat{P}$ are used to construct a general procedure for a hypothesis test concerning p.

Large-Sample Hypothesis Tests Concerning a Population Proportion

Given a random sample of size n, a large-sample hypothesis concerning the population proportion p with significance level α has the form

$H_0: p = p_0$

$H_a: p > p_0, \quad p < p_0, \quad$ or $\quad p \neq p_0,$

TS: $Z = \dfrac{\hat{P} - p_0}{\sqrt{\dfrac{p_0(1-p_0)}{n}}}$

RR: $Z \geq z_\alpha, \quad Z \leq -z_\alpha, \quad$ or $\quad |Z| \geq z_{\alpha/2}$

This is the template for a hypothesis test about a population proportion. The hypothesized value is p_0.

As usual, use only one (appropriate) alternative hypothesis and the corresponding rejection region.

A CLOSER LOOK

1. This test is valid as long as $np_0 \geq 5$ and $n(1-p_0) \geq 5$ (the nonskewness criterion).
2. The critical values for this test are from the standard normal distribution Z (as in a hypothesis test about a population mean when σ is known).

Proportion Test Example

The next example illustrates this hypothesis test procedure.

EXAMPLE 9.17 Is This a Lug Wrench?

Americans have a love–hate relationship with their cars. However, many feel they take a calculated risk every time they drive and, in fact, a recent survey showed that almost 70% of cars have at least one thing wrong with them. In addition, 36% of Americans indicated that they did not know how to fix a flat tire.[55] A roadside assistance company is concerned that the percentage of drivers who do not know how to fix a flat tire is increasing and is contributing to higher costs. A random sample of 2500 American drivers was obtained, and 935 indicated that they did not know how to fix a flat tire. Is there any evidence to suggest that the proportion of American drivers who do not know how to fix a flat tire has increased? Use $\alpha = 0.01$.

Solution

We need to conduct a large-sample hypothesis test concerning a population proportion. Check the nonskewness criterion. Use the template for a one-sided, right-tailed test about p. Use $\alpha = 0.01$ to find the critical value, compute the value of the test statistic, and draw a conclusion.

STEP 1 The given information:

Sample size: $n = 2500$

Number of people with the specific characteristic, do not know how to fix a flat tire: $x = 935$

Sample proportion: $\hat{p} = \dfrac{x}{n} = \dfrac{935}{2500} = 0.3740$

The assumed value of the population proportion is 0.36 ($= p_0$), and the significance level is $\alpha = 0.01$.

STEP 2 Check the nonskewness criterion.

$np_0 = (2500)(0.36) = 900 \geq 5$

$n(1 - p_0) = (2500)(0.64) = 1600 \geq 5$

Both inequalities are satisfied, so $\hat{P}$ is approximately normal, and we can use the large-sample hypothesis test concerning a population proportion.

STEP 3 The four parts of the hypothesis test are

$H_0: p = 0.36$

$H_a: p > 0.36$

TS: $Z = \dfrac{\hat{P} - p_0}{\sqrt{\dfrac{p_0(1 - p_0)}{n}}}$

RR: $Z \geq z_\alpha = z_{0.01} = 2.3263$

STEP 4 The value of the test statistic is

$z = \dfrac{\hat{p} - p_0}{\sqrt{\dfrac{p_0(1 - p_0)}{n}}} = \dfrac{0.3740 - 0.36}{\sqrt{\dfrac{(0.36)(0.64)}{2500}}} = 1.4583$

STEP 5 The value of the test statistic does not lie in the rejection region. Equivalently, the p value is $p = 0.0724 > 0.01$. We do not reject the null hypothesis. There is no evidence to suggest that the true population proportion of American drivers who do not know how to fix a flat tire is greater than 0.36 or has increased.

Figure 9.29 shows a technology solution, and **Figure 9.30** illustrates the p value associated with the hypothesis test.

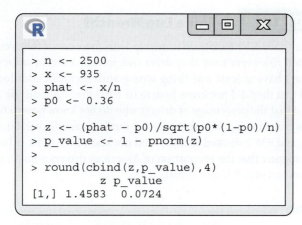
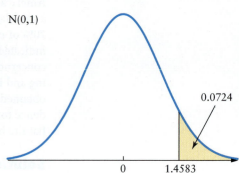

Figure 9.29 R calculations to find the value of the test statistic and p value.

Figure 9.30 p value illustration: $p = (Z \geq 1.4583) = 0.0724 > 0.01 = \alpha$.

Just a reminder: The p in H_0 and H_a represents the population proportion. The symbol p is also used later in this problem to represent the p value associated with the hypothesis test. ∎

TRY IT NOW Go to Exercises 9.191 and 9.195

Hypothesis Test and p Value

Because a large-sample test concerning a population proportion is based on a standard normal distribution Z, the p value is computed the same way as in a hypothesis test concerning a population mean when σ is known (which is also based on a standard normal distribution). The following example includes a p value computation.

EXAMPLE 9.18 Drug Shortages

Drug shortages have become a way of life for pharmacists in Canada, who often do not learn about a shortage until they attempt to fill a prescription. Each year, approximately 1200 drugs go into short supply, and there are usually 700–1000 drugs out of stock at any given time. These shortages occur for various reasons, including the supply chain and shipping delays. However, 58% of all drug shortages are caused by a disruption in manufacturing.[56] A random sample of 400 drugs in short supply was obtained, and 204 of the shortages were due to a disruption in manufacturing.

(a) Is there any evidence to suggest that the proportion of drugs in short supply due to a disruption in manufacturing is different from 0.58? Use $\alpha = 0.05$.

(b) Find the p value associated with this hypothesis test.

Solution

(a) The sample size is $n = 400$ and $x = 204$.

The sample proportion is $\hat{p} = \dfrac{x}{n} = \dfrac{204}{400} = 0.51$.

The assumed value of the population proportion is $p_0 = 0.58$ and $\alpha = 0.05$.

Check the nonskewness criterion:

$np_0 = (400)(0.58) = 232 \geq 5$

$n(1 - p_0) = (400)(0.42) = 168 \geq 5$.

Both inequalities are satisfied, so $\hat{P}$ is approximately normal, and we can use the large-sample hypothesis test concerning a population proportion.

We are looking for any difference from $p_0 = 0.58$, so this is a two-sided hypothesis test.

$H_0: p = 0.58$
$H_a: p \neq 0.58$

TS: $Z = \dfrac{\hat{P} - p_0}{\sqrt{\dfrac{p_0(1-p_0)}{n}}}$

RR: $|Z| \geq z_{\alpha/2} = z_{0.025} = 1.96$

The value of the test statistic is

$z = \dfrac{\hat{p} - p_0}{\sqrt{\dfrac{p_0(1-p_0)}{n}}} = \dfrac{0.51 - 0.58}{\sqrt{\dfrac{(0.58)(0.42)}{400}}} = -2.8365$

The value of the test statistic lies in the rejection region, so we reject the null hypothesis. There is evidence to suggest that the proportion of drugs in short supply due to a disruption in manufacturing is different from 0.58.

(b) Because this is a two-sided test and the value of the test statistic ($z = -2.8365$) is negative, $p/2$ is a left-tail probability.

$p/2 = P(Z \leq -2.8365) \approx P(Z \leq -2.84)$ *Definition of p value for $H_a: p \neq p_0$.*

$= 0.0023$ *Appendix Table 3.*

$p = 2(0.0023) = 0.0046 \ (\leq 0.05 = \alpha)$ *Solve for p.*

Because $p \leq \alpha$, we reject the null hypothesis.

Figure 9.31 shows a technology solution and **Figure 9.32** illustrates the p value associated with the hypothesis test.

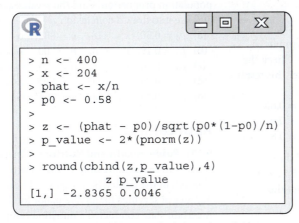

Figure 9.31 R calculations to find the value of the test statistic and the p value.

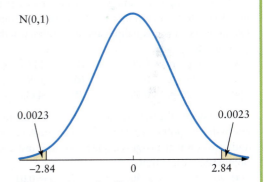

Figure 9.32 p value illustration: $p = 2P(Z \leq -2.84) = 0.0046 \leq 0.05 = \alpha$.

TRY IT NOW Go to Exercises 9.193 and 9.203

CHAPTER APP **Heavy Metal**

Is there any evidence to suggest that less than 68% of baby foods contain toxic chemicals? Use $\alpha = 0.05$ and the p value associated with this hypothesis test to draw a conclusion.

Section 9.6 Exercises

Concept Check

9.179 True or False A large-sample hypothesis test concerning a population proportion is based on a standard normal distribution.

9.180 True or False In a large-sample hypothesis test concerning a population proportion, the hypothesized proportion, p_0, cannot be close to 0 or 1.

9.181 True or False In a large-sample hypothesis test concerning a population proportion, the number of trials must be at least 250.

9.182 True or False In a large-sample hypothesis test concerning a population proportion, there are only two possible alternative hypotheses.

9.183 Short Answer In a large-sample hypothesis test concerning a population proportion, what happens if the nonskewness criterion is not satisfied?

9.184 Short Answer In a large-sample hypothesis test concerning a population proportion, explain how the null hypothesis could be rejected even if the sample proportion is within 0.01 of p_0.

Practice

9.185 The sample size and p_0 are given for a large-sample hypothesis test concerning a population proportion. Check the two nonskewness inequalities and determine whether this test is appropriate.
(a) $n = 276$, $p_0 = 0.30$ (b) $n = 1158$, $p_0 = 0.60$
(c) $n = 645$, $p_0 = 0.03$ (d) $n = 159$, $p_0 = 0.97$
(e) $n = 322$, $p_0 = 0.38$ (f) $n = 443$, $p_0 = 0.82$

9.186 Assume the null hypothesis is $H_0: p = p_0$ and the alternative hypothesis is $H_a: p < p_0$. Use the values of p_0, x, n, and α to conduct a large-sample hypothesis test about the population proportion.
(a) $p_0 = 0.14$, $x = 40$, $n = 317$, $\alpha = 0.01$
(b) $p_0 = 0.275$, $x = 98$, $n = 404$, $\alpha = 0.05$
(c) $p_0 = 0.52$, $x = 250$, $n = 546$, $\alpha = 0.025$
(d) $p_0 = 0.78$, $x = 2710$, $n = 3580$, $\alpha = 0.001$
(e) $p_0 = 0.605$, $x = 1102$, $n = 1862$, $\alpha = 0.05$

9.187 Assume the null hypothesis is $H_0: p = p_0$ and the alternative hypothesis is $H_a: p \ne p_0$. Use the values of p_0, x, n, and α to conduct a large-sample hypothesis test about the population proportion.
(a) $p_0 = 0.28$, $x = 88$, $n = 377$, $\alpha = 0.025$
(b) $p_0 = 0.46$, $x = 130$, $n = 243$, $\alpha = 0.02$
(c) $p_0 = 0.337$, $x = 120$, $n = 414$, $\alpha = 0.05$
(d) $p_0 = 0.71$, $x = 865$, $n = 1250$, $\alpha = 0.005$
(e) $p_0 = 0.93$, $x = 1515$, $n = 1600$, $\alpha = 0.01$

9.188 Assume the null hypothesis is $H_0: p = p_0$ and the alternative hypothesis is $H_a: p > p_0$. Use the values of p_0, x, n, and α to conduct a large-sample hypothesis test about the population proportion.
(a) $p_0 = 0.15$, $x = 60$, $n = 356$, $\alpha = 0.10$
(b) $p_0 = 0.62$, $x = 298$, $n = 450$, $\alpha = 0.05$
(c) $p_0 = 0.743$, $x = 1035$, $n = 1360$, $\alpha = 0.05$
(d) $p_0 = 0.94$, $x = 795$, $n = 825$, $\alpha = 0.01$
(e) $p_0 = 0.32$, $x = 960$, $n = 2750$, $\alpha = 0.001$

9.189 Assume the null hypothesis is $H_0: p = p_0$ and the alternative hypothesis is $H_a: p < p_0$. Use the values of p_0, x, n, and α to conduct a large-sample hypothesis test about the population proportion. Find the p value associated with the test, and use it to draw a conclusion.
(a) $p_0 = 0.54$, $x = 145$, $n = 301$, $\alpha = 0.05$
(b) $p_0 = 0.39$, $x = 180$, $n = 460$, $\alpha = 0.005$
(c) $p_0 = 0.07$, $x = 35$, $n = 566$, $\alpha = 0.01$
(d) $p_0 = 0.64$, $x = 449$, $n = 747$, $\alpha = 0.025$
(e) $p_0 = 0.47$, $x = 395$, $n = 925$, $\alpha = 0.005$

9.190 Assume the null hypothesis is $H_0: p = p_0$ and the alternative hypothesis is $H_a: p \ne p_0$. Use the values of p_0, x, n, and α to conduct a large-sample hypothesis test about the population proportion. Find the p value associated with the test, and use it to draw a conclusion.
(a) $p_0 = 0.50$, $x = 418$, $n = 882$, $\alpha = 0.01$
(b) $p_0 = 0.19$, $x = 158$, $n = 700$, $\alpha = 0.05$
(c) $p_0 = 0.90$, $x = 445$, $n = 520$, $\alpha = 0.001$
(d) $p_0 = 0.75$, $x = 1095$, $n = 1400$, $\alpha = 0.025$
(e) $p_0 = 0.45$, $x = 2386$, $n = 5525$, $\alpha = 0.01$

Applications

9.191 Public Health and Nutrition In a recent research study concerning nutrition, it was reported that only 2% of children in the United States ages 6–10 eat the recommended number of servings for all five major food groups each day. A certain state conducted a nutrition campaign through elementary schools, newspapers, and television. Following the campaign, a random sample of 500 children in this age group was obtained, and 16 were found to eat the recommended number of servings each day.
(a) Identify n, x, and p_0.
(b) Check the nonskewness criterion. Is a large-sample hypothesis test about p appropriate?
(c) Conduct the appropriate hypothesis test to determine whether the proportion of children who eat the recommended number of servings each day has increased. Use $\alpha = 0.05$.
(d) Compute the p value associated with this test.

9.192 Psychology and Human Behavior It is very common to complain about one's boss. There are many

examples of bullying and harassment in TV shows and movies, and now, especially and unfortunately, in real life. In a recent survey, 20% of workers reported that a manager had hurt their career. A random sample of 500 workers in the retail industry was obtained, and 115 indicated that a boss had hurt their career.
(a) Identify n, x, and p_0.
(b) Check the nonskewness criterion. Is a large-sample hypothesis test about p appropriate?
(c) Conduct the appropriate hypothesis test to determine whether the proportion of workers in the retail industry who believe a boss has hurt their career is different from p_0. Use $\alpha = 0.05$.
(d) Compute the p value associated with this test.

9.193 Education and Child Development Homeschooling has become very popular in the United States, and many colleges try to attract students from this group. Evidence suggests that approximately 90% of all homeschooled children attend college. To check this claim, a random sample of 225 homeschooled children was obtained, and 189 of them were found to have attended college.
(a) Identify n, x, and p_0.
(b) Check the nonskewness criterion. Is a large-sample hypothesis test about p appropriate?
(c) Conduct the appropriate hypothesis test to determine whether the proportion of homeschooled children who attend college is different from 0.90. Use $\alpha = 0.05$.
(d) Compute the p value associated with this test.

9.194 Public Policy and Political Science It has always been difficult to attract general physicians to rural areas. Small-town doctors do not have the opportunity to take much time off, and their salaries are relatively low. However, the quality of life is often appealing. Past records indicate that 52% of all physicians in rural areas leave after one year. The federal government decided to offer more incentives for doctors to stay in rural areas, for example, loan forgiveness and housing allowances. A few years after this program was implemented, a random sample of rural-area physician positions showed 62 of 130 left after one year.
(a) Identify n, x, and p_0.
(b) Check the nonskewness criterion. Is a large-sample hypothesis test about p appropriate?
(c) Conduct the appropriate hypothesis test to determine whether the turnover rate of rural doctors has decreased. Use $\alpha = 0.01$.
(d) Compute the p value associated with this test.

9.195 Medicine and Clinical Studies According to a recent study, taller women have a significantly higher risk of developing some form of cancer. The American Cancer Society reports that approximately 38% of all women develop some form of cancer in their lifetime. A long-term Canadian study was conducted involving 1250 tall women between the ages of 50 and 79, and 502 developed some form of cancer. Is there any evidence to suggest that the proportion of taller women who develop cancer is greater than 0.38? Use $\alpha = 0.01$.

9.196 Public Policy and Political Science Soon after Edward Snowden disclosed information about the surveillance programs of the U.S. National Security Agency, a survey of Americans revealed that 56% were worried that the United States would go too far in violating privacy rights. Aides to Senator Richard Burr selected a random sample of 575 North Carolina residents, and 358 said they were worried about violations of their privacy rights. Is there any evidence to suggest that the true proportion of North Carolina residents worried about violations of privacy rights is greater than 0.58? Use $\alpha = 0.05$ and use the p value to draw a conclusion.

9.197 Psychology and Human Behavior Many urban legends center on a full moon and human behavior. Research at the University of Basel in Switzerland suggests that sleep is associated with the lunar cycle. In a new sleep study, 120 random adults were selected and studied during a full moon phase. Melatonin levels were used to determine whether each person experienced a deep sleep. In total, 76 experienced low levels of melatonin and, therefore, had trouble sleeping during the full moon. Is there any evidence to suggest that more than half of all people have trouble sleeping during a full moon? Use $\alpha = 0.05$.

9.198 Marketing and Consumer Behavior A recent study suggests that Canadians are getting richer but are also among the most indebted people in the world.[57] The Bank of Canada suggests that no household should have debt in excess of 160% of annual disposable income. A random sample of 1500 households in Canada's wealthiest cities (Vancouver, Calgary, and Toronto) was obtained, and 235 were found to be too much in debt. Is there any evidence to suggest that the true proportion of Canadian households that have too much debt is greater than 14%? Use $\alpha = 0.05$.

9.199 Public Policy and Political Science Oakland Mayor Libby Schaaf is considering campaigning for another term as mayor. She will enter the race if evidence suggests that more than 50% of all city residents are satisfied with her job performance. A random sample of 375 residents is obtained, and 225 indicate they are satisfied with her job performance. Do you think Mayor Libby should run for reelection? Justify your answer. Use $\alpha = 0.01$.

9.200 Physical Sciences It's known as the Hum, a steady, droning sound heard in places all around the world. This low-frequency, annoying, throbbing, unexplained sound seems to be louder at night and indoors, and approximately 2% of the people living in any given Hum-prone area can hear the sound.[58] To determine if more people are affected by the Hum, a random sample of 1300 adults in Taos, New Mexico, was obtained, and 24 reported that they regularly hear the noise. Is there any evidence to suggest that the proportion of adults in this area who hear the Hum has changed? Use $\alpha = 0.05$ and use the p value associated with this hypothesis test to draw a conclusion.

9.201 Public Policy and Political Science A local planning board must consider whether to require all new projects with five or more apartments to designate some of the units as rent-controlled. A random sample of 100 apartments in Cheyenne,

Wyoming, was obtained, and 9 were found to be rent-controlled. Is there any evidence to suggest that the proportion of rent-controlled apartments is less than 10%? Let $\alpha = 0.05$, and use the p value associated with this hypothesis test to draw a conclusion.

9.202 Public Policy and Political Science Early in 2018, a political report indicated that 75% of Americans disapprove of the job Congress is doing. In a poll of 1000 Americans during late summer 2018 by NBC News, 780 said they disapprove of the job Congress is doing.[59] Is there any evidence to suggest that the proportion of Americans who disapprove of the job Congress is doing has increased? Use $\alpha = 0.01$.

9.203 Technology and the Internet Technology is constantly improving and increasing the speed of communication. Indeed, many software programs on tablets, laptops, and phones now anticipate our next words or phrases. Docmail, a U.K.-based printing and mailing company, conducted a study and concluded that one of every three adults had not been required to produce something in handwriting for more than a year. To check this claim in the United States, a random sample of 656 adults was obtained, and 235 said they had not been required to produce something in handwriting for more than a year. Is there any evidence to suggest that the proportion of Americans not using handwriting is different in the United States? Use $\alpha = 0.05$ and find the p value associated with this test.

Extended Application

9.204 Education and Child Development College study-abroad programs have great educational value, compel students to become more globally aware, and invite participants to learn about different nations and cultures. Some colleges sponsor several programs and have high participation rates. Even so, on a national scale, only 10% of all college students participate in a study-abroad program.[60] Recent world events may have made more students leery of travel and life in another country. In a random sample of 1200 graduating college students, 109 said they had participated in a study-abroad program. Is there any evidence to suggest that the proportion of students participating in these programs has decreased? Use $\alpha = 0.05$.

9.205 Physical Sciences Interstate Batteries guarantees its lawn tractor battery will last at least three years. To increase sales, the company is planning to offer a new replacement warranty, as long as 95% of all batteries do indeed last at least three years. A random sample of 200 customers was contacted, and 183 had batteries that lasted at least three years.
(a) Is there any evidence to suggest that the proportion of tractor batteries that last at least three years is less than 0.95? Use $\alpha = 0.05$.
(b) Find the p value associated with this hypothesis test.
(c) Based on your results, would you recommend implementing the new replacement warranty? Justify your answer.

9.206 Medicine and Clinical Studies A certain puzzle task is designed to measure spatial reasoning performance. Forty-five percent of all people attempting the puzzle complete the task within the allotted time. A researcher decided to test the theory that classical music increases brain activity and improves the ability to perform such tasks. A random sample of 400 people was obtained, and each listened to 15 minutes of classical music and attempted the task. A total of 211 completed the task within the allotted time.
(a) Is there any evidence to suggest that the proportion of people who complete the puzzle within the allotted time has increased? Use $\alpha = 0.001$.
(b) Compute the p value associated with this test.
(c) Carefully sketch a graph indicating the critical value from part (a) and the p value from part (b).

9.207 Sports and Leisure Over the past five years, the number of Canadian campers who camp at least three times per year has increased dramatically. Camping continues to be a desirable lifestyle attribute among Canadians, and a recent study suggests that 65% of all Canadian households include someone who camps.[61] A random sample of 350 Canadian households was obtained, and 245 included someone who camps.
(a) Is there any evidence to suggest that the proportion of Canadian households with a camper has increased? Use $\alpha = 0.05$.
(b) Compute the p value associated with this hypothesis test.
(c) Carefully sketch a graph that shows the critical value from part (a) and the p value from part (b).

Challenge Problems

9.208 Public Policy and Political Science The high cost of health care has forced some people to choose between medical treatment and other necessities such as food, heat, or electricity. A recent survey concluded that 44% of America's working-age adults did not see a doctor or access other medical services during the past year because of the cost.[62] Following a recent advertising campaign urging adults to seek medical assistance when needed, physicians' groups hope this percentage has decreased. A random sample of 1000 working Americans was obtained, and 390 did not seek medical assistance when needed, because of the cost.
(a) Is there any evidence to suggest that the proportion of American workers who did not seek medical assistance has decreased? Use $\alpha = 0.01$.
(b) Find the probability of a type II error in this hypothesis test for $p_a = 0.39$; that is, find $\beta(0.39)$.
(c) Find a value of p_a such that the probability of a type II error is 0.1; that is, solve $\beta(p_a) = 0.1$ for p_a.
(d) Find the sample size necessary such that the probability of a type II error for $p_a = 0.36$ is 0.025; that is, solve $\beta(0.35) = 0.025$ for n.

9.209 Demographics and Population Statistics A large-sample hypothesis test concerning a population proportion is based on a normal approximation and the nonskewness criterion. If n is small (and the nonskewness criterion fails), the test statistic in an exact hypothesis test concerning p is based on the number of successes in the sample, X. If $H_0: p = p_0$ is true, then X has a binomial distribution with n trials and probability of success p_0.

The four parts of the hypothesis test are

$H_0: p = p_0$
$H_a: p > p_0$, $p < p_0$,   or $p \ne p_0$
TS: X = the number of successes in n trials
RR $X \geq x_\alpha$, $X \leq x'_\alpha$,   or $X \leq x'_{\alpha/2}$ or $X \geq x'_{\alpha/2}$

The critical values x_α, x'_α, $x_{\alpha/2}$, and $x'_{\alpha/2}$ are obtained from the binomial distribution with parameters n and p_0 to yield the desired significance level α. For example, in a one-sided, right-tailed test, the critical value x'_α is found such that $P(X \geq x'_\alpha) \approx \alpha$.

A recent poll indicated that 51% of all students admit to cheating. A sociologist conducting research believes the actual percentage is much lower. In a random sample of 25 students from Long Island public schools, 9 admitted to cheating.

(a) Write the four parts of a small-sample (exact), one-sided, left-tailed test concerning the population proportion of Long Island students who admit to cheating, based on a binomial distribution. Use $\alpha = 0.01$.

(b) Is there any evidence to suggest that the true proportion is less than 0.51?

(c) Find the p value associated with this hypothesis test.

9.7 Hypothesis Tests Concerning a Population Variance or Standard Deviation

Test for One Variance

Many real-world, practical decisions involve variability, or a population variance. For example, road inspectors make sure that asphalt pavement mix meets design specifications and that variability in asphalt properties is small. This improves the quality, safety, and lifetime of the road surface. The dose of an anticancer drug is usually determined by the patient's body surface area (BSA). However, there is still large variability in drug exposure, the amount of the drug absorbed into the bloodstream. Researchers continue to search for better methods of calculating the dose in order to decrease the variability in exposure from patient to patient.

As in Section 8.5, the sample variance S^2 is used as an estimator for the population variance, σ^2. The hypothesis test procedure is based on a theorem given in Section 8.5, which outlines a standardization, or transformation, to a chi-square random variable. Recall that if S^2 is the sample variance of a random sample of size n from a normal distribution with variance σ^2, then the random variable $(n-1)S^2/\sigma^2$ has a chi-square distribution with $n-1$ degrees of freedom. This result is used to construct a general procedure for a hypothesis test concerning σ^2.

Hypothesis Test Concerning a Population Variance

Given a random sample of size n from a normal population with variance σ^2, a hypothesis test concerning the population variance σ^2 with significance level α has the form

$H_0: \sigma^2 = \sigma_0^2$
$H_a: \sigma^2 > \sigma_0^2$, $\sigma^2 < \sigma_0^2$,   or $\sigma^2 \ne \sigma_0^2$

TS: $X^2 = \dfrac{(n-1)S^2}{\sigma_0^2}$

RR: $X^2 \geq \chi^2_{\alpha, n-1}$, $X^2 \leq \chi^2_{1-\alpha, n-1}$,   or $X^2 \leq \chi^2_{1-\alpha/2, n-1}$ or $X^2 \geq \chi^2_{\alpha/2, n-1}$

This is the template for a hypothesis test about a population variance. The hypothesized value is σ_0^2.

Use only one (appropriate) alternative hypothesis and the corresponding rejection region. Remember: The two-sided alternative rejection region cannot be written with an absolute value symbol. A chi-square distribution is not symmetric (about zero).

A CLOSER LOOK

1. X (the Greek capital letter chi) is a random variable. χ (the Greek lowercase letter chi) is a specific value.
2. This test is valid for any sample size, as long as the underlying population is normal.
3. Appendix Table 6 presents selected critical values associated with various chi-square distributions. These critical values were used to construct confidence intervals in Section 8.5.
4. **Figure 9.33**, **Figure 9.34**, and **Figure 9.35** illustrate the rejection region for each alternative hypothesis.

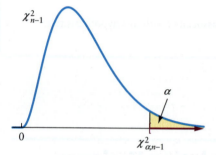

Figure 9.33 Rejection region for $H_a: \sigma^2 > \sigma_0^2$.

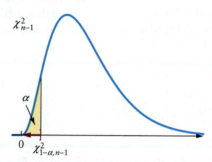

Figure 9.34 Rejection region for $H_a: \sigma^2 < \sigma_0^2$.

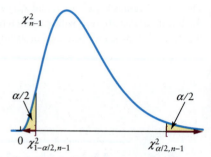

Figure 9.35 Rejection region for $H_a: \sigma^2 \ne \sigma_0^2$.

Variance Test Example

EXAMPLE 9.19 Permafrost Measurements

One measure of permafrost is the volumetric liquid water content (VWC), a unitless quantity. This measurement is used to study climate changes in the Arctic ice cap, greenhouse gases, and vegetation. Along the northern part of the Trans-Alaskan Pipeline, variation in permafrost was an important consideration in planning and construction of the pipeline. Suppose a random sample of 15 locations was obtained, and the VWC was measured at each location during the winter months. The sample mean was $\bar{x} = 0.225$ and the sample variance was $s^2 = 0.0025$. Is there any evidence to suggest the population variance is greater than 0.002? Use $\alpha = 0.05$.

Solution

We will assume the underlying distribution is normal and use the template for a one-sided, right-tailed test about σ^2. Use $\alpha = 0.05$ to find the critical value. Compute the value of the test statistic, and draw a conclusion.

STEP 1 Here is the given, relevant information.

Sample size: $n = 15$

Sample variance: $s^2 = 0.0025$

The assumed value of the population variance is 0.002 $(= \sigma_0^2)$, and the significance level is $\alpha = 0.05$.

STEP 2 The four parts of the hypothesis test are

$H_0: \sigma^2 = 0.002$

$H_a: \sigma^2 > 0.002$

TS: $X^2 = \dfrac{(n-1)S^2}{\sigma_0^2}$

RR: $X^2 \ge \chi^2_{\alpha, n-1} = \chi^2_{0.05, 14} = 23.6848$

STEP 3 The value of the test statistic is

$$\chi^2 = \frac{(n-1)s^2}{\sigma_0^2} = \frac{(14)(0.0025)}{0.002} = 17.5$$

STEP 4 The value of the test statistic does not lie in the rejection, so we do not reject the null hypothesis. There is no evidence to suggest that the true population variance in volumetric liquid water content is greater than 0.002.

Figure 9.36 shows a technology solution.

```
> n <- 15
> s_var <- 0.0025
> sigmasq_0 <- 0.002
>
> crit_value <- qchisq(.95,n-1)
> chi_sq <- (n-1)*s_var/sigmasq_0
> p_value <- 1 - pchisq(chi_sq,n-1)
>
> round(cbind(crit_value,chi_sq,p_value),4)
     crit_value chi_sq p_value
[1,]    23.6848   17.5  0.2305
```

Figure 9.36 R calculations to find the critical value, the value of the test statistic, and the *p* value.

TRY IT NOW Go to Exercises 9.229 and 9.231

p Value Bounds

A CLOSER LOOK

The table listing critical values for various chi-square distributions is very limited. When using this table, the best we can do is bound the *p* value associated with a hypothesis test, as for a *t* test.

EXAMPLE 9.20 Don't Forget the Sprinkles

ICECRM

The Ice Cream Store in Rehoboth Beach, Delaware, has more than 100 flavors available at any time, often including Boston Cream Pie, Cake Batter Cookie Dough, and Fruity Pebbles. The mean number of gallons of ice cream sold during a summer day is 354 gal, with a standard deviation of 25.7 gal. This information is used to plan production schedules and order supplies. A random sample of summer days was obtained, and the amount of ice cream sold each day (in gallons) is given in the following table.

360	372	347	358	346	387	347	359	338	343
370	372	362	369	356	349	330	344	339	334

(a) Is there any evidence to suggest that the true population variance in ice cream purchased per day is different from 660.49 gal^2? Assume normality and use $\alpha = 0.01$.

(b) Find bounds on the *p* value associated with this test.

Solution

(a) Assume the underlying distribution is normal. Use the template for a two-sided test about σ^2, and use $\alpha = 0.01$ to find the critical values.

STEP 1 The assumed population variance is $25.7^2 = 660.49 \ (= \sigma_0^2)$; $n = 20$ and $\alpha = 0.01$.

We would like to know whether the population variance is different from 660.49. The relevant alternative hypothesis is two-sided.

The underlying population is assumed to be normal. A hypothesis test concerning a population variance can be used.

STEP 2 The four parts of the hypothesis test are

$$H_0: \sigma^2 = 660.49$$
$$H_a: \sigma^2 \neq 660.49$$
$$\text{TS: } X^2 = \frac{(n-1)S^2}{\sigma_0^2}$$
$$\text{RR: } X^2 \leq \chi^2_{1-\alpha/2, n-1} = \chi^2_{0.995, 19} = 6.8440 \quad \text{or}$$
$$X^2 \geq \chi^2_{\alpha/2, n-1} = \chi^2_{0.005, 19} = 38.5823$$

STEP 3 The sample variance is

$$s^2 = \frac{1}{n-1}\left[\sum x_i^2 - \frac{1}{n}\left(\sum x_i\right)^2\right]$$ Computational formula, sample variance.

$$= \frac{1}{19}\left[2{,}511{,}964 - \frac{1}{20}(7082)^2\right] = 222.52$$ Use given data.

The value of the test statistic is

$$\chi^2 = \frac{(n-1)s^2}{\sigma_0^2} = \frac{(19)(222.52)}{660.49} = 6.4010$$

The value of the test statistic lies in the rejection region ($\chi^2 = 6.4010 \leq 6.8840$). We reject the null hypothesis at the $\alpha = 0.01$ level of significance. There is evidence to suggest that the true population variance is different from 660.49. **Figure 9.37** shows a technology solution.

(b)

> The initial bounds are on $p/2$ because the hypothesis test is two-sided.

$\chi^2 = 6.4010$.

Using Appendix Table 6, row $n - 1 = 20 - 1 = 19$, place 6.4010 in the ordered list of critical values.

$$5.4068 \leq 6.4010 \leq 6.8440$$
$$\chi^2_{0.999, 19} \leq 6.4010 \leq \chi^2_{0.999, 19}$$
$$\chi^2_{1-0.001, 19} \leq 6.4010 \leq \chi^2_{1-0.005, 19}$$

Therefore, $\quad 0.001 \leq \; p/2 \; \leq 0.005$

and $\quad\quad\quad 0.002 \leq \;\; p \;\; \leq 0.010$ See **Figure 9.38**

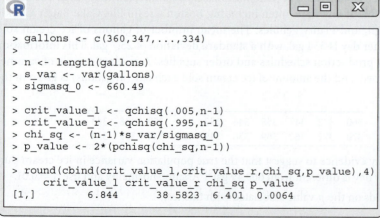

```
> gallons <- c(360,347,...,334)
>
> n <- length(gallons)
> s_var <- var(gallons)
> sigmasq_0 <- 660.49
>
> crit_value_l <- qchisq(.005,n-1)
> crit_value_r <- qchisq(.995,n-1)
> chi_sq <- (n-1)*s_var/sigmasq_0
> p_value <- 2*(pchisq(chi_sq,n-1))
>
> round(cbind(crit_value_l,crit_value_r,chi_sq,p_value),4)
     crit_value_l crit_value_r chi_sq p_value
[1,]        6.844      38.5823  6.401  0.0064
```

Figure 9.37 R calculations to find the critical values, the value of the test statistic, and the p value.

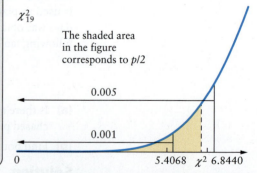

χ^2_{19}

The shaded area in the figure corresponds to $p/2$

Figure 9.38 The left tail of a chi-square distribution with 19 degrees of freedom. This illustrates the computations to find the bounds on the p value. ■

TRY IT NOW Go to Exercises 9.239 and 9.241

Section 9.7 Exercises

Concept Check

9.210 True or False A chi-square distribution with ν degrees of freedom is symmetric about ν.

9.211 True or False In a hypothesis test concerning a population variance based on a chi-square distribution, there is no assumption about the distribution of the underlying population.

9.212 True or False In a hypothesis test concerning a population variance, there are $n-1$ degrees of freedom associated with the test statistic.

9.213 True or False In a hypothesis test concerning a population variance based on a chi-square distribution, the test statistic can always be adjusted to use a right-tail rejection region.

9.214 True or False The hypothesis test concerning a population variance based on a chi-square distribution can also be used to conduct a hypothesis test concerning a population standard deviation.

9.215 True or False In a hypothesis test concerning a population variance based on a chi-square distribution, the p value could be greater than 0.5.

Practice

9.216 Consider a hypothesis test concerning a population variance from a normal population with $H_0: \sigma^2 = 27.2$ and $H_a: \sigma^2 > 27.2$.
(a) Write the appropriate test statistic.
(b) Find the rejection region for the value of n and α.
  (i) $n = 12$, $\alpha = 0.05$ (ii) $n = 19$, $\alpha = 0.025$
  (iii) $n = 23$, $\alpha = 0.01$ (iv) $n = 29$, $\alpha = 0.10$
  (v) $n = 6$, $\alpha = 0.001$ (vi) $n = 15$, $\alpha = 0.0001$

9.217 Consider a hypothesis test concerning a population variance from a normal population with $H_0: \sigma^2 = 352.98$ and $H_a: \sigma^2 < 352.98$.
(a) Write the appropriate test statistic.
(b) Find the rejection region for the value of n and α.
  (i) $n = 17$, $\alpha = 0.001$ (ii) $n = 13$, $\alpha = 0.05$
  (iii) $n = 37$, $\alpha = 0.0001$ (iv) $n = 27$, $\alpha = 0.10$
  (v) $n = 33$, $\alpha = 0.05$ (vi) $n = 8$, $\alpha = 0.005$

9.218 Consider a hypothesis test concerning a population variance from a normal population with $H_0: \sigma^2 = 43.8$ and $H_a: \sigma^2 \neq 43.8$.
(a) Write the appropriate test statistic.
(b) Find the rejection region for the value of n and α.
  (i) $n = 40$, $\alpha = 0.05$ (ii) $n = 31$, $\alpha = 0.01$
  (iii) $n = 22$, $\alpha = 0.001$ (iv) $n = 16$, $\alpha = 0.02$
  (v) $n = 29$, $\alpha = 0.002$ (vi) $n = 5$, $\alpha = 0.20$

9.219 Consider a hypothesis test concerning a population variance from a normal population with $H_0: \sigma^2 = 1.28$ and $H_a: \sigma^2 > 1.28$. Find the significance level (α) corresponding to the value of n and the rejection region.
(a) $n = 21$, $\chi^2 \geq 31.4140$ (b) $n = 33$, $\chi^2 \geq 56.3281$
(c) $n = 41$, $\chi^2 \geq 29.8195$ (d) $n = 29$, $\chi^2 \geq 59.3000$

9.220 Consider a hypothesis test concerning a population variance from a normal population with $H_0: \sigma^2 = 48.92$ and $H_a: \sigma^2 < 48.92$. Find the significance level (α) corresponding to the value of n and the rejection region.
(a) $n = 25$, $\chi^2 \leq 7.4527$ (b) $n = 36$, $\chi^2 \leq 18.5089$
(c) $n = 12$, $\chi^2 \leq 3.8157$ (d) $n = 51$, $\chi^2 \leq 27.9907$

9.221 Consider a hypothesis test concerning a population variance from a normal population with $H_0: \sigma^2 = 15.667$ and $H_a: \sigma^2 \neq 15.667$. Find the significance level (α) corresponding to the value of n and the rejection region.
(a) $n = 9$, $\chi^2 \leq 1.3444$ or $x^2 \geq 21.9550$
(b) $n = 21$, $\chi^2 \leq 8.2604$ or $x^2 \geq 37.5662$
(c) $n = 38$, $\chi^2 \leq 15.0202$ or $x^2 \geq 73.3512$
(d) $n = 18$, $\chi^2 \leq 7.5642$ or $x^2 \geq 30.1910$

9.222 Consider a hypothesis test concerning a population variance from a normal population with $H_0: \sigma^2 = 11.4$ and $H_a: \sigma^2 > 11.4$. Find bounds on the p value for the value of n and the test statistic.
(a) $n = 32$, $\chi^2 = 50.05$ (b) $n = 7$, $\chi^2 = 11.99$
(c) $n = 19$, $\chi^2 = 38.62$ (d) $n = 24$, $\chi^2 = 60.15$

9.223 Consider a hypothesis test concerning a population variance from a normal population with $H_0: \sigma^2 = 404.7$ and $H_a: \sigma^2 < 404.7$. Find bounds on the p value for the value of n and the test statistic.
(a) $n = 11$, $\chi^2 = 1.36$ (b) $n = 17$, $\chi^2 = 1.97$
(c) $n = 31$, $\chi^2 = 14.05$ (d) $n = 51$, $\chi^2 = 32.85$

9.224 Consider a hypothesis test concerning a population variance from a normal population with $H_0: \sigma^2 = 232$ and $H_a: \sigma^2 \neq 232$. Find bounds on the p value for the value of n and the test statistic.
(a) $n = 25$, $\chi^2 = 41.67$ (b) $n = 28$, $\chi^2 = 8.12$
(c) $n = 5$, $\chi^2 = 0.5005$ (d) $n = 16$, $\chi^2 = 38.88$

9.225 Consider a random sample of size 21 from a normal population with hypothesized variance 16.7.
(a) Write the four parts for a one-sided, right-tailed hypothesis test concerning the population variance with $\alpha = 0.01$.
(b) Suppose $s^2 = 28$. Find the value of the test statistic, and draw a conclusion about the population variance.
(c) Find bounds on the p value associated with this hypothesis test, and carefully sketch a graph to illustrate this value.

9.226 Consider a random sample of 16 observations, given in the table, from a normal population with hypothesized variance 36.8. 📊 EX9.226

233.1	226.1	220.3	247.6	232.9	232.8	235.9
232.4	249.4	207.4	231.8	232.1	220.7	229.6
242.5	229.3					

(a) Write the four parts for a two-sided hypothesis test concerning the population variance with $\alpha = 0.05$.
(b) Compute the sample variance and the value of the test statistic. Draw a conclusion about the population variance.
(c) Find bounds on the p value associated with this hypothesis test.

9.227 Consider a random sample of size 40 from a normal population with hypothesized variance 75.6.
(a) Write the four parts for a one-sided, left-tailed hypothesis test concerning the population variance with $\alpha = 0.001$.
(b) Suppose $s^2 = 48.5$. Find the value of the test statistic, and draw a conclusion about the population variance.
(c) Find bounds on the p value associated with this hypothesis test.

Applications

9.228 Physical Sciences A water-droplet generator is used to produce simulated rain or fog and, for example, to test emission drift rates from nuclear-power-plant cooling towers. A piezoelectric water-droplet generator is designed to produce 10-microliter (μL) mist with a variance of 0.25. A random sample of 35 water drops was obtained, and the amount of water in each (in μL) was measured. The summary statistics were $\bar{x} = 10.004$ and $s = 0.558$. Is there any evidence to suggest that the variance is larger than specified? Assume normality and use $\alpha = 0.05$.

9.229 Physical Sciences Energy companies assess potential basins and formations for recoverable shale oil and gas resources. This information is used to forecast production levels and plan further exploration. A random sample of world basins and formations was obtained, and the technically recoverable wet shale gas (in trillion cubic feet, tcf) for each site was recorded.[63] Is there any evidence to suggest that the population variance in technically recoverable wet shale gas is greater than 16,900 tcf^2? Assume normality and use $\alpha = 0.01$. 📊 SHALE

9.230 Manufacturing and Product Development A manufactured glass rod is tested for the stress required for fracture. Experiments suggest that a flame-polished rod has a higher mean fracture stress than a rod with an abraded surface. In both cases, the variance in stress is designed to be at most 324 MPa^2. A random sample of 12 polished glass rods was obtained, and the fracture stress for each was measured. The sample standard deviation in fracture stress was $s = 21.56$. Is there any evidence to refute the manufacturer's claim? Assume the underlying distribution is normal and use $\alpha = 0.01$.

9.231 Public Health and Nutrition The Italian Peoples Bakery in Pennington, New Jersey, advertises an original-recipe cream puff packed with 3 oz of filling. To produce a consistent product, the variance in cream filling is carefully monitored. In a random sample of 23 cream puffs, the sample variance in filling was $s^2 = 0.105$ oz^2. Is there any evidence to suggest that the variance in filling is greater than 0.09 oz^2? Assume normality and use $\alpha = 0.025$.

9.232 Manufacturing and Product Development In an effort to improve efficiency, Samuel Adams brewing company would like to produce yeast slurry with variance no greater than 62.5. A new delivery pump was installed, and a random sample of 10 slurry mixtures was obtained. Each mixture was measured in billion cells/mL and the sample variance was $s^2 = 70.1$. Is there any evidence to suggest that the variance is greater than the desired value? Assume normality and use $\alpha = 0.01$.

9.233 Sports and Leisure The specified weight of a 22-mm die used at casinos in Las Vegas, Nevada, is 10.4 g. The variability in die weight must be negligible (at most 0.04 g^2) for patrons to believe that games involving dice are fair. A random sample of 25 new 22-mm dice was obtained, and each die was carefully weighed. The sample mean weight was $\bar{x} = 10.38$ g and the sample standard deviation was $s = 0.244$ g. Conduct the relevant hypothesis test to determine whether there is any evidence that the variance in die weight is greater than 0.04. Use $\alpha = 0.01$ and assume normality.

9.234 Manufacturing and Product Development SoLux produces special light bulbs for museums, photography studios, automotive paint finishing, and machine vision. The SoLux 4100 Kelvin 50-watt, 12-volt 10° bulb is designed to produce a beam with a diameter of 1.75 ft from 10 ft away. In a random sample of 26 bulbs from the production line, the standard deviation was 0.342 ft. Is there any evidence to suggest that the standard deviation in beam diameter at 10 ft is greater than 0.3? Assume normality and use $\alpha = 0.05$.

9.235 Business and Management Many farmers craft unique corn mazes in their fields after the growing season. These huge puzzles challenge both children and adults, and increase revenue for farmers. Suppose a farmer has constructed a corn maze so that the mean time for completion is 1.5 hr with variance 0.57 hr^2. A random sample of 18 people entering the maze was selected, and the time (in hours) to complete the puzzle was recorded for each. The sample standard deviation was $s = 0.34$. Is there any evidence that the completion-time variance is different from the intended variance? Use $\alpha = 0.05$ and assume the underlying distribution is normal.

9.236 Sports and Leisure Bull riding has become a very popular rodeo sport—the action is fast and dangerous. A bull ride is scored by two judges; one actually rates the bull, while the other evaluates the rider. The variance in bull-riding time tends to be large, approximately 22.5 sec^2. A random sample of 37 bull rides was selected over an entire rodeo season. The sample variance in riding times was $s^2 = 15.6$ sec^2. Smaller variability in bull-riding time translates to a more monotonous, unexciting rodeo. Is there any evidence that bull riding has become less exciting? Assume normality and use $\alpha = 0.05$.

9.237 Manufacturing and Product Development Phillips Mushroom farms is the largest special mushroom grower in the United States. Its crimini mushrooms are tan with a buttery texture and have caps with a mean size of 2 in. in diameter. Packages of crimini mushrooms look and sell better if the mushrooms look uniform, meaning they have small variance in cap diameter. A random sample of Phillips crimini mushrooms was obtained and the cap diameter (in inches) for each was recorded. Is there any evidence to suggest that the true cap variance is greater than 0.25 in.^{2}? Assume normality and use $\alpha = 0.05$. CRIMINI

Extended Applications

9.238 Marketing and Consumer Behavior The wait time before speaking to a customer service representative is an important measure of call center metrics. A recent survey found that, on average, customers will hang up after waiting only 90 sec for a customer service representative. Capital One is reported to offer some of the best service for customers, and its representatives attempt to answer all waiting calls within 60 sec with small variance. A random sample of customers on hold was obtained, and the wait time (in seconds) for each was recorded. WAIT

(a) Is there any evidence to suggest that the standard deviation in wait time is greater than 10 sec? Assume normality and use $\alpha = 0.05$.

(b) Find bounds on the p value associated with this hypothesis test, and carefully sketch a graph to illustrate this value.

9.239 Manufacturing and Product Development Bittersharp apple cider is made from apples that contain more than 0.45% malic acid. It is important for the variance in malic acid content to be small, because large variability may cause the resulting cider to be too sharp. In preparation for making a batch of cider, a random sample of 20 Foxwhelp Bittersharp apples was obtained, and the malic acid content in each was measured. The sample variance was $s^2 = 0.42$.

(a) Is there any evidence to suggest that the population variance in malic acid is greater than 0.36? Assume the underlying distribution is normal and use $\alpha = 0.05$.

(b) Find bounds on the p value associated with this hypothesis test.

9.240 Business and Management Researchers studying the European Union have suggested that small variability in minimum wage among countries leads to more economic stability and growth. A random sample of European countries was obtained and the gross monthly minimum wage (in euros) was recorded for each.[64] EUWAGE

(a) Is there any evidence to suggest that the standard deviation in minimum wage is greater than 500 euros? Assume normality and use $\alpha = 0.05$.

(b) Find bounds on the p value associated with this hypothesis test.

9.241 Manufacturing and Product Development Cell phone companies attract customers by offering a wide coverage area and a consistent, strong signal. Signal strength is measured in decibels above or below 1.0 milliwatt (dBm). A random sample of locations in a company's coverage area was obtained, and the signal strength (in dBm) at each spot was measured. CELL

(a) Suppose a consistent signal means a variance of at most 230 mW^2. Is there any evidence of an inconsistent signal in this company's coverage area? Use $\alpha = 0.05$ and assume normality.

(b) Find bounds on the p value associated with this hypothesis test.

9.242 Manufacturing and Product Development The Estes Pro Series II Sahara model rocket weighs approximately 13 oz and can reach altitudes of more than 700 ft.[65] Suppose the company would like the standard deviation in height to be approximately 50 ft. If the standard deviation is any larger, then more rockets are lost (because they soar too high and drift away) or customers become angry because more launches do not achieve the advertised height. A random sample of rockets was obtained, and the height (in feet) achieved on each launch was measured. ROCKET

(a) Is there any evidence to suggest that the population variance in height is greater than the company's desired value? Assume normality and use $\alpha = 0.01$.

(b) Find bounds on the p value for this hypothesis test.

9.243 Biology and Environmental Science A monarch butterfly may fly as many as 2000 mi during its migration to Mexico. These amazing creatures actually vary dramatically in weight, color, and even flying behavior. Previous research suggests that the mean wingspan for a monarch butterfly is 50 mm, with a standard deviation of 2.75 mm. Some biologists speculate that changes in the environment (e.g., pollution and climate) have caused greater variability in the butterfly's wingspan. A random sample of monarch butterflies was obtained, and the wingspan of each (in millimeters) is given in the table. MNARCH

49.3	51.9	52.9	52.4	51.6	47.8	43.3	47.0
49.4	51.2	50.7	56.8	55.1	47.2	47.0	48.4
55.7	51.5	41.6	50.3	54.9	48.3		

(a) Is there any evidence that the variability in wingspan of the monarch butterfly has increased? Assume normality and use $\alpha = 0.05$.

(b) Find bounds on the p value associated with this hypothesis test.

9.244 Manufacturing and Product Development Following a recent night-time helicopter crash, the air ambulance service in Ontario is considering purchasing night-vision goggles for all pilots. One measure of the effectiveness of these goggles is the resolution, in line pairs per millimeter (lp/mm). A random sample of 18 ATN PS15-4 Gen 4 high-performance night-vision goggles was obtained, and the resolution was measured for each. The sample variance was $s^2 = 5.230$.

(a) Is there any evidence to suggest that the standard deviation in the resolution for these goggles is different from 2? Assume normality and use $\alpha = 0.01$.

(b) Find bounds on the p value associated with this hypothesis test.

9.245 Manufacturing and Product Development Oxford Paper Company manufactures boxboard for folding cartons made entirely from recycled material. One specific product is designed to have thickness 375 μm with standard deviation 7 μm. A random sample of 21 pieces of boxboard was obtained from the assembly line.
  a. Assume the thickness distribution is normal and let $\alpha = 0.05$. Consider a two-sided hypothesis test to determine whether the boxboard is being manufactured with the designed variance in thickness.
  b. Find the critical values in terms of the sample variance. Work backward to determine two values such that if $S^2 \leq s_L^2$ or $S^2 \geq s_H^2$, then the null hypothesis is rejected.
  c. Suppose $s^2 = 56$. Use the critical values from part (b) to draw a conclusion about the population variance.
  d. Suppose $s^2 = 15.6$. Use the critical values from part (b) to draw a conclusion about the population variance.

Challenge Problems

9.246 Economics and Finance An approximate test concerning a population variance is based on a normal distribution. Recall that if S^2 is the sample variance of a random sample of size n from a normal distribution with variance σ^2, the random variable $X = (n-1)S^2/\sigma^2$ has a chi-square distribution with $n-1$ degrees of freedom. If n is large, the random variable X is approximately normal with mean $n-1$ and variance $2(n-1)$. An approximate hypothesis test is based on standardizing X to a Z random variable.

The four parts of the hypothesis test are

$H_0: \sigma^2 = \sigma_0^2$

$H_a: \sigma^2 > \sigma_0^2, \quad \sigma^2 < \sigma_0^2, \quad$ or $\quad \sigma^2 \neq \sigma_0^2$

TS: $Z = \dfrac{S^2 - \sigma_0^2}{\sqrt{2\sigma_0^2}/\sqrt{n-1}}$

RR: $Z \geq z_\alpha, \quad Z \leq -z_\alpha, \quad$ or $\quad |Z| \geq z_{\alpha/2}$

The historic variance in the exchange rate for the Japanese yen against the U.S. dollar is approximately 1.56. A random sample of 48 closing exchange rates was obtained, and the summary statistics were $\bar{x} = 103.74$ and $s^2 = 2.2$.
  (a) Write the four parts of a large-sample, two-sided test concerning the variance in exchange rate, based on a normal approximation. Use $\alpha = 0.05$.
  (b) Is there any evidence to suggest that the variance in exchange rate has changed?
  (c) Find the p value associated with this hypothesis test.
  (d) Conduct an exact hypothesis test based on the chi-square distribution. Compute the p value, and compare it with your answer to part (c).

Chapter 9 Summary

Concept	Page	Notation / Formula / Description
Hypothesis	390	A claim about the value of a specific population parameter.
Null hypothesis	390	H_0, the claim about a parameter assumed to be true, or the hypothesis to be tested.
Alternative hypothesis	391	H_a, the possible values of the parameter if H_0 is false.
Test statistic	391	A rule related to the null hypothesis, involving information in the sample. The value of the test statistic is used to determine which hypothesis, H_0 or H_a, is more likely.
Rejection region	391	An interval or set of numbers determined such that if the value of the test statistic lies in the rejection region, then the null hypothesis is rejected.
One-sided alternatives	392	$H_a: \theta > \theta_0$ (right-tailed), $H_a: \theta < \theta_0$ (left-tailed).
Two-sided alternative	392	$H_a: \theta \neq \theta_0$.
Type I error	397	H_0 is rejected (because the value of the test statistic lies in the rejection region), but H_0 is really true.
Type II error	397	H_0 is not rejected (because the value of the test statistic does not lie in the rejection region), but H_a is really true.
Significance level	397	The probability of a type I error, denoted by P(type I error) $= \alpha$.
p value	415	The smallest significance level for which the null hypothesis can be rejected.

Parameter	Assumptions	Alternative hypothesis	Test statistic	Rejection region
μ	n large, σ known, or normality, σ known	$\mu > \mu_0$ $\mu < \mu_0$ $\mu \neq \mu_0$	$Z = \dfrac{\overline{X} - \mu_0}{\sigma/\sqrt{n}}$	$Z \geq z_\alpha$ $Z \leq -z_\alpha$ $|Z| \geq z_{\alpha/2}$
μ	normality, σ known	$\mu > \mu_0$ $\mu < \mu_0$ $\mu \neq \mu_0$	$T = \dfrac{\overline{X} - \mu_0}{S/\sqrt{n}}$	$T \geq t_{\alpha, n-1}$ $T \leq -t_{\alpha, n-1}$ $|T| \geq t_{\alpha/2, n-1}$
p	n large, nonskewness criteria	$p > p_0$ $p < p_0$ $p \neq p_0$	$Z = \dfrac{\hat{P} - p_0}{\sqrt{\dfrac{p_0(1-p_0)}{n}}}$	$Z \geq z_\alpha$ $Z \leq -z_\alpha$ $|Z| \geq z_{\alpha/2}$
σ^2	normality	$\sigma^2 > \sigma_0^2$ $\sigma^2 < \sigma_0^2$ $\sigma^2 \neq \sigma_0^2$	$X^2 = \dfrac{(n-1)S^2}{\sigma_0^2}$	$X^2 \geq \chi^2_{\alpha, n-1}$ $X^2 \leq \chi^2_{1-\alpha, n-1}$ $X^2 \leq \chi^2_{1-\alpha/2, n-1}$ or $X^2 \geq \chi^2_{\alpha/2, n-1}$

Chapter 9 Exercises

Applications

9.247 Manufacturing and Product Development Vacuum-packed coffee stays fresh longer, but the package tends to be bumpy and unappealing. In a nitrogen-flushed package, all the oxygen is pushed out by heavier nitrogen, producing a smoother, more attractive package that also stays fresh. A machine used to produce a nitrogen-flushed package should dispense 1.6 mol of nitrogen for a 1-lb package of coffee. In a random sample of 23 packings, the sample mean amount of nitrogen dispensed was $\overline{x} = 1.78$ mol. Assume the underlying distribution is normal, with $\sigma = 0.5$ mol, and use $\alpha = 0.01$.
(a) Is there any evidence to suggest that the machine is malfunctioning?
(b) Compute the p value associated with this hypothesis test.

9.248 Manufacturing and Product Development Shanghai Creative Material Co. Ltd. makes plating tape for use on circuit boards, designed to be 4 mil thick. (A mil is equal to 0.001 in.: a milli-inch.) As part of quality control, every hour a random sample of 40 pieces of tape are carefully measured. If any evidence indicates that the thickness is different from 4 mil, then the entire process is stopped and the machinery checked and cleaned. Assume $\sigma = 0.05$ mil and use $\alpha = 0.01$.
(a) Suppose the sample mean is $\overline{x} = 4.014$. Should the process be stopped?
(b) Suppose the sample mean is $\overline{x} = 3.979$. Should the process be stopped?
(c) Comment on the statistical and practical differences between the decisions in parts (a) and (b).

9.249 Physical Sciences CubeSats are tiny satellites that began as an educational experiment but are now included in many commercial and government space launches. A CubeSat is a 10-cm cube that weighs up to 1.3 kg.[66] A random sample of CubeSats set for launch in 2019 was obtained and the weight of each was measured (in kilograms). CUBESAT
(a) Each CubeSat has optimal launch characteristics if the weight is close to 1 kg. Is there any evidence to suggest that the mean weight of CubeSats scheduled for launch is different from 1? Assume the distribution is normal, $\sigma = 0.1$, and use $\alpha = 0.05$.
(b) Compute the p value for this hypothesis test.

9.250 Physical Sciences The U.S. Army Corps of Engineers is dredging the ship channel into Port Corpus Christi to accommodate tankers and bulk carriers. The plans call for the channel to be 530 ft wide.[67] Following completion, a random sample of 18 locations along the channel was selected, and the width at each location was measured. The sample mean was $\overline{x} = 523.72$. Assume the distribution of channel widths is normal and $\sigma = 12.4$.
(a) Is there any evidence to suggest the mean width of the channel is less than 530 ft? Use $\alpha = 0.05$.
(b) Find the p value associated with this hypothesis test.

9.251 Manufacturing and Product Development A piston in a particular 12-cylinder diesel engine is manufactured to have a diameter of 13 mm. Any larger or smaller diameter will cause immediate, costly damage to the engine. Every half-hour, 10 finished pistons are selected and the diameter of each (in

millimeters) is carefully measured. If there is any evidence that the mean diameter is different from 13 mm, the manufacturing process is stopped and the machinery inspected. The quality control inspector uses a significance level of $\alpha = 0.05$.
(a) Suppose the sample mean is $\bar{x} = 12.89$ and the sample standard deviation is $s = 0.96$. Should the manufacturing process be stopped?
(b) Suppose the sample mean is $\bar{x} = 13.04$ and the sample standard deviation is $s = 0.045$. Should the manufacturing process be stopped?
(c) If you were buying these pistons, would you like the manufacturer to use a smaller or a larger significance level? Justify your answer.

9.252 Medicine and Clinical Studies According to a recent study, a visit to a primary care physician lasts less than 5 min for half of the world's population.[68] In the United States, appointments with a primary care physician last, on average, 20 min. Residents at Geisinger Medical Center are being trained to spend more time with their patients, to listen carefully, and to develop a relationship with each patient. A random sample of patient visits at Geisinger was obtained, and the time spent with the primary care physician was recorded.
(a) Is there any evidence to suggest that the true population mean time spent with each patient is greater than 20 min? Assume normality and use $\alpha = 0.05$.
(b) Find bounds on the p value associated with this hypothesis test. CARE

9.253 Business and Management According to a report issued by the management consulting firm McKinsey & Company, 20% of all companies require genetic or family medical history information from employees or job applicants. A random sample of 1500 companies was obtained, and 345 said they require this information from employees or job applicants.
(a) Identify p_0 and n, and compute $\hat{p}$.
(b) Check the nonskewness criterion.
(c) Is there any evidence to suggest that the true proportion of companies that require this information is greater than 0.20? Use $\alpha = 0.01$.
(d) Compute the p value associated with this hypothesis test.

9.254 Public Policy and Political Science Despite a recent Supreme Court ruling, a candidate for district attorney claims that 75% of all residents in Elko County, Nevada, favor granting police the right to track a cell phone without a warrant. To check this claim, a random sample of 560 residents was selected, and 392 said they were in favor of tracking without a warrant.
(a) Is there any evidence to suggest that the true proportion of residents who favor tracking without a warrant is less than 0.75? Use $\alpha = 0.01$.
(b) Compute the p value associated with this hypothesis test.

9.255 Public Health and Nutrition Vitamin D coming from sunlight and proper nutrition is an essential nutrient that contributes primarily to strong bones. A recent report suggests that 42% of the U.S. population is vitamin D deficient.[69] In a random sample of 300 people older than 65, 142 were found to have a vitamin D deficiency.
(a) Is there any evidence to suggest that the proportion of people in this population, older than 65, is greater than 0.42? Use $\alpha = 0.05$.
(b) Find the p value associated with this hypothesis test.

9.256 Public Health and Nutrition Many remember actor Michael J. Fox as conservative Republican Alex Keaton on the TV show *Family Ties*. After more than 10 years, he returned to television as a news anchor diagnosed with Parkinson's disease. The show reflects much of his own frustrating experience with this disease. Most cases of Parkinson's disease are linked to environmental or genetic factors. However, it has been reported that approximately 15% of people with Parkinson's disease have a family history of the disease.[70] A random sample of 420 Parkinson's patients was obtained, and 78 had a family history of the disease. Is there any evidence to suggest that the proportion of Parkinson's patients with a family history of the disease is different from 0.15? Use $\alpha = 0.05$.

9.257 Public Policy and Political Science It was recently disclosed that soil testing conducted in certain Winnipeg neighborhoods more than 10 years ago showed potentially dangerous levels of lead. However, Canada's New Democratic Party (NDP) government withheld the test results from residents. Data were collected on the amount of lead found at 23 random locations (in $\mu g/g$).[71] SOIL
(a) Is there any evidence to suggest that the mean amount of lead in the soil is greater than the Canadian Council of Ministers of the Environment (CCME) residential guidelines for human health of 140 $\mu g/g$? Assume normality and use $\alpha = 0.01$.
(b) Find bounds on the p value for this hypothesis test.

9.258 Medicine and Clinical Studies The diameter of a virus is approximately 0.3 μm. There is some speculation that new virus strains exhibit greater variability in diameter. A medical research lab obtained a random sample of 15 new virus strains and measured the diameter (in micrometers) of each. The sample mean diameter was 0.323 and the sample variance was $s^2 = 0.0026$. Is there any evidence to suggest that the true population variation in the diameter of viruses has increased from 0.0015? Assume normality and use $\alpha = 0.05$.

9.259 Manufacturing and Product Development In going from green to oven dry, paper birch wood experiences radial shrinkage of approximately 6.3%. A new process has been developed to decrease the variability in shrinkage. In a random sample of 21 pieces of paper birch, the sample variance in shrinkage was 0.39. Is there any evidence to suggest that the population variance in shrinkage is less than 0.50? Assume normality and use $\alpha = 0.10$.

9.260 Medicine and Clinical Studies The normal blood platelet count for an adult ranges from 150,000 to 400,000 per milliliter of blood, with a standard deviation of approximately 62,500. A researcher has speculated that increased exposure to pollutants has increased the variability in numbers of blood platelets. A random sample of 37 adults was obtained, and the blood platelet count was measured in each person. The sample standard deviation was $s = 65,268$. Is there any evidence to suggest that the population variance in blood platelet count has increased? Assume normality and use $\alpha = 0.001$.

9.261 Manufacturing and Product Development Several new hotels and rental properties in New York City have been constructed using fabricated structural components, known as modules. Each module is constructed in a factory, transported to the construction site, stacked, and assembled. Carmel Place consists of 55 studio modules, each designed to be 24 ft wide. A random sample of 23 modules was obtained on site, and the width of each was carefully measured. The sample variance was $s^2 = 0.056$.
(a) Is there any evidence to suggest that the true population variance in width is greater than 0.04, the accepted tolerance for connecting each module? Assume normality and use $\alpha = 0.01$.
(b) Find bounds on the p value for this hypothesis test.

Extended Applications

9.262 Physical Sciences Workers at the Daivik diamond mine in northern Canada extract approximately 1800 dry metric tons (DMT) of ore per day, which is sifted and examined for diamonds. New machinery has just been installed that is designed to increase the amount of ore extracted per day. A random sample of 36 days was selected, and the amount of ore extracted each day was recorded. The sample mean was $\bar{x} = 1852$ DMT. Assume $\sigma = 202$ DMT.
(a) Is there any evidence to suggest that the new machinery has improved production? Use $\alpha = 0.05$.
(b) What is the probability of a type II error if the true mean amount of ore extracted has changed to 1875 DMT; that is, what is $\beta(1875)$? Find the probability of a type II error if the true mean is 1925 DMT.

9.263 Economics and Finance For those Australian taxpayers who claimed "other" work-related expenses (WRE) in 2017, the mean amount claimed was $1179.[72] The Australian Taxation Office (ATO) decided to focus on WRE by sending letters and conducting audits. As a result, it quickly collected adjustments of approximately $100 million. The following year, a random sample of taxpayers who claimed other WRE was obtained and the amount was recorded for each. **AUSWRE**
(a) Is there any evidence to suggest that the mean other WRE has decreased? Assume normality and use $\alpha = 0.05$.
(b) Find bounds on the p value for this hypothesis test.

9.264 Physical Sciences The design specifications for a new gymnasium at a local high school call for the lights to produce at least 40 footcandles (fc; a measure of brightness). Before the new facility was opened to students, the contractor collected a sample of 23 brightness measurements at random locations in the gym. The sample mean was $\bar{x} = 38.63$ fc, and the sample standard deviation was $s = 5.6$ fc.
(a) Is there any evidence to suggest that the mean brightness in the gym is less than the design specifications? Assume normality and use $\alpha = 0.01$.
(b) Find bounds on the p value associated with this hypothesis test.

9.265 Business and Management The largest above-ground storage tanks in Bayonne, New Jersey, have the capacity to hold 6,000,000 gal of oil. For safety reasons, managers prefer the mean amount stored at any given time to be no greater than 4,500,000 gal. A random sample of nine large tanks was selected, and the amount of oil stored in each (in gallons) was recorded. The summary statistics were $\bar{x} = 4,675,250$ and $s = 482,556$.
(a) Is there any evidence to suggest that the mean amount of oil stored in the large tanks exceeds the safety level? Assume normality and use $\alpha = 0.025$.
(b) Find bounds on the p value associated with this hypothesis test.

9.266 Public Health and Nutrition Cast iron is an extremely durable cookware material, good for searing and blackening foods. However, a cast-iron pan, for example, can be full of bacteria and can lend unwanted flavors to food. A company trying to promote alternative ceramic cookware asked members of a community to bring their favorite cast-iron cookware to its store for bacteria testing. The company manager claimed that at least 60% of all cast-iron cookware contains harmful bacteria. A random sample of 120 pans was selected, and 57 were found to contain harmful bacteria.
(a) Is there any evidence to refute the manager's claim? Use $\alpha = 0.01$.
(b) Find the p value associated with this hypothesis test.
(c) Do you believe the sample is really random? Why or why not?

9.267 Marketing and Consumer Behavior An assisted-living home is an alternative to a nursing home and a bridge between a skilled-care facility and a patient's residence. A recent report indicated that 92% of all patients in assisted-living homes are satisfied with the facility and the care. An insurance company believes that the actual percentage is much lower (due to health care violations and strict for-profit motives). A random sample of 5000 patients in assisted-living homes around the country was obtained, and 4576 said they were satisfied with the facility and the care.
(a) Is there any evidence to suggest that the true proportion of assisted-living patients who are satisfied is less than 0.92? Use $\alpha = 0.01$.
(b) Find the p value associated with this hypothesis test.
(c) Carefully sketch a graph illustrating the critical value, value of the test statistic, and p value.

9.268 Biology and Environmental Science New Bullards Bar Reservoir in Yuba County, California, provides water for irrigation, hydroelectricity production, and recreational activities. A random sample of days during summer 2018 was obtained, and the capacity of the reservoir on each day (in thousands of acre feet) was measured.[73] **RSRVOIR**

(a) Suppose the historic mean capacity during the summer is 662 thousand acre feet. Is there any evidence to suggest that the true population mean capacity has decreased? Assume normality and use $\alpha = 0.05$.

(b) Find the p value associated with this hypothesis test.

9.269 Manufacturing and Product Development A single-sided Cast Stone Stretcher piece is designed to weigh 50 lb.[74] This stone piece is part of a wall system for residential and commercial retaining walls. Large variability in weight of these stones could cause structural deficiencies in a wall. A random sample of 17 standard units was obtained, and the sample variance was 7.75 lb^2.

(a) Is there any evidence to suggest that the population variance is greater than 4 lb^2? Assume normality and use $\alpha = 0.05$.

(b) Find bounds on the p value for this hypothesis test.

9.270 Business and Management According to Catalyst, a nonprofit organization working to provide more opportunities for women and business, women currently hold 5% of S&P 500 CEO positions.[75] Some economic researchers believe that women own a larger proportion of small businesses. A random sample of 550 small businesses in the United States was obtained, and 37 were owned by women.

(a) Is there any evidence to suggest that the proportion of small businesses owned by women is greater than 5%? Use $\alpha = 0.01$.

(b) Find the p value for this hypothesis test.

apichart609@gmail.com/Deposit Photos

CHAPTER APP

9.271 Heavy Metal The results from a Consumer Reports study indicated that 68% of common baby food products contained levels of toxic chemicals that food safety experts suggest are worrisome. Following this study, the Food and Drug Administration (FDA) issued new guidelines to manufacturers for processing baby foods in an attempt to lower this percentage. Later, another study was conducted to determine whether the true proportion of baby foods that contain toxic chemicals had decreased. Of the 125 baby food products inspected, 75 were found to contain high levels of toxic chemicals.

(a) State the null hypothesis and the alternative hypothesis in terms of p, the true proportion of baby foods that contain toxic chemicals.

(b) Discuss the consequences of the decision to reject or not to reject the null hypothesis for each truth assumption.

(c) Is there any evidence to suggest that less than 68% of baby foods contain toxic chemicals? Use $\alpha = 0.05$ and the p value associated with this hypothesis test to draw a conclusion.

Confidence Intervals and Hypothesis Tests Based on Two Samples or Treatments

10

ASphoto777/Deposit Photos

◀ **Looking Back**
- Recall the formal, four-part hypothesis test procedure: **Section 9.1**.
- Remember the specific inference procedures concerning a single population parameter: μ, p, or σ^2: **Sections 9.3, 9.5, 9.6, and 9.7**.

Looking Forward ▶
- Adapt and extend single-sample confidence interval and hypothesis test procedures: **Chapter 10**.
- Learn how to compare two population means: **Section 10.1 and Section 10.2**.
- Learn the procedure for comparing two population means when the data are paired: **Section 10.3**.
- Learn the procedure for comparing two population proportions: **Section 10.4**.
- Learn the procedure for comparing two population variances: **Section 10.5**.

CHAPTER **APP**

Prescription Errors

We would like to think that pharmacists have some magical power that allows them to decipher even the worst handwriting on a doctor's prescription. Well, they don't. Many people just assume that doctors write illegibly but pharmacists usually know what is written. Unfortunately, bad handwriting can lead to a serious medical issue. Many examples of prescription errors can be cited, including substituting the wrong drug or indicating an incorrect dosage.

An experiment was designed to compare the proportion of pharmacist errors from computerized physician order entry (CPOE) prescriptions with hand-written prescriptions (HWP). Random samples of 2429 CPOE prescriptions and 1036 HWP prescriptions were obtained. When each prescription was carefully checked for pharmacist error, the researcher found 117 errors in the CPOE prescriptions and 69 in the HWP prescriptions.

The hypothesis test procedures presented in this chapter will be used to compare parameters (proportions, means, or variances, for example) from two different populations. In this case, we will compare the population proportion of CPOE prescription errors with the population proportion of HWP prescription errors. These tests are constructed using methods of standardization similar to those in Chapter 9.

Notation

To conduct a hypothesis test to compare two (similar) population parameters, we will simply modify the single-sample procedures presented in the previous chapter. Perhaps the most burdensome aspect of these procedures is the notation. The following table summarizes the notation used to represent similar parameters associated with two different populations.

	Population parameters			
	Mean	Variance	Standard deviation	Proportion
Population 1	μ_1	σ_1^2	σ_1	p_1
Population 2	μ_2	σ_2^2	σ_2	p_2

The next table summarizes the notation used to represent values of summary statistics associated with samples from two different populations.

Note: We do not necessarily use every summary statistic associated with a sample in every problem. For example, we may need only the sample size and proportion in one case, but use the sample size, mean, and standard deviation in another problem.

	Sample statistics				
	Sample size	Mean	Variance	Standard deviation	Proportion
Sample from population 1	n_1	$\bar{x}_1$	s_1^2	s_1	$\hat{p}_1$
Sample from population 2	n_2	$\bar{x}_2$	s_2^2	s_2	$\hat{p}_2$

To compare two population parameters to see whether any evidence indicates that they are different, we often consider a difference. For example, to compare two population means μ_1 and μ_2, we consider the difference $\mu_1 - \mu_2$. In searching for evidence that p_1 is larger than p_2, we look at the difference $p_1 - p_2$.

We may consider a difference for two important reasons.

1. A typical relationship between two population parameters can be written in terms of a difference. For example, suppose we need to compare the means from two populations, μ_1 and μ_2.

Standard notation		Difference notation
$\mu_1 = \mu_2$	is equivalent to	$\mu_1 - \mu_2 = 0$
$\mu_1 > \mu_2$	is equivalent to	$\mu_1 - \mu_2 > 0$
$\mu_1 < \mu_2$	is equivalent to	$\mu_1 - \mu_2 < 0$

Therefore, a statistical test with null hypothesis $H_0: \mu_1 - \mu_2 = 0$ corresponds to a test of $H_0: \mu_1 = \mu_2$. Note that $H_a: \mu_1 - \mu_2 > 0$ is equivalent to $H_a: \mu_1 > \mu_2$.

The hypothesized difference between two means may be nonzero. The null hypothesis $H_0: \mu_1 = \mu_2 + 5$ written using a difference is equivalent to $H_0: \mu_1 - \mu_2 = 5$.

2. A difference (for example, $\mu_1 - \mu_2$) is itself a single population parameter. We can use a natural, intuitive statistic, $\overline{X}_1 - \overline{X}_2$, to estimate the value of this parameter. In this chapter, we will use the properties of $\overline{X}_1 - \overline{X}_2$ to develop a test statistic.

As in the statistical tests presented in the last chapter, we usually make certain assumptions in any two-sample hypothesis test. The assumptions associated with the hypothesis tests in this chapter include a statement concerning the selection of individuals or objects from two different populations.

Definition

1. Two samples are **independent** if the process of selecting individuals or objects in sample 1 has no effect on, or no relation to, the selection of individuals or objects in sample 2. If the samples are not independent, they are **dependent**.

2. A **paired** data set is the result of matching each individual or object in sample 1 with a similar individual or object in sample 2. Paired data are often obtained in experiments that involve a before measurement and an after measurement on each individual or object. Each before observation is matched, or paired, with an after observation.

Similar means the individuals or objects share some common, fundamental characteristic. They may even be the same individual or object!

The notation, the idea of using differences, and the extra assumptions just described are all used in the next sections to construct hypothesis tests for comparing various characteristics of two populations.

10.1 Inference: Two Independent Samples, Population Variances Known

General Hypothesis Test

As in Chapter 8 and Chapter 9, the first hypothesis test presented here, for comparing two population means, is instructive but not very practical. Because $\overline{X}_1$ is a good estimator for μ_1 and $\overline{X}_2$ is a good estimator for μ_2, it is reasonable to use the estimator $\overline{X}_1 - \overline{X}_2$ to estimate the parameter $\mu_1 - \mu_2$. To develop a hypothesis test, we need to know the properties of the estimator, or the distribution of the random variable, $\overline{X}_1 - \overline{X}_2$.

Properties of $\overline{X}_1 - \overline{X}_2$

Interpreting the difference between two means.

Suppose

1. $\overline{X}_1$ is the mean of a random sample of size n_1 from a population with mean μ_1 and variance σ_1^2.
2. $\overline{X}_2$ is the mean of a random sample of size n_2 from a population with mean μ_2 and variance σ_2^2.
3. The samples are independent.

If the distributions of both populations are normal, then the random variable $\overline{X}_1 - \overline{X}_2$ has the following properties.

1. $E(\overline{X}_1 - \overline{X}_2) = \mu_{\overline{X}_1 - \overline{X}_2} = \mu_1 - \mu_2$

$\overline{X}_1 - \overline{X}_2$ is an unbiased estimator of the parameter $\mu_1 - \mu_2$. The distribution is centered at $\mu_1 - \mu_2$.

2. $\text{Var}(\overline{X}_1 - \overline{X}_2) = \sigma^2_{\overline{X}_1 - \overline{X}_2} = \dfrac{\sigma_1^2}{n_1} + \dfrac{\sigma_2^2}{n_2}$ and the standard deviation is

$$\sigma_{\overline{X}_1 - \overline{X}_2} = \sqrt{\dfrac{\sigma_1^2}{n_1} + \dfrac{\sigma_2^2}{n_2}}.$$

3. The distribution of $\overline{X}_1 - \overline{X}_2$ is normal.

Can you see the standardization coming?

If the underlying distributions are not known, but both n_1 and n_2 are large, then $\overline{X}_1 - \overline{X}_2$ is approximately normal (by the Central Limit Theorem).

Because the distribution of $\overline{X}_1 - \overline{X}_2$ is (approximately) normal, we can use the usual standardization to obtain a Z random variable. The resulting hypothesis test has a very typical form.

Hypothesis Test Concerning Two Population Means When Population Variances Are Known

For ease of reference, we'll call these the two-sample Z-test assumptions.

Given two independent random samples, the first of size n_1 from a population with mean μ_1 and the second of size n_2 from a population with mean μ_2, assume that:

1. The underlying populations are normal and/or both sample sizes are large.
2. The population variances, σ_1^2 and σ_2^2, are known.

A hypothesis test concerning two population means, in terms of the difference in means $\mu_1 - \mu_2$, with significance level α, has the following form:

This is the template for a hypothesis test concerning population means when the variances are known, sometimes called a two-sample Z test.

$H_0: \mu_1 - \mu_2 = \Delta_0$

$H_a: \mu_1 - \mu_2 > \Delta_0, \quad \mu_1 - \mu_2 < \Delta_0, \quad \text{or} \quad \mu_1 - \mu_2 \neq \Delta_0$

TS: $Z = \dfrac{(\overline{X}_1 - \overline{X}_2) - \Delta_0}{\sqrt{\dfrac{\sigma_1^2}{n_1} + \dfrac{\sigma_2^2}{n_2}}}$

RR: $Z \geq z_\alpha, \quad Z \leq -z_\alpha, \quad \text{or} \quad |Z| \geq z_{\alpha/2}$

A CLOSER LOOK

1. The value Δ_0 is the fixed, hypothesized difference in means. Usually $\Delta_0 = 0$, that is, the means are assumed to be equal. The null hypothesis is then $H_0: \mu_1 - \mu_2 = 0$, which is equivalent to $H_0: \mu_1 = \mu_2$. However, Δ_0 may be some nonzero value. For example, two population means may historically differ by 12 so that $H_0: \mu_1 - \mu_2 = 12 \,(= \Delta_0)$. We may conduct a test to see whether there is any change in this difference, with $H_a: \mu_1 - \mu_2 \neq 12$.

2. Just a reminder: Use only one (appropriate) alternative hypothesis and the corresponding rejection region. The z critical values come from the standard normal distribution.

3. We can use this hypothesis test procedure only if we know both population variances. If they are unknown but both sample sizes are large, some statisticians substitute s_1^2 for σ_1^2 and s_2^2 for σ_2^2. This produces an approximate test statistic. Section 10.2 presents an exact procedure for comparing population means (under certain assumptions) when the population variances are unknown.

Example: Known Variances

The next example illustrates this two-sample Z-test hypothesis test procedure.

EXAMPLE 10.1 Resting Heart Rate

The makers of the health management device Fitbit have collected at least 150 billion hours of heart rate data from millions of people all over the world. These data suggest that the resting heart rates for men and women are different at every age.[1] Independent random samples of 50-year-old men and women were obtained, and the resting heart rate was measured (in beats per minute, bpm) for each. The summary statistics and known variances are given in the table.

Group	Sample size	Sample mean	Population variance
Men	$n_1 = 18$	$\bar{x}_1 = 66.11$	$\sigma_1^2 = 7.29$
Women	$n_2 = 24$	$\bar{x}_2 = 68.17$	$\sigma_2^2 = 10.24$

Is there any evidence to suggest that the resting heart rate for 50-year-old men is less than the resting heart rate for 50-year-old women? Use $\alpha = 0.05$ and assume that each underlying distribution of resting heart rate is normal.

Solution

This example involves a hypothesis test concerning two population means when the variances are known. The samples are random and independent, the underlying populations are normal, and the population variances are known. Use a one-sided alternative hypothesis and the corresponding rejection region, find the value of the test statistic, and draw a conclusion.

STEP 1 Arbitrarily, let the 50-year-old men be population 1, and the 50-year-old women be population 2.

The current state, or assumption, is that the two population mean resting heart rates are equal:

$$\mu_1 = \mu_2 \Rightarrow \mu_1 - \mu_2 = 0 \,(= \Delta_0)$$

The sample sizes, sample means, and the population variances are given.

We are trying to find evidence that the mean resting heart rate for 50-year-old men is less than the mean resting heart rate for 50-year-old women: $\mu_1 < \mu_2$, which is the same as $\mu_1 - \mu_2 < 0$. Therefore, the alternative hypothesis is one-sided, left-tailed.

STEP 2 The four parts of the hypothesis test are

$H_0: \mu_1 - \mu_2 = 0$

$H_a: \mu_1 - \mu_2 < 0$

TS: $Z = \dfrac{(\bar{X}_1 - \bar{X}_2) - \Delta_0}{\sqrt{\dfrac{\sigma_1^2}{n_1} + \dfrac{\sigma_2^2}{n_2}}}$

RR: $Z \leq -z_\alpha = -z_{0.05} = -1.6449$

STEP 3 The value of the test statistic is

$$z = \frac{(\bar{x}_1 - \bar{x}_2) - \Delta_0}{\sqrt{\dfrac{\sigma_1^2}{n_1} + \dfrac{\sigma_2^2}{n_2}}} = \frac{66.11 - 68.17}{\sqrt{\dfrac{7.29}{18} + \dfrac{10.24}{24}}} = -2.2589 \,(\leq -1.6449)$$

STEP 4 Because -2.2589 lies in the rejection region, we reject the null hypothesis at the $\alpha = 0.05$ significance level. There is evidence to suggest that the mean resting heart rate of 50-year-old men is less than the mean resting heart rate of 50-year-old women.

The p value for this hypothesis test is

$p = P(Z \leq -2.2589) \approx P(Z \leq -2.26) = 0.0119 \,(\leq 0.05)$ Appendix Table 3.

Because $p \leq \alpha$, we reject the null hypothesis.

Figure 10.1 is a visualization of the p value, and Figure 10.2 shows a technology solution.

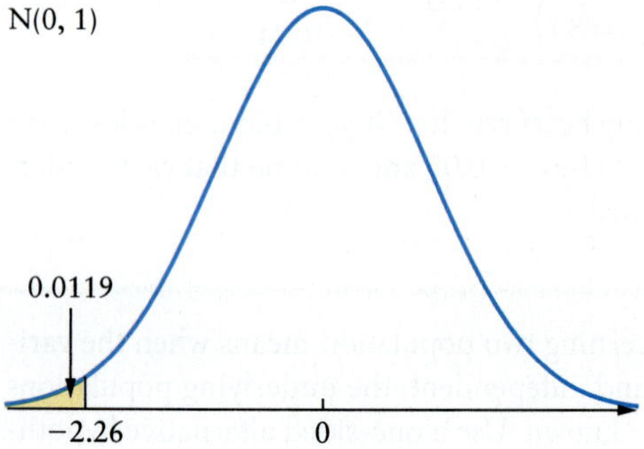

```
> n_1 <- 18
> n_2 <- 24
>
> xbar_1 <- 66.11
> xbar_2 <- 68.17
>
> var_1 <- 7.29
> var_2 <- 10.24
>
> z_crit <- qnorm(0.05)
> z <- (xbar_1 - xbar_2)/sqrt(var_1/n_1+var_2/n_2)
> p_value <- pnorm(z)
>
> round(cbind(z_crit,z,p_value),4)
       z_crit       z p_value
[1,]  -1.6449 -2.2589  0.0119
```

Figure 10.1 p value illustration: $p = P(Z \leq -2.26) = 0.0119 \leq 0.05 = \alpha$.

Figure 10.2 R calculations to find the critical value, the value of the test statistic, and the p value.

Nonzero Difference ($\Delta_0 \neq 0$)

The next example involves a hypothesis test with a nonzero value for the hypothesized difference in means Δ_0.

EXAMPLE 10.2 Low-Carb Ice Cream

LOCARB   Low-carbohydrate foods are very popular, as many Americans try to avoid the sugar and starch combination that they believe causes weight gain. An advertisement for a low-carb ice cream claims that the product has 16 fewer grams of carbohydrates per serving than the leading store brand. To check this claim, independent random samples of each type of ice cream were obtained, and the amount of carbohydrates in each serving was measured. The data are given (in grams) in the table.

Store brand (1)									
15.4	20.4	21.0	24.3	23.3	18.7	19.8	22.5	18.9	22.8
25.4	25.1	20.3	24.1	16.6	22.6	22.1	19.4	16.6	24.4
17.8	18.6	14.9	24.6	19.1	17.9	18.7	20.1	26.3	18.4
21.8	17.1	21.5	19.6	22.9	22.2	21.5	18.3		

Low-carb brand (2)									
3.7	3.9	4.5	4.3	3.2	3.6	3.7	3.6	3.7	4.0
4.1	3.1	4.3	3.4	3.4	3.5	4.4	4.9	3.7	3.8
4.1	4.7	3.7	4.2	3.1	4.4	4.2	3.4	4.8	3.6
3.2	3.4	4.2	3.0	3.9					

The variance in carbohydrates per serving is known to be 8.5 for the store brand and 0.253 for the low-carb brand. Is there any evidence to suggest that the difference in population means of carbohydrates per serving is different from 16 g? Use $\alpha = 0.01$.

Solution

This example involves a hypothesis test concerning two population means. The samples are random, independent, and large, and the population variances are known. Use the two-sided alternative hypothesis and the corresponding rejection region, find the value of the test statistic, and draw a conclusion.

STEP 1 The hypothesized difference is $\mu_1 - \mu_2 = 16 \, (= \Delta_0)$.

The sample sizes are $n_1 = 38$ and $n_2 = 35$.

The known population variances are $\sigma_1^2 = 8.5$ and $\sigma_2^2 = 0.253$. The significance level is $\alpha = 0.01$.

We are testing for any difference in population means other than 16 g of carbohydrates. This is a two-sided test.

The samples are random and independent, and the population variances are known. The underlying population distributions are unknown, but both sample sizes are large (≥ 30). We can conduct a hypothesis test concerning two population means when variances are known.

STEP 2 The four parts of the hypothesis test are

$H_0: \mu_1 - \mu_2 = 16$

$H_a: \mu_1 - \mu_2 \neq 16$

TS: $Z = \dfrac{(\overline{X}_1 - \overline{X}_2) - \Delta_0}{\sqrt{\dfrac{\sigma_1^2}{n_1} + \dfrac{\sigma_2^2}{n_2}}}$

RR: $|Z| \geq z_{\alpha/2} = z_{0.005} = 2.5758$

STEP 3 The sample means are

$\overline{x}_1 = \dfrac{1}{38}(15.4 + 20.4 + \cdots + 18.3) = 20.6579$

$\overline{x}_2 = \dfrac{1}{35}(3.7 + 3.9 + \cdots + 3.9) = 3.8486$

The value of the test statistic is

$z = \dfrac{(\overline{x}_1 - \overline{x}_2) - \Delta_0}{\sqrt{\dfrac{\sigma_1^2}{n_1} + \dfrac{\sigma_2^2}{n_2}}} = \dfrac{(20.6579 - 3.8486) - 16}{\sqrt{\dfrac{8.5}{38} + \dfrac{0.253}{35}}} = 1.6842$

STEP 4 The value of the test statistic, $z = 1.6842$, does not lie in the rejection region. Thus, we do not reject the null hypothesis. There is no evidence to suggest that the difference in population mean carbohydrates is different from 16 g at the $\alpha = 0.01$ significance level.

This is a two-sided test and the value of the test statistic is positive, so $p/2$ is a right-tail probability.

$p/2 = P(Z \geq 1.6842) \approx P(Z \geq 1.68)$ *Definition of p value for a two-sided test.*

$= 1 - P(Z \leq 1.68)$ *The Complement Rule.*

$= 1 - 0.9535 = 0.0465$ *Appendix Table 3.*

$p = 2(0.0465) = 0.0930$ *Solve for p.*

Because $p = 0.0930 > 0.01 (= \alpha)$, we do not reject the null hypothesis.

Figure 10.3 is a visualization of the p value, and **Figure 10.4** shows a technology solution.

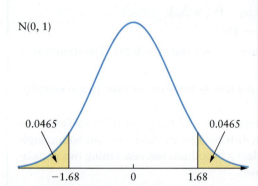

Figure 10.3 *p* value illustration:
$p = 2P(Z \geq 1.68) = 0.0930 \geq 0.01 = \alpha$.

```
> brand_1 <- c(15.4,20.4,...,18.3)
> brand_2 <- c(3.7,3.9,...,3.9)
>
> n_1 <- length(brand_1)
> n_2 <- length(brand_2)
>
> xbar_1 <- mean(brand_1)
> xbar_2 <- mean(brand_2)
>
> var_1 <- 8.5
> var_2 <- 0.253
>
> z_crit <- abs(qnorm(0.005))
> z <- ((xbar_1 - xbar_2)-16)/sqrt(var_1/n_1+var_2/n_2)
> p_value <- 2*(1-pnorm(z))
>
> round(cbind(z_crit,z,p_value),4)
     z_crit      z p_value
[1,] 2.5758 1.6842  0.0921
```

Figure 10.4 R calculations to find the critical value, the value of the test statistic, and the *p* value.

TRY IT NOW Go to Exercises 10.15 and 10.17

Confidence Interval for $\mu_1 - \mu_2$

Given the two-sample Z test assumptions and the properties of the random variable $\overline{X}_1 - \overline{X}_2$, we can construct a confidence interval (CI) for the (difference) parameter $\mu_1 - \mu_2$.

▶ As usual, to find a general CI, start with an appropriate symmetric interval about 0 such that the probability that Z lies in this interval is $1 - \alpha$.

$$P\left[-z_{\alpha/2} < \underbrace{\frac{(\overline{X}_1 - \overline{X}_2) - (\mu_1 - \mu_2)}{\sqrt{\frac{\sigma_1^2}{n_1} + \frac{\sigma_2^2}{n_2}}}}_{Z} < z_{\alpha/2}\right] = 1 - \alpha \quad (10.1)$$

Manipulate the inequality in Equation 10.1 to sandwich the parameter $\mu_1 - \mu_2$. We obtain the following probability statement.

$$P\left[(\overline{X}_1 - \overline{X}_2) - z_{\alpha/2} \cdot \sqrt{\frac{\sigma_1^2}{n_1} + \frac{\sigma_2^2}{n_2}} < \mu_1 - \mu_2 < (\overline{X}_1 - \overline{X}_2) + z_{\alpha/2} \cdot \sqrt{\frac{\sigma_1^2}{n_1} + \frac{\sigma_2^2}{n_2}}\right] = 1 - \alpha$$

This leads to the general expression for a confidence interval. ◀

How to Find a $100(1-\alpha)\%$ Confidence Interval for $\mu_1 - \mu_2$ When Variances Are Known

Given the two-sample Z test assumptions, a $100(1-\alpha)\%$ confidence interval for $\mu_1 - \mu_2$ has the following values as endpoints:

$$(\overline{x}_1 - \overline{x}_2) \pm z_{\alpha/2} \cdot \sqrt{\frac{\sigma_1^2}{n_1} + \frac{\sigma_2^2}{n_2}} \quad (10.2)$$

EXAMPLE 10.3 No Anchovies

Pizza stones designed for home use help cooks produce baked goods with brick-oven qualities, such as a crusty loaf of bread or crispy-crust pizza. However, pizza stones can be very heavy and take up a lot of space in a traditional residential oven. Independent random samples of two similar types of round pizza stones were obtained, and the weight (in pounds) of each was recorded. The summary statistics and known variances are given in the table.

Pizza stone	Sample size	Sample mean	Population variance
Kitchen Depot (1)	$n_1 = 35$	$\bar{x}_1 = 6.21$	$\sigma_1^2 = 2.1$
Head Chef (2)	$n_2 = 31$	$\bar{x}_2 = 7.08$	$\sigma_2^2 = 3.5$

Find a 95% confidence interval for the difference in population mean pizza-stone weights.

Solution

STEP 1 Sample sizes, sample means, and known variances are given.

The underlying weight distributions are unknown, but the sample sizes are both large (≥ 30).

$1 - \alpha = 0.95 \Rightarrow \alpha = 0.05 \Rightarrow \alpha/2 = 0.025$ *Find $\alpha/2$.*

$z_{\alpha/2} = z_{0.025} = 1.96$ *Find the z critical value.*

STEP 2 Use Equation 10.2.

$$(\bar{x}_1 - \bar{x}_2) \pm z_{\alpha/2} \cdot \sqrt{\frac{\sigma_1^2}{n_1} + \frac{\sigma_2^2}{n_2}}$$ *Equation 10.2.*

$$= (6.21 - 7.08) \pm (1.96)\sqrt{\frac{2.1}{35} + \frac{3.5}{31}}$$ *Use summary statistics and critical value.*

$$= -0.87 \pm 0.8150$$ *Simplify.*

$$= (-1.6850, -0.0550)$$ *Compute endpoints.*

The interval $(-1.6850, -0.0550)$ is a 95% confidence interval for the difference in population mean weights (in pounds) of the pizza stones, $\mu_1 - \mu_2$. This interval represents a set of very plausible values for the difference in population mean weights.

Figure 10.5 shows a technology solution.

```
> n_1 <- 35
> n_2 <- 31
> xbar_1 <- 6.21
> xbar_2 <- 7.08
> var_1 <- 2.1
> var_2 <- 3.5
> 
> z_crit <- abs(qnorm(0.025))
> left <- (xbar_1 - xbar_2) - z_crit * sqrt(var_1/n_1+var_2/n_2)
> right <- (xbar_1 - xbar_2) + z_crit * sqrt(var_1/n_1+var_2/n_2)
> 
> round(cbind(z_crit,left,right),4)
     z_crit   left  right
[1,]   1.96 -1.685 -0.055
```

Figure 10.5 R calculations to find the critical value and the endpoints of the confidence interval.

TRY IT NOW Go to Exercise 10.23

Section 10.1 Exercises

Concept Check

10.1 True or False A paired data set often involves a before measurement and an after measurement on each individual or object.

10.2 True or False In a two-sample Z test, the hypothesized difference in means must be 0.

10.3 True or False The two-sample Z test can be used only if both population variances are known.

10.4 True or False In a two-sample Z test, both sample sizes must be large.

10.5 True or False In a two-sample Z test, the observations may be dependent.

10.6 Short Answer Given the two-sample Z test assumptions, a $100(1-\alpha)\%$ confidence interval for $\mu_1 - \mu_2$ has the values _____ as endpoints.

10.7 Short Answer Given the two-sample Z test assumptions, explain how a $100(1-\alpha)\%$ confidence interval for $\mu_1 - \mu_2$ can be used to determine if there is evidence that $\mu_1 \neq \mu_2$.

Practice

10.8 In each of the following problems, rewrite the standard-notation hypothesis concerning two population means in terms of a difference, $\mu_1 - \mu_2$.
(a) $\mu_1 = \mu_2$
(b) $\mu_1 < \mu_2$
(c) $\mu_1 \neq \mu_2 + 7$
(d) $\mu_1 > \mu_2 - 4$
(e) $\mu_1 \neq \mu_2$
(f) $\mu_1 - 10 = \mu_2$

10.9 The values for $\mu_1, \mu_2, \sigma_1, \sigma_2, n_1$, and n_2 are given. Assume the underlying distributions are normal. Find the mean, variance, and standard deviation of the random variable $\overline{X}_1 - \overline{X}_2$, and carefully sketch the probability density function.

	μ_1	μ_2	σ_1	σ_2	n_1	n_2
(a)	12.0	9.0	3.0	7.0	15	11
(b)	25.6	37.8	7.5	10.5	10	25
(c)	125.3	250.6	15.6	25.6	8	12
(d)	3.1	2.2	0.5	0.75	21	21

10.10 Given the two-sample Z test assumptions, consider the table of sample sizes, sample means, and known standard deviations.

Group	Sample size	Sample mean	Population standard deviation
One	18	17.5	1.5
Two	26	16.2	2.6

(a) Write the four parts of a hypothesis test of $H_0: \mu_1 - \mu_2 = 0$ versus $H_a: \mu_1 - \mu_2 > 0$. Use $\alpha = 0.05$.
(b) Compute the value of the test statistic and draw a conclusion.
(c) Find the p value associated with this hypothesis test.

10.11 Given the two-sample Z test assumptions, consider the table of sample sizes, sample means, and known variances.

Group	Sample size	Sample mean	Population variance
One	25	186	14.7
Two	24	190	23.8

(a) Write the four parts of a hypothesis test of $H_0: \mu_1 - \mu_2 = 2$ versus $H_a: \mu_1 - \mu_2 < 2$. Use $\alpha = 0.01$.
(b) Compute the value of the test statistic and draw a conclusion.
(c) Carefully sketch a graph to illustrate the p value associated with this hypothesis test. Compute the p value.

10.12 Given the two-sample Z test assumptions, consider the table of sample sizes, sample means, and known variances. EX10.12

Group	Sample size	Sample mean	Population variance
One	37	1025.6	225.3
Two	42	1031.3	107.6

(a) Write the four parts of a hypothesis test of $H_0: \mu_1 - \mu_2 = 0$ versus $H_a: \mu_1 - \mu_2 \neq 0$. Use $\alpha = 0.001$.
(b) Compute the value of the test statistic and draw a conclusion.
(c) Is the normality assumption necessary to conduct this hypothesis test? Justify your answer.

10.13 Two random samples were obtained independently. Assume both populations are normal, with $\sigma_1 = 8$ and $\sigma_2 = 12$. EX10.13
(a) Find a 95% confidence interval for the true difference in means $\mu_1 - \mu_2$.
(b) Using the confidence interval in part (a), is there any evidence to suggest that the two population means are different? Justify your answer.

10.14 Suppose a random sample of size 15 is taken from a normal population with mean 25 and standard deviation 5; and a second, independent random sample of size 21 is taken from a normal population with mean 10 and standard deviation 4.
(a) Describe the distribution of the difference in sample means $\overline{X}_1 - \overline{X}_2$ in terms of the type of distribution, mean, variance, and standard deviation.
(b) Carefully sketch the probability distribution for $\overline{X}_1 - \overline{X}_2$.
(c) Find $P(\overline{X}_1 - \overline{X}_2 \geq 17)$.
(d) Find $P(13.5 < \overline{X}_1 - \overline{X}_2 < 14.5)$.
(e) Find $P(\overline{X}_1 < \overline{X}_2 + 14)$.

Applications

10.15 Manufacturing and Product Development The efficiency of an electric toothbrush is often judged by the rotation speed, measured in revolutions per minute (rpm). Two brands were selected for comparison, and independent random samples of each electric toothbrush were obtained. The rotation speed for each toothbrush was measured, and the summary statistics and known variances are given in the table.

Electric toothbrush	Sample size	Sample mean	Population variance
Sonicare Elite	23	7992.2	1260.25
Oral-B	25	7988.2	1697.44

(a) Is there any evidence to suggest that the Sonicare Elite has a greater population mean rotation speed than the Oral-B? Assume normality and use $\alpha = 0.05$.
(b) Find the p value associated with this hypothesis test.

10.16 Business and Management Gift cards have become a popular present. Retailers like these cards because they are easier to process than paper gift certificates and more difficult to counterfeit. Customers appreciate the convenience; the cards make great stocking stuffers and are easy to mail. Independent random samples of gift card purchases from two merchants were obtained, and the purchased value (in dollars) of each was recorded. The summary statistics and known variances are given in the table.

Store	Sample size	Sample mean	Population variance
Nordstrom	41	24.07	16.81
Macy's	38	26.61	10.24

Is there any evidence to suggest that the true mean Nordstrom gift card purchased value is different from the true mean Macy's gift card purchased value? Use $\alpha = 0.01$.

10.17 Manufacturing and Product Development The energy rating, water consumption, and noise level of an electric dishwasher are all-important selling features. Suppose the makers of the Bosch 55N claim that this model has a lower noise-level rating than any other comparable dishwasher. Independent random samples of the Bosch 55N and of a similar Miele dishwasher were obtained, and the noise level (in decibels) was measured for each. Is there any evidence to suggest that the population mean noise level for the Bosch dishwasher is less than the population mean noise level for the Miele? Assume the underlying distributions are normal, with $\sigma_1 = 3.75$ and $\sigma_2 = 4.14$. Use $\alpha = 0.05$. **NOISE**

10.18 Business and Management As of June 2018, Hawaii had the highest mean residential electricity rate (per kWh) and Louisiana had the lowest.[2] Independent random samples of electricity rates (dollars per kWh) were obtained from two of Louisiana's neighboring states, and the summary statistics and known variances are given in the table.

State	Sample size	Sample mean	Population variance
Texas (1)	14	11.56	0.09
Oklahoma (2)	16	10.41	0.07

(a) Assume the underlying distributions are normal and find a 95% confidence interval for the true difference in population mean electricity rates.
(b) Use the confidence interval to determine whether there is any evidence that the mean electricity rate is different for the two states.

10.19 Psychology and Human Behavior New research from the National Trust suggests that today's children play outside a lot less than their parents did.[3] Independent random samples of children and parents (the previous generation) were obtained. The number of hours each child played outside in a week was recorded, and each parent was asked to estimate the number of hours played outside each week. Is there any evidence to suggest that the mean outside playing time for children is less than the mean outside playing time for their parents? Assume normality, with $\sigma_1 = 1.2$ and $\sigma_2 = 2.5$, and use $\alpha = 0.01$. **PLAY**

10.20 Manufacturing and Product Development The total weight (with the case) of a portable sewing machine is an important consideration. Suppose Singer claims to have the lightest machine by 5 lb. Independent random samples of a Singer machine and a comparable Simplicity machine were obtained, and the weight (in pounds) of each was recorded. The summary statistics and known variances are given in the table.

Sewing machine	Sample size	Sample mean	Population variance
Simplicity	42	17.99	2.89
Singer	38	13.26	2.25

(a) Is there any evidence to refute the claim made by Singer? Use $\alpha = 0.01$.
(b) Find the p value associated with this hypothesis test.
(c) Is the normality assumption necessary in this problem? Why or why not?

10.21 Medicine and Clinical Studies Many people consume protein shakes to help build muscle mass and eliminate body fat. In a recent study, the amount of protein in two competing drinks was compared. Independent random samples were obtained, and the protein content (in grams) in each drink was measured. The summary statistics and known variances are given in the table.

Protein drink	Sample size	Sample mean	Population variance
Met-Rx	12	39.38	5.06
Pure Gro	24	39.01	6.01

Is there any evidence to suggest that the mean amount of protein is different in these two products? Use $\alpha = 0.01$ and assume normality.

10.22 Travel and Transportation The recommended tire pressure for an off-road bicycle depends on the weight of the rider. As you would expect, the greater the rider's weight, the greater the recommended tire pressure. At a well-used bicycle trail in the Black River State Forest in Wisconsin, independent random samples were obtained from two different weight groups. The front tire pressure (in pounds per square inch, psi) was measured for each person, and the summary statistics and known variances are given in the table.

Weight group	Sample size	Sample mean	Population variance
≈ 150 lb	18	38.91	2.25
≈ 180 lb	23	41.99	6.25

Is there any evidence to suggest that the difference between the 180-lb riders' mean tire pressure and the 150-lb riders' mean tire pressure is greater than 3 psi? Assume normality and use $\alpha = 0.05$.

10.23 Manufacturing and Product Development Several factors determine how well a ceiling fan cools a room, including blade pitch, height from the ceiling, and revolutions per minute. Independent random samples of two types of ceiling fans were obtained, and the revolutions per minute (on high) for each was measured. The summary statistics and known variances are given in the table.

Ceiling fan	Sample size	Sample mean	Population variance
Hampton	34	295.05	11.55
Altura	35	300.38	6.25

(a) Find a 99% confidence interval for the true mean difference in revolutions per minute, $\mu_1 - \mu_2$.

(b) Using the interval in part (a), is there any evidence to suggest that the mean revolutions per minute for the Altura ceiling fan is greater than the mean revolutions per minute for the Hampton ceiling fan? Justify your answer.

10.24 Manufacturing and Product Development Despite their name, hand-held leaf blowers are often used to sweep patios, clean driveways, and even move light snow. Leaf blowers are often compared based on their weight, noise, and airspeed. Independent random samples of two types of leaf blowers were obtained, and the airspeed (in mph) was measured for each. The summary statistics and the known variances are given in the table.

Leaf blower	Sample size	Sample mean	Population variance
Craftsman	18	200.28	24.5
Echo	19	196.74	35.7

(a) Assume the underlying distributions are normal. Is there any evidence to suggest that the population mean airspeeds are different? Use $\alpha = 0.05$.

(b) Find the p value associated with this hypothesis test.

10.25 Medicine and Clinical Studies The time it takes for general anesthesia to work (time to induction) is an important consideration during an emergency and for scheduled surgeries. Recently, a study was conducted to compare the mean induction time of similar drugs administered via inhalation and intravenously. Independent random samples of patients requiring general anesthesia were obtained, and the induction times (in minutes) were measured. Assume the variance in induction time for inhalation administration is 0.0625 and for intravenous administration is 0.1225. Is there any evidence to suggest that the mean time to induction for intravenous administration is less than the mean time to induction for inhalation administration? Use $\alpha = 0.05$. ANESTH

10.26 Psychology and Human Behavior Over the last 35 years, the mean size of a house (in square feet) in Canada has increased by approximately 70%. However, even though people in other countries dream of owning a larger home, a recent survey suggests that the majority of Canadians are content with the size of their homes.[4] Independent random samples of homes in the United States and Canada were obtained and the square footage of each was measured. The summary statistics and the known variances are given in the table.

Country	Sample size	Sample mean	Population variance
Canada	12	1792	255
United States	15	1901	497

Is there any evidence to suggest that the difference between Canadian home size and U.S. home size is different from 100 ft^2? Assume normality and use $\alpha = 0.05$.

Extended Applications

10.27 Public Health and Nutrition Magnesium is used by every cell in your body, is required for more than 300 biochemical reactions, and helps muscles and nerves function properly. According to the U.S. Department of Agriculture National Nutritional Database, 1/2 cup of vegetarian baked beans and one medium baked potato without the skin contain the same amount of magnesium (40 mg). To check this claim, independent random samples of baked beans and potatoes were obtained, and the amount of magnesium in each serving was recorded (in milligrams). The summary statistics and known variances are given in the table.

Food	Sample size	Sample mean	Population variance
Vegetarian baked beans (1)	18	39.58	2.47
Medium potato (2)	18	40.12	0.87

(a) Assume the underlying distributions are normal. Is there any evidence to refute the claim? Use $\alpha = 0.01$.

(b) Suppose that, instead, the sample sizes are $n_1 = n_2 = 38$. Now, is there any evidence to refute the claim? Find the p value for this hypothesis test.

(c) How large would the sample sizes ($n_1 = n_2$) have to be for the hypothesis test to be significant at the $\alpha = 0.01$ level?

10.28 Physical Sciences The manufacturer of an LG Electronics residential stove can order parts from two different suppliers: The Repair Clinic and The Parts Pros. The small burners, or elements, are designed to produce 7.4 kW at 240 V. To decide which supplier to use, independent random samples from each supplier were obtained, and each element's output (in kilowatts) was carefully measured. For The Repair Clinic, $n_1 = 12$, $\bar{x}_1 = 7.361$, and $\sigma_1^2 = 0.81$; for The Parts Pros, $n_2 = 15$, $\bar{x}_2 = 7.307$, and $\sigma_2^2 = 0.64$. Assume the underlying distributions are normal and use $\alpha = 0.05$ for the following.

(a) Is there any evidence to suggest that the population mean output of elements from The Repair Clinic is different from 7.4?

(b) Is there any evidence to suggest that the population mean output of elements from The Parts Pros is different from 7.4?

(c) Is there any evidence to suggest that μ_1 is different from μ_2?

(d) Using the results from parts (a), (b), and (c), which supplier should the manufacturer use?

10.29 Travel and Transportation Some people experience immense joy in riding a motorcycle: The open road and nature beckon. However, motorcycle insurance is required in most U.S. states, and the mean cost is $519 per year.[5] Independent random samples of motorcycle owners were obtained from two states, and the annual insurance cost was recorded for each. The summary statistics and known variances are given in the table.

State	Sample size	Sample mean	Population variance
Arizona	65	673	1225
California	70	665	1849

(a) Is there any evidence to suggest that the mean motorcycle insurance cost is different in the two states? Use $\alpha = 0.05$.

(b) Find the p value associated with this hypothesis test.

(c) The sample means, 673 and 665, seem fairly close together. Can you find a value $n = n_1 = n_2$ such that the hypothesis test is significant at the $\alpha = 0.05$ level?

10.30 Flood Insurance According to the U.S. National Flood Insurance Program, there were approximately 167,000 paid losses after Hurricane Katrina, and there are more than 4 million homes currently insured with insurance in force of more than $1 billion. Florida still has the most flood insurance policies, followed by Texas and then Louisiana. A random sample of flood insurance policies in these three states was obtained, and the insurance in force for each policy was recorded (in dollars). The summary statistics and known standard deviations are given in the table.[6]

State	Sample size	Sample mean	Population standard deviation
Louisiana	18	261,390	7899
Florida	22	264,660	8370
Texas	26	267,184	9580

Assume the underlying populations are normal. Is there any evidence to suggest that any pairs of population mean insurance in force are different? That is, conduct three separate hypothesis tests to consider $\mu_1 - \mu_2$, $\mu_1 - \mu_3$, and $\mu_2 - \mu_3$. Use $\alpha = 0.05$ in each case.

Challenge Problems

10.31 Sample Size Calculation Suppose a $100(1-\alpha)\%$ confidence interval is needed for the difference in two population means $\mu_1 - \mu_2$. In addition, suppose the underlying populations are normal, the population variances σ_1^2 and σ_2^2 are known, and the samples sizes are equal, $n_1 = n_2 = n$.

(a) Find an expression for the sample size necessary (from each population) in order for the resulting confidence interval to have a bound on the error of estimation B (half the width of the confidence interval).

(b) How large a sample size is necessary if $\sigma_1 = 12.7$, $\sigma_2 = 9.5$, $B = 5$, and the confidence level is 95%?

(c) Use the sample size in part (b) with $\bar{x}_1 = 57.3$ and $\bar{x}_2 = 48.6$ to construct a 95% confidence interval for $\mu_1 - \mu_2$. Compute the exact bound on the error of estimation. How does this compare with $B = 5$?

10.2 Inference: Two Independent Samples, Normal Populations

Assumptions and Properties

In Section 10.1, the hypothesis tests concerning the difference between two population means (or for comparing two population means) were based on the standard normal, or Z, distribution. These tests are valid only if both population variances are known (and with normality and/or large samples, and independent random samples). It is unrealistic to assume that the population variances are known. As in Chapter 9, we will assume that the underlying populations are normal. Even so, one additional assumption is necessary to construct a similar two-sample t test.

For reference, these are the two-sample t test assumptions.

Suppose that

1. $\overline{X}_1$ is the mean of a random sample of size n_1 from a normal population with mean μ_1.
2. $\overline{X}_2$ is the mean of a random sample of size n_2 from a normal population with mean μ_2.
3. The samples are independent.

A test for equality of population variances is discussed in Section 10.5.

4. The two population variances are unknown but equal. The common variance is denoted $\sigma^2 \, (= \sigma_1^2 = \sigma_2^2)$.

The last assumption is new and implies we are comparing populations with the same variability. If we do not assume equal variances, there is no nice test procedure. More on this later.

Properties of $\overline{X}_1 - \overline{X}_2$

If the two-sample t test assumptions are true, then the estimator $\overline{X}_1 - \overline{X}_2$ has the following properties.

1. $E(\overline{X}_1 - \overline{X}_2) = \mu_{\overline{X}_1 - \overline{X}_2} = \mu_1 - \mu_2$

   $\overline{X}_1 - \overline{X}_2$ is still an unbiased estimator of the parameter $\mu_1 - \mu_2$.

2. $\text{Var}(\overline{X}_1 - \overline{X}_2) = \sigma^2_{\overline{X}_1 - \overline{X}_2} = \dfrac{\sigma_1^2}{n_1} + \dfrac{\sigma_2^2}{n_2} = \dfrac{\sigma^2}{n_1} + \dfrac{\sigma^2}{n_2} = \sigma^2\left(\dfrac{1}{n_1} + \dfrac{1}{n_2}\right)$

   and the standard deviation is

   $$\sigma_{\overline{X}_1 - \overline{X}_2} = \sqrt{\sigma^2\left(\dfrac{1}{n_1} + \dfrac{1}{n_2}\right)}$$

In the previous section, we used the known population variances, standardized, and constructed a test based on the Z distribution. Here, we need an estimate of the common variance σ^2. The appropriate standardization results in a t distribution.

S_1^2 and S_2^2 are separate estimators for the common variance, but using only one of them means ignoring additional, useful information. Because σ^2 is the variance for both underlying populations, an estimator for this common variance should depend on both samples. Yet, it also seems reasonable for the estimator to rely more on the larger sample. Therefore, an estimate of the common variance uses both S_1^2 and S_2^2 in a weighted average.

Definition

The **pooled estimator** of the common variance σ^2, denoted S_p^2, is

$$S_p^2 = \dfrac{(n_1 - 1)S_1^2 + (n_2 - 1)S_2^2}{n_1 + n_2 - 2} \tag{10.3}$$

$$= \left(\dfrac{n_1 - 1}{n_1 + n_2 - 2}\right)S_1^2 + \left(\dfrac{n_2 - 1}{n_1 + n_2 - 2}\right)S_2^2$$

The pooled estimator of the common standard deviation σ is $S_p = \sqrt{S_p^2}$.

A CLOSER LOOK

1. S_p^2 is a weighted average. This estimator can be written in the form

   $$S_p^2 = \lambda S_1^2 + (1 - \lambda)S_2^2 \quad \text{where} \quad 0 \leq \lambda \leq 1$$

   If $n_1 = n_2$, then $\lambda = \dfrac{1}{2}$ and $S_p^2 = \dfrac{1}{2}S_1^2 + \dfrac{1}{2}S_2^2$. If $n_1 \neq n_2$, then more weight is given to the larger sample.

2. The constants in Equation 10.3 are related to the number of degrees of freedom. S_1^2 contributes $n_1 - 1$ degrees of freedom and S_2^2 contributes $n_2 - 1$ degrees of freedom. Consequently, a total of $(n_1 - 1) + (n_2 - 1) = n_1 + n_2 - 2$ degrees of freedom is associated with the estimator S_p^2.

General Hypothesis Test

The hypothesis test procedure is based on the following theorem.

Theorem
If the two-sample t test assumptions are true, then the random variable

$$T = \frac{(\overline{X}_1 - \overline{X}_2) - (\mu_1 - \mu_2)}{\sqrt{S_p^2 \left(\frac{1}{n_1} + \frac{1}{n_2} \right)}}$$

has a t distribution with $n_1 + n_2 - 2$ degrees of freedom.

As in a two-sample Z test, the null and alternative hypotheses are stated in terms of the difference $\mu_1 - \mu_2$. The critical values are from the appropriate t distribution.

Hypothesis Tests Concerning Two Population Means When Variances Are Unknown But Equal

Given the two-sample t test assumptions, a hypothesis test concerning two population means in terms of the difference in means $\mu_1 - \mu_2$, with significance level α, has the following form:

$H_0: \mu_1 - \mu_2 = \Delta_0$
$H_a: \mu_1 - \mu_2 > \Delta_0, \quad \mu_1 - \mu_2 < \Delta_0, \quad$ or $\quad \mu_1 - \mu_2 \neq \Delta_0$

TS: $T = \dfrac{(\overline{X}_1 - \overline{X}_2) - \Delta_0}{\sqrt{S_p^2 \left(\dfrac{1}{n_1} + \dfrac{1}{n_2} \right)}}$

RR: $T \geq t_{\alpha, n_1 + n_2 - 2}, \quad T \leq -t_{\alpha, n_1 + n_2 - 2}, \quad$ or $\quad |T| \geq t_{\alpha/2, n_1 + n_2 - 2}$

Examples

EXAMPLE 10.4 Surgical Wait Times

Frequently, patients must wait a long time for elective surgery. For those with chronic pain, this wait can be unbearable. Suppose researchers investigated the wait time for patients needing a knee replacement at two hospitals in British Columbia. They obtained independent random samples of patients and recorded the waiting time for each (in weeks). The resulting summary statistics are given in the table.

Hospital	Sample size	Sample mean	Sample variance
Abbotsford Regional (1)	15	17.4	34.81
Earl Ridge (2)	17	12.1	46.24

(a) Is there any evidence to suggest that there is a difference in the population mean waiting time for a knee replacement between Abbotsford Regional Hospital and Earl Ridge Hospital? Use $\alpha = 0.05$ and assume the underlying distributions are normal, with equal variances.

(b) Find bounds on the p value associated with this hypothesis test.

Solution

This example involves a hypothesis test concerning two population means. The samples are random and independent, the underlying distributions are normal, and the population variances are unknown but equal. Use the two-sided alternative hypothesis and the corresponding rejection region.

(a)

STEP 1 Let Abbotsford Regional Hospital be population 1 and Earl Ridge Hospital be population 2.

The null hypothesis is that the two population means are equal, with the same waiting time for a knee replacement:

$$\mu_1 = \mu_2 \Rightarrow \mu_1 - \mu_2 = 0 (= \Delta_0)$$

The summary statistics are given, and the samples were obtained independently. The population variances are unknown but assumed equal. A two-sample t test is appropriate.

We are looking for any difference in population means, so this is a two-sided test.

STEP 2 The four parts of the hypothesis test are

$H_0: \mu_1 - \mu_2 = 0$

$H_a: \mu_1 - \mu_2 \neq 0$

TS: $T = \dfrac{(\overline{X}_1 - \overline{X}_2) - 0}{\sqrt{S_p^2\left(\dfrac{1}{n_1} + \dfrac{1}{n_2}\right)}}$

RR: $|T| \geq t_{\alpha/2,\, n_1 + n_2 - 2} = t_{0.025, 30} = 2.0423$

STEP 3 The pooled estimate of the common population variance is

$$s_p^2 = \dfrac{(n_1 - 1)s_1^2 + (n_2 - 1)s_2^2}{n_1 + n_2 - 2} = \dfrac{(14)(34.81) + (16)(46.24)}{30} = 40.906$$

The value of the test statistic is

$$t = \dfrac{(\overline{x}_1 - \overline{x}_2) - 0}{\sqrt{s_p^2\left(\dfrac{1}{n_1} + \dfrac{1}{n_2}\right)}} = \dfrac{17.4 - 12.1}{\sqrt{(40.906)\left(\dfrac{1}{15} + \dfrac{1}{17}\right)}} = 2.3393 \;(\geq 2.0423)$$

The value of the test statistic, $t = 2.3393$, lies in the rejection region, so we reject the null hypothesis at the $\alpha = 0.05$ significance level. There is evidence to suggest that the mean waiting time for a knee replacement is different at Abbotsford Regional Hospital and at Earl Ridge Hospital.

(b)

Because of the nature of the table of critical values for t distributions, we can only bound the p value (when using the table).

$|t| = |2.3393| = 2.3393$

In Appendix Table 5, row $n_1 + n_2 - 2 = 15 + 17 - 2 = 30$, place 2.3393 in the ordered list of critical values.

$$2.0423 \leq 2.3393 \leq 2.4573$$

$$t_{0.025, 30} \leq 2.3393 \leq t_{0.01, 30}$$

Therefore, $\quad 0.01 \leq\; p/2\; \leq 0.025$

and $\quad\quad\quad\; 0.02 \leq\;\; p\;\; \leq 0.05$

Figure 10.6 is a visualization of the p value, and **Figure 10.7** shows a technology solution.

10.2 Inference: Two Independent Samples, Normal Populations

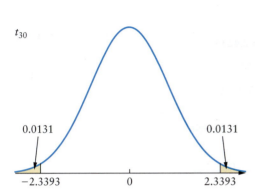

Figure 10.6 *p* value illustration:
$p = 2P(T \geq 2.3393) = 0.0262 \leq 0.05 = \alpha$.

```
> n_1 <- 15
> n_2 <- 17
> xbar_1 <- 17.4
> xbar_2 <- 12.1
> var_1 <- 34.81
> var_2 <- 46.24
>
> df <- n_1 + n_2 - 2
> t_crit <- abs(qt(0.025,df))
> sp_2 <- ((n_1-1)*var_1 + (n_2-1)*var_2)/df
> t <- (xbar_1 - xbar_2)/sqrt(sp_2*(1/n_1+1/n_2))
> p_value <- 2*(1-pt(t,df))
>
> round(cbind(t_crit,t,p_value),4)
     t_crit      t p_value
[1,] 2.0423 2.3393  0.0262
```

Figure 10.7 R calculations to find the critical value, the estimate of the pooled variance, the value of the test statistic, and the *p* value.

TRY IT NOW Go to Exercises 10.49 and 10.51

EXAMPLE 10.5 Weight of Aluminum Cans

ALUMWT

Aluminum cans are made from huge solid ingots pressed under high-pressure rollers and are cut like cookies from thin sheets. Aluminum is an ideal material for cans because it is lightweight, strong, and recyclable. A company claims that a new manufacturing process decreases the amount of aluminum needed to make a can and, therefore, decreases the weight. Independent random samples of aluminum cans made by the old and new processes were obtained, and the weight (in ounces) of each is given in the table.

Old process (1)										
0.52	0.50	0.49	0.51	0.47	0.51	0.47	0.50	0.48	0.53	0.52
0.49	0.55	0.51	0.49	0.52	0.52	0.51	0.50	0.51	0.50	

New process (2)										
0.51	0.51	0.51	0.50	0.50	0.48	0.48	0.51	0.47	0.44	0.49
0.48	0.46	0.47	0.46	0.50	0.52	0.51	0.50	0.48	0.48	

Why do you suppose a small significance level is important here?

Is there any evidence that the new-process aluminum cans have a smaller population mean weight? Assume the populations are normal, with equal variances, and use $\alpha = 0.01$.

Solution

The example involves a hypothesis test concerning two population means. The samples are random and independent, the underlying distributions are normal, and the population variances are unknown but assumed equal. Use a one-sided alternative hypothesis and the corresponding rejection region.

STEP 1 The null hypothesis is that the two mean weights are the same: $\mu_1 - \mu_2 = 0$. We are looking for evidence that the new-process cans have a smaller mean weight. Thus, the alternative hypothesis is $\mu_1 - \mu_2 > 0$.

The underlying populations are assumed to be normal with equal variances, and the samples were obtained independently. A two-sample *t* test is relevant.

STEP 2 The four parts of the hypothesis test are

$H_0: \mu_1 - \mu_2 = 0$

$H_a: \mu_1 - \mu_2 > 0$

TS: $T = \dfrac{(\overline{X}_1 - \overline{X}_2) - 0}{\sqrt{S_p^2 \left(\dfrac{1}{n_1} + \dfrac{1}{n_2}\right)}}$

RR: $T \geq t_{\alpha,\, n_1 + n_2 - 2} = t_{0.01, 40} = 2.4233$

STEP 3 Here are the summary statistics.

$\bar{x}_1 = \dfrac{1}{21}(0.50 + 0.50 + \cdots + 0.50) = 0.5048$

$\bar{x}_2 = \dfrac{1}{21}(0.51 + 0.51 + \cdots + 0.48) = 0.4886$

$s_1^2 = \dfrac{1}{20}\left[5.3580 - \dfrac{1}{21}(10.6)^2\right] = 0.0003762$

$s_2^2 = \dfrac{1}{20}\left[5.0216 - \dfrac{1}{21}(10.26)^2\right] = 0.0004429$

The pooled estimate of the common population variances:

$s_p^2 = \dfrac{(20)(0.0003762) + (20)(0.0004429)}{40} = 0.0004095$

The value of the test statistic:

$t = \dfrac{(\bar{x}_1 - \bar{x}_2) - 0}{\sqrt{s_p^2 \left(\dfrac{1}{n_1} + \dfrac{1}{n_2}\right)}} = \dfrac{0.5048 - 0.4886}{\sqrt{(0.0004095)\left(\dfrac{1}{21} + \dfrac{1}{21}\right)}} = 2.5925$

STEP 4 The value of the test statistic lies in the rejection region ($t = 2.5925 \geq 2.4233$; $p = 0.0066 \leq 0.01$; see **Figure 10.8** and **Figure 10.9**). We reject the null hypothesis at the $\alpha = 0.01$ significance level. There is evidence to suggest that new-process aluminum cans have a smaller mean weight.

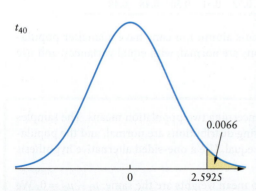

Figure 10.8 p value illustration: $p = P(T \geq 2.5925) = 0.0066 \leq 0.01 = \alpha$.

```
> old <- c(0.52,0.50,...,0.50)
> new <- c(0.51,0.51,...,0.48)
>
> t.test(old,new,alternative="greater",var.equal=TRUE)

        Two Sample t-test

data:  old and new
t = 2.5925, df = 40, p-value = 0.006622
alternative hypothesis: true difference in means is
greater than 0 95 percent confidence interval:
   0.005674524      Inf
sample estimates:
mean of x mean of y
0.5047619 0.4885714
```

Figure 10.9 The R function `t.test()` can be used to conduct both one-sample and two-sample t tests on vectors of data.

TRY IT NOW Go to Exercises 10.49 and 10.53

Confidence Interval for $\mu_1 - \mu_2$

Using the assumptions presented in this section and the technique presented in Section 10.1, we can derive a confidence interval for $\mu_1 - \mu_2$. Start with a symmetric interval about 0 such that the probability that T lies in this interval is $1 - \alpha$. Manipulate the inequality to sandwich the parameter $\mu_1 - \mu_2$.

This methodology has been used several times, beginning in Chapter 8.

How to Find a $100(1-\alpha)\%$ Confidence Interval for $\mu_1 - \mu_2$ When Variances Are Unknown But Equal

Given the two-sample t test assumptions, a $100(1-\alpha)\%$ confidence interval for $\mu_1 - \mu_2$ has the following values as endpoints:

$$(\bar{x}_1 - \bar{x}_2) \pm t_{\alpha/2,\, n_1+n_2-2} \cdot \sqrt{s_p^2\left(\frac{1}{n_1} + \frac{1}{n_2}\right)} \qquad (10.4)$$

EXAMPLE 10.6 Iron Man

Iron is an essential mineral. It is used by the body to carry oxygen to the cells, and even a slight deficiency can cause fatigue, weakness, and anemia. Certain kinds of mollusks are very high in iron content, including clams, mussels, and oysters. Independent random samples of 3-oz servings of clams and oysters were obtained, and the iron content (in milligrams) was measured in each. The summary statistics are given in the table.[7]

Mollusk	Sample size	Sample mean	Sample variance
Claims (1)	12	6.94	2.25
Oysters (2)	15	7.82	1.86

Assume the populations are normal and the variances are equal. Find a 99% confidence interval for the difference in population mean iron content.

Solution

STEP 1 The summary statistics are given, the underlying distributions are assumed to be normal, and the population variances are assumed to be equal. We can use Equation 10.4 to construct a confidence interval for the difference $\mu_1 - \mu_2$.

$1 - \alpha = 0.99 \Rightarrow \alpha = 0.01 \Rightarrow \alpha/2 = 0.005$ Find $\alpha/2$.

$t_{\alpha/2,\, n_1+n_2-2} = t_{0.005, 25} = 2.7874$ Find the t critical value.

STEP 2 Find the pooled estimate of the common variance.

$$s_p^2 = \frac{(n_1-1)s_1^2 + (n_2-1)s_2^2}{n_1+n_2-2} = \frac{(11)(2.25)+(14)(1.86)}{25} = 2.0316$$

STEP 3 Use Equation 10.4.

$$(\bar{x}_1 - \bar{x}_2) \pm t_{\alpha/2,\, n_1+n_2-2} \cdot \sqrt{s_p^2\left(\frac{1}{n_1}+\frac{1}{n_2}\right)} \qquad \text{Equation 10.4.}$$

$$= (6.94 - 7.82) \pm (2.7874)\sqrt{(2.0316)\left(\frac{1}{15}+\frac{1}{12}\right)} \qquad \text{Use summary statistics and critical value.}$$

$$= -0.88 \pm 1.5387 \qquad \text{Simplify.}$$

$$= (-2.4188, 0.6588) \qquad \text{Compute endpoints.}$$

The interval $(-2.4188, 0.6588)$ is a 99% confidence interval for the difference (in milligrams) in population mean iron content, $\mu_1 - \mu_2$. Note that because 0 is included in, or captured by, this interval, there is no evidence to suggest the mean iron content is different in 3-oz servings of clams and oysters.

Figure 10.10 shows a technology solution.

Figure 10.10 R calculations to find the critical value, the estimate of the pooled variance, and the endpoints of the confidence interval.

```
> n_1 <- 12
> n_2 <- 15
> xbar_1 <- 6.94
> xbar_2 <- 7.82
> var_1 <- 2.25
> var_2 <- 1.86
>
> df <- n_1 + n_2 - 2
> t_crit <- abs(qt(0.005,df))
> sp_2 <- ((n_1-1)*var_1 + (n_2-1)*var_2)/df
>
> left <- (xbar_1 - xbar_2) - t_crit * sqrt(sp_2 * (1/n_1 + 1/n_2))
> right <- (xbar_1 - xbar_2) + t_crit * sqrt(sp_2 * (1/n_1 + 1/n_2))
>
> round(cbind(t_crit,sp_2,left,right),4)
       t_crit   sp_2    left   right
[1,]   2.7874  2.0316  -2.4188  0.6588
```

TRY IT NOW Go to Exercise 10.63

Unequal Variances

The underlying distributions might be only approximately normal, or the population variances might not be exactly the same.

The hypothesis test procedure and the confidence interval formula presented in this section are robust. That is, if the assumptions aren't entirely true, the hypothesis test and the confidence interval are still very reliable. Even if the population variances are very different, as long as the underlying populations are normal and $n_1 = n_2$, the results will still be very reliable.

If the underlying populations are normal, the population variances are unequal, and the sample sizes are different, there is no convenient test procedure for $\mu_1 - \mu_2$ (or confidence interval for $\mu_1 - \mu_2$). In such a case, it might seem reasonable to use each sample variance as an approximation for the corresponding population variance. However, the resulting logical standardization produces only an approximate test statistic. If the sample sizes are small and the underlying populations are not normal, then we must use a nonparametric test.

Convenient means a reasonable standardization to produce a common random variable.

Hypothesis Tests and Confidence Interval Concerning Two Population Means When Variances Are Unknown and Unequal

This is the template for an approximate two-sample t test.

Given the modified two-sample t test assumptions (population variances unknown and assumed unequal), an approximate hypothesis test concerning two population means in terms of the difference $\mu_1 - \mu_2$, with significance level α, has the following form:

$H_0: \mu_1 - \mu_2 = \Delta_0$

$H_a: \mu_1 - \mu_2 > \Delta_0, \quad \mu_1 - \mu_2 < \Delta_0, \quad \text{or} \quad \mu_1 - \mu_2 \neq \Delta_0$

TS: $T' = \dfrac{(\bar{X}_1 - \bar{X}_2) - \Delta_0}{\sqrt{\dfrac{S_1^2}{n_1} + \dfrac{S_2^2}{n_2}}}$

RR: $T' \geq t_{\alpha,\nu}, \qquad T' \leq -t_{\alpha,\nu}, \qquad \text{or} \qquad |T'| \geq t_{\alpha/2,\nu}$

where $\nu \approx \dfrac{\left(\dfrac{s_1^2}{n_1} + \dfrac{s_2^2}{n_2}\right)^2}{\dfrac{(s_1^2/n_1)^2}{n_1 - 1} + \dfrac{(s_2^2/n_2)^2}{n_2 - 1}}$

The formula for ν is the Satterthwaite approximation for the number of degrees of freedom.

An approximate $100(1-\alpha)\%$ confidence interval for $\mu_1 - \mu_2$ has the following values as endpoints:

$$(\bar{x}_1 - \bar{x}_2) \pm t_{\alpha/2,\nu} \cdot \sqrt{\dfrac{s_1^2}{n_1} + \dfrac{s_2^2}{n_2}} \qquad (10.5)$$

A CLOSER LOOK

1. The random variable T' has an approximate t distribution with ν degrees of freedom.
2. It is likely that the value of ν will not be an integer. To be conservative, always round down (to the nearest integer).
3. Section 10.5 describes a test for equality of population variances. This hypothesis test is often used to determine whether equal population variances is a reasonable assumption.

EXAMPLE 10.7 Poker Chip Weights

The clay-composite poker chips used in Las Vegas and Atlantic City casinos weigh between 8.5 and 10 g each, and last between 3 and 6 years. In a recent study of poker-chip weights, a casino obtained independent random samples of $100 and $500 chips. The weight of each chip (in grams) is given in the table.

$100 chips (1)									
9.17	9.21	9.25	9.29	9.16	9.08	9.39	9.23	9.15	9.14
9.34	9.26	9.08	9.11						

$500 chips (1)								
9.37	9.98	9.04	8.74	9.58	9.45	9.08	9.96	9.69

Is there any evidence to suggest that the population mean weight of $100 chips is different from that of $500 chips? Assume both populations are normal, and use $\alpha = 0.05$.

Solution

This example involves a hypothesis test concerning two population means. The samples are random and independent, the underlying distributions are normal, and the population variances are unknown and unequal. Use the two-sided alternative hypothesis, the corresponding rejection region, and the test statistic to draw a conclusion.

STEP 1 The null hypothesis is that the two mean weights are the same, and the alternative hypothesis is two-sided. The underlying populations are assumed to be normal, and the samples were obtained independently. However, there is no assumption of equal variances. The approximate two-sample t test is appropriate.

STEP 2 Here are the summary statistics.

$$\bar{x}_1 = \frac{1}{14}(9.17 + 9.21 + \cdots + 9.11) = 9.2043$$

$$\bar{x}_2 = \frac{1}{9}(9.37 + 9.98 + \cdots + 9.69) = 9.4322$$

$$s_1^2 = \frac{1}{13}\left[1186.1804 - \frac{1}{14}(128.86)^2\right] = 0.008934$$

$$s_2^2 = \frac{1}{8}\left[802.1295 - \frac{1}{9}(84.89)^2\right] = 0.1785$$

The approximate number of degrees of freedom is

$$\nu \approx \frac{\left(\frac{0.008934}{14} + \frac{0.1785}{9}\right)^2}{\frac{(0.008934/14)^2}{13} + \frac{(0.1785/9)^2}{8}} = 8.5176$$

We round down and use $\nu = 8$.

STEP 3 The four parts of the hypothesis test are

$$H_0: \mu_1 - \mu_2 = 0$$
$$H_a: \mu_1 - \mu_2 \neq 0$$
$$\text{TS: } T' = \frac{(\overline{X}_1 - \overline{X}_2) - 0}{\sqrt{\frac{S_1^2}{n_1} + \frac{S_2^2}{n_2}}}$$
$$\text{RR: } |T'| \geq t_{\alpha/2,\nu} = t_{0.025,8} = 2.3060$$

STEP 4 The value of the test statistic is

$$t' = \frac{(\overline{x}_1 - \overline{x}_2) - 0}{\sqrt{\frac{s_1^2}{n_1} + \frac{s_2^2}{n_2}}} = \frac{9.2043 - 9.4322}{\sqrt{\frac{0.008934}{14} + \frac{0.1785}{9}}} = -1.5930$$

The value of the test statistic does not lie in the rejection region. Equivalently, $p = 0.1475 > 0.05$, as shown in **Figure 10.11**. We do not reject the null hypothesis at the $\alpha = 0.05$ significance level. There is no evidence to suggest that the mean weight of $100 chips is different from the mean weight of $500 chips.

```
> chips_1 <- c(9.17,9.21,...,9.11)
> chips_2 <- c(9.37,9.98,...,9.69)
>
> t.test(chips_1,chips_2,alternative="two.sided")

        Welch Two Sample t-test

data:  chips_1 and chips_2
t = -1.593, df = 8.5176, p-value = 0.1475
alternative hypothesis: true difference in means is not equal to 0
95 percent confidence interval:
 -0.55443591  0.09856289
sample estimates:
mean of x mean of y
 9.204286  9.432222
```

Figure 10.11 The R function `t.test()` can be used to conduct both one-sample and two-sample t tests on vectors of data. By default, the population variances are assumed unequal.

TRY IT NOW Go to Exercise 10.61

Section 10.2 Exercises

Concept Check

10.32 Short Answer State the two-sample t test assumptions.

10.33 Short Answer Under the two-sample t test assumptions, $\text{Var}(\overline{X}_1 - \overline{X}_2) = $ _____.

10.34 True or False In a two-sample t test, the pooled estimator for the common variance is the sample variance associated with the larger sample.

10.35 Short Answer The two-sample t test is robust. Explain what this means in practice.

10.36 Short Answer If the population variances are unequal, there is no nice test procedure to compare the population means. Explain what *nice* means in terms of statistics.

10.37 True or False Consider a hypothesis test or confidence interval concerning two population means when the variances are unknown and unequal. If the approximate number of degrees of freedom is a decimal, you would round it down to the nearest integer.

10.38 True or False In a hypothesis test or confidence interval concerning two population means when the variances are unknown and unequal, the hypothesized difference in means must always be 0.

10.39 Short Answer Under the two-sample t test assumptions, explain how a confidence interval for $\mu_1 - \mu_2$ can be used to test the null hypothesis $H_0: \mu_1 - \mu_2 = 0$ versus the alternative hypothesis $H_a: \mu_1 - \mu_2 \neq 0$.

Practice

10.40 Given the two-sample t test assumptions, consider the table of summary statistics.

Group	Sample size	Sample mean	Sample variance
One	14	49.6	134.56
Two	16	50.2	243.36

(a) Conduct a hypothesis test of $H_0: \mu_1 - \mu_2 = 0$ versus $H_a: \mu_1 - \mu_2 < 0$. Use $\alpha = 0.01$.
(b) Find bounds on the p value associated with this test.

10.41 Given the two-sample t test assumptions, consider the table of summary statistics.

Group	Sample size	Sample mean	Sample standard deviation
One	10	156.5	26.5
Two	11	132.6	21.5

(a) Conduct a hypothesis test of $H_0: \mu_1 - \mu_2 = 0$ versus $H_a: \mu_1 - \mu_2 > 0$. Use $\alpha = 0.05$.
(b) Find bounds on the p value associated with this test.

10.42 Given the two-sample t test assumptions, consider the independent random samples from two different populations. EX10.42
(a) Conduct a hypothesis test of $H_0: \mu_1 - \mu_2 = 0$ versus $H_a: \mu_1 - \mu_2 \neq 0$. Use $\alpha = 0.05$.
(b) Find bounds on the p value associated with this test.

10.43 Given the two-sample t test assumptions, consider the table of summary statistics.

Group	Sample size	Sample mean	Sample standard deviation
One	23	49.03	9.24
Two	23	49.57	8.15

(a) Find a 95% confidence interval for the difference in population means $\mu_1 - \mu_2$.
(b) Using the confidence interval in part (a), is there any evidence to suggest that the two population means are different? Justify your answer.

10.44 The values for n_1, n_2, s_1 and s_2 are given. Assume normal underlying distributions, independent random samples, and unknown, unequal variances. Find the approximate number of degrees of freedom ν associated with the critical value of an approximate two-sample t test.
(a) $n_1 = 12$, $n_2 = 15$, $s_1 = 11.7$, $s_2 = 16.7$
(b) $n_1 = 8$, $n_2 = 23$, $s_1 = 5.46$, $s_2 = 6.78$
(c) $n_1 = 18$, $n_2 = 26$, $s_1 = 57.8$, $s_2 = 49.9$
(d) $n_1 = 32$, $n_2 = 34$, $s_1 = 5.51$, $s_2 = 5.03$

10.45 Consider the table of summary statistics.

Group	Sample size	Sample mean	Sample variance
One	8	173.9	320.41
Two	9	150.3	655.36

Assume normal underlying distributions, independent random samples, and unknown, unequal variances.
(a) Conduct a hypothesis test of $H_0: \mu_1 - \mu_2 = 0$ versus $H_a: \mu_1 - \mu_2 > 0$. Use $\alpha = 0.05$.
(b) Find bounds on the p value associated with this test.

10.46 Consider the table of summary statistics.

Group	Sample size	Sample mean	Sample standard deviation
One	7	76.83	3.30
Two	16	66.80	14.00

Assume the underlying distributions are normal and the random samples were obtained independently.
(a) Suppose the population variances are assumed equal. Conduct a hypothesis test of $H_0: \mu_1 - \mu_2 = 0$ versus $H_a: \mu_1 - \mu_2 \neq 0$. Use $\alpha = 0.05$.
(b) Suppose the population variances are assumed to be unequal. Conduct a hypothesis test of $H_0: \mu_1 - \mu_2 = 0$ versus $H_a: \mu_1 - \mu_2 \neq 0$. Use $\alpha = 0.05$.
(c) Which of the two tests do you think is more appropriate here? Justify your answer.

10.47 Independent random samples from two normal populations were obtained. The summary statistics were $n_1 = 17, \bar{x}_1 = 32.3, s_1 = 12.9, n_2 = 19, \bar{x}_2 = 43.8, s_2 = 14.9$.
(a) Assume the population variances are unequal. Find a 99% confidence interval for the difference in population means $\mu_1 - \mu_2$.
(b) Using the confidence interval in part (a), is there any evidence to suggest that the two population means are different? Justify your answer.

Applications

10.48 Fuel Consumption and Cars The durability and flatness of the front rotors on an automobile are important for braking and for a smooth ride. The flatness of a rotor can be determined by a special optical measuring device that measures the largest deviation from perfect flatness in microinches. Suppose independent random samples of rotors from two different manufacturers were obtained. The largest deviation from perfect flatness of each rotor was measured, and the resulting summary statistics are given in the table.

Manufacturer	Sample size	Sample mean	Sample standard deviation
Power Stop	11	26.74	8.31
Hawk Sector	14	29.53	6.85

Assume the underlying distributions are normal and the population variances are equal.
(a) Is there any evidence to suggest that there is a difference in population mean deviations from perfect flatness for these two rotor brands? Use $\alpha = 0.05$.
(b) Find bounds on the p value associated with this hypothesis test.

10.49 Manufacturing and Product Development The mean weight of an ordinary key is an important consideration, as most Americans carry a pocketful of keys. A manufacturer claims that a new process produces a lighter and more durable key. Independent random samples of both types of keys were obtained, and each key was carefully weighed and its weight (in ounces) was recorded. The resulting summary statistics are given in the table.

Key type	Sample size	Sample mean	Sample variance
Old process	10	0.321	0.0137
New process	10	0.199	0.0202

Assume the underlying distributions are normal and the population variances are equal.

(a) Is there any evidence that the population mean weight of a new-process key is less than the population mean weight of an old-process key? Use $\alpha = 0.05$.

(b) Find bounds on the p value associated with this hypothesis test.

10.50 Biology and Environmental Science The tiny zebra and quagga mussels have invaded at least 600 bodies of water in the United States, including all of the Great Lakes, and are causing many environmental issues. These mollusks disrupt the natural food chain, clog pipes, cling to machinery, and foul water-delivery systems. Random samples of quagga mussels were obtained from Lake Texoma and Lake Mead, and the size of each (in centimeters) was carefully measured. Is there any evidence to suggest that the population mean size of quagga mussels is larger in Lake Texoma than in Lake Mead? Use $\alpha = 0.05$, and assume the underlying distributions are normal, with equal population variances. **MUSSEL**

10.51 Manufacturing and Product Development Shelf Safe Milk does not need to be refrigerated until it is opened. Although it is convenient, some concern has arisen that this Grade A milk contains less protein than regular milk. Independent random samples of 8-oz servings of Shelf Safe Milk and regular milk were obtained, and the amount of protein (in grams) in each was measured. The summary statistics are given in the table.

Milk	Sample size	Sample mean	Sample standard deviation
Shelf Safe	25	13.95	3.93
Regular	23	19.09	5.91

Is there any evidence to suggest that the population mean amount of protein in Shelf Safe Milk is less than the population mean amount of protein in regular milk? Use $\alpha = 0.01$, and assume the underlying distributions are normal, with equal variances.

10.52 Medicine and Clinical Studies A study was conducted to determine standard reference values for musculoskeletal ultrasonography in healthy adults. Independent random samples of men and women were obtained, and the sagittal diameter (in millimeters) of the biceps tendon was measured in each subject. The resulting summary statistics are given in the table.

Group	Sample size	Sample mean	Sample standard deviation
Women	54	2.5	0.49
Men	48	2.8	0.49

Assume the underlying populations are normal, with equal variances.

(a) Is there any evidence to suggest that the population mean sagittal diameter of women's biceps tendons is different from that of men's biceps tendons? Use $\alpha = 0.01$.

(b) Construct a 95% confidence interval for the difference in population mean sagittal diameters $\mu_1 - \mu_2$.

10.53 Manufacturing and Product Development A company that produces hospital furniture has two assembly lines dedicated to cutting and drilling wood for medical cabinets. Each computer-controlled process is designed to drill holes in a certain cabinet part with depth 12.7 mm. Independent random samples of drilled holes were obtained from the two assembly lines, and the resulting hole depths (in millimeters) were recorded. Assume the underlying populations are normal, with equal variances. **MEDCAB**

(a) Is there any evidence to suggest that Line 2 is producing holes with a greater population mean depth than Line 1? Use $\alpha = 0.05$.

(b) Find bounds on the p value associated with this hypothesis test.

10.54 Travel and Transportation During September 2018, United Airlines introduced a new two-line boarding process intended to decrease gate congestion, passenger anxiety, and the time to board the plane.[8] Independent random samples of similar cross-country flights for United Airlines and American Airlines were obtained and the boarding time (in minutes) for each was recorded. The summary statistics are given in the table.

Airline	Sample size	Sample mean	Sample standard deviation
United	21	44.5	12.3
American	26	50.7	15.5

Is there any evidence to suggest that the new United Airlines boarding process is faster than the American Airlines boarding procedure? Use $\alpha = 0.05$, and assume the populations are normal, with equal variances.

10.55 Manufacturing and Product Development A recent study was conducted to determine the curing efficiency (time to harden) of dental composites (resins for the restoration of damaged teeth) using two different types of lights. Independent random samples of lights were obtained, and a certain composite was cured for 40 sec. The depth of each cure (in millimeters) was measured using a penetrometer. The summary statistics for the Halogen light were $n_1 = 10$, $\bar{x}_1 = 5.35$, and $s_1 = 0.7$. The summary statistics for the LuxOMax light were $n_2 = 10$, $\bar{x}_2 = 3.90$, and $s_2 = 0.8$. Assume the underlying populations are normal, with equal variances.

(a) The maker of the Halogen light claims that it produces a larger cure depth after 40 sec than LuxOMax lights. Is there any evidence to support this claim? Use $\alpha = 0.01$.
(b) Construct a 99% confidence interval for the difference in population mean cure depths.

10.56 Manufacturing and Product Development Certain masonry ties used in residential construction receive a hot-dipped galvanized finish to increase their strength and protection against moisture. Independent random samples of masonry ties from two competing companies were obtained. The amount of coating on one side of each tie was measured (in g/m^2). The resulting summary statistics are given in the table.

Company	Sample size	Sample mean	Sample variance
Fero	10	331.4	201.64
Cintex	18	298.7	1190.25

Assume the underlying populations are normal, with unequal variances.
(a) Managers at Fero claim that their product has a larger mean coating than Cintex. Is there any evidence to support this claim? Use $\alpha = 0.01$.
(b) Find a 95% confidence interval for the difference in population mean coatings.

10.57 Sports and Leisure The curve in a hockey stick is measured by first placing the face of the blade against a flat surface. The curvature of the stick is restricted so that the perpendicular distance from any point at the heel to the end of the blade is at most 3/4 inch.[9] Independent random samples of hockey sticks used by players on the Toronto Maple Leafs and Montreal Canadiens teams were obtained. The curve in each stick was measured (in inches), and the resulting data are summarized in the table.

Team	Sample size	Sample mean	Sample standard deviation
Toronto	10	0.361	0.122
Montreal	20	0.425	0.051

Assume the underlying distributions are normal, with unequal variances. Is there any evidence to suggest that the mean curve in the Toronto sticks differs from the mean curve in the Montreal sticks? Use $\alpha = 0.001$.

10.58 Manufacturing and Product Development The tear strength, tensile strength, backing, and thickness all contribute to the durability of vinyl wallpaper. A new company (Aries Wallcoverings) claims to sell the thickest vinyl wallpaper currently on the market. Independent random samples of Aries wallpaper and all of its competitors were obtained. The thickness of each wallpaper (in inches) was measured. Assume the underlying distributions are normal, with unequal variances. Is there any evidence to support the claim made by Aries Wallcoverings? Use $\alpha = 0.01$. ARIES

10.59 Manufacturing and Product Development Avid video-game players are always searching for the best graphics card. Overall performance is improved and everything just runs more smoothly with a better graphics card, creating a more enjoyable game experience. Independent random samples of two video cards were obtained and the frame rate (in frames per second, fps) was measured for each. The summary statistics are given in the table.[10]

Card	Sample size	Sample mean	Sample standard deviation
RTX 2080	12	60.1	13.3
GTX 1080 Ti	14	53.4	11.6

Assume the underlying populations are normal and the population variances are equal. Is there any evidence to suggest that the mean frame rates for the two cards are different? Use $\alpha = 0.05$.

10.60 Biology and Environmental Science Many studies conducted around the world suggest that bee populations are declining.[11] Causes of the decline may include mites, fungi, and colony collapse disorder. In the United States, some states have been more affected than others. To compare the effect on honey production, independent random samples of hives in South Dakota and California were selected and the amount of honey harvested (in pounds) from each was recorded. Assume the underlying populations are normal and the population variances are equal. HONEY
(a) Is there any evidence to suggest that the mean honey harvest per hive is different in the two states? Use $\alpha = 0.05$.
(b) Find bounds on the p value for the hypothesis test in part (a).

10.61 Sports and Leisure In this fast-paced, short-attention-span world, many golfers complain that a typical 18-hole round takes too long, taking time away from family or work. Independent random samples of foursomes at two of America's best golf courses were obtained and the time (in hours) to finish an 18-hole round was recorded for each. Assume the underlying populations are normal, with unequal variances. GOLF
(a) Is there any evidence to suggest that the population mean time to complete a round of golf is different at these two courses? Use $\alpha = 0.05$.
(b) Find bounds on the p value for this hypothesis test.

10.62 Psychology and Human Behavior The voluntary reporting system for tracking cargo theft in Canada suggests that the number of thefts has been steadily rising. Independent random samples of theft reports were obtained from Ontario for two product types, and the amount (in thousands of dollars) of each theft was recorded. The summary statistics are given in the table.

Product	Sample size	Sample mean	Sample standard deviation
Auto & Parts	10	111.3	15.9
Metals	8	104.5	12.3

Assume the underlying populations are normal, with unequal variances.

(a) Is there any evidence to suggest that the mean amount of cargo theft of Auto & Parts is greater than the mean amount of cargo theft for Metals? Use $\alpha = 0.05$.
(b) Find bounds on the p value for this hypothesis test.

Extended Applications

10.63 Physical Sciences Tinted residential windows have become popular because they help a home absorb solar energy, keep out harmful ultraviolet rays, and add privacy. Two independent random samples of tinted windows were obtained, where each window was produced by applying a thin film of a specified color and density. The shading coefficient of each tinted window (a unitless quantity) was measured, and the summary statistics are given in the table.

Tinted window	Sample size	Sample mean	Sample standard deviation
Silver	8	0.601	0.113
Neutral	11	0.741	0.077

Assume the underlying populations are normal and the population variances are equal.
(a) Is there any evidence to suggest that the population mean shading coefficients are different? Use $\alpha = 0.01$.
(b) Construct a 99% confidence interval for the difference in population mean shading coefficients $\mu_1 - \mu_2$.
(c) Use the confidence interval in part (b) to determine whether any evidence suggests that the shading coefficients are different. Does your answer agree with part (a)? If so, why? If not, why not?

10.64 Economics and Finance The U.S. Bureau of Engraving and Printing produces $1, $5, $10, $20, $50, and $100 bills. The $2 banknote is still legal tender and some remain in circulation, but this bill is currently not in production. Each bill is designed to have the same width, but many people perceive larger-denomination bills to be larger in size. Independent random samples of newly minted $1 and $20 bills were obtained, and the width of each (in millimeters) was recorded. The summary statistics are given in the table.

Bill	Sample size	Sample mean	Sample variance
$1	23	66.5990	0.0132
$20	24	66.6924	0.0057

Assume the underlying distributions are normal.
(a) If the population variances are assumed to be equal, is there any evidence to suggest that the mean width of a $20 bill is greater than the mean width of a $1 bill? Use $\alpha = 0.01$.
(b) If the population variances are assumed unequal, is there any evidence to suggest that the mean width of a $20 bill is greater than the mean width of a $1 bill? Use $\alpha = 0.01$.
(c) Why do both tests lead to the same conclusion (with very similar p values)?

10.65 Marketing and Consumer Behavior Many homeowners use tiki torches for outside decoration and to burn special oil to repel insects. Independent random samples of two types of oil were obtained, and the burn time for 3 oz of each was recorded (in hours). The summary statistics are given in the table.

Oil	Sample size	Sample mean	Sample variance
Citronella Torch Fuel	21	6.25	1.04
Black Flag Mosquito Control	28	5.98	0.77

Assume the underlying populations are normal.
(a) Do you think the assumption of equal variances is reasonable? Why or why not?
(b) Based on your answer to part (a), conduct the appropriate hypothesis test to determine whether any evidence suggests that the mean burn time is different for these two brands. Use $\alpha = 0.01$.
(c) Find bounds on the p value associated with the hypothesis test in part (b).

10.66 Physical Sciences A pressure-relief valve (PRV) is installed on a residential hot-water heater to protect against overheating and, of course, high pressure. Independent random samples of PRVs from different companies were obtained. Each value was tested by recording the pressure (in pounds per square inch, psi) required to cause the valve to open. The summary statistics are given in the table.

Company	Sample size	Sample mean	Sample variance
Delta	30	147.6	7.09
Gamma	35	147.8	13.70

Assume the underlying populations are normal, with unequal variances.
(a) Is there any evidence to suggest that the mean pressure required to open each valve is different? Use $\alpha = 0.05$.
(b) Find a 95% confidence interval for the difference in population mean pressure required to open each valve. Is this confidence interval consistent with the results in part (a)? Explain your answer.

10.67 Physical Sciences The lifetime of a fuel rod in a commercial light-water nuclear reactor is related to the internal pressure. Typically, a fuel rod lasts for 36 months, and one-third of all fuel rods are replaced each year during a plant shutdown. A new type of fuel rod includes a gas-relief capsule and is designed to last longer. Independent random samples of the two types of fuel rods were obtained, and the lifetime of each (in months) was recorded. The summary statistics are given in the table.

Fuel-rod design	Sample size	Sample mean	Sample standard deviation
Old	11	34.91	3.20
New	11	39.55	3.55

Assume the underlying populations are normal and the variances are equal.
(a) Conduct the relevant hypothesis test to determine whether there is evidence to suggest that the new fuel rod lasts longer than the old version. Use $\alpha = 0.01$.
(b) Construct a 99% confidence interval for the difference in population mean lifetimes. Does this confidence interval support the hypothesis test conclusion in part (a)? Explain your answer.

10.68 Manufacturing and Product Development Root beer was originally made using the sarsaparilla root. Eventually, the oil from this root was shown to be carcinogenic (cancer-causing). Since then, many varieties have been made with cane sugar, herbs, spices, and vanilla. Independent random samples of 12-oz cans of A&W root beer and Barq's root beer were obtained, and the amount of sugar (in grams) was measured in each. Assume the underlying populations are normal, with unequal variances. **RBEER**
(a) Is there any evidence to suggest that the population mean amount of sugar in A&W root beer is greater than that in Barq's root beer? Use $\alpha = 0.05$.
(b) Construct a 95% confidence interval for the difference in the population mean sugar amounts. Does this confidence interval support the hypothesis test conclusion in part (a)? Explain your answer.

Challenge Problems

10.69 Robust Statistics The two-sample t test for comparing population means when the variances are equal is a robust statistical procedure. If the population variances are different, as long as the underlying populations are normal and the sample sizes are equal, then the hypothesis test is still very reliable.

Generate a random sample of size 25 from a normal distribution with mean $\mu_1 = 100$ and standard deviation $\sigma_1 = 5$. Generate a second random sample of size 25 from a normal distribution with mean $\mu_2 = 100$ and standard deviation $\sigma_2 = 5$. Conduct a two-sided, two-sample t test for comparing population means, assuming the population variances are equal and with $\alpha = 0.05$. Do this 100 times and record the number of times you reject the null hypothesis.

Repeat the same procedure but use $\sigma_2 = 7$. Record the number of times you reject the null hypothesis. Repeat the same procedure for $\sigma_2 = 10, 15, 20, 25, 30, 50$. Use your results to explain the robust nature of this hypothesis test.

10.3 Paired Data

Definition and Properties

When comparing population means in the previous two sections, one necessary assumption was that the samples were obtained independently. The n_1 observations from the first population and the n_2 observations from the second population were unrelated. Many experiments, however, involve only n individuals, or objects, where two observations are made of each individual. A classic example involves a diet-and-exercise program designed to help people lose weight. A random sample of n individuals is selected and each is weighed. Each person follows the regimen for a specified time period and is weighed again at the end of the experiment. There are two observations of each individual, a before weight and an after weight. These data are used to determine whether the diet-and-exercise program is effective.

Trial-and-error learning or memory experiments in animals present another good example. In a typical psychology experiment, a random sample of rats is obtained and each is timed as it maneuvers through a maze. After several weeks of training, each rat is timed again. This produces two observations of each animal, a before time and an after time. Such an experiment might be designed to determine whether animals can learn the correct path through a maze and hence decrease the mean time needed to traverse the course.

The difference between the experiments just described and those in the previous two sections is that here, the paired observations are dependent. We are still interested in comparing population means (the before mean μ_1 and the after mean μ_2), and, therefore, we are still interested in the difference $\mu_1 - \mu_2$. However, the sample means $\overline{X}_1$ and $\overline{X}_2$ are not independent. The standardizations used previously are not applicable, because the variance of $\overline{X}_1 - \overline{X}_2$ is more complicated. For this reason, we must use another method, one that addresses the dependence and yet still considers the difference $\mu_1 - \mu_2$.

The variance of $\overline{X}_1 - \overline{X}_2$ must account for the dependence.

For reference, these are the two-sample paired t test assumptions.

Suppose that

1. There are n individuals or objects, or n pairs of individuals or objects, that are related in an important way or share a common characteristic.
2. There are two observations of each individual or object. The population of first observations is normal, and the population of second observations is also normal.

▶ Let X_1 represent a randomly selected first observation and let X_2 represent the corresponding second observation on the same individual or object. Consider the random variable $D = X_1 - X_2$, the difference in the observations, and the n observed differences $d_i = (x_1)_i - (x_2)_i$, $i = 1, 2, \ldots, n$. X_1 and X_2 are both normal, so D is also normal. More important, the differences are independent. A hypothesis test concerning $\mu_1 - \mu_2$ is based on the sample mean of the differences $\bar{D}$. This random variable has the following properties.

Properties of $\bar{D}$

1. $E(\bar{D}) = \mu_1 - \mu_2$. $\bar{D}$ is an unbiased estimator of the difference in means $\mu_1 - \mu_2$.
2. The variance of $\bar{D}$ is unknown, but it can be estimated using the sample variance of the differences.
3. Both underlying populations are normal, so D is normal and, therefore, $\bar{D}$ is also normal. ◀

General Hypothesis Test

Here's what all of these results mean for us. To compare population means μ_1 and μ_2 when the data are paired, we focus on the difference $\mu_1 - \mu_2$. As in earlier two-sample tests, the null hypothesis $H_0: \mu_1 = \mu_2$ is equivalent to $H_0: \mu_1 - \mu_2 = 0$. A test to determine whether the underlying population means of two paired samples are equal is equivalent to a test to determine whether the population mean of the paired differences is zero. We compute the differences, $d_1, d_2, \ldots, d_n$, and conduct a one-sample t test (with $n-1$ degrees of freedom) using the differences.

Hypothesis Tests Concerning Two Population Means When Data Are Paired

This is the template for a hypothesis test concerning two population means when data are paired: a paired t test.

Given the two-sample paired t test assumptions, a hypothesis test concerning the two population means in terms of the difference $\mu_D = \mu_1 - \mu_2$, with significance level α, has the following form:

$H_0: \mu_D = \mu_1 - \mu_2 = \Delta_0$

$H_a: \mu_D > \Delta_0, \quad \mu_D < \Delta_0, \quad$ or $\quad \mu_D \neq \Delta_0$

TS: $T = \dfrac{\bar{D} - \Delta_0}{S_D / \sqrt{n}}$

RR: $T \geq t_{\alpha, n-1}, \quad T \leq -t_{\alpha, n-1}, \quad$ or $\quad |T| \geq t_{\alpha/2, n-1}$

🔍 A CLOSER LOOK

1. Δ_0 is the hypothesized difference in the population means. Usually $\Delta_0 = 0$: The null hypothesis is that the two population means are equal. However, Δ_0 may be nonzero. For example, the null hypothesis $H_0: \mu_1 - \mu_2 = 5 = \Delta_0$ specifies that the difference in population means is 5.

2. A paired t test is valid even if the underlying population variances are unequal, that is, even if $\sigma_1^2 \neq \sigma_2^2$. The sample variance of the differences S_D^2 is a good estimator of $\text{Var}(\bar{X}_1 - \bar{X}_2)$ when the observations are paired.

3. If a paired t test is appropriate, the test statistic is based on $n-1$ degrees of freedom. A two-sample t test (incorrect here) would be based on a test statistic with $n+n-2=2n-2$ degrees of freedom. Therefore, the correct analysis is based on a distribution with greater variability and is therefore, correctly, more conservative.

Examples

EXAMPLE 10.8 Relaxing Music

RELAX

There is no direct scientific measure of stress, but some physical properties of the body that are believed to be related to stress include pulse rate, blood pressure, breathing rate, brain waves, muscle tension, skin resistance, and body temperature. Some researchers claim that music can be relaxing and, therefore, can reduce stress. Twelve patients who claim to be suffering from job-related stress were selected at random. An initial resting pulse rate (in beats per minute, bpm) was obtained, and each person participated in a month-long music-listening, relaxation-therapy program. A final resting pulse rate was taken at the end of the experiment. The data are given in the following table.

Subject	1	2	3	4	5	6
Initial pulse rate	67	71	67	83	70	75
Final pulse rate	61	72	70	76	58	61
Difference	6	−1	−3	7	12	14
Subject	7	8	9	10	11	12
Initial pulse rate	71	68	72	88	78	70
Final pulse rate	74	59	61	64	71	77
Difference	−3	9	11	24	7	−7

Is there any evidence to suggest that the music-listening, relaxation-therapy program reduced the mean pulse rate and, therefore, the stress level? Assume the underlying distributions of initial and final pulse rate are normal, and use $\alpha = 0.05$.

Solution

The data here are certainly paired: There is a before measurement and an after measurement for each individual, and each population is assumed to be normal. So, this example involves a hypothesis test concerning two population means when data are paired. Compute the differences, use a one-sided alternative hypothesis and the corresponding rejection region, find the value of the test statistic, and draw a conclusion.

STEP 1 Typically, the before measurements are designated as population 1 and the after measurements are designated as population 2.

The null hypothesis is that the two population means are equal (i.e., the music-listening, relaxation-therapy program has no effect):

$$\mu_1 = \mu_2 \Rightarrow \mu_1 - \mu_2 = \mu_D = 0 \,(= \Delta_0).$$

Each population is assumed to be normal, and there are two observations for each individual. A paired t test is appropriate.

The therapy program is designed to reduce stress, so the alternative hypothesis is

$$\mu_1 > \mu_2 \Rightarrow \mu_1 - \mu_2 = \mu_D > 0.$$

This is a one-sided, right-tailed test.

STEP 2 The four parts of the hypothesis test are

$H_0: \mu_D = 0$

$H_a: \mu_D > 0$

TS: $T = \dfrac{\bar{D} - \Delta_0}{S_D/\sqrt{n}}$

RR: $T \geq t_{\alpha, n-1} = t_{0.05, 11} = 1.7959$

STEP 3 In anticipation of a paired t test, the differences were given in the table of data. The sample mean of the differences is

$$\bar{d} = \dfrac{1}{12}[6 + (-1) + \cdots + (-7)] = 6.3333$$

The sample variance of the differences is

$$s_D^2 = \dfrac{1}{11}\left[1320 - \dfrac{1}{12}(76)^2\right] = 76.2424$$

The sample standard deviation of the differences is

$s_D = \sqrt{76.2424} = 8.7317$

The value of the test statistic is

$$t = \dfrac{\bar{d} - 0}{s_D/\sqrt{n}} = \dfrac{6.3333}{8.7317/\sqrt{12}} = 2.5126\ (\geq 1.7959)$$

STEP 4 The value of the test statistic, $t = 2.5126$, lies in the rejection region, so we reject the null hypothesis at the $\alpha = 0.05$ significance level. There is evidence to suggest that the music-listening, relaxation-therapy program does reduce a person's resting pulse rate (and therefore the stress level).

Figure 10.12 shows a technology solution.

```
> before <- c(67,71,67,83,70,75,71,68,72,88,78,70)
> after <- c(61,72,70,76,58,61,74,59,61,64,71,77)
> 
> t.test(before,after,alternative="greater",mu=0,paired=TRUE)

        Paired t-test

data:  before and after
t = 2.5126, df = 11, p-value = 0.01443
alternative hypothesis: true difference in means is greater than 0
95 percent confidence interval:
 1.806587      Inf
sample estimates:
mean of the differences
               6.333333
```

Figure 10.12 The R function `t.test()` can be used to conduct a paired t test on vectors of data.

Note that we can use Appendix Table 5 to find bounds on the p value. However, R returns the exact p value, here $p = 0.01443 \leq 0.05 = \alpha$. ∎

TRY IT NOW Go to Exercises 10.85 and 10.87

EXAMPLE 10.9 Coal Ash Fear

COALASH

In September 2018, Hurricane Florence caused extensive flooding in North Carolina. A toxic coal ash basin near the Duke Energy L. V. Sutton power plant overflowed, sending toxic pollutants into the Cape Fear River. Duke Energy officials maintained that the coal ash had not caused any contamination downstream. Environmentalists routinely measure the concentration of mercury (in μg/L) at eight different locations in the Cape Fear River. The data given in the table represent these measurements before and after the flooding.

Location	1	2	3	4	5	6	7	8
Before	1.3	1.3	0.9	0.8	1.1	1.3	0.7	0.7
After	1.2	1.1	1.6	1.0	1.3	1.6	0.8	2.5

Is there any evidence to suggest that the population mean mercury concentration increased after the flooding? Assume the underlying populations are normal, use $\alpha = 0.05$, and find bounds on the p value associated with this hypothesis test.

Solution

This data set includes two measurements from each location; the data are paired, and each population is assumed to be normal. This example involves a hypothesis test concerning two population means when data are paired. Compute the differences; use the one-sided, left-tailed alternative hypothesis and the corresponding rejection region; find the value of the test statistic; and draw a conclusion.

STEP 1 Let population 1 be the mercury concentrations before the flooding, and let population 2 be the mercury concentrations after the flooding.

The null hypothesis is that the two population means are equal, meaning that the mercury concentrations in the river are the same at both times. Using the usual notation:

$$\mu_1 = \mu_2 \Rightarrow \mu_1 - \mu_2 = \mu_D = 0$$

Each population is assumed to be normal, and there are two observations at each location. A paired t test is appropriate.

We would like to determine if the coal ash contamination increased the mercury concentrations. Therefore, the alternative hypothesis is

$$\mu_1 < \mu_2 \Rightarrow \mu_1 - \mu_2 < \mu_D = 0$$

STEP 2 The four parts of the hypothesis test are

$H_0: \mu_D = 0$

$H_a: \mu_D < 0$

TS: $T = \dfrac{\bar{D} - \Delta_0}{S_D / \sqrt{n}}$

RR: $T \leq -t_{\alpha, n-1} = -t_{0.05, 7} = -1.8946$

STEP 3 The summary statistics for the differences are

$\bar{d} = -0.375, \quad s_D = 0.6364$

The value of the test statistic is

$$t = \frac{\bar{d} - 0}{s_D / \sqrt{n}} = \frac{-0.375}{0.6364 / \sqrt{8}} = -1.6667$$

STEP 4 The value of the test statistics, $t = -1.6667$, does not lie in the rejection region. Therefore, we do not reject the null hypothesis. There is no evidence to suggest that the population mean mercury concentration increased after the flooding.

STEP 5 Using Appendix Table 5, we can only bound the p value. In row $n - 1 = 8 - 1 = 7$ of the table, place $|-1.6667| = 1.6667$ in the ordered list of critical values.

$$1.4149 \leq 1.6667 \leq 1.8946$$

$$t_{0.10, 7} \leq 1.6667 \leq t_{0.05, 7}$$

Therefore, $\quad 0.05 \leq \ p \ \leq 0.10 \quad$ (See **Figure 10.13**.)

Figure 10.14 shows a technology solution.

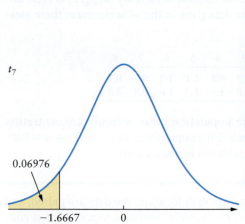

Figure 10.13 p value illustration: $p = P(T \leq -1.6667) = 0.06976$.

```
> before <- c(1.3,1.3,0.9,0.8,1.1,1.3,0.7,0.7)
> after <- c(1.2,1.1,1.6,1.0,1.3,1.6,0.8,2.5)
>
> t.test(before,after,alternative="less",mu=0,paired=TRUE)

        Paired t-test

data:  before and after
t = -1.6667, df = 7, p-value = 0.06976
alternative hypothesis: true difference in means is
less than 0 95 percent confidence interval:
       -Inf 0.05128019
sample estimates:
mean of the differences
         -0.375
```

Figure 10.14 The R function `t.test()` can be used to conduct a paired t test on vectors of data.

TRY IT NOW Go to Exercises 10.83 and 10.89

Confidence Intervals

The random variable T here is

$$T = \frac{\bar{D} - \mu_D}{S_D/\sqrt{n}}$$

The usual technique can be used to construct a confidence interval for the difference in means $\mu_D = \mu_1 - \mu_2$ when the observations are paired. Start with a symmetric interval about 0 such that the probability that T lies in this interval is $1 - \alpha$. Manipulate the inequality to sandwich μ_D.

How to Find a 100(1−α)% Confidence Interval for μ_D

Given the paired t test assumptions, a $100(1-\alpha)\%$ confidence interval for μ_D has the following values as endpoints:

$$\bar{d} \pm t_{\alpha/2,\, n-1} \cdot \frac{s_D}{\sqrt{n}} \tag{10.6}$$

EXAMPLE 10.10 Improved Mileage

A local automotive repair shop advertises a special maintenance package, including tire balancing, new spark plugs, engine oil additive, and a front-end alignment, that it promises will improve gas mileage. To check this claim, 18 cars (and drivers) were randomly selected. Each car was driven on a specially designed route and the miles per gallon for each car was recorded. Following the maintenance package, each driver took the same route, and the miles per gallon were measured again. The summary statistics for the differences (before-maintenance mpg − after-maintenance mpg) were $\bar{d} = -1.28$, $s_D = 5.62$. Assuming normality, find a 99% confidence interval for the true difference in mean miles per gallon.

Solution

STEP 1 The sample size and summary statistics are given, and the underlying distributions (before-maintenance mpg and after-maintenance mpg) are assumed to be normal.

The observations are paired, so we can use Equation 10.6.

$1 - \alpha = 0.99 \Rightarrow \alpha = 0.01 \Rightarrow \alpha/2 = 0.005$ *Find $\alpha/2$.*

$t_{\alpha/2,\, n-1} = t_{0.005, 17} = 2.8982$ *Find the t critical value with $\nu = 17$.*

STEP 2 Use Equation 10.6.

$$\bar{d} \pm t_{\alpha/2,\, n-1} \cdot \frac{s_D}{\sqrt{n}}$$ 　　　Equation 10.6.

$$= -1.28 \pm (2.8982)\frac{5.62}{\sqrt{18}}$$ 　　　Use summary statistics and critical value.

$$= -1.28 \pm 3.8391$$ 　　　Simplify.

$$= (-5.1191,\, 2.5591)$$ 　　　Compute endpoints.

The interval $(-5.1191, 2.5591)$ is a 99% confidence interval for the true mean difference in miles per gallon μ_D.

This confidence interval includes 0, so we do not have any evidence to suggest that μ_D is different from 0. Thus, there is no evidence to suggest that the maintenance program improves mileage.

Figure 10.15 shows a technology solution.

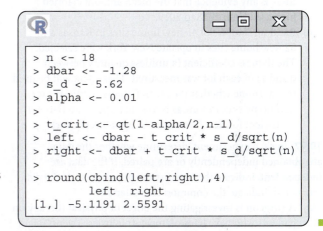

Figure 10.15 R calculations to find the critical value and the endpoints of the confidence interval.

TRY IT NOW Go to Exercise 10.94

Section 10.3 Exercises

Concept Check

10.70 Short Answer State the two-sample paired t test assumptions.

10.71 True or False In a two-sample paired t test, the before and after sample sizes must be the same.

10.72 True or False In a two-sample paired t test, the sample mean of the differences $\bar{D}$ is an unbiased estimator for $\mu_1 - \mu_2$.

10.73 True or False A paired t test is valid only if the underlying population variances are equal.

10.74 Short Answer Explain how a confidence interval for μ_D can be used to make an inference concerning μ_1 and μ_2.

10.75 Short Answer Suppose a paired t test is appropriate. How does this reference distribution compare to the distribution based on a two-sample t test?

Practice

10.76 In each experiment, determine whether the data are obtained independently or are paired. If the data are independent, indicate the two distinct populations. If the data are paired, indicate the common characteristic.

(a) School board members believe that adding a teacher's aide to each K–4 class will improve classroom management and increase instruction time. Twenty-six elementary classrooms are selected at random, and the daily instruction time for each is recorded. A teacher's aide is then added to each classroom, and the daily instruction time is recorded again. These data will be used to determine whether there is any evidence that adding a teacher's aide increases the mean daily instruction time.

(b) A researcher investigating home-insurance costs obtained a random sample of homes in the northeast and another random sample of homes in the south. The yearly insurance cost for each home was recorded. These data will be used to determine whether there is any difference in the mean yearly home-insurance costs between the northeast and the south.

(c) Officials at the transit authority of a large city would like to compare the route times during the morning and evening rush hours. Eleven routes are selected at random. Morning and evening route completion times are recorded for each. These data will be used to determine whether the mean evening route time is less than the mean morning route time.

(d) A provider of supplementary health insurance is investigating changes in claim patterns. A random sample of 32 policyholders was selected. The total amounts claimed in 2018 and in 2019 were recorded for each person. These data will be used to determine whether there is any evidence that the mean amount claimed increased from 2018 to 2019.

(e) Random samples of 45 new home sites in Kansas and 52 new home sites in upstate New York were selected. The flatness coefficient (a unitless quantity between 0 and 1) of each lot was measured. These data will be used to determine whether the mean flatness coefficient for new home sites in Kansas is less than the mean flatness coefficient for new home sites in upstate New York.

10.77 In each experiment, determine whether the data are obtained independently or are paired. If the data are independent, indicate the two distinct populations. If the data are paired, indicate the common characteristic.

(a) A surgeon is investigating the effect of physical therapy on patients who have had rotator-cuff injuries. Twenty-eight patients were selected at random. The range of motion in the affected arm was measured in each patient prior to starting physical therapy. After three weeks of consistent therapy, the range of motion in each patient was measured again. These data will be used to determine whether physical therapy has increased the mean range of motion.

(b) A woodwork manufacturer has several lathes used for shaving parts. The accuracy of each is determined by measuring the production width of a designed 1-mm wood strip. Nine lathes were selected at random, and the accuracy of each was measured. Then the gib screws on each lathe were adjusted, and the width of another 1-mm wood strip was measured. These data will be used to determine if the mean width before adjustment differs from the mean width after adjustment.

(c) A random sample of sixty 20-year-old males and a random sample of forty-five 70-year-old males were obtained. The length of each person's right ear was measured (in inches). These data will be used to determine whether the mean ear length of a 70-year-old male is greater than the mean ear length of a 20-year-old male.

(d) Modular homes are built in a closed, factory setting, indoors, and are not subject to adverse weather conditions. The quality control is often better than for on-site construction, but the cost may also differ. Ten different styles of homes were selected, and the cost of each built as a modular home and through on-site construction was estimated. These data will be used to determine whether the mean cost of modular homes is greater than for on-site construction.

(e) Random samples of 25 frequent flyers on United Airlines and on Delta Airlines were obtained. The total number of accumulated frequent-flyer miles for each person was recorded. These data will be used to determine whether the mean number of frequent-flyer miles is greater for United Airlines passengers than for Delta Airlines passengers.

10.78 In each experiment, determine whether the data are obtained independently or are paired. If the data are independent, indicate the two distinct populations. If the data are paired, indicate the common characteristic.

(a) A random sample of soccer players on the Brazil national team and a random sample of soccer players on the Argentina national team were obtained. The amount of time (in minutes) each played during a World Cup match were obtained. These data will be used to determine whether the mean time differs for the two teams.

(b) Random samples of bananas from Colombia and Costa Rica were obtained. The amount of fiber in the peel of each banana was carefully measured and recorded. These data will be used to determine whether there is a difference in the mean amount of fiber in the banana peel by country.

(c) A new stent has been designed for use in patients with diseased arteries. A random sample of patients in need of a stent was selected. The intrasaccular pressure of each was carefully measured, and then a stent was surgically placed in the diseased artery. The intrasaccular pressure was measured again following surgery. These data will be used to determine whether the stent placement reduced intrasaccular pressure.

(d) An automobile manufacturer claims that using a lighter-weight engine oil can actually increase a car's miles per gallon (of gasoline). A random sample of cars (and drivers) was selected, all of which currently use a heavy-weight oil in their engines. The miles per gallon for each car was recorded. The engine oil was drained, a lighter-weight oil was used, and the miles per gallon was recorded again. These data will be used to determine if the mean miles per gallon has increased.

(e) A new process has been developed that theoretically improves the nutritional value of barley for use in fish feed. A random sample of traditional barley and another random sample of the new barley were obtained. The percentage of protein in each grain was carefully measured. These data will be used to determine whether the new barley has a greater mean percentage of protein.

10.79 The following summary statistics were obtained in a paired-data study: $\bar{d} = 15.68$, $s_D = 33.55$, and $n = 17$. Assume normality and conduct a test of $H_0: \mu_D = 0$ versus $H_a: \mu_D > 0$. Use $\alpha = 0.05$.

10.80 Consider the following paired data.

Subject	1	2	3	4	5
Before treatment	332.5	289.3	288.2	268.0	278.0
After treatment	317.3	302.5	312.9	325.4	267.3

(a) Assume normality and conduct a test of $H_0: \mu_D = 0$ versus $H_a: \mu_D < 0$. Use $\alpha = 0.01$.

(b) Find bounds on the p value associated with this hypothesis test.

10.81 Consider the paired data given on the text website. EX10.81

(a) Assume normality and conduct a test of $H_0: \mu_D = 0$ versus $H_a: \mu_D \neq 0$. Use $\alpha = 0.01$.

(b) Find bounds on the p value associated with this hypothesis test.

Applications

10.82 Technology and the Internet Twenty-one computer programmers from IT firms around the country were selected at random. Each was asked to write code in Python and in Java for a specific application. The runtime (in seconds) for each program, by computer language, was recorded. CODE

(a) What is the common characteristic that makes these data paired?

(b) Assume normality. Conduct the appropriate hypothesis test to determine whether there is any evidence that the mean runtime for Java programs is greater than the mean runtime for Python programs. Use $\alpha = 0.001$.

(c) Find bounds on the p value associated with this hypothesis test.

10.83 Physical Sciences A consultant working for a State Police barracks contends that service weapons will fire with a higher muzzle velocity if the barrel is properly cleaned. A random sample of Glock 9-mm handguns was obtained, and the muzzle velocity (in feet per second) of a single shot from each gun was measured. Each gun was professionally cleaned, and the muzzle velocity of a second shot (with the same bullet type) was measured. The data are given in the table. MUZZLE

Gun	1	2	3	4	5	6
Before	1505	1419	1504	1494	1510	1506
After	1625	1511	1459	1441	1472	1521

(a) What is the common characteristic that makes these data paired?

(b) Assume normality. Conduct the appropriate hypothesis test to determine whether there is any evidence that a clean gun fires with a higher muzzle velocity. Use $\alpha = 0.01$.

(c) Find bounds on the p value associated with this hypothesis test.

10.84 Sports and Leisure In 1969, the height of Major League Baseball pitching mounds was lowered from 15 to 10 in. This change decreased a pitcher's leverage and presumably the speed of a typical fastball. Fifteen Major League pitchers were selected at random, and each threw his best fastball from a 15-in. mound and from a 10-in. mound. The speed of each pitch (in mph) was recorded. MOUND

(a) Assume normality. Construct a 99% confidence interval for the true mean difference in fastball speeds from a 15-in. mound and a 10-in. mound.

(b) Using the confidence interval in part (a), is there any evidence to suggest that pitching speed is, on average, faster from a higher mound? Justify your answer.

10.85 Public Health and Nutrition A filtration system made for small businesses is designed to remove particulate matter from the air. To test the new device, a random sample of businesses was obtained. The concentration of particulate matter was measured (in $\mu g/m^3$). The filtration system was then allowed to run for 24 hr, and the concentration of particulate matter was measured again. The differences (before filtration − after filtration) were recorded. Assume normality. Is there any evidence to suggest that the new filtration system improves air quality by removing particulate matter? Use $\alpha = 0.05$. PARTMAT

10.86 Biology and Environmental Science The best hurricane forecasting models use global data and take hours to run on the world's fastest supercomputers. A random sample of 26 tropical storms or hurricanes that passed within 25 mi of Miami were selected. The European Center for Medium-Range Weather Forecasting (ECMWF) and the Global Forecast System (GFS) models were used to predict storm surge (in feet). The summary statistics for the differences (ECMWF − GFS) were $\bar{d} = 1.4923$ and $s_D = 2.8097$.

(a) Is there any evidence to suggest that the ECMWF model predicts higher storm surges? Assume normality and use $\alpha = 0.05$.

(b) Find bounds on the p value associated with this hypothesis test.

10.87 Physical Sciences It is important to maintain a low ammonia-ion concentration in freshwater aquariums to ensure healthy fish (and plants). An ammonia neutralizer is advertised to almost instantly detoxify ammonia (i.e., reduce the concentration of ammonia ions) so as to protect fish. Fourteen untreated 20-gal aquariums were selected at random, and the ammonia-ion concentration (in parts per million, ppm) in each was measured. One hour after the directed amount of the neutralizer was used, the ammonia ion concentration was measured again. Assuming normality, is there any evidence to suggest that the neutralizer decreases the mean ammonia-ion concentration? Use $\alpha = 0.025$. IONS

10.88 Public Health and Nutrition Beef boullion generally has a high salt content, which can cause health problems. A food columnist for a local newspaper suggested simmering boullion with slices of raw potato to remove salt. To check this claim, 10 different boullion brands were selected at random, and the salt content in each was measured (in milligrams per cup of water). Five potato slices were then added to each broth and the mixtures were left to simmer for 15 min. Following this procedure, the salt content was measured again. The difference between the initial salt content and the final salt content was computed for each boullion brand. The data are given in the table. 📊 POTATO

| −169 | −222 | 431 | 110 | −168 | 353 | −207 | 68 | 25 | 203 |

Assume normality. Is there evidence to suggest that simmering with raw potatoes decreases the mean salt content in beef boullion? Use $\alpha = 0.05$.

10.89 Psychology and Human Behavior A recent study suggests that the speed of a person's step is related to the genre of music being listened to while walking. A random sample of 18 adults was obtained. Each walked in a circular path while listening to two different genres of music, hard rock and ballads. The steps per minute for each person for each genre were recorded. 📊 WALK

(a) What is the common characteristic that makes these data paired?
(b) Assume normality. Conduct the appropriate hypothesis to determine whether there is any evidence to suggest that the population mean steps per minute while listening to hard rock is greater than the population mean steps per minute while listening to ballads. Use $\alpha = 0.01$.

10.90 Psychology and Human Behavior The cost of long-term health care continues to rise each year, but it varies considerably from state to state. Many older adults who must enter a nursing home may select either a semi-private or private room. Several states were selected at random, and the cost of each type of room was recorded. 📊 ROOMS

(a) Assume normality and conduct the appropriate hypothesis test to determine whether the population mean cost of a private room is $15 greater than the population mean cost of a semi-private room. Use $\alpha = 0.001$.
(b) What characteristic of the differences suggests that the hypothesis test in part (a) will be significant?
(c) Find bounds on the p value associated with this hypothesis test.

10.91 Manufacturing and Product Development The porosity of a concrete block is a measure (as a percentage) of the amount of empty space in the block. In residential homes with concrete-block foundations, a larger porosity leads to damper, colder basements. A contractor recommends pretreatment of concrete blocks with a product designed to decrease the porosity. A random sample of concrete blocks was obtained, and the porosity of each was measured. The clear, paintlike product was applied to each block, and the porosity was measured again. The differences (before treatment − after treatment) in porosities were recorded. Assume normality. Is there any evidence to suggest that the new product decreases the mean porosity of concrete blocks? Use $\alpha = 0.001$. 📊 BLOCKS

10.92 Psychology and Human Behavior Each employee hired at an electronics parts assembly line in Edmonton, Alberta, is given a general intelligence test. To determine which method of training is more effective, eight pairs of new hires were matched according to their exam scores. One set of employees was asked to read appropriate training manuals, while the other group watched interactive training videos. Each employee was then asked to assemble a part used in a locater-beacon transmitter, and the time (in minutes) to completion was recorded. The data are given in the table. 📊 MANUAL

Employee pair	1	2	3	4	5	6	7	8
Written manual	4.9	4.6	5.3	4.9	4.9	5.4	5.5	5.0
Interactive video	3.1	4.1	4.4	4.9	3.6	3.9	6.5	5.3

(a) What is the common characteristic that makes these data paired?
(b) Is there any evidence to suggest that the true mean time difference μ_D differs from 0? Assume normality and use $\alpha = 0.05$.

10.93 Public Health and Nutrition Americans love hamburgers, but the high fat content in some cooked patties presents a severe health threat. Certain electric grills are designed to drain fat away from the patty, resulting in a healthier, although perhaps less tasty, meal. A random sample of ground beef packages was obtained (with various fat contents). Two patties were made from each package. One was cooked in an electric grill, while the other was prepared in a frying pan on top of a stove. The fat content (as a percentage) in each cooked patty was measured. 📊 FAT

(a) Conduct the appropriate hypothesis test to determine whether the true mean fat content in hamburgers cooked on an electric grill is less than the true mean fat content of hamburgers cooked in a frying pan. Assume normality and use $\alpha = 0.001$.
(b) Find bounds on the p value associated with this hypothesis test.

10.94 Psychology and Human Behavior The World Happiness Report presents happiness levels for 156 countries around the world. This unitless number summarizes GDP, social support, healthy life expectancy, and other variables. A random sample of countries was obtained, and the happiness levels for the years 2015 and 2018 were recorded for each. 📊 HAPPY

(a) Conduct the appropriate hypothesis test to determine whether the true mean world happiness level changed from 2015 to 2018. Assume normality and use $\alpha = 0.05$.
(b) Construct a 95% confidence interval for the true mean difference in happiness levels.
(c) Using the confidence interval in part (b), is there any evidence to suggest a difference in the true mean world happiness level from 2015 to 2018? Does this result agree with your answer in part (a)? Why or why not?

Extended Applications

10.95 Medicine and Clinical Studies A new drug designed to reduce fever (and relieve aches and pains) is being tested for efficacy and side effects. Ten patients entering a hospital with a high fever were selected at random. The temperature (in degrees Fahrenheit) of each patient was measured, the drug was administered, and two hours later the temperature was measured again. The data are given in the table. 📊 DRUG

Patient	1	2	3	4	5
Before drug	102.6	99.2	102.3	101.1	102.7
After drug	99.8	98.8	97.5	100.3	99.6
Patient	6	7	8	9	10
Before drug	102.6	100.5	103.5	105.7	104.3
After drug	102.8	99.0	101.8	97.1	99.2

(a) What is the common characteristic that makes these data paired?
(b) Assume normality. Conduct the appropriate hypothesis test to determine whether any evidence suggests that the new drug reduces the mean patient temperature after two hours. Use $\alpha = 0.05$.
(c) Find bounds on the p value associated with this hypothesis test.
(d) What characteristic of the differences suggests that a hypothesis test will be significant?

10.96 Fuel Consumption and Cars Biodiesel fuel has a cloud point, the temperature at which the fuel becomes cloudy, of approximately 13°C. This clouding can lead to poor engine performance and can even cause an engine to stop completely. An industrial chemical company produces an additive designed to lower the cloud point of this type of fuel. A random sample of six different biodiesel fuels was obtained, and the cloud point was measured for each. One ounce of the chemical additive was mixed in with every fuel sample, and the cloud point was measured again. The resulting data are given in the table (temperatures in degrees Celsius). CLOUDPT

Fuel	1	2	3	4	5	6
Before additive	11.7	12.9	14.2	12.7	11.3	12.4
After additive	10.3	10.7	14.1	10.0	11.2	12.1

(a) Assume normality, and conduct the appropriate hypothesis test to determine whether the additive lowers the mean cloud point in biodiesel fuel. Use $\alpha = 0.05$.

(b) Conduct an *inappropriate* two-sample t test to compare the population mean cloud point before treatment with the population mean cloud point after treatment. Assume the population variances are unequal and use $\alpha = 0.05$.
(c) Compare the conclusions in parts (a) and (b). How are the test statistics the same, and how do they differ?

10.97 Biology and Environmental Science Recently, Vancouver Coastal Health warned swimmers that the coliform count at East False Creek was approximately twice the safe level. The contamination was attributed to boats, birds, geese, and warm weather. A random sample of locations along the creek was obtained, and the coliform count (bacteria per 100 mL of water) was measured in early August and again following several rain storms. COLICT
(a) What is the common characteristic that makes these data paired?
(b) Use a one-sided paired t test to determine whether the rain caused a decrease in the coliform count. Assume normality and use $\alpha = 0.001$.
(c) Find the safe level of coliform. Do you think it was safe to swim in the creek after the rain? Why or why not?

10.98 Travel and Transportation For anyone planning to travel, either for vacation or on business, the pricing plans of airlines remain a guarded mystery. A study by CheapAir.com suggested that the cheapest fares are found 49 days before a flight. However, according to Travelers Today, the best prices are offered 21 days before a flight. The Washington, D.C.-to-Los Angeles route was selected as a test case. A random sample of days was selected, and the best price for a one-way ticket was recorded 21 and 49 days prior to the flight. CHPTIX
(a) Assume normality. Is there any evidence to suggest that the price of a ticket on this route differs if the ticket is purchased 21 versus 49 days in advance? Use $\alpha = 0.05$.
(b) Find bounds on the p value associated with this hypothesis test.
(c) Find a 95% confidence interval for the difference in population mean cost per ticket. Does this confidence interval support your conclusion in part (a)? Why or why not?

10.4 Comparing Two Population Proportions Using Large Samples

Notation and Properties

The methods presented in this section can be used to compare two population proportions. For example, a social scientist may conduct an experiment to determine whether the true proportion of men who favor legalized sports gambling is the same as the true proportion of women. Or an advertising agency might be interested in comparing the true proportions of children who saw a certain television commercial in two different regions of the country.

Here is a quick review of the notation (presented at the beginning of this chapter) associated with populations 1 and 2 and samples 1 and 2.

Population proportion of successes: p_1, p_2
Sample sizes: n_1, n_2
Number of successes: x_1, x_2
  Corresponding random variables: X_1, X_2
Sample proportion of successes: $\hat{p}_1 = x_1/n_1$, $\hat{p}_2 = x_2/n_2$
  Corresponding random variables: $\hat{P}_1 = X_1/n_1$, $\hat{P}_2 = X_2/n_2$

CHAPTER APP   **Prescription Errors**

Suppose a success is a prescription error. Find the sample proportion of successes for the CPOE prescriptions and the HWP prescriptions.

The general null hypothesis is stated (as usual) in terms of a difference, $H_0: p_1 - p_2 = \Delta_0$. However, there are two cases to consider: (1) $\Delta_0 = 0$ and (2) $\Delta_0 \neq 0$. In both cases, a reasonable estimator for $p_1 - p_2$ is the difference between the sample proportions, $\hat{P}_1 - \hat{P}_2$. The following properties are used to construct a hypothesis test (and confidence interval) concerning the difference between two population proportions.

Properties of the Sampling Distribution of $\hat{P}_1 - \hat{P}_2$

Interpreting the difference between two proportions.

1. The mean of $\hat{P}_1 - \hat{P}_2$ is the true difference between population proportions, $p_1 - p_2$. That is,
$$E(\hat{P}_1 - \hat{P}_2) = \mu_{\hat{P}_1 - \hat{P}_2} = p_1 - p_2$$

2. The variance of $\hat{P}_1 - \hat{P}_2$ is
$$\text{Var}(\hat{P}_1 - \hat{P}_2) = \sigma^2_{\hat{P}_1 - \hat{P}_2} = \frac{p_1(1-p_1)}{n_1} + \frac{p_2(1-p_2)}{n_2}$$
The standard deviation of $\hat{P}_1 - \hat{P}_2$ is
$$\sigma_{\hat{P}_1 - \hat{P}_2} = \sqrt{\frac{p_1(1-p_1)}{n_1} + \frac{p_2(1-p_2)}{n_2}}$$

3. If (a) both n_1 and n_2 are large,
   (b) $n_1 p_1 \geq 5$ and $n_1(1-p_1) \geq 5$, and
   (c) $n_2 p_2 \geq 5$ and $n_2(1-p_2) \geq 5$,
then the distribution of $\hat{P}_1 - \hat{P}_2$ is approximately normal.

In symbols: $\hat{P}_1 - \hat{P}_2 \overset{\bullet}{\sim} N\left[p_1 - p_2, \frac{p_1(1-p_1)}{n_1} + \frac{p_2(1-p_2)}{n_2}\right]$

General Hypothesis Test: $\Delta_0 = 0$

The appropriate standardization will result in an approximate Z distribution. The estimate of the standard deviation $\sigma_{\hat{P}_1 - \hat{P}_2}$ is determined by the value of Δ_0.

Case 1: $H_0: p_1 - p_2 = 0$ or $p_1 = p_2$ ($\Delta_0 = 0$)

If this null hypothesis is true, there is one common value for the two population proportions, denoted p ($= p_1 = p_2$). The variance of $\hat{P}_1 - \hat{P}_2$ becomes

$$\sigma^2_{\hat{P}_1 - \hat{P}_2} = \frac{p(1-p)}{n_1} + \frac{p(1-p)}{n_2} = p(1-p)\left(\frac{1}{n_1} + \frac{1}{n_2}\right) \qquad (10.7)$$

Using the properties of $\hat{P}_1 - \hat{P}_2$, the random variable

$$Z = \frac{(\hat{P}_1 - \hat{P}_2) - 0}{\sqrt{p(1-p)\left(\frac{1}{n_1} + \frac{1}{n_2}\right)}} \qquad (10.8)$$

is approximately standard normal. As for the common variance in Section 10.2, an estimator for the common proportion p is obtained by using information from both samples.

The pooled or **combined estimate of the common population proportion** is

$$\hat{P}_c = \frac{X_1 + X_2}{n_1 + n_2} = \left(\frac{n_1}{n_1 + n_2}\right)\hat{P}_1 + \left(\frac{n_2}{n_1 + n_2}\right)\hat{P}_2 \qquad (10.9)$$

The general hypothesis test procedure is based on the standardization in Equation 10.8 with $\hat{P}_c$ as an estimate of p.

Hypothesis Tests Concerning Two Population Proportions When $\Delta_0 = 0$

> This is the template for a large-sample hypothesis test concerning two population proportions when $\Delta_0 = 0$.

Given two random samples of sizes n_1 and n_2, a large-sample hypothesis test concerning two population proportions in terms of the difference $p_1 - p_2$ (with $\Delta_0 = 0$) with significance level α has the following form:

$H_0: p_1 - p_2 = 0$

$H_a: p_1 - p_2 > 0, \quad p_1 - p_2 < 0, \quad \text{or} \quad p_1 - p_2 \neq 0$

$$\text{TS: } Z = \frac{\hat{P}_1 - \hat{P}_2}{\sqrt{\hat{P}_c(1 - \hat{P}_c)\left(\frac{1}{n_1} + \frac{1}{n_2}\right)}}$$

> Remember: Use only one (appropriate) alternative hypothesis and the corresponding rejection region.

RR: $Z \geq z_\alpha, \quad Z \leq -z_\alpha, \quad \text{or} \quad |Z| \geq z_{\alpha/2}$

There is no confidence interval for the difference in population proportions in this case. If we assume $p_1 = p_2$, then there is no reason to construct a confidence interval for the difference $p_1 - p_2 = 0$.

A CLOSER LOOK

1. This test is valid as long as the nonskewness criterion holds for both samples. Use the estimates $\hat{p}_1$ and $\hat{p}_2$ to check the inequalities.
2. Just as a reminder, the z critical values for this test are from the standard normal distribution.
3. We can also determine whether to reject or not to reject the null hypothesis by comparing the p value associated with the value of the test statistic to the significance level α.

Example: Case 1

The next example illustrates the Case 1 hypothesis test procedure.

EXAMPLE 10.11   Living with Mom and Dad

A high percentage of the millennial generation, adults approximately 24 to 36 years old, are living in their parents' homes. This may be due to unemployment, lower marriage rates, and/or the economy.[12] In a random sample of 300 male millennials, 81 lived with their parents; in a random sample of 400 female millennials, 84 lived with their parents. Is there any evidence to suggest that the true proportion of male millennials living with their parents is greater than the true proportion of female millennials living with their parents? Use $\alpha = 0.05$.

Solution

This example involves a large-sample hypothesis test concerning two population proportions when $\Delta_0 = 0$. Check the large-sample assumptions. Use the template for a one-sided, right-tailed test concerning $p_1 - p_2$ when $\Delta_0 = 0$. Use $\alpha = 0.05$ to find the critical value, compute the value of the test statistic, and draw a conclusion.

STEP 1 This is a one-sided test in which we are looking for evidence that a greater proportion of males than females are living with their parents. Therefore, $\Delta_0 = 0$, and case 1 is appropriate. Arbitrarily, let male millennials be population 1 and female millennials be population 2. Here's the given information.

	Male millennials	Female millennials
Sample size	$n_1 = 300$	$n_2 = 400$
Number of successes	$x_1 = 81$	$x_2 = 84$
Sample proportion	$\hat{p}_1 = 81/300 = 0.27$	$\hat{p}_2 = 84/400 = 0.21$

STEP 2 Check the nonskewness criterion using estimates for p_1 and p_2.

$n_1 \hat{p}_1 = (300)(0.27) = 81 \geq 5 \quad n_1(1-\hat{p}_1) = (300)(0.73) = 219 \geq 5$
$n_2 \hat{p}_2 = (400)(0.21) = 84 \geq 5 \quad n_2(1-\hat{p}_2) = (400)(0.79) = 316 \geq 5$

All of the inequalities are satisfied, so $\hat{P}_1 - \hat{P}_2$ is approximately normal, and we can use the large-sample hypothesis test concerning population proportions.

STEP 3 The four parts of the hypothesis test are

$H_0: p_1 - p_2 = 0$
$H_a: p_1 - p_2 > 0$

$$\text{TS: } Z = \frac{\hat{P}_1 - \hat{P}_2}{\sqrt{\hat{P}_c(1-\hat{P}_c)\left(\frac{1}{n_1} + \frac{1}{n_2}\right)}}$$

RR: $Z \geq z_\alpha = z_{0.05} = 1.6449$

STEP 4: The estimate of the common population proportion is

$$\hat{p}_c = \frac{x_1 + x_2}{n_1 + n_2} = \frac{81 + 84}{300 + 400} = 0.2357$$

The value of the test statistic is

$$z = \frac{\hat{P}_1 - \hat{P}_2}{\sqrt{\hat{P}_c(1-\hat{P}_c)\left(\frac{1}{n_1} + \frac{1}{n_2}\right)}} = \frac{0.27 - 0.21}{\sqrt{(0.2357)(0.7643)\left(\frac{1}{300} + \frac{1}{400}\right)}} = 1.8509$$

STEP 5 The value of the test statistic lies in the rejection region. Equivalently, for $z \approx 1.85$, $p = 0.0322 \leq 0.05$, as illustrated in **Figure 10.16**. We reject the null hypothesis at the $\alpha = 0.05$ significance level. There is evidence to suggest that the true proportion of male millennials living with their parents is greater than the true proportion of female millennials living with their parents.

Figure 10.17 shows a technology solution.

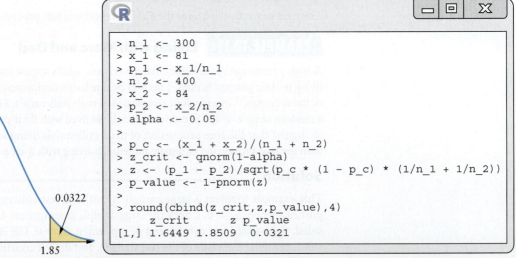

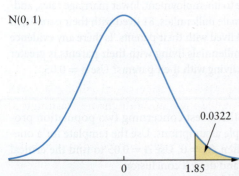

Figure 10.16 p value illustration:
$p = P(Z \geq 1.85) = 0.0322 \leq 0.05 = \alpha$.

Figure 10.17 R calculations to find the estimate of the common population proportion, the critical value, the value of the test statistic, and the p value.

TRY IT NOW Go to Exercises 10.113 and 10.115

> **CHAPTER APP** **Prescription Errors**
>
> Is there any evidence to suggest that the proportion of errors in CPOE prescriptions is less than the proportion of errors in HWP prescriptions? Use $\alpha = 0.05$.

General Hypothesis Test: $\Delta_0 \neq 0$

Case 2: $H_0: p_1 - p_2 = \Delta_0 \neq 0$

Case 2, with $\Delta_0 \neq 0$, is less common. Because p_1 and p_2 are assumed to be unequal, there is no hypothesized common population proportion. However, the hypothesis test follows routinely from the properties of $\hat{P}_1 - \hat{P}_2$.

Hypothesis Tests Concerning Two Population Proportions When $\Delta_0 \neq 0$

Given two random samples of sizes n_1 and n_2, a large-sample hypothesis test concerning two population proportions in terms of the difference $p_1 - p_2$ (with $\Delta_0 \neq 0$) with significance level α has the following form:

$H_0: p_1 - p_2 = \Delta_0$

$H_a: p_1 - p_2 > \Delta_0, \quad p_1 - p_2 < \Delta_0, \quad \text{or} \quad p_1 - p_2 \neq \Delta_0$

TS: $Z = \dfrac{(\hat{P}_1 - \hat{P}_2) - \Delta_0}{\sqrt{\dfrac{\hat{P}_1(1-\hat{P}_1)}{n_1} + \dfrac{\hat{P}_2(1-\hat{P}_2)}{n_2}}}$

RR: $Z \geq z_\alpha, \qquad Z \leq -z_\alpha, \qquad \text{or} \qquad |Z| \geq z_{\alpha/2}$

This is the template for a large-sample hypothesis test concerning two population proportions when $\Delta_0 \neq 0$. The nonskewness criterion must also be met.

Suppose two random samples of sizes n_1 and n_2 are obtained, and the nonskewness criterion is satisfied. Using the properties of $\hat{P}_1 - \hat{P}_2$, we can derive a confidence interval for $p_1 - p_2$ (p_1 and p_2 are assumed unequal). Start with a symmetric interval about 0 such that the probability that Z lies in this interval is $1 - \alpha$. As usual, manipulate the inequality to sandwich the parameter $p_1 - p_2$.

How to Find a $100(1-\alpha)\%$ Confidence Interval for $p_1 - p_2$

Given two (large) random samples of sizes n_1 and n_2, a $100(1-\alpha)\%$ confidence interval for $p_1 - p_2$ has the following values as endpoints:

$$(\hat{p}_1 - \hat{p}_2) \pm z_{\alpha/2} \cdot \sqrt{\dfrac{\hat{p}_1(1-\hat{p}_1)}{n_1} + \dfrac{\hat{p}_2(1-\hat{p}_2)}{n_2}} \qquad (10.10)$$

Example: Case 2

EXAMPLE 10.12 Make Room for Canadian Fliers

More Canadians are using smaller U.S. airports near the border when they travel to cities in the United States. Reasons include less crowding and higher taxes and fees on air travel in Canada. In a random sample of 500 fliers at Buffalo Niagara International Airport, 245 were Canadian; in a random sample of 400 fliers at Bellingham (Washington) International Airport, 160 were Canadian.

(a) Conduct the appropriate hypothesis test to determine whether there is evidence that the true proportion of Canadian fliers at Niagara is more than 0.05 greater than the true proportion of Canadian fliers at Bellingham. Use $\alpha = 0.01$.

(b) Find the p value associated with this hypothesis test.

Solution

This example involves a large-sample hypothesis test concerning two population proportions when $\Delta_0 \neq 0$. Check the large-sample assumptions. Use the template for a one-sided test concerning $p_1 - p_2$ when $\Delta_0 \neq 0$. Use $\alpha = 0.01$ to find the critical value, compute the value of the test statistic, and draw a conclusion.

(a)

STEP 1 Let Niagara fliers be population 1 and Bellingham fliers be population 2. We are looking for evidence that the difference $p_1 - p_2$ is greater than $0.05 = \Delta_0 \neq 0$. Therefore, case 2 is appropriate.

Here's the given information.

	Niagara fliers	Bellingham fliers
Sample size	$n_1 = 500$	$n_2 = 400$
Number of successes	$x_1 = 245$	$x_2 = 160$
Sample proportion	$\hat{p}_1 = 245/500 = 0.49$	$\hat{p}_2 = 160/400 = 0.40$

STEP 2 Check the nonskewness criterion using estimates for p_1 and p_2.

$n_1 \hat{p}_1 = (500)(0.49) = 245 \geq 5 \qquad n_1(1 - \hat{p}_1) = (500)(0.51) = 255 \geq 5$
$n_2 \hat{p}_2 = (400)(0.40) = 160 \geq 5 \qquad n_2(1 - \hat{p}_2) = (400)(0.60) = 240 \geq 5$

All of the inequalities are satisfied, so $\hat{P}_1 - \hat{P}_2$ is approximately normal, and we can use the large-sample hypothesis test concerning population proportions.

STEP 3 The four parts of the hypothesis test are

$H_0: p_1 - p_2 = 0.05$
$H_a: p_1 - p_2 > 0.05$

$$\text{TS: } Z = \frac{(\hat{P}_1 - \hat{P}_2) - 0.05}{\sqrt{\frac{\hat{P}_1(1 - \hat{P}_1)}{n_1} + \frac{\hat{P}_2(1 - \hat{P}_2)}{n_2}}}$$

RR: $Z \geq z_\alpha = z_{0.01} = 2.3263$

STEP 4 The value of the test statistic is

$$z = \frac{(\hat{p}_1 - \hat{p}_2) - 0.05}{\sqrt{\frac{\hat{p}_1(1 - \hat{p}_1)}{n_1} + \frac{\hat{p}_2(1 - \hat{p}_2)}{n_2}}}$$

$$= \frac{(0.49 - 0.40) - 0.05}{\sqrt{\frac{(0.49)(0.51)}{500} + \frac{(0.40)(0.60)}{400}}} = 1.2062$$

The value of the test statistic does not lie in the rejection region, so we do not reject the null hypothesis. At the $\alpha = 0.01$ significance level, there is no evidence to suggest that the population proportion of Canadian fliers at Niagara is more than 0.05 greater than the population proportion of Canadian fliers at Bellingham.

(b)

This is a one-sided, right-tailed test, so the p value is a right-tail probability.

$p = P(Z \geq 1.2062) \approx P(Z \geq 1.21)$ Definition of p value for $H_a: p_1 - p_2 > 0.05$.

$= 1 - P(Z \leq 1.21)$ The Complement Rule.

$= 1 - 0.8869 = 0.1131 \, (> 0.05 = \alpha)$ Use Appendix Table 3.

Figure 10.18 illustrates this p value, and **Figure 10.19** shows a technology solution.

Figure 10.18 p value illustration:
$p = P(Z \geq 1.21) = 0.1131 > 0.05 = \alpha$.

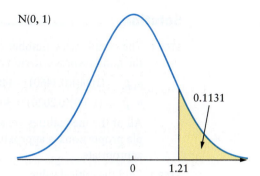

```
> n_1 <- 500
> x_1 <- 245
> p_1 <- x_1/n_1
> n_2 <- 400
> x_2 <- 160
> p_2 <- x_2/n_2
> alpha <- 0.01
>
> z_crit <- qnorm(1-alpha)
> z <- ((p_1 - p_2)-0.05)/sqrt(p_1*(1-p_1)/n_1 + p_2*(1-p_2)/n_2)
> p_value <- 1-pnorm(z)
>
> round(cbind(z_crit,z,p_value),4)
       z_crit      z p_value
[1,]   2.3263 1.2062  0.1139
```

Figure 10.19 R calculations to find the critical value, the value of the test statistic, and the p value.

TRY IT NOW Go to Exercise 10.128

Example: Confidence Interval

The computations for a confidence interval for $p_1 - p_2$ are illustrated in the next example.

EXAMPLE 10.13 Storm Watch

The Weather Channel (TWC) is one of the most popular cable TV networks. Who hasn't seen Jim Cantore struggling in the wind in the middle of some wild weather? However, the number of viewers varies greatly according to geographic region and the current weather conditions. Random samples of cable TV viewers in the northeast (population 1) and in the west (population 2) were obtained. The number of viewers who watched TWC in the past week was recorded. The data are given in the table.

	Northeast	West
Sample size	$n_1 = 1000$	$n_2 = 1500$
Number of successes	$x_1 = 446$	$x_2 = 303$
Sample proportion	$\hat{p}_1 = 446/1000 = 0.4460$	$\hat{p}_2 = 303/1500 = 0.2020$

Construct a 99% confidence interval for the true difference in proportions of cable TV viewers who watched TWC in the past week.

Solution

STEP 1 The sample sizes, number of successes, and sample proportions are given. Check the nonskewness criterion using estimates for p_1 and p_2.

$n_1 \hat{p}_1 = (1000)(0.4460) = 446 \geq 5 \quad n_1(1-\hat{p}_1) = (1000)(0.5540) = 554 \geq 5$
$n_2 \hat{p}_2 = (1500)(0.2020) = 303 \geq 5 \quad n_2(1-\hat{p}_2) = (1500)(0.7980) = 1197 \geq 5$

All of the inequalities are satisfied, so the distribution of the difference in sample proportions is approximately normal. A large-sample confidence interval is appropriate.

STEP 2 Find the critical value.

$1 - \alpha = 0.99 \Rightarrow \alpha = 0.01 \Rightarrow \alpha/2 = 0.005$ Find $\alpha/2$.

$z_{\alpha/2} = z_{0.005} = 2.5758$ Common critical value.

STEP 3 Use Equation 10.10.

$$(\hat{p}_1 - \hat{p}_2) \pm z_{\alpha/2} \cdot \sqrt{\frac{\hat{p}_1(1-\hat{p}_1)}{n_1} + \frac{\hat{p}_2(1-\hat{p}_2)}{n_2}}$$ Equation 10.10.

$$= (0.4460 - 0.2020) \pm (2.5758)\sqrt{\frac{(0.4460)(0.5540)}{1000} + \frac{(0.2020)(0.7980)}{1500}}$$

Use summary statistics and critical value.

$= 0.2440 \pm 0.0485$ Simplify,

$= (0.1955, 0.2925)$ Compute endpoints.

The interval (0.1955, 0.2925) is a 99% confidence interval for the difference in the proportion of cable TV viewers who watched TWC in the past week in the northeast and in the west, $p_1 - p_2$. This interval does not include 0, so there is evidence to suggest that the two proportions are different.

Figure 10.20 shows a technology solution.

```
> n_1 <- 1000
> x_1 <- 446
> p_1 <- x_1/n_1
> n_2 <- 1500
> x_2 <- 303
> p_2 <- x_2/n_2
> alpha <- 0.01
>
> z_crit <- qnorm(1-alpha/2)
> b <- z_crit * sqrt(p_1*(1-p_1)/n_1 + p_2*(1-p_2)/n_2)
> left <- (p_1 - p_2) - b
> right <- (p_1 - p_2) + b
>
> round(cbind(p_1,p_2,left,right),4)
         p_1   p_2   left   right
[1,]   0.446 0.202 0.1955 0.2925
```

Figure 10.20 R calculations to find the critical value and the endpoints of the confidence interval.

TRY IT NOW Go to Exercise 10.127

Section 10.4 Exercises

Concept Check

10.99 Short Answer State the nonskewness criterion.

10.100 True or False The estimate of $\sigma_{\hat{P}_1-\hat{P}_2}$ is determined by the value of Δ_0.

10.101 True or False The pooled estimate of the common population proportion is based on the largest sample.

10.102 True or False If p_1 and p_2 are assumed unequal, there is no hypothesized common population proportion.

10.103 True or False In a hypothesis test concerning two population proportions, Δ_0 can be negative.

10.104 True or False A confidence interval for $p_1 - p_2$ can be used to draw a conclusion about the equality of the two population proportions.

10.105 True or False In a hypothesis test concerning two population proportions, population 1 should always correspond to the sample with the larger size.

Practice

10.106 The sample size and the number of individuals or objects with a certain characteristic are given for samples from two populations. Check the two nonskewness inequalities for both samples, and determine whether a large-sample test concerning two population proportions is appropriate.
(a) $n_1 = 303$, $x_1 = 175$, $n_2 = 463$, $x_2 = 250$
(b) $n_1 = 560$, $x_1 = 140$, $n_2 = 530$, $x_2 = 125$
(c) $n_1 = 160$, $x_1 = 155$, $n_2 = 185$, $x_2 = 170$
(d) $n_1 = 1020$, $x_1 = 700$, $n_2 = 1277$, $x_2 = 950$
(e) $n_1 = 842$, $x_1 = 319$, $n_2 = 755$, $x_2 = 280$
(f) $n_1 = 4375$, $x_1 = 237$ $n_2 = 5005$ $x_2 = 245$

10.107 The sample sizes and population proportions are given. Find the mean, variance, and standard deviation of the estimator $\hat{P}_1 - \hat{P}_2$, and compute each probability.
(a) $n_1 = 645$, $p_1 = 0.24$, $n_2 = 650$, $p_2 = 0.26$,
$P(\hat{P}_1 - \hat{P}_2 \geq 0.045)$
(b) $n_1 = 250$, $p_1 = 0.37$, $n_2 = 270$, $p_2 = 0.33$,
$P(\hat{P}_1 - \hat{P}_2 \leq -0.04)$
(c) $n_1 = 144$, $p_1 = 0.87$, $n_2 = 156$, $p_2 = 0.86$,
$P(-0.05 < \hat{P}_1 - \hat{P}_2 < 0.05)$
(d) $n_1 = 520$, $p_1 = 0.65$, $n_2 = 480$, $p_2 = 0.72$,
$P(\hat{P}_1 - \hat{P}_2 > -0.10)$
(e) $n_1 = 1200$, $p_1 = 0.73$, $n_2 = 1150$, $p_2 = 0.85$,
$P(\hat{P}_1 - \hat{P}_2 < -0.06)$
(f) $n_1 = 500$, $p_1 = 0.645$, $n_2 = 525$, $p_2 = 0.604$,
$P(-0.02 \leq \hat{P}_1 - \hat{P}_2 \leq 0.10)$

10.108 A hypothesis test concerning two population proportions is described in each problem. Identify each population, and determine the appropriate null and alternative hypotheses in terms of p_1 and p_2.
(a) A study was conducted to determine whether there is any difference in the proportion of people who listen to satellite radio in California versus in Tennessee.
(b) Two random samples, comprising men and women who tried to talk their way out of a traffic ticket, were obtained. The number who said they missed a street sign was recorded. These data will be used to determine whether the proportion of women who say they missed a street sign is greater than the proportion of men who say so.
(c) Random samples of teens ages 13–19 from two different school districts were obtained. The number of teens who own wireless headphones was recorded. These data will be used to determine whether there is any difference in the proportion of teens who own wireless headphones in the two districts.
(d) Random samples of Americans who received an income tax refund were obtained. All were classified by income level (low versus high) and asked whether they intended to use their refund to pay outstanding bills. These data will be used to determine whether the proportion of high-income Americans who pay bills with tax refunds is 0.10 greater than the proportion of low-income Americans who pay bills with tax refunds.
(e) Independent random samples of videos on U.S. and Canadian newspaper websites were obtained. The number of videos that were advertisements was recorded. These data will be used to determine whether the proportion of videos that are advertisements in the United States is higher than that in Canada.

10.109 Use the summary statistics to conduct the appropriate hypothesis test concerning two population proportions, find the p value, and state your conclusion.
(a) $n_1 = 500$, $x_1 = 400$, $n_2 = 525$, $x_2 = 405$,
$H_0: p_1 - p_2 = 0$, $H_a: p_1 - p_2 > 0$, $\alpha = 0.05$
(b) $n_1 = 646$, $x_1 = 280$, $n_2 = 680$, $x_2 = 330$,
$H_0: p_1 - p_2 = 0$, $H_a: p_1 - p_2 < 0$, $\alpha = 0.01$
(c) $n_1 = 255$, $x_1 = 81$, $n_2 = 266$, $x_2 = 110$,
$H_0: p_1 - p_2 = 0$, $H_a: p_1 - p_2 \neq 0$, $\alpha = 0.025$
(d) $n_1 = 1440$, $x_1 = 907$, $n_2 = 1562$, $x_2 = 970$,
$H_0: p_1 - p_2 = 0$, $H_a: p_1 - p_2 \neq 0$, $\alpha = 0.001$

10.110 Use the summary statistics to conduct the appropriate hypothesis test concerning two population proportions, find the p value, and state your conclusion.
(a) $n_1 = 200$, $x_1 = 100$, $n_2 = 300$, $x_2 = 165$,
$H_0: p_1 - p_2 = 0.05$, $H_a: p_1 - p_2 < 0.05$, $\alpha = 0.01$

(b) $n_1 = 480$, $x_1 = 384$, $n_2 = 490$, $x_2 = 367$,
$H_0: p_1 - p_2 = 0.02$, $H_a: p_1 - p_2 > 0.02$, $\alpha = 0.05$
(c) $n_1 = 610$, $x_1 = 450$, $n_2 = 675$, $x_2 = 470$,
$H_0: p_1 - p_2 = 0.10$, $H_a: p_1 - p_2 \neq 0.10$, $\alpha = 0.01$
(d) $n_1 = 2500$, $x_1 = 710$, $n_2 = 3100$, $x_2 = 770$,
$H_0: p_1 - p_2 = 0.07$, $H_a: p_1 - p_2 \neq 0.07$, $\alpha = 0.001$

10.111 Use the summary statistics and confidence level to construct a confidence interval for the difference of two population proportions $p_1 - p_2$.
(a) $n_1 = 388$, $x_1 = 230$, $n_2 = 402$, $x_2 = 250$, 95%
(b) $n_1 = 528$, $x_1 = 475$, $n_2 = 530$, $x_2 = 497$, 95%
(c) $n_1 = 180$, $x_1 = 92$, $n_2 = 194$, $x_2 = 100$, 99%
(d) $n_1 = 2300$, $x_1 = 1705$, $n_2 = 2404$, $x_2 = 1690$, 90%

Applications

10.112 Public Health and Nutrition Over the last decade, many Americans have been able to stop smoking. However, a recent survey suggests that asthmatic children are more likely to be exposed to second-hand smoke than are children without asthma. In a random sample of 300 children without asthma, 132 were regularly exposed to second-hand smoke; in a random sample of 325 children with asthma, 177 were regularly exposed to second-hand smoke. Is there any evidence to suggest that the proportion of children with asthma who are exposed to second-hand smoke is greater than the corresponding proportion of children without asthma? Use $\alpha = 0.01$.

10.113 Public Policy and Political Science A recent survey suggested that more than one-fourth of registered U.S. voters believe an armed revolution might be necessary to protect our liberties. In a random sample of 250 voters in western states, 73 indicated that an armed revolution might be necessary; in a random sample of 275 voters in eastern states, 85 said that an armed revolution might be necessary. Is there any evidence to suggest that the proportion of voters in the west who believe that an armed revolution might be necessary is different from the corresponding proportion of voters in the east? Use $\alpha = 0.05$.

10.114 Technology and Internet The number of mobile phone users in the world is predicted to reach almost 5 billion by the year 2020. Most of this increase is due to the popularity of smartphones. Random samples of people in Brazil and Russia were obtained, and the number of people who use a mobile phone was recorded. The data are given in the table.

Country	Sample size	Number of mobile phone users
Brazil	326	195
Russia	390	240

Is there any evidence to suggest that the proportion of mobile phone users is greater in Russia than in Brazil? Use $\alpha = 0.01$.

10.115 Marketing and Consumer Behavior Many critics have been known to say, "They sure don't make movies like they used to." To assess Americans' opinions of movies, a random sample of people was obtained and each was asked about the quality of movies. The data are given in the table.

Age group	Sample size	Number who said movies are getting better
18–29	347	238
30–49	387	221

(a) Conduct the appropriate hypothesis test to determine whether there is any evidence to suggest that the true proportion of 18- to 29-year-olds who believe movies are getting better is greater than the proportion of 30- to 49-year-olds who hold this belief. Use $\alpha = 0.001$.
(b) Find the p value associated with the hypothesis test in part (a).

10.116 Public Health and Nutrition Several years ago, most doctors believed that it was not necessary to take any dietary supplement. Now, because many Americans do not eat a healthy, balanced diet, many physicians recommend a once-a-day multivitamin. A random sample of people was obtained and asked whether they regularly take a multivitamin. The data are given in the table.

Group	Sample size	Number who take a multivitamin
Men	490	181
Women	428	214

Is there any evidence that the proportion of women who take a multivitamin is greater than the proportion of men who take a multivitamin? Use $\alpha = 0.005$.

10.117 Travel and Transportation Many people who commute to work by car in New York City every day use either the George Washington Bridge or the Lincoln Tunnel. Random samples of commuters who use one of these two routes were obtained, and each was asked whether he or she carpooled to work. The data are given in the table.

Commuting route	Sample size	Number who carpool
Bridge	1055	530
Tunnel	1663	825

(a) Verify that the nonskewness criterion inequalities are satisfied.
(b) Is there any evidence to suggest that the proportion of carpoolers crossing the George Washington Bridge is greater than the proportion of carpoolers using the Lincoln Tunnel? Use $\alpha = 0.01$.
(c) Find the p value associated with the hypothesis test in part (b).

10.118 Conspiracy Theory Research suggests that half the American public believes in at least one conspiracy theory.[13] Some of the more popular theories involve a secret group controlling the world, a faked moon landing, and Area 51 and aliens. You might consider investigating the theory about lizard people who control our society. The research results are often very different according to political affiliation. A random sample of Democrats and Republicans were asked if they believe pharmaceutical companies invent new diseases to make money. The resulting data are given in the table.

Political affiliation	Sample size	Number who believe pharmaceutical companies invent diseases
Democrats	788	137
Republicans	866	105

Is there any evidence to suggest that the proportion of voters who believe pharmaceutical companies invent diseases to make money is different for Democrats and Republicans? Use $\alpha = 0.01$.

10.119 Medicine and Clinical Studies According to WebMD, approximately 30% of U.S. adults have allergies.[14] Approximately 8% suffer from hay fever. A random sample of people who suffer from hay fever was obtained, and each was treated with either a conventional antihistamine or butterbur extract. The number of subjects who experienced relief from hay fever was recorded for each group. The resulting data are given in the table.

Treatment	Sample size	Number who experienced relief
Antihistamine	255	71
Butterbur extract	237	55

(a) Compute the sample proportion of people who experienced relief for each treatment.
(b) Conduct the appropriate hypothesis test to determine whether the proportion of people who experienced relief due to the antihistamine is different from the proportion of people who experienced relief from butterbur extract. Use $\alpha = 0.01$.

10.120 Public Health and Nutrition A survey was conducted concerning physical activity of adults in two states. Random samples of adults were obtained from Arizona and from West Virginia, and all participants were asked whether they consider themselves physically inactive. The data are given in the table.

State	Sample size	Number who are physically inactive
Arizona	1122	163
West Virginia	1181	205

Is there any evidence to suggest that the proportion of adults who consider themselves physically inactive is greater in West Virginia than in Arizona? Use $\alpha = 0.001$ and find the p value.

10.121 Manufacturing and Product Development Blenko Specs has two different processes for the manufacture of optical lenses supplied to the military. A random sample of finished lenses was obtained from each process, and each lens was carefully inspected for defects. Of the 106 lenses from Process A, 8 were defective, whereas 12 of the 121 lenses from Process B were defective.

(a) Compute the sample proportion of defective lenses for each process.
(b) Check the nonskewness criterion and verify that the inequalities are satisfied.
(c) Conduct a hypothesis test to determine whether there is any evidence that the sample proportion of defective lenses is different for Process A and for Process B. Use $\alpha = 0.05$.

10.122 Sports and Leisure A major league sports franchise can contribute a great deal to the local economy and unite an entire region. In a recent survey, a random sample of adults in the Portland, Oregon, area were asked if they would support a National Football League team. The data are given in the table.

County	Sample size	Number who would support an NFL team
Clackamas	469	117
Multnomah	1985	337

Is there any evidence to suggest that the proportion of residents who would support an NFL team is different in the two counties? Use $\alpha = 0.01$.

10.123 Psychology and Human Behavior In a recent survey, residents of Reston, Virginia, were asked about their quality of life, characteristics of the community, child care, and crime. Respondents were asked to rate how safe or unsafe they felt in their neighborhood. The data are given in the table.

District	Sample size	Number who felt safe in their neighborhood after dark
Hunter Woods	213	141
Lake Anne	218	155

Is there any evidence to suggest that the true proportion of residents who feel safe after dark is different in the two districts? Use $\alpha = 0.05$.

10.124 Technology and the Internet The percentage of Canadians who use technology is very high, but a recent survey suggests that they greatly overrate their tech savviness. Approximately 60% of Canadians rated themselves as B or better for tech savviness, but a large proportion of respondents do not know how to use search engines, do not back up files regularly, and do not use email properly. Random samples of Canadians from two regions were obtained and asked to rate their tech savviness. The results are given in the table.

Region	Sample size	Number who rated their tech savviness as B or better
Edmonton	566	345
Thunder Bay	617	330

Is there any evidence to suggest that the true proportion of Canadians who rate their tech savviness as B or better is greater in Edmonton than in Thunder Bay? Use $\alpha = 0.01$.

10.125 Demographics and Population Statistics Over the last 15 years, employment in the STEM fields (science, technology, engineering, and mathematics) has outpaced overall job growth. Women make up approximately half of all

U.S. workers in STEM professions, but this percentage varies widely across specific occupations. Random samples of workers in two occupational groups were obtained and the number of women workers was recorded. The data are given in the table.

Occupational group	Sample size	Number of women
Computer science	500	114
Engineering	328	62

Is there any evidence to suggest that the true proportion of women in computer science occupations is greater than the proportion of women in engineering occupations? Use $\alpha = 0.05$.

Extended Applications

10.126 Marketing and Consumer Behavior Americans have many sources for daily news, including local television shows, public radio, social media, and national newspapers. A random sample of Americans was obtained and classified by age. Each person was asked whether he or she obtained news every day from three specific sources. The data are given in the table.

	Age group			
	18- to 29-year-olds		30- to 49-year-olds	
News source	Sample size	Number who obtained news every day	Sample size	Number who obtained news every day
Nightly network news	570	103	462	120
Cable news networks	450	108	520	182
Internet	546	197	568	239

(a) Conduct the appropriate hypothesis test to determine whether there is evidence that the true proportion of 18- to 29-year-olds who obtain news every day from nightly network news shows is less than the true proportion of 30- to 49-year-olds who obtain news every day from nightly network news shows. Use $\alpha = 0.05$. Find the p value associated with this hypothesis test.
(b) Conduct the appropriate hypothesis test to determine whether there is evidence that the true proportion of 18- to 29-year-olds who obtain news every day from cable news networks is less than the true proportion of 30- to 49-year-olds who obtain news every day from cable news networks shows. Use $\alpha = 0.01$. Find the p value associated with this hypothesis test.
(c) Conduct the appropriate hypothesis test to determine whether there is evidence that the true proportion of 18- to 29-year-olds who obtain news every day from the Internet is different from the true proportion of 30- to 49-year-olds who obtain news every day from the Internet. Use $\alpha = 0.005$. Find the p value associated with this hypothesis test.

10.127 Manufacturing and Product Development Two different machines in a manufacturing facility are designed to fill cans with 280 g of Tang orange-flavored drink mix. A random sample of filled cans from each machine was obtained, and each can was carefully weighed. Of the 134 cans from machine A, 10 were underfilled, whereas 7 of 114 cans from machine B were underfilled.

(a) Compute the sample proportion of underfilled cans for each machine.
(b) Verify the nonskewness criterion.
(c) Find a 95% confidence interval for the true difference in the proportion of underfilled cans for machines A and B.
(d) Using the confidence interval in part (c), is there any evidence to suggest that the proportion of underfilled cans is different for the two machines? Justify your answer.

10.128 Psychology and Human Behavior Historically, the three most popular home-improvement projects have been, in order, interior decorating, landscaping, and expansion. A random sample of homeowners and condominium owners was obtained and asked whether they planned any landscaping within the next year. The data are given in the table.

Residence	Sample size	Number who are planning to landscape
Homeowner	261	90
Condominium owner	303	65

Is there any evidence to suggest that the proportion of homeowners planning a landscaping project is more than 0.10 greater than the proportion of condominium owners planning a landscaping project? Use $\alpha = 0.01$.

10.129 Medicine and Clinical Studies Young children usually get 5–10 colds each year. To ease cold symptoms, such as a runny nose or sore throat, some parents give their children over-the-counter cough and cold medicines. However, many of these medicines are not effective and can cause serious side effects in young children. A random sample of parents of young children was obtained and asked whether they give their children cough or cold medicine. The data are given in the table.

Parent	Sample size	Number who give cold medicine
Male	376	142
Female	428	183

(a) Is there any evidence to suggest that the true population proportion of males who give their children cough medicine is different from the true proportion of females who give this treatment? Use $\alpha = 0.05$.
(b) Find the p value associated with this hypothesis test.
(c) Find a 95% confidence interval for the difference in the proportion of males who give their children cold medicine and the proportion of females who give their children cold medicine. Does this confidence interval support your conclusion in part (a)? Why or why not?
(d) What must be true of the respondents for the hypothesis test in part (a) to be valid?

10.5 Comparing Two Population Variances or Standard Deviations

The F Distribution

Many practical business decisions are based on a comparison of variability, or manufacturing precision. For example, a company that produces a certain drug via fermentation would like to maintain a very small variability in yield. One fermentation process may be more reliable and less variable than another. A hardware store wants very little variability in paint color from gallon to gallon. One paint mixer may be more precise (less variable) than another. Even food manufacturers strive for small differences in product taste from one batch to the next. Finally, we may compare population variances to decide which two-sample t test is appropriate, the pooled test or the approximate test.

S_1^2 and S_2^2 are good (unbiased) estimators for the population variances σ_1^2 and σ_2^2, respectively. However, a hypothesis test for comparing σ_1^2 and σ_2^2 is based on a new standardization and an F distribution, introduced in this section.

An F distribution has positive probability only for non-negative values. The probability density function for an F random variable is 0 for $x < 0$. Once again, it is important to focus on the properties of an F distribution and the method for finding critical values associated with this distribution.

Properties of an F Distribution

> The numerator and denominator designations will make more sense as you read on.

1. An F distribution is completely determined by two parameters, the number of degrees of freedom in the numerator and the number of degrees of freedom in the denominator, given in that order. Both values must be positive integers $(1, 2, 3, \ldots)$. There is, of course, a different F distribution for every combination.

2. If X has an F distribution with ν_1 and ν_2 degrees of freedom ($X \sim F_{\nu_1, \nu_2}$), then

$$\mu_X = \frac{\nu_2}{\nu_2 - 2}, \; \nu_2 \geq 3 \quad \text{and} \quad \sigma_X^2 = \frac{2\nu_2^2(\nu_1 + \nu_2 - 2)}{\nu_1(\nu_2 - 2)^2(\nu_2 - 4)}, \; \nu_2 \geq 5$$

> ✓ Why are these restrictions on ν_2 necessary? What do you suppose the mean is if $\nu_2 = 2$?

3. Suppose $X \sim F_{\nu_1, \nu_2}$. The density curve for X is positively skewed (not symmetric), and gets closer and closer to the x-axis but never touches it. As both degrees of freedom increase, the density curve becomes taller and more compact. See **Figure 10.21**.

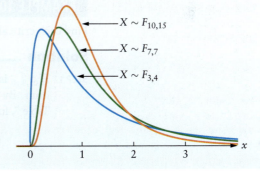

Figure 10.21 Density curves for several F distributions.

F Critical Values

The definition and notation for an F critical value are analogous to those for Z, t, and χ^2 critical values.

Definition

F_{α, ν_1, ν_2} is a critical value related to an F distribution with ν_1 and ν_2 degrees of freedom. If $X \sim F_{\nu_1, \nu_2}$, then $P(X \geq F_{\alpha, \nu_1, \nu_2}) = \alpha$.

A CLOSER LOOK

Yes, there are three subscripts, but don't panic. The notation looks more complicated than it is. F_{α,ν_1,ν_2} is simply a consistent, concise notation to represent a critical value related to an F distribution.

1. F_{α,ν_1,ν_2} is a value on the measurement axis in an F world such that there is α of the area (probability) to the right of F_{α,ν_1,ν_2}. As for a chi-square distribution, there is no symmetry in F critical values.

2. Critical values are defined in terms of right-tail probability, and the F distribution is not symmetric, so the notation here is similar to that for chi-square critical values. It will be necessary to find critical values denoted $F_{1-\alpha,\nu_1,\nu_2}$, where $1-\alpha$ is large. By definition, $P(X \geq F_{1-\alpha,\nu_1,\nu_2}) = 1-\alpha$, and by the Complement Rule, $P(X \leq F_{1-\alpha,\nu_1,\nu_2}) = \alpha$. See **Figure 10.22**.

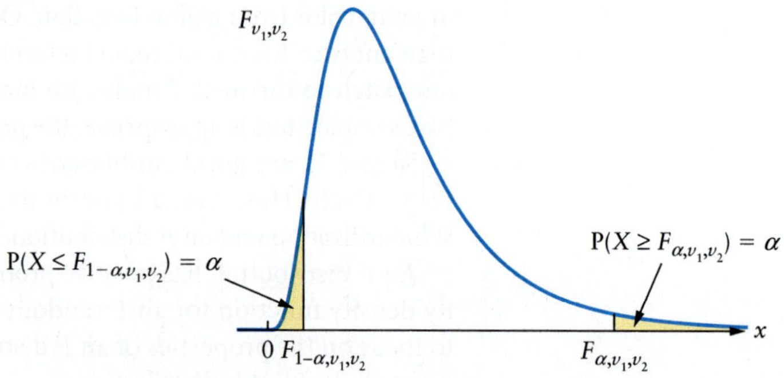

Figure 10.22 An illustration of F distribution critical values.

3. F critical values are related according to the equation

Notice how the degrees of freedom switch positions.

$$F_{1-\alpha,\nu_1,\nu_2} = \frac{1}{F_{\alpha,\nu_2,\nu_1}} \tag{10.11}$$

Appendix Table 7 presents selected critical values associated with various F distributions and right-tail probabilities. The degrees of freedom in the numerator are given in the top row and the degrees of freedom in the denominator are given in the left column. In the body of the table, F_{α,ν_1,ν_2} is at the intersection of the appropriate row and column. Left-tail probabilities are found using Equation 10.11. The next example illustrates the use of Appendix Table 7 for finding critical values associated with an F distribution.

EXAMPLE 10.14 Critical Value Look-ups

Find each critical value. **(a)** $F_{0.05,8,10}$ **(b)** $F_{0.99,9,15}$

Solution

STEP 1 $F_{0.05,8,10}$ is a critical value related to an F distribution with 8 and 10 degrees of freedom. By definition, if $X \sim F_{8,10}$, then $P(X \geq F_{0.05,8,10}) = 0.05$. Using Appendix Table 7, for $\alpha = 0.05$, find the intersection of the $\nu_1 = 8$ column and the $\nu_2 = 10$ row.

$\alpha = 0.05$

		ν_1					
ν_2		6	7	8	9	10	
⋮	⋮	⋮	⋮	⋮	⋮	⋮	⋮
8	⋯	3.58	3.50	3.44	3.39	3.35	⋯
9	⋯	3.37	3.29	3.23	3.18	3.14	⋯
10	⋯	3.22	3.14	3.07	3.02	2.98	⋯
11	⋯	3.09	3.01	2.95	2.90	2.85	⋯
12	⋯	3.00	2.91	2.85	2.80	2.75	⋯
⋮	⋮	⋮	⋮	⋮	⋮	⋮	⋮

Therefore, $F_{0.05,8,10} = 3.07$. This value is confirmed using technology, as shown in **Figure 10.23**. If the random variable $X \sim F_{8,10}$, then $P(X \geq 3.07) = 0.05$, as illustrated in **Figure 10.24**.

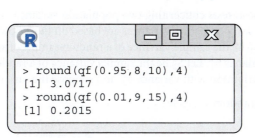

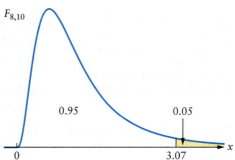

Figure 10.23 Use the R function `qf()` to find F critical values.

Figure 10.24 Visualization of the critical value $F_{0.05,8,10}$.

STEP 2 $F_{0.99,9,15}$ is a critical value related to an F distribution with 9 and 15 degrees of freedom. By definition, if $X \sim F_{9,15}$, then $P(X \geq F_{0.99,9,15}) = 0.99$. Because $F_{0.99,9,15}$ is in the left tail of the distribution, use Equation 10.11.

$$F_{0.99,9,15} = F_{1-0.01,9,15} = \frac{1}{F_{0.01,15,9}}$$

Using Appendix Table 7, for $\alpha = 0.01$, find the intersection of the $\nu_1 = 15$ column and the $\nu_2 = 9$ row.

$\alpha = 0.01$

				ν_1			
ν_2		9	10	15	20	30	
⋮	⋮	⋮	⋮	⋮	⋮	⋮	⋮
7	…	6.72	6.62	6.31	6.16	5.99	…
8	…	5.91	5.81	5.52	5.36	5.20	…
9	…	5.35	5.26	4.96	4.81	4.65	…
10	…	4.94	4.85	4.56	4.41	4.25	…
11	…	4.63	4.54	4.25	4.10	3.94	…
⋮	⋮	⋮	⋮	⋮	⋮	⋮	⋮

Therefore, $F_{0.99,9,15} = 1/4.96 = 0.202$. This value is confirmed using technology, as shown in Figure 10.23. If the random variable $X \sim F_{9,15}$, then $P(X \geq 0.202) = 0.99$ and $P(X \leq 0.202) = 0.01$, as illustrated in **Figure 10.25**.

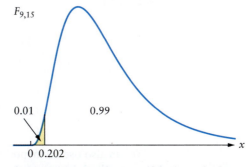

Figure 10.25 Visualization of the critical value $F_{0.99,9,15}$.

TRY IT NOW Go to Exercise 10.135

For reference, we'll call these the two-sample F test assumptions.

This is yet another kind of standardization, a transformation to an F distribution.

If σ_1^2 and σ_2^2 are equal, they cancel out.

Interpreting the ratio of two variances.

This is the template for a hypothesis test concerning two population variances, sometimes called a two-sample F test.

Remember that Appendix Table 7 is very limited. It includes only three values for α and a limited number of values for ν_1 and ν_2. Technology can be used to find other critical values.

General Hypothesis Test and CI

Hypothesis tests concerning two population variances and a confidence interval for the ratio of two population variances are based on the following results.

Let S_1^2 be the sample variance of a random sample of size n_1 from a normal distribution with variance σ_1^2. Let S_2^2 be the sample variance of a random sample of size n_2 from a normal distribution with variance σ_2^2, and suppose the samples are independent.

1. The random variable

$$F = \frac{S_1^2/\sigma_1^2}{S_2^2/\sigma_2^2} \qquad (10.12)$$

has an F distribution with $n_1 - 1$ (from the numerator) and $n_2 - 1$ (from the denominator) degrees of freedom.

2. If the null hypothesis is $H_0: \sigma_1^2 = \sigma_2^2$, then the random variable simplifies to $F = \frac{S_1^2/\sigma_1^2}{S_2^2/\sigma_2^2} = \frac{S_1^2}{S_2^2}$. This simple ratio is the test statistic for comparing two population variances.

Hypothesis Tests Concerning Two Population Variances

Given the two-sample F test assumptions, a hypothesis test concerning two population variances with significance level α has the following form:

$H_0: \sigma_1^2 = \sigma_2^2$

$H_a: \sigma_1^2 > \sigma_2^2, \qquad \sigma_1^2 < \sigma_2^2, \qquad \text{or} \qquad \sigma_1^2 \neq \sigma_2^2$

TS: $F = \dfrac{S_1^2}{S_2^2}$

RR: $F \geq F_{\alpha, n_1-1, n_2-1}, \quad F \leq F_{1-\alpha, n_1-1, n_2-1}, \quad$ or

$F \leq F_{1-\alpha/2, n_1-1, n_2-1} \quad$ or $\quad F \geq F_{\alpha/2, n_1-1, n_2-1}$

Using the same assumptions, we can derive a confidence interval for the ratio of two population variances. Let $X \sim F_{n_1-1, n_2-1}$ and find an interval that captures $1 - \alpha$ of the probability in the middle of this F distribution. Manipulate the inequality to sandwich the ratio σ_1^2/σ_2^2.

How to Find a $100(1-\alpha)\%$ Confidence Interval for the Ratio of Two Population Variances

Given the two-sample F test assumptions, a $100(1-\alpha)\%$ confidence interval for σ_1^2/σ_2^2 is given by

$$\left(\frac{s_1^2}{s_2^2} \frac{1}{F_{\alpha/2, n_1-1, n_2-1}}, \frac{s_1^2}{s_2^2} \frac{1}{F_{1-\alpha/2, n_1-1, n_2-1}} \right) \qquad (10.13)$$

Using Equation 10.11, the confidence interval can be written as

$$\left(\frac{s_1^2}{s_2^2} \frac{1}{F_{\alpha/2, n_1-1, n_2-1}}, \frac{s_1^2}{s_2^2} F_{\alpha/2, n_2-1, n_1-1} \right) \qquad (10.14)$$

We can also use the hypothesis test procedure described here to compare two population standard deviations. In addition, we can take the square root of each endpoint of Equation 10.13 to find a $100(1-\alpha)\%$ confidence interval for the ratio of two population standard deviations. The next example illustrates the hypothesis test procedure.

Example: Hypothesis Test

EXAMPLE 10.15 Long-Term-Care Cost

The cost of long-term care in a nursing home varies considerably by region and may be as much as $60,000 per year. Two independent samples of nursing homes in Connecticut and in Colorado were obtained, and the cost per day for each was recorded. The data are given in the table.

Connecticut (1)									
270	294	174	180	314	274	160	210	255	187	271

Colorado (2)										
161	150	164	109	168	172	133	148	94	157	138	120
166	116	98	168	153	118	138	116	120			

(a) Conduct the appropriate hypothesis test to determine whether there is any evidence that the population variance in cost per day is different in Connecticut and in Colorado. Assume the underlying populations for the costs per day are normal and use $\alpha = 0.02$.

(b) Find bounds on the p value for the hypothesis test in part (a).

Solution

(a)

STEP 1 The null hypothesis is that the two population variances are equal. We are looking for any difference in the variances, so the alternative hypothesis is two-sided. The underlying populations are assumed to be normal and the samples were obtained independently. A two-sample F test is appropriate.

STEP 2 The four parts of the hypothesis test are

$H_0: \sigma_1^2 = \sigma_2^2$
$H_a: \sigma_1^2 \neq \sigma_2^2$
TS: $F = \dfrac{S_1^2}{S_2^2}$
RR: $F \leq F_{1-\alpha/2, n_1-1, n_2-1} = F_{0.99, 10, 20} = 1/4.41 = 0.2268$ or
$F \geq F_{\alpha/2, n_1-1, n_2-1} = F_{0.01, 10, 20} = 3.37$

STEP 3 The summary statistics are

$$s_1^2 = \frac{1}{10}\left[638{,}819 - \frac{1}{11}(2589)^2\right] = 2946.25$$

$$s_2^2 = \frac{1}{20}\left[414{,}601 - \frac{1}{21}(2907)^2\right] = 609.46$$

The value of the test statistic is

$$f = \frac{s_1^2}{s_2^2} = \frac{2946.25}{609.46} = 4.83 \; (\geq 3.37)$$

The value of the test statistic lies in the rejection region. Therefore, we reject the null hypothesis at the $\alpha = 0.02$ significance level. There is evidence to suggest that the two population variances are different.

(b)

Because the tables of critical values for F distributions are very limited, we can only bound the p value. Place the value of the test statistic, $f = 4.83$, in an ordered list of critical values with degrees of freedom 10 and 20.

$$3.37 \leq 4.83 \leq 5.08$$
$$F_{0.01,10,20} \leq 4.83 \leq F_{0.001,10,20}$$
Therefore, $\quad 0.001 \leq p/2 \leq 0.01$
and, $\quad\quad\quad 0.002 \leq p \leq 0.02$

The p value is illustrated in **Figure 10.26**, and a technology solution with the exact p value is shown in **Figure 10.27**.

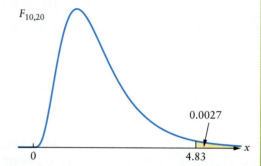

Figure 10.26 p value illustration:
$p = 2P(Z \geq 4.83) = 0.0027 \leq 0.02 = \alpha$.

```
> con <- c(270,294,174,180,314,274,160,210,255,187,271)
> col <- c(161,150,164,109,168,172,133,148,94,157,138,120,
+           166,116,98,168,153,118,138,116,120)
>
> var_1 <- var(con)
> var_2 <- var(col)
> f_crit_low <- qf(0.01,10,20)
> f_crit_high <- qf(0.99,10,20)
> f <- var_1/var_2
> p_value <- 2* (1 - pf(f,10,20))
>
> round(cbind(var_1,var_2,f_crit_low,f_crit_high,f,p_value),4)
         var_1    var_2 f_crit_low f_crit_high      f p_value
[1,]  2946.255 609.4571      0.227      3.3682 4.8342  0.0027
```

Figure 10.27 R calculations to find the sample variances, the critical values, the value of the test statistic, and the p value.

TRY IT NOW Go to Exercises 10.145 and 10.147

Example: Confidence Interval

The next example involves constructing a confidence interval for the ratio of two population variances.

EXAMPLE 10.16 Importance of Coral Reefs

Coral reefs are diverse and valuable ecosystems. These vast ocean environments support approximately 4000 species of fish, may potentially lead to medical advances, and contribute to local economies through tourism. Independent random samples of brain corals were obtained from two Caribbean reefs. The diameter of each polyp was carefully measured (in millimeters). The data are summarized in the table.

Navassa Island	$n_1 = 16$,	$s_1^2 = 0.267$
U.S. Virgin Islands	$n_2 = 21$,	$s_2^2 = 0.172$

Construct a 90% confidence interval for the ratio of population variances in diameter of brain coral polyps.

Solution

STEP 1 The samples are independent; the sample sizes and sample variances are given. A confidence interval for the ratio of two population variances is appropriate.

STEP 2 Find the critical values.

$1 - \alpha = 0.90 \Rightarrow \alpha = 0.10 \Rightarrow \alpha/2 = 0.05$ Find $\alpha/2$.

$F_{\alpha/2, n_1-1, n_2-1} = F_{0.05, 15, 20} = 2.20$ Critical value, left endpoint. Use Appendix Table 7.

$F_{\alpha/2, n_2-1, n_1-1} = F_{0.05, 20, 15} = 2.33$ Critical value, right endpoint. Use Appendix Table 7.

STEP 3 Use Equation 10.14.

$$\left(\frac{s_1^2}{s_2^2} \frac{1}{F_{\alpha/2, n_1-1, n_2-1}}, \frac{s_1^2}{s_2^2} F_{\alpha/2, n_2-1, n_1-1} \right)$$ Equation 10.14.

$$= \left(\frac{0.267}{0.172} \frac{1}{2.20}, \frac{0.267}{0.172} (2.33) \right)$$ Use sample variances and critical values.

$$= (0.7046, 3.6131)$$ Simplify.

The interval (0.7046, 3.6131) is a 90% confidence interval for the ratio of the population variances. **Figure 10.28** shows a technology solution.

```
> n_1 <- 16
> var_1 <- 0.267
> n_2 <- 21
> var_2 <- 0.172
>
> f_crit_low <- qf(0.95,n_1-1,n_2-1)
> f_crit_high <- qf(0.95,n_2-1,n_1-1)
> left <- (var_1/var_2)*(1/f_crit_low)
> right <- (var_1/var_2)*f_crit_high
>
> round(cbind(f_crit_low,f_crit_high,left,right),4)
     f_crit_low f_crit_high   left  right
[1,]     2.2033      2.3275 0.7046 3.6131
```

Figure 10.28 R calculations to find the critical values and the endpoints of the confidence interval.

TRY IT NOW Go to Exercise 10.149

Suppose a two-sample t test will be used to compare two population means. The hypothesis test presented in this section is often used first to compare the population variances. The results and conclusion suggest the appropriate hypothesis test concerning population means from Section 10.2, according to whether there is evidence that the two population variances are unequal.

Section 10.5 Exercises

Concept Check

10.130 True or False Every F distribution is symmetric about its mode.

10.131 True or False An F distribution has positive probability only for non-negative values.

10.132 True or False $F_{\alpha, \nu_1, \nu_2} = F_{1-\alpha, \nu_2, \nu_1}$

10.133 True or False A hypothesis test to compare two population variances can also be used to test for equality of population standard deviations.

10.134 Short Answer S_1^2 and S_2^2 are unbiased estimators for the population variances σ_1^2 and σ_2^2, respectively. Explain the meaning of this statement.

10.135 Find the critical value.
(a) $F_{0.05, 7, 19}$ (b) $F_{0.05, 30, 25}$ (c) $F_{0.01, 6, 19}$
(d) $F_{0.001, 40, 40}$ (e) $F_{0.95, 17, 15}$ (f) $F_{0.95, 12, 10}$
(g) $F_{0.99, 21, 30}$ (h) $F_{0.999, 11, 8}$

10.136 Find the critical value.
(a) $F_{0.01, 20, 60}$ (b) $F_{0.01, 15, 19}$ (c) $F_{0.05, 6, 8}$
(d) $F_{0.001, 8, 24}$ (e) $F_{0.99, 12, 9}$ (f) $F_{0.99, 6, 8}$
(g) $F_{0.95, 10, 10}$ (h) $F_{0.999, 23, 20}$

10.137 Consider a hypothesis test concerning two population variances with null hypothesis $H_0: \sigma_1^2 = \sigma_2^2$. The alternative hypothesis, the sample sizes, and the value of the test statistic are given. Find bounds on the p value associated with the hypothesis test.
(a) $H_a: \sigma_1^2 > \sigma_2^2$, $n_1 = 16$, $n_2 = 17$, $f = 2.89$
(b) $H_a: \sigma_1^2 < \sigma_2^2$, $n_1 = 11$, $n_2 = 16$, $f = 0.34$
(c) $H_a: \sigma_1^2 \neq \sigma_2^2$, $n_1 = 7$, $n_2 = 10$, $f = 7.36$
(d) $H_a: \sigma_1^2 > \sigma_2^2$, $n_1 = 31$, $n_2 = 26$, $f = 7.26$

10.138 Consider independent random samples of sizes $n_1 = 31$ and $n_2 = 25$ from normal populations.
(a) Write the four parts for a one-sided, right-tailed hypothesis test concerning the population variances with $\alpha = 0.05$.
(b) Suppose $s_1^2 = 44.89$ and $s_2^2 = 17.64$. Find the value of the test statistic, and draw a conclusion about the population variances.
(c) Find bounds on the p value associated with this hypothesis test and carefully sketch a graph to illustrate this value.

10.139 Consider the two independent random samples from normal distributions given in the table. **EX10.139**

Sample 1							
89.6	56.5	76.1	61.1	72.2	73.2	83.7	72.4
74.2	77.1	60.6	58.2	50.4	72.3	82.4	71.6
79.0	70.3						

Sample 2							
73.7	76.2	73.7	37.8	82.7	41.5	76.2	88.9
76.9	64.7	60.3	55.0	74.8	34.4	76.0	75.8
74.5	67.9	68.0	61.6	100.2			

(a) Write the four parts for a one-sided, left-tailed hypothesis test concerning the two population variances with $\alpha = 0.05$.
(b) Compute each sample variance, find the value of the test statistic, and draw a conclusion.
(c) Find bounds on the p value associated with this hypothesis test.

10.140 Consider independent random samples of sizes $n_1 = 10$ and $n_2 = 16$ from normal populations.
(a) Write the four parts for a two-sided hypothesis test concerning the population variances with $\alpha = 0.01$.
(b) Suppose $s_1^2 = 426.42$ and $s_2^2 = 88.36$. Find the value of the test statistic, and draw a conclusion about the population variances.
(c) Find bounds on the p value associated with this hypothesis test.

10.141 Suppose independent random samples were obtained, and the sample sizes and the confidence level are given here. Find the appropriate F critical values for use in constructing a confidence interval for the ratio of the population variances.
(a) $n_1 = 10$, $n_2 = 10$, 90%
(b) $n_1 = 21$, $n_2 = 10$, 98%
(c) $n_1 = 9$, $n_2 = 7$, 98%
(d) $n_1 = 41$, $n_2 = 31$, 99.8%

10.142 The sample sizes, the sample variances, and the confidence level are given here. Assume the underlying populations are normal and the samples were obtained independently. Find the associated confidence interval for the ratio of the population variances.
(a) $n_1 = 10$, $s_1^2 = 17.2$, $n_2 = 9$, $s_2^2 = 15.6$ 90%
(b) $n_1 = 16$, $s_1^2 = 54.1$, $n_2 = 16$, $s_2^2 = 32.6$ 98%
(c) $n_1 = 16$, $s_1^2 = 3.35$, $n_2 = 31$, $s_2^2 = 4.59$ 98%
(d) $n_1 = 31$, $s_1^2 = 126.8$, $n_2 = 41$, $s_2^2 = 155.3$ 99.8%

10.143 Use Appendix Table 7 and linear interpolation (see Chapter 6, page 267) to approximate the critical value. Verify the approximation using technology.
(a) $F_{0.05, 25, 15}$ (b) $F_{0.99, 30, 32}$ (c) $F_{0.01, 10, 56}$
(d) $F_{0.025, 15, 20}$ (e) $F_{0.995, 10, 7}$ (f) $F_{0.05, 35, 35}$

Applications

10.144 Biology and Environmental In a recent study conducted by the National Oceanic and Atmospheric Administration (NOAA), the aerosol light absorption coefficient was measured (in Mm^{-1}) at randomly selected locations in Africa and in South America. The resulting data are summarized in the table.

Country	Sample size	Sample variance
Africa	10	243.36
South America	21	51.84

Is there any evidence that the population variance in the aerosol light absorption coefficient is greater in Africa than in South America? Use $\alpha = 0.05$ and assume normality.

10.145 Medicine and Clinical Studies A study in the *British Medical Journal* suggested that children who received a diagnostic computed tomography (CT) scan using ionizing radiation were more likely to develop a cancer 10 years after radiation exposure. Newer CT scanners use less radiation. Therefore, the increased risk for children today may be decreased. Independent random samples of hospital CT scans in the United States and England were obtained, and the amount of radiation for each was recorded (in millisieverts, mSV). For the United States, $n_1 = 10$ and $s_1^2 = 1.075$; for England, $n_2 = 16$ and $s_2^2 = 2.786$. Is there any evidence to suggest that the variability in CT radiation per scan is different for these two countries? Use $\alpha = 0.05$ and assume normality.

10.146 Travel and Transportation Most airlines now charge passengers to check a bag and impose a surcharge if the weight of the bag is exceeds 50 lb. Independent random samples of checked luggage on Delta and American Airlines flights were obtained, and the weight (in pounds) of each was recorded. For Delta, $n_1 = 25$, and $s_1^2 = 96.23$; for American Airlines, $n_2 = 21$ and $s_2^2 = 194.02$. Is there any evidence to suggest that the variability in checked baggage weight is different for these two airlines? Use $\alpha = 0.05$ and assume normality.

10.147 Sports and Leisure Many basketball purists believe that the three-point shot (a shot from behind the three-point line, 22 ft from the basket) has dramatically changed the game, for the worse. And some even think Larry Bird couldn't compete with today's stars. Independent random samples of attempted shots from National Basketball Association (NBA) games played in 1975 (prior to the three-point shot) and in 2018 were obtained. The shot distance (in feet) was recorded for each attempt. The summary statistics are given in the table.

Year	Sample size	Sample variance
1975	61	12.25
2018	61	26.01

Is there any evidence to suggest that the variability in shot distance is greater in the year 2018 than it was in 1975? Use $\alpha = 0.01$ and assume normality. (Why do you suppose there is greater variability in shot distance with a three-point shot?)

10.148 Sports and Leisure A Laurel Downs racetrack official believes there is less variability in winning times for a race in which the purse, or total amount of money paid out to owners of horses, is at least $10,000, called a stakes race. Independent random samples of ordinary races and stakes races were obtained, and the winning time (in seconds) for each race was recorded. The summary statistics were as follows: ordinary race, $n_1 = 26$ and $s_1^2 = 110.25$; stakes race, $n_2 = 26$ and $s_2^2 = 38.44$.
(a) Write the four parts for a hypothesis test to check for evidence of the official's assertion. Use $\alpha = 0.01$, assume normality, and find the critical value using technology. Conduct the hypothesis test and draw a conclusion.
(b) Construct a 98% confidence interval for the ratio of population variances.

10.149 Sports and Leisure A study was conducted to compare the variability in times for men and women involved in collegiate swimming events. Independent random samples of 800-m freestyle competitors were obtained, and the time (in minutes) was recorded for each swimmer. The data are summarized in the table.

Group	Sample size	Sample variance
Men	11	0.1025
Women	12	0.1241

(a) Find the critical values necessary to construct a 95% confidence interval for the ratio of population variances.
(b) Construct the confidence interval.
(c) Use the confidence interval in part (b) to determine whether there is any evidence that the variability in times differs for men and women.

10.150 Take the Stairs The World Summit Wing Hotel in China is Beijing's tallest hotel and each year hosts a Vertical Run. Competitors must run up 81 floors, 330 m, and 2041 steps to reach the roof. Independent random samples of men's and women's times (in minutes) from the 2018 run were obtained.[15] **VERTRUN**
(a) Is there any evidence of a difference in variability of times for men and women who finished this vertical run? Use $\alpha = 0.02$ and assume normality.
(b) Find bounds on the p value associated with this hypothesis test.

10.151 Manufacturing and Product Development A sailboat manufacturer has two machines for constructing main mast poles with a diameter designed to be 76.2 mm. Maintaining small variability in production is very important to ensure boat control and safety. Independent random samples of mast poles produced on each machine were obtained, and the diameter of each was carefully measured. The summary statistics were as follows: Machine A, $n_1 = 7$, $s_1^2 = 0.0231$; Machine B, $n_2 = 8$, $s_2^2 = 0.0096$. Conduct the appropriate hypothesis test to determine whether there is any evidence of a difference in variability of mast-pole diameter between the two machines. Assume normality, use $\alpha = 0.02$, find the p value associated with this test, and use this value to draw a conclusion.

10.152 Economics and Finance Tower cranes along a city skyline often indicate the success of economic development efforts. In 2017, there were 62 tower cranes in operation; by comparison, at the beginning of 2018 there were 30 such cranes being used. Independent random samples of items lifted by cranes at two different sites were obtained, and the weight (in tons) of each item was recorded for each. The summary statistics are given in the table.

Site	Sample size	Sample variance
Moxy Hotel	31	109.86
Old Town Park	31	339.43

Is there any evidence to suggest that the variability in item weight at Moxy Hotel is greater than at Old Town Park? Use $\alpha = 0.01$ and assume normality.

10.153 Sports and Leisure The Pikes Peak International Hill Climb is also known as The Race to the Clouds. This 12.42-mi race up Pikes Peak in Colorado features more than 156 turns, grades of 7%, and a finish line at 14,110 ft altitude. Independent random samples of two classes of cars in the 2018 race were obtained, and the speed (in mph) for each was recorded. The summary statistics are given in the table.

Class	Sample size	Sample variance
Exhibition Powersports	9	44.05
Heavyweight Supermoto	9	9.76

Is there any evidence to suggest that the variability in speed for the Exhibition Powersports class is greater than the variability in speed for the Heavyweight Supermoto class? Use $\alpha = 0.01$ and assume normality.

10.154 Manufacturing and Product Development The Akashi–Kaikyo bridge in Japan is the longest suspension bridge in the world, with a main span of 1991 m. Two million workers took 10 years to construct this bridge using 181,000 tons of steel and 1.4 million m³ of concrete. Independent random samples of suspension bridges in China and the United States were obtained, and the span (in meters) of each was recorded. For China, $n_1 = 14$ and $s_1^2 = 66{,}096.8$; for the United States, $n_2 = 10$ and $s_2^2 = 59{,}524.5$. Is there any evidence to suggest a difference in the variability of the span of suspension bridges in China and the United States? Use $\alpha = 0.02$ and assume normality.

10.155 Sports and Leisure Formula One, or Grand Prix, racing events occur all over the world and usually take place over three days, including practice and qualifying sessions. Independent random samples of racers during events in Italy and Canada were obtained, and a pit stop time for each was recorded (in seconds).[16] Is there any evidence to suggest a difference in the variability of the pit stop times at races in Italy and in Canada? Use $\alpha = 0.02$ and assume normality. **PITSTOP**

Extended Applications

10.156 Physical Sciences Crude oil pumped from ocean wells contains salt that must be removed before the oil is refined. Otherwise, equipment would erode quickly. Independent random samples of unrefined crude oil from two ocean wells were obtained, and the percentage of salt in each sample was recorded. The summary statistics are given in the table.

Oil well	Sample size	Sample variance
North Sea	21	56.40
Antarctica	31	82.42

(a) Conduct the appropriate test to determine whether there is evidence of any difference in variability of salt content between these two wells. Use $\alpha = 0.10$ and assume normality.

(b) Use Appendix Table 7 to find bounds on the p value for this hypothesis test. Use technology to find an exact p value.

10.157 Public Health and Nutrition Saccharin is a low-calorie sweetener used in sugar-free foods and beverages. According to the U.S. Food and Drug Administration, the acceptable daily intake (ADI) of saccharin is 5 mg for a person with a body weight of 60 kg. If saccharin is used as an additive, it must be included on the food label and cannot exceed certain limits. Independent random samples of 12-oz bottles of iced tea from two different manufacturers were obtained, and the amount of saccharin in each drink was measured (in milligrams). The summary statistics were as follows: Fishing Creek, $n_1 = 20$, $s_1^2 = 7.84$; Honest Tea, $n_2 = 15$, $s_2^2 = 2.89$.

(a) Conduct the appropriate test to determine whether there is any difference in the population variance of saccharin amounts. Use $\alpha = 0.02$.

(b) Find bounds on the p value associated with this hypothesis test.

10.158 Fuel Consumption and Cars The length of time that brake pads last in an automobile varies depending on driving style and the type of car. Brake pads are made from organic, semimetallic, metallic, or synthetic materials, and typically last between 30,000 and 70,000 mi. Independent random samples of cars in for an inspection at dealerships and private garages were obtained, and the width (in millimeters) of the brake pad on the front driver's side was measured for each. The summary statistics are given in the table.

Location	Sample size	Sample variance
Dealership	41	1.056
Private garage	41	2.771

(a) Is there any evidence to suggest that the variance in brake-pad widths of cars in for inspection is greater at private garages than at dealerships? Use $\alpha = 0.01$ and assume normality.

(b) Find bounds on the p value associated with this hypothesis test.

Challenge Problems

10.159 Physical Sciences When comparing two population variances, if both sample sizes n_1 and n_2 are large and the null hypothesis $H_0: \sigma_1^2 = \sigma_2^2$ is true, then the test statistic $F = S_1^2/S_2^2$ is approximately normal with

$$\mu_F = \frac{n_2 - 1}{n_2 - 3} \quad \text{and} \quad \sigma_F^2 = \frac{2(n_2 - 1)^2 (n_1 + n_2 - 4)}{(n_1 - 1)(n_2 - 3)^2 (n_2 - 5)}$$

An approximate hypothesis test is based on standardizing F to a Z random variable.

The four parts of the hypothesis test are

$H_0: \sigma_1^2 = \sigma_2^2$

$H_a: \sigma_1^2 > \sigma_2^2, \quad \sigma_1^2 < \sigma_2^2, \quad \text{or} \quad \sigma_1^2 \neq \sigma_2^2$

$$\text{TS: } Z = \frac{(S_1^2/S_2^2) - [(n_2 - 1)/(n_2 - 3)]}{\sqrt{\dfrac{2(n_2 - 1)^2 (n_1 + n_2 - 4)}{(n_1 - 1)(n_2 - 3)^2 (n_2 - 5)}}}$$

RR: $Z \geq z_\alpha, \quad Z \leq -z_\alpha, \quad \text{or} \quad |Z| \geq z_{\alpha/2}$

The National Wind Energy Assessment contains data from 975 stations and includes measurements of wind speed and wind power density. Suppose independent random samples of wind power density (in watts per m²) during the winter were obtained from two stations. The data are summarized in the table.

Location	Sample size	Sample mean	Sample variance
Chanute	31	207	95.35
Dodge City	31	283	53.68

(a) Write the four parts of a large-sample, two-sided, approximate test based on the standard normal distribution to determine whether any evidence suggests that the two population variances are different. Conduct the test using $\alpha = 0.05$. Find the p value associated with this hypothesis test.

(b) Conduct an exact hypothesis test based on the F distribution. Use technology to find the p value associated with this hypothesis test. Compare your answers to part (a).

Chapter 10 Summary

Concept	Page	Notation / Formula / Description
Independent samples	453	Two samples are independent if the process of selecting individuals or objects in sample 1 has no effect on the selection of individuals or objects in sample 2.
Paired data set	453	The result of matching each individual or object in sample 1 with a similar individual or object in sample 2.
Pooled estimator for the common variance	464	$S_p^2 = \dfrac{(n_1-1)S_1^2 + (n_2-1)S_2^2}{n_1 + n_2 - 2}$
Combined estimate of the common population proportion	489	$\hat{P}_c = \dfrac{X_1 + X_2}{n_1 + n_2}$

Summary of Confidence Intervals

Parameter	Assumptions	$100(1-\alpha)\%$ Confidence Interval
$\mu_1 - \mu_2$	Either n_1, n_2 large, independence, σ_1^2, σ_2^2 known, or normality, independence, σ_1^2, σ_2^2 known	$(\bar{x}_1 - \bar{x}_2) \pm z_{\alpha/2} \cdot \sqrt{\dfrac{\sigma_1^2}{n_1} + \dfrac{\sigma_2^2}{n_2}}$
$\mu_1 - \mu_2$	Normality, independence, σ_1^2, σ_2^2 unknown but equal	$(\bar{x}_1 - \bar{x}_2) \pm t_{\alpha/2, n_1+n_2-2} \cdot \sqrt{s_p^2\left(\dfrac{1}{n_1} + \dfrac{1}{n_2}\right)}$ $s_p^2 = \dfrac{(n_1-1)s_1^2 + (n_2-1)s_2^2}{n_1 + n_2 - 2}$
$\mu_1 - \mu_2$	Normality, independence, σ_1^2, σ_2^2 unknown and unequal	$(\bar{x}_1 - \bar{x}_2) \pm t_{\alpha/2, \nu} \cdot \sqrt{\dfrac{s_1^2}{n_1} + \dfrac{s_2^2}{n_2}}$ $\nu = \dfrac{\left(\dfrac{s_1^2}{n_1} + \dfrac{s_2^2}{n_2}\right)^2}{\dfrac{(s_1^2/n_1)^2}{n_1 - 1} + \dfrac{(s_2^2/n_2)^2}{n_2 - 1}}$
$\mu_D = \mu_1 - \mu_2$	Normality, n pairs, dependence	$\bar{d} \pm t_{\alpha/2, n-1} \cdot \dfrac{s_D}{\sqrt{n}}$
$p_1 - p_2$	n_1, n_2 large, nonskewness, independence	$(\hat{p}_1 - \hat{p}_2) \pm z_{\alpha/2} \cdot \sqrt{\dfrac{\hat{p}_1(1-\hat{p}_1)}{n_1} + \dfrac{\hat{p}_2(1-\hat{p}_2)}{n_2}}$
$\dfrac{\sigma_1^2}{\sigma_2^2}$	Normality, independence	$\left(\dfrac{s_1^2}{s_2^2} \dfrac{1}{F_{\alpha/2, n_1-1, n_2-1}}, \dfrac{s_1^2}{s_2^2} \dfrac{1}{F_{1-\alpha/2, n_1-1, n_2-1}}\right)$

Summary of Hypothesis Tests

Null hypothesis	Assumptions	Alternative hypothesis	Test statistic	Rejection region
$\mu_1 - \mu_2 = \Delta_0$	Either n_1, n_2 large, independence, σ_1^2, σ_2^2 known, or normality, independence, σ_1^2, σ_2^2 known	$\mu_1 - \mu_2 > \Delta_0$ $\mu_1 - \mu_2 < \Delta_0$ $\mu_1 - \mu_2 \neq \Delta_0$	$Z = \dfrac{(\bar{X}_1 - \bar{X}_2) - \Delta_0}{\sqrt{\dfrac{\sigma_1^2}{n_1} + \dfrac{\sigma_2^2}{n_2}}}$	$Z \geq z_\alpha$ $Z \leq -z_\alpha$ $\|Z\| \geq z_{\alpha/2}$
$\mu_1 - \mu_2 = \Delta_0$	Normality, independence, σ_1^2, σ_2^2 unknown, $\sigma_1^2 = \sigma_2^2$	$\mu_1 - \mu_2 > \Delta_0$ $\mu_1 - \mu_2 < \Delta_0$ $\mu_1 - \mu_2 \neq \Delta_0$	$T = \dfrac{(\bar{X}_1 - \bar{X}_2) - \Delta_0}{\sqrt{S_p^2\left(\dfrac{1}{n_1} + \dfrac{1}{n_2}\right)}}$ $S_p^2 = \dfrac{(n_1-1)S_1^2 + (n_2-1)S_2^2}{n_1 + n_2 - 2}$	$T \geq t_{\alpha, n_1+n_2-2}$ $T \leq -t_{\alpha, n_1+n_2-2}$ $\|T\| \geq t_{\alpha/2, n_1+n_2-2}$
$\mu_1 - \mu_2 = \Delta_0$	Normality, independence, σ_1^2, σ_2^2 unknown, $\sigma_1^2 \neq \sigma_2^2$	$\mu_1 - \mu_2 > \Delta_0$ $\mu_1 - \mu_2 < \Delta_0$ $\mu_1 - \mu_2 \neq \Delta_0$	$T' = \dfrac{(\bar{X}_1 - \bar{X}_2) - \Delta_0}{\sqrt{\dfrac{S_1^2}{n_1} + \dfrac{S_2^2}{n_2}}}$ $\nu = \dfrac{\left(\dfrac{s_1^2}{n_1} + \dfrac{s_2^2}{n_2}\right)^2}{\dfrac{(s_1^2/n_1)^2}{n_1-1} + \dfrac{(s_2^2/n_2)^2}{n_2-1}}$	$T' \geq t_{\alpha, \nu}$ $T' \leq -t_{\alpha, \nu}$ $\|T'\| \geq t_{\alpha/2, \nu}$
$\mu_D = \Delta_0$	Normality, n pairs, dependence	$\mu_D > \Delta_0$ $\mu_D < \Delta_0$ $\mu_D \neq \Delta_0$	$T = \dfrac{\bar{D} - \Delta_0}{S_D/\sqrt{n}}$	$T \geq t_{\alpha, n-1}$ $T \leq -t_{\alpha, n-1}$ $\|T\| \geq t_{\alpha/2, n-1}$
$p_1 - p_2 = 0$	n_1, n_2 large, nonskewness, independence	$p_1 - p_2 > 0$ $p_1 - p_2 < 0$ $p_1 - p_2 \neq 0$	$Z = \dfrac{\hat{P}_1 - \hat{P}_2}{\sqrt{\hat{P}_c(1-\hat{P}_c)\left(\dfrac{1}{n_1} + \dfrac{1}{n_2}\right)}}$ $\hat{P}_c = \dfrac{X_1 + X_2}{n_1 + n_2}$	$Z \geq z_\alpha$ $Z \leq -z_\alpha$ $\|Z\| \geq z_{\alpha/2}$
$p_1 - p_2 = \Delta_0$	n_1, n_2 large, nonskewness, independence	$p_1 - p_2 > \Delta_0$ $p_1 - p_2 < \Delta_0$ $p_1 - p_2 \neq \Delta_0$	$Z = \dfrac{(\hat{P}_1 - \hat{P}_2) - \Delta_0}{\sqrt{\dfrac{\hat{P}_1(1-\hat{P}_1)}{n_1} + \dfrac{\hat{P}_2(1-\hat{P}_2)}{n_2}}}$	$Z \geq z_\alpha$ $Z \leq -z_\alpha$ $\|Z\| \geq z_{\alpha/2}$
$\sigma_1^2 = \sigma_2^2$	Normality, independence	$\sigma_1^2 > \sigma_2^2$ $\sigma_1^2 < \sigma_2^2$ $\sigma_1^2 \neq \sigma_2^2$	$F = \dfrac{S_1^2}{S_2^2}$	$F \geq F_{\alpha, n_1-1, n_2-1}$ $F \leq F_{1-\alpha, n_1-1, n_2-1}$ $F \leq F_{1-\alpha/2, n_1-1, n_2-1}$ or $F \geq F_{\alpha/2, n_1-1, n_2-1}$

Chapter 10 Exercises

Applications

10.160 Travel and Transportation The U.S. Department of Transportation requires vehicles transporting hazardous materials to use special placards indicating the type of cargo they are carrying. Many other regulations focus on the containers used, separation of various materials, and gross weight. Independent random samples of trucks carrying corrosive materials were stopped on highways in North Carolina and in Virginia, and the weight (in kilograms) of the hazardous material was recorded. The summary statistics and known variances are given in the table.

State	Sample size	Sample mean	Population variance
North Carolina	22	835.6	3192.25
Virginia	25	884.2	3956.41

Is there any evidence to suggest that the mean amount of corrosive material carried by trucks in North Carolina differs from the mean amount of corrosive material carried by trucks in Virginia? Use $\alpha = 0.01$ and assume each underlying distribution of weight is normal.

10.161 Taser Effectiveness Tasers are nonlethal weapons used by police to subdue dangerous people. This electroshock weapon uses an electrical current to disrupt control of an individual's muscles. The police forces in England maintain careful records concerning the use of Tasers. The table shows data from last year.

Police force	Lancashire	West Mercia
Number of Taser uses	186	138
Number of chest hits	120	62

Is there any evidence to suggest that the true proportion of chest hits is greater in Lancashire than in West Mercia? Use $\alpha = 0.01$.

10.162 Biology and Environmental Science Soybeans are an important source of oil and protein and are also used to produce many food additives. The leading producers of soybeans are the United States, Brazil, Argentina, and China. The first genetically modified (GM) soybeans were grown in the United States in 1996, and now GM soybeans are grown in at least nine countries. Independent random samples of soybean farmers in the United States and Brazil were obtained, and the number growing GM soybeans was recorded. The data are given in the table.

Country	Sample size	Number of GM soybean farmers
United States	238	202
Brazil	162	104

(a) Find the sample proportion of GM soybean farmers for each country. Verify the nonskewness criterion.
(b) Conduct the appropriate hypothesis test to determine whether there is any evidence that the true proportion of GM soybean farmers in the United States is 0.15 greater than the proportion in Brazil. Use $\alpha = 0.05$.
(c) Find the p value associated with this hypothesis test.

10.163 Biology and Environmental Maple syrup producers in New York and Vermont collect sweet-water sap from sugar maples and black maples in early spring. It takes approximately 30–50 gal of sap to yield, through boiling and evaporation, 1 gal of maple syrup. Independent random samples of maple trees in both states were obtained, and the amount of sap collected from each tree was recorded. Assume the underlying populations are normal, with equal variances. Is there any evidence to suggest that the population mean amount of sap from trees in New York is different from the population mean amount of sap from trees in Vermont? Use $\alpha = 0.01$. SAP

10.164 Physical Sciences Recycling of aluminum, glass, newspapers, and magazines is good for the environment and the economy. San Francisco has one of the highest recycling rates in the United States, and Germany has the best recycling rate in the world (recycling rate = tons collected for recycling/tons of all waste generated). Despite efforts to make the process easier, many people still do not recycle. Independent random samples of residents in Ohio and in Florida were obtained and asked whether they recycle newspapers. Of the 909 Ohio residents, 700 said they recycled newspapers, whereas 691 of the 923 Florida residents said they recycled newspapers.
(a) Is there any evidence to suggest that the population proportion of residents in Ohio who recycle newspapers is greater than the population proportion of residents who do so in Florida? Use $\alpha = 0.01$.
(b) Find the p value for this hypothesis test.

10.165 Sports and Leisure Archery target shooters use a variety of arrows made from wood, carbon, aluminum, or even platinum. One measure of the quality of an arrow (and bow) is the speed of the arrow when shot. A random sample of archers was obtained, and each was asked to shoot a carbon arrow and a similarly made aluminum arrow. The speed (in feet per second) of each arrow was measured. ARROW
(a) What is the common characteristic that makes these data paired?
(b) Assume normality. Conduct the appropriate hypothesis test to determine whether there is any evidence that the aluminum arrow flies faster. Use $\alpha = 0.05$.
(c) Find bounds on the p value associated with this hypothesis test.

10.166 Manufacturing and Product Development Raytheon Aircraft is now manufacturing business jets with a molded carbon fiber fuselage instead of aluminum. This construction reduces the overall weight of the plane, speeds production time, and increases cabin space. The total wall thickness of a carbon fiber fuselage is 0.81 in. versus 3 in. for aluminum, and the variability in thickness is theoretically much smaller as well. Independent random samples of the two fuselage types were obtained, and the thickness of each (in inches) was measured. The data are summarized in the table.

Fuselage type	Sample size	Sample variance
Aluminum	9	0.0196
Carbon fiber	11	0.0025

Is there any evidence to suggest that the variability in fuselage thickness is less for carbon fiber fuselages? Use $\alpha = 0.01$.

10.167 Biology and Environmental Science Piers on public beaches are usually supported by widely spread piles or pillars and can extend a thousand feet into the ocean. Many piers are extensions of boardwalks, and visitors frequently fish or simply sightsee along these walkways. Longer piers tend to be more susceptible to wind and storm damage. Independent random samples of concrete and wooden piers on public beaches along the California and Florida coasts were obtained and the length of each (in feet) was recorded. Assume the underlying populations are normal, with equal variances. Is there any evidence to suggest that the population mean pier length in California is greater than the population mean pier length in Florida? Use $\alpha = 0.01$. **PIERS**

10.168 Public Health and Nutrition In case you missed it, the United Nations declared 2008 to be the International Year of the Potato. Seriously, potatoes are a good source of carbohydrates, protein, fiber, and potassium. However, the amount of each element present varies depending on where the potato is grown. Independent random samples of medium-sized potatoes from Russia and China were obtained, and the amount of potassium (in milligrams) was measured in each. The data are summarized in the table.

Location	Sample size	Sample mean	Sample standard deviation
Russia	25	896.8	92.9
China	30	866.0	120.0

Assume normality and equal population variances. Is there any evidence to suggest that the population mean potassium level is different for a medium-sized potato in Russia and in China? Use $\alpha = 0.05$.

10.169 Biology and Environmental Science The moisture content in bulk grain is important, because high values can encourage the development of fungi. Potential buyers want to know how much water they are buying along with their grain. Two direct methods for measuring the moisture content are by means of a chemical reaction (with iodine in the presence of sulfur dioxide) and by distillation. A random sample of bulk grain was obtained, and the moisture content of each grain sample was measured as a percentage of water using each method. Assuming normality, conduct the appropriate hypothesis test to determine whether there is any difference in the population mean moisture content of bulk grain measured by chemical reaction and by distillation. Use $\alpha = 0.05$. **GRAIN**

10.170 Psychology and Human Behavior Two recent studies suggest that people who drive really nice cars exhibit some very bad habits. In one study, as a car approached a crosswalk, a person stepped into the road, and the driver's reaction was recorded. In another, similar study, independent random samples of drivers were selected, and their behavior was observed at a four-way intersection. For luxury-car drivers, $n_1 = 217$ and 130 cut ahead in the usual four-way rotation. For ordinary-car drivers, $n_2 = 182$ and 82 violated the four-way-intersection rotation rule. Is there any evidence to suggest that the proportion of luxury-car drivers with insufferable driving habits is greater than the proportion of ordinary-car drivers with similar habits? Use $\alpha = 0.01$. (Note: The largest group of driving-rule etiquette violators are generally men, ages 35–50, with blue BMWs.)

10.171 Public Policy and Political Science California law requires fuel outlets to install special catch basins designed to contain gasoline leaks in underground storage tanks. Owners who do not comply can face stiff fines and other penalties. Independent random samples of gasoline stations around Los Angeles and around San Francisco were obtained, and each station was inspected for catch basins. Sixteen of 140 stations near Los Angeles had no catch basins, and 12 of 126 in San Francisco were not complying with the law.

(a) Find the sample proportion of stations without catch basins near each city. Verify the nonskewness criterion.
(b) Is there any evidence that the population proportion of stations in noncompliance with the law is different near Los Angeles and near San Francisco? Use $\alpha = 0.01$.

10.172 Manufacturing and Product Development During the first quarter of 2018, the Food and Drug Administration issued at least four recalls for dog foods that may have been contaminated with *Salmonella*. While pets can become ill from eating contaminated foods, the Centers for Disease Control and Prevention also reminded people to wash their hands thoroughly after handling pet food. Independent random samples of Procter & Gamble cat foods and dog foods were obtained, and each was tested for *Salmonella*. For cat food, $n_1 = 1250$ and 50 were contaminated; for dog food, $n_2 = 1448$ and 87 were contaminated. Is there any evidence to suggest that the true proportion of contaminated cat food is different from the true proportion of contaminated dog food? Use $\alpha = 0.05$.

Extended Applications

10.173 Economics and Finance The U.S. Internal Revenue Service (IRS) estimates that the average taxpayer takes approximately 6 hr to complete Form 1040. A study was conducted to examine the amount of time it takes to complete this dreaded form, by income level. Independent random samples of federal filers in two income ranges were obtained, and the length of time (in hours) to complete Form 1040 was recorded for each. The summary statistics are given in the table.

Income level (in dollars)	Sample size	Sample mean	Sample variance
50,000 to < 100,000	17	4.56	1.5625
100,000 to < 200,000	14	6.58	15.0544

Assume the underlying populations are normal.
(a) Conduct an F test to determine whether there is any evidence that the two population variances are different. Use $\alpha = 0.02$.
(b) Using your conclusion from part (a), conduct the appropriate test for evidence that the mean time to complete Form 1040 for the lower-income level is less than the mean time for the higher-income level. Use $\alpha = 0.05$. State your conclusion and find bounds on the p value.

10.174 Public Policy and Political Science In many states, lawyers are encouraged to do *pro bono* work by both their firms and judicial advisory councils. However, in recent years lawyers have been devoting more time to their paying clients and less time to *pro bono* legal aid. Independent random samples of lawyers from two large firms were obtained, and the number of *pro bono* hours for the past year was recorded for each lawyer. The summary statistics are given in the table.

Law firm	Sample size	Sample mean	Sample variance
Dewey, Cheatum, & Howe	26	75.1	5.92
Fine, Howard, & Fine	26	80.9	5.65

Assume the underlying populations are normal, with equal variances.
(a) Is there any evidence to suggest that the mean number of yearly *pro bono* hours is different at these two law firms? Use $\alpha = 0.01$.
(b) Construct a 99% confidence interval for the difference in mean *pro bono* hours.
(c) Does the confidence interval in part (b) support your conclusion in part (a)? Explain your answer.

10.175 Economics and Finance Online investing has grown with the Internet and with the founding of companies like E*TRADE and Charles Schwab. Independent random samples of investors were obtained and asked whether they traded online within the past year. The data are given in the table, by portfolio size.

Portfolio	Sample size	Number of online traders
Less than $100,000	348	132
At least $100,000	226	65

(a) Compute the sample proportions and verify the nonskewness criterion.
(b) Construct a 95% confidence interval for the difference in population proportions of online investors.
(c) Using the interval in part (b), is there any evidence to suggest that the population proportion of online investors is different for these two portfolio classifications? Justify your answer.

10.176 Look, Up in the Sky The American Meteor Society maintains a running fireball-tracking system. All reports are analyzed and grouped according to several variables. In 2017, there were 5469 fireball reports, and 1851 of these were confirmed by at least two witnesses. As of September 2018, there were 3971 reports and 1263 were confirmed by at least two witnesses.[17] Assume the samples are independent. Is there any evidence to suggest that the proportion of fireball reports that are confirmed by at least two witnesses is different for these two time periods? Use $\alpha = 0.05$.

10.177 Biology and Environmental Science Benzene, toluene, ethylbenzene, *m*-xylene, *p*-xylene, and *o*-xylene are volatile organic compounds that are found in residential environments and can cause severe health problems, including dizziness, tremors, eye, ear, and throat irritation. The Canadian Health Measures Survey was administered to more than 5000 individuals in an effort to predict the presence of these compounds. Suppose a random sample of these respondents was selected, and the concentration of toluene was measured in each residence (in $\mu g/m^3$). The results by type of dwelling are given in the table.

Dwelling	Sample size	Sample mean	Sample standard deviation
Single detached	15	8.68	6.32
Double/duplex	15	13.16	9.20
Apartment	16	23.70	16.52

Assume all three underlying populations are normal, with equal variances. Conduct the appropriate hypothesis tests to determine which pairs of population mean toluene levels are different. Use $\alpha = 0.01$ in each test.

10.178 Physical Sciences Independent random samples of ore taken from two high-grade gold mines were obtained, and the gold value (in grams/tonne) was measured for each. The summary statistics are given in the table.

El Aguila mine	$n_1 = 8$	$\bar{x}_1 = 15.6$	$s_1 = 5.2$
Dolaucothi mine	$n_2 = 11$	$\bar{x}_2 = 26.8$	$s_2 = 21.6$

Assume normality and unequal variances.
(a) Conduct the appropriate hypothesis test to determine whether there is any evidence that the population mean gold value at the El Aguila mine is less than that of the Dolaucothi mine. Use $\alpha = 0.05$.
(b) Your conclusion in part (a) should be that there is no evidence to suggest a difference. Explain why this result is correct even though the sample means are very far apart.

ASphoto777/Deposit Photos

CHAPTER APP

10.179 Prescription Errors An experiment was designed to compare the proportion of pharmacist errors from computerized physician order entry (CPOE) prescriptions with hand-written prescriptions (HWP). Random samples of 2429 CPOE prescriptions and 1036 HWP prescriptions were obtained. When each prescription was carefully checked for pharmacist error, there were 117 errors in the CPOE prescriptions and 69 in the HWP prescriptions.

(a) Suppose a success is a prescription error. Find the proportion of successes for the CPOE prescriptions and the HWP prescriptions.
(b) Is there any evidence to suggest that the proportion of errors in CPOE prescriptions is less than the proportion of errors in HWP prescriptions? Use $\alpha = 0.05$. Use technology to find the *p* value associated with this hypothesis test.

The Analysis of Variance

11

◀ Looking Back
- Recall the statistical tests to compare two population means: **Sections 10.1 and 10.2**.
- Remember the assumptions and four parts of a two-sample *t* test: **Section 10.2**.

Looking Forward ▶
- Understand the assumptions for a one-way, or single-factor, analysis of variance (ANOVA) test and learn how to compare $k\ (>2)$ population means: **Section 11.1**.
- Learn methods to isolate group differences if an ANOVA test is significant: **Section 11.2**.
- Understand the assumptions, how the total variation is decomposed, and the hypothesis tests in a two-way ANOVA: **Section 11.3**.

MarinaP/Shutterstock

CHAPTER APP

TRACESE

Selenium Concentration

Selenium is an important trace mineral. Although the human body needs only small amounts, selenium helps prevent free radicals from damaging the body's cells and works to recycle antioxidants. A person who doesn't get enough selenium could experience nerve or heart problems. Fortunately, most people in the world get enough selenium through the natural food supply.

Cereals, grains, and Brazil nuts are rich in selenium. However, the selenium concentration in grain, for example, is determined by the amount of selenium in the soil in which it is grown. The four largest grain-producing states in the United States are Kansas, North Dakota, Washington, and Montana. Random samples of wheat farms in each state were obtained, and the selenium concentration in a soil sample was measured (in $\mu g/g$). The summary statistics are shown in the table.

State	Sample size	Sample mean	Sample standard deviation
Kansas	15	3.420	1.112
North Dakota	15	5.413	1.262
Washington	15	5.187	0.966
Montana	15	4.060	1.234

The statistical techniques presented in this chapter can be used to determine whether any two population mean selenium concentrations are different. If at least two means are different, then we need to identify which pairs of means are contributing to an overall difference.

11.1 One-Way ANOVA

The Question and Notation

A one-way, or single-factor, ANOVA (ANalysis Of VAriance) involves the analysis of data sampled from more than two populations. The only difference among the populations is a single factor. For example, consider a study in which random samples of the amount of carbon dioxide in underground train tunnels are obtained in four different cities. The data may be used to determine whether there is any difference in the mean amount of carbon dioxide in train tunnels among the four cities. The single factor that varies among the populations is the city.

Or suppose a researcher is investigating techniques for controlling the amount of electricity lost during transmission over utility lines. Experimental results may be used to determine whether there is any difference in the mean amount of electricity lost for five differently designed lines. The single factor here is the design of the electricity line. In theory, everything else is the same among the five populations—for example, the initial amount of electricity transmitted, the distance the electricity is transmitted, and the weather conditions.

Suppose we obtain three random samples, and we construct a histogram using all of the data. **Figure 11.1** shows the resulting graph. We can use analysis of variance to determine whether the data came from a single population, or whether at least two samples came from populations with different means.

The notation used in a one-way ANOVA is similar to, and is an extension of, the notation used in Chapter 10.

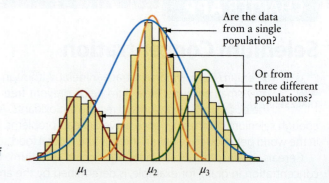

Figure 11.1 A visualization of a typical ANOVA problem.

ANOVA Notation

k = the number of populations under consideration.

Population	1	2	...	i	...	k
Population mean	μ_1	μ_2	...	μ_i	...	μ_k
Population variance	σ_1^2	σ_2^2	...	σ_i^2	...	σ_k^2
Sample size	n_1	n_2	...	n_i	...	n_k
Sample mean	$\bar{x}_{1.}$	$\bar{x}_{2.}$	...	$\bar{x}_{i.}$	...	$\bar{x}_{k.}$
Sample variance	s_1^2	s_2^2	...	s_i^2	...	s_k^2

$$n = n_1 + n_2 + \cdots + n_k$$
$$= \text{the total number of observations in the entire data set}$$

The null and alternative hypotheses are stated in terms of the population means.

$H_0: \mu_1 = \mu_2 = \cdots = \mu_k$ (All k population means are equal.)

$H_a: \mu_i \neq \mu_j$ for some $i \neq j$ (At least two of the k population means differ.)

For reference, these are the one-way ANOVA assumptions.

The assumptions for this test procedure are similar to those for a two-sample t test.

1. The k population distributions are normal.
2. The k population variances are equal; that is, $\sigma_1^2 = \sigma_2^2 = \cdots = \sigma_k^2$.
3. The samples are selected randomly and independently from the respective populations.

To denote observations, we use a single letter with two subscripts. The first subscript indicates the sample number and the second subscript denotes the observation number within that sample. In general,

x_{ij} = the jth measurment taken from the ith population
X_{ij} = the corresponding random variable

RAPID REVIEW *Double subscripts and summation notation.*

We place a comma between i and j in case of any ambiguity. For example, $x_{1,23}$ is the 23rd observation in the first sample, but $x_{12,3}$ is the 3rd observation in the 12th sample.

The mean of the observations in the ith sample is

$$\bar{x}_{i.} = \frac{1}{n_i} \sum_{j=1}^{n_i} x_{ij} = \frac{1}{n_i}(x_{i1} + x_{i2} + \cdots + x_{in_i}) \tag{11.1}$$

The dot in the second subscript is used to indicate a sum over that subscript (j) while the first subscript (i) is held fixed.

The mean of all the observations, called the **grand mean**, is the sum of all the observations divided by n.

$$\bar{x}_{..} = \frac{1}{n} \sum_{i=1}^{k} \sum_{j=1}^{n_i} x_{ij} \tag{11.2}$$

Just a little more notation will make some of the calculations easier.

A more theoretical development of a one-way ANOVA test includes the corresponding random variables $\bar{X}_{i.}$, $\bar{X}_{..}$, $T_{i.}$, and $T_{..}$.

$$t_{i.} = \sum_{j=1}^{n_i} x_{ij} = \text{sum of the observations in the } i\text{th sample}$$

$$t_{..} = \sum_{i=1}^{k} \sum_{j=1}^{n_i} x_{ij} = \text{sum of all the observations}$$

Total Variation Decomposition

Here's where the analysis of *variance* plays a role. The total variation in the data (the total sum of squares) is decomposed into a sum of *between-samples* variation (the sum of squares due to factor) and *within-samples* variation (the sum of squares due to error). The total variation is the variability of individual observations from the grand mean. The between-samples (or between-factor) variation is the variability in the sample means; this tells us how different the sample means are from each other. The within-samples variation is the variability of the observations from their sample mean; this is just like the sample variance we've used previously. The three sums of squares are defined in the following fundamental identity, which shows the decomposition of the total variation in the data.

One-Way ANOVA Identity

Let SST = total sum of squares, SSA = sum of squares due to factor, and SSE = sum of squares due to error.

$$\underbrace{\sum_{i=1}^{k}\sum_{j=1}^{n_i}(x_{ij}-\bar{x}_{..})^2}_{\text{SST}} = \underbrace{\sum_{i=1}^{k}n_i(\bar{x}_{i.}-\bar{x}_{..})^2}_{\text{SSA}} + \underbrace{\sum_{i=1}^{k}\sum_{j=1}^{n_i}(x_{ij}-\bar{x}_{i.})^2}_{\text{SSE}} \quad (11.3)$$

A CLOSER LOOK

1. SSA is used to denote the sum of squares due to factor instead of SSF because in a two-way ANOVA there is a factor A and a factor B.
2. The sample size is used as a weight in the expression for SSA.

If the null hypothesis is true, then each observation comes from the same population with mean μ and variance σ^2. Therefore, the sample means, the $\bar{x}_{i.}$'s, should all be about the same and should all be close to the grand mean $\bar{x}_{..}$. Also, SSA should be (relatively) small. If at least two population means are different, then at least two $\bar{x}_{i.}$'s should be very different, and these values will be far away from $\bar{x}_{..}$. In this case, SSA should be (relatively) large.

The one-way ANOVA test statistic is based on two separate estimates for the common variance σ^2, computed using SSA and SSE.

Definition

The **mean square due to factor**, MSA, is SSA divided by $k-1$:

$$\text{MSA} = \frac{\text{SSA}}{k-1} \quad (11.4)$$

The **mean square due to error**, MSE, is SSE divided by $n-k$:

$$\text{MSE} = \frac{\text{SSE}}{n-k} \quad (11.5)$$

The Test Procedure

If the null hypothesis is true, then (the random variable) MSA is an unbiased estimator of σ^2. If H_a is true, then MSA tends to overestimate σ^2. The mean square due to error, MSE, is an unbiased estimator of the common variance whether H_0 or H_a is true.

Consider the ratio $F = \text{MSA}/\text{MSE}$. If the value of F is close to 1, then the two estimates of σ^2, or sources of variation, are approximately the same. There is no evidence to suggest that the population means are different. If the value of F is much greater than 1, then the variation between samples is greater than the variation within samples. This suggests that the alternative hypothesis is true.

If the one-way ANOVA assumptions are satisfied and H_0 is true, then the statistic $F = \text{MSA}/\text{MSE}$ has an F distribution with $k-1$ and $n-k$ degrees of freedom. Because large values of F suggest H_a is true, the rejection region is only in the right tail of the distribution.

> One-way ANOVA does not mean a one-sided statistical test, but rather indicates that there is one factor.

One-Way ANOVA Test Procedure

Given the one-way ANOVA assumptions, the test procedure with significance level α is as follows:

$H_0: \mu_1 = \mu_2 = \cdots = \mu_k$

$H_a: \mu_i \neq \mu_j$ for some $i \neq j$

TS: $F = \dfrac{\text{MSA}}{\text{MSE}}$

RR: $F \geq F_{\alpha, k-1, n-k}$

> The critical values for the F distribution are found in Appendix Table 7.

If the value of F is smaller than the critical value, then there is no evidence from the data to reject H_0. If the value of F is in the rejection region, we say that the F test is significant and that there is a statistically significant difference among the population means. Recall that we can also use the p value to conduct an equivalent test: If $p \leq \alpha$, then we reject H_0.

If any of the one-way ANOVA assumptions are violated, then the conclusion is unreliable. Several methods to check for normality were presented in Section 6.3, and statistical procedures are also available for testing equality of variances. To use these procedures, the samples must be selected randomly and independently from the appropriate populations.

The following computational formulas (rather than the definitions) are used to find first the sums of squares, and then the mean squares.

Computational Formulas

> These equations are used for the same reasons that the computational formula for the sample variance is used: They are easier, faster, and more accurate.

$$\text{SST} = \underbrace{\sum_{i=1}^{k} \sum_{j=1}^{n_i} (x_{ij} - \bar{x}_{..})^2}_{\text{definition}} = \underbrace{\left(\sum_{i=1}^{k} \sum_{j=1}^{n_i} x_{ij}^2 \right) - \dfrac{t_{..}^2}{n}}_{\text{computational formula}}$$

$$\text{SSA} = \underbrace{\sum_{i=1}^{k} n_i (\bar{x}_{i.} - \bar{x}_{..})^2}_{\text{definition}} = \underbrace{\left(\sum_{i=1}^{k} \dfrac{t_{i.}^2}{n_i} \right) - \dfrac{t_{..}^2}{n}}_{\text{computational formula}}$$

$$\text{SSE} = \underbrace{\sum_{i=1}^{k} \sum_{j=1}^{n_i} (x_{ij} - \bar{x}_{i.})^2}_{\text{definition}} = \underbrace{\text{SST} - \text{SSA}}_{\text{computational formula}}$$

One last detail: One-way ANOVA calculations are often presented in an **analysis of variance table**, also called simply an ANOVA table. The values included in this table are associated with the three sources of variation and the calculation of the F statistic.

Here is a template for the ANOVA table:

One-way ANOVA summary table

Source of variation	Sum of squares	Degrees of freedom	Mean square	F	p Value
Factor	SSA	$k-1$	$\text{MSA} = \dfrac{\text{SSA}}{k-1}$	$\dfrac{\text{MSA}}{\text{MSE}}$	p
Error	SSE	$n-k$	$\text{MSE} = \dfrac{\text{SSE}}{n-k}$		
Total	SST	$n-1$			

Burger/Phanie/Age Fotostock

Example

The following example illustrates all the computations involved in a one-way ANOVA and the process of making an inference based on the value of the test statistic.

EXAMPLE 11.1 Take with a Grain of Salt

A recent research study suggests that a high-salt diet in older women increases the risk of breaking a bone. Although the biological mechanism is still unclear, there appears to be an association between excessive sodium intake and bone fragility. One measure of bone health is the vitamin D blood level. Based on their responses to food questionnaires, independent random samples of older women in four different salt intake categories were obtained. The vitamin D blood level (in nmol/L) was measured in each. The data are given in the table.

Sample	Observations				
Very high	91.5	77.5	94.5	77.5	92.0
High	89.0	92.0	98.2	80.0	86.7
Moderate	92.5	100.7	94.0	93.3	106.3
Low	100.1	98.0	99.1	103.9	97.6

Is there any evidence to suggest that at least two of the population mean vitamin D blood levels are different? Use $\alpha = 0.05$.

Solution

VITD

This example involves a one-way ANOVA test. Compute the summary statistics needed to complete the ANOVA table. Find the critical value, and draw the appropriate conclusion.

STEP 1 Assume the one-way ANOVA assumptions are true. There are $k = 4$ samples, or groups, and $n = 5 + 5 + 5 + 5 = 20$ total observations. Some summary statistics are given in the table.

Sample	Sample size	Sample total	Sample mean	Sample variance
Very high	$n_1 = 5$	$t_{1.} = 433.0$	$\bar{x}_{1.} = 86.60$	$s_1^2 = 70.30$
High	$n_2 = 5$	$t_{2.} = 445.9$	$\bar{x}_{2.} = 89.18$	$s_2^2 = 44.94$
Moderate	$n_3 = 5$	$t_{3.} = 486.8$	$\bar{x}_{3.} = 97.36$	$s_3^2 = 35.62$
Low	$n_4 = 5$	$t_{4.} = 498.7$	$\bar{x}_{4.} = 99.74$	$s_4^2 = 6.36$

The summary statistics table suggests a possible violation of the equal-variances assumption. There are formal statistical procedures for testing equality of (several) variances, but the ANOVA test statistic is robust. That is, even if the population variances are a little different, the test still provides reliable results. In addition, because the sample sizes are small, it is more difficult to detect a real difference in population variances.

The sum of all the observations, or grand total, is
$$t_{..} = 433.0 + 445.9 + 486.8 + 498.7 = 1864.4$$

STEP 2 The four parts of the hypothesis test are given here:

$H_0: \mu_1 = \mu_2 = \mu_3 = \mu_4$ (all four population means are equal)
$H_a: \mu_i \neq \mu_j$ for some $i \neq j$ (at least two population means differ)

TS: $F = \dfrac{\text{MSA}}{\text{MSE}}$

RR: $F \geq F_{\alpha, k-1, n-k} = F_{0.05, 3, 16} = 3.24$

STEP 3 Find the total sum of squares.

$$\text{SST} = \left(\sum_{i=1}^{k} \sum_{j=1}^{n_i} x_{ij}^2 \right) - \frac{t_{..}^2}{n}$$
Computational formula for SST.

$$= (91.5^2 + 77.5^2 + \cdots + 97.6^2) - \frac{1864.4^2}{20}$$
Apply the formula.

$$= 175{,}027.24 - 173{,}799.368 = 1227.872$$
Simplify.

Find the sum of squares due to factor.

$$SSA = \left(\sum_{i=1}^{k} \frac{t_{i\cdot}^2}{n_i}\right) - \frac{t_{\cdot\cdot}^2}{n}$$ 　　Computational formula for SSA.

$$= \left(\frac{433.0^2}{5} + \frac{445.9^2}{5} + \frac{486.8^2}{5} + \frac{498.7^2}{5}\right) - \frac{1864.4^2}{20}$$ 　　Use sample totals and grand total.

$$= 174{,}398.348 - 173{,}799.368 = 598.98$$ 　　Simplify.

Use these two values to find the sum of squares due to error.

$$SSE = SST - SSA = 1227.872 - 598.98 = 628.892$$

STEP 4 Compute MSA and MSE.

$$MSA = SSA/(k-1)$$ 　　Definition.
$$= 598.98/(4-1) = 199.66$$ 　　Use SSA and $k = 4$ groups.
$$MSE = SSE/(n-k)$$ 　　Definition.
$$= 628.892/(20-4) = 39.3058$$ 　　Use SSE, $n = 20$ total observations, and $k = 4$.

STEP 5 The value of the test statistic is

$$f = \frac{MSA}{MSE} = \frac{199.66}{39.3058} = 5.08 \; (\geq 3.24)$$

The value of the test statistic lies in the rejection region, so we reject the null hypothesis. There is evidence to suggest that at least two population means are different.

> The next reasonable question is, "Which pairs of means are contributing to this overall difference?" We'll address this issue in Section 11.2.

STEP 6 Recall that because the tables of critical values for F distributions are limited, we can only bound the p value (using the tables). Place the value of the test statistic, $f = 5.08$, in an ordered list of critical values with degrees of freedom 3 and 16.

$$3.24 \quad \leq 5.08 \leq 5.29$$
$$F_{0.05,3,16} \leq 5.08 \leq F_{0.01,3,16}$$

Therefore, $\quad 0.01 \quad \leq p \leq 0.05$

The exact p value is shown in the technology solution and is illustrated in **Figure 11.2**.

STEP 7 Here's how all of these calculations are presented in an ANOVA table.

One-way ANOVA summary table

Source of variation	Sum of squares	Degrees of freedom	Mean square	F	p Value
Factor	598.980	3	199.66	5.08	0.0117
Error	628.892	16	39.31		
Total	1227.872	19			

Figure 11.3 shows a technology solution.

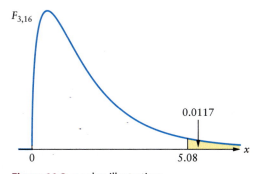

Figure 11.2 p value illustration: $p = P(X \geq 5.08) = 0.0117 \leq 0.05 = \alpha$

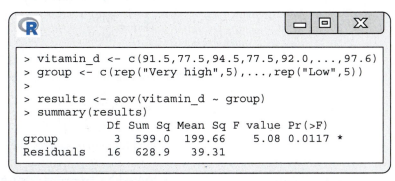

Figure 11.3 Use the R command `aov()` to produce the ANOVA summary table.

TRY IT NOW Go to Exercises 11.16 and 11.27

CHAPTER APP Selenium Concentration

Is there any evidence to suggest that at least two of the population mean selenium concentrations are different?

Focus on Interpretation

In the next example, we show fewer detailed calculations. Instead, we present the ANOVA table and focus on the inference process and interpretation of the results.

EXAMPLE 11.2 Mercury Rising

MERCURY

Although mercury is useful in many ways, recent research suggests that overexposure can cause severe health problems. The use of mercury has increased worldwide in industry and in mining, and traces of mercury are emitted during power generation. The National Atmospheric Deposition Program maintains a long-term record of total mercury concentration in precipitation in the United States and Canada. Independent random samples from five sites were obtained and the mercury concentrations (in ng/L) in precipitation are given in the table.

Site	Observations					
CA 20	6.59	2.78	7.85	5.99	4.27	5.25
CA 75	4.48	6.82	3.44	7.66	2.76	5.85
CA 94	6.08	8.81	3.11	3.16	2.57	3.87
CO 96	2.62	4.92	4.78	7.45	3.43	3.01
NV 99	3.75	2.45	3.31	8.62	4.16	3.12

Is there any evidence to suggest that the population mean mercury concentration in precipitation differs at these five sites? Use $\alpha = 0.05$.

Solution

This example involves a one-way ANOVA test. Compute the summary statistics needed to complete the ANOVA table. Find the critical value, and draw the appropriate conclusion.

STEP 1 Assume the one-way ANOVA assumptions are true. There are $k = 5$ groups and $n = 6+6+6+6+6 = 30$ total observations. We would like to know whether the data suggest that any two of the population mean mercury concentrations are different.

The four parts of the hypothesis test are as follows:

$H_0: \mu_1 = \mu_2 = \mu_3 = \mu_4 = \mu_5$

$H_a: \mu_i \neq \mu_j$ for some $i \neq j$

TS: $F = \dfrac{\text{MSA}}{\text{MSE}}$

RR: $F \geq F_{\alpha, k-1, n-k} = F_{0.05, 4, 25} = 2.76$

STEP 2 Find the total sum of squares.

$$\text{SST} = \left(\sum_{i=1}^{k}\sum_{j=1}^{n_i} x_{ij}^2\right) - \dfrac{t_{..}^2}{n} = \cdots = 110.5687$$

Find the sum of squares due to factor.

$$\text{SSA} = \left(\sum_{i=1}^{k} \dfrac{t_{i.}^2}{n_i}\right) - \dfrac{t_{..}^2}{n} = \cdots = 6.6255$$

Use these two values to find the sum of squares due to error.

SSE = SST − SSA = 110.5687 − 6.6255 = 103.9433

STEP 3 Compute MSA and MSE.

$$MSA = SSA/(k-1)$$ *Definition.*
$$= 6.6255/(5-1) = 1.6564$$ *Use SSA and $k = 5$ groups.*
$$MSE = SSE/(n-k)$$ *Definition.*
$$= 103.9432/(30-5) = 4.1577$$ *Use SSE, $n = 30$ total observations, and $k = 5$.*

STEP 4 The value of the test statistic is

$$f = \frac{MSA}{MSE} = \frac{1.6564}{4.1577} = 0.3984$$

The value of the test statistic does not lie in the rejection region. Equivalently, $p = 0.8079 > 0.05$ as shown in the technology solution and illustrated in **Figure 11.4**. Do not reject the null hypothesis. There is no evidence to suggest that any two population mean mercury concentrations are different.

STEP 5 Here is the ANOVA summary table.

One-way ANOVA summary table

Source of variation	Sum of squares	Degrees of freedom	Mean square	F	p Value
Factor	6.6255	4	1.6564	0.39	0.8079
Error	103.9433	25	4.1577		
Total	110.5687	29			

Figure 11.5 shows a technology solution.

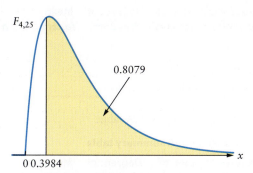

Figure 11.4 p value illustration: $p = P(X \geq 0.3984) = 0.8079 > 0.05 = \alpha$

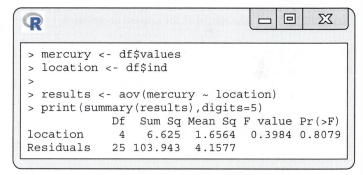

Figure 11.5 Use the R command aov() to produce the ANOVA summary table.

TRY IT NOW Go to Exercises 11.17 and 11.25

Section 11.1 Exercises

Concept Check

11.1 True or False In a one-way ANOVA, the k population variances are assumed to be equal.

11.2 True or False In a one-way ANOVA, all sample sizes must be the same in order for the test statistic to be valid.

11.3 True or False If we reject the null hypothesis in a one-way ANOVA, there is evidence to suggest that all pairs of population means are unequal.

11.4 True or False If we reject the null hypothesis in a one-way ANOVA, there is evidence to suggest that the underlying population variances are unequal.

11.5 True or False In a one-way ANOVA, we reject the null hypothesis for large values of F or for values of F close to 0.

11.6 Short Answer If the null hypothesis is true in a one-way ANOVA, then MSA is an unbiased estimator for _____.

11.7 Short Answer In a one-way ANOVA, if the value of F is much greater than 1, then the variation _____ samples is greater than the variation _____ samples.

11.8 Short Answer In a one-way ANOVA, why do we use the computation formulas rather than the definitions to find SSA, SSE, and SST?

11.9 Short Answer If we reject the null hypothesis in a one-way ANOVA, what is the next reasonable question to consider?

Practice

11.10 Consider these data in the context of a one-way ANOVA. **EX11.10**

Group	Observations
1	33 27 27 32 27 31 23 26 34
2	27 35 32 28 35 39 33
3	30 36 33 35 33 28

(a) Find n_i, $i = 1, 2, 3$ (the number of observations in each sample), and n (the total number of observations).
(b) Find $t_{i.}$, $i = 1, 2, 3$ (the total for each sample), and $t_{..}$ (the grand total).
(c) Find $\sum_{i=1}^{3}\sum_{j=1}^{n_i} x_{ij}^2$ (the sum of all the squared observations).

11.11 Consider these data in the context of a one-way ANOVA. **EX11.11**

		Group			
1	2	3	4	5	6
2.8	4.8	6.3	6.8	6.4	6.3
5.6	4.2	3.9	3.3	5.8	4.7
3.7	4.4	5.4	5.7	6.8	5.9
4.4	2.8	4.4	4.0	5.7	3.4
5.8	5.1	3.0	6.1	5.7	5.7
6.1	3.2	6.3	7.5	4.3	2.3
5.0	4.6	6.9	2.9	2.5	4.5
5.9	2.3	2.2	5.1	6.8	4.9
5.2	6.3	6.8	3.8		3.6
6.8		3.9	5.5		5.6

(a) Find n_i, $i = 1, 2, \ldots, 6$ (the number of observations in each sample), and n (the total number of observations).
(b) Find $t_{i.}$, $i = 1, 2, \ldots, 6$ (the total for each sample), and $t_{..}$ (the grand total).
(c) Find $\sum_{i=1}^{6}\sum_{j=1}^{n_i} x_{ij}^2$ (the sum of all the squared observations).

11.12 Consider these data in the context of a one-way ANOVA. **EX11.12**

Factor	Observations				
1	162	155	157	144	157
2	168	147	163	131	136
3	135	138	155	172	168
4	144	162	150	140	157

(a) Find $t_{i.}$ for $i = 1, 2, 3, 4$. Find $t_{..}$.
(b) Find $\sum_{i=1}^{4}\sum_{j=1}^{n_i} x_{ij}^2$.
(c) Find SST (the total sum of squares), SSA (the sum of squares due to factor), and SSE (the sum of squares due to error).
(d) Find MSA (the mean square due to factor), and MSE (the mean square due to error).
(e) Compute the value of the test statistic.

11.13 Consider the corresponding data in the context of a one-way ANOVA. **EX11.13**

(a) Find SST, SSA, and SSE.
(b) Find MSA and MSE.
(c) Compute the value of the test statistic.
(d) If $\alpha = 0.05$, would you reject the null hypothesis? Justify your answer.

11.14 Complete the ANOVA table.

One-way ANOVA summary table

Source of variation	Sum of squares	Degrees of freedom	Mean square	F	p Value
Factor		4			
Error	12062.1				
Total	12646.2	51			

11.15 Complete the ANOVA table.

One-way ANOVA summary table

Source of variation	Sum of squares	Degrees of freedom	Mean square	F	p Value
Factor			4.522		
Error		62	0.988		
Total		65			

(a) State the null hypothesis and the alternative hypothesis in terms of the population means.
(b) Find the rejection region for $\alpha = 0.01$.
(c) Conduct the hypothesis test. Is there any evidence to suggest that at least two population means are different? Justify your answer.

Applications

11.16 Public Health and Nutrition A study was conducted to compare the amount of salt in potato chips. Random samples of four varieties were obtained and the amount of salt in each 1-oz portion of potato chips was recorded (in milligrams of sodium). The data are given in the table. CHIPS

Variety	Observations					
BBQ	338	155	239	184	185	261
Cheese-flavored	235	238	251	229	233	232
Olestra-based	164	197	136	214	148	230
Baked	290	343	294	373	306	357

Conduct an analysis of variance test to determine whether there is any evidence that the population mean amount of salt per serving is different for at least two varieties. Use $\alpha = 0.05$.
(a) State the four parts of the hypothesis test.
(b) Complete an ANOVA table.
(c) Draw the appropriate conclusion.

11.17 Manufacturing and Product Development Breitling sells men's gold, silver, and titanium watchbands. A random sample of each type was obtained (in similar styles), and the weight of each watchband (in grams) was recorded. The data are given in the table. WATCH

Watchband	Observations							
Gold	7.9	7.2	7.8	8.1	7.9	8.3	9.9	
Silver	9.5	7.0	8.7	7.6	7.5	9.3	7.3	6.9
Titanium	6.7	7.1	6.5	7.1	5.5	6.7	4.9	3.9

(a) Conduct an analysis of variance test to determine whether there is any evidence that the mean weights of any two watchband types are different. Include an ANOVA table. Use $\alpha = 0.01$.
(b) Compute the sample mean weight for each sample. Given your conclusion in part (a), which pair(s) of population means do you think are different?

11.18 Travel and Transportation According to a recent study, U.S. drivers spend approximately 42 hours each year just sitting in traffic. Los Angeles drivers experience the worst traffic congestion; drivers spend approximately 102 hours in congested traffic each year.[1] Random drivers in some of the most congested cities were asked how long (in minutes) it normally takes them to get through the traffic congestion during morning rush hour. The cities and summary statistics are given in the table.

City	Sample size	Sample total	Sum of squared observations
Chicago	14	463	16185
Honolulu	18	620	22458
New York	20	506	13686
Seattle	15	356	8964
Washington, D.C.	19	586	18840

Conduct an analysis of variance test to determine whether there is any evidence that the population mean times through some pair of traffic bottlenecks are different. Use $\alpha = 0.05$.

11.19 Sports and Leisure A study was conducted to determine the tension needed by various types of guitar strings to produce the proper frequency. A high E was used for comparison, and tension was measured in newtons. The guitar string brands and summary statistics are given in the table.

String brand	Sample size	Sample total	Sum of squared observations
Darco Acoustic	8	458.5	26347.8
Ernie Ball	9	508.8	28947.4
Martin	9	554.5	34242.7
Gibson	8	523.9	34373.9

Conduct the appropriate test to check the hypothesis of no difference in population mean tensions. Use $\alpha = 0.01$.

11.20 Marketing and Consumer Behavior The deli department in a Publix Supermarket conducted a survey to compare orders on certain items. A random sample of sliced ham, roast beef, and turkey orders was obtained, and the weight of each order (in pounds) was recorded. Conduct an analysis of variance test to determine whether there is any evidence to suggest that at least two of the population mean weights are different. Use $\alpha = 0.01$. DELI

11.21 Public Health and Nutrition Nondairy creamers (for coffee and tea) contain vegetable fat, corn-syrup solids, casein, and other ingredients. Independent random samples of various nondairy creamers were obtained, and the percentage of fat in a serving size was measured. Is there any evidence to suggest that at least two nondairy creamers have a different population mean percentage of fat per serving? Use $\alpha = 0.05$. CREAMER

11.22 Biology and Environmental Science Sulfur dioxide is a toxic gas that has a distinct, pungent odor; it contributes to acid rain and changes in climate. It is usually released into the atmosphere by volcanoes, but sulfur dioxide is also emitted from coal-fired electricity-generating plants. Independent random times at four coal-fired power plants in the United States were selected, and the sulfur dioxide emission rate was measured (in lb/MWh) for each. Summary statistics are given in the table. Use this information with $\alpha = 0.01$ to test the null hypothesis of no difference in the population mean sulfur dioxide emission rates.

Coal-fired plant	Sample size	Sample total	Sum of squared observations
Bowen	18	574.0	18731.76
Seminole	21	584.1	16493.11
Gibson	15	409.0	11169.80
Amos	16	421.4	11455.32

11.23 Medicine and Clinical Studies The sweetener sorbitol has fewer calories than sucrose, adds texture to food, is non-carcinogenic, and is very useful for people with diabetes. Surprisingly, this sweetener is also found in children's cough syrups. Independent random samples of four cough syrups were obtained, and the amount of sorbitol per teaspoon (in grams) was measured. Conduct the appropriate test to determine whether there is any evidence that at least two population mean levels of sorbitol are different. Use $\alpha = 0.01$ and include an ANOVA table. 📊 COUGH

11.24 Fuel Consumption and Cars Biodiesel is a specific type of biofuel. It is renewable, burns cleanly, and helps to reduce our dependence on imported diesel fuel. Biodiesel is made from a mixture of agricultural oils, cooking oil, and animal fats. Independent random samples of daily production (in gallons) were obtained from five facilities near Houston, Texas. Use these data to test the hypothesis of no difference in the population mean daily production of biodiesel at these five plants. Use $\alpha = 0.05$. 📊 BIODIES

11.25 Manufacturing and Product Development Lumber and forestry are important to the Canadian economy. More than 180,000 people are employed in this industry and its supporting activities, which include shipping and manufacturing lumber. Independent random samples of monthly hardwood shipments (in cubic meters) from three provinces were obtained. The summary statistics are given in the table. Conduct an analysis of variance test to determine whether there is evidence that some pair of population mean monthly shipments are different. Use $\alpha = 0.01$ and include an ANOVA table.

Province	Sample size	Sample total	Sum of squared observations
New Brunswick	5	992.6	198016.20
Ontario	6	1275.5	272977.23
Nova Scotia	6	425.7	30443.45
British Columbia Coast	5	1285.8	336878.00

11.26 Biology and Environmental Science A study was conducted to examine the thaw depths in late August to early September in the floodplain sites of the Tanana River southwest of Fairbanks, Alaska. The data in the table are a subset of the thaw depth measurements (in centimeters) from four locations. 📊 THAW

Study site	Observations						
253	45	47	44	45	50	38	
1133	47	58	46	50	55	46	
254	69	34	37	39	40	62	
1113	44	36	38	41	43	55	

Is there any evidence to suggest that at least two population mean thaw depths are different? Use $\alpha = 0.05$.

11.27 Biology and Environmental Science Dissolved organic compounds (DOCs) consist of a variety of organic material in water systems. These dissolved particles are mostly the result of decaying organic matter and highly organic soils. The DOC concentration is one measure of the quality of a water system. Independent random samples of water in the Clackamas River in northwestern Oregon at four locations were obtained, and the DOC was measured for each (in mg/L). Is there any evidence to suggest that at least two population mean DOC concentrations are different? Use $\alpha = 0.01$. 📊 DOC

11.28 Kerosene Purity Many people who live in northern states supplement their main heating system with kerosene heaters. These devices are generally inexpensive, but a kerosene flame emits a noticeable ash and the fumes must be well ventilated. The flash point of kerosene is related to purity, ash, fumes, and safety. A low flash point is a fire hazard and may be an indication of contamination. Independent random samples of kerosene from five dealers in Maine were obtained and the flash point of each was measured (in degrees Celsius). Is there any evidence to suggest that at least two of the population mean flash points are different? Use $\alpha = 0.05$. 📊 FLASH

11.29 Expense IQ A publication by Concur provided a summary of travel expenses for both small and medium-sized businesses (SMBs) and large-scale companies. A random sample of business trips to some of the most visited international cities was obtained, and the amount spent on ground transportation was recorded (in dollars). Is there any evidence to suggest that at least two of the population mean amounts spent on ground transportation are different? Use $\alpha = 0.01$. 📊 EXPENSE

11.30 Public Health and Nutrition There seems to be a never-ending debate about whether eggs should be included in your diet. However, a typical egg has lots of nutrients, protein, and vitamins. Independent random samples of different size eggs was obtained and the amount of protein was measured in each (in grams). Is there any evidence to suggest that at least two of the population mean amounts of protein in eggs are different? Use $\alpha = 0.05$. 📊 EGGS

Extended Applications

11.31 Manufacturing and Product Development Many cities, towns, and college campuses use a rotary broom to sweep debris and snow from sidewalks. The pressure created by the broom is one measure of how effective the machine will be in removing debris. Random samples of rotary brooms from four manufacturers were obtained, and the pressure of each broom was measured (in pounds per square inch, psi). The data are given in the table. 📊 BROOMS

Company	Observations			
Ditch Witch	2081	1980	2210	2297
	2204	2765	2327	
Schwarze	2567	1799	2422	2437
	2367	2244	2245	
Elgin	2228	2581	2364	2375
	2066	2091	2543	
Holder	2905	2695	2503	2931
	2657	2591	2138	

(a) Conduct the appropriate test to determine whether there is any evidence that at least two population mean pressures are different. Use $\alpha = 0.05$.

(b) Which manufacturer would you recommend and why?

11.32 Fuel Consumption and Cars Automobile service clubs offer free maps, trip-interruption protection, payment for some legal fees, and roadside assistance. Independent random samples of roadside service calls were selected for three different clubs. The time required (in minutes) for a tow truck to arrive was recorded for each call. The summary statistics are given in the table.

Club	Sample size	Sample mean	Sample variance
AAA	15	36.8	10.18
Discover	17	43.8	9.97
Executive	20	34.8	12.15

(a) Use the definitions to find SSA and SSE. Use these two values to find SST.
(b) Complete an ANOVA table, and use this information to determine whether there is any evidence to suggest that at least two population mean waiting times are different. Use $\alpha = 0.01$.

11.33 Physical Sciences During the summer months, grocery stores, convenience stores, and gas stations sell bags of ice. Independent random sample of bags were obtained from various locations, and the weight (in pounds) of each was recorded. The summary statistics are given in the table.

Location	Sample size	Sample total	Sum of squared observations
Giant	10	80.8	654.38
Sheetz	10	88.8	789.20
Star Market	14	119.9	1028.31
Unimart	12	104.1	903.57
Weis	15	139.7	1303.41

(a) Do the data suggest that the population mean weight of bags of ice is the same at these five locations? Use $\alpha = 0.001$.
(b) If each store sells these bags of ice for approximately the same price, where would you make your purchase? Justify your answer.

11.34 Sports and Leisure The ESPN Home Run Tracker computes a variety of measurements for every home run hit in Major League Baseball, including true distance, speed off bat, elevation angle, and apex. Independent random samples of home runs were obtained from six different ballparks, and the true distance was recorded for each. HMRUNS
(a) Is there any evidence to suggest that at least two population mean home-run distances are different? Use $\alpha = 0.05$.
(b) Do you think there are any violations in the one-way ANOVA assumptions? If so, why?
(c) Do the data suggest that the population mean home-run distance is greatest in one ballpark? Why do you suppose the distances tend to be greater in that ballpark?

Challenge Problems

11.35 Public Policy and Political Science Every city has design specifications for streets, including minimum right of way, minimum vertical grade, and minimum centerline radii on curves. Independent random samples of city streets were obtained from Washington, D.C., and New York City, and the width (in feet) of a randomly selected section was recorded. The data are given in the table. STREETS

City	Observations
New York City	28.2 32.9 34.6 31.5 31.6 29.5 30.3 29.2 25.6 28.6 28.8 31.5
Washington, D.C.	32.5 36.3 34.3 33.0 31.0 36.5 29.8 30.0 30.2 34.7 31.9 30.0

(a) Conduct a two-sample t test to determine whether there is any evidence to suggest that the population mean street widths are different. Assume the population variances are equal and use $\alpha = 0.05$. Find the exact p value associated with this test.
(b) Conduct a one-way analysis of variance test to determine whether there is any difference among the $k = 2$ population mean street widths due to city. Use $\alpha = 0.05$. Find the exact p value associated with this test.
(c) What is the relationship between the value of the test statistic in part (a) and the value of the test statistic in part (b)? How are the p values related? Why do these relationships make sense?

11.36 The Effect Size Generate a random sample of size $n_1 = 20$ from a normal distribution with mean $\mu_1 = 50$ and variance $\sigma_1^2 = 100$. Generate a second random sample of size $n_2 = 20$ from a normal distribution with mean $\mu_2 = 50$ and variance $\sigma_2^2 = 100$. Generate a third random sample of size $n_3 = 20$ from a normal distribution with mean $\mu_3 = 52$ and variance $\sigma_3^2 = 100$. Conduct a one-way ANOVA test with $\alpha = 0.05$ to determine whether there is any evidence to suggest that the population means differ.

Repeat this process 100 times and record the proportion of times you reject the null hypothesis, p_r.

Let $\mu_T = (\mu_1 + \mu_2 + \mu_3)/3$ and compute the *effect size*

$$e = \sqrt{\frac{\sum_{i=1}^{3}(\mu_i - \mu_T)^2/3}{100}}$$

Plot the point (e, p_r).

Repeat this process for various values of $\mu_1, \mu_2,$ and μ_3, and therefore for various effect sizes. Plot the points (e, p_r).

Explain the resulting graph. What concept related to an ANOVA test does this graph illustrate?

11.2 Isolating Differences

The Bonferroni Procedure

If we fail to reject the null hypothesis in a one-way ANOVA, there is no evidence to suggest any difference among population means. The statistical analysis stops there. However, if we reject the null hypothesis, then we have evidence to suggest an overall difference among means. The next logical step is to try and isolate the difference(s), so as to determine which pair(s) of means are contributing to the overall (significant) difference. Several multiple comparison procedures have been developed for isolating differences, including the two methods presented in this section.

If two population means are being compared, then a t test (or Z test) is usually appropriate. However, if we are comparing three or more population means, the analysis requires a little more finesse. Here's why. Suppose that, in a one-way ANOVA with three groups, we reject H_0. It seems reasonable to conduct a test on every possible pair of means (μ_1 versus μ_2, μ_1 versus μ_3, and μ_2 versus μ_3). However, we cannot simply set the significance level in each individual hypothesis test. The probability of making a type I error (a mistake) is set in each test under the assumption that only one test is conducted per experiment. Therefore, as we conduct more tests, the chance of making an error also increases.

Think about a waiter totaling a customer's bill. Suppose the probability that the waiter makes a mistake on any single bill is 0.10. If the waiter must total three bills, the probability that he makes a mistake on at least one bill is 0.2710. The more bills and totals, the greater the chance that the waiter will make at least one error. The same principle applies to hypothesis tests. Suppose three hypothesis tests (using data from the same experiment) are conducted, each with significance level 0.05. The probability of making at least one mistake is 0.1426. The more hypothesis tests (or the more comparisons) we conduct, the more likely we are to make at least one error. As we have seen, the probability of making at least one error is more than two times as big as α.

We would really like to control the (overall) probability of making at least one mistake. For example, we might want the probability of making at least one mistake in three hypothesis tests to be 0.10. We typically set this overall error probability and work backward to compute the individual error probabilities associated with each individual test. The procedures presented here provide methods for capping the probability of making at least one mistake in all of the comparisons.

Although we could actually conduct hypothesis tests, usually we construct multiple confidence intervals for the difference between population means. Recall that, if a confidence interval for $\mu_1 - \mu_2$ contains 0, there is no evidence to suggest that the population means are different. However, if the confidence interval does not include 0, there is evidence to suggest that the two population means are different. If we want to find a $100(1-\alpha)\%$ confidence interval for all possible paired comparisons (i.e., so that the overall probability is α), we must make the intervals much wider than those for individual differences of means.

In a one-way ANOVA with three groups, suppose there is evidence to suggest that at least one pair of means is different and a multiple comparison procedure produces the following confidence intervals.

Difference	Confidence interval
$\mu_1 - \mu_2$	$(-1.21,\ 9.00)$
$\mu_1 - \mu_3$	$(\ 3.09,\ 13.31)$
$\mu_2 - \mu_3$	$(-0.81,\ 9.41)$

The confidence intervals for both the differences $\mu_1 - \mu_2$ and $\mu_2 - \mu_3$ contain 0. There is no evidence to suggest that these pairs of population means are different. The confidence

---

If we reject H_0, then we would like to know which means are different.

Recall that a type I error means rejecting H_0 when H_0 is true.

Let X = the number of bills in which there is a mistake.
$X \sim B(3, 0.10)$
$P(X \geq 1) = 1 - P(X = 0)$
$= 1 - 0.7290 = 0.2710$

interval for $\mu_1 - \mu_3$ does not include 0. There is evidence to suggest that μ_1 is different from μ_3.

The general form of each **Bonferroni confidence interval** is very similar to a confidence interval for the difference between two means based on a t distribution using a pooled estimate of the common variance (introduced in Section 10.2). Here, we use the mean square due to error (MSE) as an estimate of the common variance. In addition, we use a t critical value to achieve a simultaneous, or family-wise, confidence level of $100(1-\alpha)\%$.

The Bonferroni Multiple Comparison Procedure

In a one-way analysis of variance, suppose there are k groups, $n = n_1 + n_2 + \cdots + n_k$ total observations, and H_0 is rejected.

1. There are $c = \binom{k}{2} = \dfrac{k(k-1)}{2}$ pairs of population means to compare.

2. The c simultaneous $100(1-\alpha)\%$ **Bonferroni confidence intervals** have the following values as endpoints:

$$(\bar{x}_{i.} - \bar{x}_{j.}) \pm t_{\alpha/(2c), n-k} \cdot \sqrt{MSE} \cdot \sqrt{\frac{1}{n_i} + \frac{1}{n_j}} \quad \text{for all } i \neq j \qquad (11.6)$$

Bonferroni Example

EXAMPLE 11.3 Small-Scale Farming

Many people believe most farms are huge, corporate-like enterprises covering thousands of acres. However, in Canada, most of the country's farms are small-scale, family-owned and -operated businesses. Only a small portion of the farms in Canada earn more than $1 million in revenue. Independent random samples of farms were obtained from four Canadian provinces, with 10 observations in each group. The size (in acres) was recorded for each farm. The resulting sample means and the ANOVA table are shown here.

Group number	Province (factor)	Sample mean
1	Prince Edward Island	402.3
2	New Brunswick	421.1
3	Quebec	326.1
4	British Columbia	314.3

One-way ANOVA summary table

Source of variation	Sum of squares	Degrees of freedom	Mean square	F	p Value
Factor	86185.9	3	28728.6	6.26	0.0016
Error	165086.0	36	4585.7		
Total	251271.9	39			

The ANOVA test is significant at the $p = 0.0016$ level. There is evidence to suggest that at least one pair of population means is different (an overall difference). Construct the Bonferroni 95% confidence intervals and use them to isolate the pair(s) of means contributing to this overall experiment difference.

Solution

STEP 1 The number of pairwise comparisons needed is

$$c = \binom{4}{2} = \frac{4 \cdot 3}{2} = 6$$

$$95\% = 100(1-\alpha)\% \quad \Rightarrow \quad \alpha = 0.05 \quad \Rightarrow \quad \frac{\alpha}{2c} = \frac{0.05}{2 \cdot 6} = 0.0042$$

This (infrequent) right-tail probability is not specified in Appendix Table 5; however, we can use linear interpolation or technology, with $n - k = 40 - 4 = 36$ degrees of freedom, to find $t_{\alpha/(2c),36} = t_{0.0042,36} = 2.7888$.

STEP 2 The Bonferroni confidence interval for the difference $\mu_1 - \mu_2$ is

$$(\bar{x}_{1.} - \bar{x}_{2.}) \pm t_{\alpha/(2c),\, n-k} \cdot \sqrt{\text{MSE}} \cdot \sqrt{\frac{1}{n_1} + \frac{1}{n_2}}$$

$$= (\bar{x}_{1.} - \bar{x}_{2.}) \pm t_{0.0042,\, 36} \cdot \sqrt{\text{MSE}} \cdot \sqrt{\frac{1}{n_1} + \frac{1}{n_2}}$$

$$= (402.3 - 421.1) \pm (2.7888)\sqrt{4585.7}\sqrt{\frac{1}{10} + \frac{1}{10}}$$

$$= (-103.4,\, 65.8)$$

STEP 3 The remaining five confidence intervals are found in the same manner. Each Bonferroni confidence interval and its corresponding conclusion are shown in the table.

Difference	Bonferroni confidence interval	Conclusion
$\mu_1 - \mu_2$	$(-103.4,\ 65.8)$	0 in CI. μ_1 and μ_2 are not significantly different.
$\mu_1 - \mu_3$	$(-8.4,\ 160.8)$	0 in CI. μ_1 and μ_3 are not significantly different.
$\mu_1 - \mu_4$	$(3.4,\ 172.6)$	0 **not** in CI. Evidence to suggest that $\mu_1 \neq \mu_4$.
$\mu_2 - \mu_3$	$(10.4,\ 179.6)$	0 **not** in CI. Evidence to suggest that $\mu_2 \neq \mu_3$.
$\mu_2 - \mu_4$	$(22.2,\ 191.4)$	0 **not** in CI. Evidence to suggest that $\mu_2 \neq \mu_4$.
$\mu_3 - \mu_4$	$(-72.8,\ 96.4)$	0 in CI. μ_3 and μ_4 are not significantly different.

Finally, the initial ANOVA test indicates that there is an overall difference among the four population means. The simultaneous 95% Bonferroni confidence intervals suggest that this overall difference is due to a difference between μ_1 and μ_4, μ_2 and μ_3, and μ_2 and μ_4. **Figure 11.6** and **Figure 11.7** show technology solutions.

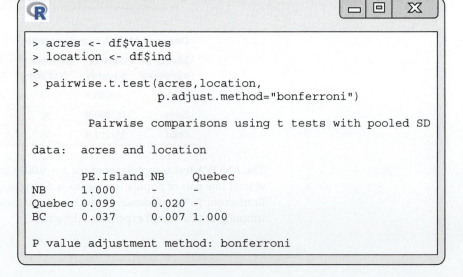

Figure 11.6 The R function `pairwise.t.test()` can be used to make pairwise comparisons. The *p* values reported that are less than or equal to 0.05 indicate a significant difference in means.

Figure 11.7 The R function `PostHocTest` in the package `DescTools` can be used to construct the simultaneous Bonferroni confidence intervals. Note that the differences are reversed in the R output.

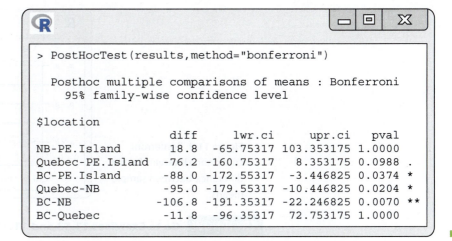

CHAPTER APP **Selenium Concentration**

Construct the Bonferroni 95% confidence intervals and use them to isolate the pair(s) of means contributing to this overall experiment difference.

Another common, compact, graphical method can be used to summarize the results of a multiple comparison procedure. Write the sample means in order from smallest to largest. Use the results from a multiple comparison procedure to draw a horizontal line under the groups of means that are not significantly different.

Here is the graphical summary for Example 11.3.

1. The sample means in order:

	$\bar{x}_{4.}$	$\bar{x}_{3.}$	$\bar{x}_{1.}$	$\bar{x}_{2.}$
Sample mean	314.3	326.1	402.3	421.1

2. There is no significant difference between μ_1 and μ_2. Draw a horizontal line under the sample means from populations 1 and 2.

	$\bar{x}_{4.}$	$\bar{x}_{3.}$	$\bar{x}_{1.}$	$\bar{x}_{2.}$
Sample mean	314.3	326.1	402.3	421.1

3. There is no significant difference between μ_1 and μ_3. Draw a horizontal line under the sample means from populations 1 and 3. There is no significant difference between μ_3 and μ_4. Draw a horizontal line under the sample means from populations 3 and 4.

	$\bar{x}_{4.}$	$\bar{x}_{3.}$	$\bar{x}_{1.}$	$\bar{x}_{2.}$
Sample mean	314.3	326.1	402.3	421.1

Those pairs of means not connected by a horizontal line are significantly different.

Figure 11.8 shows an R plot of the confidence intervals for the difference between sample means, another way to visualize which pair(s) of mean(s) is(are) different.

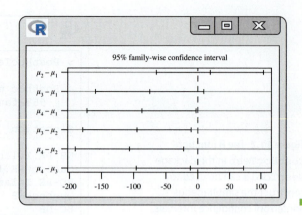

Figure 11.8 Bonferroni confidence intervals for the difference between sample means.

TRY IT NOW Go to Exercises 11.53 and 11.55

CHAPTER APP **Selenium Concentration**

Construct a graph to summarize the results from the Bonferroni multiple comparison procedure.

Tukey's Procedure

Tukey's multiple comparison procedure also yields simultaneous confidence intervals for all pairwise differences. The form of the confidence intervals is similar to Equation 11.6, but it uses a Q critical value from the **Studentized range distribution**. This distribution is completely characterized by two parameters: the degrees of freedom in the numerator and denominator, m and ν, respectively.

Using the usual notation, let $Q_{\alpha,m,\nu}$ denote the right-tail critical value of the Studentized range distribution with m and ν degrees of freedom. Appendix Table 8 presents selected critical values associated with various Studentized range distributions. The degrees of freedom in the numerator are given in the top row and the degrees of freedom in the denominator are given in the left column. In the body of the table, $Q_{\alpha,m,\nu}$ is at the intersection of column m and row ν.

Tukey's Multiple Comparison Procedure

In a one-way analysis of variance, suppose there are k groups, $n = n_1 + n_2 + \cdots + n_k$ total observations, and H_0 is rejected. The set of $c = \binom{k}{2}$ simultaneous $100(1-\alpha)\%$ confidence intervals have the following values as endpoints:

$$(\bar{x}_{i.} - \bar{x}_{j.}) \pm \frac{1}{\sqrt{2}} \cdot Q_{\alpha,k,n-k} \cdot \sqrt{\text{MSE}} \cdot \sqrt{\frac{1}{n_i} + \frac{1}{n_j}} \quad \text{for all } i \neq j \quad (11.7)$$

If all pairwise comparisons are considered, the Bonferroni procedure produces wider confidence intervals compared to the Tukey procedure. However, if only a subset of all pairwise comparisons is needed, then the Bonferroni method may be better. Other methods for comparing population means following an ANOVA test are also available. No single comparison method is uniformly best.

In the next example, Tukey's procedure is used to isolate pairwise differences contributing to an overall significant ANOVA test.

Tukey Example

EXAMPLE 11.4   People on the Go, Go to Wawa

A recent study suggests that convenience store shopping is affected by geographic region. For example, shoppers in the northeastern United States are more likely to frequent a convenience store during late night hours, and those in the western United States are the most likely to utilize a DVD rental kiosk. Independent random samples of shoppers at convenience stores were selected in four regions, and the total amount spent was recorded (in dollars). The summary statistics and the ANOVA table are shown in the tables.

Group number	1	2	3	4
Region	Northeast	South	Midwest	West
Sample size	10	11	12	11
Sample mean	13.43	16.42	9.95	15.48

One-way ANOVA summary table

Source of variation	Sum of squares	Degrees of freedom	Mean square	F	p Value
Factor	285.4	3	95.14	5.79	0.0022
Error	657.1	40	16.43		
Total	942.5	43			

The ANOVA test is significant at the $p = 0.0022$ level. There is evidence to suggest that at least one pair of means is different (an overall difference). Construct the Tukey 95% simultaneous confidence intervals and use them to isolate the pair(s) of means contributing to this overall experiment difference.

Solution

STEP 1 The number of pairwise comparisons needed is $\binom{4}{2} = 6$ and $\alpha = 0.05$.

The Studentized range critical value (Appendix Table 8) is $Q_{\alpha, k, n-k} = Q_{0.05, 4, 40} = 3.791$.

STEP 2 The first confidence interval, for the difference $\mu_1 - \mu_2$, using Tukey's procedure is

$$(\bar{x}_{1.} - \bar{x}_{2.}) \pm \frac{1}{\sqrt{2}} \cdot Q_{0.05, k, n-k} \cdot \sqrt{MSE} \cdot \sqrt{\frac{1}{n_1} + \frac{1}{n_2}}$$

$$= (13.43 - 16.42) \pm \frac{1}{\sqrt{2}} \cdot (3.791) \cdot \sqrt{16.34} \cdot \sqrt{\frac{1}{10} + \frac{1}{11}}$$

$$= -2.99 \pm 4.75 = (-7.74, 1.75)$$

STEP 3 The remaining confidence intervals and conclusions are given in the table.

Difference	Tukey confidence interval	Conclusion
$\mu_1 - \mu_2$	$(-7.74, 1.75)$	0 in CI. μ_1 and μ_2 are not significantly different.
$\mu_1 - \mu_3$	$(-1.18, 8.13)$	0 in CI. μ_1 and μ_3 are not significantly different.
$\mu_1 - \mu_4$	$(-6.80, 2.70)$	0 in CI. μ_1 and μ_4 are not significantly different.
$\mu_2 - \mu_3$	$(1.93, 11.00)$	0 **not** in CI. Evidence to suggest that $\mu_2 \neq \mu_3$.
$\mu_2 - \mu_4$	$(-3.69, 5.58)$	0 in CI. μ_2 and μ_4 are not significantly different.
$\mu_3 - \mu_4$	$(-10.06, -0.99)$	0 **not** in CI. Evidence to suggest that $\mu_3 \neq \mu_4$.

Here are the results presented in graphical form:

	$\bar{x}_{3.}$	$\bar{x}_{1.}$	$\bar{x}_{4.}$	$\bar{x}_{2.}$
Sample mean	9.95	13.43	15.48	16.42

The simultaneous 95% confidence intervals constructed using Tukey's procedure suggest that the overall significance is due to a difference between μ_2 and μ_3, and between μ_3 and μ_4.

Figure 11.9 shows a technology solution, and **Figure 11.10** shows an R plot of the confidence intervals for the difference between sample means.

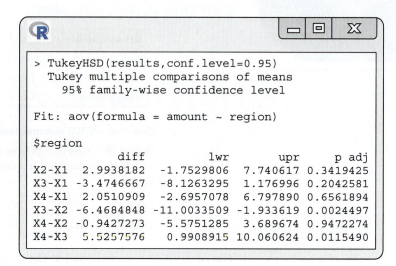

Figure 11.9 The R function `TukeyHSD()` is used to construct simultaneous Tukey confidence intervals.

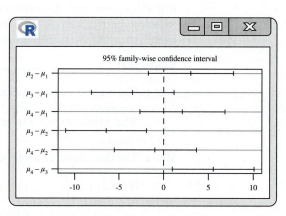

Figure 11.10 Tukey confidence intervals for the difference between sample means.

TRY IT NOW Go to Exercises 11.51 and 11.59

CHAPTER APP Selinium Concentration

Construct the Tukey 95% confidence intervals and use them to isolate the pair(s) of means contributing to this overall experiment difference. Construct a graph to summarize the results from this multiple comparison procedure.

The Bonferroni multiple comparison procedure is conservative; that is, the overall resulting confidence level is probably higher than the specified value. To compute Bonferroni intervals, we simply divide the overall confidence level evenly among all pairwise comparisons. If we are interested in only some of the pairwise comparisons, we divide the overall confidence level accordingly. In this case, the resulting confidence level may be closer to the specified value. Tukey's procedure is based on a special distribution and tends to result in more accurate confidence levels. Therefore, if all pairwise comparisons are necessary, Tukey's multiple comparison procedure is usually better.

Consider the Bonferroni confidence intervals for Example 11.4.

Difference	Bonferroni confidence interval	Conclusion
$\mu_1 - \mu_2$	$(-7.91, 1.92)$	0 in CI. μ_1 and μ_2 are not significantly different.
$\mu_1 - \mu_3$	$(-1.34, 8.29)$	0 in CI. μ_1 and μ_3 are not significantly different.
$\mu_1 - \mu_4$	$(-6.97, 2.86)$	0 in CI. μ_1 and μ_4 are not significantly different.
$\mu_2 - \mu_3$	$(1.77, 11.16)$	0 **not** in CI. Evidence to suggest that $\mu_2 \neq \mu_3$.
$\mu_2 - \mu_4$	$(-3.85, 5.74)$	0 in CI. μ_2 and μ_4 are not significantly different.
$\mu_3 - \mu_4$	$(-10.22, -0.83)$	0 **not** in CI. Evidence to suggest that $\mu_3 \neq \mu_4$.

Notice that each confidence interval is wider than the corresponding Tukey confidence interval. A wider, more conservative, Bonferroni confidence interval could result in a different conclusion concerning the difference between two population means. For example, compare the Tukey and Bonferroni confidence intervals for $\mu_2 - \mu_3$: The left endpoint of the Bonferroni CI is closer to 0.

Section 11.2 Exercises

Concept Check

11.37 True or False The Bonferroni multiple comparison procedure is conservative.

11.38 True or False The Bonferroni multiple comparison procedure can be used even if we are interested in only some of the pairwise comparisons.

11.39 True or False For Tukey's multiple comparison procedure to be valid, all sample sizes must be the same.

11.40 True or False In a one-way ANOVA with k groups, suppose we reject the null hypothesis. Using a multiple comparison procedure, there must be at least one pairwise comparison in which there is evidence to suggest that the population means are significantly different.

11.41 Short Answer In a one-way ANOVA with k groups, suppose we reject the null hypothesis. What is the total number of possible pairwise comparisons?

11.42 Short Answer Explain why we use a multiple comparison procedure rather than several individual hypothesis tests or confidence intervals.

11.43 Short Answer In a one-way ANOVA, suppose we do not reject the null hypothesis. Explain why there is no need for a multiple comparison procedure.

Practice

11.44 The number of groups k, the total number of observations n, and the overall confidence level for the Bonferroni multiple comparison procedure are given. Find the total number of comparisons c and the t critical value used in the calculation of each Bonferroni confidence interval.
(a) $k = 3$, $n = 30$, 95%
(b) $k = 3$, $n = 43$, 99%
(c) $k = 4$, $n = 32$, 99%
(d) $k = 5$, $n = 55$, 90%
(e) $k = 6$, $n = 30$, 95%

11.45 The number of observations in each group and the overall confidence level for Tukey's multiple comparison procedure are given. Find the critical value from the Studentized range distribution to be used in the calculation of each Tukey confidence interval.
(a) $n_1 = 6$, $n_2 = 8$, $n_3 = 7$, 95%
(b) $n_1 = n_2 = 11$, $n_3 = 12$, $n_4 = 10$, 95%
(c) $n_1 = 14$, $n_2 = 12$, $n_3 = 10$, $n_4 = 18$, 99%
(d) $n_1 = n_2 = n_3 = n_4 = n_5 = 11$, 99%
(e) $n_1 = n_2 = n_3 = 5$, $n_4 = n_5 = 6$, $n_6 = 9$, 99.9%

11.46 Use the summary statistics and the Bonferroni confidence intervals to construct the corresponding graph indicating which pairs of means are and are not significantly different.

(a)

Sample means	Difference	Confidence interval
$\bar{x}_{1.} = 52.8$	$\mu_1 - \mu_2$	$(-6.00, 17.85)$
$\bar{x}_{2.} = 46.9$	$\mu_1 - \mu_3$	$(-9.50, 14.35)$
$\bar{x}_{3.} = 50.4$	$\mu_2 - \mu_3$	$(-15.42, 8.42)$

(b)

Sample means	Difference	Confidence interval
$\bar{x}_{1.} = 4.82$	$\mu_1 - \mu_2$	$(-5.69, -1.13)$
$\bar{x}_{2.} = 8.23$	$\mu_1 - \mu_3$	$(-4.49, 0.07)$
$\bar{x}_{3.} = 7.03$	$\mu_2 - \mu_3$	$(-1.08, -3.48)$

(c)

Sample means	Difference	Confidence interval
$\bar{x}_{1.} = 16.09$	$\mu_1 - \mu_2$	$(-12.16, -1.07)$
$\bar{x}_{2.} = 22.71$	$\mu_1 - \mu_3$	$(-7.98, 3.11)$
$\bar{x}_{3.} = 18.53$	$\mu_1 - \mu_4$	$(-1.37, 9.72)$
$\bar{x}_{4.} = 16.33$	$\mu_2 - \mu_3$	$(-5.78, 5.31)$
	$\mu_2 - \mu_4$	$(0.83, 11.92)$
	$\mu_3 - \mu_4$	$(-3.34, 7.75)$

(d)

Sample means	Difference	Confidence interval
$\bar{x}_{1.} = 201.7$	$\mu_1 - \mu_2$	$(9.98, 56.20)$
$\bar{x}_{2.} = 168.6$	$\mu_1 - \mu_3$	$(13.20, 59.42)$
$\bar{x}_{3.} = 165.4$	$\mu_1 - \mu_4$	$(-19.89, 26.33)$
$\bar{x}_{4.} = 219.7$	$\mu_2 - \mu_3$	$(-41.10, 5.12)$
	$\mu_2 - \mu_4$	$(-74.19, -27.98)$
	$\mu_3 - \mu_4$	$(-77.41, 31.20)$

11.47 Use the summary statistics and the Tukey confidence intervals to construct the corresponding graph indicating which pairs of means are and are not significantly different.

(a)

Sample means	Difference	Confidence interval
$\bar{x}_{1.} = -33.44$	$\mu_1 - \mu_2$	$(-37.57, 0.36)$
$\bar{x}_{2.} = -14.83$	$\mu_1 - \mu_3$	$(-52.88, -14.95)$
$\bar{x}_{3.} = 0.48$	$\mu_1 - \mu_4$	$(-56.71, -18.77)$
$\bar{x}_{4.} = 4.30$	$\mu_2 - \mu_3$	$(-37.28, 3.66)$
	$\mu_2 - \mu_4$	$(-38.10, -0.17)$
	$\mu_3 - \mu_4$	$(-22.79, 15.14)$

(b)

Sample means	Difference	Confidence interval
$\bar{x}_{1.} = 1.62$	$\mu_1 - \mu_2$	$(0.086, 0.320)$
$\bar{x}_{2.} = 1.41$	$\mu_1 - \mu_3$	$(0.197, 0.432)$
$\bar{x}_{3.} = 1.30$	$\mu_1 - \mu_4$	$(-0.006, 0.229)$
$\bar{x}_{4.} = 1.50$	$\mu_2 - \mu_3$	$(0.001, 0.235)$
	$\mu_2 - \mu_4$	$(-0.203, 0.320)$
	$\mu_3 - \mu_4$	$(-0.314, -0.079)$

(c)

Sample means	Difference	Confidence interval
$\bar{x}_{1.} = 64.35$	$\mu_1 - \mu_2$	$(0.91, 19.38)$
$\bar{x}_{2.} = 54.21$	$\mu_1 - \mu_3$	$(-5.68, 12.79)$
$\bar{x}_{3.} = 60.80$	$\mu_1 - \mu_4$	$(3.20, 21.66)$
$\bar{x}_{4.} = 51.92$	$\mu_1 - \mu_5$	$(-9.73, 8.73)$
$\bar{x}_{5.} = 64.85$	$\mu_2 - \mu_3$	$(-15.83, 2.64)$
	$\mu_2 - \mu_4$	$(-6.95, 11.52)$
	$\mu_2 - \mu_5$	$(-19.88, -1.41)$
	$\mu_3 - \mu_4$	$(-0.36, 18.11)$
	$\mu_3 - \mu_5$	$(-13.29, 5.18)$
	$\mu_4 - \mu_5$	$(-22.16, -3.70)$

11.48 The results from a multiple comparison procedure are shown graphically. Use each illustration to identify all pairs of population means that are significantly different.

(a)

	$\bar{x}_{3.}$	$\bar{x}_{1.}$	$\bar{x}_{2.}$	$\bar{x}_{4.}$
Sample mean	12.69	14.64	15.94	16.21

(b)

	$\bar{x}_{1.}$	$\bar{x}_{2.}$	$\bar{x}_{3.}$	$\bar{x}_{4.}$
Sample mean	29.98	32.59	32.62	37.25

(c)

	$\bar{x}_{1.}$	$\bar{x}_{2.}$	$\bar{x}_{3.}$	$\bar{x}_{4.}$
Sample mean	39.01	41.40	50.83	51.62

(d)

	$\bar{x}_{2.}$	$\bar{x}_{3.}$	$\bar{x}_{5.}$	$\bar{x}_{1.}$	$\bar{x}_{4.}$
Sample mean	4.67	6.08	6.23	6.30	7.20

11.49 Suppose data collected from an experiment resulted in a significant ANOVA test. Using the summary statistics and MSE $= 29.83$, construct the Bonferroni 99% confidence intervals for all pairwise comparisons. Use these intervals to identify the significantly different population means.

Group	1	2	3	4
Sample size	20	20	20	20
Sample mean	19.59	21.58	16.50	25.76

11.50 Suppose data collected from an experiment resulted in a significant ANOVA test. Using the summary statistics and MSE $= 8111.8$, construct the Tukey 95% confidence intervals for all pairwise comparisons. Use these intervals to identify the significantly different population means.

Group	1	2	3	4
Sample size	16	18	21	9
Sample mean	476.3	450.2	698.2	597.6

Applications

11.51 Medicine and Clinical Studies Some people believe that unequal Medicare spending across the United States is due to waste and unnecessary tests. However, recent research suggests that health care costs do indeed differ by region.[2] Independent random samples of adults in three states were obtained, and the Medicare payments to each were recorded. The resulting ANOVA test was significant at the $p = 0.0190$

level. The summary statistics are given in the table. If MSE = 136,146, construct the Bonferroni 95% confidence intervals to isolate the population mean payments that are significantly different.

State	Sample size	Sample mean
Connecticut	18	3489.22
Georgia	22	3774.50
Mississippi	25	3794.76

11.52 Demographics and Population Statistics A study was conducted to compare the population mean age of women at the time of their first marriage. Independent random samples of size 11 were obtained from weddings in four different years. The resulting ANOVA test was significant at the $p = 0.002$ level. The sample means are given in the table.[3] If MSE = 6.834, construct the Tukey 95% confidence intervals to isolate the population mean ages that are significantly different.

Year	1985	1995	2003	2017
Sample mean	22.2	23.6	24.9	27.9

11.53 Manufacturing and Product Development An experiment was conducted to compare the true cost of microwave popcorn brands by examining the percentage of popped kernels. Independent random samples of six packages of buttered microwave popcorn (all weighing the same) were obtained for each of four different brands. Each bag was popped, and the percentage of popped kernels was measured. An ANOVA test resulted in the test statistic $f = 7.69$ and MSE = 2.72. The sample means are given in the table.

Microwave popcorn	Pop Secret	Best Choice	Act II	Orville Redenbacher's
Sample mean	90.8	86.9	90.9	89.2

(a) Find bounds on the p value associated with this test statistic. State the conclusion of the ANOVA test.
(b) Construct the Bonferroni 99% simultaneous confidence intervals to determine which pairs of population mean percentages of popped kernels are significantly different.

11.54 Biology and Environmental Science A person's metabolic rate certainly depends on level of activity. However, a recent study suggests that metabolic rate is also related to personality type. For example, the performer, who is people oriented and living for the moment, might have a higher resting metabolic rate, while the artist, who is quiet, sensitive, and kind, could have a lower resting metabolic rate. Suppose independent random samples of different personality types were obtained and the resting metabolic rate (in kcal/day) was measured for each using a metabolic chamber. An ANOVA test was conducted to determine whether there was any difference in population mean resting metabolic rates ($f = 5.83$ and MSE = 52,025.4). The summary statistics by personality type are given in the table.

Personality type	Sample size	Sample mean
Mechanic	14	1962.7
Nurturer	16	1757.4
Protector	20	2057.8
Idealist	18	2019.5

(a) Show that there is evidence of an overall difference in population mean resting metabolic rate. Use $\alpha = 0.01$ and justify your answer.
(b) Construct the Tukey 99% confidence intervals and indicate which pairs of means are contributing to the overall difference.

11.55 Biology and Environmental Science Soil permeability, the rate at which water can flow through the soil, is an important measure prior to construction. Four different potential county development sites in Wisconsin were selected for testing. Independent random samples of locations within each development were selected, and the soil permeability was measured at each location (in inches per hour). An ANOVA test was conducted to determine whether there was any difference in population mean permeability rates ($f = 9.10$ and MSE = 0.02504). The summary statistics are given in the table.

Development	Sample size	Sample mean
Grant	10	1.3164
Green	12	0.2567
Dane	11	0.5601
Rock	11	0.9206

(a) Find bounds on the p value associated with this (significant) F statistic.
(b) Construct the Bonferroni 95% confidence intervals for all pairwise comparisons and draw a graph to represent the results.

11.56 Sports and Leisure The shot clock in men's NCAA basketball games was changed to 30 seconds in the 2015–2016 season. A team must shoot before the shot clock expires, or it turns the ball over to the opposing team. A team rarely uses the entire 30 seconds, and some teams shoot as quickly as possible. A study was conducted to compare the mean time to take a shot in five different athletic conferences. Independent random samples of possessions were obtained, and the time (in seconds) to take a shot was recorded for each. An ANOVA test was conducted to determine whether there was any overall difference in population mean shot times ($f = 4.06$ and MSE = 50.9). The summary statistics are given in the table.

Conference	Sample size	Sample mean
ACC	20	17.90
Big East	26	24.65
Pac-12	23	18.39
Sun Belt	25	22.92
WAC	26	19.77

Construct the Bonferroni 99% confidence intervals for all pairwise comparisons and draw a graph to represent the results. Identify the pairs of athletic conference population means that are significantly different.

11.57 Biology and Environmental Science Most honey comes from European honey bee colonies cultivated in the United States. Africanized killer bees tend to be aggressive and produce less honey. A study was conducted to compare the mean amount of honey produced by colonies in four parts of the country. Independent random samples of colonies were obtained in each area, and the amount of honey (in pounds) per year per colony was recorded. HONEY
(a) Conduct an analysis of variance test to show that there is evidence of an overall difference in population mean honey production. Use $\alpha = 0.05$.
(b) Construct the Bonferroni 95% confidence intervals to isolate the pairs of means contributing to the overall difference.
(c) Draw a graph to represent the results of the multiple comparison procedure in part (b).

11.58 Olive Oil Purity Olive oil is used in a variety of cooking methods and has been shown to provide some health benefits. However, the chemical composition and purity of olive oil can also vary and depends on the supplier region and climate. Independent random samples of domestic regular olive oils were obtained and the peroxide value (in mEq O_2/kg oil) was measured for each. OLIVEOIL
(a) Conduct an analysis of variance test to show that there is evidence of an overall difference in population mean peroxide value. Use $\alpha = 0.01$.
(b) Construct the Tukey 99% confidence intervals to isolate the pairs of means contributing to the overall difference.
(c) Draw a graph to represent the results of the multiple comparison procedure in part (b).
(d) Based on these data, which olive oil appears to be the purest? Why?

11.59 Manufacturing and Product Development A stun gun delivers a high-voltage, low-amperage electrical shock to an attacker. In certain states, cities, and countries, the use of this self-protection device is restricted or even illegal. Independent random samples of stun gun owners in cities, suburban areas, and rural areas were obtained. Each stun gun was examined, and the voltage (in thousands of volts) on the device was carefully measured in a single test. STUN
(a) Conduct an analysis of variance test to show that there is evidence of an overall difference in population mean stun gun voltage. Use $\alpha = 0.05$.
(b) Construct the Tukey 95% confidence intervals to isolate the pairs of means contributing to the overall difference.
(c) Draw a graph to represent the results of the multiple comparison procedure in part (b).

11.60 Public Health and Nutrition A simple gelatin is known today as "America's most famous dessert." Although it was patented in 1845, Jell-O sales were minimal until the early 1900s. While primarily made from processed collagen, gelatin also contains sugar, artificial flavors, and coloring. Independent random samples of three gelatin brands were obtained, and the amount of sugar (in grams) in one serving of each was measured. GELATIN
(a) Conduct an analysis of variance test to show that there is evidence of an overall difference in population mean sugar content in gelatin servings. Use $\alpha = 0.01$.
(b) Construct the Bonferroni 99% confidence intervals to isolate the pairs of means contributing to the overall difference.
(c) Draw a graph to represent the results of the multiple comparison procedure in part (b).

11.61 Sports and Leisure The outside dimensions and playing lines of a tennis court are standard and well established. However, the distance between courts, or sidelines, varies. Many construction guidelines recommend 24 feet between courts, but to conserve space, builders plan for only at least 12 feet. Independent random samples of public courts were obtained in five different cities, and the distance (in feet) between courts was recorded. The data are given in the table. TENNIS

City	Observations				
Atlanta	14.0	14.2	13.0	15.2	15.0
Boston	14.2	16.8	18.6	15.5	16.6
Los Angeles	14.3	14.9	16.5	15.1	14.4
Miami	14.3	17.3	17.3	14.9	16.4
New York	22.0	18.3	19.3	20.5	18.5

(a) Conduct an analysis of variance test to show that there is evidence of an overall difference in population mean distances between tennis courts in public parks. Use $\alpha = 0.01$.
(b) Construct the Bonferroni 95% confidence intervals to isolate the pairs of means contributing to the overall difference.
(c) Draw a graph to represent the results of the multiple comparison procedure in part (b).

11.62 Public Policy and Political Science The Social Security Administration issues decisions about disability and supplemental security income. However, any individual who is denied benefits may request an appeal. If the person disagrees with the reconsideration decision, then a hearing is held before an administrative law judge. Independent random samples of hearings were obtained in four different cities, and the time (in months) until the hearing was held was recorded for each. HEARING
(a) Conduct an analysis of variance test to show that there is evidence of an overall difference in population mean time until hearings were held. Use $\alpha = 0.01$.
(b) Construct the Tukey 95% confidence intervals to isolate the pairs of means contributing to the overall difference.
(c) Draw a graph to represent the results of the multiple comparison procedure in part (b).

Extended Applications

11.63 Biology and Environmental Science Hatcheries around the country provide chicks to poultry farms for growing into eventual roasters and broilers. Independent random samples of chicks from different hatcheries were obtained, and the shipping weight (in grams) of each chick was measured. The data are given in the table. 📊 CHICK

Hatchery	Observations				
Bedwell Farms	47.1	48.8	52.6	49.1	53.3
Clinton Chicks	63.0	52.0	56.4	56.2	55.3
Sunny Creek	51.6	52.9	54.9	52.8	54.6
Wild Wings	57.4	57.4	55.2	56.0	56.0

(a) Conduct an analysis of variance test to show that there is evidence of an overall difference in population mean chick weights. Use $\alpha = 0.05$.

(b) Construct the Bonferroni 95% confidence intervals to isolate the pairs of means contributing to the overall difference.

(c) Construct the Tukey 95% confidence intervals to isolate the pairs of means contributing to the overall difference.

(d) Are your answers to parts (b) and (c) the same? If so, did you expect this to happen? If not, why not?

11.64 Public Health and Nutrition Butter and margarine contain saturated fat, which can increase the *bad* cholesterol in your blood. To advise her clients, a dietitian examined four types of margarine as determined by the percentage of fat listed on the product label. A random sample of each type was obtained, and the amount of saturated fat (in grams) per serving was recorded. 📊 SATFAT

(a) Conduct an analysis of variance test to show that there is evidence of an overall difference in population mean saturated fat per serving. Use $\alpha = 0.01$.

(b) Construct the Tukey 99% confidence intervals to isolate the pairs of means contributing to the overall difference.

(c) Are the differences found in part (b) consistent with the percentage of fat indicated on each product label? If not, can you explain any discrepancy?

11.65 Washer Water Usage According to the U.S. Environmental Protection Agency, a typical American family washes about 400 loads of laundry each year. Front-loading washing machines have become popular because they have a small carbon footprint, use less electricity, and tend to get clothes cleaner. Independent random samples of front-loading washing machines were obtained and the amount of water used (in gallons) in a regular load was measured for each. 📊 WASH

(a) Conduct an analysis of variance test to show that there is evidence of an overall difference in population mean water usage. Use $\alpha = 0.05$.

(b) Construct the Bonferroni 95% confidence intervals to isolate the pairs of means contributing to the difference.

(c) Construct the Tukey 95% confidence intervals to isolate the pairs of means contributing to the overall difference.

(d) Are the results the same in parts (b) and (c)? If so, why? If not, why not?

Challenge Problems

11.66 Manufacturing and Product Development An experiment was conducted to compare the acoustic properties of a control plastic (styrene) with four alternative treatment plastics. Independent random samples were selected, and the attenuation value of each piece was measured (in dB/mm at 5 MHz). 📊 PLASTIC

(a) Conduct a one-way analysis of variance to test for an overall difference among the five population means. Use $\alpha = 0.05$.

(b) Construct 95% Bonferroni confidence intervals only for the differences between each treatment plastic population mean and the control, styrene, population mean. Distribute the confidence level among the four comparisons (1 versus 2, 1 versus 3, 1 versus 4, and 1 versus 5) so that the simultaneous confidence level is 95%. Which treatment plastic attenuation means are significantly different from the control?

11.3 Two-Way ANOVA

Two Factors: Total Variation Decomposition

In the previous two sections, we considered the effect of a single factor on a response variable. In this section, we will consider experiments in which two factors may contribute to the overall variability in response. A two-way ANOVA is designed to compare the means of populations that can be classified in two different ways. For example, suppose we are interested in the miles on a charge for electric cars. The miles may vary by the motor and the battery configuration.

Without as much detail as Section 11.1, suppose there are a levels of factor A, b levels of factor B, and n observations for each combination of levels, for a total of abn observations. Using the electric car example, there could be $a = 5$ levels of factor A, 5 different motors; and $b = 3$ levels of factor B, 3 different battery configurations; and $n = 6$ observations for each combination of motor and battery configuration, for a total of $abn = (5)(3)(6) = 90$ observations.

In **Table 11.1**, there are $a = 3$ levels of factor A, $b = 4$ levels of factor B, and $n = 2$ observations per cell. Let x_{ijk} represent the kth observation for the ith level of factor A and the jth level of factor B. For example, in Table 11.1, $x_{132} = 71$, the 2nd observation for the 1st level of factor A and the 3rd level of factor B.

Table 11.1 Presentation of data in a two-way ANOVA.

		\multicolumn{8}{c}{Factor B}

		\multicolumn{2}{c	}{1}	\multicolumn{2}{c	}{2}	\multicolumn{2}{c	}{3}	\multicolumn{2}{c	}{4}
Factor A	1	83	84	84	77	84	71	100	102
	2	71	68	78	66	114	90	108	114
	3	78	89	95	104	115	119	119	126

An interaction effect is significant if one level of factor A and one level of factor B interact differently, or inconsistently, from other factor combinations.

The total variation in the data, sum of squares (**SST**), is decomposed into the sum of squares due to factor A (**SSA**), the sum of squares due to factor B (**SSB**), the sum of squares due to interaction [**SS(AB)**], and the sum of squares due to error (**SSE**).

Consistent with previous notation, dots in the subscript of $\bar{x}$ and t indicate the mean and the sum of x_{ijk}, respectively, over the appropriate subscript(s). For example,

$$\bar{x}_{\cdot j \cdot} = \frac{1}{an} \sum_{i=1}^{a} \sum_{k=1}^{n} x_{ijk} \quad \text{and} \quad t_{\cdots} = \sum_{i=1}^{a} \sum_{j=1}^{b} \sum_{k=1}^{n} x_{ijk}$$

Two-Way ANOVA Identity

Let SST = total sum of squares, SSA = sum of squares due to factor A, SSB = sum of squares due to factor B, SS(AB) = sum of squares due to interaction, and SSE = sum of squares due to error.

$$\text{SST} = \underbrace{\sum_{i=1}^{a} \sum_{j=1}^{b} \sum_{k=1}^{n} (x_{ijk} - \bar{x}_{\cdots})^2}_{\text{definition}} = \underbrace{\left(\sum_{i=1}^{a} \sum_{j=1}^{b} \sum_{k=1}^{n} x_{ijk}^2\right) - \frac{t_{\cdots}^2}{abn}}_{\text{computational formula}}$$

$$\text{SSA} = \underbrace{bn \sum_{i=1}^{a} (\bar{x}_{i \cdot \cdot} - \bar{x}_{\cdots})^2}_{\text{definition}} = \underbrace{\frac{\sum_{i=1}^{a} t_{i \cdot \cdot}^2}{bn} - \frac{t_{\cdots}^2}{abn}}_{\text{computational formula}}$$

$$\text{SSB} = \underbrace{an \sum_{j=1}^{b} (\bar{x}_{\cdot j \cdot} - \bar{x}_{\cdots})^2}_{\text{definition}} = \underbrace{\frac{\sum_{j=1}^{b} t_{\cdot j \cdot}^2}{an} - \frac{t_{\cdots}^2}{abn}}_{\text{computational formula}}$$

$$\text{SS(AB)} = \underbrace{n \sum_{i=1}^{a} \sum_{j=1}^{b} (\bar{x}_{ij \cdot} - \bar{x}_{i \cdot \cdot} - \bar{x}_{\cdot j \cdot} + \bar{x}_{\cdots})^2}_{\text{definition}}$$

$$= \underbrace{\frac{\sum_{i=1}^{a} \sum_{j=1}^{b} t_{ij \cdot}^2}{n} - \frac{\sum_{i=1}^{a} t_{i \cdot \cdot}^2}{bn} - \frac{\sum_{j=1}^{b} t_{\cdot j \cdot}^2}{an} + \frac{t_{\cdots}^2}{abn}}_{\text{computational formula}}$$

$$\text{SSE} = \underbrace{\sum_{i=1}^{a} \sum_{j=1}^{b} \sum_{k=1}^{n} (x_{ijk} - \bar{x}_{ij \cdot})^2}_{\text{definition}} = \underbrace{\text{SST} - \text{SSA} - \text{SSB} - \text{SS(AB)}}_{\text{computational formula}}$$

$$\text{SST} = \text{SSA} + \text{SSB} + \text{SS(AB)} + \text{SSE}$$

The assumptions for a two-way ANOVA are stated in terms of each cell, which is considered a population.

For reference, these are the two-way ANOVA assumptions.

1. The ab population distributions are normal.
2. The ab population variances are equal.
3. The samples are selected randomly and independently from the respective populations.

ANOVA Table and Test Procedure

Two-way ANOVA calculations are usually presented in a summary table. The values in this table are associated with the five sources of variation, and the F statistics are used to conduct appropriate hypothesis tests.

Two-way ANOVA summary table

Source of variation	Sum of squares	Degrees of freedom	Mean square	F	p Value
Factor A	SSA	$a-1$	$MSA = \dfrac{SSA}{a-1}$	$F_A = \dfrac{MSA}{MSE}$	p_A
Factor B	SSB	$b-1$	$MSB = \dfrac{SSB}{b-1}$	$F_B = \dfrac{MSB}{MSE}$	p_B
Interaction	SS(AB)	$(a-1)(b-1)$	$MS(AB) = \dfrac{SS(AB)}{(a-1)(b-1)}$	$F_{AB} = \dfrac{MS(AB)}{MSE}$	p_{AB}
Error	SSE	$ab(n-1)$	$MSE = \dfrac{SSE}{ab(n-1)}$		
Total	SST	$abn-1$			

There are three hypothesis tests associated with a two-way ANOVA.

Two-Way ANOVA Hypothesis Tests

Test 1: Test for an interaction effect.

H_0: There is no interaction effect.

H_a: There is an effect due to interaction.

TS: $F_{AB} = \dfrac{MS(AB)}{MSE}$

RR: $F_{AB} \geq F_{\alpha,(a-1)(b-1),ab(n-1)}$

Test 2: Test for an effect due to factor A.

H_0: There is no effect due to factor A.

H_a: There is an effect due to factor A.

TS: $F_A = \dfrac{MSA}{MSE}$

RR: $F_A \geq F_{\alpha,a-1,ab(n-1)}$

Test 3: Test for an effect due to factor B.

H_0: There is no effect due to factor B.

H_a: There is an effect due to factor B.

TS: $F_B = \dfrac{MSB}{MSE}$

RR: $F_B \geq F_{\alpha,b-1,ab(n-1)}$

The hypothesis test for an interaction effect is usually considered first. An interaction effect is present when the relationship between the two factors is not linear, or additive, for at least one combination of levels. For example, suppose the total production of corn depends on two factors, the amount of water and the amount of fertilizer. Intuitively, one might expect the total production of corn to increase as the amount of water increases and/or as the amount of fertilizer increases. Each change in water and/or fertilizer results in a predictable, additive change in the total amount of corn produced. However, consider a very high level of water and a high level of fertilizer. The additive model predicts a huge total production in corn. Yet there could be an interaction or inconsistent effect associated with these two levels. Too much water and too much fertilizer might combine to produce very little corn production. An interaction plot is a scatter plot of each cell sample mean versus factor A level. Connect the points in the graph corresponding to the same factor B levels. Parallel lines suggest no evidence of interaction.

Case 1: If the null hypothesis is not rejected, then the other two hypothesis tests can be conducted as usual, to see whether there are effects due to either (or both) of the factors.

Case 2: If the null hypothesis is rejected, then there is evidence of a significant interaction. Interpretation of the other two hypothesis tests is tricky.

1. If we reject a null hypothesis of no effect due to factor A (and/or factor B), then the effect due to factor A (and/or factor B) is probably significant.

2. If we do not reject a null hypothesis of no effect due to factor A (and/or factor B), then the effect due to factor A (and/or factor B) is inconclusive.

Examples

EXAMPLE 11.5 Satellite TV Quality

The quality of the picture on a home satellite TV depends on the strength of the signal. Professional installation is often necessary to properly align a dish and tune in a satellite. A study was conducted to determine whether the signal strength was related to the satellite company and/or the geographic region. For each company and region, a random sample of satellite TV users was obtained, and the signal strength of each system was measured (in dBμV). The data are given in the table.

		Geographic region							
		Northeast		Southeast		Midwest		West	
Company	DIRECTV	65.9	73.9	64.0	61.4	55.9	64.2	54.7	69.9
		70.9	74.9	55.0	52.1	67.2	74.3	62.2	65.5
	DISH	71.3	76.8	64.3	64.0	69.4	68.6	65.6	67.8
		71.4	65.2	61.3	68.2	61.2	64.9	55.2	66.0
	Glorystar	64.0	64.2	60.8	62.9	58.3	64.4	62.4	62.9
		76.7	65.1	57.0	69.6	69.3	73.1	64.5	71.3

Conduct a two-way analysis of variance to determine whether signal strength is affected by company and/or geographic region. Use $\alpha = 0.05$.

Solution

STEP 1 Assume the two-way ANOVA assumptions are true. There are $a = 3$ levels of factor A (satellite TV company), $b = 4$ levels of factor B (geographic region), and $n = 4$ observations in each cell. Sample totals are given in the table.

		Geographic region				
		Northeast	Southeast	Midwest	West	
Company	DIRECTV	$t_{11.} = 285.6$	$t_{12.} = 232.5$	$t_{13.} = 261.6$	$t_{14.} = 252.3$	$t_{1..} = 1032.0$
	DISH	$t_{21.} = 284.7$	$t_{22.} = 257.8$	$t_{23.} = 264.1$	$t_{24.} = 254.6$	$t_{2..} = 1061.2$
	Glorystar	$t_{31.} = 270.0$	$t_{32.} = 250.3$	$t_{33.} = 265.1$	$t_{34.} = 261.1$	$t_{3..} = 1046.5$
		$t_{.1.} = 840.3$	$t_{.2.} = 740.6$	$t_{.3.} = 790.8$	$t_{.4.} = 768.0$	$t_{...} = 3139.7$

STEP 2 Find the sums of squares.

$$SST = \left(\sum_{i=1}^{3}\sum_{j=1}^{4}\sum_{k=1}^{4} x_{ijk}^2\right) - \frac{t_{...}^2}{(3)(4)(4)}$$ Computational formula.

$$= (65.9^2 + 73.9^2 + \cdots + 71.3^2) - \frac{3139.7^2}{48}$$ Apply the formula.

$$= 206{,}990.75 - 205{,}369.09 = 1621.66$$ Simplify.

$$SSA = \frac{\sum_{i=1}^{3} t_{i..}^2}{(4)(4)} - \frac{t_{...}^2}{(3)(4)(4)}$$ Computational formula.

$$= \frac{1032.0^2 + 1061.2^2 + 1046.5^2}{16} - \frac{3139.7^2}{48}$$ Apply the formula.

$$= 205{,}395.73 - 205{,}369.09 = 26.65$$ Simplify.

$$SSB = \frac{\sum_{j=1}^{4} t_{.j.}^2}{(3)(4)} - \frac{t_{...}^2}{(3)(4)(4)}$$ Computational formula.

$$= \frac{840.3^2 + 740.6^2 + 790.8^2 + 768.0^2}{12} - \frac{3139.7^2}{48}$$ Apply the formula.

$$= 205{,}815.09 - 205{,}369.09 = 446.01$$ Simplify.

$$SS(AB) = \frac{\sum_{i=1}^{3}\sum_{j=1}^{4} t_{ij.}^2}{4} - \frac{\sum_{i=1}^{3} t_{i..}^2}{(4)(4)} - \frac{\sum_{j=1}^{4} t_{.j.}^2}{(3)(4)} + \frac{t_{...}^2}{(3)(4)(4)}$$ Computational formula.

$$= \frac{285.6^2 + 232.5^2 + \cdots + 261.1^2}{4}$$

$$\quad - 205{,}395.73 - 205{,}815.09 + 205{,}369.09$$ Apply the formula.

$$= 108.18$$ Simplify.

$$SSE = SST - SSA - SSB - SS(AB)$$ Computational formula.

$$= 1621.66 - 26.65 - 446.01 - 108.18 = 1040.83$$ Simplify.

STEP 3 Check for an interaction effect first.

H_0: There is no interaction effect.

H_a: There is an effect due to interaction.

TS: $F_{AB} = \dfrac{MS(AB)}{MSE}$

RR: $F_{AB} \geq F_{\alpha,(a-1)(b-1),ab(n-1)} = F_{0.05,6,36} = 2.37$

$$MS(AB) = \frac{SS(AB)}{(a-1)(b-1)} = \frac{108.18}{6} = 18.0302$$

$$MSE = \frac{SSE}{ab(n-1)} = \frac{1040.83}{36} = 28.9120$$

Remember that intermediate calculations are shown rounded but that final numerical answers are computed and presented using greater accuracy.

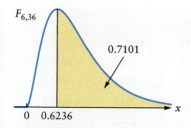

Figure 11.11 p value illustration:
$p = P(X \geq 0.6236)$
$= 0.7101 > 0.05 = \alpha$

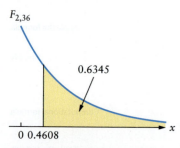

Figure 11.12 p value illustration:
$p = P(X \geq 0.4608)$
$= 0.6345 > 0.05 = \alpha$

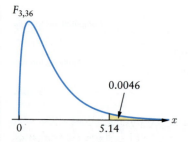

Figure 11.13 p value illustration:
$p = P(X \geq 5.14)$
$= 0.0046 \leq 0.05 = \alpha$

The value of the test statistic is

$$f_{AB} = \frac{\text{MS(AB)}}{\text{MSE}} = \frac{18.0302}{28.9120} = 0.6236$$

The value of the test statistic does not lie in the rejection region. Equivalently, $p = 0.7101 > 0.05$, as illustrated in **Figure 11.11**. There is no evidence of an interaction effect. The tests for factor effects can be conducted as usual.

STEP 4 Check for an effect due to satellite company (factor A).

H_0: There is no effect due to factor A.

H_a: There is an effect due to factor A.

TS: $F_A = \dfrac{\text{MSA}}{\text{MSE}}$

RR: $F_A \geq F_{\alpha, a-1, ab(n-1)} = F_{0.05, 2, 36} = 3.26$

The value of the test statistic is

$$f_A = \frac{\text{MSA}}{\text{MSE}} = \frac{\text{SSA}/(a-1)}{\text{SSE}/ab(n-1)} = \frac{26.65/2}{1040.83/36} = 0.4608$$

The value of the test statistic does not lie in the rejection region. Equivalently, $p = 0.6345 > 0.05$, as illustrated in **Figure 11.12**. There is no evidence to suggest that satellite company has an effect on the signal strength.

STEP 5 Check for an effect due to geographic region (factor B).

H_0: There is no effect due to factor B.

H_a: There is an effect due to factor B.

TS: $F_B = \dfrac{\text{MSB}}{\text{MSE}}$

RR: $F_B \geq F_{\alpha, b-1, ab(n-1)} = F_{0.05, 3, 36} = 2.87$

The value of the test statistic is

$$f_B = \frac{\text{MSB}}{\text{MSE}} = \frac{\text{SSB}/(b-1)}{\text{SSE}/ab(n-1)} = \frac{446.01/3}{1040.83/36} = 5.14 \ (\geq 2.87)$$

The value of the test statistic lies in the rejection region. Equivalently, $p = 0.0046 \leq 0.05$, as illustrated in **Figure 11.13**. There is evidence to suggest that signal strength is affected by geographic region.

STEP 6 Finally, here is the ANOVA summary table.

One-way ANOVA summary table

Source of variation	Sum of squares	Degrees of freedom	Mean square	F	p Value
Factor A	26.65	2	13.32	0.46	0.6345
Factor B	446.01	3	148.67	5.14	0.0046
Interaction	108.18	6	18.03	0.62	0.7101
Error	1040.83	36	28.91		
Total	1621.66	47			

Figure 11.14 shows a technology solution. **Figure 11.15** shows a profile plot of the means (with confidence intervals) to help visualize and detect an interaction effect. An interaction effect is suggested by lines that do not run in parallel.

11.3 Two-Way ANOVA

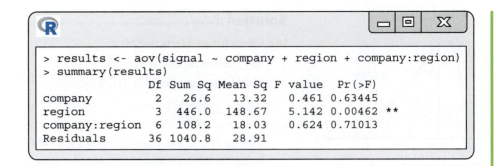

Figure 11.14 Use the R command `aov()` to produce the ANOVA summary table.

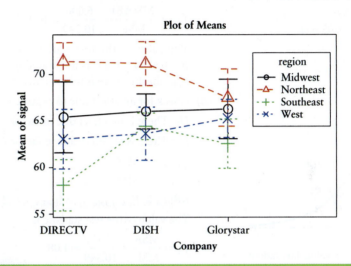

Figure 11.15 Profile plot of the means.

TRY IT NOW Go to Exercises 11.79 and 11.81

EXAMPLE 11.6 Snail Invasion

r_silver/Shutterstock

Giant African land snails have invaded Florida. These invasive creatures can be very destructive, eating plants, stucco, and plaster in their never-ending search for calcium. The snails are difficult to eliminate because they have no natural predators. However, specially trained dogs are used to detect and help capture the snails. A study was conducted to determine whether the size of these snails is related to the capture mode (factor A: bait, chemical treatments, trap, Labrador retriever, human) and/or location (factor B: commercial, residential). For each of the five capture modes ($=5$) and locations (b$=2$), independent random samples of size n$=3$ were obtained, and the length of each shell (in centimeters) was measured. Here is the ANOVA summary table.

One-way ANOVA summary table

Source of variation	Sum of squares	Degrees of freedom	Mean square	F	p Value
Factor A	44.690	4	11.172	1.08	0.3929
Factor B	122.533	1	122.533	11.84	0.0026
Interaction	24.383	4	6.096	0.59	0.6744
Error	206.973	20	10.349		
Total	398.579	29			

(a) Is there any evidence of interaction? Use $\alpha = 0.05$.

(b) Is there any evidence that capture mode affects the length of the shell? Use $\alpha = 0.05$.

(c) Is there any evidence that location affects the length of the shell? Use $\alpha = 0.05$.

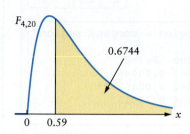

Figure 11.16 p value illustration:
$p = P(X \geq 0.59)$
$= 0.6744 > 0.05 = \alpha$

Solution

(a) Check for an interaction effect.

H_0: There is no interaction effect.

H_a: There is an effect due to interaction.

TS: $F_{AB} = \dfrac{MS(AB)}{MSE}$

RR: $F_{AB} \geq F_{\alpha,(a-1)(b-1),ab(n-1)} = F_{0.05,4,20} = 2.87$

Using the ANOVA summary table, the value of the test statistic is

$f_{AB} = \dfrac{MS(AB)}{MSE} = \dfrac{6.096}{10.349} = 0.59$

The value of the test statistic does not lie in the rejection region. Equivalently, $p = 0.6744 > 0.05$, as illustrated in **Figure 11.16**. Therefore, we do not reject the null hypothesis. There is no evidence of an interaction effect.

(b) Check for an effect due to capture mode (factor A).

H_0: There is no effect due to factor A.

H_a: There is an effect due to factor A.

TS: $F_A = \dfrac{MSA}{MSE}$

RR: $F_A \geq F_{\alpha,a-1,ab(n-1)} = F_{0.05,4,20} = 2.87$

The value of the test statistic is

$f_A = \dfrac{MSA}{MSE} = \dfrac{11.172}{10.349} = 1.08$

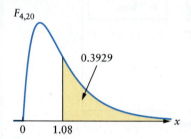

Figure 11.17 p value illustration:
$p = P(X \geq 1.08)$
$= 0.3929 > 0.05 = \alpha$

The value of the test statistic does not lie in the rejection region. Equivalently, $p = 0.3929 > 0.05$, as illustrated in **Figure 11.17**. Therefore, we do not reject the null hypothesis. There is no evidence of an effect on snail shell size due to capture mode.

(c) Check for an effect due to location (factor B).

H_0: There is no effect due to factor B.

H_a: There is an effect due to factor B.

TS: $F_B = \dfrac{MSB}{MSE}$

RR: $F_B \geq F_{\alpha,b-1,ab(n-1)} = F_{0.05,1,20} = 4.35$

The value of the test statistic is

$f_B = \dfrac{MSB}{MSE} = \dfrac{122.533}{10.349} = 11.84$

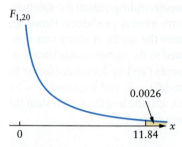

Figure 11.18 p value illustration:
$p = P(X \geq 11.84)$
$= 0.0026 \leq 0.05 = \alpha$

The value of the test statistic lies in the rejection region. Equivalently, $p = 0.0026 \leq 0.05$, as illustrated in **Figure 11.18**. Therefore, we reject the null hypothesis. There is evidence to suggest an effect on snail shell size due to location.

Figure 11.19 shows a technology solution.

```
> results <- aov(length ~ factor(capture) + factor(location) +
+                factor(capture):factor(location))
> print(summary(results),digits=5)
                                 Df  Sum Sq  Mean Sq  F value    Pr(>F)
factor(capture)                   4   44.690  11.172   1.0796  0.392945
factor(location)                  1  122.533 122.533  11.8405  0.002585 **
factor(capture):factor(location)  4   24.383   6.096   0.5890  0.674400
Residuals                        20  206.973  10.349
```

Figure 11.19 Use the R command aov() to produce the ANOVA summary table.

TRY IT NOW Go to Exercises 11.83 and 11.85

Section 11.3 Exercises

Concept Check

11.67 True or False In a two-way ANOVA, a significant interaction means there is evidence to suggest that the relationship between the two factors is not additive.

11.68 True or False In a two-way ANOVA, it is only possible to identify one of two factors as significant.

11.69 True or False In a two-way ANOVA, if there is evidence of a significant interaction, then the analysis stops.

11.70 Short Answer In a two-way ANOVA, list the degrees of freedom associated with the F critical value for each test.
(a) Test for an interaction effect.
(b) Test for an effect due to factor A.
(c) Test for an effect due to factor B.

11.71 Short Answer In a two-way ANOVA, explain why this analysis fails if the number of observations per cell is $n = 1$.

Practice

11.72 Consider the data in the table in the context of a two-way ANOVA. EX11.72

			Factor B			
		1		2		3	
Factor A	1	5	7	9	8	9	14	14	12	11
	2	8	13	7	10	11	11	16	17	15

(a) Find a, the number of levels for factor A; b, the number of levels for factor B; and n, the number of observations for each combination of levels.
(b) Find $t_{ij\cdot}$ for $i = 1, 2$ and $j = 1, 2, 3$.
(c) Find $t_{i\cdot\cdot}$ for $i = 1, 2$. Find $t_{\cdot j\cdot}$ for $j = 1, 2, 3$. Find $t_{\cdot\cdot\cdot}$.

11.73 Consider the data in the table in the context of a two-way ANOVA. EX11.73

				Factor B			
			1			2		
Factor A	1	1.5	2.5	0.7	1.8	3.1	4.4	4.1	3.9
	2	3.6	2.8	2.8	3.4	5.7	4.7	7.0	6.0
	3	1.0	2.1	1.5	1.6	4.6	3.2	2.0	4.5
	4	3.8	2.6	3.3	1.3	4.4	4.1	3.6	4.8

(a) Find $t_{i\cdot\cdot}$ for $i = 1, 2, 3, 4$, the sample total for each level of factor A. Find $t_{\cdot j\cdot}$ for $j = 1, 2$, the sample total for each level of factor B. Find $t_{ij\cdot}$, the sample total for each cell. Finally, find $t_{\cdot\cdot\cdot}$, the grand total.
(b) Find $\sum_{i=1}^{4}\sum_{j=1}^{2}\sum_{k=1}^{4} x_{ijk}^2$.
(c) Compute SST, SSA, SSB, SS(AB), and SSE.

11.74 Consider the data in the table in the context of a two-way ANOVA. EX11.74

			Factor B				
		1		2			3	
Factor A	1	11.4	16.4	9.6	4.9	9.6	12.6	17.7	16.8	14.9
	2	8.9	8.5	11.7	8.6	12.7	8.5	6.5	17.3	9.0
	3	8.7	6.4	11.2	9.4	6.6	9.2	8.4	10.0	11.7
	4	14.8	13.9	9.3	13.8	11.7	15.6	19.8	15.3	16.2

(a) Compute SST, SSA, SSB, SS(AB), and SSE.
(b) Compute MSA, MSB, MS(AB), and MSE.
(c) Compute f_A, f_B, and f_{AB}.
(d) Conduct the hypothesis tests to check for effects due to interaction, factor A, and factor B. Use $\alpha = 0.05$ for each test. State your conclusions.

11.75 Consider the data in the table in the context of a two-way ANOVA. EX11.75

			Factor B				
		1		2		3	
Factor A	1	34.2	33.3	36.3	25.9	29.7	37.6
		30.4	27.9	28.5	34.7	38.9	35.6
	2	45.5	39.7	37.5	37.9	33.2	45.0
		38.3	35.9	42.4	37.5	40.1	44.5
	3	31.0	35.1	38.3	35.7	41.4	39.8
		29.5	39.1	36.3	33.5	39.0	36.7

(a) Complete an ANOVA summary table.
(b) Conduct the hypothesis tests to check for effects due to interaction, factor A, and factor B. Use $\alpha = 0.01$ for each test. State your conclusions.

11.76 Complete the summary ANOVA table.

Source of variation	Sum of squares	Degrees of freedom	Mean square	F	p Value
Factor A		4	40.22		
Factor B		2			
Interaction	144.23				
Error	1206.87	75			
Total	1670.28	89			

Conduct the hypothesis tests to check for effects due to interaction, factor A, and factor B. Use $\alpha = 0.05$ for each test. State your conclusions.

11.77 Complete the summary ANOVA table.

Source of variation	Sum of squares	Degrees of freedom	Mean square	F	p Value
Factor A	121.25			1.57	
Factor B	91.12	4			
Interaction					
Error	1446.41	75			
Total	1833.33	99			

Conduct the hypothesis tests to check for effects due to interaction, factor A, and factor B. Use $\alpha = 0.05$ for each test. State your conclusions.

Applications

11.78 Public Health and Nutrition The food editor for *The Boston Herald* conducted a study to examine the quality of wine (measured by age only) at local restaurants. Independent random samples of three types of wines (red, white, and rosé) were obtained from four exclusive restaurants (The Federalist, Top of the Hub, Capital Grille, and Azure). The age (in years) of each wine was recorded. Consider the partial ANOVA summary table.

Source of variation	Sum of squares	Degrees of freedom	Mean square	F	p Value
Type	90.14	2			
Restaurant	266.49	3			
Interaction	56.59				
Error		48			
Total	1128.82	59			

(a) Complete the ANOVA table.
(b) What was the total number of observations?
(c) Conduct the hypothesis test for any effect due to interaction. Use $\alpha = 0.05$. What does your conclusion imply about the other two hypothesis tests?
(d) Conduct the hypothesis tests for factor effects. Use $\alpha = 0.05$. State your conclusions.

11.79 Business and Management A temp agency recently conducted a study to determine the effects of job type and gender on length of employment. Independent random samples of employees who worked in service, technology, sales, security, and labor were obtained. The length (in weeks) of each assignment was recorded. Consider the partial ANOVA summary table.

Source of variation	Sum of squares	Degrees of freedom	Mean square	F	p Value
Gender	16.33	1			
Job type	184.39	4			
Interaction		4			
Error	202.42	50			
Total	422.48				

(a) Complete the ANOVA table.
(b) Is there any evidence of interaction? Use $\alpha = 0.01$.
(c) Is there any evidence that gender or job type affects the length of employment? Use $\alpha = 0.01$.

11.80 Biology and Environmental Science Many factors affect feed intake of chickens and, therefore, egg production. For example, lighting, noise, and water may all affect the efficiency of poultry production. A study was conducted to determine whether the feed intake of chickens (in kg/day) is affected by flock size and/or temperature. **FEED**

(a) Construct a two-way ANOVA summary table.
(b) Use $\alpha = 0.05$ to test the following hypotheses: (i) there is no effect due to interaction; (ii) there is no effect due to flock size; and (iii) there is no effect due to temperature.

11.81 Sports and Leisure Tailgate parties before college football games have become long and lavish, and some present an added security risk. A study was conducted to determine whether the length of a tailgate party is affected by the outside temperature and/or by the college team. The data presented in the table are the lengths (in hours) of randomly selected tailgate parties (before a game) by school and temperature (C, cold; M, moderate; H, hot). **TAILGATE**

		School							
		Michigan		Miami		Ohio St		UCLA	
Temperature	C	5.1	5.0	4.5	7.0	3.4	4.6	2.9	2.5
		6.0	2.2	6.6	2.8	0.2	3.6	3.1	3.6
	M	5.9	5.5	4.7	3.4	3.1	5.7	6.7	7.4
		4.4	4.9	5.4	7.9	5.0	7.2	5.3	5.8
	H	4.8	2.8	6.5	5.1	6.3	5.8	4.4	5.8
		4.7	3.4	7.5	5.3	8.0	8.5	9.8	7.2

Use $\alpha = 0.05$ to test the following hypotheses: (a) there is no effect due to interaction; (b) there is no effect due to temperature; and (c) there is no effect due to school. Include the ANOVA summary table.

11.82 Public Policy and Political Science Austin, Texas, has special procedures in place to streamline the review of commercial and residential building proposals. A random sample of proposals was obtained and classified by type of development and by the city office conducting the review. The length of the review process (in business days) for each proposal is given in the table. **REVIEW**

		Type of development						
		Commercial			Residential			
Office	1	5	18	32	32	30	27	27	30
	2	20	23	23	28	31	26	31	30
	3	33	24	24	20	31	31	26	34

Use $\alpha = 0.05$ to test the following hypotheses: (a) there is no effect due to interaction; (b) there is no effect due to type of development; and (c) there is no effect due to office. Include the ANOVA summary table.

11.83 Travel and Transportation Many *snowbirds*, people who live in the northeast during the summer and in Florida during the winter, ship their vehicles back and forth. However, automobile transportation costs are affected by a number of factors. In a study conducted by Corsia Logistics, a random sample of automobiles shipped from Boston to Miami was obtained for each combination of type of car (compact, midsize, full-size, SUV, pickup) and transport type (open, enclosed). The cost for each (in dollars) was recorded. Conduct an analysis of variance with $\alpha = 0.01$ to test for interaction and factor effects. State your conclusions. **SNOWBIRD**

11.84 Biology and Environmental Science Wetlands, sometimes called the *nurseries of life*, are home to thousands of species of plants and animals and are found on every continent except Antarctica. For example, Canada has approximately 130 million hectares of wetlands, representing approximately 14% of the land area in Canada. Suppose that in a recent Department of Agriculture study, a random sample of wetlands in the United States was obtained for each combination of state and cover type. The cover type classifications are defined by the U.S. Fish and Wildlife service and are determined by the representative plant species. The area covered by water was measured (in hectares), and the data are given in the table. WETLAND

		\multicolumn{8}{c}{State}
		\multicolumn{2}{c}{LA}	\multicolumn{2}{c}{MS}	\multicolumn{2}{c}{CA}	\multicolumn{2}{c}{AR}
Cover type	AB3	0.8	1.1	1.2	0.7	1.2	1.8	3.3	0.2
	FO1	2.0	1.7	0.5	0.1	2.4	3.0	1.4	1.5
	OW	2.6	0.6	2.7	1.6	1.3	1.8	2.7	1.8
	SS1	2.7	2.3	1.1	0.8	1.3	2.4	1.8	1.3

(a) Construct the ANOVA summary table. Is there any evidence of an effect due to interaction of state and cover type? Use $\alpha = 0.01$.
(b) Is there any evidence of an effect due to state or cover type? Use $\alpha = 0.01$ for both hypothesis tests. State your conclusions.

11.85 Sports and Leisure The Genesis Diving Institute of Florida certifies scuba divers using several different systems of education. This organization recently studied the time spent underwater by scuba divers exploring caves and in open water. Each diver was also classified by age group. Independent dives were selected at random for each combination of age group and dive type. The data (in minutes) are given in the table. DIVER

		\multicolumn{7}{c}{Dive type}
		\multicolumn{3}{c}{Cave}	\multicolumn{4}{c}{Open water}
Age group	20–30	39	38	41	39	42	40	40	39
		41	42	40	40	37	45	36	40
	30–40	43	40	38	36	45	39	43	43
		41	46	40	37	47	50	46	44
	40–50	42	41	50	47	37	40	40	38
		43	38	46	47	34	45	34	41
	≥ 50	38	34	35	36	33	37	42	38
		40	38	42	38	30	39	41	39

Construct the two-way ANOVA summary table. Interpret the results. Use $\alpha = 0.001$ for each hypothesis test.

11.86 Public Health and Nutrition The PieCaken is an all-in-one holiday dessert. The original PieCaken consists of a pecan pie, a pumpkin pie, and an apple upside-down cake layered together and topped with cinnamon buttercream frosting. You can imagine the calories in a single slice, and that there are now many imitations. Independent slices of PieCakens were obtained for various combinations of pies and cakes. The number of calories in a single slice was recorded for each. Conduct an analysis of variance with $\alpha = 0.01$ to test for interaction and factor effects. State your conclusions. PIECAKEN

Extended Applications

11.87 Travel and Transportation The Swedish Highway Department conducted a study to examine the speed of cars on major highways. Three regions were selected and rated using four different quality scores (based on retroreflective properties) for road markings. (The four classes of road markings in Sweden are K1, certainly approved; K2, probably approved; K3, probably rejected; and K4, certainly rejected.) The speed of randomly selected cars (in km/hr) was recorded. The sample totals for each cell are given in the table ($n = 4$), and SST = 1680.66. SEHWY

		\multicolumn{4}{c}{Road marking quality}
		K1	K2	K3	K4
Region	North	402.4	393.4	386.6	353.3
	Mälardalen	398.6	418.3	381.6	374.2
	Mitt	419.4	425.3	385.2	386.3

Use $\alpha = 0.05$ to test the following hypotheses: (a) there is no effect due to interaction; (b) there is no effect due to region; and (c) there is no effect due to road-marking quality.

11.88 Psychology and Human Behavior The FBI studied the amount of money stolen (in thousands of dollars) during randomly selected bank robberies at different types of banks and in different locations. The sample totals for each combination of bank type and location are given in the table ($n = 3$) and $\sum_{i=1}^{3}\sum_{j=1}^{4}\sum_{k=1}^{3} x_{ijk}^2 = 2514.28$. ROBBERY

		\multicolumn{4}{c}{Bank type}
		Commercial	Savings	Savings & loan	Credit union
Location	City	25.2	23.7	26.8	24.6
	Suburban	32.2	20.2	24.4	31.6
	Rural	20.5	22.5	16.0	22.0

Use $\alpha = 0.01$ to test the following hypotheses: (a) there is no effect due to interaction; (b) there is no effect due to location; and (c) there is no effect due to bank type. Construct the ANOVA summary table.

11.89 Medicine and Clinical Studies A low hemoglobin level may be an indication of anemia, an iron deficiency, or even kidney disease. A high hemoglobin level could indicate lung disease or extreme physical exercise. In a study conducted by a new testing facility, a random sample of adults was obtained for each combination of gender and altitude. The hemoglobin level (in grams per liter) for each person was recorded. HEMOGLOB
(a) Construct the summary ANOVA table. Is there any evidence of an effect due to interaction of gender and altitude? Use $\alpha = 0.05$.
(b) Is there any evidence of an effect due to gender or altitude? Use $\alpha = 0.05$ for both hypothesis tests.

11.90 Public Health and Nutrition Cerner Corporation recently presented its vision of future hospital rooms, which includes more technology to improve medical efficiency and decrease the chance of human error. In addition to a system that scans staff IDs and a monitor that displays all relevant patient medical information, another consideration is the size of the room. Suppose a random sample of existing hospital rooms was obtained, and the amount of square feet per patient in each room was measured. Each room was classified by location and type, and the data are given in the table. **HOSPITAL**

		Room type					
		Private		Semiprivate two beds		Semiprivate four beds	
Location	City	206	190	210	244	182	183
		263	212	233	201	142	240
		217	205	185	213	177	176
		206	203	228	192	143	200
	Suburban	198	191	204	216	170	195
		197	179	178	200	124	187

(a) Construct the ANOVA summary table. Is there any evidence of an effect due to interaction of location and hospital room type? Use $\alpha = 0.05$.

(b) Is there any evidence of an effect due to location or hospital room type? Use $\alpha = 0.05$ for both hypothesis tests. State your conclusions.

(c) Does this analysis suggest there is one combination of location and room type that is, on average, the smallest in square feet per patient? If so, what is the location and room type?

11.91 Medicine and Clinical Studies A research study in Nova Scotia, Canada, suggested that several factors affect the age at diagnosis of autism spectrum disorder (ASD). Suppose an additional study was conducted to determine whether age at ASD diagnosis is affected by county and a diagnosis of attention-deficit/hyperactivity disorder (ADHD). A random sample of newly diagnosed ASD children was obtained for each combination of county and ADHD diagnosis (1 = no; 2 = yes), and the age of each child was recorded. **ASD**

(a) Construct the ANOVA summary table. Is there any evidence of an effect due to interaction of county and ADHD diagnosis? Use $\alpha = 0.05$.

(b) Is there any evidence of an effect due to county or ADHD diagnosis? Use $\alpha = 0.05$ for both hypothesis tests. State your conclusions.

(c) Early diagnosis of ASD is important to begin intervention services. Is diagnosis of ADHD associated with a decrease or increase in age at diagnosis for ASD? Justify your answer.

Chapter 11 Summary

One-Way ANOVA

Assumptions

1. The k population distributions are normal.
2. The k population variances are equal; that is, $\sigma_1^2 = \sigma_2^2 = \cdots = \sigma_k^2$.
3. The samples are selected randomly and independently from the respective populations.

A one-way, or single-factor, analysis of variance is a statistical technique used to determine whether there is any difference among k population means.

The sums of squares

The fundamental identity is SST = SSA + SSE.

$$\text{SST} = \text{total sum of squares} = \sum_{i=1}^{k}\sum_{j=1}^{n_i}(x_{ij} - \bar{x}_{..})^2 = \left(\sum_{i=1}^{k}\sum_{j=1}^{n_i}x_{ij}^2\right) - \frac{t_{..}^2}{n}$$

$$\text{SSA} = \text{sum of squares due to factor} = \sum_{i=1}^{k}n_i(\bar{x}_{i.} - \bar{x}_{..})^2 = \left(\sum_{t=1}^{k}\frac{t_{i.}^2}{n_i}\right) - \frac{t_{..}^2}{n}$$

$$\text{SSE} = \text{sum of squares due to error} = \sum_{i=1}^{k}\sum_{j=1}^{n_i}(x_{ij} - \bar{x}_{i.})^2 = \text{SST} - \text{SSA}$$

One-way ANOVA summary table

Source of variation	Sum of squares	Degrees of freedom	Mean square	F	p Value
Factor A	SSA	$k-1$	$\text{MSA} = \dfrac{\text{SSA}}{k-1}$	$\dfrac{\text{MSA}}{\text{MSE}}$	p
Error	SSE	$n-k$	$\text{MSE} = \dfrac{\text{SSE}}{n-k}$		
Total	SST	$n-1$			

Hypothesis test

$H_0: \mu_1 = \mu_2 = \cdots = \mu_k$ (all k population means are equal)

$H_a: \mu_i \neq \mu_j$ for some $i \neq j$ (at least two population means differ)

TS: $F = \dfrac{\text{MSA}}{\text{MSE}}$

RR: $F \geq F_{\alpha, k-1, n-k}$

Bonferroni multiple comparison procedure

The c simultaneous $100(1-\alpha)\%$ Bonferroni confidence intervals have the following values as endpoints:

$$(\bar{x}_{i.} - \bar{x}_{j.}) \pm t_{\alpha/(2c), n-k} \cdot \sqrt{\text{MSE}} \cdot \sqrt{\frac{1}{n_i} + \frac{1}{n_j}} \quad \text{for all } i \neq j$$

Tukey multiple comparison procedure

The c simultaneous $100(1-\alpha)\%$ Tukey confidence intervals have the following values as endpoints:

$$(\bar{x}_{i.} - \bar{x}_{j.}) \pm \frac{1}{\sqrt{2}} \cdot Q_{\alpha, k, n-k} \cdot \sqrt{\text{MSE}} \cdot \sqrt{\frac{1}{n_i} + \frac{1}{n_j}} \quad \text{for all } i \neq j$$

Two-Way ANOVA

A two-way analysis of variance is a statistical procedure used to determine the effect of two factors on a response variable.

The sums of squares

The fundamental identity is $\text{SST} = \text{SSA} + \text{SSB} + \text{SS(AB)} + \text{SSE}$.

SST = total sum of squares

$$= \sum_{i=1}^{a} \sum_{j=1}^{b} \sum_{k=1}^{n} (x_{ijk} - \bar{x}_{...})^2 = \left(\sum_{i=1}^{a} \sum_{j=1}^{b} \sum_{k=1}^{n} x_{ijk}^2 \right) - \frac{t_{...}^2}{abn}$$

SSA = sum of squares due to factor A

$$= bn \sum_{i=1}^{a} (\bar{x}_{i..} - \bar{x}_{...})^2 = \frac{\sum_{i=1}^{a} t_{i..}^2}{bn} - \frac{t_{...}^2}{abn}$$

SSB = sum of squares due to factor B

$$= an \sum_{j=1}^{b} (\bar{x}_{.j.} - \bar{x}_{...})^2 = \frac{\sum_{j=1}^{b} t_{.j.}^2}{an} - \frac{t_{...}^2}{abn}$$

SS(AB) = sum of squares due to interaction

$$= n \sum_{i=1}^{a} \sum_{j=1}^{b} (\bar{x}_{ij.} - \bar{x}_{i..} - \bar{x}_{.j.} + \bar{x}_{...})^2 = \frac{\sum_{i=1}^{a} \sum_{j=1}^{b} t_{ij.}^2}{n} - \frac{\sum_{i=1}^{a} t_{i..}^2}{bn} - \frac{\sum_{j=1}^{b} t_{.j.}^2}{an} + \frac{t_{...}^2}{abn}$$

SSE = sum of squares due to error

$$= \sum_{i=1}^{a}\sum_{j=1}^{b}\sum_{k=1}^{n}(x_{ijk}-\bar{x}_{ij.})^2 = \text{SST} - \text{SSA} - \text{SSB} - \text{SS(AB)}$$

Two-way ANOVA summary table

Source of variation	Sum of squares	Degrees of freedom	Mean square	F	p Value
Factor A	SSA	$a-1$	$\text{MSA} = \dfrac{\text{SSA}}{a-1}$	$F_A = \dfrac{\text{MSA}}{\text{MSE}}$	p_A
Factor B	SSB	$b-1$	$\text{MSB} = \dfrac{\text{SSB}}{b-1}$	$F_B = \dfrac{\text{MSB}}{\text{MSE}}$	p_B
Interaction	SS(AB)	$(a-1)(b-1)$	$\text{MS(AB)} = \dfrac{\text{SS(AB)}}{(a-1)(b-1)}$	$F_{AB} = \dfrac{\text{MS(AB)}}{\text{MSE}}$	p_{AB}
Error	SSE	$ab(n-1)$	$\text{MSE} = \dfrac{\text{SSE}}{ab(n-1)}$		
Total	SST	$abn-1$			

Hypothesis tests

1. Test for an interaction effect

   H_0: There is no interaction effect.

   H_a: There is an effect due to interaction.

   TS: $F_{AB} = \dfrac{\text{MS(AB)}}{\text{MSE}}$

   RR: $F_{AB} \geq F_{\alpha,(a-1)(b-1),ab(n-1)}$

2. Test for an effect due to factor A

   H_0: There is no effect due to factor A.

   H_a: There is an effect due to factor A.

   TS: $F_A = \dfrac{\text{MSA}}{\text{MSE}}$

   RR: $F_A \geq F_{\alpha,a-1,ab(n-1)}$

3. Test for an effect due to factor B

   H_0: There is no effect due to factor B.

   H_a: There is an effect due to factor B.

   TS: $F_B = \dfrac{\text{MSB}}{\text{MSE}}$

   RR: $F_B \geq F_{\alpha,b-1,ab(n-1)}$

Chapter 11 Exercises

Applications

11.92 Sports and Leisure Backing a trailer down a boat ramp to launch a boat can be awkward and tricky. A steep ramp angle often makes this task even more intimidating, especially for new boat owners. Independent random samples of boat ramps in various regions of Florida were obtained, and the angle of each was measured (in degrees). Consider the partial ANOVA table.

One-way ANOVA summary table

Source of variation	Sum of squares	Degrees of freedom	Mean square	F	p Value
Factor		4			
Error	470.355	45			
Total	502.650	50			

(a) Complete the ANOVA summary table.
(b) How many regions in Florida were considered? How many total observations were obtained?
(c) Is there any evidence that the population mean boat-ramp angle differs among these populations? Use $\alpha = 0.05$.

11.93 Physical Sciences It can get very hot for actors on stage. Not only do they have to remember all of their lines, but theater spotlights are also usually bright and intense. Independent random samples of spotlights were obtained from four different Broadway theaters (Cort Theater, Imperial Theater, Majestic Theater, and New Amsterdam Theater). The wattage of each light was measured, and a partial ANOVA summary table is shown.

One-way ANOVA summary table

Source of variation	Sum of squares	Degrees of freedom	Mean square	F	p Value
Factor	106568	3			
Error		20			
Total	258565				

Complete the table and use this information to determine whether there is any evidence that the population mean wattage of spotlights is different among the four theaters. Use $\alpha = 0.05$.

11.94 Biology and Environmental Science Ethiopia is building the $5 billion Grand Ethiopian Renaissance Dam (GERD) on the Blue Nile. When construction is complete, it will be the largest dam in Africa, will generate approximately 600 mW of electricity per year, and could cause a power shift in the region where water is a lifeline. Independent random samples of locations in four Nile River countries were obtained and the depth of the river was measured at each (in feet). Construct the one-way ANOVA table and test the hypothesis of no difference in population mean depths due to country. Use $\alpha = 0.01$. DAM

11.95 Manufacturing and Product Development Microlithography is a process that includes baking a semiconductor wafer. An experiment was conducted to study the temperature of a 200-mm-diameter wafer at different locations on the wafer during a new baking process. Independent random samples of wafers were selected, and a temperature sensor was placed on each wafer at one of four distances from the center of the wafer. The temperature was recorded (in degrees Celsius) 80 seconds into the process. The summary statistics are given in the table.

Location	T1	T2	T3	T4
Sample size	10	10	10	10
Sample mean	106.7	104.6	104.3	98.6

A one-way analysis of variance test was significant at the $p = 0.009$ level. Use MSE = 26.92 to find the Bonferroni 95% confidence intervals for all pairwise differences. Draw a graph to represent the results.

11.96 Technology and Internet An experiment was conducted to compare the accuracy of five different computer algorithms for translating Chinese into English. Each algorithm was tested 25 times on randomly selected passages, and an accuracy score (between 0 and 1) was recorded for each trial. A one-way analysis of variance test was significant at the $p = 0.003$ level. Use MSE = 0.0276 and the summary statistics to find the Tukey 95% confidence intervals for all pairwise differences. Draw a graph to represent the results.

Algorithm	OO	CP	SLCP	AC	T
Sample mean	0.5566	0.7020	0.6190	0.7023	0.7115

11.97 Manufacturing and Product Development Using a new atomic imaging technique, a study was conducted to measure randomly selected thin films of NaCl on a metal substrate. Four different growth modes were used, and the step height was measured (in nanometers) in each case. NACL
(a) Conduct a one-way analysis of variance test to determine whether there is an overall difference in population mean step heights. Use $\alpha = 0.05$.
(b) Construct the Bonferroni 99% confidence intervals to isolate any population means contributing to an overall difference, and draw a graph to represent the results.

11.98 Public Health and Nutrition A root canal is an endodontic procedure to remove the infected or damaged nerve tissue of a tooth. The American Dental Association conducted a study to determine whether the major canal diameter of a tooth is related to the tooth type and/or the patient's race. Independent random samples of adult male root canal patients were selected for each combination of tooth type and race. The canal diameters (in millimeters) are given in the table. ROOT

		Tooth type		
Group		Maxillary		Mandibular	
White	0.97	1.00	1.09	1.39
	0.75	1.06	1.17	1.19
African American	1.17	1.00	1.21	1.48
	1.09	0.88	0.94	1.16
Native American	0.86	1.03	1.10	1.45
	0.77	0.88	1.58	1.18
Hispanic	1.13	0.81	0.90	1.01
	0.78	0.86	1.57	1.42

Construct the two-way ANOVA summary table. Interpret the results. Use $\alpha = 0.05$ for each hypothesis test.

11.99 Travel and Transportation Many factors affect the number of automobile accidents on major highways—for example, increased traffic, weather conditions, and driver fatigue. A recent study was conducted to determine whether the frequency of median crashes is related to the number of lanes and/or the differential elevation of opposite travel lanes. Independent random samples of U.S. highways were obtained for each combination of lanes and differential elevation. The number of median crashes over a 6-month time period was recorded for each. Construct a two-way ANOVA table. Interpret the results. Use $\alpha = 0.05$ for each hypothesis test. CRASH

11.100 Medicine and Clinical Studies A recent study was conducted to estimate the effects of socioeconomic status—for example, place of residence, household wealth, and marital status—on body mass index (BMI).[4] Suppose independent random samples of 15- to 49-year-old women in the United States were obtained. The highest degree earned and the BMI (in kg/m^2) for each was measured. DEGREES
(a) Conduct an analysis of variance test to show that there is evidence of an overall difference in population mean BMI. Use $\alpha = 0.01$. Find the p value.
(b) Construct the Tukey 95% confidence intervals to isolate the pairs of means contributing to the overall difference.
(c) Draw a graph to represent the results of the multiple comparison procedure in part (b).

Extended Applications

11.101 Physical Sciences There are approximately 400 nuclear power plants in operation around the world, with an estimated power production of 390 GW per year.[5] Independent random samples of nuclear power plants in three countries were obtained, and the net generating capacity per year (in MW) of each was obtained. Summary statistics are given in the table.

Country	Sample size	Sample total	Sum of squared observations
United States	5	5848	6861942
France	6	6470	7623250
Germany	5	5212	5726814

Conduct the appropriate test to check the hypothesis of no difference in population mean net generating capacity due to country. Use $\alpha = 0.01$.

11.102 Sports and Leisure The biggest catfish caught in the Catfish Chasers Tournament in Burwell, Nebraska, in August 2018 was a record 180.3 lb.[6] In preparation for the next catfish tournament, a fisherman (familiar with statistics) randomly selected anglers from five different locations and recorded the weight (in pounds) of their last catfish caught. The data (in pounds) are given in the table. CATFISH

Location	Observations					
Hill's Landing	22.7	14.1	9.6	5.9	21.5	28.7
	45.7	17.4				
Eagle Nest	38.7	44.6	34.6	19.9	59.3	23.2
	28.3	35.5	38.8	34.5	39.1	
Santee	37.5	53.2	47.0	31.8	19.9	35.9
	50.0	27.6	30.6	22.4		
Campground	23.7	32.1	38.9	31.1	34.2	37.2
	49.8	56.6	52.2	34.2	44.5	38.9
Rock Hill	42.0	41.6	63.2	41.6	25.5	28.4
	33.8					

(a) Conduct a one-way analysis of variance test to determine whether there is an overall difference in population mean catfish weights. Use $\alpha = 0.01$.
(b) Construct the Tukey 99% confidence intervals to isolate any population means contributing to an overall difference, and draw a graph to represent the results. Which site would you recommend for the fisherman to have a good chance at winning the tournament? Why?

11.103 Biology and Environmental Science We all expect hospitals to be clean, almost spotless. However, according to the U.S. Centers for Disease Control and Prevention, approximately 1 million people contract an infection directly from a hospital visit each year. A study suggested that hospital lobbies may have high levels of airborne particles, including bacteria and fungi. Independent random samples of hospital lobbies were obtained for each combination of season (1, winter; 2, spring; 3, summer; 4, fall) and lobby size (1, small; 2, medium; 3, large). The level of airborne bacteria was measured in each (in CFU/m^3). BACTERIA
(a) Construct the two-way ANOVA summary table. Interpret the results. Use $\alpha = 0.05$ for each hypothesis test.
(b) Suppose more airborne bacteria in the lobby means there is a greater chance of contracting an infection. Is there one season in which we should avoid going to a hospital? Why?
(c) Similarly, is there a specific size of hospital lobby to avoid? Why?

11.104 Marketing and Consumer Behavior A real-estate agent conducted a study to compare the effect of location and season on weekly time-share costs (in dollars). The sample totals for each combination of island (A, Aruba; M, Martinique;

SK, St. Kitts; SL, St. Lucia) and season are given in the table ($n = 6$) and SST $= 487,902,980.64$.

		Season			
		Spring	Summer	Fall	Winter
Island	A	18045	12168	19894	26214
	M	14495	1925	13890	17538
	SK	18075	32887	24398	18834
	SL	25457	20757	27505	36428

Use $\alpha = 0.01$ to test the following hypotheses: (a) there is no effect due to interaction; (b) there is no effect due to island; and (c) there is no effect due to season.

11.105 Biology and Environmental Science A random sample of male baby California sea lions was captured, tagged, and monitored for several years. The weight of each sea lion was measured (in pounds) once, and each sea lion was classified by species ID and age (in years). The sample totals are given in the table ($n = 4$) and SST $= 103,152.38$.

		Age		
		1	2	3
Island	50011	1768	1735	1752
	50023	1794	1563	1517
	50037	1750	1742	1328

Use $\alpha = 0.05$ to test the following hypotheses: (a) there is no effect due to interaction; (b) there is no effect due to species; and (c) there is no effect due to age.

11.106 Free Radicals Glutathione is an antioxidant found in every cell and is important to our health because it neutralizes free radicals. This antioxidant is believed to contribute to our immune system and is found naturally in many foods. However, some research suggests that common, over-the-counter pain medications may lower the body's glutathione level. Independent random samples of healthy male adults and of those taking medication for ordinary aches were obtained. The glutathione level in each was measured (in micrograms/10(10) erythrocytes). GLUTATH
(a) Conduct a one-way analysis of variance test to determine whether there is an overall difference in population mean glutathione levels. Use $\alpha = 0.05$.
(b) Construct the Tukey 95% confidence intervals to isolate any population means contributing to an overall difference. Draw a graph to represent the results.
(c) Because we are interested mainly in whether pain relievers affect glutathione levels in healthy males, find the Bonferroni 95% confidence intervals only for the differences between each pain reliever mean and the healthy adult mean. Based on these results, do you believe pain relievers lower glutathione levels? Justify your answer.

11.107 Public Health and Nutrition A consumer group recently studied the amount of partially hydrogenated oils (trans fat) in various peanut butter brands. Four brands were selected in smooth and chunky varieties. The amount of trans fat was measured (in grams) per serving (two tablespoons) in each randomly selected jar, and the data are given in the table. TRANSFAT

	Peanut butter brand							
	Jif		Peter Pan		Skippy		Smucker's	
Smooth	0.51	0.72	0.89	0.69	0.75	0.85	0.73	0.71
Chunky	0.46	0.63	0.88	0.81	0.76	0.69	0.73	0.76

(a) Construct the two-way ANOVA summary table. Interpret the results. Use $\alpha = 0.05$ for each hypothesis test.
(b) If a consumer wants to avoid trans fat as much as possible, which combination of brand and style would you recommend? Why?

ASphoto777/Deposit Photos

CHAPTER APP

11.108 Selenium Concentration The four largest grain producing U.S. states are Kansas, North Dakota, Washington, and Montana. Random samples of wheat farms in each state were obtained, and the selenium concentration in a soil sample was measured (in $\mu g/g$). The summary statistics are shown in the table. TRACESE

State	Sample size	Sample mean	Sample standard deviation
Kansas	15	3.420	1.112
North Dakota	15	5.413	1.262
Washington	15	5.187	0.966
Montana	15	4.060	1.234

(a) Is there any evidence to suggest that at least two of the population mean selenium concentrations are different?
(b) Construct the Bonferroni 95% confidence intervals and use them to isolate the pair(s) of means contributing to this overall experiment difference.
(c) Construct a graph to summarize the results from the Bonferroni multiple comparison procedure.
(d) Construct the Tukey 95% confidence intervals and use them to isolate the pair(s) of means contributing to this overall experiment difference.
(e) Do your results in parts (b) and (d) agree? If so, why? If not, why not?

Correlation and Linear Regression

Rob Arnold/Alamy

◀ Looking Back
- Recall the properties and the methods to compute probabilities associated with a normal distribution: **Section 6.3**.
- Remember the procedures for conducting hypothesis tests and constructing confidence intervals based on a *t* distribution: **Sections 8.3 and 9.5**.
- Remember the procedures for conducting hypothesis tests based on an *F* distribution: **Sections 10.5, 11.1 and 11.3**.

Looking Forward ▶
- Learn the concepts of simple linear regression, the principle of least squares, and how to test for a significant regression: **Section 12.1**.
- Learn the hypothesis tests associated with simple linear regression and how to measure the strength of a linear relationship between two variables: **Section 12.2**.
- Learn how to make an inference concerning the mean value and an observed value in the context of simple linear regression: **Section 12.3**.
- Construct and interpret diagnostic graphs associated with the simple linear regression assumptions: **Section 12.4**.
- Adapt and extend estimation procedures and hypothesis tests to multiple linear regression: **Section 12.5**.

CHAPTER APP

Wind Turbine Noise

The Waleny Extension, the world's largest wind farm, is located in the Irish Sea, 12 miles off the coast of England. The 87 turbines, each approximately 640 ft tall, occupy an area bigger than San Francisco and have a generating capacity of 659 megawatts (MW). They produce clean, renewable energy and are part of a growing trend of investment in offshore wind farms.

Despite the environmental appeal and favorable wind conditions, wind farms cannot be constructed near many towns because of noise-level regulations. Some evidence also suggests that noise caused by wind farms can cause illness, or even *wind farm syndrome.* For example, a typical wind turbine produces approximately 50 decibels (dB) of sound 100 m away. This is about the same amount of noise produced by a nearby home air conditioner.

Suppose a small coastal community is considering the construction of a wind turbine to generate electricity for the town hall. An experiment is conducted to measure the wind turbine's noise level (in decibels) at various distances (in meters) from the proposed site. The data are summarized in the table.

Distance	10	50	75	120	150	160	200	250	400	500
Noise level	105	95	80	52	58	77	56	57	28	40

We can use the techniques presented in this chapter to determine whether a significant linear relationship exists between distance and noise level. We will see how regression analysis can be used to predict a value of the noise level for a given distance from the wind turbine.

12.1 Simple Linear Regression

Probabilistic Model, Notation, and Example

A *deterministic* relationship between two variables x and y is one in which the value of y is completely determined by the value of x. In general, $y = f(x)$ is a deterministic relationship between x and y. The value of y *depends* on the value of x. The *independent* variable is x and the *dependent* variable is y. We are free to choose x, but y depends on the value selected. For example, $y = x^2 + 2x + 5$ is a deterministic relationship between x and y. If $x = 1$, then y is completely determined and, in this example, $y = 1^2 + 2(1) + 5 = 8$.

Function notation, evaluation.

Linear functions.

Here, the Greek letter β is not related to a type II error, as discussed in Section 9.3.

One of the simplest deterministic relationships between x and y is a linear relationship: $y = \beta_0 + \beta_1 x$, where β_0 and β_1 are constants. The set of ordered pairs (x, y) such that $y = \beta_0 + \beta_1 x$ forms a straight line with slope β_1 and y intercept β_0. For example, the graph of $y = 3 + 7x$ is a straight line with slope 7 and y intercept 3. (Could you sketch the line described by this equation?)

A *probabilistic model* is an extension of a deterministic relationship. For a fixed value x, the value of the second variable is randomly distributed. For example, suppose we are investigating the relationship between a one-way airfare (in dollars) from New York City to Los Angeles and the number of tickets sold on the 10:00 A.M. flight. If we select $x = 200$, then the number of tickets sold (for $x = 200$) is a random variable Y. On one particular day, the observed value of Y associated with $x = 200$ may be $y = 135$ tickets.

There is often a need to predict the value of the dependent variable for a given value of the independent variable. For example, if the price is 250, predict the number of tickets that will be sold. Simple linear regression allows us to find the best prediction equation.

The independent variable, fixed by the experimenter, is usually denoted by x. For a fixed value x, the second variable is randomly distributed. This random variable is the dependent variable and is usually denoted by Y. Consistent with the previous notation, an observed value of Y is denoted by y.

A general additive probabilistic model includes a deterministic part and a random part. The value of Y differs from $f(x)$ (the deterministic part) by a random amount. The model can be written as

$$Y = (\text{deterministic function of } x) + (\text{random deviation})$$
$$= f(x) + E \qquad (12.1)$$

where E is a random variable, called the random error.

A value of the random variable E is denoted by e. Here, e is not the base of the natural logarithm.

Suppose we fix a value of x, say $x = x_0$, and observe a value of Y for this value of x, denoted y. If the value of the random variable E is positive, then $y > f(x_0)$. Similarly, if $e < 0$, then $y < f(x_0)$. And if $e = 0$, then $y = f(x_0)$. Geometrically, the value of y lies

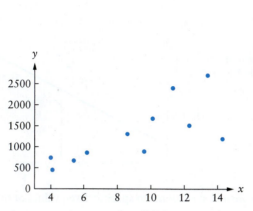

Figure 12.2 A scatter plot of TDS versus sodium level.

Figure 12.3 An R scatter plot of TDS versus sodium level.

TRY IT NOW Go to Exercise 12.13

CHAPTER APP **Wind Turbine Noise**

Construct a scatter plot for the data. What does the plot suggest about the relationship between the variables?

Simple Linear Regression Model

In a **simple linear regression model**, the deterministic function $f(x)$ in Equation 12.1 is assumed to be linear [i.e., $f(x) = \beta_1 + \beta_1 x$]. The graph of this regression equation is a straight line that describes how a dependent variable y changes as an independent variable x changes. **Figure 12.4** shows the graph of a possible deterministic straight line added to the scatter plot in Figure 12.2. Notice that the data points lie close to the line; the vertical distance between a point and the line depends on the value of the random error. Also, the line has positive slope, which conveys the relationship between the two variables: As the sodium level rises, so does the TDS.

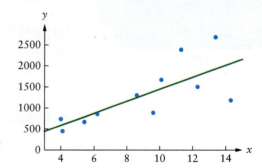

Figure 12.4 TDS versus sodium level scatter plot with a (deterministic) line added.

In just a few pages, we'll define what is meant by *best*.

This section presents a method for finding the best deterministic straight line for a given set of data. We will use hypothesis tests to determine whether there is a significant linear relationship between two variables. As with all other hypothesis tests, certain assumptions must be met for the related statistical procedures to be valid.

either above, below, or on the graph of $y = f(x)$. **Figure 12.1** illustrates the case where $e > 0$.

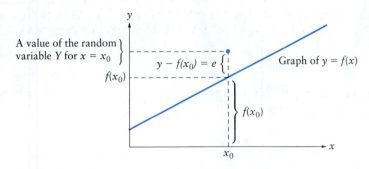

Figure 12.1 An illustration of a probabilistic model. An observed value y is composed of a deterministic part, $f(x_0)$, and a random error e.

Suppose there are n observations on specific, fixed values of the independent variable. Here is the notation we use:

1. The observed values of the independent variable are denoted $x_1, x_2, \ldots, x_n$.
2. Y_i and y_i are the random variable and the observed value of the random variable associated with x_i, for $i = 1, 2, \ldots, n$. Here's the interpretation: For each x_i, there is a corresponding random variable Y_i. So there are really n random variables in these problems.
3. The data set consists of n ordered pairs: $(x_1, y_1), (x_2, y_2), \ldots, (x_n, y_n)$.

Let's use some of the new notation in this example and learn to visualize the relationship between two variables.

EXAMPLE 12.1 Tee Time

Stacey Stout/Shutterstock

GROUNDS

Golf courses in Arizona are watered regularly to keep the fairways and greens pleasing and playable. However, high levels of dissolved salts in the water can harm the course. A groundskeeper is investigating the relationship between the amount of sodium in the water and the total dissolved salts (TDS). A random sample of Arizona golf courses was obtained and a water sample from each was used to measure the amount of sodium (in mEq/liter) and TDS (in mg/L). The data are given in the table.

Sodium, x	4.1	5.4	4.0	6.2	8.6	9.6	14.3	10.1	12.3	13.4	11.3
TDS, y	450	672	740	860	1307	891	1190	1676	1512	2700	2398

(a) Identify the independent and dependent variables.
(b) List the ordered pairs in the data set.
(c) Construct a scatter plot for these data. What does the plot suggest about the relationship between the variables?

Solution

(a) The independent variable is sodium, and the dependent variable is TDS. The groundskeeper believes that there is a relationship between these two variables and hopes to predict the TDS as a function of the sodium level.

(b) The data set consists of 11 ordered pairs: (4.1, 450), (5.4, 672), ..., (11.3, 2398). For each (fixed) value of sodium, there is a corresponding observation on a random variable for TDS.

(c) **Figure 12.2** is a scatter plot of TDS (y) versus sodium (x). This plot suggests that as the sodium level increases, the TDS also increases, so the relationship might be positive linear. **Figure 12.3** shows a technology solution, a similar scatter plot produced using R.

Simple Linear Regression Model

Let $(x_1, y_1), (x_2, y_2), \ldots, (x_n, y_n)$ be n pairs of observations such that y_i is an observed value of the random variable Y_i. We assume that there exist constants β_0 and β_1 such that

$$Y_i = \beta_0 + \beta_1 x_i + E_i$$

where $E_1, E_2, \ldots, E_n$ are independent, normal random variables with mean 0 and variance σ^2. That is,

1. The E_i's are normally distributed, which implies that the Y_i's are normally distributed.
2. The expected value of E_i is 0, which implies that $E(Y_i) = \beta_0 + \beta_1 x_i$.
3. $\text{Var}(E_i) = \sigma^2$, which implies that $\text{Var}(Y_i) = \sigma^2$.
4. The E_i's are independent, which implies that the Y_i's are independent.

A CLOSER LOOK

1. The E_i's are the **random deviations** or **random error terms**.
2. $y = \beta_1 + \beta_1 x$ is the **true regression line**. Each point (x_i, y_i) lies near the true regression line, depending on the value of the random error term e_i.
3. The four assumptions in the simple linear regression model definition can be stated compactly in terms of the random error term: $E_i \overset{\text{ind}}{\sim} N(0, \sigma^2)$.

Just a little more notation is necessary to understand the simple linear regression model and the resulting properties.

Recall, that $Y|x_i$ means "Y given x_i."

Consider each random variable $Y_i = Y|x_i$.

$\mu_{Y|x_i} = E(Y|x_i)$ is the expected value of Y for a fixed value x_i, and $\sigma^2_{Y|x_i}$ is the variance of Y for a fixed value x_i.

The simple linear regression model assumptions imply that

$$\mu_{Y|x_i} = E(\beta_0 + \beta_1 x_i + E_i) = \beta_0 + \beta_1 x_i + E(E_i) = \beta_0 + \beta_1 x_i$$

$$\sigma^2_{Y|x_i} = \text{Var}(\beta_0 + \beta_1 x_i + E_i) = \sigma^2$$

$Y|x_i$ is normal.

Therefore, the mean value of Y is a linear function of x. The true regression line passes through the line of mean values.

The variability in the distribution of Y is the same for every value of x (this is called **homogeneity of variance**).

Figure 12.5 illustrates the model assumptions and resulting properties. Each Y_i has a normal distribution, centered at $\beta_0 + \beta_1 x_i$. All the distributions have the same width, or variance.

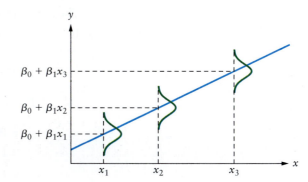

Figure 12.5 The true regression line connects the mean values $\beta_0 + \beta_1 x_i$.

Example: Use the True Regression Line

EXAMPLE 12.2 Get the Lead Out

According to the U.S. Environmental Protection Agency, children are very susceptible to the effects of lead exposure. Children who are exposed to high levels of lead may experience developmental, behavioral, and learning problems.[1] For six-year-old children, suppose that IQ level (y) is related to blood lead level (x, measured in μg/dL) and that the true regression line is $y = 96.8 - 0.45x$.

(a) Find the expected IQ for a 6-year-old child with a blood lead level of 30 μg/dL.

(b) How much change in IQ is expected if the blood lead level increases by 10 μg/dL? What if it decreases by 20 μg/dL?

(c) Suppose $\sigma = 8$ μg/dL. Find the probability that an observed IQ is greater than 100 when the blood lead level is 20 μg/dL.

Solution

(a) We need the expected value of Y for the value $x = 30$.

$E(Y \mid x) = \beta_0 + \beta_1 x$ Simple linear regression model implication.
$E(Y \mid 30) = 96.8 - 0.45(30)$ Use the true regression line with $x = 30$.
$= 83.3$ Simplify.

The expected IQ for a 6-year-old child with a blood level of 30 μg/dL is 83.3. **Figure 12.6** illustrates this result.

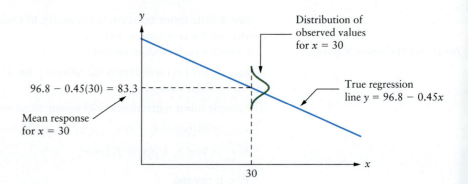

Figure 12.6 The true regression line, the mean response for $x = 30$, and the distribution of observed values around the mean response.

(b) The slope of the true regression line, $\beta_1 = -0.45$, is the change in IQ associated with a 1 μg/dL change in blood lead level.

If the blood lead level increases by 10 μg/dL, the expected change in IQ is (β_1)·(change in x) = $(-0.45)(10) = -4.5$.

If the blood lead level decreases by 20 μg/dL, the expected change in IQ is (β_1)·(change in x) = $(-0.45)(-20) = 9.0$.

The change in x is negative because the blood level is decreasing.

(c) If $\sigma = 8$, then for $x = 20$ the random variable Y is normally distributed with mean $96.8 - 0.45(20) = 87.8$ and variance $\sigma^2 = 8^2 = 64$: $Y \sim N(87.8, 64)$.

Find the probability that Y exceeds 100.

$P(Y > 100) = P\left(\dfrac{Y - 87.8}{8} > \dfrac{100 - 87.8}{8}\right)$ Standardize
$= P(Z > 1.525) \approx P(Z > 1.53)$ Equation 6.8; simplification
$= 1 - P(Z \leq 1.53)$ Use cumulative probability
$= 1 - 0.9370 = 0.0630$ Appendix Table

Figure 12.7 shows a technology solution. For a blood lead level of 20 µg/dL, the probability of observing an IQ greater than 100 in a 6-year-old child is 0.0630. Figure 12.8 illustrates this result.

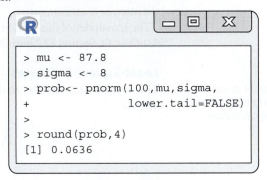

Figure 12.7 Probability calculation using R.

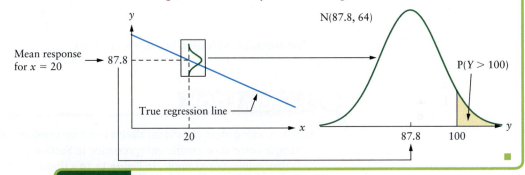

Figure 12.8 The distribution of Y for x = 20.

TRY IT NOW Go to Exercises 12.19 and 12.21

CHAPTER APP Wind Turbine Noise

For distance (x) and noise (y), suppose the true regression line is $y = 90 - 0.1x$.
(a) Find the expected noise level when the distance is 300 m.
(b) Suppose $\sigma = 14$, and find the probability that the noise level is less than 50 dB at a distance of 300 m.

Least-Squares Estimates

Suppose two variables are related via a simple linear regression model. The parameters β_0 and β_1 are usually unknown. However, if we assume that the observations $(x_1, y_1), (x_2, y_2), \ldots, (x_n, y_n)$ are independent, then we can use these sample data to estimate the model parameters β_0 and β_1.

To obtain the **line of best fit**, or **estimated regression line**, we use the **principle of least squares**. Figure 12.9 illustrates this concept: Minimize the sum of the squared

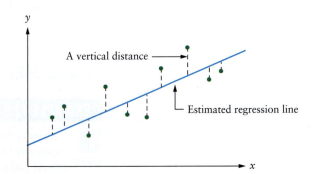

Figure 12.9 An illustration of the principle of least squares.

deviations, or vertical distances from the observed points to the line. Consider the vertical distances from the points $(x_1, y_1), (x_2, y_2), \ldots, (x_n, y_n)$ to the line. The principle of least squares produces an estimated regression line such that the sum of all squared vertical distances is a minimum.

The remainder of this section focuses on the interpretations associated with finding the line of best fit. Section 12.2 introduces several hypothesis tests utilizing these various statistics.

Least-Square Estimates

$\hat{\beta}_0$ and $\hat{\beta}_1$ are estimates of β_0 and β_1, respectively, but $\hat{\beta}_1$ is found first, and its value is used in the calculation of $\hat{\beta}_0$.

The least-square estimates of the y intercept β_0 and the slope β_1 of the true regression line are

$$\hat{\beta}_1 = \frac{n\sum x_i y_i - (\sum x_i)(\sum y_i)}{n\sum x_i^2 - (\sum x_i)^2} \quad (12.2)$$

and

$$\hat{\beta}_0 = \frac{\sum y_i - \hat{\beta}_1 \sum x_i}{n} = \bar{y} - \hat{\beta}_1 \bar{x} \quad (12.3)$$

The estimated regression line is $y = \hat{\beta}_0 + \hat{\beta}_1 x$.

A CLOSER LOOK

1. Before using these equations, always consider creating a scatter plot and compute the sample correlation coefficient (presented in Section 12.2) to make sure a linear model is reasonable. For example, in **Figure 12.10** a linear model seems reasonable; the relationship between x and y appears to be negative linear. In **Figure 12.11**, a linear model is not reasonable; the relationship between the variables appears to be quadratic.

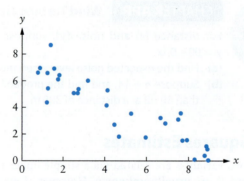

Figure 12.10 A scatter plot of data in which the relationship appears linear.

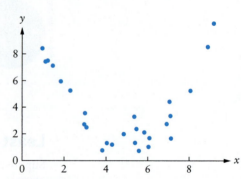

Figure 12.11 A scatter plot in which the relationship does not appear to be linear.

The predictor variable is the independent variable, and the response variable is the dependent variable.

2. If x^* is a specific value of the independent, or predictor, variable x, let $y^* = \hat{\beta}_0 + \hat{\beta}_1 x^*$.
   (a) y^* is an estimate of the mean value of Y for $x = x^*$, denoted $\mu_{Y|x^*}$.
   (b) y^* is also an estimate of an observed value of Y for $x = x^*$.

Example: Estimating Coefficients

EXAMPLE 12.3 Snow Cover

SNOW

Climate change and snow depth may affect permafrost and soil composition, especially in Canada. Researchers conducted a study to examine how snow cover affects soil temperature across northern latitudes. A random sample of locations was obtained, and the snow depth (in centimeters) and the difference between the soil

meunierd/Shutterstock

temperature and the air temperature (in °C) (surface offset) was measured for each. The data are given in the table.[2]

Depth	6.7	41.1	20.9	58.9	18.2	44.6	53.8	23.1	47.2	48.2
Offset	7.0	22.8	8.5	20.9	15.1	14.7	21.8	11.5	17.8	21.3

(a) Find the estimated regression line.

(b) Estimate the true mean surface offset for a snow depth of 45 cm.

Solution

Use the equations to find the least-squares estimates $\hat{\beta}_0$ and $\hat{\beta}_1$. Compute $y^* = \hat{\beta}_0 + \hat{\beta}_1(45)$ as an estimate of the mean value of Y for $x = 20$.

(a) Snow depth (x) is the independent, or predictor, variable, and surface offset (y) is the dependent, or response variable. Find the necessary summary statistics for the $n = 10$ pairs of observations, and use Equations 12.2 and 12.3.

$$\sum x_i = 362.7 \qquad \sum y_i = 161.4 \qquad \sum x_i y_i = 6628.39$$

$$\sum x_i^2 = 15{,}939.65 \qquad \sum y_i^2 = 2900.02$$

$$\hat{\beta}_1 = \frac{n\sum x_i y_i - (\sum x_i)(\sum y_i)}{n\sum x_i^2 - (\sum x_i)^2} \qquad \text{Equation 12.2.}$$

$$= \frac{(10)(6628.39) - (362.7)(161.4)}{(10)(15{,}939.65) - (362.7)^2} \qquad \text{Use summary statistics.}$$

$$= \frac{7744.12}{27{,}845.21} = 0.2781 \qquad \text{Simplify.}$$

$$\hat{\beta}_0 = \frac{\sum y_i - \hat{\beta}_1 \sum x_i}{n} \qquad \text{Equation 12.3.}$$

$$= \frac{161.4 - (0.2781)(362.7)}{10} \qquad \text{Use summary statistics and } \hat{\beta}_1.$$

$$= 6.0528 \qquad \text{Simplify.}$$

The estimated regression line is $y = 6.0528 + 0.2781x$. **Figure 12.12** shows a scatter plot of the data and the graph of the estimated regression line.

(b) To find the estimated true mean surface offset temperature for a snow depth of 45 cm, substitute $x^* = 45$ into the estimated regression line equation.

$$y = 6.0528 + 0.2781x \qquad \text{Estimated regression line equation.}$$

$$= 6.0528 + (0.2781)(45) = 18.57 \qquad \text{Substitute and simplify.}$$

An estimate for the expected surface offset temperature for a snow depth of 45 cm is 18.57°C. **Figure 12.13** illustrates this calculation.

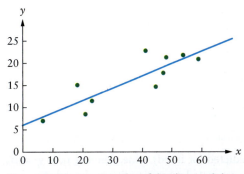

Figure 12.12 A scatter plot of the data and the graph of the estimated regression line.

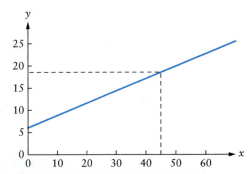

Figure 12.13 The expected surface offset temperature for a snow depth of 45 cm.

Figure 12.14 shows a technology solution.

```
> x <- c(6.7,41.1,20.9,58.9,18.2,44.6,53.8,23.1,47.2,48.2)
> y <- c(7,22.8,8.5,20.9,15.1,14.7,21.8,11.5,17.8,21.3)
>
> results <- lm(y ~ x)
> summary(results)

Coefficients:
            Estimate Std. Error t value Pr(>|t|)
(Intercept)   6.0528     2.3873   2.535  0.03496 *
x             0.2781     0.0598   4.651  0.00164 **
---
Residual standard error: 3.155 on 8 degrees of freedom
Multiple R-squared:  0.73,    Adjusted R-squared:  0.6963
F-statistic: 21.63 on 1 and 8 DF,  p-value: 0.001642
```

Figure 12.14 The estimated regression line and relevant hypothesis tests concerning the regression coefficients (explanations to follow in Section 12.2).

TRY IT NOW Go to Exercises 12.25 and 12.27

CHAPTER APP **Wind Turbine Noise**

Find the estimated regression line. Estimate the true mean noise level for a distance of 175 m.

Rob Arnold/Alamy

ANOVA Table and Coefficient of Determination

The variance σ^2 is a measure of the underlying variability in the simple linear regression model. A large value of σ^2 means the data will vary widely from the true regression line. A small value of σ^2 implies that the observed values will lie close to the true regression line.

An estimate of σ^2 is used to conduct hypothesis tests and to construct confidence intervals related to simple linear regression. This estimate is based on the deviations of the observed values from the estimated regression line. Section 12.2 provides more details on hypothesis tests and confidence intervals.

One method to assess the accuracy of a simple linear regression model involves an analysis of variance (ANOVA) table, similar to those we constructed in Chapter 11. To explain the ANOVA table entries and to make computations easier, a little more notation is necessary.

For computational purposes, define

$$S_{xx} = \underbrace{\sum(x_i - \bar{x})^2}_{\text{definition}} = \underbrace{\sum x_i^2 - \frac{1}{n}(\sum x_i)^2}_{\text{computational formula}}$$

$$S_{yy} = \underbrace{\sum(y_i - \bar{y})^2}_{\text{definition}} = \underbrace{\sum y_i^2 - \frac{1}{n}(\sum y_i)^2}_{\text{computational formula}}$$

$$S_{xy} = \underbrace{\sum(x_i - \bar{x})(y_i - \bar{y})}_{\text{definition}} = \underbrace{\sum x_i y_i - \frac{1}{n}(\sum x_i)(\sum y_i)}_{\text{computational formula}}$$

The ith predicted (or fitted) value, denoted $\hat{y}_i$, is $\hat{y}_i = \hat{\beta}_0 + \hat{\beta}_1 x_i$. This is simply the equation for the estimated regression line evaluated at x_i.

The ith residual is $y_i - \hat{y}_i$. This difference measures how far away the observed value of Y is from the estimated value of Y.

The total variation in the data (the **total sum of squares**, denoted **SST**) is decomposed into a sum of the variation explained by the model (the **sum of squares due to regression**, denoted **SSR**) and the variation about the regression line (the **sum of squares due to error**, denoted **SSE**). These three sums are defined in the following identity, which shows the decomposition of the total variation in the data.

Sum of Squares

$$\underbrace{\sum(y_i - \bar{y})^2}_{\text{SST}} = \underbrace{\sum(\hat{y}_i - \bar{y})^2}_{\text{SSR}} + \underbrace{\sum(y_i - \hat{y}_i)^2}_{\text{SSE}} \qquad (12.4)$$

Here are the computational formulas for these sums of squares:

$$\text{SST} = S_{yy} \qquad \text{SSR} = \hat{\beta}_1 S_{xy} \qquad \text{SSE} = \text{SST} - \text{SSR}$$

> You can probably guess that a large F statistic suggests a significant regression. The formal hypothesis test is presented in Section 12.2.

Regression computations are often summarized in an analysis of variance table, as shown next. As in Chapter 11, mean squares are the corresponding sums of squares divided by the corresponding degrees of freedom. The F test for a significant regression and the associated p value are discussed in Section 12.2.

ANOVA summary table for simple linear regression

Source of variation	Sum of squares	Degrees of freedom	Mean square	F	p Value
Regression	SSR	1	$\text{MSR} = \dfrac{\text{SSR}}{1}$	$\dfrac{\text{MSR}}{\text{MSE}}$	p
Error	SSE	$n-2$	$\text{MSE} = \dfrac{\text{SSE}}{n-2}$		
Total	SST	$n-1$			

Coefficient of Determination

The **coefficient of determination**, denoted r^2, is a measure of the proportion of the variation in the data that is explained by the regression model:

$$r^2 = \text{SSR}/\text{SST} \qquad (12.5)$$

Because $0 \leq \text{SSR} \leq \text{SST}$, the coefficient of determination r^2 is always a number between 0 and 1 (inclusive). The higher r^2 is, the better the model is. Many statistical software packages report $100r^2$, the percentage of variation explained by the regression model.

Example: Estimated Regression Line and ANOVA Table

The next example illustrates the computations necessary to produce the estimated regression line and the ANOVA table.

EXAMPLE 12.4 Turn Down the Noise

NOISE A recent study suggests that both noise pollution and air pollution can increase the risk of cardiovascular disease. Suppose a random sample of adults who have lived in the same home for at least five years was obtained. The nighttime noise level was measured in each home (in decibels) and an Agatston score for each person was found using electron-beam computed tomography. This score indicates the level of coronary calcification, where a higher Agatston score suggests a greater cardiovascular risk. The data are given in the table.

Noise	58	95	102	105	23	68	58	89	54
Agatston	72	135	82	188	108	192	40	61	89

Noise	15	105	21	61	20	72	113	80	27
Agatston	23	137	38	106	108	75	170	69	40

(a) Find the estimated regression line and explain the meaning of the estimated coefficient $\hat{\beta}_1$.

(b) Complete the ANOVA table (without the p value), and find and interpret the coefficient of determination.

Solution

Find the necessary summary statistics. Use Equation 12.2 to find $\hat{\beta}_1$ and Equation 12.3 to find $\hat{\beta}_0$. Compute the ANOVA summary table and compute r^2.

(a) Noise level (x) is the independent variable and Agatston score (y) is the dependent variable. Here are the summary statistics for the $n = 18$ pairs of observations:

$$\sum x_i = 1166 \qquad \sum y_i = 1733 \qquad \sum x_i y_i = 128{,}564$$

$$\sum x_i^2 = 94{,}066 \qquad \sum y_i^2 = 211{,}775$$

$$\hat{\beta}_1 = \frac{n\sum x_i y_i - (\sum x_i)(\sum y_i)}{n\sum x_i^2 - (\sum x_i)^2} \qquad \text{Equation 12.2.}$$

$$= \frac{(18)(128{,}564) - (1166)(1733)}{(18)(94{,}066) - (1166)^2} \qquad \text{Use summary statistics.}$$

$$= \frac{293{,}474}{333{,}632} = 0.8796 \qquad \text{Simplify.}$$

$$\hat{\beta}_0 = \frac{\sum y_i - \hat{\beta}_1 \sum x_i}{n} \qquad \text{Equation 12.3.}$$

$$= \frac{1733 - (0.8796)(1166)}{18} \qquad \text{Use summary statistics.}$$

$$= 39.3 \qquad \text{Simplify.}$$

The estimated regression line is $y = 39.3 + 0.8796x$. The value $\hat{\beta}_1 = 0.8796$ suggests that an increase of 1 dB leads to an increase of approximately 0.9 in the Agatston score. **Figure 12.15** shows a scatter plot of the data and the graph of the estimated regression line.

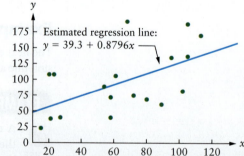

Figure 12.15 A scatter plot of the data and the graph of the estimated regression line.

(b) Complete the ANOVA table, except for the p value ($n = 18$).

$$\text{SST} = S_{yy} = \sum y_i^2 - \frac{1}{18}(\sum y_i)^2 \qquad \text{Computational formula.}$$

$$= 211{,}775 - \frac{1}{18}(1733)^2 = 44{,}925.6 \qquad \text{Use summary statistics.}$$

$$SSR = \hat{\beta}_1 S_{xy} = \hat{\beta}_1 \left[\sum x_i y_i - \frac{1}{18}(\sum x_i)(\sum y_i) \right]$$ Computational formula.

$$= 0.8796 \left[128{,}564 - \frac{1}{18}(1166)(1733) \right]$$ Use summary statistics.

$$= 14{,}341.6$$ Simplify.

$$SSE = SST - SSR$$ Computational formula.

$$= 44{,}925.6 - 14{,}341.6 = 30{,}584.0$$

Compute the mean squares, MSR and MSE.

$$MSR = SSR/1$$ Definition.

$$= 14{,}341.6/1 = 14{,}341.6$$ Use SSR.

$$MSE = SSE/(n-2)$$ Definition.

$$= 30{,}584.0/16 = 1911.5$$ Use SSE and $n = 18$.

The value of the test statistic is

$$f = \frac{MSR}{MSE} = \frac{14{,}341.6}{1911.5} = 7.50$$

Here are all of these calculations presented in an ANOVA table.

ANOVA summary table

Source of variation	Sum of squares	Degrees of freedom	Mean square	F
Regression	14341.6	1	14341.6	7.50
Error	30584.0	16	1911.5	
Total	44925.6	17		

The coefficient of determination is

$$r^2 = SSR/SST = 14{,}341.6/44{,}925.6 = 0.3192$$

Approximately 0.3192, or 32%, of the variation in the data is explained by the regression model.

Figure 12.16 shows a technology solution.

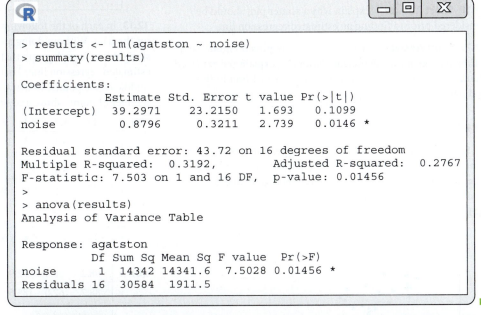

Figure 12.16 The estimated regression line, relevant hypothesis tests, ANOVA table, and r^2. The default R ANOVA table does not include a Total row, and these calculations are more accurate, not subject to round-off error.

TRY IT NOW Go to Exercise 12.33

CHAPTER APP Wind Turbine Noise

Complete the ANOVA table (without the *p* value). Find and interpret the coefficient of determination.

Section 12.1 Exercises

Concept Check

12.1 Short Answer Name the two parts in a general additive probabilistic model.

12.2 Short Answer Name the four assumptions in a simple linear regression model.

12.3 Fill in the Blank In a simple linear regression model, the variability in the distribution of Y is the same for every value of x. This property is called _____.

12.4 Fill in the Blank The principle of least squares produces an estimated regression line such that _____ is a minimum.

12.5 Fill in the Blank In a simple linear regression model, if x^* is a specific value of the independent variable, then $y^* = \hat{\beta}_0 + \hat{\beta}_1 x^*$ is an estimate of _____ and _____.

12.6 Fill in the Blank An estimate of _____ is used to conduct hypothesis tests and to construct confidence intervals related to simple linear regression.

12.7 Short Answer State the sum of squares identity in a simple linear regression model.

12.8 Fill in the Blank The coefficient of determination is defined as _____ and measures _____.

12.9 True or False In a simple linear regression model, all of the points lie on the estimated regression line.

12.10 Short Answer Explain why a scatter plot should be considered prior to finding an estimated regression line.

12.11 Short Answer In a simple linear regression model with estimated regression coefficients $\hat{\beta}_0$ and $\hat{\beta}_1$, explain the expected change in the response variable for a change of 1 unit in the predictor variable.

Practice

12.12 Decide whether a simple linear regression model is appropriate for each of the following graphs. If it is, indicate whether the slope ($\hat{\beta}_1$) of the estimated regression line is positive, negative, or zero. If it is not, state why.

(a)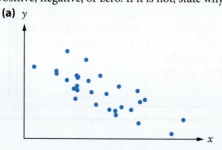

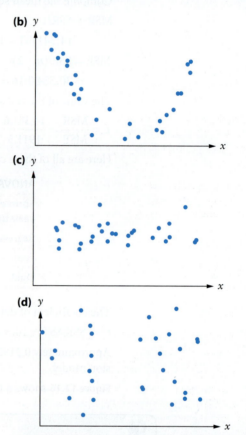

12.13 In each of the following problems, construct a scatter plot for the data. Without doing any calculations for $\hat{\beta}_0$ and $\hat{\beta}_1$, add a straight line to each graph that you believe would be close to the estimated regression line. That is, try to draw a line through the middle of the points, so that there are approximately the same number of points above and below the line.

(a)
x	8.3	1.9	8.4	8.6	2.3	4.6	2.4
y	11.6	3.4	9.3	10.1	5.6	7.6	4.3
x	7.7	8.0	5.6	6.9	1.8	4.8	9.4
y	10.5	11.5	6.6	9.8	2.9	6.8	12.8
x	6.8	4.9	7.8	5.3	1.1	0.3	
y	10.2	4.9	9.4	8.2	3.6	3.2	

EX12.13a

(b)
x	98.2	56.4	52.9	99.3	73.5	82.7
y	88.7	167.0	185.5	72.5	149.5	113.0
x	54.1	68.2	70.3	71.2	53.4	90.2
y	163.0	135.4	135.8	130.1	168.4	82.0

x	53.8	99.4	85.5	72.6	92.7	89.6
y	184.2	90.1	130.3	132.9	93.2	103.6

x	80.9	76.4	54.5	97.3	58.7	92.6
y	121.0	115.7	174.1	79.1	159.3	86.4

EX12.13b

12.14 Suppose the true regression line relating the variables x and y, for values of x between 100 and 200, is $y = 75.0 + 3.6x$.
(a) Find the expected value of Y when $x = 150$, $E(Y|150)$.
(b) How much change in the dependent variable is expected when x increases by 1 unit? Justify your answer.
(c) Suppose $\sigma = 6$. Find the probability that an observed value of Y is less than 500 when $x = 120$.

12.15 Suppose the true regression line relating the variables x and y, for values of x between 25 and 35, is $y = -35.1 - 7.2x$.
(a) Find the expected value of Y when $x = 27.4$.
(b) How much change in the dependent variable is expected when x decreases by 5 units? Justify your answer.
(c) Suppose $\sigma = 1.5$. Find the probability that an observed value of Y is between -252 and -250 when $x = 30$.

12.16 Suppose a simple linear regression model is used to explain the relationship between x and y. A random sample of $n = 12$ values for the independent variable was selected, and the corresponding values of the dependent variable were observed. The summary statistics were

$\sum x_i = 460.53$ $\sum y_i = -6349.7$

$\sum x_i^2 = 17,875.1$ $\sum y_i^2 = 3,421,892$

$\sum x_i y_i = -246,677$

(a) Find the estimated regression line.
(b) Estimate the true mean value of Y when $x = 41$.

12.17 Suppose a simple linear regression model is used to explain the relationship between x and y. A random sample of $n = 15$ values for the independent variable was selected, and the corresponding values of the dependent variable were observed. The data are given in the table. EX12.17

x	4.9	9.9	0.3	3.8	9.8	4.6	9.8	5.4
y	49.1	53.4	21.7	40.4	78.6	60.4	73.9	37.4

x	3.1	3.0	4.2	4.0	1.7	1.2	7.0
y	41.6	29.6	46.7	28.5	23.8	15.2	72.3

(a) Construct a scatter plot for these data. Is a simple linear regression model reasonable? Justify your answer.
(b) Find the estimated regression line. Add a graph of this line to your scatter plot.
(c) What is the predicted value of Y when $x = 5.7$?

12.18 Suppose a simple linear regression model is used to explain the relationship between x and y. A random sample of $n = 21$ values for the independent variable was selected, and the corresponding values of the dependent variable were observed. EX12.18
(a) Construct a scatter plot for these data. Is a simple linear regression model reasonable? Justify your answer.
(b) Find the estimated regression line. Add a graph of this line to your scatter plot.
(c) Estimate the true mean value of Y when $x = 62$.
(d) Construct an ANOVA summary table, except for the p value.

Applications

12.19 Biology and Environmental Science Agricultural research suggests that the final corn yield in bushels per acre (y) is linearly related to the number of inches between rows (x). Suppose the true regression line is $y = 197.5 - 6.1x$.
(a) Find the expected yield when there are 15 in. between rows.
(b) How much change in yield is expected if the distance between rows decreases by 2 in.?
(c) Suppose $\sigma = 3.2$ bushels per acre. Find the probability that an observed value of yield is between 90 and 95 bushels per acre when rows are 17 in. apart.

12.20 Physical Sciences Once a polyvinyl chloride (PVC) pipe joint is assembled using solvent cement, a certain amount of time must be allowed for the new joint to set. For 4- to 8-in.-diameter pipes, suppose the setting time (y, measured in hours) is linearly related to the ambient temperature (x, measured in °F). The true regression line is assumed to be $y = 8.3 - 0.09x$.
(a) Find an estimate of an observed setting time for an ambient temperature of 65°F.
(b) How much less time does a joint take to set if the temperature rises by 10°F?
(c) If $\sigma = 15$ min, find the probability that an observed set time is less than 1 hr if the ambient temperature is 80°F.

12.21 Sports and Leisure A recent study investigated the association between heading a soccer ball and subclinical evidence of traumatic brain injury. Amateur soccer players were selected, and each completed a questionnaire to determine the number of times (x) they headed the ball in the last 12 months. In addition, each player was subject to diffusion-tensor magnetic resonance to measure the fractional anisotropy (FA, y). Suppose the estimated regression line was $y = -0.00026x + 0.6877$.
(a) Find the expected value of Y if the number of headers is 1000.
(b) Find an estimate of an observed Y value for $x = 1200$.
(c) Suppose $\sigma = 0.06$. Find the probability that an observed Y value is more than 0.5 when the number of headers is 750.

12.22 Manufacturing and Product Development As soon as a bottle of soda is opened, it begins to lose its carbonation. Fourteen 12-oz bottles of cola were obtained, and each was assigned a randomly selected time period (in hours). Each bottle was opened and allowed to stand at room temperature. The carbonation (y) in each bottle was measured (in volumes) after the prescribed time period (x). The summary statistics are shown.

$\sum x_i = 8.6$ $\sum y_i = 37.4$ $\sum x_i y_i = 19.61$

$\sum x_i^2 = 7.26$ $\sum y_i^2 = 110.24$

(a) Find the estimated regression line.
(b) Estimate the true mean carbonation after 1 hr 15 min.

12.23 Biology and Environmental Science Some research suggests that higher levels of carbon dioxide (CO_2) increase the rate at which plants and trees grow. In fact, even without ideal conditions (e.g., normal precipitation and temperature), some trees grow more in elevated levels of CO_2. Suppose that, in a research study, data were collected on oak trees in northern U.S. forests. For randomly selected levels of CO_2 (x, measured in parts per million, ppm), the most recent tree-ring growth (y, in centimeters) was measured. The summary statistics are given for $n = 10$.

$\sum x_i = 5035.0$ $\sum y_i = 13.7$ $\sum x_i y_i = 7012.5$
$\sum x_i^2 = 2{,}607{,}575$ $\sum y_i^2 = 19.25$

(a) Find the estimated regression line.
(b) Find an estimate of an observed value of tree-ring growth if the level of CO_2 is 600 ppm.

12.24 Business and Management A recent study considered the effects of innovation on employment in Latin America. It seems reasonable that as more firms produce new products, they would need more workers, and employment would rise. For small firms in Argentina, let y be the yearly percentage of employment growth and let x be the percentage of small firms that are product or process innovators. Assume the estimated regression line is $y = -5.399 + 5.790x$.
(a) Find the expected percentage of employment growth if the percentage of product or process innovators is 2%.
(b) Find an estimate of an observed value for Y for $x = 3.5$.
(c) Suppose $\sigma = 1.5$. Find the probability that an observed value of Y is more than 19 when $x = 4$.
(d) Suppose $x = 0.9\%$. Find the expected value of Y and interpret this result.

12.25 Sports and Leisure Sailboat enthusiasts believe that the wind speed (x, in miles per hour, mph) is linearly related to (downwind) boat speed (y, in knots). For wind speeds between 10 and 30 mph and Hobie catamarans, the data in the table were recorded. **SAIL**

x	26.2	11.8	27.3	24.3	20.4	24.6	20.7	14.0	15.5	20.3
y	16.3	7.7	17.3	18.2	16.9	10.3	12.7	13.3	10.0	6.8

(a) Construct a scatter plot for these data.
(b) Find the estimated regression line. Add a graph of the estimated regression line to the scatter plot in part (a).
(c) Estimate the mean speed of a Hobie catamaran for a wind speed of 18 mph.

12.26 Psychology and Human Behavior A development officer at a large university believes that the amount of money donated to the general fund by alumni each year is linearly related to the football team's winning percentage. A random sample of Division I football teams was selected. The winning percentage (x) and subsequent alumni donation amount (y, in millions of dollars) are given in the table. **DONATE**

x	92	67	92	25	83	83	58	8	75
y	15.8	11.1	13.5	7.0	17.4	10.8	11.2	3.9	23.7
x	100	92	58	92	17	25	75	92	33
y	27.8	23.3	11.3	6.6	7.3	10.3	15.8	12.0	6.8

(a) Construct a scatter plot for these data.
(b) Find the estimated regression line. Add a graph of the estimated regression line to the scatter plot in part (a).
(c) Find an estimate of an observed value of the donation amount for a football team with a winning percentage of 75%.

12.27 Public Health and Nutrition Evidence suggests that for children, sugar-sweetened soft drink consumption is related to the amount of salt intake (and obesity).[3] A random sample of children in Great Britain was obtained. The amount of salt intake (x, in grams per day) and sugar-sweetened soft drink consumption (y, in milliliters per day) was measured for each. **DRINK**
(a) Construct a scatter plot for these data. What does this plot suggest about the sign of the estimated regression coefficient $\hat{\beta}_1$?
(b) Find the estimated regression line. Add a graph of the estimated regression line to the scatter plot in part (a).
(c) Find an estimate for the mean value of Y (soft drink consumption) for a salt intake of 2 g.

Extended Applications

12.28 Public Health and Nutrition A recent study suggests that the number of steps walked per day is strongly associated with good health. A person's heart rate indicates how hard the heart is working to circulate blood throughout the body and is a measure of health. A lower resting heart rate may reduce the risk of heart attack and stroke, as well as increase endurance. A random sample of adults between 35 and 40 years old was obtained, and each person wore a pedometer for a day. The number of steps taken, x, was recorded, and the next day the resting pulse rate (beats per minute), y, for each person was also measured. **STEPS**
(a) Find the estimated regression line and explain the meaning of the estimated coefficient $\hat{\beta}_1$.
(b) Complete the ANOVA table (without the p value), and find and interpret the coefficient of determination.

12.29 Medicine and Clinical Studies Many automobile accidents are caused by tired drivers. A typical test for alertness involves measuring the speed and degree to which a person's eyes respond to certain stimuli. A random sample of 25 drivers was obtained, and the oscillations in pupil size (x, in millimeters per second) was measured using a pupillograph. Each person's tiredness (y) was also recorded using the pupil unrest index (PUI). The summary statistics are shown here.

$\sum x_i = 7.1$ $\sum y_i = 192.0$ $\sum x_i y_i = 49.22$
$\sum x_i^2 = 2.1064$ $\sum y_i^2 = 2094.0$

(a) Find the estimated regression line.
(b) Find the expected PUI for $x = 0.3$ mm/sec.

(c) Suppose a driver is considered too tired to drive if the PUI score is 15 (or higher). What value of x yields an expected PUI score of 15?

12.30 Biology and Environmental Science Virginia Tech provides yield data and other performance statistics on barley as well as a summary of each growing season. A random sample of barley varieties was obtained and the 2018 yield (y, in bushels per acre) and the date headed (x, number of days since the beginning of the calendar year at which 50% of the heads have emerged from the plants) were recorded for each.[4] BARLEY

(a) Construct a scatter plot of these data. What does the scatter plot suggest about the sign of the estimated regression coefficient $\hat{\beta}_1$?
(b) Find the estimated regression line.
(c) Complete the ANOVA table (without the p value).
(d) Do you believe that the date headed is a good predictor of yield? Why or why not?

12.31 Medicine and Clinical Studies Ultrasound measurements are often used to predict the birth weight of a newborn. However, there is some evidence to suggest that pre-pregnancy body mass index (BMI) is related to birth weight.[5] A random sample of Caucasian pregnant women experiencing no complications was obtained. The pre-pregnancy BMI (x, in kg/m^2) of each was measured, and the newborn weight (y, in grams) was also recorded. BIRTH

(a) Construct a scatter plot for these data.
(b) Find the estimated regression line.
(c) Estimate the expected weight of a newborn if the mother's pre-pregnancy BMI is 25 kg/m^2.
(d) If ultrasound is not available, do you believe that BMI is a good predictor of birth weight? Why or why not?

12.32 Physical Sciences Deep-water (more than 300 m) wave forecasts are important for large cargo ships. One method of prediction suggests that the wind speed (x, in knots) is linearly related to the wave height (y, in feet). A random sample of buoys was obtained, and the wind speed and wave height were measured at each. The data are given in the table.[6] BUOY

| Wind speed | 19.4 | 1.9 | 3.9 | 15.5 | 5.5 | 15.5 | 2.9 | 11.7 |
| Wave height | 13.1 | 8.1 | 6.2 | 12.2 | 7.9 | 7.5 | 6.1 | 7.2 |

| Wind speed | 4.0 | 19.4 | 13.0 | 7.5 | 6.4 | 10.1 | 9.5 | 14.3 |
| Wave height | 4.9 | 10.1 | 8.5 | 8.3 | 7.1 | 9.8 | 6.5 | 11.1 |

| Wind speed | 14.0 | 17.3 | 17.5 | 8.9 | 10.5 | 12.2 | 13.6 | 15.5 |
| Wave height | 9.8 | 10.5 | 11.5 | 7.5 | 9.4 | 9.7 | 7.9 | 9.9 |

(a) Find the estimated regression line.
(b) Complete the ANOVA table (without the p value).
(c) Find the coefficient of determination. Interpret this value.
(d) Suppose a 10-ft wave is considered to be the storm threshold. What wind speed yields an expected storm threshold?

12.33 Manufacturing and Product Development Good evidence suggests that the depth of a bounce on a certain circular trampoline is linearly related to the stiffness of the springs around the edges, or spring constant k. A random sample of production trampolines was obtained, and the spring constant of each was recorded (in pounds per inch). The bounce was measured (in feet) with a testing weight of 200 lb released from a height of 2 ft. The data are given in the table. BOUNCE

| Spring constant | 0.43 | 0.77 | 0.63 | 0.73 | 0.55 | 0.42 |
| Bounce | 1.1 | 2.6 | 2.3 | 3.0 | 3.0 | 1.9 |

| Spring constant | 0.38 | 0.39 | 0.20 | 0.73 | 0.56 | 0.57 |
| Bounce | 1.4 | 1.4 | 1.3 | 2.2 | 2.0 | 1.2 |

(a) Find the estimated regression line.
(b) Complete the ANOVA table (without the p value).
(c) Find the coefficient of determination. Interpret this value.
(d) Suppose the trampoline sits 4 ft above the ground. For what value of k can a 200-lb person expect to hit the ground?

12.34 Public Health and Nutrition According to Map the Meal Gap 2018, some Americans are at risk of hunger in every county and congressional district in the United States.[7] A random sample of counties was obtained, and the cost of a typical meal (x, in dollars) and the food insecurity (y, a percentage of the population, adults and children, who are food insecure) was recorded for each. HUNGER

(a) Construct a scatter plot for these data. Based on this plot, do you believe there is a significant linear relationship between the cost of a typical meal and food insecurity? Why or why not?
(b) Find the estimated regression line. Add a graph of the estimated regression line to the scatter plot in part (a).
(c) Complete the ANOVA table (without the p value).
(d) Do you think the value of the F statistic in the ANOVA table supports your conclusion in part (a)? Why or why not?

Challenge Problems

12.35 Medicine and Clinical Studies Some physicians use the cholesterol ratio (CR = total cholesterol/high-density lipoprotein [HDL] cholesterol) as a measure of a patient's risk of heart disease. In addition, the triglyceride concentration (TG) is associated with coronary artery disease in many patients. In a study of the relationship between these two variables, a random sample of adults was obtained, and the triglyceride level (x_1; mg/dL) and cholesterol ratio (y) were obtained for each person. CHOLEST

(a) Construct a scatter plot of these data (y versus x_1). The relationship does not appear to be linear. Can you describe this relationship between y and x_1?
(b) Compute the natural logarithm of each difference $(x_1 - 129)$. That is, find the values of a new predictor variable, $x_2 = \ln(x_1 - 129)$.
(c) Construct a scatter plot of y versus the new predictor variable x_2. Describe this relationship.
(d) Find the estimated regression line, and complete the ANOVA table using x_2 as the predictor variable.

12.2 Hypothesis Tests and Correlation

Hypothesis Test for a Significant Linear Regression

Suppose theoretical or empirical evidence indicates that two variables are linearly related. The mean squares (in the regression ANOVA table) are used to compute a value of a test statistic to determine whether the linear relationship is statistically significant. The null hypothesis states that the variation in Y is completely random and is independent of the value of x; knowing the value of x provides no additional information about the value of Y. In this case, a scatter plot would have no discernible pattern, as points would be scattered randomly in the plane (**Figure 12.17**).

For simple linear regression, a test of significance is equivalent to testing the hypothesis $H_0: \beta_1 = 0$. If H_0 is true, then the model assumptions imply that the mean value of Y for any value of x is the same. That is, $\mu_{Y|x_i} = \beta_0$. Therefore, the values of Y vary around the horizontal line $y = \beta_0$, and knowing the value of x adds no additional information. See **Figure 12.18**.

Even though Figure 12.18 shows a true regression line, there is really no significant linear relationship. The true regression line is of absolutely no use for predicting or estimating values of Y.

Remember: β_1 is the slope of the true regression line. A horizontal line has slope β_1 equal to 0.

Figure 12.17 Scatter plot of data showing no linear pattern.

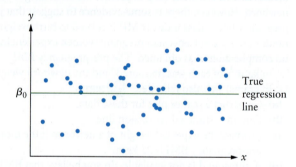

Figure 12.18 If $\beta_1 = 0$, the values of Y vary randomly around the true regression line $y = \beta_0$.

Here is a summary of an F test for a significant regression with significance level α.

Hypothesis Test for a Significant Linear Regression

H_0: There is no significant linear relationship ($\beta_1 = 0$).
H_a: There is a significant linear relationship ($\beta_1 \neq 0$).

TS: $F = \dfrac{\text{MSR}}{\text{MSE}}$

RR: $F \geq F_{\alpha,\,1,\,n-2}$

A CLOSER LOOK

1. We reject the null hypothesis only for large values of the test statistic. The associated p value is a right-tail probability. The F ratio will be larger when $\beta_1 \neq 0$ than when $\beta_1 = 0$.
2. This is often called a **model utility test**. In general, if we reject H_0, then the value of r^2 is large.
3. Alternatively, we can conduct this hypothesis test by using the p value. Recall that if H_0 is true, the p value is the probability of obtaining a value of the test statistic at least as large as the observed value. If $p \leq \alpha$, we reject the null hypothesis. If $p > \alpha$, we do not reject the null hypothesis.

Hypothesis Tests and Confidence Intervals Concerning the Regression Coefficients

Consider the random variable B_1, an estimator for β_1, and $S^2 = \text{MSE} = \text{SSE}/(n-2)$, an estimator for the underlying variance σ^2. If the simple linear regression assumptions are true, then S^2 is an unbiased estimator for σ^2, and the estimator B_1 has the following properties.

Recall: $\hat\beta_1$ is a specific value of B_1.

1. B_1 is an unbiased estimator for β_1: $E(B_1) = \mu_{B_1} = \beta_1$.
2. The variance of B_1 is $\text{Var}(B_1) = \sigma^2_{B_1} = \sigma^2/S_{xx}$. If we use s^2 as an estimate of σ^2, then an estimate of the variance of B_1 is $s^2_{B_1} = s^2/S_{xx}$.
3. The random variable B_1 has a normal distribution.

The hypothesis test procedure concerning β_1 is based on the next theorem.

Theorem
If the simple linear regression assumptions are true, then the random variable

$$T = \frac{B_1 - \beta_1}{S/\sqrt{S_{xx}}} = \frac{B_1 - \beta_1}{S_{B_1}}$$

has a t distribution with $n-2$ degrees of freedom.

No proof is given here that the two tests are equivalent, but Example 12.5 provides some numerical evidence.

The null and alternative hypotheses are stated in terms of the parameter β_1. The most common test has $H_0: \beta_1 = 0$. If $\beta_1 = 0$, there is no significant linear relationship between the two variables. The true regression line is $y = \beta_0$ (a horizontal line), and knowing the value of x does not help us predict the value of $Y|x$. For simple linear regression, the following hypothesis test (with $\beta_{10} = 0$) is equivalent to an F test for a significant regression with significance level α.

β_{10} is the hypothesized value of β_1.

Hypothesis Test and Confidence Interval Concerning β_1

$H_0: \beta_1 = \beta_{10}$
$H_a: \beta_1 > \beta_{10}, \quad \beta_1 < \beta_{10}, \quad$ or $\quad \beta_1 \neq \beta_{10}$

TS: $T = \dfrac{B_1 - \beta_{10}}{S_{B_1}}$

RR: $T \geq t_{\alpha, n-2}, \quad T \leq -t_{\alpha, n-2}, \quad$ or $\quad |T| \geq t_{\alpha/2, n-2}$

A $100(1-\alpha)\%$ confidence interval for β_1 has the following values as endpoints:

$$\hat\beta_1 \pm t_{\alpha/2, n-2} \cdot s_{B_1} \tag{12.6}$$

The properties, hypothesis test procedure, and confidence interval are similar for the simple linear regression parameter β_0. The estimator B_0 has the following properties.

1. B_0 is an unbiased estimator for β_0: $E(B_0) = \mu_{B_0} = \beta_0$.
2. The variance of B_0 is $\text{Var}(B_0) = \sigma^2_{B_0} = \dfrac{\sigma^2 \sum x_i^2}{n S_{xx}}$. If we use s^2 as an estimate of σ^2, then an estimate of the variance of B_0 is $s^2_{B_0} = \dfrac{s^2 \sum x_i^2}{n S_{xx}}$
3. The random variable B_0 has a normal distribution.

Theorem
If the simple linear regression assumptions are true, then the random variable

$$T = \frac{B_0 - \beta_0}{S\sqrt{\sum x_i^2 / n S_{xx}}} = \frac{B_0 - \beta_0}{S_{B_0}}$$

has a t distribution with $n-2$ degrees of freedom.

β_{00} is the hypothesized value of β_0.

> What does $\beta_0 = 0$ imply about the true regression line?

Hypothesis Test and Confidence Interval Concerning β_0

$H_0: \beta_0 = \beta_{00}$
$H_a: \beta_0 > \beta_{00}, \quad \beta_0 < \beta_{00}, \quad \text{or} \quad \beta_0 \ne \beta_{00}$

TS: $T = \dfrac{B_0 - \beta_{00}}{S_{B_0}}$

RR: $T \ge t_{\alpha, n-2}, \quad T \le -t_{\alpha, n-2}, \quad \text{or} \quad |T| \ge t_{\alpha/2, n-2}$

A $100(1-\alpha)\%$ confidence interval for β_0 has the following values as endpoints:

$$\hat{\beta}_0 \pm t_{\alpha/2,\, n-2} \cdot s_{B_0} \qquad (12.7)$$

Example: Hypothesis Tests

EXAMPLE 12.5 Ready to Operate

The Pulsar Corporation sells a large sterilizer with four extendable shelves for medical tools. Company engineers believe that the time to reach the operating temperature from a cold start (y, measured in minutes) is linearly related to the thickness of insulation (x, in inches). A random sample of $n = 12$ thicknesses was selected, and the time to reach operating temperature was recorded for each. The data and the summary statistics are given.

x	1.3	1.8	0.9	1.6	2.6	1.5	2.1	3.0	0.8	2.4	2.5	2.6
y	8.0	6.9	8.1	7.0	6.3	6.5	6.4	5.8	8.3	8.3	6.6	6.6

$\sum x_i = 23.1 \qquad \sum y_i = 84.8 \qquad \sum x_i y_i = 158.5$
$\sum x_i^2 = 50.13 \qquad \sum y_i^2 = 607.66$

(a) Find the estimated regression line.

(b) Complete the ANOVA table and conduct an F test for a significant regression. Use a significance level of 0.05.

(c) Conduct a t test concerning β_1 for a significant regression. Use a significance level of 0.05.

(d) Interpret your results.

Solution

(a) Use the summary statistics and Equations 12.2 and 12.3 to find $\hat{\beta}_1$ and $\hat{\beta}_0$.

$\hat{\beta}_1 = \dfrac{n \sum x_i y_i - (\sum x_i)(\sum y_i)}{n \sum x_i^2 - (\sum x_i)^2}$ *Equation 12.2.*

$= \dfrac{(12)(158.5) - (23.1)(84.8)}{(12)(50.13) - (23.1)^2}$ *Use summary statistics.*

$= \dfrac{-56.88}{67.95} = -0.8371$ *Simplify.*

$\hat{\beta}_0 = \dfrac{\sum y_i - \hat{\beta}_1 \sum x_i}{n}$ *Equation 12.3.*

$= \dfrac{84.8 - (-0.8371)(23.1)}{12}$ *Use summary statistics and $\hat{\beta}_1$.*

$= 8.6781$ *Simplify.*

The estimated regression line is $y = 8.6781 - 0.8371x$.

(b) Here are the calculations for the ANOVA table and the F test for a significant regression.

$$\text{SST} = S_{yy} = \sum y_i^2 - \frac{1}{n}(\sum y_i)^2 \quad \text{Computational formula.}$$

$$= 607.66 - \frac{1}{12}(84.8)^2 = 8.4067 \quad \text{Use summary statistics.}$$

$$\text{SSR} = \hat{\beta}_1 S_{xy} = \hat{\beta}_1 \left[\sum x_i y_i - \frac{1}{n}(\sum x_i)(\sum y_i)\right] \quad \text{Computational formula.}$$

$$= -0.8371\left[158.5 - \frac{1}{12}(23.1)(84.8)\right] \quad \text{Use summary statistics.}$$

$$= 3.9678 \quad \text{Simplify.}$$

$$\text{SSE} = \text{SST} - \text{SSR} \quad \text{Computational formula.}$$

$$= 8.4067 - 3.9678 = 4.4389$$

Compute the mean squares.

$$\text{MSR} = \text{SSR}/1 = 3.9678/1 = 3.9678$$

$$\text{MSE} = \text{SSE}/(n-2) = 4.4389/10 = 0.4439$$

Here is the F test for a significant regression with significance level 0.05.

H_0: There is no significant linear relationship.
H_a: There is a significant linear relationship.

$$\text{TS: } F = \frac{\text{MSR}}{\text{MSE}}$$

RR: $F \geq F_{\alpha,1,n-2} = F_{0.05,1,10} = 4.96$

The value of the test statistic is

$$f = \frac{\text{MSR}}{\text{MSE}} = \frac{3.9678}{0.4439} = 8.9387 \ (\geq 4.96)$$

Because f lies in the rejection region, there is evidence to suggest that insulation thickness is linearly related to time to reach the operating temperature.

Recall that we can use the tables of critical values for F distributions to bound the p value. Place the value of the test statistic in an ordered list of critical values with $\nu_1 = 1$ and $\nu_2 = 10$.

$$4.96 \leq 8.9387 \leq 10.04$$

$$F_{0.05,1,10} \leq 8.9387 \leq F_{0.01,1,10}$$

Therefore, $\quad 0.01 \leq p \leq 0.05$

Figure 12.19 shows the calculation and illustrates the exact p value.

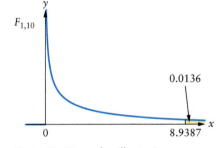

Figure 12.19 p value illustration:
$p = P(F \geq 8.9387)$
$= 0.0136 \leq 0.05 = \alpha$

Here are all the calculations presented in the ANOVA table, with some help from technology (Figure 12.21) to find the exact p value.

ANOVA summary table

Source of variation	Sum of squares	Degrees of freedom	Mean square	F	p Value
Regression	3.9678	1	3.9678	8.9387	0.0136
Error	4.4389	10	0.4439		
Total	8.4067	11			

(c) The t test for a significant regression concerning β_1 is two-sided and has $\beta_{10} = 0$. Use a significance level of 0.05.

H_0: $\beta_1 = 0$
H_a: $\beta_1 \neq 0$

TS: $T = \dfrac{B_1}{S_{B_1}}$

RR: $|T| \ge t_{\alpha/2, n-2} = t_{0.025, 10} = 2.2281$

$S_{xx} = \sum x_i^2 - \dfrac{1}{n}(\sum x_i)^2$ *Computational formula.*

$= 50.13 - \dfrac{1}{12}(23.1)^2 = 5.6625$ *Use summary statistics.*

$s_{B_1}^2 = \dfrac{s^2}{S_{xx}} = \dfrac{0.4439}{5.6625} = 0.0784$

The value of the test statistic is

$t = \dfrac{\hat{\beta}_1}{s_{B_1}} = \dfrac{-0.8371}{\sqrt{0.0784}} = -2.9898; \qquad |-2.9898| = 2.9898 \ge 2.2281$

Because $|t|$ lies in the rejection region, there is evidence to suggest that $\beta_1 \ne 0$, so the regression is significant. The tables of critical values for t distributions can be used to bound the p value. Using technology,

$\dfrac{p}{2} = P(T \le -2.9898) = 0.0068 \Rightarrow p = 0.0136$

Figure 12.20 shows this calculation and an illustration of the exact p value.

Note: Comparing the two tests for a significant regression, notice that $t^2 \approx f$. In fact, for simple linear regression, $t^2 = f$ in every case, subject to round-off error. The p values associated with these two tests are also always the same. It seems reasonable that these two tests should always lead to the same conclusion, because both are overall tests of a significant regression.

(d) This analysis suggests that insulation thickness is linearly related to the time to reach the operating temperature. Because $\hat{\beta}_1 = -0.8371 < 0$, this suggests that for each 1-in. increase in thickness, the time to reach the operating temperature decreases by 0.8371 min.

Figure 12.21 shows a technology solution.

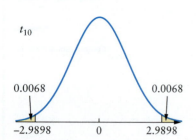

Figure 12.20 p value illustration:
$p = 2P(T \le -2.9898)$
$= 0.0136 \le 0.05 = \alpha$

Why have two hypothesis tests that are essentially the same? Multiple linear regression may include many (more than one) hypothesized explanatory variables. An overall F test for a significant regression is conducted first. If the regression is significant, then a t test is conducted for each variable to see whether it makes a significant contribution to the variability in y.

```
> results <- lm(y ~ x)
> summary(results)

Coefficients:
            Estimate Std. Error t value Pr(>|t|)
(Intercept)   8.6781     0.5723   15.16 3.15e-08 ***
x            -0.8371     0.2800   -2.99   0.0136 *

Residual standard error: 0.6662 on 10 degrees of freedom
Multiple R-squared:  0.472,     Adjusted R-squared:  0.4192
F-statistic: 8.939 on 1 and 10 DF,  p-value: 0.01358
>
> anova(results)
Analysis of Variance Table

Response: y
          Df Sum Sq Mean Sq F value  Pr(>F)
x          1 3.9678  3.9678  8.9387 0.01358 *
Residuals 10 4.4389  0.4439
```

Figure 12.21 The estimated regression line, relevant hypothesis tests, ANOVA table, and r^2.

TRY IT NOW Go to Exercises 12.47 and 12.63

CHAPTER APP Wind Turbine Noise

Complete the ANOVA table and conduct an F test for a significant regression. Conduct a t test concerning β_1 for a significant regression. Use $\alpha = 0.05$ in both tests. Interpret your results.

Correlation

Remember: Correlation, or an association, between two variables does not imply causation.

Correlation is a statistical term indicating a relationship between two variables. For example, the temperature is correlated with the number of cars that will not start in the morning. Similarly, the amount of lead exposure as a child is correlated with IQ level. In each case, a change in one variable is associated with a steady, consistent change in the other variable. As the temperature decreases, for example, the number of cars that will not start in the morning increases.

The **sample correlation coefficient** is a measure of the strength of a linear relationship between two continuous variables, x and y. Suppose we have n pairs of observations $(x_1, y_1), (x_2, y_2), \ldots, (x_n, y_n)$. If large values of x are associated with large values of y, or if as x increases the corresponding value of y tends to increase, then x and y are positively related. If small values of x are associated with large values of y, or if as x increases the corresponding value of y tends to decrease, then x and y are negatively related.

To understand the formula for the sample correlation coefficient, along with why it is a measure of a linear relationship, consider the scatter plots in **Figure 12.22** and **Figure 12.23** and the quantity

$$S_{xy} = \sum (x_i - \bar{x})(y_i - \bar{y})$$

The horizontal line $y = \bar{y}$ and the vertical line $x = \bar{x}$ divide the plane region into four parts. The signs, positive (+) or negative (−), on each graph indicate the sign of the product $(x_i - \bar{x})(y_i - \bar{y})$ in each part.

For example, for any ordered pair in the top right corner of Figure 12.22, $x_i > \bar{x}$ and $y_i > \bar{y}$. Therefore, $x_i - \bar{x} > 0$ and $y_i - \bar{y} > 0$, and the product is positive. Similarly, consider any ordered pair in the bottom right corner of Figure 12.23, $x_i > \bar{x}$ and $y_i < \bar{y}$. In this case, $x_i - \bar{x} > 0$ and $y_i - \bar{y} < 0$, and the product is negative.

If x and y are positively related, as in Figure 12.22, then most of the products $(x_i - \bar{x})(y_i - \bar{y})$ are positive. And if x and y are negatively related, as in Figure 12.23, then most of the products $(x_i - \bar{x})(y_i - \bar{y})$ are negative. Therefore, it seems reasonable to find the sum of all of these products, S_{xy}, and use this sum as a measure of the linear relationship between the two variables. Large positive values of S_{xy} should indicate a positive linear relationship, whereas large negative values should suggest a negative linear relationship.

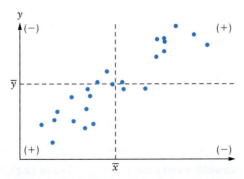

Figure 12.22 If x and y have a positive linear relationship, then most of the products $(x_i - \bar{x})(y_i - \bar{y})$ are positive.

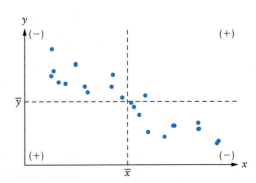

Figure 12.23 If x and y have a negative linear relationship, then most of the products $(x_i - \bar{x})(y_i - \bar{y})$ are negative.

This approach is intuitive, but it must be modified slightly. The magnitude of S_{xy} depends on the units of x and y. For example, if we multiply every value of x by 10, the inherent linear relationship between x and y should not change, but the value of S_{xy} increases. The sample correlation coefficient adjusts S_{xy} so that it is unit independent.

Sample Correlation Coefficient

> The sample correlation coefficient r estimates the population correlation coefficient ρ.

Suppose there are n pairs of observations $(x_1, y_1), (x_2, y_2), \ldots, (x_n, y_n)$. The **sample correlation coefficient** for these n pairs is

$$r = \frac{S_{xy}}{\sqrt{S_{xx} S_{yy}}} = \frac{\sum x_i y_i - \frac{1}{n}(\sum x_i)(\sum y_i)}{\sqrt{\left[\sum x_i^2 - \frac{1}{n}(\sum x_i)^2\right]\left[\sum y_i^2 - \frac{1}{n}(\sum y_i)^2\right]}} \tag{12.8}$$

A CLOSER LOOK

1. The value of r does not depend on the order of the variables and is independent of units (unitless).
2. The value of r is always between -1 and $+1$, that is, $-1 \leq r \leq +1$. r is exactly $+1$ if and only if all of the ordered pairs lie on a straight line with positive slope. r is exactly -1 if and only if all of the ordered pairs lie on a straight line with negative slope.
3. The square of the sample correlation coefficient is the coefficient of determination in a simple linear regression model. Also, because $-1 \leq r \leq +1$, $0 \leq r^2 \leq 1$.
4. r is a measure of the strength of a linear relationship. If r is near 0, there is no evidence of a linear relationship, but x and y may be related in another way.
5. Suppose there is a horizontal line ($y = \beta_0$) with zero slope, and all the data points lie very close to this line. There is no association between the variables; the correlation is close to 0. Intuitively, some of the products are positive and some are negative, and they tend to cancel out one another.

Figure 12.24, **Figure 12.25**, **Figure 12.26** and **Figure 12.27** illustrate the approximate value of r corresponding to each scatter plot. The following general rule is used to describe the linear relationship between two variables, based on the value of the sample correlation coefficient.

1. If $0 \leq |r| \leq 0.5$, then there is a weak linear relationship.
2. If $0.5 \leq |r| \leq 0.8$, then there is a moderate linear relationship.
3. If $|r| > 0.8$, then there is a strong linear relationship.

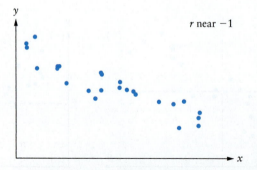

Figure 12.24 The variables x and y are strongly negatively related. The points all fall near a straight line with negative slope. The sample correlation coefficient is near -1.

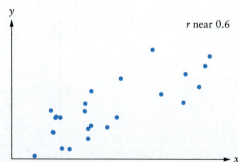

Figure 12.25 The variables x and y are moderately positively related. As the values of x increase, the values of y tend to increase. The sample correlation coefficient is near 0.6.

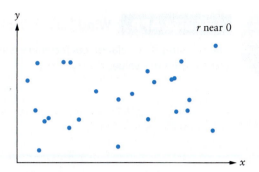

Figure 12.26 The scatter plot shows no apparent linear relationship between the variables *x* and *y*. The sample correlation coefficient is near 0.

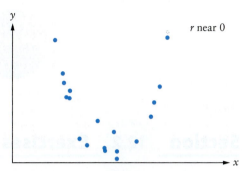

Figure 12.27 The variables *x* and *y* appear to be related, but not linearly. There is a pattern in the scatter plot, but because the sample correlation coefficient measures linear association, *r* is near 0.

Example: Correlation

EXAMPLE 12.6 Credit Scores and Income

A credit score is a numerical indication of a borrowers' credit risk, and this score can be vital when trying to secure a loan for a car or home. Credit scores are based on debt payment history, level of indebtedness, and other credit utilization. Although income is not used in calculating a credit score, some evidence suggests that income is indeed related to credit score. A random sample of adult consumers was obtained, and their credit score and yearly income level (in hundreds of thousands of dollars) were recorded. The data are given in the table.

Income, *x*	1.3	1.1	0.8	1.2	1.4	0.9	0.9	1.4	1.2	1.0
Credit score, *y*	756	728	635	599	760	722	743	726	694	726

Find the sample correlation coefficient between credit score and yearly income, and interpret this value.

Solution

STEP 1 Compute the summary statistics needed to find the sample correlation coefficient.

$$S_{xx} = \sum x_i^2 - \frac{1}{n}(\sum x_i)^2 = 12.96 - \frac{1}{10}(11.2)^2 = 0.416$$

$$S_{yy} = \sum y_i^2 - \frac{1}{10}(\sum y_i)^2 = 5{,}050{,}267 - \frac{1}{10}(7089)^2 = 24{,}874.9$$

$$S_{xy} = \sum x_i y_i - \frac{1}{10}(\sum x_i)(\sum y_i) = 7968.1 - \frac{1}{10}(11.2)(7089) = 28.42$$

STEP 2 Using Equation 12.8, the sample correlation coefficient is

$$r = \frac{S_{xy}}{\sqrt{S_{xx} S_{yy}}} = \frac{28.42}{\sqrt{(0.416)(24{,}874.9)}} = 0.2794$$

Because $r = 0.2794 < 0.5$, there is a weak positive linear relationship between income level and credit score. This suggests that as a person's income level increases, so does the credit score. **Figure 12.28** shows a technology solution. ■

TRY IT NOW Go to Exercises 12.49 and 12.55

```
> x <- c(1.3,1.1,...,1.0)
> y <- c(756,728,...,726)
>
> round(cor(x,y),4)
[1] 0.2794
```

Figure 12.28 Use the R command `cor()` to find the sample correlation coefficient.

Remember: *r* is a measure of association, not causation. Two variables may be strongly related, but that does not mean one variable causes the other. In children, for example, shoe size and height are associated, but probably both are caused by a third variable, age.

CHAPTER APP Wind Turbine Noise

Find the sample correlation coefficient between distance and noise level, and interpret this value.

Section 12.2 Exercises

Concept Check

12.36 Short Answer In a simple linear regression model, suppose the variation in Y is independent of the value of x. Describe the scatter plot of the data.

12.37 Short Answer For simple linear regression, state an equivalent test to the F test for a significant regression.

12.38 Short Answer Using the model utility test for a significant regression, the test statistic is _____.

12.39 True or False When using the model utility test for a significant linear regression, we will reject the null hypothesis for small and large values of the test statistic.

12.40 True or False The sample correlation coefficient is a measure of the strength of a linear relationship between two continuous variables.

12.41 Short Answer What are the bounds on the value of the sample correlation coefficient?

12.42 Short Answer Suppose there is a strong positive linear relationship between the variables x and y. Describe the products $(x_i - \bar{x})(y_i - \bar{y})$.

12.43 True or False Suppose the sample correlation coefficient for the variables x and y is close to 0. This suggests there is no relationship between the two variables.

Practice

12.44 Consider the following (partial) ANOVA summary table from a simple linear regression analysis.

ANOVA summary table

Source of variation	Sum of squares	Degrees of freedom	Mean square	F	p Value
Regression	11691.9				
Error		23			
Total	116064.0				

(a) Complete the ANOVA summary table.
(b) Conduct an F test for a significant regression. Use $\alpha = 0.05$.
(c) Find the coefficient of determination.
(d) Use your answer in part (c) to find the sample correlation coefficient.

12.45 Consider the following (partial) ANOVA summary table from a simple linear regression analysis.

ANOVA summary table

Source of variation	Sum of squares	Degrees of freedom	Mean square	F	p Value
Regression	2772.93				
Error	12988.70				
Total		35			

(a) Complete the ANOVA summary table.
(b) Conduct an F test for a significant regression. Use $\alpha = 0.01$.
(c) Find the coefficient of determination.
(d) Use your answer in part (c) to find the sample correlation coefficient. Is the relationship between these two variables positive or negative? Justify your answer.

12.46 Consider the following summary statistics for data obtained in a simple linear regression analysis.

$$n = 23 \qquad MSE = 561.088$$
$$S_{xx} = 151.086 \qquad \hat{\beta}_1 = 4.4285$$

(a) Conduct a hypothesis test for a significant regression based on the value of β_1 ($H_0: \beta_1 = 0$ versus $H_a: \beta_1 \neq 0$). Use a significance level of 0.05.
(b) Find a 95% confidence interval for the slope of the true regression line β_1.
(c) Using your confidence interval in part (b), is there any evidence to suggest that $\beta_1 \neq 0$? Justify your answer. How does your conclusion compare with your answer in part (a)?

12.47 Consider the data obtained in a simple linear regression study. EX12.47

x	3.27	1.26	4.55	0.86	4.07	4.79	3.25
y	16.67	19.93	14.65	17.48	18.18	13.58	15.70

(a) Find the estimated regression line, and complete the ANOVA table.
(b) Conduct the hypothesis test $H_0: \beta_0 = 0$ versus $H_a: \beta_0 \neq 0$ with significance level 0.01. Is there any evidence to suggest that the true regression line does not pass through the origin?
(c) Find a 99% confidence interval for β_0.
(d) Interpret the value of $\hat{\beta}_0$.

12.48 Consider the summary statistics for data obtained in a simple linear regression analysis.

$$n = 16 \quad SSR = 1155.9 \quad SSE = 3912.82$$
$$S_{xx} = 1980.24 \quad \hat{\beta}_1 = 0.7640$$

(a) Conduct an F test for a significant regression. Use $\alpha = 0.05$. Bound the p value and use technology to find the exact p value.
(b) Conduct a t test for a significant regression. Use $\alpha = 0.05$. Bound the p value and use technology to find the exact p value.
(c) Square the value of the test statistic in part (b). Compare this value with the test statistic in part (a).
(d) How do the exact p values compare in parts (a) and (b)? Explain this result.

12.49 Consider the summary statistics for data obtained in a study of the linear relationship between two variables.

$$S_{xx} = 199.418 \quad S_{yy} = 81.430 \quad S_{xy} = 111.774$$

(a) Find the sample correlation coefficient.
(b) Use the value of r from part (a) to describe the relationship between the two variables.

12.50 Consider the data obtained in a study of the relationship between two variables. EX12.50

x	40.4	44.8	40.7	31.7	41.3	38.1
y	−320.3	−303.9	−264.1	−197.0	−311.8	−280.3

(a) Find the estimated regression line and complete the ANOVA table.
(b) Conduct an F test for a significant regression. Use $\alpha = 0.05$.
(c) Find the coefficient of determination.
(d) Find the sample correlation coefficient. Use this value to describe the relationship between the two variables. How does the estimate $\hat{\beta}_1$ support this answer?
(e) Square the sample correlation coefficient. Verify that this value is the coefficient of determination.

Applications

12.51 Sports and Leisure Many factors affect the length of a professional football game, for example the number of running plays versus the number of passing plays. A study was conducted to determine the relationship between the total number of penalty yards (x) and the time required to complete a game (y, in hours). The data are given in the table. PLAYS

x	196	164	167	35	111	78	150	121	40
y	4.2	4.1	3.5	3.2	3.2	3.6	4.0	3.1	1.9

(a) Find the estimated regression line, and complete the ANOVA table.
(b) Conduct a t test (concerning β_1) for a significant regression. Use a significance level of 0.05.
(c) What proportion of the observed variation in game length can be explained by this simple linear regression model?

12.52 Biology and Environmental Science In a recent study, the weight of an orange (x, in pounds) was compared with the amount of fresh-squeezed juice from the orange (y, in ounces). A random sample of $n = 15$ oranges was obtained. Each was carefully weighed and squeezed. The summary statistics are given.

$$\sum x_i = 9.09 \quad \sum y_i = 50.44 \quad \sum x_i y_i = 30.72$$
$$\sum x_i^2 = 5.73 \quad \sum y_i^2 = 173.21$$

(a) Find the estimated regression line, and complete the ANOVA table. Interpret your results.
(b) Conduct an F test for a significant regression. Use a significance level of 0.01.
(c) Find a 95% confidence interval for the regression parameter β_0. Does this interval suggest that β_0 is different from 0? Justify your answer.

12.53 Physical Sciences The temperature of the upper layer of ocean water is affected by sunlight and wind. There is often a very sharp difference in temperature between the surface zone and the more stationary deep zone. The thermocline layer marks the abrupt drop-off in temperature. The data in the table were obtained in a study of temperature (x, measured in °C) versus depth (y, measured in meters) above the thermocline layer in the Mediterranean Sea. WATER

x	76	54	146	7	91	130	131	117
y	11.0	16.9	8.0	24.7	17.4	16.0	11.1	16.5

(a) Complete the ANOVA summary table for simple linear regression.
(b) Conduct an F test for a significant regression and find bounds on the p value associated with this test. Interpret your results.
(c) Find a 95% confidence interval for the true value of β_1.

12.54 Physical Sciences Permafrost is soil or rock that remains at or below 0°C for at least two years. Suppose an Arctic study analyzed the observed changes in mean annual air temperature and the depth of the freezing layer. The depth of the upper surface of the permafrost layer (y, in feet) was measured for $n = 26$ mean annual air temperatures (x, in °F). The summary statistics are given.

$$S_{xx} = 8.6786 \quad S_{yy} = 2.2771 \quad S_{xy} = -2.8600$$

(a) Find the sample correlation coefficient.
(b) Use the value of r from part (a) to describe the relationship between mean annual temperature and the depth of the permafrost layer.

12.55 Public Health and Nutrition Crimini mushrooms are more common than white mushrooms, and they contain a relatively large amount of copper, which is an essential element according to the U.S. Food and Drug Administration. A study was conducted to determine whether the weight of a mushroom is linearly related to the amount of copper it contains. A random sample of crimini mushrooms was obtained, and the weight (in grams) and the total copper content (in milligrams) were measured for each. CRIMINI

(a) Construct a scatter plot of the data.
(b) Find the sample correlation coefficient.
(c) Use your results in parts (a) and (b) to describe the relationship between crimini mushroom weight and copper content.
(d) To conduct a simple linear regression analysis, which independent and dependent variables would you use?

12.56 Sports and Leisure Many factors determine the outcome in NASCAR races, including car design and mechanics, the weather, and the efficiency of the pit crew. A recent study suggests that, for NASCAR racers, starting position is strongly related to finish position. A random sample of NASCAR racers was obtained, and the average starting position (x) and average finish position (y) for the 2018 season were recorded.[8] NASCAR
(a) Construct a scatter plot of the data. Describe the relationship between average starting position and average finish position.
(b) Find the sample correlation coefficient. Is the value of r consistent with your interpretation in part (a)? Why or why not?
(c) Find the estimated regression line, and complete the ANOVA table.
(d) Using your results from part (c), what would the average starting position have to be for the average finish position to be 11?

12.57 Fuel Consumption and Cars An investigative reporter believes that certain automobile service stations that offer state vehicle inspections routinely charge for unnecessary repair work. Preliminary data suggest that the cost of the repair work may be related to the age of the car. A random sample of automobiles inspected at these stations was obtained, and the age (in years) and the cost of the repairs (in dollars) were recorded for each vehicle. The data are given in the table. REPAIR

Age	Repair cost	Age	Repair cost
9.1	1882	3.2	17
3.8	193	3.3	1268
5.0	368	6.7	1126
1.7	1047	3.9	646
1.9	315	9.7	955
5.3	1631	2.0	801
5.9	652	5.4	973
6.4	475		

(a) Find the sample correlation coefficient, and describe the linear relationship.
(b) Find the estimated regression line for age (x) and repair cost (y). Conduct an F test for a significant regression, and find the bounds on the p value for this test.
(c) Using your results in parts (a) and (b), do you believe the reporter's claim? Justify your answer.

12.58 Physical Sciences As the temperature of air increases, it has a greater capacity to hold moisture. However, humidity can increase only if there is a supply of moisture. For the area of southern Florida, a random sample of days was obtained, and for random times the relative humidity (x, a percentage) and daily temperature (y, in °F) were recorded for each.[9] RELHUM
(a) Construct a scatter plot of the data.
(b) Find the sample correlation coefficient.
(c) Based on the value of r in part (b), predict the sign of $\hat{\beta}_1$ in the estimated regression line.
(d) Find the estimated regression line. Do the results support your answer to part (c)? Why or why not?

12.59 Biology and Environmental Science A recent report suggests that airplanes flying into and out of Australia contribute approximately 3% of global aviation emissions.[10] An area over the Pacific Ocean, approximately 1000 km to the east of the Solomon Islands, is very sensitive to aircraft emissions. A random sample of aircraft flying from Australia was obtained, and the amount of aircraft emissions (x, oxides, in kilograms) and the amount of ozone created (y, in kilograms) were measured for each. OZONE
(a) Construct a scatter plot of these data. Describe the relationship.
(b) Compute the sample correlation coefficient, and use this value to describe the linear relationship.
(c) Find the estimated regression line. Conduct an F test for a significant regression.
(d) Conduct the appropriate test to determine if there is any evidence that the coefficient β_0 is different from 0. Do your results seem practical? Why or why not?

12.60 Biology and Environmental Science The Arctic Oscillation (AO) has two distinct phases (positive and negative) that vary in intensity. The phase and intensity affect the weather patterns of North America and especially eastern North America. Some research suggests that the snow cover in the month of October in Siberia/Eurasia is associated with the AO and, therefore, with weather patterns in North America. The Eurasian snow cover (x, in millions of square kilometers) and AO anomaly (y) were measured for randomly selected Octobers.[11] AO
(a) Construct a scatter plot of these data. Describe the relationship between the variables x and y.
(b) Compute the sample correlation coefficient. Interpret this value.
(c) Suppose high AO anomalies are associated with fierce, cold winters, and the Eurasian snow cover for October 2018 was approximately 12 million km². Find the estimated regression line and use it to predict the AO anomaly for the 2018–2019 winter.

12.61 Economics and Finance Apple is one of the world's most valuable companies and its stock is an important component of many major benchmark financial indexes. Recent research suggests that Apple's stock price is correlated with the general stock market: As Apple goes, so goes the Dow Jones Industrial Average (DJIA). A random sample of days was obtained, and the closing price of Apple stock (x, in dollars) and the DJIA (y) were recorded for each.[12] APPLE

(a) Construct a scatter plot of these data. Describe the relationship between the variables x and y.
(b) Compute the sample correlation coefficient. Interpret this value.
(c) Find the estimated regression line. Conduct an F test for a significant regression. Find the expected DJIA for a closing price of $250 for Apple stock.

Extended Applications

12.62 Business and Management The owner of a small ice cream stand believes that total weekly revenue (y, in dollars) during the summer months is related to money spent on advertising (x, dollars per week). A random sample of summer weeks was selected, and the resulting data are given in the table. ICECRM

x	30	300	380	275	350	190	85
y	957	1125	1202	1028	1134	1124	1062

(a) Find the estimated regression line, and complete the ANOVA table.
(b) What proportion of the observed variation in weekly revenue is explained by this regression model?
(c) Find a 99% confidence interval for the regression parameter β_1.
(d) Conduct a hypothesis test of $H_0: \beta_0 = 0$ versus $H_a: \beta_0 > 0$. Interpret the results. (What happens if the owner spends nothing on advertising in a week?)

12.63 Manufacturing and Product Development It seems reasonable that U.S. timber and lumber sales and production are related to the total number of housing starts. However, it has taken a long time for demand for soft lumber to recover since the 2008 recession. A random sample of years was obtained, and the number of housing starts (x in thousands) and the lumber production in the United States (y in million board feet) were recorded for each.[13] LUMBER

(a) Find the estimated regression line, and complete the ANOVA table.
(b) Conduct an F test for a significant regression. Find bounds on the p value for this test. Use technology to find the exact p value.
(c) Conduct the hypothesis test $H_0: \beta_1 = 17$ versus $H_a: \beta_1 \neq 17$. Use a significance level of 0.05.
(d) Find a 99% confidence interval for the true value of β_1.

12.64 Biology and Environmental Science Environmental scientists routinely measure the water quality in many lakes and rivers by tracking the turbidity, a measure of the total amount of suspended solids in the water. High turbidity, measured in nephelometric turbidity units (NTUs), suggests murky water. Heavy rains tend to raise river levels and increase turbidity. A sample of weekly rainfall total (x, in inches) and turbidity (y, in NTUs) was obtained for the Wide Waters site at Owasco Lake in New York. The data are given in the table. TURBID

x	0.67	0.08	0.64	0.98	0.34	0.37	0.85
y	5.29	3.21	5.68	1.77	3.66	3.38	3.19

x	0.75	0.07	0.68	0.64	0.26	0.09	1.44
y	2.45	2.22	2.44	2.26	2.55	1.66	8.41

(a) Find the estimated regression line.
(b) Conduct a test of $H_0: \beta_1 = 0$ versus $H_a: \beta_1 \neq 0$ with a significance level of 0.05.
(c) Find an estimate of the expected turbidity if 0.55 in. of rain has fallen within the past week.
(d) What proportion of the observed variation in turbidity is explained by this regression model? How do you think this model could be improved?

12.65 Biology and Environmental Science At least 700 distinct species of reptiles and amphibians reside in the Caribbean Islands.[14] A recent research study suggested that the number of distinct species is related to the island area. A random sample of Carribean Islands was obtained and the area (x, in square kilometers) and the number of distinct species of reptiles and amphibians (y) were recorded for each. SPECIES

(a) Construct a scatter plot for these data. Describe the relationship between the variables x and y. Do you think a linear model is appropriate? Why or why not?
(b) Find the sample correlation coefficient. Interpret this value.
(c) Create a new set of observations, $w_i = \ln(x_i)$, where ln is the natural logarithm function. Construct a scatter plot for the variables y and w. Do you think a linear model is appropriate? Why or why not?
(d) Find the sample correlation coefficient for the variables y and w. Interpret this value.

12.66 Physical Sciences The rate of evaporation at the surface of the water in a swimming pool (kg/hr) is believed to be related to the air velocity (m/sec) or the relative humidity (measured as a percentage). A random sample of swimming pools was obtained, and the evaporation rate, air velocity, and relative humidity were recorded for each. EVAP

(a) Find the estimated regression line, and complete the ANOVA table for evaporation rate (y) and air velocity (x).
(b) Find the estimated regression line and complete the ANOVA table for evaporation rate (y) and relative humidity (x).
(c) Which of these two models do you think is better? Justify your answer.

12.67 Physical Sciences The U.S. Geological Survey recently completed a national assessment of geologic carbon dioxide storage resources. A random sample of formations was obtained, and the area of each formation (x, in thousands of acres) and the depth from the surface (y, in thousands of feet) was obtained for each. USGS

(a) Construct a scatter plot of these data. Describe the relationship between these two variables.
(b) Compute the sample correlation coefficient. Does this value support your answer in part (a)? Why or why not?
(c) Using the results in parts (a) and (b), do you believe the relationship between these two variables is significant? Why or why not?

(d) Find the estimated regression line and conduct an F test for a significant regression. Use $\alpha = 0.05$. Does your conclusion support your answer in part (c)?

Challenge Problems

12.68 Sports and Leisure Let ρ be the population correlation coefficient between two variables. A hypothesis test for a correlation different from zero is based on the sample correlation coefficient, R (the random variable), and involves the t distribution. **CLIMB**

$H_0: \rho = 0$

$H_a: \rho > 0, \qquad \rho < 0, \qquad$ or $\qquad \rho \neq 0$

TS: $T = \dfrac{R\sqrt{n-2}}{\sqrt{1-R^2}}$

RR: $T \geq t_{\alpha, n-2}, \quad T \leq -t_{\alpha, n-2}, \quad$ or $\quad |T| \geq t_{\alpha/2, n-2}$

A new mountain climber believes that there is a linear relationship between the diameter of a single rope and its dry weight. A random sample of climbing ropes was obtained. The diameter of each was measured (x, in millimeters) and the weight of each (y, in grams per meter) was recorded. The data are given in the table.

x	10.07	9.54	10.34	10.85	9.85
y	71.20	64.50	67.20	73.50	64.50
x	9.45	9.95	10.63	9.73	10.34
y	65.20	68.20	70.90	67.80	72.00

(a) Find the sample correlation coefficient, and conduct a two-sided hypothesis test with $H_0: \rho = 0$. Use a significance level of 0.05.

(b) Find the estimated regression line, and conduct a test of $H_0: \beta_1 = 0$ versus $H_a: \beta_1 \neq 0$. Use a significance level of 0.05.

(c) Explain the similarities between the values of the test statistics in parts (a) and (b). Explain why this relationship makes sense.

12.3 Inferences Concerning the Mean Value and an Observed Value of Y for $x = x^*$

Mean Value of Y for $x = x^*$

Recall from Section 12.1: Suppose x^* is a specific value of the independent, or predictor, variable x and $y = \hat{\beta}_0 + \hat{\beta}_1 x$ is the estimated regression line. Then the value $y^* = \hat{\beta}_0 + \hat{\beta}_1 x^*$ is

1. An estimate of the mean value of Y for $x = x^*$, and
2. An estimate of an observed value of Y for $x = x^*$.

Remember: $E(Y|x^*)$ is the mean of all values of Y for which $x = x^*$, and y^* is a single observation for $x = x^*$.

The error in estimating the mean value of Y is less than the error in estimating an observed value of Y. In the first case, y^* is used to estimate a single value, the mean. However, an estimate of an observed value of Y is a guess at the next value selected from an entire distribution; the error in estimation must be greater. In this section, we will first consider a hypothesis test and confidence interval concerning the mean value of Y for $x = x^*$. Then we will consider a prediction interval for an observed value of Y if $x = x^*$.

Suppose the simple linear regression model assumptions are true. For $x = x^*$, the random variable $B_0 + B_1 x^*$ has a normal distribution with expected value

$$E(B_0 + B_1 x^*) = \beta_0 + \beta_1 x^* \qquad (12.9)$$

and variance

$$\text{Var}(B_0 + B_1 x^*) = \sigma^2 \left[\dfrac{1}{n} + \dfrac{(x^* - \bar{x})^2}{S_{xx}} \right] \qquad (12.10)$$

The standard deviation is the square root of the expression in Equation 12.10, and an estimate of the standard deviation is obtained by using s as an estimate for σ.

The numerator $(x^* - \bar{x})^2$ is 0 when $x = \bar{x}$.

The variance of $B_0 + B_1 x^*$ is smallest when $x = \bar{x}$. The further x^* is from $\bar{x}$, the greater the squared difference $(x^* - \bar{x})^2$ and the greater the variance. Therefore, the estimator $B_0 + B_1 x^*$ for the mean value of Y is most precise near $\bar{x}$. Also, a confidence interval for the mean value of Y would be narrower for values of x^* near $\bar{x}$. Intuitively, think about

12.3 Inferences Concerning the Mean Value and an Observed Value of Y for x = x*

predicting the weather. The further into the future we try to predict the weather, the more inaccurate our forecast is likely to be.

The following theorem is used to conduct a hypothesis test and construct a confidence interval for the mean value of Y given $x = x^*$.

Theorem
If the simple linear regression assumptions are true, then the random variable

$$T = \frac{(B_0 + B_1 x^*) - (\beta_0 + \beta_1 x^*)}{S\sqrt{(1/n) + [(x^* - \bar{x})^2 / S_{xx}]}}$$

has a t distribution with $n - 2$ degrees of freedom.

The null and alternative hypotheses are stated in terms of the parameter $y^* = \beta_0 + \beta_1 x^*$.

Hypothesis Test and Confidence Interval Concerning the Mean Value of Y for x = x*

▶ H_0: $y^* = y_0^*$

H_a: $y^* > y_0^*$, $\quad y^* < y_0^*$, $\quad$ or $\quad y^* \neq y_0^*$

TS: $T = \dfrac{(B_0 + B_1 x^*) - y_0^*}{S\sqrt{(1/n) + [(x^* - \bar{x})^2 / S_{xx}]}}$

RR: $T \geq t_{\alpha, n-2}$, $\quad T \leq -t_{\alpha, n-2}$, $\quad$ or $\quad |T| \geq t_{\alpha/2, n-2}$ ◀

A $100(1 - \alpha)\%$ confidence interval for $\mu_{y|x^*}$, the mean value of Y for $x = x^*$, has the following values as endpoints:

$$(\hat{\beta}_0 + \hat{\beta}_0 x^*) \pm t_{\alpha/2,\, n-2} \cdot s \cdot \sqrt{\frac{1}{n} + \frac{(x^* - \bar{x})^2}{S_{xx}}} \tag{12.11}$$

EXAMPLE 12.7 Health Care Expenditures

Research suggests that a country's total expenditures on health care are related to the gross domestic product (GDP). A random sample of countries was obtained, and the GDP per capita x and health expenditures per capita y (each in thousands of U.S. dollars) were obtained for each. The data are given in the table.[15]

x	41.364	26.701	43.378	50.012	46.607	50.012	37.270
y	4.600	2.101	4.033	4.376	4.840	4.376	2.822

x	17.953	58.792	27.058	30.606	30.460	36.318	42.622
y	1.080	6.647	1.798	2.734	2.150	3.248	4.192

The estimated regression line is $y = -1.441 + 0.1283x$. The summary ANOVA table is shown here:

valphoto/Deposit Photos

ANOVA summary table

Source of variation	Sum of squares	Degrees of freedom	Mean square	F	p Value
Regression	26.6218	1	26.6218	143.39	0.0000
Error	2.2279	12	0.1857		
Total	28.8498	13			

In addition, $s = \sqrt{MSE} = \sqrt{0.1857} = 0.4309$, $\bar{x} = 38.5109$, and $S_{xx} = 1617.07$.

(a) ▶ For $x = 30$ thousand dollars per capita, conduct a hypothesis test to determine whether there is any evidence that the mean expenditure for health care per capita is greater than 2.2 thousand dollars. Use $\alpha = 0.05$. ◀

(b) Construct a 95% confidence interval for the true mean health care expenditure in thousands of dollars for a country with a GDP of 48 thousand dollars per capita.

Solution

The hypothesized value of the mean expenditures for a GDP of 30 is $y_0^* = 2.2$. Use this value and the summary statistics to conduct a one-sided, right-tail hypothesis test concerning the mean value of Y for $x = x^* = 30$.

(a) ▶ The hypothesis test is one-sided, right-tail with $y_0^* = 2.2$. Use a significance level of 0.05.

$H_0: y^* = 2.2$
$H_a: y^* > 2.2$

TS: $T = \dfrac{(B_0 + B_1 x^*) - y_0^*}{S\sqrt{(1/n) + [(x^* - \bar{x})^2 / S_{xx}]}}$

RR: $T \geq t_{\alpha, n-2} = t_{0.05, 12} = 1.7823$

The value of the test statistic is

$t = \dfrac{(\hat{\beta}_0 + \hat{\beta}_1 x^*) - y_0^*}{s\sqrt{(1/n) + [(x^* - \bar{x})^2 / S_{xx}]}}$

$= \dfrac{[-1.441 + 0.1283(30)] - 2.2}{0.4309\sqrt{(1/14) + [(30 - 38.5109)^2 / 1617.07]}} = 1.4144$

The value of the test statistic does not lie in the rejection region. Equivalently, $p = 0.0913 > 0.05$, as illustrated in **Figure 12.29**. At the $\alpha = 0.05$ significance level, there is no evidence to suggest that the mean health care expenditure per capita is greater than 2.2 thousand dollars for a country with a GDP of 30 thousand dollars per capita. ◀

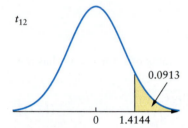

Figure 12.29 p value illustration:
$p = P(T \geq 1.4144)$
$= 0.0913 > 0.05 = \alpha$

(b) To construct a confidence interval, first find the appropriate critical value.

$1 - \alpha = 0.95 \quad \Rightarrow \quad \alpha = 0.05 \quad \Rightarrow \quad \alpha/2 = 0.025$ Find $\alpha/2$.

$t_{\alpha/2, n-2} = t_{0.025, 12} = 2.1788$ Use Appendix Table 5 to find the critical value.

Now use Equation 12.11.

$(\hat{\beta}_0 + \hat{\beta}_1 x^*) \pm t_{\alpha/2, n-2} \cdot s \cdot \sqrt{\dfrac{1}{n} + \dfrac{(x^* - \bar{x})^2}{S_{xx}}}$ Equation 12.11.

$= (\hat{\beta}_0 + \hat{\beta}_1 x^*) \pm t_{0.025, 12} \cdot s \cdot \sqrt{\dfrac{1}{n} + \dfrac{(x^* - \bar{x})^2}{S_{xx}}}$ Use the values of α and n.

$= [-1.441 + 0.1283(48)] \pm (2.1788)(0.4309)\sqrt{\dfrac{1}{14} + \dfrac{(48 - 38.5109)^2}{1617.07}}$

Use summary statistics, values for s and $t_{0.025, 12}$.

$= 4.7173 \pm 0.3347$ Simplify.

$= (4.3826, 5.0520)$ Compute endpoints.

The interval (4.3826, 5.0520) is a 95% confidence interval for the true mean health care expenditure in a country with a GDP of 48 thousand dollars per capita.

Note: Figure 12.30 shows a scatter plot of the data, the estimated regression line, and the 95% confidence bands for the true mean health expenditure for each GDP. The confidence bands allow us to visualize 95% confidence intervals for the true mean value of Y for any value of $x = x^*$. The top confidence band connects the right, or upper, endpoint of each 95% confidence interval, and the bottom confidence band connects the left, or lower, endpoint of each 95% confidence interval. To estimate the 95% confidence interval for the true mean value of Y for $x = 48$ from the graph, draw a vertical line at $x = 48$. The intersection of the line and the confidence bands yields the endpoints of the interval. As x^* moves away from $\bar{x}$, the width of the confidence interval increases.

Figure 12.31 shows a technology solution.

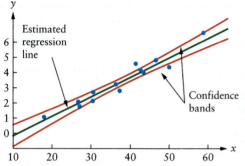

Figure 12.30 Scatter plot, estimated regression line, and confidence bands for the health care expenditure–GDP data.

```
> results <- lm(y ~ x)
> x_star <- data.frame(x = 48)
> round(predict(results,x_star,interval='confidence'),4)
    fit    lwr   upr
1 4.7173 4.3826 5.052
```

Figure 12.31 Use the R command `predict()` to find a confidence interval for the true mean health care expenditure in a country with a GDP of 48.

TRY IT NOW Go to Exercises 12.81 and 12.85

CHAPTER APP   **Wind Turbine Noise**

(a) For $x = 180$ m, conduct a hypothesis test to determine whether there is any evidence to suggest that the mean noise level is greater than 70 dB.

(b) Construct a 95% confidence interval for the true mean noise level for a distance of 225 m.

Observed Value of Y for $x = x^*$

An investigator may be interested in constructing an interval of possible values for an observed value of Y if $x = x^*$. For example, suppose temperature is used to predict the amount of expansion (in inches) of a certain type of vinyl siding. We may need an interval of possible values of expansion when the temperature is 90°F.

An observed value of Y (for $x = x^*$) is a value of a random variable, not a fixed parameter. This fact helps explain why the error of estimation for an observed value is larger than the error of estimation for a single mean value of Y. The interval of possible values is called a **prediction interval**.

We construct a confidence interval for the mean value of Y for $x = x^*$, and we construct a prediction interval for an observed value of Y when $x = x^*$.

The random variable $(B_0 + B_1 x^*) - (\beta_0 + \beta_1 x^* + E^*)$ is used to derive a prediction interval for Y. Using the properties of this random variable, we can construct a prediction interval.

The only difference between this prediction interval and the confidence interval in Equation 12.11 is the extra 1 underneath the square-root symbol. This extra 1 is reasonable because the random variable here includes an extra term, E^*, with variance s^2.

Prediction Interval for an Observed Value of Y

A $100(1-\alpha)\%$ prediction interval for an observed value of Y when $x = x^*$ has the following values as endpoints:

$$(\hat{\beta}_0 + \hat{\beta}_1 x^*) \pm t_{\alpha/2,\, n-2} \cdot s \cdot \sqrt{1 + \frac{1}{n} + \frac{(x^* - \bar{x})^2}{S_{xx}}} \qquad (12.12)$$

EXAMPLE 12.8 Health Care Expenditures (Continued)

Use the health care expenditure–GDP data presented in Example 12.7. Suppose a country with a GDP of 60 thousand dollars per capita is selected at random. (This was the approximate U.S. GDP in 2017.) Find a 90% prediction interval for the health care expenditures per capita.

Solution

STEP 1 Find the appropriate critical value.

$$1 - \alpha = 0.90 \Rightarrow \alpha = 0.10 \Rightarrow \alpha/2 = 0.05 \qquad \text{Find } \alpha/2.$$

$$t_{\alpha/2, n-2} = t_{0.05, 12} = 1.7823 \qquad \text{Use Appendix Table 5 to find the critical value.}$$

STEP 2 Use Equation 12.12.

$$(\hat{\beta}_0 + \hat{\beta}_1 x^*) \pm t_{\alpha/2, n-2} \cdot s \cdot \sqrt{1 + \frac{1}{n} + \frac{(x^* - \bar{x})^2}{S_{xx}}} \qquad \text{Equation 12.12.}$$

$$(\hat{\beta}_0 + \hat{\beta}_1 x^*) \pm t_{0.05, 12} \cdot s \cdot \sqrt{1 + \frac{1}{n} + \frac{(x^* - \bar{x})^2}{S_{xx}}} \qquad \text{Use the values of } \alpha \text{ and } n.$$

$$= [-1.441 + 0.1283(60)] \pm (1.7823)(0.4309)\sqrt{1 + \frac{1}{14} + \frac{(60 - 38.5109)^2}{1617.09}}$$

Use summary statistics, values for s and $t_{0.05,12}$.

$$= 6.257 \pm 0.8946 \qquad \text{Simplify.}$$

$$= (5.3624, 7.1516) \qquad \text{Compute endpoints.}$$

The interval (5.3624, 7.1516) is a 90% prediction interval for an observed value of health care expenditure per capita in a country with a GDP of 60 thousand dollars per capita.

Figure 12.32 shows a technology solution.

```
> results <- lm(y ~ x)
> x_star <- data.frame(x = 60)
> round(predict(results,x_star,interval='prediction',level=0.90),4)
    fit    lwr    upr
1 6.257 5.3624 7.1516
```

Figure 12.32 Use the R command `predict()` to find a prediction interval for an observed value of health care expenditure in a country with a GDP of 60.

✓ Can you explain why the prediction interval is wider the further x^* is from $\bar{x}$? Look carefully at Equation 12.12.

A CLOSER LOOK

1. A confidence interval for the mean value of Y and a prediction interval for an observed value of Y (for $x = x^*$) are centered at the same value. Compare the endpoints for these two intervals in Equations 12.11 and 12.12. The only difference is a 1 underneath the radical in the prediction interval.

2. As x^* moves farther away from $\bar{x}$, the width of the corresponding prediction interval increases.

TRY IT NOW Go to Exercises 12.87 and 12.89

CHAPTER APP

Find a 95% prediction interval for the noise level for a distance of 100 m.

Section 12.3 Exercises

Concept Check

12.69 Short Answer The value $y^* = \hat{\beta}_0 + \hat{\beta}_1 x^*$ is used to estimate _____ and _____.

12.70 True or False For $x = x^*$ and a fixed confidence level, a prediction interval for an observed value Y is wider than a confidence interval for the mean value of Y.

12.71 True or False For a fixed confidence level, the width of a confidence interval for the mean value of Y is the same for any value of x^*.

12.72 True or False For $x = x^*$, a confidence interval for the mean value of Y and a prediction interval for an observed value of Y are centered at the same value.

12.73 Short Answer When is the variance of the estimator for the mean value of Y the smallest?

12.74 Short Answer For $x = x^*$, explain how the width of a confidence interval for the mean value of Y and a prediction interval for an observed value of Y change as the sample size n increases. Justify your answer.

Practice

12.75 Consider the summary statistics for data obtained in a simple linear regression analysis.

$\hat{\beta}_0 = -34.38 \quad n = 18 \quad \text{MSE} = 103.27$
$\hat{\beta}_1 = 3.38 \quad \bar{x} = 15.367 \quad S_{xx} = 138.14$

(a) Conduct a hypothesis test to determine whether there is any evidence that the true mean value of Y for $x = 16.2$ is greater than 20. Use a significance level of 0.05.
(b) Conduct a hypothesis test to determine whether there is any evidence that the true mean value of Y for $x = 11.5$ is different from 5. Use a significance level of 0.01.

12.76 Consider the summary statistics for data obtained in a simple linear regression analysis.

$\hat{\beta}_0 = 38.86 \quad n = 21 \quad \text{MSE} = 21.23$
$\hat{\beta}_1 = -4.318 \quad \bar{x} = 30.891 \quad S_{xx} = 6.298$

(a) Find a 95% confidence interval for the true mean value of Y when $x = 31.5$. Find the width of the resulting interval.
(b) Find a 95% confidence interval for the true mean value of Y when $x = 31.9$. Find the width of the resulting interval.
(c) Explain why the width of the confidence interval in part (b) is greater than the width of the confidence interval in part (a).

12.77 Consider the summary statistics for data obtained in a simple linear regression analysis.

$\hat{\beta}_0 = 23.69 \quad n = 7 \quad \text{MSE} = 277.0$
$\hat{\beta}_1 = 1.452 \quad \bar{x} = 19.1 \quad S_{xx} = 15.131$

(a) Find a 99% prediction interval for an observed value of Y when $x = 19.25$. Find the width of the resulting interval.
(b) Find a 99% prediction interval for an observed value of Y when $x = 18.10$. Find the width of the resulting interval.
(c) Explain why the width of the prediction interval in part (b) is greater than the width of the prediction interval in part (a).

12.78 Consider the data in the table obtained in a simple linear regression analysis. EX12.78

x	5.6	5.1	7.6	3.9	6.5	6.7	5.1	2.5
y	5.7	2.8	10.5	1.6	7.7	8.8	3.4	6.1

(a) Find the estimated regression line, and complete the summary ANOVA table. Conduct an F test for a significant regression.
(b) Find s, an estimate of the standard deviation.
(c) Conduct a hypothesis test to determine whether there is any evidence that the true mean value of Y for $x = 6$ is greater than 4. Use a significance level of 0.05.
(d) Find a 99% confidence interval for the true mean value of Y when $x = 8.5$. Give an interpretation of this interval.

12.79 Consider the data obtained in a simple linear regression analysis. EX12.79

(a) Find the estimated regression line, and complete the summary ANOVA table. Conduct an F test for a significant regression. Use a significance level of 0.01.
(b) Find s, an estimate of the standard deviation.
(c) Find a 95% prediction interval for an observed value of Y when $x = 37$. Based on this interval, do you think it is likely that an observed value of Y (for $x = 37$) will be greater than 170? Justify your answer.

Applications

12.80 Psychology and Human Behavior The Linguistics Department at the University of Massachusetts at Amherst recently studied the relationship between external facial movements and the acoustics of speech sounds. Facial motions were captured using infrared markers and summarized using a unitless measure (x). A special unidirectional microphone was used for acoustic recording and to measure the sound level of speech (y, in decibels). A random sample of individuals was selected, and each was asked to read a certain word. The facial expression and sound level were recorded for each person, and the summary statistics are given.

$\hat{\beta}_0 = 11.669 \quad n = 16 \quad \text{MSE} = 65.42$
$\hat{\beta}_1 = 28.009 \quad \bar{x} = 0.4269 \quad S_{xx} = 0.8615$

(a) Find an estimate of the true mean decibel level for a facial expression value of $x = 0.40$.
(b) Find a 95% confidence interval for the true mean decibel level for a facial expression value of $x = 0.40$.

(c) Suppose the facial expression value is 0.55. Conduct a hypothesis test to determine whether there is any evidence that the true mean decibel level is greater than 30. Use a significance level of 0.01.

12.81 Physical Sciences A new solar collector is being tested for use in charging batteries that can provide electricity for an entire home. A random sample of days was selected, and the amount of solar radiation was measured (x, in langleys) for each. The total battery charge was measured as a proportion (y, between 0 and 1). The summary statistics are given.

$$\hat{\beta}_0 = 0.2007 \quad n = 21 \quad \text{MSE} = 0.06135$$
$$\hat{\beta}_1 = 0.00446 \quad \bar{x} = 103.095 \quad S_{xx} = 12{,}335.8$$

(a) What proportion of a charge can you expect the batteries to take if the amount of solar radiation is 100 langleys?
(b) Find a 95% confidence interval for the true mean battery charge proportion if the amount of solar radiation is 130 langleys.
(c) A value of 80 langleys indicates a typical cloudy day. On such a day, is there any evidence to suggest that the true mean charge proportion is greater than 0.06 (the proportion needed to ensure a home will have sufficient energy until the next day)? Use a significance level of 0.01.

12.82 Fuel Consumption and Cars An automobile mechanic believes the weight of a tire is related to the overall diameter of the tire. A random sample of automobile tires was obtained and the diameter (x, in inches) and the weight (y, in pounds) were measured for each. The estimated regression line was $y = -37.8316 + 2.46657x$ with $n = 19$, SSR $= 87.1673$, SSE $= 258.8706$, $\bar{x} = 25.3474$, $\sum x_i^2 = 12{,}221.6504$, and $S_{xx} = 14.3274$.

(a) Complete the ANOVA summary table, and conduct an F test for a significant regression. Explain the relationship between tire weight and tire diameter.
(b) Find an estimate of the observed tire weight for a tire diameter of 25.2 in.
(c) Find a 99% prediction interval for an observed tire weight if the tire diameter is 24.8 in.
(d) Conduct the hypothesis test $H_0: \beta_0 = 0$ versus $H_a: \beta_0 \neq 0$ with a significance level of 0.01. Is there any evidence to suggest that the true regression line does not pass through the origin? Does this result make practical sense? Why or why not?

12.83 Biology and Environmental Science A study was conducted to investigate the relationship between asthma prevalence and annual rainfall in countries around the world. A random sample of 11 countries was obtained. The total annual rainfall (x, in millimeters) was recorded for each country, as well as the percentage of adults, 22–44 years old, treated for asthma (y). The summary statistics are given.

$$S_{xx} = 178{,}661 \quad S_{yy} = 49.2655 \quad S_{xy} = 380.955$$
$$\bar{x} = 792.636 \quad \bar{y} = 5.26$$

(a) Find the estimated regression line.
(b) Complete the ANOVA summary table, and conduct an F test for a significant regression. Does annual rainfall help to explain the variation in asthma prevalence? Justify your answer.
(c) Find a 95% prediction interval for an observed asthma prevalence if the total annual rainfall is 1000 mm. Comment on anything odd about this interval.

12.84 Public Health and Nutrition The European Food Safety Authority recently issued a scientific opinion on the public health risks related to mechanically separated meat (MSM). The analysis suggested that calcium could be used to distinguish between MSM and non-MSM products. A random sample of MSM poultry was obtained, and the deboner head pressure (x, in lb/in^2) and the amount of calcium (y, in ppm) were measured for each. The data are given in the table. MSM

Pressure	Calcium	Pressure	Calcium
51	573	112	577
95	654	76	600
104	581	143	666
143	709	93	616
77	560	87	514
109	629	70	586
102	623	49	584
72	560	142	634
120	598	132	632

(a) Find the estimated regression line.
(b) Find an estimate for the true mean calcium for a pressure of 100 lb/in^2.
(c) Suppose a deboner is set at a pressure of 120 lb/in^2. Is there any evidence to suggest that the true mean calcium is greater than 625 ppm? Find the p value associated with this test.
(d) Suppose certain hand-deboned poultry has, on average, 600 ppm calcium. What should the pressure be for the mean calcium level for MSM poultry to be the same value?

12.85 Travel and Transportation Highway engineers have long argued that roads designed with high skid resistance help prevent accidents, especially in wet conditions. A random sample of two-lane highways was selected from across the United States, and the skid resistance was measured (in skid numbers) using a Skid Resistance Tester (SRT). The accident rate (per 10,000 vehicles) was computed for 25-mi sections of each highway during wet conditions. SKID

(a) Identify the independent variable and the dependent variable.
(b) Construct a scatter plot for these data, and describe the relationship between the two variables.
(c) Find the estimated regression line.
(d) Suppose the skid resistance is 0.50. Conduct a hypothesis test to determine whether there is any evidence that the true mean accident rate is less than 0.60. Use a significance level of 0.05.

12.86 Biology and Environmental Science A recent study suggests that the onset of Alzheimer's disease (AD) is related to exposure to microorganisms. A greater exposure to microorganisms may actually protect against AD. A random sample of countries was obtained, and a measure of contemporary parasite stress (x) and AD burden (y) was recorded for each. The data are given in the table. **AD**

Parasite stress	AD burden	Parasite stress	AD burden
3.81	5.14	4.99	3.48
3.84	4.59	4.91	3.59
4.45	4.27	3.79	5.06
4.86	4.46	4.36	4.76
3.95	4.28	4.07	4.31
4.19	4.40	3.89	4.33
4.05	4.49	4.59	4.80
4.46	4.43	3.54	4.77
4.42	4.28	4.51	4.96
4.27	4.99	4.18	5.44

(a) Find the estimated regression line. Explain the relevance of the sign on the estimate of β_1 in the context of this problem.
(b) Conduct an F test for a significant regression. Find the p value associated with this test.
(c) Suppose the United States has a parasite stress of approximately 5.6. Find a 95% confidence interval for the true mean AD burden in the United States.

12.87 Psychology and Human Behavior Several studies suggest that the number of violent crimes is related to temperature. Suppose the FBI investigated this relationship between temperature and the number of violent crimes in several large cities in the United States. A random sample of days was obtained, and the average temperature (x, in °F) and the number of violent crimes per 100,000 people (y) were recorded for each day. **CRIME**
(a) Construct a scatter plot for these data, and describe the relationship between the two variables.
(b) Find the estimated regression line.
(c) Find a 95% prediction interval for an observed number of violent crimes if the average temperature is 80°F.
(d) Find a 95% prediction interval for an observed number of violent crimes if the average temperature is 60°F.

12.88 Demographics and Population Statistics A recent study suggests that there is a positive correlation between the linguistic diversity in a community and the number of traffic accidents in that community. A random sample of U.S. cities was obtained, and the annual road fatalities per 1000 people (y) and the linguistic diversity (x, Greenberg diversity index) were recorded for each. **LINGDIV**
(a) Construct a scatter plot for these data, and describe the relationship between the two variables. Does your answer support the conclusions of the recent study? Why or why not?
(b) Find the estimated regression line. Explain how the sign of $\hat{\beta}_1$ supports the conclusions of the recent study.
(c) The Greenburg diversity index in the United States is approximately 0.333. Find a 95% confidence interval for the true mean number of accidents per 1000 people in the United States.

12.89 Public Health and Nutrition Walking is perceived as a simple way to contribute to a healthy lifestyle. A Walk Score, a number between 0 and 100, is based on a patented system and measures the walkability of any area, neighborhood, or city. For example, Boston has a Walk Score of 81. Recent research suggests that the length of a leisurely walk is related to the neighborhood Walk Score. A random sample of people walking in neighborhoods near Vancouver, British Columbia, was obtained, and the neighborhood Walk Score (x) and the length of time spent walking (y, in minutes) were recorded for each. **WALK**
(a) Construct a scatter plot for these data, and describe the general relationship between the two variables.
(b) Find the estimated regression line. Explain how the sign of $\hat{\beta}_1$ supports your observation in part (a).
(c) Conduct an F test for a significant regression. Find the p value associated with this test.
(d) Find a 95% confidence interval for the true mean time spent walking in a Vancouver neighborhood with a Walk Score of 75.
(e) Find a 95% prediction interval for an observed time spent walking in a Vancouver neighborhood with a Walk Score of 55.

Extended Applications

12.90 Technology and the Internet A recent study suggests a relationship between Twitter use and TV ratings. A random sample of prime-time television shows was obtained. During the month of September, the Twitter volume (x, in millions) of tweets on the day of each show and the Nielsen TV rating (y) was recorded for each show. The summary statistics are given.

$$\hat{\beta}_0 = -0.54 \quad n = 25 \quad MSE = 0.2042$$
$$\hat{\beta}_1 = 0.006795 \quad \bar{x} = 407.6 \quad S_{xx} = 21{,}612.0$$

(a) Find an estimate of the true mean TV rating if there are 400 million tweets on the day of the show.
(b) Suppose the number of tweets during a day is 450 million. Is there any evidence that the true mean TV rating will exceed 2.3? Use a significance level of 0.05.
(c) Construct a 99% confidence interval for the true mean TV rating when $x = 375$. On the basis of this interval, do you think the mean TV rating is less than 2.3? Justify your answer.

12.91 Public Policy and Political Science In a global study conducted by Transparency International, researchers found that a country's environmental performance is related to political corruption. A random sample of 15 countries was selected. Transparency International's Corruption Perceptions Index (CPI, x), a measure of perceived corruption among public officials and politicians, was computed for each country. In addition, the Environmental Sustainability Index (ESI, y), a measure of overall progress toward environmental sustainability, was computed for each country. The estimated regression line was $y = 26.432 + 4.546x$, with SSR = 853.50, SSE = 1234.23, $\bar{x} = 5.4333$, and $S_{xx} = 41.2933$.

(a) Complete the ANOVA summary table, and conduct an F test for a significant regression. Explain the relationship between CPI and ESI.

(b) Find an estimate of the observed ESI for a CPI score of 6.7.

(c) Find a 95% prediction interval for an observed ESI if the CPI is 8.2. Based on this interval, do you think a randomly selected country with CPI 8.2 will have an ESI of 90 or greater? Justify your answer.

12.92 Physical Sciences The frequency of air exchange is believed to influence the indoor climate in unheated historic buildings. This relationship is important because many unheated historic buildings hold collections of artifacts, and the indoor climate is related to preservation of these items. Skokloster Castle in Sweden is a masonry building without active climate control that houses many artifacts. A random sample of days during the year was obtained. The air changes per hour (ACH, x) and the climate fluctuations transmittance related to relative humidity (CFT, y) were measured in similar rooms. **AIREX**

(a) Construct a scatter plot for these data, and describe the relationship between the two variables.

(b) Find the estimated regression line. Does the estimate of β_1 agree with your description in part (a)? Why or why not?

(c) Conduct an F test for a significant regression. Find the p value associated with this test.

(d) Find a 99% confidence interval for the true mean CFT for an ACH of 0.5.

(e) Using your confidence interval in part (d), is there any evidence that the mean CFT is different from 0.75?

12.93 Medicine and Clinical Studies Golden Rule medical insurance company recently investigated the relationship between the number of patients per registered nurse in a hospital and the patient's length of stay. A random sample of hospitals was selected, and the number of patients per registered nurse was computed (x). A patient was randomly selected for each hospital, and the length of stay was recorded (y, in hours). **NURSE**

(a) Construct a scatter plot for these data, and describe the relationship between the two variables.

(b) Find the estimated regression line.

(c) Find a 99% prediction interval for an observed length of stay if the number of patients per nurse is 3.7.

(d) Find a 99% prediction interval for an observed length of stay if the number of patients per nurse is 3.3.

(e) Why is the prediction interval in part (c) wider than the prediction interval in part (d)?

12.94 Physical Sciences Doxylamine succinate (DS) is an over-the-counter antihistamine and is used in many nighttime sleep-aid products. Unfortunately, because this drug is so accessible, its use leads to frequent overdoses. To treat these patients properly, it is important to know the amount of DS ingested. However, many overdose patients are unconscious when they arrive at a hospital emergency room. A study was conducted to determine if the DS amount ingested could be reliably predicted from the plasma drug concentration upon arrival at the hospital. A random sample of DS overdose patients was obtained from various hospitals. The plasma drug concentration (x, in μg/mL) was measured for each patient upon arrival at the hospital. Following recovery, each patient was interviewed and the amount of DS ingested (y, in milligrams) was also recorded.

(a) Construct a scatter plot for these data, and describe the relationship between the plasma concentration and the amount ingested.

(b) Find the estimated regression line.

(c) Find a 95% confidence interval for the true mean amount of DS ingested for a plasma concentration of 6 μg/mL.

(d) Suppose the lethal dose of DS for a typical teenager is 2000 mg and that an overdose patient has a plasma concentration of 5 μg/mL. Do you believe that this patient has taken a lethal dose of DS? Why or why not?

12.4 Regression Diagnostics

Check the Residuals

Recall that the assumptions in a simple linear regression model are stated in terms of the random deviations, $E_i, i = 1, 2, \ldots, n$. It is assumed that the E_i's are independent, normal random variables with mean 0 and (constant) variance σ^2. If any one of these assumptions is violated, then the results and subsequent inferences are in doubt.

If the true regression line were known, then we could compute the set of actual random errors,

$$e_1 = y_1 - (\beta_0 + \beta_1 x_1)$$
$$e_2 = y_2 - (\beta_0 + \beta_1 x_2)$$
$$\vdots$$
$$e_n = y_n - (\beta_0 + \beta_1 x_n)$$

and use them to check the assumptions. However, we usually do not know the values for β_0 and β_1, so we use the **residuals**, or deviations from the estimated regression line,

$$\hat{e}_1 = y_1 - (\hat{\beta}_0 + \hat{\beta}_1 x_1)$$
$$\hat{e}_2 = y_2 - (\hat{\beta}_0 + \hat{\beta}_1 x_2)$$
$$\vdots$$
$$\hat{e}_n = y_n - (\hat{\beta}_0 + \hat{\beta}_1 x_n)$$

to check for assumption violations. In practice, these estimates of the random errors are used in a variety of diagnostic checks. This section presents several preliminary graphical procedures used to reveal assumption violations.

Recall from Section 6.3 that a normal probability plot is a scatter plot of each observation versus its corresponding expected value from a Z distribution, or normal score. For observations from a normal distribution, the points will fall close to a straight line. The data axis can be horizontal or vertical. If the scatter plot exhibits a nonlinear pattern, there is evidence to suggest that the data did not come from a normal distribution.

We can use a normal probability plot of the residuals to check the normality assumption. We can also use a simple histogram or stem-and-leaf plot to reveal departures from normality. Finally, we can consider the (backward) Empirical Rule and the ratio IQR/s.

A scatter plot of the residuals versus the independent variable values is also used to check the simple linear regression assumptions. The ordered pairs in this plot are $(x_i, \hat{e}_i)$. **Figure 12.33** illustrates the definition of a residual. This is a scatter plot of y versus x with the estimated regression line. **Figure 12.34** shows a scatter plot of the resulting residuals versus the independent variable.

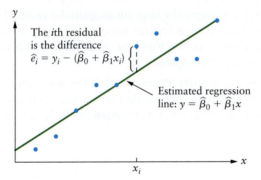

Figure 12.33 An illustration of the definition of a residual.

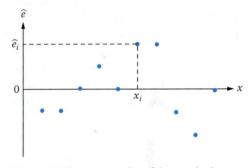

Figure 12.34 A scatter plot of the residuals versus the independent variable.

If there are no violations in assumptions, a scatter plot of the residuals versus the independent variable should look like a horizontal band around zero with randomly distributed points and no discernible pattern. **Figure 12.35** provides an example of such a scatter plot. There should be no obvious relation or pattern between the residuals and the predictor variable.

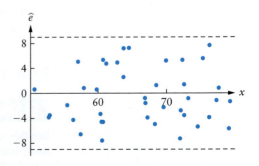

Figure 12.35 A desirable residual plot: no discernible pattern in a horizontal band.

The patterns in a residual plot may sometimes indicate a possible violation in assumptions.

1. Check for a distinct curve in the plot, either mound or bowl-shaped (parabolic), as in **Figure 12.36**. A curved plot suggests that an additional (or different) predictor variable may be necessary. A linear model is not appropriate.
2. Check for a nonconstant spread, as shown in **Figure 12.37**. If no uniform horizontal band is apparent, or if the spread of the residuals varies outside this band, this suggests that the variance is not constant.

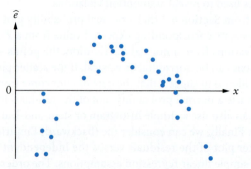

Figure 12.36 A curved residual plot. This suggests that a linear model is not appropriate.

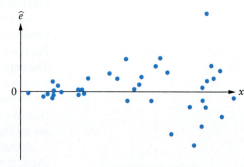

Figure 12.37 A residual plot with a nonconstant spread. This suggests that the variance is not the same for each value of x.

3. Check for any unusually large (in magnitude) residual (**Figure 12.38**). This suggests that one observation is very far away from the rest. The data may have been recorded or entered incorrectly. Often, the offending point is omitted and a new estimated regression line is computed.
4. Check for any outliers (**Figure 12.39**). If the observation is correct, an outlying residual suggests that one observation has an unusually large influence on the estimated regression line. This point is also often omitted, and a new line is computed.

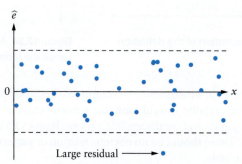

Figure 12.38 A residual plot with an unusually large residual. This suggests that an observation is far away from the rest.

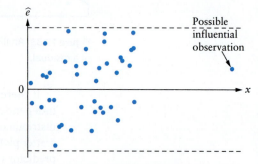

Figure 12.39 A residual plot with an outlying residual. The observation is very influential when the estimated regression line is computed.

Example: Residual Analysis

EXAMPLE 12.9 Pillbugs and Red Clover

Isopods, also called pillbugs, use moss, ferns, and even poison ivy for shelter. However, some scientists believe that their preferred natural shelter is red clover and that the density of pillbugs can be predicted from the density of red clover. A random sample of fields

was obtained. The number of red clover plants per square meter (x) and the number of pillbugs per square meter (y) were measured for each field. The data are given in the table, and the estimated regression line is $y = -12.360 + 14.213x$. Compute the residuals, construct a normal probability plot of the residuals, and carefully sketch a graph of the residuals versus the predictor variable values.

Solution

STEP 1 Use the estimated regression line to find the predicted value, $\hat{y}_i$, for each x value: $\hat{y}_i = -12.36 + 14.213x_i$. Then compute each residual: $\hat{e}_i = y_i - \hat{y}_i$. The normal scores for $n = 20$ are given in Appendix Table 4.

x_i	y_i	$\hat{y}_i$	$\hat{e}_i$	Normal score
5.9	108	71.5	36.5	0.92
4.4	12	50.2	−38.2	−0.92
3.6	86	38.8	47.2	1.13
9.1	136	117.0	19.0	0.45
8.8	103	112.7	−9.7	−0.19
6.3	99	77.2	21.8	0.59
11.7	136	153.9	−17.9	−0.31
10.9	94	142.6	−48.6	−1.13
6.3	9	77.2	−68.2	−1.87
5.7	145	68.7	76.3	1.87
5.0	4	58.7	−54.7	−1.40
8.4	124	107.0	17.0	0.31
4.5	68	51.6	16.4	0.19
13.3	177	176.7	0.3	0.06
4.9	91	57.3	33.7	0.74
14.2	259	189.5	69.5	1.40
10.4	101	135.5	−34.5	−0.74
10.2	101	132.6	−31.6	−0.45
5.0	56	58.7	−2.7	−0.06
3.6	7	38.8	−31.8	−0.59

STEP 2 Figure 12.40 shows the normal probability plot of the residuals. The points lie close to a straight line. There is no evidence to suggest that the normality assumption is violated.

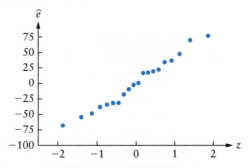

Figure 12.40 The normal probability plot of the residuals.

STEP 3 Figure 12.41 shows the residual plot ($\hat{e}_i$ versus x_i). There is no discernible pattern in this graph. One point in the upper right, corresponding to $x = 14.2$, may be an

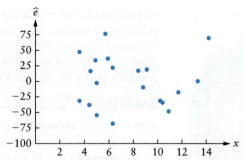

Figure 12.41 The scatter plot of the residuals versus the predictor variable values.

outlier, but the graphical evidence is not convincing enough. These two graphs suggest that there is no violation in the simple linear regression assumptions.

Figure 12.42 and **Figure 12.43** show technology solutions.

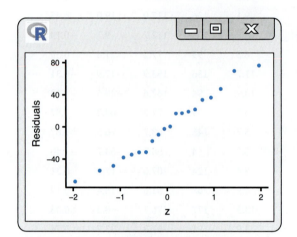

Figure 12.42 R normal probability plot of the residuals.

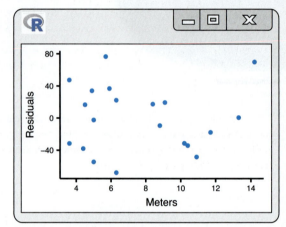

Figure 12.43 R scatter plot of the residuals versus the predictor variable, meters.

TRY IT NOW Go to Exercises 12.113 and 12.115

CHAPTER APP **Wind Turbine Noise**

Compute the residuals. Sketch a normal probability plot of the residuals and a graph of the residuals versus the predictor variable. Is there any evidence that the simple linear regression assumptions are violated?

A CLOSER LOOK

1. The sum of the residuals is always 0, subject to round-off error. We'll check a few examples empirically.
2. Other diagnostic plots include a plot of the residuals versus the predicted values $\hat{y}_i$. Any distinct pattern suggests a violation of the simple linear regression assumptions.
3. Other types of residuals are also used to check for violations in assumptions. **Standardized residuals** utilize the standard deviation of each residual and are useful for identifying residuals with large magnitudes. **Studentized residuals** are also standardized but by using a model without the current observation.

Section 12.4 Exercises

Concept Check

12.95 Short Answer Write a mathematical expression for the ith residual, $\hat{e}_i$.

12.96 Short Answer What are the simple linear regression assumptions in terms of the random deviations, E_i, $i = 1, 2, \ldots, n$.

12.97 Short Answer If the simple linear regression assumptions are satisfied, describe a normal probability plot of the residuals.

12.98 Fill in the Blank The sum of the residuals is _____.

12.99 Short Answer Suppose the plot of the residuals versus the predictor variable shows a nonconstant spread. What does this suggest about the simple linear regression assumptions?

12.100 Short Answer Suppose the plot of the residuals versus the predictor variable shows a distinct curved pattern. What does this suggest about the simple linear regression assumptions?

12.101 Short Answer In addition to a normal probability plot of the residuals and a plot of the residuals versus the predictor variable, name some other techniques to check the residuals for evidence of non-normality.

Practice

12.102 Consider the data in the table used in a simple linear regression analysis. EX12.102

x	11.9	11.3	10.4	12.4	18.5	12.5	15.7	10.2
y	45.2	33.1	34.4	25.1	−4.4	37.6	27.5	52.6

The estimated regression line is $y = 97.63 - 5.15x$.
(a) Find the residuals.
(b) Find the sum of the residuals.

12.103 In a study of the relationship between x and y, the estimated regression line is $y = 4.7245 + 3.3717x$. EX12.103
(a) Find the residuals.
(b) Find the sum of the residuals.

12.104 Examine each normal probability plot of residuals. Is there any evidence to suggest that the random error terms are not normal?

(a) $\hat{e}$

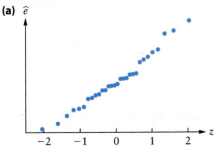

(b) $\hat{e}$

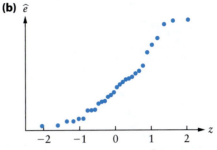

(c) $\hat{e}$

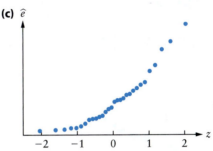

(d) $\hat{e}$

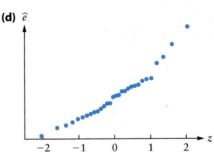

12.105 The residuals in the table were obtained in a simple linear regression analysis. EX12.105

−22.2	−67.9	−1.1	−6.3	−12.4	−33.0	−24.2
35.0	−13.2	33.7	42.3	14.2	13.6	−10.7
−1.6	23.7	38.0	0.4	−20.3	12.1	

(a) Construct a normal probability plot for these residuals.
(b) Is there any evidence to suggest that the random errors are not normal? Justify your answer.

12.106 The residuals for a data set were obtained in a simple linear regression analysis. EX12.106
(a) Construct a normal probability plot for these residuals.
(b) Is there any evidence to suggest that the random errors are not normal? Justify your answer.

12.107 Examine each residual plot (residuals versus predictor variable). Is there any evidence to suggest a violation of the simple linear regression model assumptions? If the graph suggests a violation, indicate which assumption is in doubt. The residuals were obtained in a simple linear regression analysis.

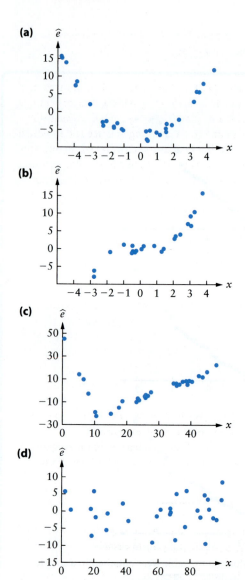

12.108 A simple linear regression analysis was conducted to study the relationship between x and y. The estimated regression line is $y = 91.74 - 3.7384x$. EX12.108
(a) Find the residuals.
(b) Sketch a graph of the residuals versus the predictor variable x. Is there any evidence of a violation of the regression assumptions? Justify your answer.

12.109 A simple linear regression analysis was conducted to study the relationship between x and y. EX12.109
(a) Find the estimated regression line.
(b) Find the residuals.
(c) Sketch a graph of the residuals versus the predictor variable x. Is there any evidence of a violation of the regression assumptions? Justify your answer.

Applications

12.110 Physical Sciences The Atlantic Elevator Company conducted a study to examine the relationship between the total weight of passengers on an elevator (x, in pounds) and the total energy (y, in kilowatts) required to lift the passengers one floor in an office building. A simple linear regression analysis was conducted, and the residuals are given in the table. ELEVAT

−1.2	−1.1	1.5	−8.5	−13.1	−3.3	−17.4
7.5	−6.0	8.2	15.6	6.9	−4.3	−19.2
18.9	−9.4	16.3	0.8	−3.0	10.9	

(a) Sketch a normal probability plot for the residuals.
(b) Is there any evidence that the random error terms are not normal? Justify your answer.

12.111 Manufacturing and Product Development Suppose the research department for Century Tire Company is developing a snow tire based on a new chemical formulation. A study was conducted to examine the relationship between the amount of the new additive per liter (x, in milliliters) and the amount of wear after 50,000 indoor test miles (y, in millimeters). A simple linear regression analysis was conducted, and the residuals are given in the table. TIRES

0.7	−0.2	0.5	−1.1	−0.8	−0.2	0.2
−0.1	0.7	0.3	−0.4	0.5	0.3	0.5
0.8	0.2	0.6	0.1	−0.8	−0.3	0.4
−0.7	−0.4	−0.7	−0.1			

(a) Sketch a normal probability plot for the residuals.
(b) Is there any evidence that the random error terms are not normal? Justify your answer.

12.112 Medicine and Clinical Studies A family counselor believes that there is a relationship between number of years married and blood pressure. For each married man in a random sample, the number of years married (x) and the systolic blood pressure (y, in millimeters of mercury, mm Hg) were recorded. The data are given in the table. BPRESS

Years married	Blood pressure	Years married	Blood pressure
4.9	133	3.4	94
10.3	157	3.1	114
11.9	129	12.0	144
13.0	141	12.9	147
13.0	166	7.1	124

The estimated regression line is $y = 97.95 + 4.03x$.
(a) Compute the residuals, and verify that they sum to 0.
(b) Sketch a normal probability plot for the residuals. Is there any evidence that the random error terms are not normal? Justify your answer.

12.113 Manufacturing and Product Development An automotive component manufacturer believes that the speed of the milling machine is related to the surface roughness of aluminum components. A random sample of aluminum components was obtained. Each piece had a 50×50 mm cross section and was 100 mm in length. The speed of the milling machine (x, in revolutions per minute, rpm)

and the surface roughness (y, in micrometers) was measured for each piece. **MILLING**
(a) Find the estimated regression line and compute the residuals.
(b) Sketch a normal probability plot of the residuals and a plot of the residuals versus the predictor variable.
(c) Do these graphs provide any evidence that the simple linear regression assumptions are invalid? Justify your answer.

12.114 Biology and Environmental Science Agriculture researchers recently investigated the relationship between wheat yield and wind speed. A random sample of wheat-field plots was obtained. The mean daily wind speed over the growing period (x, in meters per second) was recorded for each field, along with the yield (y, in bushels per acre). The estimated regression line is $y = 50.052 - 1.103x$. **WHEAT**
(a) Compute the residuals, and verify that they sum to 0.
(b) Sketch a graph of the residuals versus the predictor variable (wind speed). Is there any evidence that the simple linear regression assumptions are violated? Justify your answer.

12.115 Business and Management Many economists believe that the number of people employed in an area is strongly related to the number of businesses, or establishments, in that area. During the first quarter of 2018, a random sample of states was obtained. The number of establishments (x, in thousands) and the number of people employed (y, in thousands) were recorded for each.[16] **EMPLOY**
(a) Find the estimated regression line and compute the residuals.
(b) Sketch a normal probability plot of the residuals and a graph of the residuals versus the predictor variable (establishments). Is there any evidence that the simple linear regression assumptions are violated? Justify your answer.

12.116 Physical Sciences The solar wind, or the flow of charged particles from the Sun, affects communications, navigation, and the Earth's static pressure (the force exerted by the atmosphere on the surface of the Earth). A random sample of days was selected, and the solar wind density was measured (x, in protons/cm^3) using a special NASA satellite. For each day selected, the static pressure (y, in megapascals) was also recorded. The data are given in the table. **SOLAR**

Solar wind density	Static pressure	Solar wind density	Static pressure
5.6	0.099	9.9	0.117
7.4	0.080	9.4	0.110
8.0	0.115	11.0	0.138
8.7	0.110	6.7	0.104
10.1	0.110	6.9	0.094

(a) Find the estimated regression line, and compute the residuals.
(b) Sketch a normal probability plot of the residuals and a plot of the residuals versus the predictor variable.
(c) Do these graphs provide any evidence that the simple linear regression assumptions are invalid? Justify your answer.

12.117 Medicine and Clinical Studies A health researcher recently investigated the relationship between balance and bone density. A random sample of 25-year-old women was obtained. Each woman took a standing balance test in which she held her arms out horizontally and balanced on one foot. The length of time (x, in seconds) each person remained in that position was recorded. The bone density (y, in mg/cm^2) for each person was measured using the DEXA (dual-energy X-ray absorptiometry) technique. The data are given in the table. **DEXA**

Balance time	Bone density	Balance time	Bone density
29.8	1127	20.9	1094
26.3	1105	27.5	1115
26.0	1187	39.8	1228
37.2	1334	10.7	871
30.1	1067	35.6	1377

(a) Find the estimated regression line, and compute the residuals.
(b) Sketch a normal probability plot of the residuals and a plot of the residuals versus the predictor variable.
(c) Do these graphs provide any evidence that the simple linear regression assumptions are invalid? Justify your answer.

12.118 Medicine and Clinical Studies A coal miner's job is extremely hazardous and can cause serious health problems, including black lung disease and vibration-induced white finger. A study was conducted to examine the hearing loss of coal miners exposed to years of vibration and cool temperatures underground. A random sample of coal miners was selected, and the length of time on the job (x, in years) was recorded for each person. Each miner took a test in which the hearing threshold level (HTL, y, in decibels) was measured. **MINER**
(a) Find the estimated regression line, and compute the residuals.
(b) Sketch a normal probability plot of the residuals and a plot of the residuals versus the predictor variable.
(c) Do these graphs provide any evidence that the simple linear regression assumptions are invalid? Justify your answer. How do you think this regression model could be improved?

12.119 Biology and Environmental Science A study was conducted to predict the width of mountain streams from the height, or stage, of the stream. The research site was a stream near the Dome Glacier in Jasper National Park, Alberta, Canada.[17] A random sample of days and times was obtained, and the height (x, in meters) and the width (y, in meters) were obtained for each. **STREAM**
(a) Find the estimated regression line, and compute the residuals.

(b) Sketch a normal probability plot of the residuals and a plot of the residuals versus the predictor variable.
(c) Do these graphs provide any evidence that the simple linear regression assumptions are invalid? Justify your answer.

Extended Applications

12.120 Sports and Leisure Many studies have sought to determine the most important characteristic that enables golfers to produce long drives. A recent investigation considered grip strength. A random sample of golfers was obtained and a handgrip dynamometer was used to measure grip strength (x, in newtons) for each person. Each golfer then drove a ball, and the length of the drive was measured (y, in yards). DRIVE
(a) Find the estimated regression line, and compute the residuals.
(b) Conduct an F test for a significant regression. Use a significance level of 0.01.
(c) Sketch a normal probability plot of the residuals and a plot of the residuals versus the predictor variable.
(d) Do these graphs provide any evidence that the simple linear regression assumptions are invalid? Justify your answer. How do you think this regression model could be improved?

12.121 Business and Management Suppose the Human Resources director at NVR, a large construction company, believes that the number of sick hours per year taken by an employee is related to the commuting distance. A random sample of employees was obtained, and each person's travel distance to work (x, in miles) was recorded. Personnel records were used to obtain the number of hours off for sickness in a year (y) for each employee included in the study. COMMUTE
(a) Find the estimated regression line, and compute the residuals.
(b) Conduct an F test for a significant regression. Use a significance level of 0.10. Do you believe that there is a relationship between commuting distance and sick hours? Why or why not?
(c) Sketch a normal probability plot of the residuals and a plot of the residuals versus the predictor variable.
(d) Do these graphs provide any evidence that the simple linear regression assumptions are invalid? Justify your answer.

12.122 Physical Sciences Researchers at a marine institute investigated the relationship between the dissolved barium concentration and the silicate/nitrate ratio in the northern Pacific Ocean. The silicate/nitrate ratio (x) and the dissolved barium concentration (y, in nmol/kg) were measured at randomly selected locations. OCEAN
(a) Find the estimated regression line, and compute the residuals.
(b) Conduct an F test for a significant regression. Use a significance level of 0.01.
(c) Sketch a normal probability plot of the residuals and a plot of the residuals versus the predictor variable.
(d) Do these graphs provide any evidence that the simple linear regression assumptions are invalid? Justify your answer.

12.123 Biology and Environmental Science Research suggests that the electrical conductivity (EC) of soil is related to plant growth. Nutrients in the soil take the form of ions. These ions in the soil become dissolved in water and carry an electrical charge, which then determines the soil EC. Too little EC indicates a lack of nutrients, and a high EC level suggests a salinity problem. A random sample of corn fields in Iowa was obtained and the soil EC was measured in each (x, in μS/cm). The resulting crop yield was also recorded (y, in bushels per acre). The data are given in the table. SOILEC

EC	Yield	EC	Yield	EC	Yield
1087	193	711	165	534	90
399	104	672	159	661	163
709	165	931	171	789	152
916	134	636	122	482	155
941	151	921	165	841	157
877	191	707	131	533	131
687	177	1045	140	737	176

(a) Find the estimated regression line.
(b) Conduct an F test for a significant regression ($\alpha = 0.05$). Explain your results in the context of this problem. Find bounds on the p value associated with this test. Use technology to verify your answer.
(c) Sketch a normal probability plot of the residuals and a plot of the residuals versus the predictor variable. Do these graphs provide any evidence that the simple linear regression assumptions are invalid? Justify your answer.

12.124 Physical Sciences In a study conducted in Scandinavia, certain physiological parameters of endurance horses were found to be related to performance. A random sample of horses and races was obtained. The total fluid intake pre-race (x, in liters) and the speed of the horse during the race (y, in kilometers per hour) was measured for each. HORSE
(a) Find the estimated regression line, and compute the residuals.
(b) Conduct an F test for a significant regression. Explain your results in the context of this problem.
(c) Sketch a plot of the residuals versus the predictor variable and a normal probability plot of the residuals. Do these graphs provide any evidence that the simple linear regression assumptions are invalid? Justify your answer.

Challenge Problems

12.125 Sum of the Residuals In a simple linear regression analysis, show that the sum of the residuals is always 0. *Hint:* In the definition of the ith residual, use the definition of $\hat{y}_i$, and then the formula for $\hat{\beta}_0$.

12.5 Multiple Linear Regression

Multiple Linear Regression Model

Although simple linear regression analysis has practical uses, many real-world applications involve a model with a dependent variable Y and at least two independent variables, $x_1, x_2, \ldots, x_k$. For example, the number of days it takes to complete construction of a new home (Y) might be predicted by the square footage of the home (x_1) and the total linear feet of wiring in the home (x_2). Or the temperature of an automobile engine (Y) might be modeled (predicted) by the amount of coolant (x_1), the amount of engine oil (x_2), and the rate of air flow (x_3). In this section, we extend the simple linear regression model to cases involving k (≥ 2) predictor variables. The formal model and assumptions are given below. The procedure is similar to that used for the simple linear regression model.

Multiple Linear Regression Model

Let $(x_{11}, x_{21}, \ldots, x_{k1}, y_1), (x_{12}, x_{22}, \ldots, x_{k2}, y_2), \ldots, (x_{1n}, x_{2n}, \ldots, x_{kn}, y_n)$ be n sets of observations such that y_i is an observed value of the random variable Y_i. We assume there exist constants $\beta_0, \beta_1, \ldots, \beta_k$ such that

$$Y_i = \beta_0 + \beta_1 x_{1i} + \beta_2 x_{2i} + \cdots + \beta_k x_{ki} + E_i \qquad (12.13)$$

where $E_1, E_2, \ldots, E_n$ are independent, normal random variables with mean 0 and variance σ^2.

1. The E_i's are normally distributed
   (which means that the Y_i's are normally distributed).
2. The expected value of E_i is 0
   (which implies that $E(Y_i) = \beta_0 + \beta_1 x_{1i} + \beta_2 x_{2i} + \cdots + \beta_k x_{ki}$).
3. $\mathrm{Var}(E_i) = \sigma^2$ (which implies that $\mathrm{Var}(Y_i) = \sigma^2$).
4. The E_i's are independent (which implies that the Y_i's are independent).

A CLOSER LOOK

1. The double subscript notation on x is necessary to indicate both the variable and the observation. For example, x_{21} is the value of the variable x_2 that corresponds to the observed value of y_1. Similarly, x_{12} is the value of the variable x_1 that corresponds to the observed value of y_2.

2. In the multiple linear regression model, the E_i's again represent the **random deviations** or **random error terms**.

3. The **true regression equation** is $y = \beta_0 + \beta_1 x_1 + \beta_2 x_2 + \cdots + \beta_k x_k$. Notice that this equation is a linear function of the unknown parameters $\beta_0, \beta_1, \ldots, \beta_k$. The graph of the true regression equation is, in general, a surface, not a line. However, we often refer to this equation as the true regression line.

4. The unknown constants $\beta_0, \beta_1, \ldots, \beta_k$ are called partial regression coefficients. Recall that in Section 12.1, β_1 represented the mean amount of change in y for every 1-unit increase in x_1. In a multiple linear regression model, β_i represents the mean change in y for every increase of one unit in x_i if the values of all the other predictor variables are kept fixed.

This section focuses on the method for finding the best deterministic linear model—that is, finding estimates of the unknown parameters $\beta_0, \beta_1, \ldots, \beta_k$, and interpreting the results. Hypothesis tests are used to determine whether the overall model explains a significant amount of the variability in the dependent variable and to evaluate the contribution of each independent variable. Once again, we can use the principle of least squares to minimize the sum of the squared deviations between the observations and the estimated values (the sum of squares due to error).

Although we can perform hand calculations if necessary, it is much more efficient to use technology to compute the estimates $\hat{\beta}_0, \hat{\beta}_1, \ldots, \hat{\beta}_k$ for the true regression parameters, or coefficients, $\beta_0, \beta_1, \ldots, \beta_k$. For specific values of the independent variables, $(x_1^*, x_2^*, \ldots, x_k^*) = x^*$, let $Y^* = \hat{\beta}_0 + \hat{\beta}_1 x_1^* + \hat{\beta}_2 x_2^* + \cdots + \hat{\beta}_k x_k^*$. The value y^* is an estimate of the true mean value of Y for $x = x^*$, denoted $\mu_{Y|x^*}$, and also an estimate of an observed value of Y for $x = x^*$.

Multiple Linear Regression Example

EXAMPLE 12.10 MLB Ticket Prices

MLBTIX

The cost of a Major League Baseball (MLB) ticket has increased rapidly over the last decade. Add in the price of a few soft drinks, hot dogs, a program, and parking, and the total cost of seeing a game can exceed $300 for a family of four. A sports writer believes that the cost of a typical ticket is affected by the team's opening-day payroll and the cost of a hot dog in the stadium. To test this theory, data were obtained for each of the 30 MLB teams during the 2018 season.[18] Some of the data are given in the table, with payroll in millions of dollars.

Team	Ticket	Payroll	Hot Dog
Boston Red Sox	56.97	235.65	5.25
Los Angels Dodgers	41.13	186.14	6.50
New York Yankees	47.62	168.54	3.00
Chicago Cubs	58.57	183.46	6.50
Houston Astros	40.25	160.04	5.25
⋮	⋮	⋮	⋮

(a) Find the estimated regression line.
(b) Estimate the true mean ticket price for a payroll of $150 million and a $5 hot dog.

Solution

Use technology to find the estimates of the regression coefficients. Compute $y^* = \hat{\beta}_0 + \hat{\beta}_1(150) + \hat{\beta}_2(5)$ as an estimate of the mean value of Y for $x = (150, 5)$.

(a) Ticket price (y) is the dependent, or response, variable. Payroll (x_1) and hot dog price (x_2) are the independent, or predictor, variables. Use technology to find the estimated regression coefficients, shown in **Figure 12.44**.

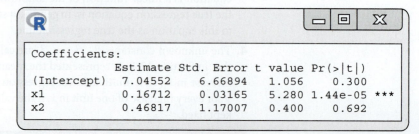

Figure 12.44 Estimated regression coefficients.

The sign of $\hat{\beta}_i$ represents the way that the surface is tilted along the x_i axis.

The estimated regression equation is $y = 7.0455 + 0.1671x_1 + 0.4682x_2$. That is, $\hat{\beta}_0 = 7.0455$, $\hat{\beta}_1 = 0.1671$, and $\hat{\beta}_2 = 0.4682$. Note that the estimated regression coefficients on both payroll and the price of a hot dog are positive. This suggests that as payroll and/or the price of a hot dog increase, ticket prices also increase. The signs on both estimated regression coefficients seem reasonable.

(b) To find the estimated true mean ticket price, substitute $x_1 = x_1^* = 150$ and $x_2 = x_2^* = 5$ into the estimated regression equation.

$y = 7.0455 + 0.1671x_1 + 0.4682x_2$ Estimated regression equation.
$ = 7.0455 + 0.1671(150) + 0.4682(5)$ Use values for x_1^* and x_2^*.
$ = 34.45$ Simplify.

An estimate of the expected ticket price for a $150 million payroll and a $5 hot dog is approximately $34.45.

In this example, the graph of $y = 7.0455 + 0.1671x_1 + 0.4682x_2$ is a plane in three dimensions. **Figure 12.45** is a three-dimensional scatter plot of the data and the graph of the estimated regression equation. Note that some of the points are behind the plane and are not visible in this particular view. **Figures 12.46, 12.47,** and **12.48** illustrate how y depends on each predictor variable separately. In addition to the graph of the estimated regression equation, Figure 12.46 includes the graphs of two lines (in three dimensions): one for $x_1 = 150$ held constant, and one for $x_2 = 5.00$ held constant. Figure 12.47 and Figure 12.48 each represent a slice through the three-dimensional plane at a particular value of the other predictor variable. In Figure 12.47, $\hat{\beta}_1 = 0.1671$ is the slope of the line when $x_2 = 5.00$, and in Figure 12.48, $\hat{\beta}_2 = 0.4682$ is the slope of the line when $x_1 = 150$.

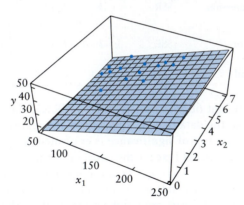

Figure 12.45 A scatter plot of the data and the graph of the estimated regression equation.

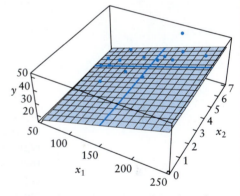

Figure 12.46 A scatter plot of the data, the graph of the estimated regression equation, and two lines: for $x_1 = 150$ and for $x_2 = 5$.

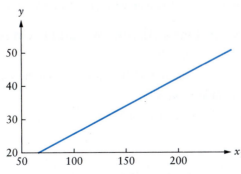

Figure 12.47 A graph of y versus x_1 when $x_2 = 5$.

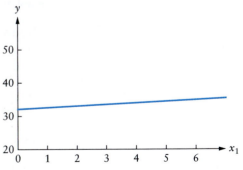

Figure 12.48 A graph of y versus x_2 when $x_1 = 150$.

TRY IT NOW Go to Exercise 12.143

ANOVA Table and Model Utility Test

The following concepts from simple linear regression also apply to multiple linear regression.

1. The variance σ^2 is a measure of the underlying variability in the model. An estimate of σ^2 is used to conduct hypothesis tests and to construct confidence intervals related to multiple linear regression.
2. The ith predicted value is $\hat{y}_i = \hat{\beta}_0 + \hat{\beta}_1 x_{1i} + \cdots + \hat{\beta}_k x_{ki}$.
3. The ith residual is $y_i - \hat{y}_i$.
4. The total sum of squares, the sum of squares due to regression, and the sum of squares due to error have the same definitions. The sum of squares identity, Equation 12.4, is also true. The total variation in the data (SST) is decomposed into a sum of the variation explained by the model (SSR) and the variation about the regression equation (SSE).

We can use an analysis of variance table to summarize the computations in multiple linear regression. The differences occur in the degrees of freedom associated with the sum of squares due to regression and the sum of squares due to error. As usual, the mean squares are the sums of squares divided by the corresponding degrees of freedom.

ANOVA summary table

Source of variation	Sum of squares	Degrees of freedom	Mean square	F	p Value
Regression	SSR	k	$MSR = \dfrac{SSR}{k}$	$\dfrac{MSR}{MSE}$	p
Error	SSE	$n-k-1$	$MSE = \dfrac{SSE}{n-k-1}$		
Total	SST	$n-1$			

The coefficient of determination, $r^2 = SSR/SST$, is a measure of the proportion of variation in the data that is explained by the regression model. For multiple linear regression, a test of significance is equivalent to testing the hypothesis that all regression coefficients (except the constant term) are zero. That is, the null hypothesis is $H_0: \beta_1 = \beta_2 = \cdots = \beta_k = 0$, and none of the predictor variables helps explain any variation in the dependent variable. Suppose $k = 2$ and H_0 is true. Graphically, this means the values of Y vary around the plane (in three dimensions) $y = \beta_0$ with no discernible pattern.

A test for a significant multiple linear regression model is based on the ratio of the mean square due to regression and the mean square due to error. Here is a summary of an F test for a significant regression with significance level α.

Hypothesis Test for a Significant Multiple Linear Regression

$H_0: \beta_1 = \beta_2 = \cdots = \beta_k = 0$

(None of the predictor variables helps explain the variation in y.)

$H_a: \beta_i \neq 0$ for at least one i

(At least one predictor variable helps explain the variation in y.)

TS: $F = \dfrac{MSR}{MSE}$

RR: $F \geq F_{\alpha, k, n-k-1}$

ANOVA Table and Model Utility Test Example

EXAMPLE 12.11   Price at the Pump

A study was conducted to investigate the effects of the cost of crude oil (x_1, in dollars per barrel), the state tax on gasoline (x_2, in dollars per gallon), and the number of publicly owned automobiles in a state (x_3, in thousands) on the price of gasoline (in dollars per gallon). Data from 29 randomly selected days were used to produce the multiple linear regression equation.

$$y = 2.59 + 0.008x_1 + 0.191x_2 + 0.0146x_3$$

In addition, SSR $= 0.3517$ and SSE $= 0.5085$.

(a) Complete the summary ANOVA table and conduct an F test for a significant regression. Use a significance level of 0.05.

(b) Compute r^2 and interpret this value.

Solution

Use the sums of squares, number of predictor variables, and number of observations to complete the ANOVA summary table for multiple linear regression. Use the template for a hypothesis test for a significant regression with $\alpha = 0.05$. The sum of squares due to regression and the total sum of squares are used to compute r^2.

(a) Use the sum of squares identity to compute the total sum of squares.

$$\text{SST} = \text{SSR} + \text{SSE} = 0.3517 + 0.5085 = 0.8602 \qquad \text{Use Equation 12.4.}$$

There are $n = 29$ observations and $k = 3$ predictor variables. Use these values to compute the mean squares.

$$\text{MSR} = \frac{\text{SSR}}{k} = \frac{0.3517}{3} = 0.1172$$

$$\text{MSE} = \frac{\text{SSE}}{n-k-1} = \frac{\text{SSE}}{29-3-1} = \frac{0.5085}{25} = 0.0203$$

The F test for a significant regression with $\alpha = 0.05$ is

$H_0: \beta_1 = \beta_2 = \beta_3 = 0$

$H_a: \beta_i \neq 0$ for at least one i

TS: $F = \dfrac{\text{MSR}}{\text{MSE}}$

RR: $F \geq F_{\alpha, k, n-k-1} = F_{0.05, 3, 25} = 2.99$

The value of the test statistic is

$$f = \frac{\text{MSR}}{\text{MSE}} = \frac{0.1172}{0.0203} = 5.7637 \quad (\geq 2.99)$$

Because f lies in the rejection region, equivalently, $p = 0.0039 \leq 0.05$, there is evidence to suggest that at least one of the regression coefficients is different from zero. At least one of the predictor variables can be used to explain a significant amount of variation in the price per gallon of gasoline. The next reasonable step is to determine which of the three predictors is really contributing to this overall significant regression.

Recall that when using tables of critical values for F distributions, we can only bound the p value. Place the value of the test statistic in an ordered list of critical values with $\nu_1 = 3$ and $\nu_2 = 25$.

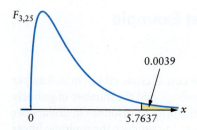

Figure 12.49 *p* value illustration:
$p = P(X \geq 5.7637)$
$= 0.0039 \leq 0.05 = \alpha$

$$4.68 \leq 5.7637 \leq 7.45$$
$$F_{0.01,3,25} \leq 5.7637 \leq F_{0.001,3,25}$$

Therefore, $0.001 \leq p \leq 0.01$

Here are all the calculations presented in the ANOVA table, with some help from technology to find the exact *p* value (illustrated in **Figure 12.49**).

ANOVA summary table

Source of variation	Sum of squares	Degrees of freedom	Mean square	F	p Value
Regression	0.3517	3	0.1172	5.7637	0.0039
Error	0.5085	25	0.0203		
Total	0.8602	28			

(b) The coefficient of determination is

$$r^2 = \frac{\text{SSR}}{\text{SST}} = \frac{0.3517}{0.8602} = 0.4089$$

Approximately 0.4089, or 41%, of the variation in the price per gallon of gasoline is explained by this regression model.

TRY IT NOW Go to Exercise 12.151

Hypothesis Test and CI Concerning β_i

If the overall multiple linear regression F test is significant, then there is evidence to suggest that at least one of the independent variables can be used to predict the value of Y. We can use hypothesis tests (or confidence intervals) to determine whether x_i helps to predict the value of $Y|x$ or, equivalently, whether β_i is different from 0.

Consider the random variable B_i, an estimator for β_i, and $S^2 = \text{MSE} = \text{SSE}/(n-k-1)$, an estimator for the underlying variance σ^2. If the multiple linear regression assumptions are true, then S^2 is an unbiased estimator for σ^2 and is used in constructing a hypothesis test concerning β_i.

Hypothesis Test and Confidence Interval Concerning β_i

$H_0: \beta_i = \beta_{i0}$
$H_a: \beta_i > \beta_{i0}, \quad \beta_i < \beta_{i0}, \quad \text{or} \quad \beta_i \neq \beta_{i0}$

TS: $T = \dfrac{B_i - \beta_{i0}}{S_{B_i}}$

RR: $T \geq t_{\alpha, n-k-1}, \quad T \leq -t_{\alpha, n-k-1}, \quad \text{or} \quad |T| \geq t_{\alpha/2, n-k-1}$

A $100(1-\alpha)\%$ confidence interval for β_i has the following values as endpoints:

$$\hat{\beta}_i \pm t_{\alpha/2, n-k-1} \cdot s_{B_i} \quad (12.14)$$

β_{i0} can be any constant. We can conduct a test for evidence that the regression coefficient β_i is different from any value. However, usually $\beta_{i0} = 0$ and $H_a: \beta_i \neq 0$. This means we are looking for any evidence to suggest that the value of x_i helps to predict the value of $Y|x$.

EXAMPLE 12.12 Milk Alternatives

ALTMILK For those people who want to or need to stay away from drinking cow's milk, a number of non-dairy milk alternatives are available. A random sample of almond, soy, cashew, and coconut milks was obtained, and the number of calories per serving (y) was measured in each. In addition, the total fat (x_1, in grams) and sugar (x_2, in grams) were measured.[19] Multiple linear regression was used to investigate the effects of total fat and sugar on the number of calories. **Figure 12.50** shows the R results from this linear model.

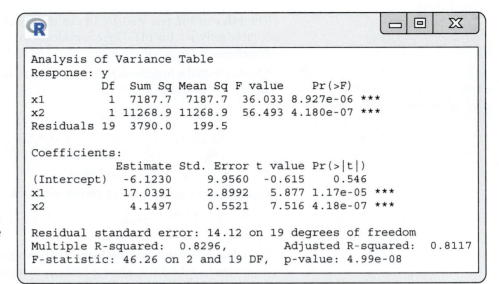

Figure 12.50 R regression analysis results. Note that the ANOVA table presents the sum of squares due to regression attributed to each predictor variable.

(a) Verify that the multiple linear regression is significant at the $\alpha = 0.05$ level.
(b) Conduct separate hypothesis tests to determine whether each predictor variable contributes to the overall significant regression. Use $\alpha = 0.05$ in each test.

Solution

Use the values in the output provided to conduct a formal F test for a significant regression. Test the hypotheses $H_0: \beta_1 = 0$ and $H_0: \beta_2 = 0$.

(a) There are $k = 2$ predictor variables and, from the ANOVA table, there are $n = 22$ observations ($n - 1 = 21$).

The F test for a significant regression with $\alpha = 0.05$ is

$H_0: \beta_1 = \beta_2 = 0$
$H_a: \beta_i \neq 0$ for at least one i

TS: $F = \dfrac{\text{MSR}}{\text{MSE}}$

RR: $F \geq F_{\alpha, k, n-k-1} = F_{0.05, 2, 19} = 3.52$

As given in Figure 12.50, the value of the test statistic is

$$f = \frac{\text{MSR}}{\text{MSE}} = \frac{(7187.7 + 11{,}268.9)/2}{199.5} = 46.26 \; (\geq 3.52)$$

Because f lies in the rejection region (or equivalently, $p < 0.0001$), there is evidence to suggest that at least one of the regression coefficients is different from 0.

(b) To test whether x_1, total fat, is a significant predictor, conduct a hypothesis test concerning the regression coefficient β_1.

$H_0: \beta_1 = 0$
$H_a: \beta_1 \neq 0$

TS: $T = \dfrac{B_1 - 0}{S_{B_1}}$

RR: $|T| \geq t_{\alpha/2, n-k-1} = t_{0.025, 19} = 2.0930$

Using the R output, the value of the test statistic is

$$t = \frac{\hat{\beta}_1 - 0}{s_{B_1}} = \frac{17.0391}{2.8992} = 5.877$$

The value of the test statistic lies in the rejection region, $|t| = 5.877 \geq 2.0930$ (equivalently, $p < 0.0001$). There is evidence to suggest that the predictor total fat contributes to the overall regression effect and to the variability in y.

Conduct a similar hypothesis test concerning the regression coefficient β_2.

$H_0: \beta_2 = 0$

$H_a: \beta_2 \neq 0$

TS: $T = \dfrac{B_2 - 0}{S_{B_2}}$

RR: $|T| \geq t_{\alpha/2,\, n-k-1} = t_{0.025,\,19} = 2.0930$

Using the R output, the value of the test statistic is

$$t = \dfrac{\hat{\beta}_2 - 0}{s_{B_2}} = \dfrac{4.1497}{0.5521} = 7.516$$

The value of the test statistic lies in the rejection region, $|t| = |7.516| = 7.516 \geq 2.0930$ (equivalently, $p < 0.0001$). There is evidence to suggest that the regression coefficient β_2 is different from 0. This implies that the predictor variable sugar also contributes to the overall regression effect and contributes significantly to the variability in y. ∎

TRY IT NOW Go to Exercise 12.159

A CLOSER LOOK

1. The R output in Example 12.12 also includes the results of a hypothesis test concerning the constant term β_0, with $H_0: \beta_0 = 0$. If we fail to reject this null hypothesis, a model without a constant term may be more appropriate.

2. If a model utility test is significant in a multiple linear regression model, then we must consider $k \geq 2$ hypothesis tests to isolate those variables contributing to the overall effect. Recall from Chapter 11 that we cannot set the significance level in each individual hypothesis test. The probability of making a type I error (a mistake) is set in each test under the assumption that only one test is conducted per experiment. Therefore, the more tests that are conducted, the greater the chance of making an error. To control the probability of making at least one mistake, we can use the Bonferroni technique (presented in Chapter 11) applied to simultaneous hypothesis tests or confidence intervals.

   (a) If k hypothesis tests are conducted, then the significance level in each case is α/k.

   (b) The k simultaneous $100(1-\alpha)\%$ confidence intervals have the values $\hat{\beta}_i \pm t_{\alpha/(2k),\, n-k-1} \cdot s_{B_i}$ as endpoints.

3. Many different statistical procedures can be used to select the best regression model. For now, the most reasonable method is simply to include in the model only those variables that have regression coefficients significantly different from zero. Eliminate the others, and calculate a new, reduced model for prediction.

If a confidence interval for β_i contains 0, there is no evidence to suggest that the ith predictor variable is significant. However, if the confidence interval does not include 0, there is evidence to suggest that the ith regression coefficient is different from 0.

Additional analyses in multiple linear regression include a partial F test, stepwise regression, and best subsets regression.

Mean Value of Y and Observed Value of Y

Two very useful inferences in multiple linear regression involve an estimate of the mean value of Y for $x = x^*$ and an estimate of an observed value of Y for $x = x^*$. Recall from the simple linear regression case that the error in estimating the mean value of Y is less than the error in estimating an observed value of Y. The difference is due to estimating a single value versus estimating the next value from an entire distribution.

We can conduct a test of $y^* = y_0^*$ using either a formal hypothesis test or a confidence interval. Most statistical software packages produce a confidence interval rather than conduct a hypothesis test.

If the multiple linear regression assumptions are true, then the hypothesis test and confidence interval concerning the mean value of Y for $x = x^*$ are based on the t distribution. Calculating the standard deviation by hand is challenging. An appropriate symbol is used in the discussion that follows, and we will rely on technology to provide the necessary calculations. The null and alternative hypotheses are stated in terms of the parameter $y^* = \beta_0 + \beta_1 x_1^* + \beta_2 x_2^* + \cdots + \beta_k x_k^*$. The random variable Y^* is used as an estimate of y^*.

Hypothesis Test and Confidence Interval Concerning the Mean Value of Y for $x = x^*$

$H_0: y^* = y_0^*$

$H_a: y^* > y_0^*, \quad y^* < y_0^*, \quad$ or $\quad y^* \neq y_0^*$

TS: $T = \dfrac{(B_0 + B_1 x_1^* + \cdots + B_k x_k^*) - y_0^*}{S_{Y^*}}$

RR: $T \geq t_{\alpha, n-k-1}, \quad T \leq -t_{\alpha, n-k-1}, \quad$ or $\quad |T| \geq t_{\alpha/2, n-k-1}$

A $100(1-\alpha)\%$ confidence interval for $\mu_{Y|x^*}$, the mean value of Y for $x = x^*$, has the following values as endpoints:

$$(\hat{\beta}_0 + \hat{\beta}_1 x_1^* + \cdots + \hat{\beta}_k x_k^*) \pm t_{\alpha/2, n-k-1} \cdot s_{Y^*} \tag{12.15}$$

Recall: $s = \sqrt{MSE}$

Prediction Interval for an Observed Value of Y

A $100(1-\alpha)\%$ prediction interval for an observed value of Y when $x = x^*$ has the following values as endpoints:

$$(\hat{\beta}_0 + \hat{\beta}_1 x_1^* + \cdots + \hat{\beta}_k x_k^*) \pm t_{\alpha/2, n-k-1} \cdot \sqrt{s^2 + s_{Y^*}^2} \tag{12.16}$$

EXAMPLE 12.13 Shear Stress

SHEAR

The total weight of an automobile contributes to its fuel consumption and, therefore, to its CO_2 emissions. Recent automobile design research has focused on lighter materials to reduce total weight. However, the strength and durability of any new product must also be considered. Data were obtained from several new chassis designs and used to fit a multiple linear regression equation of the form $y = \beta_0 + \beta_1 x_1 + \beta_2 x_2 + \beta_3 x_3$, where

$y =$ shear stress, in megapascals, MPa

$x_1 =$ thickness of web, in millimeters

$x_2 =$ thickness of upper flange, in millimeters

$x_3 =$ thickness of lower flange, in millimeters

(a) Construct a 95% confidence interval for the mean chassis shear stress when $x_1 = 3$, $x_2 = 5$, and $x_3 = 7$. Use this confidence interval to determine whether there is any evidence that the mean chassis shear stress for these values is different from 125 MPa.

(b) Construct a 95% prediction interval for an observed value of chassis shear stress when $x_1 = 3$, $x_2 = 5$, and $x_3 = 7$.

Solution

(a) The R command `lm()` was used to find the best linear model. **Figure 12.51** shows the ANOVA table and the parameter estimates, and **Figure 12.52** shows the confidence interval and prediction interval.

```
Analysis of Variance Table
Response: y
          Df Sum Sq Mean Sq F value   Pr(>F)
x1         1 1529.9 1529.93 11.2755 0.002429 **
x2         1 1717.0 1717.02 12.6543 0.001466 **
x3         1  671.1  671.14  4.9463 0.035043 *
Residuals 26 3527.8  135.69

Coefficients:
            Estimate Std. Error t value Pr(>|t|)
(Intercept)  207.354     17.468  11.871 5.33e-12 ***
x1            -7.797      1.806  -4.317 0.000204 ***
x2            -5.665      1.886  -3.004 0.005826 **
x3            -5.041      2.266  -2.224 0.035043 *

Residual standard error: 11.65 on 26 degrees of freedom
Multiple R-squared:  0.5262,     Adjusted R-squared:  0.4715
F-statistic: 9.625 on 3 and 26 DF,  p-value: 0.0001898
```

Figure 12.51 Analysis of variance table and parameter estimates.

```
> x_star <- data.frame(x1=3,x2=5,x3=7)
> predict(results,x_star,interval='confidence',level=0.95)
       fit      lwr      upr
1 120.3546 108.6389 132.0704
> predict(results,x_star,interval='prediction',level=0.95)
       fit      lwr     upr
1 120.3546 93.69827 147.011
```

Figure 12.52 Confidence interval and prediction interval.

The estimated regression equation is

$$y = 207.354 - 7.797x_1 - 5.665x_2 - 5.041x_3$$

The model utility test (using the ANOVA table) is significant at the $p = 0.0002$ level, and each predictor variable is also significant. These results together with $r^2 = 0.5262$ suggest that the model is effective in predicting the chassis shear stress.

Using Figure 12.52, a 95% confidence interval for the true mean chassis shear stress when $x_1 = 3$, $x_2 = 5$, and $x_3 = 7$ is (108.64, 132.07). Because 125 is included in this interval, there is no evidence to suggest that the mean shear stress is different from 125.

(b) Using Figure 12.52, a 95% prediction interval for a single observation of the chassis shear stress when $x_1 = 3$, $x_2 = 5$, and $x_3 = 7$ is (93.70, 147.01). Notice that this 95% prediction interval is larger (wider) than the corresponding confidence interval. ∎

TRY IT NOW   Go to Exercise 12.161

Checking Assumptions

The assumptions in a multiple linear regression model are also given in terms of the random deviations, E_i, $i = 1, 2, \ldots, n$. As in a simple linear regression model, it is assumed that the E_i's are independent, normal random variables with mean 0 and constant variance σ^2. If any of these assumptions is violated, then the resulting analysis is unreliable. The residuals $\hat{e}_i = y_i - (\hat{\beta}_0 + \hat{\beta}_1 x_{1i} + \cdots + \hat{\beta}_k x_{ki})$, $i = 1, 2, \ldots, n$ are estimates of the random errors and are used to check the regression assumptions.

Here are some graphical procedures, developed for a simple linear regression model, that can be extended to a multiple linear regression model.

1. Construct a histogram, stem-and-leaf plot, scatter plot, and/or normal probability plot of the residuals. These graphs may all be used to check the normality assumption.
2. Construct a scatter plot of the residuals versus each independent variable. For example, the first scatter plot has the ordered pairs $(x_{1i}, \hat{e}_i)$, the second has the ordered pairs $(x_{2i}, \hat{e}_i)$, and so on. If the assumptions are not violated, each scatter plot should appear as a horizontal band around 0. There should be no recognizable pattern. Typical patterns in a residual plot that suggest a violation of the assumptions include a distinct curve, a nonconstant spread, an unusually large residual, or an outlier.

EXAMPLE 12.14 Hurricane Damage

DAMAGE Research by the insurance industry suggests that the building damage due to a hurricane is related to several vulnerability indicators. Suppose several buildings damaged due to hurricanes in 2017 and 2018 were selected at random. These data were recorded for each building: the damage ratio, the quotient of the insurance payout and the building's appraised value times 100, y; the maximum sustained wind speed, in meters per second, x_1; the building age, in years, x_2; and the distance from the shoreline, in 1000 m, x_3. Figure 12.53 shows the estimated regression coefficients. Construct a normal probability plot of the residuals and sketch a graph of the residuals versus each predictor variable. Discuss any indication of violations of the regression assumptions.

```
Analysis of Variance Table
Response: y
          Df Sum Sq Mean Sq F value    Pr(>F)
x1         1  6.922   6.922  4.7298   0.04499 *
x2         1 73.218  73.218 50.0282 2.637e-06 ***
x3         1  0.415   0.415  0.2837   0.60160
Residuals 16 23.416   1.464


Coefficients:
            Estimate Std. Error t value Pr(>|t|)
(Intercept)  5.71076    2.87183   1.989   0.0641 .
x1           0.20847    0.07851   2.655   0.0173 *
x2           0.22578    0.03183   7.093 2.55e-06 ***
x3          -0.13504    0.25353  -0.533   0.6016

Residual standard error: 1.21 on 16 degrees of freedom
Multiple R-squared:  0.7748,    Adjusted R-squared:  0.7326
F-statistic: 18.35 on 3 and 16 DF,  p-value: 1.976e-05
```

Figure 12.53 R regression analysis.

Solution

STEP 1 Using technology, the normal probability plot of the residuals is shown in **Figure 12.54**. The points appear to lie close to an approximately straight line. There is no overwhelming evidence to suggest that the normality assumption is violated.

STEP 2 **Figure 12.55**, **Figure 12.56**, and **Figure 12.57** show the residual plots: $\hat{e}_i$ versus x_1, $\hat{e}_i$ versus x_2, and $\hat{e}_i$ versus x_3, respectively. There is no obvious pattern in any of these graphs.

Together, these four graphs suggest that there is no violation of the multiple linear regression model assumptions.

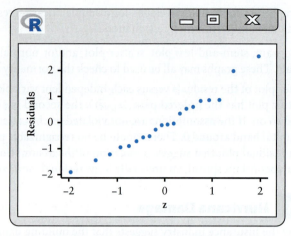

Figure 12.54 Normal probability plot of the residuals. **Figure 12.55** Plot of the residuals versus the first predictor.

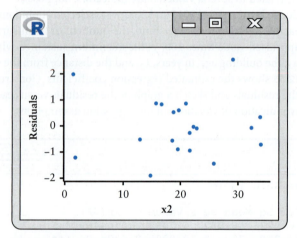 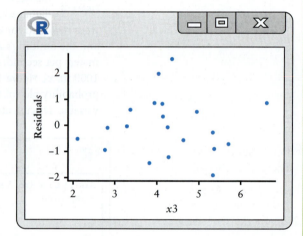

Figure 12.56 Plot of the residuals versus the second predictor. **Figure 12.57** Plot of the residuals versus the third predictor.

TRY IT NOW   **Go to Exercise 12.155**

Many other models can also be considered to explain the variation in a dependent variable y. For example, a quadratic model with one predictor variable has the form

$$Y_i = \beta_0 + \beta_1 x_i + \beta_2 x_i^2 + E_i$$

A more general kth-degree polynomial model with one predictor has the form

$$Y_i = \beta_0 + \beta_1 x_i + \beta_2 x_i^2 + \cdots + \beta_k x_i^k + E_i$$

A model with two predictor variables and an interaction term has the form

$$Y_i = \beta_0 + \beta_1 x_{1i} + \beta_2 x_{2i} + \beta_3 x_{1i} x_{2i} + E_i$$

These models are still considered linear regression models because each is a linear combination of the regression coefficients, that is, a sum of terms of the form $\beta_i \nu_i$, where ν_i is a function of one or more predictor variables.

Some other models do not appear to be linear but are fundamentally linear. These models can be transformed into linear models, and the techniques presented in this chapter can be used to estimate the regression coefficients. For example, the exponential model

$$Y_i = \beta_0 e^{\beta_1 x_i} E_i$$

is essentially linear. However, the general growth model

$$Y_i = \beta_0 + \beta_1 e^{\beta_2 x_i} + E_i$$

cannot be made into a linear model. The power model and the general logistic model are other commonly used models. No matter which model you decide is best, you should use technology to estimate the unknown parameters.

Section 12.5 Exercises

Concept Check

12.126 Short Answer State the four assumptions in a multiple linear regression model in terms of the error terms.

12.127 Fill in the Blank The true regression equation $y = \beta_0 + \beta_1 x_1 + \beta_2 x_2 + \cdots + \beta_k x_k$ is a linear function of _____.

12.128 Fill in the Blank In the regression equation $y = \beta_0 + \beta_1 x_1 + \beta_2 x_2 + \cdots + \beta_k x_k$, β_i represents the mean change in y for every increase of _____ if the values of all the other predictor variables are held constant.

12.129 Fill in the Blank In a multiple linear regression model, _____ is used to minimize the sum of squares due to error.

12.130 True or False There can be no more than five predictor variables in a multiple linear regression model.

12.131 True or False The graph of the estimated regression equation $y = \hat{\beta}_0 + \hat{\beta}_1 x_1 + \hat{\beta}_2 x_2$ is a line in space.

12.132 Short Answer A test for a significant multiple linear regression model is based on what ratio?

12.133 Short Answer In a hypothesis test for a significant multiple linear regression, suppose we reject the null hypothesis. What does this suggest?

12.134 True or False Suppose a test for a significant multiple linear regression model is significant. Every predictor in the model helps determine the value of the dependent variable.

12.135 Short Answer In a multiple linear regression model with k predictor variables, we should not set the significance level in each individual hypothesis test concerning β_i. Why?

12.136 Short Answer Suppose a test for a significant multiple linear regression model is significant. What is the most reasonable way to select the best model?

12.137 Fill in the Blank In a multiple linear regression model, the value of Y for $x = x^*$ can be used as an estimate of _____ and _____.

12.138 Short Answer Describe two type of plots used to check the assumptions in a multiple linear regression model.

12.139 True or False For a fixed confidence level, a prediction interval for an observed value of Y given $x = x^*$ is always wider than the associated confidence interval for the mean value of Y given $x = x^*$.

Practice

12.140 Suppose the true regression equation relating the variables x_1, x_2, and y for values of x_1 between 50 and 100 and for values of x_2 between 4.5 and 11 is $y = 32.0 + 7.5 x_1 - 5.3 x_2$.

(a) Find the expected value of Y when $x_1 = 65$ and $x_2 = 7$.
(b) Use the sign of the coefficient of x_1 to explain the relationship between the predictor variable x_1 and y.
(c) How much of a change in the dependent variable is expected when x_2 increases by 1 unit?
(d) Suppose $\sigma = 3.7$. Find the probability that an observed value of Y is less than 554 when $x_1 = 77$ and $x_2 = 9.8$.

12.141 Suppose the true regression equation relating the variables x_1, x_2, x_3, x_4, and y is
$y = -10.7 + 5 x_1 - 14 x_2 - 23 x_3 + 6.7 x_4$.

(a) Find the expected value of Y when $x_1 = 1$, $x_2 = 2$, $x_3 = 2$, and $x_4 = 4.5$ [or $x = (1, 2, 2, 4.5)$].
(b) How much change in the dependent variable is expected when x_2 increases by 2 units and x_4 decreases by 5 units?
(c) Suppose $\sigma = 24$. Find the probability that an observed value of Y is between -460 and 400 when $x = (10, 12.5, 15, 7)$.

12.142 Suppose a multiple linear regression model is used to explain the relationship between y, x_1, and x_2. A random sample of $n = 12$ pairs for the independent variables was selected, and the corresponding values of the dependent variable were observed. The data are given in the table. EX12.142

y	196.7	203.4	227.0	221.3	154.8	185.1
x_1	7.8	8.7	4.0	2.6	1.9	3.1
x_2	22.5	24.0	27.4	27.8	19.0	21.8

y	198.7	200.9	220.8	193.6	206.2	214.1
x_1	8.3	7.8	2.2	2.4	3.9	6.4
x_2	22.1	23.4	26.6	24.6	24.3	25.6

(a) Find the estimated regression equation.
(b) What is the predicted value of Y when $x_1 = 3.3$ and $x_2 = 20$?

12.143 An experiment resulted in observations on a single dependent variable and three independent variables. EX12.143
(a) Construct three separate scatter plots of y versus x_1, y versus x_2, and y versus x_3. Describe the relationship in each plot.
(b) Estimate the regression coefficients in the model $Y_i = \beta_0 + \beta_1 x_{1i} + \beta_2 x_{2i} + \beta_3 x_{3i} + E_i$. Explain the relationship between the sign of each estimated regression coefficient and the scatter plots in part (a).
(c) Estimate the true mean value of Y when $x_1 = -10$, $x_2 = 1.35$, and $x_3 = 52.6$.

12.144 Data from an observational study were used to fit a multiple linear regression model, and the summary statistics are given:

$n = 23 \qquad SSE = 80.75 \qquad SST = 152.5 \qquad MSE = 4.75$

(a) Complete the ANOVA summary table.
(b) How many predictor variables are in the multiple linear regression model?
(c) Conduct a model utility test with $\alpha = 0.05$. Find bounds on the p value associated with this test. Use technology to find the exact p value and to support your answer.
(d) Find the value of r^2 and explain the meaning of this value.

12.145 An experiment was conducted and the data obtained were used to fit the multiple linear regression model $Y_i = \beta_0 + \beta_1 x_{1i} + \beta_2 x_{2i} + \beta_3 x_{3i} + \beta_4 x_{4i} + E_i$. R was used to estimate the regression coefficients, and the results are shown in the figure. EX12.145

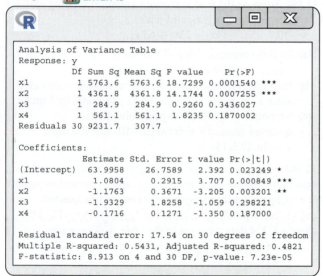

(a) Is the overall regression significant? Conduct the appropriate hypothesis test and justify your answer.
(b) Conduct four hypothesis tests with $H_0: \beta_i = 0$ for $i = 1, 2, 3, 4$ and $\alpha = 0.05$ in each test. Which regression coefficients are significantly different from 0, and therefore which predictor variables are significant?
(c) Conduct four hypothesis tests with $H_0: \beta_i = 0$ for $i = 1, 2, 3, 4$ with an overall type I error of $\alpha = 0.05$. Using these results, which predictor variables are significant? Compare your answers in parts (b) and (c).

12.146 An experiment resulted in the observations on a single dependent variable and three independent variables given in the table. EX12.146

y	x_1	x_2	x_3
−31.56	3.76	0.94	16.5
−44.48	4.47	0.23	17.2
−26.28	1.97	0.52	15.7
−27.54	1.16	0.45	15.8
−22.68	0.73	0.48	16.7
−20.34	0.82	0.63	16.5
−46.19	5.84	0.27	18.2
−29.00	2.69	0.61	16.4
−31.60	3.08	0.46	16.3
−34.31	5.38	0.73	17.5

(a) Estimate the regression coefficients in the model $Y_i = \beta_0 + \beta_1 x_{1i} + \beta_2 x_{2i} + \beta_3 x_{3i} + E_i$.
(b) Conduct an F test for a significant regression. Use $\alpha = 0.05$. Use technology to find the exact p value.
(c) Find the coefficient of determination.
(d) Which predictor variable(s) is (are) significant in explaining the variation in the dependent variable? Conduct the appropriate hypothesis tests.
(e) Construct a 95% confidence interval for the constant regression coefficient β_0. Use this interval to determine whether there is evidence to suggest the constant regression coefficient is different from 0.

12.147 An experiment resulted in observations on a single dependent variable and four independent variables. EX12.147
(a) Estimate the regression coefficients in the model $Y_i = \beta_0 + \beta_1 x_{1i} + \beta_2 x_{2i} + \beta_3 x_{3i} + \beta_4 x_{4i} + E_i$. Find the coefficient of determination.
(b) Conduct the appropriate hypothesis tests to determine which predictor variables are significant (equivalently, which regression coefficients are significantly different from 0).
(c) Using the results from part (b), write the reduced multiple linear regression model (including only the significant predictor variables). Estimate the regression coefficients and find the coefficient of determination in this reduced model.
(d) Compare the value of r^2 in the two models. Which model do you think is better? Why?

12.148 An experiment was conducted to determine whether the freezing point of a certain solution can be predicted from the concentrations of three chemicals. Twenty combinations of chemical concentrations were studied. EX12.148
(a) Estimate the regression coefficients in the model $Y_i = \beta_0 + \beta_1 x_{1i} + \beta_2 x_{2i} + \beta_3 x_{3i} + E_i$.
(b) Compute the residuals and construct a normal probability plot for the residuals. Is there any evidence to suggest that the random error terms are not normal? Justify your answer.
(c) Construct a graph of the residuals versus each predictor variable. Is there any evidence of a violation of the regression assumptions? If so, what modification(s) could you make to improve the model?

12.149 The amount of sludge buildup in an automobile engine can severely affect performance. In an article in *Vanagon Maintenance Guide*, a technical editor discussed the advantages of synthetic lubricants. Suppose an experiment was conducted to determine whether the sludge buildup in an automobile engine can be predicted by the oil viscosity, oxidation inhibitors, detergents, dispersants, and/or anti-wear additives. The data obtained were used to fit the multiple linear regression model $Y_i = \beta_0 + \beta_1 x_{1i} + \beta_2 x_{2i} + \beta_3 x_{3i} + \beta_4 x_{4i} + \beta_5 x_{5i} + E_i$. R was used to estimate the regression coefficients, and the results are shown here. SLUDGE

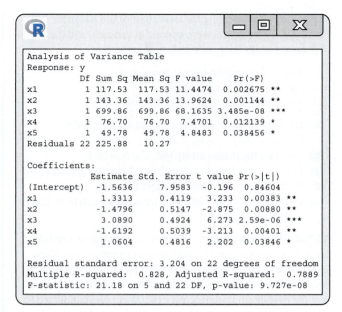

Consider the value $x^* = (6.1, 5.5, 8.3, 7.2, 6.5)$ and suppose an estimate of the standard deviation of Y^* is $s_{Y^*} = 1.973$.

(a) Find a 95% confidence interval for the true mean value of Y when $x = x^*$. Give an interpretation of this interval.

(b) Find a 95% prediction interval for an observed value of Y when $x = x^*$. Give an interpretation of this interval.

Applications

12.150 Economics and Finance A financial analyst believes that the U.S. 6-month Treasury bill yield (y) can be predicted from the crude oil price per barrel (x_1), the price of gold (x_2), and the M1 money supply (x_3, in billions of dollars). A random sample of days was selected and the data were recorded for those days.[20] YIELD

(a) Find the estimated regression line. Conduct an F test for a significant regression with $\alpha = 0.05$.

(b) Find an estimate of the true mean 6-month Treasury bill yield for $x_1 = 50$, $x_2 = 1200$, and $x_3 = 3200$.

(c) Which of these variables do you think is the most important predictor for the 6-month Treasury bill yield? Why?

12.151 Business and Management A study was conducted to determine if the rental price of an apartment in a college town is related to the distance from campus and the apartment size. A random sample of apartments in State College, Pennsylvania, was obtained, and the monthly rental price (y, in dollars), distance from campus (x_1, in miles), and square feet (x_2) were recorded for each. RENT

(a) Estimate the regression coefficients in the model $Y_i = \beta_0 + \beta_1 x_{1i} + \beta_2 x_{2i} + E_i$. Interpret the value of each estimated coefficient.

(b) Do these predictor variables explain a significant amount of variation in Y? Conduct the appropriate model utility test using $\alpha = 0.05$.

(c) Find an estimate of an observed value of Y for $x_1 = 1.5$ and $x_2 = 1200$.

12.152 Physical Sciences An experiment was conducted to test the effect of temperature (x_1, °F), contact area (x_2, cm²), and wood density (x_3, g/cm³) on the bonding strength (y) of a certain (wood) glue. Each piece of wood was glued to the same control block, and the entire fixture was then placed in an industrial oven. The bonding strength was measured by recording the time (in minutes) until the test piece of wood separated from the control block. The multiple linear regression model equation is $Y_i = \beta_0 + \beta_1 x_{1i} + \beta_2 x_{2i} + \beta_3 x_{3i} + E_i$. R was used to estimate the regression coefficients, and a portion of the results is shown here. WOOD

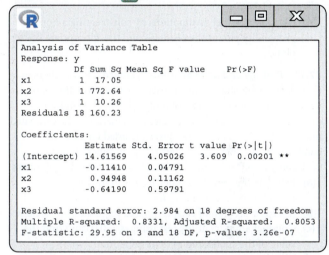

(a) Complete the ANOVA table and conduct a model utility test.

(b) Interpret the sign of each estimated regression coefficient.

(c) Conduct three hypothesis tests with $H_0: \beta_i = 0$ and $\alpha = 0.05$ in each test. Find bounds on the p value associated with each test. Use these results to determine which predictor variables are the most important in determining the bonding strength of glue.

12.153 Physical Sciences A Daniel cell uses zinc and copper solutions to produce electricity. An experiment was conducted to determine whether voltage (y) is affected by the temperature of the solutions (x_1, in °F), the concentration of the solutions (x_2, in molarity, M), and/or the surface area of the electrodes (x_3, in cm²). DANIEL

(a) Estimate the regression coefficients in the model $Y_i = \beta_0 + \beta_1 x_{1i} + \beta_2 x_{2i} + \beta_3 x_{3i} + E_i$.

(b) Conduct a model utility test and the other appropriate tests to determine which variables are the most important predictors of voltage.

(c) Construct a normal probability plot of the residuals. Use this plot to determine whether there is any evidence of violation of the multiple linear regression assumptions.

12.154 Biology and Environmental Science To keep golf courses attractive in the arid southwest, groundskeepers carefully monitor the salt content of the water used for irrigation. An excessively high salt concentration can damage plants and grass. A random sample of golf courses in the southwest was obtained and a sample of water used for irrigating was obtained for each. The following variables were measured for each water sample: dissolved salt (y, in dS/m); total dissolved solids (x_1, in mg/L); and sodium, calcium, magnesium, and bicarbonate ($x_2, x_3, x_4,$ and x_5, all in mEq/L). COURSE

(a) Estimate the regression coefficients in the model
$$Y_i = \beta_0 + \beta_1 x_{1i} + \beta_2 x_{2i} + \beta_3 x_{3i} + \beta_4 x_{4i} + \beta_5 x_{5i} + E_i.$$
(b) Construct the ANOVA table and conduct the model utility test.
(c) Conduct the necessary hypothesis tests to determine the most important variables in predicting dissolved salt. What do you think is the single most important predictor variable? Why?
(d) Check the model assumptions by constructing a normal probability plot of the residuals and the appropriate scatter plots.

12.155 Biology and Environmental Science Researchers in Canada have investigated the impacts on birds from the increased use of wind turbines and loss of nesting habitat. A random sample of Canadian bird habitats was obtained, and the estimated number of nests (y, per hectare) was recorded for each. In addition, the number of wind farms (x_1), number of turbines (x_2), and area of the habitat (x_3, in ha) were recorded. NESTS

(a) Estimate the regression coefficients in the model
$$Y_i = \beta_0 + \beta_1 x_{1i} + \beta_2 x_{2i} + \beta_3 x_{3i} + E_i.$$
(b) Interpret the sign of each estimated regression coefficient.
(c) Conduct a model utility test using $\alpha = 0.05$ and find the value of the coefficient of determination.
(d) Which predictor variable do you believe is most important? Why?
(e) Check the model assumptions by constructing a normal probability plot of the residuals and the appropriate scatter plots.

12.156 Physical Sciences It is important to correctly characterize boiler efficiency in order to maintain peak performance. However, traditional methods for computing boiler efficiency are time consuming and very expensive. A study was conducted to investigate a quicker, more economical method to compute boiler efficiency by considering the relationship between boiler efficiency (y), loss due to dry flue gas (x_1), loss due to hydrogen in the fuel (x_2), and loss due to moisture (x_3), all measured as percentages.[21] BOILER

(a) Estimate the regression coefficients in the model
$$Y_i = \beta_0 + \beta_1 x_{1i} + \beta_2 x_{2i} + \beta_3 x_{3i} + E_i.$$
(b) Interpret the sign of each estimated regression coefficient.
(c) Conduct a model utility test using $\alpha = 0.05$. Find the value of the coefficient of determination and interpret this value.
(d) Check the model assumptions by constructing a normal probability plot of the residuals and the appropriate scatter plots.

Extended Applications

12.157 Biology and Environmental Science Some evidence suggests that variation in the size of penguin colonies is related to sea ice extent. Suppose a study was conducted to investigate the effect of sea ice extent and stormy weather on the size of penguin colonies along the west coast of the Antarctic Peninsula. Fifteen years were selected at random, and the values of these variables were recorded. PENGUIN

y = size of the penguin colony population
x_1 = sea ice extent (as a percentage)
x_2 = number of stormy days (yearly total)

(a) Estimate the regression coefficients in the model
$$Y_i = \beta_0 + \beta_1 x_{1i} + \beta_2 x_{2i} + E_i.$$
(b) Conduct the model utility test. Use $\alpha = 0.01$.
(c) Interpret the sign of each estimated regression coefficient.
(d) Find the value of r^2 and interpret this value.
(e) Estimate the mean value of Y for a sea ice extent of 12.5% and 35 stormy days.

12.158 Physical Sciences Potassium ferricyanide crystals are used for etching and in eloctroplating applications. Many factors affect the crystal growth and, therefore, the size of these crystals. These factors include the solubility of the compound and the evaporation process. An experiment was conducted to determine whether the weight of the final crystal (y, in grams) is affected by the amount of initial solid (x_1, in grams), the amount of water (x_2, in milliliters), and/or the temperature of the water (x_3, in °F). R was used to estimate the regression coefficients in the model $Y_i = \beta_0 + \beta_1 x_{1i} + \beta_2 x_{2i} + \beta_3 x_{3i} + E_i$, and a portion of the results is shown here. CRYSTAL

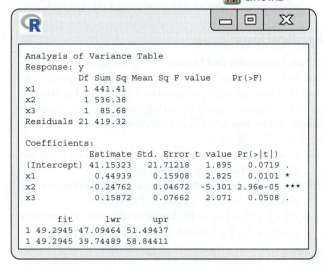

```
Analysis of Variance Table
Response: y
          Df  Sum Sq  Mean Sq  F value   Pr(>F)
x1         1  441.41
x2         1  536.38
x3         1   85.68
Residuals 21  419.32

Coefficients:
             Estimate Std. Error t value Pr(>|t|)
(Intercept)  41.15323   21.71218   1.895   0.0719 .
x1            0.44939    0.15908   2.825   0.0101 *
x2           -0.24762    0.04672  -5.301 2.96e-05 ***
x3            0.15872    0.07662   2.071   0.0508 .

       fit      lwr      upr
1  49.2945 47.09464 51.49437
1  49.2945 39.74489 58.84411
```

(a) Complete a summary ANOVA table for multiple linear regression and conduct a model utility test.
(b) Find the value of r^2 and interpret this value.
(c) Use the R output to determine which of the three predictor variables contributes to the overall significant regression.
(d) Two intervals are shown in the R output. One is a confidence interval and one is a prediction interval for $x^* = (93, 200, 100)$. Identify each interval and justify your answer.

12.159 Biology and Environmental Science In many rural areas in countries around the world, it is very costly to construct an efficient water supply system. Therefore, groundwater is the principal source of water for drinking and other activities. Prior to use, the groundwater must be tested for safety and quality. Chloride in water affects the taste of many foodstuffs, and large amounts may lead to hypertension. However, little is known about the long-term effects. A recent multiple linear regression model was used to explain the variation in chloride in groundwater (y). The model included five predictor variables. The independent variables are x_1, total dissolved solids (mg/L); x_2, iron (mg/L); x_3, nitrate (mg/L); x_4, total hardness (ppm); and x_5, depth of sample (feet). CHLOR

A portion of the analysis associated with this model is shown here.

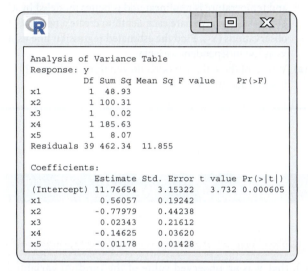

(a) Complete a summary ANOVA table for multiple linear regression and conduct a model utility test. Find the p value associated with this test. Find the value of r^2. Use the results to describe the overall significance of the model.

(b) Find the value of the test statistic and the p value associated with each hypothesis test for a significant regression coefficient. Use these results to determine the most important predictor variables in the model. For each of these variables, explain the effect on chloride in groundwater. Justify your answers.

12.160 Economics and Finance A recent study investigated the variables that can be used to predict economic growth in China, as measured by gross domestic product (GDP, y, in hundred millions of U.S. dollars). A sample of years was obtained and the following measurements were recorded: capital stock (x_1, in hundred millions of U.S. dollars); labor force (x_2, in ten thousands); and energy consumption (x_3, in ten thousand tons standard coal). GDP

(a) Estimate the regression coefficients in the model
$Y_i = \beta_0 + \beta_1 x_{1i} + \beta_2 x_{2i} + \beta_3 x_{3i} + E_i$.

(b) Conduct the model utility test. Use $\alpha = 0.05$.

(c) Find the value of r^2 and interpret this value. Explain this remarkable value in the context of this problem.

(d) Conduct the appropriate hypothesis tests to determine which regression coefficient(s) is (are) significantly different from 0. Use $\alpha = 0.05$ in each test.

(e) Do you believe there is any way to improve this model? Why or why not?

12.161 Public Health and Nutrition A recent experiment was designed to determine whether the percentage of lactic acid (y) in cultured buttermilk (produced from whole milk) can be predicted by the temperature to which the milk is heated (x_1, in °C), the temperature to which the milk is cooled before the started culture is added (x_2, in °C), the percentage of started culture added (x_3), and/or the fermentation time (x_4, in hours). LACTIC

(a) Estimate the regression coefficients in the model
$Y_i = \beta_0 + \beta_1 x_{1i} + \beta_2 x_{2i} + \beta_3 x_{3i} + \beta_4 x_{4i} + E_i$.

(b) Conduct a model utility test using $\alpha = 0.05$ and find the value of the coefficient of determination.

(c) Conduct the appropriate hypothesis tests to determine which predictor(s) is (are) significant.

(d) Conduct a hypothesis test to determine whether there is any evidence to suggest that β_3 is less than 0.93. Use $\alpha = 0.05$.

(e) Find a 95% confidence interval for the true mean amount of lactic acid in cultured buttermilk when $x_1 = 95$, $x_2 = 22$, $x_3 = 1.5$, and $x_4 = 20$.

12.162 The Rainy Season Hawaii is a popular tourist destination during the winter months, and Honolulu offers many outdoor attractions. However, the months of December through February are the rainy season, and records indicate that the most rain falls during the month of January. Using data from random years, a multiple linear regression model was used to explain the total January rainfall (y) that consisted of four predictor variables, one of which was a dummy variable (a variable with a value of either 0 or 1). The independent variables were monthly rainfall total in the previous December, January, and February (x_1, x_2, and x_3, respectively) and a dummy variable (x_4) that indicated whether there was a moderate to weak El Niño during the previous year. Rainfall totals were recorded in inches. HAWAII

(a) Construct the ANOVA table and conduct a model utility test. Use technology to find the p value associated with this test. Find the value of r^2. Use these results to describe the overall significance of the model.

(b) Find the value of the test statistic and the p value associated with each hypothesis test for a significant regression coefficient. Use these results to determine the most important predictor variables in this model.

(c) Consider a reduced model with the two most important predictor variables. Conduct a model utility test. Describe the significance of this model compared to the full model in part (a).

(d) Find the most recent rainfall totals for Honolulu and records on El Niño. Use this information to find a 95% prediction interval for the most recent January rainfall total in Honolulu. Does the actual value lie in this interval? What does this suggest about the reduced model?

Challenge Problems

12.163 Geothermal Gradient The temperature naturally increases as one digs deeper into the continental crust—a phenomenon called the geothermal gradient. A random sample of continental crust depths (x, in kilometers) was obtained, and the temperature at each depth was recorded (y, in °C). The data are given in the table. **GEOGRAD**

Depth	192	306	75	375	96	120	399
Temperature	212	682	15	855	93	49	943
Depth	469	439	209	224	313	379	296
Temperature	1456	1156	271	336	604	911	606
Depth	37	316	41	242	21	230	
Temperature	35	576	31	462	10	274	

(a) Sketch a scatter plot of these data.
(b) Find the estimated regression line, and complete the ANOVA summary table.
(c) Some evidence indicates that the rate of change in temperature differs according to the depth zone. For example, at small depths, the temperature changes only slightly for each kilometer. However, at large depths, the temperature change is much greater for each kilometer. Use your scatter plot from part (a) to divide the data into two depth zones. Find the estimated regression line and the ANOVA summary table for each zone.
(d) Some researchers believe the relationship between depth and temperature is nonlinear and is better modeled by a quadratic curve. Square each depth to create a new list of observations (x_2). Find the estimated regression line using x_2 as the independent variable.
(e) Which of these three models do you think is the best? Justify your answer.

Chapter 12 Summary

Simple Linear Regression Model

Let $(x_1, y_1), (x_2, y_2), \ldots, (x_n, y_n)$ be n pairs of observations such that y_i is an observed value of the random variable Y_i. We assume that there exist constants β_0 and β_1 such that

$$Y_i = \beta_0 + \beta_1 x_i + E_i$$

where $E_1, E_2, \ldots, E_n$ are independent, normal random variables with mean 0 and variance σ^2. That is,

1. The E_i's are normally distributed, which implies that the Y_i's are normally distributed.
2. The expected value of E_i is 0, which implies that $E(Y_i) = \beta_0 + \beta_1 x_i$.
3. $\text{Var}(E_i) = \sigma^2$, which implies that $\text{Var}(Y_i) = \sigma^2$.
4. The E_i's are independent, which implies that the Y_i's are independent.

The E_i's are the random deviations or random error terms. $y = \beta_0 + \beta_1 x$ is the true regression line.

Principle of least squares

The estimated regression line is obtained by minimizing the sum of the squared deviations, or vertical distances from the observed points to the line.

Least-squares estimates

$$\hat{\beta}_1 = \frac{n \sum x_i y_i - (\sum x_i)(\sum y_i)}{n \sum x_i^2 - (\sum x_i)^2} \qquad \hat{\beta}_0 = \frac{\sum y_i - \hat{\beta}_1 \sum x_i}{n} = \bar{y} - \hat{\beta}_1 \bar{x}$$

The estimated regression line is $y = \hat{\beta}_0 + \hat{\beta}_1 x$.
The ith predicted (fitted) value is $\hat{y}_i = \hat{\beta}_0 + \hat{\beta}_1 x_i$ ($i = 1, 2, \ldots, n$).
The ith residual is $\hat{e}_i = y_i - \hat{y}_i$.

The sum of squares

$$\underbrace{\sum (y_i - \bar{y})^2}_{\text{SST}} = \underbrace{\sum (\hat{y}_i - \bar{y})^2}_{\text{SSR}} + \underbrace{\sum (y_i - \hat{y}_i)^2}_{\text{SSE}}$$

Computational formulas

$$SST = S_{yy} = \sum(y_i - \bar{y})^2 = \sum y_i^2 - \frac{1}{n}(\sum y_i)^2$$

$$SSR = \hat{\beta}_1 S_{xy} = \hat{\beta}_1 \sum(x_i - \bar{x})(y_i - \bar{y}) = \hat{\beta}_1 \left[\sum x_i y_i - \frac{1}{n}(\sum x_i)(\sum y_i)\right]$$

$$SSE = SST - SSR$$

ANOVA summary table for simple linear regression

Source of variation	Sum of squares	Degrees of freedom	Mean square	F	p Value
Regression	SSR	1	$MSR = \frac{SSR}{1}$	$\frac{MSR}{MSE}$	p
Error	SSE	$n-2$	$MSE = \frac{SSE}{n-2}$		
Total	SST	$n-1$			

Coefficient of determination

$$r^2 = \frac{SSR}{SST}$$

Estimate of variance

$$s^2 = \frac{SSE}{n-2}$$

F test for a significant linear regression

H_0: There is no significant linear relationship ($\beta_1 = 0$).
H_a: There is a significant linear relationship ($\beta_1 \neq 0$).

TS: $F = \dfrac{MSR}{MSE}$

RR: $F \geq F_{\alpha, 1, n-2}$

Hypothesis test and confidence interval concerning β_1

H_0: $\beta_1 = \beta_{10}$
H_a: $\beta_1 \ne \beta_{10}$, $\quad \beta_1 < \beta_{10}$, $\quad$ or $\quad \beta_1 \neq \beta_{10}$

TS: $T = \dfrac{B_1 - \beta_{10}}{S/\sqrt{S_{xx}}} = \dfrac{B_1 - \beta_{10}}{S_{B_1}}$

RR: $T \geq t_{\alpha, n-2}, \quad T \leq -t_{\alpha, n-2}, \quad$ or $\quad |T| \geq t_{\alpha/2, n-2}$

A $100(1-\alpha)\%$ confidence interval for β_1 has as endpoints $\hat{\beta}_1 \pm t_{\alpha/2, n-2} \cdot s_{B_1}$.

Hypothesis test and confidence interval concerning β_0

H_0: $\beta_0 = \beta_{00}$
H_a: $\beta_0 > \beta_{00}$, $\quad \beta_0 < \beta_{00}$, $\quad$ or $\quad \beta_0 \neq \beta_{00}$

TS: $T = \dfrac{B_0 - \beta_{00}}{S\sqrt{\sum x_i^2/(nS_{xx})}} = \dfrac{B_0 - \beta_{00}}{S_{B_0}}$

RR: $T \geq t_{\alpha, n-2}, \quad T \leq -t_{\alpha, n-2}, \quad$ or $\quad |T| \geq t_{\alpha/2, n-2}$

A $100(1-\alpha)\%$ confidence interval for β_0 has as endpoints $\hat{\beta}_0 \pm t_{\alpha/2, n-2} \cdot s_{B_0}$.

Sample correlation coefficient

$$r = \frac{S_{xy}}{\sqrt{S_{xx}S_{yy}}} = \frac{\sum x_i y_i - \frac{1}{n}(\sum x_i)(\sum y_i)}{\sqrt{\left[\sum x_i^2 - \frac{1}{n}(\sum x_i)^2\right]\left[\sum y_i^2 - \frac{1}{n}(\sum y_i)^2\right]}}$$

Hypothesis test and confidence interval concerning the mean value of Y for $x = x^*$

$H_0: y^* = y_0^*$
$H_a: y^* > y_0^*,$ $y^* < y_0^*,$ or $y^* \neq y_0^*$
TS: $T = \dfrac{(\hat{\beta}_0 + \hat{\beta}_1 x^*) - y_0^*}{S\sqrt{(1/n) + [(x^* - \bar{x})^2 / S_{xx}]}}$
RR: $T \geq t_{\alpha, n-2},$ $T \leq -t_{\alpha, n-2},$ or $|T| \geq t_{\alpha/2, n-2}$

A $100(1 - \alpha)\%$ confidence interval for $\mu_{Y|x^*}$, the mean of Y for $x = x^*$, has as endpoints

$$(\hat{\beta}_0 + \hat{\beta}_1 x^*) \pm t_{\alpha/2, n-2} \cdot s \cdot \sqrt{\frac{1}{n} + \frac{(x^* - \bar{x})^2}{S_{xx}}}$$

Prediction interval for an observed value of Y

A $100(1 - \alpha)\%$ prediction interval for an observed value of Y when $x = x^*$ has as endpoints

$$(\hat{\beta}_0 + \hat{\beta}_1 x^*) \pm t_{\alpha/2, n-2} \cdot s \cdot \sqrt{1 + \frac{1}{n} + \frac{(x^* - \bar{x})^2}{S_{xx}}}$$

Regression diagnostics

1. Normal probability plot of the residuals. The points should lie close to an approximately straight line.
2. Scatter plot of residuals versus predictor variable. There should be no distinct pattern.

Multiple Linear Regression Model

Let $(x_{11}, x_{21}, \ldots, x_{k1}, y_1), (x_{12}, x_{22}, \ldots, x_{k2}, y_2), \ldots, (x_{1n}, x_{2n}, \ldots, x_{kn}, y_n)$ be n sets of observations such that y_i is an observed value of the random variable Y_i. We assume there exist constants $\beta_0, \beta_1, \ldots, \beta_k$ such that

$$Y_i = \beta_0 + \beta_1 x_{1i} + \beta_2 x_{2i} + \cdots + \beta_k x_{ki} + E_i$$

where $E_1, E_2, \ldots, E_n$ are independent, normal random variables with mean 0 and variance σ^2. That is,

1. The E_i's are normally distributed (which means that the Y_i's are normally distributed).
2. The expected value of E_i is 0 (which implies that $E(Y_i) = \beta_0 + \beta_1 x_{1i} + \beta_2 x_{2i} + \cdots + \beta_k x_{ki}$).
3. $\text{Var}(E_i) = \sigma^2$ (which implies that $\text{Var}(Y_i) = \sigma^2$).
4. The E_i's are independent (which implies that the Y_i's are independent).

The E_i's are the random deviations or random error terms.

$y = \beta_0 + \beta_1 x_1 + \beta_2 x_2 + \cdots + \beta_k x_k$ is the true regression line.

Principle of least squares

The estimated regression equation is obtained by minimizing the sum of the squared deviations between the observations and the estimated values.

The estimated regression equation is $y = \hat{\beta}_0 + \hat{\beta}_1 x_1 + \hat{\beta}_2 x_2 + \cdots + \hat{\beta}_k x_k$.
The ith predicted (fitted) value is $\hat{y}_i = \hat{\beta}_0 + \hat{\beta}_1 x_{1i} + \hat{\beta}_2 x_{2i} + \cdots + \hat{\beta}_k x_{ki}$ ($i = 1, 2, \ldots, n$).
The ith residual is $\hat{e}_i = y_i - \hat{y}_i$.

The sum of squares

$$\underbrace{\sum(y_i - \bar{y})^2}_{\text{SST}} = \underbrace{\sum(\hat{y}_i - \bar{y})^2}_{\text{SSR}} + \underbrace{\sum(y_i - \hat{y}_i)^2}_{\text{SSE}}$$

ANOVA summary table for multiple linear regression

Source of variation	Sum of squares	Degrees of freedom	Mean square	F	p Value
Regression	SSR	k	$\text{MSR} = \dfrac{\text{SSR}}{k}$	$\dfrac{\text{MSR}}{\text{MSE}}$	p
Error	SSE	$n-k-1$	$\text{MSE} = \dfrac{\text{SSE}}{n-k-1}$		
Total	SST	$n-1$			

Coefficient of determination

$$r^2 = \frac{\text{SSR}}{\text{SST}}$$

Estimate of variance

$$s^2 = \frac{\text{SSE}}{n-k-1}$$

F test for a significant multiple linear regression

$H_0: \beta_1 = \beta_2 = \cdots = \beta_k = 0$
  (None of the predictor variables helps explain the variation in y.)

$H_a: \beta_i \neq 0$ for at least one i
  (At least one predictor variable helps explain the variation in y.)

TS: $F = \dfrac{\text{MSR}}{\text{MSE}}$

RR: $F \geq F_{\alpha, k, n-k-1}$

Hypothesis test and confidence interval concerning β_i

$H_0: \beta_i = \beta_{i0}$
$H_a: \beta_i > \beta_{i0}, \qquad \beta_i < \beta_{i0}, \qquad \text{or} \qquad \beta_i \neq \beta_{i0}$

TS: $T = \dfrac{B_i - \beta_{i0}}{S_{B_i}}$

RR: $T \geq t_{\alpha, n-k-1}, \quad T \leq -t_{\alpha, n-k-1}, \quad \text{or} \quad |T| \geq t_{\alpha/2, n-k-1}$

A $100(1-\alpha)\%$ confidence interval for β_i has as endpoints

$\hat{\beta}_i \pm t_{\alpha/2, n-k-1} \cdot s_{B_i}$

Hypothesis test and confidence interval concerning the mean value of Y for $x = x^*$

$H_0: y^* = y_0^*$
$H_a: y^* > y_0^*, \qquad y^* < y_0^*, \qquad \text{or} \qquad y^* \neq y_0^*$

TS: $T = \dfrac{(B_0 + B_1 x_1^* + \cdots + B_k x_k^*) - y_0^*}{S_{Y^*}}$

RR: $T \geq t_{\alpha, n-k-1}, \quad T \leq -t_{\alpha, n-k-1}, \quad \text{or} \quad |T| \geq t_{\alpha/2, n-k-1}$

A $100(1-\alpha)\%$ confidence interval for $\mu_{Y|x^*}$, the mean value of Y for $x = x^*$, has as endpoints

$$(\hat{\beta}_0 + \hat{\beta}_1 x_1^* + \cdots + \hat{\beta}_k x_k^*) \pm t_{\alpha/2, n-k-1} \cdot s_{Y^*}$$

Prediction interval for an observed value of Y

A $100(1-\alpha)\%$ prediction interval for an observe value of Y when $x = x^*$ has as endpoints

$$(\hat{\beta}_0 + \hat{\beta}_1 x_1^* + \cdots + \hat{\beta}_k x_k^*) \pm t_{\alpha/2, n-k-1} \cdot \sqrt{s^2 + s_{Y^*}^2}$$

Regression diagnostics
1. Construct a histogram, stem-and-leaf plot, scatter plot, and/or normal probability plot of the residuals. These graphs are all used to check the normality assumption.
2. Construct a scatter plot of the residuals versus each independent variable. If there are no violations in assumptions, each scatter plot should appear as a horizontal band around 0. There should be no recognizable pattern.

Chapter 12 Exercises

Applications

12.164 Medicine and Clinical Studies Officials at the U.S. Environmental Protection Agency worked with hospital physicians in Anaheim, California, to examine the relationship between environmental pollution and illness. A random sample of summer days was selected, and at 2:00 P.M. on each day the air quality was assessed by measuring the concentration of carbon monoxide (x, in ppm). Illness was measured by counting the number of new patients (y) seen for respiratory problems at the hospital that day. Suppose the true regression line is $y = -5.1 + 0.9x$.
(a) Find the expected number of respiratory patients at the hospital on a day when the carbon monoxide concentration is 12 ppm.
(b) What is the expected change in the number of patients per day if the carbon monoxide concentration decreases by 5 ppm?
(c) Suppose $\sigma = 1.2$. Find the probability that an observed number of patients is between 5 and 7 when the carbon monoxide concentration is 13 ppm.

12.165 Economics and Finance An economist believes that a credit card company sets the spending limits for customers based on loyalty, that is, the number of years the customer has been a cardholder. A random sample of 22 cardholders was selected. The number of years since obtaining the card (x) was recorded as well as the spending limit (y, in thousands of dollars) on the card. The summary statistics are shown.

$\sum x_i = 155.2$ $\sum y_i = 392.0$ $\sum x_i y_i = 3655.2$
$\sum x_i^2 = 1508.8$ $\sum y_i^2 = 10{,}090.0$

(a) Find the estimated regression line.
(b) Estimate the true mean spending limit for a customer who has had this credit card for 10 years.
(c) Find an estimate of an observed value of the spending limit for a customer who has had this credit card for two years.

12.166 Biology and Environmental Science Farmers in northern Sweden sell one of the most expensive cheeses in the world, which is made from moose milk. Suppose a random sample of female moose was obtained, and the weight of each was measured (x, in kilograms). The amount of milk produced by each moose in one day was also measured (y, in liters). MOOSE
(a) Construct a scatter plot for these data.
(b) Find the estimated regression line. Add a graph of the estimated regression line to the scatter plot in part (a).
(c) Complete the ANOVA summary table, and conduct an F test for a significant regression. Use a significance level of 0.05.
(d) Remove the first observation, a possible outlier, from the data set. Find the new estimated regression line. Is this new regression significant at the 0.05 level?

12.167 Public Health and Nutrition Fragments of protein waste, known as beta amyloids, may be a reliable marker of Alzheimer's disease (AD). Recent studies suggest that these protein fragments accumulate in the brain and form hard plaques, which are found in many AD patients. However, as omega-3 intake increases, there is some evidence to suggest that the beta amyloid blood level decreases. Suppose a study was conducted to investigate this relationship. A random sample of adults was obtained, the amount of omega-3 that each consumed in a day (x, in milligrams) was estimated, and the beta amyloids level (y, in picograms per milliliter) was measured for each. OMEGA3
(a) Find the estimated regression line, and complete the ANOVA table.
(b) Conduct an F test for a significant regression. Use $\alpha = 0.01$.
(c) Find the coefficient of determination, r^2.
(d) Interpret your results. Do you believe that omega-3 intake is a good predictor of beta amyloid blood level? Why or why not?

12.168 Travel and Transportation Some evidence suggests that as Americans spend more time online, they spend less time on many routine duties and tasks, that is, less time working. A random sample of adults ages 25–55 was obtained, and each completed a survey concerning work and leisure activities. The amount of time spent online each day (x, in hours) and the amount of time spent working each day (y, in hours) for each are given in the table. ONLINE

Time online	Time working	Time online	Time working
3.1	7.6	5.8	7.0
5.9	10.1	0.0	9.9
2.2	11.9	5.8	8.7
1.0	11.1	5.0	10.2
3.9	4.0	3.0	7.3
2.8	11.0	2.2	8.5
3.9	11.4	1.0	14.5
0.5	10.0	5.6	9.2
2.7	10.3	5.0	10.2
2.4	11.9	5.6	6.1
0.5	13.9	0.8	13.0
5.2	10.0	0.9	8.9

(a) Sketch a scatter plot of these data.
(b) Find the sample correlation coefficient, and describe the relationship between time spent online and time spent working.
(c) Find the estimated regression line. Explain why the estimated regression coefficient $\hat{\beta}_1$ supports your answer in part (b).

12.169 Biology and Environmental Science In a study by the U.S. Forestry Service, researchers investigated the relationship between certain chemical elements and oak tree defoliation. A random sample of oak trees was obtained, and the concentration of phosphorus (x, in g/kg) in the leaves was measured for each tree. The percentage of defoliation (y) for each tree was carefully estimated at the end of the growing season. OAK
(a) Sketch a scatter plot of these data. Describe the relationship between these two variables.
(b) Find the estimated regression line. Is this regression significant? Conduct an appropriate test at the 0.05 level. Add a graph of this line to your scatter plot in part (a).
(c) Find the residuals, and sketch a graph of the residuals versus the predictor variable.
(d) How could this model be improved? Justify your answer.

12.170 Medicine and Clinical Studies In a study of hair regeneration, 30 men older than age 45 years who had used minoxidil for six months were randomly selected. The daily dose of minoxidil (x, in milligrams) and the hair density (y, in hairs per square millimeter) were recorded for each man. The summary statistics are shown.

$$S_{xx} = 21{,}423.0 \quad S_{yy} = 80.119 \quad S_{xy} = 165.593$$
$$\bar{x} = 49.033 \quad \bar{y} = 2.207$$

(a) Find the estimated regression line.
(b) Complete the ANOVA table, and conduct an F test for a significant regression. Does the dosage of minoxidil help explain variation in hair density? Justify your answer.
(c) Find a 95% confidence interval for the mean hair density for a man taking a daily 50-mg dose of minoxidil.

12.171 Economics and Finance A statistician (and cigar smoker) recently studied the relationship between *Cigar Aficionado*'s blind ratings of cigars and price. The purpose of this study was to determine whether premium cigars are really more expensive. A random sample of 50 cigars was selected, and the rating (x) and price per cigar (y, in dollars) were recorded for each. The summary statistics are shown.

$$S_{xx} = 185.725 \quad S_{yy} = 1481.83 \quad S_{xy} = 111.877$$
$$\bar{x} = 5.6084 \quad \bar{y} = 11.7654$$

(a) Find the estimated regression line.
(b) Complete the ANOVA table.
(c) Conduct a test of $H_0: \beta_1 = 0$ versus $H_a: \beta_1 \neq 0$ with a significance level of 0.05.
(d) Do you believe the saying "You get what you pay for" applies to cigars? Justify your answer.

12.172 Manufacturing and Product Development The Consumer Federation of America recently studied the relationship between tensile strength and amount of nickel in household stainless steel products. The percentage of nickel by weight (x) and the tensile strength (y, in megapascals) were measured for each product. The data are given in the table. NICKEL

x	5.9	2.8	5.3	4.6	4.6	3.5	3.8	5.0	5.3	4.2
y	948	859	921	909	915	876	828	964	964	900

(a) Find the estimated regression line, and complete the ANOVA table.
(b) Compute the residuals.
(c) Sketch a normal probability plot of the residuals, and interpret the graph.
(d) Sketch a scatter plot of the residuals versus the predictor variable. Is there any indication of a violation of the regression model assumptions?

12.173 Marketing and Consumer Behavior Black Friday, the Friday after Thanksgiving, is traditionally the beginning of the holiday shopping season and the busiest shopping day of the year. An economist believes there is a relationship between the time of day at which a consumer starts shopping on Black Friday and the amount of money spent that day. Suppose a random sample of Black Friday shoppers was obtained, and the shopping start time (x, in hours after 6:00 A.M.) and the total amount spent (y, in hundreds of dollars) were recorded for each. FRIDAY
(a) Find the estimated regression line, and complete the ANOVA table.
(b) Describe the relationship between shopping start time and total amount spent on Black Friday. What would you recommend to retailers to increase business?
(c) Find a 95% prediction interval for an observed value of shopping start time of 8:00 A.M.

(d) Suppose the shopping start time is 9:00 A.M. Is there any evidence to suggest that the true mean amount spent is greater than $725? Use a significance level of 0.05.

12.174 Economics and Finance Scientists at the U.S. Mint believe that the number of years a quarter is in circulation is linearly related to the condition of the coin. To test this theory, a random sample of quarters was obtained. The number of years in circulation (x) was recorded for each coin, and each quarter was assessed for wear using the Official American Numismatic Association Grading Standards for U.S. Coins (y, a 70-point scale). A simple linear regression analysis was performed, and the partial ANOVA table was obtained.

ANOVA summary table

Source of variation	Sum of squares	Degrees of freedom	Mean square	F	p Value
Regression					
Error	7671.0				
Total	11148.4	54			

(a) Complete the ANOVA table.
(b) Find an estimate of the variance of the random error terms.

12.175 Biology and Environmental Science Every 13 or 17 years, a brood of cicadas, sometimes in biblical proportions, emerges from underground in the northeastern part of the United States. Although harmless to humans, cicadas cause damage to small trees and shrubs, and they make a piercing, irritating sound. Some research suggests that the density of cicadas is related to the density of moles, natural predators of cicada. Plots were randomly selected, and the density of moles per acre (x) and the density of cicadas per acre (y, in millions) were carefully estimated. The data are given in the table. CICADA

x	4	6	3	0	2	4	0	5	4
y	1.5	0.8	1.2	1.4	1.5	1.5	1.7	1.2	1.1

(a) Sketch a scatter plot of the data. Describe the relationship between the density of moles and the density of cicadas.
(b) Find the estimated regression line. Conduct a hypothesis test of $H_0: \beta_1 = 0$ versus $H_a: \beta_1 \neq 0$. Use a significance level of 0.05.
(c) Construct a normal probability plot of the residuals and a scatter plot of the residuals versus the predictor variable. Is there any evidence that the regression assumptions are not satisfied?

12.176 Travel and Transportation Since airlines began charging for checked luggage, more passengers have used carry-on bags to avoid the extra fee. For those who check luggage, the wait at the baggage carousel while jostling with other passengers can be long and infuriating. A study was conducted to determine whether there is a relationship between the flight time (x, in hours) and the amount of time for a bag to arrive from a plane to the carousel (y, in minutes). A random sample of arriving flights at the Denver International Airport was obtained, and the flight time and baggage time were recorded for each. BAGGAGE

(a) Construct a scatter plot for these data. Describe the relationship between flight time and the amount of time for a bag to arrive from a plane to the carousel.
(b) Find the estimated regression line, complete the ANOVA table, and conduct an F test for a significant regression. Use $\alpha = 0.05$.
(c) Based on your results in part (b), does flight time affect the time it takes for a bag to arrive at the carousel? What are some other variables that might affect the time it takes for a bag to arrive at the carousel?

12.177 Manufacturing and Product Development In structural tests of material to be used in an airplane's fuselage, rivets are cycled (tapped) until they crack. The number of cycles (x, in thousands) is recorded along with the detectable crack length (y, in millimeters). The data are given in the table. RIVET

Cycles	Length	Cycles	Length	Cycles	Length
128.0	0.193	118.1	0.165	199.0	0.191
246.4	0.167	100.7	0.191	100.8	0.177
188.8	0.209	205.9	0.208	131.1	0.207
152.3	0.180	107.6	0.153	145.8	0.193
129.9	0.153	182.5	0.169	229.1	0.191
101.5	0.172	132.8	0.187		

(a) Sketch a scatter plot of these data. Does there appear to be a linear relationship between the number of cycles and the length of the crack? Explain your answer.
(b) Find the estimated regression line, and complete the ANOVA table.
(c) How does the F test for a significant regression support your answer to part (a)?

12.178 Medicine and Clinical Studies Twenty-one adults were selected at random for a study of the effect of castor oil on the immune system. Participants took a certain amount of castor oil (x, in milliliters) each day for three months. At the end of this regimen, the white blood cell count was measured (y, in thousands of cells per microliter of blood) for each person. Analyze these data using any appropriate methods to determine whether there is a significant linear relationship between the amount of castor oil consumed and the number of white blood cells. Check the relevant assumptions. Assume that the more white blood cells, the stronger the immune system is, and draw a conclusion about the effect of castor oil on the immune system. CASTOR

12.179 Sports and Leisure Despite the rising cost of equipment, the number of fishing licenses issued in many states continues to increase. Officials are trying to predict the number of participants in freshwater fishing so that they can hire and deploy the appropriate number of fish and game wardens. They obtained data for the percentage of freshwater (x) and the freshwater fishing participation rate (y) by state. (*Note:* No data are given for Hawaii.) Analyze these data using any appropriate methods to determine whether there is a significant linear relationship between the percentage of freshwater and the participation rate. Check the relevant assumptions. Does the sign of $\hat{\beta}_1$ seem appropriate? Why or why not? FISH

Extended Applications

12.180 Manufacturing and Product Development A study was conducted to determine the effect of thread count (x_1) and the percentage of cotton (x_2) on the lifetime of standard twin bed sheets (y, in years). A machine was constructed to simulate continued usage and to allow a measurement of the lifetime. COTTON

(a) Estimate the regression coefficients in the model
$Y_i = \beta_0 + \beta_1 x_{1i} + \beta_2 x_{2i} + E_i$.

(b) Construct an ANOVA table and conduct a model utility test. Use technology to find the p value associated with this test. Find the value of r^2. Explain the meaning of this value.

(c) Conduct the necessary hypothesis tests to determine whether both predictor variables are significant. Use an overall significance level of $\alpha = 0.05$.

(d) Find a 99% prediction interval for an observed value of Y for $x_1 = 320$ and $x_2 = 100$.

(e) Suppose a certain brand of bed sheets is sold with 50% cotton and various thread counts. What thread count would guarantee an estimate of the mean lifetime of at least 15 years?

12.181 Travel and Transportation Many of the highways in Pennsylvania are paved with concrete rather than tar. Suppose an experiment was conducted to determine whether the compression strength in cured concrete pavement (y, psi) is affected by the water-to-cement ratio (x_1), the sand-to-cement ratio (x_2), the percentage of fly ash (x_3), and/or the ambient temperature during curing (x_4, °F). HWY

(a) Estimate the regression coefficients in the model
$Y_i = \beta_0 + \beta_1 x_{1i} + \beta_2 x_{2i} + \beta_3 x_{3i} + \beta_4 x_{4i} + E_i$. Use the sign of each estimated regression coefficient to explain the effect of each predictor variable on the compression strength of concrete.

(b) Construct an ANOVA table and conduct a model utility test. Use technology to find the p value associated with this test.

(c) Conduct the appropriate hypothesis tests to determine the significant predictor variables.

(d) Construct a normal probability plot of the residuals and the plots of the residuals versus each predictor variable. Is there any evidence of a violation of the multiple linear regression assumptions?

12.182 Public Health and Nutrition Some evidence suggests that exposure to airborne fungal products may be associated with adverse health effects, including respiratory tract infections. A random sample of Canadian elementary schools were obtained and the following data were obtained for each: mesophilic fungi (y, in cfu/m³), building age (x_1, in years), air exchange rate (x_2, per hour), indoor CO_2 (x_3, in ppm), outdoor temperature (x_4, in °C), and a dummy variable, signs of moisture in the room (x_5, 0 for no signs, 1 for signs). FUNGI

(a) Estimate the regression coefficients in the model
$Y_i = \beta_0 + \beta_1 x_{1i} + \beta_2 x_{2i} + \beta_3 x_{3i} + \beta_4 x_{4i} + \beta_5 x_{5i} + E_i$.

(b) Determine the most important predictor variables. Find the estimated regression line for this reduced model.

(c) Use the sign of each estimated regression coefficient in the reduced model to explain the effect of each predictor variable on the volume of mesophilic fungi.

(d) Check the relevant assumptions for the reduced model.

CHAPTER APP

Rob Arnold/Alamy

12.183 Wind Turbine Noise Suppose a small coastal community is considering the construction of a wind turbine to generate electricity for the town hall. An experiment is conducted to measure the wind turbine's noise level (in decibels) at various distances (in meters) from the proposed site. The data are given in the table. WIND

Distance	10	50	75	120	150	160	200	250	400	500
Noise level	105	95	80	52	58	77	56	57	28	40

(a) Construct a scatter plot for the data. What does the plot suggest about the relationship between the variables?

(b) For distance (x) and noise (y), suppose the true regression line is $y = 90 - 0.1x$.
 (i) Find the expected noise level when the distance is 300 m.
 (ii) Suppose $\sigma = 14$, and find the probability that the noise level is less than 50 dB.

(c) Find the estimated regression line. Estimate the true mean noise level for a distance of 175 m.

(d) Complete the ANOVA table. Find and interpret the coefficient of determination.

(e) Conduct an F test for a significant regression. Conduct a t test concerning β_1 for a significant regression. Use $\alpha = 0.05$ in both tests. Interpret your results.

(f) Find the sample correlation coefficient between distance and noise level, and interpret this value.

(g) (i) For $x = 180$ m, conduct a hypothesis test to determine whether there is any evidence to suggest that the mean noise level is greater than 70 dB.
 (ii) Construct a 95% confidence interval for the true mean noise level for a distance of 225 m.

(h) Find a 95% prediction interval for the noise level for a distance of 100 m.

(i) Compute the residuals. Sketch a normal probability plot of the residuals and a graph of the residuals versus the predictor variable. Is there any evidence that the simple linear regression assumptions are violated?

13 Categorical Data and Frequency Tables

Matt Leung/Shutterstock

◀ Looking Back

- Remember how univariate categorical data are identified: non-numerical observations that may be placed in categories: **Section 2.1**.
- Recall how bivariate categorical data are obtained and characterized: two non-numerical observations on each individual or object: **Section 2.1**.
- Remember the natural summary measures for categorical data: frequency and relative frequency: **Sections 2.2 and 3.1**.
- Remember the chi-square distribution and hypothesis tests based on the chi-square distribution: **Sections 8.5 and 9.7**.

Looking Forward ▶

- Learn the background, computations, and interpretations of a goodness-of-fit test concerning the true population proportions: **Section 13.1**.
- Learn how to conduct and interpret a test for homogeneity or independence: **Section 13.2**.

CHAPTER APP

Traffic Composition

In 2018, after nine years of construction and an estimated $127 billion yuan, China opened the Hong Kong–Zhuhai–Macau Bridge, the world's longest sea crossing bridge. The 34.2-mi bridge cuts across the mouth of the Pearl River and includes a 4.2-mi tunnel between two human-made islands. This creates a channel above the tunnel for large cargo ships.

The new bridge significantly cuts the commuting time between three cities: Hong Kong, Zhuhai, and Macau. Before the bridge opened, it took 4 hours to drive between Hong Kong and Zhuhai. Now the commute is 45 minutes. Chinese officials expect approximately 29,000 vehicles to use the bridge daily.

A random sample of vehicles using the bridge at various times was obtained and each was classified by vehicle type. The resulting frequencies, or counts, are given in the two-way, or contingency, table.

	Time of day			
Vehicle type	Morning	Afternoon	Evening	Overnight
Trucks	33	51	28	13
Buses	30	44	31	23
Shuttles	35	47	18	41
Cars	23	42	23	23

The techniques presented in this chapter can be used to determine whether the two categorical variables, vehicle type and time of day, are dependent. That is, in the context of this problem, we would like to know whether the traffic composition on the bridge is different at various time periods.

13.1 Univariate Categorical Data, Goodness-of-Fit Tests

Goodness-of-Fit Tests

Categorical data are often displayed in a frequency distribution. In this section, we will focus only on the number of observations in each category, and display these results in a one-way frequency table. This table lists each possible category and the number of times each category occurred (the observed count or frequency for each category). For example, four types of limited liability companies (LLCs) are available to real estate investors. Suppose the Internal Revenue Service (IRS) selects 200 real estate investor LLCs at random and classifies each. The total in each category is given in the following one-way frequency table.

LLC type	Single member	Married couple	Multi-member	Series
Frequency	70	55	45	30

The hypothesis test procedure presented in this section is designed to compare a set of hypothesized proportions with the set of true proportions, to check the **goodness of fit (GOF)**. For example, the IRS might use the data in the preceding table to determine whether there is any evidence that the true proportions are different from the following hypothesized proportions: 0.4 (for single member), 0.3 (for married couple), 0.2 (for multi-member), and 0.1 (for series).

The notation used here is an extension of the notation used to represent sample and population proportions. Suppose each observation falls into one of k categories.

	True proportion	Hypothesized proportion
Category 1	p_1	p_{10}
Category 2	p_2	p_{20}
⋮	⋮	⋮
Category i	p_i	p_{i0}
⋮	⋮	⋮
Category k	p_k	p_{k0}

The subscript in the hypothesized proportion p_{10} is read as "one zero," not "ten." The first number in the subscript (1) denotes the category, and the second number (0) indicates that this is a hypothesized value.

The true proportion and the hypothesized proportion are both population proportions. The sum of the hypothesized proportions (like the sum of the true proportions) must be 1. That is, $p_{10} + p_{20} + \cdots + p_{k0} = 1$.

13.1 Univariate Categorical Data, Goodness-of-Fit Tests

The goodness-of-fit test is used to determine whether there is any evidence (from the observed, or sample, counts) that the true population proportions differ from the hypothesized population proportions. The null hypothesis and the alternative hypothesis are stated in terms of the true and hypothesized category proportions.

H_0: $p_1 = p_{10}, p_2 = p_{20}, \ldots, p_k = p_{k0}$
(Each true category proportion is equal to a specified hypothesized value.)

H_a: $p_i \neq p_{i0}$ for at least one i.
(At least one true category proportion is not equal to the corresponding specified hypothesized value.)

> H_0 is a composite null hypothesis. It involves several parameters and is true only if all the equalities hold.

Suppose a random sample of size n is selected; let n_i ($i = 1, 2, \ldots, k$) be the number of observations falling into each category. To decide whether the sample data fit the hypothesized proportions, the observed cell counts (the n_i's) are compared with the expected cell counts. If the null hypothesis is true, then the expected frequency, or count, for category 1, or in cell 1, is $e_1 = np_{10}$. For cell 2, the expected frequency is $e_2 = np_{20}$, and so on. For example, if the sample size is $n = 100$ and $p_{10} = 0.25$, then we expect the count for category 1 to be, on average, $np_{10} = (100)(0.25) = 25$.

> In this context, a cell is simply a category.

The test statistic is a measure of how far away the observed cell counts are from the expected cell counts. If the null hypothesis is true, then the random variable

$$X^2 = \sum_{i=1}^{k} \frac{(\text{observed cell count} - \text{expected cell count})^2}{\text{expected cell count}} = \sum_{i=1}^{k} \frac{(n_i - e_i)^2}{e_i}$$

> The X in X^2 is the uppercase Greek letter chi (χ is the lowercase form).

has approximately a chi-square distribution with $k - 1$ degrees of freedom. This approximation is good if $e_i = np_{i0} \geq 5$ for all i, that is, if all expected cell counts are at least 5.

If the observed cell counts are close to the expected cell counts, then the value of X^2 will be small. If the observed cell counts differ substantially from the expected cell counts, then the value of X^2 will be large. Therefore, the null hypothesis is rejected only for large values of the test statistic. Now we have all of the pieces for the complete hypothesis test.

Goodness-of-Fit Test

Let n_i be the number of observations falling into the ith category ($i = 1, 2, \ldots, k$), and let $n = n_1 + n_2 + \cdots + n_k$. A hypothesis test about the true category population proportions with significance level α has the form

H_0: $p_1 = p_{10}, p_2 = p_{20}, \ldots, p_k = p_{k0}$

H_a: $p_i \neq p_{i0}$ for at least one i

TS: $X^2 = \sum_{i=1}^{k} \frac{(n_i - e_i)^2}{e_i}$ where $e_i = np_{i0}$

RR: $X^2 \geq \chi^2_{\alpha, k-1}$

Just a reminder: This test is appropriate if all expected cell counts are at least 5 ($e_i = np_{i0} \geq 5$ for all i). The chi-square distribution was introduced in Section 8.5, and a hypothesis test based on this distribution was discussed in Section 9.7.

Examples: Goodness of Fit

EXAMPLE 13.1   Sweet as Tupelo Honey

For beekeepers looking to harvest more honey, there are four possibilities for obtaining more bees: package bees, nucs, established colonies, or swarms. A university agricultural sciences department obtained a random sample of recent bee purchases, and each was classified into one of the four categories. Use the following one-way frequency table to test the hypothesis that the four possible bee purchases occur with equal frequency. Use $\alpha = 0.05$.

Bee purchase	Package bees	Nucs	Colonies	Swarms
Frequency	31	36	26	20

Solution

Find the expected cell counts, and use the goodness-of-fit test procedure to determine whether there is any evidence that any one of the true population proportions differs from 0.25.

STEP 1 If the bee purchase types are equally likely, the proportion of purchases falling into each category is $1/k = 1/4 = 0.25$.

The four parts of the hypothesis test are

H_0: $p_1 = 0.25$, $p_2 = 0.25$, $p_3 = 0.25$, $p_4 = 0.25$
H_a: $p_i \neq p_{i0}$ for at least one i

TS: $X^2 = \sum_{i=1}^{k} \dfrac{(n_i - e_i)^2}{e_i}$

RR: $X^2 \geq \chi^2_{\alpha, k-1} = \chi^2_{0.05, 3} = 7.8147$

STEP 2 There are $n = 31 + 36 + 26 + 20 = 113$ total observations. The expected counts are given in the table.

Cell	Category	Observed cell count	Expected cell count
1	Package bees	31	$e_1 = np_{10} = (113)(0.25) = 28.25$
2	Nucs	36	$e_2 = np_{20} = (113)(0.25) = 28.25$
3	Colonies	26	$e_3 = np_{30} = (113)(0.25) = 28.25$
4	Swarms	20	$e_4 = np_{40} = (113)(0.25) = 28.25$

All four expected cell counts are greater than 5. The chi-square goodness-of-fit test is appropriate.

STEP 3 The value of the test statistic is

$$\chi^2 = \sum_{i=1}^{4} \dfrac{(n_i - e_i)^2}{e_i}$$

$$= \dfrac{(31-28.25)^2}{28.25} + \dfrac{(36-28.25)^2}{28.25} + \dfrac{(26-28.25)^2}{28.25} + \dfrac{(20-28.25)^2}{28.25}$$

Use observed and expected cell counts.

$$= 0.2677 + 2.1261 + 0.1792 + 2.4093 = 4.9823 \; (<7.8147)$$

Simplify.

STEP 4 The value of the test statistic does not lie in the rejection region or, equivalently, $p = 0.1731 > 0.05$. **Figure 13.1** shows the calculation and an illustration of the exact p value. At the $\alpha = 0.05$ significance level, there is no evidence to suggest that any of the true population proportions differs from 0.25.

Figure 13.2 shows a technology solution.

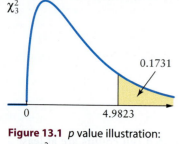

Figure 13.1 p value illustration:
$p = P(X^2 \geq 4.9823)$
$= 0.1731 > 0.05 = \alpha$

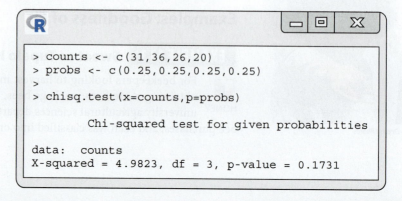

Figure 13.2 Use the R command `chisq.test()` to conduct a goodness-of-fit test.

Recall that, using Appendix Table 6, we can only bound the p value for this hypothesis test. R and other technology solutions can be used to find the exact p value for this test.

TRY IT NOW Go to Exercises 13.13 and 13.15

EXAMPLE 13.2   Thanksgiving Traditions

On Thanksgiving, many families traditionally gather for a wonderful meal, lively conversation, and, in some cases, a spirited street hockey game. A random sample of adults older than age 18 was obtained and asked to name their favorite Thanksgiving food. The data and the proportions from a previous survey are given in the table.

Favorite food	Frequency	Previous proportions
Turkey	250	0.38
Stuffing	148	0.26
Mashed potatoes	98	0.17
Yams	55	0.10
Green bean casserole	30	0.05
Cranberry sauce	42	0.04

Is there evidence to suggest that any of the true cell proportions differ from the previous proportions? Use $\alpha = 0.05$.

Solution

Find the expected cell counts, and use the goodness-of-fit test procedure to determine whether there is evidence that any of the true population proportions differs from the previous proportions.

STEP 1 A goodness-of-fit test is appropriate to determine whether there is evidence that any of the true category population proportions has changed. There are $k = 6$ categories.

The four parts of the hypothesis test are

H_0: $p_1 = 0.38$, $p_2 = 0.26$, $p_3 = 0.17$, $p_4 = 0.10$, $p_5 = 0.05$, $p_6 = 0.04$

H_a: $p_i \neq p_{i0}$ for at least one i

TS: $X^2 = \sum_{i=1}^{k} \frac{(n_i - e_i)^2}{e_i}$

RR: $X^2 \geq \chi^2_{\alpha, k-1} = \chi^2_{0.05, 5} = 11.0705$

STEP 2 There are $n = 250 + 148 + 98 + 55 + 30 + 42 = 623$ total observations. The expected cell counts are given in the table.

Cell	Category	Observed cell count	Expected cell count
1	Turkey	250	$e_1 = np_{10} = (623)(0.38) = 236.74$
2	Stuffing	148	$e_2 = np_{20} = (623)(0.26) = 161.98$
3	Mashed potatoes	98	$e_3 = np_{30} = (623)(0.17) = 105.91$
4	Yams	55	$e_4 = np_{40} = (623)(0.10) = 62.30$
5	Green bean casserole	30	$e_5 = np_{50} = (623)(0.05) = 31.15$
6	Cranberry sauce	42	$e_6 = np_{60} = (623)(0.04) = 24.92$

All expected cell counts are greater than 5. The chi-square goodness-of-fit test is appropriate.

STEP 3 The value of the test statistic is

$$\chi^2 = \sum_{i=1}^{6} \frac{(n_i - e_i)^2}{e_i}$$

$$= \frac{(250-236.74)^2}{236.74} + \frac{(148-161.98)^2}{161.98} + \frac{(98-105.91)^2}{105.91}$$

$$+ \frac{(55-62.30)^2}{62.30} + \frac{(30-31.15)^2}{31.15} + \frac{(42-24.92)^2}{24.92}$$

Use observed and expected cell counts.

$$= 0.74 + 1.21 + 0.59 + 0.86 + 0.04 + 11.71$$

$$= 15.144 \; (\geq 11.0705)$$

Simplify.

STEP 4 The value of the test statistic ($\chi^2 = 15.144$) lies in the rejection region or, equivalently, $p = 0.0098 \leq 0.05$. At the $\alpha = 0.05$ significance level, there is evidence to suggest that at least one population proportion has changed from its previous value.

Note: We can use Appendix Table 6 to bound the p value for this hypothesis test. In row $k - 1 = 5$ (degrees of freedom), place 15.144 in the ordered list of critical values.

$$15.0863 \leq 15.144 \leq 16.7496$$

$$\chi^2_{0.01,5} \leq 15.144 \leq \chi^2_{0.005,5}$$

Therefore, $\quad 0.005 \leq p \leq 0.01$

Figure 13.3 shows the calculation and an illustration of the exact p value.
Figure 13.4 shows a technology solution.

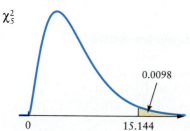

Figure 13.3 p value illustration:
$p = P(X^2 \geq 15.144)$
$= 0.0098 \leq 0.05 = \alpha$

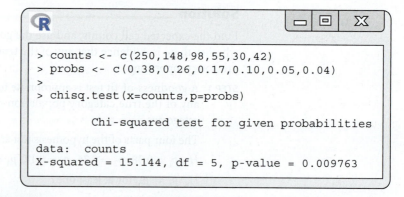

Figure 13.4 Use the R command `chisq.test()` to conduct a goodness-of-fit test. Use the observed counts and the hypothesized proportions.

TRY IT NOW Go to Exercises 13.25 and 13.29

Section 13.1 Exercises

Concept Check

13.1 True or False In a goodness-of-fit test, the hypothesized proportions are always assumed to be equal.

13.2 Short Answer Explain how to compute each expected cell count in a goodness-of-fit test.

13.3 Fill in the Blank If the observed cell counts are close to the expected cell counts in a goodness-of-fit test, then the value of X^2 should be _____.

13.4 Fill in the Blank In a goodness-of-fit test with k categories, the test statistic is compared to a critical value from a chi-square distribution with _____ degrees of freedom.

13.5 Fill in the Blank The goodness-of-fit test is appropriate if all expected cell counts are _____.

13.6 True or False In a goodness-of-fit test, the sum of the hypothesized proportions must be 1.

13.7 Short Answer In a goodness-of-fit test, what is the sum of the expected cell counts?

Practice

13.8 Use the hypothesized proportions in the frequency table to find the expected count for each category. EX13.8

Category	1	2	3	4
Frequency	42	58	62	56
p_{i0}	0.25	0.40	0.20	0.15

13.9 Use the hypothesized proportions in the frequency table to find the expected count for each category. EX13.9

Category	1	2	3	4	5
Frequency	125	150	300	220	205
p_{i0}	0.15	0.14	0.31	0.23	0.17

13.10 Consider the one-way frequency table. EX13.10

Category	1	2	3	4
Frequency	115	85	70	30

(a) Suppose the hypothesized proportions are $p_{10} = 0.4$, $p_{20} = 0.3$, $p_{30} = 0.2$, and $p_{40} = 0.1$. Write the four parts of the appropriate goodness-of-fit test. Use $\alpha = 0.01$.
(b) Find the expected count for each cell. Verify that each is at least 5.
(c) Find the value of the test statistic. State your conclusion, and justify your answer.

13.11 Consider the one-way frequency table. EX13.11

Category	1	2	3	4	5	6
Frequency	90	82	75	95	96	36

(a) Write the four parts of a goodness-of-fit test with $p_{10} = 0.175$, $p_{20} = 0.171$, $p_{30} = 0.162$, $p_{40} = 0.225$, $p_{50} = 0.202$, $p_{60} = 0.065$. Use $\alpha = 0.05$.
(b) Find the value of the test statistic. State your conclusion and justify your answer.
(c) Find bounds on the p value.

13.12 Consider the one-way frequency table. EX13.12

Category	1	2	3	4	5
Frequency	140	135	155	152	168

Conduct a goodness-of-fit test to check the hypothesis that all five categories are equally likely. Use $\alpha = 0.05$. Find bounds on the p value.

Applications

13.13 Psychology and Human Behavior A random sample of adult males was obtained and asked whether they belonged to a fraternal organization. The results are given in the one-way frequency table. ELKS

Lodge	Elks	Moose	Rotary	None
Frequency	150	60	60	230

Past research indicated that 25% of all men belonged to an Elks lodge, 15% belonged to a Moose lodge, and 10% belonged to a Rotary organization. Conduct a goodness-of-fit test to determine whether there is evidence that at least one of the population proportions associated with the four categories in the table has changed. Use $\alpha = 0.05$.

13.14 Marketing and Consumer Behavior In a random sample of Caribou coffee-shop customers who purchased flavor shots, 185 bought almond, 180 purchased vanilla, 230 chose raspberry, 220 opted for caramel, and 185 went with hazelnut. Use a goodness-of-fit test to determine whether there is any evidence that customers who purchase flavor shots prefer one choice over the others. Use $\alpha = 0.05$. CARIBOU

13.15 Economics and Finance The AAII Sentiment Survey is administered every week to determine the percentage of investors who are bullish, bearish, or neutral on stock prices in the next six months. The long-term, or historical, percentages are bullish 41.3%, neutral 27.5%, and bearish 31.2%.[1] The data in the table were obtained in a random sample of investors for the week ending December 14, 2018. AAII

Sentiment	Bullish	Neutral	Bearish
Frequency	420	301	288

Is there any evidence that the percentages during this week are different from the long-term percentages? Use $\alpha = 0.01$.

13.16 Biology and Environmental Science In a recent study of adult eye color, the hypothesized distribution was given as blue 25%, green 10%, brown 50%, and black 15%. A random sample of adults was obtained, and the resulting eye colors are summarized in the one-way frequency table. EYE

Eye color	Blue	Green	Brown	Black
Frequency	121	65	242	72

Conduct a goodness-of-fit test to determine whether these data provide any evidence that the true proportions differ from the hypothesized proportions. Use $\alpha = 0.05$.

13.17 Marketing and Consumer Behavior Original Designs sells home plans in five different styles. A random sample of purchases was obtained, and the number falling into each category was recorded. The data are given in the table. HOME

Style	A-frame	Cape Cod	Colonial	Log home	Ranch
Frequency	75	180	268	58	385

Conduct a goodness-of-fit test to determine whether there is evidence that any of the true proportions differ from the hypothesized values of 0.10, 0.20, 0.25, 0.05, and 0.40. Use $\alpha = 0.05$.

13.18 Marketing and Consumer Behavior A random sample of West Bay Yacht Club members was obtained and asked to identify their dinner event preference. The resulting data, along with the historical proportions, are given in the table. DINNER

Dinner event preference	Frequency	p_{i0}
Pot luck	120	0.33
Catered $8	95	0.18
Catered $12	100	0.26
Catered $15	36	0.10
Catered $18	61	0.13

Conduct a goodness-of-fit test to determine whether there is evidence that any of the true proportions differs from the historical value. Use $\alpha = 0.05$.

13.19 Marketing and Consumer Behavior The manager at a Publix grocery store is trying to determine which types of lettuce to order and sell to customers. A random sample of shoppers was obtained and asked to select their favorite lettuce type. The data are given in the table. LETTUCE

Lettuce	Frequency
Crisphead	45
Butterhead	17
Romaine	130
Leafy	90
Celtuce	28

Test the fit of these data to the hypothesized proportions 0.14, 0.06, 0.37, 0.30, and 0.13. Use $\alpha = 0.05$.

13.20 Public Health and Nutrition The U.S. Centers for Disease Control and Prevention collects data on the number of reported flu cases. Suppose the table shows the number of reported H1N1 flu cases by region for the week ending December 15, 2018, and the hypothesized proportions from 2017. FLU

Surveillance region	Frequency	p_{i0}
1	40	0.014
2	125	0.047
3	145	0.061
4	501	0.210
5	345	0.132
6	380	0.151
7	140	0.051
8	501	0.221
9	164	0.066
10	110	0.047

Conduct a goodness-of-fit test to determine whether there is evidence that any of this year's true population proportions differs from the 2017 proportion. Use $\alpha = 0.05$.

13.21 Demographics and Population Statistics A random sample of employed adults in Canada was obtained and classified by occupation. The resulting data and the hypothesized proportions are given in the following table.[2] EMPLOY

Occupation	Frequency	p_{i0}
Management	2122	0.11
Business, finance	3133	0.16
Sciences	1348	0.07
Health	1322	0.07
Education	2328	0.11
Art, culture, sport	659	0.05
Sales, service	4795	0.22
Trades, transport	2877	0.14
Natural resources, agriculture	483	0.03
Manufacturing, utilities	884	0.04

Conduct a goodness-of-fit test to determine whether there is evidence that any of the true proportions differs from the hypothesized value. Use $\alpha = 0.01$.

13.22 Public Health and Nutrition Many insurance companies have adopted a tiered plan that allows the user to select a generic or preferred drug at a reduced cost or copayment instead of the named or specialty drug at a higher copayment. Medicare Blue RX has four tiers for prescription medication, each with different copayments. Past records indicate that the proportions of users who select each tier are as follows: Tier 1, 0.35; Tier 2, 0.28; Tier 3, 0.12; Tier 4, 0.25. Suppose a random sample of policyholders who purchased medication was obtained and the frequency associated with each tier is given in the table. MEDINS

Tier	1	2	3	4
Frequency	152	107	36	76

(a) Use a goodness-of-fit test to show that there is a shift in the proportion of users by tiers. Use $\alpha = 0.05$.
(b) Use your results in part (a) to explain the shift in the proportions, that is, which tiers are more/less utilized.

13.23 Sports and Leisure As soon as the gates at an amusement park open, there is a mad rush to the most popular rides. A random sample of people waiting in line was obtained and asked which ride they were headed to first. The results and the hypothesized proportions are given in the table. RIDES

Ride	Frequency	p_{i0}
Tower of Terror	63	0.40
Rockin' Roller Coaster	35	0.30
Star Tours	16	0.15
Studios Backlot Tour	14	0.15

Conduct a goodness-of-fit test to determine whether any of the true population proportions differ from the hypothesized proportions. Use $\alpha = 0.01$.

13.24 Marketing and Consumer Behavior Americans love to eat out. In fact, according to a recent survey, diners eat out approximately 5 times per week.[3] Consequently, customers are quick to voice dissatisfaction when their dining experience is unpleasant. A random sample of diners was obtained and asked to name their most common restaurant complaint. The results and the hypothesized proportions are given in the table. DINEOUT

Complaint	Frequency	p_{i0}
Food quality	45	0.21
Cleanliness	32	0.19
Service speed	50	0.28
Prices	66	0.32

Conduct a goodness-of-fit test to determine whether any of the true population proportions differs from the hypothesized proportion. Use $\alpha = 0.05$.

13.25 Marketing and Consumer Behavior In a recent survey by the National Association of Realtors, participants were asked about their preferred housing type. A random sample of adults in Florida was obtained and asked to select their housing type preference. The results are given in the table. NAR

Housing type	Florida frequencies	Realtors' proportions
Single-family detached home, large yard	144	0.52
Single-family detached home, small yard	101	0.24
Apartment or condominium	74	0.14
Single-family home or townhouse	25	0.06
Something else	21	0.04

(a) Conduct a goodness-of-fit test to show that there is evidence the true proportions in Florida differ from those found in the national survey. Use $\alpha = 0.05$.
(b) Use technology to find the exact p value associated with this hypothesis test.
(c) Use your results in part (a) to explain which proportions are different. Why would you expect these proportions to be different in Florida?

13.26 Demographics and Population Statistics The mining industry in Canada employs approximately 400,000 people accounting for roughly one in every 45 Canadian jobs.[4] A random sample of people working in the mining industry in Manitoba was obtained and each was classified according to job. The results and the proportions from 2017 are given in the table. MINING

Job	Frequency	p_{i0}
Engineering	31	0.14
Operation roles	20	0.08
Construction	62	0.38
Mechanics	50	0.24
Exploration	34	0.16

Conduct a goodness-of-fit test to determine if the true proportion of workers in each job has changed from 2017. Use $\alpha = 0.05$.

Extended Applications

13.27 Psychology and Human Behavior Berkeley Breathed stopped writing his Pulitzer Prize–winning comic strip *Bloom County* in 1989. Since then, he has written several books and animations, and has been involved in writing a children's movie and two sequel comic strips, *Outland* and *Opus*, featuring some of the old *Bloom County* gang. A random sample of people who read the original *Bloom County* comic strip was obtained and asked to name their favorite character. The results are given in the table. BLOOM

Character	Frequency
Opus	250
Michael Binkley	210
Oliver Wendell Jones	205
Milo Bloom	190
Bill the Cat	260
Cutter John	195
Steve Dallas	201
Portnoy	206
Hodge Podge	185
Rosebud	204

Is there any evidence to suggest that one (or more) character(s) are more popular than the others? Find bounds on the p value associated with this test.

13.28 Biology and Environmental Science Maine is the largest producer of wild blueberries in the United States—in fact, it is the largest producer in the world.[5] Blueberry farms use several methods of pest management and pruning practices to increase the harvest. The table contains the total number of farms that used each pruning practice in 2017 and corresponding totals from a random sample in 2018. PRUNING

Pruning practice	2017 totals	2018 sample
Straw burn	33	44
Oil burn	44	40
Mow	87	110
Prune every other year	93	94

(a) Compute the proportion associated for each pruning practice based on the 2017 totals.
(b) Conduct a goodness-of-fit test to determine whether there is any evidence that the true 2017 proportions of pruning practice changed in 2018. Use $\alpha = 0.01$.

13.29 Technology and the Internet Apple and Samsung together control more than half of the world mobile phone market.[6] A random sample of mobile phone owners was obtained and each was classified by phone vendor. The results and the 2017 market share are given in the table. MOBILE

Vendor	Frequency	p_{i0}
Samsung	315	0.3371
Apple	252	0.1937
Unknown	75	0.0742
Huawei	52	0.0485
LG	36	0.0382
Xiaomi	35	0.0359
Other	275	0.2724

(a) Conduct a goodness-of-fit test to show that there is evidence the true market share proportions differ from those in 2017. Use $\alpha = 0.05$.
(b) Use technology to find the p value associated with this test.
(c) Which vendor proportion(s) do you believe has (have) changed and why?

13.30 Travel and Transportation Capital Bikeshare is a Washington, D.C., organization sponsored by several agencies that offers short-term use of more than 4300 bicycles to registered members. There are approximately 500 bicycle stations in the District of Columbia, Arlington County, and the city of Alexandria, Virginia.[7] To distribute bicycles appropriately to stations, a survey is conducted each year to determine the home location and work location of members. The table contains the historical proportions associated with each home location and a summary of a random sample of members in 2018. BIKE

Home location	2018 sample	Historical proportions
District of Columbia	3750	0.78
Arlington County (VA)	530	0.11
Montgomery County (MD)	185	0.04
Fairfax County (VA)	101	0.02
Prince Georges County (MD)	52	0.01
Alexandria City (VA)	98	0.02
Other	92	0.02

Conduct a goodness-of-fit test to determine whether there is evidence that any of the historical proportions have changed. Use $\alpha = 0.05$.

13.31 Travel and Transportation As incomes have risen and credit has become more available, China has become the largest car market and the largest automobile producer in the world. The best-selling models include cars made by Dongfeng Motor Corporation and the Changan Automobile group. The automobile production totals for each country are recorded for 2016 and 2017. AUTOS

(a) Compute the proportion of automobiles produced for each country based on the 2016 totals.
(b) Conduct a goodness-of-fit test to determine whether there is any evidence to suggest that the true 2016 proportions of automobiles produced in each country changed in 2017. Use $\alpha = 0.05$.
(c) Find the p value associated with this test.

Challenge Problems

13.32 Business and Management In Britain, a small baker does not have a fully automatic plant and sells the majority of his or her production on-site or from vehicles. According to the Birmingham City Council trading standards, the law states that the average weight of one loaf type must be 400 g. Suppose the Small Baker's Association (SBA) claims that the weight of each loaf of bread of this type is approximately normal, with mean 400 g and standard deviation 15 g. To check this claim, a random sample of small-baker loaves was obtained, and each was carefully weighed. The observed frequency of weights in each specified interval (in grams) is summarized in the one-way table. BAKER

Interval	Frequency
< 370	18
370–385	67
385–400	175
400–415	184
415–430	75
≥ 430	14

(a) Assume the SBA claim is true. Find the probability that a randomly selected loaf of bread falls into each interval.
(b) Use the probabilities computed in part (a) as the hypothesized population proportions associated with each interval. Conduct a goodness-of-fit test to determine whether the observed weights fit the hypothesized distribution. Use $\alpha = 0.05$. State your conclusion.

Note: This application of the goodness-of-fit test provides another formal test for normality. It complements the methods used to check for normality in Section 6.3.

13.2 Bivariate Categorical Data: Tests for Homogeneity and Independence

Test for Homogeneity

If two categorical observations are made on the same individual or object, the data set is bivariate. This type of data arises in two common ways.

If any two observations are made on an individual or object, the data set is bivariate. For example, one observation might be categorical and the other numerical, or both observations could be numerical.

1. Random samples are obtained from two or more populations, and each individual is classified by values of a categorical variable.
2. Suppose there are two categorical variables of interest. In a (single) random sample, a value of each variable is recorded for each individual.

The test for homogeneity applies to the first kind of data (samples from two or more populations), and the test for independence applies to the second kind of data (data from a single sample, with two categorical variables).

Let's focus on the first type of bivariate categorical data.

Suppose random samples of home insurance claims are obtained from three companies, and the type of claim is recorded for each. The insurance companies are the populations and the claim type is the categorical variable. The data are bivariate because there are two values (company and claim type) for, or associated with, each claim. The data may be recorded in the following manner.

Trade	Firm	Futures type
1	State Farm	Human-incited property damage
2	Nationwide	Nature-induced property damage
3	Allstate	Burglary/theft
4	Nationwide	Fire/lightning
⋮	⋮	⋮
300	State Farm	Fire/lightning

The natural summary for this type of bivariate data set (in which each observation is categorical) is the number of observations in each category combination.

For example, we can compute the number of claim types (or frequency) reported by State Farm *and* involving human-incited property damage, the number of claim types reported by State Farm *and* involving nature-induced property damage, and so on. This summary information can be displayed in a 3×4 two-way frequency table, or contingency table, as shown in **Table 13.1**. Each cell in this table contains the number of observations (observed count or frequency) in a category combination, or pairing.

Two-way tables are described by the number of rows and the number of columns. For example, a 2×6 contingency table has 2 rows and 6 columns.

		Homeowner insurance claim type			
		Burglary	Fire	Human	Nature
Company	Allstate	15	25	30	22
	Nationwide	22	24	15	30
	State Farm	32	25	20	40

Table 13.1 A two-way frequency table for the data obtained from insurance claims

The columns of the two-way table in Table 13.1 correspond to claim type, and the rows correspond to insurance companies. Each cell in the body of the table contains a frequency, or observed cell count. For example, there were 15 burglary/theft claims reported by Allstate and 40 nature-induced property damage claims reported by State Farm.

If we consider a single insurance company, say Allstate, then we can use a bar chart to represent the distribution of homeowner insurance claim types (**Figure 13.5**). We can also construct a bar chart for each insurance company. For example, **Figure 13.6** is another bar chart associated with these data, a summary of claim types for Nationwide.

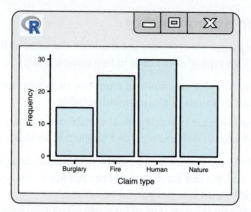
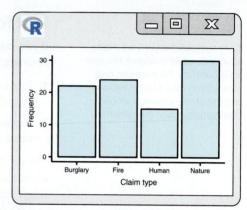

Figure 13.5 A bar chart showing the frequency of claim types for Allstate.

Figure 13.6 A bar chart showing the frequency of claim types for Nationwide.

We can use a side-by-side or stacked bar chart to compare categorical data from two or more sources, or populations. **Figure 13.7** shows a side-by-side bar chart of claim types grouped by insurance company. **Figure 13.8** shows a stacked bar chart of claim types grouped by insurance company.

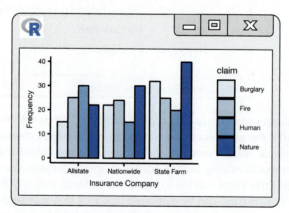
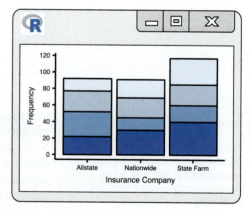

Figure 13.7 A side-by-side bar chart showing the frequency of claim types, by insurance company.

Figure 13.8 A stacked bar chart showing the frequency of claim types, by insurance company.

Although these graphs are useful for suggesting possible differences among populations, a more precise statistical test would be helpful. The practical problem to consider is this: Are all of the true category proportions the same for each population? This is a test for **homogeneity** of populations. Homogeneity is the state of having identical properties or values; in this case, it refers to populations having identical true category proportions. In the insurance company and claim type example, it seems reasonable to ask whether the proportion of claim types is the same for each company. The statistical procedure used to analyze this problem is based on the observed and expected cell counts (as in Section 13.1). Under the null hypothesis that the populations have the same category proportions, the test statistic also has a chi-square distribution.

13.2 Bivariate Categorical Data: Tests for Homogeneity and Independence

Suppose there are I rows and J columns in a two-way frequency table. In this table, dot notation in a subscript (introduced in Chapter 11) indicates a sum over that subscript while the other subscript is held fixed.

n_{ij} = observed cell count, or frequency, in the (ij) cell (the intersection of the ith row and the jth column)

$$n_{i.} = \sum_{j=1}^{J} n_{ij}$$

= ith row total, the sum of the cell counts, or observed frequencies, in the ith row

$$n_{.j} = \sum_{i=1}^{I} n_{ij}$$

= jth column total, the sum of the cell counts, or observed frequencies, in the jth column

$$n = \sum_{i=1}^{I} \sum_{j=1}^{J} n_{ij}$$

= grand total, the total of all cell counts, or observed frequencies

Table 13.2 is a visualization of this notation in an $I \times J$ two-way frequency table.

Table 13.2 Notation used in an $I \times J$ two-way frequency table

		Category					Row total
		1	2	...	j	...	J	
Population	1	n_{11}	n_{12}	...	n_{1j}	...	n_{1J}	$n_{1.}$
	2	n_{21}	n_{22}	...	n_{2j}	...	n_{2J}	$n_{2.}$
	⋮	⋮	⋮	⋮	⋮	⋮	⋮	⋮
	i	n_{i1}	n_{i2}	...	n_{ij}	...	n_{iJ}	$n_{i.}$
	⋮	⋮	⋮	⋮	⋮	⋮	⋮	⋮
	I	n_{I1}	n_{I2}	...	n_{Ij}	...	n_{IJ}	$n_{I.}$
Column total		$n_{.1}$	$n_{.2}$	...	$n_{.j}$	...	$n_{.J}$	n

The row and column totals are used to compute the expected cell counts. The insurance company and claim type data will be used to illustrate these calculations. **Table 13.3** is a modified two-way table containing the observed cell counts, the row and column totals, and the grand total.

Table 13.3 A modified two-way frequency table including the row and column totals and the grand total for the insurance company data

		Homeowner insurance claim type				Row total
		Burglary	Fire	Human	Nature	
Company	Allstate	15	25	30	22	92
	Nationwide	22	24	15	30	91
	State Farm	32	25	20	40	117
	Column total	69	74	65	92	300

Here, as in Section 13.1, the expected cell count may not be an integer.

There were 300 insurance claims in this study, and 69 involved burglary. The proportion of all burglary claims in the data set is $69/300 = 0.23$. Suppose there is no difference in the proportion of burglary claims among the insurance companies. Then, we expect 23% of the claim types from Allstate to involve burglary. This expected frequency in the (11) cell is denoted e_{11} and is computed by

The e here is used to denote expected frequency and is not connected in any way with the random errors or residuals in Chapter 12.

$$e_{11} = 0.23 \times 92 = \frac{69}{300} \times 92 = \frac{(92)(69)}{300}$$

$$= \frac{\text{(1st row total)(1st column total)}}{\text{grand total}} = \frac{n_{1.} \times n_{.1}}{n} = 21.16$$

Similarly, we expect 23% of the insurance claims from Nationwide to involve burglary. The expected cell count in the (21) cell is

$$e_{21} = 0.23 \times 91 = \frac{69}{300} \times 91 = \frac{(91)(69)}{300}$$

$$= \frac{(\text{2nd row total})(\text{1st column total})}{\text{grand total}} = \frac{n_{2.} \times n_{.1}}{n} = 20.93$$

The expected counts in column 2 are computed in a similar manner. There are 74 insurance claims that involve fire. The proportion of all claims involving fire is $74/300 = 0.2467$. If there is no difference in the proportion of fire claims, then we expect 24.67% of the claims or

$$e_{12} = 0.2467 \times 92 = \frac{74}{300} \times 92 = \frac{(92)(74)}{300}$$

$$= \frac{(\text{1st row total})(\text{2nd column total})}{\text{grand total}} = \frac{n_{1.} \times n_{.2}}{n} = 22.69$$

to be the number of fire claims from Allstate. We continue in the same manner to compute all the expected counts. **Table 13.4** shows the expected counts in parentheses beneath the corresponding observed counts.

Table 13.4 A modified two-way frequency table including the row and column totals, the grand total, and the expected cell counts for the insurance company data

		Homeowner insurance claim type				
		Burglary	Fire	Human	Nature	Row total
Company	Allstate	15 (21.16)	25 (22.69)	30 (19.93)	22 (28.21)	92
	Nationwide	22 (20.93)	24 (22.45)	15 (19.72)	30 (27.91)	91
	State Farm	32 (26.91)	25 (28.86)	20 (25.35)	40 (35.88)	117
	Column total	69	74	65	92	300

The computations in this example suggest an easy formula for finding the expected frequencies. In an $I \times J$ two-way table, the expected count, or frequency, in the (ij) cell can be written as

$$e_{ij} = \frac{(i\text{th row total})(j\text{th column total})}{\text{grand total}} = \frac{n_{i.} \times n_{.j}}{n}$$

The test statistic is a measure of how far away the observed cell counts are from the expected cell counts. If there is no difference in category proportions among populations, the random variable

$$X^2 = \sum_{\text{all cells}} \frac{(\text{observed cell count} - \text{expected cell count})^2}{\text{expected cell count}} = \sum_{i=1}^{I} \sum_{j=1}^{J} \frac{(n_{ij} - e_{ij})^2}{e_{ij}}$$

has approximately a chi-square distribution with $(I-1)(J-1)$ degrees of freedom. This approximation is good if $e_{ij} \geq 5$ for all i and j, that is, if all expected cell counts are at least 5.

If the observed cell counts are close to the expected cell counts, then the value of X^2 will be small. If the observed cell counts are considerably different from the expected cell counts, then the value of X^2 will be large. As in Section 13.1, the null hypothesis is rejected only for large values of the test statistic.

For the insurance company data, the value of the test statistic is

$$\chi^2 = \sum_{i=1}^{3}\sum_{j=1}^{4}\frac{(n_{ij}-e_{ij})^2}{e_{ij}}$$

$$= \frac{(15-21.16)^2}{21.16} + \frac{(25-22.69)^2}{22.69} + \cdots + \frac{(40-35.88)^2}{35.88} = 13.009$$

This test is significant at the $\alpha = 0.05$ level, because the critical value is

$$\chi^2_{\alpha,(I-1)(J-1)} = \chi^2_{0.05,(2)(3)} = \chi^2_{0.05,6} = 12.5916.$$

Here is the formal test procedure.

Test for Homogeneity of Populations

In an $I \times J$ two-way frequency table, let n_{ij} be the observed count in the (ij) cell and let e_{ij} be the expected count in the (ij) cell. A hypothesis test for homogeneity of populations with significance level α has the following form.

H_0: The true category proportions are the same for all populations (homogeneity of populations).

H_a: The true category proportions are not the same for all populations.

TS: $X^2 = \sum_{i=1}^{I}\sum_{j=1}^{J}\frac{(n_{ij}-e_{ij})^2}{e_{ij}}$

where $e_{ij} = \dfrac{(\text{ith row total})(\text{jth column total})}{\text{grand total}} = \dfrac{n_{i.} \times n_{.j}}{n}$

RR: $X^2 \geq \chi^2_{\alpha,(I-1)(J-1)}$

A reminder: This test is appropriate if all expected cell counts are at least 5 ($e_{ij} \geq 5$ for all i and j). Note that this procedure is called a test of homogeneity and this property is stated in the null hypothesis. However, we are really testing for evidence of inhomogeneity. We cannot prove homogeneity, but can only test for evidence of inhomogeneity.

EXAMPLE 13.3 Hospital Readmissions

PHIS One measure of the quality of health care is patient readmissions. In particular, some children with chronic illnesses are frequently readmitted to a hospital. A random sample of children admitted to a Pediatric Health Information System hospital was obtained. The number of times each child was readmitted within one year was recorded. The observed frequencies are given in the table. Is there any evidence to suggest that the true category proportions of readmissions are different for males and females? Use $\alpha = 0.05$.

		Number of times readmitted				
		0	1	2	3	≥ 4
Gender	Males	346	56	19	15	24
	Females	454	66	28	18	16

Solution

We need to compare the proportions of the number of readmissions for males and females. Find the expected cell counts, and use the test for homogeneity of populations procedure to compute the value of the test statistic and draw the appropriate conclusion.

STEP 1 There are two populations, males and females, and five categories for readmissions ($I = 2, J = 5$). We would like to know whether the true proportions of readmission categories are the same for each population.

STEP 2 The four parts of the hypothesis test are as follows:

H_0: The true readmission category proportions are the same for males and females.
H_a: The true readmission category proportions are not the same.

TS: $X^2 = \sum_{i=1}^{2}\sum_{j=1}^{5}\frac{(n_{ij}-e_{ij})^2}{e_{ij}}$

RR: $X^2 \geq \chi^2_{\alpha,(I-1)(J-1)} = \chi^2_{0.05,4} = 9.4877$

STEP 3 Find each expected cell count. Here is one calculation.

$$e_{11} = \frac{n_{1.} \times n_{.1}}{n} = \frac{(460)(800)}{1042} = 353.17$$

All of the expected cell counts are given in **Table 13.5**.

Table 13.5 The two-way table for the hospital readmissions example, including the row and column totals, the grand total, and the expected cell counts

		\multicolumn{5}{c	}{Number of times readmitted}	
		0	1	2	3	≥4	Row total
Gender	Males	346 (353.17)	56 (53.86)	19 (20.75)	15 (14.57)	24 (17.66)	460
	Females	454 (446.83)	66 (68.14)	28 (26.25)	18 (18.43)	16 (22.34)	582
		800	122	47	33	40	1042

STEP 4 The value of the test statistic is

$$X^2 = \sum_{i=1}^{2}\sum_{j=1}^{5}\frac{(n_{ij}-e_{ij})^2}{e_{ij}}$$

$$= \frac{(346-353.17)^2}{353.17} + \frac{(56-53.86)^2}{53.86} + \cdots + \frac{(16-22.34)^2}{22.34}$$

$$= 4.7772\ (<9.4877)$$

STEP 5 The value of the test statistic does not lie in the rejection region, or equivalently, $p = 0.3109 > 0.05$. **Figure 13.9** shows the calculation and illustration of the exact p value. At the $\alpha = 0.05$ significance level, there is no evidence to suggest that the true readmission category proportions are different for males and females. **Figure 13.10** shows a technology solution.

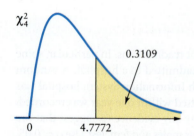

Figure 13.9 p value illustration:
$p = P(X \geq 4.7772)$
$= 0.3109 > 0.05 = \alpha$

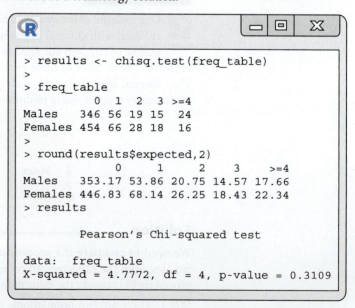

Figure 13.10 Use the R command `chisq.test()` to conduct a test for homogeneity of populations.

TRY IT NOW Go to Exercises 13.47 and 13.49

Test for Independence

Suppose bivariate data arise from a single random sample in which the values of two categorical variables are recorded for each individual or object. For example, in a random sample of individuals who stop for gasoline, the grade of gasoline and the make of car might be recorded. Or suppose a random sample of employed people is obtained, and the occupation and the employee classification (full time, part time, or temporary) are recorded for each person. In each instance, the data are bivariate, and there are observations on two categorical variables.

Given this type of bivariate data, it seems reasonable to ask whether the values of one variable affect the values of the other—that is, are the variables dependent? Or are they independent? In that case, knowing the value of one variable suggests nothing special about the value of the other. For instance, suppose that for those people living in a city, the proportions for transportation type are as follows: 0.23 drive their car to work, 0.51 use public transportation, and 0.26 walk to work. Now, suppose we know a person lives in a rural area. Does this change the proportions? If so, then the variables (residence and work commute) are dependent. If knowing the place of residence does not alter the commuting proportions, then the variables are independent.

We use the same notation as in the test for homogeneity. In an $I \times J$ two-way table, there are I categories for the first variable (instead of I populations), and J categories for the second variable. The test for independence is based on observed and expected counts once again. The test statistic is exactly the same as in the test for homogeneity. Here's why.

Recall from Chapter 4 that if two events A and B are independent, then the probability of A and B is the product of the corresponding probabilities; that is,

As in the test for homogeneity, this is actually a test for dependence. We cannot prove independence, but can only test for evidence of dependence.

$P(A \cap B) = P(A) \cdot P(B)$ if A and B are independent

Suppose the two categorical variables are independent and an individual or object falls into the (ij) cell. Consider the following probability:

P[an individual falls in the (ij) cell]

$= P\begin{pmatrix} \text{an individual responds with the } i\text{th value of the first variable} \\ \text{and the } j\text{th value of the second variable} \end{pmatrix}$

$= P[(i\text{th value for first variable}) \cap (j\text{th value for the second variable})]$ *And* means intersection.

$= P(i\text{th value for first variable}) \cdot P(j\text{th value for second variable})$

Independent events; multiply corresponding probabilities.

$= \dfrac{n_{i.}}{n} \cdot \dfrac{n_{.j}}{n}$ Using the notation introduced in this section, the probability of falling into the ith row times the probability of falling into the jth column.

Because there is a total of n individuals, the expected count, or frequency, in the (ij) cell is

e_{ij} = expected count in the (ij) cell

$= \begin{bmatrix} \text{sample} \\ \text{size} \end{bmatrix} \cdot \begin{bmatrix} \text{probability an individual falls} \\ \text{into the } (ij) \text{ cell} \end{bmatrix}$

$= n \cdot \left(\dfrac{n_{i.}}{n} \cdot \dfrac{n_{.j}}{n} \right)$

$= \dfrac{n_{i.} \times n_{.j}}{n}$ Simplify; cancel an n.

$= \dfrac{(i\text{th row total})(j\text{th column total})}{\text{grand total}}$ Symbol translation.

This is identical to the expression we used before, so the test statistic is the same as in the test for homogeneity.

646 CHAPTER 13 Categorical Data and Frequency Tables

> **CHAPTER APP** Traffic Composition
>
> Suppose vehicle type and time of day are independent. Find the expected count for each cell.

The test statistic is again a measure of how far away the observed cell counts are from the expected cell counts. Here is the formal hypothesis test.

Test for Independence of Two Categorical Variables

In a random sample of n individuals, suppose the values of two categorical variables are recorded. In the resulting $I \times J$ two-way frequency table, let n_{ij} be the observed count in the (ij) cell and let e_{ij} be the expected count in the (ij) cell. A hypothesis test for independence of the two categorical variables with significance level α has the following form.

H_0: The two variables are independent.

H_a: The two variables are not independent.

TS: $X^2 = \sum_{i=1}^{I} \sum_{j=1}^{J} \frac{(n_{ij} - e_{ij})^2}{e_{ij}}$

where $e_{ij} = \dfrac{(i\text{th row total})(j\text{th column total})}{\text{grand total}} = \dfrac{n_{i.} \times n_{.j}}{n}$

RR: $X^2 \geq \chi^2_{\alpha,(I-1)(J-1)}$

This test is appropriate if all expected cell counts are at least 5 ($e_{ij} \geq 5$ for all i and j).

EXAMPLE 13.4 Rest Stop Preferences

The Pilot Travel Center in Pennsylvania is located at the intersection of Routes 80 and 81 so that travelers on each road have easy-on/easy-off access in both directions. Research is being conducted to summarize food preferences and to attract vendors. A random sample of people who purchased food at this plaza was obtained, and the traveling direction and the food vendor were recorded for each person. The observed frequencies are given in the table. Is there any evidence to suggest that traveling direction and food vendor are dependent? Use $\alpha = 0.01$.

Michael Doolittle/Alamy

		Food vendor				
		Auntie Anne's	Pizza Hut	Taco Bell	Mrs. Fields	Hot Dog Company
Traveling direction	North	25	30	17	38	56
	South	40	22	25	45	41
	East	34	24	20	43	48
	West	28	27	25	31	32

Solution

Compute the expected cell counts, and use the test for independence of two categorical variables procedure to find the value of the test statistic and draw the appropriate conclusion.

STEP 1 The two categorical variables are traveling direction and food vendor. There are four possible values for traveling direction and five possible responses for food vendor ($I = 4$ and $J = 5$).

STEP 2 The four parts of the hypothesis test are

H_0: Traveling direction and food vendor are independent.

H_a: Traveling direction and food vendor are not independent.

TS: $X^2 = \sum_{i=1}^{I} \sum_{j=1}^{J} \frac{(n_{ij} - e_{ij})^2}{e_{ij}}$

RR: $X^2 \geq \chi^2_{\alpha,(I-1)(J-1)} = \chi^2_{0.01,12} = 26.2170$

13.2 Bivariate Categorical Data: Tests for Homogeneity and Independence

STEP 3 Find each expected cell count. Here is one calculation.

$$e_{11} = \frac{n_{1.} \times n_{.1}}{n} = \frac{(166)(127)}{651} = 32.38$$

All of the expected cell counts are given in **Table 13.6**.

Table 13.6 The two-way table for the travel plaza example, including the row and column totals, the grand total, and the expected counts

		Food vendor					Row Total
		Auntie Anne's	Pizza Hut	Taco Bell	Mrs. Fields	Hot Dog Company	
Traveling direction	North	25 (32.38)	30 (26.26)	17 (22.18)	38 (40.03)	56 (45.13)	166
	South	40 (33.75)	22 (27.37)	25 (23.12)	45 (41.72)	41 (47.04)	173
	East	34 (32.97)	24 (26.74)	20 (22.59)	43 (40.76)	48 (45.95)	169
	West	28 (27.90)	27 (22.63)	25 (19.11)	31 (34.49)	32 (38.88)	143
	Column total	127	103	87	157	177	651

STEP 4 The value of the test statistic is

$$\chi^2 = \sum_{i=1}^{4} \sum_{j=1}^{5} \frac{(n_{ij} - e_{ij})^2}{e_{ij}}$$

$$= \frac{(25 - 32.38)^2}{32.38} + \frac{(30 - 26.26)^2}{26.26} + \cdots + \frac{(32 - 38.88)^2}{38.88}$$

$$= 14.598 \ (<26.2170)$$

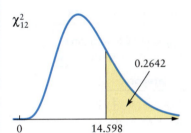

Figure 13.11 p value illustration: $p = P(X \geq 14.598) = 0.2642 > 0.01 = \alpha$

STEP 5 The value of the test statistic does not lie in the rejection region or, equivalently, $p = 0.2642 > 0.01$. **Figure 13.11** shows the calculation and illustration of the exact p value. At the $\alpha = 0.01$ significance level, there is no evidence to suggest that the two categorical variables are dependent.

Figure 13.12 shows a technology solution.

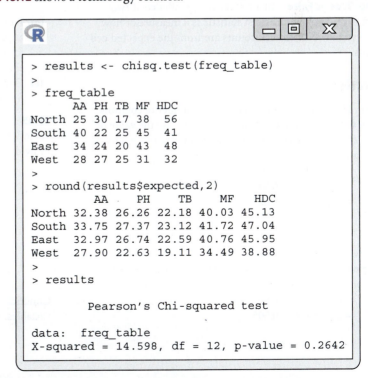

Figure 13.12 Use the R command `chisq.test()` to conduct a test for independence.

TRY IT NOW Go to Exercises 13.53 and 13.55

CHAPTER APP Traffic Composition

Is there any evidence to suggest that vehicle type and time of day are dependent?

Section 13.2 Exercises

Concept Check

13.33 Short Answer Explain two common ways to obtain bivariate data.

13.34 Short Answer Name two types of graphs that may be used to compare categorical data from two or more groups.

13.35 Short Answer State the null hypothesis in a test for homogeneity of populations.

13.36 True or False If we reject the null hypothesis in the test for homogeneity of populations, there is evidence to suggest that the category proportions are all the same.

13.37 Short Answer In a test for independence of two categorical variables, write the equation for the expected number of observations in the (ij) cell.

13.38 True or False If we reject the null hypothesis in the test for independence of two categorical variables, there is evidence to suggest that the two variables are dependent.

13.39 True or False In a test for independence of two categorical variables, we reject the null hypothesis only for large values of the test statistic.

13.40 True or False In a test for independence of two categorical variables, the test statistic is a measure of how far away the observed cell counts are from the expected cell counts.

Practice

13.41 The number of rows (I) and the number of columns (J) for a two-way frequency table are given. Use the value of α to determine the critical value in a test for homogeneity of populations.
(a) $I = 3, J = 4, \alpha = 0.05$
(b) $I = 2, J = 6, \alpha = 0.01$
(c) $I = 4, J = 3, \alpha = 0.025$
(d) $I = 5, J = 3, \alpha = 0.001$

13.42 The number of rows (I) and the number of columns (J) for a two-way frequency table are given. Use the value of α to determine the critical value in a test for independence of two categorical variables.
(a) $I = 6, J = 4, \alpha = 0.05$
(b) $I = 2, J = 8, \alpha = 0.005$

(c) $I = 3, J = 7, \alpha = 0.0005$
(d) $I = 4, J = 6, \alpha = 0.0001$

13.43 Find the missing observed cell counts and the row and column totals in the two-way frequency table.

		Category				Row total
		1	2	3	4	
Population	1	18	14	18		65
	2	25			12	
	3		33	26	28	119
Column total			68	60		258

13.44 Consider the two-way frequency table with three populations and three values of a categorical variable. EX13.44

		Category		
		1	2	3
Population	1	58	62	33
	2	65	55	40
	3	70	60	25

(a) Find the row and column totals and the grand total.
(b) Suppose the true category proportions are the same for each population. Find the expected count for each cell.
(c) Conduct a test for homogeneity of populations. Use $\alpha = 0.025$. State your conclusion.

13.45 Consider the two-way frequency table, in which the values of two categorical variables were recorded for each individual. EX13.45

		Variable 2			
		1	2	3	4
Variable 1	1	235	267	245	386
	2	241	264	280	305
	3	228	254	270	394
	4	219	235	263	363

Conduct a test for independence of the two categorical variables. Use $\alpha = 0.05$. State your conclusion.

Applications

13.46 Sports and Leisure Random samples of gamblers at four Las Vegas casinos were obtained, and each person was asked which game he or she played most. The results are given in the two-way frequency table. CASINO

	Game			
Casino	Blackjack	Poker	Roulette	Slots
Bellagio	22	20	38	66
Caesar's	30	38	22	69
Golden Nugget	28	25	21	81
Harrah's	38	25	29	84

Conduct a test for homogeneity of populations. Is there any evidence to suggest that the true proportion of gamblers at each game is not the same for all casinos? Use $\alpha = 0.05$.

13.47 Marketing and Consumer Behavior A marketing manager obtained random samples of shoppers from three different grocery stores and asked each shopper to name his or her favorite Tastykake product. The results are summarized in the two-way frequency table. KAKE

	Product			
Store	Krimpets	Cupcakes	Kandy Kakes	Creamies
Giant	90	80	95	92
Shaw	81	66	87	56
Weis	94	83	92	55

Is there any evidence to suggest that the true proportion of each favorite is not the same for all populations? Use $\alpha = 0.01$.

13.48 Marketing and Consumer Behavior Random samples of customers at two different office-supply stores were obtained, and each customer was asked which type of writing implement he or she prefers. The results are summarized in the two-way frequency table. WRITE

	Writing implement		
Store	Traditional pencil	Mechanical pencil	Pen
Office Max	183	164	480
Staples	130	202	420

Conduct a test for homogeneity of populations. Use $\alpha = 0.01$. State your conclusion, justify your answer, and find bounds on the p value associated with this test.

13.49 Public Policy and Political Science CBC/Radio Canada is Canada's public broadcaster and airs regional and cultural programs in several languages. A survey was conducted concerning the funding level for this broadcasting corporation. A random sample of Canadians was selected, and each was asked whether funding should be increased, decreased, or maintained at the current levels. The data are summarized in the two-way table. FUNDING

	Funding for CBC			
Region	Decrease	Maintain	Increase	Unsure
Atlantic	17	58	36	9
Quebec	27	167	65	37
Ontario	59	191	79	44
West	68	212	106	44

Conduct a test for homogeneity of populations. Use $\alpha = 0.05$. State your conclusion, and justify your answer.

13.50 Education and Child Development Random samples of boys and girls in elementary and secondary schools were obtained, and their parents were asked how often they helped with homework. The data are summarized in the two-way frequency table. HMWK

	Number of times spent helping with homework per week			
	Zero times	1 or 2 times	3 or 4 times	5 or more times
Girls	259	369	254	118
Boys	274	335	262	129

Is there any evidence that the number of times spent helping with homework is different for boys and girls? Use $\alpha = 0.01$. Find bounds on the p value associated with this test.

13.51 Psychology and Human Behavior While food-and-wine pairing is subjective and an inexact science, traditionally red wine goes with red meat, and white wine goes with fish and poultry. A random sample of diners at four-star restaurants was obtained, and each diner was classified according to the food and wine ordered. Here is the resulting two-way frequency table. WINEPR

	Wine	
Food	Red	White
Red meat	86	46
Fish or poultry	50	64

Is there any evidence that food and wine choices are dependent? Test the relevant hypothesis with $\alpha = 0.05$. Do these data suggest that diners are still following the traditional food-and-wine pairings?

13.52 Sports and Leisure The Whitewater Ski Resort in British Columbia offers downhill skiing, cross-country skiing, snowboarding, and snow tubing. The manager is planning to develop a new, targeted advertising campaign. A random sample of customers was obtained, and each was classified by age group and activity. The data are summarized in the two-way table. WINTER

	Resort activity			
Age group	Downhill skiing	Cross-country skiing	Snow-boarding	Snow tubing
<16	24	35	23	48
16–20	50	25	40	32
20–30	21	27	27	32
30–40	29	39	29	29
≥40	26	37	35	30

Is there any evidence to suggest that resort activity is dependent on age group? Conduct the appropriate hypothesis test with $\alpha = 0.01$.

13.53 Sports and Leisure Amid the hype and media coverage, the National Football League (NFL) draft takes place every spring. NFL teams jostle for position in an attempt to draft a college player who will help their team win a Super Bowl. A random sample of players selected in the draft was obtained, and each was classified by round and position (offense, defense, special teams).[8] The data are summarized in the two-way table. NFL

	Position		
Round	Offense	Defense	Special teams
1	1003	928	6
2	927	932	12
3	972	999	30
4	983	936	26

(a) Is there any evidence to suggest that position selected is dependent on draft round? Conduct the appropriate hypothesis test with $\alpha = 0.01$.
(b) Interpret your results in part (a). For example, when is a special teams player most likely to be drafted?

13.54 Medicine and Clinical Studies Some research suggests that college student athletes who are subject to high stress levels (due to academic demands, personal problems, or other sources) are more likely to be injured during a game. A random sample of student athletes was obtained, and a questionnaire was used to determine the stress level of each prior to a game. Following the game, the injury status of each athlete was also recorded. The data are summarized in the table. STRESS

	Stress level		
Injury	Low	Medium	High
Yes	12	17	25
No	335	292	288

Is there any evidence to suggest that stress level and injury are dependent? Conduct the appropriate hypothesis test with $\alpha = 0.05$.

13.55 Public Health and Nutrition The National Advisory Committee on Immunization in Canada provides the public with specific health recommendations and summary reports. A random sample of patients who tested positive for influenza during the 2017–2018 flu season was obtained. Each person was classified by age group and influenza type. The data are summarized in the table.[9] FLUAGE

	Influenza type			
Age group	A(H1)	A(H3)	A(UnS)	B
0–4	40	325	264	140
5–19	38	387	339	462
20–44	55	676	632	468
45–64	47	682	622	610
65+	17	2165	1406	961

Is there any evidence to suggest that the age group and influenza type are dependent? Conduct the appropriate hypothesis test with $\alpha = 0.01$.

13.56 Travel and Transportation Interstate 20 in Georgia is a four-lane highway that runs from the Alabama state line to the South Carolina state line. The Georgia Department of Transportation monitors this highway and collects vehicular traffic data. Five major interchanges by county along I-20 were selected. A random sample of days was obtained and the number of vehicles using each I-20 ramp was obtained. The data are given in the two-way table. GAHWY

	I-20 ramp			
County	On east	Off east	On west	Off west
Carroll	607	625	488	525
Fulton	635	635	501	490
DeKalb	768	787	603	605
Greene	845	860	665	681
Columbia	470	467	375	420

Conduct a test for homogeneity of (county interchange) populations. Use $\alpha = 0.01$. Find the exact p value. State your conclusions and justify your answer.

Extended Applications

13.57 Public Policy and Political Science Public swimming pools are routinely inspected by city health officials. In a random sample of pool code violations from cities in the Unites States, the type of pool and the violation class were recorded. The resulting two-way frequency table is shown. SWIM

	Violation		
Pool type	Serious, pool closed	Water chemistry	Policy/ management
Hotel/motel	1525	5462	2914
Condo/apartment	3282	13,227	7542
School/university	90	550	390
Private club	367	1508	863
Child care	6	32	20
Water park	33	111	62
Hospital	18	72	35
Municipal	84	368	259
Campground	32	134	91
Camp	20	227	192

Is there any evidence to suggest that the type of violation and the type of pool are dependent? Conduct the appropriate hypothesis test with $\alpha = 0.01$ and find bounds on the p value associated with this test.

13.58 Psychology and Human Behavior The Greenville, South Carolina, Police Department maintains a comprehensive analysis of traffic collisions on all roadways in the city. Most collisions occur on Friday, and the most common time of day for these events is between 5:00 P.M. and 6:00 P.M.[10] A random sample of collisions from 2018 was obtained, and each was classified by day of the week and time of day. The data are summarized in the two-way table. CRASH

	Time of day			
Day of the week	Dawn	Morning	Afternoon	Evening
Sunday	34	138	241	80
Monday	31	201	415	173
Tuesday	70	265	517	166
Wednesday	52	259	471	179
Thursday	47	262	538	205
Friday	60	264	532	193
Saturday	52	186	346	125

(a) Is there any evidence that day of the week and time of day for collisions are dependent? Use $\alpha = 0.05$.
(b) Use your results in part (a) to suggest the day and time of day when collisions are much different than expected.

13.59 Public Policy and Political Science The United Nations (UN) is more than 75 years old. However, many people in countries all over the world are still unfamiliar with the UN and do not understand its function. A survey concerning familiarity with the UN was conducted in six nations. The data are summarized in the two-way frequency table. Is there any evidence to suggest that familiarity with the UN and the country are dependent? Conduct the appropriate hypothesis test with $\alpha = 0.01$. THEUN

	Response			
Country	Very familiar	Somewhat familiar	Not that familiar	Not at all familiar
U.S.	340	1020	553	213
U.K.	33	402	457	196
France	31	261	669	84
Italy	32	547	400	74
Spain	60	423	403	121
Germany	21	238	610	165

13.60 Psychology and Human Behavior Homeopathic medicine is based on the theory that the body has the ability to heal itself. However, homeopathic health practitioners do use pills or liquids with a small amount of an active ingredient to treat diseases. These treatments are loosely regulated, and homeopathic practitioners hold various degrees. A random sample of homeopathic practitioners was obtained in the United States and Canada, and the academic degree of each was recorded. The data are summarized in the table. HOMEO

	Country	
Degree	United States	Canada
MD	78	30
ND	22	31
RN	15	14
DO	12	5
NP	11	4
DVM	10	18
Other	21	20

Is there any evidence to suggest that the true proportions for type of degree held are different for each country? Use $\alpha = 0.01$.

Challenge Problems

13.61 Marketing and Consumer Behavior For most customers shopping online, shipping options are important when making buying decisions. Online customers often have several shipping options, including overnight delivery, 3-day ground service, and free, usually slow, shipping. Suppose that, in a random sample of 500 adults in California, 135 said they prefer free shipping. In a random sample of 600 adults in New Jersey, 204 indicated they prefer free shipping.
(a) Conduct a hypothesis test concerning two population proportions to determine whether there is any evidence that the proportions of adults who prefer free shipping are different in California and in New Jersey. Find the p value associated with this test.
(b) Consider California and New Jersey as populations, and prefer and not prefer as categories. Using the data given in this problem, construct a two-way frequency table and conduct a test for homogeneity of populations. Use technology to find the p value for this test.
(c) What is the relationship between the value of the test statistic in part (a) and the value of the test statistic in part (b)? How are the p values related? Why do these relationships make sense?

13.62 Public Health and Nutrition The American Dental Association recently conducted a survey regarding dental hygiene habits. A random sample of elderly people was obtained, and each person was classified according to brushing and flossing frequency. The following codes and categories were used for each variable. **DENTAL**

Code	Floss/brush frequency
1	Never
2	Once per month
3	A few times per month
4	Once per week
5	A few times per week
6	Once per day
7	More than once per day

The survey results are summarized in the two-way frequency table.

		Floss						
		1	2	3	4	5	6	7
Brush	1	35	37	48	52	55	80	101
	2	36	38	42	47	54	75	97
	3	38	43	46	51	58	77	94
	4	33	51	42	42	51	52	86
	5	76	79	85	87	93	102	115
	6	81	41	46	78	107	103	116
	7	98	94	126	136	142	198	252

Is there any evidence to suggest that flossing frequency and brushing frequency are dependent? Conduct the appropriate hypothesis test with $\alpha = 0.01$. Find the p value associated with this test.

Chapter 13 Summary

One-Way Frequency Table
A one-way frequency table is a method for summarizing a univariate categorical data set. The table lists each possible category and the number of times each category occurred (the observed count or frequency for each category).

Goodness-of-Fit Test
Let n_i be the number of observations falling into the ith category ($i = 1, 2, \ldots, k$), and let $n = n_1 + n_2 + \cdots + n_k$. A hypothesis test about the true category population proportions with significance level α has the form

$H_0: p_1 = p_{10}, p_2 = p_{20}, \ldots, p_k = p_{k0}$
$H_a: p_i \neq p_{i0}$ for at least i

TS: $X^2 = \sum_{i=1}^{k} \frac{(n_i - e_i)^2}{e_i}$ where $e_i = np_{i0}$

RR: $X^2 \geq \chi^2_{\alpha, k-1}$

This test is appropriate if all expected cell counts are at least 5 ($np_{i0} \geq 5$ for all i).

Two-Way Frequency, or Contingency, Table
A two-way frequency table is a method for summarizing a bivariate categorical data set. Each cell contains the number of observations (observed count or frequency) in a category combination, or pairing.

Test for Homogeneity of Populations
In an $I \times J$ two-way frequency table, let n_{ij} be the observed count in the (ij) cell and let e_{ij} be the expected count in the (ij) cell. A hypothesis test for homogeneity of populations with significance level α has the form

H_0: The true category proportions are the same for all populations (homogeneity of populations).

H_a: The true category proportions are not the same for all populations.

TS: $X^2 = \sum_{i=1}^{I} \sum_{j=1}^{J} \frac{(n_{ij} - e_{ij})^2}{e_{ij}}$

where $e_{ij} = \frac{(i\text{th row total})(j\text{th column total})}{\text{grand total}} = \frac{n_{i.} \times n_{.j}}{n}$

RR: $X^2 \geq \chi^2_{\alpha, (I-1)(J-1)}$

This test is appropriate if all expected cell counts are at least 5 ($e_{ij} \geq 5$ for all i and j).

Test for Independence of Two Categorical Variables

In a random sample of n individuals, suppose the values of two categorical variables are recorded. In the resulting $I \times J$ two-way frequency table, let n_{ij} be the observed count in the (ij) cell and let e_{ij} be the expected count in the (ij) cell. A hypothesis test for independence of the two categorical variables with significance level α has the form

H_0: The two variables are independent.

H_a: The two variables are not independent.

TS: $X^2 = \sum_{i=1}^{I} \sum_{j=1}^{J} \frac{(n_{ij} - e_{ij})^2}{e_{ij}}$

where $e_{ij} = \frac{(i\text{th row total})(j\text{th column total})}{\text{grand total}} = \frac{n_{i.} \times n_{.j}}{n}$

RR: $X^2 \geq \chi^2_{\alpha, (I-1)(J-1)}$

This test is appropriate if all expected cell counts are at least 5 ($e_{ij} \geq 5$ for all i and j).

Chapter 13 Exercises

Applications

13.63 Marketing and Consumer Behavior Although coffee consumers are very familiar with the flavor pumps used to make their favorite caramel, vanilla, or hazelnut latte, several new flavors are now available in gourmet shops. A random sample of coffee consumers was asked to taste and select their favorite flavor from among five unique offerings.[11] The data are given in the table. **FLAVOR**

Flavor	Pecan	Maple	Tumeric	Lavender	Donut
Frequency	49	67	45	45	70

Is there any evidence to suggest that one flavor is preferred over another? Use $\alpha = 0.05$.

13.64 Psychology and Human Behavior The Rescue Pet Store sells five breeds of dogs, and the owner is trying to determine whether one breed is preferred over the others. A random sample of recent dog sales was obtained, and the number of each breed purchased is given in the table. **DOGS**

Dog breed	Frequency
American bulldog	46
Collie	54
Golden retriever	32
German shepherd	30
Yorkshire terrier	46

Is there evidence to suggest that the true population proportion of sales for any breed is different from 0.20? Use $\alpha = 0.05$.

13.65 Marketing and Consumer Behavior The manager of a CVS drugstore in Madison, Wisconsin, obtained a random sample of customers who purchased adhesive bandages. The brands and frequencies are given in the table. **BANDAGE**

Brand	Band-Aid	Curad	Nexcare	Generic
Frequency	220	215	95	510

Historical records indicate that the population proportions are Band-Aid 0.2, Curad 0.2, Nexcare 0.1, and Generic 0.5. Conduct a goodness-of-fit test to determine whether there is any evidence to suggest that the data are not consistent with the past proportions. Use $\alpha = 0.01$.

13.66 Public Policy and Political Science A Lubbock County, Texas, government official discovered some extra money in the budget that must be spent by the end of the fiscal year. A random sample of county residents was obtained, and each was asked how the money should be spent. The data and the hypothesized population proportions (from past county referendum votes) are summarized in the one-way frequency table. LUBBOCK

Project	Frequency	p_{i0}
Road construction	103	0.3
Road resurfacing	119	0.4
Bicycle paths	40	0.1
New sidewalks	35	0.1
Park improvements	20	0.1

Is there evidence to suggest that any of the true population proportions are different from the hypothesized proportions? Use $\alpha = 0.05$.

13.67 Education and Child Development The Canadian University Survey Consortium regularly conducts research into university students' satisfaction and adjustment. As part of the survey, each first-year student is asked to select the most important reason for attending a specific university. The table summarizes the results associated with this question from students surveyed at Carleton University and the proportions from the school's comparison group. CUSC

Response	Carleton frequency	Comparison group
To prepare for a specific job or career	568	0.43
To get a good job	455	0.26
To increase my knowledge in an academic field	179	0.08
To get a good general education	146	0.07
To prepare for graduate/professional school	130	0.07
To develop a broad base of skills	65	0.04
To meet parental expectations	32	0.02
Other	32	0.02
To meet new friends	16	0.01

Is there evidence to suggest that any of the Carleton true population proportions are different from the comparison group proportions? Use $\alpha = 0.01$.

13.68 Economics and Finance Many financial institutions offer customers with an individual retirement account (IRA) four different plans for automatic transfer of funds from a checking or savings account to their IRA. Random samples of IRA customers from each of five different companies were obtained, and the transfer plan was recorded for each. The data are summarized in the two-way table (Bear Sterns, BS; Commonfund, CF; Lincoln, LI; Prudential, PR; Ultimus: UL). IRA

		Transfer plan			
		Monthly	Quarterly	Semi-annually	Annually
Company	BS	71	70	67	59
	CF	90	51	56	61
	LI	75	82	70	60
	PR	69	57	78	69
	UL	93	92	77	91

Conduct a test for homogeneity of populations with $\alpha = 0.05$. Is there any evidence to suggest that the true proportions associated with transfer plans are different for any of the populations? Justify your answer.

13.69 Psychology and Human Behavior The Ohio State Bar Association regularly surveys the legal community about the economics of law practice. The resulting report includes information about attorney demographics, experience, and prevailing hourly billing rates. In addition, each survey participant is asked to indicate job satisfaction and attorney category. Suppose these data are summarized in the two-way table. ATTY

		Attorney category		
		Private practice	House counsel	Government
Current satisfaction	A great deal	109	62	101
	Some	97	56	44
	Very little	19	7	5

(a) Is there any evidence to suggest that the current satisfaction level is associated with the attorney category? Use $\alpha = 0.01$
(b) Find the p value associated with the hypothesis test in part (a).
(c) Which attorneys tend to be most satisfied with their job? Justify your answer.

13.70 Marketing and Consumer Behavior One of the most common home remodeling jobs involves the kitchen. In almost every house, this room is heavily used and often needs to be expanded to accommodate personal tastes. Three building-supply stores were selected, and a random sample of individuals purchasing kitchen countertops was obtained from each. The type of countertop was recorded for each person, and the data are summarized in the two-way frequency table (Home Depot, HD; Lowe's, LO; True Value, TV). COUNTER

		Countertop			
		Concrete	Corian	Marble	Granite
Store	HD	52	24	37	90
	LO	76	36	36	87
	TV	53	43	31	78

Is there any evidence to suggest that the true proportion of each type of countertop purchased is not the same for all supply stores? Use $\alpha = 0.05$.

13.71 Demographics and Population Statistics The Scottish Household Survey is designed to gather information on the composition, characteristics, attitudes, and behavior of both Scottish households and individuals. As part of the survey, all adults are asked about the number of days they could survive on stored food supplies in an emergency. The responses are summarized in the two-way table by tenure of household. 📊 SCOTTISH

		Household tenure		
		Owner occupied	Social rented	Private rented
Days	0	18	18	8
	1–2	140	89	51
	3–5	473	212	95
	6–9	665	171	86
	10–15	298	71	19
	16–25	88	18	8
	≥26	70	12	3

Is there any evidence to suggest that the number of days they could survive is associated with household tenure? Use $\alpha = 0.01$. Interpret your results in the context of this problem.

13.72 Psychology and Human Behavior Some research suggests that background music influences how much time people spend browsing in a store. To investigate this theory, a random sample of customers at a Paramount retail store was obtained (over a long period of time), and each was classified by the amount of time spent shopping (in minutes) and the type of music played during the day. The results are given in the table. 📊 MUSIC

		Type of music			
		Classical	Easy listening	Rock	Country
Time	<15	49	35	17	14
	15–30	51	27	19	41
	30–60	67	41	41	22
	≥60	47	26	26	33

Is there any evidence of an association between music type and time spent shopping? Conduct the appropriate hypothesis test with $\alpha = 0.01$.

13.73 Physical Sciences Recent research suggests that the *Titanic* luxury liner broke into three sections, causing it to sink faster than was previously believed. The collision with an iceberg was a terrifying event, and some experts believe the chance of survival was associated with location aboard the vessel. The table presents the class and survival status of passengers aboard the *Titanic*. 📊 TITANIC

		Survival status	
		Died	Survived
Class	First	122	203
	Second	167	118
	Third	528	178
	Crew	673	212

Is there any evidence of an association between class and survival status? Conduct the appropriate hypothesis test with $\alpha = 0.05$.

Extended Applications

13.74 Marketing and Consumer Behavior A new radio station in Fort Myers, Florida, WINK, plays a wide variety of music from 10 different genres. To assess customer preferences and narrow its focus, the station obtained a random sample of listeners, who were asked to indicate their favorite type of music. The data are summarized in the one-way frequency table. 📊 RADIO

Genre	Frequency	Genre	Frequency
Country	50	Jazz	44
Hits	62	Dance	36
Christian	62	Latin	38
Rock	68	World	42
Urban	70	Classical	35

Is there any evidence to suggest that one music genre is most preferred? Conduct the appropriate hypothesis test with $\alpha = 0.01$. Find bounds on the p value associated with this test.

13.75 Biology and Environmental Science In a recent national litter survey, three cities were selected and similar areas were inspected. The number of discarded paper bags found at each site was recorded, along with the source. The data are summarized in the table. 📊 LITTER

		City		
		San Francisco	Washington, D.C.	Oakland
Bag source	Take-out food	35	60	82
	Conv. store	40	30	78
	Pharmacy	5	6	4
	Grocery	60	5	5
	Other	7	6	12

Is there any evidence to suggest that the proportion of discarded paper bags by source is not the same for all cities? Use $\alpha = 0.05$.

13.76 Marketing and Consumer Behavior The Silicon Valley Bank Wine Report suggests that there is a relationship between wine sales and age group. A random sample of adults purchasing wine was obtained from various states. Each person was classified by generation and retail bottle price. The summary data are given in the two-way table. 📊 WINEAGE

	Generation			
Bottle price	Millennial	Gen X	Boomers	Matures
<$15	31	47	55	23
$15–$19	29	73	102	39
$20–$29	44	132	147	44
$30–$39	22	59	85	36
$40–$69	12	41	67	26
>$69	4	22	45	18

(a) Is there any evidence to suggest that generation and retail bottle price of the wine purchases are dependent? Use $\alpha = 0.05$.
(b) Find bounds on the p value. Use technology to confirm your answer and to find the exact p value.
(c) Suppose you are marketing wine in a predominantly millennial area. Using the table and the hypothesis test results as guides, how would you allocate your wine stock according to bottle price? Justify your answer.

13.77 Travel and Transportation A random sample of New York City commuters was obtained, and each was classified by resident borough and the type of transportation used to get to work. The resulting frequencies are given in the two-way table. 📊 COMMUTE

	Type of transportation			
	Bus	Subway	Ferry	Drive
Bronx	40	45	19	15
Brooklyn	45	52	14	8
Manhattan	30	54	17	11
Queens	47	48	15	13
Staten Island	30	43	20	14

(Borough on left axis)

Is there any evidence to suggest that borough residence and type of transportation are dependent? Use $\alpha = 0.05$.

CHAPTER APP

Matt Leung/Shutterstock

13.78 Traffic Composition A random sample of vehicles using the Hong Kong–Zhuhai–Macau Bridge at various times was obtained, and each was classified by vehicle type. The resulting frequencies, or counts, are given in the two-way, or contingency, table. 📊 VEHICLE

	Time of day			
Vehicle type	Morning	Afternoon	Evening	Overnight
Trucks	33	51	28	13
Buses	30	44	31	23
Shuttles	35	47	18	41
Cars	23	42	23	23

(a) Suppose vehicle type and time of day are independent. Find the expected count for each cell.
(b) Is there any evidence to suggest that vehicle type and time of day are dependent? Use $\alpha = 0.05$.

Nonparametric Statistics

14

◀ Looking Back
- Recall the parametric methods concerning population parameters: Statistical techniques based on the normality assumption: **Chapters 9, 10, and 11**.
- Remember that if any assumptions are violated, the conclusions may be invalid: **Chapters 9, 10, and 11**.

Looking Forward ▶
- Learn how to use the sign test concerning a population median and how to compare two population medians with this test: **Section 14.1**.
- Learn how to use the signed-rank test concerning a population median and how to compare two population medians with this test: **Section 14.2**.
- Learn how to use the rank-sum test to compare two population medians: **Section 14.3**.
- Learn how to perform a nonparametric analysis of variance of ranks and use the Kruskal–Wallis test to compare populations: **Section 14.4**.
- Use the runs test to determine whether a sequence of observations is random: **Section 14.5**.
- Learn how to compute and interpret Spearman's rank correlation coefficient: **Section 14.6**.

CHAPTER APP

Image Lagoon/Shutterstock

Keyboard Angles

CARPAL Carpal tunnel syndrome is a condition involving the median nerve in the wrist. Certain job or recreational activities may compress this nerve and cause pain, numbness, and tingling in the arm and hand.[1] One measure of the severity of carpal tunnel syndrome is the cross-sectional area of the median nerve. Generally, affected wrists have a larger median nerve cross-sectional area.

For those people who work with a computer, recent research suggests that the slope of a keyboard may have an effect on the median nerve. For each of two keyboard

657

slopes, 0° and 20°, a random sample of individuals who regularly type at a computer was obtained. The right wrist median nerve cross-sectional area was measured (in mm²) for each person. The data are given in the table.

0°	6.5	5.7	6.2	5.3	6.6	5.9	5.3	6.0	6.6	5.2	
20°	11.6	9.2	11.3	8.3	9.2	10.6	10.5	11.2	8.5	9.2	8.7	7.1

It seems reasonable to conduct a two-sample t test for a difference in population means to compare these two groups. However, we know that the median nerve cross-sectional areas are not normally distributed. Because the normality assumption is violated, a two-sample t test is not valid.

The methods presented in this chapter allow for comparisons of continuous distributions, with few assumptions necessary. These nonparametric procedures are handy when very little is known about the underlying distributions. A statistical test based on ranks can be used to compare the median nerve cross-sectional areas for these two keyboard angles.

14.1 The Sign Test

Nonparametric Methods and the Sign Test

Each of the hypothesis tests presented in the previous chapters depends on a set of assumptions. If any of the assumptions is violated, the conclusions may be invalid. Most of the statistical procedures include a normality assumption: The random sample(s) is (are) drawn from a normal distribution. Statistical techniques based on this assumption are called **parametric methods**. This chapter presents some alternative statistical techniques called **nonparametric**, or **distribution-free**, **procedures**. These techniques usually require very few assumptions about the underlying population(s).

The normality assumption is usually very reasonable because almost every distribution is normal or approximately normal.

Usually, very few assumptions are necessary for a nonparametric test to be valid. In addition, many statisticians consider the formula, or rule, for computing the test statistic in a nonparametric procedure to be more intuitive and easier to apply than a comparable parametric test. For example, the test statistic in a **sign test** is simply a count of the number of observations that are greater than the hypothesized median.

Nevertheless, nonparametric tests have some disadvantages. Because they require few assumptions, these procedures usually do not utilize all of the information captured in a sample. Nonparametric tests may ignore certain inherent information, for simplicity or ease of use. Consequently, the chance of making an error increases when you use a nonparametric test. Therefore, if you have a choice between using a parametric test or a nonparametric test, you should usually opt for the parametric procedure. Nonparametric tests are most useful when we cannot assume normality, or for analyzing certain nonnumerical data sets.

Remember: The population median divides the distribution in half.

Suppose a random sample is obtained from a continuous (non-normal) distribution. Consider a test concerning the population median with null hypothesis $H_0: \tilde{\mu} = \tilde{\mu}_0$. If the null hypothesis is true, then approximately half of the observations should be greater than $\tilde{\mu}_0$, and the other half should be less than $\tilde{\mu}_0$.

To use the sign test, we replace each observation above $\tilde{\mu}_0$ with a plus sign and each observation below $\tilde{\mu}_0$ with a minus sign. If H_0 is true, then the number of plus signs and the number of minus signs should be about the same. A problem arises if an observation is equal to $\tilde{\mu}_0$. Because the underlying distribution is continuous, theoretically the probability of obtaining an observation exactly equal to $\tilde{\mu}_0$ is zero. In practice, however, it is common to obtain such an observation. Any observations equal to $\tilde{\mu}_0$ are excluded from the analysis.

The test statistic X is a count of the number of plus signs (or number of observations greater than $\tilde{\mu}_0$). If the null hypothesis is true, then the probability of a plus sign (an

The binomial distribution was defined in Section 5.4.

observation is greater than $\tilde{\mu}_0$) is 1/2. Therefore, if H_0 is true, the random variable X has a binomial distribution with number of trials equal to n, the number of observations included in the test, and $p = 1/2$: $X \sim B(n, 0.5)$. We should reject the null hypothesis for very large or very small values of X.

The Sign Test Concerning a Population Median

Suppose a random sample is obtained from a continuous distribution. A hypothesis test concerning a population median $\tilde{\mu}_0$ with significance level α has the form

H_0: $\tilde{\mu} = \tilde{\mu}_0$
H_a: $\tilde{\mu} > \tilde{\mu}_0$, $\tilde{\mu} < \tilde{\mu}_0$, or $\tilde{\mu} \neq \tilde{\mu}_0$
TS: X = the number of observations greater than $\tilde{\mu}_0$
RR: $X \geq c_1$, $X \leq c_2$, or $X \geq c$ or $X \leq n - c$

The critical values c_1, c_2, and c are obtained from Appendix Table 1, with parameters n and $p = 0.5$, to yield a significance level of approximately α, that is, so that $P(X \geq c_1) \leq \alpha$, $P(X \leq c_2) \leq \alpha$, and $P(X \geq c) \leq \alpha/2$.

Observations equal to $\tilde{\mu}_0$ are excluded from the analysis, and the sample size is reduced accordingly.

A CLOSER LOOK

1. If the underlying (continuous) distribution is symmetric, then $\mu = \tilde{\mu}$ and we can use the sign test to test a hypothesis about a population mean.

2. We really do not need to literally replace observations with plus or minus signs. Discard any observations equal to $\tilde{\mu}_0$, and simply count the number of observations greater than $\tilde{\mu}_0$.

3. Because the binomial distribution is discrete, we usually cannot find critical values to yield the exact significance level α. Use critical values such that the significance level is as close to α as possible but not greater than α.

Example: The Sign Test

EXAMPLE 14.1 Unemployment Duration

UEDUR

The duration of unemployment is often used as an indicator of the state of the U.S. economy. During October 2018, the median duration of unemployment was approximately 10 weeks.[2] Almost one-third of unemployed people were able to find a new job within 5 weeks, but one-fourth took at least 27 weeks to find new employment.

Suppose a random sample of 15 unemployed adults in the United States was obtained in January 2019. The number of weeks each was unemployed is given in the table.

| 6 | 4 | 3 | 14 | 8 | 18 | 22 | 15 | 16 | 6 | 13 | 7 | 15 | 27 | 12 |

Is there any evidence to suggest that the median duration of unemployment is greater than 10 weeks? Use a significance level of $\alpha = 0.10$.

Solution

We cannot assume anything about the shape of the underlying distribution of unemployment duration. Since the normality assumption is not justified and this is a test concerning the median, the (nonparametric) sign test is appropriate.

STEP 1 The assumed median is $\tilde{\mu} = 10 \, (= \tilde{\mu}_0)$, the sample size is $n = 15$, and we will use $\alpha = 0.10$.

We are looking for any evidence that the median unemployment duration is greater than 10 weeks. Therefore, the relevant alternative hypothesis is one-sided, right-tailed.

STEP 2 The four parts of the hypothesis test are

$H_0: \tilde{\mu} = 10$

$H_a: \tilde{\mu} > 10$

TS: X = the number of observations greater than 10

RR: $X \geq c_1 = 11$

To find the critical value c_1:

(a) There are no observations equal to 10, the hypothesized median. Therefore, no values are excluded from the analysis.

If the null hypothesis is true, X is a binomial random variable with $n = 15$ and $p = 0.5$: $X \sim B(15, 0.5)$. We need to find a value c_1 such that $P(X \geq c_1)$ is as close to $\alpha = 0.10$ as possible without exceeding this value.

Therefore, find the smallest c_1 such that $P(X \geq c_1) \leq 0.10$.

(b) Use the Complement Rule applied to this known discrete random variable to convert this equation to cumulative probability.

$P(X \geq c_1) = 1 - P(X < c_1) = 1 - P(X \leq c_1 - 1) \leq 0.10$

Or, find the smallest c_1 such that $P(X \leq c_1 - 1) \geq 0.90$.

(c) Using Appendix Table 1, with $n = 15$ and $p = 0.5$, $c_1 - 1 = 10$. Therefore, $c_1 = 11$. The actual significance level using this critical value is

$P(X \geq 11) = 1 - P(X < 11) = 1 - P(X \leq 10) = 1 - 0.9408 = 0.0592$

STEP 3 Using signs, classify each observation as either greater than or less than the hypothesized median.

Observation	6	4	3	14	8	18	22	15	16	6	13	7	15	27	12
Sign	−	−	−	+	−	+	+	+	+	−	+	−	+	+	+

The value of the test statistic is the number of plus signs, or the number of observations greater than $\tilde{\mu}_0 = 10$. Therefore, $x = 9$.

STEP 4 The value of the test statistic does not lie in the rejection region. At the $\alpha = 0.10$ significance level, there is no evidence to suggest that the population median unemployment duration is greater than 10 weeks.

STEP 5 It is unlikely that a critical value will yield the exact desired significance level, so it is often more appropriate to find a p value when using the sign test.

$p = P(X \geq 9)$ *Definition of p value.*

$= 1 - P(X \leq 8)$ *The Complement Rule; binomial random variable.*

$= 1 - 0.6964 = 0.3036$ *Appendix Table 1; $n = 15$, $p = 0.5$.*

Using the p value to draw a conclusion, because $p = 0.3036 > 0.10 = \alpha$, we do not reject the null hypothesis. There is no evidence to suggest that the population median unemployment duration is greater than 10 weeks.

Figure 14.1 shows a technology solution.

A portion of Appendix Table 1, $n = 15$.

x	...	p 0.50	...
⋮		⋮	
8		0.6964	
9		0.8491	
10	...	0.9408	
11		0.9824	
12		0.9963	
⋮		⋮	

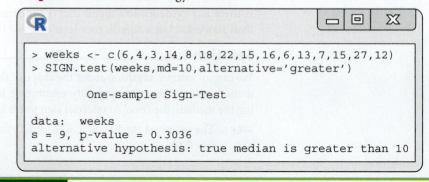

Figure 14.1 Use the R command SIGN.test() from the BSDA package to conduct a sign test.

TRY IT NOW Go to Exercises 14.15 and 14.17

CHAPTER APP **Keyboard Angles**

Use the sign test to determine whether there is any evidence that the cross-sectional area for the 0° group is greater than 5.8, a typical median nerve cross-section area in people unaffected by carpal tunnel syndrome.

The Sign Test to Compare Two Population Medians

This is a nonparametric counterpart to a paired t test.

If the data are paired (two observations on each individual or object) and the underlying distributions are not normal, then we can use the sign test to compare population medians. Compute each pairwise difference, disregard the magnitude of the difference, and consider only the sign of the difference. Replace each positive difference with a plus sign and each negative difference with a minus sign. Use these signs and the test procedure described earlier. This analysis is appropriate for comparing two population medians $\tilde{\mu}_1$ and $\tilde{\mu}_2$, with hypothesized difference $\Delta_0 = 0$.

The Sign Test to Compare Two Population Medians

Suppose we have n independent pairs of observations such that the population of first observations is continuous and the population of second observations is also continuous. A hypothesis test concerning the two population medians in terms of the difference $\tilde{\mu}_D = \tilde{\mu}_1 - \tilde{\mu}_2$, with significance level α, has the form

H_0: $\tilde{\mu}_D = \Delta_0$

H_a: $\tilde{\mu}_D > \Delta_0$, $\tilde{\mu}_D < \Delta_0$, or $\tilde{\mu}_D \neq \Delta_0$

TS: $X = $ the number of pairwise differences greater than Δ_0

RR: $X \geq c_1$, $X \leq c_2$, or $X \geq c$ or $X \leq n - c$

The critical values c_1, c_2, and c are obtained from Appendix Table 1, with parameters n and $p = 0.5$, to yield a significance level of approximately α, that is, so that $P(X \geq c_1) \leq \alpha$, $P(X \leq c_2) \leq \alpha$, and $P(X \geq c) \leq \alpha/2$.

Differences equal to Δ_0 are excluded from the analysis, and the sample size is reduced accordingly.

EXAMPLE 14.2 The Incredible Egg

EGG

Numerous studies have investigated the benefits and possible harmful effects of eating eggs. A single egg is packed with protein and nutrients, especially lutein, which can reduce the risk of chronic eye diseases.[3] A random sample of healthy male adults, ages 25–45, was obtained. The dynamic visual acuity (DVA, logMAR scale) was used to measure each subject's ability to detect objects while moving his head. A higher DVA score indicates better perception. Each person was put on a diet that included one egg per day. After four weeks, each subject was tested again for dynamic visual acuity. The data are given in the table.

Subject	1	2	3	4	5	6	7	8	9	10
Before	0.58	0.78	0.76	0.54	0.60	0.73	0.56	0.45	0.59	0.45
After	0.60	0.49	0.41	0.74	0.80	0.77	0.52	0.58	0.69	0.71

Subject	11	12	13	14	15	16	17	18	19	20
Before	0.42	0.62	0.76	0.61	0.67	0.49	0.48	0.56	0.57	0.52
After	0.58	0.70	0.73	0.43	0.52	0.70	0.47	0.44	0.64	0.68

Suppose other research suggests that the underlying DVA score populations are not normal. Use a sign test to compare the median DVA score before the egg diet with the median DVA score after the egg diet. Is there any evidence to suggest that eggs help to improve eyesight, as measured by DVA? Use a significance level of 0.01.

Solution

This study obtained before and after measurements for each individual; the data are paired. Compute each difference, and use the sign test to compare two medians, find the value of the test statistic, and draw the appropriate conclusion.

STEP 1 The data are certainly paired: There are before and after measurements for each individual. Typically, the before measurements are considered population 1 and the after measurements are population 2. The underlying populations are not normal, so a paired t test is not appropriate.

STEP 2 The null hypothesis is that the two population medians are equal: The egg-a-day diet has no effect on visual acuity, or

$$\tilde{\mu}_1 = \tilde{\mu}_2 \Rightarrow \tilde{\mu}_1 - \tilde{\mu}_2 = \tilde{\mu}_D = 0 \ (= \Delta_0)$$

We are searching for evidence that the egg diet improves visual acuity, so the alternative hypothesis is $\tilde{\mu}_1 < \tilde{\mu}_2 \Rightarrow \tilde{\mu}_1 - \tilde{\mu}_2 = \tilde{\mu}_D < 0$. This is a one-sided, left-tailed test.

STEP 3 The four parts of the hypothesis test are

H_0: $\tilde{\mu}_D = 0$

H_a: $\tilde{\mu}_D < 0$

TS: $X =$ the number of differences greater than 0

RR: $X \leq c_2 = 4$

To find the critical value c_2:

(a) There are no differences equal to Δ_0, the hypothesized difference, as shown in the following table. Therefore, no pairs are excluded from the analysis.

If the null hypothesis is true, X has a binomial distribution with $n = 20$ and $p = 0.5$: $X \sim B(20, 0.5)$. Find the largest value c_2 such that $P(X \leq c_2) \leq 0.01 = \alpha$.

(b) Using Appendix Table 1, with $n = 20$ and $p = 0.5$, the critical value is $c_2 = 4$. The actual significance level using this critical value is $P(X \leq 4) = 0.0059 \ (\leq 0.01)$.

STEP 4 Compute each pairwise difference (before − after) and determine the signs.

A portion of Appendix Table 1, $n = 20$

x	...	p 0.50	...
⋮		⋮	
2		0.0002	
3		0.0013	
4	...	0.0059	
5		0.0207	
6		0.0577	
⋮		⋮	

Subject	1	2	3	4	5	6	7	8	9	10
Before	0.58	0.78	0.76	0.54	0.60	0.73	0.56	0.45	0.59	0.45
After	0.60	0.49	0.41	0.74	0.80	0.77	0.52	0.58	0.69	0.71
Difference	−0.02	0.29	0.35	−0.20	−0.20	−0.04	0.04	−0.13	−0.10	−0.26
Sign	−	+	+	−	−	−	+	−	−	−

Subject	11	12	13	14	15	16	17	18	19	20
Before	0.42	0.62	0.76	0.61	0.67	0.49	0.48	0.56	0.57	0.52
After	0.58	0.70	0.73	0.43	0.52	0.70	0.47	0.44	0.64	0.68
Difference	−0.16	−0.08	0.03	0.18	0.15	−0.21	0.01	0.12	−0.07	−0.16
Sign	−	−	+	+	+	−	+	+	−	−

STEP 5 The value of the test statistic is the number of plus signs, or the number of differences greater than $\Delta_0 = 0$. Therefore, $x = 8$.

STEP 6 The value of the test statistic does not lie in the rejection region. At the $\alpha = 0.01$ significance level, there is no evidence to suggest that the egg diet increased the median DVA score.

STEP 7 The p value for this hypothesis test is

$p = P(X \le 8)$ *Definition of p value.*

$= 0.2517$ *Cumulative probability; Appendix Table 1, n = 20, p = 0.5.*

Using the p value to draw a conclusion, because $p = 0.2517 > 0.01 = \alpha$, we cannot reject the null hypothesis. There is no evidence to suggest that the egg diet increased the median DVA score.

Figure 14.2 shows a technology solution.

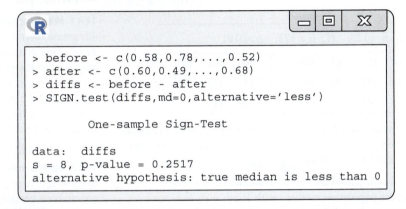

Figure 14.2 Find the pairwise differences and use the R command `SIGN.test()` from the BSDA package to conduct a sign test.

TRY IT NOW Go to Exercises 14.21 and 14.23

Section 14.1 Exercises

Concept Check

14.1 Short Answer A nonparametric test is also called a _____ procedure.

14.2 Short Answer Name two advantages to using a nonparametric statistical test.

14.3 True or False In a case where either a nonparametric test or a parametric test can be used, it is usually better to use the parametric procedure.

14.4 True or False The sign test concerning a population median is based on the number of positive values (observations greater than 0) and the number of negative values (observations less than 0).

14.5 Short Answer Under what conditions can the sign test be used to test a hypothesis about a population mean?

14.6 True or False The sign test is based on the binomial distribution with n trials and probability of a success $p = 0.50$.

14.7 Short Answer Explain why we usually cannot find critical values in a sign test to yield an exact, specified significance level α.

14.8 Short Answer In a sign test to compare two population medians, how do we treat a difference equal to Δ_0?

14.9 Short Answer If the null hypothesis is true in a sign test, explain why approximately half the observations should be greater than $\tilde{\mu}_0$ and the other half should be less than $\tilde{\mu}_0$.

Practice

14.10 A data set and a null hypothesis (concerning the population median) are given. Assume that the sign test will be used, and find the value of the test statistic x. **EX14.10**
(a) {20, 20, 13, 20, 16, 19, 19, 11, 20, 14} $H_0: \tilde{\mu} = 15$
(b) {66, 90, 77, 68, 70, 56, 56, 75, 57, 65, 65, 56, 66, 70, 54} $H_0: \tilde{\mu} = 70$
(c) {46.4, 42.6, 50.9, 49.2, 46.4, 47.6, 50.7, 49.3, 59.3, 51.5, 45.3, 52.9, 54.5, 51.1, 50.7, 41.3, 57.8, 59.5, 52.3, 47.2} $H_0: \tilde{\mu} = 51.5$
(d) {−4, −7, 8, −7, 3, 9, 1, 1, 3, 10, −9, −7, −5, 8, −1, −9, −2, 6, 7, −6, −7, −9, 7, 9, 3} $H_0: \tilde{\mu} = 0$

14.11 Use the random sample to conduct a sign test with the indicated null hypothesis, alternative hypothesis, and significance level. Find the exact p value for each test. **EX14.11**
(a) $H_0: \tilde{\mu} = 16$, $H_a: \tilde{\mu} < 16$, $\alpha = 0.05$

13.1	17.4	13.6	18.4	11.2	15.9	14.6
13.6	13.6	14.5				

(b) $H_0: \tilde{\mu} = -25$, $H_a: \tilde{\mu} > -25$, $\alpha = 0.01$

−19	−20	−25	−28	−17	−25	−16	−11
−12	−28	−23	−11	−11	−25	−28	

(c) $H_0: \tilde{\mu} = 8$, $H_a: \tilde{\mu} \neq 8$, $\alpha = 0.05$

8.90	7.68	9.41	9.47	8.98	8.41	9.51	7.40
9.66	8.17	5.32	9.06	7.67	7.87	8.16	8.77
7.01	9.22	7.41	8.57				

(d) $H_0: \tilde{\mu} = 125$, $H_a: \tilde{\mu} \neq 125$, $\alpha = 0.01$

196	126	187	168	111	196	164	181	110
113	187	136	159	154	177	132	103	151
187	191	192	139	187	190	112		

14.12 Use the paired data given to conduct a sign test to compare population medians with $H_0: \tilde{\mu}_1 - \tilde{\mu}_2 = 0$, $H_a: \tilde{\mu}_1 - \tilde{\mu}_2 > 0$, and significance level $\alpha = 0.05$. EX14.12

14.13 Use the paired data given to conduct a sign test to compare population medians with $H_0: \tilde{\mu}_1 - \tilde{\mu}_2 = 3$, $H_a: \tilde{\mu}_1 - \tilde{\mu}_2 \neq 3$, and significance level $\alpha = 0.01$. Find the exact p value for this test. EX14.13

Applications

14.14 Medicine and Clinical Studies Studies suggest that Parkinson's disease (PD) affects nerve endings in the heart that produce the chemical norepinephrine. Low levels of norepinephrine decrease a person's ability to control muscle movement, producing the shaking symptoms associated with PD. A random sample of women with PD was obtained, and the concentration of norepinephrine (in nmol/L) was measured in each. The data are given in the table. PD

1.36	1.29	1.05	1.25	1.34	1.50	1.33	1.31
1.08	1.11	1.29	1.04	1.09	1.38	1.44	

Assume that the underlying distribution is continuous. Use the sign test to determine whether there is any evidence that the median concentration of norepinephrine in women with PD is less than 1.38 nmol/L (the normal level). Use $\alpha = 0.05$.

14.15 Fuel Consumption and Cars Disc brake pads are critical safety features on an automobile and should be replaced when the thickness has worn to approximately 2 mm. A random sample of automobiles entering a New Jersey State Inspection Station was obtained, and the brake-pad thickness on the left front tire was measured on each. Assume that the underlying distribution of brake-pad thicknesses is continuous. Use the sign test to determine whether there is any evidence that the median brake-pad thickness is less than 2 mm. Use $\alpha = 0.10$ for this test and find the exact p value. BRAKE

14.16 Manufacturing and Product Development Modular homes are constructed in sections in a factory, with no delays due to weather conditions. Sections are transported to a home site on truck beds, attached together, and placed on a premade foundation. A random sample of modular-home deliveries was obtained, and the mileage from the factory to the home site was recorded for each. Assume that the underlying distribution of mileage is continuous. Use the sign test to determine whether there is any evidence that the median mileage is greater than 100. Use $\alpha = 0.05$ for this test and find the exact p value. MODHOME

14.17 Manufacturing and Product Development Grade 40s grease mohair must have a fiber diameter of approximately 24 microns. A manufacturer of mohair jackets received a large shipment of 40s-grade raw material and obtained 30 random measurements of fiber diameter. The manufacturer can only assume that the underlying distribution of fiber diameters is continuous. Is there any evidence to suggest that the median diameter is different from 24 microns? Use a significance level of $\alpha = 0.05$. MOHAIR

14.18 Sports and Leisure There has been increased attention lately to sports-related head injuries. An increase in NFL concussions produced a *call to action* from the (NFL) chief medical officer.[4] Physicians have become more sensitive to recognizing and treating head trauma in younger children, and recent studies suggest that there may be a gender difference in memory impairment after a concussion. A random sample of female soccer players who sustained a concussion was obtained. A computer test to measure memory skills was administered, and the time (in seconds) needed to complete the test was recorded for each. Assume that the underlying distribution of time needed to complete this memory test is continuous. Is there any evidence to suggest that the median time to complete this test for women is greater than 40 seconds? Use $\alpha = 0.05$. TRAUMA

14.19 Travel and Transportation According to a report from Kelley Blue Book, the typical transaction price for a new light vehicle in the United States in September 2018 was $35,742.[5] A transaction price of a vehicle is the out-the-door-price, which includes the price of the vehicle, discounts, add-ons, taxes, and license fees. A random sample of new vehicles purchased in January 2019 was obtained, and the transaction price of each was recorded. Assume that the underlying distribution of new-vehicle transaction prices is continuous. CARPRC

(a) Is there any evidence to suggest that the median new light vehicle transaction price has increased (from $35,742)?
(b) Why is the median, rather than the mean, new light vehicle transaction price a better measure of the center of this distribution?

14.20 Biology and Environmental Science Chlorophyll is an important part of photosynthesis in plants and is also present in microscopic algae and other phytoplankton. A random sample of 20 locations along the western coastline of the United States was selected. Using fluorometry, the quantity of chlorophyll in the surface water was measured (in mg/m³) in early April and in late August. The pairwise differences (April measurement − August measurement) are given in the table. **CHLOR**

−10.96	6.51	6.24	9.52	5.39	5.88	−9.03
4.58	−8.16	6.99	4.92	−9.35	4.88	−1.99
4.96	3.90	−8.13	7.55	12.61	7.68	

The distribution of chlorophyll in surface water is assumed to be continuous but not normal. Is there any evidence to suggest that the median chlorophyll amount in surface water is different in April than in August? Use a significance level of $\alpha = 0.05$.

14.21 Travel and Transportation Even though the number of vehicles driving into Manhattan's central business district on a typical workday has declined since 2004, traffic is still a nightmare. The average speed of vehicles in midtown Manhattan is approximately 4.7 mph. The worst traffic days are now called gridlock alert days. Drivers can choose to ignore these alerts, but construction is suspended and special traffic agents are placed on duty when an alert is issued. One measure of better traffic flow is the speed of a taxi ride. After carefully reviewing taxi records, a random sample of similar trips on gridlock alert days and non-gridlock alert days was obtained. The average speed (in mph) of each trip was recorded. Use the sign test with $\alpha = 0.05$ to determine whether there is any evidence that the median speed of a taxi trip is faster on non-gridlock alert days. Find the p value associated with this test. **TAXI**

14.22 Biology and Environmental Science In 2018, the Manua Loa Observatory in Hawaii reported the highest monthly average of atmospheric carbon dioxide in recorded history, approximately 410 parts per million.[6] Despite this ominous report, many countries have actually reduced their total CO_2 emissions. A random sample of countries was obtained and the total CO_2 emissions (in million tonnes) was recorded for 2017 and 2018.[7] Use the sign test with $\alpha = 0.05$ to determine whether there is any evidence that the median CO_2 emissions have decreased. Find the p value associated with this test. **EMISS**

Extended Applications

14.23 Biology and Environmental Science Jet-engine emissions at high altitudes cause vapor trails that contribute to cloud cover and pollution, and may even affect weather patterns. An experiment was conducted to measure the concentration of volatile organic compounds (VOCs) directly behind an operating engine. A random sample of jet engines was obtained. The VOC concentration was measured (in mg/m^3) before and after a special exhaust scrubber was installed. The data are given in the table. **VOC**

Before	13.0	14.9	14.0	13.4	9.8	14.9	12.0	11.2
After	11.0	13.6	12.3	9.9	10.3	10.7	9.6	9.5
Before	14.9	15.0	9.7	13.1	14.4	11.3	14.6	
After	11.8	13.1	9.2	10.0	11.9	13.7	12.3	

Use the sign test with $\alpha = 0.05$ to determine whether there is any evidence that the median VOC concentration is smaller when the scrubber is installed. Find the p value associated with this test.

14.24 Medicine and Clinical Studies Calcium channel blockers have many effects on a person's heart rate. These medications are most often used to slow the heart rate in patients with atrial fibrillation. This class of medications causes dilation of the arteries, which in turn, increases the supply of blood and oxygen to the heart.[8] Suppose an experiment was conducted to determine whether calcium channel blockers affect heart rhythms in patients with arrhythmias. A random sample of patients was obtained, and their resting pulse rate was measured (in beats per minute). After a two-week regimen of a calcium blocker, each person's resting pulse rate was measured again. Suppose no assumptions can be made about the shape of the continuous distributions of before and after pulse rates. **HEART**

(a) Conduct a sign test to determine whether there is any evidence that the median pulse rate before use of the calcium channel blocker medication is different from the median pulse rate after use of the medication. Use a significance level of $\alpha = 0.01$.

(b) Do you believe the assumptions for the sign test are valid in this case? Why or why not?

14.25 Medicine and Clinical Studies Lung transplants have become more common and more successful over the past few years, but they remain one of the riskiest transplant operations. Approximately 90% of lung transplant patients survive after the first year, but only about 50% survive after five years. The median survival rate is approximately 5.8 years.[9] The University of California, San Francisco Medical Center (UCSF) claims to have a survival rate higher than the national average. Suppose a random sample of single-lung transplant patients from UCSF was obtained. The survival (in years) after the transplant was recorded for each. Suppose no assumptions can be made about the shape of the underlying continuous distribution. **LUNG**

(a) Conduct a sign test to determine whether there is any evidence to suggest that the median survival at UCSF is greater than 5.8 years. Use $\alpha = 0.05$.

(b) Conduct a one-sample t test concerning a population mean to determine whether there is any evidence to suggest that the mean survival at UCSF is greater than 5.8 years. Use $\alpha = 0.05$. How do your results compare with part (a)? Which analysis do you believe is more appropriate? Why?

14.2 The Signed-Rank Test

The Wilcoxon Signed-Rank Test

This test was developed in 1945 by Frank Wilcoxon and is usually called the Wilcoxon signed-rank test.

The sign test concerning a population median or to compare two population medians uses only the signs (plus or minus) of the relevant differences. The signed-rank test also utilizes the magnitude of each difference and, of course, the ranks. The test statistic does not have a common distribution, but if n is sufficiently large, a normal approximation may be used.

Suppose we obtain a random sample of size n from a continuous, symmetric distribution, and the null hypothesis is H_0: $\tilde{\mu} = \tilde{\mu}_0$.

1. Subtract $\tilde{\mu}_0$ from each observation; that is, compute the differences $x_1 - \tilde{\mu}_0$, $x_2 - \tilde{\mu}_0, \ldots, x_n - \tilde{\mu}_0$.
2. Consider the magnitude, or absolute value, of each difference; compute $|x_1 - \tilde{\mu}_0|$, $|x_2 - \tilde{\mu}_0|, \ldots, |x_n - \tilde{\mu}_0|$.

The rank of an observation is its position in the ordered list.

3. Place the absolute values in increasing order, and assign a rank to each, from smallest (rank 1) to largest (rank n).
4. Equal absolute values are assigned the mean rank of their positions in the ordered list. For example, if the fifth, sixth, and seventh absolute values were all equal, then each would be assigned the rank $(5+6+7)/3 = 6$.
5. Add the ranks associated with positive differences.

If the null hypothesis is true, then approximately half of the observations should be greater than the median and approximately half less than the median. Because the distribution is assumed to be symmetric, for every positive difference, there should be a corresponding negative difference of approximately the same magnitude. Therefore, the sum of the ranks associated with the positive differences should be approximately equal to the sum of the ranks associated with the negative differences. If the sum of the ranks associated with the positive differences is very large or very small, this outcome suggests that the population median is different from $\tilde{\mu}_0$. The dot plots in **Figure 14.3** and **Figure 14.4** provide a visual interpretation of this intuitive concept.

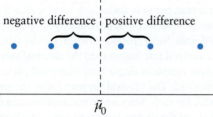

Figure 14.3 If H_0 is true, for every positive difference there should be a negative difference of about the same magnitude.

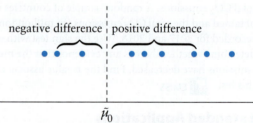

Figure 14.4 If $\tilde{\mu} > \tilde{\mu}_0$, then there should be more positive differences of large magnitude. The sum of the ranks associated with the positive differences would be large.

The Wilcoxon Signed-Rank Test

Suppose a random sample is obtained from a continuous, symmetric distribution. A hypothesis test concerning a population median with significance level α has the form

H_0: $\tilde{\mu} = \tilde{\mu}_0$

H_a: $\tilde{\mu} > \tilde{\mu}_0$, $\quad \tilde{\mu} < \tilde{\mu}_0$, $\quad$ or $\quad \tilde{\mu} \neq \tilde{\mu}_0$

Rank the absolute differences $|x_1 - \tilde{\mu}_0|$, $|x_2 - \tilde{\mu}_0|, \ldots, |x_n - \tilde{\mu}_0|$.

Equal absolute values are assigned the mean rank for their positions.

TS: $T_+ =$ sum of the ranks corresponding to the positive differences $x_i - \tilde{\mu}_0$

RR: $T_+ \geq c_2$, $\quad T_+ \leq c_1$, $\quad$ or $\quad T_+ \geq c$ or $T_+ \leq n(n+1) - c$

The critical values c_1, c_2, and c are obtained from Appendix Table 9 such that $P(T_+ \geq c_2) \approx \alpha$, $P(T_+ \leq c_1) \approx \alpha$, and $P(T_+ \geq c) \approx \alpha/2$.

Differences equal to 0 $(x_i - \tilde{\mu}_0 = 0)$ are excluded from the analysis, and the sample size is reduced accordingly.

The Normal Approximation: As n increases $(n \geq 20)$, the statistic T_+ approaches a normal distribution with mean and variance

$$\mu_{T_+} = \frac{n(n+1)}{4} \quad \text{and} \quad \sigma^2_{T_+} = \frac{n(n+1)(2n+1)}{24}$$

Therefore, the random variable $Z = \dfrac{T_+ - \mu_{T_+}}{\sigma_{T_+}}$ has approximately a standard normal distribution. In this case, $(n \geq 20)$, Z is the test statistic and the rejection region is

RR: $Z \geq z_\alpha$, $Z \leq -z_\alpha$, or $|Z| \geq z_{\alpha/2}$

We assume the underlying population is symmetric (when using the signed-rank test), so the median is equal to the mean. Therefore, we can use this procedure to test a hypothesis concerning a population mean, when the underlying population is not normal but is symmetric.

Example: The Signed-Rank Test

EXAMPLE 14.3 Sweet River Water

Diet foods and beverages contain artificial sweeteners that are not processed by the body. Consequently, these sweeteners—for example, sucralose, cyclamate, saccharin, and ace-sulfame—flow into water-treatment facilities. Most of these chemicals remain unaffected by the treatment process and flow into rivers and streams. The Grand River in Ontario is believed to have high concentrations of artificial sweeteners. Suppose a random sample of Grand River water was obtained from 10 different locations, and the saccharin concentration (in μg/L) was measured for each. The data are given in the table.

| 5.2 | 7.2 | 3.5 | 4.7 | 5.9 | 3.6 | 2.5 | 5.8 | 4.4 | 3.3 |

Assume the underlying distribution of saccharin concentration is continuous and symmetric. Use a signed-rank test to determine whether there is any evidence that the median saccharine concentration is greater than 4 μg/L, with a significance level of 0.05.

magnez2/iStock/Getty Images

Solution

Rank the absolute differences, and find the value of the test statistic. Use this value to draw the appropriate conclusion.

STEP 1 The distribution of saccharin concentration is assumed to be continuous and symmetric. Because this problem involves the population median, the signed-rank test is appropriate. We are looking for evidence that the population median is greater than 4; the hypothesis test is one-sided, right-tailed. The number of observations is $n = 10$, (< 20); the test statistic is T_+.

STEP 2 Here are the four parts of the hypothesis test.

H_0: $\tilde{\mu} = 4$

H_a: $\tilde{\mu} > 4$

TS: $T_+ =$ the sum of the ranks corresponding to the positive differences $x_i - \tilde{\mu}_0$

RR: $T_+ \geq c_2 = 44$

We need a value for c_1 such that $P(T_+ \geq c_1) \approx 0.05$. Using Appendix Table 9, $P(T_+ \geq 44) = 0.0527$.

STEP 3 The next table shows the data, each difference $x_i - 4$, the absolute value of each difference $|x_i - 4|$, the rank associated with each absolute difference, and the signed rank. The positive or negative sign from the pairwise difference has been attached to the rank to create the signed rank. Because there are no zero differences, no observations are excluded from the analysis.

Observation	5.2	7.2	3.5	4.7	5.9	3.6	2.5	5.8	4.4	3.3
Difference	1.2	3.2	−0.5	0.7	1.9	−0.4	−1.5	1.8	0.4	−0.7
Absolute difference	1.2	3.2	0.5	0.7	1.9	0.4	1.5	1.8	0.4	0.7
Rank	6.0	10.0	3.0	4.5	9.0	1.5	7.0	8.0	1.5	4.5
Signed rank	+6.0	+10.0	−3.0	+4.5	+9.0	−1.5	−7.0	+8.0	+1.5	−4.5

To assign each rank, consider the ordered list of absolute differences and their position in the list.

Absolute difference	0.4	0.4	0.5	0.7	0.7	1.2	1.5	1.8	1.9	3.2
Position	1	2	3	4	5	6	7	8	9	10
Rank	1.5	1.5	3.0	4.5	4.5	6.0	7.0	8.0	9.0	10.0

The absolute differences in positions 1 and 2 are equal. Each is assigned the mean rank, $(1+2)/2 = 1.5$.

The absolute difference 0.5 is in the third position. It is assigned the rank 3.0.

The absolute differences in positions 4 and 5 are equal. Each is assigned the mean rank, $(4+5)/2 = 4.5$.

The absolute difference 1.2 is in the sixth position. It is assigned the rank 6.0. Similarly, 1.5, 1.8, 1.9, and 3.2 are assigned the ranks 7.0, 8.0, 9.0, and 10.0.

STEP 4 The value of the test statistic is the sum of the positive signed ranks.
$$t_+ = 1.5 + 4.5 + 6.0 + 8.0 + 9.0 + 10.0 = 39 \; (< 44)$$

STEP 5 The value of the test statistic does not lie in the rejection region. There is no evidence to suggest that the median saccharin concentration in the Grand River is greater than 4 μg/L.

Figure 14.5 shows a technology solution.

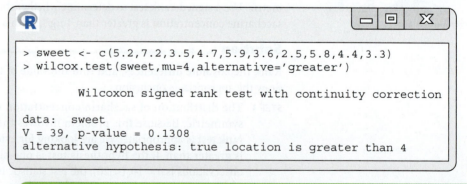

Figure 14.5 Use the R command `wilcox.test()` to conduct a signed-rank test.

TRY IT NOW Go to Exercises 14.35 and 14.37

CHAPTER APP **Keyboard Angles**

Use the signed-rank test to determine whether there is any evidence that the cross-sectional area for the 0° group is greater than 5.8, a typical median nerve cross-section area in people unaffected by carpal tunnel syndrome.

The Signed-Rank Test to Compare Two Population Medians

Two kinds of differences are present here: The pairwise differences are $d_i = x_i - y_i$, and the differences used in the calculation of the test statistic are $d_i - \Delta_0$.

The signed-rank test may also be used to compare population medians when the data are paired and the underlying distributions are not normal. We assume that the distribution of the pairwise differences is continuous and symmetric, and that each pair of values is independent of all the other pairs. If the null hypothesis is $H_0: \tilde{\mu}_1 - \tilde{\mu}_2 = \tilde{\mu}_D = \Delta_0$, we apply the one-sample test procedure to the pairwise differences.

1. Subtract Δ_0 from each pairwise difference.
   That is, compute $d_i - \Delta_0$ $(i = 1, 2, \ldots, n)$.
2. Rank the absolute differences $|d_i - \Delta_0|$ $(i = 1, 2, \ldots, n)$.
3. Find the sum of the ranks associated with the positive differences t_+.

EXAMPLE 14.4 Night-Shift Hazards

Studies suggest that people who work night shifts are at greater risk of developing obesity and diabetes, and these conditions may lead to heart disease and strokes.[10] The body clock of those people who work evening, night, and rotating shifts becomes disrupted, which may be associated with high blood pressure, high cholesterol, and diabetes. Suppose a study was conducted to compare the cholesterol level in workers on the night shift and workers on the traditional day shift in a food-processing facility. Random samples of employees on the night and day shifts were selected and paired according to gender, age, and weight. The total cholesterol level was measured (in mg/dL) in each, and the data are given in the table.

Night shift	215	185	175	185	211	158	171	199	190	151	154	233
Day shift	203	199	168	177	176	153	173	179	187	165	145	212
Night shift	180	202	180	160	195	171	152	210	212	196	188	213
Day shift	178	190	174	167	189	156	148	200	197	204	190	224

Assume that the underlying distribution of the pairwise differences is continuous and symmetric. Use the Wilcoxon signed-rank test to determine whether there is any evidence that the median total cholesterol level is greater in night-shift workers. Use $\alpha = 0.05$.

Solution

Compute each pairwise difference and rank the absolute differences. Find the value of the test statistic, and draw the appropriate conclusion.

STEP 1 The underlying distributions are not assumed to be normal, but the distribution of the differences is assumed to be continuous and symmetric. The assumptions for the Wilcoxon signed-rank test are met. Note that we could also use the sign test to compare two medians, but the Wilcoxon signed-rank test is a more reliable test because it uses more of the information in the sample.

We are searching for evidence that night-shift workers have higher median total cholesterol, so the hypothesis test is one-sided, right-tailed, with $\Delta_0 = 0$.

Let population 1 be the total cholesterol level in night-shift workers and population 2 be the total cholesterol level in day-shift workers.

Because $\Delta_0 = 0$, we do not need to modify the pairwise differences. No pairwise differences are equal to zero, so no pairs are excluded from the analysis.

There are $n = 24 \geq 20$ observations (pairwise differences), so we will use the normal approximation. The four parts of the hypothesis test are

$H_0: \tilde{\mu}_1 - \tilde{\mu}_2 = \tilde{\mu}_D = \Delta_0 = 0$

$H_a: \tilde{\mu}_1 - \tilde{\mu}_2 > \tilde{\mu}_D = \Delta_0 = 0$

TS: $Z = \dfrac{T_+ - \mu_{T_+}}{\sigma_{T_+}}$

RR: $Z \geq z_\alpha = z_{0.05} = 1.6449$

STEP 2 Compute the absolute value of each pairwise difference, and rank these numbers. Equal values are assigned the mean rank for their positions. The following table shows the original data, the pairwise differences, the absolute value of each pairwise difference, the rank associated with each pairwise difference, and the signed rank.

Night shift	Day shift	Pairwise difference	Absolute difference	Rank	Signed rank
215	203	12	12	16.5	16.5
185	199	−14	14	18.5	−18.5
175	168	7	7	9.5	9.5
185	177	8	8	11.5	11.5
211	176	35	35	24.0	24.0
158	153	5	5	6.0	6.0
171	173	−2	2	2.0	−2.0
199	179	20	20	22.0	22.0
190	187	3	3	4.0	4.0
151	165	−14	14	18.5	−18.5
154	145	9	9	13.0	13.0
233	212	21	21	23.0	23.0
180	178	2	2	2.0	2.0
202	190	12	12	16.5	16.5
180	174	6	6	7.5	7.5
160	167	−7	7	9.5	−9.5
195	189	6	6	7.5	7.5
171	156	15	15	20.5	20.5
152	148	4	4	5.0	5.0
210	200	10	10	14.0	14.0
212	197	15	15	20.5	20.5
196	204	−8	8	11.5	−11.5
188	190	−2	2	2.0	−2.0
213	224	−11	11	15.0	−15.0

STEP 3 The sum of the positive signed ranks is

$t_+ = 2.0 + 4.0 + 5.0 + \cdots + 24.0 = 223.0$

The mean and the variance of the random variable T_+ are

$\mu_{T_+} = \dfrac{n(n+1)}{4} = \dfrac{(24)(25)}{4} = 150$

$\sigma_{T_+}^2 = \dfrac{n(n+1)(2n+1)}{24} = \dfrac{(24)(25)(49)}{24} = 1225.0$

The value of the test statistic is

$$z = \frac{t_+ - \mu_{T_+}}{\sigma_{T_+}} = \frac{223.0 - 150.0}{\sqrt{1225.0}} = 2.0857$$

STEP 4 The value of the test statistic lies in the rejection region; we reject the null hypothesis. At the $\alpha = 0.05$ significance level, there is evidence to suggest that the median total cholesterol level of night-shift workers is greater than the median total cholesterol level of day-shift workers.

To find the p value associated with this test, find the right-tail probability.

$p = P(Z \geq 2.0857) \approx P(Z \geq 2.09)$ Definition of p value.

$ = 1 - P(Z \leq 2.09)$ Complement Rule.

$ = 1 - 0.9817 = 0.0183$ Cumulative probability; use Appendix Table 3.

Using the p value to draw a conclusion, because $p = 0.0183 \leq 0.05 = \alpha$, we reject the null hypothesis. There is evidence to suggest that the median total cholesterol level in night-shift workers is greater than the median total cholesterol level in day-shift workers.

Figure 14.6 shows a technology solution.

```
> night <- c(215,185,175,...,213)
> day <- c(203,199,168,...,224)
>
> results <- wilcox.test(night,day,paired=TRUE,alternative='greater')
>
> t_plus <- results$statistic
> n <- length(night)
> mu <- n*(n+1)/4
> var <- n*(n+1)*(2*n+1)/24
>
> z <- (t_plus-mu)/sqrt(var)
> p <- pnorm(z,lower.tail=FALSE)
>
> round(cbind(t_plus,z,p),4)
  t_plus      z      p
V    223 2.0857 0.0185
```

Figure 14.6 Use the sum of the positive signed ranks t_+ to find the value of the test statistic z.

TRY IT NOW Go to Exercise 14.44

Section 14.2 Exercises

Concept Check

14.26 Fill in the Blank The signed-rank test is based on the _____ and _____ of each difference.

14.27 True or False The test statistic for a signed-rank test has a chi-square distribution with $n-1$ degrees of freedom.

14.28 Short Answer What are the assumptions for a signed-rank test?

14.29 True or False The signed-rank test can be used to test a hypothesis about a population mean.

14.30 Short Answer If both the sign test and the signed-rank test are appropriate, which would you use, and why?

14.31 True or False In a signed-rank test, the normal approximation can be used only if the underlying population(s) is(are) assumed to be normal.

Practice

14.32 For the null hypothesis and data set, assume a Wilcoxon signed-rank test will be used to test H_0. Find (i) the differences, (ii) the absolute differences, and (iii) the rank associated with each absolute difference. 📊 EX14.32

(a) $H_0: \tilde{\mu} = 60$

41	66	36	72	33	22	24	36	47
53	28	77	31	34	38	59	60	45

(b) $H_0: \tilde{\mu} = 20$

21.4	20.3	18.5	18.8	21.5	21.6	20.8	21.9
19.8	19.6	19.4	20.8	19.6	18.7	20.4	20.2
21.6	19.2	20.1	18.5	18.5	19.7	19.7	

(c) $H_0: \tilde{\mu} = -5$

−2	−8	−9	−8	−7	−2	−2	−1	−5	−4
−9	−7	−9	−7	−5	−5	−8	−8	−3	−1
−4	−6	−7	−7	−2					

(d) $H_0: \tilde{\mu} = 305.4$

296	303	271	263	288	312	260	305
250	308	315	264	254	258	274	314
279	267	312	310	309	273	269	293
264	278	255	285	272	307		

14.33 Use the random sample to conduct a Wilcoxon signed-rank test with the indicated null hypothesis, alternative hypothesis, and significance level. Find the p value associated with each test. 📊 EX14.33

(a) $H_0: \tilde{\mu} = 70$, $H_a: \tilde{\mu} \neq 70$, $\alpha = 0.10$

67.1	63.5	70.1	62.5	72.2	63.7	63.7	67.7
79.1	61.6	79.7	69.5	72.2	64.5	69.3	

(b) $H_0: \tilde{\mu} = 0.7$, $H_a: \tilde{\mu} < 0.7$, $\alpha = 0.05$

0.072	0.348	0.319	0.502	0.733	0.603	0.052
0.493	0.721	0.762	0.166	0.965	0.616	0.904
0.882	0.773	0.771	0.506	0.735		

(c) $H_0: \tilde{\mu} = -45$, $H_a: \tilde{\mu} > -45$, $\alpha = 0.02$

−50	−35	−35	−31	−41	−32	−32
−40	−46	−35	−38	−31	−36	−32

(d) $H_0: \tilde{\mu} = 450$, $H_a: \tilde{\mu} \neq 450$, $\alpha = 0.001$

461	424	436	485	476	457	463	409	424	450
406	402	435	457	420	410	402	438	428	450
410	418	423	466	497	492	411	426		

14.34 Use the paired data to conduct a Wilcoxon signed-rank test to compare population medians with $H_0: \tilde{\mu}_1 - \tilde{\mu}_2 = 0$, $H_a: \tilde{\mu}_1 - \tilde{\mu}_2 \neq 0$, and significance level $\alpha = 0.02$. Find the p value associated with this test. 📊 EX14.34

Applications

14.35 Medicine and Clinical Studies A normal adult brain oxidizes approximately 120 g of glucose per day.[11] Some researchers believe that the rate of metabolism is higher in adults whose jobs require them to make many decisions daily. A random sample of basketball referees was obtained, and the brain oxidation rate of each was measured. The data are given in the table. 📊 BRAIN

120	121	119	122	121	122	121	119	121
122	124	121	123	122	123	119	120	119

The underlying distribution of oxidation rates is assumed to be continuous and symmetric.

(a) Use the Wilcoxon signed-rank test to determine whether there is any evidence that the mean oxidation rate for basketball referees is greater than 120 g per day. Use a significance level of $\alpha = 0.05$. Find the p value for this test.

(b) Why can the signed-rank test be used in this case to test a hypothesis concerning the population mean (rather than the population median)?

14.36 Medicine and Clinical Studies A random sample of patients with a certain kidney disorder was obtained, and the renal blood-flow rate was measured (in L/min) for each. In healthy patients, the median renal blood-flow rate is known to be 3.00 L/min, and the distribution of flow rates is assumed to be continuous and symmetric. Use the Wilcoxon signed-rank test to determine whether there is any evidence that the median renal blood-flow rate in patients with this kidney disorder is different from 3.00 L/min. Use $\alpha = 0.01$, and find the p value for this test. 📊 KIDNEY

14.37 Public Policy and Political Science Regulations for a new office building specify that the median strength for nonmetallic, nonshrink grout should be 6000 psi when it is supporting concrete. A random sample of grout from various locations in an office building project was obtained, and the strength of each batch (in psi) was determined using the grout-cube test at 28 days. Assume the underlying distribution of grout strength is continuous and symmetric. Use the Wilcoxon signed-rank test with $\alpha = 0.01$ to determine whether there is any evidence that the median grout strength is different from 6000 psi. Find the p value for this test. 📊 GROUT

14.38 Economics and Finance Near the end of 2018, the annual percentage rate (APR) on a typical credit card was approximately 13.64%.[12] Usually, the APR on airline credit cards is slightly higher. To check this claim, a random sample of airline credit cards was obtained and the annual percentage rate was recorded for each. Assume the distribution of APR on airline credit cards is continuous and symmetric. Use the Wilcoxon signed-rank test with $\alpha = 0.01$ to determine whether there is any evidence that the median APR on airline credit cards is greater than 13.64%. 📊 AIRCRED

14.39 Travel and Transportation The American Automobile Association (AAA) estimated that 54.3 million Americans traveled at least 50 mi away from home during Thanksgiving weekend 2018 and 48.5 million Americans took a Thanksgiving road trip.[13] The price of gasoline was higher in 2018 than in the previous year, which may have contributed to the trend toward shorter road trips. A random sample of Thanksgiving road trip travelers was obtained and the distance to their destination was recorded for each. Assume the distribution of travel distance is continuous and symmetric. Use the Wilcoxon signed-rank test with $\alpha = 0.05$ to determine whether there is any evidence that the median travel distance is less than 50 mi. THANKS

14.40 Economics and Finance Two counties in Pennsylvania handle delinquent property-tax payments in different manners. The first county has very strict regulations that include interest on the unpaid balance and eventually property seizure. The second county uses a more personal touch. Someone calls or visits the property owner to discuss the overdue bill and, if necessary, creates a payment schedule. A random sample of delinquent payments in each county was obtained, and the data were paired according to the size of the original property-tax bill. The amount of time (in months) until full payment was made was recorded for each property. Assume the distribution of pairwise differences is continuous and symmetric. Use the Wilcoxon signed-rank test to determine whether there is any evidence of a difference in median collection time due to the method of handling delinquent payments. Use $\alpha = 0.10$, and find the p value associated with this test. PROPTAX

14.41 Biology and Environmental Science The 154-day weights (in pounds) of 16 randomly selected Rambouillet lambs on a farm in Nebraska were recorded. The data are given in the table. LAMB

118	119	120	115	114	113	119	119
118	117	120	116	113	115	118	119

Assume the underlying distribution of weights is continuous and symmetric. Use the Wilcoxon signed-rank test to determine whether there is any evidence that the median 154-day weight is less than 118 lb. Use a significance level of 0.05.

14.42 Public Health and Nutrition Nearly 8 million acres of farmland in China is reportedly too polluted to use for growing food. In these areas, the soil contains a high concentration of heavy metals and other chemicals, and some of these pollutants have already made it into the food chain through rice and other crops. Cadmium is a carcinogenic metal that can cause kidney damage, is absorbed by rice, and is of particular concern. A random sample of farmland near Guangzhou was identified, and samples of soil were taken from each. The concentration of cadmium in each sample was measured. Assume that the distribution of the concentration of cadmium is continuous and symmetric. Use the Wilcoxon signed-rank test to determine whether there is any evidence that the median cadmium concentration is greater than 1 ppm. Use $\alpha = 0.05$ and find the p value for this test. FARM

14.43 Public Health and Nutrition Ottawa city officials recently installed generators near the Correctional Service of Canada offices to facilitate emergency repairs. However, nearby residents and business owners have been complaining about the nonstop drone from the generators, 27 formal complaints have been filed, and several people have indicated that they cannot sleep due to the noise.[14] A random sample of locations in neighborhoods within two blocks of the generators was obtained and the noise level was recorded (in weighted decibels, dBA). Assume that the distribution of noise level is continuous and symmetric. Use the Wilcoxon signed-rank test to determine whether there is any evidence that the median noise level is greater than 65 dBA, the noise level of normal conversation. Use $\alpha = 0.01$ and find the p value for this test. NOISE

Extended Applications

14.44 Travel and Transportation A study was conducted to assess two methods to estimate vehicle speed on interstate highways. A random sample of vehicles on Interstate 75 along Alligator Alley in Florida was obtained. The speed of each vehicle was measured using a loop detector and an electronic toll transponder. The data (in mph) are given in the table. SPEED

Vehicle	Loop detector	Toll transponder
1	66	65
2	66	66
3	70	67
4	82	80
5	74	68
6	70	67
7	63	65
8	63	61
9	80	78
10	71	68
11	82	81
12	66	65
13	72	68
14	70	68
15	65	60
16	63	62
17	66	65
18	79	79

Assume that the distribution of pairwise differences is continuous and symmetric.
(a) Use the Wilcoxon signed-rank test with $\alpha = 0.05$ to determine whether there is any evidence that the difference in median speed estimates is different from zero.
(b) Based on your conclusion in part (a), which method for estimating speed do you think the state police should use, and why?

14.45 Physical Sciences In a study of several Chesapeake Bay tributaries, a random sample of nontidal freshwater locations on the Patuxent River was obtained. The total arsenic concentration was measured at each location (in $\mu g/L$), and the data are given in the table. 📊 **ARSENIC**

0.67	0.22	0.56	0.27	0.17	0.57	0.55	0.55
0.53	0.09	0.61	0.45	0.49	0.04	0.44	0.19

Suppose the distribution of arsenic concentrations is continuous and symmetric and that the safe level of arsenic concentration is $0.30\ \mu g/L$.

(a) Conduct a sign test to determine whether there is any evidence that the median arsenic concentration in freshwater locations is greater than $0.30\ \mu g/L$. Use $\alpha = 0.05$.

(b) Conduct a signed-rank test to determine whether there is any evidence that the median arsenic concentration in freshwater locations is greater than $0.30\ \mu g/L$. Use $\alpha = 0.05$.

(c) Compare your conclusions in parts (a) and (b). Which test do you think is more accurate? Why?

14.46 Manufacturing and Product Development Modern dishwashers are reported to be more energy efficient and use less water. Most dishwashers reviewed by Consumer Reports earn an Excellent rating for energy efficiency and use between 4 and 6 gal of water per load.[15] A random sample of comparable, paired loads was obtained. One load was washed by hand and the other in a modern dishwasher. The amount of water used by each was carefully measured (in gallons). Assume the distribution of pairwise differences is continuous and symmetric. 📊 **WASH**

(a) Conduct a sign test to determine whether there is any evidence that the median amount of water used by a dishwasher is different from the median amount of water used when washing by hand. Use $\alpha = 0.05$.

(b) Conduct a signed-rank test to determine whether there is any evidence that the median amount of water used by a dishwasher is different from the median amount of water used when washing by hand. Use $\alpha = 0.05$.

(c) Compare your conclusions in parts (a) and (b). Which test do you think is more accurate? Why?

(d) If you want to save water, which washing method would you prefer? Why?

14.3 The Rank-Sum Test

The Wilcoxon Rank-Sum Test

The nonparametric Wilcoxon rank-sum test is used to compare two population medians. Suppose two independent random samples are obtained from continuous, non-normal distributions. Assume the first sample has size n_1, the second sample has size n_2, and $n_1 \leq n_2$.

Combine all of the data, for a total of $n_1 + n_2$ observations. Place these combined data in increasing order, and assign a rank to each from smallest (rank 1) to largest (rank $n_1 + n_2$). Equal values are assigned the mean rank of their positions in the ordered list (just as in the signed-rank test). Let w be the sum of the ranks associated with observations from the first sample.

Suppose, for example, that the sample sizes are equal. If $\tilde{\mu}_1 < \tilde{\mu}_2$, then the observations in the first sample will tend to be smaller than the observations in the second sample, and w will be small. If $\tilde{\mu}_1 > \tilde{\mu}_2$, then the observations in the first sample will tend to be larger than the observations in the second sample, and w will be large. There is also a normal approximation, which is valid when both n_1 and n_2 are large.

> Can you figure out the smallest and the largest possible value of w?

The Wilcoxon Rank-Sum Test

Suppose two independent random samples of sizes n_1 and n_2 ($n_1 \leq n_2$) are obtained from continuous distributions. A hypothesis test concerning the two population medians with significance level α has the form

H_0: $\tilde{\mu}_1 - \tilde{\mu}_2 = \Delta_0$

H_a: $\tilde{\mu}_1 - \tilde{\mu}_2 > \Delta_0$, $\tilde{\mu}_1 - \tilde{\mu}_2 < \Delta_0$, or $\tilde{\mu}_1 - \tilde{\mu}_2 \neq \Delta_0$

Subtract Δ_0 from each observation in the first sample. Combine these differences and the observations in the second sample, and rank all of these values. Equal values are assigned the mean rank for their positions.

TS: $W = $ the sum of the ranks corresponding to the differences from the first sample

RR: $W \geq c_2$ $\quad\quad W \leq c_1 \quad\quad$ or $\quad W \geq c$ or $W \leq n_1(n_1+n_2+1) - c$

The critical values c_1, c_2, and c are obtained from Appendix Table 10 such that $P(W \geq c_2) \approx \alpha$, $P(W \leq c_1) \approx \alpha$, and $P(W \geq c) \approx \alpha/2$.

The Normal Approximation: As n_1 and n_2 increase, the statistic W approaches a normal distribution with

$$\mu_W = \frac{n_1(n_1 + n_2 + 1)}{2} \quad \text{and} \quad \sigma_W^2 = \frac{n_1 n_2(n_1 + n_2 + 1)}{12}$$

Therefore, the random variable $Z = \dfrac{W - \mu_W}{\sigma_W}$ has approximately a standard normal distribution. The normal approximation is appropriate when both n_1 and n_2 are greater than 8. In this case, Z is the test statistic and the rejection region is

RR: $Z \geq z_\alpha$, $Z \leq -z_\alpha$, or $|Z| \geq z_{\alpha/2}$

A CLOSER LOOK

1. This general test procedure allows for any hypothesized difference between the two population medians, Δ_0. Usually, $\Delta_0 = 0$, and the null hypothesis is that the two population medians are equal.
2. If $\Delta_0 \neq 0$, we subtract this quantity from each observation in the first sample. Intuitively, if H_0 is true, this shifts sample 1 so that the two samples have the same median.
3. Like other nonparametric tests, this procedure is appropriate when the underlying distributions are non-normal, or when the data consist of ranks.
4. Suppose both underlying populations are symmetric. We can use the Wilcoxon rank-sum test to compare population means, because in each population the mean is equal to the median. In this case, we might also consider the two-sample t test, discussed in Section 10.2. However, if the underlying populations are not normal, then the t test results are not valid.
5. Sometimes, a slightly different test statistic is used in this test procedure. The Mann–Whitney U statistic is a function of W and also approaches a normal distribution as n_1 and n_2 increase. The two statistics lead to similar conclusions.

Many software packages call this the Mann–Whitney rank-sum test.

Example: The Wilcoxon Rank-Sum Test

EXAMPLE 14.5 Icebreakers

According to the U.S. Coast Guard, a large icebreaker program is very important to maintain America's presence in the Arctic. However, as of 2018 the polar icebreaking fleet included only one heavy ship, the *Polar Star*, and one medium ship, the *Healy*. Several countries around the world also maintain a fleet of icebreakers. These ships are designed with a strong hull, powerful engines, and a wide beam. Independent random samples of icebreakers from the United States and Canada were obtained, and the beam width of each was measured (in meters). The data are given in the table.

United States	19.60	25.45	24.40	25.00	18.30	
Canada		24.38	17.80	19.50	24.45	19.84	19.15

U.S. Coast Guard Photo/Alamy

Suppose the underlying distributions of icebreaker beam widths are continuous. Conduct a Wilcoxon rank-sum test to determine whether there is any evidence to suggest a difference in the median population icebreaker beam widths. Use $\alpha = 0.05$.

Solution

Combine the samples, and rank the ordered observations. Find the value of the test statistic W and draw the appropriate conclusion.

STEP 1 The samples are independent, and the populations are assumed to be continuous. We are searching for any difference in the medians, so $\Delta_0 = 0$.

The sample sizes are $n_1 = 5$ and $n_2 = 6$. Because n_1 and n_2 are both less than 8, the test statistic is W.

The four parts of the hypothesis test are

H_0: $\tilde{\mu}_1 - \tilde{\mu}_2 = \Delta_0 = 0$

H_a: $\tilde{\mu}_1 - \tilde{\mu}_2 \neq \Delta_0 = 0$

TS: W = the sum of the ranks corresponding to the differences from the first sample

$\Delta_0 = 0$, so W is the sum of the ranks corresponding to the first (smaller) sample

RR: $W \geq 41$ or $W \leq 5(5+6+1) - 41 = 19$

We need a value for such that $P(W \geq c) \approx 0.05/2 = 0.025$. From Appendix Table 10, $P(W \geq 41) = 0.026$. Therefore, the actual significance level for this test is $2(0.026) = 0.052 \approx 0.05$.

STEP 2 The table shows the combined samples in order and the rank associated with each value. The shaded columns correspond to observations from the first sample.

Observation	17.80	18.30	19.15	19.50	19.60	19.84	24.38	24.40	24.45	25.00	25.45
Rank	1	2	3	4	5	6	7	8	9	10	11

STEP 3 The value of the test statistic is the sum of the ranks corresponding to the observations from the first sample.

$w = 2 + 5 + 8 + 10 + 11 = 36$

The value of the test statistic does not lie in the rejection region. At the $\alpha = 0.05$ significance level, there is no evidence to suggest that the population median beam widths are different.

Figure 14.7 shows a technology solution.

```
> us <- c(19.60,25.45,24.40,25.00,18.30)
> canada <- c(24.38,17.80,19.50,24.45,19.84,19.15)
>
> wilcox.test(us,canada,md=0,paired=FALSE,alternative='two.sided')

        Wilcoxon rank sum test

data:  us and canada
W = 21, p-value = 0.329
alternative hypothesis: true location shift is not equal to 0
```

Figure 14.7 Use the R command wilcox.test() to conduct a Wilcoxon rank-sum test. Note that R returns an adjusted test statistic by subtracting the minimum possible value, $n_1(n_1+1)/2$.

TRY IT NOW Go to Exercises 14.61 and 14.63

The Wilcoxon Rank-Sum Test: Normal Approximation

Here is one more example that involves the normal approximation in the Wilcoxon rank-sum test.

EXAMPLE 14.6 Economic Mobility

The Opportunity Index (OI) is a unique measure of the economic mobility in counties and states. Factors that influence the kinds of opportunities people have include gender, race and ethnicity, family background, income level, and the quality of schools. A higher OI suggests more social mobility and greater economic security. A random sample of counties in the northeast and the west were obtained, and the OI was recorded for each.[16] The data are given in the table.

Northeast	67.0	58.3	58.7	60.5	56.9	60.4	58.9
	57.6	61.8	54.1	51.0	56.1		
West	47.5	56.3	56.0	40.7	58.4	45.9	48.8
	58.1	58.7	46.3	48.0	46.7	43.2	44.5

Assume the underlying distributions of the OI in each area of the country are continuous. Use a Wilcoxon rank-sum test to determine whether there is any evidence to suggest a difference in the population median OIs. Use $\alpha = 0.01$.

Solution

Combine the samples, and rank the ordered observations. Find the value of the test statistic Z, and draw the appropriate conclusion.

STEP 1 The samples are independent, and the populations are assumed to be continuous. We are searching for any difference in the medians, so $\Delta_0 = 0$.

The sample sizes are $n_1 = 12$ and $n_2 = 14$. Both sample sizes are greater than 8, so we will use the normal approximation.

The four parts of the hypothesis test are

$H_0: \tilde{\mu}_1 - \tilde{\mu}_2 = \Delta_0 = 0$
$H_a: \tilde{\mu}_1 - \tilde{\mu}_2 \neq \Delta_0 = 0$
TS: $Z = \dfrac{W - \mu_W}{\sigma_W}$
RR: $|Z| \geq z_{\alpha/2} = z_{0.005} = 2.5758$

STEP 2 The table shows the combined samples in order and the rank associated with each value. The shaded columns correspond to observations from the first sample.

Observation	40.7	43.2	44.5	45.9	46.3	46.7	47.5	48.0	48.8
Rank	1.0	2.0	3.0	4.0	5.0	6.0	7.0	8.0	9.0

Observation	51.0	54.1	56.0	56.1	56.3	56.9	57.6	58.1	58.3
Rank	10.0	11.0	12.0	13.0	14.0	15.0	16.0	17.0	18.0

Observation	58.4	58.7	58.7	58.9	60.4	60.5	61.8	67.0
Rank	19.0	20.5	20.5	22.0	23.0	24.0	25.0	26.0

Recall that equal values are assigned the mean rank for their position.

STEP 3 The value w is the sum of the ranks corresponding to the observations from the first sample.

$w = 10.0 + 11.0 + 13.0 + 15.0 + 16.0 + 18.0 + 20.5 + 22.0$
$\quad + 23.0 + 24.0 + 25.0 + 26.0 = 223.50$

The mean and variance of the random variable W are

$\mu_W = \dfrac{n_1(n_1 + n_2 + 1)}{2} = \dfrac{12(12 + 14 + 1)}{2} = 162.0$

$\sigma_W^2 = \dfrac{n_1 n_2 (n_1 + n_2 + 1)}{12} = \dfrac{(12)(14)(12 + 14 + 1)}{12} = 378.0$

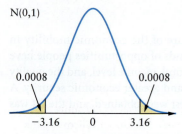

Figure 14.8 *p* value illustration:
$p = 2P(Z \geq 3.16)$
$= 0.0016 \leq 0.01 = \alpha$

The value of the test statistic is

$$z = \frac{w - \mu_W}{\sigma_W} = \frac{223.5 - 162.0}{\sqrt{378.0}} = 3.1632 \approx 3.16$$

The value of the test statistic lies in the rejection region. Equivalently, $p = 0.0016 \leq 0.01$. The *p* value calculation and illustration are shown in **Figure 14.8**. At the $\alpha = 0.01$ significance level, there is evidence to suggest that the population median OIs are different.

Figure 14.9 shows a technology solution.

```
> northeast <- c(67.0,58.3,58.7,...,56.1)
> west <- c(47.5,56.3,56.0,...,44.5)
> n_1 <- length(northeast)
> n_2 <- length(west)
>
> results <- wilcox.test(northeast,west,md=0,
                        paired=FALSE,alternative='two.sided')
>
> w <- results$statistic + n_1 * (n_1+1)/2
> mu <- n_1 * (n_1 + n_2 + 1)/2
> var <- n_1 * n_2 *(n_1 + n_2 + 1)/12
>
> z <- (w - mu)/sqrt(var)
> p <- 2*pnorm(z,lower.tail=FALSE)
>
> round(cbind(w,z,p),4)
        w      z      p
W   223.5 3.1632 0.0016
```

Figure 14.9 Use the R command `wilcox.test()` to find the value *w*. Find the value of the test statistic and the *p* value.

TRY IT NOW Go to Exercise 14.67

CHAPTER APP **Keyboard Angles**

Conduct a Wilcoxon rank-sum test to determine whether there is any evidence to suggest a difference in the population median, median nerve cross-sectional areas.

Section 14.3 Exercises

Concept Check

14.47 True or False The sample sizes must be equal for the Wilcoxon rank-sum test to be valid.

14.48 Short Answer In a Wilcoxon rank-sum test, explain how ranks are assigned to equal values.

14.49 True or False The Wilcoxon rank-sum test allows for any hypothesized difference between two population medians.

14.50 Fill in the Blank The Wilcoxon rank-sum test is most appropriate when the underlying distributions are _____.

14.51 True or False The Wilcoxon rank-sum test can be used to compare two population means.

14.52 Short Answer If both the Wilcoxon rank-sum test and the two-sample *t* test are appropriate, which would you use, and why?

14.53 True or False The normal approximation in the Wilcoxon rank-sum test is used when the underlying distributions are normal.

Practice

14.54 Two independent samples are given. Combine all of the observations, and find the rank associated with each value. EX14.54

(a)

| Sample 1 | 37 | 21 | 46 | 29 | 34 |
| Sample 2 | 45 | 42 | 22 | 41 | 24 | 39 |

(b)

Sample 1	4.5	1.8	3.4	1.4	2.2	2.1	1.5	3.7
Sample 2	4.6	6.6	1.2	6.0	2.4	2.3	2.2	0.4
	2.5	2.7

(c) Sample 1

| 820 | 872 | 814 | 825 | 876 | 858 | 841 | 892 | 882 |
| 809 | 826 | 887 | 884 | 862 | 846 | 801 | 871 | 803 |

Sample 2

850	842	888	879	832	827	831	810	871
840	870	816	821	865	899	830	875	800
813	839	822	865	818	818

14.55 Assume a Wilcoxon rank-sum test will be used to compare population medians. Two independent random samples and the value of Δ_0 are given. Find w, the value of the test statistic. EX14.55

(a) $\Delta_0 = 0$

| Sample 1 | 90 | 90 | 87 | 87 | 81 |
| Sample 2 | 83 | 84 | 80 | 84 | 87 |

(b) $\Delta_0 = 0$

Sample 1	58	69	52	53	57	58	60	56	59
Sample 2	53	52	60	55	70	66	65	69	51
	69	66

(c) $\Delta_0 = 5$
Sample 1

| 36.2 | 33.5 | 32.5 | 33.2 | 32.5 | 34.6 | 39.0 | 39.7 |
| 32.1 | 39.5 | 38.3 | 30.6 | 32.5 | 38.4 | 38.4 |

Sample 2

31.7	34.4	32.3	25.6	27.3	32.7	33.0	28.4
32.6	28.6	27.8	30.5	30.6	33.1	25.4	32.4
31.5	33.5

14.56 Assume a Wilcoxon rank-sum test will be used to compare population medians. Use the sample sizes and alternative hypothesis to find the best rejection region for the given significance level. Report the exact significance level for your rejection region.

(a) $n_1 = 3$, $n_2 = 5$, $H_a: \tilde{\mu}_1 < \tilde{\mu}_2$, $\alpha = 0.05$
(b) $n_1 = 4$, $n_2 = 4$, $H_a: \tilde{\mu}_1 > \tilde{\mu}_2$, $\alpha = 0.05$
(c) $n_1 = 4$, $n_2 = 10$, $H_a: \tilde{\mu}_1 \neq \tilde{\mu}_2$, $\alpha = 0.05$
(d) $n_1 = 6$, $n_2 = 8$, $H_a: \tilde{\mu}_1 < \tilde{\mu}_2$, $\alpha = 0.01$
(e) $n_1 = 7$, $n_2 = 9$, $H_a: \tilde{\mu}_1 \neq \tilde{\mu}_2$, $\alpha = 0.10$
(f) $n_1 = 8$, $n_2 = 8$, $H_a: \tilde{\mu}_1 > \tilde{\mu}_2$, $\alpha = 0.01$

14.57 Assume a Wilcoxon rank-sum test with the normal approximation will be used to compare population medians. Find the mean, the variance, and the standard deviation for the test statistic W.

(a) $n_1 = 15$, $n_2 = 21$
(b) $n_1 = 18$, $n_2 = 18$
(c) $n_1 = 11$, $n_2 = 27$
(d) $n_1 = 12$, $n_2 = 16$
(e) $n_1 = 23$, $n_2 = 24$
(f) $n_1 = 25$, $n_2 = 30$

14.58 Two independent random samples from continuous distributions are given in the table. Conduct a Wilcoxon rank-sum test of $H_0: \tilde{\mu}_1 = \tilde{\mu}_2$ versus $H_a: \tilde{\mu}_1 > \tilde{\mu}_2$ using a significance level of 0.01. EX14.58

| Sample 1 | 51.8 | 55.6 | 67.6 | 58.7 | 66.2 | 63.1 |
| Sample 2 | 57.2 | 51.2 | 58.2 | 56.9 | 57.1 | 63.3 | 59.7 | 60.6 |

14.59 Two independent random samples are obtained from continuous distributions. Conduct a Wilcoxon rank-sum test (using the normal approximation) of $H_0: \tilde{\mu}_1 = \tilde{\mu}_2$ versus $H_a: \tilde{\mu}_1 \neq \tilde{\mu}_2$ using a significance level of 0.05. EX14.59

Applications

14.60 Medicine and Clinical Studies Many complications may accompany type 2 diabetes, including heart disease, eye problems, and nerve damage. Other studies suggest that type 2 diabetes also affects bone strength.[17] Suppose a random sample of patients at the Mayo Clinic aged 50 to 80 with and without type 2 diabetes was obtained. Researchers measured the bone material strength in each person's tibia using micro indentation testing. The data are given in the tables. DIABET

With type 2 diabetes

| 1.09 | 1.14 | 1.38 | 1.17 | 1.29 | 1.32 |

Without type 2 diabetes

| 1.06 | 1.78 | 1.72 | 1.48 | 1.68 | 1.87 | 1.40 |

The indentation data are unitless, and a larger value indicates less bone strength. Use the Wilcoxon rank-sum test with $\alpha = 0.05$ to determine whether there is any evidence that patients with type 2 diabetes have less bone strength. Assume that the underlying distributions are continuous.

14.61 Technology and the Internet Users surfing the Internet generally leave webpages very quickly, often within 10–20 sec. Some research suggests that the first 10 sec of a page visit is critical to a decision to read further. A random sample of users visiting similar webpages, one with advertisements and one without, was obtained. The time (in seconds) for each visit was recorded, and the data are given in the tables. SURF

With advertisements

| 14.5 | 10.1 | 9.7 | 11.6 | 8.2 |

Without advertisements

| 10.4 | 12.0 | 13.2 | 11.7 | 10.9 | 13.2 | 14.5 |

Use the Wilcoxon rank-sum test with $\alpha = 0.05$ to determine whether there is any evidence to suggest that the population median visit times are different. Assume the underlying distributions are continuous.

14.62 Manufacturing and Product Development Better World Technologies claims to have produced a revolutionary new jackhammer with less vibration, reduced noise and energy input, fewer moving parts, and greater impact strength. Independent random samples of conventional jackhammers and new jackhammers were obtained. The impact strength of each, using heavy-duty concrete breakers, was measured (in ft-lb) at 75 psi. The data are given in the tables. JACK

Conventional jackhammer

95	94	102	100	93	93

New jackhammer

119	130	99	126	114	130	101

(a) Use the Wilcoxon rank-sum test with $\alpha = 0.01$ to determine whether there is any evidence to suggest that the population median impact strength is higher for the new jackhammer than for the conventional jackhammer. Assume the underlying distributions are continuous.

(b) Find the p value for this hypothesis test.

14.63 Sports and Leisure Many hockey players use a slapshot—a short, quick, powerful swing with the stick—to try to score a goal. The speed of this shot makes it very difficult for a goalie to react and make a save. The fastest slapshot recorded was 108.8 mph in 2012 by Zdeno Chara.[18] A National Hockey League (NHL) scout believes that defensemen have faster slapshots than forwards do. To test this claim, independent random samples from each group of players were obtained. Each player took a slapshot from the blue line, and the speed of the puck was recorded (in mph) using a radar gun. The data are given in the tables. SLAP

Defensemen

94	98	97	94	94	91	100	98	96

Forwards

92	88	86	94	90	85	86	95	89	85

Use the Wilcoxon rank-sum test with $\alpha = 0.01$ to determine whether there is any evidence to suggest that the population median slapshot speed of NHL defensemen is greater than the population median slapshot speed of NHL forwards. Assume the underlying distributions are continuous.

14.64 Business and Management Kroll's South restaurant serves dairy products in two separate buffet lines. Independent random samples of the dairy products holding temperature (in °F) in each line were obtained. Use the Wilcoxon rank-sum test to determine whether there is any evidence to suggest that the population median holding temperatures are different. Use $\alpha = 0.05$, and find the p value associated with this test. DAIRY

14.65 Terrestrial Radio Radio listeners have several options other than traditional terrestrial radio, including, Internet radio, music services, satellite radio, and individual playlists. Despite the loss of listeners, traditional terrestrial radio stations still run lots of commercials. A list of traditional music and talk radio stations was obtained, and a random hour during the day was selected for each. The total time for commercials during the hour (in minutes) was recorded for each. RADIO

(a) Use the Wilcoxon rank-sum test to determine whether there is any evidence to suggest that the population median commercial time per hour is greater on talk radio stations than on music stations. Use $\alpha = 0.05$ and find the p value associated with this test.

(b) Suppose the underlying distributions are approximately normal. Use a two-sample t test assuming equal variances to determine whether there is any evidence to suggest that the population mean commercial time per hour is greater on talk radio stations than on music stations. Use $\alpha = 0.05$ and find the p value associated with this test.

(c) Which test do you think is more appropriate? Why?

(d) Summarize your results: Which type of station plays more commercials?

14.66 Sports and Leisure Professional surf competitions are organized by the World Surf League (WSL). The world's best surfers compete in some very remote and exotic locations around the world. A random sample of surfers was obtained from three competitions, and the wave 1 score for each was recorded.[19] WSL

(a) Assume the underlying distributions are continuous. Use the Wilcoxon rank-sum test (three times) to determine whether there is any evidence to suggest that the population median wave score is different between any two sites. Use $\alpha = 0.05$.

(b) Higher wave scores are generally associated with more exciting tournaments. Summarize your results from part (a). Which tournament do you believe would be the most exciting?

Extended Applications

14.67 Public Health and Nutrition Independent random samples of almonds from two suppliers in the United Kingdom were obtained, and the amount of protein (in grams) per 100 g of edible portion was measured for each portion. SBP claims that its almonds contain more protein than any other brand. Use the Wilcoxon rank-sum test with $\alpha = 0.01$ to compare the population median amounts of protein in SBP and WTL International almonds. Do you believe the SBP claim? Why or why not? Find the p value for this test. ALMOND

14.68 Fuel Consumption and Cars New cars sold in the European Union in 2021 are targeted to have an emission rate of 95 g CO_2/km.[20] The emission rates were stagnant in 2017, so reduction rates will have to increase to meet the 2021 target. A random sample of new cars sold in Germany and Denmark in 2018 was obtained, and the CO_2 emission rate was measured for each. Assume the underlying distributions are continuous and symmetric. ERATE

(a) Use the Wilcoxon signed-rank test to determine whether there is any evidence that the median CO_2 emission rate for new cars in Germany is less than 110 g (the 2018 target rate). Use $\alpha = 0.01$.

(b) Use the Wilcoxon signed-rank test to determine whether there is any evidence that the median CO_2 emission rate for new cars in Denmark is less than 110 g. Use $\alpha = 0.01$.

(c) Use the Wilcoxon rank-sum test to compare the population median CO_2 emission rates in Germany and Denmark.

(d) Summarize your results and each country's progress toward the 2021 target emission rate.

Challenge Problems

14.69 Biology and Environmental Science The indoor air quality (the concentration of particles less than 2.5 microns in diameter) in 15 bars, restaurants, and other public venues in Atlanta was measured prior to the introduction of new smoking regulations. Two months after all public venues were required to be smoke-free, the air quality was measured again. The data (in $\mu g/m^3$) are given in the table. **AIRQ**

Venue	1	2	3	4	5	6	7	8
Before	353	386	104	198	597	62	412	273
After	56	35	28	21	83	10	27	34

Venue	9	10	11	12	13	14	15
Before	38	156	35	87	105	101	324
After	27	31	13	26	26	18	25

Suppose the underlying before and after distributions of air quality are continuous and symmetric.

(a) Conduct a sign test with $\alpha = 0.05$ to determine whether there is any evidence that the median amount of particulates decreased after the smoking regulations went into effect.

(b) Conduct a Wilcoxon signed-rank test with $\alpha = 0.05$ to determine whether there is any evidence that the median amount of particulates was greater before the smoking regulations went into effect than after they were implemented.

(c) Use the Wilcoxon rank-sum test with $\alpha = 0.05$ to determine whether there is any evidence that the median amount of particulates decreased after the smoking regulations went into effect.

(d) Compare the results of these three statistical tests. Which test(s) is (are) appropriate and why?

14.4 The Kruskal–Wallis Test

Background and Test Procedure

The Kruskal–Wallis test is a nonparametric analysis of variance of ranks. This procedure is an alternative to the analysis of variance F test, which does not require assumptions concerning normality or equal population variances. Although there is a table of critical values associated with this procedure, even for small sample sizes the test statistic has approximately a chi-square distribution.

Suppose that $k > 2$ independent random samples are obtained from continuous distributions. Assume the first sample has size n_1, the second sample has size n_2, and so on, such that the kth sample has size n_k.

Combine all of the data, for a total of $n = n_1 + n_2 + \cdots + n_k$ observations. Place these combined data in increasing order and assign a rank to each from smallest (rank 1) to largest (rank n). Equal values are assigned the mean rank for their positions in the ordered list (just as in the signed-rank test). Let r_i be the sum of the ranks associated with observations from sample i.

If all of the populations have identical distributions, then the sums of the ranks associated with each sample should be approximately equal. We expect the sum of the ranks associated with a sample from a different population to be distinct and separate from the rest. The test statistic in the Kruskal–Wallis test assesses the differences in the sums of the ranks.

The Kruskal–Wallis test

Suppose $k > 2$ independent random samples of sizes $n_1, n_2, \ldots, n_k$ are obtained from continuous distributions. A hypothesis test concerning the general populations with significance level α has the following form:

H_0: The k samples are from identical populations.

H_a: At least two of the populations are different.

Combine all observations, and rank these values from smallest (1) to largest (n). Equal values are assigned the mean rank for their positions. Let R_i be the sum of the ranks associated with the ith sample.

TS: $H = \left[\dfrac{12}{n(n+1)} \sum \dfrac{R_i^2}{n_i} \right] - 3(n+1)$

Critical values for the Kruskal–Wallis test statistic are available. However, if H_0 is true and either

1. $k = 3$, $n_i \geq 6$, $(i = 1, 2, 3)$ or
2. $k > 3$, $n_i > 5$, $(i = 1, 2, \ldots, k)$

then H has an approximate chi-square distribution with $k - 1$ degrees of freedom.

RR: $H \geq \chi^2_{\alpha, k-1}$

A CLOSER LOOK

1. We reject the null hypothesis only for large values of the test statistic H. If the sample sizes are large enough and we use the approximate chi-square distribution, the rejection region is always in the right tail of the appropriate chi-square distribution.
2. If we reject the null hypothesis, there is evidence to suggest that at least two populations are different. Further analysis is needed to determine which pairs of populations are different, and how they differ; the means, medians, variances, shapes of the distributions, or other characteristics could be dissimilar.

Example: Kruskal–Wallis Test

EXAMPLE 14.7 Shipping Distances

SHIP

During the year-end holiday season, shipping delays in delivering packages for Christmas often occuur due to increased volume and bad weather. However, some shipping companies claim that shipping distances contribute to delivery delays. Suppose a study was conducted to compare typical package shipping distances by company. A random sample of packages shipped within the United States from each of three major companies was obtained, and the shipping distance (in miles) from the point of origin was recorded for each. The data are given in the table; the sample numbers are in parentheses.

Airborne Express (1)	UPS (2)	Federal Express (3)
834	245	1617
2954	600	1538
2845	915	1298
1889	998	1580
2006	284	1968
1318	325	1526
1675	558	1002
1959	493	

The underlying populations are assumed to be continuous. Use the Kruskal–Wallis test to determine whether there is any evidence that the package shipping distance populations are different. Use a significance level of $\alpha = 0.05$.

Solution

The underlying distributions are assumed to be continuous. Use the Kruskal–Wallis test procedure; combine and rank all of the observations. Find the value of the test statistic, and draw the appropriate conclusion.

STEP 1 There are $k=3$ independent random samples of sizes $n_1 = 8 \geq 6$, $n_2 = 8 \geq 6$, and $n_3 = 7 \geq 6$, so the Kruskal–Wallis test statistic has an approximate chi-square distribution with $k - 1 = 3 - 1 = 2$ degrees of freedom.

STEP 2 The four parts of the hypothesis test are

H_0: The three samples are from identical populations.

H_a: At least two of the populations are different.

TS: $H = \left[\dfrac{12}{n(n+1)} \sum \dfrac{R_i^2}{n_i}\right] - 3(n+1)$

RR: $H \geq \chi^2_{\alpha, k-1} = \chi^2_{0.05, 2} = 5.9915$

STEP 3 There are $n = n_1 + n_2 + n_3 = 8 + 8 + 7 = 23$ total observations. The table shows all 23 observations, sorted in ascending order, with the associated rank. Each sample is color-coded to facilitate a visual comparison of the ranks.

White corresponds to sample 1, blue to sample 2, and green to sample 3.

Observation	245	284	325	493	558	600	834	915
Rank	1	2	3	4	5	6	7	8

Observation	998	1002	1298	1318	1526	1538	1580	1617
Rank	9	10	11	12	13	14	15	16

Observation	1675	1889	1959	1968	2006	2845	2954
Rank	17	18	19	20	21	22	23

STEP 4 If the populations are identical, we expect the ranks to be evenly distributed among the three samples. The shaded cells suggest that there are more low ranks associated with sample 2, middle ranks associated with sample 3, and high ranks associated with sample 1; this suggests that the populations are different.

All the observations, the associated ranks, and the rank sums are given in the table.

Airborne Express (1)	Rank	UPS (2)	Rank	Federal Express (3)	Rank
834	7	245	1	1617	16
2954	23	600	6	1538	14
2845	22	915	8	1298	11
1889	18	998	9	1580	15
2006	21	284	2	1968	20
1318	12	325	3	1526	13
1675	17	558	5	1002	10
1959	19	493	4		
Rank sum	139		38		99

STEP 5 The value of the test statistic is

$h = \left[\dfrac{12}{n(n+1)} \sum \dfrac{r_i^2}{n_i}\right] - 3(n+1)$

$= \left[\dfrac{12}{(23)(24)} \left(\dfrac{139^2}{8} + \dfrac{38^2}{8} + \dfrac{99^2}{7}\right)\right] - 3(24) = 14.8645 \ (\geq 5.9915)$

The value of the test statistic ($h = 14.8645$) lies in the rejection region. Equivalently, $p = 0.0006 \leq 0.05$. At the $\alpha = 0.05$ significance level, there is evidence to suggest that the package shipping distance populations are different.

Figure 14.10 shows a technology solution and the exact p value.

```
> distance <- c(834,2954,2845,1889,2006,1318,1675,1959,
+              245,600,915,998,284,325,558,493,
+              1617,1538,1298,1580,1968,1526,1002)
>
> company <- c(rep(1,length(air)),rep(2,length(ups)),
+             rep(3,length(fedex)))
>
> kruskal.test(distance ~ company)

        Kruskal-Wallis rank sum test

data: distance by company
Kruskal-Wallis chi-squared = 14.865, df = 2, p-value = 0.0005918
```

Figure 10 Use the R command `kruskal.test()` to conduct a Kruskal–Wallis test.

TRY IT NOW Go to Exercises 14.83 and 14.85

Section 14.4 Exercises

Concept Check

14.70 Fill in the Blank The Kruskal–Wallis test is a nonparametric alternative to ____________.

14.71 Short Answer If there are k groups in a Kruskal–Wallis test, what is the approximate distribution of the test statistic?

14.72 Fill in the Blank In a Kruskal–Wallis test, suppose the k populations have identical distributions. Then the sum of the ranks associated with each sample should be ____________.

14.73 True or False In a Kruskal–Wallis test, we reject the null hypothesis only for large values of the test statistic.

14.74 True or False In a Kruskal–Wallis test, if we reject the null hypothesis, there is evidence to suggest that all k population means are different.

14.75 True or False In a Kruskal–Wallis test, the underlying population variances are assumed to be equal.

14.76 Short Answer Suppose $k > 2$ independent random samples are obtained from normal distributions. Explain why the Kruskal–Wallis test is appropriate to compare the general populations instead of a one-way ANOVA.

Practice

14.77 Three independent random samples are given in the tables. Assume a Kruskal–Wallis test will be used to compare populations. Find the rank sum associated with each sample. EX14.77

Sample 1

88	94	94	79	91	76	77	79

Sample 2

19	72	77	79	89	72	87	74	82	90

Sample 3

85	96	95	96	93	85	95	93	87	95	93	93

14.78 Five independent random samples are obtained; assume a Kruskal–Wallis test will be used to compare populations. Use the sample sizes and the rank sums in the table to conduct this test at the $\alpha = 0.01$ level of significance.

Sample	Size	Rank sum
1	$n_1 = 12$	$r_1 = 395.5$
2	$n_2 = 14$	$r_2 = 428.5$
3	$n_3 = 12$	$r_3 = 287.0$
4	$n_4 = 16$	$r_4 = 620.0$
5	$n_5 = 20$	$r_5 = 1044.0$

14.79 Four independent random samples are obtained. Conduct a Kruskal–Wallis test to compare populations. Use $\alpha = 0.05$. EX14.79

14.80 Three independent random samples are obtained. Use the Kruskal–Wallis test with $\alpha = 0.05$ to test the hypothesis that all three populations are identical. EX14.80

Applications

14.81 Travel and Transportation A marketing manager at a rental-car agency believes that the number of miles a customer drives per week is related to the type of car rented. Independent random samples of week-long rental reservations were obtained, and the number of miles driven was recorded for each car. The summary statistics are given in the table.

Car classification	Sample size	Rank sum
Compact car	$n_1 = 10$	$r_1 = 201.0$
Standard car	$n_2 = 10$	$r_2 = 227.0$
Luxury car	$n_3 = 12$	$r_3 = 100.0$

Use a Kruskal–Wallis test to compare these three populations. Is there any evidence to suggest that the populations are different?

Use $\alpha = 0.05$, and assume the underlying populations are continuous.

14.82 Physical Sciences Raytheon is testing different propellants for the Tomahawk missile used by the U.S. Navy. Independent random samples of missiles were obtained, and each was totally fueled with one of four types of solid propellant. Each missile was fired at the Navy test range near Point Mugu, California, and the total distance traveled was recorded (in kilometers). The summary statistics are given in the table.

Solid propellant	Sample size	Rank sum
1	$n_1 = 14$	$r_1 = 525.0$
2	$n_2 = 15$	$r_2 = 355.5$
3	$n_3 = 12$	$r_3 = 373.0$
4	$n_4 = 16$	$r_4 = 399.5$

Use the Kruskal–Wallis test to determine whether there is any evidence to suggest that at least two distance-traveled populations are different. Use $\alpha = 0.05$, and assume the underlying populations are continuous.

14.83 Biology and Environmental Science The National Snow and Ice Data Center collects information on glaciers from around the world. Some of the parameters include the number of basins, the mean depth, and the primary class. A random sample of glaciers in Western Europe was obtained, and the total area (in square kilometers) of each was recorded.[21] GLACIER
(a) Use the Kruskal–Wallis test with $\alpha = 0.05$ to test the hypothesis that all four glacier-area populations are identical. Assume the underlying populations are continuous.
(b) Interpret your results in part (a). Which region has on average the largest glaciers by area?

14.84 Psychology and Human Behavior Our emotional reaction to different colors affects advertising, product design, and even architecture. Three subway-tunnel walkways of identical length were painted different colors. A random sample of adults was selected in each tunnel and secretly timed (in seconds) as they walked through the tunnel. The data are given in the table. COLOR

Red	Orange	Yellow	Black
12 19	21 14	29 23	22 23
16 23	27 12	18 24	11 27
15 22	26 24	25 30	10 24
18 25	22 30	25 12	24 14

Use the Kruskal–Wallis test with $\alpha = 0.01$ to determine whether there is any evidence that the tunnel-walking-time populations are different. Assume the underlying populations are continuous.

14.85 Public Health and Nutrition Independent random samples of four types of 16-oz steaks were obtained, and the amount of fat (in grams) in each was measured. Assume the underlying populations are continuous. Is there any evidence to suggest that the fat populations are different? Use $\alpha = 0.025$. STEAK

14.86 Public Policy and Political Science The U.S. Product Safety Commission issues playground specifications and guidelines to help communities build safe playgrounds. One concern is the uncompressed depth of wood chips and other loose-fill material (used as shock absorbers). Independent random samples of playgrounds in various cities were obtained, and the uncompressed depth of wood chips (in inches) was measured for each playground. The data are given in the table. CHIPS

Atlanta	Dallas	Denver
10.5 12.8 12.9	11.8 11.6 11.3	10.9 11.8 11.2
10.2 11.5 10.5	11.6 10.0 10.2	12.4 12.0 11.6

Assume the underlying distributions of uncompressed depths are continuous. Is there any evidence to suggest that the populations are different? Use $\alpha = 0.05$, and find bounds on the p value associated with this test.

14.87 Sports and Leisure The New York City Marathon is one of the largest marathons held in the world, with more than 52,000 finishers in 2018. The annual race takes runners through the five boroughs of New York City over 26.219 mi. A random sample of the top 100 men finishers from five countries in 2018 was obtained, and the finish time for each was recorded (in minutes).[22] RUN
(a) Is there any evidence to suggest that the populations are different? Use $\alpha = 0.01$.
(b) Using the ranks, which of these countries has the fastest runners? Why?

14.88 Travel and Transportation Airline credit cards offer all kinds of perks to attract customers, such as free checked bags, airport lounge access, and priority boarding. In addition, consumers accumulate miles toward free flights by using the card to make purchases. However, these miles may disappear if not used within a certain time and often it takes a huge number of miles to earn an overseas trip or seat upgrade. A random sample of users from four airline credit cards was obtained and the accumulated miles was recorded for each (in thousands). PERKS
(a) Is there any evidence to suggest that the populations are different? Use $\alpha = 0.01$.
(b) Suppose a small number of accumulated miles means the card is not regularly used to make purchases. Based on your analysis in part (a), which card is the least used? Justify your answer.

Extended Applications

14.89 Business and Management One indication of morale in county government positions is the length of service in the current position. Independent random samples of county employees were obtained, and the length of service (in years) was recorded for each person, by job classification. SERVICE
(a) Assume the underlying distributions are continuous. Use the Kruskal–Wallis test with $\alpha = 0.05$ to show that there is evidence to suggest that at least two populations are different.
(b) Which pair(s) of populations do you think is (are) different? Why?

14.90 Physical Sciences In 2018, various observation posts around the world reported detecting more than 6100 supernovae. When possible, these exploding stars were classified by their light curves and absorption lines. A random sample of these Type I and Type II Supernovae were obtained, and the magnitude of each was recorded.[23] STARS

(a) Assume the underlying distributions are continuous. Use the Kruskal–Wallis test with $\alpha = 0.01$ to determine whether there is any evidence that the populations are different. Find the p value associated with this test.

(b) Use the normal approximation in the Wilcoxon rank-sum test to determine whether there is any evidence to suggest the population median magnitudes are different. Use $\alpha = 0.01$. Find the p value associated with this test.

(c) What is the relationship between the p values in parts (a) and (b)?

(d) Consider the value of the test statistics found in parts (a) and (b). Compare z^2 and h. How are these values related?

14.5 The Runs Test

Background and Test Procedure

Some practical applications of the runs test involve the sequence of positive or negative gains of a certain stock, the strength of signals from an object in space, and monitoring stream pollution.

In all of the inference procedures presented in this text, it is very important for the sample, or samples, to be selected randomly from the underlying population(s). Otherwise, the results are not valid. Although it is not a test for verifying that a sample has been randomly selected, the procedure described in this section can be used to examine the order in which observations were drawn from a population. The runs test is used to assess only whether there is evidence that the sequence of observations is not random.

Suppose an usher at a Broadway play records the seating section for the next 15 patrons, O for the orchestra section and B for the balcony. If all 15 people sat in the balcony or all sat in the orchestra section, we would certainly conclude that the order of patrons entering the theater was not random. Similarly, if the first 10 sat in the balcony, and the last 5 sat in the orchestra section, we would still question whether the order was random.

To determine whether the order of observations is random, we first separate the entire sequence into smaller subsequences in which the observations are the same. Consider the sequence of theater patrons in the following table, grouped by seating section.

O O	B B B B	O	B B	O O O O	B B

The grouped subsequences of similar observations, or symbols, are called **runs**. The sequence of observations for the Broadway play contains six runs. The test for randomness is based on the total number of runs.

Definition
A **run** is a series, or subsequence, of one or more identical observations.

A CLOSER LOOK

1. The runs test is appropriate for testing whether a sequence of observations is not random. It can be used if the data can be divided into two mutually exclusive categories—for example, defective or satisfactory, working or retired, or pass or fail. This test may also be used for quantitative data that can be classified into one of two categories—for example, above or below the median, or dangerous versus acceptable temperature.

A run can have length 1.

2. The smallest possible number of runs in any sample is 1. This will occur if every observation in the sample falls into the same category or has the same attribute. The largest possible number of runs in a sample depends on the number of observations in each category.

3. The statistical test is based on the total number of runs. It seems reasonable that if the order of observations is random, then the total number of runs should not be very large or very small. A table of critical values is available that uses the exact distribution for the number of runs. In addition, a normal approximation can be used if the number of observations in each category is large.

The runs test is a nonparametric procedure because no assumptions are made about the underlying population.

The Runs Test

Suppose a sample is obtained in which each observation is classified into one of two mutually exclusive categories. Assume there are m observations in one category and n observations in the other.

H_0: The sequence of observations is random.

H_a: The sequence of observations is not random.

TS: $V =$ the total number of runs

RR: $V \geq v_1$ or $V \leq v_2$

The critical values v_1 and v_2 are obtained from Appendix Table 11 such that $P(V \geq v_1) \approx \alpha/2$ and $P(V \leq v_2) \approx \alpha/2$.

The Normal Approximation: As m and n increase, the statistic V approaches a normal distribution with

$$\mu_V = \frac{2mn}{m+n} + 1 \quad \text{and} \quad \sigma_V^2 = \frac{2mn(2mn-m-n)}{(m+n)^2(m+n-1)}$$

Therefore, the random variable $Z = \dfrac{V - \mu_V}{\sigma_V}$ has approximately a standard normal distribution. The normal approximation is appropriate when both m and n are greater than 10. In this case, Z is the test statistic and the rejection region is

RR: $|Z| \geq z_{\alpha/2}$

Example: The Runs Test

EXAMPLE 14.8   Lithium Mining

An Australian mining company has proposed to build an open-air lithium mine near the town of La Motte, Quebec. Although the project will create new jobs in the tiny community, many are concerned about the possible effects on the water supply. About 30 km north of La Motte, Amos has won awards for the best tap water in North America. Residents there are naturally worried that the new mine will damage the Saint-Mathieu-Lac-Berry esker, a geological formation that naturally filters rain and snow.[24] A random sample of residents in the area of the proposed new mine was obtained, and each was classified as in favor of (F) or against (A) the lithium mine. The sequence of responses is given in the table.

F	F	A	A	F	A	F	F	A	A	A	A	A	F

Use the runs test with $\alpha = 0.05$ to determine whether there is any evidence that the order in which the sample was selected was not random.

Solution

Use the runs test procedure: Compute the total number of runs, and draw the appropriate conclusion.

STEP 1 Each observation is classified into one of two mutually exclusive categories, in favor of (F) or against (A) the lithium mine. We are looking for evidence to suggest that the sequence of observations was not random. Therefore, the runs test is appropriate. There are $m = 6$ in favor of observations and $n = 8$ against observations.

STEP 2 The four parts of the hypothesis test are

H_0: The sequence of observations is random.

H_a: The sequence of observations is not random.

TS: $V =$ the the total number of runs

RR: $V \geq v_1 = 12$ or $V \leq v_2 = 4$

If $v_1 = 11$, then $P(V \geq 11) = 0.0629$, which leads to a much higher significance level.

Find the value v_1 such that $P(V \geq v_1) \approx 0.05/2 = 0.025$. From Appendix Table 11, with $m = 6$ and $n = 8$,

$$P(V \geq 12) = 1 - P(V \leq 11) = 1 - 0.9837 = 0.0163 \approx 0.025$$

Therefore, $v_1 = 12$.

In addition, $P(V \leq 4) = 0.0280 \approx 0.025$. Therefore $v_2 = 4$.

STEP 3 The table shows the original sequence of observations separated into runs.

F F	A A	F	A	F F	A A A A A	F

There are a total of seven runs. The value of the test statistic, $v = 7$, does not lie in the rejection region. At the $\alpha = 0.05$ significance level, there is no evidence to suggest that the order of observations was not random.

Figure 14.11 shows a technology solution.

```
> mine <- c(1,1,0,0,1,0,1,1,0,0,0,0,0,1)
> runs.test(mine,alternative='two.sided',threshold=.5)

        Runs Test

data:  mine
statistic = -0.48765, runs = 7, n1 = 6, n2 = 8, n = 14, p-value = 0.6258
alternative hypothesis: nonrandomness
```

Figure 14.11 Use the R command `runs.test()` in the package randtests to conduct a runs test. The input must be a numeric vector.

TRY IT NOW Go to Exercises 14.105 and 14.107

CHAPTER APP Keyboard Angles

Use the runs test with $\alpha = 0.05$ to determine whether there is any evidence that the order in which the 0° sample was selected was not random with respect to the population median 5.8 mm².

Example: The Runs Test, The Normal Approximation

The original data in the next example are quantitative but can be classified into two mutually exclusive categories. The number of observations in each category is large, so we will use the normal approximation in the runs test.

EXAMPLE 14.9 The Modern Kitchen

MODKIT Modern kitchens are often equipped with Internet-connected appliances, a kitchen island, beverage centers, and commercial ranges. According to Pi Consulting, the size of the average American kitchen is 225 ft².[25] A sample of single-family homes across the United States was obtained and the area of each kitchen was measured (in square feet). The data are given in the table in the order in which the kitchens were selected, from left to right.

236	241	202	291	218	228	236	231	196	206
245	218	213	281	203	282	205	270	189	253
244	200	250	255	229	191	231	211	254	279

Is there any evidence to suggest that the order of observations is not random with respect to the average kitchen size? Use $\alpha = 0.05$.

Solution

Use the runs test procedure: Compute the total number of runs, find the value of the test statistic, and draw the appropriate conclusion.

STEP 1 Replace each observation above the average value with an A and each observation below the average value with a B. Any observation equal to 225 will be excluded from the analysis and the sample size will be reduced. Each observation now falls into one of two mutually exclusive categories: above or below the average value.

STEP 2 The runs test will be used to determine whether there is evidence to suggest that the sequence of observations is not random. There are $m = 18$ observations above 225 and $n = 12$ observations below 225.

STEP 3 Because both m and n are greater than 10, we will use the normal approximation. The four parts of the hypothesis test are

H_0: The sequence of observations is random.

H_a: The sequence of observations is not random.

TS: $Z = \dfrac{V - \mu_V}{\sigma_V}$

RR: $|Z| \geq z_{\alpha/2} = z_{0.025} = 1.96$

STEP 4 Compute the number of runs using the original sequence of observations.

236	241	202	291	218	228	236	231	196	206
A	A	B	A	B	A	A	A	B	B

$\Rightarrow \quad\quad\quad \Rightarrow \quad\quad\quad \Rightarrow \quad\quad\quad \Rightarrow$

245	218	213	281	203	282	205	270	189	253
A	B	B	A	B	A	B	A	B	A

$\Rightarrow \quad\quad\quad \Rightarrow \quad\quad\quad \Rightarrow \quad\quad\quad \Rightarrow$

244	200	250	255	229	191	231	211	254	279
A	B	A	A	A	B	A	B	A	A

There are $v = 21$ runs.

STEP 5 The mean and the variance of the random variable V are

$$\mu_V = \frac{2mn}{m+n} + 1 = \frac{2(18)(12)}{18+12} + 1 = 15.4$$

$$\sigma_V^2 = \frac{2mn(2mn - m - n)}{(m+n)^2(m+n-1)} = \frac{2(18)(12)[2(18)(12) - 18 - 12]}{(18+12)^2(18+12-1)} = 6.6538$$

The value of the test statistic is

$$z = \frac{v - \mu_V}{\sigma_V} = \frac{21 - 15.4}{\sqrt{6.6538}} = 2.171$$

STEP 6 Because $|z| = |2.171| = 2.171 \geq 1.96$, the value of the test statistic lies in the rejection region. Equivalently, $p = 0.03 \leq 0.05$. At the $\alpha = 0.05$ significance level, there is evidence to suggest that the order of observations is not random.

To find the p value associated with this test, find the right-tail probability and multiply by 2.

$p/2 = P(Z \geq 2.171) \approx P(Z \geq 2.17)$ Definition of p value.

$= 1 - P(Z \leq 2.17)$ Complement Rule.

$= 1 - 0.9850 = 0.0150$ Cumulative probability; Appendix Table 3.

$p = 2(0.0150) = 0.03$ Solve for p.

Figure 14.12 shows a technology solution and the exact p value.

```
> area <- c(236,241,202,...,279)
> runs.test(area,alternative='two.sided',threshold=225)

        Runs Test

data:  area
statistic = 2.171, runs = 21, n1 = 18, n2 = 12, n = 30, p-value = 0.02993
alternative hypothesis: nonrandomness
```

Figure 14.12 Use the R command `runs.test()` with a threshold of 225.

TRY IT NOW Go to Exercises 14.109 and 14.111

Section 14.5 Exercises

Concept Check

14.91 True or False The runs test can be used to conclude that a sample has been randomly selected.

14.92 True or False The runs test can be used if numerical data can be divided into two mutually exclusive categories.

14.93 Fill in the Blank The smallest number of runs in a sample of size n is _____.

14.94 Short Answer What assumptions are made about the underlying distribution if a runs test is to be valid?

14.95 Short Answer When is the normal approximation in the runs test appropriate?

14.96 Short Answer What is a run?

14.97 True or False In a runs test, the population median can be used to divide each observation into two mutually exclusive categories.

Practice

14.98 Find the number of runs in each sequence of observations. EX14.98
(a) A A B B B A B A B B A B
(b) G B B B B B G G B B G B G G B
(c) F F S F F S F F S S F S F F S S S F F F
(d) + − + − − + − − − + + − + −
+ + − − + − − +

14.99 Use the values for m and n to find the critical values in a runs test with approximate significance level α. Find the exact significance level for your choice of critical values.

(a) $m = 4$, $n = 7$, $\alpha = 0.05$
(b) $m = 5$, $n = 8$, $\alpha = 0.01$
(c) $m = 6$, $n = 6$, $\alpha = 0.05$
(d) $m = 8$, $n = 9$, $\alpha = 0.01$

14.100 Use the population median to classify each observation in the ordered sample as either greater than or less than the median. Find the number of runs in each sequence of observations. The order of observations is left to right, then down. EX14.100

(a) $\tilde{\mu} = 20.0$

22.4	15.1	22.0	25.6	19.0	25.1	18.7	21.2
11.5	19.6						

(b) $\tilde{\mu} = 4.00$

4.14	3.26	3.23	3.12	4.80	5.52	5.81	5.71
3.77	3.68	3.79	3.03	4.36	5.29		

(c) $\tilde{\mu} = 150$

115	169	139	101	102	138	123	195	107
175	178	151	181	107	110	174	196	167

(d) $\tilde{\mu} = 0.44$

0.07	0.19	0.10	0.12	0.44	0.08	0.26	0.03
0.10	0.14	0.41	0.20	0.01	0.38	0.28	0.03
0.15	0.03	0.00	0.04	0.19	0.04		

14.101 Use the values for m and n to find the expected number of runs μ_V and the variance of the number of runs σ_V^2.
(a) $m = 10$, $n = 15$ (b) $m = 15$, $n = 21$
(c) $m = 2$, $n = 23$ (d) $m = 26$, $n = 26$

14.102 A sample was obtained, and each observation was classified into one of two mutually exclusive categories. The sequence of observations is given in the table. **EX14.102**

B	A	B	A	A	B	B	A	B	A
B	B	B	B	A	A	B	A	A	B

Conduct a runs test with $\alpha = 0.05$. Is there any evidence to suggest that the order of observations is not random?

14.103 A sample was obtained from a population with hypothesized median $\tilde{\mu} = 10.0$ and the sequence of observations recorded. **EX14.103**
(a) Conduct a runs test using the normal approximation with $\alpha = 0.01$. Is there any evidence to suggest that the order of observations is not random?
(b) Find the p value associated with this test.

Applications

14.104 Public Policy and Political Science During freshmen move-in day at the University of Michigan, the College Republicans staffed an information table in an effort to solicit new members. A sample of consecutive new members was obtained, and each was classified by gender (M or F). The ordered observations are given in the table. **MOVEIN**

M	M	M	M	M	M	M	F	F
M	F	M	M					

Use the runs test with a $\alpha \approx 0.05$ to determine whether there is any evidence to suggest that the order of observations is not random with respect to gender.

14.105 Business and Management In November 2018, General Motors (GM) announced plans to cut approximately 14,000 jobs and close some assembly plants in an effort to decrease its costs. Factories being shut down were located in Michigan, Maryland, and Ontario, Canada. A sample of GM workers who lost their jobs was obtained, and each was classified as either management (M) or blue collar (B). The ordered observations are given in the table. **WORK**

B	B	B	B	M	M	B	M	B	B
B	B	M	B	B	B	B	M	M	B

Is there any evidence to suggest that the order of observations is not random? Use the runs test with $\alpha \approx 0.05$.

14.106 Marketing and Consumer Behavior Staples sells photo paper in either glossy (G) or matte (M) finish. A sample of online customers who purchased photo paper was obtained, and each purchase was classified according to finish. The ordered observations are given in the table. **PHOTO**

M	M	G	M	G	M	G	M	G
M	G	M	M	G	M	M	G	M
G	G	G	M	M	M	G	M	M
G	G	G	M	M	M	G	G	

Is there any evidence to suggest that the order of observations is not random? Use the runs test with the normal approximation and a 0.02 level of significance.

14.107 Public Policy and Political Science A recent survey showed that approximately half of Canadians think it is acceptable, in certain cases, for the government to read private email and monitor other online activities. A sample of adults in Ottawa was obtained and asked if they believe it is acceptable for the government to monitor private email. Each response was classified as yes (Y) or no (N). The sequence of responses is given in the table. **PRIVACY**

Y	Y	Y	N	Y	Y	Y	N	N	N	N	N	Y	N	N

(a) Use the runs test with $\alpha \approx 0.05$ to determine whether there is any evidence that the order of observations is not random.
(b) Find the p value associated with this test.

14.108 Elvis Has Left the Building More than 40 years after the death of Elvis Presley, over 500,000 fans still travel to Graceland every year. Despite definitive reports from doctors and a coroner, many Americans believe that Elvis faked his own death and is still alive, in hiding. A sample of visitors to Graceland was obtained and asked if they believe Elvis is still alive. There responses were classified as alive (L) or dead (D). Is there any evidence to suggest that the order of observations is not random? Use the runs test with the normal approximation and $\alpha = 0.05$. **ELVIS**

14.109 Psychology and Human Behavior In a recent survey of Americans, U.S. Bank reported that people are not carrying a lot of cash these days. Approximately 76% of respondents said they carry less than $50 with them and almost half walk around with less than $20.[26] A random sample of customers at a convenience store in Philadelphia was obtained, and each was classified as paying with cash (C) or a credit/debit card (D). Is there any evidence to suggest that the order of observations is not random? Use the runs test with the normal approximation and $\alpha = 0.05$. **CASH**

14.110 Manufacturing and Product Development Composite decks are made of recycled wood and plastic, require little maintenance and are quite sturdy, and are designed to last much longer than traditional wood decks. A manufacturer of composite decks routinely checks the surface wear as part of quality control. A sample of planks was obtained, and each was measured for surface wear using a Taber tester. The ordered observations (in Taber units) are given in the table. **DECK**

204	211	188	208	190	203	203	193	204
214	176	209						

The planks are manufactured to have a median wear index of 200 units. Use the runs test with $\alpha \approx 0.02$ to determine whether there is any evidence that the order of observations is not random with respect to the median.

14.111 Education and Child Development Most applicants to the Kellogg School of Management at Northwestern University do not need to submit GMAT scores. However, for those in the class of 2018 who did include a GMAT score, the *average* was 732.[27] A sample of students in this MBA program was obtained, and the GMAT score for each was recorded. Use the runs test with the normal approximation and $\alpha = 0.05$ to determine whether there is any evidence that the order of GMAT scores is not random. 📊 GMAT

14.112 Medicine and Clinical Studies The School of Medicine and Public Health at the University of Wisconsin Madison reports that approximately 13% of American males and 7% of American females will develop a kidney stone at some time in their life.[28] Some research suggests that drinking lots of sugar-sweetened soda may lead to an increased risk of developing a kidney stone. A sample of patients who were admitted to Geisinger Medical Center for a kidney stone was obtained. Low-dose computed tomography was used to measure the size (in milliters) of each stone, and the ordered observations were recorded. Use the runs test with the normal approximation and $\alpha = 0.05$ to determine whether there is any evidence that the order of kidney stone sizes is not random with respect to the median size, 3.0 mL. Find the p value associated with this test. 📊 STONE

14.113 Sports and Leisure The Nike Zoom Vaporfly 4% Flynit running shoe is expensive, costing approximately $250, but is lightweight, soft, and designed to provide as much as 85% energy return. That is, it will, in theory, help an individual run faster. A sample of runners competing in the Bank of America Chicago Marathon who wear this running shoe was obtained. Each was asked if they believe the Nike running shoe makes them run faster, yes (Y) or no (N). Use the runs test with the normal approximation and $\alpha = 0.05$ to determine whether there is any evidence that the order of responses is not random. Find the p value associated with this test. 📊 SHOE

Challenge Problems

14.114 Random Number Generators The purpose of this exercise is to determine whether a random number generator really produces a random sequence of observations. Using R, or your favorite statistical software, or a graphing calculator:

(a) Generate 100 observations from a standard normal distribution.

(b) Classify each observation as either above or below the mean (and median) 0.

(c) Conduct a runs test using the normal approximation to determine whether there is any evidence to suggest that the sequence of observations is not random with respect to the mean. Use $\alpha = 0.05$.

Do this 100 times. How many times did you reject the null hypothesis H_0: The sequence is random? Draw a conclusion about the sequence of observations produced by your random number generator.

14.6 Spearman's Rank Correlation

Background and Test Procedure

The sample correlation coefficient, r, was introduced in Chapter 12 as a measure of the strength of the linear relationship between two continuous variables. Spearman's rank correlation coefficient is a nonparametric alternative and is computed using ranks. Without any assumptions about the underlying populations, each observation is converted to a rank, and the sample correlation coefficient is computed using the ranks in place of the actual observations.

Spearman's Rank Correlation Coefficient

Suppose we have n pairs of observations $(x_1, y_1), (x_2, y_2), \ldots, (x_n, y_n)$. Rank the observations on each variable separately, from smallest to largest. Let u_i be the rank of the ith observation on the first variable, and let v_i be the rank of the ith observation on the second variable. **Spearman's rank correlation coefficient** r_S is the sample correlation coefficient between the ranks and is computed using the equation

$$r_S = 1 - \frac{6 \sum d_i^2}{n(n^2 - 1)} \tag{14.1}$$

where $d_i = u_i - v_i$.

A CLOSER LOOK

1. As usual, equal values within each variable are assigned the mean rank of their positions in the ordered list. Equation 14.1 is not exact when there are tied observations within either variable. In this case, we should compute r_S by finding the sample correlation coefficient between the ranks.

2. Because r_S is really a sample correlation coefficient, the value is always between -1 and $+1$. Values near -1 indicate a strong negative linear relationship, and values near $+1$ suggest a strong positive linear relationship between the ranks.

3. Remember: Correlation does not imply causation. That is, r_S is a measure of the linear association between the ranks. In addition, a strong linear relationship between the ranks does not imply that the relationship between the original variables is also linear.

The sample correlation coefficient was defined in Section 12.2.

Example: Spearman's Rank Correlation

EXAMPLE 14.10 Driving and the Economy

Several studies suggest that driving habits are related to certain economic indicators. For example, as the price of gas increases, it seems reasonable that people will drive fewer miles. Measures of housing starts and the stock market may also be related to driving miles. Normally, an economic recession and high unemployment also affect miles traveled. A random sample of months was selected, and the total miles traveled by Americans (in millions of miles) and the unemployment rate (a percentage) was recorded for each. The data are given in the table.

Month	1	2	3	4	5	6	7	8	9	10	11	12
Miles	212.74	226.30	227.70	227.90	233.66	233.28	261.62	179.54	263.06	256.37	186.83	205.98
Unemp rate	9.7	7.6	3.8	3.9	5.8	4.9	4.8	6.7	4.7	6.6	6.5	7.1

Compute Spearman's rank correlation coefficient and interpret this value.

Solution

STEP 1 Let x represent the miles (in millions) driven per month, and let y represent the unemployment rate. For each variable, order the observations from smallest to largest and assign a rank to each value. Compute the difference between each pair of ranks.

The table shows each observation, its associated rank, and each difference.

Month i	Miles x_i	Rank u_i	Unemp rate y_i	Rank v_i	Difference d_i
1	212.74	4	9.7	12	−8
2	226.30	5	7.6	11	−6
3	227.70	6	3.8	1	5
4	227.90	7	3.9	2	5
5	233.66	9	5.8	6	3
6	233.28	8	4.9	5	3
7	261.62	11	4.8	4	7
8	179.54	1	6.7	9	−8
9	263.06	12	4.7	3	9
10	256.37	10	6.6	8	2
11	186.83	2	6.5	7	−5
12	205.98	3	7.1	10	−7

STEP 2 There are $n=12$ pairs and no ties within either variable. Use Equation 14.1 to compute Spearman's rank correlation coefficient.

$$r_S = 1 - \frac{6\Sigma d_i^2}{n(n^2-1)}$$

Equation 14.1.

$$= 1 - \frac{6\left[(-8)^2 + (-6)^2 + \cdots + (-7)^2\right]}{12(12^2-1)}$$

Use the d_i's.

$$= 1 - \frac{2640}{1716} = -0.5385$$

STEP 3 Because $r_S = -0.5385$, there is a moderate negative linear relationship between the miles-traveled ranks and the unemployment ranks. This suggests that there is a moderate negative correlation between miles traveled and unemployment rate; high unemployment rates are associated with fewer miles traveled.

Figure 14.13 shows a technology solution.

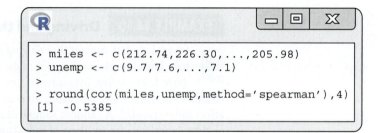

```
> miles <- c(212.74,226.30,...,205.98)
> unemp <- c(9.7,7.6,...,7.1)
>
> round(cor(miles,unemp,method='spearman'),4)
[1] -0.5385
```

Figure 14.13 Use the R command `cor()` to find Spearman's rank correlation coefficient.

TRY IT NOW Go to Exercises 14.123 and 14.125

Section 14.6 Exercises

Concept Check

14.115 Short Answer What are the assumptions about the underlying populations for Spearman's rank correlation coefficient?

14.116 True or False If there are equal values of a variable, you cannot compute Spearman's rank correlation coefficient.

14.117 True or False The values of r_S are always between -1 and $+1$.

14.118 True or False A strong linear relationship between ranks implies a strong linear relationship between the original variables.

14.119 True or False If the underlying populations are normally distributed, the correlation coefficient between the original variables and Spearman's rank correlation coefficient will be the same.

Practice

14.120 Random samples of pairs of observations are given. Rank the observations in each sample separately, from smallest to largest, and find each difference between the ranks, d_i. EX14.120

(a)
Observation	1	2	3	4	5	6
Sample 1	54	17	28	69	13	49
Sample 2	113	114	139	173	145	121

(b)
Observation	1	2	3	4	5	6
Sample 1	57	40	32	56	33	60
Sample 2	35	50	51	57	38	45

Observation	7	8	9
Sample 1	51	35	53
Sample 2	44	52	32

(c)
Observation	1	2	3	4	5	6
Sample 1	22.5	27.0	22.8	26.5	29.9	20.8
Sample 2	27.0	30.5	21.6	27.9	33.0	22.9

Observation	7	8	9	10	11	12
Sample 1	19.3	28.3	19.5	20.3	28.2	27.9
Sample 2	24.7	22.2	24.3	26.2	25.1	29.4

Observation	13	14	15
Sample 1	27.8	31.6	25.3
Sample 2	22.6	24.8	20.3

(d)

Observation	1	2	3	4	5	6
Sample 1	49.0	51.2	46.7	54.9	53.6	48.9
Sample 2	78.1	60.8	70.4	72.2	64.9	70.8

Observation	7	8	9	10	11	12
Sample 1	46.8	46.2	55.8	40.8	46.8	55.7
Sample 2	74.3	68.4	66.5	75.9	65.9	70.3

Observation	13	14	15	16	17	18
Sample 1	54.9	45.0	55.1	43.2	43.8	53.9
Sample 2	71.4	75.6	72.0	69.4	71.7	78.3

14.121 Consider the data obtained in a study of the relationship between two variables, find Spearman's rank correlation coefficient, and use it to describe the relationship between the two variables. EX14.121

(a)

x	156	268	262	206	162	166	148
y	131	258	235	189	180	262	207

(b)

x	5.39	5.60	5.12	4.53	6.01	5.00
y	−15.3	−15.4	−16.5	−16.3	−14.2	−13.6

x	5.62	5.04	4.65	6.19
y	−14.9	−14.4	−15.6	−15.2

(c)

x	25.51	25.80	24.10	24.19	25.52	26.16
y	24.45	26.98	25.72	27.20	28.56	25.69

x	24.79	23.95	25.14	25.09	25.35	26.64
y	27.38	25.84	24.04	25.87	25.51	26.94

x	25.63	23.84
y	26.64	26.53

(d)

x	y	x	y	x	y
418	754	411	755	414	776
484	725	378	759	453	781
448	712	425	749	478	745
458	718	436	733	465	785
476	756	443	727	438	775
421	772	442	716	459	735
480	728	435	763	490	751
461	753	440	752	407	744

14.122 Consider the data obtained in a study of the relationship between two variables. EX14.122

x	y	x	y
1.5	7.3	1.4	7.4
1.8	6.8	1.3	7.2
1.5	7.4	1.2	7.2
1.6	7.5	1.4	7.3
1.7	6.7	1.2	6.6

(a) Rank the observations in each sample separately, and find each difference between the ranks, d_i.
(b) Find the sample correlation coefficient between the ranks.
(c) Find Spearman's rank correlation coefficient using the differences d_i.
(d) Explain why the values in parts (b) and (c) are different.

Applications

14.123 Travel and Transportation Dynasty Travel offers cruises to the Caribbean and often discounts unsold tickets as the sailing date approaches. These last-minute deals were studied by comparing the price of an inside stateroom (x, in dollars) with the number of days before sailing (y). A random sample of last-minute cruise deals was obtained, and the data are given in the table. CRUISE

x	998	956	763	1313	1071	1091
y	9	3	4	13	10	6

Find Spearman's rank correlation coefficient, and interpret this value. Do these data suggest that cruise prices are reduced at the last minute? Justify your answer.

14.124 Biology and Environmental Science Farmers consider heavy molasses to be a high-energy food for cattle. In a recent study, the percentage of digestible energy (x) was compared with the percentage of protein (y) in heavy molasses. Random samples of various brands of molasses were obtained. The percentages of digestible energy and of protein were carefully measured. Compute Spearman's rank correlation coefficient. Use this value to describe the relationship between the percentage of digestible energy and the percentage of protein in heavy molasses. ENERGY

14.125 Biology and Environmental Science The bulk density of soil (x, in g/cm^3) affects plant growth, especially how easily roots can penetrate the soil. A group of carrot growers conducted a study to compare soil bulk density with soil texture (y), measured on a scale from 1 to 10, with 1 being very sandy soil and 10 corresponding to clay. A random sample of plots was obtained. Compute Spearman's rank correlation coefficient, and use this value to describe the relationship between these two variables. SOIL

14.126 Biology and Environmental Science A recent study examined the relationship between measures of egg albumen height and pH. A random sample of eggs from Brown Leghorn hens was obtained, and the albumen height (x, in millimeters) and pH (y) was measured for each. Compute Spearman's rank correlation coefficient. What does this value suggest about the relationship between albumen height and pH? **EGGS**

14.127 Nature Deficit Only a few generations ago, children played much more outside, participated in spontaneous play, communicated face to face, and learned how to compromise. Today, out of fear, many parents are reluctant to let their children play outdoors alone. As a result, many children have very little contact with nature. Some research suggests that children who play regularly in natural environments tend to be healthier. A random sample of middle school children was obtained, and the number of hours per week each child played outside and the number of days of school missed due to illness was recorded. The data are given in the table. **PLAY**

Hours	3.6	6.3	5.0	7.9	8.7	4.9	6.2	10.6	6.8	8.2
Days	15	3	6	0	0	11	2	45	3	14

Compute Spearman's rank correlation coefficient. Does this value support the theory that healthier children spend more time outside? Justify your answer.

14.128 Economics and Finance For those people trying to buy a new home, the mortgage process can be daunting, and long. There's the preapproval, the home appraisal, and then finally the approval of the actual loan. The entire process takes, on average, 30–60 days. A consumer group believes that the time to mortgage approval is related to the amount borrowed. A random sample of mortgage applications was obtained and the time to approval (in days) and borrowed amount (in thousands of dollars) was recorded for each. Compute Spearman's rank correlation coefficient. What does the value suggest about the relationship between mortgage approval time and borrowed amount? **MTGAPP**

14.129 Medicine and Clinical Studies Some research suggests that heart rate (or heart rate variability) is related to intelligence. A random sample of healthy adult women was obtained. Their resting heart rate was measured, and each was given an IQ test. Compute Spearman's rank correlation coefficient. What does this suggest about the relationship between heart rate and intelligence in women? **HEART**

Extended Applications

14.130 Sports and Leisure The owner of a snow tubing resort in the Pocono Mountains believes that the total snowfall (x, in feet) during a winter is related to the number of customers (y, in thousands). A random sample of winters was obtained. **TUBING**
(a) Construct a scatter plot for these data, and describe the relationship between the two variables.
(b) Compute Spearman's rank correlation coefficient, and use this value to describe the relationship between the two variables.
(c) Explain any differences in your answers to parts (a) and (b).

14.131 Physical Sciences A certain solder joint on the undercarriage of a city bus is being tested for shear strength. A sample of solder joints was randomly selected, and the shear strength for each was measured (in newtons, N) before (x) and after (y) temperature-cycle stress. The data are given in the table. **SOLDER**

x	71	69	66	69	62	67	69	64	73	69
y	64	56	65	62	65	64	57	63	57	63

(a) Rank the observations in each sample separately, and find each difference between the ranks, d_i.
(b) Find the sample correlation coefficient between the ranks.
(c) Find Spearman's rank correlation coefficient using the d_i's.
(d) Explain why the values in parts (b) and (c) are different, and use these results to explain the relationship between the two variables.

14.132 Psychology and Human Behavior Recent studies suggest that students who take music classes tend to perform better in all of their other courses. To examine this relationship, a random sample of college students who regularly practice on a musical instrument was obtained, and the time spent practicing each week (in hours) and their semester GPA were recorded. **MUSIC**
(a) Compute Spearman's rank correlation coefficient.
(b) Does the value of r_S support the claim that students taking music classes perform better in college? Why or why not?

Challenge Problems

14.133 Fuel Consumption and Cars Suppose there are n pairs of observations and ρ_S is the population correlation between ranks. The four parts of a hypothesis test concerning ρ_S based on a normal approximation are

$H_0: \rho_S = 0$ (no population correlation between ranks)

$H_a: \rho_S > 0$, $\rho_S < 0$, or $\rho_S \neq 0$

TS: $Z = R_S \sqrt{n-1}$

R_S is Spearman's rank correlation coefficient.

RR: $Z \geq z_\alpha$, $Z \leq -z_\alpha$, or $|Z| \geq z_{\alpha/2}$

A turbocharger is a compact way to add more power to an automobile, by increasing the amount of air going into the cylinders. A study was conducted to examine the relationship between the added pressure (x, in pounds per square inch, psi) and power (y, percentage change in horsepower) provided by a certain turbocharger. A variety of automobiles were selected, and a randomly selected turbocharger was installed in each. The additional air pressure and power measurements were recorded. **TURBO**
(a) Compute Spearman's rank correlation coefficient, and use this value to describe the relationship between added air pressure and change in engine power.
(b) Conduct a hypothesis test to determine whether there is any evidence that the true population correlation between ranks is greater than 0. Use $\alpha = 0.01$.

Chapter 14 Summary

Concept	Page	Notation / Formula / Description
Sign test	658	A nonparametric test concerning a population median, based on the number of observations greater than $\tilde{\mu}_0$. It can also be used to compare two population medians, based on the number of pairwise differences greater than Δ_0.
Wilcoxon signed-rank test	666	A nonparametric test concerning a population median, based on the sum of the ranks corresponding to the positive differences $x_i - \tilde{\mu}_0$. It can also be used to compare two population medians, based on the sum of the ranks corresponding to the positive differences $d_i - \Delta_0$.
Wilcoxon rank-sum test	674	A nonparametric test concerning the difference of two population medians, based on the sum of the ranks corresponding to the differences $x_i - \Delta_0$.
Kruskal–Wallis test	681	A nonparametric test to determine whether at least two of k samples are from different populations, based on the ranks of all observations combined.
Run	686	A series, or subsequence, of one or more identical observations.
Runs test	686	A nonparametric test to determine whether there is evidence that a sequence of observations is not random, based on the total number of runs.
Spearman's rank correlation coefficient	692	A nonparametric alternative to the sample correlation coefficient. Each observation is converted to a rank, and the correlation coefficient is computed using the ranks in place of the actual observations.

A comparison of parametric and nonparametric procedures:

Case	Nonparametric procedure		Parametric procedure	
	Null hypothesis	Statistical test	Null hypothesis	Statistical test
One sample	$\tilde{\mu} = \tilde{\mu}_0$	Sign test	$\mu = \mu_0$	One-sample t test
	$\tilde{\mu} = \tilde{\mu}_0$	Wilcoxon signed-rank test	$\mu = \mu_0$	One-sample t test
	Sequence of observations is random	Runs test	No comparable parametric test	
Two independent samples	$\tilde{\mu}_1 - \tilde{\mu}_2 = \Delta_0$	Wilcoxon rank-sum test	$\mu_1 - \mu_2 = \Delta_0$	Two-samples t test
Paired data	$\tilde{\mu}_D = \Delta_0$	Sign test	$\mu_D = \Delta_0$	Paired t test
	$\tilde{\mu}_D = \Delta_0$	Wilcoxon signed-rank test	$\mu_D = \Delta_0$	Paired t test
		Spearman's rank correlation coefficient		Sample correlation coefficient
k independent samples	k samples are from identical populations	Kruskal–Wallis test	$\mu_1 = \cdots = \mu_k$	One-way ANOVA

Note: Nonparametric procedures are applicable in a very broad range of situations. However, if the underlying population is normal, then the corresponding parametric procedure is more efficient.

Chapter 14 Exercises

Applications

14.134 Biology and Environmental Science Lake Powell in Colorado was created after the construction of the Glen Canyon Dam and has more than 2000 miles of shoreline. It is one of the largest human-made reservoirs in the United States and provides water to Arizona, Nevada, and California. A random sample of outflow (in cubic feet per second, cfs) per day from Lake Powell was obtained, and the data are given in the table.[29] POWELL

| 8212 | 8219 | 8265 | 11833 | 10921 | 8533 | 8245 | 8254 |
| 8473 | 7173 | 8199 | 7820 | 20703 | 13961 | 11020 | 8182 |

Suppose that to maintain an adequate water supply, the outflow from Lake Powell should be approximately 8500 cfs per day. Assume the underlying distribution of outflow is continuous. Use the sign test with $\alpha = 0.05$ to determine whether there is any evidence that the median outflow per day is less than 8500 cfs.

14.135 Biology and Environmental Science Farms across the country have old agricultural chemicals stored in deteriorating containment vessels that pose a threat to humans. The U.S. Department of Agriculture has started a program to clean up and properly dispose of these unusable chemicals. A sample of farms was randomly selected, and the amount of DDT still stored on each farm was recorded. The data (in kilograms) are given in the table. DDT

| 3.5 | 4.3 | 7.0 | 3.0 | 6.0 | 2.1 | 2.5 | 4.9 | 5.9 | 2.4 |
| 3.5 | 8.6 | 0.8 | 1.3 | 2.9 | 2.3 | 2.1 | 1.0 | 5.5 | 1.7 |

Assume the underlying distribution of stored DDT is continuous.
(a) Use the sign test with $\alpha = 0.01$ to determine whether there is any evidence that the median amount of stored DDT on farms is greater than 2.0 kg.
(b) Find the p value associated with this test.

14.136 Economics and Finance Insurance companies recommend a personal umbrella policy for most customers, even if they already have a home and an automobile policy. Basic liability coverage may not be adequate as lawsuits become more common and jury awards escalate. A random sample of adults with umbrella policies was obtained, and the amount of coverage was recorded for each (in millions of dollars). Assume the underlying distribution of umbrella coverage amounts is continuous. Use the sign test with $\alpha = 0.05$ to determine whether there is any evidence that the median coverage amount is different from \$2.00 million. UMBRELL

14.137 Public Health and Nutrition Nutella is a chocolate and hazelnut spread made by an Italian company; this popular condiment is used on breakfast toast, on pancakes, and with desserts. Consumers often perceive Nutella as relatively low in calories, sugar, and fat, but these numbers appear small on the label due to the small serving size. A random sample of Nutella servings was obtained and the amount of sugar was carefully measured in each (in grams). Assume the underlying distribution of sugar content is continuous. Use the Wilcoxon signed-rank test with $\alpha = 0.05$ to determine whether there is any evidence that the median amount of sugar in a serving of Nutella is greater than 21 g, the figure on the label. Find the p value for this test. NUTELLA

14.138 Physical Sciences The fountain beneath the Gateway Arch in St. Louis is designed to spray water 630 ft straight up, the same height as the Arch. A random sample of days was selected, and on each day the height of the spray was measured (in feet) during the fountain show. The data are given in the table. ARCH

| 603 | 633 | 612 | 619 | 624 | 627 | 622 | 619 | 630 | 609 |
| 630 | 626 | 630 | 615 | 630 | 643 | 639 | 620 | 606 | |

The underlying distribution is assumed to be continuous.
(a) Use the Wilcoxon signed-rank test with $\alpha = 0.01$ to determine whether there is any evidence that the median spray height is less than 630 ft.
(b) Find the p value associated with this test.
(c) Why can't this procedure be used in a test concerning the mean spray height?

14.139 Biology and Environmental Science Pelicans have a long flat bill and expandable pouch, and use a spectacular dive-bomb-type plunge into the water to capture fish. A random sample of pelicans at Tigertail Beach on Marco Island, Florida, was obtained. The height of a plunge was measured (in feet) for each bird, and the data are given in the table. PELICAN

| 41 | 43 | 48 | 46 | 44 | 40 | 43 | 47 | 40 |

The underlying distribution is assumed to be continuous. Use the Wilcoxon signed-rank test with $\alpha = 0.05$ to determine whether the median plunge height is different from 42 ft. Find the p value associated with this test.

14.140 Swim at Your Own Risk Some research suggests that exposure to disinfection by-products (DBPs) in swimming pools may increase the risk of some cancers. A study was conducted to determine whether exposure to DBPs in pool water increased toxic biomarkers. A random sample of adults was obtained, and the concentration of four trihalomethanes (THMs, in μ/m^3) in exhaled breath was measured before and one hour after swimming. Assume that the distribution of pairwise differences is continuous and symmetric. Use the Wilcoxon signed-rank test with $\alpha = 0.05$ to determine whether there is any evidence that the median THM concentration is greater after swimming. DBP

14.141 Travel and Transportation A random sample of Georgia drivers renewing their vehicle registration was obtained. Using motor vehicle records, the number of miles driven in the past year was recorded for each. The data (in thousands of miles) were classified by whether the driver had an organ donor card and are given in the tables. **DONOR**

Organ donor card				
13.5	13.8	14.1	13.3	14.5

No organ donor card						
13.1	12.7	12.9	13.4	14.1	12.3	13.0

Assume the underlying distributions are continuous. Is there any evidence to suggest that the median number of miles driven is different for people who carry an organ donor card and for those who do not? Use the Wilcoxon rank-sum test with $\alpha = 0.05$.

14.142 Public Health and Nutrition Irradiation is a method of food preservation and is used to reduce bacteria and microorganisms in meat, poultry, and spices. The dose of an ionizing energy source varies according to the food type and is measured in grays (Gy), the amount of radiation absorbed. In a recent study, two types of radiation sources were compared. A random sample of spices was selected, and each irradiation machine was set for the maximum allowed dose of ionizing energy. The absorbed radiation was measured for each food sample (in kGy), and the data are given in the table. **IONENER**

X-ray generator							
22.6	25.9	26.3	22.2	23.5	24.9	27.2	22.9

Electron accelerator							
27.5	26.3	22.1	19.5	26.7	21.6	21.2	21.4

Assume the underlying distributions are continuous.
(a) Is there any evidence to suggest that there is a difference in the median absorbed radiation by machine? Use the Wilcoxon rank-sum test with $\alpha = 0.10$.
(b) Find the p value associated with this test.

14.143 Manufacturing and Product Development The pressure for brewing espresso is necessary to force hot water through grounds and to distinguish this drink from strong drip coffee. Random samples of espressos from each of two commercial machines were obtained, and the pressure (in atmospheres) while brewing each cup was recorded. Assume the underlying distributions are continuous. Is there any evidence to suggest that the espresso setting on each machine produces a different median pressure? Use the Wilcoxon rank-sum test with $\alpha = 0.05$, and find the p value associated with this test. **BREW**

14.144 Medicine and Clinical Studies Auriculotherapy is a branch of alternative medicine that involves pressure points on the ears. Some research suggests that stimulating certain spots on the ears can affect other parts of the body. This medical technique is used by some practitioners to promote weight loss. Patients wear special tiny biomagnetic ear magnets, which are attached to the ear with tape. Suppose an experiment was conducted to determine if these devices and the associated method promote weight loss. A random sample of adults was obtained, and each wore a set of biomagnetic ear magnets for two months. Their before and after weights (in pounds) were recorded. Assume the underlying distribution of pairwise differences is continuous and symmetric. Use a Wilcoxon signed-rank test with $\alpha = 0.05$ to determine if there is any evidence that the median weight of the participants was greater before wearing the device than after doing so. **EARMAG**

14.145 Sports and Leisure In the sport of speed skiing, racers ski downhill in a straight line as fast as possible. These skiers use special equipment and can reach speeds of 200 km/hr. A random sample of men speed skiers was obtained from two events during 2018, and the speed was recorded for each.[30] Assume the underlying distributions are continuous. Use the Wilcoxon rank-sum test to determine if there is any evidence to suggest that the median speed is different at these two events. Use $\alpha = 0.05$. **SPDSKI**

14.146 Manufacturing and Product Development The slate in a pool table may be one piece, but because it is prone to fracturing during transporting, it is often split into three slabs. Slate slabs from three pool-table manufacturers were randomly selected. The weight of each slab (in kilograms) was recorded, and the summary statistics are given in the table. **SPDSKI**

Company	Sample size	Rank sum
Brunswick	$n_1 = 12$	$r_1 = 263.0$
AMF	$n_2 = 14$	$r_2 = 185.0$
Olhausen	$n_3 = 18$	$r_3 = 542.0$

Assume the underlying weight populations are continuous. Use the Kruskal–Wallis test with $\alpha = 0.05$ to determine whether there is any evidence to suggest that at least two slate weight populations are different.

14.147 Physical Sciences Amateur radio operators (hams) communicate with one another all over the world, and many help coordinate relief efforts during natural disasters. Unlike the 5-watt power limit on a CB radio, ham radios can have as much as 1500 W of power. A random sample of amateur radio operators was obtained from four states. The power on each transmitter was measured (in watts), and the summary statistics are given in the table.

State	Sample size	Rank sum
New Hampshire	$n_1 = 15$	$r_1 = 611.0$
Alabama	$n_2 = 16$	$r_2 = 605.5$
Texas	$n_3 = 18$	$r_3 = 609.0$
California	$n_4 = 20$	$r_4 = 589.5$

Assume the underlying transmitter power distributions are continuous. Use the Kruskal–Wallis test with $\alpha = 0.025$ to determine whether there is any evidence to suggest that at least two transmitter power populations are different. Find the p value associated with this test.

14.148 Steam Clean Many steam cleaners advertise a very high internal boiler temperature that produces a dry steam vapor and a chemical-free cleaning system. Four types of home floor steam cleaners were identified, and a random sample of each type was obtained. The temperature of the steam (in °F) was measured at the nozzle for each device. Is there any evidence to suggest that at least two of the steam cleaner nozzle temperature population distributions are different? Assume that the underlying populations are continuous, and use the Kruskal–Wallis test with $\alpha = 0.01$. STEAM

14.149 Manufacturing and Product Development The global economic conditions in 2018 contributed to a short-term increased demand for steel. China produces the most steel, by far, followed by Japan and India.[31] Four other countries in the top 10 production list were selected, and a random sample of their crude steel production per month (in thousand tonnes) was obtained. The data are given in the table. STEEL

Russia		Germany		Brazil		Italy	
87	81	78	37	42	50	52	49
50	102	59	56	55	52	53	46
82	109	74	84	60	49	52	20
106	31	84	88	54	60	56	54

Assume the underlying distributions of crude steel production per month are continuous. Use the Kruskal–Wallis test with $\alpha = 0.05$ to determine whether there is any evidence to suggest that at least two production populations are different.

14.150 Sky-High Luxury The world's tallest hotel is the Gevora. This Dubai hotel has 528 rooms, 75 floors, and two luxury restaurants. A sample of guests checking in who were planning to have dinner in the hotel was obtained, and each was classified by restaurant, Level Twelve (L) or Highest View (H). The ordered observations are given in the table. GEVORA

| L | H | H | L | H | H | L | L | H | L | H | L | H | H | H |

(a) Use the runs test with $\alpha \approx 0.05$ to determine whether there is any evidence to suggest that the order of observations is not random with respect to restaurant.
(b) Find the p value associated with this test.

14.151 Fuel Consumption and Cars A sample of automobiles entering a parking garage in Monterey, California, before 6:00 A.M. on a Monday morning was obtained, and each was classified as either foreign (F) or domestic (D). The ordered observations are given in the table. PKING

| D | F | D | F | F | D | F | F | D | F | D | F | D | D | F |
| D | F | D | F | F | F | F | D | F | D | F | F | D | D | F |

Is there any evidence to suggest that the order of automobiles entering the parking garage is not random? Use the runs test with the normal approximation and a 0.05 level of significance.

14.152 Public Health and Nutrition The Mayflower Health Insurance Company is conducting a survey of customers to estimate the number who have received a routine physical exam during the past year. A sample of customers was contacted by phone during the evening, and each was classified as exam (E) or no exam (N). The sequence of observations is given in the table. EXAM

E	E	N	E	E	E	N	E	E	N	E	N	E	E	N
N	E	N	N	N	E	E	E	N	E	E	E	N	E	E
N	N	E	N	E	N	E	N	E	N					

Is there any evidence to suggest that the order of observations is not random? Use the runs test with the normal approximation and $\alpha = 0.01$. Find the p value associated with this test.

14.153 Public Policy and Political Science The Canadian Senate fisheries committee recently released a report indicating that an emergency distress beacon should be mandatory on all commercial fishing vessels. This kind of walkie-talkie-like device emits a distress signal (when engaged), and many models include a GPS system. Only about half of all Canadian fishing vessels have this safety device installed.[32] A sample of small- and medium-sized Canadian fishing vessels was obtained, and each was classified as either having an emergency distress beacon (B) or not having one (N). The sequence of observations is given in the table. BEACON

N	N	B	N	B	N	B	B	B	B	N	N	B	B	B
N	B	N	N	B	B	B	N	N	N	N	N	N	N	N
N	B	N	N	B	N	N	N	B	B	B	N	N	N	N

Is there any evidence to suggest that the order of observations is not random? Use the runs test with the normal approximation and $\alpha = 0.05$. Find the p value associated with this test.

14.154 Travel and Transportation Interstate 95 is a major north/south highway along the eastern part of the United States. Travelers on this road have a wide variety of hotel options at most of the major exits. A random sample of hotels at the intersection of I-95 and Highway 870 in Florida was obtained. The distance (in miles) from I-95 and the starting room price (in dollars) for each were recorded. The data are given in the table. HOTEL

| Miles | 0.2 | 0.3 | 0.8 | 1.1 | 2.7 | 1.5 | 1.4 | 2.3 |
| Price | 67 | 72 | 163 | 129 | 125 | 161 | 144 | 253 |

Compute Spearman's rank correlation coefficient and use this value to describe the relationship between these two variables.

14.155 Public Health and Nutrition A study was conducted to examine the relationship between the volume and the quality of health care. A random sample of patients admitted to various hospitals in Indianapolis, Indiana, was obtained. After being discharged, each patient was asked to complete a questionnaire regarding the quality of care, and the results were evaluated to yield a quality score (x) between 1 (bad) and 50 (good). The total number of people in the hospital when the patient was admitted (y) was also recorded. Compute Spearman's rank correlation coefficient. What does this value suggest about the relationship between the volume and the quality of health care? CAREQ

14.156 Biology and Environmental Science An agricultural researcher investigated the relationship between the mineral concentrations in pre-bloom leaves and the nutrients in blackcurrant. A random sample of pre-bloom leaves was obtained and the concentration of boron was measure in each (in mg/kg).

The resulting blackcurrant was examined, and the amount of vitamin C in each was measured (in milligrams). Compute Spearman's rank correlation coefficient. Use this value to explain the relationship between the concentration of boron in pre-bloom leaves and the amount of vitamin C in the resulting blackcurrant. LEAVES

14.157 Medicine and Clinical Studies The traditional method for people to test their blood glucose level is to prick a finger with a short needle, place a drop of blood on a test strip, and then use a special measuring device. The process can be painful. A group of scientists has developed a noninvasive, painless approach to measuring glucose level using infrared laser light. A random sample of healthy adults was obtained. The glucose level was measured in each using both the traditional method and the laser light approach (both measurements in milligrams per deciliter). Assume the underlying distribution of pairwise differences is continuous and symmetric. GLUCOSE
(a) Use the Wilcoxon signed-rank test to determine whether there is any evidence that the median glucose levels are different for the two procedures. Use $\alpha = 0.01$.
(b) Using your answer in part (a), do you believe the new approach is accurate and should be recommended for people with diabetes who test their glucose level regularly? Justify your answer.

14.158 Technology and the Internet Research suggests that it takes, on average, 22 seconds for a mobile landing page to fully load, but most users will impatiently leave the page if it takes more than 3 seconds.[33] A fully loaded page includes all the associated ads and scripts. A random sample of websites in the United States was obtained and the speed index (the time until most information is loaded) was measured for each (in seconds). The data were classified by industry. SPINDEX
(a) Assume the underlying speed index populations are continuous. Use the Kruskal–Wallis test with $\alpha = 0.05$ to determine whether there is any evidence to suggest that at least two speed index populations are different.
(b) Find the p value associated with this test.
(c) Which industry websites tend to have the fastest speed indices? Justify your answer.

Extended Applications

14.159 Manufacturing and Product Development Most home air conditioners built before 2010 used Freon gas to provide cooling in a typical evaporation cycle. However, Freon, which contains chlorine, was phased out of new equipment by 2010 and will not be used at all by 2020. R-410A, a mixture of difluoromethane and pentafluoroethane, is now the most commonly used refrigerant in the United States. Although an air conditioner is, theoretically, a closed system, units typically lose refrigerant and must be recharged every few years. A random sample of 10,000-BTU air conditioners was obtained, and the amount of refrigerant (in pounds) in each was carefully measured. Each unit was recharged by a technician, and the amount of refrigerant in each was measured after the service. Suppose no assumptions can be made about the shape of the continuous distributions of before and after refrigerant weights. RECHG
(a) Conduct a sign test to determine whether there is any evidence that the median refrigerant weight before service is less than the median refrigerant weight after service. Use a significance level of $\alpha = 0.01$.
(b) Based on your results, do you believe recharging an air conditioner really results in more refrigerant in the system? Why or why not?

14.160 Manufacturing and Product Development A nail gun is a handy tool, especially if you need to install a wood floor, replace a roof, or attach a Venetian blind to a metal support, or if you tend to hit your thumb a lot with a hammer. These devices propel a nail at incredible speeds and can save time and energy. A random sample of nail guns from four companies was obtained, and the speed of the nail was measured for each gun (in feet per second, fps). Assume the underlying speed distributions are continuous. NAILGUN
(a) Use the Kruskal–Wallis test with $\alpha = 0.05$ to determine whether there is any evidence that at least two of the nail-gun speed population distributions are different.
(b) Find the p value associated with this test.
(c) Suppose you would like to purchase the brand of nail gun that fires a nail at the highest speed. Which brand would you choose, and why?

14.161 Marketing and Consumer Behavior A home-building company offers a variety of styles in only two exterior finishes: vinyl siding (V) or brick (B). Immediately following an advertising campaign explaining the advantages of vinyl siding, a sample of consecutive customers was obtained, and each was classified by the exterior finish chosen. The ordered observations are given in the table. FINISH

B	V	V	V	V	B	V	V	V	V	B	V	B	B

(a) Conduct a runs test with $\alpha \approx 0.05$ to determine whether there is any evidence to suggest that the order of observations is not random with respect to exterior finish.
(b) Using this sample, do you believe the advertising campaign was successful? Why or why not?

14.162 Psychology and Human Behavior The manager of concessions at Fenway Park believes that the number of runs scored by the Boston Red Sox (x) is related to the number of Fenway Franks consumed by fans (y). A random sample of nine-inning games was obtained. The hot dog and run totals were recorded. HOTDOG
(a) Construct a scatter plot for these data, and describe the relationship between the two variables.
(b) Rank the observations in each sample separately, and find each difference between the ranks d_i.
(c) Find the sample correlation coefficient between the ranks.
(d) Find Spearman's rank correlation coefficient using the d_i's.
(e) Explain why the values in parts (c) and (d) are different.
(f) Suppose the manager of concessions would like to sell as many hot dogs as possible. How may runs would he or she like the Red Sox to score? Why?

14.163 Travel and Transportation SunPass is the prepaid toll program in Florida. A SunPass transponder allows drivers to pay the lowest toll, pay for parking at most airports in Florida, and, of course, zip through toll plazas by using electronic express lanes. A random sample of vehicles with and without SunPass traveling along Alligator Alley in Florida, from Miami to Fort Myers, was obtained, and the travel time was recorded (in minutes). Assume the underlying distributions are continuous. **SUNPASS**

(a) Use a Wilcoxon rank-sum test to determine whether there is any evidence to suggest the median time to traverse Alligator Alley for SunPass users is less than the median time it takes for non-SunPass users. Use $\alpha = 0.05$. Explain this result in the context of the problem.

(b) Use a two-sample t test to determine whether there is any evidence to suggest the mean time to traverse Alligator Alley for SunPass users is less than the mean time it takes for non-SunPass users. Use $\alpha = 0.05$. Explain this result in the context of the problem.

(c) Which test do you believe is more appropriate? Why?

14.164 Technology and the Internet In 2018, Marriott International announced possibly the largest consumer data breach ever, in which as many as 500 million guests may have had reservation information stolen, such as their phone number, passport number, and Starwood Preferred Guest (SPG) account information. A sample of SPG members who had information stolen was obtained, and the number of SPG reward points at the time of the breach was recorded for each. Is there any evidence to suggest that the order of observations is not random with respect to the median number of SPG points, 35,000? Use $\alpha = 0.01$. **BREACH**

14.165 Boarding Pain The process of boarding a flight in any airport can be slow and agonizing. Passengers with priority boarding often crowd the gate area, and passengers with carry-on bags jostle for position in line. Because faster boarding time saves airlines money and eases anxiety, airlines often experiment with more efficient ways to seat passengers and prepare for departure. Consider the following boarding methods.

> Block boarding: Boarding in blocks starting at the rear of the airplane.
>
> Random seating: Passengers board the airplane at random, in the order in which they line up.
>
> Wilma: Boarding by window seats first, then middle seats, followed by aisle seats.
>
> Steffen's method: Boarding in a very precise order, beginning with the window seat passengers on the right side of the plane, then the left side, then the middle seats, and so on.

A random sample of similar flights and airplanes using one of these four methods was selected from London's Heathrow Airport. The boarding time (in minutes) for each flight was recorded. Assume the underlying boarding time distributions are continuous. **BOARD**

(a) Use the Kruskal–Wallis test with $\alpha = 0.05$ to determine whether there is any evidence that at least two of the boarding time populations are different.

(b) Find the exact p value for this test.

(c) What other statistical test might be appropriate in this case? Check the assumptions and conduct this test. Compare your results with part (a).

(d) Which of these boarding methods do you think is the most efficient? Why?

Image Lagoon/Shutterstock

CHAPTER APP

14.166 Keyboard Angles For each of two keyboard slopes, 0° and 20°, a random sample of individuals who regularly type at a computer was obtained. The right wrist median nerve cross-sectional area was measured (in mm^2) for each person. The data are given in Section 14.1. **CARPAL**

(a) Use the sign test to determine if there is any evidence that the cross-sectional area for the 0° group is greater than 5.8, a typical median nerve cross-section area in people unaffected by carpal tunnel syndrome. Assume the underlying distribution is continuous.

(b) Use the signed-rank test to determine if there is any evidence that the cross-sectional area for the 0° group is greater than 5.8, a typical median nerve cross-section area in people unaffected by carpal tunnel syndrome. Assume the underlying distribution is continuous and symmetric.

(c) Conduct a Wilcoxon rank-sum test to determine whether there is any evidence to suggest a difference in the population median, median nerve cross-sectional areas. Assume the underlying distributions are continuous.

(d) Use the runs test with $\alpha = 0.05$ to determine whether there is any evidence that the order in which the 0° sample was selected was not random with respect to the population median 5.8 mm^2.

(e) Suppose that for each of two additional keyboard slopes, 5° and 10°, random samples of individuals who regularly type at a computer were obtained. The right wrist median nerve cross-sectional area was measured (in mm^2) for each person. The underlying populations are assumed to be continuous. Use the Kruskal–Wallis test to determine whether there is any evidence that the populations are different. Use a significance level of $\alpha = 0.05$. Assume the underlying distributions are continuous.

Tables Appendix

Table 1	Binomial Distribution Cumulative Probabilities	T-2
Table 2	Poisson Distribution Cumulative Probabilities	T-4
Table 3	Standard Normal Distribution Cumulative Probabilities	T-7
Table 4	Standardized Normal Scores	T-9
Table 5	Critical Values for the t Distribution	T-10
Table 6	Critical Values for the Chi-Square Distribution	T-11
Table 7	Critical Values for the F Distribution	T-13
Table 8	Critical Values for the Studentized Range Distribution	T-16
Table 9	Critical Values for the Wilcoxon Signed-Rank Statistic	T-19
Table 10	Critical Values for the Wilcoxon Rank-Sum Statistic	T-22
Table 11	Critical Values for the Runs Test	T-25
Table 12	Greek Alphabet	T-27

Table 1 Binomial Distribution Cumulative Probabilities

Let X be a binomial random variable with parameters n and p: $X \sim B(n, p)$. This table contains cumulative probabilities:

$$P(X \leq x) = \sum_{k=0}^{x} P(X = k) = P(X = 0) + P(X = 1) + P(X = 2) + \cdots + P(X = x).$$

$n = 5$								p							
x	0.01	0.05	0.10	0.20	0.25	0.30	0.40	0.50	0.60	0.70	0.75	0.80	0.90	0.95	0.99
0	0.9510	0.7738	0.5905	0.3277	0.2373	0.1681	0.0778	0.0313	0.0102	0.0024	0.0010	0.0003	0.0000		
1	0.9990	0.9774	0.9185	0.7373	0.6328	0.5282	0.3370	0.1875	0.0870	0.0308	0.0156	0.0067	0.0005	0.0000	
2	1.0000	0.9988	0.9914	0.9421	0.8965	0.8369	0.6826	0.5000	0.3174	0.1631	0.1035	0.0579	0.0086	0.0012	0.0000
3		1.0000	0.9995	0.9933	0.9844	0.9692	0.9130	0.8125	0.6630	0.4718	0.3672	0.2627	0.0815	0.0226	0.0010
4			1.0000	0.9997	0.9990	0.9976	0.9898	0.9688	0.9222	0.8319	0.7627	0.6723	0.4095	0.2262	0.0490

$n = 10$								p							
x	0.01	0.05	0.10	0.20	0.25	0.30	0.40	0.50	0.60	0.70	0.75	0.80	0.90	0.95	0.99
0	0.9044	0.5987	0.3487	0.1074	0.0563	0.0282	0.0060	0.0010	0.0001	0.0000					
1	0.9957	0.9139	0.7361	0.3758	0.2440	0.1493	0.0464	0.0107	0.0017	0.0001	0.0000	0.0000			
2	0.9999	0.9885	0.9298	0.6778	0.5256	0.3828	0.1673	0.0547	0.0123	0.0016	0.0004	0.0001	0.0000		
3	1.0000	0.9990	0.9872	0.8791	0.7759	0.6496	0.3823	0.1719	0.0548	0.0106	0.0035	0.0009	0.0000		
4		0.9999	0.9984	0.9672	0.9219	0.8497	0.6331	0.3770	0.1662	0.0473	0.0197	0.0064	0.0001	0.0000	
5		1.0000	0.9999	0.9936	0.9803	0.9527	0.8338	0.6230	0.3669	0.1503	0.0781	0.0328	0.0016	0.0001	
6			1.0000	0.9991	0.9965	0.9894	0.9452	0.8281	0.6177	0.3504	0.2241	0.1209	0.0128	0.0010	0.0000
7				0.9999	0.9996	0.9984	0.9877	0.9453	0.8327	0.6172	0.4744	0.3222	0.0702	0.0115	0.0001
8				1.0000	1.0000	0.9999	0.9983	0.9893	0.9536	0.8507	0.7560	0.6242	0.2639	0.0861	0.0043
9						1.0000	0.9999	0.9990	0.9940	0.9718	0.9437	0.8926	0.6513	0.4013	0.0956

$n = 15$								p							
x	0.01	0.05	0.10	0.20	0.25	0.30	0.40	0.50	0.60	0.70	0.75	0.80	0.90	0.95	0.99
0	0.8601	0.4633	0.2059	0.0352	0.0134	0.0047	0.0005	0.0000							
1	0.9904	0.8290	0.5490	0.1671	0.0802	0.0353	0.0052	0.0005	0.0000						
2	0.9996	0.9638	0.8159	0.3980	0.2361	0.1268	0.0271	0.0037	0.0003	0.0000					
3	1.0000	0.9945	0.9444	0.6482	0.4613	0.2969	0.0905	0.0176	0.0019	0.0001	0.0000				
4		0.9994	0.9873	0.8358	0.6865	0.5155	0.2173	0.0592	0.0093	0.0007	0.0001	0.0000			
5		0.9999	0.9978	0.9389	0.8516	0.7216	0.4032	0.1509	0.0338	0.0037	0.0008	0.0001			
6		1.0000	0.9997	0.9819	0.9434	0.8689	0.6098	0.3036	0.0950	0.0152	0.0042	0.0008			
7			1.0000	0.9958	0.9827	0.9500	0.7869	0.5000	0.2131	0.0500	0.0173	0.0042	0.0000		
8				0.9992	0.9958	0.9848	0.9050	0.6964	0.3902	0.1311	0.0566	0.0181	0.0003	0.0000	
9				0.9999	0.9992	0.9963	0.9662	0.8491	0.5968	0.2784	0.1484	0.0611	0.0022	0.0001	
10				1.0000	0.9999	0.9993	0.9907	0.9408	0.7827	0.4845	0.3135	0.1642	0.0127	0.0006	
11					1.0000	0.9999	0.9981	0.9824	0.9095	0.7031	0.5387	0.3518	0.0556	0.0055	0.0000
12						1.0000	0.9997	0.9963	0.9729	0.8732	0.7639	0.6020	0.1841	0.0362	0.0004
13							1.0000	0.9995	0.9948	0.9647	0.9198	0.8329	0.4510	0.1710	0.0096
14								1.0000	0.9995	0.9953	0.9866	0.9648	0.7941	0.5367	0.1399

Table 1 Binomial Distribution Cumulative Probabilities (Continued)

n = 20

x	0.01	0.05	0.10	0.20	0.25	0.30	0.40	0.50	0.60	0.70	0.75	0.80	0.90	0.95	0.99
0	0.8179	0.3585	0.1216	0.0115	0.0032	0.0008	0.0000								
1	0.9831	0.7358	0.3917	0.0692	0.0243	0.0076	0.0005	0.0000							
2	0.9990	0.9245	0.6769	0.2061	0.0913	0.0355	0.0036	0.0002							
3	1.0000	0.9841	0.8670	0.4114	0.2252	0.1071	0.0160	0.0013	0.0000						
4		0.9974	0.9568	0.6296	0.4148	0.2375	0.0510	0.0059	0.0003						
5		0.9997	0.9887	0.8042	0.6172	0.4164	0.1256	0.0207	0.0016	0.0000					
6		1.0000	0.9976	0.9133	0.7858	0.6080	0.2500	0.0577	0.0065	0.0003	0.0000				
7			0.9996	0.9679	0.8982	0.7723	0.4159	0.1316	0.0210	0.0013	0.0002	0.0000			
8			0.9999	0.9900	0.9591	0.8867	0.5956	0.2517	0.0565	0.0051	0.0009	0.0001			
9			1.0000	0.9974	0.9861	0.9520	0.7553	0.4119	0.1275	0.0171	0.0039	0.0006			
10				0.9994	0.9961	0.9829	0.8725	0.5881	0.2447	0.0480	0.0139	0.0026	0.0000		
11				0.9999	0.9991	0.9949	0.9435	0.7483	0.4044	0.1133	0.0409	0.0100	0.0001		
12				1.0000	0.9998	0.9987	0.9790	0.8684	0.5841	0.2277	0.1018	0.0321	0.0004		
13					1.0000	0.9997	0.9935	0.9423	0.7500	0.3920	0.2142	0.0867	0.0024	0.0000	
14						1.0000	0.9984	0.9793	0.8744	0.5836	0.3828	0.1958	0.0113	0.0003	
15							0.9997	0.9941	0.9490	0.7625	0.5852	0.3704	0.0432	0.0026	
16							1.0000	0.9987	0.9840	0.8929	0.7748	0.5886	0.1330	0.0159	0.0000
17								0.9998	0.9964	0.9645	0.9087	0.7939	0.3231	0.0755	0.0010
18								1.0000	0.9995	0.9924	0.9757	0.9308	0.6083	0.2642	0.0169
19									1.0000	0.9992	0.9968	0.9885	0.8784	0.6415	0.1821

n = 25

x	0.01	0.05	0.10	0.20	0.25	0.30	0.40	0.50	0.60	0.70	0.75	0.80	0.90	0.95	0.99
0	0.7778	0.2774	0.0718	0.0038	0.0008	0.0001	0.0000								
1	0.9742	0.6424	0.2712	0.0274	0.0070	0.0016	0.0001								
2	0.9980	0.8729	0.5371	0.0982	0.0321	0.0090	0.0004	0.0000							
3	0.9999	0.9659	0.7636	0.2340	0.0962	0.0332	0.0024	0.0001							
4	1.0000	0.9928	0.9020	0.4207	0.2137	0.0905	0.0095	0.0005	0.0000						
5		0.9988	0.9666	0.6167	0.3783	0.1935	0.0294	0.0020	0.0001						
6		0.9998	0.9905	0.7800	0.5611	0.3407	0.0736	0.0073	0.0003						
7		1.0000	0.9977	0.8909	0.7265	0.5118	0.1536	0.0216	0.0012	0.0000					
8			0.9995	0.9532	0.8506	0.6769	0.2735	0.0539	0.0043	0.0001					
9			0.9999	0.9827	0.9287	0.8106	0.4246	0.1148	0.0132	0.0005	0.0000				
10			1.0000	0.9944	0.9703	0.9022	0.5858	0.2122	0.0344	0.0018	0.0002	0.0000			
11				0.9985	0.9893	0.9558	0.7323	0.3450	0.0778	0.0060	0.0009	0.0001			
12				0.9996	0.9966	0.9825	0.8462	0.5000	0.1538	0.0175	0.0034	0.0004			
13				0.9999	0.9991	0.9940	0.9222	0.6550	0.2677	0.0442	0.0107	0.0015			
14				1.0000	0.9998	0.9982	0.9656	0.7878	0.4142	0.0978	0.0297	0.0056	0.0000		
15					1.0000	0.9995	0.9868	0.8852	0.5754	0.1894	0.0713	0.0173	0.0001		
16						0.9999	0.9957	0.9461	0.7265	0.3231	0.1494	0.0468	0.0005		
17						1.0000	0.9988	0.9784	0.8464	0.4882	0.2735	0.1091	0.0023	0.0000	
18							0.9997	0.9927	0.9264	0.6593	0.4389	0.2200	0.0095	0.0002	
19							0.9999	0.9980	0.9706	0.8065	0.6217	0.3833	0.0334	0.0012	
20							1.0000	0.9995	0.9905	0.9095	0.7863	0.5793	0.0980	0.0072	0.0000
21								0.9999	0.9976	0.9668	0.9038	0.7660	0.2364	0.0341	0.0001
22								1.0000	0.9996	0.9910	0.9679	0.9018	0.4629	0.1271	0.0020
23									0.9999	0.9984	0.9930	0.9726	0.7288	0.3576	0.0258
24									1.0000	0.9999	0.9992	0.9962	0.9282	0.7226	0.2222

Table 2 Poisson Distribution Cumulative Probabilities

Let X be a Poisson random variable with parameter λ. This table contains cumulative probabilities:

$$P(X \leq x) = \sum_{k=0}^{x} P(X = k) = P(X = 0) + P(X = 1) + \cdots + P(X = x).$$

					λ					
x	0.05	0.10	0.15	0.20	0.25	0.30	0.35	0.40	0.45	0.50
0	0.9512	0.9048	0.8607	0.8187	0.7788	0.7408	0.7047	0.6703	0.6376	0.6065
1	0.9988	0.9953	0.9898	0.9825	0.9735	0.9631	0.9513	0.9384	0.9246	0.9098
2	1.0000	0.9998	0.9995	0.9989	0.9978	0.9964	0.9945	0.9921	0.9891	0.9856
3		1.0000	1.0000	0.9999	0.9999	0.9997	0.9995	0.9992	0.9988	0.9982
4				1.0000	1.0000	1.0000	1.0000	0.9999	0.9999	0.9998
5								1.0000	1.0000	1.0000

					λ					
x	0.55	0.60	0.65	0.70	0.75	0.80	0.85	0.90	0.95	1.00
0	0.5769	0.5488	0.5220	0.4966	0.4724	0.4493	0.4274	0.4066	0.3867	0.3679
1	0.8943	0.8781	0.8614	0.8442	0.8266	0.8088	0.7907	0.7725	0.7541	0.7358
2	0.9815	0.9769	0.9717	0.9659	0.9595	0.9526	0.9451	0.9371	0.9287	0.9197
3	0.9975	0.9966	0.9956	0.9942	0.9927	0.9909	0.9889	0.9865	0.9839	0.9810
4	0.9997	0.9996	0.9994	0.9992	0.9989	0.9986	0.9982	0.9977	0.9971	0.9963
5	1.0000	1.0000	0.9999	0.9999	0.9999	0.9998	0.9997	0.9997	0.9995	0.9994
6			1.0000	1.0000	1.0000	1.0000	1.0000	1.0000	0.9999	0.9999
7									1.0000	1.0000

					λ					
x	1.1	1.2	1.3	1.4	1.5	1.6	1.7	1.8	1.9	2.0
0	0.3329	0.3012	0.2725	0.2466	0.2231	0.2019	0.1827	0.1653	0.1496	0.1353
1	0.6990	0.6626	0.6268	0.5918	0.5578	0.5249	0.4932	0.4628	0.4337	0.4060
2	0.9004	0.8795	0.8571	0.8335	0.8088	0.7834	0.7572	0.7306	0.7037	0.6767
3	0.9743	0.9662	0.9569	0.9463	0.9344	0.9212	0.9068	0.8913	0.8747	0.8571
4	0.9946	0.9923	0.9893	0.9857	0.9814	0.9763	0.9704	0.9636	0.9559	0.9473
5	0.9990	0.9985	0.9978	0.9968	0.9955	0.9940	0.9920	0.9896	0.9868	0.9834
6	0.9999	0.9997	0.9996	0.9994	0.9991	0.9987	0.9981	0.9974	0.9966	0.9955
7	1.0000	1.0000	0.9999	0.9999	0.9998	0.9997	0.9996	0.9994	0.9992	0.9989
8			1.0000	1.0000	1.0000	1.0000	0.9999	0.9999	0.9998	0.9998
9							1.0000	1.0000	1.0000	1.0000

Table 2 Poisson Distribution Cumulative Probabilities (Continued)

					λ					
x	2.1	2.2	2.3	2.4	2.5	2.6	2.7	2.8	2.9	3.0
0	0.1225	0.1108	0.1003	0.0907	0.0821	0.0743	0.0672	0.0608	0.0550	0.0498
1	0.3796	0.3546	0.3309	0.3084	0.2873	0.2674	0.2487	0.2311	0.2146	0.1991
2	0.6496	0.6227	0.5960	0.5697	0.5438	0.5184	0.4936	0.4695	0.4460	0.4232
3	0.8386	0.8194	0.7993	0.7787	0.7576	0.7360	0.7141	0.6919	0.6696	0.6472
4	0.9379	0.9275	0.9162	0.9041	0.8912	0.8774	0.8629	0.8477	0.8318	0.8153
5	0.9796	0.9751	0.9700	0.9643	0.9580	0.9510	0.9433	0.9349	0.9258	0.9161
6	0.9941	0.9925	0.9906	0.9884	0.9858	0.9828	0.9794	0.9756	0.9713	0.9665
7	0.9985	0.9980	0.9974	0.9967	0.9958	0.9947	0.9934	0.9919	0.9901	0.9881
8	0.9997	0.9995	0.9994	0.9991	0.9989	0.9985	0.9981	0.9976	0.9969	0.9962
9	0.9999	0.9999	0.9999	0.9998	0.9997	0.9996	0.9995	0.9993	0.9991	0.9989
10	1.0000	1.0000	1.0000	1.0000	0.9999	0.9999	0.9999	0.9998	0.9998	0.9997
11				1.0000	1.0000	1.0000	1.0000	1.0000	0.9999	0.9999
12									1.0000	1.0000

					λ					
x	3.1	3.2	3.3	3.4	3.5	3.6	3.7	3.8	3.9	4.0
0	0.0450	0.0408	0.0369	0.0334	0.0302	0.0273	0.0247	0.0224	0.0202	0.0183
1	0.1847	0.1712	0.1586	0.1468	0.1359	0.1257	0.1162	0.1074	0.0992	0.0916
2	0.4012	0.3799	0.3594	0.3397	0.3208	0.3027	0.2854	0.2689	0.2531	0.2381
3	0.6248	0.6025	0.5803	0.5584	0.5366	0.5152	0.4942	0.4735	0.4532	0.4335
4	0.7982	0.7806	0.7626	0.7442	0.7254	0.7064	0.6872	0.6678	0.6484	0.6288
5	0.9057	0.8946	0.8829	0.8705	0.8576	0.8441	0.8301	0.8156	0.8006	0.7851
6	0.9612	0.9554	0.9490	0.9421	0.9347	0.9267	0.9182	0.9091	0.8995	0.8893
7	0.9858	0.9832	0.9802	0.9769	0.9733	0.9692	0.9648	0.9599	0.9546	0.9489
8	0.9953	0.9943	0.9931	0.9917	0.9901	0.9883	0.9863	0.9840	0.9815	0.9786
9	0.9986	0.9982	0.9978	0.9973	0.9967	0.9960	0.9952	0.9942	0.9931	0.9919
10	0.9996	0.9995	0.9994	0.9992	0.9990	0.9987	0.9984	0.9981	0.9977	0.9972
11	0.9999	0.9999	0.9998	0.9998	0.9997	0.9996	0.9995	0.9994	0.9993	0.9991
12	1.0000	1.0000	1.0000	0.9999	0.9999	0.9999	0.9999	0.9998	0.9998	0.9997
13				1.0000	1.0000	1.0000	1.0000	1.0000	0.9999	0.9999
14									1.0000	1.0000

Table 2 Poisson Distribution Cumulative Probabilities (Continued)

					λ					
x	4.1	4.2	4.3	4.4	4.5	4.6	4.7	4.8	4.9	5.0
0	0.0166	0.0150	0.0136	0.0123	0.0111	0.0101	0.0091	0.0082	0.0074	0.0067
1	0.0845	0.0780	0.0719	0.0663	0.0611	0.0563	0.0518	0.0477	0.0439	0.0404
2	0.2238	0.2102	0.1974	0.1851	0.1736	0.1626	0.1523	0.1425	0.1333	0.1247
3	0.4142	0.3954	0.3772	0.3594	0.3423	0.3257	0.3097	0.2942	0.2793	0.2650
4	0.6093	0.5898	0.5704	0.5512	0.5321	0.5132	0.4946	0.4763	0.4582	0.4405
5	0.7693	0.7531	0.7367	0.7199	0.7029	0.6858	0.6684	0.6510	0.6335	0.6160
6	0.8786	0.8675	0.8558	0.8436	0.8311	0.8180	0.8046	0.7908	0.7767	0.7622
7	0.9427	0.9361	0.9290	0.9214	0.9134	0.9049	0.8960	0.8867	0.8769	0.8666
8	0.9755	0.9721	0.9683	0.9642	0.9597	0.9549	0.9497	0.9442	0.9382	0.9319
9	0.9905	0.9889	0.9871	0.9851	0.9829	0.9805	0.9778	0.9749	0.9717	0.9682
10	0.9966	0.9959	0.9952	0.9943	0.9933	0.9922	0.9910	0.9896	0.9880	0.9863
11	0.9989	0.9986	0.9983	0.9980	0.9976	0.9971	0.9966	0.9960	0.9953	0.9945
12	0.9997	0.9996	0.9995	0.9993	0.9992	0.9990	0.9988	0.9986	0.9983	0.9980
14	0.9999	0.9999	0.9998	0.9998	0.9997	0.9997	0.9996	0.9995	0.9994	0.9993
15	1.0000	1.0000	1.0000	0.9999	0.9999	0.9999	0.9999	0.9999	0.9998	0.9998
16				1.0000	1.0000	1.0000	1.0000	1.0000	0.9999	0.9999
17									1.0000	1.0000

					λ					
x	5.5	6.0	6.5	7.0	7.5	8.0	8.5	9.0	9.5	10.0
0	0.0041	0.0025	0.0015	0.0009	0.0006	0.0003	0.0002	0.0001	0.0001	0.0000
1	0.0266	0.0174	0.0113	0.0073	0.0047	0.0030	0.0019	0.0012	0.0008	0.0005
2	0.0884	0.0620	0.0430	0.0296	0.0203	0.0138	0.0093	0.0062	0.0042	0.0028
3	0.2017	0.1512	0.1118	0.0818	0.0591	0.0424	0.0301	0.0212	0.0149	0.0103
4	0.3575	0.2851	0.2237	0.1730	0.1321	0.0996	0.0744	0.0550	0.0403	0.0293
5	0.5289	0.4457	0.3690	0.3007	0.2414	0.1912	0.1496	0.1157	0.0885	0.0671
6	0.6860	0.6063	0.5265	0.4497	0.3782	0.3134	0.2562	0.2068	0.1649	0.1301
7	0.8095	0.7440	0.6728	0.5987	0.5246	0.4530	0.3856	0.3239	0.2687	0.2202
8	0.8944	0.8472	0.7916	0.7291	0.6620	0.5925	0.5231	0.4557	0.3918	0.3328
9	0.9462	0.9161	0.8774	0.8305	0.7764	0.7166	0.6530	0.5874	0.5218	0.4579
10	0.9747	0.9574	0.9332	0.9015	0.8622	0.8159	0.7634	0.7060	0.6453	0.5830
11	0.9890	0.9799	0.9661	0.9467	0.9208	0.8881	0.8487	0.8030	0.7520	0.6968
12	0.9955	0.9912	0.9840	0.9730	0.9573	0.9362	0.9091	0.8758	0.8364	0.7916
13	0.9983	0.9964	0.9929	0.9872	0.9784	0.9658	0.9486	0.9261	0.8981	0.8645
14	0.9994	0.9986	0.9970	0.9943	0.9897	0.9827	0.9726	0.9585	0.9400	0.9165
15	0.9998	0.9995	0.9988	0.9976	0.9954	0.9918	0.9862	0.9780	0.9665	0.9513
16	0.9999	0.9998	0.9996	0.9990	0.9980	0.9963	0.9934	0.9889	0.9823	0.9730
17	1.0000	0.9999	0.9998	0.9996	0.9992	0.9984	0.9970	0.9947	0.9911	0.9857
18		1.0000	0.9999	0.9999	0.9997	0.9993	0.9987	0.9976	0.9957	0.9928
19			1.0000	1.0000	0.9999	0.9997	0.9995	0.9989	0.9980	0.9965
20					1.0000	0.9999	0.9998	0.9996	0.9991	0.9984
21						1.0000	0.9999	0.9998	0.9996	0.9993
22							1.0000	0.9999	0.9999	0.9997
23								1.0000	0.9999	0.9999
24									1.0000	1.0000

Table 3 Standard Normal Distribution Cumulative Probabilities

Let Z be a standard normal random variable: $\mu = 0$ and $\sigma = 1$.
This table contains cumulative probabilities: $P(Z \leq z)$.

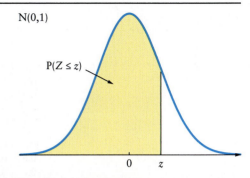

z	.00	.01	.02	.03	.04	.05	.06	.07	.08	.09
−3.4	0.0003	0.0003	0.0003	0.0003	0.0003	0.0003	0.0003	0.0003	0.0003	0.0002
−3.3	0.0005	0.0005	0.0005	0.0004	0.0004	0.0004	0.0004	0.0004	0.0004	0.0003
−3.2	0.0007	0.0007	0.0006	0.0006	0.0006	0.0006	0.0006	0.0005	0.0005	0.0005
−3.1	0.0010	0.0009	0.0009	0.0009	0.0008	0.0008	0.0008	0.0008	0.0007	0.0007
−3.0	0.0013	0.0013	0.0013	0.0012	0.0012	0.0011	0.0011	0.0011	0.0010	0.0010
−2.9	0.0019	0.0018	0.0018	0.0017	0.0016	0.0016	0.0015	0.0015	0.0014	0.0014
−2.8	0.0026	0.0025	0.0024	0.0023	0.0023	0.0022	0.0021	0.0021	0.0020	0.0019
−2.7	0.0035	0.0034	0.0033	0.0032	0.0031	0.0030	0.0029	0.0028	0.0027	0.0026
−2.6	0.0047	0.0045	0.0044	0.0043	0.0041	0.0040	0.0039	0.0038	0.0037	0.0036
−2.5	0.0062	0.0060	0.0059	0.0057	0.0055	0.0054	0.0052	0.0051	0.0049	0.0048
−2.4	0.0082	0.0080	0.0078	0.0075	0.0073	0.0071	0.0069	0.0068	0.0066	0.0064
−2.3	0.0107	0.0104	0.0102	0.0099	0.0096	0.0094	0.0091	0.0089	0.0087	0.0084
−2.2	0.0139	0.0136	0.0132	0.0129	0.0125	0.0122	0.0119	0.0116	0.0113	0.0110
−2.1	0.0179	0.0174	0.0170	0.0166	0.0162	0.0158	0.0154	0.0150	0.0146	0.0143
−2.0	0.0228	0.0222	0.0217	0.0212	0.0207	0.0202	0.0197	0.0192	0.0188	0.0183
−1.9	0.0287	0.0281	0.0274	0.0268	0.0262	0.0256	0.0250	0.0244	0.0239	0.0233
−1.8	0.0359	0.0351	0.0344	0.0336	0.0329	0.0322	0.0314	0.0307	0.0301	0.0294
−1.7	0.0446	0.0436	0.0427	0.0418	0.0409	0.0401	0.0392	0.0384	0.0375	0.0367
−1.6	0.0548	0.0537	0.0526	0.0516	0.0505	0.0495	0.0485	0.0475	0.0465	0.0455
−1.5	0.0668	0.0655	0.0643	0.0630	0.0618	0.0606	0.0594	0.0582	0.0571	0.0559
−1.4	0.0808	0.0793	0.0778	0.0764	0.0749	0.0735	0.0721	0.0708	0.0694	0.0681
−1.3	0.0968	0.0951	0.0934	0.0918	0.0901	0.0885	0.0869	0.0853	0.0838	0.0823
−1.2	0.1151	0.1131	0.1112	0.1093	0.1075	0.1056	0.1038	0.1020	0.1003	0.0985
−1.1	0.1357	0.1335	0.1314	0.1292	0.1271	0.1251	0.1230	0.1210	0.1190	0.1170
−1.0	0.1587	0.1562	0.1539	0.1515	0.1492	0.1469	0.1446	0.1423	0.1401	0.1379
−0.9	0.1841	0.1814	0.1788	0.1762	0.1736	0.1711	0.1685	0.1660	0.1635	0.1611
−0.8	0.2119	0.2090	0.2061	0.2033	0.2005	0.1977	0.1949	0.1922	0.1894	0.1867
−0.7	0.2420	0.2389	0.2358	0.2327	0.2296	0.2266	0.2236	0.2206	0.2177	0.2148
−0.6	0.2743	0.2709	0.2676	0.2643	0.2611	0.2578	0.2546	0.2514	0.2483	0.2451
−0.5	0.3085	0.3050	0.3015	0.2981	0.2946	0.2912	0.2877	0.2843	0.2810	0.2776
−0.4	0.3446	0.3409	0.3372	0.3336	0.3300	0.3264	0.3228	0.3192	0.3156	0.3121
−0.3	0.3821	0.3783	0.3745	0.3707	0.3669	0.3632	0.3594	0.3557	0.3520	0.3483
−0.2	0.4207	0.4168	0.4129	0.4090	0.4052	0.4013	0.3974	0.3936	0.3897	0.3859
−0.1	0.4602	0.4562	0.4522	0.4483	0.4443	0.4404	0.4364	0.4325	0.4286	0.4247
−0.0	0.5000	0.4960	0.4920	0.4880	0.4840	0.4801	0.4761	0.4721	0.4681	0.4641

Table 3 Standard Normal Distribution Cumulative Probabilities (Continued)

z	.00	.01	.02	.03	.04	.05	.06	.07	.08	.09
0.0	0.5000	0.5040	0.5080	0.5120	0.5160	0.5199	0.5239	0.5279	0.5319	0.5359
0.1	0.5398	0.5438	0.5478	0.5517	0.5557	0.5596	0.5636	0.5675	0.5714	0.5753
0.2	0.5793	0.5832	0.5871	0.5910	0.5948	0.5987	0.6026	0.6064	0.6103	0.6141
0.3	0.6179	0.6217	0.6255	0.6293	0.6331	0.6368	0.6406	0.6443	0.6480	0.6517
0.4	0.6554	0.6591	0.6628	0.6664	0.6700	0.6736	0.6772	0.6808	0.6844	0.6879
0.5	0.6915	0.6950	0.6985	0.7019	0.7054	0.7088	0.7123	0.7157	0.7190	0.7224
0.6	0.7257	0.7291	0.7324	0.7357	0.7389	0.7422	0.7454	0.7486	0.7517	0.7549
0.7	0.7580	0.7611	0.7642	0.7673	0.7704	0.7734	0.7764	0.7794	0.7823	0.7852
0.8	0.7881	0.7910	0.7939	0.7967	0.7995	0.8023	0.8051	0.8078	0.8106	0.8133
0.9	0.8159	0.8186	0.8212	0.8238	0.8264	0.8289	0.8315	0.8340	0.8365	0.8389
1.0	0.8413	0.8438	0.8461	0.8485	0.8508	0.8531	0.8554	0.8577	0.8599	0.8621
1.1	0.8643	0.8665	0.8686	0.8708	0.8729	0.8749	0.8770	0.8790	0.8810	0.8830
1.2	0.8849	0.8869	0.8888	0.8907	0.8925	0.8944	0.8962	0.8980	0.8997	0.9015
1.3	0.9032	0.9049	0.9066	0.9082	0.9099	0.9115	0.9131	0.9147	0.9162	0.9177
1.4	0.9192	0.9207	0.9222	0.9236	0.9251	0.9265	0.9279	0.9292	0.9306	0.9319
1.5	0.9332	0.9345	0.9357	0.9370	0.9382	0.9394	0.9406	0.9418	0.9429	0.9441
1.6	0.9452	0.9463	0.9474	0.9484	0.9495	0.9505	0.9515	0.9525	0.9535	0.9545
1.7	0.9554	0.9564	0.9573	0.9582	0.9591	0.9599	0.9608	0.9616	0.9625	0.9633
1.8	0.9641	0.9649	0.9656	0.9664	0.9671	0.9678	0.9686	0.9693	0.9699	0.9706
1.9	0.9713	0.9719	0.9726	0.9732	0.9738	0.9744	0.9750	0.9756	0.9761	0.9767
2.0	0.9772	0.9778	0.9783	0.9788	0.9793	0.9798	0.9803	0.9808	0.9812	0.9817
2.1	0.9821	0.9826	0.9830	0.9834	0.9838	0.9842	0.9846	0.9850	0.9854	0.9857
2.2	0.9861	0.9864	0.9868	0.9871	0.9875	0.9878	0.9881	0.9884	0.9887	0.9890
2.3	0.9893	0.9896	0.9898	0.9901	0.9904	0.9906	0.9909	0.9911	0.9913	0.9916
2.4	0.9918	0.9920	0.9922	0.9925	0.9927	0.9929	0.9931	0.9932	0.9934	0.9936
2.5	0.9938	0.9940	0.9941	0.9943	0.9945	0.9946	0.9948	0.9949	0.9951	0.9952
2.6	0.9953	0.9955	0.9956	0.9957	0.9959	0.9960	0.9961	0.9962	0.9963	0.9964
2.7	0.9965	0.9966	0.9967	0.9968	0.9969	0.9970	0.9971	0.9972	0.9973	0.9974
2.8	0.9974	0.9975	0.9976	0.9977	0.9977	0.9978	0.9979	0.9979	0.9980	0.9981
2.9	0.9981	0.9982	0.9982	0.9983	0.9984	0.9984	0.9985	0.9985	0.9986	0.9986
3.0	0.9987	0.9987	0.9987	0.9988	0.9988	0.9989	0.9989	0.9989	0.9990	0.9990
3.1	0.9990	0.9991	0.9991	0.9991	0.9992	0.9992	0.9992	0.9992	0.9993	0.9993
3.2	0.9993	0.9993	0.9994	0.9994	0.9994	0.9994	0.9994	0.9995	0.9995	0.9995
3.3	0.9995	0.9995	0.9995	0.9996	0.9996	0.9996	0.9996	0.9996	0.9996	0.9997
3.4	0.9997	0.9997	0.9997	0.9997	0.9997	0.9997	0.9997	0.9997	0.9997	0.9998

Special critical values: $P(Z \geq z_\alpha) = \alpha$

α	0.10	0.05	0.025	0.01	0.005	0.001	0.0005	0.0001
z_α	1.2816	1.6449	1.9600	2.3263	2.5758	3.0902	3.2905	3.7190

α	0.00009	0.00008	0.00007	0.00006	0.00005	0.00004	0.00003	0.00002	0.00001
z_α	3.7455	3.7750	3.8082	3.8461	3.8906	3.9444	4.0128	4.1075	4.2649

Table 4 Standardized Normal Scores

This table contains the standardized normal scores, z_i, for selected values of n.

i	n=10	20	25	30	40	50
1	−1.55	−1.87	−1.96	−2.04	−2.16	−2.24
2	−1.00	−1.40	−1.52	−1.61	−1.75	−1.85
3	−0.66	−1.13	−1.26	−1.36	−1.51	−1.62
4	−0.38	−0.92	−1.06	−1.18	−1.34	−1.46
5	−0.12	−0.74	−0.90	−1.02	−1.20	−1.33
6	0.12	−0.59	−0.76	−0.89	−1.08	−1.22
7	0.38	−0.45	−0.64	−0.78	−0.98	−1.12
8	0.66	−0.31	−0.52	−0.67	−0.88	−1.03
9	1.00	−0.19	−0.41	−0.57	−0.79	−0.95
10	1.55	−0.06	−0.30	−0.47	−0.71	−0.87
11		0.06	−0.20	−0.38	−0.63	−0.80
12		0.19	−0.10	−0.29	−0.56	−0.73
13		0.31	0.00	−0.21	−0.49	−0.67
14		0.45	0.10	−0.12	−0.42	−0.61
15		0.59	0.20	−0.04	−0.35	−0.55
16		0.74	0.30	0.04	−0.28	−0.49
17		0.92	0.41	0.12	−0.22	−0.44
18		1.13	0.52	0.21	−0.16	−0.38
19		1.40	0.64	0.29	−0.09	−0.33
20		1.87	0.76	0.38	−0.03	−0.28
21			0.90	0.47	0.03	−0.23
22			1.06	0.57	0.09	−0.18
23			1.26	0.67	0.16	−0.13
24			1.52	0.78	0.22	−0.07
25			1.96	0.89	0.28	−0.02
26				1.02	0.35	0.02
27				1.18	0.42	0.07
28				1.36	0.49	0.13
29				1.61	0.56	0.18
30				2.04	0.63	0.23
31					0.71	0.28
32					0.79	0.33
33					0.88	0.38
34					0.98	0.44
35					1.08	0.49
36					1.20	0.55
37					1.34	0.61
38					1.51	0.67
39					1.75	0.73
40					2.16	0.80
41						0.87
42						0.95
43						1.03
44						1.12
45						1.22
46						1.33
47						1.46
48						1.62
49						1.85
50						2.24

Table 5 Critical Values for the t Distribution

This table contains critical values associated with the t distribution, $t_{\alpha,\nu}$, defined by α and the degrees of freedom, ν.

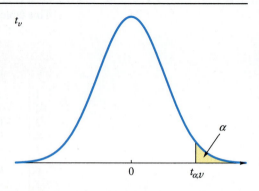

ν	0.20	0.10	0.05	0.025	0.01	0.005	0.001	0.0005	0.0001
1	1.3764	3.0777	6.3138	12.7062	31.8205	63.6567	318.3088	636.6192	3183.0988
2	1.0607	1.8856	2.9200	4.3027	6.9646	9.9248	22.3271	31.5991	70.7001
3	0.9785	1.6377	2.3534	3.1824	4.5407	5.8409	10.2145	12.9240	22.2037
4	0.9410	1.5332	2.1318	2.7764	3.7469	4.6041	7.1732	8.6103	13.0337
5	0.9195	1.4759	2.0150	2.5706	3.3649	4.0321	5.8934	6.8688	9.6776
6	0.9057	1.4398	1.9432	2.4469	3.1427	3.7074	5.2076	5.9588	8.0248
7	0.8960	1.4149	1.8946	2.3646	2.9980	3.4995	4.7853	5.4079	7.0634
8	0.8889	1.3968	1.8595	2.3060	2.8965	3.3554	4.5008	5.0413	6.4420
9	0.8834	1.3830	1.8331	2.2622	2.8214	3.2498	4.2968	4.7809	6.0101
10	0.8791	1.3722	1.8125	2.2281	2.7638	3.1693	4.1437	4.5869	5.6938
11	0.8755	1.3634	1.7959	2.2010	2.7181	3.1058	4.0247	4.4370	5.4528
12	0.8726	1.3562	1.7823	2.1788	2.6810	3.0545	3.9296	4.3178	5.2633
13	0.8702	1.3502	1.7709	2.1604	2.6503	3.0123	3.8520	4.2208	5.1106
14	0.8681	1.3450	1.7613	2.1448	2.6245	2.9768	3.7874	4.1405	4.9850
15	0.8662	1.3406	1.7531	2.1314	2.6025	2.9467	3.7328	4.0728	4.8800
16	0.8647	1.3368	1.7459	2.1199	2.5835	2.9208	3.6862	4.0150	4.7909
17	0.8633	1.3334	1.7396	2.1098	2.5669	2.8982	3.6458	3.9651	4.7144
18	0.8620	1.3304	1.7341	2.1009	2.5524	2.8784	3.6105	3.9216	4.6480
19	0.8610	1.3277	1.7291	2.0930	2.5395	2.8609	3.5794	3.8834	4.5899
20	0.8600	1.3253	1.7247	2.0860	2.5280	2.8453	3.5518	3.8495	4.5385
21	0.8591	1.3232	1.7207	2.0796	2.5176	2.8314	3.5272	3.8193	4.4929
22	0.8583	1.3212	1.7171	2.0739	2.5083	2.8188	3.5050	3.7921	4.4520
23	0.8575	1.3195	1.7139	2.0687	2.4999	2.8073	3.4850	3.7676	4.4152
24	0.8569	1.3178	1.7109	2.0639	2.4922	2.7969	3.4668	3.7454	4.3819
25	0.8562	1.3163	1.7081	2.0595	2.4851	2.7874	3.4502	3.7251	4.3517
26	0.8557	1.3150	1.7056	2.0555	2.4786	2.7787	3.4350	3.7066	4.3240
27	0.8551	1.3137	1.7033	2.0518	2.4727	2.7707	3.4210	3.6896	4.2987
28	0.8546	1.3125	1.7011	2.0484	2.4671	2.7633	3.4082	3.6739	4.2754
29	0.8542	1.3114	1.6991	2.0452	2.4620	2.7564	3.3962	3.6594	4.2539
30	0.8538	1.3104	1.6973	2.0423	2.4573	2.7500	3.3852	3.6460	4.2340
40	0.8507	1.3031	1.6839	2.0211	2.4233	2.7045	3.3069	3.5510	4.0942
50	0.8489	1.2987	1.6759	2.0086	2.4033	2.6778	3.2614	3.4960	4.0140
60	0.8477	1.2958	1.6706	2.0003	2.3901	2.6603	3.2317	3.4602	3.9621
70	0.8468	1.2938	1.6669	1.9944	2.3808	2.6479	3.2108	3.4350	3.9257
80	0.8461	1.2922	1.6641	1.9901	2.3739	2.6387	3.1953	3.4163	3.8988
90	0.8456	1.2910	1.6620	1.9867	2.3685	2.6316	3.1833	3.4019	3.8780
100	0.8452	1.2901	1.6602	1.9840	2.3642	2.6259	3.1737	3.3905	3.8616
200	0.8434	1.2858	1.6525	1.9719	2.3451	2.6006	3.1315	3.3398	3.7891
500	0.8423	1.2832	1.6479	1.9647	2.3338	2.5857	3.1066	3.3101	3.7468
∞	0.8416	1.2816	1.6449	1.9600	2.3263	2.5758	3.0902	3.2905	3.7190

Table 6 Critical Values for the Chi-Square Distribution

This table contains critical values associated with the chi-square distribution, $\chi^2_{\alpha,\nu}$, defined by α and the degrees of freedom, ν.

ν	α							
	0.9999	0.9995	0.999	0.995	0.99	0.975	0.95	0.90
1	0.0^{7}157	0.0^{6}393	0.0^{5}157	0.0^{4}393	0.0002	0.0010	0.0039	0.0158
2	0.0002	0.0010	0.0020	0.0100	0.0201	0.0506	0.1026	0.2107
3	0.0052	0.0153	0.0243	0.0717	0.1148	0.2158	0.3518	0.5844
4	0.0284	0.0639	0.0908	0.2070	0.2971	0.4844	0.7107	1.0636
5	0.0822	0.1581	0.2102	0.4117	0.5543	0.8312	1.1455	1.6103
6	0.1724	0.2994	0.3811	0.6757	0.8721	1.2373	1.6354	2.2041
7	0.3000	0.4849	0.5985	0.9893	1.2390	1.6899	2.1673	2.8331
8	0.4636	0.7104	0.8571	1.3444	1.6465	2.1797	2.7326	3.4895
9	0.6608	0.9717	1.1519	1.7349	2.0879	2.7004	3.3251	4.1682
10	0.8889	1.2650	1.4787	2.1559	2.5582	3.2470	3.9403	4.8652
11	1.1453	1.5868	1.8339	2.6032	3.0535	3.8157	4.5748	5.5778
12	1.4275	1.9344	2.2142	3.0738	3.5706	4.4038	5.2260	6.3038
13	1.7333	2.3051	2.6172	3.5650	4.1069	5.0088	5.8919	7.0415
14	2.0608	2.6967	3.0407	4.0747	4.6604	5.6287	6.5706	7.7895
15	2.4082	3.1075	3.4827	4.6009	5.2293	6.2621	7.2609	8.5468
16	2.7739	3.5358	3.9416	5.1422	5.8122	6.9077	7.9616	9.3122
17	3.1567	3.9802	4.4161	5.6972	6.4078	7.5642	8.6718	10.0852
18	3.5552	4.4394	4.9048	6.2648	7.0149	8.2307	9.3905	10.8649
19	3.9683	4.9123	5.4068	6.8440	7.6327	8.9065	10.1170	11.6509
20	4.3952	5.3981	5.9210	7.4338	8.2604	9.5908	10.8508	12.4426
21	4.8348	5.8957	6.4467	8.0337	8.8972	10.2829	11.5913	13.2396
22	5.2865	6.4045	6.9830	8.6427	9.5425	10.9823	12.3380	14.0415
23	5.7494	6.9237	7.5292	9.2604	10.1957	11.6886	13.0905	14.8480
24	6.2230	7.4527	8.0849	9.8862	10.8564	12.4012	13.8484	15.6587
25	6.7066	7.9910	8.6493	10.5197	11.5240	13.1197	14.6114	16.4734
26	7.1998	8.5379	9.2221	11.1602	12.1981	13.8439	15.3792	17.2919
27	7.7019	9.0932	9.8028	11.8076	12.8785	14.5734	16.1514	18.1139
28	8.2126	9.6563	10.3909	12.4613	13.5647	15.3079	16.9279	18.9392
29	8.7315	10.2268	10.9861	13.1211	14.2565	16.0471	17.7084	19.7677
30	9.2581	10.8044	11.5880	13.7867	14.9535	16.7908	18.4927	20.5992
31	9.7921	11.3887	12.1963	14.4578	15.6555	17.5387	19.2806	21.4336
32	10.3331	11.9794	12.8107	15.1340	16.3622	18.2908	20.0719	22.2706
33	10.8810	12.5763	13.4309	15.8153	17.0735	19.0467	20.8665	23.1102
34	11.4352	13.1791	14.0567	16.5013	17.7891	19.8063	21.6643	23.9523
35	11.9957	13.7875	14.6878	17.1918	18.5089	20.5694	22.4650	24.7967
36	12.5622	14.4012	15.3241	17.8867	19.2327	21.3359	23.2686	25.6433
37	13.1343	15.0202	15.9653	18.5858	19.9602	22.1056	24.0749	26.4921
38	13.7120	15.6441	16.6112	19.2889	20.6914	22.8785	24.8839	27.3430
39	14.2950	16.2729	17.2616	19.9959	21.4262	23.6543	25.6954	28.1958
40	14.8831	16.9062	17.9164	20.7065	22.1643	24.4330	26.5093	29.0505
50	21.0093	23.4610	24.6739	27.9907	29.7067	32.3574	34.7643	37.6886
60	27.4969	30.3405	31.7383	35.5345	37.4849	40.4817	43.1880	46.4589
70	34.2607	37.4674	39.0364	43.2752	45.4417	48.7576	51.7393	55.3289
80	41.2445	44.7910	46.5199	51.1719	53.5401	57.1532	60.3915	64.2778
90	48.4087	52.2758	54.1552	59.1963	61.7541	65.6466	69.1260	73.2911
100	55.7246	59.8957	61.9179	67.3276	70.0649	74.2219	77.9295	82.3581

Table 6 Critical Values for the Chi-Square Distribution (Continued)

ν	α 0.10	0.05	0.025	0.01	0.005	0.001	0.0005	0.0001
1	2.7055	3.8415	5.0239	6.6349	7.8794	10.8276	12.1157	15.1367
2	4.6052	5.9915	7.3778	9.2103	10.5966	13.8155	15.2018	18.4207
3	6.2514	7.8147	9.3484	11.3449	12.8382	16.2662	17.7300	21.1075
4	7.7794	9.4877	11.1433	13.2767	14.8603	18.4668	19.9974	23.5127
5	9.2364	11.0705	12.8325	15.0863	16.7496	20.5150	22.1053	25.7448
6	10.6446	12.5916	14.4494	16.8119	18.5476	22.4577	24.1028	27.8563
7	12.0170	14.0671	16.0128	18.4753	20.2777	24.3219	26.0178	29.8775
8	13.3616	15.5073	17.5345	20.0902	21.9550	26.1245	27.8680	31.8276
9	14.6837	16.9190	19.0228	21.6660	23.5894	27.8772	29.6658	33.7199
10	15.9872	18.3070	20.4832	23.2093	25.1882	29.5883	31.4198	35.5640
11	17.2750	19.6751	21.9200	24.7250	26.7568	31.2641	33.1366	37.3670
12	18.5493	21.0261	23.3367	26.2170	28.2995	32.9095	34.8213	39.1344
13	19.8119	22.3620	24.7356	27.6882	29.8195	34.5282	36.4778	40.8707
14	21.0641	23.6848	26.1189	29.1412	31.3193	36.1233	38.1094	42.5793
15	22.3071	24.9958	27.4884	30.5779	32.8013	37.6973	39.7188	44.2632
16	23.5418	26.2962	28.8454	31.9999	34.2672	39.2524	41.3081	45.9249
17	24.7690	27.5871	30.1910	33.4087	35.7185	40.7902	42.8792	47.5664
18	25.9894	28.8693	31.5264	34.8053	37.1565	42.3124	44.4338	49.1894
19	27.2036	30.1435	32.8523	36.1909	38.5823	43.8202	45.9731	50.7955
20	28.4120	31.4104	34.1696	37.5662	39.9968	45.3147	47.4985	52.3860
21	29.6151	32.6706	35.4789	38.9322	41.4011	46.7970	49.0108	53.9620
22	30.8133	33.9244	36.7807	40.2894	42.7957	48.2679	50.5111	55.5246
23	32.0069	35.1725	38.0756	41.6384	44.1813	49.7282	52.0002	57.0746
24	33.1962	36.4150	39.3641	42.9798	45.5585	51.1786	53.4788	58.6130
25	34.3816	37.6525	40.6465	44.3141	46.9279	52.6197	54.9475	60.1403
26	35.5632	38.8851	41.9232	45.6417	48.2899	54.0520	56.4069	61.6573
27	36.7412	40.1133	43.1945	46.9629	49.6449	55.4760	57.8576	63.1645
28	37.9159	41.3371	44.4608	48.2782	50.9934	56.8923	59.3000	64.6624
29	39.0875	42.5570	45.7223	49.5879	52.3356	58.3012	60.7346	66.1517
30	40.2560	43.7730	46.9792	50.8922	53.6720	59.7031	62.1619	67.6326
31	41.4217	44.9853	48.2319	52.1914	55.0027	61.0983	63.5820	69.1057
32	42.5847	46.1943	49.4804	53.4858	56.3281	62.4872	64.9955	70.5712
33	43.7452	47.3999	50.7251	54.7755	57.6484	63.8701	66.4025	72.0296
34	44.9032	48.6024	51.9660	56.0609	58.9639	65.2472	67.8035	73.4812
35	46.0588	49.8018	53.2033	57.3421	60.2748	66.6188	69.1986	74.9262
36	47.2122	50.9985	54.4373	58.6192	61.5812	67.9852	70.5881	76.3650
37	48.3634	52.1923	55.6680	59.8925	62.8833	69.3465	71.9722	77.7977
38	49.5126	53.3835	56.8955	61.1621	64.1814	70.7029	73.3512	79.2247
39	50.6598	54.5722	58.1201	62.4281	65.4756	72.0547	74.7253	80.6462
40	51.8051	55.7585	59.3417	63.6907	66.7660	73.4020	76.0946	82.0623
50	63.1671	67.5048	71.4202	76.1539	79.4900	86.6608	89.5605	95.9687
60	74.3970	79.0819	83.2977	88.3794	91.9517	99.6072	102.6948	109.5029
70	85.5270	90.5312	95.0232	100.4252	104.2149	112.3169	115.5776	122.7547
80	96.5782	101.8795	106.6286	112.3288	116.3211	124.8392	128.2613	135.7825
90	107.5650	113.1453	118.1359	124.1163	128.2989	137.2084	140.7823	148.6273
100	118.4980	124.3421	129.5612	135.8067	140.1695	149.4493	153.1670	161.3187

Table 7 Critical Values for the F Distribution

This table contains critical values associated with the F distribution, F_{α,ν_1,ν_2}, defined by α and the degrees of freedom ν_1 and ν_2.

$\alpha = 0.05$

ν_2 \ ν_1	1	2	3	4	5	6	7	8	9	10	15	20	30	40	50	60	100
1	161.45	199.50	215.71	224.58	230.16	233.99	236.77	238.88	240.54	241.88	245.95	248.01	250.10	251.14	251.77	252.20	253.04
2	18.51	19.00	19.16	19.25	19.30	19.33	19.35	19.37	19.38	19.40	19.43	19.45	19.46	19.47	19.48	19.48	19.49
3	10.13	9.55	9.28	9.12	9.01	8.94	8.89	8.85	8.81	8.79	8.70	8.66	8.62	8.59	8.58	8.57	8.55
4	7.71	6.94	6.59	6.39	6.26	6.16	6.09	6.04	6.00	5.96	5.86	5.80	5.75	5.72	5.70	5.69	5.66
5	6.61	5.79	5.41	5.19	5.05	4.95	4.88	4.82	4.77	4.74	4.62	4.56	4.50	4.46	4.44	4.43	4.41
6	5.99	5.14	4.76	4.53	4.39	4.28	4.21	4.15	4.10	4.06	3.94	3.87	3.81	3.77	3.75	3.74	3.71
7	5.59	4.74	4.35	4.12	3.97	3.87	3.79	3.73	3.68	3.64	3.51	3.44	3.38	3.34	3.32	3.30	3.27
8	5.32	4.46	4.07	3.84	3.69	3.58	3.50	3.44	3.39	3.35	3.22	3.15	3.08	3.04	3.02	3.01	2.97
9	5.12	4.26	3.86	3.63	3.48	3.37	3.29	3.23	3.18	3.14	3.01	2.94	2.86	2.83	2.80	2.79	2.76
10	4.96	4.10	3.71	3.48	3.33	3.22	3.14	3.07	3.02	2.98	2.85	2.77	2.70	2.66	2.64	2.62	2.59
11	4.84	3.98	3.59	3.36	3.20	3.09	3.01	2.95	2.90	2.85	2.72	2.65	2.57	2.53	2.51	2.49	2.46
12	4.75	3.89	3.49	3.26	3.11	3.00	2.91	2.85	2.80	2.75	2.62	2.54	2.47	2.43	2.40	2.38	2.35
13	4.67	3.81	3.41	3.18	3.03	2.92	2.83	2.77	2.71	2.67	2.53	2.46	2.38	2.34	2.31	2.30	2.26
14	4.60	3.74	3.34	3.11	2.96	2.85	2.76	2.70	2.65	2.60	2.46	2.39	2.31	2.27	2.24	2.22	2.19
15	4.54	3.68	3.29	3.06	2.90	2.79	2.71	2.64	2.59	2.54	2.40	2.33	2.25	2.20	2.18	2.16	2.12
16	4.49	3.63	3.24	3.01	2.85	2.74	2.66	2.59	2.54	2.49	2.35	2.28	2.19	2.15	2.12	2.11	2.07
17	4.45	3.59	3.20	2.96	2.81	2.70	2.61	2.55	2.49	2.45	2.31	2.23	2.15	2.10	2.08	2.06	2.02
18	4.41	3.55	3.16	2.93	2.77	2.66	2.58	2.51	2.46	2.41	2.27	2.19	2.11	2.06	2.04	2.02	1.98
19	4.38	3.52	3.13	2.90	2.74	2.63	2.54	2.48	2.42	2.38	2.23	2.16	2.07	2.03	2.00	1.98	1.94
20	4.35	3.49	3.10	2.87	2.71	2.60	2.51	2.45	2.39	2.35	2.20	2.12	2.04	1.99	1.97	1.95	1.91
21	4.32	3.47	3.07	2.84	2.68	2.57	2.49	2.42	2.37	2.32	2.18	2.10	2.01	1.96	1.94	1.92	1.88
22	4.30	3.44	3.05	2.82	2.66	2.55	2.46	2.40	2.34	2.30	2.15	2.07	1.98	1.94	1.91	1.89	1.85
23	4.28	3.42	3.03	2.80	2.64	2.53	2.44	2.37	2.32	2.27	2.13	2.05	1.96	1.91	1.88	1.86	1.82
24	4.26	3.40	3.01	2.78	2.62	2.51	2.42	2.36	2.30	2.25	2.11	2.03	1.94	1.89	1.86	1.84	1.80
25	4.24	3.39	2.99	2.76	2.60	2.49	2.40	2.34	2.28	2.24	2.09	2.01	1.92	1.87	1.84	1.82	1.78
30	4.17	3.32	2.92	2.69	2.53	2.42	2.33	2.27	2.21	2.16	2.01	1.93	1.84	1.79	1.76	1.74	1.70
40	4.08	3.23	2.84	2.61	2.45	2.34	2.25	2.18	2.12	2.08	1.92	1.84	1.74	1.69	1.66	1.64	1.59
50	4.03	3.18	2.79	2.56	2.40	2.29	2.20	2.13	2.07	2.03	1.87	1.78	1.69	1.63	1.60	1.58	1.52
60	4.00	3.15	2.76	2.53	2.37	2.25	2.17	2.10	2.04	1.99	1.84	1.75	1.65	1.59	1.56	1.53	1.48
100	3.94	3.09	2.70	2.46	2.31	2.19	2.10	2.03	1.97	1.93	1.77	1.68	1.57	1.52	1.48	1.45	1.39

Table 7 Critical Values for the F Distribution (Continued)

$\alpha = 0.01$

ν_2	ν_1 = 1	2	3	4	5	6	7	8	9	10	15	20	30	40	50	60	100
2	98.50	99.00	99.17	99.25	99.30	99.33	99.36	99.37	99.39	99.40	99.43	99.45	99.47	99.47	99.48	99.48	99.49
3	34.12	30.82	29.46	28.71	28.24	27.91	27.67	27.49	27.35	27.23	26.87	26.69	26.50	26.41	26.35	26.32	26.24
4	21.20	18.00	16.69	15.98	15.52	15.21	14.98	14.80	14.66	14.55	14.20	14.02	13.84	13.75	13.69	13.65	13.58
5	16.26	13.27	12.06	11.39	10.97	10.67	10.46	10.29	10.16	10.05	9.72	9.55	9.38	9.29	9.24	9.20	9.13
6	13.75	10.92	9.78	9.15	8.75	8.47	8.26	8.10	7.98	7.87	7.56	7.40	7.23	7.14	7.09	7.06	6.99
7	12.25	9.55	8.45	7.85	7.46	7.19	6.99	6.84	6.72	6.62	6.31	6.16	5.99	5.91	5.86	5.82	5.75
8	11.26	8.65	7.59	7.01	6.63	6.37	6.18	6.03	5.91	5.81	5.52	5.36	5.20	5.12	5.07	5.03	4.96
9	10.56	8.02	6.99	6.42	6.06	5.80	5.61	5.47	5.35	5.26	4.96	4.81	4.65	4.57	4.52	4.48	4.41
10	10.04	7.56	6.55	5.99	5.64	5.39	5.20	5.06	4.94	4.85	4.56	4.41	4.25	4.17	4.12	4.08	4.01
11	9.65	7.21	6.22	5.67	5.32	5.07	4.89	4.74	4.63	4.54	4.25	4.10	3.94	3.86	3.81	3.78	3.71
12	9.33	6.93	5.95	5.41	5.06	4.82	4.64	4.50	4.39	4.30	4.01	3.86	3.70	3.62	3.57	3.54	3.47
13	9.07	6.70	5.74	5.21	4.86	4.62	4.44	4.30	4.19	4.10	3.82	3.66	3.51	3.43	3.38	3.34	3.27
14	8.86	6.51	5.56	5.04	4.69	4.46	4.28	4.14	4.03	3.94	3.66	3.51	3.35	3.27	3.22	3.18	3.11
15	8.68	6.36	5.42	4.89	4.56	4.32	4.14	4.00	3.89	3.80	3.52	3.37	3.21	3.13	3.08	3.05	2.98
16	8.53	6.23	5.29	4.77	4.44	4.20	4.03	3.89	3.78	3.69	3.41	3.26	3.10	3.02	2.97	2.93	2.86
17	8.40	6.11	5.18	4.67	4.34	4.10	3.93	3.79	3.68	3.59	3.31	3.16	3.00	2.92	2.87	2.83	2.76
18	8.29	6.01	5.09	4.58	4.25	4.01	3.84	3.71	3.60	3.51	3.23	3.08	2.92	2.84	2.78	2.75	2.68
19	8.18	5.93	5.01	4.50	4.17	3.94	3.77	3.63	3.52	3.43	3.15	3.00	2.84	2.76	2.71	2.67	2.60
20	8.10	5.85	4.94	4.43	4.10	3.87	3.70	3.56	3.46	3.37	3.09	2.94	2.78	2.69	2.64	2.61	2.54
21	8.02	5.78	4.87	4.37	4.04	3.81	3.64	3.51	3.40	3.31	3.03	2.88	2.72	2.64	2.58	2.55	2.48
22	7.95	5.72	4.82	4.31	3.99	3.76	3.59	3.45	3.35	3.26	2.98	2.83	2.67	2.58	2.53	2.50	2.42
23	7.88	5.66	4.76	4.26	3.94	3.71	3.54	3.41	3.30	3.21	2.93	2.78	2.62	2.54	2.48	2.45	2.37
24	7.82	5.61	4.72	4.22	3.90	3.67	3.50	3.36	3.26	3.17	2.89	2.74	2.58	2.49	2.44	2.40	2.33
25	7.77	5.57	4.68	4.18	3.85	3.63	3.46	3.32	3.22	3.13	2.85	2.70	2.54	2.45	2.40	2.36	2.29
30	7.56	5.39	4.51	4.02	3.70	3.47	3.30	3.17	3.07	2.98	2.70	2.55	2.39	2.30	2.25	2.21	2.13
40	7.31	5.18	4.31	3.83	3.51	3.29	3.12	2.99	2.89	2.80	2.52	2.37	2.20	2.11	2.06	2.02	1.94
50	7.17	5.06	4.20	3.72	3.41	3.19	3.02	2.89	2.78	2.70	2.42	2.27	2.10	2.01	1.95	1.91	1.82
60	7.08	4.98	4.13	3.65	3.34	3.12	2.95	2.82	2.72	2.63	2.35	2.20	2.03	1.94	1.88	1.84	1.75
100	6.90	4.82	3.98	3.51	3.21	2.99	2.82	2.69	2.59	2.50	2.22	2.07	1.89	1.80	1.74	1.69	1.60

Table 7 Critical Values for the F Distribution (Continued)

$\alpha = 0.001$

ν_2	ν_1=1	2	3	4	5	6	7	8	9	10	15	20	30	40	50	60	100
2	998.50	999.00	999.17	999.25	999.30	999.33	999.36	999.37	999.39	999.40	999.43	999.45	999.47	999.47	999.48	999.48	999.49
3	167.03	148.50	141.11	137.10	134.58	132.85	131.58	130.62	129.86	129.25	127.37	126.42	125.45	124.96	124.66	124.47	124.07
4	74.14	61.25	56.18	53.44	51.71	50.53	49.66	49.00	48.47	48.05	46.76	46.10	45.43	45.09	44.88	44.75	44.47
5	47.18	37.12	33.20	31.09	29.75	28.83	28.16	27.65	27.24	26.92	25.91	25.39	24.87	24.60	24.44	24.33	24.12
6	35.51	27.00	23.70	21.92	20.80	20.03	19.46	19.03	18.69	18.41	17.56	17.12	16.67	16.44	16.31	16.21	16.03
7	29.25	21.69	18.77	17.20	16.21	15.52	15.02	14.63	14.33	14.08	13.32	12.93	12.53	12.33	12.20	12.12	11.95
8	25.41	18.49	15.83	14.39	13.48	12.86	12.40	12.05	11.77	11.54	10.84	10.48	10.11	9.92	9.80	9.73	9.57
9	22.86	16.39	13.90	12.56	11.71	11.13	10.70	10.37	10.11	9.89	9.24	8.90	8.55	8.37	8.26	8.19	8.04
10	21.04	14.91	12.55	11.28	10.48	9.93	9.52	9.20	8.96	8.75	8.13	7.80	7.47	7.30	7.19	7.12	6.98
11	19.69	13.81	11.56	10.35	9.58	9.05	8.66	8.35	8.12	7.92	7.32	7.01	6.68	6.52	6.42	6.35	6.21
12	18.64	12.97	10.80	9.63	8.89	8.38	8.00	7.71	7.48	7.29	6.71	6.40	6.09	5.93	5.83	5.76	5.63
13	17.82	12.31	10.21	9.07	8.35	7.86	7.49	7.21	6.98	6.80	6.23	5.93	5.63	5.47	5.37	5.30	5.17
14	17.14	11.78	9.73	8.62	7.92	7.44	7.08	6.80	6.58	6.40	5.85	5.56	5.25	5.10	5.00	4.94	4.81
15	16.59	11.34	9.34	8.25	7.57	7.09	6.74	6.47	6.26	6.08	5.54	5.25	4.95	4.80	4.70	4.64	4.51
16	16.12	10.97	9.01	7.94	7.27	6.80	6.46	6.19	5.98	5.81	5.27	4.99	4.70	4.54	4.45	4.39	4.26
17	15.72	10.66	8.73	7.68	7.02	6.56	6.22	5.96	5.75	5.58	5.05	4.78	4.48	4.33	4.24	4.18	4.05
18	15.38	10.39	8.49	7.46	6.81	6.35	6.02	5.76	5.56	5.39	4.87	4.59	4.30	4.15	4.06	4.00	3.87
19	15.08	10.16	8.28	7.27	6.62	6.18	5.85	5.59	5.39	5.22	4.70	4.43	4.14	3.99	3.90	3.84	3.71
20	14.82	9.95	8.10	7.10	6.46	6.02	5.69	5.44	5.24	5.08	4.56	4.29	4.00	3.86	3.77	3.70	3.58
21	14.59	9.77	7.94	6.95	6.32	5.88	5.56	5.31	5.11	4.95	4.44	4.17	3.88	3.74	3.64	3.58	3.46
22	14.38	9.61	7.80	6.81	6.19	5.76	5.44	5.19	4.99	4.83	4.33	4.06	3.78	3.63	3.54	3.48	3.35
23	14.20	9.47	7.67	6.70	6.08	5.65	5.33	5.09	4.89	4.73	4.23	3.96	3.68	3.53	3.44	3.38	3.25
24	14.03	9.34	7.55	6.59	5.98	5.55	5.23	4.99	4.80	4.64	4.14	3.87	3.59	3.45	3.36	3.29	3.17
25	13.88	9.22	7.45	6.49	5.89	5.46	5.15	4.91	4.71	4.56	4.06	3.79	3.52	3.37	3.28	3.22	3.09
30	13.29	8.77	7.05	6.12	5.53	5.12	4.82	4.58	4.39	4.24	3.75	3.49	3.22	3.07	2.98	2.92	2.79
40	12.61	8.25	6.59	5.70	5.13	4.73	4.44	4.21	4.02	3.87	3.40	3.14	2.87	2.73	2.64	2.57	2.44
50	12.22	7.96	6.34	5.46	4.90	4.51	4.22	4.00	3.82	3.67	3.20	2.95	2.68	2.53	2.44	2.38	2.25
60	11.97	7.77	6.17	5.31	4.76	4.37	4.09	3.86	3.69	3.54	3.08	2.83	2.55	2.41	2.32	2.25	2.12
100	11.50	7.41	5.86	5.02	4.48	4.11	3.83	3.61	3.44	3.30	2.84	2.59	2.32	2.17	2.08	2.01	1.87

Table 8 Critical Values for the Studentized Range Distribution

This table contains critical values associated with the Studentized range distribution, $Q_{\alpha, k, \nu}$, defined by α, and the degrees of freedom k and ν, where k is the number of degrees of freedom in the numerator (the number of treatment groups) and ν is the number of degrees of freedom in the denominator.

$\alpha = 0.05$

ν	2	3	4	5	6	7	8	9	10	11	12	13	14	15	16	17	18	19	20
2	6.085	8.331	9.798	10.881	11.734	12.434	13.027	13.538	13.987	14.387	14.747	15.076	15.375	15.650	15.905	16.143	16.365	16.573	16.769
3	4.501	5.910	6.825	7.502	8.037	8.478	8.852	9.177	9.462	9.717	9.946	10.155	10.346	10.522	10.686	10.838	10.980	11.114	11.240
4	3.926	5.040	5.757	6.287	6.706	7.053	7.347	7.602	7.826	8.027	8.208	8.373	8.524	8.664	8.793	8.914	9.027	9.133	9.233
5	3.635	4.602	5.218	5.673	6.033	6.330	6.582	6.801	6.995	7.167	7.324	7.465	7.596	7.716	7.828	7.932	8.030	8.122	8.208
6	3.460	4.339	4.896	5.305	5.629	5.895	6.122	6.319	6.493	6.649	6.789	6.917	7.034	7.143	7.244	7.338	7.426	7.509	7.587
7	3.344	4.165	4.681	5.060	5.359	5.606	5.815	5.997	6.158	6.302	6.431	6.550	6.658	6.759	6.852	6.939	7.020	7.097	7.169
8	3.261	4.041	4.529	4.886	5.167	5.399	5.596	5.767	5.918	6.053	6.175	6.287	6.389	6.483	6.571	6.653	6.729	6.801	6.870
9	3.199	3.948	4.415	4.755	5.024	5.244	5.432	5.595	5.738	5.867	5.983	6.089	6.186	6.276	6.359	6.437	6.510	6.579	6.644
10	3.151	3.877	4.327	4.654	4.912	5.124	5.304	5.460	5.598	5.722	5.833	5.935	6.028	6.114	6.194	6.269	6.339	6.405	6.467
11	3.113	3.820	4.256	4.574	4.823	5.028	5.202	5.353	5.486	5.605	5.713	5.811	5.901	5.984	6.062	6.134	6.202	6.265	6.325
12	3.081	3.773	4.199	4.508	4.750	4.950	5.119	5.265	5.395	5.510	5.615	5.710	5.797	5.878	5.953	6.023	6.089	6.151	6.209
13	3.055	3.734	4.151	4.453	4.690	4.884	5.049	5.192	5.318	5.431	5.533	5.625	5.711	5.789	5.862	5.931	5.995	6.055	6.112
14	3.033	3.701	4.111	4.407	4.639	4.829	4.990	5.130	5.253	5.363	5.463	5.554	5.637	5.714	5.785	5.852	5.915	5.973	6.029
15	3.014	3.673	4.076	4.367	4.595	4.782	4.940	5.077	5.198	5.306	5.403	5.492	5.574	5.649	5.719	5.785	5.846	5.904	5.958
16	2.998	3.649	4.046	4.333	4.557	4.741	4.896	5.031	5.150	5.256	5.352	5.439	5.519	5.593	5.662	5.726	5.786	5.843	5.896
17	2.984	3.628	4.020	4.303	4.524	4.705	4.858	4.991	5.108	5.212	5.306	5.392	5.471	5.544	5.612	5.675	5.734	5.790	5.842
18	2.971	3.609	3.997	4.276	4.494	4.673	4.824	4.955	5.071	5.173	5.266	5.351	5.429	5.501	5.567	5.629	5.688	5.743	5.794
19	2.960	3.593	3.977	4.253	4.468	4.645	4.794	4.924	5.037	5.139	5.231	5.314	5.391	5.462	5.528	5.589	5.647	5.701	5.752
20	2.950	3.578	3.958	4.232	4.445	4.620	4.768	4.895	5.008	5.108	5.199	5.282	5.357	5.427	5.492	5.553	5.610	5.663	5.714
25	2.913	3.523	3.890	4.153	4.358	4.526	4.667	4.789	4.897	4.993	5.079	5.158	5.230	5.297	5.359	5.417	5.471	5.522	5.570
30	2.888	3.487	3.845	4.102	4.301	4.464	4.601	4.720	4.824	4.917	5.001	5.077	5.147	5.211	5.271	5.327	5.379	5.429	5.475
40	2.858	3.442	3.791	4.039	4.232	4.388	4.521	4.634	4.735	4.824	4.904	4.977	5.044	5.106	5.163	5.216	5.266	5.313	5.358
50	2.841	3.416	3.758	4.002	4.190	4.344	4.473	4.584	4.681	4.768	4.847	4.918	4.983	5.043	5.098	5.150	5.199	5.245	5.288
100	2.806	3.365	3.695	3.929	4.109	4.256	4.379	4.484	4.577	4.659	4.733	4.800	4.862	4.918	4.971	5.020	5.066	5.108	5.149
200	2.789	3.339	3.664	3.893	4.069	4.212	4.332	4.435	4.525	4.605	4.677	4.742	4.802	4.857	4.908	4.955	4.999	5.041	5.080
300	2.783	3.331	3.654	3.881	4.056	4.198	4.317	4.419	4.508	4.587	4.659	4.723	4.782	4.837	4.887	4.934	4.978	5.019	5.057
400	2.780	3.327	3.649	3.875	4.050	4.191	4.309	4.411	4.500	4.578	4.649	4.714	4.772	4.826	4.876	4.923	4.967	5.007	5.046
500	2.779	3.324	3.645	3.872	4.046	4.187	4.305	4.406	4.494	4.573	4.644	4.708	4.766	4.820	4.870	4.917	4.960	5.001	5.039

Table 8 Critical Values for the Studentized Range Distribution (Continued)

$\alpha = 0.01$

ν	2	3	4	5	6	7	8	9	10	11	12	13	14	15	16	17	18	19	20
2	14.035	19.019	22.293	24.717	26.628	28.199	29.528	30.677	31.687	32.585	33.395	34.129	34.802	35.421	35.995	36.529	37.028	37.496	37.937
3	8.260	10.616	12.169	13.324	14.240	14.997	15.640	16.198	16.689	17.128	17.524	17.884	18.214	18.519	18.802	19.065	19.311	19.543	19.761
4	6.511	8.118	9.173	9.958	10.582	11.099	11.539	11.925	12.264	12.566	12.840	13.089	13.318	13.530	13.726	13.909	14.081	14.242	14.394
5	5.702	6.976	7.806	8.421	8.913	9.321	9.669	9.971	10.239	10.479	10.695	10.893	11.075	11.243	11.399	11.544	11.681	11.809	11.930
6	5.243	6.331	7.033	7.556	7.974	8.318	8.611	8.869	9.097	9.300	9.485	9.653	9.808	9.951	10.084	10.208	10.325	10.434	10.538
7	4.948	5.919	6.543	7.006	7.373	7.678	7.940	8.167	8.368	8.548	8.711	8.859	8.996	9.124	9.242	9.353	9.456	9.553	9.645
8	4.745	5.635	6.204	6.625	6.960	7.238	7.475	7.681	7.864	8.028	8.177	8.312	8.437	8.552	8.659	8.760	8.854	8.942	9.026
9	4.595	5.428	5.957	6.347	6.658	6.915	7.134	7.326	7.495	7.647	7.785	7.910	8.026	8.133	8.233	8.326	8.413	8.495	8.573
10	4.482	5.270	5.769	6.136	6.428	6.669	6.875	7.055	7.214	7.356	7.485	7.603	7.712	7.813	7.906	7.994	8.076	8.153	8.226
11	4.392	5.146	5.621	5.970	6.247	6.476	6.671	6.842	6.992	7.127	7.250	7.362	7.465	7.560	7.649	7.732	7.810	7.883	7.952
12	4.320	5.046	5.502	5.836	6.101	6.321	6.507	6.670	6.814	6.943	7.060	7.167	7.265	7.356	7.441	7.520	7.594	7.664	7.731
13	4.261	4.964	5.404	5.727	5.981	6.192	6.372	6.528	6.666	6.791	6.903	7.006	7.100	7.188	7.269	7.345	7.417	7.484	7.548
14	4.210	4.895	5.322	5.634	5.881	6.085	6.258	6.409	6.543	6.664	6.772	6.871	6.962	7.047	7.125	7.199	7.268	7.333	7.394
15	4.167	4.836	5.252	5.556	5.796	5.994	6.162	6.309	6.438	6.555	6.660	6.757	6.845	6.927	7.003	7.074	7.141	7.204	7.264
16	4.131	4.786	5.192	5.488	5.722	5.915	6.079	6.222	6.348	6.461	6.564	6.658	6.743	6.824	6.897	6.967	7.032	7.093	7.151
17	4.099	4.742	5.140	5.430	5.659	5.847	6.007	6.147	6.270	6.380	6.480	6.572	6.656	6.733	6.806	6.873	6.937	6.997	7.053
18	4.071	4.703	5.094	5.379	5.603	5.787	5.944	6.081	6.201	6.309	6.407	6.496	6.579	6.655	6.725	6.791	6.854	6.912	6.967
19	4.046	4.669	5.054	5.333	5.553	5.735	5.888	6.022	6.141	6.246	6.342	6.430	6.510	6.585	6.654	6.719	6.780	6.837	6.891
20	4.024	4.639	5.018	5.293	5.509	5.687	5.839	5.970	6.086	6.190	6.285	6.370	6.449	6.523	6.591	6.654	6.714	6.770	6.823
25	3.942	4.527	4.884	5.143	5.346	5.513	5.654	5.777	5.885	5.982	6.070	6.150	6.223	6.291	6.355	6.414	6.469	6.521	6.571
30	3.889	4.454	4.799	5.048	5.242	5.401	5.536	5.653	5.756	5.848	5.932	6.008	6.078	6.142	6.202	6.258	6.311	6.360	6.407
40	3.825	4.367	4.695	4.931	5.114	5.265	5.392	5.502	5.599	5.685	5.764	5.835	5.900	5.961	6.017	6.069	6.118	6.165	6.208
50	3.787	4.316	4.634	4.863	5.040	5.185	5.308	5.414	5.507	5.590	5.665	5.734	5.796	5.854	5.908	5.958	6.005	6.050	6.092
100	3.714	4.216	4.516	4.730	4.896	5.031	5.144	5.242	5.328	5.405	5.474	5.537	5.594	5.648	5.697	5.743	5.786	5.826	5.864
200	3.714	4.216	4.516	4.730	4.896	5.031	5.144	5.242	5.328	5.405	5.474	5.537	5.594	5.648	5.697	5.743	5.786	5.826	5.864
300	3.666	4.152	4.440	4.645	4.803	4.931	5.039	5.132	5.213	5.286	5.351	5.410	5.464	5.514	5.560	5.603	5.644	5.682	5.717
400	3.661	4.144	4.431	4.634	4.791	4.919	5.026	5.118	5.199	5.271	5.335	5.394	5.448	5.498	5.543	5.586	5.626	5.664	5.699
500	3.657	4.139	4.425	4.628	4.784	4.911	5.018	5.110	5.190	5.262	5.327	5.385	5.438	5.488	5.533	5.576	5.616	5.653	5.688

Table 8 Critical Values for the Studentized Range Distribution (Continued)

α = 0.001 ν	2	3	4	5	6	7	8	9	10	11	12	13	14	15	16	17	18	19	20
2	44.666	60.323	70.586	78.162	84.127	89.022	93.650	97.285	100.480	103.325	105.886	108.211	110.340	112.300	114.115	115.805	117.385	118.867	120.263
3	18.275	23.298	26.609	29.075	31.030	32.645	34.016	35.327	36.389	37.338	38.194	38.974	39.688	40.347	40.959	41.529	42.062	42.564	43.036
4	12.174	14.965	16.798	18.225	19.333	20.253	21.037	21.719	22.323	22.862	23.350	23.795	24.204	24.581	24.932	25.259	25.566	25.854	26.126
5	9.710	11.671	12.959	13.924	14.695	15.335	15.882	16.358	16.780	17.158	17.500	17.811	18.098	18.402	18.651	18.884	19.102	19.307	19.501
6	8.431	9.955	10.965	11.719	12.322	12.824	13.254	13.629	13.961	14.260	14.530	14.777	15.004	15.215	15.411	15.593	15.765	15.927	16.079
7	7.649	8.933	9.761	10.388	10.883	11.316	11.674	11.988	12.265	12.515	12.742	12.949	13.139	13.316	13.480	13.634	13.778	13.914	14.043
8	7.130	8.252	8.980	9.523	9.948	10.317	10.625	10.894	11.133	11.347	11.559	11.740	11.906	12.060	12.203	12.337	12.463	12.582	12.694
9	7.130	8.252	8.980	9.523	9.948	10.317	10.625	10.894	11.133	11.347	11.559	11.740	11.906	12.060	12.203	12.337	12.463	12.582	12.694
10	6.486	7.411	8.007	8.451	8.805	9.100	9.353	9.574	9.770	9.954	10.106	10.245	10.387	10.512	10.629	10.737	10.840	10.936	11.027
11	6.274	7.137	7.688	8.099	8.427	8.700	8.934	9.138	9.320	9.483	9.631	9.767	9.892	10.017	10.121	10.218	10.309	10.394	10.475
12	6.106	6.917	7.442	7.820	8.128	8.383	8.602	8.793	8.963	9.116	9.254	9.381	9.498	9.607	9.708	9.803	9.892	9.976	10.055
13	5.969	6.740	7.234	7.595	7.885	8.126	8.333	8.513	8.674	8.818	8.949	9.068	9.179	9.281	9.377	9.466	9.550	9.630	9.705
14	5.855	6.593	7.070	7.410	7.692	7.914	8.111	8.282	8.434	8.571	8.696	8.810	8.915	9.012	9.103	9.188	9.268	9.343	9.414
15	5.760	6.470	6.920	7.257	7.517	7.742	7.924	8.088	8.234	8.365	8.483	8.592	8.693	8.786	8.873	8.954	9.030	9.102	9.170
16	5.678	6.365	6.799	7.125	7.377	7.585	7.769	7.923	8.063	8.189	8.303	8.407	8.504	8.593	8.676	8.754	8.828	8.897	8.963
17	5.614	6.274	6.695	7.010	7.254	7.457	7.629	7.783	7.921	8.037	8.147	8.248	8.341	8.427	8.508	8.583	8.654	8.720	8.783
18	5.550	6.201	6.609	6.909	7.147	7.343	7.511	7.658	7.781	7.908	8.017	8.116	8.199	8.283	8.361	8.433	8.502	8.566	8.628
19	5.493	6.129	6.527	6.820	7.051	7.243	7.407	7.550	7.676	7.790	7.894	7.990	8.079	8.162	8.238	8.302	8.369	8.431	8.491
20	5.444	6.065	6.455	6.741	6.967	7.154	7.314	7.453	7.577	7.687	7.788	7.880	7.966	8.046	8.121	8.190	8.256	8.318	8.376
25	5.264	5.840	6.196	6.456	6.662	6.831	6.976	7.102	7.213	7.314	7.404	7.487	7.558	7.629	7.696	7.758	7.816	7.871	7.924
30	5.154	5.698	6.033	6.277	6.469	6.628	6.763	6.880	6.984	7.077	7.161	7.238	7.309	7.375	7.436	7.494	7.548	7.598	7.646
40	5.022	5.527	5.837	6.062	6.239	6.385	6.508	6.615	6.710	6.795	6.872	6.942	7.006	7.066	7.121	7.173	7.222	7.268	7.312
50	4.946	5.426	5.725	5.939	6.107	6.245	6.361	6.463	6.552	6.632	6.705	6.771	6.832	6.888	6.940	6.989	7.035	7.078	7.119
100	4.795	5.244	5.512	5.706	5.855	5.978	6.083	6.173	6.252	6.323	6.387	6.445	6.499	6.548	6.594	6.637	6.678	6.715	6.751
200	4.723	5.151	5.408	5.596	5.738	5.854	5.952	6.038	6.110	6.178	6.237	6.292	6.342	6.388	6.431	6.471	6.509	6.544	6.577
300	4.700	5.122	5.375	5.556	5.696	5.814	5.910	5.993	6.066	6.131	6.189	6.244	6.291	6.335	6.379	6.418	6.455	6.489	6.522
400	4.688	5.107	5.358	5.538	5.677	5.791	5.890	5.972	6.044	6.108	6.166	6.219	6.267	6.312	6.355	6.393	6.427	6.460	6.494
500	4.681	5.098	5.348	5.527	5.665	5.778	5.874	5.959	6.031	6.095	6.152	6.205	6.253	6.297	6.338	6.376	6.412	6.448	6.479

Table 9 Critical Values for the Wilcoxon Signed-Rank Statistic

This table contains critical values and probabilities for the Wilcoxon signed-rank statistic T_+: n is the sample size, and c_1 and c_2 are defined by $P(T_+ \leq c_1) = \alpha$ and $P(T_+ \geq c_2) = \alpha$, respectively.

n	c_1	c_2	α
1	0	1	0.5000
2	0	3	0.2500
3	0	6	0.1250
4	0	10	0.0625
	1	9	0.1250
5	0	15	0.0313
	1	14	0.0625
	2	13	0.0938
	3	12	0.1563
6	0	21	0.0156
	1	20	0.0313
	2	19	0.0469
	3	18	0.0781
	4	17	0.1094
	5	16	0.1563
7	0	28	0.0078
	1	27	0.0156
	2	26	0.0234
	3	25	0.0391
	4	24	0.0547
	5	23	0.0781
	6	22	0.1094
	7	21	0.1484
8	0	36	0.0039
	1	35	0.0078
	2	34	0.0117
	3	33	0.0195
	4	32	0.0273
	5	31	0.0391
	6	30	0.0547
	7	29	0.0742
	8	28	0.0977
	9	27	0.1250
9	0	45	0.0020
	1	44	0.0039
	2	43	0.0059
	3	42	0.0098
	4	41	0.0137
	5	40	0.0195
	6	39	0.0273
	7	38	0.0371
	8	37	0.0488
	9	36	0.0645
	10	35	0.0820
	11	34	0.1016
	12	33	0.1250

n	c_1	c_2	α
10	0	55	0.0010
	1	54	0.0020
	2	53	0.0029
	3	52	0.0049
	4	51	0.0068
	5	50	0.0098
	6	49	0.0137
	7	48	0.0186
	8	47	0.0244
	9	46	0.0322
	10	45	0.0420
	11	44	0.0527
	12	43	0.0654
	13	42	0.0801
	14	41	0.0967
	15	40	0.1162
	16	39	0.1377
11	0	66	0.0005
	1	65	0.0010
	2	64	0.0015
	3	63	0.0024
	4	62	0.0034
	5	61	0.0049
	6	60	0.0068
	7	59	0.0093
	8	58	0.0122
	9	57	0.0161
	10	56	0.0210
	11	55	0.0269
	12	54	0.0337
	13	53	0.0415
	14	52	0.0508
	15	51	0.0615
	16	50	0.0737
	17	49	0.0874
	18	48	0.1030
	19	47	0.1201
	20	46	0.1392

n	c_1	c_2	α
12	0	78	0.0002
	1	77	0.0005
	2	76	0.0007
	3	75	0.0012
	4	74	0.0017
	5	73	0.0024
	6	72	0.0034
	7	71	0.0046
	8	70	0.0061
	9	69	0.0081
	10	68	0.0105
	11	67	0.0134
	12	66	0.0171
	13	65	0.0212
	14	64	0.0261
	15	63	0.0320
	16	62	0.0386
	17	61	0.0461
	18	60	0.0549
	19	59	0.0647
	20	58	0.0757
	21	57	0.0881
	22	56	0.1018
	23	55	0.1167
	24	54	0.1331
	25	53	0.1506

n	c_1	c_2	α
13	0	91	0.0001
	1	90	0.0002
	2	89	0.0004
	3	88	0.0006
	4	87	0.0009
	5	86	0.0012
	6	85	0.0017
	7	84	0.0023
	8	83	0.0031
	9	82	0.0040
	10	81	0.0052
	11	80	0.0067
	12	79	0.0085
	13	78	0.0107
	14	77	0.0133
	15	76	0.0164
	16	75	0.0199
	17	74	0.0239
	18	73	0.0287
	19	72	0.0341
	20	71	0.0402
	21	70	0.0471
	22	69	0.0549
	23	68	0.0636
	24	67	0.0732
	25	66	0.0839
	26	65	0.0955
	27	64	0.1082
	28	63	0.1219
	29	62	0.1367
	30	61	0.1527

n	c_1	c_2	α
14	0	105	0.0001
	1	104	0.0001
	2	103	0.0002
	3	102	0.0003
	4	101	0.0004
	5	100	0.0006
	6	99	0.0009
	7	98	0.0012
	8	97	0.0015
	9	96	0.0020
	10	95	0.0026
	11	94	0.0034
	12	93	0.0043
	13	92	0.0054
	14	91	0.0067
	15	90	0.0083
	16	89	0.0101
	17	88	0.0123
	18	87	0.0148
	19	86	0.0176
	20	85	0.0209
	21	84	0.0247
	22	83	0.0290
	23	82	0.0338
	24	81	0.0392
	25	80	0.0453
	26	79	0.0520
	27	78	0.0594
	28	77	0.0676
	29	76	0.0765
	30	75	0.0863
	31	74	0.0969
	32	73	0.1083
	33	72	0.1206
	34	71	0.1338
	35	70	0.1479
	36	69	0.1629

Table 9 Critical Values for the Wilcoxon Signed-Rank Statistic (Continued)

n	c_1	c_2	α
15	0	120	0.0000
	1	119	0.0001
	2	118	0.0001
	3	117	0.0002
	4	116	0.0002
	5	115	0.0003
	6	114	0.0004
	7	113	0.0006
	8	112	0.0008
	9	111	0.0010
	10	110	0.0013
	11	109	0.0017
	12	108	0.0021
	13	107	0.0027
	14	106	0.0034
	15	105	0.0042
	16	104	0.0051
	17	103	0.0062
	18	102	0.0075
	19	101	0.0090
	20	100	0.0108
	21	99	0.0128
	22	98	0.0151
	23	97	0.0177
	24	96	0.0206
	25	95	0.0240
	26	94	0.0277
	27	93	0.0319
	28	92	0.0365
	29	91	0.0416
	30	90	0.0473
	31	89	0.0535
	32	88	0.0603
	33	87	0.0677
	34	86	0.0757
	35	85	0.0844
	36	84	0.0938
	37	83	0.1039
	38	82	0.1147
	39	81	0.1262
	40	80	0.1384
	41	79	0.1514

n	c_1	c_2	α
16	0	136	0.0000
	1	135	0.0000
	2	134	0.0000
	3	133	0.0001
	4	132	0.0001
	5	131	0.0002
	6	130	0.0002
	7	129	0.0003
	8	128	0.0004
	9	127	0.0005
	10	126	0.0007
	11	125	0.0008
	12	124	0.0011
	13	123	0.0013
	14	122	0.0017
	15	121	0.0021
	16	120	0.0026
	17	119	0.0031
	18	118	0.0038
	19	117	0.0046
	20	116	0.0055
	21	115	0.0065
	22	114	0.0078
	23	113	0.0091
	24	112	0.0107
	25	111	0.0125
	26	110	0.0145
	27	109	0.0168
	28	108	0.0193
	29	107	0.0222
	30	106	0.0253
	31	105	0.0288
	32	104	0.0327
	33	103	0.0370
	34	102	0.0416
	35	101	0.0467
	36	100	0.0523
	37	99	0.0583
	38	98	0.0649
	39	97	0.0719
	40	96	0.0795
	41	95	0.0877
	42	94	0.0964
	43	93	0.1057
	44	92	0.1156
	45	91	0.1261
	46	90	0.1372
	47	89	0.1489

n	c_1	c_2	α
17	0	153	0.0000
	1	152	0.0000
	2	151	0.0000
	3	150	0.0000
	4	149	0.0001
	5	148	0.0001
	6	147	0.0001
	7	146	0.0001
	8	145	0.0002
	9	144	0.0003
	10	143	0.0003
	11	142	0.0004
	12	141	0.0005
	13	140	0.0007
	14	139	0.0008
	15	138	0.0010
	16	137	0.0013
	17	136	0.0016
	18	135	0.0019
	19	134	0.0023
	20	133	0.0028
	21	132	0.0033
	22	131	0.0040
	23	130	0.0047
	24	129	0.0055
	25	128	0.0064
	26	127	0.0075
	27	126	0.0087
	28	125	0.0101
	29	124	0.0116
	30	123	0.0133
	31	122	0.0153
	32	121	0.0174
	33	120	0.0198
	34	119	0.0224
	35	118	0.0253
	36	117	0.0284
	37	116	0.0319
	38	115	0.0357
	39	114	0.0398
	40	113	0.0443
	41	112	0.0492
	42	111	0.0544
	43	110	0.0601
	44	109	0.0662
	45	108	0.0727
	46	107	0.0797
	47	106	0.0871
	48	105	0.0950
	49	104	0.1034
	50	103	0.1123
	51	102	0.1217
	52	101	0.1317
	53	100	0.1421
	54	99	0.1530

n	c_1	c_2	α
18	0	171	0.0000
	1	170	0.0000
	2	169	0.0000
	3	168	0.0000
	4	167	0.0000
	5	166	0.0000
	6	165	0.0001
	7	164	0.0001
	8	163	0.0001
	9	162	0.0001
	10	161	0.0002
	11	160	0.0002
	12	159	0.0003
	13	158	0.0003
	14	157	0.0004
	15	156	0.0005
	16	155	0.0006
	17	154	0.0008
	18	153	0.0010
	19	152	0.0012
	20	151	0.0014
	21	150	0.0017
	22	149	0.0020
	23	148	0.0024
	24	147	0.0028
	25	146	0.0033
	26	145	0.0038
	27	144	0.0045
	28	143	0.0052
	29	142	0.0060
	30	141	0.0069
	31	140	0.0080
	32	139	0.0091
	33	138	0.0104
	34	137	0.0118
	35	136	0.0134
	36	135	0.0152
	37	134	0.0171
	38	133	0.0192
	39	132	0.0216
	40	131	0.0241
	41	130	0.0269
	42	129	0.0300
	43	128	0.0333
	44	127	0.0368
	45	126	0.0407
	46	125	0.0449
	47	124	0.0494
	48	123	0.0542
	49	122	0.0594
	50	121	0.0649
	51	120	0.0708
	52	119	0.0770
	53	118	0.0837
	54	117	0.0907
	55	116	0.0982
	56	115	0.1061
	57	114	0.1144
	58	113	0.1231
	59	112	0.1323
	60	111	0.1419
	61	110	0.1519

Table 9 Critical Values for the Wilcoxon Signed-Rank Statistic (Continued)

n	c_1	c_2	α	n	c_1	c_2	α	n	c_1	c_2	α	n	c_1	c_2	α
19	0	190	0.0000	19	41	149	0.0145	20	0	210	0.0000	20	41	169	0.0077
	1	189	0.0000		42	148	0.0162		1	209	0.0000		42	168	0.0086
	2	188	0.0000		43	147	0.0180		2	208	0.0000		43	167	0.0096
	3	187	0.0000		44	146	0.0201		3	207	0.0000		44	166	0.0107
	4	186	0.0000		45	145	0.0223		4	206	0.0000		45	165	0.0120
	5	185	0.0000		46	144	0.0247		5	205	0.0000		46	164	0.0133
	6	184	0.0000		47	143	0.0273		6	204	0.0000		47	163	0.0148
	7	183	0.0000		48	142	0.0301		7	203	0.0000		48	162	0.0164
	8	182	0.0000		49	141	0.0331		8	202	0.0000		49	161	0.0181
	9	181	0.0001		50	140	0.0364		9	201	0.0000		50	160	0.0200
	10	180	0.0001		51	139	0.0399		10	200	0.0000		51	159	0.0220
	11	179	0.0001		52	138	0.0437		11	199	0.0001		52	158	0.0242
	12	178	0.0001		53	137	0.0478		12	198	0.0001		53	157	0.0266
	13	177	0.0002		54	136	0.0521		13	197	0.0001		54	156	0.0291
	14	176	0.0002		55	135	0.0567		14	196	0.0001		55	155	0.0319
	15	175	0.0003		56	134	0.0616		15	195	0.0001		56	154	0.0348
	16	174	0.0003		57	133	0.0668		16	194	0.0002		57	153	0.0379
	17	173	0.0004		58	132	0.0723		17	193	0.0002		58	152	0.0413
	18	172	0.0005		59	131	0.0782		18	192	0.0002		59	151	0.0448
	19	171	0.0006		60	130	0.0844		19	191	0.0003		60	150	0.0487
	20	170	0.0007		61	129	0.0909		20	190	0.0004		61	149	0.0527
	21	169	0.0008		62	128	0.0978		21	189	0.0004		62	148	0.0570
	22	168	0.0010		63	127	0.1051		22	188	0.0005		63	147	0.0615
	23	167	0.0012		64	126	0.1127		23	187	0.0006		64	146	0.0664
	24	166	0.0014		65	125	0.1206		24	186	0.0007		65	145	0.0715
	25	165	0.0017		66	124	0.1290		25	185	0.0008		66	144	0.0768
	26	164	0.0020		67	123	0.1377		26	184	0.0010		67	143	0.0825
	27	163	0.0023		68	122	0.1467		27	183	0.0012		68	142	0.0884
	28	162	0.0027		69	121	0.1562		28	182	0.0014		69	141	0.0947
	29	161	0.0031		70	120	0.1660		29	181	0.0016		70	140	0.1012
	30	160	0.0036						30	180	0.0018		71	139	0.1081
	31	159	0.0041						31	179	0.0021		72	138	0.1153
	32	158	0.0047						32	178	0.0024		73	137	0.1227
	33	157	0.0054						33	177	0.0028		74	136	0.1305
	34	156	0.0062						34	176	0.0032		75	135	0.1387
	35	155	0.0070						35	175	0.0036		76	134	0.1471
	36	154	0.0080						36	174	0.0042		77	133	0.1559
	37	153	0.0090						37	173	0.0047				
	38	152	0.0102						38	172	0.0053				
	39	151	0.0115						39	171	0.0060				
	40	150	0.0129						40	170	0.0068				

Table 10 Critical Values for the Wilcoxon Rank-Sum Statistic

This table contains critical values and probabilities for the Wilcoxon rank-sum statistic W = the sum of the ranks of the m observations in the smaller sample: m and n are the sample sizes, and c_1 and c_2 are defined by $P(W \leq c_1) = \alpha$ and $P(W \geq c_2) = \alpha$, respectively.

m	n	c_1	c_2	α
2	3	3	9	0.1000
2	4	3	11	0.0667
		4	10	0.1333
2	5	3	13	0.0476
		4	12	0.0952
2	6	3	15	0.0357
		4	14	0.0714
		5	13	0.1429
2	7	3	17	0.0278
		4	16	0.0556
		5	15	0.1111
2	8	3	19	0.0222
		4	18	0.0444
		5	17	0.0889
		6	16	0.1333
2	9	3	21	0.0182
		4	20	0.0364
		5	19	0.0727
		6	18	0.1091
2	10	3	23	0.0152
		4	22	0.0303
		5	21	0.0606
		6	20	0.0909
		7	19	0.1364
3	3	6	15	0.0500
		7	14	0.1000
3	4	6	18	0.0286
		7	17	0.0571
		8	16	0.1143
3	5	6	21	0.0179
		7	20	0.0357
		8	19	0.0714
		9	18	0.1250
3	6	6	24	0.0119
		7	23	0.0238
		8	22	0.0476
		9	21	0.0833
		10	20	0.1310
3	7	6	27	0.0083
		7	26	0.0167
		8	25	0.0333
		9	24	0.0583
		10	23	0.0917
		11	22	0.1333

m	n	c_1	c_2	α
3	8	6	30	0.0061
		7	29	0.0121
		8	28	0.0242
		9	27	0.0424
		10	26	0.0667
		11	25	0.0970
		12	24	0.1394
3	9	6	33	0.0045
		7	32	0.0091
		8	31	0.0182
		9	30	0.0318
		10	29	0.0500
		11	28	0.0727
		12	27	0.1045
		13	26	0.1409
3	10	6	36	0.0035
		7	35	0.0070
		8	34	0.0140
		9	33	0.0245
		10	32	0.0385
		11	31	0.0559
		12	30	0.0804
		13	29	0.1084
		14	28	0.1434
4	4	10	26	0.0143
		11	25	0.0286
		12	24	0.0571
		13	23	0.1000
4	5	10	30	0.0079
		11	29	0.0159
		12	28	0.0317
		13	27	0.0556
		14	26	0.0952
		15	25	0.1429
4	6	10	34	0.0048
		11	33	0.0095
		12	32	0.0190
		13	31	0.0333
		14	30	0.0571
		15	29	0.0857
		16	28	0.1286

m	n	c_1	c_2	α
4	7	10	38	0.0030
		11	37	0.0061
		12	36	0.0121
		13	35	0.0212
		14	34	0.0364
		15	33	0.0545
		16	32	0.0818
		17	31	0.1152
		18	30	0.1576
4	8	10	42	0.0020
		11	41	0.0040
		12	40	0.0081
		13	39	0.0141
		14	38	0.0242
		15	37	0.0364
		16	36	0.0545
		17	35	0.0768
		18	34	0.1071
		19	33	0.1414
4	9	10	46	0.0014
		11	45	0.0028
		12	44	0.0056
		13	43	0.0098
		14	42	0.0168
		15	41	0.0252
		16	40	0.0378
		17	39	0.0531
		18	38	0.0741
		19	37	0.0993
		20	36	0.1301
4	10	10	50	0.0010
		11	49	0.0020
		12	48	0.0040
		13	47	0.0070
		14	46	0.0120
		15	45	0.0180
		16	44	0.0270
		17	43	0.0380
		18	42	0.0529
		19	41	0.0709
		20	40	0.0939
		21	39	0.1199
		22	38	0.1518

m	n	c_1	c_2	α
5	5	15	40	0.0040
		16	39	0.0079
		17	38	0.0159
		18	37	0.0278
		19	36	0.0476
		20	35	0.0754
		21	34	0.1111
		22	33	0.1548
5	6	15	45	0.0022
		16	44	0.0043
		17	43	0.0087
		18	42	0.0152
		19	41	0.0260
		20	40	0.0411
		21	39	0.0628
		22	38	0.0887
		23	37	0.1234
5	7	15	50	0.0013
		16	49	0.0025
		17	48	0.0051
		18	47	0.0088
		19	46	0.0152
		20	45	0.0240
		21	44	0.0366
		22	43	0.0530
		23	42	0.0745
		24	41	0.1010
		25	40	0.1338
5	8	15	55	0.0008
		16	54	0.0016
		17	53	0.0031
		18	52	0.0054
		19	51	0.0093
		20	50	0.0148
		21	49	0.0225
		22	48	0.0326
		23	47	0.0466
		24	46	0.0637
		25	45	0.0855
		26	44	0.1111
		27	43	0.1422

Table 10 Critical Values for the Wilcoxon Rank-Sum Statistic (Continued)

m	n	c_1	c_2	α	m	n	c_1	c_2	α	m	n	c_1	c_2	α	m	n	c_1	c_2	α
5	9	15	60	0.0005	6	7	21	63	0.0006	6	10	21	81	0.0001	7	8	28	84	0.0002
		16	59	0.0010			22	62	0.0012			22	80	0.0002			29	83	0.0003
		17	58	0.0020			23	61	0.0023			23	79	0.0005			30	82	0.0006
		18	57	0.0035			24	60	0.0041			24	78	0.0009			31	81	0.0011
		19	56	0.0060			25	59	0.0070			25	77	0.0015			32	80	0.0019
		20	55	0.0095			26	58	0.0111			26	76	0.0024			33	79	0.0030
		21	54	0.0145			27	57	0.0175			27	75	0.0037			34	78	0.0047
		22	53	0.0210			28	56	0.0256			28	74	0.0055			35	77	0.0070
		23	52	0.0300			29	55	0.0367			29	73	0.0080			36	76	0.0103
		24	51	0.0415			30	54	0.0507			30	72	0.0112			37	75	0.0145
		25	50	0.0559			31	53	0.0688			31	71	0.0156			38	74	0.0200
		26	49	0.0734			32	52	0.0903			32	70	0.0210			39	73	0.0270
		27	48	0.0949			33	51	0.1171			33	69	0.0280			40	72	0.0361
		28	47	0.1199			34	50	0.1474			34	68	0.0363			41	71	0.0469
		29	46	0.1489	6	8	21	69	0.0003			35	67	0.0467			42	70	0.0603
5	10	15	65	0.0003			22	68	0.0007			36	66	0.0589			43	69	0.0760
		16	64	0.0007			23	67	0.0013			37	65	0.0736			44	68	0.0946
		17	63	0.0013			24	66	0.0023			38	64	0.0903			45	67	0.1159
		18	62	0.0023			25	65	0.0040			39	63	0.1099			46	66	0.1405
		19	61	0.0040			26	64	0.0063			40	62	0.1317	7	9	28	91	0.0001
		20	60	0.0063			27	63	0.0100			41	61	0.1566			29	90	0.0002
		21	59	0.0097			28	62	0.0147								30	89	0.0003
		22	58	0.0140			29	61	0.0213	7	7	28	77	0.0003			31	88	0.0006
		23	57	0.0200			30	60	0.0296			29	76	0.0006			32	87	0.0010
		24	56	0.0276			31	59	0.0406			30	75	0.0012			33	86	0.0017
		25	55	0.0376			32	58	0.0539			31	74	0.0020			34	85	0.0026
		26	54	0.0496			33	57	0.0709			32	73	0.0035			35	84	0.0039
		27	53	0.0646			34	56	0.0906			33	72	0.0055			36	83	0.0058
		28	52	0.0823			35	55	0.1142			34	71	0.0087			37	82	0.0082
		29	51	0.1032			36	54	0.1412			35	70	0.0131			38	81	0.0115
		30	50	0.1272	6	9	21	75	0.0002			36	69	0.0189			39	80	0.0156
		31	49	0.1548			22	74	0.0004			37	68	0.0265			40	79	0.0209
6	6	21	57	0.0011			23	73	0.0008			38	67	0.0364			41	78	0.0274
		22	56	0.0022			24	72	0.0014			39	66	0.0487			42	77	0.0356
		23	55	0.0043			25	71	0.0024			40	65	0.0641			43	76	0.0454
		24	54	0.0076			26	70	0.0038			41	64	0.0825			44	75	0.0571
		25	53	0.0130			27	69	0.0060			42	63	0.1043			45	74	0.0708
		26	52	0.0206			28	68	0.0088			43	62	0.1297			46	73	0.0869
		27	51	0.0325			29	67	0.0128			44	61	0.1588			47	72	0.1052
		28	50	0.0465			30	66	0.0180								48	71	0.1261
		29	49	0.0660			31	65	0.0248								49	70	0.1496
		30	48	0.0898			32	64	0.0332										
		31	47	0.1201			33	63	0.0440										
		32	46	0.1548			34	62	0.0567										
							35	61	0.0723										
							36	60	0.0905										
							37	59	0.1119										
							38	58	0.1361										

Table 10 Critical Values for the Wilcoxon Rank-Sum Statistic (Continued)

m	n	c_1	c_2	α
7	10	28	98	0.0001
		29	97	0.0001
		30	96	0.0002
		31	95	0.0004
		32	94	0.0006
		33	93	0.0010
		34	92	0.0015
		35	91	0.0023
		36	90	0.0034
		37	89	0.0048
		38	88	0.0068
		39	87	0.0093
		40	86	0.0125
		41	85	0.0165
		42	84	0.0215
		43	83	0.0277
		44	82	0.0351
		45	81	0.0439
		46	80	0.0544
		47	79	0.0665
		48	78	0.0806
		49	77	0.0966
		50	76	0.1148
		51	75	0.1349
		52	74	0.1574
8	8	36	100	0.0001
		37	99	0.0002
		38	98	0.0003
		39	97	0.0005
		40	96	0.0009
		41	95	0.0015
		42	94	0.0023
		43	93	0.0035
		44	92	0.0052
		45	91	0.0074
		46	90	0.0103
		47	89	0.0141
		48	88	0.0190
		49	87	0.0249
		50	86	0.0325
		51	85	0.0415
		52	84	0.0524
		53	83	0.0652
		54	82	0.0803
		55	81	0.0974
		56	80	0.1172
		57	79	0.1393
8	9	36	108	0.0000
		37	107	0.0001
		38	106	0.0002
		39	105	0.0003
		40	104	0.0005
		41	103	0.0008
		42	102	0.0012
		43	101	0.0019
		44	100	0.0028
		45	99	0.0039
		46	98	0.0056
		47	97	0.0076
		48	96	0.0103
		49	95	0.0137
		50	94	0.0180
		51	93	0.0232
		52	92	0.0296
		53	91	0.0372
		54	90	0.0464
		55	89	0.0570
		56	88	0.0694
		57	87	0.0836
		58	86	0.0998
		59	85	0.1179
		60	84	0.1383
8	10	36	116	0.0000
		37	115	0.0000
		38	114	0.0001
		39	113	0.0002
		40	112	0.0003
		41	111	0.0004
		42	110	0.0007
		43	109	0.0010
		44	108	0.0015
		45	107	0.0022
		46	106	0.0031
		47	105	0.0043
		48	104	0.0058
		49	103	0.0078
		50	102	0.0103
		51	101	0.0133
		52	100	0.0171
		53	99	0.0217
		54	98	0.0273
		55	97	0.0338
		56	96	0.0416
		57	95	0.0506
		58	94	0.0610
		59	93	0.0729
		60	92	0.0864
		61	91	0.1015
		62	90	0.1185
		63	89	0.1371
		64	88	0.1577
9	9	45	126	0.0000
		46	125	0.0000
		47	124	0.0001
		48	123	0.0001
		49	122	0.0002
		50	121	0.0004
		51	120	0.0006
		52	119	0.0009
		53	118	0.0014
		54	117	0.0020
		55	116	0.0028
		56	115	0.0039
		57	114	0.0053
		58	113	0.0071
		59	112	0.0094
		60	111	0.0122
		61	110	0.0157
		62	109	0.0200
		63	108	0.0252
		64	107	0.0313
		65	106	0.0385
		66	105	0.0470
		67	104	0.0567
		68	103	0.0680
		69	102	0.0807
		70	101	0.0951
		71	100	0.1112
		72	99	0.1290
9	10	45	135	0.0000
		46	134	0.0000
		47	133	0.0000
		48	132	0.0001
		49	131	0.0001
		50	130	0.0002
		51	129	0.0003
		52	128	0.0005
		53	127	0.0007
		54	126	0.0011
		55	125	0.0015
		56	124	0.0021
		57	123	0.0028
		58	122	0.0038
		59	121	0.0051
		60	120	0.0066
		61	119	0.0086
		62	118	0.0110
		63	117	0.0140
		64	116	0.0175
		65	115	0.0217
		66	114	0.0267
		67	113	0.0326
		68	112	0.0394
		69	111	0.0474
		70	110	0.0564
		71	109	0.0667
		72	108	0.0782
		73	107	0.0912
		74	106	0.1055
		75	105	0.1214
		76	104	0.1388
10	10	55	155	0.0000
		56	154	0.0000
		57	153	0.0000
		58	152	0.0000
		59	151	0.0001
		60	150	0.0001
		61	149	0.0002
		62	148	0.0002
		63	147	0.0004
		64	146	0.0005
		65	145	0.0008
		66	144	0.0010
		67	143	0.0014
		68	142	0.0019
		69	141	0.0026
		70	140	0.0034
		71	139	0.0045
		72	138	0.0057
		73	137	0.0073
		74	136	0.0093
		75	135	0.0116
		76	134	0.0144
		77	133	0.0177
		78	132	0.0216
		79	131	0.0262
		80	130	0.0315
		81	129	0.0376
		82	128	0.0446
		83	127	0.0526
		84	126	0.0615
		85	125	0.0716
		86	124	0.0827
		87	123	0.0952
		88	122	0.1088
		89	121	0.1237
		90	120	0.1399
		91	119	0.1575

Table 11 Critical Values for the Runs Test

This table contains cumulative probabilities associated with the runs test. Let m be the number of observations in one category, n be the number of observations in the other category ($m \leq n$), and V be the number of runs. The values in this table are the probabilities $P(V \leq v)$ if the order of observations is random.

m	n	2	3	4	5	6	7	8	9
2	2	0.3333	0.6667	1.0000					
2	3	0.2000	0.5000	0.9000	1.0000				
2	4	0.1333	0.4000	0.8000	1.0000				
2	5	0.0952	0.3333	0.7143	1.0000				
2	6	0.0714	0.2857	0.6429	1.0000				
2	7	0.0556	0.2500	0.5833	1.0000				
2	8	0.0444	0.2222	0.5333	1.0000				
2	9	0.0364	0.2000	0.4909	1.0000				
2	10	0.0303	0.1818	0.4545	1.0000				
3	3	0.1000	0.3000	0.7000	0.9000	1.0000			
3	4	0.0571	0.2000	0.5429	0.8000	0.9714	1.0000		
3	5	0.0357	0.1429	0.4286	0.7143	0.9286	1.0000		
3	6	0.0238	0.1071	0.3452	0.6429	0.8810	1.0000		
3	7	0.0167	0.0833	0.2833	0.5833	0.8333	1.0000		
3	8	0.0121	0.0667	0.2364	0.5333	0.7879	1.0000		
3	9	0.0091	0.0545	0.2000	0.4909	0.7455	1.0000		
3	10	0.0070	0.0455	0.1713	0.4545	0.7063	1.0000		
4	4	0.0286	0.1143	0.3714	0.6286	0.8857	0.9714	1.0000	
4	5	0.0159	0.0714	0.2619	0.5000	0.7857	0.9286	0.9921	1.0000
4	6	0.0095	0.0476	0.1905	0.4048	0.6905	0.8810	0.9762	1.0000
4	7	0.0061	0.0333	0.1424	0.3333	0.6061	0.8333	0.9545	1.0000
4	8	0.0040	0.0242	0.1091	0.2788	0.5333	0.7879	0.9293	1.0000
4	9	0.0028	0.0182	0.0853	0.2364	0.4713	0.7455	0.9021	1.0000
4	10	0.0020	0.0140	0.0679	0.2028	0.4186	0.7063	0.8741	1.0000

Table 11 Critical Values for the Runs Test (Continued)

m	n	2	3	4	5	6	7	8	9	10	11	12	13	14	15	16	17	18	19	20
5	5	0.0079	0.0397	0.1667	0.3571	0.6429	0.8333	0.9603	0.9921	1.0000										
5	6	0.0043	0.0238	0.1104	0.2619	0.5216	0.7381	0.9113	0.9762	0.9978	1.0000									
5	7	0.0025	0.0152	0.0758	0.1970	0.4242	0.6515	0.8535	0.9545	0.9924	1.0000									
5	8	0.0016	0.0101	0.0536	0.1515	0.3473	0.5758	0.7933	0.9293	0.9837	1.0000									
5	9	0.0010	0.0070	0.0390	0.1189	0.2867	0.5105	0.7343	0.9021	0.9720	1.0000									
5	10	0.0007	0.0050	0.0290	0.0949	0.2388	0.4545	0.6783	0.8741	0.9580	1.0000									
6	6	0.0022	0.0130	0.0671	0.1753	0.3918	0.6082	0.8247	0.9329	0.9870	0.9978	1.0000								
6	7	0.0012	0.0076	0.0425	0.1212	0.2960	0.5000	0.7331	0.8788	0.9662	0.9924	0.9994	1.0000							
6	8	0.0007	0.0047	0.0280	0.0862	0.2261	0.4126	0.6457	0.8205	0.9371	0.9837	0.9977	1.0000							
6	9	0.0004	0.0030	0.0190	0.0629	0.1748	0.3427	0.5664	0.7622	0.9021	0.9720	0.9944	1.0000							
6	10	0.0002	0.0020	0.0132	0.0470	0.1369	0.2867	0.4965	0.7063	0.8636	0.9580	0.9895	1.0000							
7	7	0.0006	0.0041	0.0251	0.0775	0.2086	0.3834	0.6166	0.7914	0.9225	0.9749	0.9959	0.9994	1.0000						
7	8	0.0003	0.0023	0.0154	0.0513	0.1492	0.2960	0.5136	0.7040	0.8671	0.9487	0.9879	0.9977	0.9998	1.0000					
7	9	0.0002	0.0014	0.0098	0.0350	0.1084	0.2308	0.4266	0.6224	0.8059	0.9161	0.9748	0.9944	0.9993	1.0000					
7	10	0.0001	0.0009	0.0064	0.0245	0.0800	0.1818	0.3546	0.5490	0.7433	0.8794	0.9571	0.9895	0.9981	1.0000					
8	8	0.0002	0.0012	0.0089	0.0317	0.1002	0.2145	0.4048	0.5952	0.7855	0.8998	0.9683	0.9911	0.9988	0.9998	1.0000				
8	9	0.0001	0.0007	0.0053	0.0203	0.0687	0.1573	0.3186	0.5000	0.7016	0.8427	0.9394	0.9797	0.9958	0.9993	1.0000	1.0000			
8	10	0.0000	0.0004	0.0033	0.0134	0.0479	0.1170	0.2514	0.4194	0.6209	0.7822	0.9031	0.9636	0.9905	0.9981	0.9998	1.0000			
9	9	0.0000	0.0004	0.0030	0.0122	0.0445	0.1090	0.2380	0.3992	0.6008	0.7620	0.8910	0.9555	0.9878	0.9970	0.9996	1.0000	1.0000		
9	10	0.0000	0.0002	0.0018	0.0076	0.0294	0.0767	0.1786	0.3186	0.5095	0.6814	0.8342	0.9233	0.9742	0.9924	0.9986	0.9998	1.0000	1.0000	
10	10	0.0000	0.0001	0.0010	0.0045	0.0185	0.0513	0.1276	0.2422	0.4141	0.5859	0.7578	0.8724	0.9487	0.9815	0.9955	0.9990	0.9999	1.0000	1.0000

Table 12 Greek Alphabet

This table contains the Greek alphabet: the letter name, the lowercase letter, the variant of the lowercase letter where applicable, and the uppercase letter.

Name	Lowercase letter	Lowercase variant	Uppercase letter
Alpha	α		A
Beta	β		B
Gamma	γ		Γ
Delta	δ		Δ
Epsilon	ε	ϵ	E
Zeta	ζ		Z
Eta	η		H
Theta	θ	ϑ	Θ
Iota	ι		I
Kappa	κ		K
Lambda	λ		Λ
Mu	μ		M
Nu	ν		N
Xi	ξ		Ξ
Omicron	o		O
Pi	π	ϖ	Π
Rho	ρ	ϱ	R
Sigma	σ	ς	Σ
Tau	τ		T
Upsilon	υ		Υ
Phi	ϕ	φ	Φ
Chi	χ		X
Psi	ψ		Ψ
Omega	ω		Ω

Answers to Odd-Numbered Exercises

Chapter 0

0.1 Claim, experiment or observed outcome, likelihood, conclusion.

0.3 Currently, NOAA expects to see an average of 2 whales perish per year. This observation of 41 humpback whale deaths is extremely rare according to NOAA's claim.

0.5 Kimberly may have had an incredibly lucky series of events occur or she could have cheated. Because the tickets appear to be randomly purchased and independent of one another, the more reasonable explanation may be that she is very lucky.

0.7 Answers may vary. If there is a 50% chance of rain, this means that if the same atmospheric conditions presented themselves on 10 identical days, we would actually observe rain on approximately 5 of those days.

0.9 (a) The large number of illnesses was very rare in the short time span and locality.
(b) Claim: Very few people should get sick after eating at Chipotle in any given week. Experiment: 130 customers became sick soon after eating at a specific Chipotle in Virginia. Likelihood: Observing this many customers getting sick in a week due to chance is incredibly rare. Conclusion: Chipotle officials decide to investigate the cause of the illness because it may be attributed to the food eaten at this establishment.

0.11 Since there were only 15 burglaries in the year, it is not that unusual for none to have occurred on a Thursday. This observation may simply be due to chance.

0.13 This instance does not give reason to doubt the claim made by the NRF for several reasons. First, the NRF's claim is likely about the entire country, but this sample represents only a small region. Also, the statistic presented is likely an average, and as we will see later in this text, although 17 of those surveyed may plan to purchase 12 or fewer gifts, one person may plan to buy 100 gifts, which will affect the calculated mean.

0.15 (a) A subscript is a numerical label that is used to distinguish between several values of the variable, in this case x.
(b) A superscript in mathematics indicates the power the term below is being raised to.
(c) Summation notation is a more concise way of writing a string of terms to be added together.
(d) A function is a rule that assigns a unique output value to each input value.

Chapter 1

1.1 Statistics Today
No exercises.

1.2 Populations, Samples, Probability, and Statistics

1.1 True

1.3 True

1.5 Answers may vary.
(a) Probability question: How likely would it be to observe that fewer than 50 of a sample of 100 Americans have a social media profile? Statistics question: Within the sample of 100 Americans, only 50 report having a social media profile. What does this result suggest about the reported national percentage of social media users?
(b) Probability question: What is the probability that exactly 21 of 50 randomly selected individuals use the same password for everything? Statistics question: We observe that 21 of the selected 50 use the same password for everything. Does this suggest that the true percentage of people who use the same password for everything is actually different than 36%?
(c) Probability question: Suppose you walk through a parking lot in Norway. What is the probability that none of the first 16 cars you observe is a plug-in electric vehicle? Statistics question: Suppose you take a random sample of 35 cars in Denmark and 19 of them are plug-in electric vehicles. Does this sample provide evidence to suggest that Denmark has a higher percentage of electric cars than Norway does?
(d) Probability question: Suppose that 3 people in your neighborhood suffer snow shoveling injuries this winter. What is the likelihood that all 3 injuries are categorized as pulled muscles? Statistics question: Suppose an insurance agency sees an atypically large number of snow shoveling–related insurance claims this winter, all from the same neighborhood. If the company observes that 6 of the 10 claims are not categorized as pulled muscles, does it have evidence to suggest that these claims may be fraudulent?

1.7 (a) Descriptive statistics
(b) Inferential statistics
(c) Descriptive statistics
(d) Inferential statistics
(e) Descriptive statistics
(f) Descriptive statistics

1.9 Population: All T-shirt buyers. Sample: 50 surveyed consumers. Variable: Tag preference.

1.11 Population: All adult residents in Arizona. Sample: 500 selected Arizona residents.

1.13 (a) Population: All people who bought a dining room table in the last month. Sample: 5 selected buyers. Probability question.
(b) Population: All people who shop at this Walgreens. Sample: 25 selected shoppers. Statistics question.

(c) Population: All people who have used or will use this water slide. Sample: 50 selected riders. Probability question.
(d) Population: All automatic doors in public buildings in Henderson, Nevada. Sample: 10 randomly selected doors. Statistics question.
(e) Population: All travelers entering LAX airport. Sample: 1000 selected travelers. Statistics question.
(f) Population: All people who sold a home in the past year. Sample: 35 selected sellers. Probability question.
(g) Populations: All for-profit nursing homes and all Medicare nursing homes. Samples: The several nursing homes of each type that were selected. Statistics question.

1.15 Population: All pumpkin seeds. Sample: 50 selected seeds. Variable: Amount of zinc that the seed contains.

1.17 (a) Population: All Bounty paper towels.
(b) Sample: The single sheet tested from each of the 35 selected rolls.
(c) Variable: Amount of absorption for each paper towel.

1.19 (a) Population: All traditionally produced cheddar cheese. Sample: 20 selected cheddar cheeses from around the world.
(b) Probability problem: How likely is it that only 3 of the selected 20 cheeses were aged for fewer than 2 years? Statistics problem: If only 3 of the selected 20 cheeses were aged for fewer than 2 years, does this provide evidence to suggest that less than 75% of cheddar cheeses are aged for fewer than 2 years?

1.21 (a) All computer systems of U.S. government agencies.
(b) The computer systems of 100 U.S. government agencies.
(c) Whether the computer systems contained a trace of the Kaspersky software.
(d) Probability question: What is the likelihood that at most 20 of the selected 100 agencies' systems have traces of the Kaspersky software? Statistics question: Suppose the sample indicates that 20 of the selected systems contain traces of the Kaspersky software. Do these findings suggest that the true proportion of government agencies with traces is higher than 15%?

1.3 Experiments and Random Samples

1.23 True

1.25 False

1.27 False

1.29 Claim, experiment, likelihood, conclusion

1.31 (a) Observational study.
(b) The sample is all student responses that the volunteers were able to collect that evening.
(c) This is not a random sample. The dorm selected was close to the proposed site, and students living in that dorm may have a different opinion than those living in a dormitory farther from the site. Additionally, only students in their rooms at the time the volunteers walked through were able to participate, leaving out students at the library studying for their statistics exam.

1.33 (a) Population: All 4-oz packages of Reese's Pieces produced by the Hershey Company. Sample: Packages selected by the inspection team.
(b) This is a random sample. The inspection team has selected packages from various locations within the plant, ensuring that all packages have the same chance of being selected and measured.

1.35 (a) Observational study.
(b) Population: All Lyft drivers. Sample: 25 selected Atlanta drivers.
(c) This is not a random sample because drivers were only selected from Atlanta. Perhaps this city has drivers who travel an oddly large number of miles in a typical 8-hr shift compared to Lyft drivers in the rest of the country.

1.37 Obtain a list of customers who have recently purchased the TV and place each of their names on an equal-sized slip of paper. Put the slips into a hat and after mixing well, select a group of 25 customer names to then call and survey.

1.39 (a) Experiment.
(b) Tumor count for each rodent.

1.41 (a) Experiment.
(b) Lifetime of each rose blossom.
(c) Assign each of the 50 roses an ID number. Using a random number generator, produce 25 unique ID numbers that correspond to the roses in the "treated" group. The rest of the roses will be placed in the "untreated" group.

1.43 (a) Population: All floor tiles of this particular type. Sample: The 25 tiles that were ordered and tested.
(b) This is not a random sample. One method for obtaining a random sample of tiles would be to visit the factory production line and, using a random number generator, select one of the 4000 tiles produced each hour throughout the work day. The collection of selected tiles will then make up your sample.

1.45 (a) Observational study.
(b) Weight of each seat cushion.
(c) Assign a random sample of new seat cushions and a random sample of old seat cushions to each of the two rail cars chosen at random. After the rail cars travel the same route, measure the energy efficiency of the cars and compare the two measurements.

Chapter 2
2.1 Types of Data

2.1 False

2.3 True

2.5 (a) Numerical (or quantitative) **(b)** Categorical (or qualitative) **(c)** Discrete **(d)** Continuous

2.7 (a) Numerical, continuous **(b)** Numerical, discrete **(c)** Numerical, continuous **(d)** Numerical, continuous

(e) Numerical, discrete (f) Numerical, discrete
(g) Categorical (h) Categorical

2.9 (a) Numerical, continuous (b) Numerical, continuous
(c) Categorical (d) Categorical (e) Numerical, continuous
(f) Categorical (g) Numerical, discrete (h) Categorical

2.11 (a) Continuous (b) Continuous (c) Discrete
(d) Continuous (e) Continuous (f) Discrete
(g) Continuous (h) Discrete

2.13 (a) Continuous (b) Discrete (c) Discrete
(d) Discrete (e) Categorical (f) Categorical

2.15 (a) Discrete (b) Continuous (c) Discrete
(d) Categorical (e) Categorical (f) Discrete

2.2 Bar Charts and Pie Charts

2.17 True

2.19 True

2.21 False

2.23

Class	Frequency	Relative Frequency
Abstract	15	0.357
Surrealist	9	0.214
Realist	12	0.286
Expressionist	6	0.143
Total	42	1.000

2.25 (a)

County	Relative Frequency
Hidalgo	0.159 (= 225/1419)
Cameron	0.116 (= 165/1419)
Brazoria	0.102 (= 145/1419)
Fort Bend	0.072 (= 102/1419)
Harris	0.125 (= 177/1419)
Montgomery	0.115 (= 163/1419)
Williamson	0.085 (= 120/1419)
Bell	0.083 (= 118/1419)
Colin	0.072 (= 102/1419)
Parker	0.072 (= 102/1419)
Total	1.000 (= 1419/1419)

(b)
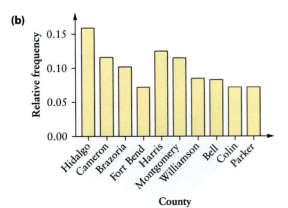

2.27 (a)

Attitude	Relative Frequency
Alarmed	0.182
Concerned	0.293
Cautious	0.242
Disengaged	0.061
Doubtful	0.121
Dismissive	0.101
Total	1.000

(b)
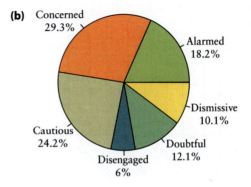

2.29 (a)

Grade	Frequency	Relative Frequency
A	10	0.067
B	43	0.287
C	56	0.373
D	26	0.173
F	15	0.100
Total	150	1.000

(b)
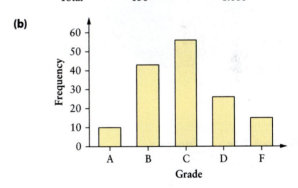

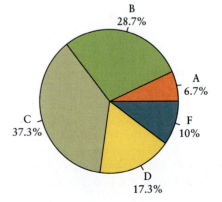

(c) 150 students, 90% (135/150) earned a passing grade.

2.31 (a)

Brand	Frequency
GE	6
LG	5
Maytag	4
Samsung	8
Whirlpool	7
Total	30

(b)

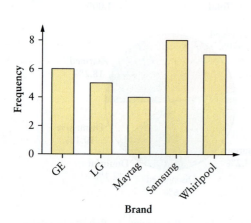

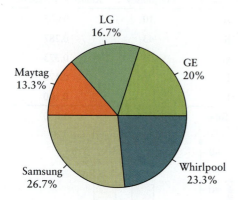

(c) 0.367
(d) 0.733

2.33 (a)

Book Type	Frequency
Education	5
Law	3
Literature	4
Medicine	7
Science	5
Technology	6
Total	30

(b)

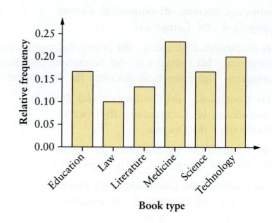

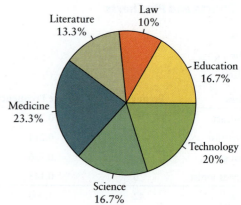

(c) There appears to be a fairly even distribution of interest in book types, suggesting that the library does not need to purchase more books of a certain type. If the library were to purchase more books, it might choose to buy more about medicine, as it was the most popular choice.

2.35 (a)

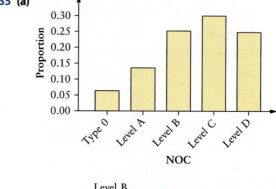

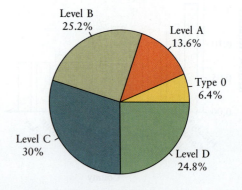

(b)

NOC	Frequency
Skill Type 0	16
Skill Level A	34
Skill Level B	63
Skill Level C	75
Skill Level D	62
Total	250

2.37 (a)

Service Provider	Relative Frequency
Netflix	0.288
Hulu	0.124
Amazon	0.113
Playstation Vue	0.138
Sling	0.192
Crackle	0.054
HBO	0.092
Total	1.000

(b)

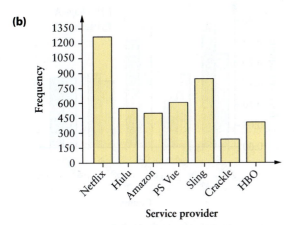

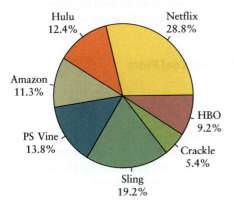

2.39

Casino	Frequency	Relative Frequency
Borgata	40	0.200
Caesars	25	0.125
Tropicana	32	0.160
Harrah's	22	0.110
Bally's	25	0.125
Golden Nugget	56	0.280
Total	200	1.000

(a) 200 people
(b) The Golden Nugget casino had the highest frequency (56) of visitors in the sample and had the largest relative frequency (0.280) of visitors, making it the most popular in the sample.

2.41 (a)

Response	Frequency	Relative Frequency
Excellent	50	0.0500
Very good	152	0.1520
Good	255	0.2550
Fair	425	0.4250
Poor	118	0.1180
Total	1000	1.0000

(b)

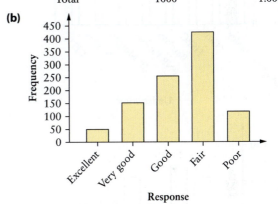

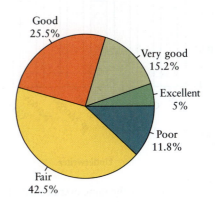

(c) 0.798

2.43 (a)

Underwriter	Relative Frequency
Morgan Stanley	0.108
Citi	0.156
HSBC	0.172
Barclays	0.118
BNP Paribas	0.156
JP Morgan	0.124
Natikis	0.118
CITIC	0.048
Total	1.000

(b)

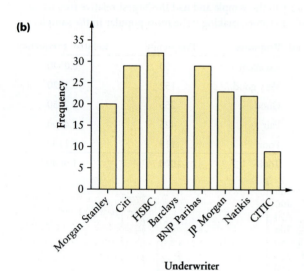

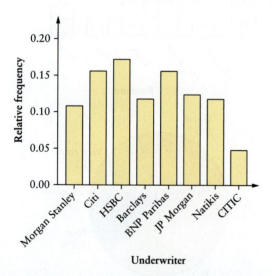

While both bar charts show the same overall shape, the chart that uses frequency on the vertical axis will help with comparing the volumes of green bonds that have been issued. The second chart, with relative frequency on the vertical axis, allows us to compare the various underwriters to one another, letting us observe the percentage of green bonds each group has issued relative to the whole.

2.45 (a)

	Package A	
City	Frequency	Relative Frequency
Winnipeg	67	0.221
Vancouver	52	0.172
Toronto	86	0.284
Montreal	48	0.158
Calgary	50	0.165
Total	303	1.000

	Package B	
City	Frequency	Relative Frequency
Winnipeg	63	0.292
Vancouver	24	0.111
Toronto	55	0.255
Montreal	38	0.176
Calgary	36	0.167
Total	216	1.000

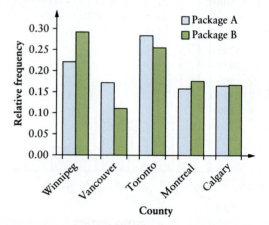

(c) Comparing frequencies for each city between packages can be misleading since the total number of packages sold is not the same for both types.

2.3 Stem-and-Leaf Plots

2.47 False

2.49 True

2.51 A stem-and-leaf plot with fewer than 5 stems will not allow us to observe trends in the data because values will be too close together, while in a stem-and-leaf plot with more than 20 stems, the data will be very spread out, making it difficult to observe the shape of the distribution.

2.53

2	7 9
3	5 6 6 6 9
4	1 1 2 7 7 9
5	0 1 1 2 4 5 7 7 8 9
6	1 4 4 4 6 8
7	1

Stem = 1, Leaf = 0.1

The center of these data appears to be between 4.5 and 5.5. More specifically, 5 appears to be a typical value in this data set.

2.55

53	0 3 4 4
53	7 9 9
54	1 1 1 3 4 4
54	5 6 6 6 7 7 7 7 7 8 8 9
55	1 1 2 3 3 4
55	6 7 7 7 7
56	0 0 2
56	9

Stem = 10, Leaf = 1

The center of these data appears to be between 545 and 550, with a typical value of 547.

2.57 **(a)** 543, 543, 549. **(b)** A typical value appears to be around 561. **(c)** The data seem to tail off more slowly at the lower values, with the majority of the observations being between 550 and 590. **(d)** There do not appear to be any outliers in this data set.

2.59 (a)

0	4
1	1 6
2	1 2 7 7 9
3	3 4 4 9
4	0 1 1 1 1 2 5 5 5 5 7 9 9
5	6 9 9
6	0 2 3 4 4 4
7	0 1 2 3 6 7 9
8	2 5 8
9	3

Stem = 1, Leaf = 0.1

(b)

0	4
1	1 6
2	1 3 7 8
3	0 3 5 5
4	0 0 1 2 2 2 2 5 6 6 6 8
5	0 0 7 9 9
6	0 3 3 4 4 5
7	0 1 3 4 7 7 9
8	2 6 9
9	3

Stem = 1, Leaf = 0.1

(c) The data in plot (b) appear to be fairly symmetric around the central values, whereas in plot (a), there are almost two clusters of data. A typical value appears to be approximately 4.7 for both sets of data.

2.61 (a)

Lower		Upper
	10	1 4
	10	
4 3 0 0	11	1
9 9 8 8 7 7 7 7 7 7 7 6 6 6 6 6 5 5 5 5	11	5 5 5 7 8
3 3 3 3 3 2 2 2 2 1 1 1 1 1 1 1 0 0 0 0 0 0 0 0	12	1 2 4 4
6 6 5	12	5 5 5 5 6 7 7 7 8 9 9
0	13	1 2 2 2 3 4
	13	5 6 7 8 8 8 9 9 9
	14	0 2 2 3 4
	14	5 8 8 8
	15	2 4 4
	15	

Stem = 10, Leaf = 1

(b) There appears to be a lot less variability in the temperatures recorded on the lower floors, where all temperatures were between 110 and 130 degrees. On the upper floors, the data had values ranging from 101 to 154 degrees. Additionally, the typical temperature for the lower floors appears to be lower than the typical temperature for the upper floors (120 degrees for the lower floors versus 130 degrees for the upper floors).

2.63

4	2 4
5	6
6	2 4 8 8
7	3 4 9 9
8	1 3 6
9	0 0 2 3 7 8 8
10	0 1 1 1 7 8
11	2 3 6 6 6 7
12	2 3 6 9
13	1
14	
15	8
16	
17	2

Stem = 10, Leaf = 1

(b) The typical time spent in the emergency room appears to be approximately 97 minutes. The observations of 158 and 172 minutes appear to be outliers.

2.65 (a)

```
2 | 1 3 5 6 6 7
3 | 1 3 4 5 6 7 7 7 7 8 8 9 9 9
4 | 0 0 0 1 1 2 2 3 5 5
5 | 0 0 2 2 4 5 5 6 6
6 |
7 |
8 | 0
```
Stem = 10, Leaf = 1

(b) The typical number of years a gas station has been in operation appears to be approximately 41 years. The station that has been open for 80 years appears to be an outlier.

2.67 (a)

```
16 | 2 3 3 3 3 3 3 3 4 4 4 4 4 4
16 | 5 5 5 5 5 5 6 6 6 6 7 8 9 9 9
17 | 0 0 0 0 1 1 2 2 2 2 3 3 3 3 4 4
17 | 5 5 5 5 5 6 7 7 8 8 9
18 | 2 2 3 4 4
18 | 7 7 7 8 8
19 | 0 2 3
19 | 5 5 7 7
20 | 0 1 4
20 | 6
21 | 0 0
21 |
22 |
22 |
23 | 2
```
Stem = 100, Leaf = 10

(b) A typical pumpkin in this data set appears to weigh approximately 1770 lb. The pumpkin that weighs 2324 lb appears to be an outlier.

2.69 (a)

With		Without
	250	2 4
	251	
8 4	252	1 4 7
3	253	7 8 8
6 3	254	4 9
8 7 4 2 2 2 1	255	0 2 3
9 8 6 6 6 5 3 1 1 0	256	1 4 9
8 7 3	257	1 2 4 8
6 5 5 3	258	2 3 3 6 7
4	259	3
	260	2 6
	261	2
	262	3

Stem = 100, Leaf = 0.1

(b) The center of both distributions appears to be approximately 256 cycles per second; however, the frequencies from pianos without a humidifier appear to have more variability. The data set with the humidifier appears to have a frequency of a middle C closer, on average, to the desired 256 cycles per second, indicating that a humidifier seems to have the ability to better keep the piano in tune.

2.71 (a)

```
3 | 3 4
3 | 8 9
4 | 0 0 0 1 1 2 2 2 3 3 3 3 3 3 3 3 4 4 4
4 | 5 5 5 5 5 5 5 5 5 5 5 6 6 6 6 6 7 7 7 8 8 8 8 8 8 9 9 9 9 9 9
5 | 0 0 1 1 1 1 1 1 2 2 4 4 4
5 | 5 6 7 9
```
Stem = 10, Leaf = 1

(b) The distribution of ages is roughly symmetric and very consistent, with ages ranging from 33 to 59 years. A typical age for loan payoff appears to be approximately 46 years.
(c) 0.0571

2.4 Frequency Distributions and Histograms

2.73 False

2.75 False

2.77 True

2.79 False

2.81 (a) A density histogram is used when class intervals do not have equal width.
(b) The sum of the areas of all rectangles must be 1.

2.83

Class	Frequency
70–80	2
80–82	4
82–84	4
84–86	4
86–88	9
88–90	6
90–92	9
92–94	2
Total	40

2.85

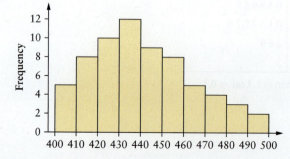

2.87

Class	Frequency	Relative Frequency	Cumulative Relative Frequency
100–150	155	0.194	0.194
150–200	120	0.150	0.344
200–250	130	0.163	0.507
250–300	145	0.181	0.688
300–350	150	0.188	0.876
350–400	100	0.125	1.001
Total	800	1.000	

*Note: The total cumulative relative frequency is not equal to 1.000 due to round-off error.

2.89

Class	Frequency	Relative Frequency	Cumulative Relative Frequency
0–25	150	0.150	0.150
25–50	200	0.200	0.350
50–75	175	0.175	0.525
75–100	150	0.150	0.675
100–125	125	0.125	0.800
125–150	100	0.100	0.900
150–175	75	0.075	0.975
175–200	25	0.025	1.000
Total	1000	1.000	

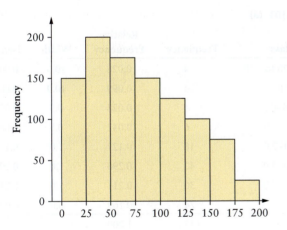

2.91 (a)

Class	Frequency
0–10	1
10–20	3
20–30	9
30–40	11
40–50	12
50–60	10
60–70	7
70–80	5
80–90	2
Total	60

(b)

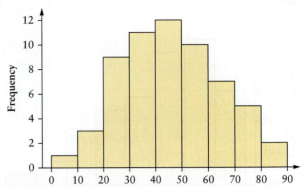

This distribution is unimodal and symmetric.
(c) $M = 43$
(d) $Q_1 = 31.75$
(e) $Q_3 = 58.25$

2.93 (a)

Class	Frequency
0–50	5
50–100	9
100–150	5
150–200	3
200–250	2
250–300	1
300–350	4
350–400	0
400–450	0
450–500	0
500–550	1
Total	30

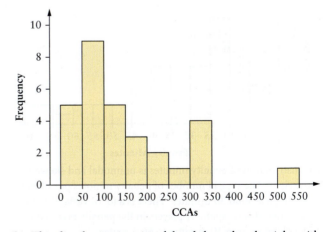

(b) This distribution is unimodal and skewed to the right, with one high outlier at 514.
(c) $M = 111.5$

2.95 (a)

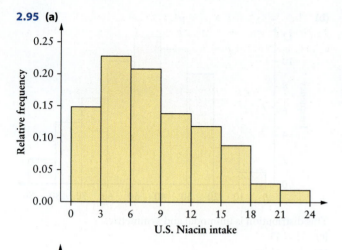

(b) The histogram for niacin intake in the United States is skewed to the right and unimodal, whereas the histogram for niacin intake in Europe is also unimodal but skewed to the left. A typical value of niacin intake in Europe appears to be higher than that in the United States.

2.97 (a)

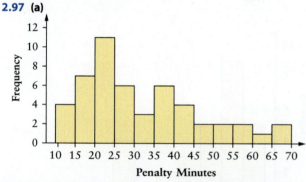

The distribution of penalty minutes is unimodal and skewed to the right. The typical number of penalty minutes per player seems to be around 30 minutes. There is a lot of variability in the amount of time spent by players in the penalty box, with values ranging from 10 minutes to 66 minutes.
(b) $m = 52$

2.99 (a)

Class	Frequency	Relative Frequency	Cumulative Relative Frequency
20.0–20.5	6	0.04	0.04
20.5–21.0	12	0.07	0.11
21.0–21.5	17	0.10	0.21
21.5–22.0	21	0.13	0.34
22.0–22.5	28	0.17	0.51
22.5–23.0	25	0.15	0.66
23.0–23.5	19	0.12	0.78
23.5–24.0	15	0.09	0.87
24.0–24.5	11	0.07	0.94
24.5–25.0	10	0.06	1.00
Total	164	1.00	

(b)

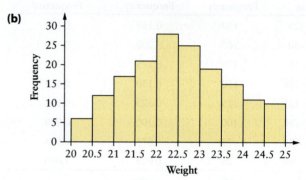

(c) $w = 20.9$

2.101 (a)

Class	Frequency	Relative Frequency	Width	Density
−20 to −10	4	0.028	10.0	0.0028
−10 to 0	14	0.099	10.0	0.0099
0–0.5	4	0.028	0.5	0.0560
0.5–1.0	6	0.042	0.5	0.0840
1.0–2.0	18	0.127	1.0	0.1270
2.0–3.0	42	0.296	1.0	0.2960
3.0–4.0	30	0.211	1.0	0.2110
4.0–10.0	24	0.169	6.0	0.0282
Total	142	1.000		

(b) Because the class sizes are not all the same width, both a frequency histogram and a relative frequency histogram would be inappropriate for these data.

(c)

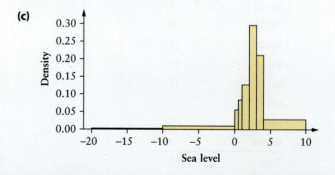

2.103 (a)

Class	Female Frequency	Female Relative Frequency	Male Frequency	Male Relative Frequency
30–45	14	0.14	0	0.0000
45–60	62	0.62	0	0.0000
60–75	24	0.24	1	0.0067
75–90	0	0.00	40	0.2667
90–105	0	0.00	67	0.4467
105–120	0	0.00	39	0.2600
120–135	0	0.00	3	0.0200

(b)

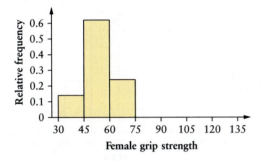

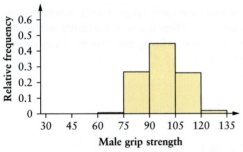

(c) Both sexes appear to have approximately symmetric distributions. The male's grip strengths are centered higher than the women's (near 97 versus 52). Also, the women's data appear to be less spread out than the grip strengths for the men.
(d) Grip strengths were provided for 150 men, but only 100 women, making relative frequency a more appropriate value to display in a histogram.

Chapter 2 Exercises

2.105 (a)

```
1  |
1  | 5 7 8 9
2  | 0 1 1 3 4
2  | 5 5 6 7 8 8 9 9 9 9
3  | 0 0 1 1 2 2 3 3 4 4
3  | 5 5 5 6 6 6 7 9
4  | 0 1 2 3 4
4  | 8 9
5  | 2
```

Stem = 0.1, Leaf = 0.01

(b) The coefficient of drag appears to have a symmetric distribution, centered at approximately 0.30. The values have a large range, varying from 0.15 to 0.52.

2.107 (a)

Class	Frequency	Relative Frequency
70–75	4	0.0816
75–80	12	0.2449
80–85	20	0.4082
85–90	6	0.1224
90–95	3	0.0612
95–100	2	0.0408
100–105	2	0.0408
Total	49	1.0000

(b)

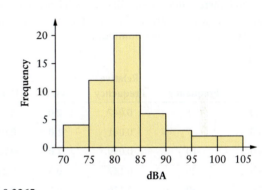

(c) 0.3265
(d) 0.1429

2.109 (a)

```
                    |  New  |    | Traditional
                    |       |  0 | 9
                    |       |  1 | 0 5
            8 7 7 6 0|       |  2 | 6
        7 6 6 4 4 3 3 1 1|   |  3 | 4 5
    9 9 8 7 5 4 4 4 3 2 2|   |  4 | 1 3 5 7 9
                2 1 0|       |  5 | 6 7
                  5 4|       |  6 | 0 1 2 2 8 8 9
                    |       |  7 | 3 3 6 6 7
                    |       |  8 | 0 3 4 6 8 9
                    |       |  9 | 2 8 8
                    |       | 10 | 1 3 4
                    |       | 11 | 3
                    |       | 12 | 0 3 4
                    |       | 13 | 0 4
                    |       | 14 | 1 7
```

Stem = 1, Leaf = 0.1

(b) Both distributions appear to be roughly symmetric. The new method is much less spread out than the traditional one, and is also centered at a lower value (approximately 4.2 min).
(c) Because it is, on average, quicker and far less variable than the traditional method, the new method appears to be better.

2.111 (a)

Defensive Backs		Safetys
8	42	
9 6 5 2 2 2	43	4
9 9 8 6 6 6 6 4 4 3	44	0 0 1 4 5 7 7
7 7 6 4 4 3 3 3 2 2 1 0 0	45	0 1 3 6 6 8 8
7 4 3 3 1 0	46	1 2 3 5 6 9
0	47	

Stem = 0.1, Leaf = 0.01

(b) Both distributions appear to be roughly symmetric, with the safetys' times perhaps a bit uniform. The defensive backs have less variability in 40-yd sprint times with a slightly lower center than the safetys.

(c) Although there is more variability in the defensive backs data set, its average is slightly smaller than the safetys' data set. This data set also has a smaller minimum.

2.113 (a)

Class	Frequency	Relative Frequency	Width	Density
30.0–32.0	8	0.047	2.0	0.0235
32.0–33.0	7	0.041	1.0	0.0410
33.0–34.0	10	0.059	1.0	0.0590
34.0–34.5	25	0.147	0.5	0.2940
34.5–35.0	30	0.176	0.5	0.3520
35.0–35.5	40	0.235	0.5	0.4700
35.5–36.0	45	0.265	0.5	0.5300
36.0–50.0	5	0.029	14.0	0.0021
Total	170	1.000		

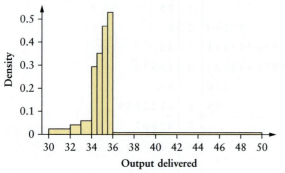

2.115 (a)

```
0  
0  9
1  0 2 3 3
1  5 7 7 8 8 9 9
2  0 0 1 1 2 2 2 2 2 3 3 3 3 3 4 4 4 4 4 4 4
2  5 5 5 5 5 5 5 6 6 6 6 6 6 6 6 6 7 7 7 7 7 7 7 7 7 7 8 8 8 8 8 8 8 8
   9 9 9 9 9 9 9 9
3  0 0 0 0 0 0 1 1 1 1 2 2 2 2 2 3 4 4 4
3  5 5 6 8 9 9 9
4  0 0 0
```
Stem = 10, Leaf = 1

Class	Frequency	Relative Frequency
5–10	1	0.01
10–15	4	0.04
15–20	7	0.07
20–25	21	0.21
25–30	39	0.39
30–35	18	0.18
35–40	7	0.07
40–45	3	0.03
Total	100	1.00

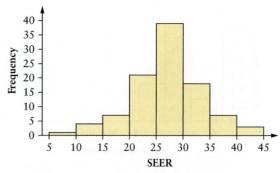

(c) These data have a very symmetric distribution, centered at approximately 26. There is a lot of variability in the SEER ratios that were measured, ranging from 9 to 40. There do not appear to be any outliers.

(d) 0.86

(e)

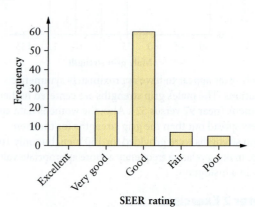

2.117 (a)

Class	Frequency	Relative Frequency	Cumulative Relative Frequency
100–110	1	0.02	0.02
110–120	2	0.04	0.06
120–130	9	0.18	0.24
130–140	7	0.14	0.38
140–150	9	0.18	0.56
150–160	9	0.18	0.74
160–170	11	0.22	0.96
170–180	1	0.02	0.98
180–190	1	0.02	1.00
Total	50	1.00	

(b)

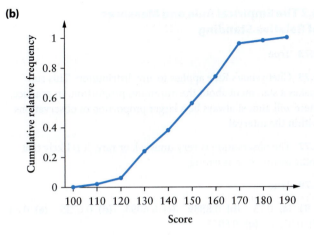

2.119

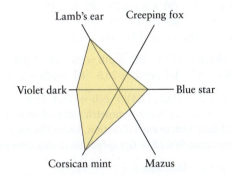

Chapter 3

3.1 Numerical Summary Measures

3.1 (a) center; variability (b) centered or clustered

3.3 (a) 70 (b) 2676 (c) 20 (d) 2101 (e) 140 (f) 140

3.5 (a) 105.7 (b) 13.1852 (c) 6.9583 (d) 0.1232
(e) −2.4933 (f) 17.7432

3.7 (a) $\bar{x} = 6.6667; \tilde{x} = 7$ (b) $\bar{x} = 6.6364; \tilde{x} = 9$
(c) $\bar{x} = 10.6889; \tilde{x} = 7.7$ (d) $\bar{x} = -107.69; \tilde{x} = -109.1$

3.9 (a) $\bar{x} < \tilde{x}$; suggests the distribution is skewed left.
(b) $\bar{x} \approx \tilde{x}$; suggests the distribution is approximately symmetric.
(c) $\bar{x} < \tilde{x}$; suggests the distribution is skewed left.
(d) $\bar{x} < \tilde{x}$; suggests the distribution is skewed left.

3.11 (a) $M = 6$ (b) $M = 0$ (c) There is no mode. Each observation occurs exactly once.

3.13 (a) $\bar{x} = 68.5238$ (b) $\tilde{x} = 67$ (c) Since $\tilde{x} < \bar{x}$, this suggests that the distribution is slightly skewed to the right.

3.15 (a) $\bar{x} = 85.53; \tilde{x} = 87.2$ (b) $\bar{x} = 84.03; \tilde{x} = 87.2$. The mean is smaller and is pulled in the direction of the new, smaller value. The median stays the same.

3.17 (a) $\bar{x} = 619.5; \tilde{x} = 620$ (b) $\bar{x}_{tr(0.05)} = 619.1667$
(c) Since all three numerical summary measures are approximately the same, this suggests the distribution is approximately symmetric.

3.19 (a) $\bar{x} = 26176.11; \tilde{x} = 21694.5$ (b) The median is a better measure of central tendency. The distribution appears to be skewed to the right, with several large values pulling the mean in their direction.

3.21 (a) $\bar{x} = 82; \tilde{x} = 83.5$ (b) $\bar{x}_{tr(0.10)} = 82.625$ (c) There are two modes: 84 and 86. Each observation occurs 5 times.

3.23 (a) City: $\bar{x}_c = 756.1; \tilde{x}_c = 755.5$
Rural: $\bar{x}_r = 684.53; \tilde{x}_r = 691$
(b) Since $\bar{x}_c > \bar{x}_r$ and $\tilde{x}_c > \tilde{x}_r$, this suggests that, on average, it is more expensive to have root canal in a city.

3.25 (a) $\bar{x} = 68; \tilde{x} = 67.5$ (b) Since $\bar{x} \approx \tilde{x}$, this suggests the shape of the distribution is approximately symmetric.
(c) $\bar{y} = 72; \bar{y} = \bar{x} + 4 = 68 + 4 = 72$

3.27 (a) $\hat{p} = 0.8438$ (b) $\bar{x} = 0.8438 = \hat{p}$ (c) It is not possible for the sample proportion of successes to be 0.9. Even if eight additional panels are successes, the total number of successes will be 35, and $\hat{p} = 35/40 = 0.8750$.

3.29 (a) $\bar{x} = 38.3250$ (b) There is no way to determine the sample median. We need to know the seventh and eighth observations in the ordered list.

3.31 $x_{16} = 67.8$

3.33 (a) $\bar{x} = 1.6238$ (b) $\bar{y} = 1.9973$ (c) $\bar{y} = 1.23 * \bar{x}$.

3.2 Measures of Variability

3.35 False

3.37 False

3.39 True

3.41 (a) $R = 3.9; s^2 = 2.1690; s = 1.4728$
(b) $R = 20.6; s^2 = 27.2812; s = 5.2231$
(c) $R = 98.32; s^2 = 1096.2665; s = 33.1099$
(d) $R = 5.7; s^2 = 2.5843; s = 1.6076$

3.43 (a) $Q_1 \Rightarrow$ depth $= 15.5; Q_3 \Rightarrow$ depth $= 45.5$
(b) $Q_1 \Rightarrow$ depth $= 10; Q_3 \Rightarrow$ depth $= 28$
(c) $Q_1 \Rightarrow$ depth $= 25.5; Q_3 \Rightarrow$ depth $= 75.5$
(d) $Q_1 \Rightarrow$ depth $= 12.5; Q_3 \Rightarrow$ depth $= 36.5$

3.45 (a) $s^2 = 430.4; s = 20.7461$ (b) $s^2 = 430.4; s = 20.7461$
The sample variance and the sample standard deviation are the same in parts (a) and (b).
(c) $s^2 = 172,160; s = 414.9217$
The sample variance is multiplied by $20^2 = 400$, and the sample standard deviation is multiplied by 20.

3.47 (a) $R = 7.31$ (b) $s^2 = 6.622; s = 2.5733$
(c) $Q_1 = 4.995; Q_3 = 9.455;$ IQR $= 4.46$

3.49 (a) $s_2 = 1,126,239.7; s = 1061.24$
(b) $Q_1 = 4726; Q_3 = 5777$ (c) IQR $= 1051;$ QD $= 525.5$

3.51 (a) $\bar{x} = \frac{1}{8}(252 + \cdots + 251) = 245.875$

$s^2 = \frac{1}{7}[(252 - 245.875)^2 + \cdots + (251 - 245.875)^2]$

$= \frac{1}{7}[750.875] = 107.2679$

(b) $s^2 = \frac{1}{7}(484,387 - \frac{1}{8}(3,869,089)) = 107.2679$
(c) The answers in parts (a) and (b) are the same.

3.53 (a) $Q_1 = 291; Q_3 = 313;$ IQR $= 22$
(b) $s^2 = 536.4889; s = 23.1622$ **(c)** IQR $= 22; s^2 = 1160.3222$
(d) IQR is the same; s^2 is larger. s^2 is more sensitive to outliers.

3.55 (a) $s^2 = 1,437,360.5; s = 1198.9$
(b) $Q_1 = 9; Q_3 = 278.5;$ IQR $= 269.5$
(c) $s^2 = 69,875.5; s = 264.34$
$Q_1 = 7; Q_3 = 200;$ IQR $= 193$
(d) The sample standard deviation is smaller in the reduced data set. This makes sense since we removed the two largest observations (outliers). The remaining data set is more compact. IQR is slightly less in the reduced data set. Since we removed the two largest observations, the middle 50% of the data shifted slightly to the left, which resulted in a smaller IQR.

3.57 (a) $Q_1 = 12; Q_3 = 21;$ IQR $= 9$
(b) The minimum observation (5) can be as large as 21 without changing IQR.
(c) CQV $= 27.27$

3.59 (a) $s^2 = 5,343,315.007; s = 2311.56$
(b) $Q_1 = 364; Q_3; Q_3 = 2422;$ IQR $= 2058$
(c) Since there is one outlier, IQR is probably a better measure of variability.
(d) $\bar{x} = 1707.778$
$\sum(x_i - \bar{x}) = -1343.778 + \cdots + (-1170.7778) = 0$

3.61 (a) $s^2 = 1,157,005.05; s = 1075.64$
(b) $s^2 = 968.08; s = 968.08$
(c) The new sample variance is $(0.9)^2$ times the original sample variance. The new sample standard deviation is 0.9 times the original sample standard deviation.

3.63 (a) $s^2 = 70.1449; s = 8.3753$
(b) $s^2 = 14.0181; s = 3.7441$
(c) The new sample variance is $(0.44704)^2$ times the original sample variance. The new sample standard deviation is 0.44704 times the original sample standard deviation.

3.65 (a) $s^2 = 3048.9333; s = 55.2171$
(b) $s^2 = 3048.9333; s = 55.2171$
(c) The sample variance and the sample standard deviation are the same in each data set.
(d) $s_y^2 = s_x^2; s_y = s_x$

3.67 $s_y^2 = a^2 s_x^2; s_y = |a| s_x$

3.69 (a) $s_x^2 = 8175.3553; s_x = 90.4177$
(b) $s_y^2 = 130,805.7; s_y = 361.6707$
(c) $s_y^2 = 4^2 \cdot s_x^2; s_y = |4| \cdot s_x$
(d) For the original data set:
$Q_1 = 341.5; Q_3 = 481.5;$ IQR $= 140$
For the modified data set:
$Q_1 = 2366; Q_3 = 2926;$ IQR $= 560$
The new quartiles are 4 times the original plus 1000.
The new IQR is 4 times the original IQR.

3.71 Answers will vary. Larger values of g_1 indicate a greater lack of symmetry, or more skewness. Larger values (in magnitude) of g_2 suggest a more defined peak in the distribution. Smaller values of g_2 suggest a flatter, more uniform distribution.

3.3 The Empirical Rule and Measures of Relative Standing

3.73 True

3.75 Chebyshev's Rule applies to any distribution. This rule makes a statement about the minimum proportion. In practice, there will almost always be a larger proportion of observations within the interval.

3.77 The observation is very unusual, or rare. It is likely that some assumption is wrong.

3.79 True

3.81 (a) 0.75   **(b)** 0.8889   **(c)** 0.6094   **(d)** 0.6735   **(e)** 0.84
(f) 0.8724   **(g)** 0.8025

3.83 (a) $z = 3.0$ **(b)** $z = -1.25$ **(c)** $z = 0.8333$
(d) $z = -1.1111$ **(e)** $z = 1.6563$ **(f)** $z = 0.4545$
(g) $z = -2.2$ **(h)** $z = 4.1143$ **(i)** $z = 1.1111$ **(j)** $z = 5.8125$

3.85 (a) depth $= 120.5$ **(b)** depth $= 90$ **(c)** depth $= 22$
(d) depth $= 30.5$ **(e)** depth $= 20.5$ **(f)** depth $= 3525$

3.87 (a) $\bar{x} \pm 2s = (18.8, 32.4); \bar{x} \pm 3s = (15.4, 35.8)$
(b) $\bar{x} \pm s = (22.2, 29.0)$ is a symmetric interval about the mean, 1 standard deviation in each direction. Using the Empirical Rule, approximately 0.68 of the speeds lie in this interval.

3.89 (a) $\bar{x} \pm s = (6.4, 8.8); \bar{x} \pm 2s = (5.2, 10)$
(b) Using Chebyshev's Rule, at least 0.75 of the times lie in this interval.
(c) Using Chebyshev's Rule, at most 0.1111 of the times are either less than 4 or greater than 11.2 years.
(d) Using the Empirical Rule, approximately 0.95 of the times lie in the interval (5.2, 10).
Using the Empirical Rule, approximately 0.003 of the times are either less than 4 or greater than 11.2 years.

3.91 (a) Using the Empirical Rule, approximately 0.68 of the depths lie in this interval.   **(b)** Approximately 0.16 of the depths are less than 14.6.   **(c)** Approximately 0.8385 of the depths are between 14.6 and 23.

3.93 (a) 89% of all scores were 513 or less, and 11% of the scores were greater than 513.
(b) The 50th percentile is the median. Half of all scores were below 500 and half were above 500.
(c) A score in the 99th percentile is greater than 99% of all scores.

3.95 (a) Claim: $\mu = 11$ ($\sigma = 2.5$, distribution approximately normal)
Experiment: $x = 13$
Likelihood: $z = (13 - 11)/2.5 = 0.80$
Conclusion: This is a reasonable z-score subject to ordinary variability. There is no evidence to suggest the manager's claim is false.
(b) Claim: $\mu = 11$ ($\sigma = 2.5$, distribution approximately normal)
Experiment: $x = 20$
Likelihood: $z = (20 - 11)/2.5 = 3.6$
Conclusion: This is a very unusual observation. Almost all observations from a normal distribution have a z-score between −3 and 3. There is evidence to suggest the manager's claim is false.

3.97 (a) Claim: $\mu = 15$ ($\sigma = 3$, distribution approximately normal)
Experiment: $x = 11$
Likelihood: $z = (11-15)/3 = -1.33$
Conclusion: This is a reasonable z-score subject to ordinary variability. There is no evidence to suggest the general physician's claim is false.
(b) Claim: $\mu = 15$ ($\sigma = 3$, distribution approximately normal)
Experiment: $x = 22$
Likelihood: $z = (22-15)/3 = 2.33$
Conclusion: This is an unusual observation. Almost all observations from a normal distribution have a z-score between -2 and 2. There is evidence to suggest the general physician is spending more than 15 minutes with each patient.

3.99 (a)

Interval	Proportion
Within 1s: (116.83, 224.79)	50/68 = 0.74
Within 2s: (52.84, 308.77)	64/68 = 0.94
Within 3s: (−11.14, 372.76)	68/68 = 1.00

(b) These proportions are close enough to the Empirical Rule proportions. There is not enough evidence to suggest that the shape of the distribution is non-normal.
(c) Frequency histogram:

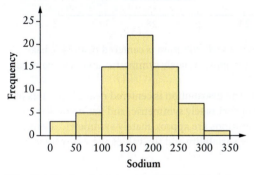

The distribution is approximately bell-shaped.

3.101 (a) $\bar{x} = 6.0248$; $s^2 = 0.6423$; $s = 0.8015$
(b) z-scores: 1.19, −1.10, 1.12, −0.22, 0.26, 0.82, 0.14, −1.90, 1.23, 0.43, −1.40, 0.23, 0.53, −0.06, 0.37, −0.57, −1.18, −1.42, −0.39, 1.89, 1.04, −0.03, −0.22, 0.23, −0.23, −0.23, −0.41, −1.53, 0.56, 1.34, −0.68, 1.27, 0.73, 1.05, 2.51, 0.16, −1.37, −0.11, 0.09, −0.98, 1.44, −0.77, 0.31, −0.72, −0.12, 0.74, −0.82, −0.48, −2.31, −0.46
(c) $p_{16} = -1.0416$, $p_{84} = 1.0858$
By the Empirical Rule, approximately 68% of the data is within 1 standard deviation of the mean. Therefore, we expect these percentiles to be close to -1 and $+1$, for a normal distribution.

3.103 Using Chebyshev's Rule with $k = 2$, at least 0.75 of the lifetimes lie in the interval
$\bar{x} \pm 2s = (48, 96)$.
A reasonable guaranteed life is 48 months.

3.4 Five-Number Summary and Box Plots

3.105 False

3.107 False

3.109 False

3.111 (a)

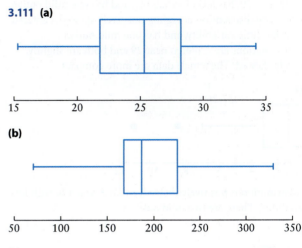

(b)

(c)

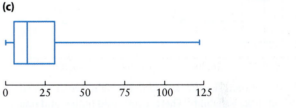

(d)

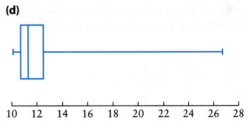

3.113 (a) $IF_L = 6.5$, $IF_H = 42.5$, $OF_L = -7.0$, $OF_H = 56.0$; $x = 35$ is within the inner fences and therefore is not an outlier.
(b) $IF_L = 448.9$, $IF_H = 548.1$, $OF_L = 411.7$, $OF_H = 585.3$; $x = 440$ is between OF_L and IF_L and therefore is a mild outlier.
(c) $IF_L = 3.47$, $IF_H = 8.03$, $OF_L = 1.76$, $OF_H = 9.74$; $x = 4.2$ is within the inner fences and therefore is not an outlier.
(d) $IF_L = 89.1$, $IF_H = 108.3$, $OF_L = 81.9$, $OF_H = 115.5$; $x = 116.5$ is greater than OF_H and therefore is an extreme outlier.
(e) $IF_L = 68.695$, $IF_H = 69.295$, $OF_L = 68.470$, $OF_H = 69.520$; $x = 68.4$ is less than OF_L and therefore is an extreme outlier.
(f) $IF_L = 36.265$, $IF_H = 209.585$, $OF_L = -28.730$, $OF_H = 274.580$; $x = 132.6$ is within the inner fences and therefore is not an outlier.

3.115

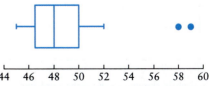

The distribution is centered near 48, is positively skewed, and has two mild outliers.

3.117 The distribution is centered near 1800, has a lot of variability, and is skewed to the right. There are three mild outliers.

3.119 Males: The distribution is slightly positively skewed, centered near 29, has lots of variability, and has one mild outlier.
Females: The distribution is slightly positively skewed, centered near 29, has little variability, and has one mild outlier.
Both distributions are centered near 29 and both are slightly positively skewed. The female data are more compact.

3.121

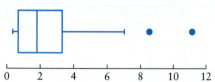

The distribution is positively skewed, entered near 1.8, with lots of variability. There are two mild outliers.

3.123

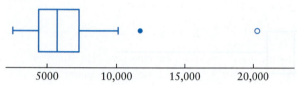

The distribution is slightly skewed to the right, centered near 5700, with little variability. There is one mild outlier and one extreme outlier. The box plot does not provide an evidence to suggest the center of the distribution is different from 6300.

3.125 (a)

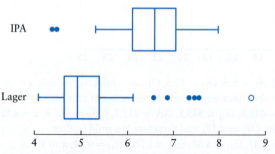

(b) The IPA distribution is slightly skewed to the left, centered near 6.5, with moderate variability. There are two mild outliers. The lager distribution is skewed to the right, centered near 4.9, with moderate variability. There are five mild outliers and one extreme outlier.
(c) The box plots suggest that, on average, the ABV value is less for IPA home brews than for lager home brews.

3.127 (a)

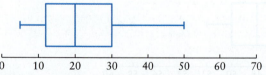

(b) The distribution is slightly skewed to the right, centered near 20, with lots of variability. There are no outliers.
(c) New box plot:

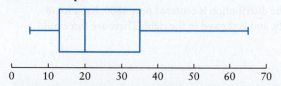

The two box plots are centered at approximately 20. Neither box plot has outliers. The new box plot is skewed to the right and appears to have more variability.

3.129 (a)

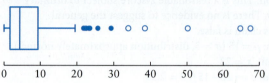

(b) The distribution is skewed to the right, centered near 4, with lots of variability. There are at least 5 mild outliers and 5 extreme outliers.

Chapter 3 Exercises

3.131 (a)

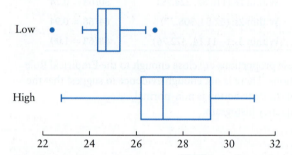

(b) Miami (low): the distribution is centered near 24.5, has little variability, is approximately symmetric, and has two mild outliers.
Denver (high): The distribution is centered near 27, has lots of variability, is approximately symmetric, and has no outliers.
(c) Both distributions are approximately symmetric. Denver (high) has more variability and the values are, on average, larger.

3.133 (a) Claim: $\mu = 12$ ($\sigma = 0.07$, distribution approximately normal)
Experiment: $x = 12.04$
Likelihood: $z = (12.04 - 12)/0.07 = 0.57$
Conclusion: This is a reasonable z-score subject to ordinary variability. There is no evidence to suggest this generating capacity is unusual.
(b) Claim: $\mu = 12$ ($\sigma = 0.07$, distribution approximately normal)
Experiment: $x = 11.8$
Likelihood: $z = (11.8 - 12)/0.07 = -2.86$
Conclusion: This is an unusual observation. Almost all observations from a normal distribution have a z-score between -2 and 2. There is evidence to suggest the generating capacity is unusual.

3.135 (a) $\bar{x} = 3.0003$; $\tilde{x} = 2.995$
(b) Since the mean is approximately equal to the median, this suggests the distribution is approximately symmetric.
(c) $\bar{x}_{tr(0.10)} = 3.0046$
A trimmed mean is not necessary. The distribution is approximately symmetric, and there are no extreme outliers.

3.137 **(a)** $\bar{x} = 88.1302$; $s^2 = 216.6583$; $s = 14.7193$

(b)

Interval	Proportion
Within $1s$: (73.41, 102.85)	$166/192 = 0.86$
Within $2s$: (58.69, 117.57)	$182/192 = 0.95$
Within $3s$: (43.97, 132.29)	$187/192 = 0.97$

Since these proportions are not close enough to the Empirical Rule proportions, this suggests the shape of the distribution is not normal.

(c)

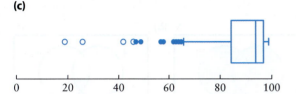

The box plot also suggests that the distribution is non-normal. The distribution is skewed to the left, centered near 90, with several mild and extreme outliers.

3.139 **(a)** $\bar{x} = 458.2083$; $s^2 = 2231.4764$; $s = 47.2385$

(b)

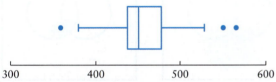

There are three mild outliers: 359, 551, 566.

(c) $p_8 = 380$; $p_9 = 409$; 400 is in the ninth percentile.

(d) At least 0.75 of the observations lie in the interval $\bar{x} \pm 2s = (363.73, 552.69)$.

At least 0.89 of the observations lie in the interval $\bar{x} \pm 3s = (316.49, 599.92)$.

3.141 **(a)** $\bar{x} = 1.5201$; $s^2 = 1.4953$; $s = 1.2228$

(b)

Interval	Proportion
Within $1s$: (0.30, 2.74)	$387/500 = 0.78$
Within $2s$: (-0.93, 3.97)	$456/500 = 0.91$
Within $3s$: (-2.15, 5.19)	$496/500 = 0.99$

The first two proportions are far enough away from the Empirical Rule proportions to suggest that the distribution is non-normal.

(c)

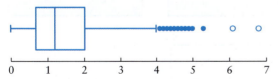

The box plot suggests that the distribution is skewed to the right. This conclusion supports part (b).

(d) $p_{40} = 0.955$; $p_{80} = 2.245$

(e) $z = (5 - 1.5201)/1.2228 = 2.85$. A magnitude 5 earthquake is 2.85 standard deviations from the mean. Therefore, this is a relatively unlikely magnitude.

3.143 Consider the number line.

3588	4038	4488	4938	5388	5838	6288
$\mu - 3\sigma$	$\mu - 2\sigma$	$\mu - \sigma$	μ	$\mu + \sigma$	$\mu + 2\sigma$	$\mu + 3\sigma$

(a) 0.95 **(b)** 0.0015 **(c)** 0.8385

(d) Claim: $\mu = 4938$ ($\sigma = 450$, distribution approximately normal)

Experiment: $x = 5300$

Likelihood: $z = (5300 - 4938)/450 = 0.80$

Conclusion: This is a reasonable z-score subject to ordinary variability. There is no evidence to suggest the mean depth of wells has changed.

3.145 **(a)** (2.76, 2.84) is a symmetric interval about the mean, $k = 2$ standard deviations in each direction. Using Chebyshev's Rule, at least 0.75 of the times lie in this interval.

(b) Claim: $\mu = 2.8$ ($\sigma = 0.02$)

Experiment: $x = 2.78$

Likelihood: $z = (2.78 - 2.8)/0.02 = -1.00$

Conclusion: This is a reasonable z-score subject to ordinary variability. There is no evidence to suggest the mean weight of a king-size package is different from 2.8 oz.

3.147 **(a)** $z = \dfrac{1 - 0.7}{0.1} = 3.0$. It is unlikely that a fisherman will catch a smallmouth bass with mercury level greater than 1 because this value is 3 standard deviations from the mean.

(b) $z = \dfrac{1 - 0.7}{0.5} = 6$. It is even more unlikely that a fisherman will catch a smallmouth bass with mercury level greater than 1 because this is 6 standard deviations from the mean.

(c) Blue: $\sigma = 0.1$; green: $\sigma = 0.05$

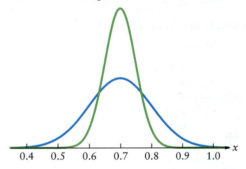

3.149 **(a)** $\bar{x} = 258.84$; $\tilde{x} = 277.5$

Since $\bar{x} < \tilde{x}$, this suggests the distribution is slightly skewed to the left.

(b) $s^2 = 13{,}407.5$; $s = 115.79$

Interval	Proportion
Within $1s$: (143.0, 374.6)	$384/600 = 0.64$
Within $2s$: (27.3, 490.4)	$588/600 = 0.98$
Within $3s$: (-88.5, 606.2)	$600/600 = 1.00$

These proportions are reasonable close to the Empirical Rule proportions. There is not enough evidence to suggest that the distribution is non-normal.

(c)

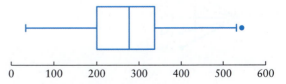

The box plot is approximately symmetric, with one mild outlier. This graph supports the conclusions in part (b). There is not enough evidence to suggest that the distribution is non-normal.

(d) $p_{98} = 491.3$; $p_{99} = 515$. Therefore, 500 lies in the 99th percentile. A person lifting 500 kg has a very good chance of winning the meet.

Chapter 4

4.1 Experiments, Sample Spaces, and Events

4.1 False

4.3 True

4.5 (a) not (b) or (c) and

4.7

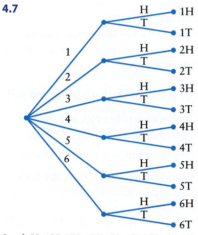

$S = \{1H, 2H, 3H, 4H, 5H, 6H, 1T, 2T, 3T, 4T, 5T, 6T\}$

4.9 There are $5 \times 5 = 25$ outcomes.

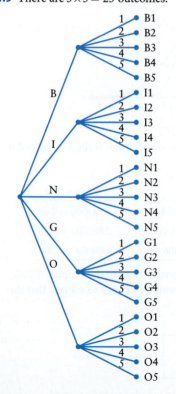

4.11 (a) $A' = \{1, 3, 5, 7, 9\}$ (b) $C' = \{5, 6, 7, 8, 9\}$
(c) $D' = \{0, 1, 2, 3, 4\}$ (d) $A \cup B = \{0, 1, 2, 3, 4, 5, 6, 7, 8, 9\} = S$
(e) $A \cup C = \{0, 1, 2, 3, 4, 6, 8\}$ (f) $A \cup D = \{0, 2, 4, 5, 6, 7, 8, 9\}$

4.13 (a) $A' = \{b, d, f, h, i, j, k\}$ (b) $C' = \{a, b, d, e, j, k\}$
(c) $D' = \{c, f, i\}$ (d) $A \cap B = \{c\}$ (e) $A \cap C = \{c, g\}$
(f) $C \cap D = \{g, h\}$

4.15 (a) $(A \cap B \cap C)' = \{a, b, d, e, f, g, h, i, j, k\}$
(b) $A \cup B \cup C \cup D = \{a, b, c, d, e, f, g, h, i, j, k\}$
(c) $(B \cup C \cup D)' = \{\ \}$ (d) $B' \cap C' \cap D' = \{\ \}$

4.17

4.19

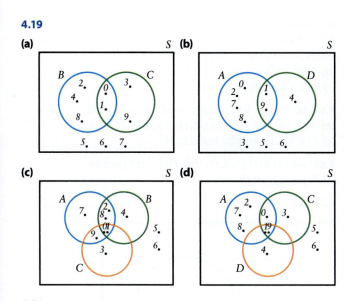

4.21

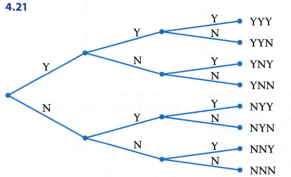

$S = \{YYY, YYN, YNY, YNN, NYY, NYN, NNY, NNN\}$

4.23 Use the following abbreviations: valid registration (V), invalid registration (I), properly insured (P), not properly insured (N), and muffler condition: none (W), standard (T), modified (M).

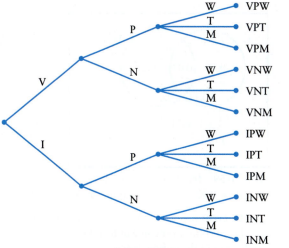

$S = \{VPW, VPT, VPM, VMW, VNT, VNM$
$\quad IPW, IPT, IPM, INW, INT, INM\}$

4.25 (a) There are 4 outcomes: S, FS, FFS, FFFS.
(b) The outcome FSFF is not possible because the experiment is over as soon as the strawberry cream is found.

4.27 For either a pass or a rush, there are 199 different yard possibilities. Therefore, the total number of outcomes is $n = 2 \times 199 = 398$.

4.29 (a) $S = \{LS, LU, LV, LP, RS, RU, RV, RP, SS, SU, SV, SP\}$
(b) $A = \{LV, RV, SV\}$,
$B = \{LS, LP, RS, RP, SS, SP\}$,
$C = \{LS, LU, LV, LP\}$,
$D = \{RS, RU, RV, RP, SS, SU, SV, SP\}$
(c) $C \cup D = S, C \cap D = \{\ \}$

4.31 (a) $S = \{D0, D1, D2, D3, D4, D5, I0, I1, I2, I3, I4, I5\}$
(b) $A =$ The passenger had 0 bags.
$B =$ The passenger was on an international flight.
$C =$ The passenger had 1 or 2 bags.
$D =$ The passenger was on an international flight and had 0 or 5 bags.
$E =$ The passenger had an odd number of bags.

4.33 (a)
$S = \{CSY, CSN, CWY, CWN, CBY, CBN, CKY, CKN,$
$\quad MSY, MSN, MWY, MWN, MBY, MBN, MKY,$
$\quad MKN, ESY, ESN, EWY, EWN, EBY, EBN, EKY, EKN\}$
(b) i. $\{CWY, CWN, MWY, MWN, EWY, EWN\}$
ii. $\{CBY, MBY, EBY\}$
iii. $\{CSY, CSN, CWY, CWN, CBY, CBN, CKY, CKN, MSN,$
$\quad MKN, ESN, EWN, EBN, EKN\}$
iv. $\{EKY, EKN\}$
v. $\{CSY, CSN, CWY, SWN, CBY, CBN, ESY, ESN, EWY, EWN,$
$\quad EBY, EBN\}$

4.35 (a) There are infinitely many outcomes in this experiment.
(b) Some of the outcomes in this experiment are H, BH, BBH, BBBH, BBBBH.

4.37 (a) $S = \{1Y, 2Y, 3Y, 4Y, 5Y, 6Y, 7Y, 1N,$
$2N, 3N, 4N, 5N, 6N, 7N\}$
(b) i. $B' = \{4Y, 5Y, 6Y, 7Y, 4N, 5N, 6N, 7N\}$
ii. $A \cup B = \{Y1, Y2, Y3, N1, N2, N3, N4, N5, N6, N7\}$
iii. $A \cap B = \{N1, N2, N3\}$
iv. $C \cap D = \{5Y, 7Y\}$
v. $A \cap B \cap D = \{1N, 3N\}$
vi. $(A \cap D)' = \{1Y, 2Y, 3Y, 4Y, 5Y, 6Y, 7Y, 2N, 4N, 6N\}$

4.39 (a)

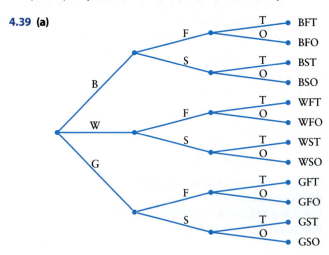

(b) $S = \{$BFT, BFO, BST, BSO, WFT, WFO, WST, WSO, GFT, GFO, GST, GSO$\}$
i. $A \cup B = \{$BFO, BFT, BSO, BST, GSO, GST, WSO, WST$\}$
ii. $B \cup C = \{$BFO, BFT, BSO, BST, GFT, GSO, GST, WFT, WSO, WST$\}$
iii. $B \cap C = \{$BSO, BST$\}$
iv. $C' = \{$GSO, GST, WSO, WST$\}$

4.2 An Introduction to Probability

4.41 True

4.43 False

4.45 True

4.47 False

4.49 Sometimes true

4.51 (a) $P(A) = 0.5$ (b) $P(B) = 0.3333$
(c) $P(C) = 0.3333$ (d) $P(D) = 0.5$

4.53 (a) $P(A \cup B) = 0.85$ (b) $P(A \cap B) = 0.15$
(c) P(just A) = $P(A) - P(A \cap B)$
$\phantom{P(\text{just }A)} = 0.55 - 0.15 = 0.40$
(d) P(just A or just B) = $P(A \cup B) - P(A \cap B)$
$\phantom{P(\text{just }A \text{ or just }B)} = 0.85 - 0.15 = 0.70$

4.55 (a) $P(A \cup B) = 0.532$ (b) $P[(A \cup B)'] = 0.468$
(c) $P(B') = 0.594$ (d) $P[(A \cap B)'] = 0.771$

4.57

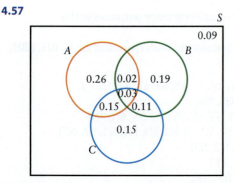

4.59 (a) $n = 10$
{HLP, HLD, HLV, HPD, HPV, HDV, LPD, LPV, LDV, PDV}
(b) P(includes health insurance) = 0.60
(c) P(life insurance and prescription plan) = 0.30

4.61 (a) P(0%, 10%, 12%, 22%) = 0.9461
(b) P(35% or 37%) = 0.0078
(c) P(not 24%) = 0.9666

4.63 (a) P(NS or NB) = 0.0476
(b) P(not BC or Yukon) = 0.8668
(c) P(not Ontario or Quebec or BC) = 0.2620
(d) $n = (35{,}151{,}728)(0.0010) \approx 35{,}152$

4.65 (a) $P(A) = 0.23$; $P(B) = 0.52$; $P(C) = 0.40$
(b) $P(A \cup B) = 0.75$
$P(A \cap B) = P(\{\ \}) = 0$
$P(B \cap C) = 0.12$

(c) $P(A') = 0.77$
$P(A' \cap C) = 0.17$
$P(A \cap B \cap C) = P(\{\ \}) = 0$
(d) $P(B' \cap C') = 0.20$
$P[(B \cup C)'] = 0.20$

4.67 (a) $P(A) = 0.513$; $P(B) = 0.925$; $P(C) = 0.105$
(b) $P(A \cap B) = 0.513$; $P(A \cup C) = 0.618$; $P(A \cap C) = 0$
(c) $P(A' \cup C) = 0.4870$; $P(A \cup B \cup C') = 1$

4.69 (a) $P(F \cup H) = 0.32$ (b) $P[(F \cup H)'] = 0.68$
(c) P(just F) = 0.18

4.71 (a)

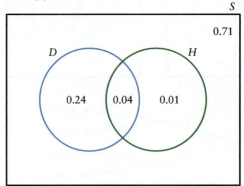

(b) $P(D \cup H) = 0.29$ (c) $P(D \cup H) - P(D \cap H) = 0.25$
(d) P(just D) = 0.24 (e) $P(H') = 0.95$

4.73 (a) There are 6 defect-free tires (G1–G6) and 2 reject tires (B1 and B2).
G_1G_2, G_1G_3, G_1G_4, G_1G_5, G_1G_6, G_1B_1, G_1B_2,
G_2G_3, G_2G_4, G_2G_5, G_2G_6, G_2B_1, G_2B_2, G_3G_4,
G_3G_5, G_3G_6, G_3B_1, G_3B_2, G_4G_5, G_4G_6, G_4B_1,
G_4B_2, G_5G_6, G_5B_1, G_5B_2, G_6B_1, G_6B_2, B_1B_2
(b) P(both defect-free) = 0.5357
(c) P(at least one defect) = 0.4643 (d) P(both defects) = 0.0357

4.75 (a)

(b) $P(C \cup W) = 0.59$ (c) $P(W \cup J) = 0.73$
(d) $P(C \cup W \cup J) = 0.90$ (e) P(some other reason) = 0.10

4.77 Let B = the event that everyone buys a breakfast sandwich with a bagel. There is only 1 outcome in B and there are $2^{10} = 1024$ outcomes in the sample space.

$P(B) = \dfrac{1}{1024} = 0.000976$

Since this probability is so small, all 10 people buying a breakfast sandwich with a bagel is a rare event. This suggests that the assumption is wrong. There is evidence to suggest that the demand for each type of sandwich is not equal.

4.3 Counting Techniques

4.79 True

4.81 True

4.83 tree diagram

4.85 combination

4.87 equally likely outcomes experiment

4.89 (a) $\binom{9}{5} = \frac{9!}{5!4!} = 126$ (b) $\binom{9}{4} = \frac{9!}{4!5!} = 126$

(c) $\binom{14}{7} = \frac{14!}{7!7!} = 3432$ (d) $\binom{10}{10} = \frac{10!}{10!0!} = 1$

(e) $\binom{10}{1} = \frac{10!}{1!9!} = 10$ (f) $\binom{10}{0} = \frac{10!}{0!10!} = 1$

(g) $\binom{12}{3} = \frac{12!}{3!9!} = 220$ (h) $\binom{16}{7} = \frac{16!}{7!9!} = 11,440$

(i) $\binom{20}{18} = \frac{20!}{18!2!} = 190$

4.91 (a) $n = 5 \times 8 \times 15 = 600$ (b) $n = 5 \times 8 \times 15 \times 7 = 4200$

4.93 $n = 6 \times 10 \times 4 = 240$

4.95 (a) $\binom{20}{6} = \frac{20!}{6!14!} = 38,760$

(b) $\binom{17}{5}\binom{3}{1} = \left(\frac{17!}{5!12!}\right)\left(\frac{3!}{1!2!}\right) = 18,564$

(c) $\binom{17}{3}\binom{3}{3} = \left(\frac{17!}{3!14!}\right)\left(\frac{3!}{3!0!}\right) = 680$

4.97 (a) $n = 64,000$ (b) P(single-digit numbers) = 0.0156
(c) $n = {}_{40}P_3 = 59,280$;
P(single-digit numbers) = 0.0121

4.99 (a) $n = 400$ (b) $n = 80$ (c) $n = 360$

4.101 (a) Number of different schedules = ${}_{20}P_8 = 5,079,110,400$
(b) P(all appointments with women) = ${}_{15}P_8 / 5,079,110,400$
= 0.0511

4.103 (a) Number of different collections = $\binom{10}{5} = 252$

(b) P(no Crazy Crawler) = 0.5
(c) P(at least one is an Excalibur) = 0.9167

4.105 (a) Number of different collections = $\binom{20}{7} = 77,520$

(b) P(all 7 by the local artist) = 0.0015
(c) Number of different ordered arrangements
= ${}_{20}P_7 = 390,700,800$

4.107 (a) Number of ways to select 8 tiles = $\binom{12}{8} = 495$

P(2 tiles with blue tint included) = 0.4242
(b) P(5 family herbs included) = 0.0707

4.109 (a) Number of faculty book collections $\binom{25}{10} = 3,268,760$

(b) P(no math faculty books) = 0.1978

(c) P(all 10 from COST faculty) = 0.0009
(d) P(no books from COST faculty) = 0.0000003
If none of the books is from COST faculty, the process was probably not random, since this probability is so small.

4.111 (a) P(all fraternal twins) = $\dfrac{\binom{14}{8}}{\binom{20}{8}} = 0.0238$

(b) P(all identical twins selected) = $\dfrac{\binom{14}{2}\binom{6}{6}}{\binom{20}{8}} = 0.0007$

(c) P(4 fraternal and 4 identical) = $\dfrac{\binom{14}{4}\binom{6}{4}}{\binom{20}{8}} = 0.1192$

(d)
P(at most 2 identical sets of twins) = $\dfrac{\binom{14}{8}}{\binom{20}{8}} + \dfrac{\binom{14}{7}\binom{6}{1}}{\binom{20}{8}} + \dfrac{\binom{14}{6}\binom{6}{2}}{\binom{20}{8}}$

= 0.5449

4.113 (a) Number of access codes = $2^{10} = 1024$
(b) P(code with exactly 1 zero) = 0.0098
(c) Number of access codes = $3^{10} = 59049$
P(code with exactly 1 zero) = 0.0867

4.115 (a) Number of possible committees = $\binom{14}{4} = 1001$
(b) P(all Republicans) = 0.0699
(c) P(all Democrats) = 0.0150
There is some evidence to suggest the selection process was not random because the probability of selecting a committee with all Democrats is so small.

4.117 (a) P(10 murals, 10 sculptures) = 0.0517
(b) P(20 murals) = 0.00008
(c) P(18 murals: codifiers of law) = 3.46×10^{-18}
(d) Note: There are now 115 possible items.
P(all panels and murals) = 3.1×10^{-34}

4.119 (a) 4 (b) 8 (c) 16 (d) 2^n

4.121 Number of ways to arrange n people at a round table = $n!/n = (n-1)!$

4.123 (a) $\binom{100}{18} = 30,664,510,802,988,208,300$

(b) P(all Democrats) = 0.000000149
(c) P(6 Democrats) = 0.0934

4.4 Conditional Probability

4.125 True

4.127 False

4.129 B

4.131 $P(A) = P(A \cap B_1) + P(A \cap B_2) + P(A \cap B_3) + P(A \cap B_4)$

4.133 (a) Conditional (b) Unconditional (c) Unconditional (d) Conditional (e) Unconditional (f) Conditional

4.135 (a) $P(A_1) = 0.118$; $P(A_2) = 0.396$; $P(A_3) = 0.486$
(b) $P(B_1) = 0.455$; $P(B_2) = 0.442$; $P(B_3) = 0.103$
(c) $P(A_1 \cap B_1) = 0.095$; $P(A_2 \cap B_2) = 0.188$; $P(A_3 \cap B_3) = 0.093$
(d) $P(A_1 \mid B_1) = 0.2088$; $P(B_1 \mid A_1) = 0.8051$; $P(A_1' \cap B_1') = 0.5220$
(e) $P(B_2 \mid A_2) = 0.4747$; $P(B_3 \mid A_3) = 0.1914$

4.137 (a)

	B_1	B_2	B_3	
A_1	178	231	406	815
A_2	123	150	244	517
A_3	165	202	335	702
	466	583	985	2034

(b) Number of people who participated in the survey is 2034.
(c) $P(A_1) = 0.4007$; $P(A_2) = 0.2542$; $P(A_3) = 0.3451$
(d) $P(B_1 \cap A_1) = 0.0875$; $P(B_2 \cap A_2) = 0.0737$; $P(B_3 \cap A_3) = 0.1647$
(e) $P(A_3 \mid B_1) = 0.3541$; $P(B_2 \mid A_2) = 0.2901$; $P(A_3 \cap B_3') = 0.1804$; $P(A_3 \mid B_1') = 0.3425$

4.139 (a) $P(A) = 0.574$; $P(B) = 0.488$; $P(C) = 0.465$
(b) $P(A \cap B) = 0.297$; $P(B \cap C) = 0.218$
(c) $P(A \mid B) = 0.6086$; $P(B \mid C) = 0.4688$; $P[(A \cap B) \mid C] = 0.3333$
(d) $P(0 \mid C') = 0.2523$; $P(7 \mid C) = 0.1355$; $P[(A \cap B) \mid C'] = 0.7477$
(e) $P(2 \mid B) = 0.2910$; $P(3 \mid B) = 0.2623$; $P(7 \mid B) = 0.1291$

4.141 Let O = participates in an office football pool and C = cheats on income taxes. $P(C \mid O) = 0.60$

4.143 (a) $P(S \mid C) = 0.75$ (b) $P(S \mid C') = 0.75$

4.145 (a) $P(AR) = 0.1074$ (b) $P(CA \cap Tech) = 0.3420$
(c) $P(FL \mid Adv) = 0.3123$ (d) $P(Nov \mid MI) = 0.0123$
(e) $P[(AR \cup CA) \mid (Adv \cup Ex)'] = 0.6926$

4.147 (a) $P(east \mid A) = 0.7311$
(b) $P(E \mid west) = 0.1671$
(c) $P(east \mid E') = 0.5921$

4.149 (a) $P(W \cap P) = 0.4825$
(b) $P(P \mid M) = 0.66874$
(c) $P(W \mid P) = 0.6433$

4.151 (a) $P(O \mid L) = 0.4765$
(b) $P(O \mid L') = 0.6522$
(c) $P(O' \mid L') = 0.3478$

4.153 (a) $P(midwest \cap fair) = 0.144$
(b) $P(excellent \mid north) = 0.4493$
(c) $P(south \mid poor) = 0.7843$
(d) $P[north \mid (good \cup fair)] = 0.1814$
$P[midwest \mid (good \cup fair)] = 0.3733$
$P[south \mid (good \cup fair)] = 0.4453$
The customer is most likely from the south.

4.155 (a)

		Bus	Car	Walk	
Lunch	Carries	625	466	142	1233
	Buys	345	122	500	967
		970	588	642	2200

Arrival mode

(b) $P(carries \cap car) = 0.2118$
(c) $P(buys \mid bus) = 0.3557$
(d) $P(carries \mid walk') = 0.7003$
(e) $P(bus \mid buys) = 0.3568$
$P(car \mid buys) = 0.1262$
$P(walk \mid buys) = 0.5171$
The student most likely walked to school.

4.157 (a) $P(Adv) = 0.1613$
(b) $P(Univ \cap thrill) = 0.0263$
(c) $P(drama \mid Sony) = 0.2394$
(d) $P(Warner \mid comedy) = 0.0667$
(e) $P(other\ genre \mid Other') = 0.2083$

4.5 Independence

4.159 $P(A)$

4.161 $P(A) \cdot P(B) \cdot P(C)$

4.163 True

4.165 The events A and B are independent if the occurrence or nonoccurrence of one has no effect on the occurrence or nonoccurrence of the other.

4.167 (a) Independent (b) Dependent (c) Dependent (d) Dependent

4.169 (a) $P(A \cap B) = 0.085$; $P(B' \mid A) = 0.66$; $P(A \cap B') = 0.165$
(b) $P(A \cap B \cap C) = 0.0527$; $P[C' \mid (A \cap B)] = 0.38$; $P(A \cap B \cap C') = 0.0323$
(c) The events A and B are dependent since $P(B) \neq P(B \mid A)$.

4.171 (a) $P(A \cap B) = 0.10$; $P(A \cap C) = 0.18$; $P(A \cap D) = 0.12$
(b) There is not enough information to determine independence or dependence.
(c) $P(B' \mid A) = 0.75$
If the event A occurs, only B, C, or D can occur because $P(B \mid A) + P(C \mid A) + P(D \mid A) = 1$.

4.173 (a) $P(A') = 0.65$; $P(C \mid A) = 0.18$; $P(B \mid A') = 0.36$
(b) $P(A \cap C) = 0.063$; $P(A' \cap B) = 0.234$;
(c) $P(D) = 0.451$

4.175 Let F_i = fisherman i is fatally injured.
(a) $P(F_1 \cap F_2) = 0.00000081$
(b) $P(F_1' \cap F_2') = 0.9982$
(c) $P(1\ fatality) = 0.0018$

4.177 Let S_i = American i owns a smart speaker.
(a) $P(S_1 \cap S_2) = 0.0256$
(b) $P(1\ American\ owns\ a\ smart\ speaker) = 0.2688$
(c) $P(at\ most\ 1) = 0.9744$

4.179 (a) P(earthquake in all four fault regions) = 0.00069696
(b) P(earthquake in none of the four fault regions) = 0.4126
(c) P(earthquake in at least one fault region) = 0.5874
(d) P(earthquake in at least one fault region) = 0.8861

4.181 Let A_i = person i has Alzheimer's disease.
(a) P(all four have Alzheimer's) = 0.00083521
(b) P(exactly one has Alzheimer's) = 0.3888
(c) P(at least 2 have Alzheimer's) = 0.1366

4.183 Let R_i = stock i will rise.
(a) P(both stocks will rise) = 0.2646
(b) $P(R_1 \cap R_2') = 0.1554$
(c) $P(R_1 \cap R_2) = 0.3402$

4.185 Let S_i = makes shot i.
(a) P(makes both shots) = 0.7531
(b) P(misses both shots) = 0.0057
(c) P(makes only 1 shot) = 0.2412

4.187 Let C_i = person i believes aliens have contacted us.
(a) P(three believe aliens have contacted us) = 0.0270
(b) P(none believes aliens have contacted us) = 0.3430
(c) P(one believes aliens have contacted us) = 0.4410

4.189 Let C_i = oyster i contains a pearl.
Let $p = 1/12000$.
(a) P(none contains a pearl) = 0.9997
(b) P(at least one contains a pearl) = 0.0003
(c) Let n = the number of oysters necessary.
P(at least one contains a pearl) = $1 - (1-p)^n \geq 0.50 \Rightarrow n \geq 8318$

4.191 (a) Let L = a patient has a penicillin allergy listed in their medical records.
Let R = a patient has a reaction to penicillin.
$P(L) = 0.10$; $P(R|L) = 0.10$; $P(R \cap L') = 0.045$
(b)

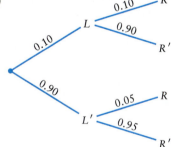

(c) $P(R \cap L) = 0.01$
(d) $P(L|R) = 0.1818$

4.193 (a) $P(A \cap L) = 0.0255$
(b) $P(L) = 0.1071$
P(passenger is on time) = 0.8929
(c) $P(A|L) = 0.2382$
$P(J|L) = 0.4044$
$P(D|L) = 0.2151$
$P(U|L) = 0.0945$
$P(V|L) = 0.0474$
The passenger most likely flew on Jet Blue.

4.195 Let C_i = person i has celiac disease.
(a) P(1 of 5 has celiac disease) = 0.0480
(b) P(first and fifth adults have celiac disease) = 0.00009703
(c) Claim: $P(C_i) = 0.01$
Experiment: All 5 have celiac disease.
Likelihood:
P(all 5 have celiac disease) = 0.0000000001
Conclusion: Since this probability is so small, there is evidence to suggest the claim concerning the percentage of adults with celiac disease is wrong.

4.197 Let L_i = professional i is considering leaving.
(a) P(all 6 are considering leaving) = 0.0055
(b) P(1 is considering leaving) = 0.1654
(c) Claim: $P(L_i) = 0.42$
Experiment: All 6 are considering leaving
Likelihood: P(all 6 considering leaving) = 0.0055
Conclusion: Since this probability is so small, there is evidence to suggest the claim concerning the percentage of professionals considering leaving the city is wrong.

4.199 $P(A|R) = 0.2830$
$P(B|R) = 0.3396$
$P(C|R) = 0.3774$
The salesperson most likely stayed at Hotel C.

Chapter 4 Exercises

4.201 (a) $\binom{20}{3} = 1140$

(b) P(2 mirrors selected will be damaged) = $\dfrac{\binom{17}{1}\binom{3}{2}}{1140} = 0.0447$

(c) P(at least 1 mirror damaged) = 1 − P(0 mirrors damaged) = 0.4035

4.203 (a)

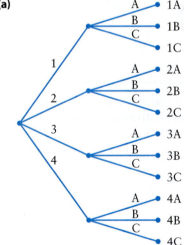

(b) $S = \{1A, 1B, 1C, 2A, 2B, 2C, 3A, 3B, 3C, 4A, 4B, 4C\}$
(c) $E = \{2A, 2B, 2C\}$
$F = \{1B, 2B, 3B, 4B\}$
$G = \{1A, 1B, 1C, 2A, 3A, 4A\}$
$H = \{3C\}$
(d) $E \cup F = \{2A, 2B, 2C, 1B, 3B, 4B\}$
$F \cap G = \{1B\}$
$H' = \{1A, 1B, 1C, 2A, 2B, 2C, 3A, 3B, 4A, 4B, 4C\}$

(e) $E \cap H' = \{2A, 2B, 2C\}$
$E \cup F \cup G = \{1A, 1B, 1C, 2A, 2B, 2C, 3A, 3B, 4A, 4B\}$
$F \cup G' = \{1B, 2B, 3B, 4B, 2C, 3C, 4C\}$

4.205 (a) $P(D \cap G) = 0.26$ (b) $P(G) = 0.80$
(c) $P(D \mid G') = 0.70$

4.207 (a) $P(S \cup H) = 0.95$ (b) $P[(S \cup H)'] = 0.05$
(c) $P(S) - P(S \cap H) = 0.25$ (d) $P(S \mid H) = 0.2143$
(e) P(customer gets only a haircut) $= P(H) - P(S \cap H) = 0.55$
P(2 customers get only a haircut) $= 0.3025$

4.209 (a) $P(S' \mid G) = 0.80$ (b) $P(S \cap G) = 0.09$
(c) $P(S' \cap G) = 0.36$

4.211 Let R_i = wedding i takes place in a rustic venue.
(a) $P(R_1 \cap R_2) = 0.0225$ (b) $P(R_1' \cap R_2') = 0.7225$
(c) P(both rustic | at least 1 rustic) $= 0.0811$

4.213 (a) $P(A' \mid U) = 0.65$ (b) $P(A \cap U) = 0.1995$
(c) $P(U_1 \cap U_2' \cap U_3') + P(U_1' \cap U_2 \cap U_3') + P(U_1' \cap U_2' \cap U_3)$
$= 0.3162$

4.215 Let N = never traveled outside of the state.
Let L = owns luggage.
(a) $P(N \cap L') = 0.0748$ (b) $P(L) = 0.7294$
(c) $P(N \mid L) = 0.0483$

4.217 (a) $P(CC \cap C') = 0.0630$
(b) $P(C') = 0.4475$ (c) $P(\text{ATM} \mid C') = 0.5520$

4.219 (a)

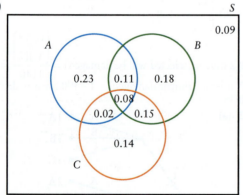

(b) P(just A) $= 0.23$ (c) $P(A' \cap B' \cap C') = 0.09$
(d) $P(A \mid C) = 0.2564$; $P[B \mid (A \cap C)] = 0.80$;
$P[(A \cap B \cap C) \mid A] = 0.1818$

4.221 (a) P(transfer) $= 0.29$
(b) P(deposit | south) $= 0.2857$
(c) P(west | cash) $= 0.2143$
(d) P(midwest | (cash)') $= 0.3056$
(e) P(balance | (north)') $= 0.2987$
(f) P(south $\cap$ cash) $= 0.04$
P(south) $\cdot$ P(cash) $= 0.0588 \neq$ P(south $\cap$ cash)
Therefore, the events south and withdraw cash are dependent.
(g) P(west $\cap$ transfer)$^2 = 0.0144$

4.223 (a) A simulation produced the following awards: 1, 2, 1, 5, 3, 1, 1, 1, 1, 1. One person won 5 or more free nights.

(b)

n	Rel Freq	n	Rel Freq	n	Rel Freq
50	0.0600	100	0.0900	150	0.0800
200	0.0950	250	0.0800	300	0.0567
350	0.0629	400	0.0775	450	0.0844
500	0.0600	550	0.0618	600	0.0733
650	0.0492	700	0.0586	750	0.0640
800	0.0713	850	0.0541	900	0.0633
950	0.0642	1000	0.0540	1050	0.0590
1100	0.0755	1150	0.0635	1200	0.0692
1250	0.0640	1300	0.0677	1350	0.0681
1400	0.0671	1450	0.0517	1500	0.0680
1550	0.0516	1600	0.0613	1650	0.0709
1700	0.0729	1750	0.0674	1800	0.0661
1850	0.0595	1900	0.0658	1950	0.0585
2000	0.0635				

(c)

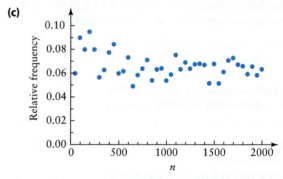

(d) An estimate of the probability of winning 5 or more free nights is 0.063.
(e) The exact probability is 0.0625.

Chapter 5

5.1 Random Variables

5.1 True

5.3 False

5.5 Function

5.7 Counting

5.9 Find all the outcomes that are mapped to 2, and add the probabilities associated with those outcomes.

5.11 (a) Discrete (b) Continuous
(c) Discrete (d) Discrete (e) Continuous
(f) Continuous

5.13 (a) Discrete (b) Continuous
(c) Discrete (d) Discrete (e) Continuous
(f) Continuous

5.15 (a) $S = \{MM, MW, MB, MG, WM, WW, WB, WG, BM, BW, BB, BG, GM, GW, GB, GG\}$
(b) Possible values for X: 0, 1, 2. X is discrete. X can assume only a finite number of values.

5.17 (a) Discrete **(b)** Continuous **(c)** Discrete **(d)** Continuous

5.19 X is continuous since it is a measurement of acceleration.

5.21 (a) Continuous **(b)** Continuous **(c)** Discrete **(d)** Continuous **(e)** Discrete **(f)** Discrete **(g)** Discrete **(h)** Continuous

5.2 Probability Distributions for Discrete Random Variables

5.23 True

5.25 False

5.27 All the possible values of X; the probability associated with each value

5.29 A probability distribution for a discrete random variable may be given by a complete listing of all values and associated probabilities, a table of values and probabilities, a probability histogram, a point representation, or a formula.

5.31 (a) $p(45) = 0.044$ **(b)** $P(Y \geq 25) = 0.608$ $P(Y > 25) = 0.424$ **(c)** $P(Y = 10, 20, 30, 50) = 0.772$
(d)

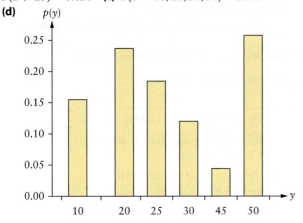

5.33 (a) Not valid because $\sum p(x) > 1$.
(b) Not valid because $p(8) < 0$.
(c) Valid because $0 \leq p(x) \leq 1$ for all x and $\sum p(x) = 1$.
(d) Not valid because $\sum p(x) > 1$.

5.35 Probability distribution:

y	1	2	3
p(y)	0.028	0.324	0.648

Probability histogram:

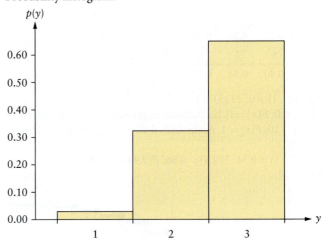

5.37 (a) $P(X = 0) = 0.900$ **(b)** $P(X \leq 2) = 0.975$
(c) $P(X_1 \geq 3 \cap X_2 \geq 3) = 0.000625$
(d) $P(2 \leq X \leq 4) = 0.049$ $P(2 < X < 4) = 0.020$

5.39 (a) $P(X = 1) = 0.247$ **(b)** $1 - P(X = 1) = 1 - 0.371$
(c) $P(X \leq 3) = 0.776$
(d) $P(X \geq 5 \mid X \geq 3) = 0.3351$
(e) $P(X_1 > 5 \cap X_2 > 5 \cap X_3 > 5) = 0.0003$

5.41 (a) Probability distribution:

y	0	1	2	3	4
p(y)	0.1897	0.3910	0.3021	0.1038	0.0134

(b) $P(Y \geq 1) = 0.8103$
(c) $P(Y = 4 \mid Y \geq 2) = 0.0320$

5.43

x	2	3	4	5	6	7	8
p(x)	0.01	0.03	0.0725	0.175	0.2125	0.25	0.25

5.45 (a) $P(X > 2) = 0.10$ **(b)** $P(X \neq 2) = 0.65$
(c) $P(X_1 = 1 \cap X_2 = 1) = 0.3025$
(d) P(total bags at least 8) = 0.0023
(e) Probability distribution:

y	50	100	150	200	250
p(y)	0.55	0.35	0.07	0.02	0.01

5.47 (a) Probability distribution:

m	100	250	500	1000
p(m)	0.0667	0.1333	0.4667	0.3333

(b) $P(M_1 = 1000 \cap M_2 = 1000) = 0.1111$

5.49 (a) Probability distribution:

y	0	1	2	3	4
p(y)	0.0031	0.0458	0.2195	0.4311	0.3005

(b) $P(Y \geq 1) = 0.9969$ **(c)** $P(Y_1 \geq 3 \cap Y_2 \geq 3) = 0.5352$

5.51 (a)

x	0	1	2	3	4	5
F(x)	0.21	0.39	0.54	0.66	0.76	0.84

x	6	7	8	9	10
F(x)	0.90	0.94	0.97	0.99	1.00

(b) $F(-1) = 0$; $F(25) = 1$
(c) $x < 0$; $F(x) = 0$. No cumulative probability until $x = 0$.
(d) $x > 10$; $F(x) = 1$. Accumulated all of the probability (1) at $x = 10$.
(e) $F(2.1) = 0.54$; $F(2.5) = 0.54$; $F(2.99) = 0.54$
(f) $F(x)$

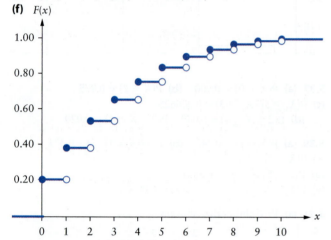

The graph of F is a step function.

5.3 Mean, Variance, and Standard Deviation for a Discrete Random Variable

5.53 True

5.55 True

5.57 False

5.59 The sum of the squared deviations about the mean

5.61

$x^2 \cdot p(x)$	x^2	$p(x)$	x	$x \cdot p(x)$
0.40	4	0.10	2	0.20
2.56	16	0.16	4	0.64
7.20	36	0.20	6	1.20
15.36	64	0.24	8	1.92
18.00	100	0.18	10	1.80
17.28	144	0.12	12	1.44
60.80				7.20

$\mu = 7.20$; $\sigma^2 = 8.96$; $\sigma = 2.9933$

5.63 (a) $\mu = 0$; $\sigma^2 = 270$; $\sigma = 16.4317$
(b) $P(\mu - 2\sigma \leq Y \leq \mu + 2\sigma)$
$= P(-32.8634 \leq Y \leq 32.8634) = 1$
(c) $P(Y \geq \mu) = P(Y \geq 0) = 0.55$
$P(Y > \mu) = P(Y > 0) = 0.45$

5.65 (a) $\mu = 7.35$; $\sigma^2 = 24.6275$; $\sigma = 4.9626$
(b) $P(X < \mu - \sigma) + P(X > \mu + \sigma)$
$= P(X < 2.3874) + P(X > 12.3126) = 0.40$
(c) $P(X \leq \mu + 2\sigma) = P(X \leq 17.2752) = 0.95$

5.67 (a) This is a valid probability distribution since $0 \leq p(x) \leq 1$ and $\sum p(x) = 1$.
(b) $\mu = 1.025$; $\sigma^2 = 1.8744$; $\sigma = 1.3691$
(c) $P(X < \mu) = P(X < 1.025) = 0.750$
(d) $P(X_1 \geq 4 \cap X_2 \geq 4) = 0.01$

5.69 (a) $\mu = 2.61$; $\sigma^2 = 1.2779$; $\sigma = 1.1304$
(b) $P(X \geq 5 \mid X \geq 3) = 0.0561$
(c) $P(X > \mu + \sigma) = P(X > 3.7404) = 0.230$

5.71 (a) $\mu = 15.55$; $\sigma^2 = 40.4475$; $\sigma = 6.3598$
(b) $P(\mu - \sigma \leq X \leq \mu + \sigma)$
$= P(9.1902 \leq X \leq 21.9098) = 0.82$
(c) $P(\mu - 2\sigma \leq X \leq \mu + 2\sigma)$
$= P(2.8304 \leq X \leq 28.2696) = 0.92$
(d) A sunlamp lasts for $(100)(60) = 6000$ min. The mean tanning session lasts 15.55 min. A sunlamp lasts $6000/15.55 = 385.85$ sessions, or approximately 386.

5.73 (a) $\mu = 7.998$; $\sigma^2 = 6.706$; $\sigma = 2.5896$
(b) $P(\mu - \sigma \leq X \leq \mu + \sigma)$
$= P(5.41 \leq X \leq 10.59) = 0.60$
(c) $1 - P(\mu - 2\sigma \leq X \leq \mu + 2\sigma)$
$= 1 - P(2.82 \leq X \leq 13.18) = 0.012$
P(both more than two standard deviations from the mean) = 0.000144
(d) P(both trips same number of riders) = 0.1120

5.75 (a) Probability distribution:

x	1	5	10	15	20
p(x)	0.4899	0.1984	0.0162	0.0364	0.2591

This is a valid probability distribution since $0 \leq p(x) \leq 1$ and $\sum p(x) = 1$.
(b) $\mu = 7.3725$; $\sigma^2 = 64.5576$; $\sigma = 8.0348$
(c) $P(X < \mu - \sigma) + P(X > \mu + \sigma)$
$= P(X < -0.6623) + P(X > 15.4073) = 0.2591$

5.77 (a) $\mu = 366.75$; $\sigma^2 = 2294.4375$; $\sigma = 47.9003$
(b) Let $F_i =$ temperature at least 400 for use i.
$P(F_i) = 0.280$; P(all 3 at least 400) = 0.022
(c) P(1 use is for 350) $= 3(0.400)(0.600)^2 = 0.432$

5.79 (a) $\mu = 3.27$; $\sigma^2 = 1.3771$; $\sigma = 1.1735$
(b) $P(X = 6 \mid X \geq 4) = 0.125$
(c) Let $B_i =$ wedding i has at least 3 bridesmaids.
$P(B_i \geq 3) = 0.74$
P(all 4 have at least 3 bridesmaids) = 0.2999

5.81 (a) $\mu = 0.6$; $\sigma^2 = 0.24$; $\sigma = 0.4899$
(b) $\mu = 0.7$; $\sigma^2 = 0.21$; $\sigma = 0.4583$
(c) $\mu = 0.8$; $\sigma^2 = 0.16$; $\sigma = 0.4000$

(d) $\mu = 0(1-p) + 1(p) = p$
$\sigma^2 = p - p^2 = p(1-p) = pq$
$\sigma = \sqrt{pq}$
(e) The variance is greatest when $p = q = 0.5$.

5.83 $E[(X-\mu)^2]$
$= \sum_{\text{all } x}(x-\mu)^2 p(x) = \sum_{\text{all } x}(x^2 - 2x\mu + \mu^2)p(x)$
$= \sum_{\text{all } x} x^2 p(x) - 2\mu \sum_{\text{all } x} xp(x) + \mu^2 \sum_{\text{all } x} p(x)$
$= E(X^2) - 2\mu\mu + \mu^2 = E(X^2) - \mu^2$

5.4 The Binomial Distribution

5.85 False

5.87 True

5.89 True

5.91 The number of successes in n trials

5.93 $0, 1, 2, 3, \ldots, n$

5.95 $P(X \le x)$

5.97 (a) $P(X \ge 12) = 0.0565$ (b) $P(X \ne 10) = 0.8829$
(c) $P(X \le 15) = 0.9997$ (d) $P(2 < X \le 8) = 0.5920$

5.99 (a) $\mu = 20; \sigma^2 = 4; \sigma = 2$
(b) $P(\mu - \sigma \le X \le \mu + \sigma)$
$= P(18 \le X \le 22) = 0.7927$
(c) $P(X < \mu - 2\sigma) + P(X > \mu + 2\sigma)$
$= P(X < 16) + P(X > 24) = 0.0211$

5.101 (a) $\mu = 9; \sigma^2 = 4.95; \sigma = 2.2249$
(b) $P(X > \mu + \sigma \mid X > \mu)$
$= P(X > 11.2249 \mid X > 9) = 0.32$
(c) $P(X < \mu - \sigma \cup X > \mu + \sigma)$
$= P(X < 6.7751) + P(X > 11.2249) = 0.2607$
(d) $P(X \le 8) = 0.4143, P(X \le 9) = 0.5914$
Therefore, $m \approx 8$ or 9

5.103 Let $X =$ the number of Gmail accounts that use 2FA. $X \sim B(20, 0.10)$.
(a) $P(X \le 2) = 0.6769$ (b) $P(X \ge 4) = 0.133$
(c) $\mu = 2$ (d) $P(X = 0 \mid X < 4) = 0.1402$

5.105 Let $X =$ the number of days on which all 50 loaves are sold. $X \sim B(30, 0.80)$.
(a) $\mu = 24.0$ (b) $P(X \ge 20) = 0.9744$
(c) $P(X \le 18) = 0.0095$

5.107 Let $X =$ the number of parents who correctly identify their child's illness. $X \sim B(50, 0.90)$.
(a) $\mu = 45; \sigma^2 = 4.5; \sigma = 2.1213$ (b) $P(X \ge 42) = 0.9421$
(c) $P(42 \le X \le 47) = 0.8304$
(d) Claim: $p = 0.9 \Rightarrow X \sim B(50, 0.9)$
Experiment: $x = 41$
Likelihood: $P(X \le 41) = 0.0579$
Conclusion: There is no evidence to suggest the claim is false.

5.109 Let $X =$ the number of students who are chronically absent. $X \sim B(35, 0.18)$.
(a) $P(X = 5) = 0.1593$ (b) $P(X \ge 8) = 0.2873$
(c) Claim: $p = 0.18 \Rightarrow X \sim B(35, 0.18), \mu = 6.3$
Experiment: $x = 4$
Likelihood: $P(X \ge 4) = 0.2196$
Conclusion: There is no evidence to suggest the claim is false.

5.111 Let $X =$ the number of data-science jobs in Washington. $X \sim B(20, 0.12)$. Let $Y =$ the number of data-science jobs in Maryland. $Y \sim B(20, 0.08)$.
(a) $P(X \le 2) = 0.5631$ (b) $P(Y = 0) = 0.1887$
(c) $P(X \ge 1 \cap Y \ge 1) = 0.7484$

5.113 Let $X =$ the number of jobs that could be completed by a robot. $X \sim B(50, 0.14)$.
(a) $P(X = 7) = 0.1606$ (b) $P(X \le 4) = 0.1528$
(c) Claim: $p = 0.14 \Rightarrow X \sim B(50, 0.14), \mu = 7$
Experiment: $x = 12$
Likelihood: $P(X \ge 12) = 0.0402$
Conclusion: There is evidence to suggest the claim is false.

5.115 Let $X =$ the number of orders that are filled correctly. $X \sim B(20, 0.75)$.
(a) $P(X = 15) = 0.2023$ (b) $P(X \le 12) = 0.1018$
(c) $P(10 \le X \le 14) = 0.3789$
(d) $P(X_i \ge 16) = 0.4148$
$P(X_1 \ge 16 \cap X_2 \ge 16) = 0.1721$

5.117 Let $X =$ the number of preschool children who can open a medication bottle. $X \sim B(10, 0.25)$.
(a) Probability histogram:

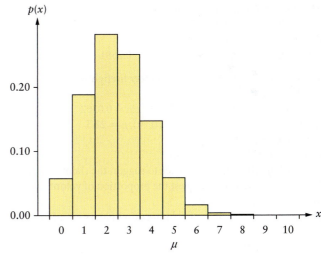

(b) $\mu = 2.5; \sigma^2 = 1.875; \sigma = 1.3693$
(c) $P(\mu - \sigma \le X \le \mu + \sigma)$
$= P(1.1307 \le X \le 3.8693)$
$= P(2 \le X \le 3) = 0.5319$

Probability visualization:

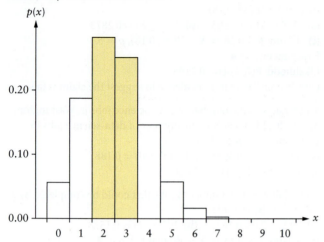

(d) Claim: $p = 0.25 \Rightarrow X \sim B(10, 0.25)$
Experiment: $x = 1$
Likelihood: $P(X \leq 1) = 0.2440$
Conclusion: There is no evidence to suggest the claim is false. There is no evidence to suggest the new cap is more effective.

5.119 Let X = the number of buyers who purchase white flip-flops. $X \sim B(30, 0.50)$.
(a) $\mu = 15$; $\sigma^2 = 7.5$; $\sigma = 2.7386$
(b) $P(\mu - 2\sigma \leq X \leq \mu + 2\sigma)$
$= P(9.5228 \leq X \leq 20.4772)$
$= P(X \leq 20) - P(X \leq 9) = 0.9572$
Using Chebyshev's Rule with $k = 2$, at least 0.75 of the values are within 2 standard deviations of the mean.
(c) Let F_i = group i has exactly 15.
$P(F_i) = P(X = 15) = 0.1445$
P(at least one group has exactly 15)
$= 1 - $ P(neither group has exactly 15) $= 0.2681$

5.121 Let X = the number of volunteer firefighters.
$X \sim B(30, 0.70)$.
(a) $P(X = 22) = 0.1501$ (b) $P(X > 25) = 0.0302$
(c) Claim: $p = 0.70 \Rightarrow X \sim B(30, 0.70)$, $\mu = 22.5$
Experiment: $x = 17$
Likelihood: $P(X \leq 17) = 0.0845$
Conclusion: There is no evidence to suggest the claim is false. There is no evidence to suggest the proportion of volunteer firefighters has decreased.
(d) Let Y_i = the number of volunteer firefighters out of 50 in group i. $Y_i \sim B(50, 0.70)$.
$P(Y_i \geq 40) = 0.0789$
$P(Y_1 \geq 40 \cap Y_2 \geq 40) = 0.0062$

5.123 Let X = the number of Mini Bites that contain undeclared almonds. $X \sim B(20, 0.008)$.
(a) $P(X \geq 1) = 0.1484$
(b) Let Y = the number of packages that contain at least one Mini Bite with undeclared almonds.
$Y \sim B(2, 0.1484)$.
$P(Y \geq 1) = 0.2748$

(c) Let W = the number of packages that contain at least one Mini Bite with undeclared almonds.
$Y \sim B(n, 0.1484)$.
$P(Y \geq 1) \geq 0.5$
$\Rightarrow 1 - P(Y = 0) \geq 0.5$
$\Rightarrow P(Y = 0) = (1 - 0.1484)^n \leq 0.5$
$\Rightarrow n \geq 4.315$ Therefore, $n = 5$ packages are needed.

5.5 Other Discrete Distributions

5.125 True

5.127 True

5.129 A Poisson random variable is typically used to model rare events. An event occurring, on average, 705 times per unit of time or volume does not appear to be rare.

5.131 Possible values: 0, 1, 2, 3, 4, 5

5.133 $\mu = 4 \Rightarrow p = 1/4 = 0.25$
(a) $P(X = 1) = 0.25$ (b) $P(3 \leq X \leq 7) = 0.4290$
(c) $P(X > \mu + 2\sigma) = P(X > 10.9282) = 0.0563$
(d) $P(X < 3 \mid X < 10) = 0.4636$

5.135 (a) $P(X > 4.5) = 0.4679$ (b) $P(X = 2) = 0.1125$
(c) $P(X = 4) + P(X = 5) = 0.3606$
(d) $P(X \leq \mu + 2\sigma) = P(X \leq 8.7426)$
$= P(X \leq 8) = 0.9597$

5.137 (a) Possible values for X: 4, 5, 6, 7, 8
(b) $\mu = 6$; $\sigma^2 = 0.80$; $\sigma = 0.8944$
(c) $P(X = 5) = 0.2462$ (d) $P(X = 8) = 0.0385$

5.139 Let X = the number of official engagements per week. X is a Poisson random variable with $\lambda = 4.75$.
(a) $P(X = 3) = 0.1545$ (b) $P(X > 7) = 0.1086$
(c) $P(X_1 \leq 2 \cap X_2 \leq 2) = 0.0217$

5.141 Let X = the number of calls until the first false alarm. X is a geometric random variable with $p = 0.17$.
(a) $P(X = 3) = 0.1171$ (b) $P(X = 12) = 0.0219$
(c) $\mu = 5.88$ (d) $P(X > 9 \mid X > 5) = 0.4746$
(e) $P(X_1 \leq 2 \cap X_2 \leq 2) = 0.0968$

5.143 Let X = the number of traffic accidents per week on Highway 401. X is a Poisson random variable with $\lambda = 4$.
(a) $P(X \leq 4) = 0.6288$
(b) $\mu = 4$, $\sigma = 2$
$P(X > \mu + 2\sigma) = P(X > 8) = 0.0214$
(c) $P(X_1 \geq 6 \cap X_2 \geq 6 \cap X_3 \geq 6 \cap X_4 \geq 6 \cap X_5 \geq 6)$
$= (0.2149)^5 = 0.000458$

5.145 Let X = the number of Americans selected until the first person cooks every day. X is a geometric random variable with $p = 0.27$.
(a) $P(X = 5) = 0.0767$ (b) $P(X > 10) = 0.0430$
(c) $\mu = 3.7$ (d) $P(X < 15) = 0.9878$

5.147 Let X = the number of oysters selected until the first is infected with norovirus. X is a geometric random variable with $p = 0.70$.

(a) $P(X = 2) = 0.2100$ **(b)** $P(X \leq 3) = 0.9730$
(c) Claim: $p = 0.70 \Rightarrow X$ is a geometric random variable with $p = 0.70$ and $\mu = 1.43$.
Experiment: $x = 10$
Likelihood: $P(X \geq 10) = 0.00002$
Conclusion: There is evidence to suggest the claim is false. There is evidence to suggest that the proportion of oysters infected with norovirus is less than 0.70.

5.149 Let $X =$ the number of job-related accidents per year. X is a Poisson random variable with $\lambda = 3.9$.
(a) $P(X = 0) = 0.0202$ **(b)** $P(2 \leq X \leq 5) = 0.7014$
(c) $P(X > \mu + 3\sigma) = P(X > 9.8245)$
$= 1 - P(X \leq 9) = 0.0069$

5.151 Let $X =$ the number of people waiting to have their license renewed. X is a hypergeometric random variable with $n = 6$, $N = 20$, and $M = 8$.
(a) $P(X = 2) = 0.3576$ **(b)** $P(X = 6) = 0.0007224$
(c) $P(X \leq 4) = 0.9819$

5.153 Let $X =$ the number of earthquakes with magnitude greater than 7 per month. X is a Poisson random variable with $\lambda = 1$.
(a) $P(X = 2) = 0.1839$ **(b)** $P(X \leq 4) = 0.9963$
(c) Claim: $\lambda = 1 \Rightarrow X$ is a Poisson random variable with $\lambda = \mu = 1$.
Experiment: $x = 4$
Likelihood: $P(X \geq 4) = 0.0190$
Conclusion: There is evidence to suggest the claim is false. There is evidence to suggest that the number of earthquakes with magnitude greater than 7 per month is greater than 1.

5.155 Let $X =$ the number of tenants who do not pay their rent on time in any given month. X is a Poisson random variable with $\lambda = 4.7$.
(a) $P(X = 0) = 0.0091$ **(b)** $P(X \geq 7) = 0.1954$
(c) $P(X_1 \leq 3 \cap X_2 \leq 3) = 0.0959$

5.157 Let $X =$ the number of people selected who multitask while driving. X is a hypergeometric random variable with $n = 4$, $N = 30$, and $M = 6$.
(a) $P(X = 1) = 0.4431$ **(b)** $P(X \leq 2) = 0.9819$
(c) Let $Y =$ the number of people selected who multitask while driving. Y is a hypergeometric random variable with $n = 4$, $N = 50$ and $M = 10$.
$P(Y = 1) = 0.4290$
$P(Y \leq 2) = 0.9782$
(d) Let $W =$ the number of people selected who multitask while driving. W is approximately a binomial random variable with $n = 4$ and $p = 0.20$.
$P(W = 1) = 0.4096$
$P(W \leq 2) = 0.9728$
(e) In a hypergeometric experiment, the probability of a success changes on each trial. As N increases with M/N constant, the hypergeometric probabilities approach the corresponding binomial probabilities.

5.159 $\mu = 0(0.1353) + \cdots + 3(0.3233) = 1.7820$
$\mu = 0(0.1353) + \cdots + 4(0.1429) = 1.9249$
$\mu = 0(0.1353) + \cdots + 5(0.0527) = 1.9775$
$E(Y)$ is converging to $\mu = 2$. This is the expected value of a Poisson random variable with $\lambda = 2$.

5.161 (a) Possible values: 2, 3, 4, ... **(b)** $P(X = 2) = p^2$
(c) $P(X = 3) = 2(1-p)p^2$
$P(X = 4) = 3(1-p)^2 p^2$
(d) $P(X = x) = (x-1)(1-p)^{x-2} p^2$

Chapter 5 Exercises

5.163 Let $X =$ the number of successful calls. $X \sim B(25, 0.80)$.
(a) $\mu = 20$; $\sigma^2 = 4$; $\sigma = 2$ **(b)** $P(X \geq 18) = 0.8909$
(c) Claim: $p = 0.8 \Rightarrow X \sim B(25, 0.8)$, $\mu = 20$
Experiment: $x = 21$
Likelihood: $P(X \geq 21) = 0.4207$
Conclusion: There is no evidence to suggest the supervisor's claim is false. There is no evidence to suggest the proportion of successful calls has changed.

5.165 Let $X =$ the number of satellites launched until the first fails to initiate the solar panels. X is a geometric random variable with $p = 0.08$.
(a) $P(X = 5) = 0.0573$ **(b)** $\mu = 12.5$
(c) $P(X \geq 21) = 0.1887$

5.167 Let $X =$ the number of people who suffered an injury involving a gas grill until the first who experiences an injury by starting the grill with the cover closed. X is a geometric random variable with $p = 0.08$.
(a) $P(X = 3) = 0.0677$ **(b)** $P(X \geq 10) = 0.472$
(c) $\mu = 12.5$
(d) Claim: $p = 0.08 \Rightarrow X$ is a geometric random variable, $\mu = 12.5$
Experiment: $x = 30$
Likelihood: $P(X \geq 30) = 0.0891$
Conclusion: There is no evidence to suggest the GGSC claim is false.

5.169 Let $X =$ the number of shark attacks at the beaches in North Carolina during a year. X is a Poisson random variable with $\lambda = 3.3$.
(a) $P(X = 0) = 0.0369$ **(b)** $P(2 \leq X \leq 5) = 0.7243$
(c) Claim: $\lambda = 3.3 \Rightarrow X$ is a Poisson random variable with $\lambda = \mu = 3.3$.
Experiment: $x = 8$
Likelihood: $P(X \geq 8) = 0.0198$
Conclusion: There is evidence to suggest the claim is false. There is evidence to suggest that the mean number of shark attacks per year has increased.
(d) According to the Florida Museum, there were 3 shark attacks at the beaches in North Carolina during 2018.
$P(X = 3) = 0.5803$
(e) Consider three additional Poisson random variables with $\lambda = 24.4$, 6.5, and 3.9, respectively.
P(no attacks in all four states)
$= P(X_{NC} = 0 \cap X_F = 0 \cap X_H = 0 \cap X_{SC} = 0)$
$= (0.0369)(2.53 \times 10^{-11})(0.0015)(0.0202)$
$= 2.84 \times 10^{-17}$ (pretty unlikely)

5.171 Let $X =$ the number of people who have taken a cruise. $X \sim B(25, 0.24)$.
(a) $P(X = 7) = 0.1578$ **(b)** $P(X \leq 5) = 0.4233$

(c) Claim: $p = 0.24 \Rightarrow X \sim B(25, 0.24)$, $\mu = 6$
Experiment: $x = 11$
Likelihood: $P(X \geq 11) = 0.0222$
Conclusion: There is evidence to suggest the claim is false. There is evidence to suggest the proportion of people who have taken a cruise has increased.

5.173 Let $X =$ the number of people who believe the barriers should be installed. $X \sim B(50, 0.20)$.
(a) $\mu = 10$; $\sigma^2 = 8$; $\sigma = 2.8284$ (b) $P(X \leq 10) = 0.5836$
(c) $P(\mu - 2\sigma \leq X \leq \mu + 2\sigma) = P(4.34 \leq X \leq 15.66)$
$= P(5 \leq X \leq 15) = 0.9507$

5.175 Let $X =$ the number of consumers who believe companies are obligated to disclose the use of AI and how they are using it. $X \sim B(50, 0.80)$.
(a) $P(X \geq 45) = 0.0480$ (b) $P(38 \leq X \leq 43) = 0.7105$
(c) Claim: $p = 0.80 \Rightarrow X \sim B(50, 0.80)$, $\mu = 40$
Experiment: $x = 36$
Likelihood: $P(X \leq 36) = 0.1106$
Conclusion: There is no evidence to suggest the claim is false. There is no evidence to suggest the proportion of people who believe companies are obligated to disclose the use of AI and how they are using it is different from 0.80.

5.177 Let $X =$ the number of people contacted until the first person indicates he or she skipped medical care due to cost. X is a geometric random variable with $p = 0.45$.
(a) $P(X = 3) = 0.1361$ (b) $P(X \leq 6) = 0.9723$
(c) Claim: $p = 0.45 \Rightarrow X$ is a geometric random variable with $p = 0.45$ and $\mu = 2.22$.
Experiment: $x = 12$.
Likelihood: $P(X \geq 12) = 0.0014$
Conclusion: There is evidence to suggest the claim is false. There is evidence to suggest that the proportion of people who skip medical care due to cost is less than 0.45, resulting in a higher mean (number of observations).

5.179 Let $X =$ the number of Americans who attempt to diagnose a medical condition online. $X \sim B(25, p)$.
(a) Assume $p = 0.35$
$P(X \geq 12) = 0.1254$
(b) Assume $p = 0.40$: $P(X \leq 11) = 0.7323$
Assume $p = 0.50$: $P(X \leq 11) = 0.3450$
(c) $p = 0.35$: $P(X \geq 13) = 0.0604$
$p = 0.40$: $P(X \leq 12) = 0.8462$
$p = 0.50$: $P(X \leq 12) = 0.5000$

5.181 Let $X =$ the number of parents who set up GPS tracking systems on their children's phones without consent. $X \sim B(40, 0.30)$.
(a) $P(X = 14) = 0.1042$ (b) $P(X \leq n) \leq 20 \Rightarrow n = 9$
(c) Claim: $p = 0.30 \Rightarrow X \sim B(40, 030)$, $\mu = 12$
Experiment: $x = 16$
Likelihood: $P(X \geq 16) = 0.1151$
Conclusion: There is no evidence to suggest the claim is false. There is no evidence to suggest that the proportion of parents who secretly set up GPS tracking systems is different from 0.30.

5.183
(a) Answers will vary.

Complete shipments	Relative frequency	$p(y)$
0	0.0000	0.0000
1	0.0000	0.0007
2	0.0060	0.0052
3	0.1600	0.0234
4	0.0790	0.0701
5	0.1300	0.1471
6	0.2020	0.2207
7	0.2640	0.2365
8	0.1850	0.1774
9	0.0830	0.0887
10	0.0320	0.0266
11	0.0030	0.0036

(b)

Complete shipments	Relative frequency	$p(y)$
0	0.0000	0.0000
1	0.0000	0.0000
2	0.0000	0.0001
3	0.0000	0.0008
4	0.0050	0.0040
5	0.0140	0.0142
6	0.0390	0.0392
7	0.0960	0.0840
8	0.1210	0.1417
9	0.1980	0.1889
10	0.1910	0.1983
11	0.1590	0.1623
12	0.1050	0.1014
13	0.0580	0.0468
14	0.0120	0.0150
15	0.0020	0.0030
16	0.0000	0.0003

(c) $Y \sim B(25, 0.60)$
$P(Y = 15) = 0.1612$
$P(Y \leq 12) = 0.1538$
$P(Y > 16) = 0.2735$
$\mu = 15$

Chapter 6

6.1 Probability Distributions for a Continuous Random Variable

6.1 False

6.3 True

6.5 True

6.7 The probability that X takes on a value between a and b $(a < b)$ is the area under the curve between a and b.

6.9 Because the uniform distribution is perfectly symmetric, the mean and the median will have the same value. Thus the mean will be the value halfway between a and b, or the average of the endpoints.

6.11 (a)

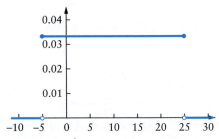

(b) $\mu = 10$, $\sigma^2 = 75$, $\sigma = 8.66$ **(c)** $P(-10 < X < -1) = 0.3$
(d) $P(X > 0) = 0.833$ **(e)** $P(X \geq 20 \mid X \geq 10) = 0.333$

6.13 (a) $\mu = 50$, $\sigma^2 = 208.33$, $\sigma = 14.43$
(b) $P(X \leq \mu - 2\sigma \text{ or } X \geq \mu + 2\sigma) = 0$ **(c)** $c = 55$
(d) $P(30 \leq X \leq 40) = 0.2$.

The probability that two randomly selected values are both between 30 and 40 is 0.04.

6.15 (a) The total area under the curve is $\frac{1}{2}(10)(0.05) + (10)(0.05) + \frac{1}{2}(10)(0.05) = 1$. Additionally, the entire density function falls on or above the x-axis.

(b)
$$f(x) = \begin{cases} 0.005x - 0.05 & \text{if } 10 \leq x \leq 20 \\ 0.05 & \text{if } 20 \leq x \leq 30 \\ -0.005x + 0.2 & \text{if } 30 \leq x \leq 40 \\ 0 & \text{otherwise} \end{cases}$$

(c) $P(X \leq 15) = 0.0625$ **(d)** $P(X > 27) = 0.4$
(e) Because this distribution is perfectly symmetric, the mean will be the center of the interval.

6.17 (a) Let W = weight of ball. $P(W \geq 5.14) = 0.2143$
(b) Let C = circumference of ball. $P(C \leq 9.03) = 0.3$
(c) $P((5.11 \leq W \leq 5.13) \cap (9.04 \leq C \leq 9.06)) = (0.2857) \cdot (0.2) = 0.0571$

6.19 Let X = drying time of paint.
(a) $P(X \leq 45) = 0.5$ **(b)** $P(40 \leq X \leq 50) = 0.3333$
(c) $P(X \leq t) = 0.75 \Rightarrow t = 37.5$ **(d)** $P(X \geq 55) = 0.1667$

6.21 Let X = length of pillow.
(a) $\mu = 24.375$, $\sigma^2 = 0.1302$, $\sigma = 0.3608$

(b) $P(24.0141 < X < 24.7358) = 0.5774$
(c) $P(X > L) = 0.25 \Rightarrow L = 24.6875$

6.23 Let X = hours the car is parked.
(a) $P(X < 2) = 0.5$ **(b)** $P(X < 1.4) = 0.08$ **(c)** $P(X > 2.6) = 0.08$
(d) $P(1.4 < X < 2.6) = 0.84$

6.25 Let X = change in price.
(a) Note that the graph of f lies entirely on or above the x-axis. Also, the total area under this curve is 1.
(b) $P(X \geq 1.00) = 0.3125$
(c) $P(-1.00 < X < 1.00) = 0.375$
(d) $P(-c < X < c) = 0.90 \Rightarrow c \approx 1.8635$

6.27

(a)

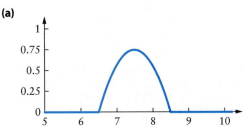

(b)

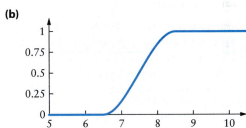

(c) $P(X < 7) = 0.1563$ **(d)** $P(X \geq 8.25) = 0.0430$
(e) $P(X > 8) = 0.1563$.
Let Y = the number of rides that take more than 8 minutes.
Then, $P(Y = 5) = (0.1563)^5 = 0.000093$.

6.29 (a) $P(X_3 < 1) = 0.3333$
(b) $\mu_{X_4} = 2$ hours. $\sigma_{X_4} = 1.1547$
$P(X_4 < -0.3094 \text{ or } X_4 > 4.30) = 0$
(c) $P(X_5 < t) = 0.25 \Rightarrow t = 3.75$
(d) $P(\text{all families drive less than 0.5 hours}) = 0.00026$
$P(\text{all 5 families drive more than 90 minutes}) = 0$

6.2 The Normal Distribution

6.31 True

6.33 $\mu = 0$ and $\sigma^2 = 1$

6.35 (a)

6.37

(a)

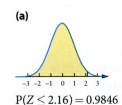

$P(Z \leq 2.16) = 0.9846$

(b)

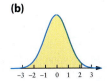

$P(Z < 2.16) = 0.9846$

(c)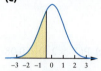
$P(Z \leq -0.47) = 0.3912$

(d)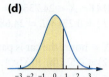
$P(0.73 > Z) = 0.7673$

(e)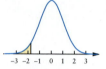
$P(-1.75 \geq Z) = 0.0401$

(f)
$P(-0.35 \leq Z \leq 0.65) = 0.3790$

(g)
$P(Z < 5) \approx 1$

(h)
$P(Z \leq -4) \approx 0$

(i)
$P(Z \leq 4) \approx 1$

(j)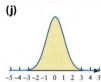
$P(Z \geq -5) = 1$

6.39 (a) $P(-1.00 \leq Z \leq 1.00) = 0.6827$
(b) $P(-2.00 \leq Z \leq 2.00) = 0.9545$
(c) $P(-3.00 \leq Z \leq 3.00) = 0.9973$
These are the probabilities given by the Empirical Rule.

6.41 (a)
$P(Z \leq b) = 0.5100$
$\Rightarrow b = 0.0251$

(b)
$P(Z > b) = 0.1080$
$\Rightarrow b = 1.2372$

(c)
$P(Z \geq b) = 0.0500$
$\Rightarrow b = 1.6449$

(d)
$P(Z \leq b) = 0.0100$
$\Rightarrow b = -2.3263$

(e)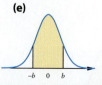
$P(-b \leq Z \leq b) = 0.8000$
$\Rightarrow b = 1.2816$

(f)
$P(-b < Z < b) = 0.6535$
$\Rightarrow b = 0.9412$

6.43 (a) $Q_1 = -0.6745$, $Q_3 = 0.6745$
(b) IQR $= 1.3490$, IF$_L = -2.6980$, IF$_H = 2.6980$
(c) $P(Z \leq -2.6980 \cup Z \geq 2.6980) = 0.0070$
(d) OF$_L = -4.7215$, OF$_H = 4.7215$
(e) $P(Z \leq -4.7215 \cup Z \geq 4.7215) = 0.0000023$

6.45
(a)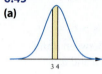
$P(3.0 \leq X \leq 4.0) = 0.1844$

(b)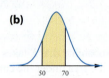
$P(50 < X < 70) = 0.6731$

(c)
$P(X \geq 45) = 0.0088$

(d)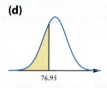
$P(X < 76.95) = 0.3085$

(e)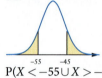
$P(X < -55 \cup X > -45) = 0.2133$

(f)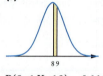
$P(8 \leq X \leq 9) = 0.1110$

6.47 (a) $Q_1 = 20.9531$, $Q_3 = 29.0470$
(b) IQR $= 8.0939$, IF$_L = 8.8122$, IF$_H = 41.1878$
(c) $P(X \leq 8.8122 \cup X \geq 41.1878) = 0.0070$
(d) OF$_L = -3.3286$, OF$_H = 53.3286$
(e) $P(X \leq -3.3286 \cup X \geq 53.3286) = 0.0000023$

6.49 Let $X =$ cost of wedding; $X \sim N(33{,}391, 1500^2)$.
(a) $P(X > 35{,}000) = 0.1417$
(b) $P(30{,}000 \leq X \leq 34{,}000) = 0.6457$
(c) $P(X < 29{,}000) = 0.0017$

6.51 Let $X =$ weight of Versa-Lok block; $X \sim N(37.19, 0.8^2)$.
(a) $P(X > 38) = 0.1556$ (b) $P(36 \leq X \leq 37) = 0.3377$
(c) $P(X < 35.5) = 0.0173$

6.53 Let $X =$ salinity of the sample; $X \sim N(35, 0.52^2)$.
(a) $P(X > 36) = 0.0272$
(b) $P(X < 33.5) = 0.0020$ (c) $P(33 \leq X \leq 35) = 0.4999$
(d) $P(\mu - b \leq X \leq \mu + b) = 0.50 \Rightarrow b = 0.3507$
$\Rightarrow P(34.6493 \leq X \leq 35.3507) = 0.50$
These are the lower and upper quartiles.

6.55 Let $X =$ percent of 2-butoxyethanol in the bottle; $X \sim N(3, 1)$.
(a) $P(X < 2.5) = 0.3085$ (b) $P(2.2 \leq X \leq 3.5) = 0.4796$
(c) $P(X > 5) = 0.0228$

6.57 Let $X =$ player's sprint speed; $X \sim N(27, 1.5^2)$.
(a) $P(X > 28.2) = 0.2119$ (b) $P(25 \leq X \leq 30) = 0.8860$
(c) $P(X < 23) = 0.0038$
Because this probability is so small, there is evidence to suggest that the mean sprint speed is less than 27 ft/sec.

6.59 Let X = lap speed; $X \sim N(150.545, 16^2)$.
(a) $P(X<135)=0.1656$ **(b)** $P(150\leq X\leq 160)=0.2363$
(c) $P(X>210.364)=0.000093$ **(d)** $P(X>170)=0.1120$
$P(\text{all 4 speeds greater than } 170)=0.00016$

6.61 Let X = tariff on imported good; $X \sim N(2.4, 0.6^2)$.
(a) $P(X>3.5)=0.0334$ **(b)** $P(2\leq X\leq 3)=0.5889$
(c) $P(X<1.5)=0.0668$
While this is an unusual observation, there is not enough evidence to suggest that the claim made by the WTO is false.

6.63 Let X = bow weight; $X \sim N(60, 3.2^2)$.
(a) $P(58\leq X\leq 62)=0.4680$ **(b)** $P(X>66.4)=0.0228$
(c) $P(X>66)=0.0304$ **(d)** $P(X\leq 55)=0.0591$
While this is an unusual observation, there is not enough evidence to suggest that the mean weight is less than 60 g.

6.65 Let X = phosphorus concentration of the sample; $X \sim N(14, 5.5^2)$.
(a) $P(X<12)=0.3581$ **(b)** $P(X<9 \cup X>19)=0.3633$
(c) $P(X<10\,|\,X<20)=0.2708$ **(d)** $P(X\geq 25)=0.0228$
This occurrence is incredibly rare. There is evidence to suggest that the mean phosphorus level has increased.

6.67 Let X = electricity generated; $X \sim N(35, \sigma^2)$.
(a) $P(X<34)=0.3540 \Rightarrow \sigma = 2.67$ **(b)** $P(X>37.8)=0.1472$
$P(\text{4 consecutive years of profit})=0.00047$
(c) $P(X\leq 33.5)=0.2871$
There is not enough evidence to suggest that the claim made by the committee is false.

6.69 Let X = diameter of the sand dollar; $X \sim N(3, 0.55^2)$.
(a) $P(X>d)=0.90 \Rightarrow d=2.2951$ **(b)** $P(X>4)=0.0345$
(c) $P(2.5\leq X\leq 3.5)=0.6367$
Let Y = number of sand dollars between 2.5 and 3.5 inches; $Y \sim \text{Binomial}(n=10, p=0.6367) \Rightarrow P(Y=7)=0.2441$.
(d) $P(X\leq 2)=0.0345$
There is evidence to suggest that the mean diameter has decreased.

6.71 Let X = diameter of the tennis ball; $X \sim N(2.5625, 0.04^2)$.
(a) $P(2.5\leq X\leq 2.625)=0.8818$
(b) Let Y = number of balls that meet specifications;
$Y \sim B(6, 0.8818) \Rightarrow P(Y=5)=0.3781$.
(c) $P(\text{1 ball less than 2.5 and 1 ball greater than } 2.63)=2(0.0591)(0.0458)=0.0054$

6.3 Checking the Normality Assumption

6.73 Graphs (histogram, stemplot, or dotplot), Backward Empirical Rule, IQR/s, normal probability plot

6.75 False

6.77 True

6.79 Order the observations from smallest to largest. The associated normal score for each value, in order, appears in Table 4 in the $n=25$ column.

6.81

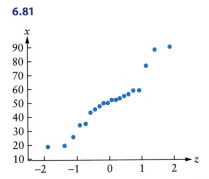

These data do not appear to be normally distributed, as the normal probability plot has a curved pattern.

6.83 Histogram:

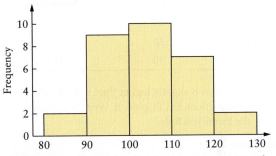

The histogram shows that the data are symmetric with no apparent outliers. There is no evidence to suggest that these data are from a non-normal distribution. Backward Empirical Rule: $\bar{x}=103.4$, $s=9.9441$

Interval	Frequency	Proportion
(93.5, 113.4)	20	0.67
(83.5, 123.3)	30	1.00
(73.6, 133.2)	30	1.00

The middle proportion is slightly higher than it should be (0.95), but the proportions are, in general, very close to those suggested by the Empirical Rule. IQR/s: IQR $=15.8$, $s=9.9441 \Rightarrow$ IQR/$s=1.586$ This ratio differs from 1.3 by only a small amount.
Normal probability plot:

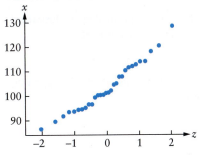

There is a relatively linear pattern in the normal probability plot, suggesting no evidence of non-normality.

6.85 (a)

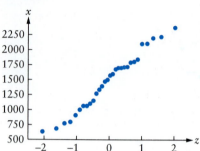

There is a relatively linear pattern in the normal probability plot, suggesting no evidence of non-normality. (b) $\bar{x} = 1480.00$, $s = 482.75$

Interval	Frequency	Proportion
(997.25, 1962.75)	20	0.67
(514.50, 2445.50)	30	1.00
(31.75, 2928.25)	30	1.00

The middle proportion is slightly higher than it should be (0.95), but the proportions are, in general, very close to those suggested by the Empirical Rule.

6.87 Histogram:

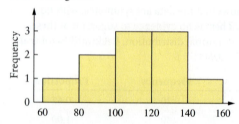

This distribution appears to be slightly skewed to the left, which shows some evidence of non-normality.
Backward Empirical Rule: $\bar{x} = 111.93$, $s = 20.78$

Interval	Frequency	Proportion
(91.14, 132.71)	6	0.60
(70.36, 153.49)	10	1.00
(49.57, 174.28)	10	1.00

These proportions differ from those specified by the Empirical Rule, suggesting some evidence of non-normality.

IQR/s: IQR = 25.85, $s = 20.78 \Rightarrow$ IQR/s = 1.24.

This ratio does not differ too significantly from 1.3, suggesting little evidence of non-normality.
Normal probability plot:

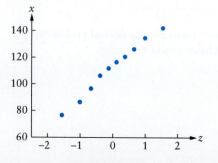

This plot shows a very linear pattern, suggesting little evidence of non-normality.

6.89 Histogram:

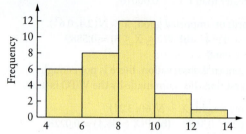

This histogram is roughly symmetric, suggesting little evidence of non-normality. Backward Empirical Rule: $\bar{x} = 7.7$, $s = 2.0$

Interval	Frequency	Proportion
(5.8, 9.7)	19	0.63
(3.8, 11.6)	29	0.97
(1.9, 13.6)	30	1.00

These proportions are very close to those specified by the Empirical Rule, suggesting almost no evidence of non-normality.

IQR/s: IQR = 2.6, $s = 2.0 \Rightarrow$ IQR/s = 1.3

This ratio shows no evidence of non-normality.
Normal probability plot:

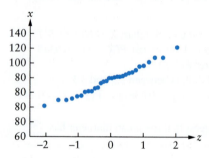

The strong linear pattern in this plot does not suggest any evidence of non-normality.

6.91 Because this plot shows a distinct curved pattern, there is evidence to suggest that these data are from a non-normal distribution.

6.93 Histogram:

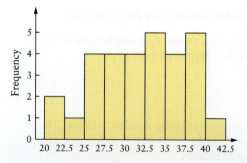

This distribution appears uniform and not unimodal, suggesting that the distribution is non-normal.
Backward Empirical Rule: $\bar{x} = 31.73$, $s = 5.4$

Interval	Frequency	Proportion
(26.3, 37.1)	19	0.63
(20.9, 42.5)	29	0.97
(15.5, 47.9)	30	1.00

The first proportion is a bit lower than it should be (0.68) according to the Empirical Rule, which suggests that this distribution is not normal. IQR/s: IQR = 7.75, s = 5.4 ⇒ IQR/s = 1.4 This ratio is very close to 1.3, providing little evidence that this distribution is non-normal.
Normal probability plot:

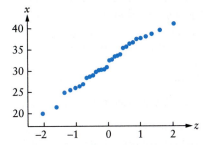

There are several unusual points in the normal probability plot that lie away from the overall pattern, providing some evidence that these data are from a non-normal distribution.

6.95 If the samples come from a standard normal distribution, then each sample should have an approximately normal distribution and, on average, will be close to those predicted by Table 4. These ordered pairs form the cumulative density function of a normal distribution with mean 50 and standard deviation 10.

6.4 The Exponential Distribution

6.97 True

6.99 Solutions will vary. Example: The amount of time (years) until a washing machine breaks

6.101 (a)

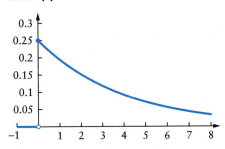

(b) $P(0.01 \leq X \leq 0.05) = 0.0099$ **(c)** $P(X > 0.06) = 0.9851$

6.103 $P(X \leq 20) = 0.7981 \Rightarrow \lambda = 0.08$

6.105 Let X = time until the carton spoils.
(a) $P(X < 4) = 0.7275$
(b) $P(2 \leq X \leq 3) = 0.1449$ **(c)** $P(X > 5) = 0.1969$

6.107 Let X = wait time.
(a) $P(X \leq 5) = 0.0488$
(b) $P(10 \leq X \leq 20) = 0.0861$ **(c)** $P(X > 30) = 0.7408$

6.109 Let X = time between tests. $\lambda = 0.0417$
(a) $P(X < 12) = 0.3937$
(b) $P(X > 8) = 0.7163$ **(c)** $P(3 \leq X \leq 4) = 0.0360$

6.111 Let X = lifetime of booster shot.
(a) $P(X > 10) = 0.6065$ **(b)** $P(X > t) = 0.10 \Rightarrow t = 46.05$
(c) P(both still effective after 5 years) = 0.6065

6.113 Let X = time the child plays with the toy.
(a) $P(X \leq 10) = 0.3935$ **(b)** $P(5 \leq X \leq 20) = 0.4109$
(c) $P(X > 35 \mid X \geq 15) = P(X > 20) = 0.3679$
(d) P(all 4 play at least 25 minutes) = 0.0067

6.115 Let X = time aroma lasts.
(a) $\mu = 30, \sigma^2 = 900, \sigma = 30$ **(b)** $P(X \geq 40) = 0.2636$
(c) $P(30 \leq X \leq 50) = 0.1790$ **(d)** $P(X \leq t) = 0.90 \Rightarrow t = 69.08$
(e) Let X_1 = time aroma lasts from batch 1, and let X_2 = time aroma lasts from batch 2,
$\Rightarrow P(X_1 \leq 35 \cap X_2 \leq 25) = 0.3893$

6.117 (a) $P(4 \leq X \leq 5) = 0.03 \Rightarrow \lambda \approx 0.035$ **(b)** $P(X \leq 3) = 0.0997$
(c) P(at least one within 3 years) = 0.6113

Chapter 6 Exercises

6.119 Let X = time until document prints.
(a) $\mu = 2.5$ **(b)** $P(X \leq 0.5) = 0.1813$ **(c)** $P(X > 5) = 0.1353$
(d) $P(X > t) = 0.02 \Rightarrow t = 9.78$

6.121 Let X = length of time for approval; $x \sim N(21, 16)$.
(a) $P(X < 14) = 0.0401$ **(b)** $P(15 \leq X \leq 19) = 0.2417$
(c) $P(20 \leq X \leq 30 \mid X \geq 20) = 0.9796$
(d) P(both longer than 30 days) = 0.00015

6.123

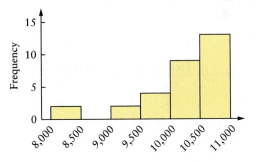

$\bar{x} = 10,252.50, s = 654.94$

Interval	Frequency	Proportion
(9597.56, 10,907.44)	26	0.87
(8942.62, 11,562.38)	28	0.93
(8287.68, 12,217.32)	30	1.00

IQR = 716, s = 654.94 ⇒ IQR/s = 1.09

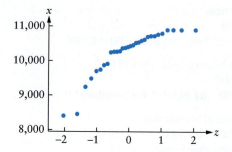

Using the four methods to check for non-normality, the histogram of these data is heavily skewed to the left, proportions of observations within intervals specified by the Empirical Rule are not consistent with those specified by the rule, the IQR/s ratio is significantly different from 1.3, and there is a very curved pattern in the normal probability plot. All of these methods provide strong evidence to suggest the data come from a non-normal distribution.

6.125 Let $X =$ time to recharge; $X \sim N(2.5, 0.75^2)$.
(a) $P(1 \leq X \leq 2) = 0.2297$ (b) $P(X > 4.5) = 0.0038$
(c) $P(X < 0.5) = 0.0038$
(d) $P(\text{all 3 cars take more than 3 hours}) = 0.0161$
(e) $P(X > 3.5) = 0.0912$
There is no evidence to suggest that the claim is false.

6.127

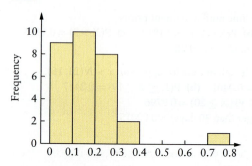

$\bar{x} = 0.17, s = 0.13$

Interval	Frequency	Proportion
(0.04, 0.30)	26	0.87
(−0.08, 0.43)	29	0.97
(−0.21, 0.56)	29	0.97

IQR $= 0.15, s = 0.13 \Rightarrow$ IQR/$s = 1.14$

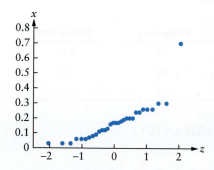

Using the four methods to check for non-normality, the histogram is skewed to the right with a possible outlier, the proportion of observations within 1, 2, and 3 standard deviations from the mean differ significantly from the proportions specified by the Empirical Rule, the IQR/s ratio differs from 1.3, and the normal probability plot does not show a linear pattern, again suggesting a high outlier. All four of these methods provide strong evidence to suggest that the data came from non-normal distribution.

6.129 Let $X =$ caffeine in the cup; $X \sim N(180, 15)$.
(a) $P(X < 170) = 0.2525$ (b) $P(160 \leq X \leq 190) = 0.6563$
(c) $P(X > 200) = 0.0912$ (d) $P(X < 157) = 0.0626$
There is no evidence to suggest that the claim is false.

6.131 Let $X =$ length of car loan; $X \sim N(68, 81)$.
(a) $P(X < 60) = 0.1870$ (b) $P(50 \leq X \leq 70) = 0.5652$
(c) $P(50 \leq X \leq 86) = 0.95$ (d) $P(X < 72) = 0.6716$
Let $Y =$ amount of the car loan.
$P(X < 72 \cap Y > 2700) = 0.0542$

6.133 Let $X =$ proportion of water in the barrel;
$X \sim N(0.12, 0.025^2)$.
(a) $P(X < 0.12) = 0.5000$
(b) $P(0.15 \leq X \leq 0.17) = 0.0923$ (c) $P(X > 0.20) = 0.00069$

6.135 (a)

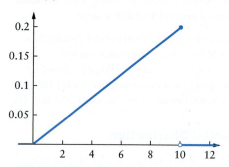

(b) $P(X < 5) = 0.25$ (c) $P(X > 8) = 0.36$
(d) $P(2 \leq X \leq 6) = 0.32$ (e) $P(X < 1 | X < 2) = 0.25$

6.137 (a) $P(X_2 < 1.5) = 0.75$
(b) $\mu_{X_1} = 0.5, \mu_{X2} = 1, \mu_{X_3} = 1.5, \mu_{X_4} = 2$
(c) $P(\text{all 4 times less than 30 seconds}) = 0.0026$
(d) $P(\text{exactly 1 person finished in less than 1 minute}) = 0.25$

6.139 Let $X =$ protein in serving; $X \sim N(12, 0.7^2)$.
(a) $P(X > 13.5) = 0.0161$
(b) $P(11 \leq X \leq 13) = 0.8469 \Rightarrow$ in control
$P(\text{out of control}) = 0.1531$
(c) $P(11.5 < X < 12.5 | X < 12.5) = 0.6885$
(d) Let $Y =$ number of cups with the desired amount;
$P(Y \geq 3) = 0.9686$

6.141 Let $X =$ time to process EMV chip; $X \sim N(13, 6.25)$.
(a) $P(X < 10) = 0.1151$ (b) $P(X \geq 20) = 0.0026$
There is evidence to suggest that the mean processing time is greater than 13 seconds.

(c) Histogram:

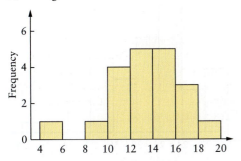

Backward Empirical Rule:
$\bar{x} = 13.41, s = 3.18$

Interval	Frequency	Proportion
(10.23, 16.58)	14	0.70
(7.05, 19.76)	19	0.95
(3.87, 22.94)	20	1.00

$IQR = 4.15, s = 3.18 \Rightarrow IQR/s = 1.3$

Normal probability plot:

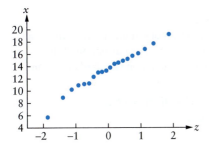

Using the four methods for assessing non-normality, the histogram of these data is roughly symmetric with one unusual observation at 5.7, the proportion of observations within the intervals 1, 2, and 3 standard deviations from the mean are close to those specified by the Empirical Rule, the IQR/s ratio is 1.3, and the normal probability plot shows a linear pattern. All of these methods show very little evidence of non-normality for this distribution.

(d) $P(X > c) = 0.01 \Rightarrow c = 18.82$

Chapter 7

7.1 Statistics, Parameters, and Sampling Distributions

7.1 population; sample

7.3 True

7.5 Take many, many samples of the same size (n) from the population, each time recording the desired statistic produced by the sample. Plot each of the recorded statistics using a histogram or other appropriate graphical display.

7.7 $\binom{N}{n} = \dfrac{N!}{n!(N-n)!}$ samples

7.9 The shape of the distribution of the sample mean appears to be approximately normal. The center of the distribution of the sample mean is approximately the mean from the original population and there is less variability in the distribution of the sample mean than in the original population.

7.11 (a) statistic **(b)** parameter **(c)** statistic **(d)** statistic **(e)** parameter

7.13 (a) $\mu = 16; \tilde{\mu} = 15$

(b) There are $\binom{5}{3} = 10$ unique samples of size 3.

Sample	$\bar{x}$	Sample	$\bar{x}$
10 12 15	12.3	10 18 25	17.7
10 12 18	13.3	12 15 18	15.0
10 12 25	15.7	12 15 25	17.3
10 15 18	14.3	12 18 25	18.3
10 15 25	16.7	15 18 25	19.3

$\mu_{\bar{x}} = 16; \sigma_{\bar{x}}^2 = 4.6; \sigma_{\bar{x}} = 2.145$

(c)

Sample	$\tilde{x}$	Sample	$\tilde{x}$
10 12 15	12	10 18 25	18
10 12 18	12	12 15 18	15
10 12 25	12	12 15 25	15
10 15 18	15	12 18 25	18
10 15 25	15	15 18 25	18

Distribution of $\tilde{x}$

$\tilde{x}$	12	15	18
$p(\tilde{x})$	0.3	0.4	0.3

$\mu_{\tilde{x}} = 15; \sigma_{\tilde{x}}^2 = 5.4; \sigma_{\tilde{x}} = 2.324$

(d) The mean of the sample means is the same as the population mean (16) and the mean of the sample medians is the same as the population median (15).

7.15 (a) $\mu_X = 65.1$

(b) There are 9 possible outcomes for the ages of two selected employees.

Sample	$\bar{x}$	Probability
64 64	64	0.01
64 65	64.5	0.07
64 66	65	0.02
65 64	64.5	0.07
65 65	65	0.49
65 66	65.5	0.14
66 64	65	0.02
66 65	65.5	0.14
66 66	66	0.04

Sampling distribution:

$\bar{x}$	64	64.5	65	65.5	66
$p(\bar{x})$	0.01	0.14	0.53	0.28	0.04

(c) $\mu_{\bar{X}} = 65.1$
The mean of the sampling distribution of $\bar{X}$ is the same as the mean of the population.

7.17 (a) There are $\binom{5}{3} = 10$ different ways to select three jaguars.

$\bar{x}$	160	163.33	166.67	170
$p(\bar{x})$	0.1	0.1	0.1	0.2

$\bar{x}$	173.33	176.67	180	183.33
$p(\bar{x})$	0.1	0.2	0.1	0.1

(b)

t	480	490	500	510	520	530	540	550
$p(t)$	0.1	0.1	0.1	0.2	0.1	0.2	0.1	0.1

7.19 (a) $\mu_X = 11.55$
(b) There are $5 \times 5 \times 5 = 125$ different groups of three committee meetings.

$\bar{x}$	10	10.5	11	11.5	12	12.5
$p(\bar{x})$	0.04	0.12	0.21	0.24	0.2	0.12

$\bar{x}$	13	13.5	14
$p(\bar{x})$	0.0525	0.015	0.0025

(c)

m	10	11	12	13	14
$p(m)$	0.2	0.24	0.15	0.03	0.0025

7.21 (a) $\mu_X = 5.75; \sigma_X^2 = 0.7875$
(b) There are $4 \times 4 = 16$ different ways to select two American Airlines flights.

s^2	0	0.5	2	4.5
$p(s^2)$	0.365	0.405	0.18	0.05

(c) $\mu_{S^2} = 0.7875$
The mean of the sampling distribution of the sample variance is equal to the population variance.

7.23 (a) Let X = sample maximum for the selected group.

x	14	15	17
$p(x)$	0.3	0.3	0.4

(b)

$\tilde{x}$	12.5	13	14	14.5	15.5	16
$p(\tilde{x})$	0.2	0.1	0.2	0.2	0.2	0.1

7.25 (a) $\mu = 25.5$
(b) Answers may vary. One possible set is shown in the histogram in part (c).

(c)

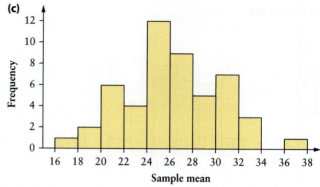

The data have a mean of 26.09, which is very close to the population mean of 25.5.

7.27 (a) $\mu_X = 2.55; \sigma_X = 1.1608$
(b) There are $5 \times 5 = 25$ different groups of 2 amounts of trapped toads.

$\bar{x}$	0	0.5	1	1.5	2
$p(\bar{x})$	0.0025	0.015	0.0475	0.105	0.1775

$\bar{x}$	2.5	3	3.5	4
$p(\bar{x})$	0.225	0.215	0.15	0.0625

(c) $\mu_{\bar{X}} = 2.55; \sigma_{\bar{X}} = 0.8208$
(d) The mean of the distribution of sample means is the same as the population mean, but the standard deviation of the sample means is less than the population standard deviation.

7.2 The Sampling Distributions of the Sample Mean and the CLT

7.29 True

7.31 False

7.33 True

7.35 Many real-world measurements are obtained by a sum of various factors and, as suggested by the Central Limit Theorem, the distribution of a sum will be approximately normal.

7.37 (a) $\bar{X} \sim N\left(17.5, \dfrac{36}{24}\right)$

(b)

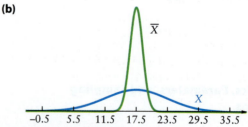

(c) $P(X \leq 14) = 0.2798$
$P(\bar{X} \leq 14) = 0.0021$
(d) $P(15 < X < 19) = 0.2602$
$P(15 < \bar{X} < 19) = 0.8690$

7.39 (a) $\bar{X} \stackrel{\bullet}{\sim} N\left(1000, \dfrac{10{,}000}{36}\right)$

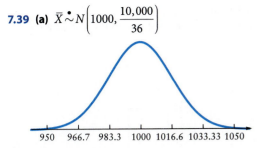

(b) $P(\bar{X} > 975) = 0.9332$
(c) $P(\bar{X} \leq 1030) = 0.9641$
(d) $P(\mu_{\bar{X}} - \sigma_{\bar{X}} \leq \bar{X} \leq \mu_{\bar{X}} + \sigma_{\bar{X}})$
$= P(983.3 \leq \bar{X} \leq 1016.7) = 0.68$
(e) According to the Empirical Rule, $c = 2\sigma_{\bar{X}}$.
Therefore, $c = 2 \times \dfrac{100}{6} = 33.3$.

7.41 The blue graph represents the density function for the random variable X, the green graph represents the sampling distribution of $\bar{X}$ for $n = 5$, and the orange graph represents the sampling distribution of $\bar{X}$ for $n = 15$. As the sample size increases, the variability of the sampling distribution decreases.

7.43 Let X = the BMI of a randomly selected Canadian male. $X \sim N(27.4, 2.25)$.
(a) $P(X > 28) = 0.3446$
(b) For $n = 10$, $\bar{X} \sim N\left(27.4, \dfrac{2.25}{10}\right)$
$P(\bar{X} > 28) = 0.103$
(c) $P(\bar{X} < 26.5) = 0.0289$
(d) $P(26 \leq \bar{X} \leq 27) = 0.1979$

7.45 (a) $\bar{X} \sim N\left(1750, \dfrac{62{,}500}{15}\right)$
(b) $P(\bar{X} > 1800) = 0.2193$
(c) $P(1650 \leq \bar{X} \leq 1850) = 0.8787$
(d) According to the Empirical Rule, 95% of samples means will lie within $2 \times \sigma_{\bar{X}}$ of the mean, 1750. Therefore, the interval is (1620.9, 1879.1).

7.47 Let X = the weight of a randomly selected chip bag. Then $\bar{X} \stackrel{\bullet}{\sim} N\left(12, \dfrac{.09}{100}\right)$:
$n = 100 > 30$, so the CLT applies.
(a) $P(\bar{X} \leq 11.9) = 0.0004$
(b) Because the standard deviation of weights of individual chip bags is so small, it will be unusual for 100 randomly selected bags to have a mean that is this small.
(c) Claim: $\mu_X = 12$
Experiment: $\bar{x} = 11.9$ oz
Likelihood: $P(\bar{X} \leq 11.9) = 0.0004$
Conclusion: Since this probability is so small, there is evidence to suggest the claim concerning the mean weight of a 12-oz bag of chips in incorrect, and the bags actually weigh less than 12 oz, on average.

7.49 Let X = the amount of rainfall in a randomly selected August. Then, $\bar{X} \stackrel{\bullet}{\sim} N\left(3.54, \dfrac{1.21}{30}\right)$:
$n = 30 \geq 30$, so the CLT applies.
(a) $P(\bar{X} < 3.3) = 0.1160$ **(b)** $P(\bar{X} > 3.9) = 0.0365$
(c) $P(\mu - c \leq \bar{X} \leq \mu + c) = 0.90 \Rightarrow c = 0.3304$
Therefore, $P(3.21 \leq \bar{X} \leq 3.87) = 0.90$.

7.51 Let X = the weight of a randomly selected Onix Pure pickleball. Since the population of weights have a normal distribution, $\bar{X} \sim N\left(25, \dfrac{0.81}{10}\right)$.
(a) $P(\bar{X} < 24.5) = 0.0395$ **(b)** $P(\bar{X} > 25.25) = 0.19$
(c) $1 - P(24.4 < \bar{X} < 25.6) = 0.035$
There is a 3.5% chance that the sample cannot be used.

7.53 Let X = the vertical leap of a randomly selected NBA player. Then, $\bar{X} \stackrel{\bullet}{\sim} N\left(28, \dfrac{49}{36}\right)$:
$n = 36 > 30$, so the CLT applies.
(a) $P(\bar{X} < 26) = 0.0432$ **(b)** $P(27.5 \leq \bar{X} \leq 28.5) = 0.3318$
(c) Claim: $\mu = 28$
Experiment: $\bar{x} = 29.75$ for $n = 50$ players
Likelihood: Now, $\bar{X} \stackrel{\bullet}{\sim} N\left(28, \dfrac{49}{50}\right)$
$P(\bar{X} \geq 29.75) = 0.0385$
Conclusion: Since this probability is so small, there is evidence to suggest that the mean vertical leap height has increased for these players.

7.55 Let X = the response time of a randomly selected call. Then, $\bar{X} \stackrel{\bullet}{\sim} N\left(101, \dfrac{324}{35}\right)$:
$n = 35 > 30$, so the CLT applies.
(a) $P(\bar{X} > 99) = 0.7445$ **(b)** $P(95 \leq \bar{X} \leq 105) = 0.8814$
(c) Claim: $\mu = 101$
Experiment: $\bar{x} = 94$
Likelihood: $P(\bar{X} \leq 94) = 0.0107$
Conclusion: Since the probability is so small, there is evidence to suggest that the mean response time has decreased.

7.57 (a) Because T is a sum and $n = 35 > 30$, it will have an approximately normal distribution.
$\mu_T = 525$ min, or 8.75 hr
$\sigma_T^2 = 140$ min^2 or 0.0389 hr^2
(b) $P(8 < T < 9) = 0.8975$ **(c)** $P(T > 9.2) = 0.0112$
(d) $P(T \geq t) = 0.01 \Rightarrow t = 9.2088$ (552.5 min)

7.59 (a) If $\mu = 0.8$, then the mean force for 25 welds $\bar{X}$ will have a normal distribution with $\mu_{\bar{X}} = 0.8$ and $\sigma_{\bar{X}}^2 = \dfrac{0.01}{25}$.
P(process is shut down)
$= 1 - P(0.75 \leq \bar{X} \leq 0.85) = 0.0124$

(b) If $\mu = 0.82$, then the mean force for 25 welds $\bar{X}$ will have a normal distribution with $\mu_{\bar{X}} = 0.82$ and $\sigma^2_{\bar{X}} = \dfrac{0.01}{25}$.

P(process continues)
$= P(0.75 \leq \bar{X} \leq 0.85) = 0.9330$

If $\mu = 0.84$, then the mean force for 25 welds $\bar{X}$ will have a normal distribution with $\mu_{\bar{X}} = 0.84$ and $\sigma^2_{\bar{X}} = \dfrac{0.01}{25}$.

P(process continues)
$= P(0.75 \leq \bar{X} \leq 0.85) = 0.6915$

(c) For $\mu = 0.8$, P(process is shut down)
$= 1 - P(0.76 \leq \bar{X} \leq 0.84) = 0.0455$
For $\mu = 0.82$, P(process continues)
$= P(0.76 \leq \bar{X} \leq 0.84) = 0.84$
For $\mu = 0.84$, $P(0.76 \leq \bar{X} \leq 0.84) = 0.50$

7.61 Let X = the flight time for a randomly selected jumper.
Then, $\bar{X} \overset{\bullet}{\sim} N\left(22.5, \dfrac{5.29}{32}\right)$.

(a) $P(\bar{X} < 22) = 0.1094$ **(b)** Claim: $\mu = 22.5$
Experiment: $\bar{x} = 23.15$
Likelihood: $P(\bar{X} \geq 23.15) = 0.0549$
Conclusion: Since this probability is small, there is some evidence to suggest that the mean flight time for the new suit is longer.

(c) $P(\bar{X} \leq w) = 0.005 \Rightarrow w = 21.45$

7.63 (a) $\bar{X} \overset{\bullet}{\sim} N\left(\mu, \dfrac{0.0169}{30}\right)$

The probabilities $P(1.95 \leq \bar{X} \leq 2.05)$ are given in the table.

μ	P(accept shipment)
1.86	0.0001
1.88	0.0016
1.90	0.0174
1.92	0.1028
1.94	0.3365
1.96	0.6634
1.98	0.8956
2.00	0.9651
2.02	0.8956
2.04	0.6634
2.06	0.3365
2.08	0.1028
2.10	0.0174
2.12	0.0016
2.14	0.0001

(b)

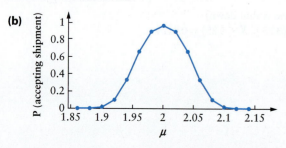

7.65 (a) $\mu = 124.85$ min
(b) Answers may vary. One set is shown in the histogram in part (c).
(c)

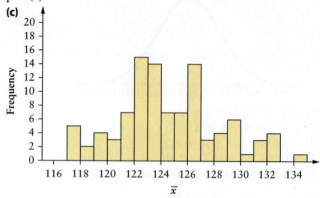

This distribution appears to be slightly skewed to the right. The mean of this distribution is 124.57 and the standard deviation is 3.841.

7.3 The Distribution of the Sample Proportion

7.67 False

7.69 True

7.71 The standard deviation of $\hat{P}$ is $\sqrt{\dfrac{p(1-p)}{n}}$. As n increases, the denominator of the standard deviation expression is growing larger, resulting in a decease of standard deviation.

7.73 Skewness criterion: $200(0.40) = 80 \geq 5$ and $200(1 - 0.40) = 120 \geq 5$
$\hat{P} \overset{\bullet}{\sim} N\left(0.40, \dfrac{(0.4)(0.6)}{200}\right); \sigma_{\hat{p}} = 0.035$

(a) $P(\hat{P} \leq 0.37) = 0.1932$
(b) $P(\hat{P} > 0.45) = 0.0742$
(c) $P(0.38 \leq \hat{P} \leq 0.42) = 0.4368$
(d) $P(\hat{P} < 0.33 \cup \hat{P} > 0.47) = 0.0430$

7.75 Skewness criterion: $80(0.35) = 28 \geq 5$ and $80(1 - 0.35) = 52 \geq 5$
$\hat{P} \overset{\bullet}{\sim} N\left(0.35, \dfrac{(0.35)(0.65)}{80}\right); \sigma_{\hat{p}} = 0.0533$

(a) $P(\hat{P} \leq a) = 0.10 \Rightarrow a = 0.2817$
(b) $P(\hat{P} > b) = 0.01 \Rightarrow b = 0.4741$
(c) According to the Empirical Rule, $c = 2 \times \sigma_{\hat{p}}$; therefore $c = 2(0.0533) = 0.1067$.

7.77 (a) Skewness criterion: $250(0.75) = 75 \geq 5$ and $250(1 - 0.7) = 75 \geq 5$
$\hat{P} \overset{\bullet}{\sim} N\left(0.7, \dfrac{(0.7)(0.3)}{250}\right); \sigma_{\hat{p}} = 0.0290$

(b) $P(\hat{P} < 0.66) = 0.0838$ **(c)** $P(\hat{P} > 0.71) = 0.3650$
(d) $P(0.68 \leq \hat{P} \leq 0.78) = 0.752$

7.79 (a) Skewness criterion: $150(0.16) = 24 \geq 5$ and $150(1 - 0.16) = 126 \geq 5$
$\hat{P} \overset{\bullet}{\sim} N\left(0.16, \dfrac{(0.16)(0.84)}{150}\right); \sigma_{\hat{p}} = 0.0299$

(b) $P(\hat{P} > 0.19) = 0.1581$
(c) $P(\hat{P} < 0.11) = 0.0474$
(d) $P(0.16 - c \leq \hat{P} \leq 0.16 + c) = 0.90$
$\Rightarrow c = 0.0492$
Therefore, $P(0.1108 \leq \hat{P} \leq 0.2092) = 0.90$

7.81 Skewness criterion: $225(0.64) = 144 \geq 5$ and $225(1 - 0.64) = 81 \geq 5$
$\hat{P} \overset{\bullet}{\sim} N\left(0.64, \dfrac{(0.64)(0.36)}{225}\right); \sigma_{\hat{p}} = 0.032$
(a) $P(\hat{P} < 0.60) = 0.1056$
(b) $P(0.65 \leq \hat{P} \leq 0.70) = 0.3469$
(c) $P(\hat{P} > m) = 0.95 \Rightarrow m = 0.5874$

7.83 Skewness criterion: $100(0.82) = 82 \geq 5$ and $100(1 - 0.82) = 18 \geq 5$
$\hat{P} \overset{\bullet}{\sim} N\left(0.82, \dfrac{(0.82)(0.18)}{100}\right); \sigma_{\hat{p}} = 0.0384$
(a) $P(\hat{P} > 0.87) = 0.0966$
(b) $P(0.75 \leq \hat{P} \leq 0.85) = 0.7483$
(c) Claim: $p = 0.82$
Experiment: $\hat{p} = 0.72$
Likelihood: $P(\hat{P} \leq 0.72) = 0.0046$
Conclusion: Since this probability is very small, there is strong evidence to suggest that the yield has decreased.

7.85 (a) Skewness criterion:
$400(0.987) = 394.8 \geq 5$ and $400(1 - 0.987) = 5.2 \geq 5$
$\hat{P} \overset{\bullet}{\sim} N\left(0.987, \dfrac{(0.987)(0.013)}{400}\right); \sigma_{\hat{p}} = 0.0057$
(b) $P(\hat{P} < 0.98) = 0.1082$
(c) $P(\hat{P} \leq f) = 0.05 \Rightarrow f = 0.9777$
(d) $P(\hat{P} > 0.995) = 0.0789$
There is a 7.9% chance that the team will earn a bonus.

7.87 Skewness criterion: $450(0.21) = 54.45 \geq 5$ and $450(1 - 0.121) = 395.55 \geq 5$
$\hat{P} \overset{\bullet}{\sim} N\left(0.121, \dfrac{(0.121)(0.879)}{450}\right); \sigma_{\hat{p}} = 0.0154$
(a) $P(\hat{P} > 0.14) = 0.1082$
(b) $P(0.10 \leq \hat{P} \leq 0.15) = 0.8844$
(c) Claim: $p = 0.121$
Experiment: $\hat{p} = \dfrac{36}{450} = 0.08$
Likelihood: $P(\hat{P} \leq 0.08) = 0.0038$
Conclusion: Since this probability is so small, there is strong evidence to suggest that the proportion of cigarette debris is less than 0.121.

7.89 Skewness criterion: $1000(0.07) = 70 \geq 5$ and $1000(1 - 0.07) = 930 \geq 5$
$\hat{P} \overset{\bullet}{\sim} N\left(0.07, \dfrac{(0.07)(0.93)}{1000}\right); \sigma_{\hat{p}} = 0.0081$
(a) $P(\text{sent back}) = P(\hat{P} > 0.09) = 0.0066$
(b) If $p = 0.08$, then $\hat{P} \overset{\bullet}{\sim} N\left(0.07, \dfrac{(0.08)(0.92)}{1000}\right)$
$P(\text{accept}) = P(\hat{P} \leq 0.09) = 0.8781$

7.91 (a) $\sigma_{\hat{p}}^2 = 0.001242 = \dfrac{(p)(1-p)}{200} \Rightarrow p \approx 0.46$
(b) $P(\hat{P} \leq r) = 0.70 \Rightarrow r = 0.4785$

7.93 (a)

p	P(accept shipment)
0.01	1.0000
0.02	1.0000
0.03	1.0000
0.04	0.9981
0.05	0.9742
0.06	0.8832
0.07	0.7103
0.08	0.5000
0.09	0.3106
0.10	0.1729
0.15	0.0028

(b)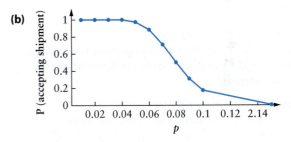

Chapter 7 Exercises

7.95 (a) $\bar{X} \overset{\bullet}{\sim} N\left(0.10, \dfrac{0.0025}{30}\right)$
(b) Claim: $\mu = 0.10$
Experiment: $\bar{x} = 0.1267$
Likelihood: $P(\bar{x} \geq 0.1267) = 0.0017$
Conclusion: Since this probability is so small, there is evidence to suggest the claim is false, that the mean amount of hydrogen peroxide in each bottle is more than 0.10 mg/m³.

7.97 (a) There are $5 \times 5 = 25$ ways to select two customers.

$\bar{x}$	1	1.5	2	2.5	3
$p(\bar{x})$	0.0100	0.1000	0.2900	0.2300	0.2000

$\bar{x}$	3.5	4	4.5	5
$p(\bar{x})$	0.1100	0.0425	0.0150	0.0025

(b) $\mu_{\bar{X}} = 2.55; \sigma_{\bar{X}}^2 = 0.5238; \sigma_{\bar{X}} = 0.7237$

7.99 (a) Because each male's weight is normally distributed, T will also have a normal distribution.
$T \sim N(782.8, 3600)$

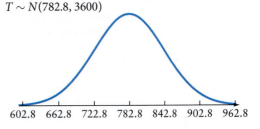

(b) P(extended ride) = P($T < 900$) = 0.9746
There is about a 97.5% chance of an extended ride.
(c) P(no takeoff) = P($T > 975$) = 0.0007
There is only an approximately 0.07% chance that the balloon will not be able to take off.

7.101 (a) Skewness criterion: $120(0.37) = 44.4 \geq 5$ and $120(1-0.37) = 75.6 \geq 5$
$\bar{X} \overset{\bullet}{\sim} N\left(0.37, \dfrac{(0.37)(0.63)}{120}\right)$

(b) P($\hat{P} < 0.30$) = 0.0561
(c) P($0.35 \leq \hat{P} \leq 0.40$) = 0.4270
(d) Claim: $p = 0.37$
Experiment: $\hat{p} = 0.42$
Likelihood: P($\hat{P} \geq 0.42$) = 0.1283
Conclusion: Since this probability is large, there is not sufficient evidence to suggest that the proportion of diners who order lobsters has increased.

7.103 (a) $\mu_X = 0.84$; $\sigma_X^2 = 1.1944$; $\sigma_X = 1.0929$
(b) There are $6 \times 6 = 36$ ways to select two machines.

t	0	1	2	3	4	5
p(t)	0.2500	0.3000	0.1900	0.1300	0.0720	0.0360

t	6	7	8	9	10
p(t)	0.0149	0.0048	0.0018	0.0004	0.0001

(c) $\mu_T = 1.68$; $\sigma_T^2 = 2.3888$; $\sigma_T = 1.5456$
(d) $2\mu_X = 2(0.84) = 1.68 = \mu_T$
$2\sigma_X^2 = 2(1.1944) = 2.3888 = \sigma_T^2$

7.105 (a) $\bar{X} \overset{\bullet}{\sim} N\left(3, \dfrac{0.0256}{40}\right)$

(b) Claim: $\mu = 3$
Experiment: $\bar{x} = 3.0338$
Likelihood: P($\bar{X} \geq 3.0338$) = 0.0908
Conclusion: Since this probability is large, there is not sufficient evidence to suggest that the average length of the stapes bone has increased.

7.107 (a) $\bar{X} \sim N\left(78, \dfrac{42.25}{10}\right)$
(b) Claim: $\mu = 78$
Experiment: $\bar{x} = 76.47$
Likelihood: P($\bar{X} \leq 76.47$) = 0.2283
Conclusion: Since this probability is quite large, there is not sufficient evidence to suggest that the study's claim is false.

7.109 (a) $\bar{X} \overset{\bullet}{\sim} N\left(70, \dfrac{25}{36}\right)$
(b) P($\bar{X} > 71$) = 0.1151
(c) Claim: $\mu = 70$
Experiment: $\bar{x} = 68.25$
Likelihood: P($\bar{X} \leq 68.25$) = 0.0179
Conclusion: Since this probability is so small, there is evidence to suggest that the mean amount of dextran is less than 70%.

7.111 (a) Skewness criterion: $150(0.5) = 75 \geq 5$ and $150(1-0.5) = 75$
$\hat{P} \overset{\bullet}{\sim} N\left(0.5, \dfrac{(0.5)(0.5)}{150}\right)$
(b) P($0.45 \leq \hat{P} \leq 0.55$) = 0.7793
(c) Claim: $p = 0.5$
Experiment: $\hat{p} = 0.59$
Likelihood: P($\hat{P} \geq 0.59$) = 0.0137
Conclusion: Since this probability is small, there is evidence to suggest that the true proportion of adults who are sensitive to mangoes has increased.

7.113 Skewness criterion: $250(0.53) = 132.5 \geq 5$ and $250(1-0.53) = 117.5 \geq 5$
$\hat{P} \overset{\bullet}{\sim} N\left(0.53, \dfrac{(0.53)(0.47)}{250}\right)$
(a) P($0.45 \leq \hat{P} \leq 0.50$) = 0.1653
(b) P($\hat{P} < b$) = 0.01 $\Rightarrow b = 0.4566$
Experiment: $\hat{p} = \dfrac{152}{250} = 0.608$
Likelihood: P($\hat{P} \geq 0.608$) = 0.0067
Conclusion: Since this probability is incredibly small, there is strong evidence to suggest that the true proportion of inpulse bookings has increased.

7.115 (a) $\bar{X} \overset{\bullet}{\sim} N\left(90, \dfrac{243.36}{40}\right)$ **(b)** P($\bar{X} < 60$) ≈ 0
(c) P($\bar{X} > 110$) ≈ 0 **(d)** P($\bar{X} < 93$) = 0.95 $\Rightarrow n = 74$

7.117 (a) Parameters **(b)** $\bar{X} \overset{\bullet}{\sim} N\left(1.75, \dfrac{0.04}{40}\right)$
(c) P($\bar{X} < 1.72$) = 0.1714
(d) Claim: $\mu = 1.75$
Experiment: $\bar{x} = 1.80$
Likelihood: P($\bar{X} \geq 1.8$) = 0.0569
Conclusion: Although this probability is small, it is still large enough that there is not sufficient evidence to suggest that the mean ride time is greater than 1.75 min.

Chapter 8

8.1 Point Estimation

8.1 True

8.3 Unbiased, small variance

8.5 Minimum variance unbiased estimator (MVUE)

8.7 Answers will vary. Example: Sample mean, $\bar{x} = \dfrac{\sum x_i}{n}$ or if $M=$ the maximum value of the data set and $N=$ the minimum value of the data set, then a statistic could be proposed as $\hat{\theta} = \dfrac{M+N}{2}$. This estimate is called a midrange.

8.9 $\hat{\theta}_3$ is the best statistic to estimate θ because it is the only unbiased statistic of the three. It also has a small variance, which is good.

8.11 Select the one with the smallest variance. This estimator will, on average, yield an estimate closer to the true value.

8.13 (a) $\bar{x} = 15.005$ (b) $\tilde{x} = 15.1$
(c) $s^2 = 7.1626$ (d) $\hat{p} = 0.45$

8.15 (a) $Q_1 = 7.2$; $Q_3 = 7.5$ (b) $p_{20} = 7.2$

8.17 (a) $\hat{p}_{QC} = \dfrac{75}{135} = 0.5556$ (b) $\hat{p}_{NS} = \dfrac{50}{107} = 0.4673$
(c) $\hat{p}_d = 0.0883$

8.19 (a) $x_{\min} = 0.866$ (b) $x_{\max} = 21.371$
(c) $Q_1 = 2.931$; $Q_3 = 12.795$
$\Rightarrow \widehat{IQR} = 12.795 - 2.931 = 9.864$ metric tons

8.2 A Confidence Interval for a Population Mean When σ Is Known

8.21 False

8.23 True

8.25 Wider

8.27 Since $B = z_{\alpha/2} \cdot \dfrac{\sigma}{\sqrt{n}}$, as n increases, the denominator of a fraction is getting larger, resulting in a smaller fraction. Therefore, if all other values remain constant, the bound B will decrease.

8.29 (a) $z_{0.10} = 1.2816$ (b) $z_{0.05} = 1.6449$
(c) $z_{0.025} = 1.9600$ (d) $z_{0.01} = 2.3263$
(e) $z_{0.005} = 2.5758$ (f) $z_{0.001} = 3.0902$
(g) $z_{0.0005} = 3.2905$ (h) $z_{0.0001} = 3.7190$

8.31 (a) $(14.0417, 21.1583)$ (b) $(127.032, 146.568)$
(c) $(310.136, 361.264)$ (d) $(-7.4259, -5.9741)$
(e) $(18.9843, 21.2357)$

8.33 (a) $n = \left[\dfrac{(7.9)(1.96)}{2.5}\right]^2 = 38.36 \Rightarrow n \geq 39$ (b) $n \geq 31$
(c) $n \geq 1{,}637{,}031$ (d) $n \geq 406$ (e) $n \geq 8189$

8.35 (a) $(136.57, 143.43)$
(b) Using the CLT, since $n = 40 \geq 30$, we know that $\bar{x} \dot\sim N\left(140, \dfrac{64}{40}\right)$. Since the sample size is large, no assumptions about the shape of the underlying distribution were necessary.

8.37 (a) $(0.6649, 0.8151)$ (b) $n \geq 75$
(c) $(0.9561, 1.1439)$ (d) $n \geq 76$

8.39 (a) $(31.2554, 36.2446)$ (b) $n \geq 113$
(c) The distribution of lighthouse heights is assumed to be normal.

8.41 (a) $(0.5834, 0.8606)$ (b) Claim: $\mu = 1$ hr
Experiment: $\bar{x} = 0.722$
Likelihood: A 95% CI or interval of likely values for μ: $(0.5834, 0.8606)$.
Conclusion: There is no evidence to suggest the mean time spent using the Internet is more than 1 hr. The CI is does not include 1 and is completely less than 1.
(c) $n \geq 97$
(d) The underlying distribution is probably not normal, but skewed to the right because most employees probably use the Internet for personal use very little, but there are likely a few employees who spend a great deal of time on the Internet during the workday.

8.43 (a) $(521.73, 599.07)$
(b) There is no evidence to suggest that the mean cruising speed is greater than 600 mph. The CI does not include 600 and is completely less than 600.

8.45 (a) $(8.3397, 8.8603)$
(b) The normality assumption seems reasonable. Even though we are counting the number of lightning strikes (which is discrete), the distribution is likely to be approximately normal looking.

8.47 (a) $(6.41, 7.78)$
(b) There is evidence to suggest that the population mean has decreased because the CI does not contain 8.2 and is entirely less than 8.2.

8.49 (a) $(122{,}495.89, 127{,}904.11)$
(b) $(144{,}999.95, 166{,}800.05)$
(c) There is evidence to suggest the mean selling price is different for the two parishes because the CIs have no overlapping values.

8.51 (a) Football: $(61.1087, 70.4313)$
Basketball: $(49.4270, 58.3730)$
Hockey: $(64.8986, 72.0014)$
(b) There is evidence to suggest the mean coping skills level is different for football and basketball players. The CIs do not overlap.
(c) Football: $n \geq 191$
Basketball: $n \geq 151$
Hockey: $n \geq 101$

8.53 (a) Cashews: (5.0591, 5.2809)
Filberts: (4.0737, 4.4063)
Pecans: (2.3367, 2.8633)
There is evidence to suggest the mean amount of protein is different for cashews and pecans because the CIs do not overlap. There is also evidence to suggest the mean amount of protein is different for filberts and pecans because the CIs do not overlap.
(b) Cashews: (4.9852, 5.3548)
Filberts: (3.9628, 4.5172)
Pecans: (2.1611, 3.0389)
There is evidence to suggest the mean amount of protein is different for cashews and pecans because the CIs do not overlap. There is also evidence to suggest the mean amount of protein is different for filberts and pecans because the CIs do not overlap.

8.55 (a) (8.513, 9.287)
(b) There is evidence to suggest that the mean mpg for the new truck is greater than 7. The CI does not contain 7 and is entirely above 7.
(c) There is evidence to suggest that the mean mpg for the new truck is not 10. The CI does not contain 10 and is entirely less than 10.

8.57 Lower bound: $\bar{x} - z_{0.01} \frac{\sigma}{\sqrt{n}} = 259.79 - (2.3263)\frac{7.5}{17} = 255.5584 \Rightarrow \mu > 255.5584$

8.3 A Confidence Interval for a Population Mean When σ Is Unknown

8.59 True

8.61 False

8.63 Confidence level $(1-\alpha)$; the degrees of freedom (ν)

8.65 As n increases, the critical value $t_{\alpha,n-1} \approx z_\alpha$. The t distribution tends toward the standard normal distribution.

8.67 (a) $t_{0.025,14} = 2.1448$ **(b)** $t_{0.01,20} = 2.5280$
(c) $t_{0.005,30} = 2.7500$ **(d)** $t_{0.0005,16} = 4.0150$
(e) $t_{0.00005,11} = 5.9212$ **(f)** $t_{0.01,3} = 4.5407$

8.69 (a) (193.6478, 228.7522) **(b)** (46.0475, 102.7925)
(c) (127.2235, 150.5765) **(d)** $(-47.643, -8.957)$
(e) (948.3804, 1080.6196)

8.71. (a) $z_{0.01} < t_{0.01,27} < t_{0.01,17} < t_{0.01,5}$
(b) $z_{0.025} < t_{0.025,45} < t_{0.025,13} < t_{0.025,11}$
(c) $t_{0.05,15} < t_{0.025,15} < t_{0.02,15} < t_{0.001,15}$
(d) $t_{0.1,21} < t_{0.05,21} < t_{0.005,21} < t_{0.0001,21}$
(e) $t_{0.10,6} < t_{0.05,17} < t_{0.001,26} < z_{0.0001}$

8.73 (a) (35.5795, 40.0205)
(b) We are 90% confident that the interval from 35.58 to 40.02 captures the true mean length of a washing machine main cycle. In repeated trials, the method used to construct this interval will successfully capture μ approximately 95% of the time.

(c) A 95% confidence interval will be larger. As confidence increases, so does the critical value, resulting in a wider interval.

8.75 (118.8967, 136.1033)
(118.8967, 136.1033) is an interval in which we are 95% confident the true mean depth of the upper mantle lies.

8.77 (a) (10.89, 17.89)
(b) There is evidence to suggest that the manufacturer's claim is false. The CI does not contain 10 and is entirely larger than 10, suggesting that the bikes are actually heavier than advertised.

8.79 (a) (1.9916, 4.9821)
(b) Frequency histogram:

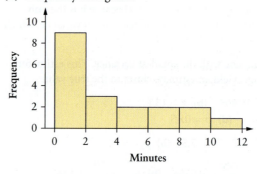

Backward Empirical Rule: $\bar{x} = 3.4868$, $s = 3.7585$

Interval	Frequency	Proportion
$(-0.27, 7.25)$	15	0.79
$(-4.02, 11.00)$	19	1.00
$(-7.79, 14.76)$	19	1.00

IQR/s = 1.8385
Normal probability plot:

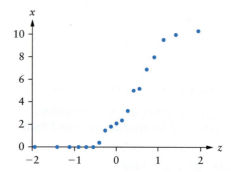

There is evidence to suggest that the data are not from a normal population. The histogram is very skewed right; the proportions 1, 2, and 3 standard deviations from the mean do not match those given by the Empirical Rule; the IQR/s ratio is not close to 1.3; and there is a distinct curve to the normal probability plot.
(c) Although it is close, there is no evidence to suggest that the mean number of minutes a bus is late is greater than 5 min. The CI does not include 5 and is completely less than 5. Because the data do not appear to come from a normal population, we should use caution in interpreting these results.

8.81 (a) Boston: (193.08, 204.71)
Dallas: (199.24, 214.96)
Detroit: (191.43, 203.17)
Edmonton: (196.18, 210.02)
Philadelphia: (187.14, 204.46)
(b) The sample from Dallas had a larger sample standard deviation.
(c) There is no evidence to suggest that the mean weights for both teams are different. Because there is overlap in the CIs, it is plausible that both teams have the same population mean weight.

8.83 (a) (826.48, 834.72)
(b) There is evidence to suggest that the mean price of platinum has increased. The 95% CI does not contain 825.75 and lies entirely above 825.75.

8.85 (a) (68.6122, 71.7878) **(b)** (71.0587, 73.1413)
(c) We have to assume that the underlying distributions are normal because the sample sizes are so small.
(d) There is no evidence to suggest that the mean thermostat settings are different since the two CIs overlap.

8.87 (a) (281.7269, 288.9397) **(b)** (249.4027, 266.0973)
(c) There is evidence to suggest that the true mean beats per minute in patients with atrial flutter is different for men and women since the CIs do not overlap.
(d) The normality assumption seems reasonable. Atrial flutter beats per minute is a measurement and probably not a skewed distribution.

8.89 (a) (11.70, 14.30) **(b)** (7.85, 9.75)
(c) Rural areas: Frequency histogram:

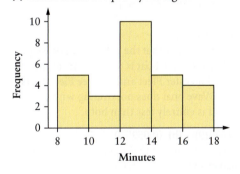

Backward Empirical Rule: $\bar{x} = 13$, $s = 2.4277$

Interval	Frequency	Proportion
(10.57, 15.43)	18	0.67
(8.14, 17.86)	27	1.00
(5.72, 20.28)	27	1.00

IQR/$s = 1.1945$
Normal probability plot:

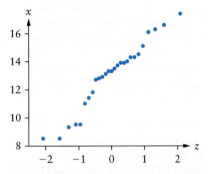

There does not appear to be much evidence of non-normality for the rural response times. The histogram is roughly symmetric; the proportions 1, 2, and 3 standard deviations away from the mean seem to follow the Empirical Rule; the IQR/s ratio is close to 1.3; and the normal probability plot is fairly linear.

City areas:
Frequency histogram:

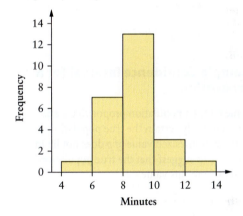

Backward Empirical Rule: $\bar{x} = 8.8$, $s = 1.6981$

Interval	Frequency	Proportion
(7.10, 10.50)	18	0.72
(5.40, 12.20)	24	0.96
(3.71, 13.89)	25	1.00

IQR/$s = 1.21$
Normal probability plot:

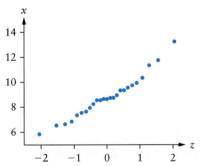

There does not appear to be much evidence of non-normality for the city response times. The histogram is roughly symmetric; the proportions 1, 2, and 3 standard deviations away from the mean seem to follow the Empirical Rule; the IQR/s ratio is close to 1.3; and the normal probability plot is fairly linear.
(d) There is evidence to suggest a difference in the mean response times for rural and city areas. The 99% CIs for each do not overlap.

8.91 (a) Men: (36.8592, 40.9408) Women: (34.1608, 37.0392)
(b) There is no evidence to suggest the mean age for men is different from the mean age for women. The CIs overlap, suggesting that it is plausible that both mean ages could be the same.
(c) (240.1594, 274.8406)
This is an interval in which we are 99% confident the true mean distance traveled lies.

8.93 (a) Upper bound: $\bar{x} + t_{\alpha, n-1} \dfrac{s}{\sqrt{n}}$

Lower bound: $\bar{x} - t_{\alpha, n-1} \dfrac{s}{\sqrt{n}}$

(b) Lower bound: $\bar{x} - t_{0.01, 29} \dfrac{s}{\sqrt{n}} \Rightarrow \mu > -6857.38$

8.4 A Large-Sample Confidence Interval for a Population Proportion

8.95 A large-sample CI for a population proportion is an interval in which we are fairly certain the true population proportion lies. If the hypothesized value *of p* does not lie in the CI, there is evidence to suggest that the true population proportion is different from this value.

8.97 False

8.99 It is possible. With a $\hat{p}$ value that is very close to 0 or 1 and a high level of confidence, the critical value, $Z_{\alpha/2}$ will be a large enough multiplier in the error of estimation term to cause the boundary on the interval to be larger than 1 or smaller than 0. Realistically the true population proportion cannot be less than 0 or greater than 1, but it is possible for our interval of likely values.

8.101 (a) (0.3868, 0.5465) **(b)** (0.2186, 0.3592)
(c) (0.9180, 0.9540) **(d)** (0.5383, 0.7881)
(e) (0.3698, 0.4350)

8.103 (a) $n = (0.45)(0.55)\left[\dfrac{1.96}{0.05}\right]^2 = 380.30 \Rightarrow n \geq 381$
(b) $n \geq 241$ **(c)** $n \geq 80$ **(d)** $n \geq 145{,}426$ **(e)** $n \geq 2065$

8.105 (a) As n increases, the square root becomes smaller and the width of the resulting CI decreases.
(b) As the confidence level decreases, the critical value decreases and the width of the resulting CI decreases.
(c) As $\hat{p}$ increases, the product $\hat{p}(1-\hat{p})$ decreases and the width of the resulting CI decreases.

8.107 (a) (0.1840, 0.2942)
(b) There is no evidence to suggest that the true proportion is less than 0.25. The CI captures 0.25, making it a plausible value for *p*.

8.109 (a) $\hat{p} = 132/952 = 0.1387$
$(952)(0.1387) = 132 \geq 5$, $(952)(0.8613) = 820 \geq 5$
The nonskewness criteria are satisfied, so $\hat{P}$ is approximately normal.
(b) (0.1167, 0.1606)
(c) There is evidence to suggest the true proportion of people who recognize the logo is greater than 0.10 since the CI does not include 0.10 and the CI is greater than 0.10.

8.111 (a) (0.0685, 0.1289) **(b)** (0.1133, 0.1823)
(c) (0.0820, 0.1352)
(d) The CI for the Canadian adults is the narrowest. This sample had the largest sample size, which leads to a smaller error of estimation.

8.113 (a) (0.3313, 0.3687) **(b)** $n \geq 423$

8.115 (a) (0.5982, 0.7618) **(b)** (0.1866, 0.3229)
(c) (0.3160, 0.4757)

8.117 (a) (0.0474, 0.0812)
(b) There is evidence that the true proportion of New England dairy farms certified as organic has changed (increased). The CI does not contain 0.03 and is greater than 0.03.

8.119 (a) (0.8736, 0.9402)
(b) There is no evidence to suggest that the TSA's claim is false. The CI captures 0.93.

8.121 (a) Gloucester Fleet: (0.5562, 0.6745)
Yankee Fleet: (0.7148, 0.8010)
Patriot Wave: (0.3486, 0.4578)
(b) There is evidence to suggest that the proportion of people who catch fish on a Patriot Wave boat is different from the proportions for both Gloucester Fleet and Yankee Fleet boats. The CI for the Patriot Wave boat does not overlap with the other two intervals and is entirely less than both.

8.123 (a) (0.0453, 0.0919) **(b)** (0.0595, 0.1147)
(c) There is no evidence to suggest that the true proportions of low-birth-weight babies in these two states are different. The two CIs overlap.

8.125 (a) Treatment: (0.1005, 0.1619) Placebo: (0.0506, 0.1442)
(b) There is no evidence to suggest the true proportion of people who suffer from headaches in the two groups is different. The CIs overlap.
(c) Treatment: (0.0398, 0.0936) Placebo: (0.0145, 0.1024)
(d) There is still no evidence to suggest the true proportion of people who suffer from a rash in the two groups is different. The CIs overlap.

8.127 (a) Upper bound: $\hat{p} + z_\alpha \sqrt{\dfrac{\hat{p}(1-\hat{p})}{n}}$

Lower bound: $\hat{p} - z_\alpha \sqrt{\dfrac{\hat{p}(1-\hat{p})}{n}}$

(b) $\hat{p} = 171/260 = 0.6577$

(c) Upper bound: $\hat{p} + z_{0.05}\sqrt{\dfrac{\hat{p}(1-\hat{p})}{n}} \Rightarrow p < 0.7061$

8.129

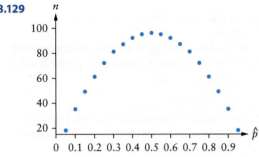

The points appear to lie along a parabola. n is largest when $\hat{p} = 0.50$.

8.5 A Confidence Interval for a Population Variance or Standard Deviation

8.131 False

8.133 False

8.135 (a) $\chi^2_{0.10,5} = 9.2364$ (b) $\chi^2_{0.001,31} = 61.0983$
(c) $\chi^2_{0.05,16} = 26.2962$ (d) $\chi^2_{0.025,21} = 35.4789$
(e) $\chi^2_{0.99,11} = 3.0535$ (f) $\chi^2_{0.95,15} = 7.2609$
(g) $\chi^2_{0.975,23} = 11.6886$ (h) $\chi^2_{0.995,9} = 1.7349$

8.137 (a) $\chi^2_{0.975,21} = 10.2829$, $\chi^2_{0.025,21} = 35.4789$
(b) $\chi^2_{0.995,36} = 17.8867$, $\chi^2_{0.005,36} = 61.5812$
(c) $\chi^2_{0.99,10} = 2.5582$, $\chi^2_{0.01,10} = 23.2093$
(d) $\chi^2_{0.95,30} = 18.4927$, $\chi^2_{0.05,30} = 43.7730$
(e) $\chi^2_{0.975,4} = 0.4844$, $\chi^2_{0.025,4} = 11.1433$
(f) $\chi^2_{0.9995,36} = 14.4012$, $\chi^2_{0.0005,36} = 70.5881$

8.139 (a) (1.3427, 11.2313) (b) (31.2940, 197.4147)
(c) (31.9769, 183.4121) (d) (3.0994, 26.5425)
(e) (18.5739, 64.0850) (f) (5.8969, 36.9707)

8.141 (a) CI for σ^2: (2.5912, 8.2250)
(b) CI for σ^2: (1.6097, 2.8679)

8.143 (a) CI for σ^2: (1.7229, 8.9829)
(b) Frequency histogram:

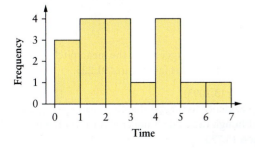

Backward Empirical Rule: $\bar{x} = 2.8294$, $s = 1.8401$

Interval	Frequency	Proportion
(0.9894, 4.6695)	12	0.67
(−0.8507, 6.5096)	17	0.94
(2.6908, 8.3497)	18	1.00

IQR/s = 1.3152
Normal probability plot:

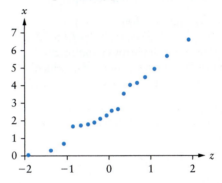

There is no strong evidence of non-normality. While the histogram shows a slight rightward skew, the proportions of values 1, 2, and 3 standard deviations from the mean follow the Empirical Rule closely; the IQR/s ratio is very close to 1.3; and the normal probability plot is fairly linear.

8.145 (a) (40.2387, 135.5132)
(b) Using the methods from Section 6.3: Frequency histogram:

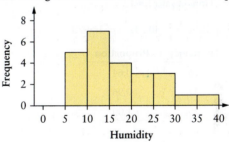

Backward Empirical Rule: $\bar{x} = 16.5417$, $s = 8.2987$

Interval	Frequency	Proportion
(8.2430, 24.8404)	14	0.58
(−0.557, −33.1391)	23	0.96
(−8.3544, 41.4378)	24	1.00

IQR/s = 1.3857
Normal Probability Plot:

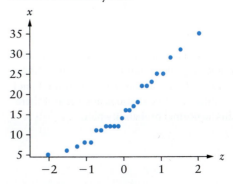

There appears to be evidence to suggest non-normality in this population. The histogram is skewed to the right; the proportions of observations 1, 2, and 3 standard deviations away from the mean do not follow the Empirical Rule; the IQR/s ratio is not 1.3; and there is a marked curve in the normal probability plot.

8.147 (a) (3.3005, 16.2993) (b) (22.2792, 76.8694)
(c) There is evidence to suggest that the two population variances are different. The two CIs do not overlap.

8.149 (a) (0.5611, 2.2381) (b) (15.5896, 71.8413)
(c) The veteran will get the starting position. The CI for the variance in yardage gained is much narrower and covers smaller values. Additionally, the CIs do not overlap, which provides evidence to suggest that the two variances are, in fact, different.

8.151 (a) (1.4239, 5.6372) (b) (1.1933, 2.3743)
Frequency histogram:

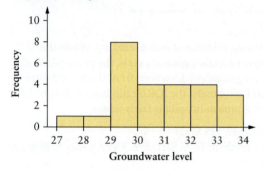

Backward Empirical Rule: $\bar{x} = 30.542$, $s = 1.5973$

Interval	Frequency	Proportion
(29.24, 32.44)	16	0.64
(27.65, 34.04)	24	0.96
(26.05, 35.63)	25	1.00

IQR/s = 1.7905
Normal probability plot:

There is evidence to suggest that the underlying distribution is not normal. The histogram is not symmetric; the proportions of values 1, 2, and 3 standard deviations from the mean do not follow the Empirical Rule; the IQR/s ratio is not 1.3; and there is a marked curve in the normal probability plot.

8.153 (a) $\dfrac{(n-1)s^2}{\chi^2_{0.025,19}} = 32,461,793,942,572,000$

$\dfrac{(n-1)s^2}{\chi^2_{0.975,19}} = 119,737.786,239,214,000$

Interval: (32.46 quadrillion, 119.74 quadrillion)
(b) (0.5842, 2.1548)
(c) There is no evidence to suggest that the variance ratio is greater than 1. The CI captures 1, making it a plausible value for the true population variance.

8.155 (a) (6.5503, 23.3373)
(b) There is no evidence to suggest that the true population variance in the height at midpoints between guard towers is less than 12 ft. The CI includes

8.157 (a) (3.6957, 10.1632)
(b) (4.7881, 14.1916)
(c) There is no evidence to suggest that the population variance is different for men and women. The CIs overlap.
(d) We needed to assume that the underlying populations are normal.

8.159 (a) CI for σ^2: (0.5064, 2.0211)
(b) CI for σ: (0.7116, 1.4217)

8.161 (a) CI for σ_C^2: (0.7665, 4.8353)
(b) CI for σ_A^2: (1.4511, 4.6060)
(c) There is no evidence to suggest that the population variance in time to relief is different for children and adults. The CIs overlap.

8.163 (a) CI for σ_{2017}^2: (25,847.8, 82,047.1)
(b) CI for σ_{1989}^2: (15,541.7, 42,740.0)
(c) There is no evidence to suggest that the variance in flow rate has increased. The CIs overlap.

8.165 (a) The CI for σ^2:

$$\left(\dfrac{s^2(n-1)}{z_{\alpha/2}\sqrt{2(n-1)}+(n-1)} < \sigma^2 < \dfrac{s^2(n-1)}{-z_{\alpha/2}\sqrt{2(n-1)}+(n-1)} \right)$$

(a) (i) (11.3763, 25.1102)
(ii) (11.6739, 26.7267)
(iii) The interval based on the normal distribution is wider since this is based on an approximate distribution.

Chapter 8 Exercises

8.167 (a) (2.096, 3.764)
(b) $n \geq 98$

8.169 (a) (90.14, 93.58)
(b) (7.6469, 39.4080)
(c) There is no evidence to suggest that the mean pitch speed has changed. The CI from part (a) captures 91.225.

8.171 (a) (12.5327, 13.2093)
(b) There is no evidence to suggest that the mean age is greater than 13.772. Although the CI does not contain 13.772, it is entirely less than 13.772.

8.173 **(a)** (0.5960, 0.6442)
(b) Use $\hat{p} = \dfrac{1670}{2693} = 0.6201 \Rightarrow n \geq 3908$

8.175 **(a)** (964.13, 1001.37)
(b) Interval for σ^2: (1240.12, 3936.41)
(c) Interval for $\sigma^2 = (35.22, 62.74)$

8.177 **(a)** (70.798, 80.002)
(b) There is no evidence to suggest that the population mean has changed. The CI includes 78.4.

8.179 **(a)** (0.1743, 0.2681) **(b)** (0.2358, 0.3246)
(c) There is no evidence to suggest a difference in the proportions of men and women who suffer from a disability. The two CIs overlap.

8.181 **(a)** (0.2832, 0.3168) **(b)** (0.3292, 0.3708)
(c) There is evidence to suggest the percentage of tannin is different in apples from fertilized and unfertilized trees. The CIs do not overlap.

8.183 **(a)** (4.5138, 4.8102) **(b)** (1.9568, 2.2012)
(c) There is no evidence that the population mean mercury concentration is greater than 5 for either group. Both CIs lie entirely below 5.
(d) There is evidence to suggest the mean mercury concentration is different for the two groups. The CIs do not overlap.

8.185 **(a)** (0.2740, 0.4294) **(b)** (0.3567, 0.4403)
(c) There is no evidence to suggest the proportion of males with heart disease is different for those who donate blood and those who do not. The CIs overlap.

8.187 **(a)** Boston: (0.3785, 0.5015)
Dallas: (0.4438, 0.5806)
Denver: (0.2227, 0.3551)
(b) There is evidence to suggest that the true proportion of job seekers who negotiate a starting salary is different in Denver than in Boston and Dallas. The CI for Denver does not overlap with the other two CIs.

Chapter 9

9.1 The Parts of a Hypothesis Test

9.1 True

9.3 False

9.5 Null hypothesis; alternative hypothesis, test statistic, rejection region

9.7 **(a)** Valid, null hypothesis
(b) Invalid, $\hat{p}$ is a statistic
(c) Invalid, s is a statistic
(d) Invalid, $\bar{x}$ is a statistic
(e) Valid, alternative hypothesis
(f) Valid, alternative hypothesis
(g) Invalid, $\tilde{x}$ is a statistic
(h) Valid, null hypothesis
(i) Valid, null hypothesis

9.9 **(a)** Valid
(b) Invalid. The alternative hypothesis should be $H_a: \mu > 9.7$.
(c) Invalid. The alternative hypothesis should involve 98.6—for example, $H_a: \sigma^2 \neq 98.6$.
(d) Invalid. The alternative hypothesis cannot be the same as the null hypothesis.

9.11 **(a)** Valid
(b) Valid
(c) Invalid. The null hypothesis should be stated so that μ (a parameter) equals a single value.
(d) Valid

9.13 Let $\mu =$ the population mean cumulative SAT score. $H_0: \mu = 1060$, $H_a: \mu > 1060$

9.15 Let $\mu =$ the population mean number of acre burned during wildfires. $H_0: \mu = 17{,}060$, $H_a: \mu < 17{,}060$

9.17 **(a)** The status quo is that the mean age is 25 years (or less). We are looking for evidence to suggest that the mean is greater than 25 years, which would support developing a new video game aimed at older players.

9.19 $H_0: p = 0.70$, $H_a: p < 0.70$

9.21 **(c)** The bus company is looking for evidence that the true proportion of parents who favor seat-belt installation is greater than 0.50.

9.23 $H_0: \tilde{\mu} = 400$, $H_a: \tilde{\mu} > 400$

9.25 $H_0: \sigma = 7$, $H_a: \sigma < 7$

9.27 $H_0: p = 0.60$, $H_a: p > 0.60$

9.29 **(c)** The City Council is looking for evidence that the true proportion of residents who favor the plan is greater than 0.80.

9.31 $H_0: \tilde{\mu} = 125.50$, $H_a: \tilde{\mu} < 125.50$

9.33 $H_0: \sigma^2 = 1.25$, $H_a: \sigma^2 < 1.25$

9.2 Hypothesis Test Errors

9.35 False

9.37 True

9.39 True

9.41 **(a)** Type I error **(b)** Correct decision
(c) Type II error **(d)** Type I error

9.43 **(a)** Type I error **(b)** Type II error
(c) Correct decision **(d)** Correct decision

9.45 There is always a chance of making a mistake in any hypothesis test because we never look at the entire population, only a sample. In addition, the value of the parameter is unknown. So it is always possible that we have incorrectly decided to reject the null hypothesis or failed to reject the null hypothesis.

9.47 Let μ = the true mean number of cars that use highway 405 per day (a) H_0: $\mu = 374{,}000$, H_a: $\mu > 374{,}000$
(b) Type I error: The true mean number of cars using highway 405 has not changed, but transportation officials conclude that it has (H_0 is rejected) and raise tolls.
Type II error: The true mean number of cars using highway 405 has increased, but transportation officials do not find evidence to conclude that there was a change (H_0 is not rejected) and do not raise tolls.
(c) The drivers are more angry. They now have to pay additional tolls for no reason.
(d) The transportation officials are more angry. They have missed out on an opportunity to make more money on a larger volume of cars.

9.49 (a) H_0: $\mu = 0.65$, H_a: $\mu > 0.65$
(b) A type I error would be canceling the race because you decide that the mean current velocity is more than 0.65 knot when it is actually less than that level. A type II error would be failing to realize that the mean current velocity is too high and running the race when it should have been canceled.
(c) A type II error is more serious for swimmers because it would result in swimming in dangerous currents.
(d) A type I error is more serious for race organizers because it would result in unnecessarily canceling the race and losing a great deal of revenue and goodwill.

9.51 (a) A type I error would be deciding the mean percentage of antioxidants in the bloodstream was more than 0.4 when it was actually the same or less. A type II error would be failing to realize the mean percentage of antioxidants in the bloodstream was more than 0.4.
(b) A type II error would be more serious for the Hershey Corporation because it would mean lost revenue from its inability to advertise the high percentages of antioxidants in its dark chocolate that were actually present.

9.53 (a) H_0: $\mu = 1200$, H_a: $\mu < 1200$
(b) Type I error: Decide $\mu < 1200$ when the true mean is really 1200 (or greater). That is, decide that the helmets do not meet the standard when they really do.
Type II error: Decide $\mu = 1200$ (or greater) when the true mean is really less than 1200. That is, decide the helmets meet the standard when they really do not

9.55 (a) H_0: $p = 0.98$, H_a: $p < 0.98$
(b) Type I error: Airport officials conclude that the scanners are not effective at a 98% detection rate (H_0 is rejected) when they really are.
Type II error: Airport officials determine that the scanners are effective at a 98% detection rate (H_0 is not rejected) when, in reality, the detection rate is less than 98%.

9.57 (a) H_0: $p = 0.20$, H_a: $p > 0.20$
(b) Type I error: Labor leaders conclude that the true proportion of contract workers is greater than 0.20 (H_0 is rejected) when it actually isn't.
Type II error: Labor leaders conclude that the true proportion of contract workers hasn't changed (H_0 is not rejected) when it has actually increased.
(c) Type II error is worse for the federal government because officials will miss out on additional taxes they should be collecting. Type I error is more serious for businesses because will have to pay additional taxes when they shouldn't.

9.59 (a) H_0: $p = 0.15$, H_a: $p > 0.15$
(b) A type I error would be to conclude that the true proportion of boats with safety violations is more than 0.15 when it is actually 0.15 or less. A type II error would be to conclude that the proportion of boats with safety violations is 0.15 (or less) when it is really more than 0.15.
(c) As the true value of p approaches 0.15 from the right, the probability of a type I error becomes smaller.

9.61 (a) H_0: $p = 0.62$, H_a: $p < 0.62$
(b) Type I error: Researchers conclude that the new implant design has a failure rate lower than 0.62, when the failure rate actually has not decreased.
Type II error: Researchers conclude that the new implant design has the same failure rate as the old design (0.62), when it actually has a lower failure rate.
(c) $\beta(0.50) < \beta(0.60)$

9.3 Hypothesis Tests Concerning a Population Mean When σ Is Known

9.63 False

9.65 False

9.67 True

9.69 (a) $Z = \dfrac{\overline{X} - 170}{15/\sqrt{38}}$
(b) (i) $Z \leq -2.3263$ (ii) $Z \leq -1.96$ (iii) $Z \leq -1.6449$
(iv) $Z \leq -1.2816$ (v) $Z \leq -3.0902$ (vi) $Z \leq -3.7190$

9.71 (a) $Z = \dfrac{\overline{X} - (-11)}{4.5/\sqrt{21}}$
(b) (i) $|Z| \geq 2.5758$ (ii) $|Z| \geq 1.2816$ (iii) $|Z| \geq 1.96$
(iv) $|Z| \geq 1.6449$ (v) $|Z| \geq 3.2905$ (vi) $|Z| \geq 3.7190$

9.73 (a) $\alpha = 0.05$ (b) $\alpha = 0.10$ (c) $\alpha = 0.005$
(d) $\alpha = 0.001$ (e) $\alpha = 0.20$ (f) $\alpha = 0.02$

9.75 (a) H_0: $\mu = 212$, H_a: $\mu > 212$
TS: $Z = \dfrac{\overline{X} - \mu_0}{\sigma/\sqrt{n}}$; RR: $Z \geq 2.3263$
(b) We assume the underlying population is normal and the population standard deviation (σ) is known.
(c) $z = \dfrac{213.5 - 212}{2.88/\sqrt{25}} = 2.6042 \geq 2.3263$. There is evidence to suggest that the population mean is greater than 212.

9.77 (a) H_0: $\mu = 365.25$; H_a: $\mu \neq 365.25$
TS: $Z = \dfrac{\overline{X} - \mu_0}{\sigma/\sqrt{n}}$; RR: $|Z| \geq 1.96$

(b) We assume the sample size is large and the population standard deviation (σ) is known.

(c) $z = \dfrac{360 - 365.25}{22.3/\sqrt{48}} = -1.6311 < 1.96$. There is no evidence to suggest the population mean is different from 365.25.

9.79 $H_0: \mu = 51{,}500$, $H_a: \mu < 51{,}500$

TS: $Z = \dfrac{\bar{X} - \mu_0}{\sigma/\sqrt{n}}$; RR: $Z \leq -2.3263$

$z = \dfrac{49{,}762 - 51{,}500}{3750/\sqrt{38}} = -2.8570 \leq -2.3263$

There is evidence to suggest the mean income per year of corporate communications workers has decreased.

9.81 $H_0: \mu = 295$, $H_a: \mu > 295$

TS: $Z = \dfrac{\bar{X} - \mu_0}{\sigma/\sqrt{n}}$; RR: $Z \geq 2.3263$

$z = \dfrac{306.3 - 295}{52/\sqrt{48}} = 1.5056 < 2.3263$

There is no evidence to suggest the mean length of international phone calls has increased.

9.83 (a) $H_0: \mu = 35$, $H_a: \mu > 35$

TS: $Z = \dfrac{\bar{X} - \mu_0}{\sigma/\sqrt{n}}$; RR: $Z \geq 2.3263$

$z = \dfrac{36.22 - 35}{\sqrt{5.7}/\sqrt{41}} = 3.2720 \geq 2.3263$

There is evidence to suggest the LOA is greater than 35 ft.
(b) The answer does not change if $\alpha = 0.10$. The critical value in this case is 1.2816, which is even smaller than 2.3263.

9.85 (a) $H_0: \mu = 2000$, $H_a: \mu < 2000$

TS: $Z = \dfrac{\bar{X} - \mu_0}{\sigma/\sqrt{n}}$; RR: $Z \leq -1.6449$

$z = \dfrac{1889 - 200}{360/\sqrt{37}} = -1.8756 \leq -1.6449$

There is evidence to suggest that the mean caloric intake of these students is below the daily energy requirement of 2000 calories.
(b) If $\alpha = 0.01$, the new critical value is -2.3263. Now our test statistic is not smaller than our critical value and we no longer have sufficient evidence to suggest that the mean caloric intake is too low.

9.87 $H_0: \mu = 40.11$, $H_a: \mu > 40.11$

TS: $Z = \dfrac{\bar{X} - \mu_0}{\sigma/\sqrt{n}}$; RR: $Z \geq 2.3263$

$z = \dfrac{42.76 - 40.11}{32/\sqrt{40}} = 0.5238 < 2.3263$

There is no evidence to suggest that the mean download speed has increased.

9.89 (a) $H_0: \mu = 12$, $H_a: \mu < 12$

TS: $Z = \dfrac{\bar{X} - \mu_0}{\sigma/\sqrt{n}}$; RR: $Z \leq -1.6449$

$z = \dfrac{11.85 - 12}{0.26/\sqrt{12}} = -1.9985 \leq -1.6449$

There is evidence to suggest that the true mean impact velocity is less than 12 m/s.
(b) If $\alpha = 0.01$, RR: $Z \leq -2.3263$
In this case we would not reject the null hypothesis but would conclude there is no evidence to suggest that the true mean impact velocity is less than 12 m/s.

9.91 $H_0: \mu = 55$, $H_a: \mu > 55$

TS: $Z = \dfrac{\bar{X} - \mu_0}{\sigma/\sqrt{n}}$; RR: $Z \geq 1.96$

$z = \dfrac{55.98 - 55}{7.1/\sqrt{43}} = 0.9051 < 1.96$

There is no evidence to suggest an increase in the mean flight time.

9.93 $H_0: \mu = 42$, $H_a: \mu \neq 42$

TS: $Z = \dfrac{\bar{X} - \mu_0}{\sigma/\sqrt{n}}$; RR: $|Z| \geq 1.96$

$z = \dfrac{43.22 - 42}{7.6/\sqrt{75}} = 1.3902 < 1.96$

There is no evidence to suggest that the true mean height of mailboxes in Des Moines is different from 42 in.

9.95 $H_0: \mu = 26.2$, $H_a: \mu > 26.2$

TS: $Z = \dfrac{\bar{X} - \mu_0}{\sigma/\sqrt{n}}$; RR: $Z \geq 1.96$

$z = \dfrac{32.9 - 26.2}{6/\sqrt{38}} = 6.8836 \geq 1.96$

There is strong evidence to suggest that the mean commuting time for people in Toronto is greater than 26.2 min.

9.97 $H_0: \mu = 0.23$, $H_a: \mu < 0.23$

TS: $Z = \dfrac{\bar{X} - \mu_0}{\sigma/\sqrt{n}}$; RR: $Z \leq -2.3263$

$z = \dfrac{0.1638 - 0.23}{0.23/\sqrt{26}} = -4.8189 \leq -2.3263$

There is evidence to suggest that the mean level of HC emissions is less than 0.23 g/hp-hr.
(b) A type I error is more important to the company: In this case, it will build the biodiesel fuel plant when the biodiesel fuel does not really decrease the mean level of HC emissions. Therefore, the company would want to use a very small significance level.

9.99 (a) $H_0: \mu = 15.5$, $H_a: \mu < 15.5$

TS: $Z = \dfrac{\bar{X} - \mu_0}{\sigma/\sqrt{n}}$; RR: $Z \leq -2.3263$

$z = \dfrac{15.45 - 15.5}{0.26/\sqrt{250}} = -3.0407 \leq -2.3263$

There is evidence to suggest that the mean weight of pretzel packages is less than 15.5 oz.
(b) The sample size is very large and σ is small. This results in a very small sampling standard deviation of $\bar{X}$.

9.101 (a) $H_0: \mu = 714$, $H_a: \mu \neq 714$

TS: $Z = \dfrac{\bar{X} - \mu_0}{\sigma/\sqrt{n}}$; RR: $|Z| \geq 1.96$

$$z = \frac{601.2 - 714}{283/\sqrt{16}} = -1.5943 > -1.96$$

There is no evidence to suggest the mean monthly water usage is different from 714 ft^3.
(b) The sample size is small and σ is large. This results in a large sampling standard deviation of $\bar{X}$.

9.103 (a) H_0: $\mu = 12.0$, H_a: $\mu > 12.0$
TS: $Z = \dfrac{\bar{X} - \mu_0}{\sigma/\sqrt{n}}$; RR: $Z \geq 2.3263$

$$z = \frac{12.3 - 12.0}{1.25/\sqrt{37}} = 1.4599 < 2.3263$$

There is no evidence to suggest that the population mean moisture content is greater than 12%.

(b) $P\left(\dfrac{\bar{X} - 12.0}{1.25/\sqrt{37}} \geq 2.3263\right) = 0.01$

$\Rightarrow P(\bar{X} \geq 12.4781) = 0.01$

$\beta(12.2) = P(\bar{X} < 12.4781) = P\left(Z < \dfrac{12.4781 - 12.2}{1.25/\sqrt{37}}\right)$

$= P(Z < 1.35) = 0.9120$

9.105 (a) H_0: $\mu = 496$, H_a: $\mu > 496$
TS: $Z = \dfrac{\bar{X} - \mu_0}{\sigma/\sqrt{n}}$; RR: $Z > 2.3263$

$$z = \frac{498.42 - 496}{46.7/\sqrt{33}} = 0.2977 < 2.3263$$

There is no evidence to suggest that the true mean weight of passengers and gear is greater than 496 lb.

(b) $P\left(\dfrac{\bar{X} - 496}{46.7/\sqrt{33}} \geq 2.3263\right) = 0.01$

$\Rightarrow P(\bar{X} \geq 514.9119) = 0.01$

$\beta(525) = P(\bar{X} < 514.9119) = P\left(Z < \dfrac{514.9119 - 525}{46.7/\sqrt{33}}\right)$

$= P(Z < -1.24) = 0.1073$

9.4 p Values

9.107 False

9.109 True

9.111 Reject the null hypothesis

9.113 (a) Do not reject H_0
(b) Reject H_0
(c) Do not reject H_0
(d) Do not reject H_0
(e) Reject H_0
(f) Do not reject H_0

9.115 (a) 0.0202   **(b)** 0.0764   **(c)** 0.0006
(d) 0.2514   **(e)** 0.000002327   **(f)** 0.5987

9.117 (a) $p = 0.0764$; do not reject H_0
(b) $p = 0.0202$; reject H_0
(c) $p = 0.0801$; reject H_0
(d) $p = 0.0009$; reject H_0
(e) $p = 0.1230$; do not reject H_0
(f) $p = 0.0188$; do not reject H_0

9.119 (a) $p = 0.2000$; do not reject H_0
(b) $p = 0.1671$; do not reject H_0
(c) $p = 0.0021$; reject H_0
(d) $p = 0.0068$; do not reject H_0
(e) $p = 0.7288$; do not reject H_0
(f) $p = 0.0094$; reject H_0

9.121 H_0: $\mu = 91.03$, H_a: $\mu \neq 91.03$
TS: $z = \dfrac{86.8 - 91.03}{11.06/\sqrt{38}} = -2.3576$; $p = 0.0184 \leq 0.05$

There is evidence to suggest the mean quarterback rating has changed.

9.123 (a) H_0: $\mu = 30$, H_a: $\mu > 30$
TS: $Z = \dfrac{\bar{X} - \mu_0}{\sigma/\sqrt{n}}$; RR: $Z \geq 2.3263$

$$z = \frac{32.2 - 30}{5.7/\sqrt{58}} = 2.9394 \geq 2.3263$$

There is evidence to suggest that the mean time to complete the safety checkup is more than 30 min.
(b) $p = 0.0016 \leq 0.01$

9.125 H_0: $\mu = 40$, H_a: $\mu \neq 40$
TS: $Z = z = \dfrac{40.9125 - 40}{2/\sqrt{8}} = 1.29$

$p = 0.1969 > 0.05$
There is no evidence to suggest that the population mean weight of the blocks is different from 40 lb.

9.127 H_0: $\mu = 42$, H_a: $\mu \neq 42$
TS: $Z = z = \dfrac{41.3 - 42}{2.5/\sqrt{32}} = -1.5839$

$p = 0.1132 > 0.05$
There is no evidence to suggest that the mean weight is different from 42 lb.

9.129 H_0: $\mu = 2500$, H_a: $\mu > 2500$
TS: $Z = z = \dfrac{2657.23 - 2500}{860/\sqrt{47}} = 1.2534$

$p = 0.1050 > 0.05$
There is no evidence to suggest that the mean depth is greater than 2500 m.

9.131 (a) H_0: $\mu = 1600$, H_a: $\mu < 1600$
TS: $Z = z = \dfrac{1595.6 - 1600}{23/\sqrt{25}} = -0.96$

$p = 0.1694 > 0.01$
There is no evidence to suggest the mean is less than 1600 lb.

(b)

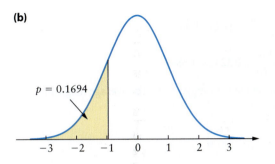

9.133 (a) $H_0: \mu = 2$, $H_a: \mu > 2$

TS: $z = \dfrac{2.16 - 2}{0.5/\sqrt{35}} = 1.8931$

$p = 0.0292 < 0.05$

There is evidence to suggest the mean sinking amount is greater than 2 cm.

(b)

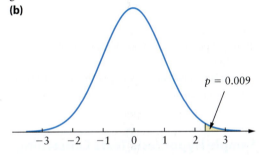

9.5 Hypothesis Tests Concerning a Population Mean When σ Is Unknown

9.135 True

9.137 True

9.139 False

9.141 (a) TS: $T = \dfrac{\bar{X} - \mu_0}{S/\sqrt{n}}$

(b) (i) $T \geq 3.3649$ **(ii)** $T \geq 2.0739$ **(iii)** $T \geq 1.7459$ **(iv)** $T \geq 1.3125$ **(v)** $T \geq 4.2968$ **(vi)** $T \geq 6.4420$

9.143 (a) TS: $T = \dfrac{\bar{X} - \mu_0}{S/\sqrt{n}}$

(b) (i) $|T| \geq 3.1058$ **(ii)** $|T| \geq 1.3304$ **(iii)** $|T| \geq 2.0595$ **(iv)** $|T| \geq 1.7033$ **(v)** $|T| \geq 5.9588$ **(vi)** $|T| \geq 2.39608$

9.145 (a) $\alpha = 0.10$ **(b)** $\alpha = 0.001$ **(c)** $\alpha = 0.005$ **(d)** $\alpha = 0.01$

9.147 (a) $0.01 \leq p \leq 0.025$ **(b)** $0.005 \leq p \leq 0.01$ **(c)** $p < 0.0001$ **(d)** $0.05 \leq p \leq 0.01$

9.149 (a) $0.05 \leq p \leq 0.10$ **(b)** $0.001 \leq p \leq 0.005$. **(c)** $p < 0.0001$ **(d)** $0.1 \leq p \leq 0.2$

9.151 (a) $H_0: \mu = 57.71$, $H_a: \mu > 57.71$

TS: $T = \dfrac{\bar{X} - \mu_0}{S/\sqrt{n}}$; RR: $T \geq 2.7638$

(b) $t = \dfrac{59.31 - 57.71}{1.6037/\sqrt{11}} = 3.3053 \geq 2.7638$

There is evidence to suggest the mean is greater than 57.71.

(c) $p = 0.0040$

9.153 (c) When the test statistic is T, the value of σ is unknown. The test statistic should be

$T = \dfrac{\bar{X} - \mu_0}{S/\sqrt{n}}$.

(b) The sample size $n = 25$ should be used in computing the value of the test statistic, not $n - 1$.

(c) This is a two-sided test, so the rejection region should be $|T| \geq 2.0639$.

(d) If the value of the test statistic is $t = 2.6732$, then the p value is between 0.01 and 0.025.

9.155 $H_0: \mu = 9.16$, $H_a: \mu > 9.16$

TS: $T = \dfrac{\bar{X} - \mu_0}{S/\sqrt{n}}$; RR: $T \geq 2.1098$

$t = \dfrac{9.68 - 9.16}{1.65/\sqrt{18}} = 1.3371 < 2.1009$

There is no evidence to suggest that the mean movie ticket cost has increased.

9.157 $H_0: \mu = 2560$, $H_a: \mu > 2560$

TS: $T = \dfrac{\bar{X} - \mu_0}{S/\sqrt{n}}$; RR: $T \geq 2.7638$

$t = \dfrac{2909.5 - 2560}{738.03/\sqrt{11}} = 1.5706 < 2.7638$

There is no evidence to suggest that the mean square footage has increased.

9.159 $H_0: \mu = 23.1$, $H_a: \mu > 23.1$

TS: $T = \dfrac{\bar{X} - \mu_0}{S/\sqrt{n}}$; RR: $T \geq 2.8965$

$t = \dfrac{24.6 - 23.1}{2.1/\sqrt{9}} = 2.1429 < 2.8965$

$p = 0.0322 > 0.01$

There is no evidence to suggest that the mean hotel room occupation rate has increased. If the test is significant, one cannot conclude the ad campaign caused the increase. This is an observational study.

9.161 $H_0: \mu = 2.3$, $H_a: \mu < 2.3$

TS: $T = \dfrac{\bar{X} - \mu_0}{S/\sqrt{n}}$; RR: $T \leq -2.4286$

$t = \dfrac{1.99 - 2.3}{1.612/\sqrt{39}} = -1.2010 > -2.4286$

$p = 0.1186$

There is no evidence to suggest that the true mean physician density is less than 2.3.

9.163 $H_0: \mu = 19.36$, $H_a: \mu \neq 19.36$

TS: $T = \dfrac{\bar{X} - \mu_0}{S/\sqrt{n}}$; RR: $|T| \geq 2.2622$

$t = \dfrac{18.66 - 19.36}{5.6715/\sqrt{10}} = -0.3903 > -2.2622$

$p = 0.7054$

There is no evidence to suggest that the mean hourly wage has changed.

9.165 (a) $H_0: \mu = 1381$, $H_a: \mu \neq 1381$

TS: $T = \dfrac{\bar{X} - \mu_0}{S/\sqrt{n}}$; RR: $|T| \geq 2.1199$

$t = \dfrac{1857 - 1381}{786/\sqrt{17}} = 2.4969 > 2.1199$

There is evidence to suggest that the true mean loss as a result of a residential break-in has changed from \$1381.
(b) $0.02 \leq p \leq 0.05$
(c)

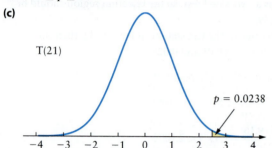

$T(21)$, $p = 0.0238$

9.167 (a) $H_0: \mu = 159{,}350$, $H_a: \mu > 159{,}350$

TS: $T = \dfrac{\bar{X} - \mu_0}{S/\sqrt{n}}$; RR: $T \geq 2.7638$

$t = \dfrac{163{,}288 - 159{,}350}{8792/\sqrt{11}} = 1.4855 < 2.7638$

There is no evidence to suggest that the mean methane output per well per day has increased.
(b) $0.05 \leq p \leq 0.10$

9.169 (a) $H_0: \mu = 14.5$, $H_a: \mu > 14.5$

TS: $T = \dfrac{\bar{X} - \mu_0}{S/\sqrt{n}}$; RR: $T \geq 2.4922$

$t = \dfrac{15.005 - 14.5}{4.039/\sqrt{24}} = 0.6125 < 2.4922$

There is no evidence to suggest an increase in the mean length of these lionfish.
(b) $p \geq 0.20$

9.171 (a) $H_0: \mu = 4.75$, $H_a: \mu < 4.75$

TS: $T = \dfrac{\bar{X} - \mu_0}{S/\sqrt{n}}$; RR: $T \leq -2.4121$

$t = \dfrac{4.66 - 4.75}{0.25/\sqrt{46}} = -2.4416 \leq -2.4121$

There is evidence to suggest that the population mean width of littlenecks has decreased.
(b) This test is statistically significant even though the sample mean (4.66) is so close to the assumed population mean (4.75) because the sample size is fairly large ($n = 46$) and the sample standard deviation is quite small ($s = 0.25$) relative to the sample mean.
(c) $0.005 \leq p \leq 0.01$

9.173 (a) $H_0: \mu = 1.0$, $H_a: \mu > 1.0$

TS: $T = \dfrac{\bar{X} - \mu_0}{S/\sqrt{n}}$; RR: $T \geq 1.6849$

$t = \dfrac{1.03 - 1.0}{0.587/\sqrt{40}} = 0.3232 < 1.6849$

There is no evidence to suggest that the average magnitude is greater than 1.0.
(b) $p \geq 0.20$

9.175 (a) $H_0: \mu = 65$, $H_a: \mu > 65$

TS: $T = \dfrac{\bar{X} - \mu_0}{S/\sqrt{n}}$; RR: $T \geq 2.6503$

$t = \dfrac{70.8571 - 65}{45.7045/\sqrt{14}} = 0.4795 < 2.6503$

There is no evidence to suggest that the mean amount of aspartame is greater than 65 mg.
(b) $p \geq 0.20$

9.177 (a) $H_0: \lambda = 4$, $H_a: \lambda < 4$
TS: X = the number of people locked out of their rooms; RR: $X \leq 0$ ($\alpha \approx 0.0183$)
(b) $x = 2 > 0$. There is no evidence to suggest the mean number of people locked out of their rooms per day is less than 4.
(c) $p = 0.2381$

9.6 Large-Sample Hypothesis Tests Concerning a Population Proportion

9.179 True

9.181 False

9.183 The hypothesis test is not valid.

9.185 (a) $276(0.3) = 82.8 \geq 5$, $276(0.7) = 193.2 \geq 5$
The test is appropriate.
(b) $1158(0.6) = 694.8 \geq 5$, $1158(0.4) = 463.2 \geq 5$
The test is appropriate.
(c) $645(0.03) = 19.35 \geq 5$, $645(0.97) = 625.65 \geq 5$
The test is appropriate.
(d) $159(0.97) = 154.23 \geq 5$, $159(0.03) = 4.77 < 5$
The test is not appropriate.
(e) $322(0.38) = 122.36 \geq 5$, $322(0.62) = 199.64 \geq 5$
The test is appropriate.
(f) $443(0.82) = 363.26 \geq 5$, $443(0.18) = 79.74 \geq 5$
The test is appropriate.

9.187

	Rejection Region	Value of TS	Conclusion
(a)	$\|Z\| \geq 2.2414$	-2.0142	Do not reject H_0
(b)	$\|Z\| \geq 2.3263$	2.3451	Reject H_0
(c)	$\|Z\| \geq 1.9600$	-2.0294	Reject H_0
(d)	$\|Z\| \geq 2.8070$	-1.4025	Do not reject H_0
(e)	$\|Z\| \geq 2.5758$	2.6455	Reject H_0

9.189

	Value of TS	p value	Conclusion
(a)	−2.0285	0.0213	Reject H_0
(b)	0.0574	0.5229	Do not reject H_0
(c)	−0.7611	0.2233	Do not reject H_0
(d)	−2.2166	0.0133	Reject H_0
(e)	−2.6187	0.0044	Reject H_0

9.191 (a) $n = 500$, $x = 16$, $p_0 = 0.02$
(b) $500(0.02) = 10 \geq 5$, $500(0.98) = 490 \geq 5$
The large sample test is appropriate.
(c) $H_0: p = 0.02$, $H_a: p > 0.02$

TS: $Z = \dfrac{\hat{P} - p_0}{\sqrt{\dfrac{p_0(1-p_0)}{n}}}$; RR: $Z \geq 1.6449$

$z = \dfrac{0.032 - 0.02}{\sqrt{\dfrac{(0.02)(0.98)}{500}}} = 1.9166 \geq 1.6449$

There is evidence to suggest that the proportion of children who eat the recommended number of servings each day has increased.
(d) $p = 0.0276$

9.193 (a) $n = 225$, $x = 189$, $p_0 = 0.90$
(b) $225(0.90) = 202.5 \geq 5$, $225(0.10) = 22.5 \geq 5$
The large-sample test is appropriate.
(c) $H_0: p = 0.90$, $H_a: p \neq 0.90$

TS: $Z = \dfrac{\hat{P} - p_0}{\sqrt{\dfrac{p_0(1-p_0)}{n}}}$; RR: $|Z| \geq 1.96$

$z = \dfrac{0.84 - 0.90}{\sqrt{\dfrac{(0.90)(0.10)}{225}}} = -3.0 \leq -1.96$

There is evidence to suggest that the proportion of home-schooled children who attend college is different from 0.90.
(d) $p = 0.0027$

9.195 $H_0: p = 0.38$, $H_a: p > 0.38$

TS: $Z = \dfrac{\hat{P} - p_0}{\sqrt{\dfrac{p_0(1-p_0)}{n}}}$; RR: $Z \geq 2.3263$

$z = \dfrac{0.4016 - 0.38}{\sqrt{\dfrac{(0.38)(0.62)}{1250}}} = 1.5733 < 2.3263$

p-value: $0.0578 > 0.01$
There is no evidence to suggest the proportion of taller women who develop cancer is greater than 0.38.

9.197 $H_0: p = 0.50$, $H_a: p > 0.50$

TS: $Z = \dfrac{\hat{P} - p_0}{\sqrt{\dfrac{p_0(1-p_0)}{n}}}$; RR: $Z \geq 1.6449$

$z = \dfrac{0.6333 - 0.50}{\sqrt{\dfrac{(0.50)(0.50)}{120}}} = 2.9212 \geq 1.6449$

p-value: $0.0017 \leq 0.05$
There is evidence to suggest that more than half of all people have trouble sleeping during a full moon.

9.199 $H_0: p = 0.5$, $H_a: p > 0.50$

TS: $Z = \dfrac{\hat{P} - p_0}{\sqrt{\dfrac{p_0(1-p_0)}{n}}}$; RR: $Z \geq 2.3263$

$z = \dfrac{0.6 - 0.5}{\sqrt{\dfrac{(0.5)(0.5)}{375}}} = 3.8730 \geq 2.3263$

p-value: $0.0001 < 0.01$ There is evidence to suggest that more than half of all city residents approve of Mayor Libby. She seems to have the support of her community to run for office.

9.201 $H_0: p = 0.10$, $H_a: p < 0.10$

TS: $Z = \dfrac{\hat{P} - p_0}{\sqrt{\dfrac{p_0(1-p_0)}{n}}}$; RR: $Z \leq -1.6449$

$z = \dfrac{0.09 - 0.10}{\sqrt{\dfrac{(0.10)(0.90)}{100}}} = -0.3333 > -1.6449$

p-value: $0.3694 > 0.05$
There is no evidence to suggest that the proportion of rent-controlled apartments is less than 0.10.

9.203 $H_0: p = 1/3$, $H_a: p \neq 1/3$

TS: $Z = \dfrac{\hat{P} - p_0}{\sqrt{\dfrac{p_0(1-p_0)}{n}}}$; RR: $|Z| \geq 1.96$

$z = \dfrac{0.3582 - 0.3333}{\sqrt{\dfrac{(0.3333)(0.6667)}{656}}} = 1.3528 < 1.96$

p-value: $0.1761 > 0.05$
There is no evidence to suggest that the proportion of Americans not using handwriting is different from 1/3.

9.205 (a) $H_0: p = 0.95$, $H_a: p < 0.95$

TS: $Z = \dfrac{\hat{P} - p_0}{\sqrt{\dfrac{p_0(1-p_0)}{n}}}$; RR: $Z \leq -1.6449$

$z = \dfrac{0.915 - 0.95}{\sqrt{\dfrac{(0.95)(0.05)}{200}}} = -2.2711 \leq -1.6449$

There is evidence to suggest that the proportion of all tractor batteries that last at least three years is less than 0.95.
(b) $p = 0.0116$
(c) Since there is evidence to suggest that less than 95% of all batteries last at least three years, the company should not implement the new warranty. It may end up losing money.

9.207 (a) $H_0: p = 0.65$, $H_a: p > 0.65$

TS: $Z = \dfrac{\hat{P} - p_0}{\sqrt{\dfrac{p_0(1-p_0)}{n}}}$; RR: $Z \geq 1.6449$

$z = \dfrac{0.7 - 0.65}{\sqrt{\dfrac{(0.65)(0.35)}{350}}} = 1.9611 > 1.6449$

There is evidence to suggest that the true proportion of households with a camper has increased.

(b) p-value: 0.0249
(c)

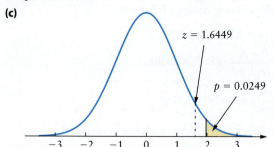

9.209 (a) $H_0: p = 0.51$, $H_a: p < 0.51$
TS: X = the number of students who admit to cheating out of the 25. RR: $X \leq 6$ ($\alpha \approx 0.0055$)
(b) $x = 9 > 6$
There is no evidence to suggest that the true proportion of Long Island students who admit to cheating is less than 0.51.
(c) Assuming $X \sim B(25, 0.51)$: $p = 0.0964$

9.7 Hypothesis Tests Concerning a Population Variance

9.211 False

9.213 False

9.215 True

9.217 (a) $X^2 = \dfrac{(n-1)S^2}{\sigma_0^2}$
(b) (i) $X^2 \leq 3.9416$ **(ii)** $X^2 \leq 5.2260$
(iii) $X^2 \leq 12.5622$ **(iv)** $X^2 \leq 17.2919$
(v) $X^2 \leq 20.0719$ **(vi)** $X^2 \leq 0.9893$

9.219 (a) $\alpha = 0.05$ **(b)** $\alpha = 0.005$ **(c)** $\alpha = 0.005$
(d) $\alpha = 0.0005$

9.221 (a) $\alpha = 0.01$ **(b)** $\alpha = 0.02$ **(c)** $\alpha = 0.001$
(d) $\alpha = 0.05$

9.223 (a) $0.0001 \leq p \leq 0.001$ **(b)** $p \leq 0.0001$
(c) $0.005 \leq p \leq 0.01$ **(d)** $0.025 \leq p \leq 0.05$

9.225 (a) $H_0: \sigma^2 = 16.7$, $H_a: \sigma^2 > 16.7$
TS: $X^2 = \dfrac{(n-1)S^2}{\sigma_0^2}$; RR: $X^2 \geq 37.5662$
(b) $\chi^2 = \dfrac{(20)(28)}{16.7} = 33.5329 < 37.5662$
There is no evidence to suggest that the population variance is greater than 16.7.

(c) $0.025 \leq p \leq 0.05$

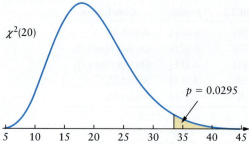

9.227 (a) $H_0: \sigma^2 = 75.6$, $H_a: \sigma^2 < 75.6$
TS: $X^2 = \dfrac{(n-1)S^2}{\sigma_0^2}$; RR: $X^2 \leq 17.2616$
(b) $\chi^2 = \dfrac{(39)(48.5)}{75.6} = 25.0198 > 17.2616$
There is no evidence to suggest that the population variance is less than 75.6.
(c) $0.025 \leq p \leq 0.05$

9.229 $H_0: \sigma^2 = 16{,}900$, $H_a: \sigma^2 > 16{,}900$
TS: $X^2 = \dfrac{(n-1)S^2}{\sigma_0^2}$; RR: $X^2 \geq 62.4281$
$\chi^2 = \dfrac{(39)(21{,}330.8)}{16{,}900} = 49.2249 < 62.4281$
There is no evidence to suggest that the population variance in technically recoverable wet shale gas is greater than 16,900.

9.231 $H_0: \sigma^2 = 0.09$, $H_a: \sigma^2 > 0.09$
TS: $X^2 = \dfrac{(n-1)S^2}{\sigma_0^2}$; RR: $X^2 \geq 36.7807$
$\chi^2 = \dfrac{(22)(0.105)}{0.09} = 25.6667 < 36.7807$
$p = 0.2663 > 0.025$
There is no evidence to suggest that the population variance in the fillings of the cream puffs is greater than 0.09 oz.

9.233 $H_0: \sigma^2 = 0.04$, $H_a: \sigma^2 > 0.04$
TS: $X^2 = \dfrac{(n-1)S^2}{\sigma_0^2}$; RR: $X^2 \geq 42.9798$
$\chi^2 = \dfrac{(24)(0.224)^2}{0.04} = 35.7216 < 42.9798$
$p = 0.0584 > 0.01$
There is no evidence to suggest that the population variance in die weight is greater than 0.04 g^2.

9.235 $H_0: \sigma^2 = 0.57$, $H_a: \sigma^2 \neq 0.57$
TS: $X^2 = \dfrac{(n-1)S^2}{\sigma_0^2}$; RR: $X^2 \leq 7.5642$ or $X^2 \geq 30.1910$
$\chi^2 = \dfrac{(17)(0.34)^2}{0.57} = 3.4477 \leq 7.5642$
$p = 0.0004 \leq 0.05$
There is evidence to suggest that the population variance is different from 0.57.

9.237 H_0: $\sigma^2 = 0.25$, H_a: $\sigma^2 > 0.25$
TS: $X^2 = \dfrac{(n-1)S^2}{\sigma_0^2}$; RR: $X^2 \geq 31.4104$
$\chi^2 = \dfrac{(20)(0.4451)^2}{0.25} = 35.6080 \geq 31.4104$
There is evidence to suggest that the true variance in cap diameter is greater than 0.25 in.

9.239 (a) H_0: $\sigma^2 = 0.36$, H_a: $\sigma^2 > 0.36$
TS: $X^2 = \dfrac{(n-1)S^2}{\sigma_0^2}$; RR: $X^2 \geq 30.1435$
$\chi^2 = \dfrac{(19)(0.42)}{0.36} = 22.1667 < 30.1435$
$p = 0.2760 > 0.05$
There is no evidence to suggest the population variance is greater than 0.36.
(b) $p > 0.10$

9.241 (a) H_0: $\sigma^2 = 230$, H_a: $\sigma^2 > 230$
TS: $X^2 = \dfrac{(n-1)S^2}{\sigma_0^2}$; RR: $X^2 \geq 38.8851$
$\chi^2 = \dfrac{(26)(194.0425)}{230} = 21.9352 < 38.8851$
$p = 0.6922 > 0.05$
There is no evidence to suggest the variance is greater than 230 mW^2.
(b) $p > 0.10$

9.243 (a) H_0: $\sigma^2 = 2.75^2$, H_a: $\sigma^2 > 2.75^2$
TS: $X^2 = \dfrac{(n-1)S^2}{\sigma_0^2}$; RR: $X^2 \geq 32.6706$
$\chi^2 = \dfrac{(21)(14.5214)}{7.5625} = 40.3239 \geq 32.6706$
$p = 0.0068 \leq 0.05$
There is evidence to suggest that the population variance in wingspan is greater than 2.75^2 mm.
(b) $0.005 \leq p \leq 0.01$

9.245 (a) H_0: $\sigma^2 = 49$, H_a: $\sigma^2 \neq 49$
TS: $X^2 = \dfrac{(n-1)S^2}{\sigma_0^2}$; RR: $X^2 \leq 9.5908$ or $X^2 \geq 34.1696$
(b) $S^2 \leq 23.4975$ or $S^2 \geq 83.7155$
(c) $23.4975 < 56 < 83.7155$; $p = 0.5917 > 0.05$
There is no evidence to suggest the population variance in thickness is different from 49 μ.
(d) $s^2 = 15.6 \leq 23.4975$; $p = 0.0034 \leq 0.05$
There is evidence to suggest the population variance in thickness is different from 49 μ.

Chapter 9 Exercises

9.247 (a) H_0: $\mu = 1.6$, H_a: $\mu \neq 1.6$
TS: $Z = \dfrac{\bar{x} - \mu_0}{\sigma/\sqrt{n}}$; RR: $|Z| \geq 2.5758$
$z = \dfrac{1.78 - 1.6}{0.5/\sqrt{23}} = 1.7265 < 2.5758$
There is no evidence to suggest the population mean is different from 1.6 mol; there is no evidence to suggest the machine is malfunctioning.
(b) $p = 0.0843$

9.249 (a) H_0: $\mu = 1$, H_a: $\mu \neq 1$
TS: $Z = \dfrac{\bar{x} - \mu_0}{\sigma/\sqrt{n}}$; RR: $|Z| \geq 1.96$
$z = \dfrac{1.044 - 1}{0.1/\sqrt{18}} = 1.8668 < 1.96$
There is no evidence to suggest that the mean weight is not 1 kg.
(b) $p = 0.062 > 0.05$

9.251 (a) H_0: $\mu = 13$, H_a: $\mu \neq 13$
TS: $T = \dfrac{\bar{X} - \mu_0}{S/\sqrt{n}}$; RR: $|T| \geq 2.2622$
$t = \dfrac{12.89 - 13}{0.96/\sqrt{10}} = -0.3623 > -2.2622$
$p = 0.7255 > 0.05$
There is no evidence to suggest the mean diameter is different from 13 mm. The process should not be stopped.
(b) $t = \dfrac{13.04 - 13}{0.045/\sqrt{10}} = 2.8109 \geq 2.2622$
$p = 0.0203 \leq 0.05$
There is evidence to suggest the mean diameter is different from 13 mm. The process should be stopped.
(c) We will want a large significance level. This would mean a smaller type II error, failing to realize the mean diameter is different from 13 mm, and less of a chance of costly engine damage.

9.253 (a) $p_0 = 0.20$, $n = 1500$, $\hat{p} = 0.23$
(b) $(1500)(0.20) = 300 \geq 5$; $(1500)(0.80) = 1200 \geq 5$
(c) H_0: $p = 0.20$, H_a: $p > 0.20$
TS: $Z = \dfrac{\hat{p} - p_0}{\sqrt{\dfrac{p_0(1-p_0)}{n}}}$; RR: $Z \geq 2.3263$
$z = \dfrac{0.23 - 0.20}{\sqrt{\dfrac{(0.20)(0.90)}{1500}}} = 2.9047 \geq 2.3263$
There is evidence to suggest that the proportion of companies that require this information is greater than 0.20.
(d) $p = 0.0018 \leq 0.01$

9.255 (a) $H_0: p = 0.42$, $H_a: p > 0.42$

TS: $Z = \dfrac{\hat{P} - p_0}{\sqrt{\dfrac{p_0(1-p_0)}{n}}}$; RR: $Z \geq 1.6449$

$z = \dfrac{0.4733 - 0.42}{\sqrt{\dfrac{(0.42)(0.58)}{300}}} = 1.8716 \geq 1.6499$

There is evidence to suggest that the proportion of people older than age 65 who are vitamin D deficient is greater than 0.42.
(b) $p = 0.0306 < 0.05$

9.257 (a) $H_0: \mu = 140$, $H_a: \mu > 140$

TS: $T = \dfrac{\bar{X} - \mu_0}{S/\sqrt{n}}$; RR: $T \geq 2.5083$

$t = \dfrac{468.2 - 140}{507.44/\sqrt{23}} = 3.1018 \geq 2.5083$

There is evidence to suggest that the mean amount of lead in the soil is above the acceptable amount.
(b) $0.001 \leq p \leq 0.01$

9.259 $H_0: \sigma^2 = 0.50$, $H_a: \sigma^2 < 0.50$

TS: $X^2 = (n-1)S^2/\sigma_0^2$; RR: $X^2 \leq 12.4426$

$\chi^2 = \dfrac{(20)(0.39)}{0.50} = 15.6000 > 12.4426$

$p = 0.2589 > 0.10$

There is no evidence to suggest the population variance in shrinkage is less than 0.50.

9.261 $H_0: \sigma^2 = 0.04$, $H_a: \sigma^2 > 0.04$

TS: $X^2 = (n-1)S^2/\sigma_0^2$; RR: $X^2 \geq 40.2894$

$\chi^2 = \dfrac{(22)(0.056)}{0.04} = 30.8 < 40.2894$

There is no evidence to suggest that the true variance in module width is greater than 0.04.
(a) $0.10 \leq p \leq 0.20$

9.263 (a) $H_0: \mu = 1179$, $H_a: \mu < 1179$

TS: $T = \dfrac{\bar{X} - \mu_0}{S/\sqrt{n}}$; RR: $T \leq -1.7396$

$t = \dfrac{1148.89 - 1179}{238.12/\sqrt{18}} = -0.5365 > -1.7396$

There is no evidence to suggest that the mean other WRE has decreased.
(b) $p \geq 0.20$

9.265 (a) $H_0: \mu = 4{,}500{,}000$, $H_a: \mu > 4{,}500{,}000$

TS: $T = \dfrac{\bar{X} - \mu_0}{S/\sqrt{n}}$; RR: $T \geq 2.3060$

$t = \dfrac{4{,}675{,}250 - 4{,}500{,}000}{482{,}556/\sqrt{9}} = 1.0895 < 2.3060$

$p = 0.1538 > 0.025$

There is no evidence to suggest that the mean amount of oil stored is above the safety level.
(b) $0.10 \leq p \leq 0.20$

9.267 $H_0: p = 0.92$, $H_a: p < 0.92$

TS: $Z = \dfrac{\hat{P} - p_0}{\sqrt{\dfrac{p_0(1-p_0)}{n}}}$; RR: $Z \leq -2.3263$

$z = \dfrac{0.9152 - 0.92}{\sqrt{\dfrac{(0.92)(0.08)}{5000}}} = -1.2511 > -2.3263$

There is no evidence to suggest that the true proportion of assisted-living patients who are satisfied is less than 0.92.
(b) p-value: 0.1055
(c)

$p = 0.1055$

$z = -2.3263$

9.269 (a) $H_0: \sigma^2 = 4$, $H_a: \sigma^2 > 4$

TS: $X^2 = \dfrac{(n-1)S^2}{\sigma_0^2}$; RR: $X^2 \geq 26.2962$

$\chi^2 = \dfrac{(16)(7.75)}{4} = 31.00 \geq 26.2962$

There is evidence to suggest that the population variance is greater than 4 lb.
(b) $0.01 \leq p \leq 0.025$

9.271 (a) $H_0: p = 0.68$, $H_a: p < 0.68$
(b) If H_0 is true:
We do not reject H_0: This is the correct test decision.
We reject H_0: This is a type I error. The FDA believes that there are fewer toxic chemicals in baby food than there really are, which might result in sick babies.
If H_a is true:
We do not reject H_0: This is a Type II error. The FDA believes that 68% of baby foods still contain toxic chemicals, when in reality the presence of chemicals in these foods has decreased. This may lead to unfair fines placed on companies or stricter and more costly regulations.
We reject H_0: This is the correct test decision.

(c) TS: $Z = \dfrac{\hat{P} - p_0}{\sqrt{\dfrac{p_0(1-p_0)}{n}}}$; RR: $Z \leq -1.6449$

$z = \dfrac{0.6 - 0.68}{\sqrt{\dfrac{(0.68)(0.32)}{125}}} = -1.9174$

p-value: $0.0276 < 0.05$

There is evidence to suggest that the true proportion of baby foods that contain toxic chemicals is less than 0.68.

Chapter 10

10.1 Inference: To Independent Samples, Population Variances Known

10.1 True

10.3 True

10.5 False

10.7 If 0 is not captured by the interval, then there is evidence to suggest that $\mu_1 \neq \mu_2$ as the interval produces a set of plausible values for the true difference and $\mu_1 - \mu_2 = 0$ would mean that $\mu_1 = \mu_2$.

10.9 (a) $\mu_{\bar{X}_1-\bar{X}_2} = 3.0, \sigma^2_{\bar{X}_1-\bar{X}_2} = 5.0545, \sigma_{\bar{X}_1-\bar{X}_2} = 2.2482$

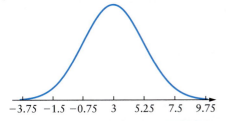

(b) $\mu_{\bar{X}_1-\bar{X}_2} = -12.2, \sigma^2_{\bar{X}_1-\bar{X}_2} = 10.035, \sigma_{\bar{X}_1-\bar{X}_2} = 3.1678$

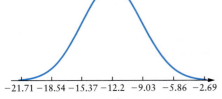

(c) $\mu_{\bar{X}_1-\bar{X}_2} = -125.3, \sigma^2_{\bar{X}_1-\bar{X}_2} = 85.0333, \sigma_{\bar{X}_1-\bar{X}_2} = 9.2214$

(d) $\mu_{\bar{X}_1-\bar{X}_2} = 0.90, \sigma^2_{\bar{X}_1-\bar{X}_2} = 0.0387, \sigma_{\bar{X}_1-\bar{X}_2} = 0.1967$

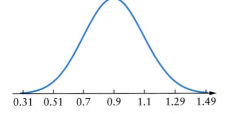

10.11 (a) $H_0: \mu_1 - \mu_2 = 2; H_a: \mu_1 - \mu_2 < 2$

TS: $Z = \dfrac{(\bar{X}_1 - \bar{X}_2) - 2}{\sqrt{\dfrac{\sigma_1^2}{n_1} + \dfrac{\sigma_2^2}{n_2}}}$; RR: $Z \leq -2.3263$

(b) $z = \dfrac{(186 - 190) - 2}{\sqrt{\dfrac{14.7}{25} + \dfrac{23.8}{24}}} = -4.7738 \leq -2.3263$

There is evidence to suggest population mean 1 is less than population mean 2 plus 2.

(c) $p = 0.0000009$

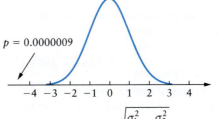

10.13 $(\bar{x}_1 - \bar{x}_2) \pm z_{0.025}\sqrt{\dfrac{\sigma_1^2}{n_1} + \dfrac{\sigma_2^2}{n_2}} = (-11.1298, 2.4492)$

(a) There is no evidence to suggest that the two population means are different since 0 is included in the CI.

10.15 $H_0: \mu_1 - \mu_2 = 0, H_a: \mu_1 - \mu_2 > 0$

TS: $Z = \dfrac{(\bar{X}_1 - \bar{X}_2) - 0}{\sqrt{\dfrac{\sigma_1^2}{n_1} + \dfrac{\sigma_2^2}{n_2}}}$; RR: $Z \geq 1.6449$

$z = \dfrac{(7992.2 - 7988.2) - 0}{\sqrt{\dfrac{1260.25}{23} + \dfrac{1697.44}{25}}} = 0.3611 \not\geq 1.6449$

There is no evidence to suggest that the mean rotation speed for the Sonicare Elite is greater than the mean rotation speed for the Oral-B.

(b) $p = 0.3590$

10.17 $H_0: \mu_1 - \mu_2 = 0, H_a: \mu_1 - \mu_2 < 0$

TS: $Z = \dfrac{(\bar{X}_1 - \bar{X}_2) - 0}{\sqrt{\dfrac{\sigma_1^2}{n_1} + \dfrac{\sigma_2^2}{n_2}}}$; RR: $Z \leq -1.6449$

$z = \dfrac{(38.300 - 39.394) - 0}{\sqrt{\dfrac{3.75^2}{16} + \dfrac{4.14^2}{18}}} = -0.8085 \not\leq -1.6449$

$p = 0.2094$

There is no evidence to suggest that the mean noise level for the Bosch dishwasher is less than the mean noise level for the Miele.

10.19 $H_0: \mu_1 - \mu_2 = 0, H_a: \mu_1 - \mu_2 < 0$

TS: $Z = \dfrac{(\bar{X}_1 - \bar{X}_2) - 0}{\sqrt{\dfrac{\sigma_1^2}{n_1} + \dfrac{\sigma_2^2}{n_2}}}$; RR: $Z \leq -2.3263$

$z = \dfrac{(4.04 - 5.88) - 0}{\sqrt{\dfrac{1.2^2}{24} + \dfrac{2.5^2}{22}}} = -3.1368 \leq -2.3263$

$p = 0.00085 < 0.01$

There is evidence to suggest that the mean outside playing time for children is less than the mean time for parents.

10.21 $H_0: \mu_1 - \mu_2 = 0, H_a: \mu_1 - \mu_2 \neq 0$

TS: $Z = \dfrac{(\bar{X}_1 - \bar{X}_2) - 0}{\sqrt{\dfrac{\sigma_1^2}{n_1} + \dfrac{\sigma_2^2}{n_2}}}$; RR: $|Z| \geq 2.5758$

$$z = \frac{(39.38 - 39.01) - 0}{\sqrt{\frac{5.06}{12} + \frac{6.01}{24}}} = 0.4513 < 2.5758$$

$p = 0.6518 > 0.01$

There is no evidence to suggest that the mean amount of protein is different in the two products.

10.23 (a) $(\bar{x}_1 - \bar{x}_2) \pm z_{0.025}\sqrt{\frac{\sigma_1^2}{n_1} + \frac{\sigma_2^2}{n_2}} = (-7.1844, -3.4756)$

(b) There is evidence to suggest that the mean number of revolutions per minute for the Altura fan is greater than the mean revolutions per minute for the Hampton fan; 0 is not included in the CI, and the CI is less than 0.

10.25 $H_0: \mu_1 - \mu_2 = 0$, $H_a: \mu_1 - \mu_2 > 0$

TS: $Z = \frac{(\bar{X}_1 - \bar{X}_2) - 0}{\sqrt{\frac{\sigma_1^2}{n_1} + \frac{\sigma_2^2}{n_2}}}$; RR: $Z \geq 1.6449$

$$z = \frac{(3.5227 - 3.4182) - 0}{\sqrt{\frac{0.0625}{44} + \frac{0.1225}{44}}} = 1.6123 < 1.6449$$

$p = 0.0534 > 0.05$

There is no evidence (although it is very close) to suggest that the mean time to induction for intravenous administration is less than the mean time to induction for inhalation administration.

10.27 $H_0: \mu_1 - \mu_2 = 0$, $H_a: \mu_1 - \mu_2 \neq 0$

TS: $Z = \frac{(\bar{X}_1 - \bar{X}_2) - 0}{\sqrt{\frac{\sigma_1^2}{n_1} + \frac{\sigma_2^2}{n_2}}}$; RR: $|Z| \geq 2.5758$

$$z = \frac{(39.58 - 40.12) - 0}{\sqrt{\frac{2.47}{18} + \frac{0.87}{18}}} = -1.2536 > -2.5758$$

There is no evidence to suggest that the population mean magnesium in each serving of baked beans and potatoes is different.

(b) $z = \frac{(39.58 - 40.12) - 0}{\sqrt{\frac{2.47}{38} + \frac{0.87}{38}}} = -1.8214 > -2.5758$

$p = 0.0685$

There is still no evidence to refute the claim.

(c) Solve for n: $\frac{(39.58 - 40.12) - 0}{\sqrt{\frac{2.47}{n} + \frac{0.87}{n}}} = -2.5758$

$\Rightarrow n = 75.99 \Rightarrow n_1 = n_2 = 76$

10.29 (a) $H_0: \mu_1 - \mu_2 = 0$, $H_a: \mu_1 - \mu_2 \neq 0$

TS: $Z = \frac{(\bar{X}_1 - \bar{X}_2) - 0}{\sqrt{\frac{\sigma_1^2}{n_1} + \frac{\sigma_2^2}{n_2}}}$; RR: $|Z| \geq 1.96$

$$z = \frac{(673 - 665) - 0}{\sqrt{\frac{1225}{65} + \frac{1849}{70}}} = 1.1891 < 1.96$$

There is no evidence to suggest that the true mean motorcycle cost is different in these two states.

(b) $p = 0.2344 > 0.05$

(c) Solve for n: $\frac{(673 - 665) - 0}{\sqrt{\frac{1225}{n} + \frac{1849}{n}}} = 1.96$

$\Rightarrow n = 184.52 \Rightarrow n_1 = n_2 = 185$

10.31 (a) Solve for n: $z_{\alpha/2}\sqrt{\frac{\sigma_1^2}{n_1} + \frac{\sigma_2^2}{n_2}} = B$

$\Rightarrow \frac{\sigma_1^2 + \sigma_2^2}{n} = \frac{B^2}{z_{\alpha/2}^2}$

$\Rightarrow n_1 = n_2 = n = \frac{(z_{\alpha/2})^2(\sigma_1^2 + \sigma_2^2)}{B^2}$

(b) $n = \frac{(1.96)^2(12.7^2 + 9.5^2)}{25} = 38.65 \Rightarrow n_1 = n_2 = 39$

(c) $(\bar{x}_1 - \bar{x}_2) \pm z_{0.05}\sqrt{\frac{\sigma_1^2}{n_1} + \frac{\sigma_2^2}{n_2}}$

$= (3.7224, 13.6776)$

$B = 4.9776 \leq 5$

10.2 Inference: Two Independent Samples, Normal Populations

10.33 $\sigma^2\left(\frac{1}{n_1} + \frac{1}{n_2}\right)$

10.35 Even if the assumptions aren't entirely true, the hypothesis test is still very reliable.

10.37 True

10.39 If 0 is not captured by the interval, then there is evidence to suggest that $\mu_1 \neq \mu_2$ as the interval produces a set of plausible values for the true difference and $\mu_1 - \mu_2 = 0$ would mean that $\mu_1 = \mu_2$.

10.41 (a) $H_0: \mu_1 - \mu_2 = 0$, $H_a: \mu_1 - \mu_2 > 0$

TS: $T = \frac{(\bar{X}_1 - \bar{X}_2) - 0}{\sqrt{S_p^2\left(\frac{1}{n_1} + \frac{1}{n_2}\right)}}$; RR: $T \geq 1.7291$

$$t = \frac{(156.5 - 132.6) - 0}{\sqrt{(575.9342)\left(\frac{1}{10} + \frac{1}{11}\right)}} = 2.2793 \geq 1.7291$$

$p = 0.0172 \leq 0.05$

There is evidence to suggest that μ_1 is greater than μ_2.

(b) $0.01 \leq p \leq 0.025$

10.43 (a) $(\bar{x}_1 - \bar{x}_2) \pm t_{0.025,44}\sqrt{s_p^2\left(\frac{1}{n_1} + \frac{1}{n_2}\right)} = (-5.718, 4.638)$

(b) Since 0 is included in the CI, there is no evidence to suggest that the population means are different.

10.45 (a) $H_0: \mu_1 - \mu_2 = 0$, $H_a: \mu_1 - \mu_2 > 0$

TS: $T' = \frac{(\bar{X}_1 - \bar{X}_2) - 0}{\sqrt{\frac{S_1^2}{n_1} + \frac{S_2^2}{n_2}}}$; RR: $T' \geq 1.7613$

$t' = \dfrac{(173.9 - 150.3) - 0}{\sqrt{\dfrac{320.41}{8} + \dfrac{655.36}{9}}} = 2.2214 \geq 1.7613$

$p = 0.0217 < 0.05$

There is evidence to suggest that the mean of population 1 is greater than the mean of population 2.

(b) $0.01 \leq p \leq 0.025$

10.47 (a) $(\bar{x}_1 - \bar{x}_2) \pm t_{0.005,33} \sqrt{\dfrac{s_1^2}{n_1} + \dfrac{s_2^2}{n_2}} = (-24.1659, 1.1659)$

(b) Since 0 is in the CI, there is no evidence to suggest that the two population means are different.

10.49 $H_0: \mu_1 - \mu_2 = 0$, $H_a: \mu_1 - \mu_2 > 0$

TS: $T = \dfrac{(\bar{X}_1 - \bar{X}_2) - 0}{\sqrt{S_p^2 \left(\dfrac{1}{n_1} + \dfrac{1}{n_2}\right)}}$; RR: $T \geq 1.7341$

$t = \dfrac{(0.321 - 0.199) - 0}{\sqrt{(0.0170)\left(\dfrac{1}{10} + \dfrac{1}{10}\right)}} = 2.0954 \geq 1.7341$

$p = 0.0253 \leq 0.05$

There is evidence to suggest that the population mean weight of a new-process key is less than the population mean weight of an old-process key.

(b) $0.025 \leq p \leq 0.05$

10.51 $H_0: \mu_1 - \mu_2 = 0$, $H_a: \mu_1 - \mu_2 < 0$

TS: $T = \dfrac{(\bar{X}_1 - \bar{X}_2) - 0}{\sqrt{S_p^2 \left(\dfrac{1}{n_1} + \dfrac{1}{n_2}\right)}}$; RR: $T \leq -2.4102$

$t = \dfrac{(13.95 - 19.09) - 0}{\sqrt{(24.7630)\left(\dfrac{1}{25} + \dfrac{1}{23}\right)}} = -3.5750 \leq -2.4102$

$p = 0.0004 \leq 0.05$

There is evidence to suggest that the population mean amount of protein in Shelf Safe Milk is less than the population mean amount of protein in regular milk.

10.53 (a) $H_0: \mu_1 - \mu_2 = 0$, $H_a: \mu_1 - \mu_2 < 0$

TS: $T = \dfrac{(\bar{X}_1 - \bar{X}_2) - 0}{\sqrt{S_p^2 \left(\dfrac{1}{n_1} + \dfrac{1}{n_2}\right)}}$; RR: $T \leq -1.6973$

$t = \dfrac{(12.562 - 12.818) - 0}{\sqrt{(0.3240)^2 \left(\dfrac{1}{16} + \dfrac{1}{16}\right)}} = -2.2393 \leq -1.6973$

$p = 0.0164 \leq 0.05$

There is evidence to suggest that the population mean depth for Line 1 is less than the population mean depth for Line 2.

(b) $0.01 \leq p \leq 0.025$

10.55 (a) $H_0: \mu_1 - \mu_2 = 0$, $H_a: \mu_1 - \mu_2 > 0$

TS: $T = \dfrac{(\bar{X}_1 - \bar{X}_2) - 0}{\sqrt{S_p^2 \left(\dfrac{1}{n_1} + \dfrac{1}{n_2}\right)}}$; RR: $T \geq 2.5524$

$t = \dfrac{(5.35 - 3.90) - 0}{\sqrt{(0.5650)\left(\dfrac{1}{10} + \dfrac{1}{10}\right)}} = 4.3135 \geq 2.5524$

$p = 0.0002 \leq 0.01$

There is evidence to suggest that the population mean Halogen cure depth is greater than the population mean LuxOMax cure depth.

(b) $(\bar{x}_1 - \bar{x}_2) \pm t_{0.005,18} \sqrt{s_p^2 \left(\dfrac{1}{n_1} + \dfrac{1}{n_2}\right)} = (0.4823, 2.3176)$

Note that 0 is not contained within the interval, providing evidence to suggest that the two population means are not equal.

10.57 $H_0: \mu_1 - \mu_2 = 0$, $H_a: \mu_1 - \mu_2 \neq 0$

TS: $T' = \dfrac{(\bar{X}_1 - \bar{X}_2) - 0}{\sqrt{\dfrac{s_1^2}{n_1} + \dfrac{s_2^2}{n_2}}}$; RR: $|T'| \geq 4.5869$

$t' = \dfrac{(0.361 - 0.425) - 0}{\sqrt{\dfrac{0.122^2}{10} + \dfrac{0.051^2}{20}}} = -1.5909 > -4.5869$

$p = 0.1427 > 0.001$

There is no evidence to suggest that the population mean curve in sticks for these two teams is different.

10.59 $H_0: \mu_1 - \mu_2 = 0$, $H_a: \mu_1 - mu_2 \neq 0$

TS: $T = \dfrac{(\bar{X}_1 - \bar{X}_2) - 0}{\sqrt{S_p^2 \left(\dfrac{1}{n_1} + \dfrac{1}{n_2}\right)}}$; RR: $|T| \geq 2.7969$

$t = \dfrac{(60.1 - 53.4) - 0}{\sqrt{(153.96)\left(\dfrac{1}{12} + \dfrac{1}{14}\right)}} = 1.3726 < 2.7969$

$p = 0.1825 > 0.01$

There is no evidence to suggest that the mean frame rates for these two cards are different.

10.61 (a) $H_0: \mu_1 - \mu_2 = 0$, $H_a: \mu_1 - \mu_2 \neq 0$

TS: $T' = \dfrac{(\bar{X}_1 - \bar{X}_2) - 0}{\sqrt{\dfrac{s_1^2}{n_1} + \dfrac{s_2^2}{n_2}}}$; RR: $|T'| \geq 2.0281$

$t' = \dfrac{(3.53 - 3.68) - 0}{\sqrt{\dfrac{0.1896^2}{18} + \dfrac{0.1925^2}{20}}} = -2.4155 \leq -2.0281$

$p = 0.0210 < 0.05$

There is evidence to suggest that the mean time to complete a round of golf is different for these two courses.

(b) $0.01 \leq p \leq 0.025$

10.63 (a) $H_0: \mu_1 - \mu_2 = 0$, $H_a: \mu_1 - \mu_2 \neq 0$

TS: $T = \dfrac{(\bar{X}_1 - \bar{X}_2) - 0}{\sqrt{S_p^2 \left(\dfrac{1}{n_1} + \dfrac{1}{n_2}\right)}}$; RR: $|T| \geq 2.8982$

$t = \dfrac{(0.601 - 0.741) - 0}{\sqrt{(0.0087)\left(\dfrac{1}{8} + \dfrac{1}{11}\right)}} = -3.2218 \leq -2.8982$

$p = 0.005 \leq 0.01$

There is evidence to suggest that the population mean shading coefficients are different.

(b) $(\bar{x}_1 - \bar{x}_2) \pm t_{0.005, 17} \sqrt{s_p^2 \left(\frac{1}{n_1} + \frac{1}{n_2}\right)} = (-0.2659, -0.0141)$

(c) Since 0 is not included in the CI, there is evidence to suggest that the population mean shading coefficients are different. This agrees with the conclusion in part (a).

10.65 (a) The assumption of equal variances seems reasonable. The sample variances are close.
(b) $H_0: \mu_1 - \mu_2 = 0$, $H_a: \mu_1 - \mu_2 \neq 0$

TS: $T = \frac{(\bar{X}_1 - \bar{X}_2) - 0}{\sqrt{S_p^2 \left(\frac{1}{n_1} + \frac{1}{n_2}\right)}}$; RR: $|T| \geq 2.6846$

$t = \frac{(6.25 - 5.98) - 0}{\sqrt{(1.04)\left(\frac{1}{21} + \frac{1}{28}\right)}} = 0.9943 < 2.6846$

$p = 0.3252$

There is no evidence to suggest that the mean burn times are different.
(c) $0.20 \leq p \leq 0.40$

10.67 (a) $H_0: \mu_1 - \mu_2 = 0$, $H_a: \mu_1 - \mu_2 < 0$

TS: $T = \frac{(\bar{X}_1 - \bar{X}_2) - 0}{\sqrt{S_p^2 \left(\frac{1}{n_1} + \frac{1}{n_2}\right)}}$; RR: $T \leq -2.5280$

$t = \frac{(34.91 - 39.55) - 0}{\sqrt{(11.4213)\left(\frac{1}{11} + \frac{1}{11}\right)}} = -3.2199 \leq -2.5280$

$p = 0.0021 \leq 0.01$

There is evidence to suggest that the population mean lifetime of the new fuel rod is greater than the population mean lifetime of the old fuel rod.

(b) $(\bar{x}_1 - \bar{x}_2) \pm t_{0.005, 20} \sqrt{s_p^2 \left(\frac{1}{n_1} + \frac{1}{n_2}\right)} = (-8.7402, -0.5398)$

Note that 0 is not captured by the CI. This supports the conclusion from part (a) that the mean lifetime of the new rod is greater than the mean lifetime of the old one.

10.69 Answers will vary. You should reject the null hypothesis roughly 5 times in 100 trials. This is true no matter what value of σ_2 is used, although as σ_2 gets farther from σ_1, you should reject the null hypothesis more often.

Robust means our results are still 95% accurate, even when the underlying assumptions are not met.

10.3 Paired Data

10.71 True

10.73 False

10.75 Both will have a t distribution, but a two-sample t test has a test statistic with a higher degrees of freedom, which yields a more conservative result.

10.77 (a) The data are paired; the common characteristic is the affected arm of the patient with rotator cuff injuries.
(b) The data are paired; the common characteristic is the lathes.
(c) The data were obtained independently; the two populations are 20-year-old males and 70-year-old males.
(d) The data are paired; the common characteristic is the style of modular home.
(e) The data were obtained independently; the two populations are frequent flyers on United Airlines and frequent flyers on Delta Airlines.

10.79 $H_0: \mu_D = 0$, $H_a: \mu_D > 0$

TS: $T = \frac{\bar{D} - 0}{S_D / \sqrt{n}}$; RR: $T \geq 1.7459$

$t = \frac{15.68 - 0}{33.55 / \sqrt{17}} = 1.9270 \geq 1.7459$

$p = 0.0360 \leq 0.05$

There is evidence to suggest that population mean 1 is greater than population mean 2.

10.81 (a) $H_0: \mu_D = 0$, $H_a: \mu_D \neq 0$

TS: $T = \frac{\bar{D} - 0}{S_D / \sqrt{n}}$; RR: $|T| \geq 2.8609$

$t = \frac{7.08 - 0}{10.07 / \sqrt{20}} = 3.1442 \geq 2.8609$

There is evidence to suggest that the means for both populations are different.
(b) $0.005 \leq p \leq 0.01$

10.83 (a) The common characteristic that makes these data paired is the gun.
(b) $H_0: \mu_D = 0$, $H_a: \mu_D < 0$

TS: $T = \frac{\bar{D} - 0}{S_D / \sqrt{n}}$; RR: $T \leq -3.3649$

$t = \frac{-15.1667 - 0}{74.8156 / \sqrt{6}} = -0.4966 > -3.3649$

There is no evidence to suggest that the population mean muzzle velocity of a clean gun is greater than the population mean muzzle velocity of a dirty gun.
(c) $p > 0.20$

10.85 $H_0: \mu_D = 0$, $H_a: \mu_D > 0$

TS: $T = \frac{\bar{D} - 0}{S_D / \sqrt{n}}$; RR: $T \geq 1.6991$

$t = \frac{3.8 - 0}{13.4354 / \sqrt{30}} = 1.5491 < 1.6991$

$p = 0.0661 > 0.05$

There is no evidence to suggest that the population mean concentration of particulate matter before filtration is greater than the population mean concentration of particulate matter after filtration.

10.87 $H_0: \mu_D = 0$, $H_a: \mu_D > 0$

TS: $T = \dfrac{\bar{D} - 0}{S_D/\sqrt{n}}$; RR: $T \geq 2.1604$

$t = \dfrac{0.0836 - 0}{0.9982/\sqrt{14}} = 0.3132$

$p = 0.3795 > 0.025$

There is no evidence to suggest that the population mean ammonia-ion concentration before treatment is greater than the population mean ammonia-ion concentration after treatment.

10.89 **(a)** The common characteristic is the adult.
(b) $H_0: \mu_D = 0$, $H_a: \mu_D > 0$

TS: $T = \dfrac{\bar{D} - 0}{S_D/\sqrt{n}}$; RR: $T \geq 2.5669$

$t = \dfrac{15.333 - 0}{22.8859/\sqrt{18}} = 2.8425 \geq 2.5669$

$p = 0.0056 \leq 0.01$
There is evidence to suggest that the population mean steps per minute listening to hard rock is greater than the population mean steps per minute listening to ballads.

10.91 $H_0: \mu_D = 0$, $H_a: \mu_D > 0$

TS: $T = \dfrac{\bar{D} - 0}{S_D/\sqrt{n}}$; RR: $T \geq 3.3962$

$t = \dfrac{4.45 - 0}{4.4720/\sqrt{30}} = 5.4503 \geq 3.3962$

$p = 0.00000363 \leq 0.001$

There is evidence to suggest that the population mean porosity before treatment is greater than the population mean porosity after treatment.

10.93 **(a)** $H_0: \mu_D = 0$, $H_a: \mu_D < 0$

TS: $T = \dfrac{\bar{D} - 0}{S_D/\sqrt{n}}$; RR: $T \leq -3.4210$

$t = \dfrac{-2.8321 - 0}{4.2912/\sqrt{28}} = -3.4924 \leq -3.4210$

$p = 0.0008 \leq 0.001$
There is evidence to suggest that the electric-grill population's mean fat content is less than the frying-pan population's mean fat content.
(b) $0.0005 \leq p \leq 0.001$

10.95 **(a)** The common characteristic that makes these data paired is the patient with high fever.
(b) $H_0: \mu_D = 0$, $H_a: \mu_D > 0$

TS: $T = \dfrac{\bar{D} - 0}{S_D/\sqrt{n}}$; RR: $T \geq 1.8331$

$t = \dfrac{2.86 - 0}{2.68/\sqrt{10}} = 3.3746 \geq 1.8331$

There is evidence to suggest that the population mean temperature before the drug is greater than the population mean temperature after the drug.

(c) $0.001 \leq p \leq 0.005$
(d) As 9 of the 10 differences are positive and most of these are more than 1 degree of difference, we should expect the hypothesis test to be significant.

10.97 **(a)** The common characteristic is the East False Creek.
(b) $H_0: \mu_D = 0$, $H_a: \mu_D > 0$

TS: $T = \dfrac{\bar{D} - 0}{S_D/\sqrt{n}}$; RR: $T \geq 3.5272$

$t = \dfrac{79.5909 - 0}{78.9096/\sqrt{22}} = 4.7309 \geq 3.5272$

$p = 0.0001 \leq 0.001$
There is evidence to suggest that the rain caused a decrease in the mean coliform count.
(c) The standard for the safe level of coliform varies by state. However, the EPA had recommended a safe level of 200/100 mL.
$H_0: \mu = 200$, $H_a: \mu > 200$

TS: $T = \dfrac{\bar{X} - 200}{S/\sqrt{n}}$; RR: $T \geq 1.7207$

$t = \dfrac{285.591 - 200}{39.85/\sqrt{22}} = 10.07 \geq 1.7207$

$p \leq 0.0001$
There is evidence to suggest that the level of coliform is still unsafe after the rain.

10.4 Comparing Two Population Proportions

10.99 If both n_1 and n_2 are large, and $n_1 p_1 \geq 5$ and $n_1(1 - p_1) \geq 5$, and $n_2 p_2 \geq 5$ and $n_2(1 - p_2) \geq 5$, then the distribution of $\hat{P}_1 - \hat{P}_2$ is approximately normal.

10.101 False

10.103 True

10.105 False

10.107

	$\mu_{\hat{P}_1 - \hat{P}_2}$	$\sigma^2_{\hat{P}_1 - \hat{P}_2}$	$\sigma_{\hat{P}_1 - \hat{P}_2}$	Probability
(a)	-0.020	0.000579	0.0241	0.0034
(b)	0.040	0.001751	0.0418	0.0280
(c)	0.010	0.001557	0.0395	0.7804
(d)	-0.070	0.000858	0.0293	0.8472
(e)	-0.120	0.000275	0.0166	0.9999
(f)	0.041	0.000914	0.0302	0.9527

10.109 TS: $Z = \dfrac{\hat{P}_1 - \hat{P}_2}{\sqrt{\hat{P}_c(1 - \hat{P}_c)\left(\frac{1}{n_1} + \frac{1}{n_2}\right)}}$

(a) RR: $Z \geq 1.6449$

$z = \dfrac{0.80 - 0.7714}{\sqrt{(0.7854)(0.2146)\left(\frac{1}{525} + \frac{1}{405}\right)}} = 1.1137 < 1.6449$

p-value: $0.1327 \Rightarrow$ Do not reject H_0.

(b) RR: $Z \leq -2.3263$

$$z = \frac{0.4334 - 0.4853}{\sqrt{(0.46)(0.54)\left(\frac{1}{646} + \frac{1}{680}\right)}} = -1.8938 > -2.3263$$

p-value: $0.0291 \Rightarrow$ Do not reject H_0.

(c) RR: $|Z| \geq 2.2414$

$$z = \frac{0.3176 - 0.4135}{\sqrt{(0.3666)(0.6334)\left(\frac{1}{255} + \frac{1}{266}\right)}} = -2.2705 \leq -2.2414$$

p-value: $0.0232 \Rightarrow$ Reject H_0.

(d) RR: $|Z| \geq 3.2905$

$$z = \frac{0.6299 - 0.6210}{\sqrt{(0.6252)(0.3748)\left(\frac{1}{1440} + \frac{1}{1562}\right)}} = 0.5012 < 3.2905$$

p-value: $0.6163 \Rightarrow$ Do not reject H_0.

10.111 $(\hat{p}_1 - \hat{p}_2) \pm z_{\alpha/2} \sqrt{\frac{\hat{p}_1(1-\hat{p}_1)}{n_1} + \frac{\hat{p}_2(1-\hat{p}_2)}{n_2}}$

(a) $(0.5928 - 0.6219)$

$\pm (1.96)\sqrt{\frac{(0.5928)(0.4072)}{388} + \frac{(0.6219)(0.3781)}{402}}$

$= (-0.0972, 0.0390)$

(b) $(0.8996 - 0.9377)$

$\pm (1.96)\sqrt{\frac{(0.8996)(0.1004)}{528} + \frac{(0.9377)(0.0623)}{530}}$

$= (-0.0710, -0.0052)$

(c) $(0.5111 - 0.5155)$

$\pm (2.5758)\sqrt{\frac{(0.5111)(0.4889)}{180} + \frac{(0.5155)(0.4845)}{194}}$

$= (-0.1376, 0.1289)$

(d) $(0.7413 - 0.7030)$

$\pm (1.6449)\sqrt{\frac{(0.7413)(0.2587)}{2300} + \frac{(0.7030)(0.2970)}{2404}}$

$= (0.0168, 0.0598)$

10.113 H_0: $p_1 - p_2 = 0$, H_a: $p_1 - p_2 \neq 0$

TS: $Z = \dfrac{\hat{P}_1 - \hat{P}_2}{\sqrt{\hat{P}_c(1-\hat{P}_c)\left(\frac{1}{n_1} + \frac{1}{n_2}\right)}}$; RR: $|Z| \geq 1.96$

$$z = \frac{0.2920 - 0.3091}{\sqrt{(0.3010)(0.6990)\left(\frac{1}{250} + \frac{1}{275}\right)}} = -0.4264 > -1.96$$

p-value: $0.6698 < 0.05$

There is no evidence to suggest that the proportion of voters in the west who believe that an armed revolution might be necessary is different from the proportion of voters in the east.

10.115 H_0: $p_1 - p_2 = 0$, H_a: $p_1 - p_2 > 0$

TS: $Z = \dfrac{\hat{P}_1 - \hat{P}_2}{\sqrt{\hat{P}_c(1-\hat{P}_c)\left(\frac{1}{n_1} + \frac{1}{n_2}\right)}}$; RR: $Z \geq 3.0902$

$$z = \frac{0.6859 - 0.5711}{\sqrt{(0.6253)(0.3747)\left(\frac{1}{347} + \frac{1}{387}\right)}} = 3.2086 \geq 3.0902$$

There is evidence to suggest that the population proportion of 18- to 29-year-olds who believe movies are getting better is greater than the population proportion of 30- to 49-year-olds who believe movies are getting better.

(a) p-value: $0.0007 < 0.001$

10.117 (a) $n_1 \hat{p}_1 = 530 \geq 5$, $n_1(1-\hat{p}_1) = 525 \geq 5$

$n_2 \hat{p}_2 = 825 \geq 5$, $n_2(1-\hat{p}_2) = 838 \geq 5$

(b) H_0: $p_1 - p_2 = 0$, H_a: $p_1 - p_2 > 0$

TS: $Z = \dfrac{\hat{P}_1 - \hat{P}_2}{\sqrt{\hat{P}_c(1-\hat{P}_c)\left(\frac{1}{n_1} + \frac{1}{n_2}\right)}}$; RR: $Z \geq 2.3263$

$$z = \frac{0.502 - 0.496}{\sqrt{(0.4985)(0.5015)\left(\frac{1}{1055} + \frac{1}{1663}\right)}} = 0.3190 < 2.3263$$

There is no evidence to suggest that the population proportion of carpoolers crossing the George Washington Bridge is greater than the population proportion of carpoolers using the Lincoln Tunnel.

(c) p-value: $0.3749 > 0.01$

10.119 (a) $\hat{p}_1 = 0.2784$, $\hat{p}_2 = 0.2321$

(b) H_0: $p_1 - p_2 = 0$, H_a: $p_1 - p_2 \neq 0$

TS: $Z = \dfrac{\hat{P}_1 - \hat{P}_2}{\sqrt{\hat{P}_c(1-\hat{P}_c)\left(\frac{1}{n_1} + \frac{1}{n_2}\right)}}$; RR: $|Z| \geq 2.5758$

$$z = \frac{0.2784 - 0.2321}{\sqrt{(0.256)(0.744)\left(\frac{1}{255} + \frac{1}{237}\right)}} = 1.1773 < 2.5758$$

p-value: $0.2391 > 0.01$

There is no evidence to suggest that the population proportion of people who obtain relief from the antihistamine is different from the population proportion of people who obtain relief from butterbur extract.

10.121 (a) $\hat{p}_1 = 0.0755$, $\hat{p}_2 = 0.0992$

(b) $n_1 \hat{p}_1 = 8 \geq 5$, $n_1(1-\hat{p}_1) = 98 \geq 5$

$n_2 \hat{p}_2 = 12 \geq 5$, $n_2(1-\hat{p}_2) = 109 \geq 5$

(c) H_0: $p_1 - p_2 = 0$, H_a: $p_1 - p_2 \neq 0$

TS: $Z = \dfrac{\hat{P}_1 - \hat{P}_2}{\sqrt{\hat{P}_c(1-\hat{P}_c)\left(\frac{1}{n_1} + \frac{1}{n_2}\right)}}$; RR: $|Z| \geq 1.96$

$$z = \frac{0.0755 - 0.0992}{\sqrt{(0.0881)(0.9119)\left(\frac{1}{106} + \frac{1}{121}\right)}} = -0.6286 > -1.96$$

p-value: $0.5296 > 0.05$

There is no evidence to suggest that the population proportion of defective lenses is different for Process A than Process B.

10.123 H_0: $p_1 - p_2 = 0$, H_a: $p_1 - p_2 \neq 0$

TS: $Z = \dfrac{\hat{P}_1 - \hat{P}_2}{\sqrt{\hat{P}_c(1-\hat{P}_c)\left(\frac{1}{n_1} + \frac{1}{n_2}\right)}}$; RR: $|Z| \geq 1.96$

$$z = \frac{0.6620 - 0.7110}{\sqrt{(0.6868)(0.3132)\left(\frac{1}{213} + \frac{1}{218}\right)}} = -1.0974 > -1.96$$

p-value: $0.2725 > 0.05$

There is no evidence to suggest that the true proportion of residents who feel safe after dark is different in the two districts.

10.125 $H_0: p_1 - p_2 = 0$, $H_a: p_1 - p_2 > 0$

TS: $Z = \dfrac{\hat{P}_1 - \hat{P}_2}{\sqrt{\hat{P}_c(1-\hat{P}_c)\left(\frac{1}{n_1}+\frac{1}{n_2}\right)}}$; RR: $Z \geq 1.6449$

$z = \dfrac{0.2280 - 0.1890}{\sqrt{(0.2126)(0.7874)\left(\frac{1}{500}+\frac{1}{328}\right)}} = 1.3408 < 1.6449$

p-value: $0.09 > 0.05$

There is no evidence to suggest that the true proportion of women with computer science occupations is different from the true proportion of women with engineering occupations.

10.127 (a) $\hat{p}_1 = 0.0746$, $\hat{p}_2 = 0.0614$
(b) $n_1\hat{p}_1 = 10 \geq 5$, $n_1(1-\hat{p}_1) = 124 \geq 5$
$n_2\hat{p}_2 = 7 \geq 5$, $n_2(1-\hat{p}_2) = 107 \geq 5$
(c) $(\hat{p}_1 - \hat{p}_2) \pm z_{0.025}\sqrt{\dfrac{\hat{p}_1(1-\hat{p}_1)}{n_1}+\dfrac{\hat{p}_2(1-\hat{p}_2)}{n_2}}$
$= (0.0746 - 0.0614)$
$\pm (1.96)\sqrt{\dfrac{(0.0746)(0.9254)}{134}+\dfrac{(0.0614)(0.9386)}{114}}$
$= (-0.0494, 0.0758)$
(d) Since 0 is included in the CI, there is no evidence to suggest any difference in the proportion of underfilled cans for the two machines.

10.129 (a) $H_0: p_1 - p_2 = 0$, $H_a: p_1 - p_2 \neq 0$

TS: $Z = \dfrac{\hat{P}_1 - \hat{P}_2}{\sqrt{\hat{P}_c(1-\hat{P}_c)\left(\frac{1}{n_1}+\frac{1}{n_2}\right)}}$; RR: $|Z| \geq 1.96$

$z = \dfrac{0.3777 - 0.4276}{\sqrt{(0.4042)(0.5958)\left(\frac{1}{376}+\frac{1}{428}\right)}} = -1.4389 > -1.96$

There is no evidence to suggest that the true population proportion of males who give their children cough medicine is different from the true proportion of females.
(b) p-value: $0.1502 > 0.05$
(c) $(\hat{p}_1 - \hat{p}_2) \pm z_{0.025}\sqrt{\dfrac{\hat{p}_1(1-\hat{p}_1)}{n_1}+\dfrac{\hat{p}_2(1-\hat{p}_2)}{n_2}}$
$= (0.3777 - 0.4276)$
$\pm (1.96)\sqrt{\dfrac{(0.3777)(0.6223)}{376}+\dfrac{(0.4276)(0.5724)}{428}}$
$= (-0.1177, 0.0179)$

This confidence interval supports the conclusion in part (a). Because 0 is in the CI, there is no evidence to suggest a difference in the proportions of men and women who give their children cold medicine.
(d) The samples must be independent. The sample should not contain solely husband and wife pairs, as the wife's answer may be affected by her husband's answer.

10.5 Comparing Two Population Variances or Standard Deviations

10.131 True

10.133 True

10.135 (a) $F_{0.05,7,19} = 2.54$ (b) $F_{0.05,30,25} = 1.92$
(c) $F_{0.01,6,19} = 3.94$ (d) $F_{0.001,40,40} = 2.73$
(e) $F_{0.95,17,15} = 0.43$ (f) $F_{0.95,12,10} = 0.36$
(g) $F_{0.99,21,30} = 0.37$ (h) $F_{0.999,11,8} = 0.12$

10.137 (a) $0.01 \leq p \leq 0.05$ (b) $0.01 \leq p \leq 0.05$
(c) $0.002 \leq p \leq 0.02$ (d) $p \leq 0.001$

10.139 (a) $H_0: \sigma_1^2 = \sigma_2^2$, $H_a: \sigma_1^2 < \sigma_2^2$
TS: $F = S_1^2/S_2^2$; RR: $F \leq 0.45$
(b) $f = 105.2708/259.8669 = 0.4051 \leq 0.45$
There is evidence to suggest that population variance 1 is less than population variance 2.
(c) $0.01 \leq p \leq 0.05$

10.141 (a) $F_{0.05,9,9} = 3.1789$, $F_{0.05,9,9} = 3.1789$
(b) $F_{0.01,20,9} = 4.808$, $F_{0.01,9,20} = 3.4567$
(c) $F_{0.01,8,6} = 8.1017$, $F_{0.01,6,8} = 6.3707$
(d) $F_{0.001,40,30} = 3.0716$, $F_{0.001,30,40} = 2.8721$

10.143 (a) $F_{0.05,25,15} = 2.28$ (b) $F_{0.99,20,32} = 0.36$
(c) $F_{0.01,10,56} = 2.66$ (d) $F_{0.025,15,20} = 2.57$
(e) $F_{0.995,10,7} = 0.16$ (f) $F_{0.05,35,35s} = 1.76$

10.145 $H_0: \sigma_1^2 = \sigma_2^2$, $H_a: \sigma_1^2 \neq \sigma_2^2$
TS: $F = S_1^2/S_2^2$; RR: $F \leq 0.27$ or $F \geq 3.12$
$f = 1.075/2.786 = 0.3859 \Rightarrow 0.27 < 0.3859 < 3.12$
$p = 0.1523 > 0.05$
There is no evidence to suggest that the variability in CT radiation per scan is different for these two countries.

10.147 $H_0: \sigma_1^2 = \sigma_2^2$, $H_a: \sigma_1^2 < \sigma_2^2$
TS: $F = S_1^2/S_2^2$; RR: $F \leq 0.5446$
$f = 12.25/26.01 = 0.4710 \leq 0.5446$
$p = 0.002 < 0.01$
There is evidence to suggest that the variability in shot distance has increased from 1975. Players are more willing to take long-range shots with the chance of making 3 points. Therefore, it seems reasonable that the variability in shot distance is greater in 2018.

10.149 (a) $F_{0.025,10,11} = 3.53$, $F_{0.025,11,10} = 3.66$
(b) $\left(\dfrac{s_1^2}{s_2^2}\dfrac{1}{F_{0.025}}, \dfrac{s_1^2}{s_2^2}F_{0.025}\right) = \left(\dfrac{0.1025}{0.1241}\dfrac{1}{3.53}, \dfrac{0.1025}{0.1241}(3.66)\right)$
$= (0.2343, 3.0270)$
(c) There is no evidence to suggest a difference in variability of times. The CI represents an interval of likely values of $\dfrac{\sigma_1^2}{\sigma_2^2}$. Since 1 is contained within the interval, it is plausible that both population variances are the same.

10.151 $H_0: \sigma_1^2 = \sigma_2^2$, $H_a: \sigma_1^2 \neq \sigma_2^2$
TS: $F = S_1^2/S_2^2$
$f = 0.0231/0.0096 = 2.4063$
$p = 0.2757 > 0.05$
There is no evidence to suggest that the population variance in mast-pole diameter in Machine A is different from the population variance in mast-pole diameter in Machine B.

10.153 $H_0: \sigma_1^2 = \sigma_2^2$, $H_a: \sigma_1^2 > \sigma_2^2$
TS: $F = S_1^2/S_2^2$; RR: $F \geq 6.03$
$f = 44.05/9.76 = 4.5133 < 6.03$
$p = 0.0238 > 0.01$
There is no evidence to suggest that the variability in speed for the Exhibition Powersports class is greater than the variability in speed for the Heavyweight Supermoto class.

10.155 $H_0: \sigma_1^2 = \sigma_2^2$, $H_a: \sigma_1^2 \neq \sigma_2^2$
TS: $F = S_1^2/S_2^2$; RR: $F \leq 0.3059$ or $F \geq 3.5561$
$f = 13.57/20.02 = 0.6778$
$\Rightarrow 0.3059 \leq 0.6778 \leq 3.5561$
$p = 0.4326 > 0.02$
There is no evidence to suggest that the variances of pit stop times are different for races in Italy and in Canada.

10.157 (a) $H_0: \sigma_1^2 = \sigma_2^2$, $H_a: \sigma_1^2 \neq \sigma_2^2$
TS: $F = S_1^2/S_2^2$; RR: $F \leq 0.31$ or $F \geq 3.53$
$f = 7.84/2.89 = 2.7128$
$\Rightarrow 0.31 < 2.7128 < 3.53$
$p = 0.0621 > 0.0$
There is no evidence to suggest that the population variance in saccharin amount for Fishing Creek is different from the population variance in saccharin amount for Honest Tea.
(b) $0.02 \leq p \leq 0.10$

10.159 (a) $H_0: \sigma_1^2 = \sigma_2^2$, $H_a: \sigma_1^2 \neq \sigma_2^2$
TS: $Z = \dfrac{(S_1^2/S_2^2) - [(n_2-1)/(n_2-3)]}{\sqrt{\dfrac{2(n_2-1)^2(n_1+n_2-4)}{(n_1-1)(n_2-3)^2(n_2-5)}}}$; RR: $|Z| \leq 1.96$

$z = \dfrac{(95.35/53.68) - (30)(28)}{\sqrt{\dfrac{2(30)^2(58)}{(30)(28)^2(26)}}} = 1.7059 < 1.96$

$p = 0.0880 > 0.05$
There is no evidence to suggest that the two population variances in wind power density are different for these two stations.
(b) TS: $F = S_1^2/S_2^2$; RR: $F \leq 0.48$ or $F \geq 2.07$
$f = 95.35/53.68 = 1.7763$
$\Rightarrow 0.48 < 1.7763 < 2.07$
$p = 0.1212$
There is no evidence to suggest that the two population variances are different. This is the same conclusion as in part (a), but the p value is larger (less significant) than the one found using a standard normal approximation.

Chapter 10 Exercises

10.161 $H_0: p_1 - p_2 = 0$, $H_a: p_1 - p_2 > 0$

TS: $Z = \dfrac{\hat{P}_1 - \hat{P}_2}{\sqrt{\hat{P}_c(1-\hat{P}_c)\left(\dfrac{1}{n_1}+\dfrac{1}{n_2}\right)}}$; RR: $|Z| \geq 2.3263$

$z = \dfrac{0.6425 - 0.4493}{\sqrt{(0.5617)(0.4383)\left(\dfrac{1}{186}+\dfrac{1}{138}\right)}} = 3.5139 \geq 2.3263$

p-value: $0.0004 \leq 0.01$
There is evidence to suggest that the true proportion of chest hits is greater in Lancashire than in West Mercia.

10.163 $H_0: \mu_1 - \mu_2 = 0$, $H_a: \mu_1 - \mu_2 \neq 0$

TS: $T = \dfrac{(\bar{X}_1 - \bar{X}_2) - 0}{\sqrt{S_p^2\left(\dfrac{1}{n_1}+\dfrac{1}{n_2}\right)}}$; RR: $|T| \geq 2.7045$

$t = \dfrac{(11.1333 - 11.0) - 0}{\sqrt{(0.3269)^2\left(\dfrac{1}{24}+\dfrac{1}{18}\right)}} = 1.3083 < 2.7045$

$p = 0.1982 > 0.01$
There is no evidence to suggest that the mean amount of sap from trees in Vermont is different from the mean amount of sap from trees in New York.

10.165 (a) The common characteristic that makes these data paired is the archer.
(b) $H_0: \mu_D = 0$, $H_a: \mu_D < 0$

TS: $T = \dfrac{\bar{D} - 0}{S_D/\sqrt{n}}$; RR: $T \leq -1.7959$

$t = \dfrac{-4.1667 - 0}{12.2016/\sqrt{12}} = -1.1829 > -1.7959$

$p = 0.1309 > 0.05$
There is no evidence to suggest that the population mean speed of a carbon arrow is less than the population mean speed of an aluminum arrow.
(c) $0.10 \leq p \leq 0.20$

10.167 $H_0: \mu_1 - \mu_2 = 0$, $H_a: \mu_1 - \mu_2 > 0$

TS: $T = \dfrac{(\bar{X}_1 - \bar{X}_2) - 0}{\sqrt{S_p^2\left(\dfrac{1}{n_1}+\dfrac{1}{n_2}\right)}}$; RR: $T \geq 2.4286$

$t = \dfrac{(1387.0 - 1101.15) - 0}{\sqrt{(427.6012)^2\left(\dfrac{1}{20}+\dfrac{1}{20}\right)}} = 2.1140 < 2.4286$

$p = 0.0206 > 0.01$

There is no evidence to suggest that the mean pier length in California is greater than the mean pier length in Florida.

10.169 $H_0: \mu_D = 0$, $H_a: \mu_D \neq 0$

TS: $T = \dfrac{\bar{D} - 0}{S_D/\sqrt{n}}$; RR: $|T| \geq 2.0930$

$t = \dfrac{0.2250 - 0}{2.03/\sqrt{20}} = 0.4957 < 2.0930$

$p = 0.6258 > 0.05$

There is no evidence to suggest that the population mean moisture content of bulk grain is different when measured by chemical reaction and by distillation.

10.171 (a) $\hat{p}_1 = 0.1143$, $\hat{p}_2 = 0.0952$

$n_1 \hat{p}_1 = 16 \geq 5$, $n_1(1-\hat{p}_1) = 124 \geq 5$,

$n_2 \hat{p}_2 = 12 \geq 5$, $n_2(1-\hat{p}_2) = 114 \geq 5$

(b) $H_0: p_1 - p_2 = 0$, $H_a: p_1 - p_2 \neq 0$

TS: $Z = \dfrac{\hat{P}_1 - \hat{P}_2}{\sqrt{\hat{P}_c(1-\hat{P}_c)\left(\frac{1}{n_1} + \frac{1}{n_2}\right)}}$; RR: $|Z| \geq 2.5758$

$z = \dfrac{0.1143 - 0.0952}{\sqrt{(0.1053)(0.8947)\left(\frac{1}{140} + \frac{1}{126}\right)}} = 0.5054 < 2.5758$

$p = 0.6133 > 0.01$

There is no evidence to suggest that the population proportion of stations in noncompliance with the law is different near Los Angeles and near San Francisco.

10.173 (a) $H_0: \sigma_1^2 = \sigma_2^2$, $H_a: \sigma_1^2 \neq \sigma_2^2$

TS: $F = S_1^2/S_2^2$; RR: $F \leq 0.29$ or $F \geq 3.78$

$f = 1.5625/15.0544 = 0.1038 \leq 0.29$

$p = 0.0001 \leq 0.02$

There is evidence to suggest that the two population variances in time are different.

(b) $H_0: \mu_1 - \mu_2 = 0$, $H_a: \mu_1 - \mu_2 < 0$

TS: $T' = \dfrac{(\bar{X}_1 - \bar{X}_2) - 0}{\sqrt{\frac{S_1^2}{n_1} + \frac{S_2^2}{n_2}}}$; RR: $T' \leq -1.7531$

$t' = \dfrac{(4.56 - 6.58) - 0}{\sqrt{\frac{1.5625}{17} + \frac{15.0544}{14}}} = -1.8697 \leq -1.7531$

$0.025 \leq p \leq 0.05$

There is evidence to suggest that the population mean time to complete Form 1040 for the lower income level is less than the population mean time to complete Form 1040 for the higher income level.

10.175 (a) $\hat{p}_1 = 0.3793$, $\hat{p}_2 = 0.2876$

$n_1 \hat{p}_1 = 132 \geq 5$, $n_1(1-\hat{p}_1) = 216 \geq 5$,

$n_2 \hat{p}_2 = 65 \geq 5$, $n_2(1-\hat{p}_2) = 161 \geq 5$

(b) $(\hat{p}_1 - \hat{p}_2) \pm z_{0.025} \sqrt{\dfrac{\hat{p}_1(1-\hat{p}_1)}{n_1} + \dfrac{\hat{p}_2(1-\hat{p}_2)}{n_2}}$

$= (0.3793 - 0.2876)$

$\pm (1.96) \sqrt{\dfrac{(0.3793)(0.6207)}{348} + \dfrac{(0.2876)(0.1724)}{226}}$

$= (0.0137, 0.1697)$

(c) There is evidence to suggest that the population proportion of online investors is different for these two portfolio classifications, as since 0 is not in the CI.

10.177 For each test:

$H_0: \mu_1 - \mu_2 = 0$, $H_a: \mu_1 - \mu_2 \neq 0$

TS: $T = \dfrac{(\bar{X}_1 - \bar{X}_2) - 0}{\sqrt{S_p^2\left(\frac{1}{n_1} + \frac{1}{n_2}\right)}}$

Single detached versus double/duplex:

RR: $|T| \geq 2.7633$

$t = \dfrac{(8.68 - 13.16) - 0}{\sqrt{(7.8925^2)\left(\frac{1}{15} + \frac{1}{15}\right)}} = -1.5545 > -2.7633$

$p = 0.1313 > 0.01$

There is no evidence to suggest that the population mean toluene levels are different.

Single detached versus apartment:

RR: $|T| \geq 2.7564$

$t = \dfrac{(8.68 - 23.7) - 0}{\sqrt{(12.6666)^2 \left(\frac{1}{15} + \frac{1}{16}\right)}} = -3.2994 \leq -2.7564$

$p = 0.0026 \leq 0.01$

There is evidence to suggest that the population mean toluene levels are different.

Double/duplex versus apartment:

RR: $|T| \geq 2.7564$

$t = \dfrac{(13.16 - 23.7) - 0}{\sqrt{(13.4915)^2 \left(\frac{1}{15} + \frac{1}{16}\right)}} = -2.1737 > -2.7564$

$p = 0.0380 > 0.01$

There is no evidence to suggest that the population mean toluene levels are different.

10.179 (a) $\hat{p}_1 = 0.0482$, $\hat{p}_2 = 0.0666$

(b) $H_0: p_1 - p_2 = 0$, $H_a: p_1 - p_2 < 0$

TS: $Z = \dfrac{\hat{P}_1 - \hat{P}_2}{\sqrt{\hat{P}_c(1-\hat{P}_c)\left(\frac{1}{n_1} + \frac{1}{n_2}\right)}}$; RR: $Z \leq -1.6449$

$z = \dfrac{0.0482 - 0.0666}{\sqrt{(0.0537)(0.9463)\left(\frac{1}{2429} + \frac{1}{1036}\right)}} = -2.2042 \leq -1.6449$

p-value: $0.0138 < 0.05$

There is evidence to suggest that the true proportion of errors in CPOE prescriptions is less than the true proportion of errors in HWP prescriptions.

Chapter 11

11.1 One-Way ANOVA

11.1 True

11.3 False

11.5 False

11.7 Between; within

11.9 We should investigate which pair(s) of means are contributing most to the overall significant difference.

11.11 (a) $n_1 = 10$, $n_2 = 9$, $n_3 = 10$, $n_4 = 10$, $n_5 = 8$, $n_6 = 10$, $n = 57$
(b) $t_{1.} = 51.3$, $t_{2.} = 37.7$, $t_{3.} = 49.1$, $t_{4.} = 50.7$, $t_{5.} = 44.0$, $t_{6.} = 46.9$, $t_{..} = 279.7$
(c) $\sum_{i=1}^{6}\sum_{j=1}^{n_i} x_{ij}^2 = 1481.97$

11.13 (a) SST = 2456.4790, SSA = 122.3315, SSE = 2334.1475
(b) MSA = 122.3315/4 = 30.5829, MSE = 2334.1475/35 = 66.6899
(c) $f = 30.5829/66.6899 = 0.4586$
(d) RR: $F \geq 2.64$
Since $f < 2.64$, do not reject the null hypothesis.

11.15

Source of variation	Sum of squares	Degrees of freedom	Mean square	F	p-value
Factor	13.566	3	4.522	4.58	0.0059
Error	61.256	62	0.988		
Total	74.822	65			

(a) $H_0: \mu_1 = \mu_2 = \mu_3 = \mu_4$, $H_a: \mu_i \neq \mu_j$ for some $i \neq j$
(b) $F \geq 4.11$
(c) $f = 4.522/0.988 = 4.58 \geq 4.11$
$p = 0.0059 \leq 0.01$
There is evidence to suggest at least two population means are different.

11.17 (a) $H_0: \mu_1 = \mu_2 = \mu_3$,
$H_a: \mu_i \neq \mu_j$ for some $i \neq j$
TS: $F = $ MSA/MSE; RR: $F \geq 5.85$

Source of variation	Sum of squares	Degrees of freedom	Mean square	F	p-Value
Factor	21.2026	2	10.6013	9.97	0.0010
Error	21.2721	20	1.0636		
Total	42.4747	22			

$f = 10.6013/1.0636 = 9.97 \geq 5.85$
$p = 0.0010 < 0.01$
There is evidence to suggest at least two of the population mean weights are different.
(b) Sample mean weights: gold: 8.1571, silver: 7.975, titanium: 6.05
The conclusion in part (a) and these sample means suggest the population mean weight for titanium is different from the population mean weights for both silver and gold.

11.19 $H_0: \mu_1 = \mu_2 = \mu_3 = \mu_4$,
$H_a: \mu_i \neq \mu_j$ for some $i \neq j$
TS: $F = $ MSA/MSE; RR: $F \geq 4.51$
SST = 826.8444, SSA = 429.2480, SSE = 397.5964
MSA = 143.0827, MSE = 13.2532
$f = 143.0827/13.2532 = 10.7961 \geq 4.51$
$p = 0.00006 \leq 0.01$
There is evidence to suggest at least two of the population mean tensions are different.

11.21 $H_0: \mu_1 = \mu_2 = \mu_3 = \mu_4$,
$H_a: \mu_i \neq \mu_j$ for some $i \neq j$
TS: $F = $ MSA/MSE; RR: $F \geq 2.83$
SST = 1573.2174, SSA = 611.3235, SSE = 961.8939
MSA = 203.7745, MSE = 22.9022
$f = 203.7745/22.9022 = 8.90 \geq 2.83$
$p = 0.0001 \leq 0.05$
There is evidence to suggest at least two nondairy creamers have a different population mean percentage of fat per serving.

11.23 $H_0: \mu_1 = \mu_2 = \mu_3 = \mu_4$,
$H_a: \mu_i \neq \mu_j$ for some $i \neq j$
TS: $F = $ MSA/MSE; RR: $F \geq 4.57$

Source of variation	Sum of squares	Degrees of freedom	Mean square	F	p-Value
Factor	0.0436	3	0.0145	0.45	0.7202
Error	0.9067	28	0.0324		
Total	0.9503	31			

$f = 0.0145/0.0324 = 0.45 < 4.57$
$p = 0.7202 > 0.01$
There is no evidence to suggest at least two population mean levels of sorbitol are different in these cough syrups.

11.25 $H_0: \mu_1 = \mu_2 = \mu_3 = \mu_4$,
$H_a: \mu_i \neq \mu_j$ for some $i \neq j$
TS: $F = $ MSA/MSE; RR: $F \geq 5.09$

Source of variation	Sum of squares	Degrees of freedom	Mean square	F	p-Value
Factor	109,187.3	3	36,395.8	70.79	<0.0001
Error	9254.2	18	514.1		
Total	118 441.5	21			

$f = 36{,}395.8/514.1 = 70.79 \geq 5.09$
$p < 0.0001$
There is strong evidence to suggest at least two of the population mean monthly shipments are different.

11.27 $H_0: \mu_1 = \mu_2 = \mu_3 = \mu_4$,
$H_a: \mu_i \neq \mu_j$ for some $i \neq j$
TS: $F = $ MSA/MSE; RR: $F \geq 4.57$
SST = 19.28, SSA = 5.1625, SSE = 14.1175
MSA = 1.7208, MSE = 0.5042

$f = 1.7208/0.5042 = 3.41 < 4.57$
$p = 0.0310 > 0.01$
There is no evidence to suggest at least two population mean DOC concentrations are different.

11.29 $H_0: \mu_1 = \mu_2 = \mu_3 = \mu_4 = \mu_5 = \mu_6$,
$H_a: \mu_i \neq \mu_j$ for some $i \neq j$
TS: $F = MSA/MSE$; RR: $F \geq 3.22$
SST $= 5370.10$, SSA $= 1405.65$, SSE $= 3964.45$
MSA $= 281.13$, MSE $= 42.18$
$f = 281.13/42.18 = 6.67 \geq 3.22$
$p < 0.0001$
There is evidence to suggest at least two population mean amounts spent on ground transportation are different.

11.31 (a) $H_0: \mu_1 = \mu_2 = \mu_3 = \mu_4$,
$H_a: \mu_i \neq \mu_j$ for some $i \neq j$
TS: $F = MSA/MSE$; RR: $F \geq 3.01$
SST $= 2,027,473.9$, SSA $= 605,145.5$,
SSE $= 1,422,327.4$
MSA $= 201,715.2$, MSE $= 59,263.6$
$f = 201,715.2/59,263.6 = 3.40 \geq 3.01$
$p = 0.0339 \leq 0.05$
There is evidence to suggest at least two of the population mean pressures are different.
(b) Recommend Holder since this broom has the highest mean pressure.

11.33 (a) $H_0: \mu_1 = \mu_2 = \mu_3 = \mu_4 = \mu_5$,
$H_a: \mu_i \neq \mu_j$ for some $i \neq j$
TS: $F = MSA/MSE$; RR: $F \geq 5.36$
SST $= 16.4292$, SSA $= 9.9652$, SSE $= 6.4640$
MSA $= 2.4913$, MSE $= 0.1154$
$f = 2.4913/0.1154 = 21.58 \geq 5.36$
$p < 0.0001$
There is evidence to suggest at least two population mean weights are different.
(b) Buy at Weis since these bags have the largest sample mean weight.

11.35 (a) $H_a: \mu_1 - \mu_2 = 0$, $H_a: \mu_1 - \mu_2 \neq 0$

TS: $T = \dfrac{(\bar{X}_1 - \bar{X}_2) - 0}{\sqrt{S_p^2\left(\frac{1}{n_1} + \frac{1}{n_2}\right)}}$; RR: $|T| \geq 2.0739$

$t = \dfrac{(30.1917 - 32.5167) - 0}{\sqrt{(5.8921)\left(\frac{1}{12} + \frac{1}{12}\right)}}; -2.3462 \leq -2.0739$

$p = 0.0284 \leq 0.05$
There is evidence to suggest the population mean widths are different.
(b) $H_0: \mu_1 = \mu_2$,
$H_a: \mu_i \neq \mu_j$ for some $i \neq j$
TS: $F = MSA/MSE$; RR: $F \geq 4.30$
SST $= 162.06$, SSA $= 32.43$, SSE $= 129.63$
MSA $= 32.43$, MSE $= 5.89$
$f = 32.43/5.89 = 5.50 \geq 4.30$
$p = 0.0284 \leq 0.05$
There is evidence to suggest the population mean widths are different.

(c) $t^2 = f$ and the p values are the same. This makes sense because these two procedures are testing the exact same hypothesis.

11.2 Isolating Differences

11.37 True

11.39 False

11.41 $\dfrac{k(k-1)}{2}$

11.43 There is no evidence to suggest any difference among population means, so there is no reason to look any further for one.

11.45 (a) $Q_{0.05,3,18} = 3.609$ **(b)** $Q_{0.05,4,40} = 3.791$
(c) $Q_{0.01,4,50} = 4.634$ **(d)** $Q_{0.01,5,50} = 4.863$
(e) $Q_{0.001,6,30} = 6.469$

11.47 (a)

	$\bar{x}_1.$	$\bar{x}_2.$	$\bar{x}_3.$	$\bar{x}_4.$
Sample mean	-33.44	-14.83	0.48	4.30

(b)

	$\bar{x}_3.$	$\bar{x}_2.$	$\bar{x}_4.$	$\bar{x}_1.$
Sample mean	1.30	1.41	1.50	1.62

(c)

	$\bar{x}_4.$	$\bar{x}_2.$	$\bar{x}_3.$	$\bar{x}_1.$	$\bar{x}_5.$
Sample mean	51.92	54.21	60.80	64.35	64.85

11.49 $c = \binom{4}{2} = 6$, $n - k = 76$, $t_{0.01/12,76} = 3.2603$

Difference	Bonferroni CI	Significantly different
$\mu_1 - \mu_2$	$(\ -7.62\ ,\quad 3.64\ )$	No
$\mu_1 - \mu_3$	$(\ -2.54\ ,\quad 8.72\ )$	No
$\mu_1 - \mu_4$	$(\ -11.80\ ,\ -0.54\ )$	Yes
$\mu_2 - \mu_3$	$(\ -0.55\ ,\quad 10.71\ )$	No
$\mu_2 - \mu_4$	$(\ -9.81\ ,\quad 1.45\ )$	No
$\mu_3 - \mu_4$	$(\ -14.89\ ,\ -3.63\ )$	Yes

11.51 $c = \binom{3}{2} = 3$, $n - k = 62$, $t_{0.05/6,62} = 2.4607$

Difference	Bonferroni CI	Significantly different
$\mu_1 - \mu_2$	$(\ -573.84\ ,\quad 3.28\ )$	No
$\mu_1 - \mu_3$	$(\ -586.20\ ,\ -24.88\ )$	Yes
$\mu_1 - \mu_4$	$(\ -285.67\ ,\quad 245.15\ )$	No

11.53 (a) $0.001 \leq p \leq 0.01$
There is evidence to suggest at least two population means are different.
(b) $c = \binom{4}{2} = 6$, $n - k = 20$, $t_{0.01/12, 20} = 3.6303$

Difference	Bonferroni CI	Significantly different
$\mu_1 - \mu_2$	(0.44 , 7.35)	Yes
$\mu_1 - \mu_3$	(−3.56 , 3.36)	No
$\mu_1 - \mu_4$	(−1.86 , 5.06)	No
$\mu_2 - \mu_3$	(−7.45 , −0.54)	Yes
$\mu_2 - \mu_4$	(−5.76 , 1.16)	No
$\mu_3 - \mu_4$	(−1.76 , 5.16)	No

11.55 (a) $0.001 \leq p \leq 0.0001$
(b) $c = \binom{4}{2} = 6$, $n - k = 40$, $t_{0.05/12, 40} = 2.7759$

Difference	Bonferroni CI	Significantly different
$\mu_1 - \mu_2$	(0.87 , 1.25)	Yes
$\mu_1 - \mu_3$	(0.56 , 0.95)	Yes
$\mu_1 - \mu_4$	(0.20 , 0.59)	Yes
$\mu_2 - \mu_3$	(−0.49 , −0.12)	Yes
$\mu_2 - \mu_4$	(−0.85 , −0.48)	Yes
$\mu_3 - \mu_4$	(−0.55 , −0.17)	Yes

	$\bar{x}_{2\cdot}$	$\bar{x}_{3\cdot}$	$\bar{x}_{4\cdot}$	$\bar{x}_{1\cdot}$
Sample mean	0.2567	0.5601	0.9206	1.3164

There are no horizontal lines under any sample means since the Bonferroni CIs suggest all population means are different.

11.57 (a) $H_0: \mu_1 = \mu_2 = \mu_3 = \mu_4$,
$H_a: \mu_i \neq \mu_j$ for some $i \neq j$
TS: $F = $ MSA/MSE; RR: $F \geq 3.10$
$f = 207.22/20.65 = 10.03 \geq 3.10$
$p = 0.0003 \leq 0.05$
There is evidence to suggest at least two population means are different.
(b) $c = \binom{4}{2} = 6$, $n - k = 20$, $t_{0.05/12, 20} = 2.9271$

Difference	Bonferroni CI	Significantly different
$\mu_1 - \mu_2$	(−15.05 , 0.31)	No
$\mu_1 - \mu_3$	(−6.53 , 8.83)	No
$\mu_1 - \mu_4$	(−18.76 , −3.40)	Yes
$\mu_2 - \mu_3$	(0.84 , 16.20)	Yes
$\mu_2 - \mu_4$	(−11.40 , 3.96)	No
$\mu_3 - \mu_4$	(−19.91 , −4.55)	Yes

(c)

	$\bar{x}_{3\cdot}$	$\bar{x}_{1\cdot}$	$\bar{x}_{2\cdot}$	$\bar{x}_{4\cdot}$
Sample mean	65.55	66.70	74.07	77.78

11.59 (a) $H_0: \mu_1 = \mu_2 = \mu_3$,
$H_a: \mu_i \neq \mu_j$ for some $i \neq j$
TS: $F = $ MSA/MSE; RR: $F \geq 3.32$
$f = 6524.76/211.50 = 30.85 \geq 3.32$
$p = 0.00000005 \leq 0.05$
There is evidence to suggest at least two population means are different.
(b) $Q_{0.05, 3, 30} = 3.486$

Difference	Tukey CI	Significantly different
$\mu_1 - \mu_2$	(32.84 , 63.58)	Yes
$\mu_1 - \mu_3$	(17.74 , 49.10)	Yes
$\mu_2 - \mu_3$	(−29.77 , 0.19)	No

(c)

	$\bar{x}_{2\cdot}$	$\bar{x}_{3\cdot}$	$\bar{x}_{1\cdot}$
Sample mean	67.80	82.59	116.01

11.61 (a) $H_0: \mu_1 = \mu_2 = \mu_3 = \mu_4 = \mu_5$
$H_a: \mu_i \neq \mu_j$ for some $i \neq j$
TS: $F = $ MSA/MSE; RR: $F \geq 4.43$
$f = 21.7904/1.6996 = 12.82 \geq 4.43$
$p = 0.00002 \leq 0.01$
There is evidence to suggest at least two population means are different.
(b) $c = \binom{5}{2} = 10$, $n - k = 20$, $t_{0.05/20, 20} = 3.1534$

Difference	Bonferroni CI	Significantly different
$\mu_1 - \mu_2$	(−4.66 , 0.54)	No
$\mu_1 - \mu_3$	(−3.36 , 1.84)	No
$\mu_1 - \mu_4$	(−4.36 , 0.84)	No
$\mu_1 - \mu_5$	(−8.04 , −2.84)	Yes
$\mu_2 - \mu_3$	(−1.30 , 3.90)	No
$\mu_2 - \mu_4$	(−2.30 , 2.90)	No
$\mu_2 - \mu_5$	(−5.98 , −0.78)	Yes
$\mu_3 - \mu_4$	(−3.60 , 1.60)	No
$\mu_3 - \mu_5$	(−7.28 , −2.08)	Yes
$\mu_4 - \mu_5$	(−6.28 , −1.08)	Yes

(c)

	$\bar{x}_{1\cdot}$	$\bar{x}_{3\cdot}$	$\bar{x}_{4\cdot}$	$\bar{x}_{2\cdot}$	$\bar{x}_{5\cdot}$
Sample mean	14.28	15.04	16.04	16.34	19.72

11.63 **(a)** $H_0: \mu_1 = \mu_2 = \mu_3 = \mu_4$,
$H_a: \mu_i \neq \mu_j$ for some $i \neq j$
TS: $F = $ MSA/MSE; RR: $F \geq 3.24$
$f = 45.5847/6.4655 = 7.05 \geq 3.24$
$p = 0.0031 \leq 0.05$
There is evidence to suggest at least two population mean chick weights are different.

(b) $c = \binom{4}{2} = 6$, $n - k = 16$, $t_{0.05/12,16} = 3.0083$

Difference	Bonferroni CI	Significantly different
$\mu_1 - \mu_2$	(-11.24 , -1.56)	Yes
$\mu_1 - \mu_3$	(-8.02 , 1.66)	No
$\mu_1 - \mu_4$	(-11.06 , -1.38)	Yes
$\mu_2 - \mu_3$	(-1.62 , 8.06)	No
$\mu_2 - \mu_4$	(-4.66 , 5.02)	No
$\mu_3 - \mu_4$	(-7.88 , 1.80)	No

(c) $Q_{0.05,4,16} = 4.046$

Difference	Tukey CI	Significantly different
$\mu_1 - \mu_2$	(-11.00 , -1.80)	Yes
$\mu_1 - \mu_3$	(-7.78 , 1.42)	No
$\mu_1 - \mu_4$	(-10.82 , -1.62)	Yes
$\mu_2 - \mu_3$	(-1.38 , 7.82)	No
$\mu_2 - \mu_4$	(-4.42 , 4.78)	No
$\mu_3 - \mu_4$	(-7.64 , 1.56)	No

(d) The numerical answers (intervals) are not the same. However, the mean pairs that are significantly different are the same. We hoped this would happen. However, the mean pairs that are significantly different might not be the same given that the Tukey CIs are smaller.

11.65 **(a)** $H_0: \mu_1 = \mu_2 = \mu_3 = \mu_4 = \mu_5$,
$H_a: \mu_i \neq \mu_j$ for some $i \neq j$
TS: $F = $ MSA/MSE; RR: $F \geq 2.50$
$f = 41.4976/1.6743 = 15.5174 \geq 2.50$
$p \leq 0.0001$
There is evidence to suggest at least two population mean water usages are different.

(b) $c = \binom{5}{2} = 10$, $n - k = 69$, $t_{0.05/20,69} = 2.9001$

Difference	Bonferroni CI	Significantly different
$\mu_1 - \mu_2$	(1.72 , 5.13)	Yes
$\mu_1 - \mu_3$	(0.76 , 4.29)	Yes
$\mu_1 - \mu_4$	(2.52 , 5.98)	Yes
$\mu_1 - \mu_5$	(1.86 , 5.38)	Yes
$\mu_2 - \mu_3$	(-2.64 , 0.83)	No
$\mu_2 - \mu_4$	(-0.88 , 2.53)	No
$\mu_2 - \mu_5$	(-1.54 , 1.93)	No
$\mu_3 - \mu_4$	(-0.04 , 3.49)	No
$\mu_3 - \mu_5$	(-0.70 , 2.89)	No
$\mu_4 - \mu_5$	(-2.39 , 1.13)	No

(c) $Q_{0.05,5,69} = 3.9615$

Difference	Tukey CI	Significantly different
$\mu_1 - \mu_2$	(1.78 , 5.08)	Yes
$\mu_1 - \mu_3$	(0.82 , 4.23)	Yes
$\mu_1 - \mu_4$	(2.58 , 5.92)	Yes
$\mu_1 - \mu_5$	(1.92 , 5.32)	Yes
$\mu_2 - \mu_3$	(-2.58 , 0.77)	No
$\mu_2 - \mu_4$	(-0.82 , 2.47)	No
$\mu_2 - \mu_5$	(-1.48 , 1.87)	No
$\mu_3 - \mu_4$	(0.02 , 3.43)	Yes
$\mu_3 - \mu_5$	(-0.64 , 2.83)	No
$\mu_4 - \mu_5$	(-2.33 , 1.07)	No

(d) One result is different concerning the difference $\mu_3 - \mu_4$. The left endpoint of the CI is very close to 0 and the Tukey CI is narrower, so this change in significance is not surprising.

11.3 Two-Way ANOVA

11.67 True

11.69 False

11.71 If $n = 1$, then MSE $= \dfrac{\text{SSE}}{ab(n-1)}$ is undefined

11.73 **(a)** $t_{1..} = 22.0$, $t_{2..} = 36.0$, $t_{3..} = 20.5$, $t_{4..} = 27.9$, $t_{.1.} = 36.3$, $t_{.2.} = 70.1$, $t_{...} = 106.4$

(b) $\sum_{i=1}^{4}\sum_{j=1}^{2}\sum_{k=1}^{4} x_{ijk}^2 = 424.82$

(c) SST $= 71.04$, SSA $= 18.5525$, SSB $= 35.7013$, SS(AB) $= 1.5563$, SSE $= 15.23$

11.75 (a)

Source of variation	Sum of squares	Degrees of freedom	Mean square	F	p-Value
Factor A	297.51	2	148.76	10.47	0.0004
Factor B	86.69	2	43.34	3.05	0.0639
Interaction	26.40	4	6.60	0.46	0.7643
Error	383.73	27	14.21		
Total	794.33	35			

(b) Interaction: $F_{0.01,4,27} = 4.11$
$f_{AB} = 0.46$; there is no evidence of interaction.
Factor A: $F_{0.01,2,27} = 5.49$
$f_A = 10.47 \geq 5.49$; there is evidence of an effect due to factor A.
Factor B: $F_{0.01,2,27} = 5.49$
$f_B = 3.05$; there is no evidence of an effect due to factor B.

11.77

Source of variation	Sum of squares	Degrees of freedom	Mean square	F	p-Value
Factor A	121.25	4	30.31	1.57	0.1906
Factor B	91.12	4	22.78	1.18	0.3260
Interaction	174.55	16	10.91	0.57	0.8996
Error	1446.41	75	19.29		
Total	1833.33	99			

Interaction: $F_{0.001,16,75} = 2.90$
$f_{AB} = 0.57$; there is no evidence of interaction.
Factor A: $F_{0.001,4,75} = 5.16$
$f_A = 1.57$; there is no evidence of an effect due to factor A.
Factor B: $F_{0.001,4,75} = 5.16$
$f_B = 1.18$; there is no evidence of an effect due to factor B.

11.79 (a)

Source of variation	Sum of squares	Degrees of freedom	Mean square	F	p-Value
Gender	16.33	1	16.33	4.03	0.0500
Job Type	184.39	4	46.10	11.39	0.0000
Interaction	19.34	4	4.84	1.19	0.3249
Error	202.42	50	4.05		
Total	422.48	59			

(b) $F_{0.01,4,50} = 3.72$
$f_{AB} = 1.19$ and $p = 0.3249 > 0.01$; there is no evidence of interaction
(c) Gender: $F_{0.01,1,50} = 7.17$
$f_A = 4.03$ and $p = 0.0500 > 0.01$; there is no evidence of an effect due to gender
Job type: $F_{0.01,4,50} = 3.72$
$f_B = 11.39$ and $p = 0.0000 < 0.01$; there is strong evidence of an effect due to job type.

11.81

Source of variation	Sum of squares	Degrees of freedom	Mean square	F	p-Value
Temperature	36.847	2	18.42	8.07	0.0013
School	6.822	3	2.27	1.00	0.4040
Interaction	37.093	6	6.18	2.71	0.0283
Error	82.138	36	2.28		
Total	162.900	47			

Interaction: $F_{0.05,6,36} = 2.36$
$f_{AB} = 2.71$ and $p = 0283 < 0.05$; there is evidence of interaction.
Temperature: $F_{0.05,2,36} = 3.26$
$f_A = 8.07$ and $p = 0.0013 < 0.05$; there is evidence of an effect due to temperature.
School: $F_{0.05,3,36} = 2.87$
$f_B = 1.00$ and $p = 0.4040 > 0.05$; the effect due to school is inconclusive.

11.83

Source of variation	Sum of squares	Degrees of freedom	Mean square	F	p-Value
Transport	4332.0	1	4332.0	1.87	0.1793
Car type	86 841.6	4	21 710.4	9.59	0.0000
Interaction	6001.6	4	1500.4	0.66	0.6217
Error	90 598.8	40	2265.0		
Total	187 674.0	49			

Interaction: $F_{0.01,4,40} = 3.83$
$f_{AB} = 0.66$ and $p = 0.6217 > 0.01$; there is no evidence of interaction.
Transport: $F_{0.01,1,40} = 7.31$
$f_A = 1.87$ and $p = 0.1793 > 0.01$; there is no evidence of an effect due to transport type.
Car type: $F_{0.01,4,40} = 3.83$
$f_B = 9.59$ and $p = 0.0000 < 0.01$; there is strong evidence of an effect due to type of car.

11.85

Source of variation	Sum of squares	Degrees of freedom	Mean square	F	p-Value
Age group	217.13	3	72.3750	6.98	0.0005
Dive type	2.25	1	2.2500	0.22	0.6409
Interaction	205.63	3	68.5417	6.61	0.0007
Error	580.75	56	10.3705		
Total	1005.75	63			

Interaction: $F_{0.001,3,56} = 6.23$
$f_{AB} = 6.61$ and $p = 0.0007 < 0.001$; there is evidence of interaction.
Age group: $F_{0.0013,56} = 6.23$
$f_A = 6.98$ and $p = 0.0005 < 0.001$; there is evidence of an effect due to age group.

Dive type: $F_{0.001,1,56} = 12.06$
$f_B = 0.22$ and $p = 0.6409$; the effect due to dive type is inconclusive.

11.87

Source of variation	Sum of squares	Degrees of freedom	Mean square	F	p-Value
Region	202.95	2	101.48	7.23	0.0023
Quality	830.49	3	276.83	19.71	0.0000
Interaction	141.62	6	23.60	1.68	0.1541
Error	505.60	36	14.04		
Total	1680.66	47			

Interaction: $F_{0.05,6,36} = 2.36$
$f_{AB} = 1.68$ and $p = 0.1541 > 0.05$; there is no evidence of interaction.
Region: $F_{0.05,2,36} = 3.26$
$f_A = 7.23$ and $p = 0.0023 < 0.05$; there is evidence of an effect due to region.
Quality: $F_{0.05,3,36} = 2.87$
$f_B = 19.71$ and $p = 0.0000$; there is evidence of an effect due to road marking quality.

11.89 (a)

Source of variation	Sum of squares	Degrees of freedom	Mean square	F	p-Value
Gender	261.33	1	261.33	10.52	0.0024
Altitude	241.17	3	80.39	3.24	0.0321
Interaction	106.17	3	35.39	1.43	0.2497
Error	993.33	40	24.83		
Total	1602.00	47			

Interaction: $F_{0.05,3,40} = 2.84$
$f_{AB} = 1.43$ and $p = 0.2497 > 0.05$; there is no evidence of interaction.
(b) Gender: $F_{0.05,1,40} = 4.08$
$f_A = 10.52$ and $p = 0.0024 \leq 0.05$; there is evidence of an effect due to gender.
Altitude: $F_{0.05,3,40} = 2.84$
$f_B = 3.24$ and $p = 0.0321 \leq 0.05$; there is evidence of an effect due to altitude.

11.91 (a)

Source of variation	Sum of squares	Degrees of freedom	Mean square	F	p-Value
ADHD	72.45	1	72.45	57.75	<0.0001
County	38.72	17	2.28	1.82	0.0290
Interaction	15.70	17	0.92	0.74	0.7627
Error	225.82	180	1.25		
Total	352.69	215			

Interaction: $F_{0.05,17,180} = 1.68$
$f_{AB} = 0.74$ and $p = 0.7627 > 0.05$; there is no evidence of interaction.
(b) ADHD: $F_{0.05,1,180} = 3.89$
$f_A = 57.75$ and $p < 0.0001 < 0.05$; there is evidence of an effect due to ADHD diagnosis.
County: $F_{0.05,17,180} = 1.68$
$f_B = 1.82$ and $p = 0.0290 < 0.05$; there is evidence of an effect due to county.
(c) Diagnosis of ADHD is associated with an increase in age at diagnosis for ASD. The mean age of all those not diagnosed with ADHD was 3.72; for those with ADHD, it was 4.30.

Chapter 11 Exercises

11.93

Source of variation	Sum of squares	Degrees of freedom	Mean square	F	p-Value
Factor	106,568	3	35,522.67	4.67	0.0124
Error	151,997	20	7599.85		
Total	258,565	23			

$H_0: \mu_1 = \mu_2 = \mu_3 = \mu_4$,
$H_a: \mu_i \neq \mu_j$ for some $i \neq j$
TS: $F = $ MSA/MSE; RR: $F \geq 3.10$
$f = 35,522.67/7599.85 = 4.67 \geq 3.10$
$p = 0.0124 < 0.05$
There is evidence to suggest at least two of the population mean wattages of spotlights are different.

11.95 $c = \binom{4}{2} = 6$, $n - k = 36$, $t_{0.05/12, 36} = 2.7920$

Difference	Bonferroni CI		Significantly different
$\mu_1 - \mu_2$	(−4.38 ,	8.58)	No
$\mu_1 - \mu_3$	(−4.08 ,	8.88)	No
$\mu_1 - \mu_4$	(1.62 ,	14.58)	Yes
$\mu_2 - \mu_3$	(−6.18 ,	6.78)	No
$\mu_2 - \mu_4$	(−0.48 ,	12.48)	No
$\mu_3 - \mu_4$	(−0.78 ,	12.18)	No

	$\bar{x}_{4.}$	$\bar{x}_{3.}$	$\bar{x}_{2.}$	$\bar{x}_{1.}$
Sample mean	98.6	104.3	104.6	106.7

11.97 (a) $H_0: \mu_1 = \mu_2 = \mu_3 = \mu_4$,
$H_a: \mu_i \neq \mu_j$ for some $i \neq j$
TS: $F = $ MSA/MSE; RR: $F \geq 3.24$
$f = 0.0623/0.0533 = 6.23 \geq 3.24$
$p = 0.0052 \leq 0.05$
There is evidence to suggest at least two of the population mean step heights are different.

(b) $c = \binom{4}{2} = 6$, $n - k = 16$, $t_{0.01/12, 16} = 3.7725$

Difference	Bonferroni CI		Significantly different
$\mu_1 - \mu_2$	(0.0139 ,	0.2892)	Yes
$\mu_1 - \mu_3$	(−0.0346 ,	0.2407)	No
$\mu_1 - \mu_4$	(−0.0276 ,	0.2477)	No
$\mu_2 - \mu_3$	(−0.1862 ,	0.0892)	No
$\mu_2 - \mu_4$	(−0.1792 ,	0.0962)	No
$\mu_3 - \mu_4$	(−0.1307 ,	0.1447)	No

	$\bar{x}_2$.	$\bar{x}_4$.	$\bar{x}_3$.	$\bar{x}_1$.
Sample mean	0.2299	0.2714	0.2784	0.3815

11.99

Source of variation	Sum of squares	Degrees of freedom	Mean square	F	p-Value
Lanes	735.06	3	245.02	16.78	< 0.0001
Diff Elev	113.56	3	37.85	2.59	0.0635
Interaction	325.81	9	36.20	2.48	0.0206
Error	701.00	48	14.60		
Total	1875.44	63			

Interaction: $F_{0.05, 9, 48} = 2.08$
$f_{AB} = 2.48$ and $p = 0.0206 < 0.05$; there is evidence of interaction.
Lanes: $F_{0.05, 3, 48} = 2.80$
$f_A = 16.78$ and $p < 0.0001 < 0.05$; there is probably an effect due to number of lanes.
Differential elevation: $F_{0.05, 3, 48} = 2.80$
$f_B = 2.59$ and $p = 0.0635 > 0.05$; since interaction is present, the effect due to differential elevation is inconclusive.

11.101 $H_0: \mu_1 = \mu_2 = \mu_3$,
$H_a: \mu_i \neq \mu_j$ for some $i \neq j$
TS: $F =$ MSA/MSE; RR: $F \geq 6.70$
$f = 21{,}660.01/74{,}029.21 = 0.29 < 6.70$
$p = 0.7530 > 0.05$
There is no evidence to suggest a difference in any of the mean generating capacities.

11.103

Source of variation	Sum of squares	Degrees of freedom	Mean square	F	p-Value
Season	97.00	3	32.33	4.32	0.0070
Lobby	90.55	2	45.28	6.05	0.0035
Interaction	45.32	6	7.55	1.01	0.4253
Error	629.12	84	7.49		
Total	862.00	95			

Interaction: $F_{0.05, 6, 84} = 2.21$
$F_{AB} = 1.01$ and $p = 0.4253 > 0.05$; there is no evidence of interaction.
Season: $F_{0.05, 3, 84} = 2.71$
$F_A = 4.32$ and $p = 0.0070 \leq 0.05$; there is evidence of an effect due to season.
Lobby: $F_{0.05, 2, 84} = 3.11$
$F_B = 6.05$ and $p = 0.0035 \leq 0.05$; there is evidence of an effect due to lobby size.
(b) Winter has the highest sample mean level of airborne particles (9.704). We should avoid sitting in a hospital lobby during winter.
(c) Large hospital lobbies have the largest sample mean level of airborne particles (9.906). We should avoid sitting in large hospital lobbies.

11.105 (a) Interaction: $F_{0.05, 4, 27} = 2.73$
$f_{AB} = 2.34$ and $p = 0.0802$; there is no evidence of interaction.
(b) Species: $F_{0.05, 2, 27} = 3.35$
$f_A = 2.36$ and $p = 0.1132 > 0.05$; there is no evidence of an effect due to species ID.
(c) Age: $F_{0.05, 2, 27} = 3.35$
$f_B = 5.48$ and $p = 0.0101 \leq 0.05$; there is evidence of an effect due to age.

11.107 (a)

Source of variation	Sum of squares	Degrees of freedom	Mean square	F	p-Value
Style	0.0011	1	0.0011	0.13	0.7318
Brand	0.1241	3	0.0414	4.94	0.0316
Interaction	0.0131	3	0.0044	0.52	0.6792
Error	0.0671	8	0.0084		
Total	0.2053	15			

Interaction: $F_{0.05, 3, 8} = 4.07$
$f_{AB} = 0.52$ and $p = 0.6792 > 0.05$; there is no evidence of interaction.
Style: $F_{0.05, 1, 8} = 5.32$
$f_A = 0.13$ and $p = 0.7318 > 0.05$; there is no evidence of an effect due to style.
Brand: $F_{0.05, 3, 8} = 4.07$
$f_B = 4.94$ and $p = 0.0316 \leq 0.05$; there is evidence of an effect due to brand.
(b) Choose Chunky Jif. This combination of style and brand has the smallest sample mean.

Chapter 12

12.1 Simple Linear Regression

12.1 Deterministic and random

12.3 Homogeneity of variance

12.5 The mean value of Y for $x = x^*$; an observed value of Y for $x = x^*$

12.7 SST = SSR + SSE

12.9 False

12.11 For each one-unit change in the independent variable, the y variable will change by approximately $\hat{\beta}_1$ units.

12.13 (a)

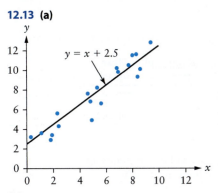

(b)

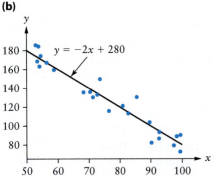

12.15 (a) $E(Y|27.4) = -232.38$
(b) Change in the dependent variable: $(-7.2)(-5) = 36$
(c) For $x = 30$, $Y \sim N(-251.1, 1.5^2)$
$P(-252 \leq Y \leq -250) = 0.4941$

12.17 (a)

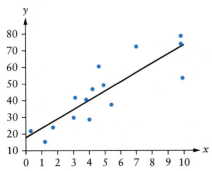

A simple linear regression model seems reasonable. The points appear to fall near a straight line.
(b) $y = 17.8373 + 5.5714x$ **(c)** $y = 49.5943$

12.19 (a) $E(Y|15) = 106$ **(b)** $(-6.1)(-2.0) = 12.2$
(c) For $x = 17$, $Y \sim N(93.8, 3.2^2)$
$P(90 \leq Y \leq 95) = 0.5287$

12.21 (a) $E(Y|1000) = 0.4277$ **(b)** $y = 0.3757$
(c) For $x = 750$, $Y \sim N(0.4927, 0.06^2)$
$P(Y > 0.5) = 0.4516$

12.23 (a) $\hat{\beta}_1 = 0.0016$, $\hat{\beta}_0 = 0.5739$
$y = 0.5739 + 0.0016x$
(b) $y = 1.5339$

12.25 (a)

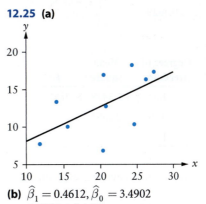

(b) $\hat{\beta}_1 = 0.4612$, $\hat{\beta}_0 = 3.4902$
$y = 3.4902 + 0.4612x$
(c) $E(Y|18) = 11.7923$

12.27 (a)

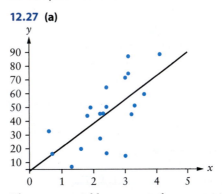

These two variables appear to have a positive association, meaning $\hat{\beta}_1$ should be positive.
(b) $\hat{\beta}_0 = 3.7427$, $\hat{\beta}_1 = 17.2603$
$y = 3.7427 + 17.2603x$
(c) $E(Y|2) = 38.2633$

12.29 (a) $\hat{\beta}_1 = -58.9778$, $\hat{\beta}_0 = 24.4297$
$y = 24.4297 - 58.9778x$
(b) $E(Y|0.3) = 6.7364$
(c) $24.4297 - 58.9778x = 15.0 \Rightarrow x = 0.1599$

12.31 (a)

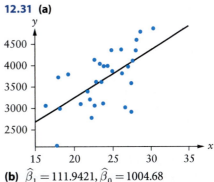

(b) $\hat{\beta}_1 = 111.9421$, $\hat{\beta}_0 = 1004.68$
$y = 1004.68 + 111.9421x$
(c) $E(Y|25) = 3803.2355$

(d) $r^2 = 0.3894$ and the points appear to fall somewhat near a straight line. While BMI is not a perfect predictor of birth weight, it seems to do a reasonably good job.

12.33 (a) $\hat{\beta}_1 = 2.8059, \hat{\beta}_0 = 0.4629$
$y = 0.4629 + 2.8059x$
(b)

Source of variation	Sum of squares	Degrees of freedom	Mean square	F
Regression	2.5478	1	2.5478	9.87
Error	2.5822	10	0.2582	
Total	5.1300	11		

(c) $r^2 = 0.497$
Approximately 50% of the variation in bounce is explained by the regression model.
(d) $0.4629 + 2.8059x = 4.0 \Rightarrow x = 1.26$

12.35 (a)

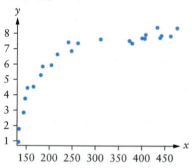

As x_1 increases, the values of y tend to level off. The relationship appears to be logarithmic.
(b) Values of x_2:
1.3863, 3.9512, 4.7185, 4.7791, 2.8904, 3.1355, 5.6240, 5.7366, 5.8493, 2.7081, 4.0254, 4.4773, 5.2040, 5.4972, 1.6094, 5.6312, 5.7462, 3.5835, 4.3175, 4.8903, 5.3083, 5.5215, 5.6021, 5.7170, 5.8081
(c)

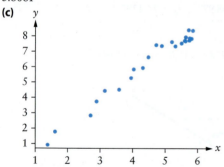

This relationship appears to be linear.
(d) $\hat{\beta}_1 = 1.5603, \hat{\beta}_0 = -0.8059$
$y = -0.8059 + 1.5603x_2$

Source of variation	Sum of squares	Degrees of freedom	Mean square	F	p Value
Regression	103.16	1	103.16	741.04	0.0000
Error	3.20	23	0.14		
Total	106.36	24			

12.2 Hypothesis Tests and Correlation

12.37 A t-test for β_1

12.39 False

12.41 $-1 \le r \le 1$

12.43 True

12.45 (a)

Source of variation	Sum of squares	Degrees of freedom	Mean square	F	p Value
Regression	2772.93	1	2772.93	7.26	0.0109
Error	12988.70	34	382.02		
Total	15761.63	35			

(b) H_0: There is no significant linear relationship.
H_a: There is a significant linear relationship.
TS: $F = $ MSR/MSE; RR: $F \ge 7.44$
$f = 7.26 < 7.44$
$p = 0.0109 > 0.01$
There is no evidence of a significant linear relationship. (It's close!)
(c) $r^2 = 0.1759$ (d) $r = 0.4194$ or -0.4194
There is no way to determine whether r is positive or negative from the ANOVA table alone.

12.47 (a) $\hat{\beta}_1 = -1.0197, \hat{\beta}_0 = 19.8108$
$y = 19.8108 - 1.0198x$

Source of variation	Sum of squares	Degrees of freedom	Mean square	F	p Value
Regression	14.9084	1	14.9084	5.65	0.0634
Error	13.1891	5	2.6378		
Total	28.0975	6			

(b) $H_0: \beta_0 = 0$, $H_a: \beta_0 \ne 0$
TS: $T = B_0/S_{B_0}$; RR: $|T| \ge 4.0321$
$t = 19.8108/1.4841 = 13.3489 \ge 4.0321$
$p < 0.0001$
There is evidence to suggest $\beta_0 \ne 0$; the true regression line does not pass through the origin.
(c) $\hat{\beta}_0 \pm t_{0.005,5} \, s_{B_0} = (13.8268, 25.7948)$
(d) $\hat{\beta}_0 = 19.8108$. If the observed value of x is 0, the predicted y-value is 19.8180.

12.49 (a) $r = 0.8771$
(b) Since $r > 0$ and close to 1, there is a strong positive relationship between the two variables.

12.51 (a) $\hat{\beta}_1 = 0.0093, \hat{\beta}_0 = 2.3220$
$y = 2.3220 + 0.0093x$

Source of variation	Sum of squares	Degrees of freedom	Mean square	F	p Value
Regression	2.2824	1	2.2824	9.55	0.0176
Error	1.6731	7	0.2390		
Total	3.9556	8			

(b) $H_0: \beta_1 = 0$, $H_a: \beta_1 \ne 0$
TS: $T = B_1/S_{B_1}$; RR: $|T| \ge 2.3646$
$t = \dfrac{0.0093}{\sqrt{0.2390/26256}} = 3.0902 \ge 2.3646$
$p = 0.0176 \le 0.05$
There is evidence to suggest that β_1 is different from 0.

(c) $r^2 = 0.5770$
Approximately 0.5770 of the variation in game length can be explained by this simple linear regression model.

12.53 (a)

Source of variation	Sum of squares	Degrees of freedom	Mean square	F	p Value
Regression	108.54	1	108.54	8.34	0.0277
Error	78.06	6	13.01		
Total	186.60	7			

(b) H_0: There is no significant linear relationship.
H_a: There is a significant linear relationship.
TS: $F = $ MSR/MSE; RR: $F \geq 5.99$
$f = 108.54/13.01 = 8.34 \geq 5.99$
$0.01 \leq p \leq 0.05$
There is evidence of a significant linear relationship.
(c) $\hat{\beta}_1 \pm t_{0.025,6} s_{B_1} = (-0.1551, -0.0128)$

12.55 (a)

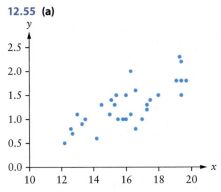

(b) $r = 0.7548$
(c) There is a moderately strong positive linear relationship between crimini mushroom weight and copper content. As mushroom weight increases, so does copper content.
(d) The independent variable would be mushroom weight and the dependent variable would be the copper content.

12.57 (a) $r = 0.4303$
(b) $\hat{\beta}_1 = 92.7141$, $\hat{\beta}_0 = 370.2037$
$y = 370.2037 + 92.7141x$
H_0: There is no significant linear relationship.
H_a: There is a significant linear relationship.
TS: $F = $ MSR/MSE; RR: $F \geq 4.67$
$f = 712{,}577.86/241{,}191.01 = 2.95 < 4.67$
$p > 0.05$
There is no evidence of a significant linear relationship.
(c) There is little evidence to support the reporter's claim. There is no evidence of a significant linear relationship.

12.59 (a)

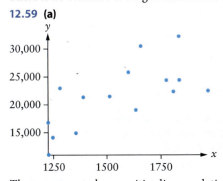

There appears to be a positive linear relationship.

(b) $r = 0.6846$
The value of r suggests there is a positive linear relationship.
(c) $y = -2734.55 + 15.601x$
H_0: There is no significant linear relationship.
H_a: There is a significant linear relationship.
TS: $F = $ MSR/MSE; RR: $F \geq 4.67$
$f = 11.47 \geq 4.67$
$p = 0.0049 \leq 0.05$
There is evidence to suggest a significant linear relationship between ozone generated and aircraft emissions.
(d) H_0: $\beta_0 = 0$, H_a: $\beta_0 \neq 0$
TS: $T = B_0/S_{B_0}$; RR: $|T| \geq 2.1604$
$t = \dfrac{-2734.55}{7262.5} = -0.3765 > -2.1604$
$p = 0.7126 > 0.05$
There is no evidence to suggest that β_0 is different from 0. This result seems practical. If there are 0 aircraft emissions, then no ozone should be generated.

12.61 (a)

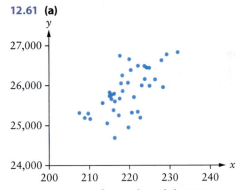

The closing price for Apple and the DJIA appear to have a moderately strong, positive, linear relationship.
(b) $r = 0.6239$
This correlation coefficient shows a moderately strong positive relationship between these two variables.
(c) $y = 12{,}256.96 + 61.99x$
H_0: There is no significant linear relationship.
H_a: There is a significant linear relationship.
TS: $F = $ MSR/MSE; RR: $F \geq 4.0847$
$f = 25.4876 \geq 4.0847$
$p \leq 0.0001$
There is very strong evidence to suggest a linear relationship between closing price for Apple and DIJA.
$E[Y \mid 250] = 27{,}754.56$

12.63 (a) $y = 22{,}310.42 + 16x$

Source of variation	Sum of squares	Degrees of freedom	Mean square	F	p Value
Regression	356,483,106.9	1	356,483,106.9	268.66	0.0000
Error	11,942,022.0	9	1,326,891.3		
Total	368,425,128.9	10			

(b) H_0: There is no significant linear relationship.
H_a: There is a significant linear relationship.
TS: $F = $ MSR/MSE; RR: $F \geq 5.1174$
$f = 268.62 \geq 5.1174$
$p = 0.000000052 \leq 0.0001$
(c) H_0: $\beta_1 = 17$, H_a: $\beta_1 \neq 17$
TS: $T = (B_1 - 17)/S_{B_1}$; RR: $|T| \geq 2.2622$

$t = -1.02 > -2.2622$
$p = 0.9994$
There is no evidence to suggest that $\hat{\beta}_1 \neq 17$.
(d) $\hat{\beta}_1 \pm t_{0.005,9} s_{B_1} = (12.8298, 19.1756)$

12.65 (a)

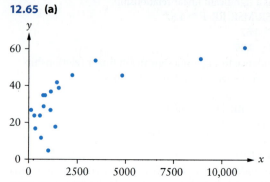

A linear model does not seem appropriate to describe the relationship between these two variables. As x increases, y appears to level off, suggesting a logarithmic model may be more appropriate.
(b) $r = 0.7365$
This correlation coefficient suggests a strong linear relationship between x and y. The scatter plot, however, suggests that the relationship is not linear, making it unreasonable to interpret the r value.
(c) Scatter plot of y versus w

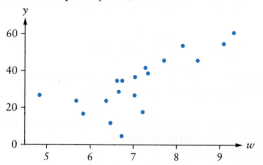

A linear model seems much more appropriate to describe these variables.
(d) $r = 0.7401$
This correlation coefficient suggests a strong positive linear relationship between these two variables.

12.67 (a)

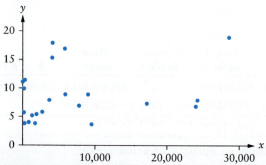

There does not appear to be any relationship between x and y. There is only very slight evidence that as x increases, y increases.
(b) $r = 0.2438$. This value supports the answer in part (a); $r > 0$ but close to 0.

(c) The relationship between the two variables does not appear to be significant. The scatter plot and r suggest a weak positive linear relationship at best.
(d) $y = 7.9620 + 0.0001x$
H_0: There is no significant linear relationship.
H_a: There is a significant linear relationship.
TS: $F = $ MSR/MSE; RR: $F \geq 4.35$
$f = 1.26 < 4.35$
$p = 0.2742 > 0.05$
There is no evidence of a significant linear relationship. This is consistent with part (c).

12.3 Inferences Concerning the Mean Value and an Observed Value of Y for $x = x^*$

12.69 The mean value of Y for $x = x^*$; an observed value of Y for $x = x^*$

12.71 False

12.73 The variance is smallest when $x^* = \bar{x}$.

12.75 (a) $H_0: y^* = 20$, $H_a: y^* > 20$
TS: $T = \dfrac{(B_0 + B_1 x^*) - y_0^*}{s\sqrt{(1/n) + [(x^* - \bar{x})^2 / S_{xx}]}}$; RR: $T \geq 1.7459$
$t = 0.1503 < 1.7459$
$p = 0.4412 > 0.05$
There is no evidence to suggest the mean value of Y for $x = 16.2$ is greater than 20.
(b) $H_0: y^* = 5$, $H_a: y^* \neq 5$
TS: $T = \dfrac{(B_0 + B_1 x^*) - y_0^*}{s\sqrt{(1/n) + [(x^* - \bar{x})^2 / S_{xx}]}}$; RR: $|T| \geq 2.9208$
$t = \dfrac{4.49 - 5.0}{(10.1622)\sqrt{(1/18) + [(11.5 - 15.3670)^2 / 138.14]}} = -0.1240 > -2.920$
$p = 0.9029 > 0.01$
There is no evidence to suggest the mean value of Y for $x = 11.5$ is different from 5.

12.77 (a) $(\hat{\beta}_0 + \hat{\beta}_1 x^*) \pm t_{0.005,5} \, s \sqrt{1 + \dfrac{1}{n} + \dfrac{(x^* - \bar{x})^2}{S_{xx}}}$
$= (-20.1474, 123.4294)$
Width $= 143.5768$
(b) $(\hat{\beta}_0 + \hat{\beta}_1 x^*) \pm t_{0.005,5} \, s \sqrt{1 + \dfrac{1}{n} + \dfrac{(x^* - \bar{x})^2}{S_{xx}}}$
$= (-23.8157, 123.7581)$
Width $= 147.5738$
(c) 18.1 is farther from the mean than 19.25.

12.79 (a) $y = 398.6420 - 8.3856x$

Source of variation	Sum of squares	Degrees of freedom	Mean square	F	p Value
Regression	11954.82	1	11954.82	13.51	0.0028
Error	11502.28	13	884.79		
Total	23457.09	14			

H_0: There is no significant linear relationship.
H_a: There is a significant linear relationship.
TS: $F = $ MSR/MSE; RR: $F \geq 9.07$
$f = 11954.82 / 884.79 = 13.51 \geq 9.07$

$p = 0.0028 \leq 0.01$
There is evidence of a significant linear relationship.
(b) $s = 29.7454$
(c) $(\hat{\beta}_0 + \hat{\beta}_1 x^*) \pm t_{0.025,13} \, s\sqrt{1 + \dfrac{1}{n} + \dfrac{(x^* - \bar{x})^2}{S_{xx}}}$
$= (16.0985, 160.6499)$
It is unlikely that an observed value of Y will be greater than 170 since this PI is completely below 170.

12.81 (a) $E(Y\,|\,100) = 0.6467$
(b) $(\hat{\beta}_0 + \hat{\beta}_1 x^*) \pm t_{0.025,19} \, s\sqrt{\dfrac{1}{n} + \dfrac{(x^* - \bar{x})^2}{S_{xx}}}$
$= (0.6115, 0.9495)$
(c) $H_0: y^* = 0.06$, $H_a: y^* > 0.06$
TS: $T = \dfrac{(B_0 + B_1 x^*) - y_0^*}{s\sqrt{(1/n) + [(x^* - \bar{x})^2/S_{xx}]}}$; RR: $T \geq 2.5395$
$t = 6.6636 \geq 2.5395$
$p < 0.0001$
There is evidence to suggest the mean value of Y for $x = 80$ is greater than 0.06.

12.83 (a) $\hat{\beta}_1 = 0.0021$, $\hat{\beta}_0 = 3.5955$
$y = 3.5955 + 0.0021x$
(b)

Source of variation	Sum of squares	Degrees of freedom	Mean square	F	p Value
Regression	0.8000	1	0.8000	0.15	0.7089
Error	48.4655	9	5.3851		
Total	49.2655	10			

H_0: There is no significant linear relationship.
H_a: There is a significant linear relationship.
TS: $F = $ MSR/MSE; RR: $F \geq 5.12$
$f = 0.80/5.3851 = 0.15 < 5.12$
$p = 0.7089 > 0.05$
There is no evidence of a significant linear relationship. Annual rainfall does not help to explain the variation in asthma prevalence.
(c) $(\hat{\beta}_0 + \hat{\beta}_1 x^*) \pm t_{0.025,9} \, s\sqrt{1 + \dfrac{1}{n} + \dfrac{(x^* - \bar{x})^2}{S_{xx}}}$
$= (-0.3622, 11.7532)$
This CI includes some negative numbers, which is odd since we cannot have a negative percentage of adults treated for asthma.

12.85 (a) The independent variable is the skid resistance. The dependent variable is the accident rate.
(b)
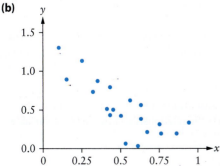
The relationship appears to be negative and linear.

(c) $\hat{\beta}_1 = -1.2285$, $\hat{\beta}_0 = 1.1570$
$y = 1.1570 - 1.2285x$
(d) $H_0: y^* = 0.60$, $H_a: y^* < 0.60$
TS: $T = \dfrac{(B_0 + B_1 x^*) - y_0^*}{s\sqrt{(1/n) + [(x^* - \bar{x})^2/S_{xx}]}}$; RR: $T \leq -1.7341$
$t = -1.1942 > -1.7341$
$p = 0.1240 > 0.05$
There is no evidence to suggest the mean value of Y for $x = 0.50$ is less than 0.60.

12.87 (a)
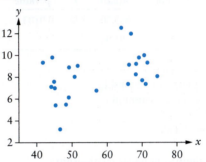
There appears to be a weak, positive, linear relationship between average temperature and number of violent crimes.
(b) $\hat{\beta}_1 = 0.0751$, $\hat{\beta}_0 = 3.8228$
$y = 3.8228 + 0.0751x$
(c) $(\hat{\beta}_0 + \hat{\beta}_1 x^*) \pm t_{0.025,23} \, s\sqrt{1 + \dfrac{1}{n} + \dfrac{(x^* - \bar{x})^2}{S_{xx}}}$
$= (5.6039, 14.0519)$
(d) $(\hat{\beta}_0 + \hat{\beta}_1 x^*) \pm t_{0.025,23} \, s\sqrt{\dfrac{1}{n} + \dfrac{(x^* - \bar{x})^2}{S_{xx}}}$
$= (4.3877, 12.2656)$

12.89 (a)
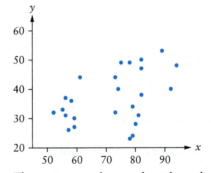
There appears to be a moderately weak, positive, linear association between Walk Score and time walking.
(b) $\hat{\beta}_0 = 14.84$, $\hat{\beta}_1 = 0.31$
$y = 14.84 + 0.31x$
Because $\hat{\beta}_1$ is positive, this supports our observation of a positive association between these two variables.
(c) H_0: There is no significant linear relationship.
H_a: There is a significant linear relationship.
TS: $F = $ MSR/MSE; RR: $F \geq 4.2793$
$f = 370.40/67.28 = 5.5058 \geq 4.2793$
$p = 0.0279 < 0.05$
There is evidence to suggest a linear relationship between Walk Score and time spent walking.

(d) $(\hat{\beta}_0+\hat{\beta}_1 x^*)\pm t_{0.025,23}\, s\sqrt{\dfrac{1}{n}+\dfrac{(x^*-\bar{x})^2}{S_{xx}}}$

$= (34.43, 41.39)$

(e) $(\hat{\beta}_0+\hat{\beta}_1 x^*)\pm t_{0.025,23}\, s\sqrt{1+\dfrac{1}{n}+\dfrac{(x^*-\bar{x})^2}{S_{xx}}}$

$= (13.84, 49.68)$

12.91 (a)

Source of variation	Sum of squares	Degrees of freedom	Mean square	F	p Value
Regression	853.50	1	853.50	8.99	0.0103
Error	1234.23	13	94.94		
Total	2087.73	14			

H_0: There is no significant linear relationship.
H_a: There is a significant linear relationship.
TS: $F = MSR/MSE$; RR: $F \geq 4.67$
$f = 835.50/94.94 = 8.99 \geq 4.67$
$p = 0.0103 \leq 0.05$
There is evidence of a significant linear relationship. As Corruption Perceptions Index increases, so does the Environmental Sustainability Index.
(b) $E(Y|6.7) = 56.8902$
(c) $(\hat{\beta}_0+\hat{\beta}_1 x^*)\pm t_{0.025,23}\, s\sqrt{1+\dfrac{1}{n}+\dfrac{(x^*-\bar{x})^2}{S_{xx}}}$

$= (40.1554, 87.2630)$
Since the prediction interval is entirely below 90, it is unlikely a randomly selected country with a CPI of 8.2 will have an ESI greater than 90.

12.93 (a)

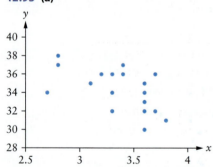

There appears to be a weak, negative, linear relationship.
(b) $\hat{\beta}_1 = -3.70$, $\hat{\beta}_0 = 46.91$
$y = 46.91 - 3.70x$
(c) $(\hat{\beta}_0+\hat{\beta}_1 x^*)\pm t_{0.005,19}\, s\sqrt{1+\dfrac{1}{n}+\dfrac{(x^*-\bar{x})^2}{S_{xx}}}$

$= (27.5775, 38.8625)$

(d) $(\hat{\beta}_0+\hat{\beta}_1 x^*)\pm t_{0.005,19}\, s\sqrt{1+\dfrac{1}{n}+\dfrac{(x^*-\bar{x})^2}{S_{xx}}}$

$= (29.2183, 40.1817)$
(e) The prediction interval in part (c) is wider than the prediction interval in part (d) because 3.7 is farther from the mean than is 3.3.

12.4 Regression Diagnostics

12.95 $\hat{e}_i = y_i - \hat{y}_i$

12.97 The points on the normal probability plot will fall along an approximate straight line.

12.99 The variance is not the same for each value of x and, therefore, not all the assumptions of linear regression are satisfied.

12.101 A histogram or stem plot of the residuals and the IQR/s ratio and Backward Empirical Rule may be used to assess the approximate normality of the residuals.

12.103 (a)

x_i	y_i	$\hat{y}_i$	$\hat{e}_i$
1.44	8.59	9.5798	−0.9898
1.98	8.40	11.4005	−3.0005
1.69	13.60	10.4227	3.1773
1.29	9.25	9.0740	0.1760
1.16	11.16	8.6357	2.5243
1.66	11.03	10.3215	0.7085
1.32	9.12	9.1752	−0.0552
1.94	13.30	11.2656	2.0344
1.42	8.77	9.5123	−0.7423
1.66	11.34	10.3215	1.0185
1.89	10.76	11.0970	−0.3370
1.31	5.58	9.1415	−3.5615
1.40	7.43	9.4449	−2.0149
1.04	8.64	8.2311	0.4089
1.32	8.89	9.1752	−0.2852
1.31	10.08	9.1415	0.9385

(b) $\sum \hat{e}_i = 0$

12.105 (a)

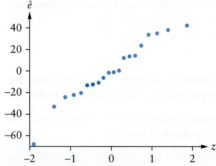

(b) There is some evidence to suggest the random error terms are not normal. There appears to be an outlier, and the points are slightly wavy.

12.107 (a) There is evidence to suggest a violation of the regression model assumptions. The obvious U-shape in the residual plot suggests the relationship between these variables is not linear.

(b) There is evidence to suggest a violation of the regression model assumptions. The curved pattern in the residual plot suggests the relationship between these variables is not linear.
(c) There is evidence to suggest a violation of the regression model assumptions. The sharp curve in the residual plot suggests the relationship between these variables is not linear.
(d) There is some evidence to suggest a violation of the regression model assumptions. The residual plot suggests the variance is not constant for each value of x because a large proportion of residuals appear to be below the line $y = 0$ (negative values) when they should be evenly distributed above and below it.

12.109 (a) $\hat{\beta}_1 = -3.5333$, $\hat{\beta}_0 = 463.3508$
$y = 463.3508 - 3.5333x$
(b)

x_i	y_i	$\hat{y}_i$	$\hat{e}_i$
71.5	51.7	210.7198	−159.0214
55.0	283.8	269.0193	14.7795
87.8	366.2	153.1271	213.0710
86.8	347.9	156.6604	191.2377
68.8	87.7	220.2598	−132.5613
68.4	57.5	221.6731	−164.1746
69.6	50.6	217.4331	−166.8346
82.1	203.7	173.2669	30.4313
89.4	424.4	147.4738	276.9243
71.1	79.0	212.1332	−133.1347
84.6	271.7	164.4336	107.2645
56.2	262.3	264.7793	−2.4806
69.6	70.1	217.4331	−147.3346
75.4	78.8	196.9400	−118.1416
57.4	212.6	260.5394	−47.9406
53.0	344.9	276.0859	68.8129
68.8	66.5	220.2598	−153.7613
51.7	403.4	280.6792	122.7197
51.5	394.9	281.3859	113.5130
57.4	216.0	260.5394	−44.5406
68.3	50.5	222.0264	−171.5279
50.7	427.6	284.2125	143.3864
81.6	184.4	175.0335	9.3647
51.0	414.8	283.1525	131.6464
54.9	284.2	269.3726	14.8262
53.4	341.3	274.6726	66.6262
51.8	397.4	280.3259	117.0730
76.3	86.6	193.7600	−107.1617
52.9	359.1	276.4392	82.6596
64.0	81.5	237.2196	−155.7210

(c)

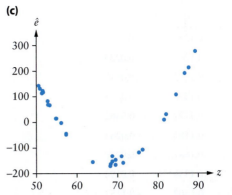

There is evidence to suggest a violation in the simple linear regression assumptions. There is a distinct curve in the residual plot.

12.111 (a)

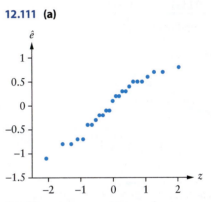

(b) There is some evidence of non-normality. Each end of the plot flattens out slightly.

12.113 (a) $\hat{\beta}_0 = 0.303575$, $\hat{\beta}_1 = -0.0000538$
$y = 0.303575 - 0.0000538x$

x_i	y_i	$\hat{y}_i$	$\hat{e}_i$
2511	0.138	0.1684	−0.0304
2817	0.157	0.1520	0.0050
3051	0.107	0.1394	−0.0324
3390	0.132	0.1211	0.0109
3191	0.124	0.1318	−0.0078
3583	0.093	0.1107	−0.0177
3402	0.117	0.1205	−0.0035
2927	0.124	0.1460	−0.0220
3083	0.218	0.1376	0.0804
3505	0.122	0.1149	0.0071
2751	0.203	0.1555	0.0475
2496	0.202	0.1692	0.0328
3043	0.194	0.1398	0.0542
2668	0.138	0.1600	−0.0220
2958	0.148	0.1444	0.0036
2536	0.167	0.1671	−0.0001
2749	0.159	0.1556	0.0034
3009	0.064	0.1416	−0.0776

x_i	y_i	$\hat{y}_i$	$\hat{e}_i$
3071	0.137	0.1383	−0.0013
3304	0.149	0.1257	0.0233
2494	0.231	0.1693	0.0617
2671	0.095	0.1598	−0.0648
3556	0.066	0.1122	−0.0462
3442	0.092	0.1183	−0.0263
2516	0.193	0.1682	0.0248
3140	0.164	0.1346	0.0294
2664	0.045	0.1602	−0.1152
3239	0.171	0.1292	0.0418
3378	0.123	0.1218	0.0012
3103	0.177	0.1366	0.0404

(b) Normal probability plot of residuals:

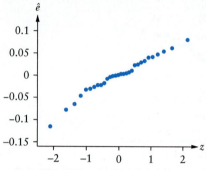

Scatter plot of residuals versus x:

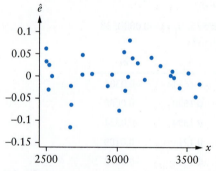

(c) There is some evidence to suggest the simple linear regression assumptions are not valid. The normal probability plot is nonlinear on the left end. There appears to be more variability in the residuals for smaller values of x suggesting that the variance is not consistent.

12.115 (a) $\hat{\beta}_0 = 43.6012$, $\hat{\beta}_1 = 15.2024$
$y = 43.6012 + 15.2024x$

x_i	y_i	$\hat{y}_i$	$\hat{e}_i$
126.2	1948.9	1962.14	−13.24
162.2	2822.5	2509.43	313.07
90.7	1211.4	1422.46	−211.06
202.6	2639.5	3123.61	−484.11
120.1	1651.9	1869.41	−217.51
41.3	770.2	671.46	98.74
281.4	4409.1	4321.56	87.54
61.2	712.6	973.99	−261.39
166.9	3018.8	2580.88	437.92
102.6	1525.8	1603.37	−77.57
88.3	1370.6	1385.97	−15.37
123.4	1873.7	1919.58	−45.88
132.3	1914.7	2054.88	−140.18
171.4	2646.9	2649.29	−2.39
257.1	3509.9	3952.14	−442.24
245.5	4289	3775.79	513.21
174.2	2823.6	2691.86	131.74
73.8	1125.9	1165.54	−39.64
207.4	2777.6	3196.58	−418.98
72.2	966	1141.21	−175.21
81.8	1351.6	1287.16	64.44
273.7	3997.6	4204.50	−206.90
59.3	813.3	945.10	−131.80
279.2	4370.6	4288.11	82.49
31.6	408.2	524.00	−115.80
297.5	5328.5	4566.32	762.18
111.3	1600.9	1735.63	−134.73
154.9	1894.3	2398.45	−504.15
358.1	5787.2	5487.58	299.62
133.6	2067.4	2074.64	−7.24
161	2950	2491.19	458.81
10.7	1458.8	206.27	1252.53
25.7	307.1	434.30	−127.20
276.4	3854.4	4245.54	−391.14
239.1	3316.1	3678.50	−362.40
50.8	684.8	815.88	−131.08
173.2	2831.7	2676.66	155.04

(b) Normal probability plot of residuals:

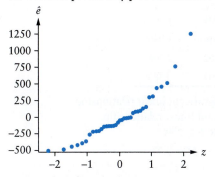

Scatter plot of residuals versus x:

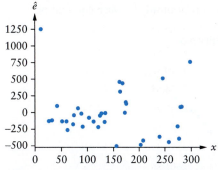

There is evidence to suggest the simple linear regression assumptions are not valid. There is a marked curve in the normal probability plot and the variance of the residuals does not appear to be constant for each x.

12.117 (a) $\hat{\beta}_1 = 14.5283$, $\hat{\beta}_0 = 738.0426$
$y = 738.0426 + 14.5283x$

x_i	y_i	$\hat{y}_i$	$\hat{e}_i$
29.8	1127	1170.9849	−43.9849
26.3	1105	1120.1359	−15.1359
26.0	1187	1115.7774	71.2226
37.2	1334	1278.4940	55.5060
30.1	1067	1175.3433	−108.3433
20.9	1094	1041.6833	52.3167
27.5	1115	1137.5698	−22.5698
39.8	1228	1316.2675	−88.2675
10.7	871	893.4950	−22.4950
35.6	1377	1255.2488	121.7512

(b) Normal probability plot of residuals:

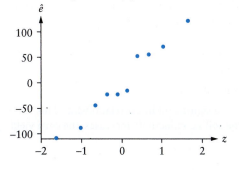

Scatter plot of residuals versus x:

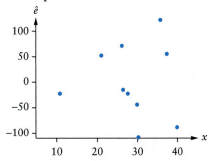

There is no overwhelming evidence that the simple linear regression assumptions are invalid. There is a possible outlier, but the number of observations is small.

12.119 (a) $\hat{\beta}_0 = 3.78$, $\hat{\beta}_1 = 15.57$
$y = 3.78 + 15.57x$

x_i	y_i	$\hat{y}_i$	$\hat{e}_i$
0.24	9.16	7.52	1.64
0.25	6.90	7.67	−0.77
0.16	6.22	6.27	−0.05
0.15	6.69	6.12	0.57
0.13	3.48	5.80	−2.32
0.23	6.64	7.36	−0.72
0.23	7.86	7.36	0.50
0.16	6.58	6.27	0.31
0.07	1.10	4.87	−3.77
0.11	6.67	5.49	1.18
0.22	8.08	7.21	0.87
0.07	3.59	4.87	−1.28
0.30	7.43	8.45	−1.02
0.11	5.47	5.49	−0.02
0.23	6.07	7.36	−1.29
0.25	6.20	7.67	−1.47
0.18	9.50	6.58	2.92
0.21	8.05	7.05	1.00
0.20	5.72	6.89	−1.17
0.14	6.46	5.96	0.50
0.11	5.82	5.49	0.33
0.20	6.67	6.89	−0.22
0.11	9.75	5.49	4.26
0.19	6.38	6.74	−0.36
0.09	7.12	5.18	1.94
0.20	9.32	6.89	2.43
0.08	2.59	5.03	−2.44
0.21	7.07	7.05	0.02
0.27	5.39	7.98	−2.59
0.29	9.31	8.30	1.01

(b) Normal probability plot of residuals:

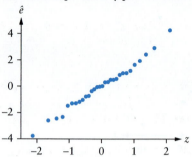

Scatter plot of residuals versus x:

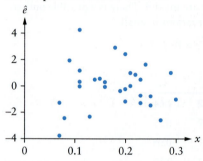

There is no evidence that the simple linear regression assumptions are invalid. The normal probability plot is relatively linear and there is no discernible pattern to the scatter plot of the residuals.

12.121 (a) $\hat{\beta}_1 = -0.0902$, $\hat{\beta}_0 = 29.8441$
$y = 29.8441 + -0.0902x$

x_i	y_i	$\hat{y}_i$	$\hat{e}_i$
28	25	27.3173	−2.3173
59	25	24.5199	0.4801
13	33	28.6709	4.3291
50	31	25.3321	5.6679
23	22	27.7685	−5.7685
16	26	28.4002	−2.4002
43	23	25.9638	−2.9638
24	32	27.6783	4.3217
48	16	25.5126	−9.5126
53	35	25.0614	9.9386
30	30	27.1369	2.8631
22	17	27.8588	−10.8588
39	37	26.3247	10.6753
39	19	26.3247	−7.3247
9	33	29.0319	3.9681
15	35	28.4905	6.5095
23	31	27.7685	3.2315
5	37	29.3929	7.6071
32	22	26.9564	−4.9564
12	19	28.7612	−9.7612
26	26	27.4978	−1.4978

x_i	y_i	$\hat{y}_i$	$\hat{e}_i$
57	23	24.7004	−1.7004
42	28	26.0540	1.9460
33	24	26.8662	−2.8662
58	25	24.6102	0.3898

(b) H_0: There is no significant linear relationship.
H_a: There is a significant linear relationship.
TS: $F = $ MSR/MSE; RR: $F \geq 2.94$
$f = 51.4402/37.8052 = 1.36 < 2.94$
$p = 0.2554 > 0.10$
There is no evidence of a significant linear relationship. There does not appear to be a relationship between commuting distance and sick hours.

(c) Normal probability plot:

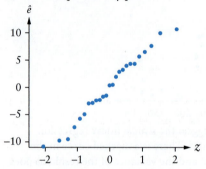

Scatter plot of residuals versus x:

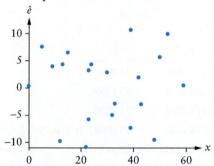

(d) The graphs do not provide any evidence that the simple linear regression assumptions are invalid. The normal probability plot is approximately linear, and the plot of the residuals versus the predictor variable exhibits no discernible pattern.

12.123 (a) $\hat{\beta}_1 = 0.0768$, $\hat{\beta}_0 = 94.1430$
$y = 94.1430 + 0.0768x$

(b) H_0: There is no significant linear relationship.
H_a: There is a significant linear relationship.
TS: $F = $ MSR/MSE; RR: $F \geq 4.38$
$f = 3997.1398/519.8347 = 7.69 \geq 4.38$
$0.01 \leq p \leq 0.05 \Rightarrow p = 0.0121$

There is evidence of a significant linear relationship. This suggests that as the soil's conductivity increases, the corn yield also increases.

(c) Normal probability plot:

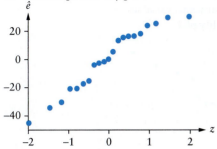

Scatter plot of residuals versus x:

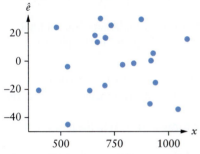

The graphs do not provide any evidence that the simple linear regression assumptions are invalid. The normal probability plot is approximately linear, and the plot of the residuals versus the predictor variable exhibits no discernible pattern.

12.125 $\sum_{i=1}^{n}(y_i - \hat{y}_i)$
$= \sum_{i=1}^{n}(y_i - (\hat{\beta}_0 + \hat{\beta}_1 x_i))$
$= \sum_{i=1}^{n}(y_i - (\bar{y} - \hat{\beta}_1 \bar{x} + \hat{\beta}_1 x_i))$
$= \sum_{i=1}^{n}(y_i - \bar{y}) - \hat{\beta}_1 \sum_{i=1}^{n}(x_i - \bar{x})$
$= n\bar{y} - n\bar{y} - \hat{\beta}_1(n\bar{x} - n\bar{x}) = 0$

12.5 Multiple Linear Regression

12.127 The unknown parameters $\beta_0, \beta_1, \ldots, \beta_k$

12.129 The principle of least squares

12.131 False

12.133 At least one of the predictor variables helps explain some of the variation in the dependent variable.

12.135 This assumes only one test is conducted.

12.137 The mean value of Y for $x = x^*$ and an estimate of an observed value of Y for $x = x^*$

12.139 True

12.141 (a) $E(Y) = -49.55$
(b) -61.5 units
(c) $E(Y) = -433.8$; $Y \sim N(-433.8, 24^2)$
$P(-460 \leq Y \leq 400) = 0.8625$

12.143 (a) y versus x_1

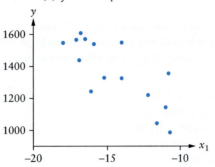

There is a slight negative linear relationship.
y versus x_2

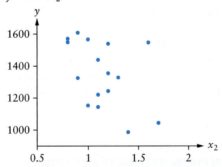

There is a very slight indication of a negative linear relationship.
y versus x_3

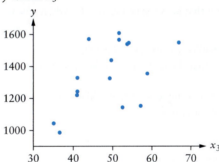

There is a slight indication of a positive linear relationship.
(b) $y = 221.9231 - 56.3497x_1 - 124.2215x_2 + 9.5798x_3$
The sign of each estimated regression coefficient reflects the relationship in each scatter plot.
(c) $E(Y) = 1121.6179$

12.145 (a) $H_0: \beta_1 = \cdots = \beta_4 = 0$
$H_a: \beta_i \neq 0$ for at least one i
TS: $F = $ MSR/MSE; RR: $F \geq 2.69$
$f = 8.91 \geq 2.69$
$p = 0.00007 \leq 0.05$
There is evidence to suggest that at least one of the regression coefficients is different from 0. The overall regression is significant.
(b) β_1 and β_2 are significantly different from 0. In these hypothesis tests to determine if the coefficient is significantly different from 0, $p \leq 0.05$. Therefore, there is evidence to suggest that x_1 and x_2 are significant predictor variables.

(c) The critical value in each test is $t_{0.05/8,30} = t_{0.0063,30} = 2.6574$.
Using the R output, β_1 and β_2 are significantly different from 0. Therefore, there is evidence to suggest that both are significant predictor variables. This result is the same as in part (b).

12.147 (a) $y = 114.4895 + 6.4722x_1 - 12.8017x_2 + 4.6091x_3 + 0.6409x_4$
$r^2 = 13{,}305.955/17{,}078.8733 = 0.7791$
(b) $\beta_1: t = 4.3402, p = 0.0015$
$\beta_2: t = -0.5592, p = 0.5883$
$\beta_3: t = 1.4814, p = 0.1693$
$\beta_4: t = 2.9202, p = 0.0153$
Using $\alpha = 0.05$, x_1 and x_4 are significant.
(c) $Y_i = \beta_0 + \beta_1 x_{1i} + \beta_4 x_{4i} + E_i$
$y = 105.4656 + 5.1074x_1 + 0.6863x_4$
$r^2 = 12{,}477.5792 / 17{,}078.8733 = 0.7306$
(d) The second model is better because it has fewer variables (is less complex) and the coefficient of determination is only slightly smaller.

12.149 (a) $y^* = 19.291$
$y^* \pm t_{0.025,22} s_{Y^*} = (15.199, 23.383)$
We are 95% confident that the true mean value of Y when $x = x^*$ lies in this interval.
(b) $y^* \pm t_{0.025,23} \sqrt{s^2 + s_{Y^*}^2}$
$= (11.487, 27.097)$
We are 95% confident that an observed value of Y when $x = x^*$ lies in this interval.

12.151 (a) $y = 749.6408 - 131.9821x_1 + 0.5227x_2$
$\hat{\beta}_1 < 0$: This suggests that the rental price is less for apartments farther from campus.
$\hat{\beta}_2 > 0$: This suggests that the rental price is higher for apartments with more square feet.
(b) $H_0: \beta_1 = \beta_2 = 0$
$H_a: \beta_i \neq 0$ for at least one i
TS: $F = $ MSR/MSE; RR: $F \geq 4.74$
$f = 227{,}338.5893/6260.4745 = 36.31 \geq 4.74$
$p = 0.0002 \leq 0.05$
There is evidence to suggest that at least one of the regression coefficients is different from 0. The overall regression is significant.
(c) $y^* = 1178.88$

12.153 (a) $y = -3.1136 + 0.0554x_1 + 0.5777x_2 + 0.0028x_3$
(b) $H_0: \beta_1 = \beta_2 = \beta_3 = 0$
$H_a: \beta_i \neq 0$ for at least one i
TS: $F = $ MSR/MSE; RR: $F \geq 3.49$
$f = 0.6564/0.1670 = 3.93 \geq 3.49$
$p = 0.0363 \leq 0.05$
There is evidence to suggest that at least one of the regression coefficients is different from 0. The overall regression is significant.
$\beta_1: t = 0.0554/0.0224 = 2.4743$
$p = 0.0293$
$\beta_2: t = 0.5777/0.2017 = 2.8634$
$p = 0.0143$
$\beta_3: t = 0.0028/0.0064 = 0.4409$
$p = 0.6671$

The temperature and concentration of the solutions are the most important (significant) variables.
(c) Normal probability plot:

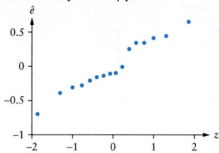

There is some evidence to suggest a violation of the multiple linear regression assumptions. The points in this plot are slightly nonlinear.

12.155 (a) $y = 3.2424 + 0.1462x_1 - 0.0104x_2 + 0.00001x_3$
(b) $\hat{\beta}_1 > 0$: As the number of wind farms increases, so does the number of nests.
$\hat{\beta}_2 < 0$: As the number of turbines increases, the number of nests decreases.
$\hat{\beta}_3 > 0$: As the area of the habitat increases, so does the number of nests.
(c) $H_0: \beta_1 = \beta_2 = \beta_3 = 0$
$H_a: \beta_i \neq 0$ for at least one i
TS: $F = $ MSR/MSE; RR: $F \geq 3.71$
$f = 15.573/8.1222 = 1.92 < 3.71$
$p = 0.1907 > 0.05$
There is no evidence to suggest that at least one of the regression coefficients is different from 0. The overall regression is not significant.
$r^2 = 46.7319/127.9543 = 0.3652$
(d) Even though the overall regression is not significant, the most important predictor appears to be x_2, the number of turbines.
$\beta_2: t = -2.0310, p = 0.0697$
(e) Normal probability plot:

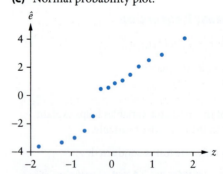

Scatter plot of residuals versus x_1:

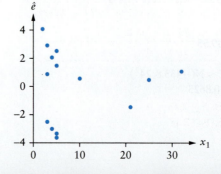

Scatter plot of residuals versus x_2:

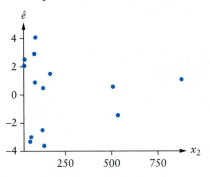

Scatter plot of residuals versus x_3:

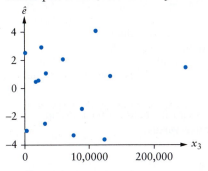

There appear to be some violations of the regression model assumptions. The points on the normal probability plot do not fall along a straight line, and the variance of residuals does not seem to be constant.

12.157 (a) $y = 2447.7015 + 51.2511x_1 - 10.8445x_2$
(b) $H_0: \beta_1 = \beta_2 = 0$
$H_a: \beta_i \neq 0$ for at least one i
TS: $F = $ MSR/MSE; RR: $F \geq 6.93$
$f = 91,242.149/11,142.8863 = 8.19 \geq 6.93$
$p = 0.0057 \leq 0.01$
There is evidence to suggest that at least one of the regression coefficients is different from 0. The overall regression is significant.
(c) $\hat{\beta}_1 > 0$: As the extent of sea ice increases, so does the size of the penguin colony population. $\hat{\beta}_2 < 0$: As the yearly total number of stormy days increases, the size of the penguin colony population decreases.
(d) $r^2 = 182,484.298/316,198.9333 = 0.5771$
Approximately 58% of the variation in y is explained by this regression model.
(e) $E(Y) = 2708.7832$

12.159 (a)

Source of variation	Sum of squares	Degrees of freedom	Mean square	F	p Value
Regression	342.970	5	68.594	5.79	0.0004
Error	462.340	39	11.855		
Total	805.310	44			

$H_0: \beta_1 = \cdots = \beta_5 = 0$
$H_a: \beta_i \neq 0$ for at least one i
TS: $F = $ MSR/MSE; RR: $F \geq 2.46$
$f = 68.594/11.855 = 5.79 \geq 2.48$
$p = 0.004 \leq 0.05$

There is evidence that at least one of the regression coefficients is different from 0.
$r^2 = 342.970/805.310 = 0.4259$
The overall model is significant. However, only approximately 43% of the variation in the data is explained by the regression model.
(b) $\beta_1: t = 2.91, p = 0.0059$
$\beta_2: t = -1.76, p = 0.0862$
$\beta_3: t = 0.11, p = 0.9130$
$\beta_4: t = -4.04, p = 0.0002$
$\beta_5: t = -0.83, p = 0.4116$
The p values suggest the most important predictor is x_4, total hardness.
Using the signs of the estimated regressions coefficients:
As the amount of total dissolved solids increases, chloride in groundwater increases. As iron increases, chloride in groundwater decreases. As nitrate increases, chloride in groundwater increases. As total hardness increases, chloride in groundwater decreases. As depth of the sample increases, chloride in groundwater decreases.

12.161 (a)
$y = 9.3465 - 0.0618x_1 - 0.0812x_2 + 0.9117x_2 - 0.0837x_4$
(b) $H_0: \beta_1 = \beta_2 = \beta_3 = \beta_4 = 0$
$H_a: \beta_i \neq 0$ for at least one i
TS: $F = $ MSR/MSE; RR: $F \geq 3.06$
$f = 3.70 \geq 3.06$
$p = 0.0275 \leq 0.05$
There is evidence that at least one of the regression coefficients is different from 0.
$r^2 = 1.4392/2.8990 = 0.4965$
(c) $\beta_1: t = -0.9881, p = 0.3388$
$\beta_2: t = -1.6693, p = 0.1158$
$\beta_3: t = 3.3806, p = 0.0041$
$\beta_4: t = -1.2554, p = 0.2286$
Only x_3 is a significant predictor variable.
(d) $H_0: \beta_3 = 0.93, H_a: \beta_1 < 0.93$
TS: $T = (B_3 - 0.93)/S_{B_3}$; RR: $T \leq -1.7531$
$t = \dfrac{0.9117 - 0.93}{0.2697} = -0.0679 > -1.7531$
$p = 0.4733 > 0.05$
There is no evidence to suggest that $\beta_3 < 0.93$.
(e) $y^* = 1.3849$
$y^* \pm t_{0.025,15} s_{Y^*} = (1.05, 1.72)$

12.163 (a)

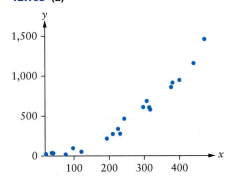

(b) $y = -215.1226 + 2.9043x$

Source of variation	Sum of squares	Degrees of freedom	Mean square	F	p Value
Regression	3,105,671.6	1	3,105,671.6	201.69	<0.0001
Error	277.162.9	18	15,397.9		
Total	3,382,834.5	19			

(c) Small depths (≤ 150):
$y = 7.4516 + 0.4828x$

Source of variation	Sum of squares	Degrees of freedom	Mean square	F	p Value
Regression	1720.68	1	1720.68	2.47	0.1915
Error	2792.15	4	698.04		
Total	4512.83	5			

Large depths (>150): $y = -598.5726 + 4.0383x$

Source of variation	Sum of squares	Degrees of freedom	Mean square	F	p Value
Regression	1,654,572.7	1	1,654,572.7	309.29	<0.0001
Error	64.194.7	12	5349.6		
Total	1,718,767.4	13			

(d) $y = 7.3195 + 0.0062x^2$
(e) The model using x^2 seems to be the best. A model with x^2 appears to fit the scatter plot, and the value of r^2 for this model is 0.9869, which is very high.

Chapter 12 Exercises

12.165 (a) $\hat{\beta}_1 = 2.1497, \hat{\beta}_0 = 2.6533$
$y = 2.6533 + 2.1497x$
(b) $E(Y|10) = 24.15$
(c) $y = 6.953$

12.167 (a) $y = 302.3637 - 0.0629x$

Source of variation	Sum of squares	Degrees of freedom	Mean square	F	p Value
Regression	44580.81	1	44580.81	21.15	0.0001
Error	48475.33	23	2107.62		
Total	93056.14	24			

(b) H_0: There is no significant linear relationship.
H_a: There is a significant linear relationship.
TS: $F = MSR/MSE$; RR: $F \geq 7.88$
$f = 44,580.81/21107.62 = 21.15 \geq 7.88$
$p = 0.0001 \leq 0.01$

There is evidence of a significant linear relationship.
(c) $r^2 = 44,580.81/93,056.14 = 0.4791$
(d) The overall regression is significant, explaining approximately 48% of the variation in the model. Omega-3 is a good predictor of blood level of beta amyloids.

12.169 (a)

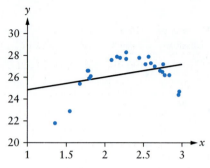

The relationship between x and y appears to be quadratic.
(b) $\hat{\beta}_1 = 1.1767, \hat{\beta}_0 = 23.6854$
$y = 23.6853 + 1.1767x$
H_0: There is no significant linear relationship.
H_a: There is a significant linear relationship.
TS: $F = MSR/MSE$; RR: $F \geq 4.30$
$f = 3.08 < 4.30$
$p = 0.0932 > 0.05$
There is no evidence of a significant linear relationship.
(c) Scatter plot of residuals versus x:

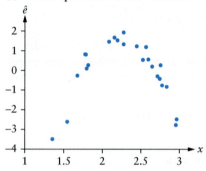

The graphs in parts (a) and (c) suggest the model could be improved by adding a quadratic term, x^2.

12.171 (a) $\hat{\beta}_1 = 0.6204, \hat{\beta}_0 = 8.3870$
$y = 8.3870 + 0.6024x$
(b)

Source of variation	Sum of squares	Degrees of freedom	Mean square	F	p Value
Regression	67.39	1	67.39	2.29	0.1370
Error	1414.44	48	29.47		
Total	1481.83	49			

(c) $H_0: \beta_1 = 0, H_a: \beta_1 \neq 0$
TS: $T = B_1/S_{B_1}$; RR: $|T| \geq 2.0106$
$t = \dfrac{0.6204}{5.4286/\sqrt{185.725}} = 1.5123 < 2.0106$
$p = 0.1370 > 0.05$
There is no evidence to suggest that $\beta_1 \neq 0$, the regression line is not significant.
(d) No. There is no significant relationship between rating and price.

12.173 (a) $\hat{\beta}_1 = 1.2584, \hat{\beta}_0 = 4.3207$
$y = 4.3207 + 1.2584x$

Source of variation	Sum of squares	Degrees of freedom	Mean square	F	p Value
Regression	104.707	1	104.707	27.31	<0.0001
Error	107.359	28	3.834		
Total	212.066	29			

The later people start shopping, the more money they spend. Retailers should entice people to start shopping later in the day.

(c) $y^* \pm t_{0.025,28}\, s\sqrt{1 + \dfrac{1}{n} + \dfrac{(x^* - \bar{x})^2}{S_{xx}}}$
$= (2.7201, 10.9526)$

(d) $H_0: y^* = 7.25,\ H_a: y^* > 7.25$
TS: $T = \dfrac{(B_0 + B_1 x^*) - y_0^*}{s\sqrt{(1/n) + [(x^* - \bar{x})^2/S_{xx}]}}$; RR: $T \geq 1.7011$

$t = \dfrac{8.096 - 7.25}{(1.9582)\sqrt{\dfrac{1}{30} + \dfrac{(2 - 3.1433)^2}{65.9737}}} = 2.3553 \geq 1.7011$

$p = 0.0129 \leq 0.05$
There is evidence to suggest the true mean spent is greater than 7.25 ($725).

12.175 (a)

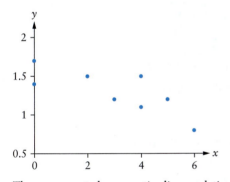

There appears to be a negative linear relationship.
(b) $\hat{\beta}_1 = -0.0924, \hat{\beta}_0 = 1.6096$
$y = 1.6096 - 0.0924x$
$H_0: \beta_1 = 0,\ H_a: \beta_1 \neq 0$
TS: $T = B_1/S_{B_1}$; RR: $|T| \geq 2.3646$
$t = -0.0924/0.0349 = -2.6441 \leq -2.3646$
$p = 0.0332 \leq 0.05$
There is evidence to suggest that $\beta_1 \neq 0$, the regression line is significant.
(c) Normal probability plot:

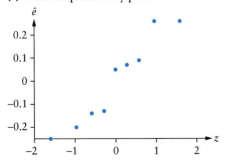

Scatter plot of residuals versus x:

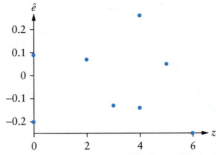

The normal probability plot suggests a violation in the normality assumption, as the graph has a slight curve. However, this is also a small sample size.

12.177 (a)

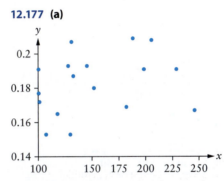

There does not appear to be a linear relationship. The scatter plot appears random.
(b) $\hat{\beta}_1 = 0.0001, \hat{\beta}_0 = 0.1673$
$y = 0.1673 + 0.0001x$

Source of variation	Sum of squares	Degrees of freedom	Mean square	F	p Value
Regression	0.0004	1	0.0004	1.15	0.3008
Error	0.0047	15	0.0003		
Total	0.0050	16			

From the ANOVA table, the F test is not significant ($p = 0.3008$). There is no evidence to suggest a significant linear relationship.

12.179 Estimated regression line:
$y = 14.8391 - 0.2436x$

Source of variation	Sum of squares	Degrees of freedom	Mean square	F	p Value
Regression	228.53	1	228.53	7.32	0.0095
Error	1466.91	47	31.21		
Total	1695.44	48			

From the ANOVA table, the F test is significant ($p = 0.0095$). There is evidence to suggest a significant linear relationship.

Normal probability plot:

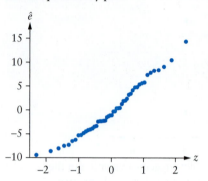

Scatter plot of residuals versus x:

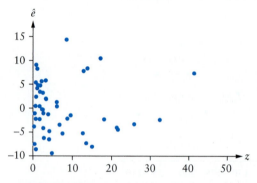

There do not appear to be any violations of the assumptions. The sign of $\hat{\beta}_1$ does not seem appropriate. One would think that as the percentage of freshwater increases, the fishing participation rate will also increase and that the sign of $\hat{\beta}_1$ will be positive.

12.181 (a) $y = 7274.5117 - 971.4403x_1 - 69.2220x_2 - 64.1724x_3 + 32.1604x_4$

$\hat{\beta}_1 < 0$: As x_1 increases, y decreases.
$\hat{\beta}_2 < 0$: As x_2 increases, y decreases.
$\hat{\beta}_3 < 0$: As x_3 increases, y decreases.
$\hat{\beta}_4 > 0$: As x_4 increases, y increases.

(b)

Source of variation	Sum of squares	Degrees of freedom	Mean square	F	p Value
Regression	4,445,366	4	1,111,342	74.81	<0.0001
Error	297,126	20	14,856		
Total	4,742,493	24			

$H_0: \beta_1 = \ldots = \beta_4 = 0$
$H_a: \beta_i \neq 0$ for at least one i
TS: $F = MSR/MSE$; RR: $F \geq 2.87$
$f = 74.81 \geq 2.87$
$p = 0.0000000000096665$

There is evidence to suggest that at least one of the regression coefficients is different from 0. The overall regression is significant.

(c) $\beta_1: t = -1.2730, p = 0.2176$
$\beta_2: t = -0.7246, p = 0.4771$
$\beta_3: t = -12.4987, p < 0.0001$
$\beta_4: t = 12.4091, p < 0.0001$

The variables x_3 and x_4 are significant predictors.

(d) Normal probability plot:

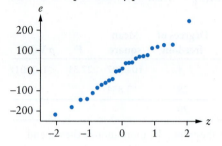

Scatter plot of residuals versus x_1:

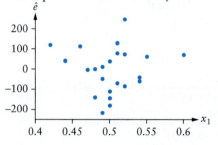

Scatter plot of residuals versus x_2:

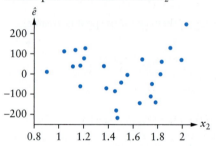

Scatter plot of residuals versus x_3:

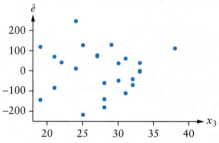

Scatter plot of residuals versus x_4:

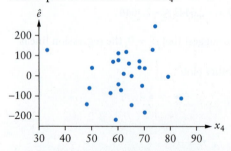

The plots suggest a possible outlier, and there is some evidence of nonconstant variance.

12.183 (a)

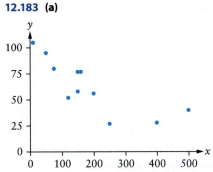

There appears to be a moderately strong, negative, linear relationship between x and y.
(b) (i) $E(Y \mid 300) = 50.57$
(ii) For $x = 300, Y \sim N(50.57, 14^2)$
$P(Y < 50) = 0.4836$
(c) $\hat{\beta}_0 = 89.9043, \hat{\beta}_1 = -0.1311$
$y = 89.9043 - 0.1311x$
$E(Y \mid 175) = 66.963$
(d)

Source of variation	Sum of squares	Degrees of freedom	Mean square	F	p Value
Regression	3720.69	1	3720.69	19.78	0.0021
Error	1504.91	8	188.11		
Total	5225.60	9			

$r^2 = 0.7120$
Approximately 71% of the variation in noise level can be explained by the regression.
(e) H_0: There is no significant linear relationship.
H_a: There is a significant linear relationship.
TS: $F = $ MSR/MSE; RR: $F \geq 5.3177$
$f = 3720.69/188.11 = 19.78 \geq 5.3177$
$p = 0.0021 \leq 0.05$
There is evidence to suggest a significant linear relationship between distance and noise level.
$H_0: \beta_1 = 0, H_a: \beta_1 \neq 0$
TS: $T = B_1/S_{B_1}$; RR: $|T| \geq 2.3060$
$t = \dfrac{-0.1311}{\sqrt{188.11/216502.5}} = -4.4473 \leq -2.3060$
$p = 0.0021$
There is evidence to suggest that $\beta_1 \neq 0$.
(f) $r = -0.8438$
Because $r < 0$, the correlation coefficient suggests a negative linear relationship between x and y
(g) (i) $H_0: y^* = 70, H_a: y^* > 70$
TS: $T = \dfrac{(B_0 + B_1 x^*) - y_0^*}{s\sqrt{(1/n) + [(x^* - \bar{x})^2/S_{xx}]}}$; RR: $T \geq 1.8595$
$t = -0.8491 < 1.85$
$p = 0.7897 > 0.05$
There is no evidence to suggest that the mean noise level is greater than 70 dB at a distance of 180 m.

(ii) $y^* \pm t_{0.025, 8}\, s\sqrt{\dfrac{1}{n} + \dfrac{(x^* - \bar{x})^2}{S_{xx}}}$
$= (50.1494, 70.6642)$
(h) $(\hat{\beta}_0 + \hat{\beta}_1 x^*) \pm t_{0.025,\, 8}\, s\sqrt{1 + \dfrac{1}{n} + \dfrac{(x^* - \bar{x})^2}{S_{xx}}}$
$= (43.0450, 110.5436)$
(i)

x_i	y_i	$\hat{y}_i$	$\hat{e}_i$
10	105	88.593	16.407
50	95	83.349	11.651
75	80	80.072	−0.072
120	52	74.172	−22.172
150	58	70.239	−12.239
160	77	68.928	8.072
200	56	63.684	−7.684
250	57	57.129	−0.129
400	28	37.464	−9.464
500	40	24.354	15.646

Normal probability plot:

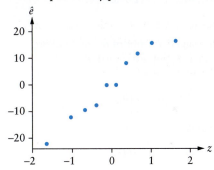

Scatter plot of residuals versus x:

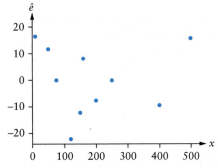

There is some evidence to suggest that the assumptions for simple linear regression are not met. There is a curve in the normal probability plot, and the scatter plot of the residuals shows that the variance of y does not remain constant.

Chapter 13
13.1 Univariate Categorical Data, Goodness-of-Fit Tests

13.1 False

13.3 Small

13.5 At least 5

13.7 The sample size, n

13.9

Category	Expected cell count
1	$(1000)(0.15) = 150$
2	$(1000)(0.14) = 140$
3	$(1000)(0.31) = 310$
4	$(1000)(0.23) = 230$
5	$(1000)(0.17) = 170$

13.11 (a) H_0: $p_1 = 0.175, p_2 = 0.171, p_3 = 0.162, p_4 = 0.225,$
$p_5 = 0.202, p_6 = 0.065$
H_a: $p_i \neq p_{i0}$ for at least one i
TS: $X^2 = \sum_{i=1}^{6}(n_i - e_i)^2/e_i$; RR: $X^2 \geq 11.0705$
(b) $X^2 = 2.7994 < 11.0705$
There is no evidence to suggest any one of the population proportions differs from its hypothesized value.
(c) $p > 0.10$

13.13 H_0: $p_1 = 0.25, p_2 = 0.15, p_3 = 0.10,$
$p_4 = 0.50$
H_a: $p_i \neq p_{i0}$ for at least one i
TS: $X^2 = \sum_{i=1}^{4}(n_i - e_i)^2/e_i$; RR: $X^2 \geq 7.8147$
$X^2 = 11.6 \geq 7.8147$
$p = 0.0089 < 0.05$
There is evidence to suggest at least one of the population proportions differs from its hypothesized value.

13.15 H_0: $p_1 = 0.413, p_2 = 0.275, p_3 = 0.312$
H_a: $p_i \neq p_{i0}$ for at least one i
TS: $X^2 = \sum_{i=1}^{4}(n_i - e_i)^2/e_i$; RR: $X^2 \geq 9.2103$
$X^2 = 4.3033 < 9.2103$
$p = 0.1163 > 0.01$
There is no evidence to suggest that any of the percentages of each type of investor is any different than historical percentages.

13.17 H_0: $p_1 = 0.10, p_2 = 0.0.20, p_3 = 0.25, p_4 = 0.05, p_5 = 0.40$
H_a: $p_i \neq p_{i0}$ for at least one i
TS: $X^2 = \sum_{i=1}^{4}(n_i - e_i)^2/e_i$; RR: $X^2 \geq 9.4877$
$X^2 = 10.5927 > 9.4877$
$p = 0.0315 < 0.05$
There is evidence to suggest that at least one of the true proportions of style is different than the hypothesized values.

13.19 H_0: $p_1 = 0.14, p_2 = 0.06, p_3 = 0.37, p_4 = 0.30, p_5 = 0.13$
H_a: $p_i \neq p_{i0}$ for at least one i
TS: $X^2 = \sum_{i=1}^{5}(n_i - e_i)^2/e_i$; RR: $X^2 \geq 9.4877$
$X^2 = 6.0884 < 9.4877$
$p = 0.1926 > 0.05$
There is no evidence to suggest any one of the population proportions differs from its hypothesized value.

13.21 H_0: $p_1 = 0.11, \ldots, p_{10} = 0.04$
H_a: $p_i \neq p_{i0}$ for at least one i
TS: $X^2 = \sum_{i=1}^{5}(n_i - e_i)^2/e_i$; RR: $X^2 \geq 21.666$
$X^2 = 203.766 \geq 21.666$
$p = 0 < 0.01$
There is strong evidence to suggest that at least one of the true proportions of occupations is different than the hypothesized value.

13.23 H_0: $p_1 = 0.40, p_2 = 0.30, p_3 = 0.15, p_4 = 0.15$
H_a: $p_i \neq p_{i0}$ for at least one i
TS: $X^2 = \sum_{i=1}^{4}(n_i - e_i)^2/e_i$; RR: $X^2 \geq 11.3449$
$X^2 = 4.9622 < 11.3449$
$p = 0.1746 > 0.01$
There is no evidence to suggest at least one of the population proportions differs from its hypothesized value.

13.25 (a) H_0: $p_1 = 0.52, p_2 = 0.24, p_3 = 0.14, p_4 = 0.06, p_5 = 0.04$
H_a: $p_i \neq p_{i0}$ for at least one i
TS: $X^2 = \sum_{i=1}^{5}(n_i - e_i)^2/e_i$; RR: $X^2 \geq 9.4877$
$X^2 = 26.6083 \geq 9.4877$
There is evidence to suggest that at least one of the true population proportions in Florida differs from those found in the national survey.
(b) $p = 0.0000238526 < 0.05$
(c) The type of housing that contributed most to the chi square statistic is the single-family detached home, large yard, where we observed fewer than expected. This makes sense, as seniors are unlikely to want to mow and maintain a large yard.

13.27 H_0: $p_1 = \cdots = p_{10} = 0.10$
H_a: $p_i \neq p_{i0}$ for at least one i
TS: $X^2 = \sum_{i=1}^{10}(n_i - e_i)^2/e_i$; RR: $X^2 \geq 16.9190$
$X^2 = 26.1369 \geq 16.9190$
$0.001 \leq p \leq 0.005$
There is evidence to suggest that at least one character is more popular than the others.

13.29 **(a)** $H_0: p_1 = 0.3371, \cdots, p_7 = 0.2724$
$H_a: p_i \neq p_{i0}$ for at least one i
TS: $X^2 = \sum_{i=1}^{10}(n_i - e_i)^2/e_i$; RR: $X^2 \geq 12.5916$
$X^2 = 17.1455 \geq 12.5916$
There is evidence to suggest that the true market share proportions have changed since 2017.
(b) $p = 0.0088 < 0.05$
(c) Apple and Samsung make the largest contributions to the test statistic. We observed more Apple shares than expected and fewer Samsung shares than expected. These appear to be the two vendor's proportions that have changed the most.

13.31 **(a)** Each country's proportion is calculated as $\dfrac{\text{oberved count}}{72{,}586{,}385}$.
(b) (a) $H_0: p_1 = 0.0033, \cdots, p_{40} = 0.0108$
$H_a: p_i \neq p_{i0}$ for at least one i
TS: $X^2 = \sum_{i=1}^{10}(n_i - e_i)^2/e_i$; RR: $X^2 \geq 54.5722$
$X^2 = 1{,}104{,}603 \leq 54.5722$
There is very strong evidence to suggest a change in the proportions of automobiles produced in at least one country has changed.
(c) $p \approx 0$

13.2 Bivariate Categorical Data: Tests for Homogeneity and Independence

13.33 Random samples are obtained from two or more populations, and each individual is classified by values of a categorical variable.
In a single random sample, a value of each of two categorical variables is recorded for each individual or object.

13.35 The true category proportions are the same for all populations.

13.37 $\dfrac{(i\text{th row total})(j\text{th column total})}{\text{grand total}}$

13.39 True

13.41 **(a)** $\chi^2_{0.05,6} = 12.5916$
(b) $\chi^2_{0.01,5} = 15.0863$
(c) $\chi^2_{0.025,6} = 14.4494$
(d) $\chi^2_{0.001,8} = 26.1245$

13.43

	1	2	3	4	Total
1	18	14	18	15	65
2	25	21	16	12	74
3	32	33	26	28	119
Total	75	68	60	55	258

13.45 H_0: Variable 1 and variable 2 are independent.
H_a: Variable 1 and variable 2 are dependent.
TS: $X^2 = \sum_{i=1}^{4}\sum_{j=1}^{4}\dfrac{(n_{ij} - e_{ij})^2}{e_{ij}}$; RR: $X^2 \geq 16.9190$

$X^2 = 16.8226 < 16.9190$
$p = 0.0516 > 0.05$
It's close, but there is no evidence to suggest the two categorical variables are dependent.

13.47 H_0: The true product proportions are the same for all stores.
H_a: The true product proportions are not the same for all stores.
TS: $X^2 = \sum_{i=1}^{3}\sum_{j=1}^{4}\dfrac{(n_{ij} - e_{ij})^2}{e_{ij}}$; RR: $X^2 \geq 16.8119$
$X^2 = 9.2674 < 16.8119$
$p = 0.1591 > 0.01$
There is no evidence to suggest the true proportion of each favorite differs by grocery store.

13.49 H_0: The true funding opinion proportions are the same for all regions.
H_a: The true funding opinion proportions are not the same for all regions.
TS: $X^2 = \sum_{i=1}^{4}\sum_{j=1}^{4}\dfrac{(n_{ij} - e_{ij})^2}{e_{ij}}$; RR: $X^2 \geq 16.9190$
$X^2 = 17.8692 \geq 16.9190$
$p = 0.0367 < 0.05$
There is evidence to suggest funding opinion differs by region.

13.51 H_0: Food and wine choice are independent.
H_a: Food and wine choice are dependent.
TS: $X^2 = \sum_{i=1}^{2}\sum_{j=1}^{2}\dfrac{(n_{ij} - e_{ij})^2}{e_{ij}}$; RR: $X^2 \geq 3.8415$
$X^2 = 11.2179 \geq 3.8415$
$p = 0.0008 < 0.05$
There is evidence to suggest that food and wine are dependent. This suggests that diners are still following the traditional food-and-wine pairings.

13.53 **(a)** H_0: Draft round and position are independent.
H_a: Draft round and position are dependent.
TS: $X^2 = \sum_{i=1}^{4}\sum_{j=1}^{3}\dfrac{(n_{ij} - e_{ij})^2}{e_{ij}}$; RR: $X^2 \geq 16.8119$
$X^2 = 23.1435 \geq 16.8119$
$p = 0.0007 < 0.01$
There is evidence to suggest that draft round and position are dependent.
(b) Using the contributions to the chi-square test statistic, the special teams have the largest discrepancy between observed players and expected players. Fewer special teams players than expected were drafted in rounds 1 and 2, suggesting that teams are more likely to wait until later rounds to draft special teams members.

13.55 H_0: Age group and influenza type are independent.
H_a: Age group and influenza type are dependent.
TS: $X^2 = \sum_{i=1}^{25}\sum_{j=1}^{34}\dfrac{(n_{ij} - e_{ij})^2}{e_{ij}}$; RR: $X^2 \geq 26.2170$
$X^2 = 383.1536 \geq 26.2170$
$p = 0 < 0.01$

There is very strong evidence to suggest that age group and influenza type are dependent.

13.57 H_0: Type of violation and type of pool are independent.
H_a: Type of violation and type of pool are dependent.
TS: $X^2 = \sum_{i=1}^{10}\sum_{j=1}^{3}\frac{(n_{ij}-e_{ij})^2}{e_{ij}}$; RR: $X^2 \geq 34.8053$
$X^2 = 127.3715 \geq 34.8053$
$p < 0.0001$
There is strong evidence to suggest that the type of violation and the type of pool are dependent.

13.59 H_0: Familiarity with the UN and country are independent.
H_a: Familiarity with the UN and country are dependent.
TS: $X^2 = \sum_{i=1}^{6}\sum_{j=1}^{4}\frac{(n_{ij}-e_{ij})^2}{e_{ij}}$; RR: $X^2 \geq 30.5779$
$X^2 = 981.9745 \geq 30.5779$
$p = 0 < 0.01$
There is strong evidence to suggest that familiarity with the UN and the country are dependent.

13.61 (a) H_0: $p_1 - p_2 = 0$, H_a: $p_1 - p_2 \neq 0$
TS: $Z = \dfrac{\hat{p}_1 - \hat{p}_2}{\sqrt{\hat{p}_c(1-\hat{p}_c)\left(\frac{1}{n_1}+\frac{1}{n_2}\right)}}$
$z = -2.5036$
p-value: $0.0123 < 0.05$
There is evidence to suggest that the proportions of adults who prefer free shipping in California and New Jersey are different.
(b)

	Prefer free shipping	Do not prefer free shipping	Total
California	135	365	500
New Jersey	204	396	600
Total	339	761	1100

H_0: The true proportions of shipping preference are the same for both states.
H_a: The true proportions of shipping preference are not the same for both states.
TS: $X^2 = \sum_{i=1}^{2}\sum_{j=1}^{2}\frac{(n_{ij}-e_{ij})^2}{e_{ij}}$
$X^2 = 6.268$
$p = 0.0123 < 0.05$
There is evidence to suggest that the proportions of shipping preferences are not the same for these two states.
(c) $z^2 = (-2.5036)^2 = 6.268 = \chi^2$
The p values are the same. These relationships make sense because we are examining the difference of two population proportions in both tests.

Chapter 13 Exercises

13.63 H_0: $p_1 = p_2 = \ldots = p_5 = 0.20$
H_a: $p_i \neq p_{i0}$ for at least one i
TS: $X^2 = \sum_{i=1}^{5}(n_i - e_i)^2/e_i$; RR: $X^2 \geq 9.4877$

$X^2 = 10.9565 \geq 9.4877$
$p = 0.0271 < 0.05$
There is evidence to suggest that at least one of these flavors is selected as a favorite at a different rate than the others. There is evidence of a preference in flavors.

13.65 H_0: $p_1 = 0.20, p_2 = 0.20, p_3 = 0.10$, $p_4 = 0.50$
H_a: $p_i \neq p_{i0}$ for at least one i
TS: $X^2 = \sum_{i=1}^{4}(n_i - e_i)^2/e_i$; RR: $X^2 \geq 11.3449$
$X^2 = 1.8990 < 11.3449$
$p = 0.5936 > 0.01$
There is no evidence to suggest the data are inconsistent with the past proportions.

13.67 H_0: $p_1 = 0.43, \ldots, p_9 = 0.01$
H_a: $p_i \neq p_{i0}$ for at least one i
TS: $X^2 = \sum_{i=1}^{9}(n_i - e_i)^2/e_i$; RR: $X^2 \geq 20.0902$
$X^2 = 56.9869 \geq 20.0902$
$p < 0.0001$
There is evidence to suggest that at least one of the Carleton true population proportions is different from the comparison group proportions.

13.69 (a) H_0: Current satisfaction level and attorney category are independent.
H_a: Current satisfaction level and attorney category are dependent.
TS: $X^2 = \sum_{i=1}^{3}\sum_{j=1}^{3}\frac{(n_{ij}-e_{ij})^2}{e_{ij}}$; RR: $X^2 \geq 13.2767$
$X^2 = 16.0672 \geq 13.2767$
There is evidence to suggest that current satisfaction level is associated with attorney category.
(b) $p = 0.0029$
(c) Attorneys who work for the government tend to be the most satisfied with their job. The cell labeled "A great deal" for the government employees contributed the most to the chi square statistic, and we observed more than were expected..

13.71 H_0: Number of days they could survive and household tenure are independent.
H_a: Number of days they could survive and household tenure are dependent.
TS: $X^2 = \sum_{i=1}^{7}\sum_{j=1}^{3}\frac{(n_{ij}-e_{ij})^2}{e_{ij}}$; RR: $X^2 \geq 26.2170$
$X^2 = 113.3284 \geq 26.2170$
$p < 0.0001$ There is very strong evidence to suggest that the number of days adults say they could survive is associated with household tenure. The table suggests that owner-occupied households could survive longer than expected.

13.73 H_0: Survival status and class are independent.
H_a: Survival status and class are dependent.
TS: $X^2 = \sum_{i=1}^{4}\sum_{j=1}^{2}\frac{(n_{ij}-e_{ij})^2}{e_{ij}}$; RR: $X^2 \geq 7.8147$

$X^2 = 190.4011 \geq 7.8147$
$p < 0.0001$
There is very strong evidence to suggest an association between class and survival status.

13.75 H_0: The true proportion of discarded paper bags by source is the same for all cities.
H_a: The true proportion of discarded paper bags by source is not the same for all cities.
TS: $X^2 = \sum_{i=1}^{5}\sum_{j=1}^{3}\frac{(n_{ij}-e_{ij})^2}{e_{ij}}$; RR: $X^2 \geq 15.5073$
$X^2 = 112.2213 \geq 15.5073$
$p < 0.0001$
There is very strong evidence to suggest that the proportion of discarded paper bags by source is not the same for all cities.

13.77 H_0: Borough and type of transportation are independent.
H_a: Borough and type of transportation are dependent.
TS: $X^2 = \sum_{i=1}^{5}\sum_{j=1}^{4}\frac{(n_{ij}-e_{ij})^2}{e_{ij}}$; RR: $X^2 \geq 21.0261$
$X^2 = 11.2908 < 21.0261$
$p = 0.5042 > 0.05$
There is no evidence to suggest that borough and type of transportation are dependent.

Chapter 14

14.1 The Sign Test

14.1 Distribution-free

14.3 True

14.5 The underlying distribution is continuous and symmetric.

14.7 Because the binomial distribution is discrete, we usually cannot find critical values to yield an exact level-α test.

14.9 The median is the 50th percentile of a data set, so if $\tilde{\mu}_0$ is the true population median, then it should be the 50th percentile of the data set.

14.11 (a) TS: $X = $ the number of observations greater than 16
$x = 2, p = 0.0547 > 0.05$
There is no evidence to suggest that the population median is less than 16.
(b) TS: $X = $ the number of observations greater than -25
$x = 9, p = 0.0730 > 0.01$
There is no evidence to suggest that the population median is greater than -25.
(c) TS: $X = $ the number of observations greater than 8
$x = 13, p = 0.2632 > 0.05$
There is no evidence to suggest that the population median is different from 8.

(d) TS: $X = $ the number of observations greater than 125
$x = 20, p = 0.0041 \leq 0.01$
There is evidence to suggest that the population median is different from 125.

14.13 TS: $X = $ the number of differences greater than 3
$x = 7, p = 0.0433 > 0.01$
There is no evidence to suggest that $\tilde{\mu}_1 - \tilde{\mu}_2 \neq 3$.

14.15 H_0: $\tilde{\mu} = 2$, H_a: $\tilde{\mu} < 2$
TS: $X = $ the number of observations greater than 2
$x = 9, p = 0.4119 > 0.10$
There is no evidence to suggest that the median thickness is less than 2 mm.

14.17 H_0: $\tilde{\mu} = 24$, H_a: $\tilde{\mu} \neq 24$
TS: $X = $ the number of observations greater than 24
$x = 9, p = 0.1221 > 0.05$
There is no evidence to suggest that the median diameter is different from 24 microns.

14.19 (a) H_0: $\tilde{\mu} = 35{,}742$, H_a: $\tilde{\mu} > 35{,}742$
TS: $X = $ the number of observations greater than $35{,}742$
$x = 18, p = 0.1808 > 0.05$
There is no evidence to suggest that the median new vehicle transaction price has increased.
(b) It seems likely that the distribution of new vehicle transaction prices is not symmetric, but rather skewed to the right. Therefore, the median is a better measure of the center of this distribution.

14.21 H_0: $\tilde{\mu}_1 - \tilde{\mu}_2 = 0$, H_a: $\tilde{\mu}_1 - \tilde{\mu}_2 > 0$
TS: $X = $ the number of observations greater than 0
$x = 10, p = 0.0494 < 0.05$
There is evidence to suggest that the the median speed of a taxi trip is faster on non-gridlock-alert days.

14.23 H_0: $\tilde{\mu}_1 - \tilde{\mu}_2 = 0$, H_a: $\tilde{\mu}_1 - \tilde{\mu}_2 > 0$
TS: $X = $ the number of observations greater than 0
$x = 13, p = 0.0037 \leq 0.05$
There is evidence to suggest that the median VOC concentration is smaller when the scrubber is installed.

14.25 (a) H_0: $\tilde{\mu} = 5.8$, H_a: $\tilde{\mu} > 5.8$
TS: $X = $ the number of observations greater than 5.8
$x = 15, p = 0.0669 > 0.05$
There is no evidence to suggest that the median survival at UCSF is greater than 5.8 years.
(b) H_0: $\mu = 5.8$, H_a: $\mu > 5.8$
TS: $T = \dfrac{\bar{X} - \mu_0}{S/\sqrt{n}}$; RR: $T \geq 1.7109$
$t = \dfrac{7.116 - 5.8}{2.453/\sqrt{25}} = 2.6824 \geq 1.7109$
$p = 0.0065 < 0.05$
There is evidence to suggest that the mean survival at UCSF is greater than 5.8 years.
The conclusion in part (b) is different than the conclusion in part (a). It seems more appropriate to use the sign test from part (a), as it is unlikely that the underlying distribution of survival times is normal.

14.2 The Wilcoxon Signed-Rank Test

14.27 False

14.29 True

14.31 False

14.33 (a) TS: T_+ = the sum of the ranks corresponding to the positive differences $x_i - 70$
$t_+ = 39, p = 0.2441$
There is no evidence to suggest that the median is different from 70.
(b) TS: T_+ = the sum of the ranks corresponding to the positive differences $x_i - 0.7$
$t_+ = 56.0, p = 0.0616$
There is no evidence to suggest that the median is less than 0.7.
(c) TS: T_+ = the sum of the ranks corresponding to the positive differences $x_i - (-45)$
$t_+ = 100.5, p = 0.0014 \leq 0.02$
There is evidence to suggest that the median is greater than -45.
(d) TS: T_+ = the sum of the ranks corresponding to the positive differences $x_i - 450$
$t_+ = 94.0, p = 2P(T+ \leq 94) = 0.040$
There is no evidence to suggest that the median is different from 450.

14.35 (a) $H_0: \tilde{\mu} = 120, H_a: \tilde{\mu} > 120$
TS: T_+ = the sum of the ranks corresponding to the positive differences $x_i - 120$
RR: $T+ \geq 101$
$t_+ = 116 \geq 101$
$p = 0.0055 \leq 0.05$
There is evidence to suggest that the mean (median) oxidation rate is greater than 120.
(b) The distribution is assumed to be symmetric, therefore $\mu = \tilde{\mu}$.

14.37 $H_0: \tilde{\mu} = 6000, H_a: \tilde{\mu} \neq 6000$
TS: $Z = \dfrac{T_+ - \mu_{T+}}{\sigma_{T+}}$; RR: $|Z| \geq 2.5758$
$z = \dfrac{113 - 162.5}{\sqrt{1381.25}} = -1.3319 > -2.5758$
$p = 0.1829 > 0.01$
There is no evidence to suggest that the median grout strength is different from 6000.

14.39 $H_0: \tilde{\mu} = 50, H_a: \tilde{\mu} < 50$
TS: T_+ = the sum of the ranks corresponding to the positive differences $x_i - 50$
RR: $T_+ \leq 53$
$t_+ = 62 > 53$
$p = 0.0978 > 0.05$
There is no evidence to suggest that the median travel distance is less than 50 mi.

14.41 $H_0: \tilde{\mu} = 118, H_a: \tilde{\mu} < 118$
TS: T_+ = the sum of the ranks corresponding to the positive differences $x_i - 118$
RR: $T_+ \leq 21$
$t_+ = 26 > 21$
$p = 0.0955 > 0.05$
There is no evidence to suggest that the median 154-day weight is less than 118 lb.

14.43 $H_0: \tilde{\mu} = 65, H_a: \tilde{\mu} > 65$
TS: $Z = \dfrac{T_+ - \mu_{T+}}{\sigma_{T+}}$; RR: $Z \geq 2.3263$
$z = \dfrac{229 - 138}{\sqrt{1081}} = 2.7678 \geq 2.3263$
$p = 0.0028 < 0.01$
There is evidence to suggest that the median noise level is greater than 65 dBA.

14.45 (a) $H_0: \tilde{\mu} = 0.30, H_a: \tilde{\mu} > 0.30$
TS: X = the number of observations greater than 0.3
$x = 10, p = 0.2272 > 0.05$
There is no evidence to suggest that the median arsenic concentration in freshwater locations is greater than 0.30 μg/L.
(b) $H_0: \tilde{\mu} = 0.30, H_a: \tilde{\mu} > 0.30$
TS: T_+ = the sum of the ranks corresponding to the positive differences $x_i - 118$
RR : $T_+ \geq 101$
$t_+ = 105.5 \geq 101$
$p = 0.0253 \leq 0.05$
There is evidence to suggest that the median arsenic concentration in freshwater locations is greater than 0.30 μg/L.
(c) The conclusions in parts (a) and (b) are not the same. The test in part (b) is probably more accurate because it uses more of the information in the sample than does the test in part (a).

14.3 The Rank Sum Test

14.47 False

14.49 True

14.51 True, if the underlying populations are symmetric

14.53 False

14.55 (a) $w = 35.0$ (b) $w = 82.5$ (c) $w = 256.5$

14.57 (a) $\mu_W = 277.5, \sigma_W^2 = 971.25, \sigma_W = 31.1649$
(b) $\mu_W = 333.0, \sigma_W^2 = 999.0, \sigma_W = 31.6070$
(c) $\mu_W = 214.5, \sigma_W^2 = 965.25, \sigma_W = 31.0685$
(d) $\mu_W = 174.0, \sigma_W^2 = 464.0, \sigma_W = 21.5407$
(e) $\mu_W = 552.0, \sigma_W^2 = 2208.0, \sigma_W = 46.9894$
(f) $\mu_W = 700.0, \sigma_W^2 = 3500.0, \sigma_W = 59.1608$

14.59 $H_0: \tilde{\mu}_1 - \tilde{\mu}_2 = 0, H_a: \tilde{\mu}_1 - \tilde{\mu}_2 \neq 0$
TS: $Z = \dfrac{W - \mu_W}{\sigma_W}$; RR: $|Z| \geq 1.96$
$\mu_W = 323.0, \sigma_W^2 = 1076.6667$
$w = 267.0$
$z = \dfrac{267.0 - 323.0}{\sqrt{1076.6667}} = -1.7067 > -1.96$

$p = 0.0879 > 0.05$
There is no evidence to suggest $\tilde{\mu}_1 \neq \tilde{\mu}_2$.

14.61 $H_0: \tilde{\mu}_1 - \tilde{\mu}_2 = 0$, $H_a: \tilde{\mu}_1 - \tilde{\mu}_2 \neq 0$
TS: $W =$ the sum of the ranks corresponding to the smaller sample
RR: $W \leq 27$ or $W \geq 57$
$w = 32.5 > 27$
$p = 0.1806 > 0.05$
There is no evidence to suggest that median visit times are different

14.63 $H_0: \tilde{\mu}_1 - \tilde{\mu}_2 = 0$, $H_a: \tilde{\mu}_1 - \tilde{\mu}_2 > 0$
TS: $W =$ the sum of the ranks corresponding to the first sample
RR: $W \geq 118$
$w = 127.5 \geq 118$
$p = 0.0012 \leq 0.01$
There is evidence to suggest that the median slapshot speed of NHL defensemen is greater than the median slapshot speed of NHL forwards.

14.65 (a) $H_0: \tilde{\mu}_1 - \tilde{\mu}_2 = 0$, $H_a: \tilde{\mu}_1 - \tilde{\mu}_2 < 0$
TS: $Z = \dfrac{W - \mu_W}{\sigma_W}$; RR: $Z \leq -1.6449$
$\mu_W = 285.0$, $\sigma_W^2 = 1045.0$
$w = 214.0$
$z = \dfrac{214.0 - 285.0}{\sqrt{1045.0}} = -2.1963 \leq -1.6449$
$p = 0.0140 \leq 0.05$
There is evidence to suggest that the median commercial time per hour is greater on talk radio.
(b) $H_0: \mu_1 - \mu_2 = 0$, $H_a: \mu_1 - \mu_2 < 0$
TS: $T = \dfrac{(\bar{X}_1 - \bar{X}_2) - 0}{\sqrt{S_P^2 \left(\frac{1}{n_1} + \frac{1}{n_2}\right)}}$; RR: $T \leq -1.6896$
$t = \dfrac{(10.62 - 12.8318) - 0}{\sqrt{(9.0786)\left(\frac{1}{15} + \frac{1}{22}\right)}} = -2.1923 \leq -1.6896$
$p = 0.0175 \leq 0.05$
There is evidence to suggest that the median commercial time per hour is greater on talk radio.
(c) The two-sample t test seems more appropriate. The underlying distribution of commercial times is probably approximately normal. The equal variance also seems reasonable.
(d) The two tests lead to the same conclusion and almost the same p value. There is evidence to suggest talk radio stations play more commercials.

14.67 $H_0: \tilde{\mu}_1 - \tilde{\mu}_2 = 0$, $H_a: \tilde{\mu}_1 - \tilde{\mu}_2 < 0$
TS: $Z = \dfrac{W - \mu_W}{\sigma_W}$; RR: $Z \leq -2.3263$
$w = 340.5$
$\mu_W = 528.0$, $\sigma_W^2 = 2200.0$
$z = \dfrac{340.5 - 528.0}{46.9042} = -3.9975 \leq -2.3263$
$p < 0.0001$
There is evidence to suggest that the population median amount of protein in SBP is greater than the population median amount of protein in WTL.

14.69 (a) $H_0: \tilde{\mu}_1 - \tilde{\mu}_2 = 0$, $H_a: \tilde{\mu}_1 - \tilde{\mu}_2 > 0$
TS: $X =$ the number of observations greater than 0
RR: $X \geq 11$
$x = 15 \geq 11$
There is overwhelming evidence to suggest that the median amount of particulates after the smoking regulations is less than the median amount before the regulations.
(b) $H_0: \tilde{\mu}_1 - \tilde{\mu}_2 = 0$, $H_a: \tilde{\mu}_1 - \tilde{\mu}_2 > 0$
TS: $T_+ =$ the sum of the ranks corresponding to the positive differences $d_i - 0$
RR: $T_+ \geq 89$
$t_+ = 120 \geq 89$
There is evidence to suggest that the median amount of particulates after the smoking regulations is less than the median amount before the regulations.
(c) $H_0: \tilde{\mu}_1 - \tilde{\mu}_2 = 0$, $H_a: \tilde{\mu}_1 - \tilde{\mu}_2 > 0$
TS: $Z = \dfrac{W - \mu_W}{\sigma_W}$; RR: $Z \geq 1.6449$
$\mu_W = 232.5$, $\sigma_W^2 = 581.25$
$w = 339.5$
$z = \dfrac{339.5 - 232.5}{\sqrt{581.25}} = 4.4382 \geq 1.6449$
There is evidence to suggest that the median amount of particulates after the smoking regulations is less than the median amount before the regulations.
(d) All three tests lead to the same conclusion. The rank sum test, however, should not be used since the samples are dependent.

14.4 The Kruskal–Wallis Test

14.71 χ^2_{k-1}

14.73 True

14.75 False

14.77 $r_1 = 113.5$, $r_2 = 86.0$, $r_3 = 265.5$

14.79 H_0: The four samples are from identical populations.
H_a: At least two of the populations are different.
TS: $H = \left[\dfrac{12}{n(n+1)} \sum \dfrac{R_i^2}{n_i}\right] - 3(n+1)$; RR: $H \geq 7.8147$
$h = 5.2723 < 7.8147$
$p = 0.1529 > 0.05$
There is no evidence to suggest that the populations are different.

14.81 H_0: The three samples are from identical populations.
H_a: At least two of the populations are different.
TS: $H = \left[\dfrac{12}{n(n+1)} \sum \dfrac{R_i^2}{n_i}\right] - 3(n+1)$; RR: $H \geq 5.9915$
$h = 14.9356 \geq 5.9915$
$p = 0.0006 < 0.05$
There is evidence to suggest that at least two of the car classification populations are different.

14.83 (a) H_0: The four samples are from identical populations.
H_a: At least two of the populations are different.
TS: $H = \left[\dfrac{12}{n(n+1)} \sum \dfrac{R_i^2}{n_i}\right] - 3(n+1)$; RR: $H \geq 7.8147$
$h = 11.0517 \geq 7.8147$
$p = 0.0114 < 0.05$
There is evidence to suggest that at least two of these glacier-area populations are different.
(b) It appears that the Danube River drainage region has the largest glaciers. This sample contributed the most to the test statistic.

14.85 H_0: The four samples are from identical populations.
H_a: At least two of the populations are different.
TS: $H = \left[\dfrac{12}{n(n+1)} \sum \dfrac{R_i^2}{n_i}\right] - 3(n+1)$; RR: $H \geq 9.3484$
$h = 4.8911 < 9.3484$
$p = 0.1799 > 0.025$
There is no evidence to suggest that the fat populations are different.

14.87 (a) H_0: The five samples are from identical populations.
H_a: At least two of the populations are different.
TS: $H = \left[\dfrac{12}{n(n+1)} \sum \dfrac{R_i^2}{n_i}\right] - 3(n+1)$
RR: $H \geq 13.2767$
$h = 16.5100 \geq 13.2767$
$p = 0.0024 < 0.01$
There is evidence to suggest that at least two of the finish time distributions are different.
(b) The smallest mean rank is associated with Kenya. Therefore, Kenya appears to have the fastest runners.

14.89 (a) H_0: The three samples are from identical populations.
H_a: At least two of the populations are different.
TS: $H = \left[\dfrac{12}{n(n+1)} \sum \dfrac{R_i^2}{n_i}\right] - 3(n+1)$; RR: $H \geq 5.9915$
$h = 22.589 \geq 5.9915$
$p < 0.0001$
There is evidence to suggest that at least two of the length-of-service populations are different.
(b) According to the sample medians and ranks sums for each group, it appears the public safety and communications populations and the support services and communications populations are different.

14.5 The Runs Test

14.91 False

14.93 1

14.95 When both m and n are greater than 10

14.97 True

14.99 (a) $v_1 = 9$, $v_2 = 3$, $\alpha = 0.0788$
(b) $v_1 = 11$, $v_2 = 3$, $\alpha = 0.0264$
(c) $v_1 = 11$, $v_2 = 3$, $\alpha = 0.0260$
(d) $v_1 = 14$, $v_2 = 4$, $\alpha = 0.0256$

14.101 (a) $\mu_V = 13.0$, $\sigma_V^2 = 5.5$
(b) $\mu_V = 18.5$, $\sigma_V^2 = 8.25$
(c) $\mu_V = 4.68$, $\sigma_V^2 = 0.4109$
(d) $\mu_V = 27.0$, $\sigma_V^2 = 12.7451$

14.103 (a) H_0: The sequence of observations is random.
H_a: The sequence of observations is not random.
TS: $Z = \dfrac{V - \mu_V}{\sigma_V}$; RR: $|Z| \geq 2.5758$
$v = 23$, $\mu_V = 15.4$, $\sigma_V^2 = 6.6538$
$z = \dfrac{23 - 15.4}{\sqrt{6.6538}} = 2.9463 \geq 2.5758$
There is evidence to suggest that the order of observations is not random.
(b) $p = 0.0032 < 0.01$

14.105 H_0: The sequence of observations is random.
H_a: The sequence of observations is not random.
TS: $Z = \dfrac{V - \mu_V}{\sigma_V}$; RR: $|Z| \geq 1.96$
$v = 9$, $\mu_V = 9.4$, $\sigma_V^2 = 3.2716$
$z = \dfrac{9 - 9.4}{\sqrt{3.2716}} = -0.2212 > -1.96$
$p = 0.825$
There is no evidence to suggest that the order of observations is not random.

14.107 (a) H_0: The sequence of observations is random.
H_a: The sequence of observations is not random.
TS: $V =$ the number of runs
RR: $V \leq 4$ or $V \geq 12$
$v = 6$
There is no evidence to suggest that the order of observations is not random.
(b) $p = 0.4522$

14.109 H_0: The sequence of observations is random.
H_a: The sequence of observations is not random.
TS: $Z = \dfrac{V - \mu_V}{\sigma_V}$; RR: $|Z| \geq 1.96$
$v = 9$, $\mu_V = 15.7333$, $\sigma_V^2 = 6.9772$
$z = \dfrac{9 - 15.7333}{\sqrt{6.9772}} = -2.5491 \leq -1.96$
$p = 0.0108 < 0.05$
There is evidence to suggest that the order of observations is not random.

14.111 H_0: The sequence of observations is random.
H_a: The sequence of observations is not random.
TS: $Z = \dfrac{V - \mu_V}{\sigma_V}$; RR: $|Z| \geq 1.96$
$v = 13$, $\mu_V = 18.3714$, $\sigma_V^2 = 8.3646$
$z = \dfrac{13 - 18.3714}{\sqrt{8.3646}} = -1.8572 > -1.96$
$p = 0.0632$
There is no evidence to suggest that the order of GMAT scores is not random.

14.113 H_0: The sequence of observations is random.
H_a: The sequence of observations is not random.
TS: $Z = \dfrac{V - \mu_V}{\sigma_V}$; RR: $|Z| \geq 1.96$
$v = 15$, $\mu_V = 20.55$, $\sigma_V^2 = 9.2988$
$z = \dfrac{15 - 20.55}{\sqrt{9.2988}} = -1.8200 > -1.96$
There is no evidence to suggest that these responses were not selected at random.

14.6 Spearman's Rank Correlation

14.115 There are no assumptions about the underlying populations.

14.117 True

14.119 False

14.121 (a) $r_S = 0.5357$. This indicates a moderate positive relationship.
(b) $r_S = 0.3333$. This indicates a weak positive relationship.
(c) $r_s = 0.0637$. This indicates there is no definitive relationship.
(d) $r_S = -0.2922$. This indicates a weak negative relationship.

14.123 $r_S = 0.7714$
There is a positive relationship between the ranks of x and y. This suggests that as the price of a stateroom increases, so does the number of days before sailing. Therefore, this suggests that cruise prices may be reduced at the last minute, although we cannot conclude this is a cause–effect relationship.

14.125 $r_S = -0.4429$
This value suggests that there is a weak to moderate negative linear relationship between density and soil texture. As bulk density increases, the soil texture tends to decrease.

14.127 $r_S = -0.4634$
This weak negative relationship suggests that the more time a child plays outdoors, the fewer school days the child misses.

14.129 $r_S = 0.1194$
This weak positive relationship suggests that a higher heart rate in a woman is associated with a higher IQ; however, this is a very weak association.

14.131 (a)

x	rank	y	rank	d_i
71	9.0	64	7.5	1.5
69	6.5	56	1.0	5.5
66	3.0	65	9.5	−6.5
69	6.5	62	4.0	2.5
62	1.0	65	9.5	−8.5
67	4.0	64	7.5	−3.5
69	6.5	57	2.5	4.0
64	2.0	63	5.5	−3.5
73	10.0	57	2.5	7.5
69	6.5	63	5.5	1.0

(b) $r = -0.5887$
(c) $r_S = -0.5212$
(d) These values are different because there are tied observations. There is a moderate negative relationship between the ranks. This suggests that as shear strength increases, temperature cycle stress decreases.

14.133 (a) $r_S = 0.5549$
There is a moderate positive linear relationship between the ranks. This suggests that as the psi increases, so does the percentage change in horsepower.
(b) $H_0: \rho_S = 0$, $H_a: \rho_S > 0$
TS: $Z = R_S \sqrt{n-1}$; RR: $Z \geq 2.3263$
$z = (0.5549)\sqrt{20-1} = 2.4187 \geq 2.3263$
$p = 0.0078 \leq 0.01$
There is evidence to suggest that the true population correlation between ranks is greater than 0.

Chapter 14 Exercises

14.135 $H_0: \tilde{\mu} = 2$; $H_a: \tilde{\mu} \neq 2$
TS: the number of observations greater than 2
RR: $X \leq 5$ or $X \geq 15$
$x = 15 \geq 15$, $p = 0.0414 \leq 0.05$
There is evidence to suggest that the median coverage amount is different from 2.

14.137 $H_0: \tilde{\mu} = 21$, $H_a: \tilde{\mu} > 21$
TS : $T_+ = $ the sum of the ranks corresponding to the positive differences $x_i - 21$
RR: $T_+ \geq 123$
$t_+ = 91 < 123$
$p > 0.1519$
There is no evidence to suggest that the median amount of sugar in a serving is greater than 21 g.

14.139 $H_0: \tilde{\mu} = 42$, $H_a: \tilde{\mu} \neq 42$
TS : $T_+ = $ the sum of the ranks corresponding to the positive differences $x_i - 42$
RR: $T_+ \leq 5$ or $T_+ \geq 40$
$t_+ = 33$
$p = 0.2500$
There is no evidence to suggest that the median plunge height is different from 42.

14.141 $H_0: \tilde{\mu}_1 - \tilde{\mu}_2 = 0$, $H_a: \tilde{\mu}_1 - \tilde{\mu}_2 \neq 0$
TS: $W = $ the sum of the ranks corresponding to the first (smaller) sample
RR: $W \leq 20$ or $W \geq 45$
$w = 45.5 \geq 45$
$p = 0.034$
There is evidence to suggest that the median number of miles driven is different for people who carry an organ donor card and for those who do not.

14.143 $H_0: \tilde{\mu}_1 - \tilde{\mu}_2 = 0$, $H_a: \tilde{\mu}_1 - \tilde{\mu}_2 \neq 0$
TS: $Z = \dfrac{W - \mu_W}{\sigma_W}$; RR: $|Z| \geq 1.96$
$w = 311.0$, $\mu_W = 232.5$, $\sigma_W^2 = 581.25$
$z = \dfrac{311 - 232.5}{\sqrt{581.25}} = 3.2560 \geq 1.96$

$p = 0.0011 < 0.05$
There is evidence to suggest that the median pressures are different.

14.145 $H_0: \tilde{\mu}_1 - \tilde{\mu}_2 = 0$, $H_a: \tilde{\mu}_1 - \tilde{\mu}_2 \neq 0$
TS: W = the sum of the ranks corresponding to the first (smaller) sample
RR: $W \leq 43$ or $W \geq 83$
$w = 65$
$p > 0.3148 > 0.05$
There is no evidence to suggest that the median speeds are different.

14.147 H_0: The four samples are from identical populations.
H_a: At least two of the populations are different.
TS: $H = \left[\dfrac{12}{n(n+1)} \sum \dfrac{R_i^2}{n_i}\right] - 3(n+1)$; RR: $H \geq 9.3484$
$h = 3.1241 < 9.3484$
$p = 0.3729 > 0.025$
There is no evidence to suggest that the transmitter power populations are different

14.149 H_0: The four samples are from identical populations.
H_a: At least two of the populations are different.
TS: $H = \left[\dfrac{12}{n(n+1)} \sum \dfrac{R_i^2}{n_i}\right] - 3(n+1)$; RR: $H \geq 7.8147$
$h = 7.0185 < 7.8147$
$p = 0.0713 > 0.05$
There is no evidence to suggest that these steel production populations are different.

14.151 H_0: The sequence of observations is random.
H_a: The sequence of observations is not random.
TS: $Z = \dfrac{V - \mu_V}{\sigma_V}$; RR: $|Z| \geq 1.96$
$v = 22$, $\mu_V = 15.7333$, $\sigma_V^2 = 6.9772$
$z = \dfrac{22 - 15.7333}{\sqrt{6.9772}} = 2.3725 \geq 1.96$
$p = 0.0177 \leq 0.05$
There is evidence to suggest that the order of automobiles entering the parking garage is not random.

14.153 H_0: The sequence of observations is random.
H_a: The sequence of observations is not random.
TS: $Z = \dfrac{V - \mu_V}{\sigma_V}$; RR: $|Z| \geq 1.96$
$v = 19$, $\mu_V = 22.9556$, $\sigma_V^2 = 10.4566$
$z = \dfrac{19 - 22.9556}{\sqrt{10.4566}} = -1.2232 > -1.96$
$p = 0.2212 > 0.05$
There is no evidence to suggest that the order of observations is not random.

14.155 $r_S = -0.4126$
This value suggests there is a weak to moderate negative linear relationship between the ranks of the quality score and the total number of people in the hospital. This suggests that as the quality score increases, the total number of people in the hospital decreases.

14.157 (a) $H_0: \tilde{\mu}_1 - \tilde{\mu}_2 = 0$, $H_a: \tilde{\mu}_1 - \tilde{\mu}_2 \neq 0$
TS: $Z = \dfrac{T_+ - \mu_T}{\sigma_T}$; RR: $|Z| \geq 2.5758$
$t_+ = 152.5$, $\mu_{T_+} = 175.5$, $\sigma_{T_+}^2 = 1550.25$
$z = \dfrac{152.5 - 175.5}{\sqrt{1550.25}} = -0.5842 < 2.5758$
$p = 0.5591 > 0.01$
There is no evidence to suggest that the median glucose levels are different for the two procedures.
(b) Since there is no difference in the median glucose levels, it seems reasonable to recommend the noninvasive, painless approach.

14.159 $H_0: \tilde{\mu}_1 - \tilde{\mu}_2 = 0$, $H_a: \tilde{\mu}_1 - \tilde{\mu}_2 < 0$
TS: X = the number of pairwise differences greater than 0
RR: $X \leq 5$
$x = 15 > 5$
There is no evidence to suggest that the median refrigerant weight before service is less than the median refrigerant weight after service.
(b) Since we cannot reject the null hypothesis, there is no evidence to suggest that recharging an air conditioner increases the amount of refrigerant in the system.

14.161 (a) H_0: The sequence of observations is random.
H_a: The sequence of observations is not random.
TS: V = the number of run; RR: $V \leq 4$ or $V \geq 11$
$v = 7$
There is no evidence to suggest that the order of observations is not random with respect to exterior finish.
(b) Using the runs test, we cannot tell if the advertising campaign was successful. Also, we don't know the historical proportion of home builders who use vinyl. Therefore, we cannot tell if this proportion has increased.

14.163 (a) $H_0: \tilde{\mu}_1 - \tilde{\mu}_2 = 0$, $H_a: \tilde{\mu}_1 - \tilde{\mu}_2 < 0$
TS: $Z = \dfrac{W - \mu_W}{\sigma_W}$; RR: $Z \leq -1.645$
$w = 1508$, $\mu_W = 1740$, $\sigma_W^2 = 13{,}340$
$z = \dfrac{1508 - 1740}{\sqrt{13.340}} = -2.0087 \leq -1.645$
$p = 0.0223 < 0.05$
There is evidence to suggest that the median time for SunPass users is less than the median travel time for non-SunPass users.
(b) $H_0: \mu_1 - \mu_2 = 0$, $H_a: \mu_1 - \mu_2 < 0$
TS: $T = \dfrac{(\bar{X}_1 - \bar{X}_2) - 0}{\sqrt{S_p^2\left(\frac{1}{n_1} + \frac{1}{n_2}\right)}}$; RR: $T \leq -1.6632$
$t = \dfrac{(133.3 - 138.2) - 0}{\sqrt{(207.5265)\left(\frac{1}{40} + \frac{1}{46}\right)}} = -7.3032$
$p = 0 \leq 0.05$
There is evidence to suggest that the mean travel time for SunPass users in less than the mean travel time for non-SunPass users.
(c) The Wilcoxon rank sum test is more appropriate since the underlying distributions are probably not normal.

14.165 H_0: The four samples are from identical populations.
H_a: At least two of the populations are different.

TS: $H = \left[\dfrac{12}{n(n+1)} \sum \dfrac{R_i^2}{n_i} \right] - 3(n+1)$; RR: $H \geq 7.8147$

$h = 11.0428 \geq 7.8147$

There is evidence to suggest that at least two of the boarding time populations are different.

(b) $p = 0.0115 < 0.05$

(c) A one-way ANOVA test can be used. Assumptions: It is reasonable to assume that all populations are normally distributed, and the population variances are equal. Each sample was selected randomly.

$H_0: \mu_1 = \mu_2 = \mu_3 = \mu_4$
$H_a: \mu_i \neq \mu_j$ for some $i \neq j$
TS: $F = $ MSA/MSE; RR: $F \geq 2.6984$

Source of variation	Sum of squares	Degrees of freedom	Mean square	F	p Value
Factor	160.3	3	53.42	3.947	0.0106
Error	1299	96	13.53		
Total	1459.3	99			

$f = 53.42/13.53 = 3.947 \geq 2.6984$
$p = 0.0106 < 0.05$

There is evidence to suggest that at least two population mean boarding times are different for these boarding methods.

(d) Steffen's method appears to be the most efficient. These data have the smallest mean, smallest median, and the smallest mean rank.

Index

A

A complement, 125, 138
A union B, 125
Actuaries, 10
Addition Rule, 140
    for two events, 138, 140–141
Additive probabilistic model, 558
Almost independent events, 180
Alternative hypothesis, 391, 631
    two-sided, 405
Alternative mean, 408
Analysis of variance (ANOVA) table, 566–569
    model utility test and, 606–608
    summary for multiple linear regression, 606, 607
    summary for simple linear regression, 567, 568
ANOVA (analysis of variance), 515–552
    one-way, 516–523
        grand mean, 517
        mean square due to error, MSE, 518
        mean square due to factor, MSA, 518
        notation, 517
        summary table, 519, 520, 521, 523, 544, 545
        test procedure, 519
        total variation decomposition, 518
    two-way, 539–546
        hypothesis tests, 541–542
        summary table, 541
        total variation decomposition, 539–541
Approximate hypothesis test, 470
Approximate two-sample test, 470
Arithmetic mean, 71
Assumptions
    in a multiple linear regression model, 612–613
    normality, 278–279
    z test, 454
Average of the squared deviations about the mean, 84
Average, weighted, 207

B

Back-to-back stem-and-leaf plots, 45–47
Backward cumulative probability, 269
Backward Empirical Rule, 97, 279
Bar charts, 35
    constructing, 35
    frequency versus relative frequency, 34
    side-by-side or stacked, 640

Bayes' Rule, 121, 176–177, 178
Bell-shaped curve, 261–263
Bernoulli variable, 77
Between-sample variation, 518
Bias, 20
    nonresponse, 20
    self-selection, 20
Biased estimator, 337
Bimodal distribution, 56
Binomial experiment, 216
    properties of, 216
Binomial probability distribution, 216, 218, 659
    normal approximation to the, 323
Binomial random variable, 216, 217, 224
    cumulative probability, 220, 221–222
    mean, variance and standard deviation of, 224
Bivariate categorical data
    test for homogeneity, 639–644
    test for independence, 645–647
Bivariate data, 645
Bivariate data set, 29
Bonferroni confidence interval, 529, 531
Bonferroni multiple comparison procedure, 528, 529, 611
Bonferroni simultaneous confidence intervals, 611
Bound on the error of estimation, 349
Box-and-whisker plot, 106
Box plot, 106
    modified, 108
    standard, 106

C

Categorical data
    bivariate, 639–644
    display in frequency distribution, 630
    univariate, 630–634
Categorical data set, 29
Categorical variables, test for independence of two, 646, 653
Cell, 631
Centers for Disease Control and Prevention (CDC), 11
Central Limit Theorem, 313–314
    inference and, 317–318
Chance, 134
Chart, 28
    bar, 35, 640
    pie, 36
Chebyshev's Rule, 94–95, 97, 211
Chi-square (χ^2) critical values, 375

Chi-square (χ^2) distribution, 375, 631
    normal approximation to, 384
    properties of, 375
Claim, questioning, in statistical inference, 21
Class, 50
Class cumulative relative frequency, 34
Class frequency, 34
Class relative frequency, 34
Coefficient(s)
    confidence, 345
    of determination, 567
        analysis of variance (ANOVA) table and, 566–567
    estimating, 564–566
    of kurtosis, 93
    of quartile variation, 91
    of skewness, 93
    of variation, 91
Combinations, 151, 153
Common variance, pooled estimator for, 466
Complement, 125
Complement Rule, 138–139, 256, 660
Composite null hypothesis, 631
Conclusion
    of hypothesis, 392
    in statistical inference, 3, 21
Conditional probability, 158–160, 159
    steps for calculating, 163
Confidence coefficient, 341, 345
Confidence intervals, 335–388
    Bonferroni simultaneous, 529, 611
    construction of, 341
    general $100(1-\alpha)\%$, 343–344
    inference using, 346–347, 360–361
    large sample, for a population proportion, 366–370
    large-sample, for p, 366–367
    for the mean value of Y for $x = x^*$, 587
    95%, 342–343
    one-sample t, 358
    for p and inference, 367–369
    for partial regression coefficient, 604
    point estimation, 336–339
    for population mean when σ is known, 341–350
    for population mean when σ is unknown, 355–361
    for population variance or standard deviation, 375–380
    for slope of regression line, 559, 561, 565

Confidence intervals (*Continued*)
    for the difference of two population means when variances are unknown and unequal, 470
    for the difference of two population proportions, 489
    for the ratio of two population variances, 502
    when variances are known, 458
    when variances are unknown but equal, 465
    for $\mu_1 - \mu_2$, 458–459, 469–470
    for σ^2 and inference, 379–380
    for σ^2 or σ, 378
Confidence level, 341
Contingency table, 161–162
Continuous data, discrete data versus, 29–30
Continuous data set, 30
Continuous probability distribution, 247–294
    checking the normality assumption in, 278–284
    for continuous random variable, 248–257
    exponential distribution, 286–291
    normal distribution, 261–273
Continuous random variable, 192
    probability distribution for, 248–257
Correlation, 579–580
Correlation coefficient, sample, 579–580
Counting numbers, 30
Counting techniques
    combinations, 153
    for events, 148
    Multiplication Rule in, 147–148
    permutations, 150–151
Critical region, 391
Critical values, 376
CrunchIt!, 5, 107
Cumulative probability, 220, 221–222, 253–256, 287–288
    backward, 269
    inverse, 269, 326
    using, 254–256
Cumulative relative frequency, 50

D

Data
    numerical, 42
    statistics, 10
    types of, 28–31
Data axis, 280
    horizontal, 280
    vertical, 280
Data collection, methods of, 10
Data mining, 10
Data sets
    categorical, 29
    continuous, 30
    discrete, 30
    n ordered pairs in, 559

numerical, 29, 49–50
range of a, 83, 106
Decomposition, total variance, 539–541
Density curve, 248
Density, finding, 55
Density function
    defined, 249
    values of, 249
Density histograms, 55
Dependent events, 171
Dependent samples, 453
Dependent value, predicting, 558
Dependent variables, 564
Descriptive statistics, 12, 28
Determination, coefficient of, 567
Deterministic relationship, 558
Diagram
    tree, 121–122
    Venn, 127
Difference in means, properties of estimator for, 466
Discrete data set, 30
Discrete data versus continuous data, 29–30
Discrete random variable, 192
    mean of, 206–208
    probability distributions for, 195–202
    properties of a valid probability distribution for, 201
Disjoint events, 125, 171
    dependent, 171
Distribution
    bimodal, 56
    binomial, 216
    center of, 42
    exponential, 286–287
    frequency, 33–35
    geometric probability, 231
    hypergeometric probability, 235
    multimodal, 56
    negatively skewed, 57
    normal, 261–263
    Poisson probability, 232, 233
    positively skewed, 57
    of the sample proportion, 323–324
    shape of, 28, 56–58
    standard normal, 263
    symmetric, 57
    uniform, 249–250
    unimodal, 56, 57
Dot plot, 83
    stacked, 83
Double subscript notation, 603

E

Empirical Rule, 96–97
    backward, 97, 279
Equally likely outcome experiment, 135–138
    finding probabilities in an, 136
Equifax, 11

Errors
    of estimation, 349, 369–370
    type I, 397
    type II, 397
Estimated regression line, 563, 567–569
Estimates, 336
    error of, 349, 369–370
    least-squares, 563–564
    point, 336, 338–339
Estimator, 336
    biased, 336
    pooled, 464
    unbiased, 84, 313, 336
Event(s), 124
    Addition Rule for two, 140–141
    dependent, 171
    disjoint, 125, 127, 171
    independent, 171
    mutually exclusive, 125
    mutually exclusive exhaustive, 164–165
    probability of an, 132, 133
    simple, 124
Excel 2019, 8
Expected value, 206–208
Experiment, 2, 120, 392
    binomial, 216
    equally likely outcome, 135–138
    geometric, 230
    hypergeometric, 234
    Poisson, 232
    random samples and, 19–20
    in statistical inference, 2, 21
Experimental study, 19
Experimentation, 10
Exponential distribution, probability density function for, 286–287
Exponential model, 615
Exponential random variables, memoryless property of, 291
Extended definitions of union, intersection, and disjoint events, 127

F

Factorial, 150
F critical values, 499
F distribution, 499
    critical values for the, 519
    properties of, 499
First (lower) quartile, 87
Five-number summary, 106–107, 108
Formal hypothesis test, 392
Frequency, 34
    class, 50
    cumulative relative, 50
    relative, 34, 50, 133
Frequency distribution, 33–35, 34
    constructing for numerical data, 50
Frequency histogram, 53, 309
F test for a significant linear regression, 607, 610, 611

G

General additive probabilistic model, 558
General growth model, 615
Generalized intersection, 127
Generalized union, 125, 127
Geometric experiment, properties of, 230
Geometric probability distribution, 231
Geometric random variable, 230–231
Goodness-of-fit test, 630–634, 652
Grand mean, 517

H

Heavy tail, 58
Histogram, 52, 53
    constructing, 52
    density, 55
    drawing by hand, 53
    frequency, 53, 309
    relative frequency, 53–54
Homogeneity, 640
Homogeneity of populations, test for, 643, 645
Homogeneity of variance, 561
Human error, 396–397
Hypergeometric experiment, 234
    properties of, 234
Hypergeometric probability distribution, 235
Hypergeometric random variable, 234–235
Hypothesis
    alternative, 391
    defined, 390
    no-change, 390
    null, 390, 392
    two-sided alternative, 405
Hypothesis test(s), 389–450
    alternative mean in, 408–409
    approximate, 470
    calculating the probability of a type II error, 408–409
    concerning a population mean when σ is known, 403–409, 405
    concerning a population mean when σ is unknown, 422–427
    concerning a population variance, 439
    concerning two population means when data are paired, 478
    concerning two population means when population variances are known, 454
    concerning two population means when variances are unknown but equal, 465
    concerning two population proportions when $\Delta_0 = 0$, 489
    concerning two population proportions when $\Delta_0 \neq 0$, 491–493
    concerning two population variances, 502
    conclusion in, 392
    and confidence interval concerning β_1, 608–610
    efficiency in, 400
    experiment in, 392
    formal procedure in, 392
    for interaction effect, 542
    large-sample, concerning a population proportion, 432–434, 432–435
    likelihood in, 392
    notation and, 452–453
    one-sample t test in, 422–423
    parts of, 390–391
    p value bounds in, 425–427
    p value in, 414–419, 434–435
    rejection region in, 414–415
    robust, 472
    significance level in, 397
    for a significant linear regression, 574
    for a significant multiple linear regression, 606
    tests for one proportion, 432
    two-sample Z test, 465
    two-way ANOVA, 541–542
Hypothesis test errors, 396–397
    significance level and, 397
    type I error as, 397
    type II error as, 397
Hypothesis tests and confidence interval concerning the mean value of Y for $x = x^*$, 611, 622
    concerning β_0, 576
    concerning β_1, 575
Hypothesized proportion, 630, 631

I

Independence, 170–171
Independent events, 171
    visualizing, 171
Independent samples, 453
    from normal populations, 463–472
    population variances known, 453–459
Independent variables, 558, 564
Index of summation, 70, 71
Inference, 11, 222–224
    Central Limit Theorem and, 317–318
    with confidence interval, 346–347, 360–361
    normal distribution and, 272–273
    $\hat{p}$ and, 326
    random variables and, 200–201
    statistical, 10–11
    with two independent samples
        normal populations, 463–472
        population variances known, 453–459
Inferences concerning the mean value and an observed value of Y for $x = x^*$, 586–590
Inferential statistics, 11, 12, 15
Interaction effect, hypothesis test for an, 542
Interpolation, 267
    linear, 267, 361
Interquartile range (IQR), 86, 87, 89, 279
Intersection, 125
    generalized, 127
Interval notation, 94
Intervals. *See also* Confidence intervals
    symmetric, 94, 96–97
Inverse cumulative probability, 266–267, 269, 326
Isolating differences, 528–535
ith deviation about the mean, 83

J

JMP Statistical Software, 6
Joint probability table, 146, 163–164

K

Known variances, 454–456
Kruskal–Wallis test, 681–684, 697
Kurtosis, coefficient of, 93

L

Large-sample confidence interval
    for p, 366–367
    for a population proportion, 366–370
Large-sample hypothesis test
    concerning a population proportion, 432–434
    concerning two population proportions when $\Delta_0 = 0$, 489
    when $\Delta_0 \neq 0$, 491
Least squares
    estimates, 563–564, 566
    principle of, 563–564
Light tails, 58
Likelihood, 11
    in statistical inference, 3, 21
Linear interpolation, 267, 361
Linear regression
    hypothesis test for a significant, 574
    hypothesis test for a significant multiple, 606
    multiple, 622–624
    simple, 558–570
Linear relationship, strength of, 580, 581
Line of best fit, 564
Line of mean values, 562
Lower tail, 57

M

Mann–Whitney rank-sum test, 675
Mann–Whitney U statistic, 675
Marginal totals, 161
Mathematica, 6
Mean, 206–208
    alternative, 408
    grand, 517
    population, 72, 74, 304
    sample, 71–73, 74, 313
    trimmed, 75

Mean square, 575
    due to error (MSE), 518, 607, 608
    due to factor (MSA), 611
    due to regression, 607, 608
Mean value of Y for $x = x^*$, 586–587
    confidence interval for, 586
    hypothesis test about, 587
Measure of central tendency, 70–78, 82
    notation in, 70–71
    sample mean as, 71–73
    sample median as, 73–74
Measure of variability, 82–89
    population standard deviation in, 84
    population variance in, 84
    quartiles in, 86–87
    sample range in, 82
    sample standard deviation in, 83, 84
    sample variance in, 83, 84, 85
Median
    population, 73, 658
    sample, 73–74
Memoryless property, 90–291
Minimum variance unbiased estimator (MVUE), 337
Minitab, 5
Mode, 76
Model utility test, 574, 575, 610
    ANOVA table and, 606–608, 611, 612
Modified box plot, 108
    constructing, 108
Modified tree diagram, 172
Modified two-way frequency table, 641, 642
Multimodal distribution, 56
Multiple linear regression, 603–614
    hypothesis test for a significant, 606
Multiple linear regression model, 622–624
    assumptions in a, 612
    partial regression coefficients, 603, 604
    probabilistic model, 558
    variability in the model, 567
Multiplication Rule, 147–148, 303
    probability, 172
    tree diagram in proving, 122
Multivariate data set, 29
Mutually exclusive events, 125
Mutually exclusive exhaustive events, 164–165

N

Negative linear relationship, 565, 580
Negatively skewed distribution, 57
News, numbers in the, 10–11
n factorial, 150
95% confidence interval, 342–343
No-change hypothesis, 390
Nonparametric procedures, comparison with parametric procedures, 697
Nonparametric statistics
    Kruskal–Wallis test, 681–684, 697

    runs test, 686–690
    sign test, 658–660, 697
    Sperman's rank correlation, 692–694, 697
    Wilcoxon rank-sum test, 674–678, 697
    Wilcoxon signed rank test, 666–668, 697
Nonparametric tests
    assumptions for validity of, 658
    disadvantages of, 658
Nonresponse bias, 20
Nonskewness criterion, 323
Nonzero differences, 456–458
Normal approximation, 667, 676–678
    concerning the variance, 677
    to the binomial distribution, 323
Normal convergence theorem, 313
Normal curve, 58
Normal probability distribution
    bell-shaped curve in, 261–263
    checking normality assumption for, 278–281
    critical values for, 344
    cumulative probability density function, 287–288
    inference and, 272–273
    inverse cumulative probability and, 266–267
    probability density function, 249–250, 262–263
    standard, 263
    Standardization Rule and, 267–268
Normality assumptions, 658
    checking, 278–284
Normality, checking with the Empirical Rule, 97
Normal probability plot, 279–281, 598, 599
    construction, 279–281
Normal score, 596
Notation
    double subscript, 603
    for events, 124
    hypothesis tests and, 452–453
    interval, 94
    for likelihood, 132
    lowercase letters in, 70
    n in, 70
    one-way ANOVA, 517
    standard set, 124
    subscript, 70, 71
    summation, 70–71
    x in, 70
Null hypothesis, 390, 392
    composite, 631
    rejection of, 682
Numbers
    counting, 30
    in the news, 10–11
    rounding of, 47
    truncation of, 47

Numerical data, 42
    constructing frequency distribution for, 50
Numerical data set, 29, 49–50
    continuous, 30
    discrete, 30
Numerical summary, 70

O

Observational sampling, 10
Observational study, 19
Observation, rank of an, 666
Observed value of Y for $x = x^*$, 589–590
    prediction interval for, 611
Odds, 11
One-sample t confidence level, 358
One-sample t test, 422
One-sided alternative hypothesis, 455
One-sided alternatives, 392
One-to-one correspondence, 30
One-way ANOVA, 516–523
    grand mean, 517
    mean square due to error, MSE, 518
    mean square due to factor, MSA, 518
    notation, 517
    summary table, 519, 520, 521, 523, 544, 545
    test procedure, 519
    total variation decomposition, 518
One-way frequency table, 630, 632, 652
Opportunity Index (OI), 677
Ordered pairs, 559, 560, 580, 596, 614
Ordered stem-and-leaf plot, 44
Outcomes, 121
    converting into real numbers, 174
Outlier, 86, 597
    extreme, 108
    mild, 108

P

Paired data, 477–483
    hypothesis tests concerning two population means with, 478
Paired t test, 478, 479
Parameter, 300
    population, 345
    versus statistic, 300–301
Parametric methods, 658
    comparison with nonparametric procedures, 697
Partial regression coefficients, 603, 604
Percentile, 101
    computing, 101
    rth, 101
Permutations, 150–151
Pie chart, 36
    constructing, 36
    drawing by hand, 37
    exploding, 37

3D, 37
  variations of, 37
p, large-sample confidence interval for, 366–367
Plot
  box
    modified, 108
    standard, 106
  dot, 83
  normal probability, 279–281
  of the residuals, 598
  scatter, 280
  stem-and-leaf, 596, 614, 625
Point estimate, 338–339
Point estimation, 336–339
Poisson probability distribution, 232, 233
Poisson random variable, 232–233
Polynomial model, 615
Pooled estimator, 464
Population
  defined, 14
  versus sample, 13–14
Population mean, 72, 74, 304
  hypothesis test concerning two
    when population variances are known, 454
    when population variances are unknown, 465
  hypothesis tests concerning
    when σ is known, 403–409, 405
    when σ is unknown, 422–427
Population median(s), 73, 658
  signed-rank test to compare two, 669–671
  sign test concerning, 659
  sign test to compare two, 661–663
Population parameter, 345
  notation for, 452–453
Population proportion
  combined estimate of the common population proportion, 489
  comparing two using large samples, 487–494
  large-sample hypothesis test concerning, 432–434
  of successes, 78
Population standard deviation, 84
Population variance(s), 84
  comparing, 499–505
  hypothesis tests concerning, 439
  hypothesis tests concerning two, 502
  hypothesis tests concerning two population means with known, 454
Positive linear relationship, 559, 580, 582
Positively skewed distribution, 57
Prediction interval for an observed value of Y when $x = x^*$, 589, 611
Predictor variables, 564

Principle of least squares, 563, 564, 622
Probabilistic model, 558
Probability, 11
  Addition Rule for events, 140–141
  of an event, 132, 133
  calculation of a type II error, 408–409
  conditional, 158–160, 159
  counting techniques in
    combinations in, 153
    Multiplication Rule, 147–148
    permutations in, 150–151
  cumulative, 220, 221–222, 253–256, 264, 287–288
    inverse, 266
  defined, 132
  dependent events in, 171
  finding, in an equally likely outcome experiment, 136
  odds and, 11
  problem solving, 15
  properties of, 134–135
  relative frequency and, 11
  versus statistic, 15–16
  unconditional, 159
Probability density function, 248, 286–287
Probability distribution
  binomial, 218
  continuous, 248
  for continuous random variable, 248–257
  for discrete random variable, 195–202
  exponential, 287
  geometric, 231
  hypergeometric, 235
  normal, 262
  Poisson, 232, 233
  properties of, 200–201
Probability mass function, 219
Probability Multiplication Rule, 121, 172–173
Problem solving, 3–4
Properties of probability, 134–135
Proportion
  hypothesized, 630
  true, 630
p value, 415
  examples of, 417
  hypothesis testing and, 414–419
p value bounds, 425–427, 441–442

Q
Quadratic model, 615
Qualitative variable, 14
Quantitative variable, 14
Quartile(s), 86–87
  computing, 87
  first (lower), 87
  second, 87
  third (upper), 87
Quartile variation, coefficient of, 93

R
R, 4, 107
Random deviations, 561, 603
Random error, 558
Random error terms, 561, 603
Randomness, 2
Random number generator, 692
Random sample, 20, 303–304
  experiments and, 19–20
  simple, 20
Random variable(s), 190–194
  binomial, 207, 216, 224
  continuous, 192
  discrete, 192
  exponential, 291
  geometric, 230–231
  hypergeometric, 234–235
  inference and, 200–201
  Poisson, 232–233
  standard normal, 263–264
  types of, 192
  variance and standard deviation of a, 208
Range
  of a data set, 106
  interquartile, 86, 87, 89, 279
  sample, 82–83
Regression coefficients, hypothesis tests and confidence intervals concerning the, 575–576
Regression diagnostics, 594–598, 622, 624
Regression equation, 561, 604, 605, 606, 607, 608
Regression line
  estimated, 563
  true, 561, 562–563
Rejection region, 391
Relative frequency, 34, 133
  cumulative, 50
  of occurrence of an event, 132
  probability and, 11
Relative frequency histogram, 53–54
Representative sample, 19–20
Residual analysis, 596–597
Residuals
  plot of the, 598
  standardized, 598
  studentized, 598
  sum of the, 598
Response variable, 564
R function PostHocTest, 531
R function t test, 472, 480
RStudio, 4
rth moment about the mean, 93
rth percentile, 101
Rules
  Addition, 140–141
  Backward Empirical, 279
  Bayes', 176–177, 178
  Chebyshev's, 94–95, 97

Rules (*Continued*)
  Complement, 138–139
  Empirical, 96–97
  Multiplication, 147–148, 303
  Probability Multiplication, 172–173
  Standardization, 267–268
Run, 686, 697
Runs test, 686–690, 697

S

Sample(s)
  comparing two population proportions using large, 487–494
  defined, 14
  dependent, 453
  independent, 453
  population versus, 13–14
  random, 303–304
    simple, 304
Sample correlation coefficient, 579–580
Sample mean, 71–73
  properties of, 313
Sample median, 73–74
Sample proportion, 77
  distribution of, 323–324
Sample range, 82–83
Sample space, 123–124
Sample standard deviation, 83, 84
Sample statistic, notation for, 452–453
Sample variance, 83, 84
  computational formula for, 85
Sampling distribution, 336
  approximating, 301–302
  Central Limit Theorem and, 313–314
  parameters in, 300–304
  of the sample mean, 309–313
  of a statistic, 301–302
Sampling variability, 309–310
SAS, 6
Satterthwaite approximation for the number of degrees of freedom, 470
Scatter plot
  nonlinear, 280
  of the residuals, 595
Second quartile, 87
Self-selection bias, 20
Side-by-side bar charts, 640
Signed-tank test, to compare two population medians, 669–671
Significance level, 397
Significant linear regression, hypothesis test for a, 574
Significant multiple linear regression, hypothesis test for a, 606
Sign test, 658–660, 697
  to compare two population medians, 661–663
  concerning a population median, 659
Simple linear regression, 558–570

ANOVA table and coefficient of determination in, 566–567
estimated regression line, 567–569
estimating coefficients in, 564–566
least squares estimates in, 563–564
model in, 560–561
negative relationship, 565, 580
positive relationship, 559, 580, 582
principle of least squares, 563, 622
probabilistic model, 558
sample correlation coefficient in, 579–580
true regression line in, 562–563
Simple linear regression model, 560–561, 620–622
Simple random sample (SRS), 20, 304
Slope of regression line, 563, 564
Small-sample test, 422
Spearman's rank correlation coefficient, 692–693, 697
SPSS, 6
Stacked bar chart, 640
Stacked dot plot, 83
Standard box plot, 106
  constructing, 106
Standard deviation, 100
  comparing, 499–505
  population, 84
  of a random variable, 208
  sample, 84
Standardization, 267–268
  visualization of the, 269
Standardization Rule, 267–268
Standardized residuals, 598
Standard normal distribution, 263, 343
Standard normal random variables, 263–264
Standard set notation, 124
Statistical inference, 10–11, 70
  claim in, 2, 21
  conclusion in, 3, 21
  experiment in, 2, 21
  likelihood in, 3, 21
  procedure for, 2–3, 21
Statistic(s)
  application of, 11–13
  defined, 10, 300
  descriptive, 12, 28
  inferential, 11, 12, 15
  versus parameter, 300
  probability, 15–16
  purpose of, 2
  reasons for studying, 1–2
  sampling distribution of, 301–302
  summary, 11, 12
  unbiased, 84
Statistics data, availability of, 10
Statistics problems, solving, 15
Stem-and-leaf plot, 42–47
  back-to-back, 45–47

constructing, 43
division of stems in, 44–45
ordered, 44
uses of, 42, 49
Studentized range distribution, 532
Studentized residuals, 598
Subscript notation, 70
Summary measures for categorical variables, 77
Summary statistics, 11, 12
Summary table, 49, 77
Summation notation, 70–71
Sum of squares due to error (SSE), 567
Sum of squares due to regression (SSR), 567
Sum of the residuals, 598
Symmetric distribution, 57
Symmetric interval, 94, 96–97

T

Table, 28
  contingency, 161–162
  joint probability, 146, 163–164
  summary, 49, 77
  two-way, 161–162, 167
Tally marks, 35
t critical values, 356
t distribution, 355
  properties of, 355–356
Technology, 4–6
Test for homogeneity, bivariate categorical data, 639–644
Test for independence, bivariate categorical data, 645–647
Test statistic, 391, 631, 642, 658–659
  value of, 632, 634, 660
Texas Instruments TI-84 Plus CE graphing calculator, 4
  TI-Nspire graphing calculator, 6
Theorem
  Central Limit, 313–314
  normal convergence, 313
Third (upper) quartile, 87
3D pie charts, 37
Tick marks, 35
TI-Nspire graphing calculator, 6
Total sum of squares, 567
Total variation decomposition, 539–541
Tree diagram, 121–122
  modified, 172, 177, 180, 183
  in proving Multiplication Rule, 122
  visualizing Probability Multiplication Rule with, 172
Trial(s), 217
  independent, 217
  single, 217
Trimming percentage, 75
True population proportion, 631
True regression equation, 603

True regression line, 561, 562–563
Truncation of numbers, 47
t test
   modified two-sample test, 470
   one-sample, 422
   paired, 478, 479
   p value bounds for, 425–427
   for a significant linear regression, 577, 578, 579, 580
   two-sample, 470
Tukey's procedure, 532–535
Two population means, comparing
   using independent samples from normal populations, 463–472
   using independent samples when variances are known, 465
   when data are paired, 478
   when variances are unknown and unequal, 470
   when variances are unknown but equal, 465
Two population variances, comparing, 499–505
Two sample F test, 502
Two-sample t test, 464, 470
Two-sample z test, 452
Two-sided alternative hypothesis, 392, 405
Two-sided confidence interval, 349
Two-way ANOVA, 539–546
   hypothesis tests, 541–542
   identity, 540–541
   summary table, 541
   total variation decomposition, 539–541
Two-way frequency table, 639
   modified, 641, 642
Two-way, or contingency table, 630
Two-way table, 161–162, 167, 639
Type I error, 397
Type II error, 397
   calculating probability of a, 408–409

U

Unbiased estimator, 84, 313, 337
Unbiased statistic, 84
Uncertainty, 2
Unconditional probability, 159
Unequal variances, 470–472
Uniform distribution, 249–250
Unimodal distribution, 56, 57
Union, 125
   generalized, 125, 127
Univariate categorical data, goodness-of-fit test for, 630–634
Univariate data classifications, 31
Univariate data set, 29
Upper tail, 57

V

Variability, sampling, 309–310
Variable(s)
   Bernoulli, 77
   defined, 14
   dependent, 564
   independent, 558, 564
   predictor, 564
   qualitative, 14
   quantitative, 14
   random, 190–194 (*See* Random variables)
   response, 564
Variance(s)
   homogeneity of, 561
   known, 454–456
   population, 84
   of a random variable, 208
   sample, 84
   unequal, 470–472
Variance F test, 681
Variation
   between-samples, 518
   coefficient of, 91
   within-simples, 518
Venn diagrams, 127, 171
Visualization of the Z standardization, 269

W

Weighted average, 207
Wilcoxon, Frank, 666
Wilcoxon rank-sum test, 674–678, 697
Wilcoxon signed rank test, 666–668, 697
Within-samples variation, 518
Word problems, solving, 3–4

Y

y-intercept of regression line, 564

Z

z critical values, 343, 454
z distribution, 279
z-scores
   negative, 100
   positive, 100
z test
   assumptions, 454
   two-sample, 452